D0821785

The Mouse in Biomedical Research

Volume II

Diseases

AMERICAN COLLEGE OF LABORATORY ANIMAL MEDICINE SERIES

Steven H. Weisbroth, Ronald E. Flatt, and Alan L. Kraus, eds.:
The Biology of the Laboratory Rabbit, 1974

Joseph E. Wagner and Patrick J. Manning, eds.:
The Biology of the Guinea Pig, 1976

Edwin J. Andrews, Billy C. Ward, and Norman H. Altman, eds.:
Spontaneous Animal Models of Human Disease, Volume I, 1979; Volume II, 1979

Henry J. Baker, J. Russell Lindsey, and Steven H. Weisbroth, eds.:
The Laboratory Rat, Volume I: Biology and Diseases, 1979; Volume II: Research Applications, 1980

Henry L. Foster, J. David Small, and James G. Fox, eds.:
The Mouse in Biomedical Research, Volume I: History, Genetics and Wild Mice, 1981; Volume II: Diseases, 1982

In preparation

Henry L. Foster, J. David Small, and James G. Fox, eds.:
The Mouse in Biomedical Research, Volume III: Husbandry

Henry L. Foster, J. David Small, and James G. Fox, eds.:
The Mouse in Biomedical Research, Volume IV: Experimental Biology

The Mouse in Biomedical Research

Volume II
Diseases

EDITED BY

Henry L. Foster
The Charles River Laboratories, Inc.
Wilmington, Massachusetts

J. David Small
Veterinary Resources Branch
Small Animal Section
National Institutes of Health
Bethesda, Maryland

James G. Fox
Division of Comparative Medicine
Massachusetts Institute of Technology
Cambridge, Massachusetts

ACADEMIC PRESS 1982
A SUBSIDIARY OF HARCOURT BRACE JOVANOVICH, PUBLISHERS
New York London
Paris San Diego San Francisco São Paulo Sydney Tokyo Toronto

ACADEMIC PRESS, INC.
111 Fifth Avenue, New York, New York 10003

United Kingdom Edition published by
ACADEMIC PRESS, INC. (LONDON) LTD.
24/28 Oval Road, London NW1 7DX

Library of Congress Cataloging in Publication Data
Main entry under title:

The Mouse in biomedical research.

(American College of Laboratory Animal Medicine series)
Includes index.
Contents: v. 1. History genetics, and wild mice -- v. 2. Diseases.
1. Mice as laboratory animals. I. Foster, Henry L. II. Small, J. David. III. Fox, James G. IV. Series. [DNLM: 1. Mice. 2. Research. 3. Animals Laboratory. QY 60.R6 M932]
QL737.R638M68 619'.93 80-70669
ISBN 0-12-262502-1 (v.2) AACR2

PRINTED IN THE UNITED STATES OF AMERICA

82 83 84 85 9 8 7 6 5 4 3 2 1

Contents

List of Contributors ix

Preface xi

Introduction xiii

List of Reviewers for Chapters in This Volume xv

Chapter 1 Bacterial and Mycotic Diseases of the Digestive System
James R. Ganaway
- I. Introduction 1
- II. Bacterial Infections 2
- III. Mycotic Infections 13
- IV. Summary and Concluding Remarks 13
- References 14
- Related References 18

Chapter 2 Mycoplasmal and Other Bacterial Diseases of the Respiratory System
J. Russell Lindsey, Gail H. Cassell, and Maureen K. Davidson
- I. Introduction 21
- II. Murine Respiratory Mycoplasmosis (MRM) 22
- III. Klebsiellosis 33
- IV. Pasteurellosis 33
- V. Chlamydial Pneumonitis 35
- VI. Corynebacteriosis 36
- VII. Miscellaneous Infections 36
- VIII. Conclusions 37
- References 37

Chapter 3 Bacterial, Mycoplasmal, Mycotic, and Immune-Mediated Diseases of the Urogenital System
Harold W. Casey and George W. Irving III
- I. Introduction 43
- II. Bacterial Diseases 44
- III. Mycoplasmal Infections 49
- IV. Mycotic Infections 49
- V. Glomerulonephritis 50
- References 52

Chapter 4 Bacterial and Mycotic Diseases of the Integumentary System
Cynthia Besch Williford and Joseph E. Wagner
- I. Introduction 55
- II. Hairy Integument 56
- III. Bacterial Diseases 57
- IV. Mycotic Diseases 65
- V. Noninfectious Skin Disorders 68
- References 72

Chapter 5 Bacterial, Mycoplasmal, and Mycotic Diseases of the Central Nervous System
Dennis F. Kohn
- I. Introduction 77
- II. *Mycoplasma neurolyticum* 78
- III. *Mycoplasma pulmonis* 79
- IV. *Pseudomonas aeruginosa* 79
- V. *Corynebacterium kutscheri* 80
- VI. *Pasteurella pneumotropica* 80
- VII. Mycoses 80
- References 80

Chapter 6 Bacterial, Mycoplasmal, and Mycotic Diseases of the Lymphoreticular, Musculoskeletal, Cardiovascular, and Endocrine Systems
John E. Harkness and Frederick G. Ferguson
- I. Introduction 83
- II. Bacterial Infections 84
- III. Mycoplasmal Infections 93
- IV. Mycotic Infections 93
- References 94

Chapter 7 Rickettsial and Chlamydial Diseases
Paul K. Hildebrandt
I. Introduction 99
II. Rickettsial Infections 99
III. Chlamydial Infections 105
References 106

Chapter 8 Viral Diseases of the Respiratory System
John C. Parker and Conrad B. Richter
I. Introduction 110
II. Sendai Virus 110
III. Pneumonia Virus of Mice (PVM) 134
IV. K Virus 142
Addendum 150
References 150

Chapter 9 Viral Diseases of the Digestive System
Lisbeth M. Kraft
I. Introduction 159
II. Epizootic (Epidemic) Diarrhea of Infant Mice (EDIM), Mouse Rotavirus Enteritis 160
III. Reovirus 3 Infection (Hepatoencephalomyelitis, ECHO 10 Virus Infection) 167
IV. Murine (Mouse) Hepatitis Virus Infection (MHV) 173
References 183

Chapter 10 Lactate Dehydrogenase-Elevating Virus
Margo A. Brinton
I. Introduction 194
II. Characteristics of Infection 194
III. Methods for Detecting the Presence of LDV 200
IV. Replication in Tissue Culture 200
V. Properties of the Virus 203
VI. Conclusion 205
References 205

Chapter 11 Mousepox
Frank Fenner
I. Introduction 209
II. Viral Agent 210
III. Summary: The Practical Problems 227
References 228

Chapter 12 Lymphocytic Choriomeningitis Virus
Fritz Lehmann-Grube
I. Introduction 231
II. Definitions 233
III. Properties of the Agent 233
IV. Infection of Mice 237
V. Epizootiology 254
VI. Diagnosis 256
VII. Control and Prevention 256
References 258

Chapter 13 Cytomegalovirus and Other Herpesviruses
June E. Osborn
I. Introduction 267
II. Murine Cytomegalovirus 268
III. Mouse Thymic Virus 288
References 289

Chapter 14 Polyomavirus
Bernice E. Eddy
I. Introduction 293
II. History 293
III. Related Viruses 295
IV. The Virus Particle 295
V. Resistance of the Virus or Its DNA to Physical and Chemical Agents 297
VI. Cultivation 298
VII. Infection of Mice and Other Animals 300
VIII. Spread of Polyomavirus and Methods for Its Prevention 304
IX. Conclusions 305
References 305

Chapter 15 Minute Virus of Mice
David C. Ward and Peter J. Tattersall
I. Introduction 313
II. Biology of MVM 314
III. Structure and Replication of MVM 322
References 333

Chapter 16 Mouse Adenovirus
James A. Otten and Raymond W. Tennant
I. Introduction 335
II. Isolations of Mouse Adenoviruses and Pathological Effects in Mice 336
III. Physical Characteristics 337
IV. Serological Reactions 338
V. Epidemiology and Prevalence 338
References 339

Chapter 17 Mouse Encephalomyelitis Virus
Wilbur G. Downs
I. Introduction 341
II. Viral Agent 341
References 350

Chapter 18 Encephalomyocarditis Virus
Thomas G. Murnane
I. Introduction 353
II. History 353
III. Etiology 354
IV. Epidemiology 354
V. Laboratory Animal Hosts 355
References 356

Chapter 19 Protozoa
Chao-Kuang Hsu
I. Introduction 359
II. Parasites of Parenteral Systems 359
III. Parasites of the Alimentary System 366
References 370

Chapter 20 Helminths
Richard B. Wescott
I. Introduction 373
II. Helminths of Major Importance 374
III. Helminths of Minor Importance 379
References 382

Chapter 21 Arthropods
Steven H. Weisbroth
I. Introduction 385
II. Ectoparasitic Ecology 387
III. Diagnosis 387
IV. Identification 388
V. Arthropod Parasities of Laboratory Mice 388
VI. Treatment, Control, and Prevention 398
References 400

Chapter 22 Zoonoses and Other Human Health Hazards
James G. Fox and James B. Brayton
I. Introduction 404
II. Viral Diseases 404
III. Rickettsial Diseases 406
IV. Bacterial Diseases 407
V. Mycoses (Ringworm) 411
VI. Protozoan Diseases (*Entamoeba coli*) 413
VII. Helminth Diseases 413
VIII. Arthropod Infestations 414
IX. Bites 416
X. Allergic Sensitivities 416
XI. Conclusion 419
References 419

Chapter 23 Selected Nonneoplastic Diseases
J. D. Burek, J. A. Molello, and S. D. Warner
I. Introduction 425
II. Occurrence 426
References 438

Index 441

List of Contributors

Numbers in parentheses indicate the pages on which the authors' contributions begin.

James B. Brayton (403), Division of Comparative Medicine, The Johns Hopkins University School of Medicine, Baltimore, Maryland 21205

Margo A. Brinton (193), The Wistar Institute, Philadelphia, Pennsylvania 19104

J. D. Burek (425), Toxicology Department, Health and Consumer Products, Dow Chemical Company, Indianapolis, Indiana 46268

Harold W. Casey[1] (43), Department of Veterinary Pathology, Armed Forces Institute of Pathology, Washington, D.C. 20306

Gail H. Cassell (21), Departments of Comparative Medicine and of Microbiology, Schools of Medicine and Dentistry, University of Alabama in Birmingham, Birmingham, Alabama 35294

Maureen K. Davidson (21), Department of Comparative Medicine, Schools of Medicine and Dentistry, University of Alabama in Birmingham, and the Veterans Administration Hospital, Birmingham, Alabama 35294

Wilbur G. Downs (341), Yale Arbovirus Research Unit, Yale University School of Medicine, New Haven, Connecticut 06510

Bernice E. Eddy[2] (293), Experimental Virology Branch, Division of Virology, Bureau of Biologics, Food and Drug Administration, Bethesda, Maryland 20034

Frank Fenner (209), John Curtin School of Medical Research, Australian National University, Canberra, Australia

Frederick G. Ferguson (83), Laboratory Animal Resources, The Pennsylvania State University, University Park, Pennsylvania 16802

James G. Fox (403), Division of Comparative Medicine, Massachusetts Institute of Technology, Cambridge, Massachusetts 02139

James R. Ganaway (1), Comparative Pathology Section, Veterinary Resources Branch, DRS, National Institutes of Health, Bethesda, Maryland 20205

John E. Harkness (83), Laboratory Animal Resources, The Pennsylvania State University, University Park, Pennsylvania 16802

Paul K. Hildebrandt (99), Tracor Jitco, Rockville, Maryland 20852

Chao-Kuang Hsu (359), School of Medicine, University of Maryland, Baltimore, Maryland 21201

George W. Irving III[3] (43), Department of Veterinary Pathology, Armed Forces Institute of Pathology, Washington, D.C. 20306

Dennis F. Kohn (77), Department of Comparative Medicine, University of Texas Medical School, Houston, Texas 77025

Lisbeth M. Kraft (159), National Aeronautics and Space Administration, Ames Research Center, Moffett Field, California 94035

Fritz Lehmann-Grube (231), Heinrich-Pette-Institut für Experimentelle Virologie und Immunologie and der Universität Hamburg, 2000 Hamburg 20, Federal Republic of Germany

J. Russell Lindsey (21), Department of Comparative Medicine, Schools of Medicine and Dentistry, University of Alabama in Birmingham, and the Veterans Administration Hospital, Birmingham, Alabama 35294

[1]Present address: Department of Pathology, Whittaker Toxigenics, Inc., Decatur, Illinois 62526

[2]Present address: 6722 Selkirk Court, Bethesda, Maryland 20817

[3]Present address: Directorate of Life Sciences, Air Force Office of Scientific Research, Bolling AFB, Washington, D.C. 20332.

J. A. Molello (425), Toxicology Department, Health and Consumer Products, Dow Chemical Company, Indianapolis, Indiana 46268

Thomas G. Murnane[4] (353), U.S. Army Veterinary Corps, Office of the Surgeon General, Washington, D.C. 20310

June E. Osborn (267), Departments of Medical Microbiology and of Pediatrics, University of Wisconsin, Madison, Wisconsin 53706

James A. Otten (335), Biology Division, Oak Ridge National Laboratory, Oak Ridge, Tennessee 37830

John C. Parker (109), Microbiological Associates, Bethesda, Maryland 20816

Conrad B. Richter[5] (109), Oak Ridge Associated Universities, Inc., Oak Ridge, Tennessee 37830

Peter J. Tattersall (313), Department of Human Genetics, Yale University School of Medicine, New Haven, Connecticut 06510

Raymond W. Tennant[6] (335), National Institute of Environmental Health Sciences, Research Triangle Park, North Carolina 27709

Joseph E. Wagner (55), College of Veterinary Medicine, University of Missouri, Columbia, Missouri 65201

David C. Ward (313), Department of Human Genetics, Yale University School of Medicine, New Haven, Connecticut 06510

S. D. Warner (425), Toxicology Department, Health and Consumer Products Department, Dow Chemical Company, Indianapolis, Indiana 46268

Steven H. Weisbroth (385), AnMed Laboratories, Inc., New Hyde Park, New York 11040

Richard B. Wescott (373), Department of Veterinary Microbiology and Pathology, Washington State University, Pullman, Washington 99164

Cynthia Besch Williford (55), College of Veterinary Medicine, University of Missouri, Columbia, Missouri 65201

[4]Present address: Apartado Postal 61-148, Mexico 6, D.F.

[5]Present address: Comparative Medicine Branch, National Institute of Environmental Health Sciences, Research Triangle Park, North Carolina 27709

[6]Present address: Cellular and Genetic Toxicology Branch, National Toxicology Program, P.O. Box 12233, Research Triangle Park, North Carolina 27709

Preface

The American College of Laboratory Animal Medicine (ACLAM) was formed in 1957 in response to the need for specialists in laboratory animal medicine. The College has promoted high standards for laboratory animal medicine by providing a structure framework to achieve certification for professional competency and by stressing the need for scientific inquiry and exchange via progressive continuing education programs. The multivolume treatise, "The Mouse in Biomedical Research," is a part of the College's effort to fulfill those goals. It is one of a series of comprehensive texts on laboratory animals developed by ACLAM over the past decade: "The Biology of the Laboratory Rabbit" was published in 1974, "The Biology of the Guinea Pig" in 1976, and a two-volume work "Biology of the Laboratory Rat" in 1979 and 1980. Also, in 1979 the College published a two-volume text on "Spontaneous Animal Models of Human Disease."

The annual use of approximately 50 million mice worldwide attests to the importance of the mouse in experimental research. In no other species of animal has such a wealth of experimental data been utilized for scientific pursuits. Knowledge of the mouse that has been accumulated is, for the most part, scattered throughout a multitude of journals, monographs, and symposia. It has been fifteen years since the publication of the second edition of "The Biology of the Laboratory Mouse" edited by E. L. Green and the scientific staff of the Jackson Laboratories. It is not the intent of this work simply to update and duplicate this earlier effort, but to build upon its framework. We are indeed fortunate to have Dr. Green and many of his colleagues at the Jackson Laboratory as contributors to this treatise. It is the intended purpose of this text to assemble established scientific data emphasizing recent information on the biology and use of the laboratory mouse. Separation of the material into multiple volumes was essential because of the number of subject areas covered.

The contents of Volume I are presented in fourteen chapters and provide information on taxonomy, nomenclature, breeding systems, and a historical perspective on the development and origins of the laboratory and wild mouse. Six chapters deal specifically with the ever-increasing diversity of inbred strains of mice, including coverage of methods of developing and the genetic monitoring and testing of these strains. The emphasis of this volume on genetics is also manifested by chapters discussing the *H-2* complex, cytogenetics, radiation genetics, and pharmacogenetics.

Because of the impact of spontaneous diseases on interpretation of, and potential for, complicating experimental research, it is of paramount importance for investigators to recognize these diseases and their effect on the mouse. Volume II, for the first time, compiles in one format a narrative detailing infectious diseases of the mouse; the chapters cover bacterial, mycotic, viral, protozoal, rickettsial, and parasitic diseases. Also, nonneoplastic and metabolic diseases are covered as well as the topic of zoonoses.

Volume III provides comprehensive coverage of selected material related to normative biology and management and care of the laboratory mouse. Developmental, anatomical, nutritional, physiological, and biochemical parameters of the mouse are compiled in several chapters and will be of great interest and an important resource for normal biological profiles. A review of the histologic features was not included because of space constraints and the availability of this information in previous texts. Environmental monitoring and disease surveillance as well as management and design of animal facilities will be particularly useful for those individuals responsible for the management of mouse colonies. The chapters on gnotobiotics and gastrointestinal flora represent the state of the art in gnotobiology. The three chapters on selected aspects of immunology in the mouse serve to highlight the explosive progress being made in immunologic techniques and instrumentation and the underlying importance of genetic differentiation.

The fourth volume includes selected applications of the mouse in research. Several chapters discuss the use of the mouse in infectious disease research, while others range from eye research to the use of the mouse in experimental embryology. The chapters devoted to the use of the mouse in oncological research follow a body system format. Research topics in other disciplines have not been included, but hopefully will be included in future editions.

This treatise was conceived with the intent to offer information suitable to a wide cross section of the scientific community. It is hoped that it will serve as a standard reference source. Students embarking on scientific careers will benefit from the broad coverage of material presented in compendia format. Certainly, specialists in laboratory animal science will benefit from these volumes; technicians in both animal care and research will find topics on surgical techniques, management, and environmental monitoring of particular value.

The editors wish to extend special appreciation to the contributors to these volumes. Authors were selected because of knowledge and expertise in their respective fields. Each individual contributed his or her time, expertise, and considerable effort to compile this resource treatise. In addition, the contributors and editors of this book, as with all volumes of the ACLAM series texts, have donated publication royalties to the American College of Laboratory Animal Medicine for the purpose of continuing education in laboratory animal science. This book could not have been completed without the full support and resources of the editors' parent institutions which allowed time and freedom to assemble this text. A special thanks is also extended to the numerous reviewers of the edited work whose suggestions helped the authors and editors present the material in a meaningful and concise manner. We acknowledge and thank Rosanne Brown and Sara Spanos for their secretarial assistance. Also, the assistance provided to us by the staff of Academic Press was greatly appreciated.

Finally, we especially acknowledge with deep appreciation the editorial assistance of Patricia Bergenheim, whose dedication and tireless commitment to this project were of immeasurable benefit to the editors in the completion of this text.

Henry L. Foster
J. David Small
James G. Fox

Introduction

Although the past several decades have seen a dramatic change in the infectious disease experience of laboratory mice, with marked improvement in animal health and quality of research results, a detailed compilation of murine infectious diseases will be of great value to users of laboratory mice. With regard to my own experience in virus research, thirty years ago it was not unusual to receive shipments from commercial sources of Swiss mice decimated by *Salmonella typhimurium*. The impact of such shipments on an experimental program and on one's own mouse stocks is easily envisioned. Ten years later *Salmonella* was no longer a problem, but when we began serologic screening of mouse colonies for indigenous viruses we found that some colonies, both commercial and institutional, contained virtually every known mouse virus while other colonies were free of almost all such agents. Not surprisingly, housing of animals from such diverse sources in common rooms led to serious morbidity in the clean stocks, particularly among infant animals. Contaminated animals have also been a serious source of introduction of extraneous agents into mouse-passage biological materials. Passenger agents in transplanted tumors and virus stocks, once accepted as an unavoidable component of complex *in vivo* systems, have become increasingly disruptive and hence unacceptable as studies of tumor biology have become more subtle and precise. Another unsavory by-product of using infested mice was the mistaken identification of a mouse virus, acquired during passage of a specimen through mouse tissues *in vivo* or even *in vitro,* as the etiologic agent of the disease being studied. The literature of the 1940s and 1950s is replete with such mistaken claims.

In recent years subclinical infections with Sendai and minute viruses have been found to disrupt sensitive assays for various parameters of immune functions, and thus constitute serious threats to immunological research programs.

Several events combined to bring about the vastly improved infectious disease experience with today's mice. Most important was the development and application of techniques for deriving stocks by cesarean section and rearing them in strictly maintained quarantine. This technology was made possible by the husbandry techniques and barrier procedures developed by the pioneers of germfree techniques, and much credit is due to them as well as to the few farsighted commercial mouse breeders who led the way in their application. A second important factor was the development and widespread application of serologic monitoring procedures for indigenous agents, combined with increased knowledge of their natural history. Definition of those agents present in a mouse colony is in many respects more important than elimination of every agent of minimal pathogenicity. Further, awareness of the problems and the wide availability of clean, defined mice have contributed greatly to improving the quality and standards of laboratory mice.

It is of the greatest importance to recognize that elimination of agents from mouse colonies *increases,* rather than reduces, the need to be informed and aware of the agents. In their natural occurrence in breeding colonies, most viruses of mice cause little disease; they infect young animals passively protected by maternal antibody, producing subclinical infections that provide immune protection for the infected animal as well as its offspring. In the absence of indigenous immunizing infection, a number of agents can induce epizootic disease with high morbidity and mortality rates. Our clean colonies are thus highly vulnerable, particularly to ectromelia, hepatitis, and Sendai virus infections. Awareness, quarantine procedures, monitoring programs, and knowledge of the natural biology of indigenous murine agents will long be required of all users of mice.

Another area in which knowledge of indigenous infections of

mice is of major importance to infectious disease research is in their value as model systems that represent naturally occurring host–parasite relationships. Antimicrobial defense mechanisms, the interplay of infectious agents with the immune system, and mechanisms of pathogenesis of various infectious diseases can be studied in mouse systems with a scale and preciseness unmatched in other host systems. The importance of naturally occurring mouse infections as models is illustrated by the fact that the original discoveries of an astounding number of major virus groups were made with the members that occur in mice. Polyomaviruses (polyoma- and K viruses), coronaviruses (mouse hepatitis), rotaviruses (epidemic diarrhea of infant mice), parainfluenza viruses (Sendai), cytomegaloviruses (MSGV), B-type retroviruses (mammary tumor virus), arenaviruses (LCM), pneumovirus (PVM), and picornaviruses (Theiler's) all fall into this category. In addition, mice carry viruses belonging to at least five other major families (C-type retrovirus, reo-, adeno-, pox-, and parvoviruses). This diversity of viral flora of the mouse provides natural models for a wide spectrum of viral diseases.

The surgical derivation and barrier-rearing techniques have also drammatically reduced bacterial and parasitic infections. However, they remain important problems, particularly in conventional colonies containing large numbers of mice from mixed sources. Mycoplasmal infection remains a common, serious health problem, and hyperplastic colitis due to *Citrobacter freundii* can also seriously threaten a colony. Pinworms and ectoparasites are by no means a thing of the past.

The detailed information on infectious agents of the mouse which has been assembled in this volume is a unique compilation that should be of much use in dealing with the practical problems of animal care and in exploiting the rich opportunities that mouse models offer for the study of the biology of infectious disease.

Retroviruses—the murine leukemia viruses, mammary tumor viruses, and A particles—are among the major model systems in viral carcinogenesis. The complexity and experimental and biochemical detail of these agents would require a book of their own. Several review books on this subject have appeared in the past few years and another major volume is in preparation.

Wallace P. Rowe
Laboratory of Viral Diseases
National Institute of Allergy and Infectious Diseases
Bethesda, Maryland 20014

List of Reviewers for Chapters in This Volume

Baker, Henry J.	University of Alabama
Barthold, Stephen W.	Yale University School of Medicine
Brennan, Patricia C.	Argonne National Laboratory
Broderson, J. Roger	Center for Disease Control
Cassell, Gail H.	University of Alabama
Cole, Gerald A.	The Johns Hopkins University
Craighead, John E.	University of Vermont
Dawe, Clyde J.	National Institutes of Health
Flynn, Robert J.	Argonne National Laboratory
Ganaway, James R.	National Institutes of Health
Garner, F. M.	Rockville, Maryland
Hoggan, M. David	National Institutes of Health
Hollander, Carel F.	Institute for Experimental Gerontology TNO
Hotchin, John	Department of Health, State of New York
Kilham, Lawrence	Dartmouth Medical School
Kraft, Lisbeth M.	National Aeronautics and Space Administration
Levine, Norman D.	University of Illinois
Lipton, Howard	Northwestern University
Marcus, Leonard	State Diagnostic Laboratory, Massachusetts
Medearis, Donald N.	Massachusetts General Hospital
Migaki, George	Armed Forces Institute of Pathology
Mims, Cedric A.	Guy's Hospital Medical School
Notkins, Abner L.	National Institutes of Health
Parker, John C.	Microbiological Associates, Inc.
Plagemann, Peter G. W.	University of Minnesota
Quimby, Fred R.	Cornell University Medical College
Richter, Conrad B.	Oak Ridge Associated Universities
Robinson, David M.	Walter Reed Army Institute of Research
Rowe, Wallace	National Institutes of Health
Tully, Joseph G.	National Institutes of Health
Weisbroth, Steven H.	AnMed Laboratories
Woode, G. N.	Iowa State University
Yunker, Conrad E.	National Institutes of Health

Chapter 1

Bacterial and Mycotic Diseases of the Digestive System

James R. Ganaway

I. Introduction 1
II. Bacterial Infections 2
A. *Salmonella* 2
B. *Bacillus* 7
C. *Citrobacter* 9
D. *Pseudomonas* 11
E. *Escherichia* 12
F. *Clostridium* 12
G. *Klebsiella* 13
III. Mycotic Infections 13
IV. Summary and Concluding Remarks 13
References 14
Related References 18

I. INTRODUCTION

The determination of which microbial agents to include under the broad subject ''Bacterial and mycotic diseases of the digestive tract of the mouse'' was not easily resolved. There are probably very few microbial pathogens of man or lower animals that colonize and cause pathology in the digestive tract *only*. Furthermore, it appears untenable to consider ''possible portal of entry'' as the criterion because the gastrointestinal tract is exposed to an infinite variety of microorganisms. Indeed, the microbiology of the gastrointestinal tract of conventional animals is extremely complex, so much so that its true character is yet to be determined. The present coverage, with the concurrence of the Editorial Committee, is

ISBN 0-12-262502-1

an attempt to include recognized pathogens which are known to colonize the gastrointestinal tract of mice.

The biological characterization and methods for identification of recognized microbial agents are not necessarily included, as these can be readily found in numerous textbooks and "Bergey's Manual of Determinative Bacteriology." Only selected references are cited in the text. A list of related references is also included.

It is hoped that constructive and critical review will impress the reader not with how much we know, but with how little we know. New findings will help to fill this void and provide a basis for correction of existing interpretations.

II. BACTERIAL INFECTIONS

A. *Salmonella*

Salmonellosis of mice is one of the most studied diseases of animals. Historically, the mouse has been used in greater numbers than any other laboratory animal, due, in part, to the comparative low cost and the ease of production and maintenance in the laboratory. Spontaneous salmonellosis in mice has been a universal problem in conventionally maintained mouse production colonies. The disease in mice has been studied experimentally because it is an excellent model of a similar disease in man, typhoid fever. Consequently, the pathogenesis and mechanism of acquired immunity have received much attention. The mouse is also used for testing the potency of vaccines destined for use in the protection of man against typhoid fever. Indeed, this enteric and progressive systemic disease in mice is seemingly so well understood that it has been used as a model to study herd immunity (Greenwood *et al.*, 1931) and the effects of nutrition (Schneider, 1956) and heredity (Gowen, 1948) upon resistance to disease.

The increasing sophistication of biomedical research has stimulated a demand for mice (and other laboratory animals) that are free of pathogenic microbiota. This is because latent or subclinical infections are often activated by experimental stresses, such as irradiation, and by the administration of corticosteroid and cytotoxic drugs and anti-lymphocyte serum. Intercurrent disease can complicate or nullify the results of experiments. Because of the recognized futility of eradicating *Salmonella* spp. (as well as a host of other agents) from conventionally maintained populations of infected mice (discussed in more detail later), a new approach was clearly needed. The cesarean derivation and production of mice under strict barrier-maintained conditions is a fairly recent and highly successful means of providing mice free of most indigenous pathogenic microbiota, including salmonellas. The disease-free state of such mice can be extended throughout the conduct of research, but only if adequate protection of the environment is maintained so as to exclude the introduction of contaminants. One must also constantly be aware of that ever-present source, the human factor. This is especially true with a zoonotic disease such as salmonellosis. The success of the cesarean-derived, barrier-maintained mouse colony is so widely acclaimed and the technology so universally practiced that there is a tendency to suggest that salmonellosis of mice no longer poses a problem. This suggestion gains favor when one considers the fact that a ready and reliable means of detection of the infection is available, and reports of spontaneous salmonellosis in mice have diminished in recent years. On the other hand, it is unrealistic to rely upon the frequency of reported occurrences in the literature as a basis for the assessment of the prevalence of a disease so common and well understood as salmonellosis. Speculation aside, the prevalence of salmonellosis in mice used in biomedical research in the United States today remains unknown. Certainly, the salmonellas are ubiquitous in nature and are encountered in research laboratories, where many species of animals, many of which are potential carriers (Fox and Beaucage, 1979), are maintained under conventional colony conditions. The only means of making this determination is to monitor for the presence of *Salmonella* spp. by isolation and identification in the laboratory.

The classification of *Salmonella* is complex and, scientifically, the present methods of nomenclature are unsatisfactory (LeMinor and Rohde, 1974). There are approximately 1600 recognized serotypes (Smith, 1977). A recent recommendation (Edwards and Ewing, 1972) would reduce the number of recognized species to three: *Salmonella cholerae-suis, S. typhi,* and *S. enteritidis*. All other previously recognized species are designated as serotypes of *S. enteritidis*. Accordingly, the most commonly isolated salmonella from mice is *S. enteritidis* ser Typhimurium (Edwards *et al.*, 1948); other reported serotypes are Poona (Franklin and Richter, 1968), Paratyphi A, Infantis, Montevideo, Oranienburg, Blockley, California, Anatum (Margard and Litchfield, 1963), Tennessee, Senftenberg (Haberman and Williams, 1958), Binza, and Bredeney (Wetmore and Hoag, 1960). Salmonellae are gram-negative, nonspore-forming, usually motile rods which do not hydrolyze urea and usually do not ferment lactose. "Bergey's Manual of Determinative Bacteriology" should be consulted for common biochemical characters. Selective media of value in isolation of salmonellae for diagnostic purposes will be discussed later.

Salmonellae are primarily intestinal parasites of vertebrates and are frequently isolated from sewage, river and sea water, and certain foods. They can survive for varying lengths of time under such conditions, but it is doubtful if they can exist indefinitely in any environment outside the animal body (Wilson and Miles, 1975). Contaminated feed and bedding are considered the usual sources of infection for mice (Carlton and Hunt, 1978; Haberman and Williams, 1958). In a study in the United

Kingdom (Stott *et al.*, 1975), *Salmonella* spp. were found in 19% of the animal feed samples tested and were associated with meat and bone meal. It was found that pelleting of the feed resulted in a 1000-fold reduction in the numbers of *Enterobacteriaceae*. Commercially available laboratory animal feeds in the United States are normally pelleted and are pasteurized or sterilized by autoclaving when fed to barrier-maintained mice. Though the usual route of infection is by ingestion, experimentally the conjunctival route is more effective than the oral route, requiring fewer organisms to establish an infection (Darlow *et al.*, 1961; Tannock and Smith, 1971; Bate and James, 1958). In the research setting, where many species of animals from varied sources are commonly maintained using conventional husbandry practice, the carrier animal is a likely source of infection to other animals in the area. In the closed or barrier-maintained colony, where food and bedding are autoclaved and vermin, birds, and feral animals are excluded, the human carrier should be considered as a possible source of infection.

Many factors strongly influence the induction and course of naturally occurring and experimentally induced *Salmonella* infection in mice: virulence, route of infection, and dose of the *Salmonella* organism; age, sex, and inheritance factors of mice which favor resistance; nutrition of the mouse; intercurrent disease, both naturally occurring and experimentally induced; other experimental stresses which suppress immunity, such as irradiation and corticosteroid drug administration; environmental factors, such as temperature fluctuation; alteration of normal gastrointestinal motility, caused by withholding food or by morphine administration; and alteration of normal gastrointestinal microflora, e.g., by the oral administration of antibiotics. Some of these factors have been studied extensively and were worthy of further comment.

Strains of *Salmonella* differ in their virulence for mice. *Salmonella gallinarium,* though related antigenically to *S. enteritidis,* is almost avirulent for the mouse (Collins *et al.*, 1966). Carter and Collins (1974a) studied five serotypes of *Salmonella* in three strains of mice and found that *S. paratyphi* A, *S. paratyphi* B, and *S. typhi* grew very poorly in mice following intravenous inoculation even after 20 serial passages. On the other hand, *S. paratyphi* C and an *S. typhi-typhimurium* hybrid produced progressive systemic infection in C57BL mice with a mean lethal dose of < 10 organisms. Tannock *et al.* (1975) have shown that mutant strains as well as the wild type of *S. typhimurium* can associate with and invade the intestinal mucosa of infected gnotobiotic mice; neither O antigen, flagella, nor pili appeared to be essential for the association with the mucosal surface of the mouse ileum. Jenkin (1962) and Bohme (1970) suggest that susceptibility of the mouse to *S. typhimurium* infection may be affected by an antigenic relationship between host and parasite. The shared antigen(s) appears to be absent or is masked in avirulent strains, whereas the virulent strain shares antigen(s) with the host, which is therefore unable to produce antibody because of "self" recognition. Others (Furness and Ferreira, 1959; Mackenzie *et al.*, 1940; Pike and Mackenzie, 1940) suggest that virulence of a strain is dependent upon the ability of the salmonella to multiply intracellularly; virulent strains do so, whereas avirulent strains fail to do so.

When considering that the natural route of infection in mice is by ingestion, it is noteworthy that a dose $\geq 10^6$ virulent *S. enteritidis* organisms is required to infect mice which have a normal gastrointestinal microflora (Miller and Bohnhoff, 1962; Collins and Carter, 1978). Experimentally, the LD_{50} of virulent *S. enteritidis* for conventional TRU:ICR mice was 2.5 $\times 10^3$ by the intraperitoneal route, 1×10^4 by the intravenous route, and 5×10^6 by the oral route; the comparable LD_{50} values for germ-free mice by similar routes were 4×10^3, 2×10^3, and 3-5, respectively (Collins and Carter, 1978). The very small number of organisms required to initiate a lethal infection in germ-free mice by the oral route is in sharp contrast to that required in the conventional counterpart and supports the suggestion of others (Miller and Bohnhoff, 1963, Margard *et al.*, 1963; Savage, 1972) that the normal intestinal microflora exerts an inhibitory effect upon the establishment of *Salmonella* infection by the oral route.

Inheritance markedly affects the resistance of mice to salmonellosis (Webster, 1937; Gowen, 1948, 1960; Collins, 1972; Bohme, 1970; Bohme *et al.*, 1959; Darlow *et al.*, 1961; Groschel *et al.*, 1970; Robson and Vas, 1972; Oakberg, 1946; Plant and Glynn, 1974, 1976). Certain studies (Robson and Vas, 1972; Plant and Glynn, 1976) indicate that BALB/c and C57BL mice are very susceptible, whereas A/J mice are quite resistant. Recognition of this inherited resistance or susceptibility of strains of mice presents a problem in the critical interpretation of the literature because many of the earlier studies used noninbred Swiss white mice, which behave similarly to the A/J strain in native and vaccine-elicited resistance to infection with *S. typhimurium* (Robson and Vas, 1972; Collins, 1972). Inherited resistance factors should not be confused with acquired immunity, as Gowen (1948) demonstrated that genetic resistant strains became increasingly resistant when immunized. The nature of inherited resistance to salmonellosis is unclear; it may be associated with glycogen metabolism in the liver (Oakberg, 1946) and/or related to the ability of the mouse to produce a good delayed type hypersensitivity reaction (Plant and Glynn, 1976; Mackaness, 1967). Apart from the factors mentioned above, weanling mice are more susceptible than older mice (Tannock and Smith, 1971), and in a single undesignated strain of mice, females were found to be more susceptible than males (Gowen, 1960). Though commercially prepared feed is readily available, universally used today, and apparently adequate, Schneider and Webster (1945; Schneider, 1956) found a resistance factor in wheat, malted barley, and dried egg white which was dializable, alcohol extractable, and not stored in the body. The resistance factor is neither a vitamin nor an antibiotic but is categorized as a member of a class of ecological ectocrines called *pacifarins* (Schneider, 1967). Nutritional iron deficiency has an attenuating effect upon *S. typhimurium* infection of mice (Puschmann and Ganzoni, 1977), whereas iron overload states appear to promote bacterial growth and inhance the virulence of *S. typhimurium* (Jones *et al.*, 1977).

Gastrointestinal motility appears to have a marked effect upon the establishment of *Salmonella* infection in the intestine (Miller and Bohnhoff, 1962; Carter and Collins, 1974b; Ruitenberg *et al.*, 1971). Miller and Bohnhoff (1962) suggested that the reason large inocula ($\geq 10^6$ virulent *S. enteritidis*) are required to infect orally mice having a normal gastrointestinal microflora is motility of the gastrointestinal tract. Using a dye marker, oral inoculum appeared in the cecum within 30–60 min and in the feces within 6 hr. Decreased resistance of mice was noted when morphine sulfate (250 mg/kg subcutaneously) was given or food was withheld for 24 hr. The ill-defined reduction in host resistance following 24-hr starvation was not related to the presence of food in the stomach at the time of experimental inoculation.

Several studies (Bohnhoff *et al.*, 1954; Miller *et al.*, 1954, 1956; Miller, 1959; Bohnhoff and Miller, 1962; Miller and Bohnhoff, 1963; Ushiba *et al.*, 1955; Meynell, 1955) indicate that alteration of the normal gastrointestinal microflora by oral administration of antibiotics markedly reduces the resistance of mice infected experimentally with *S. enteritidis* by the oral route. As previously mentioned, $\geq 10^6$ virulent *S. enteritidis* by mouth is required to initiate infection in 50% of the mice. When similar mice were given streptomycin (50 mg by mouth) within 24 hr before challenge, < 10 *S. enteritidis* organisms were required (Bohnhoff and Miller, 1962). A similar effect was noted with penicillin and, to a lesser degree, with oxytetracycline and bacitracin. Thus the extremely small inoculum (< 10 organisms) required to initiate infection in germ-free mice by the oral route (Collins and Carter, 1978) is comparable to that required to infect conventional mice following the oral administration of antibiotics and strongly favors an inhibitory role of the normal gastrointestinal microflora of mice in establishment of *Salmonella* infection in the intestine.

Considering the suggested high frequency of salmonellosis in conventionally maintained colonies of mice, it is remarkable that so few descriptions of spontaneous disease and of the epizootiology of salmonellosis in mice appear in the literature. In the absence of evidence to the contrary, it may be that high morbidity and mortality are not observed as frequently in colonies of infected mice as are commonly observed in colonies of infected guinea pigs (Ganaway, 1976). Having examined approximately 26,000 mice from various production colonies in the United States for evidence of *Salmonella* infection, Margard *et al.* (1963) suggested that obvious or clinically apparent salmonellosis was the exception rather than the rule. In an experimental study (Miller and Bohnhoff, 1962) wherein mice received 10^6 virulent *S. enteritidis* intragastrically, less than 3% of the infected mice died (usually within 7–10 days of inoculation) and infected mice (bacteremic) rarely showed signs of illness. According to Rabstein (1958), however, when salmonellosis is enzootic in the mouse production colony, there are periods of quiescence and periods of high mortality, diarrhea, anorexia, loss of weight, roughened hair coat, and small litters, both in number and in size. Conjunctivitis (Carlton and Hunt, 1978) and impaired gluconeogenesis (Moore *et al.*, 1977; Berry and Smythe, 1960) have also been observed.

The incubation period is variable depending upon one or more of the previously mentioned predisposing factors but is usually 3–6 days (Haberman and Williams, 1958). Experimental studies (Carter and Collins, 1974b) indicate that the primary site of penetration of the intestine by *S. enteritidis* administered to mice intragastrically is the distal ileum. Most of the challenge dose (99%) is excreted in the feces; only a few bacteria pass across the mucosa and invade the Peyer's patches. Here they multiply and spread to the draining mesenteric lymph nodes. The bacteria continue to multiply and spread by the lymphatics to other lymph nodes, liver, and spleen; a bacteremia ensues. After further multiplication in the liver, they pass via the bile to the intestine, where further multiplication and reinvasion of the mucosa occur. In chronic infections, bacteria are shed intermittently in the feces for months, resulting in a carrier state. Approximately 5% of such mice become carriers (Margard *et al.*, 1963; Rabstein, 1958).

Lesions seen at necropsy are highly variable depending upon whether the disease is acute or chronic (Carlton and Hunt, 1978; Carter and Collins, 1974b; Jones, 1967; Haberman and Williams, 1958). In acute cases, there may be no obvious lesions; the mucosa of the distal ileum may appear hyperemic, and the lower intestine may be empty or contain a small amount of fluid. If the infection is prolonged, yellowish-white miliary foci may be seen in the liver (Fig. 1), spleen, and lymph nodes, and varying degrees of necrosis and hemorrhage may be seen in the lower intestine. The spleen is usually enlarged (Fig. 1). In experimentally infected, germ-free mice (Collins and Carter, 1978), all mice died within 10 days of a progressive systemic infection involving the lower intestine (severe diarrhea), liver, spleen, lymph nodes, and lung.

Microscopically, lesions are found wherever there is bacterial invasion and colonization; the extent of the lesion is proportional to the course of the disease and the number of bacteria present in the tissue (Collins and Carter, 1978; Carter and Collins, 1974b; Bakken and Vogelsang, 1950; Miller and Bohnhoff, 1962). Necrotic foci may be seen in the intestine, mesenteric lymph nodes, liver (Fig. 2), and spleen. Focal ac-

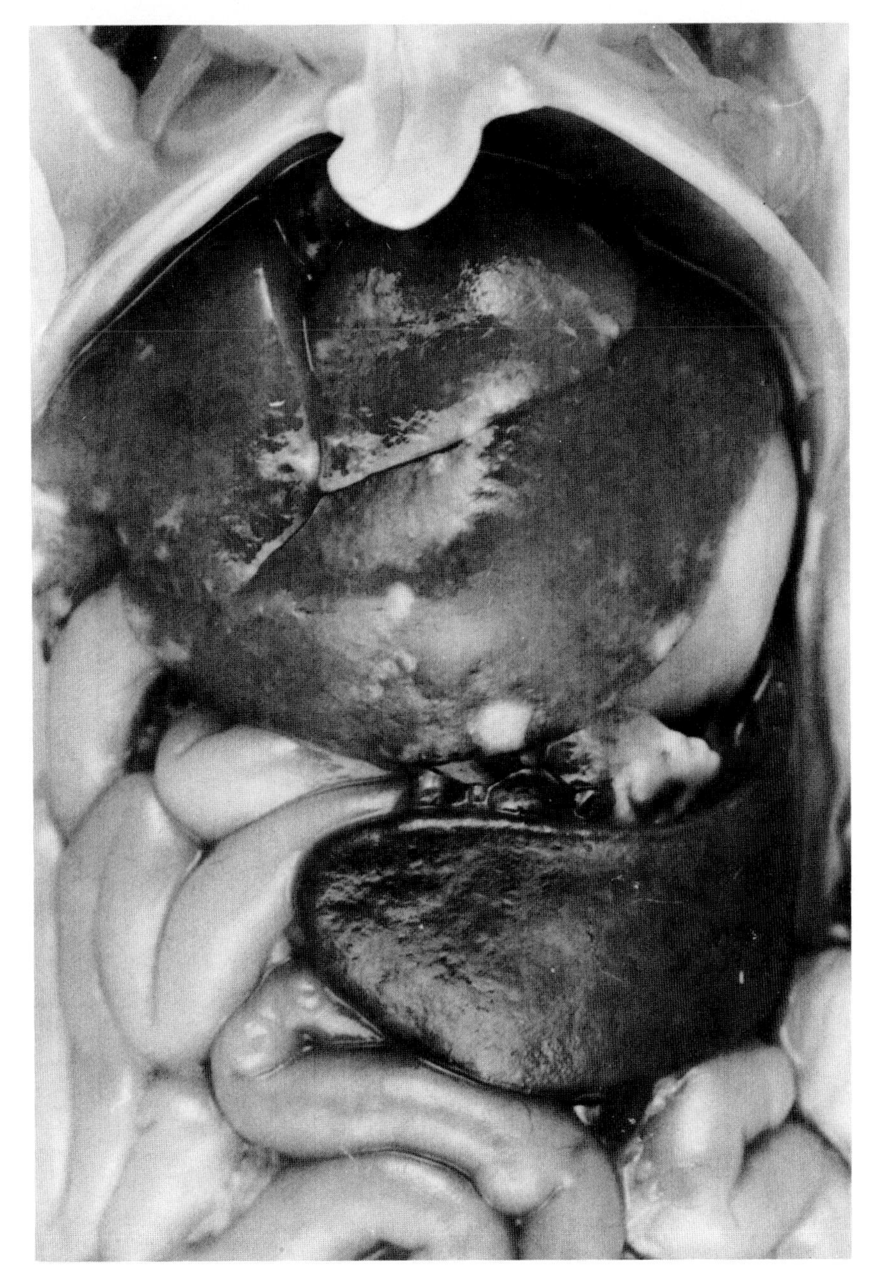

Fig. 1. Outbred Swiss mouse 7 days after peritoneal inoculation with *Salmonella enteritidis*. The spleen is enlarged, and there are small white foci in the liver. (Courtesy of Dr. Alice O'Brien, Uniformed Services University for Health Sciences.)

cumulations of polymorphonuclear leukocytes and histiocytes are found in the lymphoid follicles of the ileum, spleen, and lymph nodes. Reticuloendothelial cell hyperplasia may be so marked in the spleen as to give it a bloodless appearance. In the liver, there is venous thrombosis. Bacteria are commonly present at the periphery of the granulomatous lesions.

Probably the most intensively studied aspect of salmonellosis of mice is that which is concerned with the mechanism of acquired immunity. The principal motivation has been the need to develop an effective and safe vaccine to protect man against systemic *Salmonella* infection, especially *S. typhi,* the cause of typhoid fever. The mouse has been used not only to test the effectiveness of the numerous vaccine preparations but also as a model for studying and understanding the immune response mechanisms. It is beyond the intention of this chapter to review critically these numerous studies. However, no discussion of

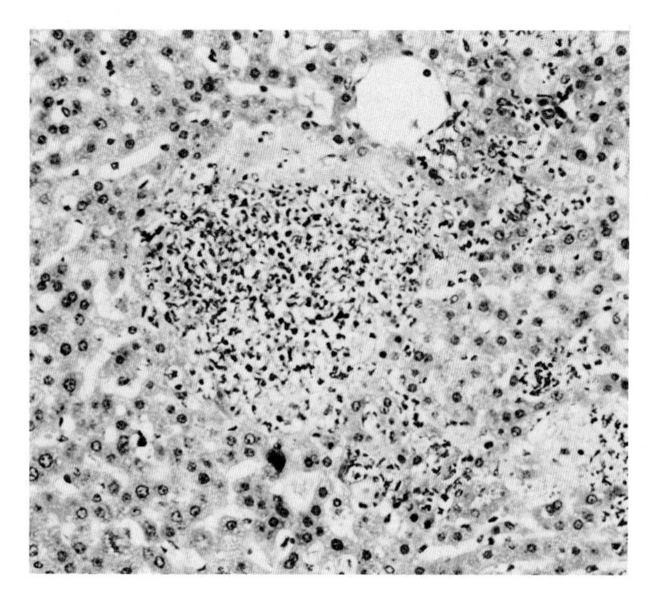

Fig. 2. N.NIH(s) mouse 7 days after intravenous inoculation with *Salmonella enteritidis*. There is focal necrosis of the liver and invasion with polymorphonuclear leukocytes. H&E stain. ×220.

salmonellosis in mice would be complete without mention of acquired immunity. Two schools of thought have been actively pursued: One suggests that killed vaccines do not afford protection and that immunity is mediated by cellular rather than humoral mechanisms (Mackaness *et al.*, 1966; Blanden *et al.*, 1966; Collins *et al.*, 1966; Collins, 1968a,b, 1969; Collins and Mackaness, 1968; Collins and Carter, 1972; Mitsuhashi *et al.*, 1961; Sato *et al.*, 1962; Germanier, 1972; Hobson, 1957a; Macleod, 1954; Howard, 1961; Venneman *et al.*, 1970; Venneman and Berry, 1971a,b); the other suggests that killed vaccines are effective and that opsonic or humoral antibody plays an important role in acquired immunity (Jenkin and Rowley, 1963, 1965; Jenkin *et al.*, 1964; Kenney and Herzberg, 1967, 1968; Turner *et al.*, 1964; Rowley *et al.*, 1964, 1968; Robson and Vas, 1972; Ushiba, 1965; Dimache *et al.*, 1976). Rowley *et al.* (1968) and Ushiba (1965) suggest that the two mechanisms are not mutually exclusive and that both opsonic antibody and phagocytic cells play vital roles. Possibly, an important point in summary for anyone contemplating the use of one of these preparations for "immunizing" or "protecting" mice from spontaneous disease in a production colony or in a research laboratory setting is that in no instance was the carrier state eliminated. If the immunogen was a live virulent or attenuated strain, the carrier state remained; if the immunogen was not alive, the challenge was alive and it persisted, producing a carrier. The results of these studies are in agreement with our understanding of the naturally occurring disease; active infection is universally recognized as the most effective means of producing immunity known, and yet, carriers persist. As immunity is a relative state, the question for users of mice in the laboratory is not one of how to protect mice from progressive systemic salmonellosis; rather, it is how to avoid the complicating effects of infection, and this is accomplished by maintaining the *Salmonella*-free status of the mice.

Diagnosis is based upon isolation and identification of *Salmonella* organisms. Isolation is readily accomplished from any of the affected tissues or blood during the septicemic stage. In mice which have experienced systemic disease, bacteria may persist in the liver and spleen for several weeks.

An important aspect of quality control of mice destined for use in biomedical research today is the detection of subclinical salmonellosis in the carrier animal. Confidence in the validity of the monitoring program is greatly facilitated by access to records of the diagnostic laboratory responsible for monitoring the health status of the production colony, whereby the cause of any illness or death in the colony is known. Lacking such information, as when purchasing mice from a vendor, one must resort to sampling procedures when the mice arrive at the using facility. Assuming a carrier rate of 5% and a desire to detect at least one infected mouse at the 95% confidence level, examination of a sample size of 58 mice would be required.* Furthermore, if a determination of the *Salmonella*-free status of the supplier is a consideration, there may be reason to question whether the sample is truly representative of the production colony population.

The detection of carriers by culture of fecal samples has been studied extensively (Banwart and Ayres, 1953; Dixon, 1961; Margard and Litchfield, 1961; Margard *et al.*, 1963). The best results of isolation from mouse feces were obtained by selective enrichment in selenite F cystine broth (18–24 hr), followed by plating on brilliant green agar and choosing suspect colonies for transfer to triple sugar iron–urea (Margard *et al.*, 1963). Confirmation is determined serologically. The investigators suggest that samples from individual mice are more effective in detection of the carrier than are pooled feces from several mice, due presumably to the presence of inhibitors in the feces of some mice. This could be an important consideration when attempting to determine the incidence of shedding within a population, as pointed out by Margard *et al.* (1963). However, to determine *only* whether *Salmonella* is present in a given population, use of the pooled sample may have merit, especially in consideration of the expected small carrier rate (5%) and the large sample size required to detect shedding in one mouse. Further study is needed.

Since antibody is not always detectable by the agglutination

**Institute of Laboratory Animal Resources News*, Vol. XIX, no. 4, 1976, p. L22. The formula for determining sample size = log 0.05/log N, where N represents the percent of *normal* animals expected in the population and the accepted confidence level is 95%. The sample should detect at least one positive case with this level of confidence provided the base population from which the sample was taken is 100 or more and the sample is random, i.e., representative of the base population.

test in culturally positive mice (Hobson, 1957b; Tannock and Smith, 1971) and serological cross-reactivity is so common among bacteria, even of different genera, the detection of *Salmonella*-infected mice (see Otis, Volume III), past and present, by serological means does not appear to be useful. However, Morello *et al.* (1964) suggest the use of a hemolytic test to detect chronic salmonellosis in mice. Further study is needed.

Salmonella-infected mice are not acceptable tools for the conduct of biomedical research. Furthermore, such mice can be a source of infection for other animals and man. Salmonellosis of mice is preventable by production methods utilizing a barrier system (see Chapter 16, this volume). The *Salmonella*-free state can be maintained in the research laboratory by adhering to strict husbandry practices (see Chapter 15, this volume, and Lang, Volume III). Neither antibiotic treatment (Seligmann and Wassermann, 1949; Hobson, 1956; Haberman and Williams, 1958; Slanetz, 1946, 1948; Rabstein, 1958) nor vaccination with a variety of preparations (Mitsuhashi *et al.*, 1958, 1959; Blanden *et al.*, 1966; Mackaness *et al.*, 1966; Rowley *et al.*, 1964; Hobson, 1957c; Hashimoto *et al.*, 1961; Germanier, 1970; Germanier and Furer, 1971; Collins and Carter, 1972; Wray *et al.*, 1977) will prevent the development of the carrier state and is therefore not recommended.

B. *Bacillus*

Ernest Tyzzer (1917) originally described this disease condition, which destroyed his colony of Japanese waltzing mice. If it seems difficult to conceive that a microbial agent would kill an entire population, consider the known characteristics of the etiologic agent: a gram-negative rod $0.5 \times 8-10\ \mu m$, motile by peritrichous flagella, spore-forming, pleomorphic, obligate intracellular parasite, very fastidious in selection of cells for metabolism and growth (epithelial cells of lower intestinal mucosa, smooth and cardiac muscle cells, nerve cells, and hepatocytes), unclassified but referred to as *Bacillus piliformis* (see reviews in Ganaway *et al.*, 1971a; Fujiwara, 1978). The literature is replete with descriptions of unsuccessful attempts to culture this interesting parasite in cell-free media. Although two reports of isolation in such media have appeared (Kanazawa and Imai, 1959; Simon, 1977), the evidence is not convincing. Other than by passage in a vertebrate host, the only presently known means of propagating *B. piliformis* is by the inoculation of embryonated hens' eggs via the yolk sac route (Craigie, 1966a; Ganaway *et al.*, 1971b; Fries, 1977a).

Tyzzer's disease was recognized in mice only until 1965, when Allen and associates (1965) described the condition in laboratory rabbits. The disease occurs worldwide, and fatal infections have since been described in a wide variety of laboratory animals, in free-living animals, and in horses (Ganaway *et al.*, 1976). In the United Kingdom, Tyzzer's disease is thought to be the greatest cause of ruined cancer research studies (Anonymous, 1961) and one of the four most important diseases of laboratory mice (Tuffery, 1956). The importance, and indeed the prevalence, of the disease in mice in the United States remain unknown.

The source of infection for colonies of mice remains unknown. Since the vegetative phase of *B. piliformis* outside the host cell is so unstable (Craigie, 1966a; Ganaway *et al.*, 1971a,b; Fujiwara, 1978), it would appear that the spore represents the interepizootic survival mechanism and the means of spread of the infection between animals. Bedding soiled by animals which die of Tyzzer's disease remains a source of infection for other animals for extended periods of time (Tyzzer, 1917; Allen *et al.*, 1965; Ganaway *et al.*, 1971a). Thus, the ingestion of feces-contaminated food seems to be the most likely means of acquiring infection. The use of animal feed that is not sterilized might explain the introduction of *B. piliformis* into cesarean-derived, barrier-type colonies of mice (Mullink, 1968; Hunter, 1971; Ganaway *et al.*, 1976). Fries (1978) recently reported infection of the mouse fetus *in utero* following an infective intravenous challenge of the pregnant dam. Such studies need to be extended to explore the possibility of vertical transmission.

Several factors that could predispose mice to Tyzzer's disease have been suggested: overcrowding (Tyzzer, 1917; Rights *et al.*, 1947), high temperature and humidity (Gard, 1944), feeding moist food (Gard, 1944), immunosuppressive therapy such as X irradiation (Tuffery, 1956) and corticosteroid drugs (Kaneko *et al.*, 1960; Fujiwara, 1978; Ganaway *et al.*, 1971a; Craigie, 1966b), and inherited susceptibility (Tyzzer, 1917; Gowen and Schott, 1933). Overcrowding, high humidity, and feeding moist food which is easily contaminated with feces may contribute to the buildup of spores in the environment to the extent that clinical disease becomes manifest. In the United Kingdom, the mortality due to Tyzzer's disease in noncompromised, noninbred mice is believed to be low (2–3%), but when a group of mice were X-irradiated, 47% died as a result of *B. piliformis* infection (Tuffery, 1956). The use of corticosteroid drugs has been a valuable aid in experimental studies whereby *B. piliformis* can be predictably maintained in passage in mice inoculated intravenously with suspensions of infected liver (Fujiwara, 1978).

The first sign of Tyzzer's disease is usually the occurrence of sudden deaths. Close examination of cagemates may reveal varying degrees of diarrhea, which, if present, is of short duration (Tyzzer, 1917; Gard, 1944; Rights *et al.*, 1947). At necropsy, conspicuous lesions are usually seen in the liver as miliary, white to yellowish-gray foci scattered throughout the parenchyma (Fig. 3). The lower intestinal tract appears normal or is slightly reddened. No other tissues appear to be affected grossly. Histologically, the miliary foci in the liver appear as areas of coagulation necrosis and are located generally in the immediate vicinity to a branch of the portal vein, indicating an

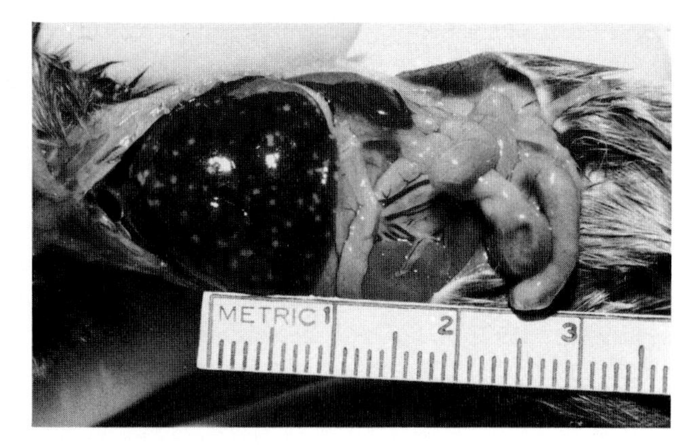

Fig. 3. CBA/N mouse 3 days after intravenous inoculation with *Bacillus piliformis* (rabbit origin). Numerous white foci are scattered throughout the liver.

embolic origin from the gastrointestinal tract. Variable numbers of neutrophils and lymphocytes surround the necrotic areas. Bundles of slender, sticklike rods, *B. piliformis,** appear randomly arranged in the cytoplasm of apparently viable hepatocytes at the border of the necrotic and normal tissue (Figs. 4 and 5). Varying stages and degrees of infection may be seen in sections of the same liver. The bacilli are numerous in early stages and absent in late stages. The bacilli are also found in the epithelial cells of the intestinal mucosa of the terminal ileum, cecum, and proximal colon. Here, as in the liver, the ease of demonstrating the bacilli depends upon the degree and stage of infection and the use of proper staining technique.

The diagnosis of mice examined within a reasonably short time after death or killed during a stage of acute disease is distinctive and easily accomplished. It is based upon the demonstration of typical bacilli within the cytoplasm of hepatocytes or intestinal epithelial cells of the mucosa.

As expected, the number of subclinical infections appears to exceed greatly the number of clinical infections. The indirect immunofluorescence antibody technique (FAT) appears to be considerably more sensitive than the complement fixation (CF) test for determining previous infection. Using the CF test, Fujiwara (1967) found antibody in individual sera of 4–10% of the mice from colonies in which Tyzzer's disease was known to be enzootic. Using FAT, Fries (1977b) found antibody in 36–83% of the mice from known infected colonies. The higher prevalence rates seem more probable for a fecal-oral-transmitted disease. Confirmation and extension of this effort are needed to provide prevalence data and to better our knowledge of the poorly understood *B. piliformis* parasite. High prevalence rates of subclinical disease may indicate that a vaccine would be useful in protecting mice (and other animals). Fujiwara *et al.* (1965) reported that formalinized infected mouse liver suspension was antigenic and, when used as a vaccine, would protect mice against a subsequent, otherwise lethal challenge (intravenous route). However, it did not prevent the development of liver lesions.

*Hemotoxylin and eosin or gram stains are unsatisfactory for visualization of the bacilli. Giemsa may be helpful, but the best method is the Warthin-Starry silver impregnation technique (Allen *et al.*, 1965; Ganaway *et al.*, 1971a).

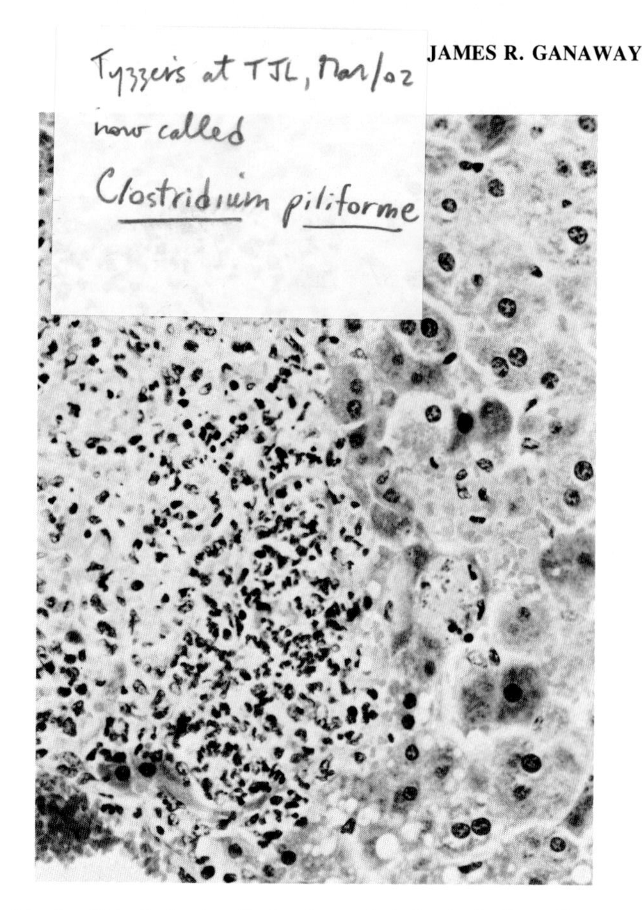

Fig. 4. Note the border of a focal necrotic lesion in the liver of a mouse with Tyzzer's disease. *Bacillus piliformis* organisms are not demonstrated with H&E stain. ×330.

The means to prevent or control Tyzzer's disease have not been adequately studied. The natural history of Tyzzer's disease, prevalence rates of infection in various animal species, the possibility of reservoirs and carrier states, and the actual mechanisms of spread of the disease among animals remain areas of speculation. As previously mentioned, either nonsterilized animal feed or vertical transmission might be the means of introducing *B. piliformis* into caesarian-derived, barrier-maintained colonies of mice. Also, as previously mentioned, a buildup of spores in the environment and/or immunosuppression might be expected to result in fatal infections. In certain instances, antibiotics might be helpful (Craigie, 1966b; Ganaway *et al.*, 1971a,b; Fujiwara, 1978). Since the diagnosis of Tyzzer's disease is based upon histopathological findings, antibiotic treatment of individual animals is not a consideration. It is conceivable that an epizootic might be averted by the oral administration of tetracycline, as

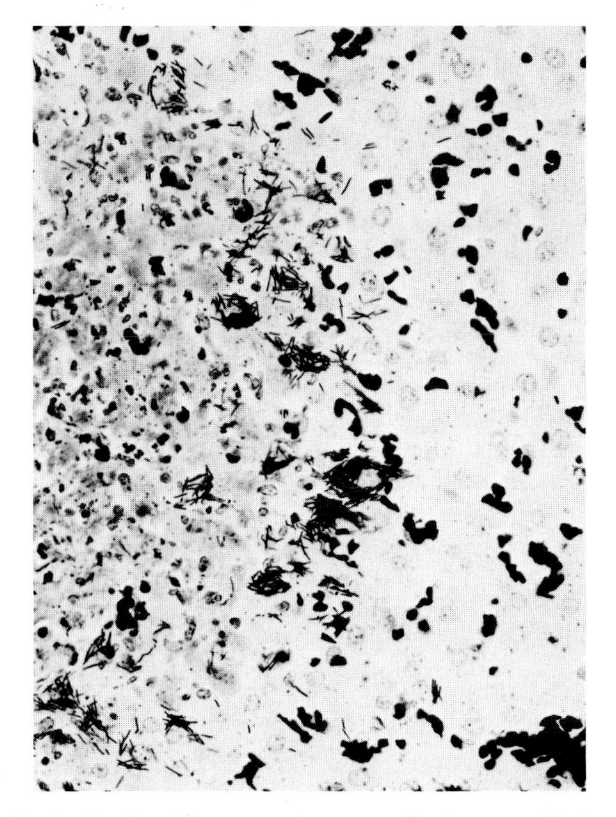

Fig. 5. *Bacillus piliformis* organisms are readily demonstrated in hepatic cells at the border of a focal necrotic lesion in the liver of a mouse with Tyzzer's disease using the Warthin-Starry silver impregnation technique. ×330.

reported by Hunter (1971). It is unlikely, however, that a spore-forming bacterium can be eliminated from a large population of animals by the use of antibiotics. A vaccine might be helpful but is yet to be developed and tested. Until convincing evidence of vertical transmission is provided, the cesarean-derived, barrier-maintained colony should offer the best means of providing Tyzzer's disease-free mice. Efforts to maintain this state in the laboratory during the conduct of research are appropriate, especially if the mice are to be used in studies involving immunosuppression.

C. *Citrobacter*

Colitis, characterized by marked mucosal hyperplasia and varying degrees of inflammation, has been observed as a naturally occurring epizootic disease of mice associated with a variant of *Citrobacter freundii*. Various terms have been used to describe the condition: neoplasia (Pullinger and Iverson, 1960), colitis cystica (Brynjolfsson and Lombard, 1969), catarrhal enterocolitis (Brennan *et al.*, 1965), colitis with rectal prolapse (Ediger *et al.*, 1974; Bieniek and Tober-Meyer, 1976), and transmissible murine colonic hyperplasia (Barthold *et al.*, 1976; Silverman *et al.*, 1979).

Citrobacter freundii, the type species of this genus, is placed in group I of the family Enterobacteriaceae (Sedlak, 1974). A normal inhabitant of the intestine of man, *C. freundii* is found in water, food, feces, and urine. It is a gram-negative rod which grows readily on ordinary media. Though most strains are motile, ferment lactose, and utilize citrate as the sole source of carbon, the atypical strains that have been associated with colonic hyperplasia of mice (Brennan *et al.*, 1965; Ediger *et al.*, 1974; Barthold *et al.*, 1977) are nonmotile and either fail to utilize citrate or do so only marginally.

Citrobacter freundii has been a cause of significant nosocomial infections associated with contamination of intravenous fluids and is an opportunistic pathogen associated with urinary tract, respiratory tract, wound, and cutaneous infections in man (Hodges *et al.*, 1978). The prevalence of typical or atypical *C. freundii* in colonies of mice, in the absence of clinical disease, remains unknown, as field survey studies have not been performed. Ediger *et al.* (1974) isolated atypical *C. freundii* from 4361 clinically normal mice from a conventional colony. Brennan *et al.* (1965) isolated atypical *C. freundii* from 43% (65 of 151) newly arrived, unaffected mice from commercial breeders. Likewise, the source of infection for the mouse colony remains unknown. For the microbiologically monitored, barrier-type colony, man might be considered as a possible source of infection.

Diet, inherited resistance and age of the mouse, and virulence of *C. freundii* strains have marked predisposing influences on the induction and severity of disease (Barthold *et al.*, 1977). In experimental studies using an atypical strain of *C. freundii*, each of four strains of mice examined (NIH Swiss, DBA/2J, C57BL/6J, and C3H/HeJ) developed colonic hyperplasia, but the degree of hyperplasia varied significantly with the mouse strain. Mortality was low or absent in DBA/2J, NIH Swiss, and C57BL/6J mice but reached 45% in C3H/HeJ mice between 2 and 3 weeks postinoculation. Both suckling mice and adults were susceptible, but moderate mortality was seen in younger mice and was rare in adults (Barthold *et al.*, 1978). On the other hand, Silverman *et al.* (1979) reported a natural outbreak in a group of 210 adult A/J mice in which they observed 50% morbidity and 25% mortality. Bieniek and Tober-Meyer (1976) observed high morbidity (37%) in a breeding colony of Han:NMRI mice, in which young adults (5–7 weeks of age) were primarily affected. Significant differences in the degree of colonic hyperplasia in affected mice were noted between groups fed four commercially available diets (Barthold *et al.*, 1977). The responsible factor(s) in the diet remains unknown. In an experimental study designed to determine the infectivity of various isolants of *C. freundii*, each of 19 isolants obtained from other animal species, including man, failed to colonize the lower intestinal tract or to cause colonic hyperplasia in otherwise susceptible mice (Barthold *et al.*, 1977).

In natural outbreaks, retarded growth, ruffled fur, soft feces,

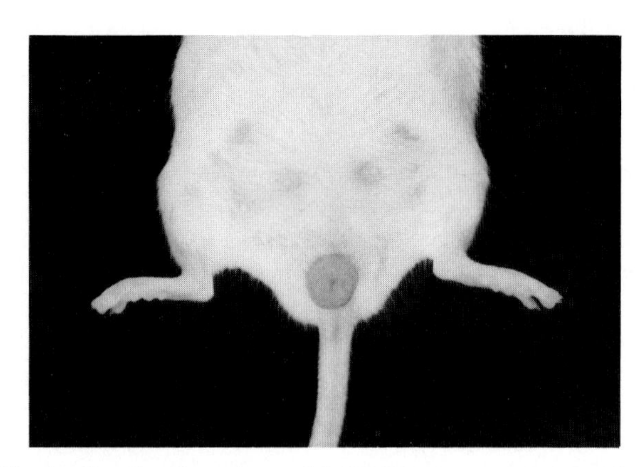

Fig. 6. Rectal prolapse in an adult N.NIH(s) mouse infected with *Citrobacter freundii* 4280. (From Barthold *et al.*, 1978, *Vet. Pathol.*, with permission of Dr. S. W. Barthold.)

occasional prolapse of the rectum (Fig. 6), and moderate mortality in late suckling and early weanling-age mice may be seen (Brennan *et al.*, 1965; Ediger *et al.*, 1974; Barthold *et al.*, 1978). Brennan *et al.* (1965) noted diarrhea with pasting or soiling of the perineum as a constant finding, whereas Ediger *et al.* (1974) did not observe diarrhea but noted prolapse of the rectum in approximately 15% of the mice with colonic hyperplasia. In a production colony, Bieniek and Tober-Meyer (1976) noted diarrhea and prolapsed rectum in almost 8% of the weanling mice. At necropsy, the distal half of the colon is thickened and appears rigid but is empty or contains semiformed feces (Fig. 7). In experimental studies, the earliest detectable change in appearance of the colon occurs 1 week postinoculation, progresses to maximal thickness between 2 and 3 weeks, and regresses thereafter (Barthold *et al.*, 1978). Histologically, there is marked mitotic activity in the mucosa of the descending colon and varying degrees of leukocytic infiltration (Fig. 8). In more severely affected young mice, varying degrees of severe mucosal necrosis, erosion, and inflammation are noted. After 2 to 3 weeks, the leukocytic infiltrate changes from neutrophilic to mononuclear, with lymphocytes and plasma cells predominating, and there is decreased mitotic activity in the mucosal cells. The regression is rapid and complete in adults but slow and more persistent in mice infected as sucklings (Barthold *et al.*, 1978). Scanning and transmission electron microscopic findings indicate that the bacteria become attached to the surface of the mucosa of the descending colon between 4 and 10 days after inoculation (Fig. 9) (Johnson and Barthold, 1979). In a sequential study, mucosal hyperplasia peaked at 16 days and regression to normal epithelium occurred by day 45. The hyperplastic epithelium throughout the crypt resembled undifferentiated crypt cells of controls. The hyperplastic response appeared to be a defense mechanism of replacing infected cells with newly migrated, uninfected epithelium.

Diagnosis is based on demonstration of mucosal hyperplasia

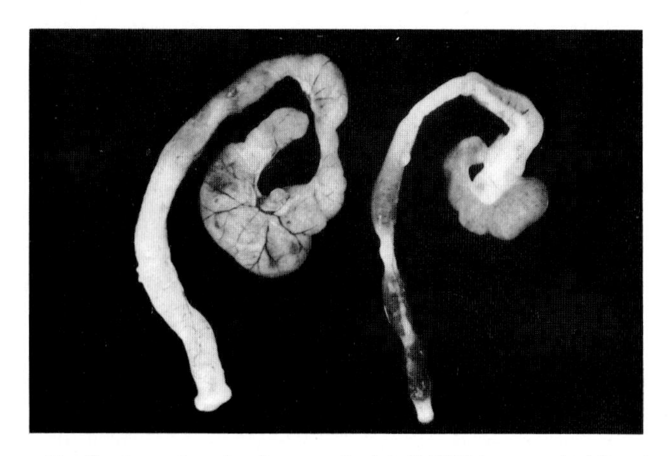

Fig. 7. Large bowels of a normal adult N.NIH(s) mouse (right) and a similar mouse infected with *Citrobacter freundii* 4280 (left). The descending colon of the affected mouse is contracted, devoid of feces, and opaque due to mucosal thickening. (From Barthold *et al.*, 1978, *Vet. Pathol.*, with permission of Dr. S. W. Barthold.)

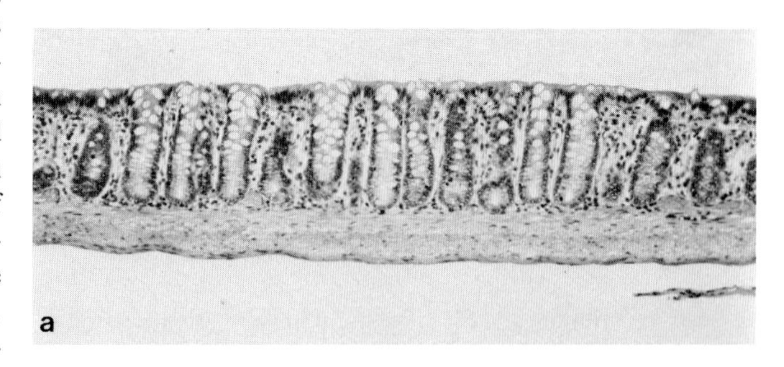

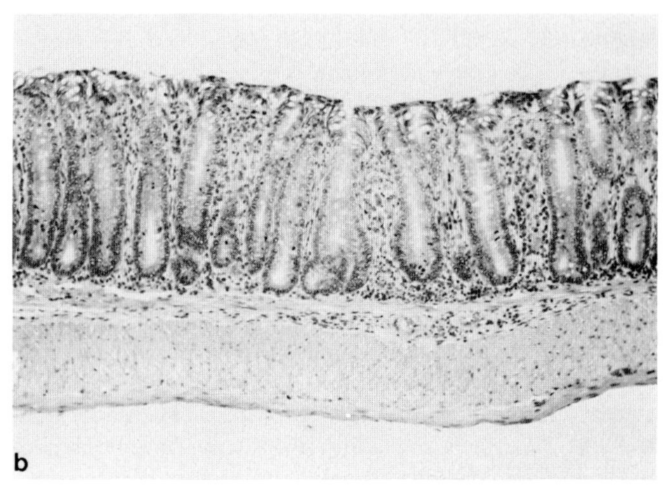

Fig. 8. Descending colon of a normal mouse (a) and an infected mouse 16 days after inoculation with *Citrobacter freundii* 4280 (b). The crypts are elongated and lined with mature hyperplastic epithelium. The lamina propria and submucosa are infiltrated with leukocytes. The outer muscular layers are thickened due to contraction. ×95. (Contributed by and with the permission of Dr. S. W. Barthold.)

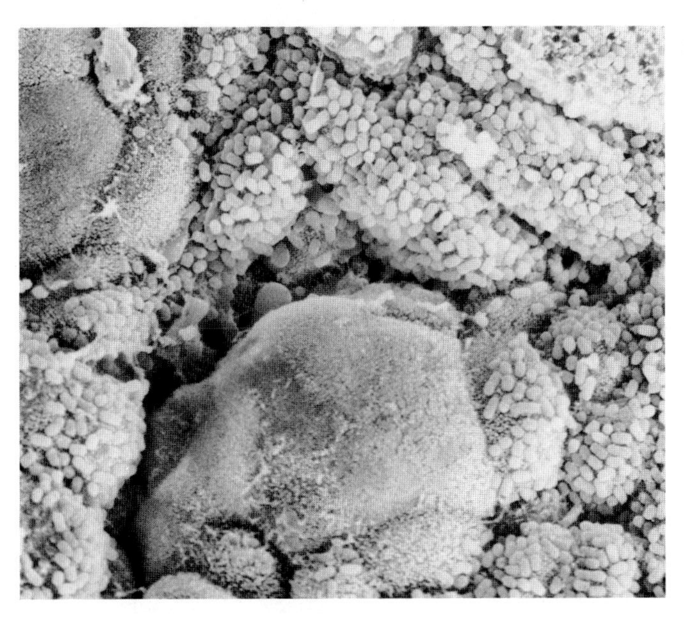

Fig. 9. Scanning electron micrograph of the descending colon of a mouse 4 days after inoculation with *Citrobacter freundii* 4280. The organisms are attached to the surface of the mucosal epithelium. ×2500. (Contributed by and with the permission of Dr. S. W. Barthold.)

of the descending colon and isolation of *C. freundii*. It has been suggested (Barthold *et al.*, 1977) that the disease may indeed be rare in mouse colonies, since an outbreak would depend upon a fortuitous combination of several predisposing factors mentioned above. Clinical signs, other than occasional prolapsed rectum, are vague, especially in adult mice. The lesions in certain strains of mice, such as C57BL/6J, could be so mild as to be overlooked by the pathologist. Sodium sulfamethazine, 0.1% of drinking water (Ediger *et al.*, 1974), tetracycline, 450 mg/liter of drinking water (Silverman *et al.*, 1979), or neomycin sulfate, 2 mg/ml of drinking water prepared daily (Barthold, 1980) are reported useful in the prevention and treatment of *C. freundii* infection.

D. *Pseudomonas*

Interest in *Pseudomonas aeruginosa* as a pathogen for mice was stimulated by investigations of acute radiation injury (Hammond *et al.*, 1954a,b, 1955; Gordon *et al.*, 1955; Wensinck, 1961). It is considered a normal inhabitant of the gastrointestinal tract of animals and man in the absence of disease. No effect was noted when *P. aeruginosa* was administered by mouth to normal mice (Hammond *et al.*, 1954b), but when mice were anesthetized (sodium pentobarbitol, 70 mg/kg), given atropine sulfate (0.45 mg/kg) to reduce intestinal motility, and administered *P. aeruginosa* ($\geqslant 5.3 \times 10^6$ colony-forming units) by mouth, they died within 72 hr of an acute systemic infection (Schook *et al.*, 1977). In conventional mouse colonies, *P. aeruginosa* may be present in drinking water devices (McPherson, 1963; Hoag *et al.*, 1965), and up to 90% of the mice may shed *P. aeruginosa* in the feces without evidence of disease (Hoag *et al.*, 1965). *Pseudomonas aeruginosa* density in the feces may increase following oral antibiotic treatment of mice (Hoag *et al.*, 1965; Urano and Maejima, 1978). Spontaneous otitis interna associated with disturbances of equilibrium has been reported in mice (Ediger *et al.*, 1971; Olson and Ediger, 1972). In other animals and man, *P. aeruginosa* is associated with wounds, burns, urinary tract infections, occasional auditory canal infection, immunosuppressive treatment for malignancies and organ transplants, and premature infant septicemia. The infections tend to be resistant to antimicrobial drug therapy.

Pseudomonas aeruginosa is a gram-negative, motile, nonspore-forming rod which is essentially ubiquitous in warm, moist habitats, being frequently isolated from soil, water, sewage, and normal human skin. Readily isolated, smooth colonies have a greenish fluorescence, a fruity odor, and are sometimes surrounded by zones of hemolysis. A distinguishing feature is the production of pigments: pyocyanin, a blue compound soluble in chloroform; and fluorescin, a greenish substance soluble in water but not in chloroform. They are the cause of blue pus. Strains producing neither pigment are extremely rare (Doudoroff and Palleroni, 1974).

When mature adult mice are exposed to 650 R of X irradiation, aplasia of the bone marrow and a profound granulocytopenia occur within 72 hr. Since the half-life of the granulocyte in the circulation is only 6–12 hr, the fall in the white cell count is precipitous. During maximum depression of granulocytes, otherwise normal flora of the nasopharyngeal (Wensinck, 1961) and/or gastrointestinal tract (Hammond *et al.*, 1954a,b, 1955; Gordon *et al.*, 1955) assume an opportunistic role and produce a bacteremia (Davis *et al.*, 1968; Flynn, 1963). Similar effects are produced by cyclophosphamide, 350 mg/kg, intraperitoneally (Brownstein, 1978). *Pseudomonas aeruginosa* infection complicates such experimental studies by causing a rapidly fatal septicemia. Virus infections which alter host defense mechanisms, such as murine cytomegalovirus, may markedly enhance the susceptibility of mice to *P. aeruginosa* (Hamilton and Overall, 1978).

Cyclophosphamide-treated mice which consume *P. aeruginosa* in the drinking water sicken on the third to fourth day and develop bilateral conjunctivitis, serosanguinous nasal discharge, generalized edema of the head, and a marked loss of weight and die on the fourth to fifth day (Brownstein, 1978). Bacteremia occurred in less than 18 hr in mice with total leukocyte counts of less than 6000/μl. Using a peroxidase-labeled antibody technique, the portal of entry of *P. aeruginosa* in such mice was specifically found to be the squamocolumnar junction of the upper respiratory tract and, in some cases, the periodontal gingiva concurrently. Invasion of the epithelium of the lower gastrointestinal tract did not occur. At necropsy, focal necrosis of the liver and spleen, diffuse fatty change of the liver, and fluid distention of the bowel were noted. The only lesions noted histologically in mice killed during the first 96 hr were lymphoid and myeloid atrophy, typical of cyclophosphamide treatment. From 96 hr on, there was rapid and progressive epithelial necrosis and ulcer formation in the nasal and oropharyngeal mucosa which coincided with *P. aeruginosa* invasion of the epithelium. The organisms spread rapidly to the regional lymph nodes, and a bacteremia ensued. Since maximum bacterial counts in the tissues are inversely related to the number of leukocytes in the circulation, the population density of bacteria in the lesions can reach 1000 times that attained in lesions of normal mice (Davis *et al.*, 1968). Depressed mobilization of the leukocytic exudate is the critical factor in the irradiation effect rather than impaired phagocytic or bacteriocidal capability.

The diagnosis of disease caused by bacterial agents is normally dependent upon isolation and identification of the bacterium. In this instance, the production of pigments by *P. aeruginosa* is so universal and characteristic that Wensinck *et al.* (1957) and Flynn (1963) have used this property to advantage in developing a simple test for determining the presence of *P. aeruginosa* in contaminated material (feces and drinking water) without resorting to the need for obtaining a pure culture and performing biochemical tests. The suspect material is added to glycerol broth (Wensinck *et al.*, 1957) and incubated at 37°C for 3–7 days. Development of a blue-green color indicates that the suspect material contained *P. aeruginosa*.

Pseudomonas aeruginosa infection in mice can be prevented by providing acidified (pH 2.5 with hydrochloric acid), chlorinated (10 ppm) drinking water (Schaedler and Dubos, 1962; Wensinck *et al.*, 1957; McPherson, 1963; Hoag *et al.*, 1965).

E. *Escherichia*

Escherichia coli appears to be of little significance as a pathogen for mice (Carlton and Hunt, 1978). However, in a survey (Schiff *et al.*, 1972) of seven strains of conventionally raised mice (C57BL/6Cr, DBA/2Cr, BDF_1, C3H/HeJ, BRVS/SrCr, BSRV/SrCr, and DBA/2Cr), *E. coli* was isolated from feces of 81% of 382 adults. Worthy of note was the observation that colibacillosis associated with clinical signs of gastroenteritis was noted in only 22 of 52 BRVS/SrCr mice. The detail of establishing this association was not given. Enteropathogenic serotypes, mainly 0119:B14, constituted 58% of the *E. coli* isolants. Similarly, Thunert (1978) found *E. coli* to be a consistent part of the gastrointestinal flora of conventional mice in the absence of associated clinical disease over an 8-year observation period. When *E. coli* of unknown serotype was intentionally introduced into a specific pathogen-free (SPF) colony of mice, no disease was noted. However, serotype 07:K1:H7 was introduced into another SPF colony, presumably by an animal technician (from whom the same strain was isolated), and clinical disease resulted. This serotype was confirmed in 700 of 4000 necropsies performed over a 5-year period and was associated with abscesses in various organs, purulent urogenital tract infections, and meningitis. In experimental studies, this serotype displaced other *E. coli* serotypes in mice and could not be displaced by superinfection with other serotypes. Thus, Thunert (1978) concluded that species specific strains of *E. coli* do exist.

Though serotypes pathogenic for one animal species are not necessarily pathogenic for another animal species, the enterotoxin plasmid can be transmitted between serotypes. Thus *E. coli* is a frequent inhabitant of the intestinal tract of conventionally maintained mice in the absence of disease and may, in ill-defined circumstances, assume the role of a pathogen. Recognition of that role would require the demonstration of pathogenicity of the isolant in the involved mouse strain. In conventionally housed mice, the source of an enteropathogenic serotype may be difficult to establish. In the barrier-maintained colony, the role of the animal handler should not be overlooked.

F. *Clostridium*

Spontaneous clostridial infection in mice is reported to be rare (Carlton and Hunt, 1978). Clapp and Graham (1970), however, reported epizootic *Clostridium perfringeus* type D infection in a cesarean-derived, barrier-sustained colony of Upj:TUC(ICR)spf mice. The condition was observed in lactating females (prevalence of 1%) and 2- to 3-week-old suckling mice, which experienced "a fairly heavy mortality" (p. 1081). New breeder stock and weaned mice in the same facility were not affected. Affected mice appeared gaunt and haunched; they had ruffled coats, pale extremities, paralysis of the hindquarters, soiled perineum, soft feces, and fecal impaction. At necropsy, the colon was enlarged and flaccid and contained frothy, whitish-yellow feces. Irregular yellowish-white foci were found in the liver of some of the mice. Unfortunately, histological description of the lesions was not provided. Test results for murine viruses, helminths, protozoa, *Salmonella* spp., and *Shigella* spp. were negative. *Cl. perfringeus* type D was isolated from the colon, blood, and liver of some of the affected mice. It was also isolated from the pasteurized diet fed to the mice and was consistently isolated from the intestine of normal mice in the barrier room. A broth culture was lethal for two of four mice inoculated intraperitoneally but was without effect in mice inoculated intravenously and intragastrically. Attempts to reproduce the disease experimentally using the intestinal content of affected mice failed. The demonstration of a toxin in the intestinal content of affected mice was not attempted. The use of a commercially available *Cl. perfringeus* type D bacterin was not effective in preventing the natural occurrence of the disease. In uncontrolled experiments, incorporating penicillin G into the diet or changing the diet prevented further occurrence of the disease.

G. *Klebsiella*

Klebsiella pneumoniae is a normal inhabitant of the intestinal tract of 30–40% of humans and animals. It is a saprophyte found in soil and water and is widespread in nature. It also occurs in areas free of fecal contamination such as nutrient-rich industrial wastes (Bagley and Seidler, 1977). *Klebsiella pneumoniae* is an opportunistic pathogen associated with urinary and respiratory tract infections in man and a variety of animals. Capsule types 1 and 2 are highly virulent for mice (Orskov, 1974). However, there are no reports to indicate that *K. pneumoniae* is a significant cause of spontaneous disease in mice (Carlton and Hunt, 1978). Flamm (1957) described throat abscesses associated with adenitis of the supraclavicular glands of white mice. A similar condition was experimentally produced by inoculating mice under the buccal mucosal membrane with the *K. pneumoniae* isolant.

III. MYCOTIC INFECTIONS

Infection of the digestive tract with pathogenic fungi does not appear to have been observed in the mouse, in contrast to that observed in birds and other animals (Smith and Austwick, 1967). Fungal infections are not likely to present a problem in cesarean-derived, barrier-maintained colonies of mice since soil is the reservoir for both animal and human infection. Certain geographic areas are known to favor enzootic and epizootic states of specific deep or systemic mycotic infections of animal and man. Such infections are usually contracted by aerosol exposure and are not believed to be transmitted from animal to animal. This situation contrasts sharply to that of the dermatomycoses, which are zoonotic infections readily transmitted by contact (Fischman *et al.*, 1976; Keep, 1977). The mouse is, nevertheless, used extensively for studying fungal infection (Patton *et al.*, 1976; Corbell and Eades, 1976, 1977; Adriano and Schwarz, 1955; Louria *et al.*, 1960; Staib *et al.*, 1976; Iwata *et al.*, 1973; Oestreicher *et al.*, 1973).

Experimental studies of *Candida albicans* infection in mice (Iwata *et al.*, 1973; Oestreicher *et al.*, 1973) attest to the contention of Corbell and Eades (1976) that infections caused by opportunistic fungi are usually associated with some predisposing condition which modifies the immunological competence of the host. Conventional mice given 10^7 to 10^8 viable *C. albicans* cells by mouth shed small numbers in the feces for only a few days. In contrast, germ-free mice given 10^3 *C. albicans* cells demonstrate colonization of all parts of the gastrointestinal tract, including the stomach; 10^7 to 10^8 cells/gm feces were obtained throughout a 90-day observation period without evidence of disease. All other organs remained culturally negative. If, however, such colonized mice were given cortisone daily, they lost weight, developed diarrhea, and died within 4 weeks of systemic infection, especially involving the kidney.

IV. SUMMARY AND CONCLUDING REMARKS

Knowledge of the spontaneous or naturally occurring diseases of laboratory animals is essential primarily because of the need to assess the impact of intercurrent disease during the conduct of biomedical research. Such knowledge is also valuable from a comparative medicine standpoint in providing animal models for studying similar disease conditions of man and other animals. The mouse being the most common and most widely used animal in biomedical research, one might expect the spontaneous diseases of this animal to be well understood. With the exception of salmonellosis, this is not true, at least with regard to bacterial pathogens of the gastrointestinal tract. Even in the case of salmonellosis, knowledge of the prevalence of this disease in mouse stocks in the United States and an accurate assessment of pooled versus individual fecal sampling for isolation purposes are lacking. Though Tyzzer's disease (*B. piliformis* infection) was recognized in 1917 as a potentially devastating disease of mice, very little is known today about the natural history of this parasite; the pathogenesis, prevalence in stocks of mice, means of introduction to closed colonies, and means of control or prevention are yet to be elucidated. *Escherichia coli, C. freundii, K. pneumoniae, Clostridium* spp., *P. aeruginosa,* and fungi can be isolated from the feces of conventionally maintained mice in the absence of clinical disease. (This is not to imply that they are universally present or that they represent part of the dominant microflora.) As such, they may revert from an opportunistic to a pathogenic role when the normal host–parasite relationship is disturbed, as for example, with oral antibiotic or immunosuppressive treatment. Antibiotics can be beneficially used in certain instances in which a causal relationship of the bacterium to clinical disease is established, sensitivity tests are performed on the isolants, and the population at risk is not exceedingly large. The benefit of antibiotics should be viewed as temporary, however, because microorganisms are not likely to be *eliminated* from conventional populations as a result of administering antibiotics. Increased sanitation, isolation or quarantine of sick mice, and reduction of recognizable stress factors should also be beneficial. Vaccines which will *prevent colonization* of the gastrointestinal tract, a goal to consider for animals that will be stressed during experimentation, have not been developed. To increase the probability of a research project being completed and reproducible results realized, cesarean-derived, barrier-maintained mice should be used. Microbiological monitoring and the provision of adequate containment facilities during the conduct of research are recommended to help assure the pathogen-free state of the mice and to minimize the devastating and complicating effect of intercurrent disease.

REFERENCES

Adriano, S. M., and Schwarz, J. (1955). Experimental monialiasis in mice. *Am. J. Pathol.* **31,** 859–874.

Allen, A. M., Ganaway, J. R., Moore, T. D., and Kinard, R. F. (1965). Tyzzer's disease syndrome in laboratory rabbits. *Am. J. Pathol.* **46,** 859–882.

Anonymous (1961). *Annu. Rep., Imp. Cancer Res. Fund, 58th* Apr., p. 19.

Bagley, S. T., and Seidler, R. J. (1977). Significance of fecal coliform-positive *Klebsiella. Appl. Environ. Microbiol.* **33,** 1141–1148.

Bakken, K., and Vogelsang, T. M. (1950). Pathogenesis of Salmonella typhimuruim infection in mice. *Acta Pathol. Microbiol. Scand.* **27,** 41–50.

Banwart, G. J., and Ayres, J. C. (1953). Effect of various enrichment broths and selection agars upon the growth of several species of Salmonella. *Appl. Microbiol.* **1,** 296–301.

Barthold, S. W. (1980). The microbiology of transmissible murine colonic hyperplasia. *Lab. Anim. Sci.* **30,** 167–173.

Barthold, S. W., Coleman, G. L., Bhatt, P. N., Osbaldiston, G. W., and Jonas, A. M. (1976). The etiology of transmissable murine colonic hyperplasia. *Lab. Anim. Sci.* **26,** 889–894.

Barthold, S. W., Osbaldiston, G. W., and Jonas, A. M. (1977). Dietary, bacterial and host genetic interactions in the pathogenesis of transmissable murine colonic hyperplasia. *Lab Anim. Sci.* **27,** 938–945.

Barthold, S. W., Coleman, G. L., Jacoby, R. O., Livstone, E. M., and Jonas, A. M. (1978). Transmissible murine colonic hyperplasia. *Vet. Pathol.* **15,** 223–236.

Bate, J. G., and James, U. (1958). *Salmonella typhimurium* infection dust-borne in a children's ward. *Lancet* **ii,** 713–715.

Berry, L. J., and Smythe, D. S. (1960). Some metabolic aspects of host-parasite interactions in the mouse typhoid model. *Ann. N.Y. Acad. Sci.* **88,** 1278–1286.

Bieniek, H., and Tober-Meyer, B. (1976). Zur atiologie der colitis und des prolapsus recti bei der maus. (Etiology of colitis and rectal prolapse in the mouse.) *Z. Versuchstierkd.* **18,** 337–348.

Blanden, R. V., Mackaness, G. B., and Collins, F. M. (1966). Mechanisms of acquired resistance in mouse typhoid. *J. Exp. Med.* **124,** 585–600.

Bohme, D. H. (1970). Resistance to salmonella infections in inbred mouse strains. *Bull. N.Y. Acad. Med.* **46,** 499–508.

Bohme, D. H., Schneider, H. A., and Lee, J. M. (1959). Some physiopathological parameters of natural resistance to infection in murine salmonellosis. *J. Exp. Med.* **110,** 9–25.

Bohnhoff, M., and Miller, C. P. (1962). Enhanced susceptibility to Salmonella infection in streptomycin-treated mice. *J. Infect. Dis.* **111,** 117–127.

Bohnhoff, M., Drake, B. L., and Miller, C. P. (1954). Effect of streptomycin on susceptibility of intestinal tract to experimental Salmonella infection. *Proc. Soc. Exp. Biol. Med.* **86,** 132–137.

Brennan, P. C., Fritz, T. E., Flynn, R. J., and Poole, C. M. (1965). *Citrobacter freundii* associated with diarrhea in laboratory mice. *Lab Anim. Care* **15,** 266–275.

Brownstein, D. G. (1978). Pathogenesis of bacteremia due to *Pseudomonas aeruginosa* in cyclophosphamide-treated mice and potentiation of virulence of endogenous streptococci. *J. Infect. Dis.* **137,** 795–801.

Brynjolfsson, G., and Lombard, L. S. (1969). Colitis cystica in mice. *Cancer (Philadelphia)* **23,** 225–229.

Carlton, W. W., and Hunt, R. D. (1978). Bacterial disease. *In* "Pathology of Laboratory Animals" (K. Benirschke, F. M. Garner, and T. C. Jones, eds.), Vol. 2, pp. 1368–1480. Springer-Verlag, Berlin and New York.

Carter, P. B., and Collins, F. M. (1974a). Growth of typhoid and partyphoid bacilli in intravenously infected mice. *Infect. Immun.* **10,** 816–822.

Carter, P. B., and Collins, F. M. (1974b). The route of enteric infection in normal mice. *J. Exp. Med.* **139,** 1189–1203.

Clapp, H. W., and Graham, W. R. (1970). An experience with *Clostridium perfringens* cesarean-derived barrier sustained mice. *Lab Anim. Care* **20,** 1081–1086.

Collins, F. M. (1968a). Recall of immunity in mice vaccinated with *Salmonella enteritidis* or *Salmonella typhimurium. J. Bacteriol.* **95,** 2014–2021.

Collins, F. M. (1968b). Cross-protection against *Salmonella enteritidis* infection in mice. *J. Bacteriol.* **95,** 1343–1349.

Collins, F. M. (1969). Effect of specific immune mouse serum on the growth *Salmonella enteritidis* in nonvaccinated mice challenged by various routes. *J. Bacteriol.* **97,** 667–675.

Collins, F. M. (1972). Salmonellosis in orally infected specific pathogen-free C57BL mice. *Infect. Immun.* **5,** 191–198.

Collins, F. M., and Carter, P. B. (1972). Comparative immunogenicity of heat-killed and living oral Salmonella vaccines. *Infect. Immun.* **6,** 451–458.

Collins, F. M., and Carter, P. B. (1978). Growth of salmonellae in orally infected germfree mice. *Infect. Immun.* **21,** 41–47.

Collins, F. M., and Mackaness, G. B. (1968). Delayed hypersensitivity and arthus reactivity in relation to host resistance in salmonella-infected mice. *J. Immunol.* **101,** 830–845.

Collins, F. M., Mackaness, G. B., and Blanden, R. V. (1966). Infection-immunity in experimental salmonellosis. *J. Exp. Med.* **124,** 601–619.

Corbell, M. J., and Eades, S. M. (1976). The relative susceptibility of New Zealand black and CBA mice to infection with opportunistic fungal pathogens. *Sabouraudia.* **14,** 17–32.

Corbell, M. J., and Eades, S. M. (1977). Examination of the effect of age and acquired immunity on the susceptibility of mice to infection with *Aspergillus fumigatus. Mycopathologia* **60,** 79–85.

Craigie, J. (1966a). *Bacillus piliformis* (Tyzzer) and Tyzzer's disease of the laboratory mouse. I. Propagation of the organism in embryonated eggs. *Proc. R. Soc. London, Ser. B* **165,** 35–60.

Craigie, J. (1966b). *Bacillus piliformis* (Tyzzer) and Tyzzer's disease of the laboratory mouse. II. Mouse pathogenicity of *B. piliformis* grown in embryonated eggs. *Proc. R. Soc. London, Ser. B* **165,** 61–77.

Darlow, H. M., Bale, W. R., and Carter, G. B. (1961). Infection of mice by the respiratory route with *Salmonella typhimurium. J. Hyg.* **59,** 303–308.

Davis, R. D., Dulbecco, R., Eisen, H. N., Ginsberg, H. S., and Wood, W. B., Jr. (1968). "Microbiology," pp. 642–644. Harper, New York.

Dimache, G., Gogulescu, L., Tarbuc, R., and Dimache, V. (1976). Intradermal vaccination against *Salmonella typhimurium*. Experimental study on mice. *Arch. Roum. Pathol. Exp. Microbiol.* **35,** 265-268.

Dixon, J. M. S. (1961). Rapid isolation of salmonella from feces. *J. Clin. Pathol.* **14,** 397-399.

Doudoroff, M., and Palleroni, N. J. (1974). Pseudomonas. *In* "Bergey's Manual of Determinative Bacteriology" (R. E. Buchanan and N. E. Gibbons, eds.), 8th ed., pp. 217-222. Williams & Wilkins, Baltimore, Maryland.

Ediger, R. D., Rabstein, M. M., and Olson, L. D. (1971). Circling in mice caused by *Pseudomonas aeruginosa*. *Lab Anim. Sci.* **21,** 845-848.

Ediger, R. D., Kovatch, R. M., and Rabstein, M. M. (1974). Colitis in mice with a high incidence of rectal prolapse. *Lab Anim. Sci.* **24,** 488-494.

Edwards, P. R., and Ewing, W. H. (1972). "Identification of Enterobacteriaceae," 3rd ed., p. 146. Burgess, Minneapolis, Minnesota.

Edwards, P. R., Bruner, D. W., and Moran, A. B. (1948). Further studies on the occurrence and distribution of salmonella types in the United States. *J. Infect. Dis.* **83,** 220-231.

Fischman, O., DeCamargo, Z. P., and Grinblat, M. (1976). *Trichophyton mentagrophytes* infection in laboratory white mice. *Mycopathologia* **59,** 113-115.

Flamm, H. (1957). Klebsiella-enzootic in einer Mäusezucht. *Schweiz. Z. Pathol. Bakteriol.* **20,** 23-27.

Flynn, R. J. (1963). The diagnosis of *Pseudomonas aeruginosa* infection of mice. *Lab Anim. Care* **13,** 126-129.

Fox, J. G., and Beaucage, C. M. (1979). The incidence of *Salmonella* in random-source cats purchased for use in research. *J. Infect. Dis.* **139,** 362-365.

Franklin, J., and Richter, C. B. (1968). The isolation of *Salmonella poona* and a nonmotile variant from laboratory mice. *Lab Anim. Care* **18,** 92-93.

Fries, A. S. (1977a). Studies on Tyzzer's disease isolation and propagation of *Bacillus piliformis*. *Lab. Anim.* **11,** 75-78.

Fries, A. S. (1977b). Studies on Tyzzer's disease: application of immunofluorescence for detection of *Bacillus piliformis* and for demonstration and determination of antibodies to it in sera from mice and rats. *Lab. Anim.* **11,** 69-73.

Fries, A. S. (1978). Demonstration of antibodies to *Bacillus piliformis* in SPF colonies and experimental transplacental infection by *Bacillus piliformis* in mice. *Lab. Anim.* **12,** 23-26.

Fujiwara, K. (1967). Complement fixation reaction and agar gel double diffusion test in Tyzzer's disease of mice. *Jpn. J. Microbiol.* **11,** 103-117.

Fujiwara, K. (1978). Tyzzer's disease. *Jpn. J. Exp. Med.* **48,** 467-480.

Fujiwara, K., Kurashina, H., Maejima, K., Tajima, Y., Takagaki, Y., and Naiki, M. (1965). Actively induced immune resistance to the experimental Tyzzer's disease of mice. *Jpn. J. Exp. Med.* **35,** 259-275.

Furness, G., and Ferreira, I. (1959). The role of macrophages in natural immunity to Salmonellae. *J. Infect. Dis.* **104,** 203-206.

Ganaway, J. R. (1976). Bacterial, mycoplasma, and rickettsial diseases. *In* "The Biology of the Guinea Pig" (J. E. Wagner and P. J. Manning, eds.), pp. 121-135. Academic Press, New York.

Ganaway, J. R., Allen, A. M., and Moore, T. D. (1971a). Tyzzer's disease. *Am. J. Pathol.* **64,** 717-732.

Ganaway, J. R., Allen, A. M., and Moore, T. D. (1971b). Tyzzer's disease of rabbits: isolation and propagation of *Bacillus piliformis* (Tyzzer) in embryonated eggs. *Infect. Immun.* **3,** 429-437.

Ganaway, J. R., McReynolds, R. S., and Allen, A. M. (1976). Tyzzer's disease in free-living cottontail rabbits in Maryland. *J. Wildl. Dis.* **12,** 545-549.

Gard, S. (1944). *Bacillus piliformis* infection in mice, and its prevention. *Acta Pathol. Microbiol. Scand.* **54,** 123-134.

Germanier, R. (1970). Immunity in experimental salmonellosis. I. Protection by rough mutants of *Salmonella typhimurium*. *Infect. Immun.* **2,** 309-315.

Germanier, R. (1972). Immunity in experimental salmonellosis. III. Comparative immunization with viable and heat inactivated cells of *Salmonella typhimurium*. *Infect. Immun.* **5,** 792-797.

Germanier, R., and Furer, E. (1971). Immunity in experimental salmonellosis II. Basis for the avirulence and protective capacity of gal E mutants of *Salmonella typhimurium*. *Infect. Immun.* **4,** 663-673.

Gordon, L. E., Ruml, D., Hahne, H. J., and Miller, C. P. (1955). Studies on susceptibility to infection following ionizing radiation. IV. The pathogenesis of the endogenous bacteremias in mice. *J. Exp. Med.* **102,** 413-424.

Gowen, J. W. (1948). Inheritance of immunity in animals. *Annu. Rev. Microbiol.* **2,** 215-254.

Gowen, J. W. (1960). Genetic effects in nonspecific resistance to infectious disease. *Bacteriol. Rev.* **24,** 192-200.

Gowen, J. W., and Schott, R. G. (1933). Genetic predisposition to *Bacillus piliformis* infection among mice. *J. Hyg.* **33,** 370-378.

Greenwood, M., Topley, W. W. C., and Wilson, J. (1931). Contribution to the experimental study of epidemiology. The effect of vaccination on herd immunity. *J. Hyg.* **31,** 257-289.

Groschel, D., Paas, C. M. S., and Rosenberg, B. S. (1970). Inherited resistance and mouse typhoid I. Some factors which effect the survival of infected mice. *J. Reticuloendothel. Soc.* **7,** 484-499.

Haberman, R. T., and Williams, F. P. (1958). Salmonellosis in laboratory animals. *J. Natl. Cancer Inst.* **20,** 933-948.

Hamilton, J. R., and Overall, J. C., Jr. (1978). Synergistic infection with cytomegalovirus and *Pseudomonas aeruginosa* in mice. *J. Infect. Dis.* **137,** 775-782.

Hammond, C. W., Tompkins, M., and Miller, C. P. (1954a). Studies on susceptibility to infection following ionizing radiation. I. The time of onset and duration of the endogenous bacteremias in mice. *J. Exp. Med.* **99,** 405-410.

Hammond, C. W., Colling, M., Cooper, D., and Miller, C. P. (1954b). Studies on susceptibility to infection following ionizing radiation. II. Its estimation by oral inoculation at different times post irradiation. *J. Exp. Med.* **99,** 411-418.

Hammond, C. W., Ruml, D., Cooper, D. B., and Miller, C. P. (1955). Studies on susceptibility to infection following ionizing radiation. III. Susceptibility of the intestinal tract to oral inoculation with *Pseudomonas aeruginosa*. *J. Exp. Med.* **102,** 403-411.

Hashimoto, H., Honda, T., Kawakami, M., and Mitsuhashi, S. (1961). Studies on the experimental Salmonellosis VI. Long-lasting immunity of mice immunized with live vaccine of *Salmonella enteritidis*. *Jpn. J. Exp. Med.* **31,** 187-190.

Hoag, W. G., Stout, H., and Meier, H. (1965). Epidemiological aspects of the control of pseudomonas infection in mouse colonies. *Lab Anim. Care* **15,** 217-225.

Hobson, D. (1956). The chemotherapy of experimental mouse typhoid with furazolidone. *Br. J. Exp. Pathol.* **37,** 20-31.

Hobson, D. (1957a). The behaviour of a mutant strain of *S. typhimurium* in experimental mouse typhoid. *J. Hyg.* **55,** 323-333.

Hobson, D. (1957b). Chronic bacterial carriage in survivors of mouse typhoid. *J. Pathol. Bacteriol.* **73,** 399-410.

Hobson, D. (1957c). Resistance to reinfection in experimental mouse typhoid. *J. Hyg.* **55,** 334-343.

Hodges, G. R., Degener, C. E., and Barnes, W. G. (1978). Clinical significance of Citrobacter isolate. *Am. J. Clin. Pathol.* **70,** 37-40.

Howard, J. G. (1961). Resistance to infection with *Salmonella paratyphi* C in mice parasitized with a relatively avirulent strain of *Salmonella typhimurium*. *Nature (London)* **191,** 87-88.

Hunter, B. (1971). Eradication of Tyzzer's disease in a colony of barrier-maintained mice. *Lab. Anim.* **5,** 271-276.

Iwata, K., Nagai, T., Ikeda, T., and Okudaira, M. (1973). Studies on the pathogenesis of *Candida albicans* by the use of germfree mice. *In* "Germfree Research, Biological Effect of Gnotobiotic Environments" (J. B. Heneghan, ed.), pp. 331-337. Academic Press, New York.

Jenkin, C. R. (1962). An antigenic basis for virulence of strains of *Salmonella typhimurium*. *J. Exp. Med.* **115,** 731-743.

Jenkin, C. R., and Rowley, D. (1963). Basis for immunity to typhoid in mice and the question of "cellular immunity." *Bacteriol. Rev.* **27,** 391-404.

Jenkin, C. R., and Rowley, D. (1965). Partial purification of the "protective" antigen of *S. typhimurium* and its distribution amongst various strains of bacteria. *Aust. J. Exp. Biol. Med. Sci.* **43,** 65-78.

Jenkin, C. R., Rowley, D., and Auzins, I. (1964). The basis of immunity to mouse typhoid I. The carrier state. *Aust. J. Exp. Biol. Med. Sci.* **42,** 215-228.

Johnson, E., and Barthold, S. W. (1979). The ultrastructure of transmissible murine colonic hyperplasia. *Am. J. Pathol.* **97,** 291-313.

Jones, R. L., Peterson, C. M., Grady, R. W., Kumbaraci, T., Cerami A., and Graziano, J. H. (1977). Effects of iron chelators and iron overload on Salmonella infection. *Nature (London)* **267,** 63-65.

Jones, T. C. (1967). Pathology of the liver of rats and mice. *In* "Pathology of Laboratory Rats and Mice" (E. Cothin and F. J. C. Roe, eds.), pp. 11-12. Davis, Philadelphia, Pennsylvania.

Kanazawa, K., and Imai, A. (1959). Pure culture of the pathogenic agent of Tyzzer's disease of mice. *Nature (London)* **184,** 1810-1811.

Kaneko, J., Fujita, H., Matsuyama, S., Kojima, H., Saskura, H., Nakamura, Y., and Kodama, T. (1960). An outbreak of Tyzzer's disease in colonies of mice. (In Jpn.) *Exp. Anim.* **9,** 148-156.

Keep, J. M. (1977). Hazards of domestic pets. Ringworm and other skin conditions. *Aust. Fam. Physician* **6,** 1527-1536.

Kenny, K., and Herzberg, M. (1967). Early antibody response in mice to either infection of immunization with *Salmonella typhimurium*. *J. Bacteriol.* **93,** 773-778.

Kenny, K., and Herzberg, M. (1968). Antibody response and protection induced by immunization with smooth and rough strains in experimental salmonellosis. *J. Bacteriol.* **95,** 406-417.

LeMinor, L., and Rohde, R. (1974). Salmonella. *In* "Bergey's Manual of Determinative Bacteriology" (R. E. Buchanan and N. E. Gibbons, eds.), 8th ed., pp. 298-318. Williams & Wilkins, Baltimore, Maryland.

Louria, D. B., Fallon, N., and Browne, H. G. (1960). The influence of cortisone on experimental fungus injections in mice. *J. Clin. Invest.* **39,** 1435-1449.

Mackaness, G. B. (1967). The relationship of delayed hypersensitivity to acquired cellular resistance. *Br. Med. Bull.* **23,** 52-54.

Mackaness, G. B., Blanden, R. V., and Collins, F. M. (1966). Host-parasite relations in mouse typhoid. *J. Exp. Med.* **124,** 573-583.

Mackenzie, G. M., Pike, R. M., and Swinney, R. D. (1940). Virulence of Salmonella typhimurium II. Studies of the polysaccharide antigens of virulent and avirulent strains. *J. Bacteriol.* **40,** 197-214.

Macleod, D. R. E. (1954). Immunity to salmonella infection in mice. *J. Hyg.* **52,** 9-17.

McPherson, C. (1963). Reduction of *Pseudomonas aeruginosa* and coliform bacteria in mouse drinking water following treatment with hydrochloric acid or chlorine. *Lab. Anim. Care* **13,** 737-744.

Margard, W. L., and Litchfield, J. H. (1961). Triple sugar iron-urea agar, a new differential tube medium for confirming salmonella in mouse fecal samples. *Bacteriol. Proc.* p. 128.

Margard, W. L., and Litchfield, J. H. (1963). Occurrence of unusual Salmonellae in laboratory mice. *J. Bacteriol.* **85,** 1451-1452.

Margard, W. L., Peters, A. C., Dorko, N., Litchfield, J. H., and Davidson, R. S. (1963). Salmonellosis in mice-diagnostic procedures. *Lab. Anim. Care* **13,** 144-165.

Meynell, C. P. (1955). Some factors affecting the resistence of mice to oral infection by *Salmonella typhrimurium*. *Proc. R. Soc. Med.* **48,** 916-918.

Miller, C. P. (1959). Protective action of the normal microflora against enteric infection: an experimental study in the mouse. *Univ. Mech. Med. Bull.* **25,** 272-279.

Miller, C. P., and Bohnhoff, M. (1962). A study of experimental *Salmonella* infection in the mouse. *J. Infect. Dis.* **111,** 107-116.

Miller, C. P., and Bohnhoff, M. (1963). Changes in the mouse's enteric microflora associated with enhanced susceptibility to *Salmonella* infection following streptomycin treatment. *J. Infect. Dis.* **113,** 59-66.

Miller, C. P., Bohnhoff, M., and Drake, B. L. (1954). The effect of antibiotic therapy on susceptibility to an experimental enteric infection. *Trans. Assoc. Am. Physicians* **67,** 156-161.

Miller, C. P., Bohnhoff, M., and Rifkind, D. (1956). The effect of an antibiotic on the susceptibility of the mouse's intestinal tract to salmonella infection. *Trans. Am. Clin. Climatol. Assoc.* **68,** 51-55.

Mitsuhashi, S., Kawakami, M., Yamaguchi, Y., and Nagai, M. (1958). Studies on the experimental typhoid I. A comparative study of living and killed vaccines against the intection of mice with *S. enteritidis*. *Jpn. J. Exp. Med.* **28,** 249-258.

Mitsuhashi, S., Kawakami, M., Goto, S., Yoshimura, T., and Hashimoto, H. (1959). Studies on the experimental typhoid Iv. The micro-organisms in the organs of infected mice. *Jpn. J. Exp. Med.* **29,** 311-321.

Mitsuhashi, S., Sato, J., and Tanaka, T. (1961). Experimental salmonellosis: Intracellular growth of *Salmonella enteritidis* ingested in mononuclear phagocytes, and cellular basis of immunity. *J. Bacteriol.* **81,** 863-868.

Moore, R. N., Johnson, B. A., and Berry, L. J. (1977). Nutritional effects of salmonellosis in mice. *Am. J. Clin. Nutr.* **30,** 1289-1293.

Morello, J. A., Digenio, T. A., and Baker, E. E. (1964). Evaluation of serological and cultural methods for the diagnosis of chronic salmonellosis in mice. *J. Bacteriol.* **88,** 1277-1282.

Mullink, J. W. M. A. (1968). Tyzzer's disease. Intestinal lesions in both S.P.F. and conventional mice. *Z. Versuchstierkd.* **10,** 271-284.

Oakberg, E. F. (1946). Constitution of liver and spleen as a physical basis for genetic resistance to mouse typhoid. *J. Infect. Dis.* **78,** 79-98.

Oestreicher, E. J., Hummell, R. P., Maley, M. P., and MacMillan, B. G. (1973). Pathogenicity of *Candida albicans* as influenced by *Escherichia coli*, gentamicin therapy, and thermal injury in the germfree mouse. *In* "Germfree Research, Biological Effect of Gnotobiotic Environments" (J. B. Heneghan, ed.), pp. 405-410. Academic Press, New York.

Olson, L. D., and Ediger, R. D. (1972). Histopathologic study of the heads of circling mice infected with *Pseudomonas aeruginosa*. *Lab. Anim. Sci.* **22,** 522-527.

Orskov, I. (1974). *Klebsiella*. *In* "Bergey's Manual of Determinative Bacteriology" (R. E. Buchanan and N. E. Gibbons, eds.), 8th ed., pp. 321-323. Williams & Wilkins, Baltimore, Maryland.

Patton, R. M., Riggs, A. R., Compton, S. B., and Chick, E. W. (1976). Histoplasmosis in purbred mice: influence of genetic susceptibility and immune depression on treatment. *Mycopathologica* **60,** 39-43.

Pike, R. M., and Mackenzie, G. M. (1940). Virulence of *Salmonella typhimurium* I. Analysis of experimental infection in mice with strains of high and low virulence. *J. Bacteriol.* **40,** 171-195.

Plant, J., and Glynn, A. A. (1974). Natural resistance to *Salmonella* infection, delayed hypersensitivity and Ir genes in different strains of mice. *Nature (London)* **248,** 345-347.

Plant, J., and Glynn, A. A. (1976). Genetics of resistance to infection with *Salmonella typhimurium* in mice. *J. Infect. Dis.* **133,** 72-78.

Pullinger, B. D., and Iverson, S. (1960). Mammary tumor incidence in relation to age and number of litters in C3H and R III mice. *Br. J. Cancer* **14,** 267-278.

Puschmann, M., and Ganzoni, A. M. (1977). Increased resistance of iron deficient mice to salmonella infection. *Infect. Immun.* **17,** 663-664.

Rabstein, M. M. (1958). The practical establishment and maintenance of Salmonella free mouse colonies. *Proc. Anim. Care Panel* **8,** 67-74.

Rights, F. L., Jackson, E. B., and Smadel, J. E. (1947). Observations on Tyzzer's disease of mice. *Am. J. Pathol.* **23,** 627-635.

Robson, H. G., and Vas, S. I. (1972). Resistance of inbred mice to *Salmonella typhimurium. J. Infect. Dis.* **126,** 378-386.

Rowley, D., Turner, K. J., and Jenkin, C. R. (1964). The basis for immunity to mouse typhoid III. Cell-bound antibody. *Aust. J. Exp. Biol. Med. Sci.* **42,** 237-248.

Rowley, D., Auzins, I., and Jenkin, C. R. (1968). Further studies regarding the question of cellular immunity in mouse typhoid. *Aust. J. Exp. Biol. Med. Sci.* **46,** 447-463.

Ruitenberg, E. J., Guinee, P. A. M., Kruyt, B. C., and Berkvens, J. M. (1971). Salmonella pathogenesis in germfree mice. A bacteriological and histological study. *Br. J. Exp. Pathol.* **52,** 192-197.

Sato, I., Tanaka, T., Saito, K., and Mitsuhashi, S. (1962). Cellular basis of immunity V. *In vitro* acquisition of immunity of the mouse mononuclear phagocytes immunized with live vaccine of *Salmonella* enteritidis. *Proc. Jpn. Acad.* **38,** 387-391.

Savage, D. C. (1972). Survival on mucosal epithelia, epithelial penetration and growth in tissue of pathogenic bacteria. *In* "Microbiol Pathogenicity in Man and Animals" (H. Smith and J. H. Pearce, eds.), Symposium of the Society for General Microbiology, Vol. 22, pp. 25-57. Cambridge Univ. Press, London and New York.

Schaedler, R. W., and Dubos, R. J. (1962). The fecal flora of various strains of mice. Its bearing on their susceptibility to endotoxin. *J. Exp. Med.* **115,** 1149-1160.

Schiff, L. J., Barbera, P. W., Port, C. S., Yamashiroya, H. M., Schefner, A. M., and Poiley, S. M. (1972). Enteropathogenic *Escherichia coli* infection: Increasing awareness of a problem in laboratory animals. *Lab. Anim. Sci.* **22,** 705-708.

Schneider, H. A. (1956). Nutritional and genetic factors in the natural resistance of mice to *Salmonella* infections. *Ann. N.Y. Acad. Sci.* **66,** 337-347.

Schneider, H. A. (1967). Ecological ectocrines in experimental epidemiology. *Science* **158,** 597-603.

Schneider, H. A., and Webster, L. T. (1945). Nutrition of the host and natural resistance to infection. I. The effect of diet on the response of several genotypes of *Mus musculus* to *Salmonella enteritidis* infection. *J. Exp. Med.* **81,** 359-384.

Schook, L. B., Carrick, L., Jr., and Berk, R. S. (1977). Experimental pulmonary infection of mice by tracheal intubation of *Pseudomonas aeruginosa:* The use of antineoplastic agents to overcome natural resistance. *Can. J. Microbiol.* **23,** 823-826.

Sedlak, J. (1974). *Citrobacter. In* "Bergey's Manual of Determinative Bacteriology" (R. E. Buchanan and N. E. Gibbons, eds.), 8th ed., pp. 296-297. Williams & Wilkins, Baltimore, Maryland.

Seligmann, E., and Wassermann, M. (1949). Action of chloromycetin on salmonella. *Proc. Soc. Exp. Biol. Med.* **71,** 253-255.

Silverman, J., Chavannes, J., Rigotty, J., and Ornaf, M. (1979). A natural outbreak of transmissable murine colonic hyperplasia in A/J mice. *Lab. Anim. Sci.* **29,** 209-213.

Simon, P. C. (1977). Isolation of *Bacillus piliformis* from rabbits. *Can. Vet. J.* **18,** 46-48.

Slanetz, C. A. (1946). Control of salmonella infection in mice by streptomycin. *Proc. Soc. Exp. Biol. Med.* **62,** 248.

Slanetz, C. A. (1948). The control of salmonella infections in colonies of mice. *J. Bacteriol.* **56,** 771-775.

Smith, A. L. (1977). "Principles of Microbiology," 8th ed., p. 369. Mosby, St. Louis, Missouri.

Smith, J. M. B., and Austwick, P. K. C. (1967). Fungal disease of rats and mice. *In* "Pathology of Laboratory Rats and Mice" (E. Cotchin and F. J. Roe, eds.), pp. 681-732. Davis, Philadelphia, Pennsylvania.

Staib, F., Gorsse, G., and Mishra, S. K. (1976). *Staphylococcus aureus* and *Candida albicans* infection (animal experiments). *Zentralbl. Bakteriol., Parasitenkd. Infektionskr. Hyg., Abt. 1: Orig., Reihe A* **234,** 450-461.

Stott, J. A., Hodgson, J. E., and Chaney, J. C. (1975). Incidence of Salmonellae in animal feed and the effect of pelleting on content of enerobacteriaciae. *J. Appl. Bacteriol.* **39,** 41-46.

Tannock, G. W., and Smith, J. M. B. (1971). A *Salmonella* carrier state involving the upper respiratory tract of mice. *J. Infect. Dis.* **123,** 502-506.

Tannock, G. W., Blumershine, R. V. H., and Savage, D. C. (1975). Association of *Salmonella typhimurium* with, and its invasion of, the ileal mucosa of mice. *Infect. Immun.* **11,** 365-370.

Thunert, A. (1978). *Escherichia coli* in laboratory animals. *Z. Versuchstierkd.* **20,** 19-27.

Tuffery, A. A. (1956). Laboratory mouse in Great Britain. IV Intercurrent infection (Tyzzer's disease). *Vet. Rec.* **68,** 511-515.

Tully, J. G. (1965). Biochemical, morphological, and serological classification of mycoplasma of murine origin. *J. Infect. Dis.* **115,** 171-185.

Turner, K. J., Jenkin, C. R., and Rowley, D. (1964). The basis of immunity to mouse typhoid II. Antibody formation during the carrier state. *Aust. J. Exp. Biol. Med. Sci.* **42,** 229-236.

Tyzzer, E. E. (1917). A fatal disease of the Japanese waltzing mouse caused by a spore-bearing bacillus (*Bacillus piliformis* N. SP.). *J. Med. Res.* **37,** 307-338.

Urano, T., and Maejima, K. (1978). Distribution of *Pseudomonas aeruginosa* in experimentally infected mice. (In Jpn.) *Exp. Anim.* **27,** 263-269.

Ushiba, D. (1965). Two types of immunity in experimental typhoid, "cellular immunity" and "humoral immunity". *Keio J. Med.* **14,** 45-61.

Ushiba, D., Yumoto, M., Ohno, S., and Sasaki, S. (1955). Infectivity in oral infection of mice with *Salmonella enteritidis:* Effect of streptomycin on the infectivity with special reference to normal intestinal flora. *Keio J. Med.* **4,** 163-173.

Venneman, M. R., and Berry, L. J. (1971a). Cell-mediated resistance induced with immunogenic preparations from *Salmonella typhimurium. Infect. Immun.* **4,** 374-380.

Venneman, M. R., and Berry, L. J. (1971b). Serum-mediated resistance induced with immunogenic preparations from *Salmonella typhimurium. Infect. Immun.* **4,** 374-380.

Venneman, M. R., Bigley, N. J., and Berry, L. J. (1970). Immunogenicity of ribonucleic acid preparations obtained from *Salmonella typhimurium. Infect. Immun.* **1,** 574-582.

Webster, L. T. (1937). Inheritance of resistance of mice to enteric bacterial and neurotropic virus infection. *J. Exp. Med.* **65,** 261-286.

Wensinck, F. (1961). The origin of induced *Pseudomonas aeruginosa* bacteriaemia in irradiated mice. *J. Pathol. Bacteriol.* **81,** 401-408.

Wensinck, F., VanBekkum, D. W., and Renaud, H. (1957). The prevention of *Pseudomonas aeruginosa* infections in irradiated mice and rats. *Radiat. Res.* **7,** 491-499.

Wetmore, P. W., and Hoag, W. G. (1960). *Salmonella binza* and *Salmonella bredeney* from laboratory mice. *J. Bacteriol.* **80,** 283.

Wilson, G. S., and Miles, A. A. (1975). "Topley and Wilson's Principles of Bacteriology and Immunity," 6th ed., p. 919. Williams & Wilkins, Baltimore, Maryland.

Wray, C., Cojka, W. J., Morris, J. A., and Brinley Morgan, W. J. (1977). The immunization of mice and calves with gal E mutants of *Salmonella typhrimurium. J. Hyg.* **79,** 17-24.

RELATED REFERENCES

Barthold, S. W. (1979). Autoradiographic cytokinetics of colonic mucosal hyperplasia in mice. *Cancer Res.* **39,** 24–29.

Barthold, S. W., and Jonas, A. M. (1977). Morphogenesis of early 1, 2-dimethylhydrazine-induced lesions and latent period reduction of colon carcinogenesis in mice by a variant of *Citrobacter freundii*. *Cancer Res.* **37,** 4352–4360.

Berry, L. J., and Mitchell, R. B. (1953). Influence of simulated altitude on resistance-susceptibility to *S. typhimurium* infection in mice. *Tex. Rep. Biol. Med.* **11,** 379–401.

Berry, L. J., Douglas, G. N., Hoops, P., and Prather, N. E. (1975). Immunization against salmonellosis. *In* "The Immune System and Infectious Disease" (E. Neter and F. Miligram, eds.), pp. 388–398. Karger, Basel.

Bjornson, A. B., and Michael, J. G. (1971). Contribution of humoral and cellular factors to the resistance to experimental infection by *Pseudomonas aeruginosa* in mice. I. Interaction between immunoglobulins, heat-labile serum factors, and phagocytic cells in the killing of bacteria. *Infect. Immun.* **4,** 462–467.

Bjornson, A. B., and Michael, J. G. (1972). Contribution of humoral and cellular factors to the resistance to experimental infection by *Pseudomonas aeruginosa* in mice. II. Opsonic, agglutinative, and protective capacities of immunoglobulin G anti-pseudomonas antibodies. *Infect. Immun.* **5,** 775–782.

Bohme, D. H. (1966). Response of inbred mice to virulent and avirulent *Salmonella typhimurium* and their endotoxins. *J. Recticuloendothel. Soc.* **3,** 18–28.

Campa, M., Ferrannini, E., Colizzi, V., and Garzelli, G. (1977). Distribution of 51Cr-labeled lymph node cells in *Pseudomonas aeruginosa*-infected mice. *J. Nucl. Med. Allied Sci.* **21,** 37–42.

Carter, P. B., and Collins, F. M. (1975). Peyer's patch responsiveness to *Salmonella* in mice. *J. Reticuloendothel. Soc.* **17,** 38–46.

Carter, P. B., and Collins, F. M. (1977). Assessment of typhoid vaccines by using the intraperitoneal route of challenge. *Infect. Immun.* **17,** 555–560.

Carter, P. B., Woolcock, J. B., and Collins, F. M. (1975). Involvement of the upper respiratory tract in orally induced salmonellosis in mice. *J. Infect. Dis.* **131,** 570–574.

Chrisp, C. E., Bookman, G. A., and Kerlan, J. T. (1971). Epizootiology and control of salmonellosis in a laboratory aviary. *Lab Anim. Sci.* **21,** 49–53.

Cole, C. R., Farrell, R. L., Chamberlain, D. M., Prior, J. A., and Saslaw, S. (1953). Histoplasmosis in animals. *J. Am. Vet. Med. Assoc.* **14,** 471–473.

Collins, F. M. (1970). Immunity to enteric infection in mice. *Infect. Immun.* **1,** 243–250.

Collins, F. M. (1979). Mucosal defenses against Salmonella infection in the mouse. *J. Infect. Dis.* **139,** 503–510.

Collins, F. M., and Milne, M. (1966). Heat-labile antigens of *Salmonella enteritidis*. *J. Bacteriol.* **92,** 549–557.

Cutler, J. E. (1976). Acute systemic candidiasis in normal and congenitally thymic-deficient (nude) mice. *J. Reticuloendothel. Soc.* **19,** 121–124.

Dubos, R. J., and Schaedler, R. W. (1962). The effect of diet on the fecal bacterial flora of mice and on their resistance to infection. *J. Exp. Med.* **115,** 1161–1172.

Edwards, P. R., and Bruner, D. W. (1943). The occurrence and distribution of salmonella types in the United States. *J. Infect. Dis.* **72,** 58–67.

Emmons, C. W., Rowley, D. A., Olsen, B. J., Mattern, C. F. T., Bell, J. A., Powell, E., and Marcey, E. A. (1955). Histoplasmosis: Occurrence of an apparent infection in dogs, cats and other animals. *Am. J. Hyg.* **61,** 40–41.

Felix, A. (1951). The preparation, testing and standardization of typhoid vaccine. *J. Hyg.* **49,** 268–287.

Fisher, M. W. (1977). A polyvalent human gamma-globulin immune to *Pseudomonas aeruginosa:* passive protection of mice against lethal infection. *J. Infect. Dis.* **136,** Suppl., S181–S185.

Folb, P. I., Timme, A., and Horowitz, A. (1977). Nocardia infections in congenitally athymic (nude) mice and in other inbred mouse strains. *Infect. Immun.* **18,** 459–466.

Fujiwara, K., Fukuda, S., Takagaki, Y., and Tajima, Y. (1963). Tyzzer's disease in mice: electron microscopy of the liver lesions. *Jpn. J. Exp. Med.* **33,** 203–212.

Fujiwara, K., Kurashina, H., Matsunuma, N., and Takahashi, R. (1968). Demonstration of peritrichous flagella of Tyzzer's disease organism. *Jpn. J. Microbiol.* **12,** 361–363.

Fujiwara, K., Takahashi, R., Kurashina, H., and Matsunuma, N. (1969). Protective serum antibodies in Tyzzer's disease of mice. *Jpn. J. Exp. Med.* **39,** 491–504.

Fujiwara, K., Hirano, N., Takenaka, S., and Sato, K. (1973). Peroral infection in Tyzzer's disease of mice. *Jpn. J. Exp. Med.* **43,** 33–42.

Gaydos, J. M., Carrick, L., Jr., and Berk, R. S. (1975). Experimental studies on mice challenged subcutaneously with *Pseudomonas aeruginosa*. *Proc. Soc. Exp. Biol. Med.* **149,** 908–914.

Gerichter, C. B. (1960). The dissemination of *Salmonella typhi, paratyphi* A and *paratyphi* B through the organs of the white mouse by oral infection. *J. Hyg.* **58,** 307–319.

Goetz, M. E., Dee, O., and Taylor, N. (1967). A naturally occurring outbreak of *Candida tropicalis* infection in a laboratory mouse colony. *Am. J. Pathol.* **50,** 361–369.

Gordon, L. E., Ruml, D., Hahne, H. J., and Miller, C. P. (1955). Studies on susceptibility to infection following ionizing radiation. IV. The pathogenesis of the endogenous bacteremias in mice. *J. Exp. Med.* **102,** 413–424.

Gorrill, R. H., and DeNavasquez, S. J. (1964). Experimental pylonephritis in the mouse produced by *Escherichia coli, Pseudomonas aeruginosa* and *Proteus mirabilis*. *J. Pathol. Bacteriol.* **87,** 79–87.

Griffin, C. A. (1952). A study of prepared feeds in relation to Salmonella infection in laboratory animals. *J. Am. Vet. Med. Assoc.* **121,** 197–200.

Hamilton, J. R., Overall, J. C., Jr., and Glosgow, L. A. (1976). Synergistic effect on mortality in mice with murine cytomegalovirus and *Pseudomonas aeruginosa, Staphylococcus aureus,* or *candida albicans*. *Infect. Immun.* **14,** 982–989.

Haranka, K., Sugane, K., and Mashimo, K. (1976). Combined therapy of antiendotoxin (OEP) antibody and gentamicin in the immuno-suppressed mice with *Pseudomonas aeruginosa* infection. *In* "Chemotherapy" (J. D. Williams and A. M. Geddes, eds.), pp. 323–350. Plenum, New York.

Hazlett, L. D., Rosen, D. D., and Berk, R. S. (1977). Pseudomonas eye infection in cyclophosphamide-treated mice. *Invest. Ophthalmol. Visual. Sci.* **16,** 649–652.

Hazlett, L. D., Rosen, D. D., and Berk, R. S. (1978). Age related susceptibility to *Pseudomonas aeruginosa* ocular infections in mice. *Infect. Immun.* **20,** 25–29.

Hochadel, J. F., and Keller, K. F. (1977). Protective effects of passively transferred immune T- or B-lymphocytes in mice infected with *Salmonella typhimurium*. *J. Infect. Dis.* **135,** 813–823.

Hoops, R., Prather, N. E., Berry, L. G., and Ravel, J. M. (1976). Evidence for an extrinsic immunogen in effective ribosomal vaccines from *Salmonella typhimurium*. *Infect. Immun.* **13,** 1184–1192.

Hubalek, Z. (1977). Mouse inoculation with various saprophytic fungi. *Mykosen* **20,** 229–234.

Hunter, P. A., Rolinson, G. N., and Witting, D. A. (1976). Effect of carbenicillin on pseudomonas infection. *In* "Chemotherapy" (J. D. Williams and A. M. Geddes, eds.), Vol. 2, pp. 289–293. Plenum, New York.

Ishibashi, T., Harada, S., Harada, Y., Kitahara, Y., Takamoto, M., and Sugiyama, K. (1978). Experimental pseudomonas infection in mice-acquired resistance against pseudomonas septicemia and altered susceptibility in BCG infected mice. *Jpn. J. Exp. Med.* **48,** 313–320.

Jensen, K. A. (1929). Immunitätsstudien. *Z. Immunitaetsforsch. Exp. Ther.* **63,** 298–326.

Johnson, J. A., Lau, B. H., Nutter, R. L., Slater, J. M., and Winter, C. E. (1978). Effect of L1210 Leukemia on the susceptibility of mice to *Candida albicans* infection. *Infect. Immun.* **19,** 146–151.

Jones, R. J., and Dyster, R. E. (1973). The role of polymorphonuclear leucocytes in protecting mice vaccinated against *Pseudomonas aeruginosa* infections. *Br. J. Exp. Pathol.* **54,** 416–421.

Kawakami, M., Osawa, N., and Mitsuhashi, S. (1966). Experimental salmonellosis VII. Comparison of the immunizing effect of live vaccine and materials extracted from *Salmonella enteritidis*. *J. Bacteriol.* **92,** 1585–1589.

Kemble, P. R. (1966). Treatment of *Salmonella typhimurium* infection. *Vet. Rec.* **79,** 410.

Kitahara, M., Kobayashi, G. S., and Medoff, G. (1976). Enhanced efficiency of amphotericin B and rifampicin combined in treatment of murine histoplasmosis and blastomycosis. *J. Infect. Dis.* **133,** 663–668.

Klinge, K. (1960). Differential techniques and methods of isolation of pseudomonas. *J. Appl. Bacteriol.* **23,** 442–462.

Koopman, J. P., Kennis, H. M., and VanDruten, J. A. M. (1978). Colonization resistance of the digestive tract and gastrointestinal transit time in SPF mice. *Lab Anim.* **12,** 223–226.

Laborde, H. F., and de Farjardo, C. L. (1965). Pseudomonas vaccine I. Propagation and assay. *J. Bacteriol.* **90,** 290–291.

Liu, P. V., and Mercer, C. B. (1963). Growth, toxigenicity, and virulence of *Pseudomonas aeruginosa*. *J. Hyg.* **61,** 485–491.

Lutz, A. (1954). Association chloramphénicol-framycétine dans le traitement de la salmonellose expérimentale de la souris. *Bull. Acad. Natl. Med. (Paris)* **138,** 28–29.

Mackaness, G. B., and Blanden, R. V. (1967). Cellular immunity. *Prog. Allergy* **11,** 89–140.

Maejima, K., Fujiwara, K., Takagaki, Y., Naiki, M., Kurashina, H., and Tajima, Y. (1965). Dietetic effects on experimental Tyzzer's disease of mice. *Jpn. J. Exp. Med.* **35,** 1–10.

Maejima, K., Urano, T., Fujiwara, K., and Homma, J. Y. (1972). Bacteriological and clinical observations of experimental infection with *Pseudomonas aeruginosa* in mice. *Jpn. J. Exp. Med.* **42,** 569–574.

Miller, C. P., Tompkins, M., Colling, M., Bohnhoff, M., and Hammond, C. W. (1952). The Gram-negative flora of the upper intestinal tract and its influence on post-irradiation bacteremia in mice. *Bacteriol. Proc.* p. 72.

Moody, M. R., Kessell, R. W. I., Young, V. M., and Fiset, P. (1978). Role of nonagglutinating antibody in the protracted immunity of vaccinated mice to *Pseudomonas aeruginosa* infection. *Infect. Immun.* **21,** 905.

Mushin, R., and Dubos, R. (1965). Colonization of the mouse intestine with *Escherichia coli*. *J. Exp. Med.* **122,** 745–757.

Naiki, M., Takagaki, Y., and Fujiwara, K. (1965). Note on the change of transaminases in the liver and the significance of the transaminase ratio in experimental Tyzzer's disease of mice. *Jpn. J. Exp. Med.* **35,** 305–309.

Nakano, M. (1962). Mutants of salmonella with unusually low toxicity for mice. *Nature (London)* **196,** 1118–1119.

Onodera, T., and Fujiwara, K. (1970). Experimental encephalopathy in Tyzzer's disease of mice. *Jpn. J. Exp. Med.* **40,** 295–323.

Onodera, T., and Fujiwara, K. (1973). Naso-encephalopathy in suckling mice inoculated intranasally with the Tyzzer's organism. *Jpn. J. Exp. Med.* **43,** 509–522.

Orskov, J., and Moltke, O. (1928). Studien über den infektions mechanisnus bei vershiedenen paratyphus-infektionen an weissen mäusen. *Z. Immunitaetsforsch. Exp. Ther.* **59,** 357–405.

Orskov, J., Jensen, K. A., and Kobayashi, K. (1928). Studien über Breslauinfektion der Mäuse Speziell mit Rüchsicht auf die Bedeutung des Retikuloendothelialgewebes. *Z. Immunitaetsforsch. Exp. Ther.* **55,** 34–68.

Ozawa, A., and Freter, R. (1964). Ecological mechanism controlling growth of *Escherichia coli* in continuous flow cultures and in the mouse intestine. *J. Infect. Dis.* **114,** 235–242.

Ozawa, A., Goto, J., Ito, Y., and Shibata, H. (1973). Histopathological and biochemical responses of germfree and conventional mice with Salmonella infection. *In* "Germfree Research, Biological Effect of Gnotobiotic Environments" (J. B. Heneghan, ed.), pp. 325–330. Academic Press, New York.

Peace, T., and Soave, O. A. (1969). Tyzzer's disease in a group of newly purchased mice. *Lab. Ahim. Dig.* **5,** 891.

Pierson, C. L., Johnson, A. G., and Feller, I. (1976). Effect of cyclophosphamide on the immune response to *Pseudomonas aeruginosa* in mice. *Infect. Immun.* **14,** 168–177.

Purnell, D. M. (1976). Enhancement of tissue invasion in murine aspergillosis by systemic administration of suspensions of killed *Corynebacterium parvum*. *Am. J. Pathol.* **83,** 547–555.

Ray, T. L., and Wuepper, K. D. (1976). Experimental cutaneous candidiosis in rodents. *J. Invest. Dermatol.* **66,** 29–33.

Rogers, T., and Balish, E. (1976). Experimental *Candida albicans* infection in conventional mice and germfree rats. *Infect. Immun.* **14,** 33–38.

Sacquet, E. (1960). Contribution a l'étude de la sensibilité de diverses lignees de souris a l'inoculation de *Salmonella typhi*. *Ann. Inst. Pasteur, Paris* **98,** 880–886.

Saito, K., Akiyama, T., Nakano, M., and Ushiba, D. (1960). Enteraction between *Salmonella enteritidis* and tissue cultured macrophages derived from immunized animals. *Jpn. J. Microbiol.* **4,** 395–407.

Saito, K., Nakano, M., Akiyama, T., and Ushiba, D. (1962). Passive transfer of immunity to typoid by macrophages. *J. Bacteriol.* **84,** 500–507.

Sato, I., Tanaka, T., Saito, K., and Mitsuhashi, S. (1961). Cellular basis for immunity II. Inhibition of the intracellular multiplication of *Salmonella enteritidis* in mouse phagocytes of the liver and subcutaneous tissue immunized with live vaccine. *Proc. Jpn. Acad.* **37,** 261–266.

Savage, N. L., and Lewis, S. H. (1972). Application of immunofluorescence to detection of Tyzzer's disease agent (*Bacillus piliformis*) in experimentally infected mice. *Am. J. Vet. Res.* **33,** 1007–1011.

Schaedler, R. W., and Dubos, R. J. (1956). Reversible changes in the susceptibility of mice to bacterial infections. II. Changes brought about by nutritional deficiencies. *J. Exp. Med.* **104,** 67–84.

Schlewinski, E., Graben, N., Funk, J., Sahm, E., and Raettig, H. (1971). Orale immunisierung mit nichtvermehrungstähigen mikroorganismen oder ihren antigenen XIII. Mitteilung: persorption und sekretion von microorganismen im tierversuch. *Zentralbl. Backteriol., Parasitenkd., Infektionskr. Hyg., Abt. 1: Orig., Reihe A* **218,** 93–104.

Schneider, H. A., and Zinder, N. D. (1956). Nutrition of the host and natural resistance to infection V. An improved assay employing genetic markers in double strain inoculation test. *J. Exp. Med.* **103,** 207–223.

Schwarz, J. (1954). The deep mycoses in laboratory animals. *Proc. Anim. Care Panel* **5,** 37–70.

Sensakovic, J. W., and Bartell, P. F. (1977). Glycolipoprotein from

Pseudomonas aeruginosa as a protective antigen against *P. aeruginosa* infection in mice. *Infect. Immun.* **18,** 304-309.

Shadomy, S. (1977). *In vitro* and *in vivo* studies on synergistic antifungal activity. *Contrib. Microbiol. Immunol.* **4,** 147-157.

Smith, R. A., and Bigley, N. J. (1972). Ribonucleic acid-protein fractions of virulent *Salmonella typhimurium* as protective immunogens. *Infect. Immun.* **6,** 377-383.

Sojka, W. J., Wray, C., Hudson, E. B., and Benson, J. A. (1975). Incidence of salmonella infection in animals in England and Wales 1968-73. *Vet. Rec.* **96,** 280-284.

Stieritz, D. D., and Holder, I. A. (1978). Experimental studies of the pathogenesis of *Pseudomonas aeruginosa* infection: evidence for the *in vivo* production of a lethal toxin. *J. Med. Microbiol.* **11,** 101-109.

Takagaki, Y., and Fujiwara, K. (1968). Bacteremia in experimental Tyzzer's disease of mice. *Jpn. J. Microbiol.* **12,** 129-143.

Takagaki, Y., Iwata, M., Fujiwara, K., and Tajima, Y. (1961). Studies on the Tyzzer's disease in mice. 1. Epizootiological observation. (In Jpn.) *Exp. Anim.* **10,** 75-82.

Takagaki, Y., Naiki, M., Ito, N., Noguchi, G., and Fujiwara, K. (1967). Checking of infections due to *Corynebacterium kutcheria* and Tyzzer's organisms among mouse breeding colonies by cortisone injection. (In Jpn.) *Exp. Anim.* **16,** 12-19.

Takeuchi, A. (1967). Electron microscope studies of experimental salmonella infection I. Penetration into the intestinal epithelium by *Salmonella typhimurium*. *Am. J. Pathol.* **50,** 109-136.

Takeuchi, A. (1971). Penetration of the intestinal epithelium by various microorganisms. *Curr. Top. Pathol.* **54,** 1-27.

Thompson, D. W., and Kaplan, W. (1977). Laboratory-acquired sporotrichosis. *Sabouraudia* **15,** 167-170.

Urano, T., Maejima, K., Okada, O., Takashina, S., Syumiya, S., and Tamura, H. (1977). Control of *Pseudomonas aeruginosa* infection in laboratory mice with gentamicin. (In Jpn.) *Exp. Anim.* **26,** 259-262.

Ushiba, D., Saito, K., Akiyama, T., Nakano, M., Sugiyama, T., and Shirono, S. (1959). Studies on experimental typhoid: Bacterial multiplication and host cell response after infection with *Salmonella enteritidis* in mice immunized with live and killed vaccines. *Jph. J. Microbiol.* **3,** 231-242.

Volkman, A., and Collins, F. M. (1968). Recovery of delayed-type hypersensitivity in mice following suppressive doses of x-radiation. *J. Immunol.* **101,** 846-859.

Webster, L. T. (1924). Microbic virulence and host susceptibility in paratyphoid-enteritidis infection of white mice III. The immunity of a surviving population. *J. Exp. Med.* **39,** 129-135.

Weitzman, I., Bonaporte, P., Guevin, V., and Crist, M. (1973). Cryptococcus in a field mouse. *Sabouraudia* **11,** 77-79.

Wensinck, F. (1961). The origin of induced *Pseudomonas aeruginosa* bacteremia in irradiated mice. *J. Pathol. Bacteriol.* **81,** 401-408.

Wensinck, O. V. F., and VanBekkum, D. W. (1959). Lesions of the tongue in irradiated mice. *Radiat. Res.* **10,** 339-346.

Winner, H. I. (1977). Recent advances in systemic candidosis. *Contrib. Microbiol. Immunol.* **4,** 64-76.

Yokoiyama, S., and Fujiwara, K. (1971). Effect of antibiotics on Tyzzer's disease. *Jpn. J. Exp. Med.* **41,** 49-57.

Chapter 2

Mycoplasmal and Other Bacterial Diseases of the Respiratory System

J. Russell Lindsey, Gail H. Cassell, and Maureen K. Davidson

I. Introduction ... 21
II. Murine Respiratory Mycoplasmosis (MRM) ... 22
A. Historical Background of MRM in the Mouse ... 22
B. Characteristics of *Mycoplasma pulmonis* ... 22
C. Experimental Disease ... 22
D. Natural Disease and Pathology ... 23
E. MRM Complicated by Other Infections ... 28
F. Immune Response ... 29
G. Diagnosis ... 30
H. Epizootiology ... 32
I. Control ... 32
J. Interference with Research ... 33
III. Klebsiellosis ... 33
IV. Pasteurellosis ... 33
A. *Pasteurella pneumotropica,* a Respiratory Tract Pathogen? ... 33
B. *Pasteurella pneumotropica,* an Opportunistic Respiratory Tract Pathogen? ... 34
C. Nonrespiratory Disease ... 34
V. Chlamydial Pneumonitis ... 35
VI. Corynebacteriosis ... 36
VII. Miscellaneous Infections ... 36
VIII. Conclusions ... 37
References ... 37

I. INTRODUCTION

The respiratory tract is the site of only a few natural bacterial and mycoplasmal infections in mice. Nevertheless, some of them are pathogens which are recognized to cause serious disease problems in contemporary mouse colonies, either through their pathogenic effects alone or through synergistic action with indigenous viruses. The purpose of this chapter is to summarize present understanding of these infections and the diseases which they cause.

ISBN 0-12-262502-1

II. MURINE RESPIRATORY MYCOPLASMOSIS (MRM)

This disease apparently was first recognized by Hektoen about 1915 in the laboratory rat (Hektoen, 1915–1916). Since that time, a voluminous literature has documented the ubiquitous distribution and supreme importance of the disease in rats (Lindsey *et al.*, 1971, 1978). Although less well known as a problem of mice, it is nevertheless an important common disease of this species. During the long controversy which surrounded the etiology of the disease until about 1970, approximately 20 terms appeared in the literature designating this clinicopathologic entity due to *Mycoplasma pulmonis* (Cassell *et al.*, 1973, 1979).

A. Historical Background of MRM in the Mouse

Much of the present understanding of MRM in the mouse derives from the foundational work of John B. Nelson, who devoted more than 40 years of his distinguished career at Rockefeller University to the study of this disease. In 1937 Nelson (1937a,b,c) published the first description of MRM (which he called *infectious catarrh*) in the mouse and attributed its etiology to a "coccobacilliform organism" (later identified as *M. pulmonis*). His studies included one of the more extensive investigations of a natural outbreak of MRM. The most consistent findings were suppurative rhinitis and otitis media, which apparently prompted Nelson to adopt the term *infectious catarrh*. Pneumonia was less frequently observed. Chattering, a sound which Nelson compared to that of a person gently clicking the teeth together, was one of the main clinical manifestations. Chronically affected mice also showed inactivity, weight loss, rough hair coats, and dyspnea. A mortality of 95% was observed over a period of 11 months (Nelson, 1937a).

In subsequent years, Nelson (1955, 1963, 1967a,b) contended that the erratic pneumonia observed in such cases was viral in origin. Because of this belief, the pneumonia was first designated *endemic pneumonia* (Nelson, 1955) and later *enzootic bronchiectasis* (Nelson, 1963). The combination of *infectious catarrh* and the alleged viral pneumonia were designated by Nelson *chronic respiratory disease* (CRD). In retrospect, the fact that Nelson's putative virus has never been isolated and characterized, and the fact that the entire syndrome has been repeatedly reproduced (i.e., Koch's postulates fulfilled) in many laboratories (Atobe and Ogata, 1974; Howard *et al.*, 1978; Jersey *et al.*, 1973; Kohn and Kirk, 1969; Lindsey and Cassell, 1973; Lindsey *et al.*, 1971; Lutsky and Organick, 1966; Organick *et al.*, 1966; Taylor *et al.*, 1977; Whittlestone *et al.*, 1972) by the inoculation of *M. pulmonis* into pathogen-free mice and rats argue convincingly in favor of the mycoplasma as the causative agent. Thus, *MRM* is clearly the disease name of choice because it connotes etiologic specificity (Lindsey *et al.*, 1978).

B. Characteristics of *Mycoplasma pulmonis*

Mycoplasma pulmonis is a bacterium of the sterol-requiring family Mycoplasmataceae. Like other members of the class Mollicutes, it lacks a cell wall and has a single outer limiting membrane. Although it exhibits extreme pleomorphism, ultrastructural studies have shown it to be predominantly spherical in shape and 600–1500 nm in diameter (Cassell and Hill, 1979). On solid medium the colonies typically have a fine granular appearance, with little tendency to grow into the medium centrally (the "fried egg" appearance). Freshly isolated strains may demonstrate motility (Andrewes and Welch, 1946; Bredt, 1973; Bredt and Radestock, 1977; Nelson, 1960). Biochemical distinctives are its ability to ferment glucose and its inability to ferment arginine. Present evidence indicates that the proteins and polysaccharides (but not lipids) of *M. pulmonis* are antigenic (Deeb and Kenny, 1967b) and that most isolates of the agent may have up to three common antigens (Ogata *et al.*, 1967).

The mechanisms by which *M. pulmonis* causes disease are incompletely understood, although a wide range of possibilities has been advanced. The organism is an extracellular pathogen which "attaches" or "adheres" to host cell membranes as an initial event in infection. In the respiratory tract, this intimate association is with the respiratory epithelium (anywhere from the anterior nasal passages down to the alveoli); it may be mediated by surface-bound glycoproteins on the organism (Schiefer *et al.*, 1974). From that vantage point the organism, like *M. pneumoniae,* may injure host cells through competition with host cells for metabolites such as carbohydrates (Hu *et al.*, 1975) and nucleic acids (Plackett, 1957; Razin *et al.*, 1964; Russell, 1966) and/or release of toxic substances such as peroxides (Brennan and Feinstine, 1969; Cole *et al.*, 1968), ciliostasis (Westerberg *et al.*, 1972), reduction in the number of cilia, ultrastructural changes (such as nuclear swelling, cytoplasmic vacuolation, and distortion of mitochondria), and desquamation (Collier and Baseman, 1973; Gabridge *et al.*, 1974; Kohn, 1971a; Organick and Lutsky, 1976).

C. Experimental Disease

Experimental infections of *M. pulmonis* in the mouse have been induced primarily in three organ systems: respiratory, genital, and joints. Each of these is instructive in understanding natural infections and will be summarized briefly.

The intranasal inoculation of *M. pulmonis* into animals under light anesthesia has served as a standard model system for many years (Cassell *et al.*, 1973; Edward, 1947; Howard *et al.*, 1978; Nelson, 1937c; Organick *et al.*, 1966). In the mouse (but not rat) the resulting disease is highly dose dependent (Lindsey and Cassell, 1973). Doses of 10^4 colony-forming units (CFU) or less result in mild, transient disease involving mostly the nasal passages, middle ears, and larynx. Animals given more than 10^4 CFU experience acute pneumonia, in addition to lesions in the upper respiratory tract, with many deaths a few days after inoculation. Many of the survivors develop chronic bronchopneumonia in which bronchiectasis and pulmonary abscesses occur (Lindsey and Cassell, 1973).

Several investigators (Barden and Tully, 1969; Cole *et al.*, 1975a; Harwick *et al.*, 1973; Nelson, 1955; Taylor and Taylor-Robinson, 1976) have shown that the intravenous inoculation of *M. pulmonis* into mice leads to arthritis. Typically, there is severe suppurative arthritis involving mainly the carpal and hock joints, reaching maximal intensity in 2–4 weeks. Tendosynovitis also occurs in some animals. Chronic inflammation supersedes the acute response within a few weeks and may persist for many months. After intravenous inoculation, abscesses may occur in the brain and spinal cord of a small percentage of mice, sometimes resulting in flaccid paralysis.

Nelson (1954) and Goeth and Appel (1974) reported that the intraperitoneal inoculation of *M. pulmonis* causes oophoritis. Such experimental infections have been found to reduce fertility (Fraser and Taylor-Robinson, 1977; Goeth and Appel, 1974; Taylor-Robinson *et al.*, 1975). Although naturally occurring genital disease due to *M. pulmonis* has not been reported in mice, the organism has been isolated from the vagina and uterus in natural infections (Hill, 1974; G. H. Cassell, unpublished observations).

Evidence (Saito *et al.*, 1978b) indicates that there are wide differences in susceptibility of different strains of mice to *M. pulmonis*. Such differences have been suggested many times in the past but usually have not been supported by controlled experiments.

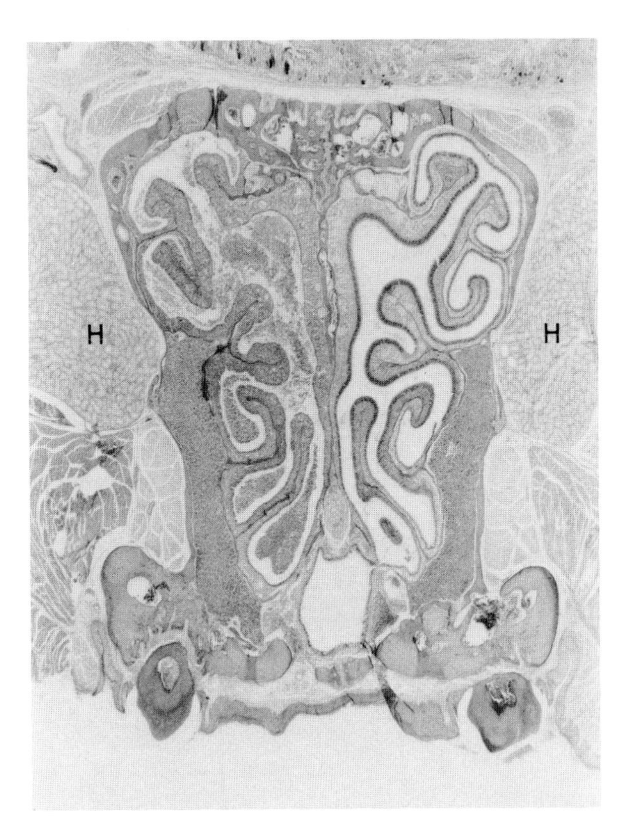

Fig. 1. Nasal passages of a mouse at the level of the eyes. The Harderian glands (H) are shown. There is unilateral suppurative rhinitis due to *Mycoplasma pulmonis* on the left side. ×15.

D. Natural Disease and Pathology

Mice from colonies experiencing natural infections of *M. pulmonis* may appear entirely normal. However, in most instances some mice show the manifestations which Nelson (1937a,b,c) described for infectious catarrh—chattering, inactivity, weight loss, rough hair coat, and dyspnea. Chattering and dyspnea presumably can be attributed to the characteristic accumulation of purulent exudate in the nasal passages, along with inflammatory thickening of the nasal mucosa (mice and rats are said to be obligate or preferential nose breathers).

Histologically, the lesions in the nasal passages are characteristic (Lindsey and Cassell, 1973; Lindsey *et al.*, 1978). (For this reason, histologic examination of the nasal passages should be mandatory in routine health monitoring of mouse colonies for *M. pulmonis*.) Early in the course of the disease, the main lesion is suppurative rhinitis. Respiratory and olfactory epithelium may undergo dramatic change to a squamoid appearance during the more acute phases of the disease (Figs. 1–4). Syncytial giant cells frequently occur in the epithelium (Figs. 5 and 6). Often there is hyperplasia of the submucosal glands suggesting increased secretion (Figs. 2 and 3). With the passage of time the epithelium may return to a more normal appearance, but there is varying infiltration of lymphoid cells in the submucosa. Chronic rhinitis persists for weeks.

Affected mice also usually have suppurative otitis media that may be bilateral. Histologic sections may reveal purulent exudate in the eustachian tubes as well. The larynx and trachea are commonly affected. The lumina contain polymorphonuclear leukocytes, which may also be present in the laryngeal glands. The mucosal epithelium becomes hyperplastic, and varying

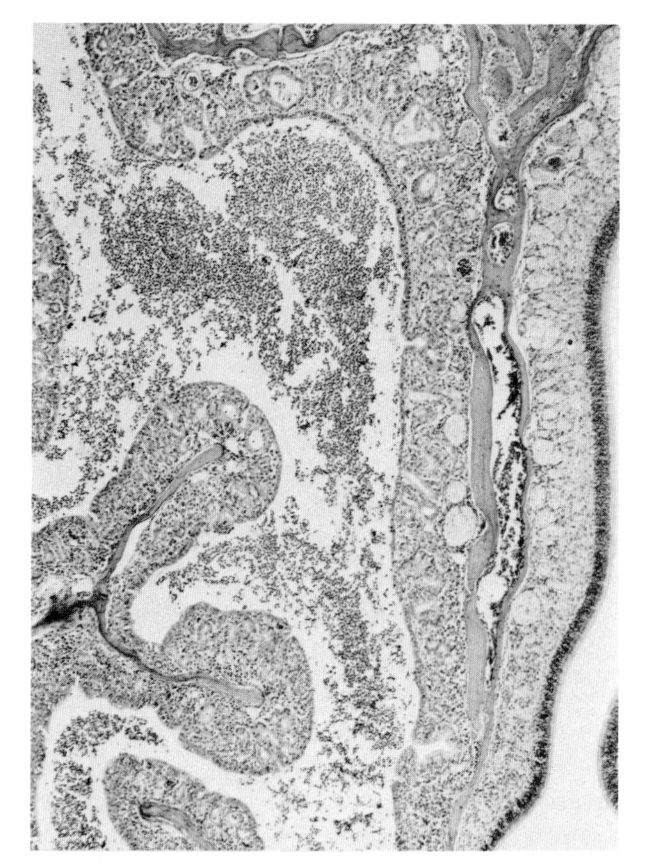

Fig. 2. Higher magnification of the nasal passages shown in Fig. 1. The mucosa throughout the left side shows replacement of olfactory epithelium by flattened epithelium, hyperplasia of the submucosal glands, and beginning infiltration of the lymphoid cells. ×60.

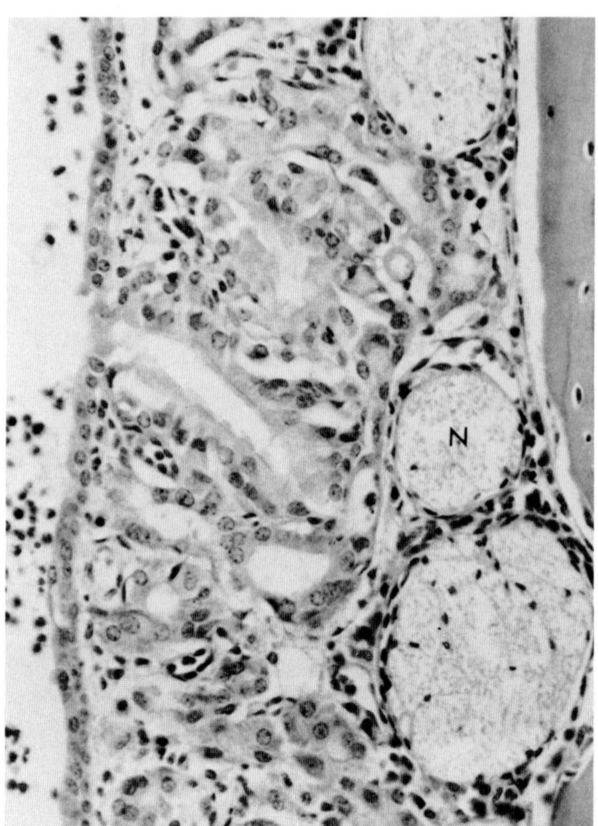

Fig. 3. Left side of the nasal septum shown in Fig. 1. Cross sections of nerve fibers (N) are present deep in the submucosa. Note the hyperplasia of the submucosal glands and squamoid appearance of the surface epithelium. ×375.

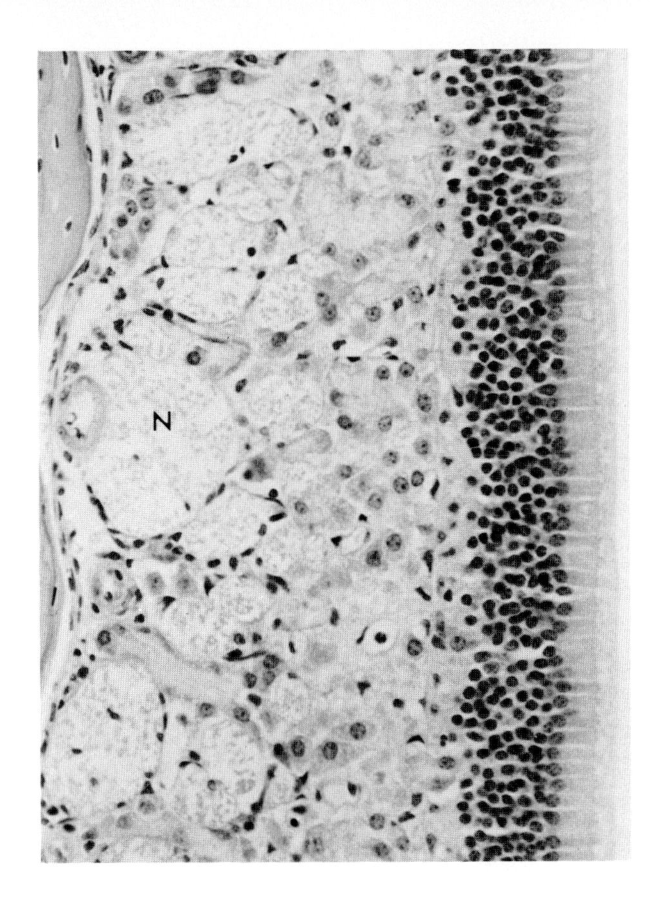

Fig. 4. Right side of the nasal septum depicted in Fig. 1, showing normal olfactory epithelium. Numerous large and small nerve fibers (N) are present near the bony septum. ×375.

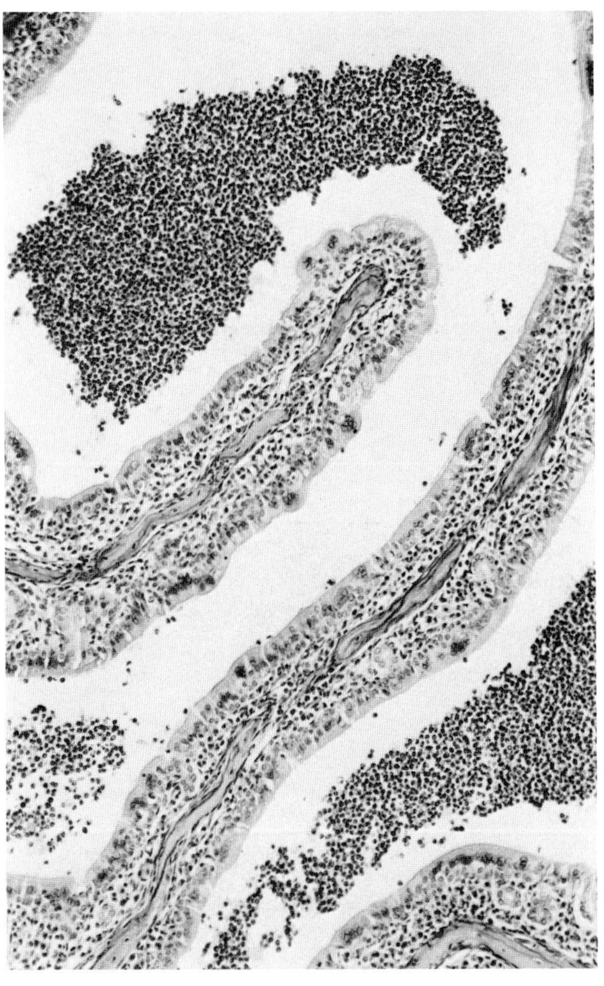

Fig. 5. Nasal passage of a mouse with suppurative rhinitis due to *Mycoplasma pulmonis*. Note the abundant purulent exudate in the lumen, lymphoid infiltrate in the submucosa of the scrolls, and syncytial giant cells in the surface epithelium. ×120.

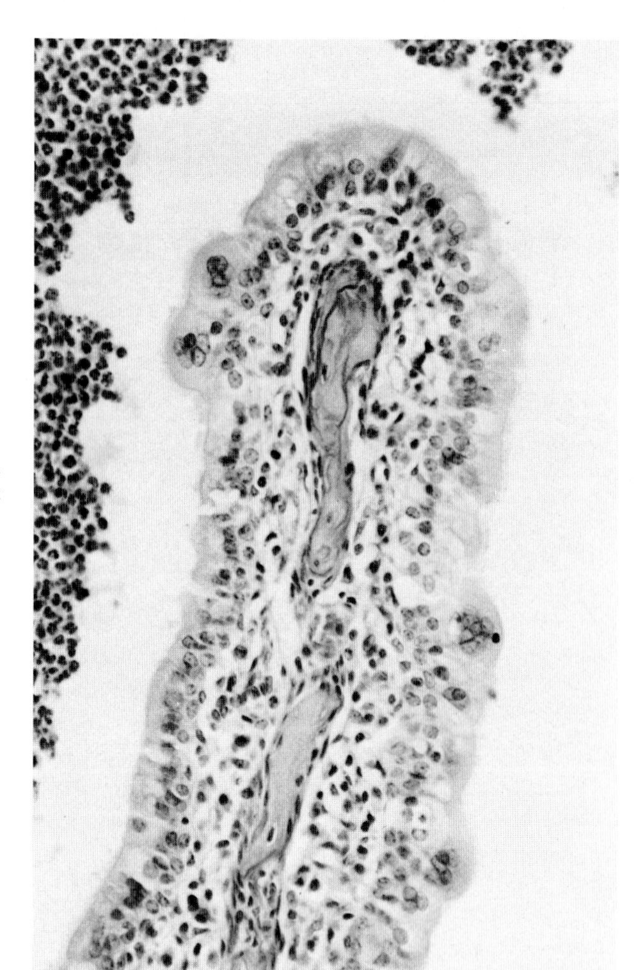

Fig. 6. Higher magnification of the tip of the nasal scroll shown in Fig. 5. Note the syncytial epithelial giant cells characteristic of *Mycoplasma pulmonis* infection. ×250.

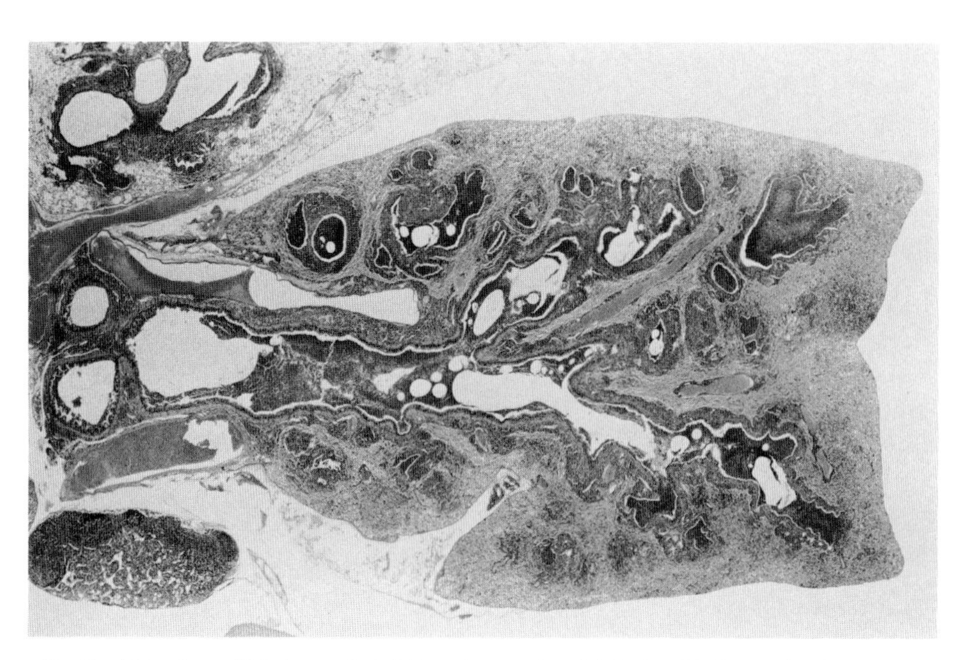

Fig. 7. Mouse lung with lesions typical of *Mycoplasma pulmonis* infection. The larger lobe shows complete atelectasis which persisted while infusion of fixative via the trachea returned more normal lobes to their normal distention. Note the massive accumulation of purulent exudate in the airways and severe peribronchial lymphoid cuffing. Early lesions are present in the small lobe above. A mediastinal node is present at the lower left. ×10.

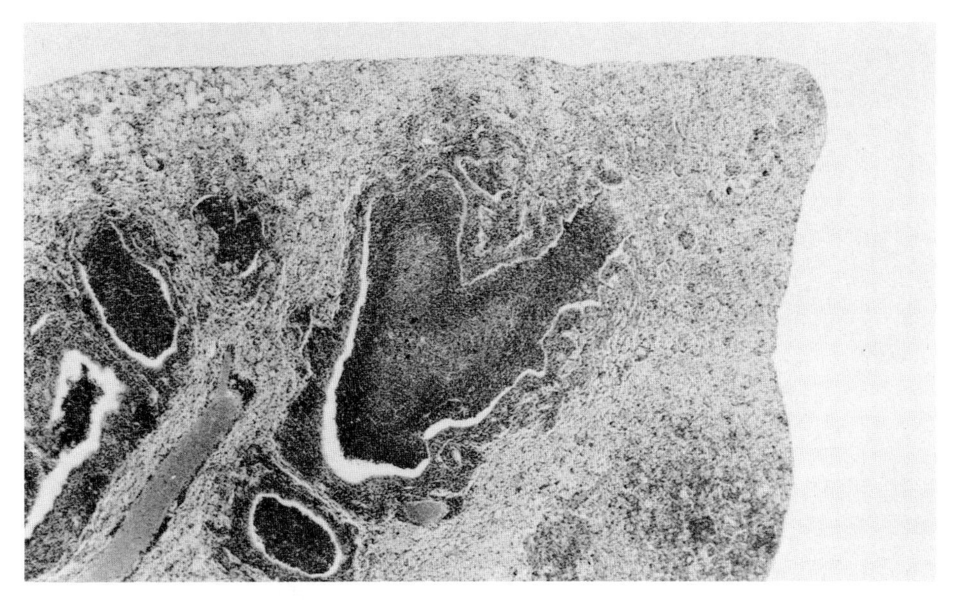

Fig. 8. Right upper tip of the large lobe of lung shown in Fig. 7. Purulent exudate distends the smaller bronchioles and adjacent alveoli, whereas macrophages predominate in the alveoli of pale more peripheral lung. ×30.

numbers of lymphoid cells infiltrate the submucosa (Lindsey and Cassell, 1973; Lindsey *et al.*, 1978).

The lung lesion is basically a chronic bronchopneumonia that tends to spread outward from the lung hilus. Varying numbers of polymorphonuclear leukocytes appear in the lumen of the bronchi and large peribronchial cuffs of lymphoid cells, primarily plasma cells, appear (Figs. 7–10). Bronchus-associated lymphoid tissue is not present in normal mouse lungs. Syncytial giant cells (Figs. 9 and 10) similar to those occurring in the nasal passages sometime appear in bronchi contain-

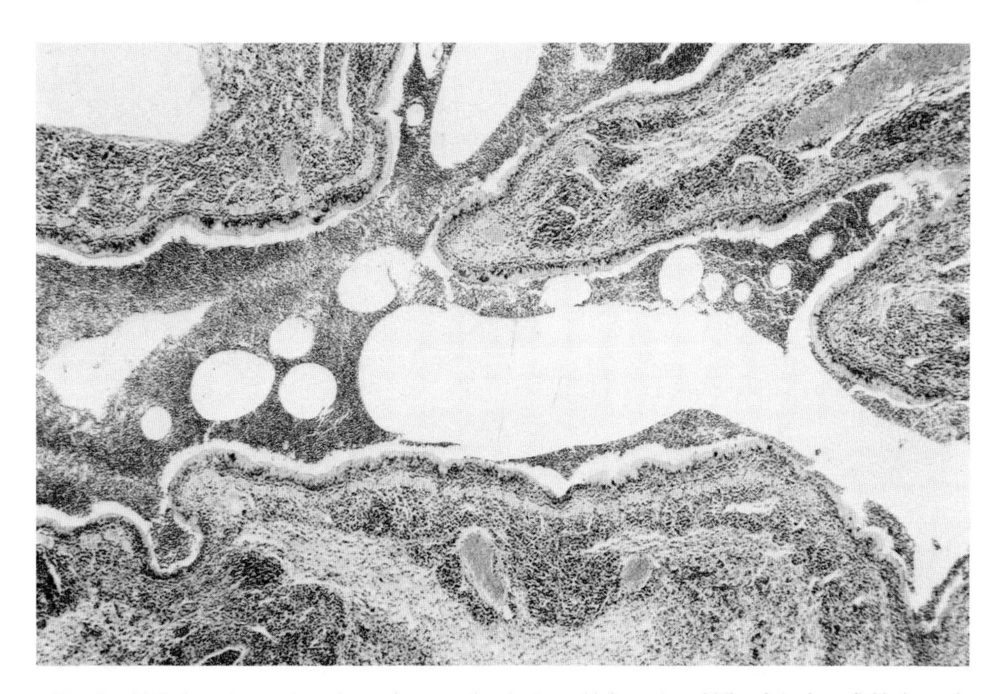

Fig. 9. Main bronchus and portions of a second-order bronchi from the middle of the lung field shown in Fig. 7. A large blood vessel occupies the upper left corner of the photograph. Note the purulent exudate in the bronchial lumen, peribronchial lymphoid cuffing, and syncytial giant cells in the hyperplastic bronchial epithelium. ×40.

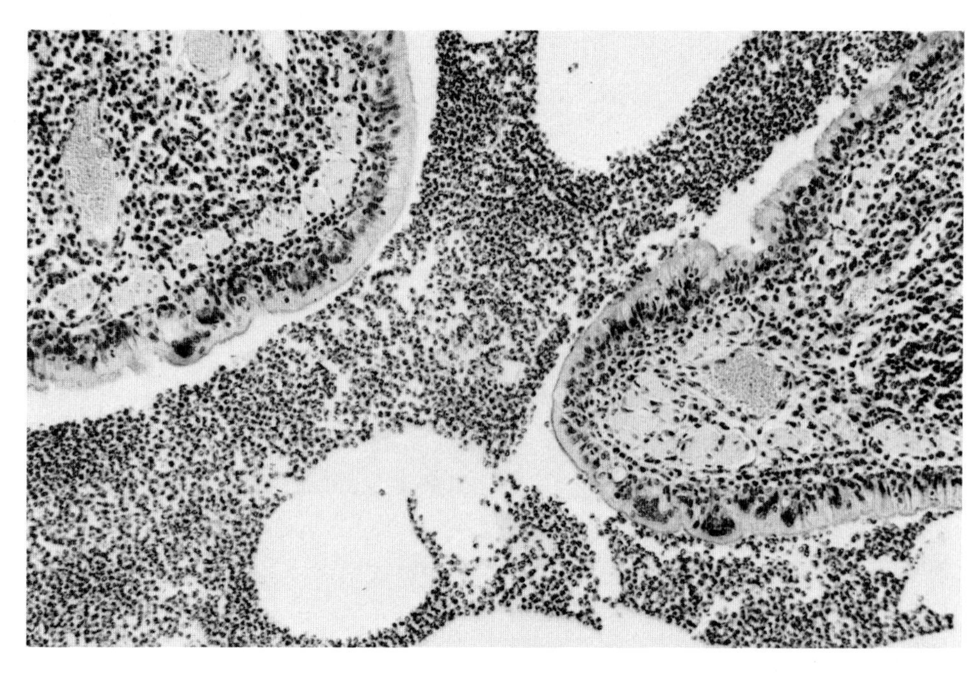

Fig. 10. Higher magnification of the bronchial bifurcation shown in Fig. 9. Syncytial giant cells are present in the bronchial epithelium. The epithelial surface frequently shows a dark, fuzzy line representing myriad *Mycoplasma pulmonis*. ×120.

ing an abundance of purulent exudate. Because of the accumulations of exudate, mainly in the larger bronchi, atelectasis, when present, appears to radiate out from hilar regions of the lung but may involve entire lobes. If the accumulation of purulent exudate is sufficiently severe, bronchiectasis, bronchiolectasis, and bronchial abscesses may develop. Cuboidal epithelium often lines the alveoli immediately surrounding severely affected airways. Pleuritis is rare (Lindsey and Cassell, 1973; Lindsey *et al.*, 1978).

E. MRM Complicated by Other Infections

Brennan *et al.* (1965, 1969a) have reported that experimental infections of mice with *M. pulmonis* and *Pasteurella pneumotropica* cause more severe respiratory disease than either agent alone. However, their work has not been confirmed, and to our knowledge, no such association has been incriminated in a natural outbreak.

As first suggested by Richter (1970), there is a synergistic effect of dual *M. pulmonis* and Sendai virus infection in the mouse. Indeed, the natural outbreaks of clinical disease with mortality involving *M. pulmonis* in mice which we have observed have actually been attributable to combined *M. pulmonis* and Sendai infections. These observations have been supported experimentally by Howard *et al.* (1978), who found that prior infection with Sendai virus greatly enhanced the growth of the mycoplasma and the severity of lung lesions following a subsequent inoculation of *M. pulmonis*. By contrast, prior inoculation of *M. pulmonis* did not enhance subsequent infection by Sendai virus in their study. Sendai virus infection has been reported to enhance pulmonary infection in the mouse through impaired bacterial killing by macrophages (Jakab and Green, 1976; Jakab *et al.*, 1980).

The pulmonary lesions of combined Sendai virus and *M. pulmonis* infections are unique. The characteristic morphologic features associated with each agent are observed, but certain aspects may be greatly exaggerated in the dual infection. The peribronchiolar proliferation of glandlike spaces due to Sendai virus (Figs. 11 and 12) becomes pronounced, with lining cells often transformed to the mucus-secreting type. Dramatic colonization of these spaces by *M. pulmonis* takes place, as evidenced by immunofluorescent demonstration of abundant antigen in their lumens and on the lining epithelium. There is an exaggerated accumulation of neutrophils in the glandlike spaces and surrounding alveoli as well as in larger conducting airways. In hemotoxylin and eosin-stained sections, purple structures, thought to represent aggregates of DNA from dead neutrophils, are seen scattered through the large pools of neutrophilic exudate (Lindsey and Cassell, 1973; Lindsey *et al.*, 1978). In uncomplicated *M. pulmonis* infections, such accumulations of neutrophils and DNA aggregates are usually confined to larger airways.

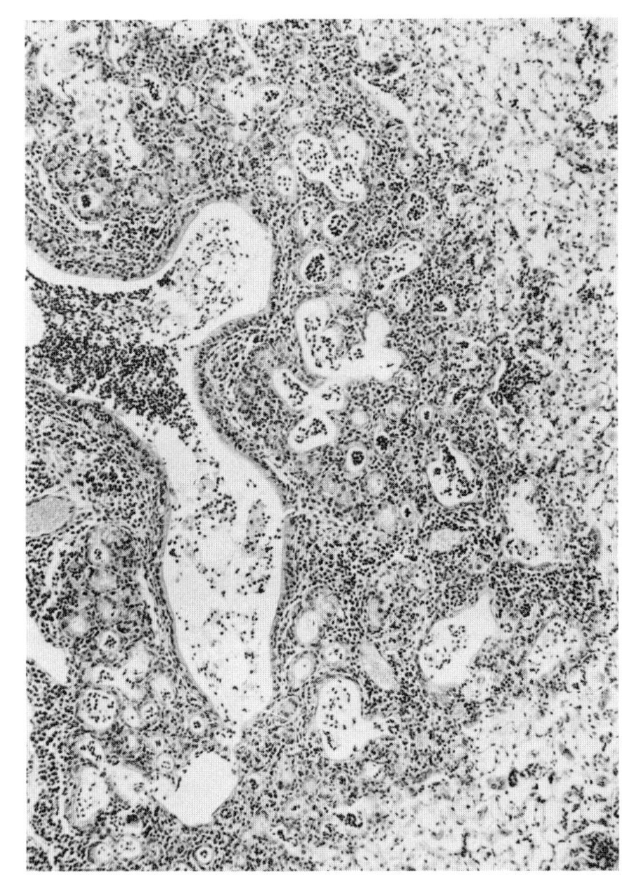

Fig. 11. Mouse lung with lesions characteristic of dual infections of Sendai virus and *Mycoplasma pulmonis*. A bronchiole enters the field from the left and bifurcates into two smaller bronchioles. The wide zone of peribronchiolar glandlike spaces is due to the virus infection, whereas the suppurative bronchiolitis and purulent inflammation in the glandlike spaces and surrounding alveoli at the right are due to *Mycoplasma pulmonis*. ×80.

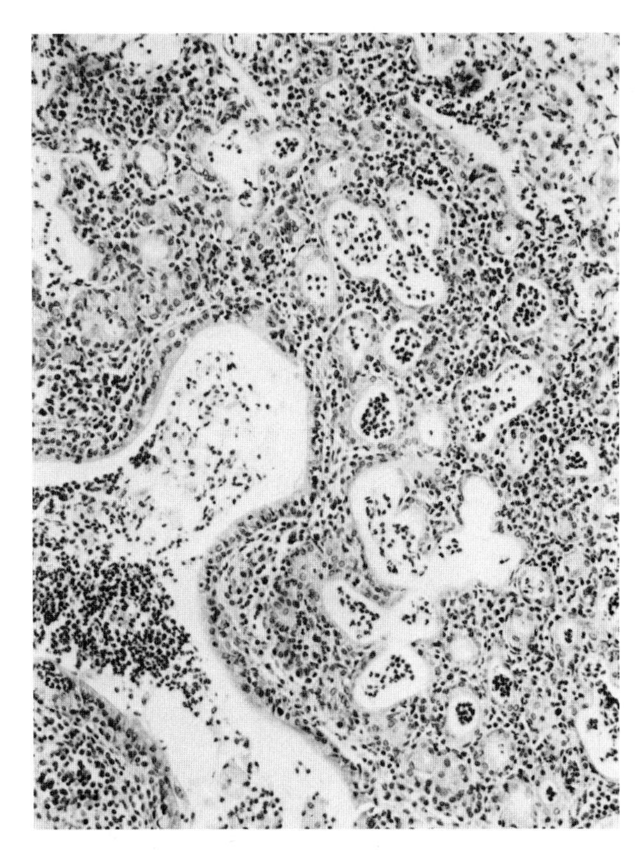

Fig. 12. Higher magnification of the lesions shown in Fig. 11. The interstitial stroma has a moderate infiltrate of lymphoid cells which, along with suppurative inflammation, is attributable to *Mycoplasma pulmonis*. ×140.

F. Immune Response

Mice can mount an effective immune response to *M. pulmonis,* as evidenced by their protection against disease after either active or passive immunization (Atobe and Ogata, 1974; Cassell *et al.*, 1973, 1974; Taylor and Taylor-Robinson, 1976; Taylor *et al.*, 1977) and by their recovery from disease induced by experimental inoculation of low numbers of organisms ($\leq 10^4$ CFU) (Lindsey and Cassell, 1973). However, mice given more than 10^4 CFU develop severe, chronic disease even in the presence of an intense immune response. The highest antibody titers occur in mice with the most severe disease, and furthermore, lymphoid cells comprise a major component of the pathologic lesions. Thus, *M. pulmonis* elicits an immune response that is at least theoretically capable of protecting the host, yet the organism is able to evade immunologic destruction, and the host becomes the victim of immunologic injury. Much of the pathogenic potential of this organism, in fact, may result from its direct mitogenic effect upon lymphocytes (Cole *et al.*, 1975; Naot *et al.*, 1977, 1979a,b).

Naturally and experimentally infected mice produce anti-*M. pulmonis* antibody of the IgM, IgG_1, IgG_2, and IgA classes, both locally in lymphoid cells in the respiratory tract and in paratracheal lymph nodes (Cassell *et al.*, 1974; Horowitz and Cassell, 1978; Taylor, 1979). Antibodies apparently are directed to similar protein antigens regardless of whether the mice are naturally infected, experimentally infected, or hyperimmunized (Asa *et al.*, 1980). Whereas complement-fixing (CF) antibodies are detected as early as 5–10 days after inoculation and begin to decline after 10 months, indirect hemagglutinating (IHA) antibodies do not appear until 1 month after infection and persist at high titers for up to 1 year. Metabolism-inhibiting (MI) antibodies reach lower titers than CF or IHA antibodies but persist for 12 months or more (Atobe and Ogata, 1974). However, the level of MI antibodies may depend upon the route of infection and the strain of mouse (Cole *et al.*, 1975). The nonspecific mitogenicity of *M. pulmonis* for mouse lymphocytes suggests that not all antibody produced is mycoplasma specific (Cole *et al.*, 1975).

The role of anti-*M. pulmonis* antibody in respiratory disease is unclear. As already mentioned, antibodies are produced earlier and in higher titers in mice with more severe disease, and the organism continues to replicate in the presence of high concentrations of specific antibody in respiratory secretions (Atobe and Ogata, 1974; Cassell *et al.*, 1974). Development of antibody does appear to correlate with a shift from acute alveolar disease to chronic bronchopneumonia. Following intranasal inoculation, the organisms initially replicate within alveoli (Cassell *et al.*, 1973; Lindsey and Cassell, 1973) but are later confined to the tracheobronchial epithelium and/or airway lumens. It has been suggested that immune phagocytosis may be important in this alveolar clearance. Antibody-mediated phagocytosis, not dependent upon complement action, can be demonstrated *in vitro* with peritoneal and pulmonary macrophages (Cole and Ward, 1973; Davis *et al.*, 1980; Jones and Hirsch, 1971; Jones and Yang, 1977; Jones *et al.*, 1972; Taylor and Howard, 1980). *M. pulmonis* apparently possesses an anti-phagocytic surface protein that is neutralized by specific antibody (Jones *et al.*, 1972).

Rabbit hyperimmune serum has more opsonic capacity than mouse hyperimmune serum with peritoneal macrophages (Taylor and Howard, 1980), and with pulmonary macrophages homologous antiserum essentially has no anti-mycoplasmal effect (Davis *et. al.*, 1980). The relatively poor opsonic activity of mouse serum may be due to the presence of high levels of IgG_1-specific antibody (Cassell *et al.*, 1974). This class of immunoglobulin, which neither promotes phagocytosis nor fixes complement, may act as a "blocking antibody," thus protecting the organism from other host defenses. It has been shown that *M. pulmonis* isolated from persistently infected mice is much more resistant to the anti-mycoplasmal effect of macrophages than are organisms grown *in vitro* (Taylor and Howard, 1980).

Mouse peritoneal macrophages exposed *in vitro* to *M. pulmonis* release large amounts of hydrolytic enzymes (Taylor-Robinson *et al.*, 1978). This release of lysosomal hydrolases by phagocytes *in vivo* may contribute to the chronic inflammatory response elicited by this organism (Lindsey and Cassell, 1973). Release of lymphokines as a result of the mitogenicity of *M. pulmonis* could also contribute (Naot *et al.*, 1977, 1979a,b).

Although ineffective in promoting recovery from disease, antibody seems capable of preventing reinfection. Mice can be protected from experimental pneumonia by immunization. Serum from immune mice inoculated intranasally (Cassell *et al.*, 1973) or intravenously (Taylor and Taylor-Robinson, 1974, 1976) consistently confers protection on normal mice. Since passively transferred antibody protects against disease but not colonization, it has been suggested that the preexisting antibody blocks the local immune response in the lung, thereby preventing development of disease (Taylor *et al.*, 1977).

Classic cell-mediated immunity does not appear to play a major role against *M. pulmonis* infection in mice, as immunity cannot be passively transferred with immune cells (Taylor and Taylor-Robinson, 1976; G. H. Cassell, unpublished observations). In addition, neither congenitally athymic (*nu*/*nu*) mice (Cassell and McGhee, 1975) nor neonatally thymectomized mice (Denny *et al.*, 1972; Taylor and Taylor-Robinson, 1974, 1975) are more susceptible to pneumonia than normal mice. However, T lymphocytes are essential to the generation of plasma cell infiltrates in the lung and to subsequent production of anti-*M. pulmonis* antibodies (Cassell and McGhee, 1975; Tanaka, 1979). In addition, the immune response generated by T lymphocytes may be necessary to limit dissemination of the mycoplasmas. Athymic mice develop arthritis following intranasal inoculation (an event which does not occur in normal mice), and increased spread to other organs has been noted in T-deficient mice (Keystone *et al.*, 1980; Tanaka, 1979).

In contrast to the mouse, rat alveolar macrophages are capable of clearing *M. pulmonis* from the alveoli even in the absence of immune factors (Cassell *et al.*, 1973) and can be shown to exert an anti-mycoplasmal effect *in vitro* in the absence of specific antibody (Davis *et al.*, 1980). The local immune response in the rat is predominantly lymphocytic as compared with the plasmacytic response in mice. In addition, immunity to respiratory disease in the rat can be transferred passively with immune spleen cells but not serum (G. H. Cassell and J. K. Davis, unpublished observations). Therefore, the two animal species provide interesting contrasts in their responses to *M. pulmonis*.

G. Diagnosis

Effective diagnostic programs must be designed to evaluate disease states properly as well as to recognize subclinical infections due to *M. pulmonis*. Despite the commonality of much of the methodology employed in accomplishing these objectives, the diagnosis of clinical and subclinical infections requires them to be treated as separate problems. Detection of subclinical infections is by far the greater diagnostic challenge.

Mice suffering from clinical respiratory disease due to *M. pulmonis* should be approached in much the same way as any other disease problem. All clinicopathologic features of the disease must be carefully correlated with the results of laboratory procedures selected to demonstrate the presence of mycoplasmas as well as other bacterial and viral agents, if present. This approach alone provides an opportunity to delineate the etiologic agent(s) responsible for each clinical outbreak and a sound basis for corrective measures. It must be remembered that MRM may be complicated by other infections, Sendai virus being the most notable example in the mouse (Richter, 1970).

A special problem exists in detecting clinically silent infections of *M. pulmonis* in samples of mice taken from colonies for routine health surveillance. In this instance few mice may be infected, few organisms may be present in individual mice, and antibody response may be meager. Newer methods, such as the ELISA (enzyme-linked immunosorbent assay) (Horowitz and Cassell, 1978), offer exceptional promise in this situation, but other methods, including histopathologic examination of all levels of the respiratory tract, may also provide invaluable clues.

At present, detection of *M. pulmonis* is best accomplished using one or more of the following methodologic approaches: (1) cultural isolation and identification, (2) immunofluorescence or peroxidase-labeled specific antibody, (3) serologic methods, and (4) histopathology. In actual practice, combinations of these methods usually prove most satisfactory, as each method has unique advantages and limitations (Cassell and Hill, 1979; Cassell *et al.*, 1979; Davidson *et al.*, 1981; Lindsey *et al.*, 1978).

1. Isolation and Identification of *Mycoplasma pulmonis*

Cultural isolation is the most widely used detection method. In general, it gives good results depending on selection of the site(s) to be cultured, appropriateness of the specimens taken, and the ability of each batch of medium to support growth. Its main disadvantages are that multiple sites have to be cultured and days or weeks may be required for detectable growth.

In natural infections *M. pulmonis* most frequently inhabits the upper respiratory tract (nasal passages, middle ears, and larynx), hence the reason these sites are most often sampled in the study of clinical cases or in routine health monitoring (Davidson *et al.*, 1981; Kappel *et al.*, 1974; Lentsch *et al.*, 1979; Lindsey *et al.*, 1978; Nelson, 1937a,b,c). Phosphate-buffered saline or mycoplasma broth are routinely used to lavage the larynx and trachea (Lentsch *et al.*, 1979) or the larynx, pharynx, and nasal passages. Unfortunately, it is not known how frequently genital tract infection occurs in the absence of respiratory infection, so that culturing the respiratory tract alone will not assure conclusive screening (Cassell and Hill, 1979).

Specimens such as exudates or secretions of body organs are preferred because they usually contain the greatest concentration of organisms. Also, tissues may contain a growth inhibitor (Tully and Rask-Nielsen, 1967), thought to be lysolecithin. Addition of lysophospholipase (Kaklamanis *et al.*, 1969; Mardh and Taylor-Robinson, 1973) or serial dilution of tissue suspensions in broth (Tully and Rask-Nielsen, 1967) abolishes this anti-mycoplasmal activity.

Mycoplasmas are more difficult to grow on artificial media than most bacteria. In keeping with their small genome size, mycoplasmas have a limited biosynthetic capacity and therefore require a wide array of precursor molecules for macromolecular synthesis (Rodwell and Mitchell, 1979). *Mycoplasma pulmonis* and *M. neurolyticum* grow aerobically at 37°C in both standard Hayflick's broth and mycoplasmal agar (Cassell and Hill, 1979; Hayflick, 1970). Various broth and agar formulations will support the growth of *M. pulmonis* (Chalquest, 1962; Frey *et al.*, 1974; Kohn, 1971b; Kohn and Kirk, 1969; Lentsch *et al.*, 1979; Olsen *et al.*, 1963).

Since different lots of medium base, serum, yeast extract, and agar can differ markedly, *it is essential that each batch of complete medium be pretested for its ability to support growth using stock strains of mycoplasmas.* Culture vessels, wash water, and medium components should be of tissue culture quality (Whittlestone, 1974). Complete broth medium may be stored at −20°C for up to 3 months. Agar plates may be stored in sealed containers for up to 3 weeks. If penicillin G is added to the medium, it is recommended that additional penicillin be added once per week to maintain adequate antibiotic activity (M. K. Davidson, unpublished observations).

Mycoplasma pulmonis usually grows within 1 week on agar or in broth, but all cultures should be incubated for at least 21 days before being discarded as negative. Blind restreaking of plates and/or rapid passages in broth may increase the isolation rate (Cassell and Hill, 1979). Because some strains may still prove difficult to isolate, negative cultures are not always conclusive. In some instances, methods such as the ELISA and immunofluorescence can serve as useful adjuncts to culture.

Identification of isolates must conform to the standards adopted by the Subcommittee on the Taxonomy of Mycoplasmatales (1972). This usually entails demonstration of "typical" colonies that do not revert to bacterial form upon removal of penicillin from the medium. However, colonial morphology is variable depending on the medium formulation and hydration of agar plates. *Mycoplasma pulmonis* shows little tendency to grow into the medium and to produce typical "fried egg" colonies but will do so under optimal growth conditions. This morphologic variation is also true of individual cells (Freundt, 1958; Maniloff *et al.*, 1965; Razin and Cosenza, 1966; Razin *et al.*, 1966, 1967); the organisms appear pleomorphic when examined by light and electron microscopy. Ultrastructural analysis reveals a single triple-layered membrane without evidence of a cell wall.

Speciation of mycoplasmas can be accomplished by immunofluorescence (Del Giudice *et al.*, 1967), immunoperoxidase staining (Hill, 1978; Polak-Vogelzang *et al.*, 1978), or growth inhibition (Clyde, 1964). A rapid presumptive test to distinguish *M. pulmonis* from the other rodent mycoplasmas is the hemadsorption test (Manchu and Taylor-Robinson, 1968), but some strains of *M. pulmonis* do not hemadsorb.

2. Immunofluorescence (IMF) and Immunoperoxidase (IMP) Methods

These methods are extremely valuable for the detection of *M. pulmonis* in tissues, exudates, or lavages of organs. Their advantages include rapidity, versatility, high specificity, and low expense.

Tissue sections may be examined by IMF or IMP as frozen sections, or after fixation in cold 95% ethanol and special processing for paraffin embedding (Sainte-Marie, 1962). Tracheobronchial and genital tract lavages can be concentrated by centrifugation and a drop placed on a slide, air-dried, fixed in methanol or 95% ethanol, and stained by IMF or IMP procedures. IMF is faster than culture and can be equally reliable, but special equipment is required. This limitation can be overcome by peroxidase labeling of smears or tissue sections and examination by light microscopy (Hill, 1978; Nakane and Kawaoi, 1974; Polak-Vogelzang *et al.*, 1978).

3. Serology

Since mycoplasmal infections are generally limited to mucosal and serosal surfaces, serum antibodies are usually present in low titer. Thus, conventional serology, i.e., complement fixation, hemagglutination inhibition, and growth inhibition, has limited value, especially in early infections or latent infections when the number of organisms is small. The more recently developed, more sensitive ELISA (Horowitz and Cassell, 1978) radioimmunosorbent assay (Brunner *et al.*, 1977), and solid phase radioimmunoassay (Taylor, 1979) may circumvent this problem. These assays are much more sensitive than most serologic procedures and can measure specific classes of antibody. The ELISA for *M. pulmonis* has been compared with culture, histopathology, and IMF (Cassell *et al.*, 1981; Davidson *et al.*, 1981; Horowitz and Cassell, 1978) and found to be a highly reproducible, sensitive, and specific assay for detection of *M. pulmonis*. Low cost, simplicity, and rapidity makes these tests extremely promising.

4. Pathology

Histopathology is extremely valuable in the diagnosis of *M. pulmonis* infection. This is particularly true for the mouse, as certain lesions of MRM in this species are quite specific. The presence of syncytial epithelial giant cells (Figs. 6, 7, 9, 10) in the nasal passages or bronchi in association with abundant purulent exudate, is in our experience, characteristic but not always present. These giant cells have been reproduced repeatedly by experimentally infecting pathogen-free and axenic mice with cultures of *M. pulmonis* (Lindsey and Cassell, 1973). We have never observed them in natural outbreaks of respiratory disease in mice except in instances in which *M. pulmonis* was present. The giant cells apparently represent a transient phenomenon, as they appear during the stage of the disease in which there is maximal accumulation of neutrophils in airway lumens. Histopathologic examinations should be made of the entire respiratory tract. Final diagnosis should correlate all morphologic and microbiologic data.

H. Epizootiology

Systematic investigations of the epizootiology of *M. pulmonis* infection are generally lacking. However, transmission between cage contacts as well as animals in adjacent cages has been reported (Hill, 1972; Nelson, 1937a,c). Thus, there is ample reason to believe that offspring of infected dams most commonly acquire the infection by aerosol transmission during the first few days of life (Klieneberger and Nobel, 1962; Lemcke, 1961). Transplacental transmission also appears possible (Ganaway *et al.*, 1973; Juhr *et al.*, 1970; Tram *et al.*, 1970). Data concerned with survival of the organism outside the host, as in contaminated food and water, are contradictory (Juhr, 1971; Vogelzang, 1975). The organism has been isolated from cotton rats (Andrewes and Niven, 1950), Syrian hamsters (Battigelli *et al.*, 1971; Hill, 1974), guinea pigs (Cassell and Hill, 1979; Juhr and Obi, 1970), and rabbits (Deeb and Kenny, 1967a), but evidence is lacking that any of these serve as a significant reservoir of infection for mice.

Accurate assessments of the prevalence of *M. pulmonis* infection in modern mouse colonies are almost nonexistent, as efficient methods for large-scale testing have not been available in the past. Based on the use of culture methods, Sparrow (1976) found *M. pulmonis* in 17% of 96 samples of mice from 15 accredited breeders in the United Kingdom, and Saito *et al.* (1978a) recovered the agent from 36% of 22 colonies in Japan. The latter authors isolated the agent from 43% of a total sample of 3270 mice studied. In a limited survey which included 734 mice, Cassell *et al.* (1981) isolated the organism from 38% of conventional mice and 8% of barrier-maintained mice in the United States.

I. Control

The most effective means of control presently is the establishment and maintenance of *M. pulmonis*-free breeding colonies and continued maintenance of this state in user colonies throughout the research process. This requires careful screening of breeding stocks (by subpopulations such as room or strain) through rigorous detection methods, such as the ELISA, culture for the organism, and histopathology, and rigid adherence to the principles of barrier maintenance (Jonas, 1976). Caesarian derivation may be used to eliminate the infection once it has become established in a colony, but rigorous testing of fetal membranes and offspring from each dam is essential because of the hazard of transplacental transmission or contamination (Ganaway *et al.*, 1973; Juhr *et al.*, 1970; Tram *et al.*, 1970). Other methods of identifying and screening uninfected breeders have apparently given some degree of success (Berg and Harmison, 1957; Nelson and Gowen, 1931). Treatment with tetracyclines is reported to suppress the disease manifestations but does not eliminate the infection (Dolowy *et al.*, 1960).

Some studies have provided evidence that mice (Atobe and Ogata, 1977; Cassell *et al.*, 1973; Taylor *et al.*, 1977) and rats (Cassell and Davis, 1978) can be vaccinated successfully against *M. pulmonis* respiratory disease. However, considerable research and testing of potential vaccines will be necessary before the feasibility and efficacy of this promising method can be fully established.

J. Interference with Research

Like many other infectious agents of mice, *M. pulmonis* poses extremely serious problems (Lindsey *et al.*, 1971) for the unwary investigator who has little appreciation of health surveillance and barrier maintenance of mice. A common experience is to obtain mice from a colony (commercial or investigator managed) of unknown disease status, only to have them develop signs of overt disease, including death, in the course of an experiment. Perhaps even more devastating to the scientific integrity of some experiments is the use of mice with clinically silent *M. pulmonis* infection. Even though rejected in principle, the latter practice is the norm in many laboratories.

Latent infections of *M. pulmonis* in mice have been activated by intranasal inoculations of various types (Edward, 1940; Lutsky and Organick, 1966; Sullivan and Dienes, 1939) and probably by inhalation of tobacco smoke (Wynder *et al.*, 1968). Transplantation of a tumor contaminated with *M. pulmonis* has been found to cause arthritis in recipient mice (Barden and Tully, 1969. Immunosuppression of mice with cyclophosphamide has been shown to enhance greatly the mortality due to *M. pulmonis* (Singer *et al.*, 1972).

Alterations in immune responsiveness of mice infected with *M. pulmonis* appear to be a distinct possibility. *Mycoplasma pulmonis* is known to be a potent mitogen for mouse and rat lymphocytes (Ginsberg and Nicolet, 1973; Naot *et al.*, 1977, 1979a,b). Thus, inappropriate immune responses might be anticipated in infected mice due to preempting helper cells or modification of immune regulatory mechanisms.

In rats, *M. pulmonis* has been reported to alter mucus secretion and ciliary function in the respiratory tract (Green, 1970; Irvani and van As, 1972; Ventura and Domaradzki, 1967) and the incidence of carcinogen-induced cancers in the respiratory tract (Schrieber *et al.*, 1972). Thus, inappropriate responses should be expected from almost any experimental procedure involving the respiratory tract of *M. pulmonis*-infected mice.

III. KLEBSIELLOSIS

Klebsiella pneumoniae is a gram-negative, capsulated bacillus which measures 0.3–1.5 × 0.6–6.0 μm and occurs in pairs or short chains. In addition to causing pneumonia in man, it has been implicated in epizootic infections of muskrats (Wyand and Hayden, 1973), agalactiae in sows (Lake and Jones, 1970), mastitis in cattle (Buntain and Field, 1953), metritis in horses (Crouch *et al.*, 1972), and respiratory infections in monkeys (Giles *et al.*, 1974; Hunt *et al.*, 1968).

According to "Bergey's Manual of Determinative Bacteriology" (Buchanan and Gibbon, 1974), *K. pneumoniae* of capsule types 1 and 2 is highly virulent for mice. *Klebsiella pneumoniae* type 6 (Schneemilch, 1976) and an untyped *K. pneumoniae* (Flamm, 1957) have been implicated in natural disease outbreaks in mice. In these outbreaks clinical signs included inappetance, hunched posture, rough hair coats, sneezing, and hyperpnea. At necropsy the observed lesions were enlargement of cervical nodes, cervical abscesses, renal and hepatic abscesses, empyema, and "granulomatous" pneumonia. The organism was recovered from most lesions in pure culture. Also, it was consistently found on all room and cage surfaces. Rats housed in the room with affected mice harbored the organism in their nasal passages but did not show clinical disease (Schneemilch, 1976). Experimentally, peritonitis was readily produced by intraperitoneal injection of the organism in one study (Schneemilch, 1976). In the other study (Flamm, 1957), it was found that inoculations into the buccal mucosa reproduced the cervical abscesses of the natural disease.

Klebsiella pneumoniae is widely disseminated in nature, soil, and water, as well as in agricultural and forest products (Duncan and Razzell, 1972), and occurs commonly in the intestinal tract of man and animals (Buchanan and Gibbon, 1974). Pathogenic serotypes apparently are much more restricted in prevalence. Control measures should attempt to identify specific source(s) responsible in a given outbreak. Cesarean derivation and strict barrier maintenance most likely would be required to eliminate the organism from infected mice.

IV. PASTEURELLOSIS

A. *Pasteurella pneumotropica,* a Respiratory Tract Pathogen?

Three decades ago, Jawetz (1948, 1950) and Jawetz and Baker (1950) reported studies which seemed to incriminate *Pasteurella pneumotropica* as a ubiquitous respiratory pathogen of major importance in mice. This view received a great deal of support several years later from Brennan *et al.* (1965, 1969a), whose experimental studies of *P. pneumotropica* and

M. pulmonis infections in mice led them to conclude that this combination of agents was a common cause of spontaneous murine pneumonia (Brennan *et al.*, 1969b). Since subsequent investigations of pneumonias and *P. pneumotropica* infections in mice have failed to confirm these claims, it seems imperative to reexamine the evidence concerning this bacterium's role as a respiratory tract pathogen.

A careful evaluation of the studies of Jawetz (1948, 1950) and Jawetz and Baker (1950) gives reason to question their claims. First, the stock of mice which they used was known to have indigenous pneumonia virus of mice (PVM) and *Chlamydia trachomatis* (the Nigg agent), and "spontaneous pulmonary consolidation" (of unknown cause) was known to occur in a small percentage of the animals (Jawetz and Baker, 1950). The stock was not tested for pathogens except *P. pneumotropica*. Second, the unusual circumstances under which their studies came about and developed must be emphasized. "In the course of attempts to adapt to Swiss mice an agent causing a minor respiratory illness in human volunteers, pneumonic lesions were encountered in animals inoculated intranasally under ether anesthesia with mouse lung material in serial passage. . . . In early passages only transient illness with lassitude, anorexia and ruffled coat occurred between the 6th and 10th day after inoculation" (Jawetz, 1950). In subsequent passages, "the proportion of deaths rapidly increased so that in the 14th passage of mouse lung material 85% of the animals died after an illness of 2 to 7 days" (p. 174). Jawetz (1950) also stated that *P. pneumotropica* "produces disease only under artificial circumstances of serial transfer of lung material or possibly drastic reduction of host resistance" (p. 181). Also, he concluded that "pathogenicity of latent *P. pneumotropica* appears to be a function of intranasal passage of mouse lung material" (p. 183). Thus, Jawetz's and Baker's data were derived from a high serial passage of human lung tissue in mice already infected with known respiratory pathogens and probably some which were unrecognized. The clinical signs and pathologic findings they described (Jawetz, 1950; Jawetz and Baker, 1950) include all the hallmarks of MRM in the mouse induced by intranasal inoculation of large doses of *M. pulmonis* (Lindsey and Cassell, 1973).

It is interesting to note that part of the work of Brennan *et al.* (1969a) was a repeat of the work of Jawetz (1948, 1950) and Jawetz and Baker (1950). Their initial effort also was to passage serially *P. pneumotropica* in mice. The result is summarized in the following statement: "When *P. pneumotropica* was serially passed in conventional mice, *M. pulmonis*, as well as *P. pneumotropica*, was recovered from mice with gross lesions" (Brennan *et al.*, 1969a, p. 337).

There are further precedents which tend to support the possibility that *M. pulmonis* was the primary pathogen responsible for disease in the mice of Jawetz (1948, 1950) and Jawetz and Baker (1950). In the study by Horsfall and Hahn (1940) of PVM, a virus that alone causes minimal or no morphologic pathology depending on the age of the mice, serial passage of the virus in conventional mice led to pneumonia compatible with that of MRM; *M. pulmonis* was isolated from the animals. Many investigators (Brennan *et al.*, 1965; Edward, 1940, 1947; Lutsky and Organick, 1966; Sullivan and Dienes, 1939) have inadvertently activated latent infections of *M. pulmonis* in mice by intranasal inoculations of sterile broth or serial passage of lung homogenates.

B. *Pasteurella pneumotropica*, an Opportunistic Respiratory Tract Pathogen?

Many questions still remain concerning the potential role of this extremely ubiquitous organism (Casillo and Blackmore, 1972; Moore *et al.*, 1973; Sparrow, 1976) as a respiratory tract pathogen. The results of experimental trials are conflicting in that a few studies have reported the production of pneumonia (Brennan *et al.*, 1969a) after intranasal inoculation of *P. pneumotropica* into pathogen-free mice, whereas others have reported negative results (Burek *et al.*, 1972; Goldstein and Green, 1967; Hoag *et al.*, 1962). In one study of natural outbreaks of pneumonias in mice (Saito *et al.*, 1978a), none could be attributed to *P. pneumotropica* alone, although it was one of the most frequently encountered agents.

Perhaps *P. pneumotropica* is an opportunist which plays a role in respiratory tract disease only under certain circumstances. A few studies of experimental *P. pneumotropica* suggest the possibility that this organism may complicate pneumonias primarily due to *M. pulmonis* (Brennan *et al.*, 1969a) or Sendai virus (Jakab, 1974). Similarly, Goldstein and Green (1967) have shown that bilateral nephrectomy of mice causes reduced pulmonary clearance of *P. pneumotropica*, suggesting that immunosuppression might predispose to respiratory tract infection by this organism. Natural infections to confirm these possibilities have not been reported, however.

C. Nonrespiratory Disease

There is abundant evidence that *P. pneumotropica* causes localized suppurative lesions in a wide variety of organs other than the respiratory tract. The reported lesions include: conjunctivitis (Needham and Cooper, 1975; Wagner *et al.*, 1969), panophthalmitis (Weisbroth *et al.*, 1969), dacryoadenitis (Needham and Cooper, 1975), subcutaneous and localized abscesses (Weisbroth *et al.*, 1969; Wilson, 1976), and infection of bulbourethral glands (Sebesteny, 1973).

For further information on *P. pneumotropica*, including bacteriology, host range, and distribution in infected hosts and pathology, see Chapter 6, this volume.

V. CHLAMYDIAL PNEUMONITIS

This disease has no known importance in contemporary laboratory mouse stocks. It is included here for historical reasons as well as the possibility that it may reappear as an indigenous disease of mice. All known reports of the natural disease emanated from American, European, and Australian laboratories in the 1940s and 1950s.

The causative agent, *Chlamydia trachomatis,* was first described by Dr. Clara Nigg (1942) at Yale. She referred to the agent as the *mouse pneumonitis virus,* but most subsequent authors called it the *Nigg agent.* Natural transmission was thought to be by inhalation, cannibalism (Karr, 1943; Nigg and Eaton, 1944), and/or the genital route (Genest, 1959). However, all studies (DeBurgh *et al.,* 1945; Gönnert, 1941, 1942; Karr, 1943; Nigg and Eaton, 1944) implicating the agent as a cause of pneumonitis came about as a consequence of passaging human tissues, particularly lung, in mice likely harboring the latent infection. Clinical signs were dyspnea, cyanosis, reluctance to move, humped posture, weight loss, ruffled fur, and chattering. Deaths occurred in 1-20 days depending on the dose of the agent (Karr, 1943; Nigg and Eaton, 1944).

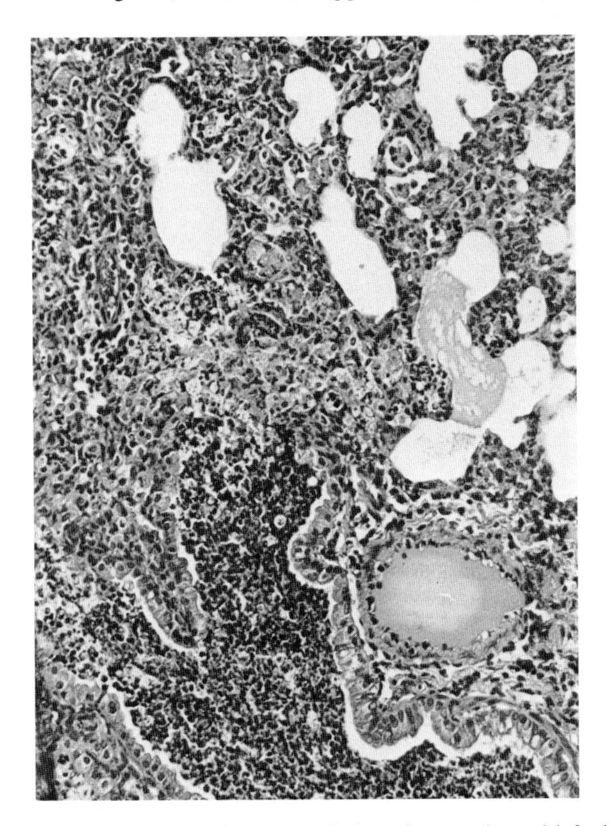

Fig. 13. Lung section with a mouse 2 days after experimental infection with a human strain of *Chlamydia trachomatis.* The field shows interstitial pneumonitis and neutrophilic exudate filling a terminal bronchiole. ×125. (Courtesy of Drs. W. Chen and C. Kuo, 1980, and the *American Journal of Pathology.*)

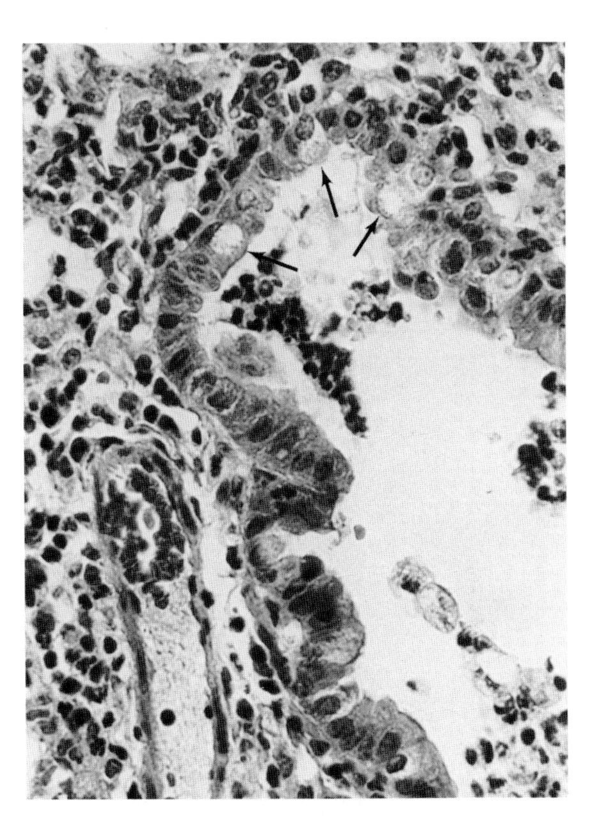

Fig. 14. Bronchus from the lung of a mouse experimentally infected with *Chlamydia trachomatis.* Note the intracytoplasmic vesicles containing inclusions (arrows) within epithelial cells. ×300. (Courtesy of Drs. W. Chen and C. Kuo, 1980, and the *American Journal of Pathology.*)

The pathology observed in the experimental disease was principally a patchy to diffuse interstitial pneumonia similar to chlamydial pneumonias of other species (Gogolak, 1953; Weiss, 1949). Alveolar macrophages and neutrophils were prominent in affected areas of the lung. The characteristic elementary bodies could be demonstrated in macrophages and bronchial epithelium by special stains (Gogolak, 1953; Gönnert, 1941; Weiss, 1949). More recent investigations (Chen and Kuo, 1980) have further characterized the dose response of experimentally infected mice (Figs. 13-15).

It is interesting to speculate about the significance of this spontaneous disease in mice in light of the recognition of chlamydial pneumonitis in human infants (Schachter and Dawson, 1978). Is it possible that this genitally transmitted disease of man due to *C. trachomatis* became established in mice of a few colonies and persisted in them as an indigenous infection for a period of time? If so, does it still persist in some colonies? Are contemporary colonies of mice at risk for contracting this infection from human adults such as animal technicians who harbor the infection in their genital tract? Similar questions can be asked about the agent of guinea pig inclusion conjunctivitis (Murray, 1964; Storz, 1964). Perhaps future studies will provide answers to these provocative questions.

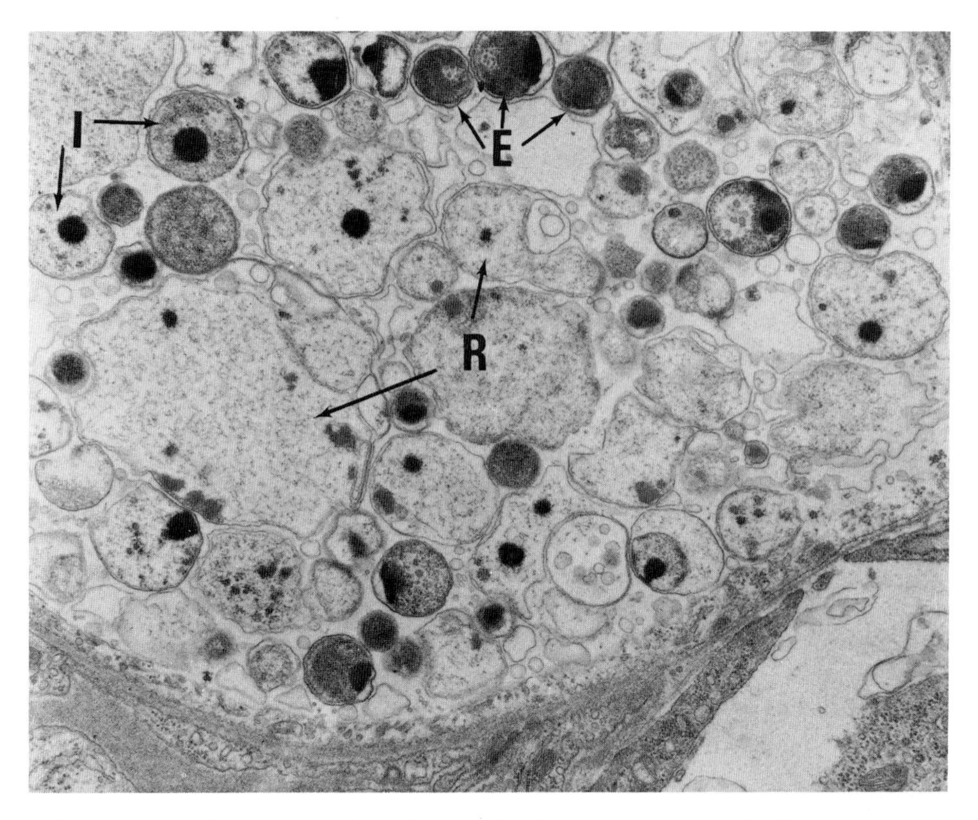

Fig. 15. Lung of a mouse experimentally infected with *Chlamydia trachomatis.* Electron micrograph showing an intracytoplasmic vesicle containing various developmental forms: reticulate bodies (R), intermediate bodies (I), and elementary bodies (E). ×10,000. (Courtesy of Drs. W. Chen and C. Kuo, 1980, and the *American Journal of Pathology.*)

VI. CORYNEBACTERIOSIS

The term *pseudotuberculosis* traditionally has referred to infections of *Corynebacterium* spp. in animals. Unfortunately, the term is often inappropriate because septicemia and acute tissue responses may predominate, depending on the species of organism and the host factors. This is particularly true of *Corynebacterium kutscheri* infection in the mouse. Latent infections are said to be common in conventional mouse colonies (Hirst and Campbell, 1977; Hirst and Olds, 1978a,b; Hirst and Wallace, 1976; Pierce-Chase *et al.,* 1964; Yokoiyama *et al.,* 1977). Active disease is precipitated by procedures which cause immunosuppression: X irradiation (Shechmeister and Adler, 1953), steroid administration (Antopol, 1950; Antopol *et al.,* 1951; Fauve *et al.,* 1964; LeMaistre and Thompsett, 1952), concurrent infections (Lawrence, 1957; Topley and Wilson, 1920), and pantothenate deficiency (Zucker, 1957). The active disease in the mouse is characteristically a massive septicemia which results in the lodgement of septic emboli in many organs, most notably in the kidney and liver and less frequently in the lung, skin, and joints (Weisbroth and Scher, 1968). Regardless of location, the septic emboli lead to formation of abscesses. If the animal survives, chronic inflammatory responses supervene. Rats appear to be more prone to develop predominantly pulmonary disease in response to *C. kutscheri* infection (Ford and Jointer, 1968; Giddens *et al.,* 1968; Weisbroth, 1979).

VII. MISCELLANEOUS INFECTIONS

The respiratory tract is constantly exposed to all the rigors and hazards contained in the ambient air. For the mouse, this no doubt includes enormous numbers of particles and aerosols from the bedding, food, cage, cagemates, and room air. Through phenomena such as sedimentation and impaction, these materials are efficiently filtered from the incoming air, particularly in the nasal passages (Brain and Valberg, 1979; Proctor, 1977; Subcommittee on Airborne Particles (1979). Thus, the nasal passages and oropharynx of the mouse are often contaminated by bacteria which normally inhabit the body surfaces and gastrointestinal tract. Presumably, this explains why mice occasionally are found to have otitis media attributable to organisms such as *Pseudomonas aeruginosa, K. pneumoniae, Proteus* spp., *E. coli,* and others (Tuffery and Innes, 1963).

Aside from occasionally causing otitis media and possibly rhinitis, there is little evidence that *P. aeruginosa* is a significant respiratory tract pathogen in the mouse. However, it is of particular interest that the nasal passages provide the most likely portal of entry for septicemia due to this organism following immunosuppression of mice which harbor the infection, usually in the nasal passages and digestive tract (Brownstein, 1978; Wensinck, 1961).

VIII. CONCLUSIONS

It should be apparent from the foregoing discussion that the subject of infectious respiratory tract diseases in the mouse has been and continues to be somewhat confusing. We conclude that far-reaching claims about the natural history of certain indigenous infections have at times been made without adequate efforts to define the total pathogenic flora of the respiratory tract in the animals being considered. Furthermore, many experimental infections of the respiratory tract have been performed using mice which were not first subjected to rigorous health surveillance evaluations to detect preexisting, often latent respiratory tract infections by other bacteria, mycoplasmas, and/or viruses. Such experiences from the past must be taken into account in performing pathologic investigations of clinical respiratory tract disease outbreaks in mice and in selecting mouse stocks for experimental studies in the future—if the natural history of indigenous respiratory infections in the mouse is ever to be understood and if the principles of the scientific method are to be observed in research using this species.

ACKNOWLEDGMENTS

Supported in part by research funds of the Veterans Administration, and USPHS Grants RR 00463, RR 00959 and HL 19741. G.H.C. is the recipient of USPHS Career Development Award 1 KO4 HL 00387.

REFERENCES

Andrewes, C. H., and Niven, J. S. F. (1950). A virus from cotton rats: Its relation to gray lung virus. *Br. J. Exp. Pathol.* **31,** 773–778.

Andrewes, C. H., and Welch, F. V. (1946). A motile organism of the pleuropneumonia group. *J. Pathol. Bacteriol.* **58,** 578–580.

Antopol, W. (1950). Anatomic changes in mice treated with excessive doses of cortisone. *Proc. Soc. Exp. Biol. Med.* **73,** 262–265.

Antopol, W., Glaubach, S., and Quittner, H. (1951). Experimental observations with massive doses of cortisone. *Rheumatism* **7,** 187–197.

Asa, P., Acton, R. T., Cassell, G. H., and Wise, K. S. (1980). Identification and characterization of protein antigens of two mycoplasma species. *J. Immunol.* **124,** 997–999.

Atobe, H., and Ogata, M. (1974). Pneumonitis in mice inoculated with *Mycoplasma pulmonis:* Production of pulmonary lesions and persistence of organisms and antibodies. *Jpn. J. Vet. Sci.* **36,** 495–503.

Atobe, H., and Ogata, M. (1977). Protective effect of killed *Mycoplasma pulmonis* vaccine against experimental infection in mice. *Jpn. J. Vet. Sci.* **39,** 39–46.

Barden, J. A., and Tully, J. G. (1969). Experimental arthritis in mice with *Mycoplasma pulmonis*. *J. Bacteriol.* **100,** 5–10.

Battigelli, M. C., Fraser, D. A., and Cole, H. (1971). Microflora of the respiratory surface of rodents exposed to "inert" particulates. *Arch. Intern. Med.* **127,** 1103–1104.

Berg, B. N., and Harmison, C. R. (1957). Growth, disease and aging in the rat. *J. Gerontol.* **12,** 370–377.

Brain, J. D., and Valberg, P. A. (1979). Deposition of aerosol in the respiratory tract. *Am. Rev. Respir. Dis.* **120,** 1325–1373.

Bredt, W. (1973). Motility of mycoplasmas. *Ann. N.Y. Acad. Sci.* **225,** 246–250.

Bredt, W., and Radestock, U. (1977). Gliding motility of *Mycoplasma pulmonis*. *J. Bacteriol.* **130,** 937–938.

Brennan, P. C., and Feinstine, R. N. (1969). Relationship of hydrogen peroxide production by *Mycoplasma pulmonis* virulence of catalase-deficient mice. *J. Bacteriol.* **98,** 1036–1040.

Brennan, P. C., Fritz, T. E., and Flynn, R. J. (1965). *Pasteurella pneumotropica:* Cultural and biochemical characteristics, and its association with disease in laboratory animals. *Lab. Anim. Care* **19,** 360–371.

Brennan, P. C., Fritz, T. E., and Flynn, R. J. (1969a). Role of *Pasteurella pneumotropica* and *Mycoplasma pulmonis* in murine pneumonia. *J. Bacteriol.* **97,** 337–349.

Brennan, P. C., Fritz, T. E., and Flynn, R. J. (1969b). Murine pneumonia: a review of etiologic agents. *Lab. Anim. Care* **19,** 360–371.

Brownstein, D. G. (1978). Pathogenesis of *Pseudomonas aeruginosa* bacteremia in cyclophosphamide treated mice and potentiation of endogenous streptococcal virulence. *J. Infect. Dis.* **137,** 795–801.

Brunner, H., Schaeg, W., Schiefer, H. G., and Bruck, U. (1977). A staphylococcal radioimmunoassay for detection of antibodies to *Mycoplasma pneumoniae*. *Med. Microbiol. Immunol.* **163,** 125–135.

Buchanan, R. E., and Gibbon, N. E., eds. (1974). "Bergey's Manual of Determinative Bacteriology," 8th ed., pp. 321–324. Williams & Wilkins, Baltimore, Maryland.

Buntain, D., and Field, H. I. (1953). An outbreak of mastitis in cattle due to infection with an organism of the Friedlander (Klebsiella) group. *Vet. Rec.* **65,** 91–93.

Burek, J. D., Jersey, G. C., Whitehair, C. K., and Carter, G. R. (1972). The pathology and pathogenesis of *Bordetella bronchiseptica* and *Pasteurella pneumotropica* in conventional and germfree rats. *Lab. Anim. Sci.* **22,** 844–849.

Casillo, S., and Blackmore, D. K. (1972). Uterine infections caused by bacteria and mycoplasma in mice and rats. *J. Comp. Pathol.* **82,** 477–482.

Cassell, G. H., and Davis, J. K. (1978). Active immunization of rats against *Mycoplasma pulmonis* respiratory disease. *Infect. Immun.* **21,** 69–75.

Cassell, G. H., and Hill, A. (1979). Murine and other small animal mycoplasmas. *In,* "The Mycoplasmas" (M. F. Barile, S. Razin, J. G. Tully, and R. F. Whitcomb, eds.), Vol. 2, pp. 235–273. Academic Press, New York.

Cassell, G. H., and McGhee, J. R. (1975). Pathologic and immunologic response of nude mice following intranasal inoculation of *Mycoplasma pulmonis*. *J. Reticuloendothel. Soc.* **18,** 342.

Cassell, G. H., Lindsey, J. R., Overcash, R. G., and Baker, H. J. (1973). Murine *Mycoplasma* respiratory disease. *Ann. N.Y. Acad. Sci.* **225,** 395–412.

Cassell, G. H., Lindsey, J. R., and Baker, H. J. (1974). Immune response of pathogen-free mice inoculated intranasally with *Mycoplasma pulmonis*. *J. Immunol.* **112,** 124–136.

Cassell, G. H., Lindsey, J. R., Baker, H. J., and Davis, J. K. (1979). Mycoplasmal and rickettsial diseases. *In* "The Laboratory Rat" (H. J. Baker, J. R. Lindsey, and S. H. Weisbroth, eds.), Vol. 1, pp. 243–269. Academic Press, New York.

Cassell, G. H., Lindsey, J. R., Davis, J. K., Davidson, M. K., Brown, M. B., and Mayo, J. G. (1981). Detection of natural *Mycoplasma pulmonis* infection in rats and mice by an enzyme linked immunosorbent assay (ELISA). *Lab. Anim. Sci.* **31,** 676–682.

Chen, W., and Kuo, C. (1980). A mouse model of pneumonitis induced by *Chlamydia trachomatis*. *Am. J. Pathol.* **100,** 365–382.

Chalquest, R. R. (1962). Cultivation of the infectious-synovitis type pleuropneumonia-like organism. *Avian Dis.* **6,** 36–43.

Clyde, W. A. (1964). *Mycoplasma* species identification based upon growth inhibition by specific antisera. *J. Immunol.* **92,** 958–965.

Cole, B. C., and Ward, J. R. (1973). Interaction of *Mycoplasma arthritidis* and other mycoplasmas with murine peritoneal macrophages. *Infect. Immun.* **7,** 691–699.

Cole, B. C., Ward, J. R., and Martin, C. H. (1968). Hemolysin and peroxide activity of *Mycoplasma* species. *J. Bacteriol.* **95,** 2022–2030.

Cole, B. C., Golightly-Rowland, L., and Ward, J. R. (1975a). Arthritis in mice induced by *Mycoplasma pulmonis:* Humoral antibody and lymphocyte responses of CBA mice. *Infect. Immun.* **12,** 1083–1092.

Collier, A. M., and Baseman, J. B. (1973). Organ culture techniques with mycoplasmas. *Ann. N.Y. Acad. Sci.* **225,** 277–289.

Crouch, J. R. F., Atherton, J. G., and Platt, H. (1972). Venereal transmission of *K. aerogenes* in a thoroughbred stud from a persistently infected stallion. *Vet. Rec.* **90,** 21–24.

Davidson, M. K., Lindsey, J. R., Brown, M. B., Schoeb, T. R., and Cassell, G. H. (1981). Comparison of methods for detection of *Mycoplasma pulmonis* in experimentally and naturally infected rats. *J. Clin. Microbiol.* **14,** 646–655.

Davis, J. K., Delozier, K. M., Asa, K. D., Minion, F. C., and Cassell, G. H. (1980). Interactions between murine alveolar macrophages and *Mycoplasma pulmonis in vitro*. *Infect. Immun.* **29,** 590–599.

DeBurgh, P., Jackson, A. V., and Williams, A. E. (1945). Spontaneous infection of laboratory mice with a psittacosis-like organism. *Aust. J. Exp. Biol. Med. Sci.* **23,** 107–110.

Deeb, B. J., and Kenny, G. E. (1967a). Characterization of *Mycoplasma pulmonis* variants isolated from rabbits. I. Identification and properties of isolates. *J. Bacteriol.* **93,** 1416–1424.

Deeb, B. J., and Kenny, G. E. (1967b). Characterization of *Mycoplasma pulmonis* variants isolated from rabbits. II. Basis for differentiation of antigenic subtypes. *J. Bacteriol.* **93,** 1425–1429.

Del Giudice, R. A., Robillard, N. F., and Carski, T. R. (1967). Immunofluorescence identification of *Mycoplasma* on agar by use of indirect incident illumination. *J. Bacteriol.* **93,** 1205–1209.

Denny, F. W., Taylor-Robinson, D., and Allison, A. C. (1972). The role of thymus-dependent immunity in *Mycoplasma pulmonis* infections of mice. *J. Med. Microbiol.* **5,** 327–336.

Dolowy, W. C., Hess, A. L., and McDonald, G. O. (1960). Oxytetracycline hydrochloride in the treatment of 590 rats in an outbreak of infectious catarrh. *Ill. Vet.* **3,** 20–24.

Duncan, D. W., and Razzell, W. E. (1972). *Klebsiella* biotypes among coliforms isolated from forest environments and farm produce. *Appl. Microbiol.* **24,** 933–938.

Edward, D. G. (1940). The occurrence in normal mice of pleuropneumonia-like organisms capable of producing pneumonia. *J. Pathol. Bacteriol.* **50,** 409–418.

Edward, D. G. (1947). Catarrh of the upper respiratory tract in mice and its association with pleuropneumonia-like organisms. *J. Pathol. Bacteriol.* **59,** 209–221.

Fauve, R. M., Pierce-Chase, C. H., and Dubos, R. (1964). Corynebacterial pseudotuberculosis in mice. II. Activation of natural and experimental latent infections. *J. Exp. Med.* **120,** 283–311.

Flamm, H. (1957). Klebsiella-Enzootie in einer Mäusezucht. *Schweiz. Z. Pathol. Bakteriol.* **20,** 23–27.

Ford, T. M., and Jointer, G. N. (1968). Pneumonia in a rat associated with *Corynebacterium pseudotuberculosis*, a case report and literature survey. *Lab. Anim. Care* **18,** 220–223.

Fraser, L. R., and Taylor-Robinson, D. (1977). The effect of *Mycoplasma pulmonis* on fertilization and preimplantation development *in vitro* of mouse eggs. *Fertil. Steril.* **28,** 488–498.

Freundt, E. A. (1958). "The Mycoplasmataceae (the Pleuropneumonia Group of Organisms)." Munksgaard, Copenhagen.

Frey, M. L., Stalheim, O. H., Ellis, E. M., Levingston, C. W., and Yedloulschnig, R. J. (1974). Diagnostic procedures in animal mycoplasma infections. *Proc. U.S. Anim. Health Assoc.* **73,** 492–499.

Gabridge, M. G., Johnson, C. K., and Cameron, A. M. (1974). Cytotoxicity of *Mycoplasma pneumoniae* membranes. *Infect. Immun.* **10,** 1127–1134.

Ganaway, J. R., Allen, A. M., Moore, T. D., and Bohner, H. J. (1973). Natural infection of germfree rats with *Mycoplasma pulmonis*. *J. Infect. Dis.* **127,** 529–537.

Genest, P. (1959). Transmission congenitale de la psittacose experimentale chez la souris blanche. *Bull. Acad. Vet. Fr.* **32,** 75–80.

Giddens, W. E., Keahey, K. K., Carter, G. R., and Whitehair, C. K. (1968). Pneumonia in rats due to infection with *Corynebacterium kutscheri*. *Pathol. Vet.* **5,** 227–237.

Giles, R. C., Jr., Hildebrandt, P. K., and Tate, C. (1974). Klebsiella air sacculitis in the owl monkey (*Aotus trivirgatus*). *Lab. Anim. Sci.* **24,** 610–616.

Ginsberg, H. S., and Nicolet, J. (1973). Extensive transformation of lymphocytes by a *Mycoplasma* organism. *Nature (London), New Biol.* **246,** 143–146.

Gönnert, R. (1941). Die Bronchopneumoniae, eine neue Viruskrankhirt der Maus. *Zentralbl. Bakteriol., Parasitenkd., Infektionskr. Hyg., Abt. 1: Orig.* **147,** 161–174.

Gönnert, R. (1942). Uber einige Eigenschaften des Bronchopneumonievirus der Maus. *Zentralbl. Bakteriol., Parasitenkd., Infektionskr. Hyg., Abt. 1: Orig.* **148,** 331–337.

Goeth, H., and Appel, K. R. (1974). Experimentelle Untersuchungen über die Beeinträchtigung der Fertilität durch Mycoplasmeninfektionen des Genital traktes. *Zentralbl. Bakteriol., Parasitenkd., Infektionskr. Hyg., Abt. 1: Orig., Reihe A* **228,** 282–289.

Gogolak, F. M. (1953). The histopathology of murine pneumonitis infection and the growth of the virus in the mouse lung. *J. Infect. Dis.* **92,** 254–272.

Goldstein, E., and Green, G. M. (1967). Alteration of the pathogenicity of *Pasteurella pneumotropica* for the murine lung caused by changes in pulmonary antibacterial activity. *J. Bacteriol.* **93,** 1651–1656.

Green, G. M. (1970). The Burns Amberson Lecture—In defense of the lung. *Am. Rev. Respir. Dis.* **102,** 691–703.

Harwick, H. J., Kalmanson, G. M., Fox, M. A., and Guze, L. B. (1973).

Arthritis in mice due to infection with *Mycoplasma pulmonis*. I. Clinical and microbiologic features. *J. Infect. Dis.* **128,** 533-540.

Hayflick, L. (1970). "The Mycoplasmatales and the L-Phase of Bacteria." Appleton, New York.

Hektoen, L. (1915-1916). Observations on pulmonary infections in rats. *Trans. Chicago Pathol. Soc.* **10,** 105-109.

Hill, A. (1972). Transmission of *Mycoplasma pulmonis* between rats. *Lab. Anim.* **6,** 331-336.

Hill, A. (1974). Mycoplasmas of small animals. *In* "Mycoplasmas of Man, Animals, Plants and Insects" (J. M. Bove and J. F. Duplan, eds.), pp. 311-316. INSERM, Paris.

Hill, A. C. (1978). Demonstration of *Mycoplasma* in tissue by the immunoperoxidase technique. *J. Infect. Dis.* **137,** 152-154.

Hirst, R. G., and Campbell, R. (1977). Mechanisms of resistance to *Corynebacterium kutscheri* in mice. *Infect. Immun.* **17,** 319-324.

Hirst, R. G., and Olds, R. J. (1978a). *Corynebacterium kutscheri* and its alleged avirulent variant in mice. *J. Hyg.* **80,** 349-356.

Hirst, R. G., and Olds, R. J. (1978b). Serological and biochemical relationships between the alleged avirulent variant of *Corynebacteriun kutscheri* and streptococci of group N. *J. Hyg.* **80,** 356-363.

Hirst, R. G., and Wallace, M. E. (1976). Inherited resistance to *Corynebacterium kutscheri* in mice. *Infect. Immun.* **14,** 475-482.

Hoag, W. G., Wetmore, P. W., Rogers, J., and Meier, H. (1962). A study of latent *Pasteurella* infection in a mouse colony. *J. Infect. Dis.* **11,** 135-140.

Horowitz, S. A., and Cassell, G. H. (1978). Enzyme-linked immunosorbent assay for detection of *Mycoplasma pulmonis* antibodies. *Infect. Immun.* **22,** 161-170.

Horsfall, F. L., Jr., and Hahn, R. G. (1940). A latent virus in normal mice capable of producing pneumonia in its natural host. *J. Exp. Med.* **71,** 391-408.

Howard, C. J., Stott, E. J., and Taylor, G. (1978). The effect of pneumonia induced in mice with *Mycoplasma pulmonis* on resistance to subsequent bacterial infection and the effect of a respiratory infection with Sendai virus on the resistance of mice to *Mycoplasma pulmonis*. *J. Gen. Microbiol.* **109,** 79-87.

Hu, P. C., Collier, A. M., and Baseman, J. B. (1975). Alterations in metabolism of hamster tracheas in organ culture after infection by virulent *Mycoplasma pneumoniae*. *Infect. Immun.* **11,** 704-710.

Hunt, D. E., Pittillo, R. F., Deneau, G. A., Schabel, F. M., Jr., and Mellett, L. B. (1968). Control of an acute *Klebsiella pneumoniae* infection in a rhesus monkey colony. *Lab. Anim. Care* **18,** 182-185.

Irvani, J., and van As, A. (1972). Mucus transport in the tracheobronchial tree of normal and bronchitic rats. *J. Pathol.* **106,** 81-93.

Jakab, G. J. (1974). Effect of sequential innoculations of Sendai virus and *Pasteurella pneumotropica* in mice. *J. Am. Vet. Med. Assoc.* **164,** 723-728.

Jakab, G. J., and Green, G. M. (1976). Defect in intracellular killing of *Staphylococcus aureus* within alveolar macrophages in Sendai virus-infected murine lungs. *J. Clin. Invest.* **57,** 1533-1539.

Jakab, G. J., Warr, G. A., and Sannes, P. L. (1980). Alveolar macrophage ingestion and phagosome-lysosome fusion defect associated with virus pneumonia. *Infect. Immun.* **27,** 960-968.

Jawetz, E. (1948). A latent pneumotropic Pasteurella of laboratory animals. *Proc. Soc. Exp. Biol. Med.* **68,** 46-48.

Jawetz, E. (1950). A pneumotropic Pasteurella of laboratory animals. I. Bacteriological and serological characteristics of the organism. *J. Infect. Dis.* **86,** 172-183.

Jawetz, E., and Baker, W. H. (1950). A pneumotropic Pasteurella of laboratory animals. II. Pathological and immunological studies with the organism. *J. Infect. Dis.* **86,** 184-196.

Jersey, G., Whitehair, C. K., and Carter, G. R. (1973). *Mycoplasma pulmonis* as the primary cause of chronic respiratory disease in rats. *J. Am. Vet. Med. Assoc.* **163,** 599-604.

Jonas, A. M., chm. (1976). Long-term holding of laboratory rodents. *ILAR News* **19,** 1-25.

Jones, T. C., and Hirsch, J. G. (1971). The interaction *in vitro* of *Mycoplasma pulmonis* with mouse peritoneal macrophages and L-cells. *J. Exp. Med.* **133,** 231-259.

Jones, T. C., and Yang, L. (1977). Attachment and ingestion of mycoplasmas by mouse macrophages. *Am. J. Pathol.* **87,** 331-346.

Jones, T. C., Yeh, S., and Hirsch, J. G. (1972). Studies on attachment and ingestion phases of phagocytosis of *Mycoplasma pulmonis* by mouse peritoneal macrophages. *Proc. Soc. Exp. Biol. Med.* **139,** 464-470.

Juhr, N. C. (1971). Untersuchungen zur chronischen murinen Pneumonie. Pathogenese der *Mycoplasma pulmonis* Infection bei der Ratte. *Z. Versuchstierkd.* **13,** 217-223.

Juhr, N. C., and Obi, S. (1970). Uterusinfektionen beim Meerschweinchen. *Z. Versuchstierkd.* **12,** 383-387.

Juhr, N. C., Obi, S., Hiller, H. H., and Eichberg, J. (1970). Mycoplasmen bei Keimfreien Ratten Mäusen. *Z. Versuchstierkd.* **12,** 318-320.

Kaklamanis, E., Thomas, L., Stavropoulos, K., Borman, I., and Boshwitz, C. (1969). Mycoplasmacidal action of normal tissue extracts. *Nature (London)* **221,** 860-862.

Kappel, H. K., Nelson, J. B., and Weisbroth, S. H. (1974). Development of a screening technique to monitor a mycoplasma-free Blu:(LE) Long-Evans rat colony. *Lab. Anim. Sci.* **24,** 768-772.

Karr, H. V. (1943). Study of latent pneumotropic virus of mice. *J. Infect. Dis.* **72,** 108-116.

Keystone, E. C., Taylor-Robinson, D., Osborn, M. F., Ling, L., Pope, C., and Fornasier, V. (1980). Effect of T-cell deficiency on the chronicity of arthritis induced in mice by *Mycoplasma pulmonis*. *Infect. Immun.* **27,** 192-196.

Klieneberger, E., and Nobel, E. (1962). "Pleuropneumonia-Like Organisms (PPLO) Mycoplasmataceae," pp. 13-15. Academic Press, New York.

Kohn, D. F. (1971a). Sequential pathogenicity of *Mycoplasma pulmonis* in laboratory rats. *Lab. Anim. Care* **21,** 849-855.

Kohn, D. F. (1971b). Bronchiectasis in rats infected with *M. pulmonis:* An electron microscopy study. *Lab. Anim. Sci.* **21,** 856-861.

Kohn, D. F., and Kirk, B. E. (1969). Pathogenicity of *Mycoplasma pulmonis* in laboratory rats. *Lab. Anim. Care* **19,** 321-330.

Lake, S. G., and Jones, J. E. T. (1970). Post-parturient disease in sows associated with *Klebsiella* infection. *Vet. Rec.* **87,** 484-485.

Lawrence, J. J. (1957). Infection of laboratory mice with *Corynebacterium murium*. *Aust. J. Sci.* **20,** 147.

LeMaistre, C., and Thompsett, R. (1952). The emergence of pseudotuberculosis in rats given cortisone. *J. Exp. Med.* **95,** 393-407.

Lemcke, R. M. (1961). Association of PPLO infection and antibody response in rats and mice. *J. Hyg.* **59,** 401-412.

Lentsch, R. H., Wagner, J. H., and Owens, D. H. (1979). Comparison of techniques for primary isolation of respiratory *Mycoplasma pulmonis* from rats. *Infect. Immun.* **26,** 590-593.

Lindsey, J. R., and Cassell, G. H. (1973). Experimental *Mycoplasma pulmonis* infection in pathogen-free mice. Models for studying mycoplasmosis of the respiratory tract. *Am. J. Pathol.* **72,** 63-90.

Lindsey, J. R., Baker, H. J., Overcash, R. G., Cassell, G. H., and Hunt, C. E. (1971). Murine chronic respiratory disease. Significance as a research complication and experimental production with *Mycoplasma pulmonis*. *Am. J. Pathol.* **64,** 675-716.

Lindsey, J. R., Cassell, G. H., and Baker, H. J. (1978). Diseases due to mycoplasmas and rickettsias. *In* "Pathology of Laboratory Animals" (K. Benirschke, F. Garner, and C. Jones, eds.), Vol. 2, pp. 1482-1550. Springer-Verlag, Berlin and New York.

Lutsky, I. J., and Organick, A. B. (1966). Pneumonia due to mycoplasma in gnotobiotic mice. I. Pathogenicity of *Mycoplasma pneumoniae, Mycoplasmas salivarium*, and *Mycoplasma pulmonis* for the lungs of conventional and gnotobiotic mice. *J. Bacteriol.* **92,** 1154-1163.

Manchu, R. J., and Taylor-Robinson, D. (1968). Haemadsorption and haemagglutination by mycoplasmas. *J. Gen. Microbiol.* **50,** 465-478.

Maniloff, J., Morowitz, H. J., and Barrnett, R. J. (1965). Ultrastructure and ribosomes of *Mycoplasma gallisepticum*. *J. Bacteriol.* **90,** 193-204.

Mardh, P. A., and Taylor-Robinson, D. (1973). New approaches to the isolation of mycoplasmas. *Med. Microbiol. Immunol.* **158,** 259-266.

Moore, T. D., Allen, A. M., and Ganaway, J. R. (1973). Latent *Pasteurella pneumotropica* infection of the intestine of gnotobiotic and barrier-held rats. *Lab. Anim. Sci.* **5,** 657-661.

Murray, E. S. (1964). Guinea pig inclusion conjunctivitis. I. Isolation and identification as a member of the psittacosis-lymphogranuoma-trachoma group. *J. Infect. Dis.* **114,** 1-12.

Nakane, P. K., and Kawaoi, A. (1974). Peroxidase labelled antibody. A new method of conjugation. *J. Histochem. Cytochem.* **22,** 1084-1091.

Naot, Y., Tully, J. G., and Ginsburg, H. (1977). Lymphocyte activation by various mycoplasma strains and species. *Infect. Immun.* **9,** 185-189.

Naot, Y., Merchav, S., Ben-David, E., and Ginsburg, H. (1979a). Mitogenic activity of *Mycoplasma pulmonis*. I. Stimulation of rat B and T lymphocytes. *Immunology* **36,** 399-406.

Naot, Y., Siman-Tov, R., and Ginsburg, H. (1979b). Mitogenic activity of *Mycoplasma pulmonis*. II. Studies on the biochemical nature of the mitogenic factor. *Eur. J. Immunol.* **9,** 149-154.

Needham, J. R., and Cooper, J. E. (1975). An eye infection in laboratory mice associated with *Pasteurella pneumotropica*. *Lab. Anim.* **9,** 197-200.

Nelson, J. B. (1937a). Infectious catarrh of mice. I. A natural outbreak of the disease. *J. Exp. Med.* **65,** 833-842.

Nelson, J. B. (1937b). Infectious catarrh of mice. II. The detection and isolation of coccobacilliform bodies. *J. Exp. Med.* **65,** 851-860.

Nelson, J. B. (1937c). Infectious catarrh of mice. III. The etiological significance of the coccobacilliform bodies. *J. Exp. Med.* **65,** 851-860.

Nelson, J. B. (1954). The selective localization of murine pleuropneumonia-like organisms in the female genital tract on intraperitoneal injection in mice. *J. Exp. Med.* **100,** 311-320.

Nelson, J. B. (1955). Chronic respiratory disease in mice and rats. *Proc. Anim. Care Panel* **6,** 9-15.

Nelson, J. B. (1960). The behavior of murine pleuropneumonia-like organisms in HeLa cultures. *Ann. N.Y. Acad. Sci.* **79,** 450-457.

Nelson, J. B. (1963). Chronic respiratory disease in mice and rats. *Lab. Anim. Care* **13,** 137-143.

Nelson, J. B. (1967a). Pathologic response of Swiss and Princeton mice to *M. pulmonis*. *Ann. N.Y. Acad. Sci.* **143,** 778-783.

Nelson, J. B. (1967b). Respiratory infections of rats and mice with emphasis on indigenous mycoplasmas. *In* "Pathology of Laboratory Rats and Mice" (E. Cotchin and F. J. C. Roe, eds.), pp. 259-289. Blackwell, Oxford.

Nelson, J. B., and Gowen, J. W. (1931). The establishment of an albino rat colony free of middle ear disease. *J. Exp. Med.* **54,** 629-636.

Nigg, C. (1942). Unidentified virus which produced pneumonia and systemic infection in mice. *Science* **95,** 49-50.

Nigg, C., and Eaton, M. D. (1944). Isolation from normal mice of a pneumotropic virus which forms elementary bodies. *J. Exp. Med.* **79,** 497-510.

Ogata, M., Ohta, T., and Atobe, H. (1967). Studies on mycoplasmas of rodent origin. Mycoplasmas from the chronic respiratory disease of rats. (In Jpn.) *Nippon Saikingaku Zasshi* **22,** 618-627.

Olsen, N. O., Kerr, K. M., and Campbell, A. (1963). Control of infectious synovitis: Preparation of an agglutination test antigen. *Avian Dis.* **7,** 310-317.

Organick, A. B., and Lutsky, I. I. (1976). *Mycoplasma pulmonis* infection in gnotobiotic and conventional mice: Aspects of pathogenicity including microbial enumeration and studies of tracheal involvement. *Lab. Anim. Sci.* **26,** 419-429.

Organick, A. B., Siegesmund, K. A., and Lutsky, I. I. (1966). Pneumonia due to *Mycoplasma* in gnotobiotic mice. II. Localization of *Mycoplasma pulmonis* in the lungs of infected gnotobiotic mice by electron microscopy. *J. Bacteriol.* **92,** 1164-1176.

Pierce-Chase, C. H., Fauve, M., and Dubos, R. (1964). Corynebacterial Pseudotuberculosis in mice. I. Comparative susceptibility of mouse strains to experimental infection with *Corynebacterium kutscheri*. *J. Exp. Med.* **120,** 267-281.

Plackett, P. (1957). Depolymerisation of ribonucleic acid by extracts of *Asterococcus mycoides*. *Biochim. Biophys. Acta* **26,** 664-665.

Polak-Vogelzang, A. A., Hagenaars, R., and Nagel, J. (1978). Evaluation of an indirect immunoperoxidase test for identification of acholeplasma and mycoplasma. *J. Gen. Microbiol.* **106,** 241-249.

Proctor, D. F. (1977). The upper airways. I. Nasal physiology and defense of the lungs. *Am. Rev. Respir. Dis.* **115,** 97-129.

Razin, S., and Cosenza, B. J. (1966). Growth phases of mycoplasma in liquid media observed with phase-contrast microscope. *J. Bacteriol.* **91,** 858-869.

Razin, S., Knyszynski, A., and Lifshitz, Y. (1964). Nucleases of mycoplasma. *J. Gen. Microbiol.* **36,** 323-332.

Razin, S., Cosenza, B. J., and Tourtellatte, M. E. (1966). Variations in mycoplasma morphology induced by long chain fatty acids. *J. Gen. Microbiol.* **42,** 139-145.

Razin, S., Cosenza, B. J., and Tourtellatte, M. E. (1967). Filamentous growth of mycoplasma. *Ann. N.Y. Acad. Sci.* **143,** 66-72.

Richter, C. B. (1970). Application of infectious agents to the study of lung cancer. *In* "Studies on the Etiology and Morphogenesis of Metaplastic Lung Lesions in Mice" (P. Nettesheim, M. V. Hanna, Jr., and J. W. Deatherage, Jr., eds.), Symposium Series, No. 21, pp. 365-382. USAEC, Oak Ridge, Tennessee.

Rodwell, A. W., and Mitchell, A. (1979). Nutrition, growth and reproduction. *In* "The Mycoplasmas" (M. F. Barile and S. Razin, eds.), Vol. 1, pp. 103-140. Academic Press, New York.

Russell, W. C. (1966). Alterations in the nucleic acid metabolism of the tissue culture (hamster fibroblast) cells infected by mycoplasmas. *Nature (London)* **212,** 1537-1540.

Sainte-Marie, G. (1962). A paraffin embedding technique for studies employing immunofluorescence. *J. Histochem. Cytochem.* **10,** 250-256.

Saito, M., Nakagawa, M., Kinoshita, K., and Imaizumi, K. (1978a). Etiological studies on natural outbreaks of pneumonia in mice. *Jpn. J. Vet. Sci.* **40,** 283-290.

Saito, M., Nakagawa, M., Muto, T., and Imaizumi, K. (1978b). Strain difference of mouse in susceptibility to *Mycoplasma pulmonis* infection. *Jpn. J. Vet. Sci.* **40,** 697-705.

Schachter, J., and Dawson, C. R. (1978). "Human Chlamydial Infections." P.S.G. Publ. Co., Littleton, Massachusetts.

Schiefer, H.-G., Gerhardt, U., Brunner, H., and Krupe, M. (1974). Studies with lectins on the surface carbohydrate structures of mycoplasma membranes. *J. Bacteriol.* **120,** 81-88.

Schneemilch, H. D. (1976). A naturally acquired infection of laboratory mice with *Klebsiella* capsule type 6. *Lab. Anim.* **10,** 305-310.

Schrieber, H., Nettesheim, P., Lijinsky, W., Richter, C. B., and Walburg, H. E., Jr. (1972). Induction of lung cancer in germ-free, specific-pathogen-free and infected rats by N-nitrosoheptamethyleneimine: Enhancement by respiratory infection. *J. Natl. Cancer Inst.* **4,** 1107-1114.

Sebesteny, A. (1973). Abscesses of the bulbourethral glands of mice due to *Pasteurella pneumotropica*. *Lab. Anim.* **7,** 315-317.

Shechmeister, J. L., and Adler, F. L. (1953). Activation of Pseudotuberculosis in mice exposed to sublethal total body radiation. *J. Infect. Dis.* **92,** 228-239.

Singer, S. H., Ford, M., and Kirschstein, R. L. (1972). Respiratory diseases in cyclophosphamide treated mice. Increased virulence of *Mycoplasma pulmonis*. *Infect. Immun.* **5,** 953-956.

Sparrow, S. (1976). The microbiological and parasitological status of laboratory animals from accredited breeders in the United Kingdom. *Lab. Anim.* **10,** 365-373.

Storz, J. (1964). Uber eine naturliche Infektion cines Meerschweinchenbestandes mit einem Erreger aus der Psittakose-Lymphogranuloma Gruppe. *Zentralbl. Bakteriol., Parasitenkd., Infektionskr. Hyg., Abt. 1: Orig.* **193,** 432-446.

Subcommittee on Airborne Particles, Assembly of Life Sciences, National Research Council (1979). Effects of inhaled particles on human and animals: Deposition, retention, and clearance. *In* "Airborne Particles," pp. 107-145. Univ. Park Press, Baltimore, Maryland.

Subcommittee on the Taxonomy of Mycoplasmatales (1972). Proposal for minimal standards for descriptions of new species of the order Mycoplasmatales. *Int. J. Syst. Bacteriol.* **22,** 184-188.

Sullivan, E. R., and Dienes, L. (1939). Pneumonia in white mice produced by a pleuropneumonia-like micro-organism. *Proc. Soc. Exp. Biol. Med.* **41,** 620-622.

Tanaka, H. (1979). *Mycoplasma pulmonis* arthritis in congenitally athymic (nude) mice. Clinical and biological features. *Microbiol. Immunol.* **23,** 1055-1065.

Taylor, G. (1979). Solid-phase micro-radioimmunoassay to measure immunoglobulin class-specific antibody to *Mycoplasma pulmonis*. *Infect. Immun.* **24,** 701-706.

Taylor, G., and Howard, C. (1980). Interaction of *Mycoplasma pulmonis* with mouse peritoneal macrophages and polymorphonuclear leucocytes. *J. Med. Microbiol.* **13,** 19-30.

Taylor, G., and Taylor-Robinson, D. (1974). Humoral and cell-mediated immune mechanisms in mycoplasma infections. *In* Mycoplasmas of Man, Animals, Plants and Insects (J. M. Bove and J. F. Duplan, eds.), pp. 331-336. INSERM, Paris.

Taylor, G., and Taylor-Robinson, D. (1975). The part played by cell-mediated immunity in mycoplasma respiratory infections. *Dev. Biol. Stand.* **28,** 195-210.

Taylor, G., and Taylor-Robinson, D. (1976). Effects of active and passive immunization on *Mycoplasma pulmonis*-induced arthritis in mice. *Ann. Rheum. Dis.* **36,** 232-238.

Taylor, G., Taylor-Robinson, D., and Slavin, G. (1974). Effect of immunosuppression on arthritis in mice induced by *Mycoplasma pulmonis*. *Ann. Rheum. Dis.* **33,** 376-382.

Taylor, G., Howard, C. J., and Gourlay, R. N. (1977). Protective effect of vaccines on *Mycoplasma pulmonis* induced respiratory disease of mice. *Infect. Immun.* **16,** 422-431.

Taylor-Robinson, D., Rassner, C., Furr, P. M., Humber, D. P., and Barnes, R. D. (1975). Fetal wastage as a consequence of *Mycoplasma pulmonis* infection in mice. *J. Reprod. Fertil.* **42,** 483-490.

Taylor-Robinson, D., Schorlemmer, H. U., Furr, P. M., and Allison, A. C. (1978). Macrophages secretion and the complement cleavage product C3a in the pathogenesis of infections by mycoplasmas and L-forms of bacteria and in immunity to these organisms. *Clin. Exp. Immunol.* **33,** 486-494.

Topley, W. W. C., and Wilson, G. S. (1920). The spread of bacterial infection. The problem of herd immunity. *J. Hyg.* **21,** 243-249.

Tram, par C., Guilon, J. C., and Chouroulinkov, I. (1970). Isolement de mycoplasmes de foetus de rats obtenus par cesarienne aseptique. *C. R. Seances Soc. Biol. Ses Fil.* **164,** 2470-2471.

Tuffery, A. A., and Innes, J. R. M. (1963). Diseases of laboratory mice and rats. *In* "Animals for Research—Principles of Breeding and Management" (W. Lane-Petter, ed.), pp. 47-108. Academic Press, New York.

Tully, J. G., and Rask-Nielsen, R. (1967). Mycoplasma in leukemic and nonleukemic mice. *Ann. N.Y. Acad. Sci.* **143,** 345-352.

Ventura, J., and Domaradzki, M. (1967). Role of mycoplasma infection in the development of experimental bronchiectasis in the rat. *J. Pathol. Bacteriol.* **93,** 342-348.

Vogelzang, A. A. (1975). The survival of *Mycoplasma pulmonis* in drinking water and in other materials. *Z. Versuchstierkd.* **17,** 240-246.

Wagner, J. E., Garrison, R. G., Johnson, D. R., and McGuire, T. J. (1969). Spontaneous conjunctivitis and dacryoadenitis of mice. *J. Am. Vet. Med. Assoc.* **155,** 1211-1217.

Weisbroth, S. H. (1979). Bacterial and mycotic diseases. *In* "The Laboratory Rat" (H. J. Baker, J. R. Lindsey, and S. H. Weisbroth, eds.), Vol. 1, pp. 214-219. Academic Press, New York.

Weisbroth, S. H., and Scher, S. (1968). *Corynebacterium kutscheri* infection in the mouse. I. Report of an outbreak, bacteriology, and pathology of spontaneous infections. *Lab. Anim. Care* **18,** 451-458.

Weisbroth, S. H., Scher, S., and Boman, I. (1969). *Pasteurella pneumotropica* abscess syndrome in a mouse colony. *J. Am. Vet. Med. Assoc.* **155,** 1206-1210.

Weiss, E. (1949). The extracellular development of agents of the psittacosis-lymphogranuloma group (Chlamydozoaceae). *J. Infect. Dis.* **84,** 125-149.

Wensinck, F. (1961). The origin of induced *Pseudomonas aeruginosa* bacteraemia in irradiated mice. *J. Pathol. Bacteriol.* **81,** 401-408.

Westerberg, S. C., Smith, C. B., Wiley, B. B., and Jensen, C. (1972). Mycoplasma-virus interrelationships in mouse tracheal organ cultures. *Infect. Immun.* **5,** 840-846.

Whittlestone, P. (1974). Isolation techniques for mycoplasmas from animal diseases. *In* "Mycoplasmas of Man, Animals, Plants and Insects" (J. B. Bove and J. F. Duplan, eds.), pp. 143-151. INSERM, Paris.

Whittlestone, P., Lemcke, R. M., and Olds, R. J. (1972). Respiratory disease in a colony of rats. II. Isolation of *Mycoplasma pulmonis* from the natural disease, and the experimental disease induced with a cloned culture of this organism. *J. Hyg.* **70,** 387-407.

Wilson, P. (1976). *Pasteurella pneumotropica* as the causal organism of abscesses in the masseter muscle of mice. *Lab. Anim.* **10,** 171-172.

Wyand, D. S., and Hayden, D. W. (1973). *Klebsiella* infection in muskrats. *J. Am. Vet. Med. Assoc.* **163,** 589-591.

Wynder, E. L., Taguchi, K. T., Baden, V., and Hoffman, D. (1968). Tobacco carcinogenesis. IX. Effect of cigarette smoke on respiratory tract of mice after passive inhalation. *Cancer (Philadelphia)* **21,** 134-153.

Yokoiyama, S., Mizuno, K., and Fujiwara, K. (1977). Antigenic heterogeneity of *Corynebacterium kutscheri* from mice and rats. (In Jpn.) *Exp. Anim.* **26,** 263-266.

Zucker, T. F. (1957). Pantothenate deficiency in rats. *Proc. Anim. Care Panel* **7,** 193-202.

Chapter 3

Bacterial, Mycoplasmal, Mycotic, and Immune-Mediated Diseases of the Urogenital System

Harold W. Casey and George W. Irving III**

I.	Introduction	43
II.	Bacterial Diseases	44
	A. Nephritis and Pyelonephritis	44
	B. Cystitis and Urethritis	47
	C. Infections of the Male Genital System	47
	D. Infections of the Female Genital Tract	49
III.	Mycoplasmal Infections	49
IV.	Mycotic Infections	49
V.	Glomerulonephritis	50
	References	52

I. INTRODUCTION

The incidence of infection and disease of the urogenital system produced by microorganisms in mice has not been well studied. Systematic studies for bacterial infections have been performed in only limited numbers, and studies utilizing appropriate culture methods for mycoplasma are essentially nonexistent. Lesions of the urogenital system are generally recognized as an expression of systemic infections. Exceptions are ascending bacterial infections of the urinary tract, which frequently occur in association with any condition that produces urinary stasis. Bacterial infections of the urogenital tract generally produce a suppurative inflammation which, in acute severe cases, may be hemorrhagic. The frequency of infections

*H. W. Casey's military designation is Colonel, USAF, VC, and that of G. W. Irving III is Lt. Col., USAF, VC. The opinions or assertions contained herein are the private views of the authors and are not to be construed as official or as reflecting the views of the Department of the Air Force or the Department of Defense.

ISBN 0-12-262502-1

of the urogenital tract varies greatly in mouse colonies, as their occurrence is closely related to husbandry and environmental conditions within the facility. Data on bacterial diseases are fragmentary, as documented reports are generally related to specific disease outbreaks. The etiology of sporadic cases is often not completely investigated. Tuffery (1966) reported the incidence of nephritis in over 2500 mice of 14 strains. In adult breeding males, the incidence was 32.9% compared to 4.8% in females. In mice on longevity studies, the incidence was 86.2% in males compared to 52.3% in females. These data probably reflect an unusually high incidence, as lesions associated with chloroform toxicity were included in data calculations. Bacterial organisms known to localize in the urinary tract include both gram-negative and gram-positive types: *Proteus, Pasteurella, Staphylococcus,* and *Corynebacterium* spp. Genital infections are less common and have been associated with *Pasteurella, Staphylococcus,* and *Streptobacillus* spp. Lesions produced by mycoplasmal and mycotic organisms are much less common, and only limited information is available on the frequency of infections with these organisms and their pathogenicity.

II. BACTERIAL DISEASES

A. Nephritis and Pyelonephritis

Grossly, bacterial lesions of the kidney generally have a focal distribution, although they may be disseminated throughout the organ. Acute lesions vary from whitish, pale foci or nodules to, in severe cases, enlarged, swollen kidneys that are whitish to red. Chronic cases are infrequently seen and are characterized grossly by irregular-shaped scars and depressions that extend from the surface to the pelvis. The microscopic appearance of the lesion depends on the microorganism involved, the severity and age of the lesion, and whether the lesion is hematogenous or ascending in origin. Nephritis of hematogenous origin generally involves the renal cortex and in the early stages may be centered in the glomeruli (Fig. 1). As the lesion increases in severity, it extends into the adjacent tubules (Fig. 2). At this point, abscessation may occur or extend down the tubules, producing pyelonephritis (Fig. 3). Ascending infections always involve the renal pelvis and secondarily spread to the renal cortex, generally in an irregular pattern (Fig. 4). Both types, ascending and descending infections, are characterized by extensive infiltration with neutrophils that may result in complete destruction of the renal architecture. Organisms that elicit a granulomatous reaction generally produce more focal lesions with cellular infiltration, with histiocytes as the predominating infiltrating cell. Diagnosis should be based on culture results accompanied by the histologic demonstration of the organisms within the lesions. Smears from fresh lesions stained for bacteria will generally permit a rapid, presumptive diagnosis.

Proteus mirabilis-induced suppurative nephritis was reported (Jones *et al.,* 1972) in 26 of 58 C3H/HeJ mice that died over a 28-month period. In addition to the suppurative nephritis, lesions were also found in other organs in eight mice, and the infection was considered septicemic. The route of the infection was not established, although investigators speculated that it was either by inhalation or by the genital tract associated with postpartum matings.

Pseudomonas aeruginosa has been isolated (Tuffery, 1966) from two sporadic cases of pyelonephritis; however, nephritic disease produced by this organism involving a significant number of animals has not been documented. Likewise, renal granulomas induced by *Staphylococcus aureus* have been described in C57BL/6Bd mice in conjunction with similar lesions in the cervical and head region (Shults *et al.,* 1973).

The kidneys are a frequent site of lesions produced by *Corynebacterium kutscheri,* as nephritis and lesions in numerous other organs are routinely found in mice with pseudotuberculosis. The disease is most frequently associated with well-defined stresses, such as steroid administration (Antopol, 1950; Fauve *et al.,* 1964), poxvirus infection (Lawrence, 1957), and radiation (Schechmeister and Adler, 1953); however, outbreaks have been reported in which specific stress factors were not identified (Weisbroth and Sher, 1968). Stress-initiated disease is associated with the activation of latent infections. A marked variation in susceptibility in different strains of mice to the organism is well documented, and avirulent strains of the organism are recognized (Fauve *et al.,* 1964). Lesions in the kidney appear to be hematogenous in origin, as bacterial thrombi can be noted in glomerular capillaries that produce a liquefactive necrosis surrounded by a zone of neutrophils. The lesion generally progresses to large abscesses and may extend down the tubules in a linear pattern. Smears of the lesion reveal a gram-positive, rod-shaped organism.

Leptospirosis has not been recognized as a naturally occurring disease in the mouse, but both laboratory and wild mice have been shown to be carriers of the organism. *Leptospira ballum* is the serotype reported in mice (Stoenner, 1957; Stoenner *et al.,* 1958; Yager *et al.,* 1953). Naturally occurring infections have been documented only in the Swiss mouse, but other laboratory strains of mice have also been shown to be lethally susceptible to experimental inoculation with *L. icterohaemorrhagiae* (Fujikura, 1965; Imamura *et al.,* 1960). Despite the absence of clinical disease in carrier mice, such animals may present a public health hazard. Four clinical cases of leptospirosis in laboratory personnel working with naturally infected carrier mice have been documented (Stoenner and Maclean, 1958).

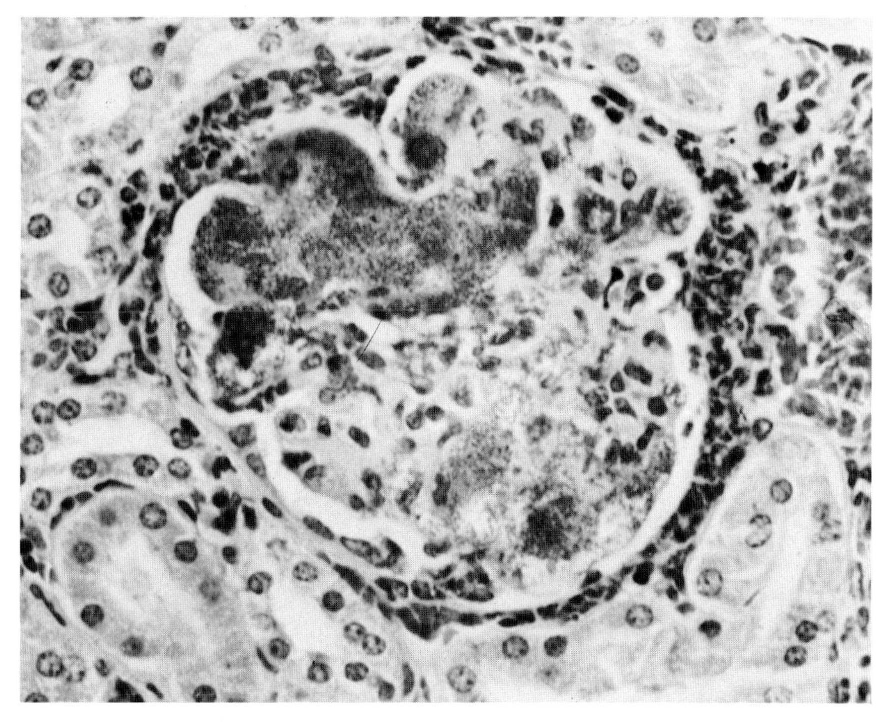

Fig. 1. Early stage of hematogenous bacterial nephritis. Bacterial colonies in glomerular capillaries and neutrophilic infiltration. H&E, AFIP Neg. 79-12912. ×450.

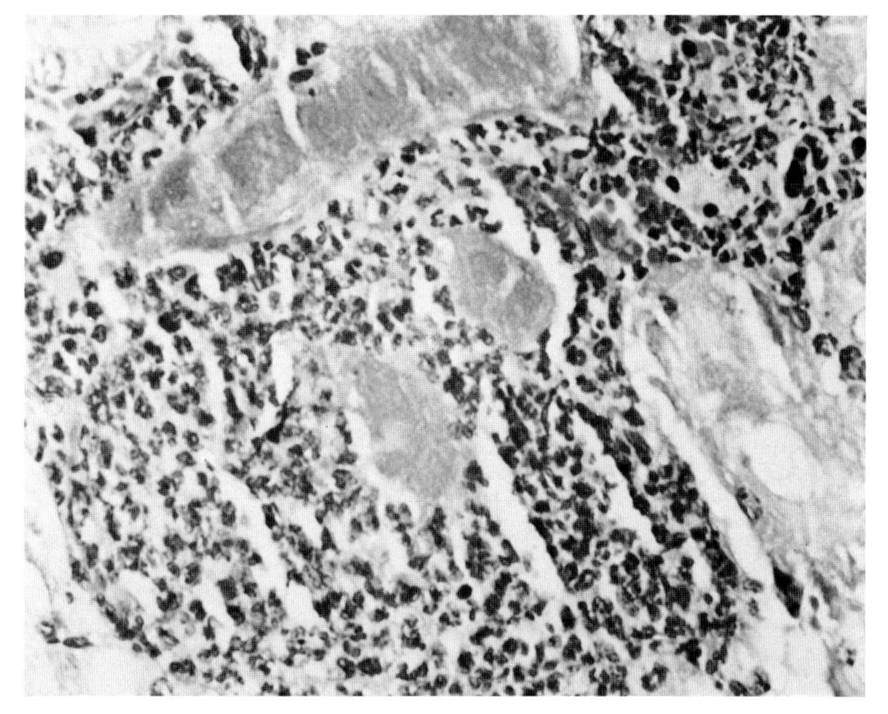

Fig. 2. Suppurative nephritis with bacterial colonies in renal tubules surrounded by an intense neutrophilic infiltrate in the interstitium. H&E, AFIP Neg. 79-12916. ×450.

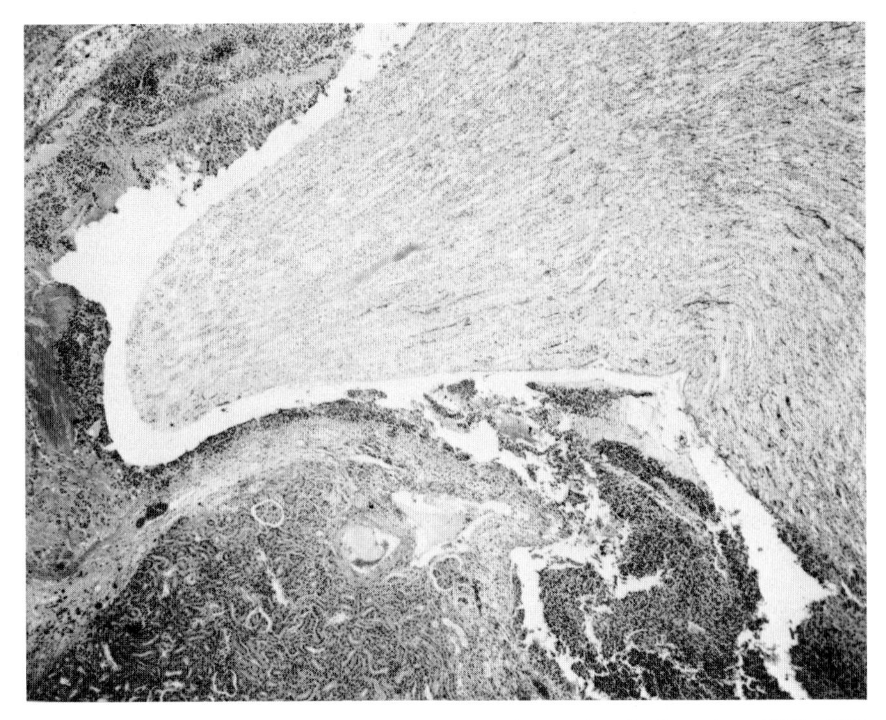

Fig. 3. Severe septic pyelonephritis with a suppurative exudate in the pelvis. H&E, AFIP Neg. 79-12915. ×45.

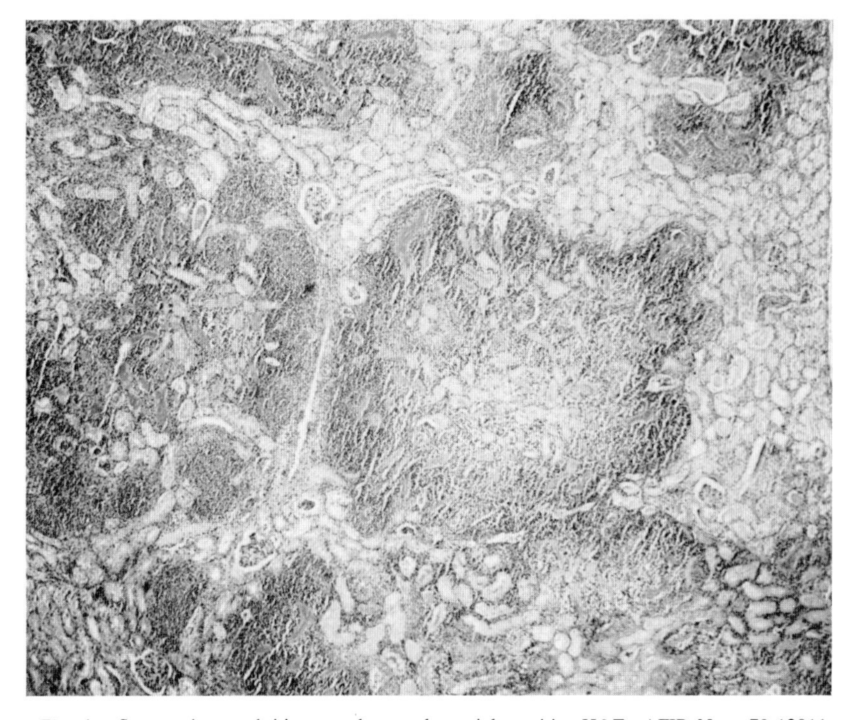

Fig. 4. Suppurative nephritis secondary to bacterial cystitis. H&E, AFIP Neg. 79-12911. ×110.

B. Cystitis and Urethritis

Bacterial infections of the lower urinary tract are generally seen in mice with conditions that produce urinary stasis or may accompany bacterial infections of the kidney. Bacterial infections may occur secondarily to urolithiasis and occur in male mice secondarily to obstructuve uropathy associated with proteinaceous urethral plugs. Organisms cultured from mice with cystitis and urethritis include *Proteus*, *Alcaligenes*, *Corynebacterium*, *Micrococcus*, and *Pasteurella* spp. (Brennan *et al.*, 1965; Sokoloff and Barile, 1962).

Although not primarily a bacterial disease, the formation of proteinaceous plugs in the penile urethra can produce significant medical problems on a colony basis due to urethral blockage and secondary bacterial infections (Sokoloff and Barile, 1962). The exact pathogenesis of the formation of the urethral plugs is not known. The disease is characterized by inflammation of the prepuce and surrounding abdomen. Proteinaceous plugs are seen within the urethral lumen. As the inflammation intensifies, urinary stasis results that produces distention of the urinary bladder, which may progress to a mild, acute hydronephrosis. Secondary bacterial infections may induce a suppurative (Fig. 5) and/or hemorrhagic component. The disease occurs principally in adult males. In one study, no cases occurred in mice under 6 months of age (Sokoloff and Barile, 1962). Genetic susceptibility to the disease has been recognized, as the STR/1N and their hybrids have a high predilection for the disease (Sokoloff and Barile, 1962). The authors have observed severe outbreaks of the disease in two separate colonies of CD-1 mice, and the disease has been reported in C57BL, AKR, and Swiss strains (Babcock and Southam, 1965). The origin of the material forming the mucous plugs is not clear. Sokoloff and Barile (1962) speculated that it represents secretions from the accessory sex glands. Prior irritation of the prepuce and urethra may be necessary for the mucous plugs to form and produce clinical signs. The finding of mucous plugs cannot always be interpreted as a lesion, as their formation has been reported as a terminal event (Rapp, 1962) depending on the manner in which the mice are killed (Rapp, 1962).

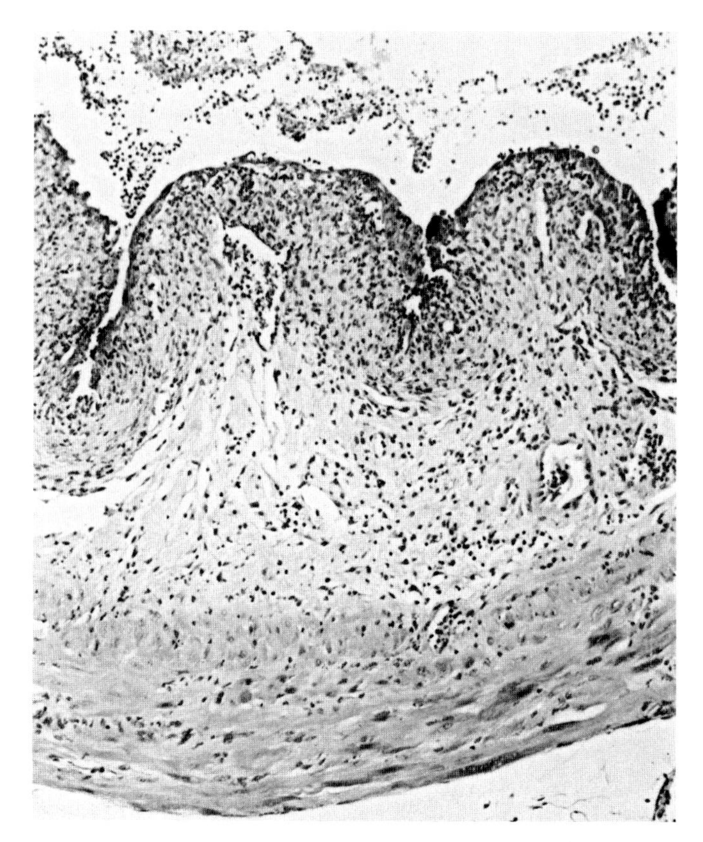

Fig. 5. Acute suppurative cystitis associated with secondary bacterial infection and the formation of mucous plugs in the penile urethra. H&E, AFIP Neg. 79-12913. ×110.

C. Infections of the Male Genital System

Bacterial diseases of the genital system of male mice are infrequently reported; however, this may be a reflection of the lack of systematic studies. In a survey of only nine animals, Ward *et al.*, (1978) cultured *P. pneumotropica* from the seminal vesicle of two animals without histologic evidence of disease, which suggests that infections may be common. Although reports of specific diseases of the testis or accessory sex glands were not located, lesions in these organs occur with systemic bacterial infections (Figs. 6 and 7). Infections of the bulbourethral gland have been reported with *S. aureus* (Needham and Cooper, 1976) and *P. pneumotropica* (Sebesteny, 1973). Mice with bacterial infections of the bulbourethral glands may range in age from 1 to 9 months. One report (Needham and Cooper, 1976) indicates that the disease has a higher frequency in breeders; however, affected animals continue to mate successfully. Diseased animals generally remain in good condition but develop unilateral or bilateral swelling adjacent to the penis in the perineal region. No particular strain susceptibility is recognized, as one report indicated that the disease was seen in eight of ten strains housed in one colony (Needham and Cooper, 1976), with an attack rate of approximately 6% in breeding males. Histologically, affected glands show suppurative inflammations that may proceed to abscessation of the entire gland. Identification of the organism involved should be made by cultural techniques.

Abscessation of the preputial gland also occurs in male mice (Franks, 1967; Tucker and Baker, 1967; Hong and Ediger, 1978). Most reports (Franks, 1967; Tucker and Baker, 1967) on the entity have been fragmentary as to the incidence and organisms involved. The disease was studied in detail by Hong

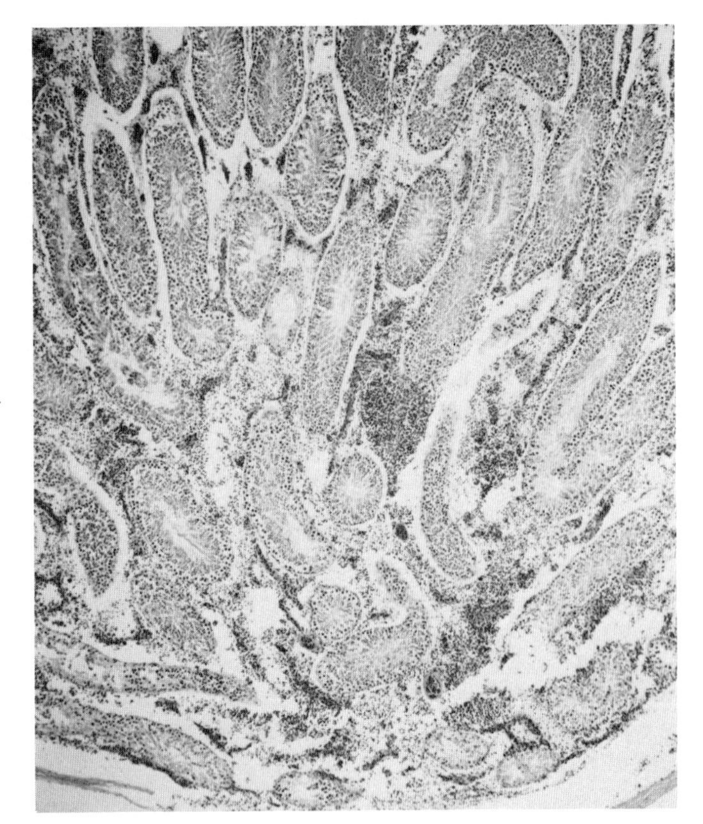

Fig. 6. Acute orchitis associated with streptococcal septicemia. H&E, AFIP Neg. 79-12909. ×45.

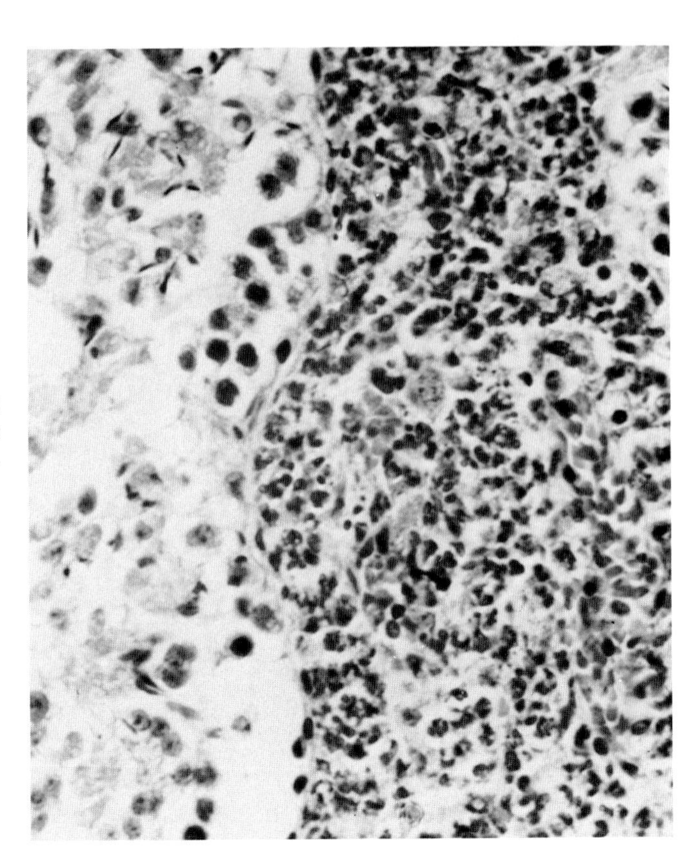

Fig. 7. Higher magnification of orchitis illustrated in Fig. 6 showing disruption of seminiferous tubules and intense neutrophilic infiltration. H&E, AFIP Neg. 79-12908. ×450.

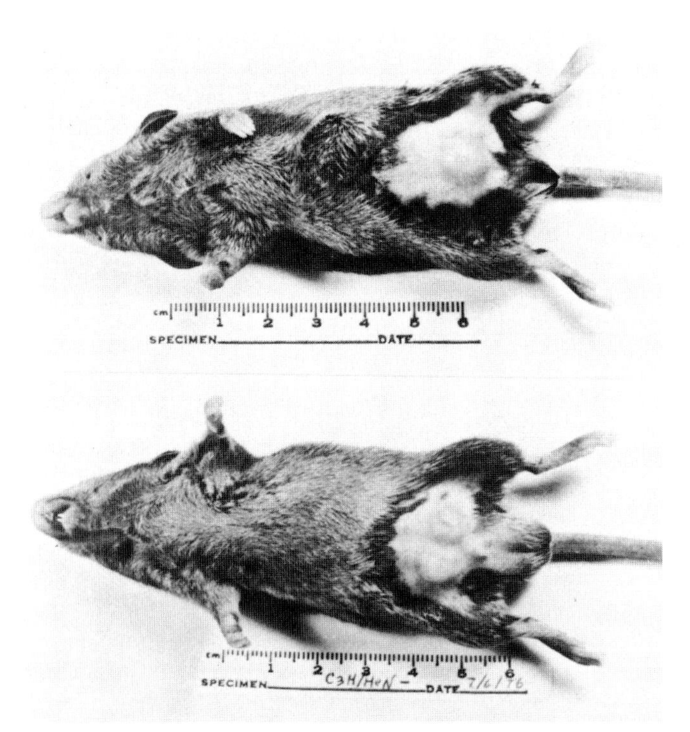

Fig. 8. Bilateral and solitary swellings in the inguinal area of mice due to preputial gland abscesses. AFIP Neg. 79-12910. (Courtesy of Hong and Ediger, 1978, and the editor of *Laboratory Animal Science*.)

and Ediger (1978) in a large colony of eight strains of mice; they reported an incidence of 0.5–8.0% in the different strains during a 1-year period in nearly 20,000 animals. The average overall incidence for all strains was 1.8%. *Staphylococcus aureus* was isolated in pure culture from 84% of the affected mice. Other isolates included *P. aeruginosa, S. epidermidis, Enterobacter cloaca, Klebsiella pneumoniae,* and *Escherichia coli*. Affected mice do not show clinical signs and present with unilateral or bilateral nodular swelling at the base of the penis (Fig. 8). Severely affected animals may exhibit ulceration and scab formation of the skin and prepuce. Histologically, the lesion is characterized by extensive suppurative inflammation and the presence of large numbers of bacteria, which can be demonstrated on impression smears. In severe cases, the infection may spread to the urethra and seminal vesicles. Treatment of the condition apparently has not been attempted.

D. Infections of the Female Genital Tract

Numerous bacteria have been isolated from the genital tract of female mice; however, lesions and disease outbreaks due to bacterial infections of the genital tract are uncommon. *Pasteurella pneumotropica* has been cultured repeatedly from the uterus (Jawetz and Baker, 1950; Hoag *et al.*, 1962; Brennan *et al.*, 1965; Flynn *et al.*, 1965; Blackmore and Casillo, 1972); however, its role in uterine disease is unclear. The organism can produce latent infections in other organs and is well recognized as a respiratory pathogen (Hoag *et al.*, 1962). Some evidence suggests that it may produce infertility and, under certain conditions, abortions. The organism was cultured by Brennan *et al.* (1965) from 11.9% of nonproductive female mice and by Hoag *et al.* (1962) from 14.6% of DBA/2J nonproductive female mice. Ward *et al.* (1978) cultured *P. pneumotropica* from 7 of 10 aborting Swiss mice which exhibited a suppurative metritis histologically. In companion studies, these investigators also cultured the organism from the uteri of 4 of 11 pregnant and nongravid mice. However, experimental inoculation of this isolate intraperitoneally and intravaginally failed to cause abortion in pregnant mice. These investigators considered *P. pneumotropica* to be a potential cause of abortion in mice when infection is combined with additional stress factors. Numerous cases of malignant lymphoma and mammary adenocarcinoma were present in the colony in which the abortions occurred, suggesting to these investigators that concurrent viral infections could have produced an immunosuppression in affected mice. *Streptobacillus moniliformis* has been recognized as a cause of abortion and stillbirths in addition to the polyarthritis and conjunctivitis normally associated with the organism. Sawicki *et al.* (1962) reported reproductive failure in approximately 279 breeding female albino mice over a 1-year period. Reproductive failures included stillbirths and abortion in the last week of pregnancy. Affected mice appeared sick; however, only one death was observed, and one animal also developed swollen feet suggestive of arthritis.

III. MYCOPLASMAL INFECTIONS

Spontaneous disease of the urogenital tract of mice has not been definitively documented as caused by mycoplasmal infections. Experimentally, these organisms have been shown capable of localizing and producing an oophoritis, salpingitis, and metritis in mice following parenteral injection (Nelson, 1954, 1970). Experimentally, *Mycoplasma pulmonis* has been shown to produce infertility and fetal wastage in mice (Goeth and Appel, 1974). As mycoplasmal organisms are common pathogens of the respiratory system and joints in many rodent colonies, investigators and clinicians should be cognizant that these organisms are also common pathogens of the genital system of numerous other species.

IV. MYCOTIC INFECTIONS

Although mice have been used extensively as an experimental animal for the study of mycotic infections, naturally occurring disease of the urogenital system has been limited to a report of one outbreak of pyelonephritis caused by *Candida tropicalis* (Goetz and Taylor, 1967). The disease occurred in female Swiss-Webster mice and resulted in approximately 10% mortality in a colony of 3000 mice during a 3-month period. The outbreak may have been precipitated by two environmental factors that occurred in the colony. The mice were on a hydrated diet designed to supply water within the food. This hydrated food proved to be highly favorable for fungal growth. Conditions also existed in the colony that could have exposed the mice to chloroform fumes. Symptoms noted were nonspecific and included a cessation of lactation, inappetence, humped posture, and a ruffled hair coat. Death occurred within 72 hr of the first signs of the disease. At necropsy the kidneys varied from enlarged and pale to small and constricted. Red streaks were present throughout the parenchyma. Microscopically, the kidneys were characterized by marked infiltration of neutrophils and tubular necrosis. Papillary necrosis was observed in one mouse. Lesions attributable to the fungi were not observed in other organs. Gridley stain sections demonstrated numerous organisms in affected areas. Diagnosis was based on smears made from affected kidneys showing "bottle-shaped," yeastlike organisms that were gram-positive and were cultured on Sabouraud's medium.

V. GLOMERULONEPHRITIS

Inclusion of glomerulonephritis in a chapter on bacterial and mycotic diseases of the urogenital system is arbitrary, as the disease in mice is frequently associated with persistent viral infections or disorders of the immune system; however, glomerulonephritis as a sequela to bacterial diseases in man is recognized as a common entity. Glomerulonephritis represents one of the most common renal diseases in the mouse. It is seen principally in adult and older animals. The incidence in the different strains of mice may range from nearly 100% in black and white hybrids of the New Zealand strain to only sporadic cases in other strains. As will be discussed later, the frequency of the disease may depend a great deal on the incidence of viral infections within the individual colony. In some of the older literature (Gude and Upton, 1960, 1962), the disease has been termed *glomerulosclerosis,* which, in chronic stages, may characterize the histologic appearance of the lesion. Mice presenting with an acute nephrotic syndrome, proteinuria, and generalized edema have been described by Kirschbaum (1944) and Kirschbaum *et al.* (1949); however, most mice will present with a "wasting" syndrome (Hotchin and Collins, 1964) without specific clinical signs, with sporadic deaths occurring in unthrifty and debilitated animals. The nature of the lesion in most mice indicates that the proteinuria occurs at some stage of the disease, even though measurements are infrequently made. Grossly, the lesions are relatively mild and are generally apparent only in severe chronic cases. Both kidneys are uniformly affected. The cortex may exhibit granularity, and the organ may be reduced in size. On a cut surface, occasional small cysts may be distinguished in the cortex.

The lesions of murine glomerulonephritis are considered immune mediated as they result from the deposition of immune complexes and complement in the capillary walls and mesangium of the glomeruli. The New Zealand black mouse (NZB) and the F_1 hybrid of the New Zealand black and white cross ($NZBW_{F1}$) develop a glomerulonephritis that is considered an autoimmune disease (Bielschowsky *et al.,* 1959; Mellors, 1965; Dixon *et al.,* 1971), and these strains of mice have been studied extensively as a model for autoimmune disease in man. The glomerular disease is relatively mild in the NZB mouse, but the B/W hybrids generally develop a fatal disease and the female mouse develops a more severe disease at an accelerated rate. Renal changes may appear as early as 3–4 months of age, but clinical signs and severe lesions generally are not present until 6–9 months of age. The lesions are progressive, becoming more severe until death. Histologically, the glomeruli exhibit extensive proteinaceous deposits within the capillaries and mesangium (Fig. 9). The deposits become so extensive that the capillary lumens are essentially occluded. In the early stages of the disease, these deposits, which are PAS positive, are more extensive in the mesangium. In the latter stages of the disease, tubular atrophy and proteinaceous casts may be seen throughout the organ. Immunofluorescent studies have shown the glomerular deposits to contain IgG and the third component of complement (Fig. 10). The immunoglobulins have been shown to develop as an autoimmune response to a number of different nuclear antigens and antigens of the murine leukemia virus (Hahn and Shulman, 1969; Abe *et al.,* 1976; Raveche *et al.,* 1978; Imamura *et al.,* 1977).

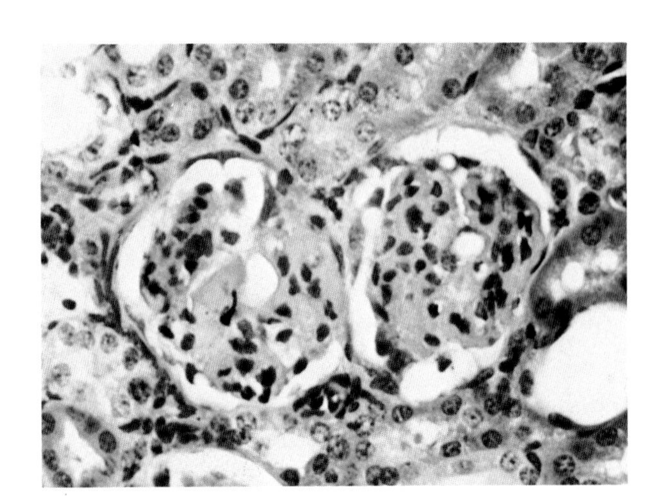

Fig. 9. Glomerulonephritis in an NZB/W mouse. Extensive eosinophilic deposits are present in the capillary walls and mesangium. H&E, AFIP Neg. 79-12919. ×450.

Fig. 10. Deposits IgG in two glomeruli of an NZB/W mouse with immune-mediated glomerulonephritis. Immunofluorescence procedure. AFIP Neg. 79-12914. ×250.

Mice infected with the lymphocytic choriomeningitis virus (LCM) may develop a glomerulonephritis that is of an immune complex nature. Development of renal disease in mice infected with this virus is associated with congenital or neonatal exposure to the virus before the mice are immunologically competent (Hotchin and Collins, 1964). These mice develop a persistent viremia that elicits a nonprotective immune response to the virus and the formation of antigen–antibody complexes that are filtered out in the renal glomeruli. Viral immune complexes are detectable in the glomeruli as early as 3 months of age, but clinical signs develop only in older mice and has been referred to as *late onset disease* (Hotchin and Collins, 1964). The appearance of the renal lesion (Oldstone and Dixon, 1969; Kajima and Pollard, 1970) is similar to that in the NZB mouse and has a similar pathogenesis. A mononuclear inflammatory cell infiltrate, lymphocytes, and plasma cells are frequently observed in the interstitial areas of the kidney, particularly around vessels. The frequency and severity of the disease vary depending on the strain of LCM virus and the strain of mouse (Cole and Johnson, 1975).

Mice infected with murine oncornaviruses may develop an immune complex glomerulonephritis. Glomerular lesions were described in mice infected with the murine leukemia viruses in 1967 (Dunn and Green, 1967), and the immune complex nature of these viral-associated lesions was later established (Pascal *et al.,* 1973; Branca *et al.,* 1971). The mouse mammary tumor virus also may induce the glomerular disease (Pascal *et al.,* 1975). The incidence of glomerulonephritis in mice infected with these viruses has not been established, but the widespread occurrence of these viruses in the different strains of laboratory mice suggests that they could be of major importance in glomerular disease.

In most cases of glomerulonephritis seen in diagnostic situations, the antigens involved in the disease process are unknown; however, Markham *et al.* (1973) showed that the deposition of immune complexes and complement in the renal glomeruli is extremely common even at an early age. In their studies of 12 common strains of laboratory mice, essentially 100% contained deposits of immunoglobulins, and up to 50% of two strains of germ-free mice also exhibited immune deposits within the glomeruli. These authors were unable to correlate the presence of immunoglobulin deposits within the glomeruli with either proteinuria or the severity of histologic changes; however, their studies were limited to animals 7–16 weeks in age, and in the authors' experience, severe glomerulonephritis generally is not seen until mice are more than 6 months old. The lesions in these spontaneous cases appear basically as enlarged sclerotic glomeruli (Fig. 11), with extensive deposition of periodic acid-Schiff-positive eosinophilic material within the glomeruli (Fig. 12). These

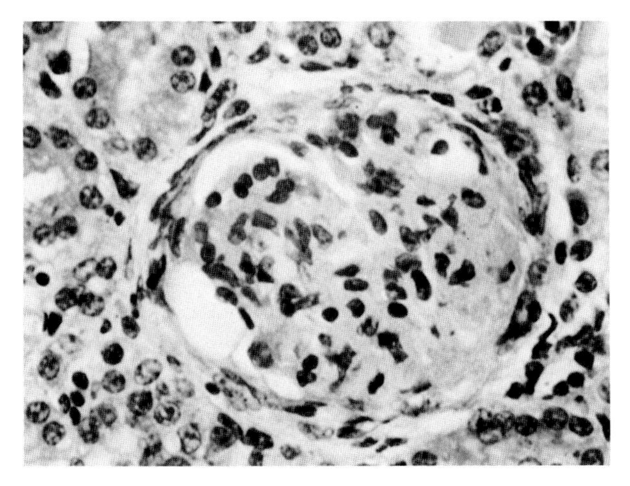

Fig. 11. Severe glomerulonephritis of unknown etiology in a CD-1 mouse that has progressed to the sclerotic stage. H&E, AFIP Neg. 79-12917. ×450.

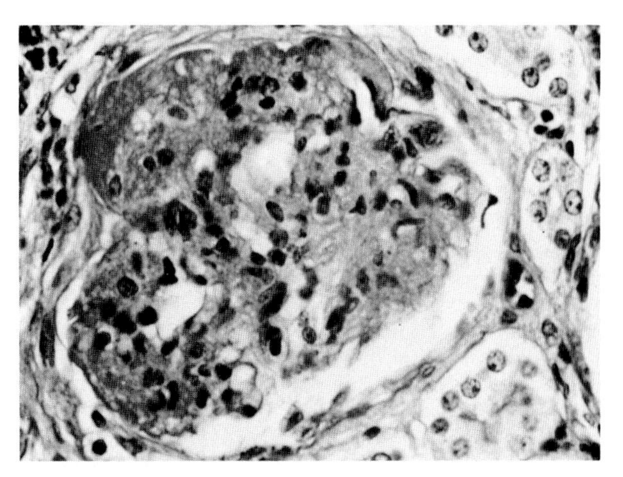

Fig. 12. Periodic acid-Schiff (PAS) procedure of the glomerular lesion illustrated in Fig. 11 showing the intense PAS-positive nature of the glomerular deposits. AFIP Neg. 79-12918. ×450.

cases probably represent an immune complex form of glomerulonephritis, but they must be differentiated from amyloidosis. Preventive methods have not been developed for the disease, but elimination of viral infections, including those that produce persistent inapparent infections, appears to be warranted given the present state of knowledge. When the frequency of the disease reaches such magnitude that it interferes with colony or experimental procedures, it is imperative that virologic studies on the colony and immunofluorescent studies on the glomerular lesion be performed to determine the pathogenesis and etiology of the lesion so that appropriate control measures can be initiated.

REFERENCES

Abe, C., Chia, D., Barnfit, E. V., Pearson, C. M., Hays, E. A., and Skiokawa, Y. (1976). Correlation of natural antibodies to nuclear substances in New Zealand and other strains of mouse. *Clin. Immunol. Immunopathol.* **6,** 369–375.

Antopol, W. (1950). Anatomic changes in mice treated with excessive doses of cortisone. *Proc. Soc. Exp. Biol. Med.* **73,** 262–265.

Babcock, V. I., and Southam, C. M. (1965). Obstructive uropathy in laboratory mice. *Proc. Soc. Exp. Biol. Med.* **120,** 580–581.

Bielschowsky, M., Helyer, B. J., and Howie, J. B. (1959). Spontaneous hemolytic anemia in mice of the NZB/B1 strain. *Proc. Univ. Otago Med. Sch.* **37,** 9–13.

Blackmore, D. K., and Casillo, S. (1972). Experimental investigation of uterine infections of mice to *Pasteurella pneumotropica. J. Comp. Pathol.* **82,** 471–475.

Branca, M., de Petris, S., Allison, A. C., Havey, J. J., and Hissch, M. S. (1971). Immune complex disease. Pathological changes in the kidneys of BALB/c mice neonatally infected with Moloney leukaemogenic and murine sarcoma viruses. *Clin. Exp. Immunol.* **9,** 853–868.

Brennan, P. C., Fritz, T. E., and Flynn, R. J. (1965). *Pasteurella pneumotropica:* Cultural and biochemical characteristics and its association with disease in laboratory animals. *Lab. Anim. Care* **15,** 307–312.

Cole, G. A., and Johnson, E. D. (1975). Immune response to LCM virus infection *in vivo* and *in vitro. Bull. W.H.O.* **52,** 465–469.

Dixon, F. J., Oldstone, M. B. A., and Tonietti, G. (1971). Pathogenesis of immune complex glomerulonephritis of New Zealand mice. *J. Exp. Med.* **134,** 65–71.

Dunn, T. B., and Green, A. W. (1967). Morphology of BALB/c mice inoculated with Rauscher virus. *J. Natl. Cancer Inst.* **36,** 987–994.

Fauve, R. M., Pierce-Chase, C. H., and Dubos, R. (1964). Corynebacterial pseudotubeculosis in mice. II. Activation of natural and experimental latent infections. *J. Exp. Med.* **120,** 283–303.

Flynn, R. J., Brennan, P. C., and Fritz, T. E. (1965). Pathogen status of commercially produced laboratory mice. *Lab. Anim. Care* **15,** 440–447.

Franks, L. M. (1967). Normal and pathological anatomy and history of the genital tract of rats and mice. *In* "Pathology of Laboratory Rats and Mice" (E. Cotchin and F. J. D. Roe, eds.), p. 480. Blackwell, Oxford.

Fujikura, T. (1965). Studies on the lethal susceptibility of laboratory mice to *Leptospira icterohaemorrhagiae. Jpn. J. Vet. Sci.* **27,** 283–287.

Goeth, H., and Appel, K. R. (1974). Experimentelle Untersuchungen uber die Beeintrachtigung der Fertilitat durch Mykoplasmeninfektionen des Genitaltraketes. *Zentralbl. Bakteriol., Hyg., Parasitenkd. Infektionskr., Abt. 1: Orig., Reihe A* **228,** 282–289.

Goetz, M. E., and Taylor, D. O. N. (1967). A naturally occurring outbreak of *Candida tropicalis* infection in a laboratory mouse colony. *Am. J. Pathol.* **50,** 361–369.

Gude, W. D., and Upton, A. C. (1960). Spontaneous glomerulonephritis in aging RF mice. *J. Gerontol.* **15,** 373–376.

Gude, W. D., and Upton, A. C. (1962). A histologic study of spontaneous glomerular lesions in aging RF mice. *Am. J. Pathol.* **40,** 699–706.

Hahn, B. H., and Shulman, L. E. (1969). Autoantibodies and nephritis in the white strain (NZW) of New Zealand mice. *Arthritis Rheum.* **12,** 355–364.

Hoag, W. G., Wetmore, P. W., Rogers, J., and Meier, H. (1962). A study of latent Pasteurella infection in a mouse colony. *J. Infect. Dis.* **111,** 135–140.

Hong, C. C., and Ediger, R. D. (1978). Preputial gland abscess in mice. *Lab. Anim. Sci.* **28,** 153–156.

Hotchin, J., and Collins, D. H. (1964). Glomerulonephritis and late onset disease of mice following neonatal virus infection. *Nature (London)* **164,** 1357–1359.

Imamura, M., Mellors, R. C., Strand, M., and August, J. T. (1977). Murine type C viral envelope glycoprotein gp69/71 and lupus-like glomerulonephritis of New Zealand mice. *Am. J. Pathol.* **86,** 375–381.

Imamura, S., Ashizawa, Y., and Nagata, Y. (1960). Studies on Leptospirosis. I. Experimental Leptospirosis of mice with jaundice, hemorrhage and high mortality. *Jpn. J. Exp. Med.* **30,** 427–431.

Jawetz, E., and Baker, W. H. (1950). A pneumotropic Pasteurella of laboratory animals. II. Pathological and immunological studies with the organism. *J. Infect. Dis.* **86,** 184–196.

Jones, J. B., Estes, P. C., and Jordan, A. E. (1972). *Proteus mirabilis* infection in a mouse colony. *J. Am. Vet. Med. Assoc.* **161,** 661–664.

Kajima, M., and Pollard, M. (1970). Ultrastructural pathology of glomerular lesions in gnotobiotic mice with congenital lymphocytic choriomeningitis (LCM) virus infection. *Am. J. Pathol.* **61,** 117–130.

Kirschbaum, A. (1944). Spontaneous glomerulonephritis in mice. *Proc. Soc. Exp. Biol. Med.* **55,** 280–281.

Kirschbaum, A., Bell, E. T., and Gordon, J. (1949). Spontaneous and induced glomerulonephritis in an inbred strain of mice. *J. Lab. Clin. Med.* **34,** 209–220.

Lawrence, J. J. (1957). Infection of laboratory mice with *Corynebacterium murium. Aust. J. Sci.* **20,** 147.

Markham, R. V., Sutherland, J. C., and Mardiney, M. R. (1973). The ubiquitous occurrence of immune complex localization in renal glomeruli of normal mice. *Lab. Invest.* **29,** 111–120.

Mellors, R. C. (1965). Autoimmune disease in NZB/B1 mice. 1. Pathology and pathogenesis of a model system of spontaneous glomerulonephritis. *J. Exp. Med.* **122,** 25–40.

Needham, J. R., and Cooper, J. E. (1976). Bulbourethral gland infections in mice associated with *Staphylococcus aureus. Lab. Anim.* **10,** 311–315.

Nelson, J. B. (1954). The selective localization of murine pleuropneumonia-like organisms in the genital tract on intraperitoneal infection in mice. *J. Exp. Med.* **100,** 311–320.

Nelson, J. B. (1970). Response of mice to *Mycoplasma pulmonis* injected alone and with an ascites tumor. *Lab. Anim. Care* **20,** 670–674.

Oldstone, M. B. A., and Dixon, F. J. (1969). Pathogenesis of chronic diseases associated with persistent lymphocytic choriomeningitis virus infection. I. Relationship of antibody production to disease in neonatally infected mice. *J. Exp. Med.* **129,** 483–505.

Pascal, R. R., Koss, M. N., and Kassel, R. L. (1973). Glomerulonephritis associated with immune complex deposit and viral particles in spontaneous murine leukemia. *Lab. Invest.* **29,** 159–165.

Pascal, R. R., Koss, Rollwagon, F. M., Harding, T. A., and Schiavone, W. A. (1975). Glomerular immune complex deposits with mouse mammary tumor. *Cancer Res.* **35,** 302–304.

Rapp, J. P. (1962). Terminal formation of urethral plugs in male mice. *Proc. Soc. Exp. Biol. Med.* **111,** 243–245.

Raveche, E. S., Steinberg, A. D., Klassen, L. W., and Tjio, J. H. (1978). Genetic studies in NZB mice. I. Spontaneous autoantibody production. *J. Exp. Med.* **147,** 1487–1502.

Sawicki, L., Bruce, H. M., and Andrews, C. H. (1962). *Streptobacillus moniliformis* infection as a probable cause of arrested pregnancy and abortion in laboratory mice. *Br. J. Exp. Pathol.* **43,** 194–197.

Schechmeister, I. L., and Adler, F. L. (1953). Activation of pseudotuberculosis in mice exposed to sublethal total body irradiation. *J. Infect. Dis.* **92,** 228–239.

Sebesteny, A. (1973). Abscesses of the bulbourethral glands of mice due to *Pasteurella pneumotropica*. *Lab. Anim.* **7,** 315–317.

Shults, F. S., Estes, P. C., Franklin, J. A., and Richter, C. B. (1973). Staphylococcal botryomycosis in a specific-pathogen-free mouse colony. *Lab. Anim. Sci.* **23,** 36–42.

Sokoloff, L., and Barile, M. F. (1962). Obstructive genitourinary disease in male STR/1N mice. *Am. J. Pathol.* **41,** 233–243.

Stoenner, H. G. (1957). The sylvatic and ecological aspects of leptospirosis. *Vet. Med.* **52,** 553–555.

Stoenner, H. G., and Maclean, D. (1958). Leptospirosis (Ballum) contracted from Swiss albino mice. *Arch. Intern. Med.* **101,** 606–610.

Stoenner, H. G., Grimes, E. F., Thrailkill, F. B., and Davis, E. (1958). Elimination of *Leptospira ballum* from a colony of Swiss albino mice by use of chlortetracycline hydrochloride. *Am. J. Trop. Med. Hyg.* **7,** 423–426.

Tucker, M. J., and Baker, S. B. (1967). Disease of specific pathogen-free mice. *In* "Pathology of Laboratory Rats and Mice" (E. Cotchin and F. J. C. Roe, eds.), p. 791. Blackwell, Oxford.

Tuffery, A. A. (1966). Urogenital lesions in laboratory mice. *J. Pathol. Bacteriol.* **9,** 301–309.

Ward, G. E., Moffath, R., and Olfert, E. (1978). Abortion in mice associated with *Pasteurella pneumotropica*. *J. Clin. Microbiol.* **8,** 177–180.

Weisbroth, S. H., and Sher, S. (1968). *Corynebacterium kutscheri* infection in the mouse. I. Report of an outbreak, bacteriology and pathology of spontaneous infections. *Lab. Anim. Care* **18,** 451–458.

Yager, R. H., Gochenour, W. S., Jr., Alexander, A. D., and Wetmore, P. W. (1953). Natural occurrence of *Leptospira ballum* in rural house mice and in an opossum. *Proc. Soc. Exp. Biol. Med.* **84,** 589–890.

Chapter 4

Bacterial and Mycotic Diseases of the Integumentary System

Cynthia Besch Williford and Joseph E. Wagner

I. Introduction 55
II. Hairy Integument 56
III. Bacterial Diseases 57
  A. *Corynebacterium* 57
  B. *Pasteurella* 58
  C. *Actinobacillus* 60
  D. *Staphylococcus* 60
  E. *Streptococcus* 62
IV. Mycotic Diseases 65
  A. General 65
  B. Trichophyton 65
  C. Other Dermatophytes 66
  D. Diagnosis 66
  E. Treatment and Control 68
V. Noninfectious Skin Disorders 68
References 72

I. INTRODUCTION

Dermatitides can be categorized into primary and secondary skin disorders. Primary skin lesions in mice can be produced by physical manipulations such as fight wounds, applications of chemical irritants, and irradiation; by physiological disorders including genetic, hormonal, and nutritional abnormalities; and by infectious agents such as poxviruses, reoviruses, ectoparasites, dermatophytes, and occasionally bacteria. Secondary skin diseases are perhaps more commonly encountered in which opportunistic bacteria or fungi colonize compromised skin or adnexa and produce or exacerbate in-

ISBN 0-12-262502-1

flammation and infection. Often the etiology of the skin disorder is overlooked or masked by the more obvious cutaneous microbial flora cultured from the skin lesion. This occurs with heavy ectoparasite infestations, which can incite a vicious cycle of irritation, pruritis, wound production, and secondary bacterial infections. The self-inflicted wounds are often more severe than the effects of the ectoparasitism. Harkness and Wagner (1975) reported that suppurative middle ear disease in mice provided sufficient irritation to cause self-mutilation of the head and neck, producing severe cutaneous excoriations from which *Staphylococcus aureus* was isolated. Clinically, the mice exhibited the scratching, mutilating behavior with no head tilt or circling; an antemortem diagnosis of staphylococcus dermatitis would have missed the clinically inapparent otitis media. These examples highlight the importance of investigating all possible causes of a skin disorder before evaluating the significance of the various microbial agents isolated from the cutaneous lesions.

The emphasis of this chapter is on the clinical and pathological manifestations of bacterial and mycotic dermatitides that have produced epizootics or significant mortality in mouse colonies. Supplemental information about strain susceptibility to infection, incidence and significance of the disease condition, and treatment or control measures has also been included.

II. HAIRY INTEGUMENT

The adnexal structures of normal mouse skin have been described in the chapter "Neoplasms of the Mouse Integument and Harderian Gland," which appears in Volume IV of this series. The natural defense barriers of the integument and the cyclic hair growth patterns will be discussed in this section.

The stratum corneum or horny layer is classically the most impermeable layer of the skin, shielding the underlying tissues from the environment and maintaining the "internal milieu." In rodents, the keratinized cells of the horny layer are stacked in orderly columns which interdigitate and laterally overlap cells of adjacent columns. There is also a gradient of cell cohesiveness which is lowest at the skin surface and increases to its maximum at the stratum granulosum–stratum corneum junction. The stacking and cohesive properties of the horny layer are perhaps the most significant contributors to the barrier role of the epidermis (Montagna and Parakkal, 1974).

Among the supportive biological functions of the dermis is the ability to protect and defend underlying tissues from mechanical and infectious injury and to repair structural defects. Hyaluronic acid, a major component of dermal collagen and amorphous dermal ground substance, deters bacterial penetration due to its biochemical composition. However, hyaluronidase-producing bacteria such as *S. aureus* can digest a pathway through the dermis and thus facilitate the spread of infection. The resiliency of the dermal connective tissue meshwork provides a mechanical cushion to minor injurious agents and assists in averting the traumatic inoculation of microorganisms. Connective tissue macrophages or histiocytes perform both phagocytic and immunological effector functions critical for initiation of the immune response. Mast cells of rats and mice synthesize and store in granules heparin, histamine, and, unlike other species, 5-hydroxytryptamine or serotonin (Parratt and West, 1957). The chemically or physically induced degranulation of the mast cells results in local inflammation with increased capillary permeability, smooth muscle contraction, and stimulation of phagocytosis in epithelioid cells. Heparin induces the formation of precursor material for the dermal ground matrix and may play a role in inhibition of hyaluronidase (Riley, 1962; Pinto, 1974).

Hair replacement in the mouse occurs in cyclic waves of growth. There is an orderly progression in which a group of hair follicles enters and leaves the growth phase, or anagen, of the hair cycle (Fig. 1). The hair follicles are grouped and form parallel rows that are circumferentially arranged around the trunk and limbs with pattern modifications on the tail and head (Hussein, 1971). The most commonly described wave pattern of hair growth in haired and nude athymic mice begins in the pectoral and axillary regions, progresses dorsally, and proceeds cephalad and/or caudad. Variations of the general wave pattern in mice occur according to age, strain, sex, and physiological condition of the animal (Borum, 1954; Johnson, 1958a,b; Chase and Eaton, 1959). Eaton (1977) later reported that C57BL/6Icr *nu/nu* mice do not have age-induced variations in hair wave patterns in animals up to 6–8 months of age.

The cyclic activity of the hair follicle plays a role in the reaction of the skin to environmental agents. In anagen, the pleuripotential matrix cells of the bulb are actively producing the cells of the hair shaft proper and the inner root sheath. The outer root sheath in mice has a prominent area of mitotic activity at the lower portion of the outer sheath. In the anagen phase of follicular activity, the outer root sheath epithelial cells and the epidermal basal cells proliferate. A break in the integrity of the epithelium in the anagen phase results in quick recovery, with both outer root sheath epithelium and epidermis participating in the reparative process (Chase and Montagna, 1951). In the active follicle, however, the "open" hair shaft can allow access of solutions as well as bacteria, fungi, and parasites to the level of the mitotically active bulb. This cyclic process should be known when evaluating the effects of topical potential carcinogens (Wolbach, 1951; Chase and Montagna, 1951). Kligman (1956) and Kostanecki *et al.* (1963) demonstrated that only in certain stages of late anagen can mice be infected with *Trichophyton mentagrophytes*. The catagen and telogen stages are the relatively quiet phases of the hair cycle.

Fig. 1. Junction of the anagen (left) and telogen (right) phases of the hair cycle in normal mouse skin.

These stages are characterized by club formation of the hair root, with decreased activity of the matrix cells, the epithelial cells of the outer root sheath, and the epidermis. Inflammation from topical irritants and from dermatophyte infections can induce early telogen from the anagen stage (Chase and Montagna, 1951). This protective transitional mechanism of the hair follicle prevents deep penetration of topical chemicals and causes the dermatophytes to be shed with the hair shaft (Kligman, 1956).

Of the glandular components of the adnexa, the sebaceous glands play a role in skin homeostasis. Sebaceous glands produce sebum, which is released via ducts into the sheath of the hair follicle or onto the skin surface in areas where hair is absent. The inguinal (clitoral or preputial) glands of mice are specialized sebaceous glands whose cells have a foamy appearance and whose ducts open into the sheath of the genital papilla (Hieronymi, 1958). The biological functions of sebum as an emollient and a pheromone are generally accepted. The enzyme and lipid components of sebum are also thought to have bacteriostatic and fungistatic properties.

III. BACTERIAL DISEASES

A. *Corynebacterium*

Murine pseudotuberculosis is a naturally occurring disease of mice caused by bacteria of the genus *Corynebacterium*. The etiological agent was first described by D. Kutscher (1894), who named it *Bacillus pseudotuberculosis murium*. The name of the organism has since been changed several times. Synonyms include *Bacterium kutscheri* (Migula, 1900, cited in Bergey, 1923), *Myobacterium pseudotuberculosis* (Chester, 1901, cited in Committee, 1923), *Corynethrix pseudotuberculosis murium* (Bongert, 1901), *Corynebacterium murium* (cited in Bergey, 1923), and *Corynebacterium kutscheri* (cited in Bergey, 1925). *Corynebacterium kutscheri* is considered an opportunistic pathogen of the mouse, causing clinically latent infections. However, spontaneously occurring clinical disease has been infrequently reported (Kutscher, 1894; Bongert, 1901; Bicks, 1957; Fauve *et al.*, 1964; Weisbroth and Scher,

1968; Soerensen *et al.*, 1975). Epizootics usually occurred in infected animals whose resistance to infection was impaired. Resistance-lowering stresses reported to have triggered clinical disease include irradiation (Schechmeister and Adler, 1953), cortisone therapy (Antopol, 1950; Antopol *et al.*, 1953; Fauve *et al.*, 1964; Caren and Rosenberg, 1966), pantothenic acid deficiency (Gundel *et al.*, 1932), and other infections (Lawrence, 1957; Wolfe, 1950).

Corynebacterium kutscheri is not a primary skin pathogen, and the classic spontaneous or induced clinical disease is manifested by embolic abscessation in highly vascular visceral organs. Transmission of the organism has been speculated to be by direct contact with overtly infected animals or by contact with infected surroundings (Bicks, 1957; Schechmeister and Adler, 1953). Also, Merkenschlager (1975) demonstrated prenatal transmission and latent infection in progeny in experimentally infected, germ-free, and conventional NMRI mice. Soerensen *et al.* (1975), however, consistently isolated *C. kutscheri* from the mouths of both clinically ill and apparently healthy mice in a breeding colony during the course of a spontaneous pseudotuberculosis epizootic. Natural transmission of the organism is dependent on the inherent susceptibility of particular strains of mice (Pierce-Chase *et al.*, 1964; Schechmeister and Adler, 1953; Weisbroth and Scher, 1968) as well as on the immunocompetence of the animal at the time of exposure. Therefore, determining the mode of transmission of *C. kutscheri* is not as critical in investigating an epizootic as is the identification of latently infected mice.

1. Clinical Manifestations

Mice exhibiting spontaneous or induced overt disease present with depression, altered gait, and a roughened hair coat prior to death. Necropsy examinations of infected mice reveal abscess formation in joints and highly vascular target organs such as the kidney, liver, and lungs. Septic emboli are spread hematogenously, accounting for the abscessation in organs responsible for blood filtration. Similarly, the lodgement of emboli and subsequent infarction in dermal vessels can produce skin necrosis. Skin lesions are occasionally observed in spontaneous outbreaks, appearing as epithelial ulcerations or fistulous tracts (Weisbroth and Scher, 1968; Giddens *et al.*, 1968). Another report describes subcutaneous nodules as the primary manifestations of a *C. kutscheri* epizootic in a breeding colony (Soerensen *et al.*, 1975).

2. Histopathology

Histologically, the skin lesions are characterized by areas of coagulative necrosis in the dermis or subcutis surrounded by intense neutrophilic infiltration that often undergoes caseation. The cutaneous ulcers and fistulous tracts are the manifestations of subcutaneous vascular thrombosis and necrosis rather than actual bacterial colonization of the skin (Weisbroth and Scher, 1968).

3. Diagnosis and Control

Diagnosis of pseudotuberculosis is dependent on isolation and identification of *C. kutscheri* from abscesses in the viscera or subcutis. Schechmeister *et al.* (1953) began to explore the possibility of serodiagnosis of *C. kutscheri* infections using agglutination techniques of pooled serum samples, comparing subsequent titer results to observations of overt disease, and finding a positive correlation. Weisbroth and Scher (1968) further investigated this by defining *C. kutscheri* as a species antigenically distinct from other closely related corynebacterial species. They also compared the sensitivity of several immunological tests (agglutination, fluorescent antibody, and agar gel immunodiffusion methods) for detection of serum antibodies and correlated serological evidence of infection with postmortem findings. It was found that agglutination reactions and immunofluorescent methods were more sensitive than the immunodiffusion technique. Also, there was excellent correlation between detection of serum antibodies and gross observation of tissue infection. However, there was no detection of latent infections serologically. From this study, it was proposed that a quick, accurate diagnosis in the event of a pseudotuberculosis epizootic could be obtained using the described immunological techniques. To identify latent carriers, the possibility of parenteral steroid challenge was suggested.

Yokoiyama *et al.* (1977) reported on antigenic heterogeneity of *C. kutscheri* from mice and rats in microtiter serum agglutinin tests. They suggested that more effective serological checking for murine *C. kutscheri* would result from the use of antigen prepared from both the CK1 (mouse origin) and FRCI (rat origin) strains. Cortisone provocation (Yokoiyama *et al.*, 1975; Utsumi *et al.*, 1969), agglutinin tests with culture, an anamnestic response (Fujiwara, 1971), and the microtiter serum agglutination tests are used in various combinations to detect *Corynebacterium* infections in Japan.

Control of the disease can best be accomplished by immediate removal of affected animals and perhaps by incorporating isolation procedures for mice that are naturally or experimentally suppressed immunologically.

B. *Pasteurella*

In 1948, Jawetz (1948) isolated a *Pasteurella*-like agent from pneumonic lesions of mice inoculated intranasally with lung homogenates of healthy mice. The bacterium was identified as *Pasteurella pneumotropica* based on biochemical reactions (Jawetz and Baker, 1950). Since its original description, *P. pneumotropica* has been considered a latent bacterium (Jawetz and Baker, 1950; Hoag *et al.*, 1962; Flynn *et al.*, 1965), a secondary invader, often acting synergistically with other bacteria to produce clinical pneumonia (Wheater, 1967; Brennan *et al.*, 1969), or an agent capable of producing subclinical infections or overt clinical disease in either the respiratory tract (Gray and Campbell, 1953; Flynn *et al.*, 1965; Wheater, 1967) or in other tissues (Brennan *et al.*, 1965; Flynn *et al.*, 1968; Weisbroth *et al.*, 1969; Wagner *et al.*, 1969; Van der Schaaf *et al.*, 1970; Casillo and Blackmore, 1972; Sebesteny, 1973; Rehbinder and Tschappat, 1974; Needham and Cooper, 1975; Ward *et al.*, 1978; Moore *et al.*, 1978).

Transmission of *P. pneumotropica* has been suggested to occur *in utero* (Casillo and Blackmore, 1972; Hong and Ediger, 1978b; Ward *et al.*, 1978), venereally (Ward *et al.*, 1978), by fecal/oral contamination (Hong and Ediger, 1978b), and by aerosolization. Conjunctivitis and orbital abscessation have commonly been associated with latent or clinical respiratory infection, with ascension of the organism from the nasal cavity to the orbit via the nasolacrimal duct (Gray and Campbell, 1953; Brennan *et al.*, 1969; Rehbinder and Tschappat, 1974; Moore *et al.*, 1978). However, Wagner *et al.* (1969) described a spontaneously occurring conjunctivitis in weanling mice in which a *Pasteurella*-like organism was isolated consistently from affected ocular tissues, with no recovery of the organism from the respiratory tract. Suppuration of the superficial skin and subcutaneous tissues (Brennan *et al.*, 1965; Weisbroth *et al.*, 1969; Van der Schaaf *et al.*, 1970), bulbourethral glands (Sebesteny, 1973; Moore and Aldred, 1978), uterus (Flynn *et al.*, 1965; Blackmore and Casillo, 1972; Ward *et al.*, 1978), masseter muscles (Wilson, 1976), and middle ears (Harkness and Wagner, 1975) caused by *P. pneumotropica* has also been associated with concurrent latent infections of the nasopharynx. These findings suggest that oral transmission of the bacteria occurs by biting, licking, and extension of the bacteria via the genital tract, the nasolacrimal duct, and the eustachian tube.

Pasteurella pneumotropica has been isolated from other laboratory animals, including dogs, hamsters, rats, guinea pigs, and kangaroo rats (Brennan *et al.*, 1965). Species harboring this agent may act as a source of infection for laboratory mice.

1. Clinical Manifestations

Although *P. pneumotropica* can produce myriad clinical disorders, only the cutaneous and subcutaneous infections will be discussed. In the spontaneous abscess syndrome described by Weisbroth *et al.* (1969), specific pathogen-free (SPF) adult breeding mice exhibited superficial abscesses of the skin in the shoulder area or lateral body regions. Unilateral and bilateral panophthalmitis with retro-orbital abscessation was also observed in the breeding colony (Fig. 2). Moore and Aldred (1978) reported the isolation of *P. pneumotropica* from preputial and orbital abscesses in a breeding colony of nude (athymic) mice. *Pasteurella pneumotropica* was isolated from sporadic cases of masseter muscle abscessation in HT mice (Wilson, 1976) and from abscesses that occurred in mice from colonies in which clinical pulmonary disease was evident (Brennan *et al.*, 1965; Wheater, 1967). Subcutaneous abscesses in the mammary region and mastitis have not been reported to occur in female mice but have been observed in SPF female strain rats (Wheater, 1967; Van der Schaaf *et al.*, 1970; Hong and Ediger, 1978b).

2. Histopathology

In all cases of cutaneous or subcutaneous infections, no evidence of penetrating wounds or foreign bodies was observed. The abscesses were, for the most part, encapsulated and filled with green-yellow thick exudate. Histologically, the lesions consisted of acute suppuration, with a center of liquefactive necrosis surrounded by a granulomatous exudate consisting of neutrophils, lymphocytes, plasma cells, and fibroblasts. As

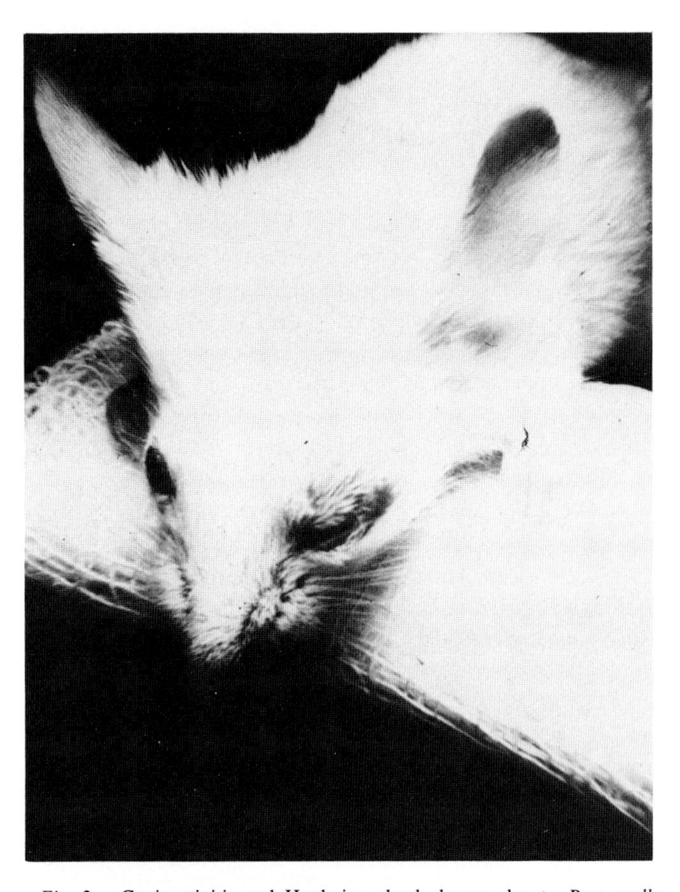

Fig. 2. Conjunctivitis and Harderian gland abscess due to *Pasteurella pneumotropica* in a juvenile mouse.

reported by Weisbroth *et al.* (1969), the superficial abscesses did not extend below the dermis. The abscesses described by Wilson (1976) were within the bulk of the masseter muscle, with no skin involvement. The ocular lesions described by Needham and Cooper (1975) were variable, ranging from a mild keratitis with edema and neutrophilic infiltration of the cornea to a chronic panophthalmitis with disruption of the cornea and the intraocular structures by fibrinonecrotic debris and fibrosis. In the article by Wagner *et al.* (1969), Harderian gland involvement was severe, with intra-acinar and intraductal neutrophilic exudate and necrosis of the glandular acini. The suppurative process appeared to extend to the lacrimal gland, causing interstitial and interlobular inflammation. Similar orbital lesions were described by Weisbroth *et al.* (1969).

3. Diagnosis and Control

Isolation and identification of *P. pneumotropica* from abscesses provide a definitive diagnosis of infection. Concurrent culture of the oropharynx, lungs, and feces of affected animals as well as from other animals in the colony delineates the distribution of the infection throughout the colony. Although attempts have been made to identify carrier animals through serological surveys, no agglutination titers were obtained in mice except those showing spontaneously occurring lesions (Hoag *et al.*, 1962; Weisbroth *et al.*, 1969). Therefore, serological techniques appear not as useful as *in vitro* microbiological examinations in identifying mice carrying *P. pneumotropica*.

Cutaneous manifestations of *P. pneumotropica* infections can occur sporadically without evidence of systemic involvement. These lesions may indicate widespread silent infection in a colony with a potential for overt disease. Moore and Aldred (1978) treated local abscesses produced by *P. pneumotropica* with antibiotic regimens and strict husbandry procedures in order to maintain their nude mouse breeding colony. They found regression of clinical signs using oral and parenteral ampicillin and tetracycline. *Pasteurella pneumotropica* was consistently isolated from the nasopharynx of oxytetracycline-treated mice but not from those treated with ampicillin. Jawetz (1948) defined the antibiotic sensitivity of the *Pasteurella* isolated and found it to be streptomycin sensitive. Treatment of experimentally infected mice with a 4-day regimen of streptomycin was reported to "cure" the mice. Owens *et al.* (1975) reported the *in vitro* antibiogram sensitivity patterns for a series of the most common pathogens of research animals. Among them were 11 isolates of *P. pneumotropica* from mice. The isolates were generally sensitive to ampicillin, cephaloglycin, chloramphenicol, gentamicin, kanamycin, neomycin, polymyxin B, and triple sulfa. Five of ten isolates were resistant to streptomycin and tetracycline. Antibiotic therapy may be beneficial for individual animals but is likely to be impractical and/or unsuccessful for elimination of infection from entire colonies.

C. *Actinobacillus*

A recent report by Simpson and Simmons (1980) described the isolation of two *Actinobacillus* species from nasopharyngeal and cecal cultures of Chinese hamsters, Syrian hamsters, and Wistar rats. One isolate was cultured from an abscessed cheek pouch in a Chinese hamster. The agents were identified as *Actinobacillus equuli* and another isolate which could not be speciated. Lentsch and Wagner (1980) reported the isolation of both *A. equuli* and *A. lignieresii* from conjunctival, oropharyngeal, and middle ear cultures from guinea pigs, rats, and mice during routine diagnostic examinations. In both papers, the authors emphasize that the physical and biochemical characteristics of actinobacilli are similar to those of *Pasteurella* species, specifically *P. pneumotropica*, and may have been misidentified in the past. Although it has been reported that laboratory animals are not susceptible hosts to *Actinobacillus* infections (Wilson and Miles, 1975), these two reports document the isolation of actinobacilli from lesions in laboratory rodents, suggesting possible pathogenicity of the organism for these species. Further, these reports raise questions about the identity of isolates classified as *P. pneumotropica* in previous papers, especially papers wherein full profiles of physical and biochemical characteristics are not listed.

D. *Staphylococcus*

Staphylococci are gram-positive, non-spore-forming cocci that characteristically grow in clusters. Staphylococci are hardy bacteria known to withstand temperatures as high as 15°C and to remain active in dried exudates for several weeks. Most pathogenic strains ferment mannitol and produce yellow pigment, hemolysins, and coagulase. Staphylococci can produce lipolytic enzymes which counter the bactericidal actions of the lipids in the skin.

The staphylococcal species common to the flora of the skin and mucous membranes of mice are the pathogenic *S. aureus* and the generally nonpathogenic *S. epidermidis*. Staphylococcal organisms isolated from suppurative infections and abscesses have generally been considered secondary invaders of devitalized, contaminated tissues. Although *S. aureus* has not been widely recognized as a primary pathogen in laboratory mice, several reports point to its increasingly important role as the etiological agent responsible for outbreaks of cutaneous, lymphatic, or mucous membrane infections (Blackmore and Francis, 1970; Shults *et al.*, 1973; Clarke *et al.*, 1978; Lenz *et al.*, 1978).

The incidence of spontaneous staphylococcal infections in mice is affected by the inherent susceptibility of various strains of mice to the organism, by the physiological condition of the mice, and by the amount of environmental contamination with the staphylococcal organism. While investigating a spontaneous *S. aureus* facial abscess epizootic in an SPF colony of mice, Shults *et al.* (1973) observed that C57BL/6Bd mice were most susceptible to staphylococcal infection. This observation was made after documenting the widespread distribution of *S. aureus* in the facility by isolating the bacteria from the feces of mouse strains C57BL/6Bd, C3Hf/Bd, DBA/2Bd, BALB/CBd, and their hybrids. These findings agree with that of others who found that the C57BL/6J, C3H/HEJ, and Swiss albino strains were most susceptible to infection (Nutini and Berberich, 1965). Festing and Blackmore (1971) reported that strain A mice had a particular susceptibility to *S. aureus* infections. In a production colony of 20,000 mice of eight different strains, Hong and Ediger (1978a) observed that spontaneous staphylococcal-induced preputial gland abscesses occurred most frequently in the C3H/HEN(MTV-) strain. Athymic nude mice, by virtue of the lack of a protective hair coat and a depleted T-cell population, quite frequently develop cutaneous and orbital staphylococcal infections even when housed in SPF barrier conditions (Cooper, 1977).

Certainly, the physiological condition of the animal affects its ability to respond to external stimuli or stress. Factors such as age, environmental conditions, and diet have been reported to influence the susceptibility of the mouse to staphylococcal infections. In experimental subcutaneous infections of neonatal CD-1 random-bred albino mice with six strains of staphylococci (strains SM1, SM6, SM9, SM10, SM14, SM15), McKay and Arbuthnott (1979) documented an age-related susceptibility to staphylococcal-induced abscesses or death in mice from birth to 5 days of age. Ash (1971) suggested that a seasonal change provoked epizootics of chronic staphylococcal dermatitis in an established SPF rat colony, with outbreaks occurring in the winter and spontaneously clearing by midsummer. The dietary influence on the susceptibility of mice challenged with virulent staphylococci has been investigated experimentally (Nutini and Berberich, 1965; Nutini *et al.*, 1968). It was initially found that random-bred BT (Boontucky) mice fed a nutritionally incomplete (deficiencies not specified), synthetic diet were more susceptible to staphylococcal disease than control mice fed a commercial chow. When the study was simulated using a nutritionally complete, semisynthetic diet versus a commercial diet fed to naturally susceptible Swiss albino mice, no difference was evident in the incidence of infection between the diet groups. Nutini *et al.* (1968) suggested that nutritionally deficient diets rather than the synthetic diet predisposed the mice to staphylococcal infections.

Prevalence of staphylococci in the environment may directly reflect the frequency of tissue infection. Staphylococcal abscesses have occurred in SPF and barrier-maintained mice, but not with the frequency of disease incidence found in conventionally housed mice (Blackmore and Francis, 1970; Clarke *et al.*, 1978; Shults *et al.*, 1973). Pathogenic staphylococci are more frequently isolated from the mucous membranes of apparently healthy conventional mice, but in SPF mouse colonies, nonclinically affected carriers of virulent staphylococci are the exception rather than the rule (Blackmore and Francis, 1970). Once virulent staphylococci contaminate the environment, natural colonization of the mucosa of the gastrointestinal tract can occur and produce a carrier state. Several articles reported that *S. aureus* isolated from cutaneous lesions of mice and rats were concurrently isolated from either the feces or oropharynx or both (Blackmore and Francis, 1970; Shults *et al.*, 1973; Wagner *et al.*, 1977; Clarke *et al.*, 1978).

Staphylococci are not proscribed organisms for SPF mice, but introduction and persistence of staphylococcal organisms in SPF colonies can be the result of faulty managerial procedures or defects in barrier equipment or its operation. Several reports suggest that pathogenic *S. aureus* can be introduced into SPF colonies through human carriers (Blackmore and Francis, 1970; Ash, 1971; Shults *et al.*, 1973). Following a microbiological survey in an animal facility, Blackmore and Francis (1970) found that all staphylococci isolated from SPF animals were of human phage types. In contrast, less than two-thirds of the staphylococci isolated from conventional animals were classified by phage types of human origin. Staphylococci of nonhuman origin are reported to be relatively species specific, with the majority of the isolates nontypable using the phage-type system developed for classification of staphylococci of human origin (Mann and Bjotvedt, 1966). Similarly, Markham and Markham (1966) reported that human phage types are generally accepted to be specific for man. However, culture and phage typing of the strains of coagulase-positive staphylococci isolated from clinically infected mice have directly correlated with the staphylococcal phage types isolated from the animal caretakers (Blackmore and Francis, 1970; Ash, 1971; Shults *et al.*, 1973). Through 4 years of bacterial surveillance and typing in an SPF facility, Lenz *et al.* (1978) reported that the phage types of *S. aureus* from healthy SPF mice often corresponded to the strains isolated from animal caretakers. The study also revealed that the most frequently occurring virulent phage type isolated from affected animals could be identified in caretakers after clinical disease was evident in the animals. Finally, the data documented the spread of a lysotype (phage type) of *S. aureus* from one SPF building to another by the transfer of an animal caretaker.

Blackmore and Francis (1970) addressed the interesting phenomenon of the different numbers of phage-typable staphylococci in SPF versus conventional mice. Although the conventionally housed mice had close contact with the human attendants, less than two-thirds of the staphylococci isolated from the mice were typable by human phages. It was suggested that conventional mice may carry indigenous strains of staphylococci that may interfere with colonization by other strains of staphylococci, with reference to colonization by staphylococci of human origin. Eichenwald (1965) showed that the normal flora of the human nasal mucosa contain several strains of *S. aureus* that impart resistance to subsequent colonization attempts by other strains of the virulent staphylococci. This "interference" phenomenon may be the reason that conventional animals carrying indigenous strains of staphylococci appear to withstand better the challenge with human strains of *S. aureus* than SPF mice that lack the indigenous organism in their natural nasopharyngeal flora.

1. Clinical Manifestations

Staphylococcal-induced infections in laboratory mice primarily produce localized disease. Cutaneous lesions are the most common manifestation of infection and usually occur secondary to a traumatic break in the integrity of the surface epithelium. Dermatitides and subcutaneous abscesses are located primarily on the face and the neck region and range in severity from superficial infections of the epidermis and hair follicles to deeper infections of the dermis, subcutis, and underlying muscles. Colonization of the conjunctiva or genital mucosa with *S. aureus* can also occur singularly or in association with other concurrent staphylococcal infections.

Superficial pyodermas most frequently appear as subcutaneous abscesses which are characterized as small, firm, raised nodules in which the overlying epidermis is intact. Rupture of the abscess can produce an exudative ulcerative dermatitis. The dermatitis is often pruritic, resulting in scratching and production of a cyclic, traumatic excoriation and autoinoculation of adjacent epithelium from scratching. A deeper staphylococcal cellulitis has been described by Shults *et al.* (1973) in which various-sized nodules were present in the mandibular region of SPF mice. These lesions were described as botryomycotic granulomas which eroded adjacent bone and often produced focal epidermal ulcerations or draining fistulous tracts.

Unilateral or bilateral purulent infections of the conjunctiva present with epiphora and crusty ocular exudate. This superficial infectious process can extend to other ocular structures, resulting in retrobulbar abscessation. Keratitis and panophthalmitis can also occur (Blackmore and Francis, 1970; Shults *et al.*, 1973).

Staphylococcal colonization of the genital mucosa of male mice has been speculated to be the source of infection in accessory sex gland abscessation (Needham and Cooper, 1976; Hong and Ediger, 1978a). Preputial gland abscesses present as raised, firm nodules in the inguinal region or at the base of the penis; these abscesses frequently rupture and spread to surrounding tissues.

2. Histopathology

Histologically, the lesions of acute ulcerative dermatitis are loss of the normal epidermis, with focal epidermal ulcerations covered with a fibrinonecrotic scab and diffuse accumulations of neutrophils in the surrounding epidermis and underlying dermis (Fig. 3). Chronic inflammatory skin lesions consist of an infiltrate with lymphocytes and macrophages with dermal fibroplasia. Deep staphylococcal infections are characterized as single or multiple coalescing granulomas with purulent, necrotic centers containing bacterial colonies. These necrotic cores are surrounded by zones of macrophages, lymphocytes, neutrophils, and fibrous connective tissue (Shults *et al.*, 1973; Clarke *et al.*, 1978). Infections of the glandular tissues of the orbit and genitalia result in acute microabscessation of the glandular acini, with cellular necrosis and neutrophilic infiltration of the affected tissues (Shults *et al.*, 1973; Needham and Cooper, 1976; Hong and Ediger, 1978a).

3. Diagnosis and Control

Diagnosis is based on isolation of gram-positive, coagulase-positive cocci that may produce β-hemolysis on blood agar and grow well on mannitol–salt agar. Phage typing of the isolant may be useful in determining the source of infection and the diversity of environmental contamination. Efforts to treat or eliminate the ubiquitous organism from the environment are, in most cases, impractical and unnecessary. However, Fox *et al.* (1977) demonstrated that trimming of the hind toenails of rats affected with chronic ulcerative dermatitis resulted in partial remission of the lesions. Similarly, Wagner *et al.* (1977) amputated the hind toes of rats with ulcerative lesions, resulting in healing of the skin lesions. Such control measures can be applied in situations in which species or strain susceptibility, SPF status, or experimental protocol requires protection from self-trauma, *S. aureus* exposure, and infection. Removal of affected animals, acidification of water, sterilization of food and bedding, frequent changing of bedding, use of filter-topped cages, and frequent bacterial monitoring of personnel may reduce or limit possible sources and transmission of the staphylococcal organism (Lenz *et al.*, 1978).

E. *Streptococcus*

Streptococci are differentiated by the Lancefield classification into groups A through O. Strains responsible for most

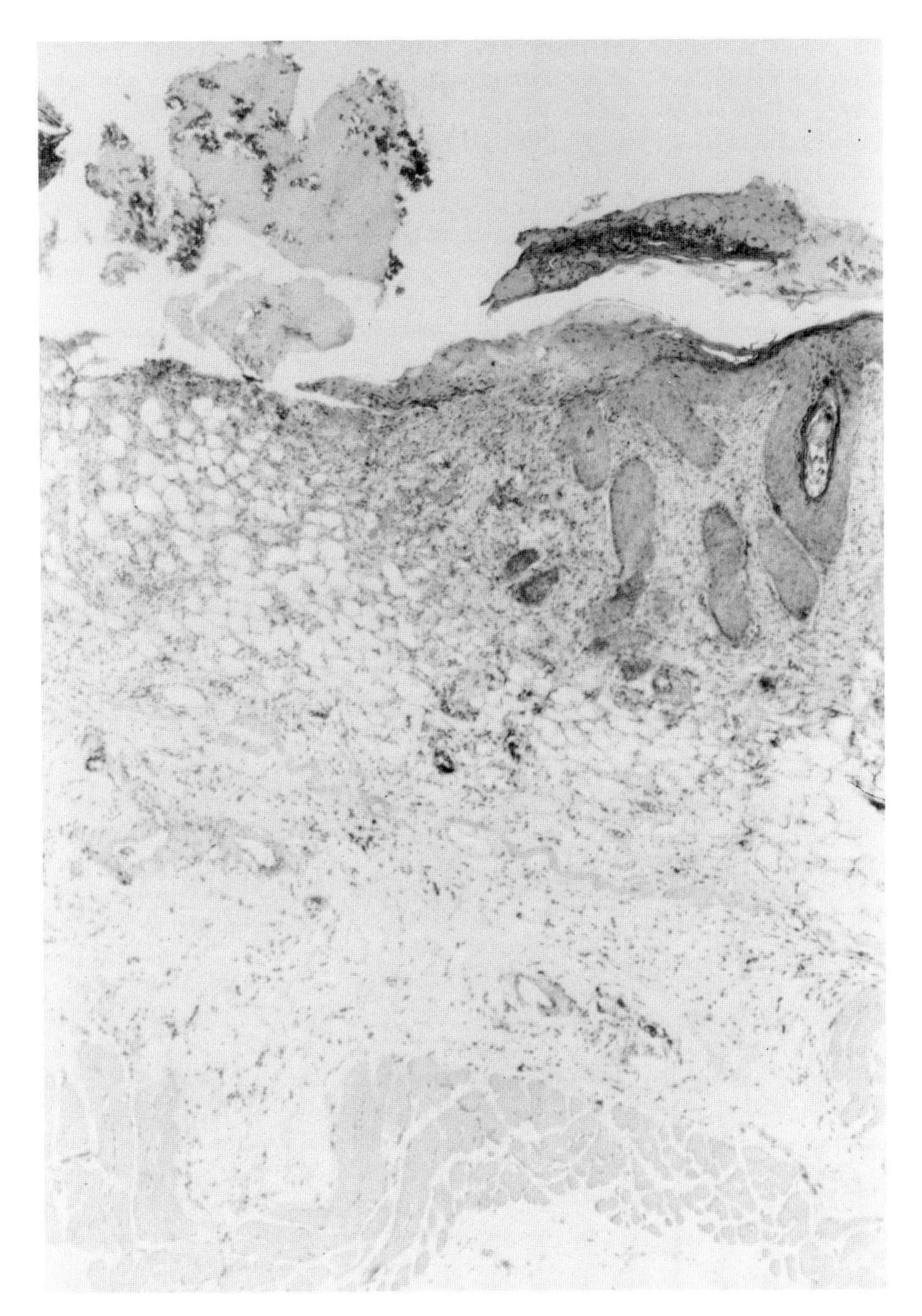

Fig. 3. An area of acute ulcerative staphylococcal dermatitis with acanthosis of the epidermal border of the ulcer, inflammatory cell infiltration of the dermis and subcutis, and bacterial colonies in the fibrinonecrotic scab.

human infections belong to group A and include *Streptococcus pyogenes*. Most strains causing suppurative infections in laboratory mice belong to group C, but epizootics caused by group A (Nelson, 1954), group B (Loewenthal, 1932), and group D (Gledhill and Rees, 1951) streptococci have occurred. Also, Stewart and associates (1975) reported isolating streptococci of Lancefield group G from skin lesions in mice.

Group A streptococcal infections in domestic animals are rare. The spontaneous occurrences of group A infections reported in laboratory mice were caused by a *Streptococcus* sp. designated group A, type 50 (Nelson, 1954; Hook *et al.*, 1960). Infected mice examined bacteriologically in separate laboratories were derived from a single commercial mouse farm (Lancefield, 1972). Attempts to identify the source of infection in mice by culturing human attendants were negative. Transmission of infection was probably by direct mouse contact and indirect exposure via contaminated fomites or aerosolization (Lancefield, 1972).

Group B streptococci are an important source of bovine mammary infections and can cause serious infections in humans. In man, this strain is most commonly isolated from the female genital tract. However, a human carrier rate of 10% has been reported with strains obtained from the throat, genital tract, and rectum (Eickhoff, 1972). Loewenthal (1932) described a spontaneous outbreak of a fatal septicemia in mice caused by a group B streptococcus. The mice that survived the systemic infection developed lymphadenitis. The source of the group B, type 1, streptococcus isolated from the mouse epizootic described by Loewenthal (1932) was not identified, but transmission was probably due to direct contact and aerosol.

Morphologically, group C streptococci closely resemble *S. pyogenes* of group A. Group C strains, however, are not fibrinolytic, ferment glycerol aerobically, and are infrequently human pathogens. Group C, *Strep. zooepidemicus,* can cause disease in the laboratory mouse and is also responsible for cervical lymphadenitis in guinea pigs. *Streptococcus zooepidemicus* colonizes the oropharynx, invading the deeper soft tissues through mucosal abrasions and lymphatics, and causes suppurative adenitis of the cervical lymph nodes. Transmission occurs by direct contact with infected fomites, oral secretions, and possibly through aerosol dissemination.

Cutaneous infections of mouse skin with group G streptococci have been reported by Stewart *et al.* (1975). The organism was further identified as group G, type 1, which is a member of the species *anginosa*. The source of the epizootic was not identified. The bacterium was consistently present in the throats and skin lesions of affected mice, suggesting transmission by direct contact (fight wounds), infected fomites, and aerosol. This bacterium is commonly isolated from the human oral, nasal, and genital mucosa, as well as from abscesses, skin, and feces (Deibel and Seeley, 1974).

1. Clinical Manifestations

Although several groups of streptococci have been identified as primary pathogens causing disease in mice, the clinical manifestations of infection can be divided into two general entities: cervical lymphadenitis with rupture and drainage through the ventral aspect of the neck and ulcerative, necrotic dermatitis.

Group A streptococci have been associated with purulent cervical lymphadenitis, pneumonia, bacteremia, and occasionally meningitis followed by death. Cervical lymph node enlargement was first observed when mice were of breeding age, even though positive throat cultures could be obtained from mice as young as weanlings. Generally, the lymph nodes remained enlarged, with occasional rupture and drainage through the skin on the neck ventrum. The hair in the fistulous area became wet and matted with yellow, thick exudate and blood. Death due to systemic streptococcal infection was common, with mortality approaching 50% (Nelson, 1954; Hook *et al.*, 1960). It is interesting to note that purulent lymphadenitis occurred only occasionally in spontaneously infected Swiss mice (Hook *et al.*, 1960) but was observed more frequently in naturally infected Princeton mice (Nelson, 1954). Animals surviving the infection often became carriers, with bacteria being consistently isolated from the oropharynx.

Spontaneous epizootics of cervical lymphadenitis resulting from group C streptococci have been infrequently reported (Neilsen, 1978). The pathogenicity of the organism for mice seems to be less than for guinea pigs. Experimental studies testing the virulence of group C strains via subcutaneous injections in mice revealed suppurative adenitis of cervical and regional lymph nodes. Death accompanied the infectious process when unstable, mucoid group C strains were injected (Seastone, 1939).

Conversely, mice infected with hemolytic streptococci from group B, type 1a, succumbed to systemic infection with high mortality (Loewenthal, 1932). In testing the virulence of the isolate, death occurred 3 or 4 days after infection by the oral or nasal route. General lymphadenitis was described to occur occasionally in those mice surviving acute bacteremia.

An epizootic of dermatitis caused by group G streptococci has been reported (Stewart *et al.*, 1975). The dermatitis was characterized by dry, necrotic lesions over the dorsal thoracic and lumbar areas. Lesions progressively spread and often involved the entire shoulder or lumbar regions or both. Underlying muscle was rarely affected, but paresis and paralysis occurred in some of the more extensively infected mice. Morbidity was 90%, with mortality approaching 35%. Microscopically, the dermatitis was characterized by epithelial ulceration covered by fibrinonecrotic debris. There was a neutrophilic infiltration of the underlying dermal collagen and subcutis. Vasculitis and lymphangitis, often resulting in vessel occlusion, were prominent features of the subcuticular inflammatory process. The subjacent muscle was mildly infiltrated with neutrophils. Gram-positive cocci were demonstrated in the dermal connective tissue, especially underlying the margin of the ulcer. Bacteria were also observed in the subcutis. Amyloidosis and extramedullary hematopoiesis were observed in the liver, spleen, and kidneys of several infected mice. The lesions were reproduced experimentally.

2. Diagnosis and Control

Diagnosis depends on isolation of hemolytic streptococci in pure culture from infected tissues. Recovery of β-hemolytic streptococci or mixed populations of bacteria that include β-hemolytic streptococci may be the result of secondary bacterial contamination of a wound rather than implicating the streptococci as a primary pathogen. Classification of isolates may be helpful in determining the prognosis of the condition, as well as in attempting to identify the source of infection.

Control of virulent streptococcal infections can be enforced and involves removal of infected animals with efforts to improve sanitation and impede transmission of the organism. These measures can include decreasing stressful stimuli, such as fighting and overcrowding, using filter tops, and perhaps identifying carriers by periodic throat cultures (Stewart *et al.*, 1975).

IV. MYCOTIC DISEASES

A. General

Dermatophytoses are superficial fungal infections of the epidermis, hair, and nails that are usually self-limiting. Dermatophytes utilize keratin, grow as branching hyphae, and divide to form rows of arthrospores. Spontaneous remissions are the rule.

Excessive environmental heat and humidity are factors that notoriously stimulate fungus infections (Dolan *et al.*, 1958); overcrowding may also contribute. Age can be a factor; in one study, 6-week-old mice were more susceptible than older animals (Dolan *et al.*, 1958). Diagnosis is based on the clinical appearance of affected mice, microscopic examinations of skin scrapings, culture of affected areas on selective agar, and histopathological examination of appropriately stained sections of skin.

Possible sources of dermatophytic infections for laboratory mice include soil (Knudtson and Robertstad, 1970), domestic (Connole, 1965; Fuentes, 1959) and wild animals (English, 1971), human contacts (Mackenzie, 1961), and other species of laboratory animals (Feuerman *et al.*, 1975), i.e., guinea pigs (Fuentes, 1959; Fischerm, 1972), rats (Povar, 1965), rabbits, hamsters, and other mice (Paveia, 1971). Buchanan (1919) credited Bennett in England in 1850 with the first record of mouse favus. He reported having "seen it on the face of a common house mouse, in which animal the same cryptogamic vegetations were to be detected as in man" (p. 97).

Gluge and d'Udekem (1857) published the first illustrated description of mouse favus, including the nature of the fungus "*Microsporon muris*" and the concept of contagion to humans.

B. *Trichophyton*

Trichophyton mentagrophytes, a worldwide, zoophilic, heterothallic fungus, has limited host species specificity and is the major cause of ringworm in mice. *Arthroderma benhamiae,* an independent species in older literature, is the perfect state of *T. mentagrophytes*. Based on successful crosses with *A. benhamiae, T.* (*Achorion*) *quinckeamum* is now believed to be either a variety of or synonymous with *T. mentagrophytes* (Ajello *et al.*, 1968; Weitzman and Padhye, 1976). A wide variety of *Trichophyton, Microsporum,* and *Achorion* species of older literature are now believed synonymus with *T. mentagrophytes*.

Classic human favus is attributed to *T. schoenleini,* with the highest incidence in southern Europe and northern Africa, in which areas it constitutes a serious public health problem. Early reports in the field of medical mycology indicated that the observed "favic type" lesions in mice were due to *T. schoenleini*. Most of these, however, were based on clinical observations, and only two such reports (DuBois, 1929) and one by Unna and Franck (cited in Patiala, 1951) were verified by Sabouraud. *Trichophyton schoenleini* has also been reported in a cat by Lebert (Richou, 1950) and a dog by Catanei (1936).

Of 14 species of *Trichophyton* experimentally inoculated into mice, only three proved pathogenic: *T. mentagrophytes, T. mentagrophytes* var. *quinckeanum,* and *T. violaceum* (Rieth, 1968).

In 1956, several field investigations of mouse favus were conducted in the lower St. Lawrence Valley (Blank, 1957). Direct examination of hairs from wild mice with skin lesions did not indicate infection; however, culture of lesions resulted in isolation of *T. mentagrophytes* var. *quinckaenum* (*Microsporum quinckeanum,* according to Blank's taxonomy) or no growth. Two of the field investigations demonstrated involvement of family members and pets from farms where mice carried the fungus. It was concluded that mice constitute a reservoir of the fungus for cats and dogs which served to transmit *T. mentagrophytes* to man.

Fifty percent of 2500 breeding stock mice were infected with *T. mentagrophytes* in England (Parish and Craddock, 1931). A survey of 222 asymptomatic 4- to 6-week-old mice from four commercial breeders in the United States by Dolan *et al.* (1958) revealed incidences of *T. mentagrophytes* in 55, 9, 0, and 66% of animals. Mackenzie (1961) examined over 800 pet shop and laboratory mice for *T. mentagrophytes*. There was a high incidence of fungus among laboratory animals (49 of 160 from breeding boxes and 104 and 149 from stock cages), whereas lesions were seen in only 2 of the 104 positive carriers. Twelve of 20 pet shop mice were infected. Absence of signs in many of the mice was of particular interest. Using a toothbrush method to collect spores, Rieth (1968) found 4.3% of 4000 apparently healthy laboratory mice infected with *T. mentagrophytes*. Others have reported natural infections of mice with *T. mentagrophytes* (DuBois, 1929; Catanei, 1942). Many authors stress the significance of the inapparent carrier state.

Booth (1952) cultured *T. mentagrophytes* from 12 mice with no lesions and demonstrated that mice which appeared healthy may be infected with the fungus. Dolan *et al.* (1958) cultured *T. mentagrophytes* from 43% of healthy domestic mice. Of these, only 2% of the mice had lesions. Similarly, Menges *et al.* (1957) reported recovering the fungus from "healthy" wild mice of Georgia, in the United States. Other dermatophyte isolations have been reported in the United States, Egypt, and New Zealand (McKeever *et al.*, 1958; Taylor *et al.*, 1964; J. M. B. Smith and M. J. Marples, 1965, unpublished observations).

Davies and Shewell (1964, 1965) found an infection rate of over 90% in a colony in which less than 1% of BALB/c and C3H/Bi mice had clinical signs. The carrier rate increased with age, and 100% of animals over 6 months of age were positive. No sex difference was noted in the carrier rate. Hair coats of long-term survivors of whole-body irradiation had higher fungal counts than untreated mice. The C3H mice developed bald patches, usually on the back and belly, but culture of tail skin scrapings was negative. The BALB/c mice developed raised, brown, circular scabs on the tail from which *T. mentagrophytes* was isolated. LaTouche (1957) considered fungal carriage on the tail to be an important factor in maintaining colony infections. In several instances, mouse lice were thought to be a possible means of transmission of *T. mentagrophytes* among mice (Booth, 1952; Rieth, 1968).

Cetin *et al.* (1965) described an epizootic of *T. mentagrophytes* (interdigital) in a colony of 400 white mice in which 20 to 25 died in 2 days. Some had signs of generalized infection characterized by sparse hair, inactivity, and anorexia. Also described were epistaxis just before death, plaque formations 4–10 mm in diameter on the face, cheek, and nose which were circular and craterlike, with a prominent periphery adherent to underlying tissue covered with a thick yellowish crust and a chalky appearance on the surface. A second type of plaque, 4–6 mm in diameter, involved the dorsal neck, back, and abdomen. The latter were circular, whitish-pink, circumscribed areas of alopecia. Similar lesions were produced experimentally in mice. The skin of animals with general alopecia was dry and cyanotic, and the hair was brittle. According to Cetin *et al.* (1965), "The tail of some animals was cut off spontaneously, whereas in some the tail was easily cut off under a slight pressure" (p. 840). Retrospectively, the influence of other intercurrent, naturally occurring diseases cannot be assessed, i.e., *Corynebacterium* spp. and ectromelia virus infections.

Trichophyton terrestre is ubiquitous in soil and is relatively nonpathogenic. It has been isolated from the hair coats of rodents (not laboratory mice) without dermatological disease (Emmons *et al.*, 1977). It has been reported to be occasionally pathogenic in man (Grimmer, 1974; Heidenbluth *et al.*, 1978), dogs and cats (Scott *et al.*, 1980), cattle (Gupta *et al.*, 1970), and experimentally in guinea pigs (Heidenbluth *et al.*, 1978). Bloch (1911) isolated a *Trichophyton* sp. of uncertain identification from a mouse with skin lesions.

C. Other Dermatophytes

Microsporum spp. infections of mice are not well documented in the literature. Simmons and Brick (1970) list *Microsporum* under the diseases of mice without describing clinical signs or citing reports of the infection in mice. In experiments with eight species of *Microsporum*, only *M. gypseum* was pathogenic for mice (Rieth, 1968). Among cats and dogs, *Microsporum* spp. account for 99 and 90% of dermatophytoses, with *T. mentagrophytes* accounting for about 1 and 10%, respectively (Scott *et al.*, 1980). Feuerman *et al.* (1975) isolated *M. gypseum* from several mice without clinical disease in Israel.

Accidental contamination of a subcutaneous inoculation site in a C3H mouse induced granuloma caused by *Absidia corymbifera* (Symeonidis and Emmons, 1955). The granulomas were transmissible to other mice by subcutaneous inoculation.

According to Bazin (1862), Draper noted the presence of lesions which resembled perrigofavosa on wild mice caught in his home. When the mice were given to a cat, a crusty lesion appeared over the cat's eye. Further, two children who had contact with the cat were later diagnosed as having perrigofavosa (most likely not human favus, considering the ease of cure of the children's condition).

For a more complete discussion of the zoonotic significance of mouse dermatophytes see Chapter 22, this volume.

D. Diagnosis

1. Direct Examination

Specimen collection is crucial to accurate diagnosis of infections. Select infected hairs from the periphery of a lesion. Infected hairs may be broken and deformed. The mass of fungal hyphae at the orifice of the hair follicle provides excellent material for diagnostic examination. Pull affected hairs with a forceps and place them on a slide in a few drops of 10% sodium or potassium hydroxide, add a coverslip, and warm (do not boil) over a small flame. Examine the preparation under reduced light at low and high magnification. Epidermal scales and hair affected with *T. mentagrophytes* rarely fluoresce under Wood's light (Ajello *et al.*, 1963). Hyphae must be differentiated from bedding fiber, food particles, and epidermal debris. In unhaired regions or animals, scrape away the upper scaly layers and discard. Collect deeper epidermal layers by scraping to the point of producing a serous exudate and

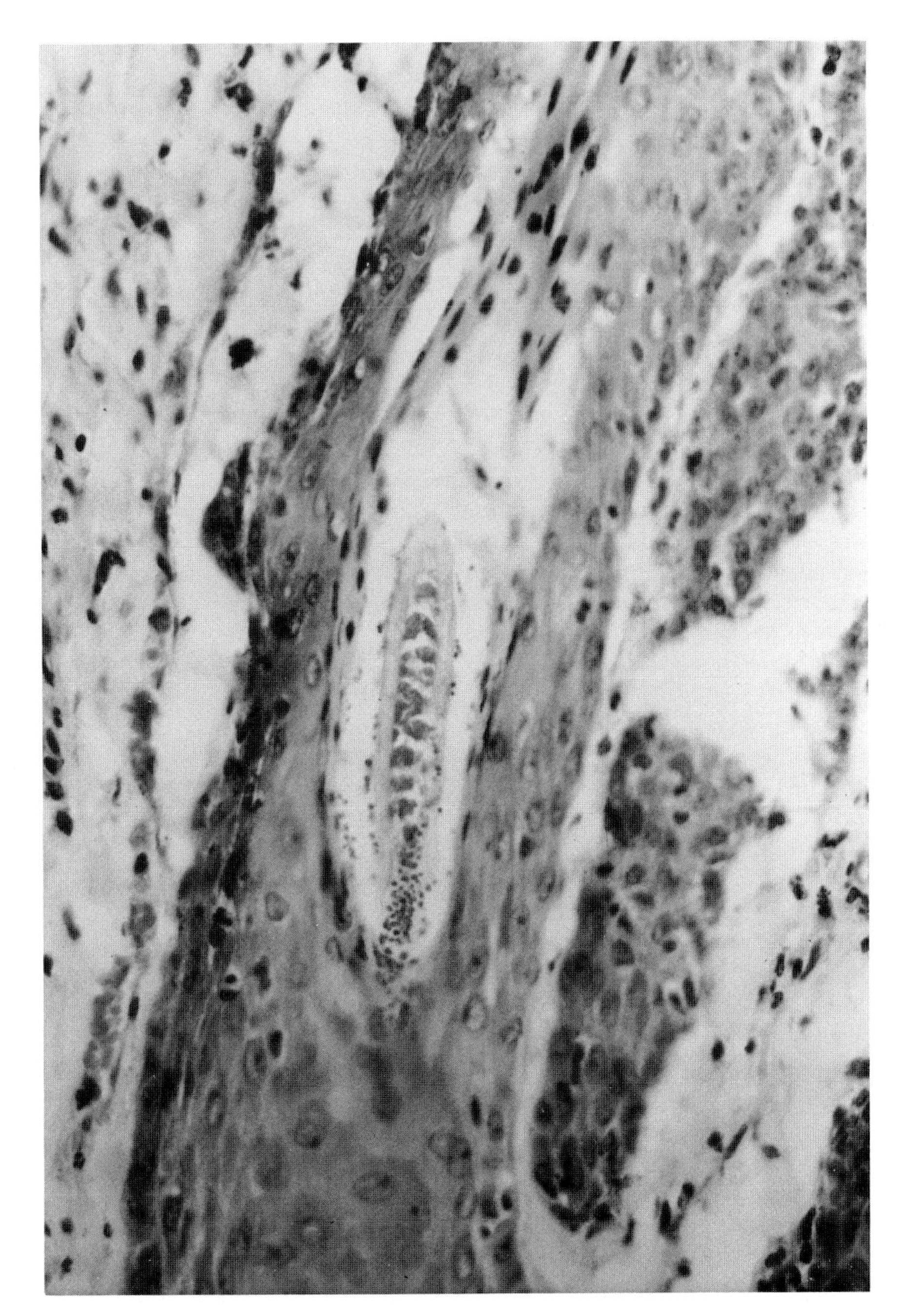

Fig. 4. Clusters of arthrospores of *Trichophyton mentagrophytes* on the hair shaft deep within the hair follicle.

prepare for examination as described above. Microscopical examination of potassium or sodium hydroxide preparations of skin scrapings may reveal chains of arthrospores and mycelia. Sheaths or isolated chains of spores (2–3 μm) may be seen on the surface of hair (Ajello *et al.,* 1963). Strains from animal sources, in particular, may invade hair follicles and hairs (Fig. 4) (Emmons *et al.,* 1977).

2. Culture

Culture is accomplished by removal of hair by clipping with sterile scissors and removing superficial debris by scraping a sterile scalpel blade across the lesion. Sterile swabs (Feuerman *et al.,* 1975), hairbrush-collected material (Mackenzie, 1961; Rosenthal and Wapnick, 1963), selected hairs, or deeper scrapings are placed directly on agar slants or in petri dishes or in a clean, sterile envelope for shipping to a diagnostic laboratory for culture. Modified Sabouraud's agar containing 0.05 mg/ml of chloramphenicol adjusted to pH 6.5-7.0 is satisfactory (Emmons *et al.,* 1977). Incubate at 22°-30°C (never 37°C) and examine daily for evidence of sporulation, which may begin within 5-10 days, and for characteristic reproductive and vegetative structures.

Trichophyton mentagrophytes of animal origin frequently produces rapid growth. Typical colonies are white and granular and are thick, flat, or heaped and irregularly folded. The surface is coarsely granular to powdery, fluffy, downy, or cottony white to cream or light buff, occasionally yellow, pink, or rose-tan. The reverse side of the colony is rose-brown or wine, occasionally yellowish-orange or deep red (Ajello *et al.,* 1963; Emmons *et al.,* 1977). Microconidia are very numerous, small, globose to slender and elongate, borne singly along hyphae or in pine-tree-like terminal clusters. Macroconidia are rare or abundant in some strains, two- to five-celled, thin-walled, slightly club-shaped, spindle-shaped, or long and nearly pencil-shaped. Spirals or coils are tightly wound, and nodular bodies may be numerous (Ajello *et al.,* 1963). Only hyphae and arthrospores are seen in skin. A small-spored ectothrix infection is seen on the hairs.

E. Treatment and Control

Pathogenic dermatophytes are not eradicated readily from experimental mice. It may be necessary to destroy the entire colony and sterilize all cages and equipment before reuse. Dolan *et al.* (1958) found that cleaning of cages with steam and detergents and regulation of temperature and humidity in animal quarters considerably decreased the incidence of clinical ringworm.

Mice that harbor dermatophytes may also be infested with ectoparasites, and fungal infections may develop more readily in irritated or abraded skin. The degree of dermatophyte infection therefore may be lessened by reducing the population of ectoparasites prior to or concurrent with specific dermatophyte treatment. Methods for treating ectoparasites are covered by Weisbroth in Chapter 21, this volume.

A moderate degree of success in treating *T. mentagrophytes* infection was achieved through dipping in an acaricide consisting of 67 gm 25% tetraethylthiuram monosulfide in industrial alcohol* made up to 1 liter in warm water at 3-week intervals (Davies and Shewell, 1964, 1965). Others have suggested killing infected mice as a control measure (Parish and Craddock, 1931).

*Tetmosol, Imperial Chemical Industries, Ltd., Wilmslow, Cheshire, England.

In other animal species and man, treatment regimens of a microcrystalline oral form of griseofulvin† and/or topical lime sulfur‡ applications are frequently used to treat dermatophytosis. However, the metabolism of griseofulvin in dogs is very rapid, and the route of liver clearance in dogs is about six times faster than in humans (Chiou and Rigelman, 1969). It may be difficult to apply dosage recommendations predictably across species; in laboratory mice, therefore, a trial-and-error approach would be needed to establish efficacy and safety. Administration of griseofulvin to breeding female mice should be avoided, since griseofulvin has been found to be embryotoxic and teratogenic when given orally to pregnant Wistar rats (Slonitskaya, 1969).

Dip solutions should be kept at 37°C and animals should be allowed to dry in a warm environment, preferably in solid cages. Care must be taken to prevent toxicities, drowning during dipping, or aspiration of dip solutions. The consequences of treatment must be considered, especially the potential effect of such treatment on interpretation of experimental results.

V. NONINFECTIOUS SKIN DISORDERS

The noninfectious disorders of the mouse skin referred to in this section include naturally occurring and accidentally induced skin lesions in which a microbial agent was not primarily incriminated.

The behavioral characteristics and social interactions of caged mice often lead to fighting, tail biting, and whisker chewing. Although fighting is not limited extensively to male mice, they tend to be more aggressive and will express dominance over other males by fighting. Bite wounds tend to be located on the head, neck, shoulders, perineal area, and tail. Often, in a cage of males, one animal is free of skin lesions; he is the aggressor or dominant animal. Removal of the unaffected male usually ends the fighting, with healing of the wounds on the subordinate animals in several weeks. However, with removal of an aggressive male, a previously submissive male may assume a dominant role, with resumption of fighting. Caging of male littermates or males paired prior to weaning tend to decrease the incidence of fighting (Les, 1972).

Les (1972) reported tail lesions resembling bite wounds which occurred in adult C3H/HeJ and C3HeB/FeJ mice of both sexes when housed in groups of 40. The lack of ectoparasites and microbial pathogens prompted behavioral investigation. The tail lesions were described as shallow, circular, cutaneous abrasions which increased in number and severity with time.

†Fulvicin U/F, Schering Corporation, Galloping Hill Road, Kenilworth, New Jersey 07033.

‡Vlem-Dome, Dome Laboratories, 400 Morgan Lane, West Haven, Connecticut 06516.

Occasionally the tail was swollen, with necrosis and loss of the distal third of it. The tail lesions healed when the mice caged in groups of 40 were divided by sex and housed in groups of five. The incidence of tail lesions was lowest when weanlings were maintained in littermate groups and housed according to sex. It was suggested that the social stress from cage density, from mixing sexes, and from combining littermates contributed to the appearance of the lesions.

Another reported manifestation of social dominance in mice is the act of hair nibbling and whisker chewing. Through the observations of Hauschka (1952), the whisker-eating habits of Swiss Webster mice as well as those of C57BL, C3H, and their hybrids were established. This activity, possibly an inherited trait, may have stemmed from an exaggerated grooming behavior, which resulted in establishment of a social hierarchy. Long (1972) also described hair chewing in C57BL and C3H mice and supported the theory of whisker chewing and hair nibbling as an expression of social dominance. Facial hair and whisker eating were activities of both C57BL and C3H strains; however, hair which was chewed along the neck and back was seen only in the C57BL mice (Fig. 5). In both reports, the dominant mice were primarily females who grasped the faces of subordinate cagemates and chewed the facial hair and whiskers down to the skin. The dominant animals usually retained their whiskers, whereas the cagemates presented with "shaved" faces. Long (1972) observed that among C57BL mice, a dominant whisker chewer occasionally was submissive to hair nibbling by another mouse, suggesting different dominance patterns in the social hierarchy. The alopecic skin was unbroken, without erythema or swelling. Thornburg *et al.* (1973) investigated the pathogenesis of hair chewing in C57BL mice. Chronic hair chewing produced histological abnormalities such as poorly formed, pigmented club hairs, resulting from early incomplete detachment of the bulb from the matrix, epidermal thickening, melanin pigmentation of the epidermis, and foreign body granulomata in the panniculus adiposes. In most cases, the mice with previous hair loss had full regrowth of hair in several weeks. In the chronically chewed mice, however, regrown hair was depigmented and sparse, with a few local areas of alopecia remaining.

In an article by Litterst (1974), muzzle alopecia in male CDF_1 (BALB/cCr female × DBA/2Cr males) mice was the result of mechanical abrasion of the nose from inserting the muzzle through ventilation holes punched in the stainless steel cage cover. Histologically, the denuded skin showed mild hyperkeratosis and acanthosis, hair shafts broken at or slightly above the skin surface, and a neutrophilic infiltration of the dermis (Fig. 6). Fungi and ectoparasites were not found, and whisker chewing was not observed. The similarities in the appearance of muzzle alopecia from drug administration and hair loss from mechanical means (both chewing and rubbing) were emphasized.

Occasionally, skin lesions have been produced during routine handling or processing of laboratory mice. When faced with unexpected deaths and skin lesions in mice on a research project, Serrano (1972) was able to link the cause of the disease with the accidental use of improperly diluted quaternary ammonium disinfectant. Forceps dipped into the disinfectant and not rinsed were used for transferring mice from cage to cage. The skin lesions were characterized by yellow staining of the hair of the scapulae and neck, loss of hair, and induration of the affected regions followed by sloughing of skin. Ulceration of the ears also occurred. The deaths and lesions were reproduced with a single topical application of 13 and 50% solutions of aqueous benzalkonium chloride. The normal dilution of aqueous benzalkonium chloride used for disinfection is 0.07%.

Permanent identification of individual mice can be accomplished by ear notching, toe amputation, hair dying, and ear tagging. A dermititis apparently resulting from an irritation to the metal ear tags and augmented by scratching by the hind feet was reported by Cover *et al.* (1979). Clinically, the tagged ears developed a moist dermatitis that progressed down the neck and shoulders. Pruritis and self-induced trauma exacerbated the irritated pinna, and the mice became moribund. On necropsy, pyogranulomas were observed in the kidneys and *S. aureus* was isolated from skin lesions, kidneys, and heart blood. It was suggested that the ear wounds acted as a portal of entry for the resulting staphylococcal septicemia.

Mice are frequently used in studies to assess the dermal toxicity or irritability of a wide variety of compounds destined for human use. These compounds are applied to the skin of mice after the hair is clipped. Irritating compounds can produce a pruritic reaction; the mouse responds by frequent scratching with its feet. Self-trauma may exacerbate the pruritic response and produce lesions that augment or exceed the severity of those resulting from the irritating nature of the test compound. The lesions of self-trauma can be reduced or eliminated by removing the first phalanx of all toes of the rear feet, a minor surgical procedure in very young animals.

Dermatophyte or ectoparasite infestations should be considered in the differential diagnoses for muzzle or body alopecia. Although the skin and adnexa are usually irritated or ulcerated in these infectious diseases (Weisbroth *et al.*, 1976), an early infection may be mild enough to resemble alopecia from hair chewing (Galton, 1963). These agents can be identified by skin scrapings and pelt examination for mites and lice, fungal culture of hair, and histopathological examination of skin sections stained with hemotoxylin and eosin, periodic acid-Schiff, Grocott's silver methenamine, and Gridley's stain for fungi.

In studies evaluating the effect of hormones on the follicular activity of rat skin, Johnson (1958b) found that the steroids produced by the ovaries and testes prolonged the resting phase or telogen, and club hairs were retained. The author also

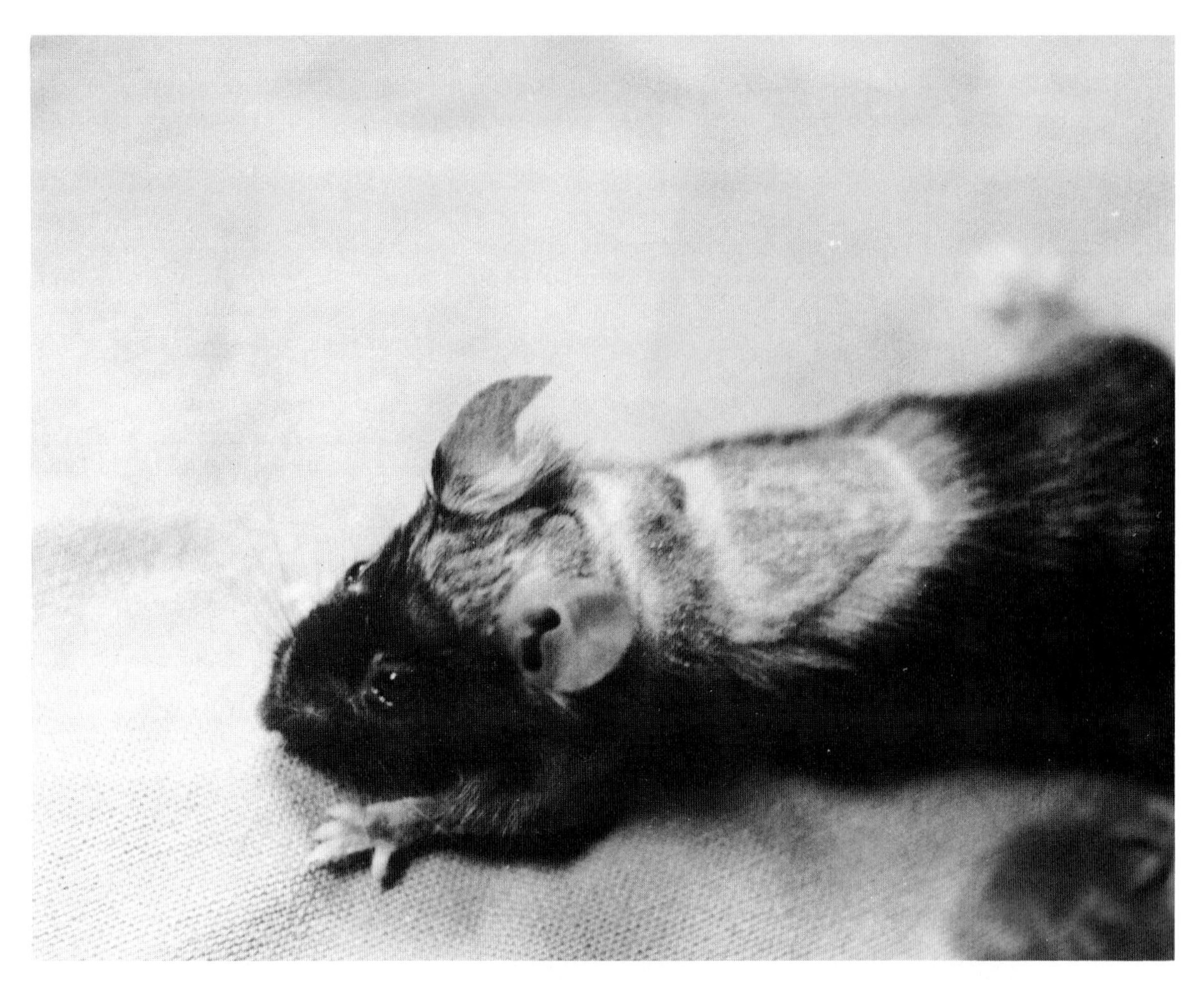

Fig. 5. Alopecia from barbaring of the neck and back of a C57BL/6 female mouse (top). Evidence of whisker chewing (bottom) in a fellow female cagemate. (Courtesy of Dr. James G. Fox.)

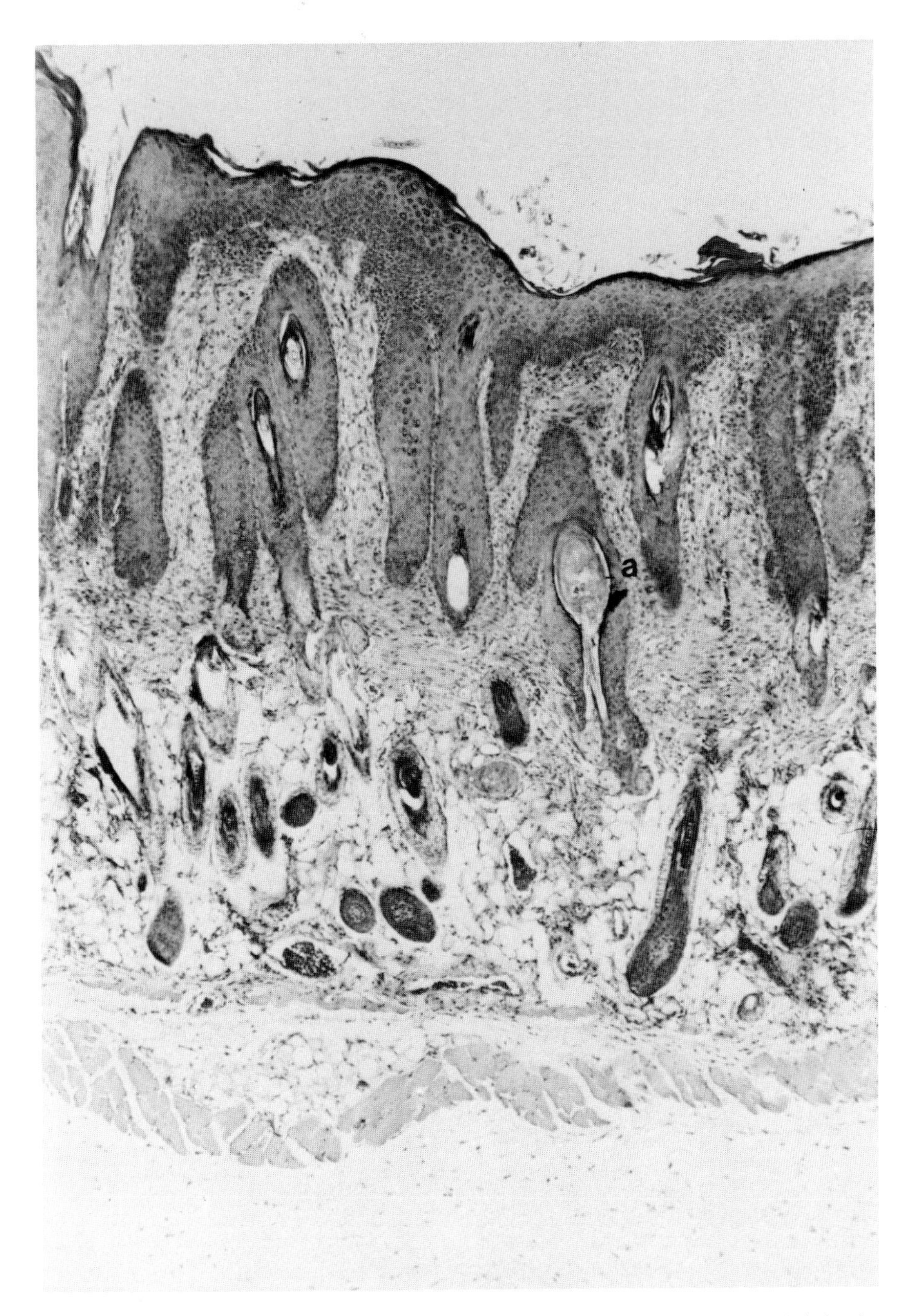

Fig. 6. Mild hyperkeratosis, acanthosis, and dermal fibroplasia associated with mechanical abrasion of the muzzle. An abnormally keratinized hair shaft is present (a).

suggested that anagen was slightly prolonged in female rats compared to male rats. The high levels of estrogen in pregnancy are thought to play a role in the postpartum shedding phenomenon. The estrogen levels drop at parturition, allowing the club hairs to be shed and releasing the anagenic follicles to progress to catagen. Davis *et al.* (1970) reported alopecia of the face and entire body in sterile female mice homozygous for the abnormal *W* gene series. These mice had nonfunctional ovaries (no follicles or ova) at 2 months of age and developed ovarian tubular tumors by 7 months of age. The tumor-bearing females began to lose hair at 2 months of age, and rarely did the hair grow back. The inhibition of hair growth was speculated to be due to compensatory steroid production by the adrenal gland and/or excessive gonadotropin release by the pituitary gland. The adrenocortical hormones act similarly to the gonadal hormones in prolonging the follicular resting phase. The thyroid hormones, however, shorten the resting phase of the hair cycle. Hypophysectomy also results in shortened resting phases, indicating that the intact gland, by virtue of its hormonal feedback relationship with the gonads and adrenal glands, can prolong the resting phase of the hair cycle (Ebling and Johnson, 1964). Theoretically, tumors of the gonads, pituitary, thymoid, and adrenal glands, whether actively secreting or not, can influence the haired state of the mouse.

ACKNOWLEDGMENTS

The authors gratefully acknowledge the clerical assistance of Ms. Sylvia Bradfield, the editorial guidance and content material provided by Dr. J. E. Harkness, and translations provided by M. Bukowski.

This work was supported in part by NIH grants RR00471 and RR07004.

REFERENCES

Ajello, L., Georg, L. K., Kaplan, W., and Kaufman, L. (1963). "Laboratory Manual for Medical Mycology." U.S. Dep. Health, Educ., Welfare, Communicable Dis. Cent., Atlanta, Georgia.

Ajello, L., Bostick, L., and Cheng, S. L. (1968). Relationship of *Trichophyton quinckeanum* to *Trichophyton mentagrophytes*. *Mycologia* **60,** 1185-1189.

Antopol, W. (1950). Anatomic changes in mice treated with excessive doses of cortisone. *Proc. Soc. Exp. Biol. Med.* **73,** 262-265.

Antopol, W., Quittner, H., and Saphra, I. (1953). "Spontaneous" infections after the administration of cortisone and ACTH. *Am. J. Pathol.* **29,** 599-600.

Ash, G. W. (1971). An epidemic of chronic skin ulceration in rats. *Lab. Anim. Care* **5,** 115-122.

Bazin, E. (1862). "Leçons Théoriques et Cliniques sur les Affections Cutanées Parasitaires," 2nd ed. Delahaye, Paris.

Bergey, D. (1923). *Corynebacteracae. In* "Bergey's Manual of Determinative Bacteriology" (Committee of the Society of American Bacteriologists, eds.), 1st ed., pp. 380-382. Williams & Wilkins, Baltimore, Maryland.

Bergey, D. (1925). *Corynebacteracae. In* "Bergey's Manual of Determinative Bacteriology" (Committee of the Society of American Bacteriologists, eds.), 2nd ed., pp. 395-405. Williams & Wilkins, Baltimore, Maryland.

Bicks, V. A. (1957). Infection of laboratory mice with *Corynebacterium murium*. *Aust. J. Sci.* **20,** 20-22.

Blackmore, D. K., and Casillo, S. (1972). Experimental investigation of uterine infections of mice due to *Pasteurella pneumotropica*. *J. Comp. Pathol.* **82,** 471-475.

Blackmore, D. K., and Francis, R. A. (1970). The apparent transmission of staphylococci of human origin to laboratory animals. *J. Comp. Pathol.* **80,** 645-651.

Blank, F. (1957). Favus of mice. *Can. J. Microbiol.* **3,** 885.

Bloch, B. (1911). Das *Acharian violaceum,* ein bisher unbekannter. *Favuspilz. Dermatol.* **2,** 815-833.

Bongert, R. (1901). *Corynethrix pseudotuberculosis murium* ein neuer pathogener Bacillus für Mäuse. *Z. Hyg. Infektionskr.* **37,** 449-475.

Booth, B. (1952). Mouse ringworm. *Arch. Dermatol.* **66,** 65-69.

Borum, K. (1954). Hair pattern and hair succession in the albino mouse. *Acta Pathol. Microbiol. Scand.* **34,** 521-541.

Brennan, P. C., Fritz, T. C., and Flynn, R. J. (1965). *Pasteurella pneumotropica:* cultural and biochemical characteristics, and its association with disease in laboratory animals. *Lab. Anim. Care* **15,** 307-312.

Brennan, P. C., Fritz, T. E., and Flynn, R. J. (1969). Role of *Pasteurella pneumotropica* in murine pneumonia. *J. Bacteriol.* **97,** 337-349.

Buchanan, R. E. (1919). Favus herpeticus or mouse favus. *J. Am. Med. Assoc.* **72,** 97-100.

Caren, L. D., and Rosenberg, L. T. (1966). The role of complement in resistance to endogenous and exogenous infections with a common mouse pathogen, *Corynebacterium kutscheri*. *J. Exp. Med.* **124,** 689-699.

Casillo, S., and Blackmore, D. K. (1972). Uterine infections caused by bacteria and mycoplasma in mice and rats. *J. Comp. Pathol.* **82,** 477-479.

Catanei, A. (1936). Premieres recherches sur les teigres du chien en Algerie. *Arch. Institut. Pasteur d'Algérie* **14,** 104-108.

Catanei, A. (1942). Les teignes de la Souris blanche à alger. *Arch. Institut. Pasteur d'Algérie.* **20,** 305-308.

Cetin, E. T., Tansinoglu, M., and Volkan, S. (1965). Epizootic of *T. mentagrophytes* (interdigitale) in white mice. *Pathol. Microbiol.* **28,** 839-846.

Chase, H. B., and Eaton, G. J. (1959). Growth of hair follicles in waves. *Ann. N.Y. Acad. Sci.* **83,** 365-368.

Chase, H. B., and Montagna, W. (1951). Relation to hair proliferation to damage induced in the mouse skin. *Proc. Soc. Exp. Biol. Med.* **76,** 35-37.

Chiou, W. L., and Riegelman, S. (1969). Disposition kinetics of griseofulvin in dogs. *J. Pharm. Sci.* **58,** 1500-1504.

Clarke, M. C., Taylor, R. J., Hall, G. A., and Jones, P. W. (1978). The occurrence in mice of facial and mandibular abscesses associated with *Staphylococcus aureus*. *Lab. Anim.* **12,** 121-123.

Connole, M. D. (1965). Keratinophilic fungi on cats and dogs. *Sabouraudia* **4,** 45-48.

Cooper, J. E. (1977). Furunculosis in the mouse (correspondence). *Vet. Rec.* **101,** 433.

Cover, C. E., Krauss, W. C., and Bettinger, G. E. (1979). Ear tag induced dermatitis in mice. *Annu. Meet. Am. Assoc. Lab. Anim. Sci.* Abstr. No. 65.

Davies, R. R., and Shewell, J. (1964). Control of mouse ringworm. *Nature (London)* **202,** 406-407.

Davies, R. R., and Shewell, J. (1965). Ringworm carriage and its control in mice. *J. Hyg.* **63,** 507-515.

Davis, R. H., McGowan, L., and Ryan, J. P. (1970). Hair loss in ovarian tumorgenic mice. *Proc. Soc. Exp. Biol. Med.* **134,** 434-436.

Deibel, R. H., and Seeley, H. W., Jr. (1974). Streptococcus. *In* "Bergey's Manual of Determinative Bacteriology" (R. E. Buchanan and N. E. Gibbons, eds.), 8th ed., pp. 490-510. Williams & Wilkens, Baltimore, Maryland.

Dolan, M. M., Kligman, A. M., Kobylinski, P. G., and Motsavage, M. A. (1958). Ringworm epizootics in laboratory mice and rats: experimental and accidental transmission of infection. *J. Invest. Dermatol.* **30,** 23-25.

DuBois (1929). Trichophytie spontanée de la souris grise avec quelques considérations sur l'évolution naturelle de l'infection. *Ann. Dermatol. Syphiligr.* **10,** 1359-1363.

Eaton, G. J. (1977). Hair growth waves and cycles in nude mice. *Proc. Int. Workshop Nude Mice* **2,** 89-93.

Ebling, F. J., and Johnson, E. (1964). The action of hormones on spontaneous hair growth cycles in the rat. *J. Endocrinol.* **29,** 193-201.

Eichenwald, H. F. (1965). Bacterial interference and staphyloccal colonization in infants and adults. *Ann. N.Y. Acad. Sci.* **128,** 365–380.

Eickhoff, T. C. (1972). Group B streptococci in human infections. *In* "Streptococci and Streptococcal Diseases" (L. W. Wannamaker and J. M. Matsen, eds.), pp. 533–541. Academic Press, New York.

Emmons, C. W., Chapman, H., Binford, J. P., and Kwon-Chung, K. J. (1977). "Medical Mycology," 3rd ed. Lea & Febiger, Philadelphia, Pennsylvania.

English, M. (1971). Ringworm in groups of wild animals. *J. Zool.* **165,** 535–544.

Fauve, R. M., Pierce-Chase, C. H., and Dubos, R. (1964). Corynebacterial pseudotuberculosis in mice II. Activation of natural and experimental latent infections. *J. Exp. Med.* **120,** 283–304.

Festing, M. R. W., and Blackmore, D. K. (1971). Lifespan of specified pathogen free (MRC Catagory 4) mice and rats. *Lab. Anim. Sci.* **5,** 179–192.

Feuerman, E., Alteras, I., Honig, E., and Lehrer, N. (1975). Saprophytic occurrence of *Trichophyton mentagrophytes* and *Microsporum gypseum* in the coats of healthy laboratory animals. (Preliminary Report.) *Mycopathologia* **55,** 13–16.

Fischerm, S. A. M. (1972). Virulence of *T. mentagrophytes* infecting steroid-treated guinea pigs. *Mycopathol. Mycol. Appl.* **47,** 121–127.

Flynn, R. J., Brennan, P. C., and Fritz, T. E. (1965). Pathogen status of commercially produced laboratory mice. *Lab. Anim. Care* **15,** 440–447.

Flynn, R. J., Simkins, R. C., and Brennan, P. C. (1968). Uterine infection in mice. *Z. Versuchstierkd.* **10,** 131–136.

Fox, J. G., Niemi, S. M., Murphy, J. C., and Quimby, F. W. (1977). Ulcerative dermatitis in the rat. *Lab. Anim. Sci.* **27,** 671–678.

Fuentes, C. A. (1959). *Trichophyton mentagrophytes* and *Microsporum gypseum* in the coats of young healthy guinea pigs and cats. *Congr. Ibero Lat. Am. Dermatol., Mem., 3rd, Mexico City, 1956* pp. 142–145.

Fujiwara, K. (1971). Problems in checking inapparent infections in laboratory mouse colonies. An attempt at serological checking by anamnestic response. *In* "Defining of the Laboratory Animals," IV Symposium, International Committee on Laboratory Animals, pp. 77–92. Natl. Acad. Sci., Washington, D.C.

Galton, M. (1963). Myobic mange in the mouse leading to skin ulceration and amyloidosis. *Am. J. Pathol.* **43,** 855–866.

Giddens, W. E., Jr., Keahey, K. K., Carter, G. R., and Whitehair, C. K. (1968). Pneumonia in rats due to infection with *Corynebacterium kutscheri. Pathol. Vet.* **5,** 227–237.

Gledhill, A. W., and Rees, R. J. W. (1951). A spontaneous enterococcal disease of mice and its enhancement by cortisone. *Br. J. Exp. Pathol.* **33,** 183–189.

Gluge, G., and d'Udekem, J. (1857). De quelques parasites végétaux developpés sur des animaux vivants. *Bull. Acad. R. Sci., Lett., Beaux-Arts Belg. [Ser. 2]* **3,** 338–352.

Gray, D. F., and Campbell, A. L. (1953). The use of chloramphenicol and foster mothers in the control of natural pasteurellosis in experimental mice. *Aust. J. Exp. Biol.* **31,** 161–166.

Grimmer, H. (1974). Mykotisches granulom durch *Trichophyton terrestre. Mykosen* **17,** 333–338.

Gundel, M., Gyorgy, P., and Pagel, W. (1932). Expeirmentelle boebachtungen zu der frage der resistenzverminderung und infektion. *Z. Hyg. Infektionskr.* **113,** 629–644.

Gupta, P. K., Singh, R. P., and Singh, I. P. (1970). A study of dermatomycoses (ringworm) in domestic animals and fowls. *Indian J. Anim. Health* **9,** 85–89.

Harkness, J. E., and Wagner, J. E. (1975). Self-mutilation in mice associated with otitis media. *Lab. Anim. Sci.* **25,** 315–318.

Hauschka, T. S. (1952). Whisker-eating mice. *J. Hered.* **43,** 77–80.

Heidenbluth, V. I., Hübner, U., and Meyer, E. (1978). Verbreitung und pathogenität von *Trichophyton terrestre. Dermatol. Monatsschr.* **164,** 432–439.

Hieronymi, E. (1958). Die Haut der Maus. *In* "Pathologie der Laboratoriumstiere" (P. Cohrs, R. Jaffe, and H. Meesen, eds.), pp. 596–597. Springer-Verlag, Berlin and New York.

Hoag, W. G., Wetmore, P. W., Rogers, J., and Meier, H. (1962). A study of Pasteurella infection in a mouse colony. *J. Infect. Dis.* **111,** 135–140.

Hong, C. C., and Ediger, R. D. (1978a). Preputial gland abscess in mice. *Lab. Anim. Sci.* **28,** 153–156.

Hong, C. C., and Ediger, R. D. (1978b). Chronic necrotizing mastitis in rats caused by *Pasteurella pneumotropica. Lab. Anim. Sci.* **28,** 317–320.

Hook, E. W., Wagner, R. R., and Lancefield, R. C. (1960). An epizootic in Swiss mice caused by a group A streotococcus, newly designated type 50. *Am. J. Hyg.* **72,** 111–119.

Hussein, M. A. F. (1971). The overall pattern of hair follicle arrangement in the rat and mouse. *J. Anat.* **109,** 307–316.

Jawetz, E. (1948). A latent pneumotropic *Pasteurella* of laboratory animals. *Proc. Soc. Exp. Biol. Med.* **68,** 46–48.

Jawetz, E., and Baker, W. H. (1950). A pneumotropic *Pasteurella* of laboratory animals. II. Pathological and immunological studies with the organism. *J. Infect. Dis.* **86,** 184–196.

Johnson, E. (1958a). Quantitative studies of hair growth in the albino rat. II. The effect of sex hormones. *J. Endocrinol.* **16,** 351–359.

Johnson, E. (1958b). Quantitative studies of hair growth in the albino rat. III. The role of the adrenal glands. *J. Endocrinol.* **16,** 360–368.

Kligman, A. M. (1956). Pathophysiology of ringworm infections in animals with skin cycle. *J. Invest. Dermatol.* **27,** 171–185.

Knudtson, W. U., and Robertstad, G. W. (1970). The isolation of keratinophilic fungi from soil and wild animals in south Dakota. *Mycopathol. Mycol. Appl.* **40,** 309–323.

Kostanecki, W., Bosse, K., and Krempl-Lamprecht, L. (1963). Der Verlauf der *Trychophyton-mentagrophytes*-Infektion bei Mausen unter der Einwirkung von 16-Epivestriol. *Arch. Klin. Exp. Dermatol.* **217,** 489–493.

Kutscher, D. (1894). Ein Beitrag zur Kenntniss der *Bacillären pseudotuberculose* der Nagethiere. *Z. Hyg. Infektionskr.* **18,** 327–342.

Lancefield, R. (1972). A streptococcal infection in animals—natural and experimental. *In* "Streptococci and Streptococcal Diseases" (L. W. Wannamaker and J. M. Matsen, eds.), pp. 316–320. Academic Press, New York.

LaTouche, C. J. (1957). The care and management of laboratory animals. *In* "The U.F.A.W. Handbook," (U.F.A.W., ed.), 2nd ed., p. 279. U.F.A.W., London.

Lawrence, J. J. (1957). Infection of laboratory mice with *Corynebacterium murium. Aust. J. Sci.* **20,** 147–154.

Lentsch, R. H., and Wagner, J. E. (1980). Isolation of *Actinobacillus lignieresii* and *Actinobacillus equuli* from laboratory rodents. *J. Clin. Microbiol.* **12,** 351–354.

Lenz, W., Thumert, A., and Brancks, H. (1978). Untersuchungen zuv Epidemiologie von Staphylokokken-infektionen in SPR-Versuchstierbeständen. *Zentralbl. Bakteriol., Parasitenkd., Infektionskr. Hyg., Abt. 1: Orig., Reihe A* **240,** 447–465.

Les, E. P. (1972). A disease related to cage population density: Tail lesions of C3H/HeJ mice. *Lab. Anim. Sci.* **22,** 56–60.

Litterst, C. I. (1974). Mechanically self induced muzzle alopecia in mice. *Lab. Anim. Sci.* **24,** 806–809.

Loewenthal, H. (1932). Beziehungenzwischen Wuchsform, Pathogenitat and Antigener Struktur Bei Streptokokken. *Z. Hyg. Infektionskr.* **114,** 379–396.

Long, S. Y. (1972). Hair-nibbling and whisker-trimming as indicators of social hierarchy in mice. *Anim. Behav.* **20,** 10–12.

McKay, S. E., and Arbuthnott, J. P. (1979). Age related susceptibility of mice to staphylococcal infection. *J. Med. Microbiol.* **12,** 99–106.

McKeever, S., Menges, R. W., Kaplan, W., and Ajello, L. (1958). Ringworm fungi of feral rodents in South-western Georgia. *Am. J. Vet. Res.* **19,** 969–972.

Mackenzie, D. W. R. (1961). *Trichophyton mentagrophytes* in mice: Infections of humans and incidence among laboratory animals. *Sabouraudia* **1,** 178–182.

Mann, P., and Bjotvedt, G. (1966). Altre informazioni riguardanti la pipizzazione batterio fagica dei ceppi di stafilococchi coagulasi positivi isolati da varie specie di animali domestici, selvatici e di laboratorio. *Arch. Vet. Ital.* **17,** 357–360.

Markham, N. P., and Markham, J. (1966). Strains of staphylococci in man and animals. *J. Comp. Pathol.* **76,** 49–56.

Menges, R. W., Georg, L. K., and Habermann, R. T. (1957). Therapeutic studies on ringworm-infected guinea pigs. *J. Invest. Dermatol.* **28,** 233–237.

Merkenschlager, M. (1975). Modellinfektion mit *Corynebacterium kutscheri* bei der Mäus. *Z. Versuchstierkd.* **17,** 129–141.

Montagna, W., and Parakkal, P. F. (1974). The pilary apparatus. *In* "The Structure and Function of Skin" (W. Montagna and P. F. Parakkal, eds.), 3rd ed., pp. 172–222. Academic Press, New York.

Moore, J., and Aldred, P. (1978). Treatment of *Pasteurella pneumotropica* abscesses in nude mice (nu/nu). *Lab. Anim.* **12,** 227–228.

Moore, T. D., Allen, A. M., and Ganaway, J. R. (1978). Latent *Pasteurella pneumotropica* infection of the gnotobiotic and barrier-held rats. *Lab. Anim. Sci.* **23,** 657–661.

Needham, J. R., and Cooper, J. E. (1975). An eye infection in laboratory mice associated with *Pasteurella pneumotropica. Lab. Anim.* **9,** 197–200.

Needham, J. R., and Cooper, J. E. (1976). Bulbourethral gland infections in mice associated with *Staphylococcus aureus. Lab. Anim.* **10,** 313–315.

Neilsen, S. W. (1978). Diseases of skin. *In* "Pathology of Laboratory Animals" (K. Benirschke, F. M. Garner, and T. C. Jones, eds.), pp. 610–614. Springer-Verlag, Berlin and New York.

Nelson, J. B. (1954). Association of group A streptococci with an outbreak of cervical lymphadenitis in mice. *Proc. Soc. Exp. Biol. Med.* **86,** 542–545.

Nutini, L. G., and Berberich, J. (1965). Effect of diet and strain difference on virulence of *Staphylococcus aureus* for mice. *Appl. Microbiol.* **13,** 614–617.

Nutini, L. G., Mukkada, A. T., and Cook, E. S. (1968). Susceptibility of Swiss albino mice to *Staphylococcus aureus:* Diet and sex factors. *Appl. Microbiol.* **16,** 815–816.

Owens, D. R., Wagner, J. E., and Addison, J. B. (1975). Antibiograms of pathogenic bacteria isolated from laboratory animals. *J. Am. Vet. Med. Assoc.* **167,** 605–609.

Parish, H. V., and Craddock, S. (1931). A ringworm epizootic in mice. *Br. J. Exp. Pathol.* **12,** 209–212.

Parratt, J. R., and West, G. B. (1957). 5-Hydroxytryptamine and tissue mast cells. *J. Physiol.* (*London*) **137,** 169–178.

Patiala, R. (1951). On fungus diseases in game. *Pap. Game Res.* (*Vanin, Saint Gaudars, Fr.*) **6,** 21–22.

Paveia, N. R. (1971). *Trichophyton mentagrophytes* in mice. *Trab. Soc. Part. Dermatol.* **29,** 145–148.

Pierce-Chase, C. H., Fauve, R. M., and Dubos, R. (1964). Corynebacterial pseudotuberculosis in mice. I. Comparative susceptibility of mouse strains to experimental infection with *Corynebacteria kutscheri. J. Exp. Med.* **120,** 267–281.

Pinto, J. (1974). 4. The dermis. *In* "The Structure and Function of Skin" (W. Montagna and P. F. Parakkal, eds.), 3rd ed., pp. 125–141. Academic Press, New York.

Povar, M. L. (1965). Ringworm (*Trichophyton mentagrophytes*). Infection in a colony of albino Norway rats. *Lab. Anim. Care* **15,** 264–265.

Rehbinder, C., and Tschappat, V. (1974). Pasteurella pneumotropica, isoliert von der konjunktivalscheimhaut gesunder laboratoriumsmause. *Z. Versuchstienkd.* **16,** 359–365.

Richou, L. (1950). Etude sur les teigres des carnivares. *Ecole Natl. Vet. Alfart, Arch. Inst. Pasteur Alger.* **14,** 104–108.

Rieth, V. H. (1968). Spontane und experimentelle pilzerkrankungen bei Mäusen. *Z. Versuchstierkd.* **10,** 75–81.

Riley, D. F. (1962). Histamine and heparin in mast-cells. Why both? *Lancet* **ii,** 40–41.

Rosenthal, S. A., and Wapnick, H. (1963). The value of MacKenzie's hair brush technique in the isolation of *Trichophyton mentagrophytes* from clinically normal guinea pigs. *J. Invest. Dermatol.* **41,** 5–6.

Schechmeister, I. L., and Adler, F. L. (1953). Activation of pseudotuberculosis in mice exposed to sublethal total body radiation. *J. Infect. Dis.* **92,** 228–239.

Scott, D. W., Kirk, R. W., and Bentinck-Smith, J. (1980). Dermatophytosis due to *Trichophyton terrestre* infection in a dog and cat. *J. Am. Anim. Hosp. Assoc.* **16,** 53–59.

Seastone, C. V. (1939). The virulence of group C hemolytic streptococci of animal origin. *J. Exp. Med.* **70,** 361–378.

Sebesteny, A. (1973). Abscesses of the bulbourethral glands of mice due to *Pasteurella pneumotropica. Lab. Anim.* **7,** 315–317.

Serrano, L. J. (1972). Dermatitis and death in mice accidentally exposed to quaternary ammonium disinfectant. *J. Am. Vet. Med. Assoc.* **161,** 652–655.

Shults, F. S., Estes, P. C., Franklin, J. A., and Richter, C. B. (1973). Staphylococcal botryomycosis in a specific-pathogen-free mouse colony. *Lab. Anim. Sci.* **23,** 36–42.

Simmons, M. L., and Brick, J. O. (1970). Animal health. *In* "The Laboratory Mouse," p. 91. Prentice-Hall, Englewood Cliffs, New Jersey.

Simpson, W., and Simmons, D. J. C. (1980). Two *Actinobacillus* species isolated from laboratory rodents. *Lab. Anim.* **14,** 15–16.

Slonitskaya, N. N. (1969). Teratogenic effect of Griseofulvin-Forte on the rat fetus. *Antibiotiki* (*Moscow*) **14,** 44–48.

Soerensen, B., Yarid, M. J. F., Neto, L. Z., and Machado, J. C. (1975). Pseudotuberculose em camundongos. Isolamento de *Corynebacterium kutscheri* de cavidade oral e da pele de animais doentes e aparentemente sãos. *Mem. Inst. Butantan, Sao Paulo* **39,** 233–238.

Stewart, D. D., Buck, G. E., McConnell, E. E., and Amster, R. L. (1975). An epizootic of necrotic dermatitis in laboratory mice caused by Lancefield group G streptococci. *Lab. Anim. Sci.* **25,** 296–302.

Symeonidis, A., and Emmons, C. W. (1955). Granulomatous growth induced in mice by *Absidia corymbifera. Arch. Pathol.* **60,** 251–258.

Taylor, W. W., Radcliffe, F., and Van Peenen, P. F. D. (1964). A survey of small Egyptian mammals for pathogenic fungi. *Sabouraudia* **3,** 140–142.

Thornburg, L. P., Stowe, H. D., and Pick, J. F. (1973). The pathogenesis of the alopecia due to hairchewing in mice. *Lab. Anim. Sci.* **23,** 843–850.

Utsumi, K., Matsui, Y., Ishikawa, T., Fukagawa, S., Tatsumi, H., and Fujiwara, K. (1969). Checking of Corynebacterial infection in rats by cortisone treatment. *Exp. Anim.* **18,** 59–67.

Van der Schaaf, A., Mullink, J. W. M. A., Nikkels, R. J., and Goudswaard, J. (1970). *Pasteurella pneumotropica* as a causal microorganism of multiple subcutaneous abscesses in a colony of Wistar rats. *Z. Versuchstierkd.* **12,** 356–362.

Wagner, J. E., Garrison, R. G., Johnson, D. R., and McGuire, T. J. (1969). Spontaneous conjunctivitis and dacryoadenitis of mice. *J. Am. Vet. Med. Assoc.* **155,** 1211–1217.

Wagner, J. E., Owens, D. R., LaRegina, M. C., and Vogler, G. A. (1977). Self trauma and *Staphylococcus aureus* in ulcerative dermatitis of rats. *J. Am. Vet. Med. Assoc.* **171,** 839–841.

Ward, G. E., Moffatt, R., and Olfert, E. (1978). Abortion in mice associated with *Pasteurella pneumotropica*. *J. Clin. Microbiol.* **8,** 177-180.

Weisbroth, S. H., and Scher, S. (1968). *Corynebacterium kutscheri* infection in the mouse. I. Report of an outbreak, bacteriology, and pathology of spontaneous infections. *Lab. Anim. Care* **18,** 451-458.

Weisbroth, S. H., Scher, S., and Boman, I. (1969). *Pasteurella pneuotropica* abscess syndrome in a mouse colony. *J. Am. Vet. Med. Assoc.* **155,** 1206-1210.

Weisbroth, S. H., Friedman, S., and Scher, S. (1976). The parasitic ecology of the rodent mite, *Myobia musculi*. III. Lesions in certain host strains. *Lab. Anim. Sci.* **26,** 725-735.

Weitzman, I., and Padhye, A. A. (1976). Is *Arthroderma simii* the perfect state of *Trichophyton quinckeanum? Sabouraudia* **14,** 65-74.

Wheater, D. F. W. (1967). The bacterial flora of an SPF colony of mice, rats and guinea pigs. *In* "Husbandry of Laboratory Animals" (M. L. Conalty, ed.), pp. 343-360. Academic Press, New York.

Wilson, G. S., and Miles, A. (1975). "Topey and Wilson's Principles of Bacteriology, Virology and Immunity," 6th ed., Chap. 14. Arnold, London.

Wilson, P. (1976). *Pasteurella pneumotropica* as the causal organism of abscesses in the masseter muscles of mice. *Lab. Anim.* **10,** 171-172.

Wolbach, S. B. (1951). Hair cycle of the mouse and its importance in the study of experimental carcinogens. *Ann. N.Y. Acad. Sci.* **53,** 517-536.

Wolfe, H. L. (1950). On some spontaneous infections observed in mice. I. *C. kutsheri* and *C. pseudotuberculosis*. *Antonie van Leeuwenhoek* **16,** 105-110.

Yokoiyama, S., Ando, T., Hayana, K., and Fujiwara, K. (1975). Latent infection with *Corynebacterium kutscheri* in mice after peroral inoculation and its provocation by cortisone. *Exp. Anim.* **24,** 103-110.

Yokoiyama, S., Mizuno, K., and Fujiwara, K. (1977). Antigenic heterogeneity of *Corynebacterium kutscheri* from mice and rats. *Exp. Anim.* **26,** 263-266.

Chapter 5

Bacterial, Mycoplasmal, and Mycotic Diseases of the Central Nervous System

Dennis F. Kohn

I.	Introduction	77
II.	*Mycoplasma neurolyticum*	78
	A. Introduction	78
	B. Incidence	78
	C. Clinical Signs	78
	D. Characteristics of Exotoxin	78
	E. Pathogenicity	78
	F. Significance	79
III.	*Mycoplasma pulmonis*	79
IV.	*Pseudomonas aeruginosa*	79
	A. Introduction	79
	B. Clinical Signs	79
	C. Pathogenicity	80
	D. Significance	80
	E. Prevention	80
V.	*Corynebacterium kutscheri*	80
VI.	*Pasteurella pneumotropica*	80
VII.	Mycoses	80
	References	80

I. INTRODUCTION

A striking characteristic about nonvirally induced infectious diseases of the central nervous system (CNS) is the rarity of their occurrence. The almost negligible number of reports is probably a function of the infrequency of clinically evident infections of the CNS combined with a lack of interest by individuals in defining cases that may appear singly and sporadically.

THE MOUSE IN BIOMEDICAL RESEARCH, VOL. II

ISBN 0-12-262502-1

II. *MYCOPLASMA NEUROLYTICUM*

A. Introduction

Mycoplasma neurolyticum is the etiological agent associated with the classic syndrome known as *rolling disease*. Sabin (1938a) and Findlay *et al.* (1938), working independently, both described the occurrence of rolling disease in 1938. In Sabin's initial report, he describes the presence of a filterable, transmissible agent that was present in *Toxoplasma*-infected mouse brains. Intracerebral inoculation of this infected tissue into mice resulted in signs not attributable to *Toxoplasma*. He found that 2–3 days after inoculation the mice began turning on the long axis of their bodies. Most of the affected mice would recover in a few days; however, some would either die or continue to have choreiform signs of neurological disease for months. He noticed that animals younger than 15 days or older than 2 months were refractory to the development of neurological signs subsequent to intracerebral inoculation of *M. neurolyticum*. Findlay *et al.* (1938) reported a similar neurological disease in mice subsequent to intracerebral inoculation of lymphocytic choriomeningitis (LCM) virus-infected brain homogenate. Sabin, in a second report (1938b), determined that sterile *M. neurolyticum* culture filtrates given intravenously to mice induced the rolling disease within hours after inoculation and that most died shortly thereafter. However, mice that recovered were immune to the effects of reinoculation.

Little appeared in the literature until 1964, when Tully and Ruchman (1964) obtained lyophilized cultures from stocks that Sabin had maintained since 1943. Most of the work done since 1964 has been performed in the laboratories of Tully or Thomas. Through their efforts and those of their colleagues, much has been learned to augment and reinforce the initial work of Sabin and Findlay.

B. Incidence

To the author's knowledge, relatively little is known on the actual incidence of the organism; however, limited surveys done on conventionally raised mice indicate that its presence in mice is not uncommon. Tully and Rask-Nielsen (1967) surveyed 42 mice from eight strains and found *M. neurolyticum* in animals from four of them. None were isolated from 12 germ-free mice surveyed. The same workers surveyed mouse recipients of several transmissible leukomogenic agents and plasma cell tumors for the presence of *M. neurolyticum*. The organism was recovered from tissues, including the brains, of some animals. Tamura (1976) surveyed 73 mice originating from five commercial rodent sources in Japan and was able to isolate *M. neurolyticum* from the brains of 11 animals.

C. Clinical Signs

Neurological signs occur 30–60 min after intravenous inoculation of *M. neurolyticum* exotoxin. The earliest signs observed by Thomas *et al.* (1966) are spasmodic hyperextension of the head and raising of one foreleg, followed in a few minutes by intermittent rolling on the long axis of the body. The rolling becomes more constant, with occasional periods in which the mice leap or move rapidly. After 1–2 hr of rolling, the animals become comatose, and nearly all die within 4 hr. Early work indicated that the majority of mice under 2 weeks and over 2 months of age were refractory to clinical manifestations of the toxin. However, later studies by Tully (1980, personal communication) indicate that mice of most ages are susceptible to virulent *M. neurolyticum*.

Intravenous, intraperitoneal, or intracerebral inoculation of viable organisms induced clinical signs identical to those associated with toxin administration. Thomas and Bilensky (1966) indicated that approximately 10^9 colony-forming units (cfu) are necessary to induce clinical signs.

D. Characteristics of Exotoxin

The exotoxin is a thermolabile protein with a molecular weight in excess of 200,000. Its neurotoxicity is expressed if it is given intravenously to mice but not if it is administered intracerebrally or intraperitoneally. Thomas *et al.* (1966) suggest that the exotoxin is fixed to receptors located in the podocytes of glial cells at their sites of attachment in the capillaries. Accordingly, it is probable that this site is accessible only via the blood, and herein lies the reason inoculation of the toxin intracerebrally does not result in disease.

Tully (1964) showed that the neurotoxicity of broth cultures is lost at 48 hr of aerobic incubation at 37°C even though the organisms within the broth remain viable long after the toxic properties of the broth are lost.

E. Pathogenicity

No remarkable histological changes are evident in those animals which succumb in less than 8 hr after inoculation of exotoxin. In animals that survive for 8 hr or more, one finds focal lesions of spongiform degeneration, with small vesicles associated with astrocytes. These lesions are most prominent in deep layers of the frontoparietal cortex and underlying white matter, along with the molecular layer of the cerebellum (Thomas, 1966). Electron microscopy confirms that the principal cellular derangement is the greatly distended astrocyte caused by an accumulation of intracellular fluid (Aleu and Thomas, 1966).

Thomas *et al.* (1966) suggest that this disruption of normal membrane regulation of fluid transport, along with the concomitant compression of neurons by the swollen astrocytes, is responsible for the neurological disease.

F. Significance

Rolling disease due to *M. neurolyticum* is considered by many to be a naturally occurring disease, whereas in fact all descriptions of it are associated with experimental inoculation of the organism or exotoxin. Experimental work indicates that 10^9 cfu are necessary to induce clinical signs; however, there is no evidence to suggest that, under natural conditions, the organism replicates in the brain to that concentration. Accordingly, rolling disease is an extremely interesting expression of an infection induced experimentally, but it has not been demonstrated to be a naturally occurring disease.

However, latent *M. neurolyticum* infections are obviously of much importance due to their impact on various investigations, such as those in which tumors or brain tissues/homogenates become contaminated. Limited surveys indicate that *M. neurolyticum* infections are not uncommon. Accordingly, investigators using mice for certain types of experimental manipulation should be cognizant of this fact. The organism is not a particularly fastidious mycoplasma, and it can be cultivated in most of the commonly used formulations. More than one species of mycoplasma may be isolated from an organ or tissue. Since most mycoplasmas vary little in their colonial morphology, it is necessary to identify isolates by some means. The epi-immunofluorescence procedure has proved to be a valuable tool in species identification and in detecting cultures containing more than one species (Barile and Delgiudice, 1972).

III. *MYCOPLASMA PULMONIS*

Mycoplasma neurolyticum has historically been the species associated with infections of the CNS; however, two reports indicate that *M. pulmonis* is probably present as often in the mouse brain as is the former organism. Since *M. pulmonis* is principally a pathogen of the respiratory and genital tracts in rodents, few studies have been made on its prevalence in the CNS.

Lemcke (1961) surveyed 40 individuals from one stock of mice and isolated *M. pulmonis* from the brains of 12 animals. More recently, the report by Tamura (1976) compared the recovery of *M. pulmonis* and *M. neurolyticum* in 73 mice from five sources. He isolated *M. pulmonis* from the brains of 14 and *M. neurolyticum* from 11. In both reports, infection with *M. pulmonis* was latent.

Accordingly, one should assume that a significant proportion of mice harboring the organism within the respiratory tract will also carry the organism latently within the brain. There is no evidence to indicate that *M. pulmonis* is responsible for an overt neurological disease in mice, even though the organism is neurotropic. However, as in any latent infection, the investigator should be aware of its possible presence and of its potential in modifying certain data.

Mycoplasma pulmonis is often the pathogen associated with otitis media in mice (Harkness and Wagner, 1975). In the rat, extension from the inner ear to the meninges occurs, and at this site the organism induces circling and torticollis. Similar involvement of the inner ear and meninges in mice probably occurs, but reports of circling due to *M. pulmonis* were not found during preparation of this chapter.

IV. *PSEUDOMONAS AERUGINOSA*

A. Introduction

Pseudomonas aeruginosa is a ubiquitous bacterium in animals and is commonly found in the oropharynx and intestinal tract of the mouse. Infections are almost always latent. However, in the immunologically compromised host, a septicemic state often results in the rapid demise of the host. Under experimental conditions, cold stress has been shown to induce septicemic deaths in *Pseudomonas*-infected mice (Halkett *et al.*, 1968). Infections associated with neurological signs are uncommon, but several reports of naturally occurring infections have been made.

B. Clinical Signs

Ediger *et al.* (1971) observed torticollis and circling in outbred Swiss-Webster mice which became progressively more severe until rolling on the longitudinal axis occurred. This author has seen clinical signs in C3H mice that are similar to those reported by Ediger *et al.* (1971). *Pseudomonas aeruginosa* was consistently isolated from the middle and inner ears and/or brains from the C3H mice displaying the rolling signs, but mycoplasmas were never recovered. In most instances, affected mice had been recently received from a commercial source. It is possible that the stress of shipping induced a shift in the host–parasite balance from a quiescent to a pathogenic mode. The clinical manifestations of the infection occurred in 10% of animals from two shipments, whereas in other cases, the occurrences were sporadic. Since other stocks/strains of mice often carry the organism, apparently without clinical signs postshipment, it is possible that C3H mice lack appropriate immunological or innate defense mechanisms against *Pseudomonas* infections of the inner ear.

C. Pathogenicity

Olson and Ediger (1972) reported on the pathology associated with spontaneous cases of *P. aeruginosa* in which neurological signs occurred. The lesions within the inner ear were limited to the cochlea and vestibular apparatus. Most often there was a chronic proliferative inflammation within the ear, and in some instances, dissolution of the bone surrounding the ear was evident. In some cases, there were abscesses within the cerebellum or cerebrum. These latter lesions were thought to be due to extension of the infection from the inner ear.

In the outbreak involving C3H mice, the lesions were similar to the above. However, the inflammatory response in the inner ear and meninges was primarily neutrophilic. Abscesses within the brain were not observed, but suppurative meningitis was present. Extension of the ear infection along the acoustic nerve to the CNS was demonstrated.

D. Significance

Although neurological manifestations due to this ubiquitous organism are apparently uncommon, *P. aeruginosa* should be considered one of the principal bacterial causes of neurological signs in the mouse.

E. Prevention

Pseudomonas can be suppressed in a closed colony by chlorination and acidification of drinking water and by subsequent removal of those mice that continue to shed the organism in the feces (Hoag *et al.,* 1965).

V. *CORYNEBACTERIUM KUTSCHERI*

Weisbroth and Scher (1968) reported an epizootic in which 55 of 120 BRVR/R (Rockefeller University) mice became overtly ill or died due to the effects of a *Corynebacterium kutscheri* bacteremia. Gross lesions were primarily restricted to the kidney and liver; however, microscopic lesions were observed in the brain. Septic emboli within the brain induced thromboses, with subsequent development of zones of necrosis and inflammation. *Corynebacterium kutscheri,* like *P. aeruginosa,* is a bacterium which is usually carried latently and induces a clinical disease only when the host is stressed.

VI. *PASTEURELLA PNEUMOTROPICA*

Pasteurella pneumotropica is usually harbored latently within the respiratory tract of the mouse, but it has been associated with pneumonia and otitis media. Hoag *et al.* (1962), working with a strain that produced an overt pulmonary disease, found that the organism was frequently carried within the brains of several inbred strains of mice. He considered the organism to be neurotropic but not responsible for any overtly evident neuropathology. Due to the lack of other reports, the prevalence and pathogenicity of *P. pneumotropica* in the mouse remain undefined.

VII. MYCOSES

Only two cases of spontaneous mycotic infections of the CNS were found in the literature and by personal correspondence. Weitzman *et al.* (1973) reported the isolation of *Cryptococcus neoformans* from the brain of a wild field mouse that had neurological signs prior to its death. F. M. Garner (1980 personal communication) found *C. neoformans*-like organisms in histological sections from the brain of a hybrid $B6C3F_1$ mouse. The animal had extensive cystic lesions in the leptomeninges and brain. The source of infection was unknown, but wood shavings used for bedding were suggested as a possible source. Rapp and McGrath (1975) have reported an outbreak of mycotic encephalitis in weanling rats. The mold was classified histologically as being a *Phycomycetes*. The authors suggest that contaminated wood chip bedding might have been the source of the opportunistic fungi. No other organs were involved, and this fact led the authors to hypothesize that the organism invaded the brain via the upper respiratory tract. Under appropriate conditions, it would appear that the mouse brain could be similarly infected.

REFERENCES

Aleu, F., and Thomas, L. (1966). Studies of PPLO infection. III. Electron microscopic study of brain lesions caused by *Mycoplasma neurolyticum* toxin. *J. Exp. Med.* **124,** 1087–1088.

Barile, M. F., and Delgiudice, R. A. (1972). Isolations of mycoplasmas and their rapid identification by plate epi-immunofluorescence. *In* "Pathogenic Mycoplasmas," pp. 165–185. Assoc. Sci. Publ., Amsterdam.

Ediger, R. D., Rabstein, M. M., and Olson, L. D. (1971). Circling in mice caused by *Pseudomonas aeruginosa*. *Lab. Anim. Sci.* **21,** 845–858.

Findlay, G. E., Klieneberger, E., MacCallum, F. O., and MacKenzie, R. D. (1938). Rolling Disease: New syndrome in mice associated with a pleuropneumonia-like organism. *Lancet* No. 235, 1511–1513.

Halkett, J. A. E., Davis, A. J., and Natsios, G. A. (1968). The effect of cold stress and *Pseudomonas aeruginosa* gavage on the survival of three-week-old Swiss mice. *Lab. Anim. Care* **18,** 94–96.

Harkness, J. E., and Wagner, J. E. (1975). Self-mutiliation in mice associated with otitis media. *Lab. Anim. Sci.* **25,** 315–318.

Hoag, W. G., Wetmore, P. W., Robers, J., and Meier, H. (1962). A study of latent *Pasteurella* infection in a mouse colony. *J. Infect. Dis.* **111,** 135–140.

Hoag, W. G., Strout, J., and Meier, H. (1965). Epidemiological aspects of the control of *Pseudomonas* infection in mouse colonies. *Lab. Anim. Care* **15,** 217–225.

Lemcke, R. M. (1961). Association of PPLO infection and antibody response in rats and mice. *J. Hyg.* **59,** 401–412.

Olson, L. D., and Ediger, R. D. (1972). Histopathologic study of the heads in circling mice infected with *Pseudomonas aeruginosa*. *Lab. Anim. Sci.* **22,** 522–527.

Rapp, J. P., and McGrath, J. T. (1975). Mycotic encephalitis in weanling rats. *Lab. Anim. Sci.* **25,** 477–480.

Sabin, A. B. (1938a). Isolation of a filtrable, transmissible agent with neurolytic properties from *Toxoplasma*-infected tissues. *Science* **88,** 189–191.

Sabin, A. B. (1938b). Identification of the filterable transmissible agent isolated from *Toxoplasma*-infected tissues as a new pleuropneumonia-like microbe. *Science* **88,** 575–576.

Tamura, H. (1976). Persistent infection of *Mycoplasma* in mouse brain and respiratory organs. *J. Jpn. Assoc. Infect. Dis.* **50,** 298–302.

Thomas, L., and Bilensky, M. W. (1966). Studies of PPLO infection. IV. The neurotoxicity of intact mycoplasmas and their production of toxin *in vivo* and *in vitro*. *J. Exp. Med.* **124,** 1089–1098.

Thomas, L., Aleu, F., Bilensky, M. W., Davidson, M., and Gesner, B. (1966). Studies of PPLO infection. II. The neurotoxin of *Mycoplasma neurolyticum*. *J. Exp. Med.* **124,** 1067–1082.

Tully, J. G. (1964). Production and biological characteristics of an extracellular neurotoxin from *Mycoplasma neurolyticum*. *J. Bacteriol.* **88,** 381–388.

Tully, J. G., and Rask-Nielsen, R. (1967). Mycoplasma in leukemic and non-leukemic mice. *Ann. N.Y. Acad. Sci.* **143,** 345–352.

Tully, J. G., and Ruchman, I. (1964). Recovery, identification and neurotoxicity of Sabin's type A and C mouse mycoplasma (PPLO) from lyophilized cultures. *Proc. Soc. Exp. Biol. Med.* **115,** 554–558.

Weisbroth, S. H., and Scher, S. (1968). *Corynebacterium kutscheri* infection in the mouse. I. Report on an outbreak, bacteriology and pathology of spontaneous infections. *Lab. Anim. Care* **18,** 451–458.

Weitzman, I., Bonaparte, P., Guevin, V., and Crist, M. (1973). Cryptococcosis in a field mouse. *Sabourandia* **11,** 77–79.

Chapter 6

Bacterial, Mycoplasmal, and Mycotic Diseases of the Lymphoreticular, Musculoskeletal, Cardiovascular, and Endocrine Systems

John E. Harkness and Frederick G. Ferguson

I. Introduction 83
II. Bacterial Infections 84
A. *Streptobacillus moniliformis* 84
B. *Corynebacterium kutscheri* 86
C. *Salmonella* spp. 86
D. *Pseudomonas aeruginosa* 87
E. *Bacillus piliformis* 88
F. *Mycobacterium lepraemurium* 88
G. *Streptococcus* spp. 89
H. *Staphylococcus aureus* 90
I. *Pasteurella pneumotropica* 91
J. *Klebsiella pneumoniae* 91
K. Miscellaneous Infections 92
III. Mycoplasmal Infections 93
IV. Mycotic Infections 93
References 94

I. INTRODUCTION

If the number of literature reports is representative of the occurrence of spontaneous disease in mice, then naturally occurring bacterial, mycoplasmal, and mycotic disease of *Mus musculus* have decreased considerably in recent years. Undoubtedly, the availability of cesarean-derived, barrier-sustained mice has been a major factor contributing to this fortunate decrease. Today mice are raised in pathogen-free environments, shipped in filtered-air containers, evaluated

ISBN 0-12-262502-1

upon receipt, and maintained under constant vigilance in appropriate facilities. As a result, unwanted infectious diseases can be eliminated for the duration of most experimental procedures.

The frequency of reported disease varies among organ systems. Bacterial, mycoplasmal, and mycotic diseases of the musculoskeletal, cardiovascular, and endocrine systems usually occur as extensions from primary foci in other organs, whereas the lymphoreticular system, because of its role as a biological filter, is frequently a primary site for disease processes. As a result, fewer of the conditions described in this chapter are associated with the heart, muscles, skeleton, and endocrine system than with the lymphoreticular organs. The paucity of reports of endocrine gland lesions with spontaneous murine disease may, however, be due to the lack of gross and histopathologic attention to these small organs.

Spontaneous bacterial, mycoplasmal, and mycotic diseases that affect the lymphoreticular, musculoskeletal, cardiovascular, and endocrine systems are discussed in this chapter under their respective etiologic agents. Detailed discussion of agents other than *Streptobacillus moniliformis* and *Mycobacterium lepraemurium* are found in other chapters in this volume. Information obtained from experimental studies is included where it contributes to knowledge of a disease process.

II. BACTERIAL INFECTIONS

A. *Streptobacillus moniliformis*

Spontaneous disease caused by *Streptobacillus moniliformis* in mice is uncommon (Levaditi *et al.*, 1932; Mackie *et al.*, 1933; Strangeways, 1933; Williams, 1941; Freundt, 1956a; Sawicki *et al.*, 1962; Yamamoto and Clark, 1966), but when it does occur, it may be either an acute, highly fatal, generalized infection or a chronic polyarthritis with a predilection for the rear limbs, spine, and tail. Mouse strain, immune status, and pregnancy may predispose to the disease (Van Rooyen, 1936; Freundt, 1959; Savage, 1972).

1. Agent

Streptobacillus moniliformis (syn: *Actinobacillus moniliformis*), an organism of uncertain taxonomic affiliation, is a gram-negative bacillus whose morphology varies with age and cultural circumstances. Organisms obtained on a blood or tissue smear from an affected host are regular, nonpleomorphic rods, but the same bacteria cultured *in vitro* vary from a nonbranching, interlacing mycelium of rods, chains, and filaments up to 200 μm long in young cultures to irregular, beaded, and clubbed bacillary and coccobacillary elements in older colonies or under unfavorable growth conditions (Van Rooyen, 1936; Freundt, 1959; Wittler and Cary, 1974; Wilson and Miles, 1975). Not only is the morphology of the bacillary form highly variable, but a reversible conversion occurs into an L-phase variant both *in vivo* and *in vitro* (Klieneberger, 1935, 1936; Dienes, 1939; Freundt, 1956a,b). This variant caused taxonomic and etiologic confusion when it was thought to be a pleuropneumonia-like symbiont rather than a variant. The variant's diminished cell wall affects its morphology, cultural characteristics, susceptibility to antibiotics, staining characteristics, and virulence. The L phase is considered nonpathogenic in mice, but it can revert *in vivo* to the virulent bacillus.

The bacterium is a nonhemolytic, facultative anaerobe requiring a serum-enriched medium for growth, and because of these fastidious characteristics, the carrier state in rats may be missed during routine bacteriologic screenings. Storage on artificial medium for periods longer than 4 weeks results in greatly diminished virulence for mice (Van Rooyen, 1936).

2. Occurrence

Streptobacillus moniliformis has been reported as a primary pathogen in man (Van Rooyen, 1936; Brown and Nunemaker, 1942), turkeys (Boyer *et al.*, 1958; Yamamoto and Clark, 1966), guinea pigs (Smith, 1941; Aldred *et al.*, 1974), rats (Strangeways, 1933; Freundt, 1959), and mice. Although the rat usually carries the organism as a harmless saphrophyte in the nasopharynx (Klieneberger, 1935), and thereby serves as a reservoir for infection to man and other animals, submaxillary abscesses have been associated with *S. moniliformis* in rats (Freundt, 1959). *Streptobacillus moniliformis* in man may cause a systemic, febrile illness with lymphadenopathy, arthritis, and rash (rat bite fever, Haverhill fever) (Brown and Nunemaker, 1942; Robbins, 1974); therefore, the public health significance of the infection must be considered.

3. Transmission

Initial infection may occur from airborne exposure, bite wounds, contaminated equipment, feed, or bedding, or exposure to infected animals; however, urine and feces are not considered significant routes for the excretion of the bacillus (Levaditi *et al.*, 1932; Mackie *et al.*, 1933; Van Rooyen, 1936; Freundt, 1959). Clinically healthy but latently infected rats housed in the same room with mice were implicated as the source of a disease outbreak in the mice (Freundt, 1956a).

4. Signs

Clinicopathologic signs of the spontaneous disease in mice vary among outbreaks. The general progression of the disease begins with a 1- to 3-day acute phase with high mortality followed by a subacute and then a chronic phase that may persist for months. The acute phase, however, does not necessarily correspond to the septicemia, which may persist for weeks (Mackie *et al.*, 1933). Specific considerations in the pathogenesis of the disease are the affinity of the organism for joints, the reversible form of the organism, the poorly defined but generally weak immune response of the hosts, and the anatomy of the hosts' vascular systems, particularly the end-vessel capillaries in the metaphyses (Savage, 1972; Lerner and Sokoloff, 1959).

Mackie *et al.* (1933) described a fatal epizootic of streptobacillary arthritis in two unrelated mouse stocks. The prominent clinical signs noted during the acute phase were a dull, damp hair coat, purulent conjunctival discharge, and death in most affected animals within 1–14 days. As the outbreak progressed, limb ulceration, swollen legs, feet, and tail, and partial to complete limb paralysis occurred. Regional lymphadenopathy and gestation arrest were also seen. The epizootic persisted for several months and resulted in an overall mortality of over 50%. Williams (1941) described a chronic, progressive polyarthritis caused by *S. moniliformis* among wild mice in Australia. Clinical signs included emaciation and bulbous, fusiform swellings of one or more wrist, ankle, or tail joints. Both death and spontaneous recovery with lesion regression occurred during the outbreak.

Anemia, diarrhea, conjunctivitis, hemoglobinuria, cyanosis, emaciation, and photophobia may be variably seen within days of exposure to *S. moniliformis* (Levaditi *et al.*, 1932; Brown and Nunemaker, 1942; Freundt, 1956a, 1959). The advent of the chronic stage is indicated by subcutaneous abscessation, weeping cutaneous ulcers, stillbirths and abortions, and limb, spine, and tail arthropathies (Van Rooyen, 1936; Lerner and Sokoloff, 1959; Sawicki *et al.*, 1962).

The first sign of joint involvement is a diffuse swelling of the limbs or tail followed in several days by more distinct joint enlargement, reddening, and immobilization. The arthritis takes a chronic course with swelling subsiding slowly over several months, leaving an ankylosed, deformed, functionally impaired joint. Ulcerative destruction of the joint and gangrenous amputation occur in some cases (Freundt, 1956a). Paralysis of the hindlimbs, urinary bladder distention, bowel incontinence, kyphosis, and priapism may occur in advanced cases when proliferative spinal lesions impinge on motor nerves (Mackie *et al.*, 1933; Van Rooyen, 1936; Brown and Nunemaker, 1942).

5. Pathology

An affinity for bones and joints is a characteristic of the organism. Experimentally, this affinity has been shown to depend on many factors, including the organism's metabolic requirements, number of bacteria inoculated, and route of exposure. Subcutaneous inoculation into a footpad results in a localized arthritis in the small joints of the foot. Intraperitoneal, intravenous, oral, and nasal exposure frequently result in a polyarthritis (Levaditi *et al.*, 1932; Mackie *et al.*, 1933; Heilman, 1940; Brown and Nunemaker, 1942; Freundt, 1956a,b, 1959). Lerner and Sokoloff (1959) have suggested that the end-artery anatomy of the vascular system in young animals is an important factor contributing to the sequestration of microorganisms in the metaphyses and ultimately the joints. In addition, based on *in vitro* studies, for some yet undetermined reason macrophages seem unable to ingest the organism effectively (Savage, 1972).

Few characteristic pathologic changes are apparent in the acute disease. In subacute infections, splenic enlargement, often up to three or four times the normal size, necrotic foci on the liver and spleen, lymphadenopathy, and petechial hemorrhages of serous membranes occur (Mackie *et al.*, 1933; Van Rooyen, 1936; Williams, 1941; Freundt, 1956a, 1959). Occasionally, in experimental infections, pericarditis, endocarditis, and myocarditis of the left ventricle have been reported (Van Rooyen, 1936; Dienes, 1939). Embolic involvement of the vessels of the left heart extends into the myocardium and occasionally to the pericardium and endocardium (Levaditi *et al.*, 1932).

The prominent lesion in chronically infected mice is a purulent polyarthritis involving the feet and legs, particularly the hindlimbs. In addition, coccygeal and vertebral involvement occurs (Mackie *et al.*, 1933). Experimentally, gross evidence of inflammatory exudation and cartilaginous and periosteal proliferation have been noted during the chronic phase of infection, when purulent or caseous exudate is present in the joints. Similar lesions may be found in the coccygeal and vertebral joints (Van Rooyen, 1936; Brown and Nunemaker, 1942; Freundt, 1956a).

Histopathologic observations of lesions in vessels, myocardium, and joints include a septic thrombosis of small vessels with accompanying mural degeneration and perivascular inflammatory cell accumulation. Neutrophilic infiltration, edema, fibrin deposition, and proliferation of adjacent tissue are followed by necrosis, abscessation, and capsule formation. Organisms remain for months within and about the lesion (Mackie *et al.*, 1933; Brown and Nunemaker, 1942; Lerner and Sokoloff, 1959).

6. Diagnosis

Disease confirmation depends on isolation and identification of the gram-negative bacillary organism from the lesions, although the agent, if present as the L-phase variant, may be more easily cultured than seen on direct smear (Brown and Nunemaker, 1942). *Streptobacillus moniliformis* has been isolated from joint fluid as long as 16 months postinfection (Freundt, 1956a). Culturing procedures involve inoculation of suspect joint fluid or blood into thioglycollate broth or brain-heart infusion semisolid and onto serum or blood agar. Cultures are incubated aerobically at 37°C for 48–72 hr, when small, fluffy colonies appear in broth and 1-3-mm, irregularly oval, gray, glistening, smooth colonies grow on agar. Giemsa and gram preparations reveal the rods and filaments of *S. moniliformis* (Carter, 1973).

Clinical signs of the acute stage may resemble those of ectromelia, Tyzzer's disease, erysipelas, and salmonellosis. The arthritic signs of the subacute and chronic stages resemble those in mice with ectromelia (Briody, 1966), other bacterial infections (Alspaugh and Van Hoosier, 1973), or traumatic wounds.

7. Treatment and Prevention

Freundt (1956b) investigated the use of bacterins and found that the L phase was antigenically incomplete and, therefore, could not provide an immune reaction against the complete bacillary form. A bacterin prepared from the bacillary phase did provide excellent protection, but this bacterin is not commercially available. Chemotherapy in mice using gold sodium thiomalate (Heilman, 1940) and penicillin (Freundt, 1956b) has been successfully attempted, although L-phase variants are highly resistant to penicillin. Penicillin and, less often, tetracyclines and streptomycin, are the antibiotics of choice for treating the human infection (Wilson and Miles, 1975).

Control of the disease is based on prevention of exposure to the organism. Mice should be separated from wild rodents, affected animals removed from the colony, and clean bedding provided (Strangeways, 1933; Klieneberger, 1935; Freundt, 1956a).

B. *Corynebacterium kutscheri*

Corynebacterium kutscheri, a gram-positive, diphtheroid bacillus, causes a relatively common spontaneous disease of rats and mice (Carlton and Hunt, 1978). This disease was first reported in mice by Kutscher (1894) in Germany and by Welch (Reed, 1902) in the United States, whereas Reed conducted the first experimental studies with the organism. Subsequent studies in mice have shown that clinical manifestations are usually related to physiologic stressors or factors that decrease immunocompetence, such as cortisone treatment (Antopol, 1950, 1951; Speirs, 1956; Fauve *et al.*, 1964), irradiation (Shechmeister and Adler, 1953), simultaneous infections (Wolff, 1950; Lawrence, 1957), and pantothenate deficiency (Zucker, 1957).

Clinical infection results in either an acute condition with high mortality or chronic disease with unknown morbidity and low mortality. Clinical signs of the infection may include inappetence, emaciation, rough hair coat, hunched posture, hyperpnea, nasal and ocular discharge, cutaneous ulceration, and swelling, stiffening, and ulceration of limbs. At necropsy raised, gray-white, friable, caseous nodules or thinly encapsulated, liquefactive abscesses up to 1 cm in diameter are found most often in the liver, lung, and kidney. Similar gross lesions have been seen in the pericardium, heart, spleen, lymph nodes, blood vessels, muscle, and joints (Kutscher, 1894; Bongert, 1901; Reed, 1902; Fischl *et al.*, 1931; Hojo, 1939; Bicks, 1957; Weisbroth and Scher, 1968). Bongert (1901) attributed the development of multiple nodular foci to hematogenous spread rather than to direct extension from a contiguous focus, as may occur in tuberculosis.

Histologically, the centers of the nodules are necrotic and surrounded by inflammatory cells, mostly neutrophils, and clumps of bacteria. Epithelioid and giant cells are not present. Portal and mesenteric veins and small vessels in the brain and heart frequently have thrombi, perivascular inflammatory cell infiltration, and necrosis (Bicks, 1957; Weisbroth and Scher, 1968). These lesions suggest an embolic "showering" following release of the organisms from a preexisting focus, probably in the intestine, and a transient septicemia.

In occasional epizootics and experimental studies joint lesions have been noted (Fischl *et al.*, 1931; Weisbroth and Scher, 1968; Juhr and Horn, 1975). These mucopurulent lesions are characterized by gross enlargement and distention of the joints, usually the carpometacarpal and tarsometatarsal joints. Acute arthritic changes are related to bacterial colonization of the synovium accompanied by synovial membrane necrosis, articular cartilage erosion, ulceration, and eventually an ankylosing panarthritis or sloughing.

C. *Salmonella* spp.

In the past, salmonellosis has been a significant disease and the causative organism a frequent isolate in mice (Loeffler, 1892; Lynch, 1922; Amoss, 1922; Olitsky and Syverton, 1934; Wolff, 1950; Erling and Rokey, 1960; Litchfield and Margard, 1963; Hoag *et al.*, 1964; Brennan *et al.*, 1965a; Jones, 1976; Shimi *et al.*, 1979), but in recent years, due to improved husbandry and elimination of carriers, few outbreaks have been reported. The species most frequently associated with the disease in mouse colonies are *Salmonella enteritidis* and *S. typhimurium,* which affect both man and animals (Read and Savage, 1913; Habermann and Williams, 1958).

The acute disease, especially severe in young mice, is characterized by anorexia, weight loss, lethargy, dull hair coat, humped posture, and, in some cases, conjunctivitis. Although gastroenteritis is a common sign of the disease, the only change in the feces may be a lighter color than normal (Olitsky and Syverton, 1934; Habermann and Williams, 1958). Animals with acute disease usually die within a week of the onset of clinical signs. In mice with subacute infections, distended abdomens due to hepatomegaly and splenomegaly are occasionally seen. Mice with the chronic disease often show no clinical signs other than anorexia and weight loss.

By monitoring the number of viable *Salmonella* in various organs, particularly in lymphatic tissues, investigators have reported that approximately 99% of ingested organisms pass rapidly through the animal. The remainder settle in the intestine (Carter and Collins, 1974). The virulence of the organism seems to depend on its ability to penetrate the intestinal wall into lymphatic tissue, multiply, disseminate, and survive systemically. The first detectable salmonellae appear within 12 hr in the ileal Peyer's patches. At 18 hr, organisms are detectable in mesenteric lymph tissue and by 36 hr in the spleen, liver, and cervical lymph nodes. In several studies, ileal Peyer's patches were the site of primary infection (Seiffert *et al.*, 1928; Carter and Collins, 1974; Carter *et al.*, 1975; Collins and Carter, 1978). Multiplication of the bacteria in the lymphatic system is followed by passage first into the thoracic duct and then into the vascular system to produce a bacteremia. The reticuloendothelial system removes bacteria from the blood, a second bacterial proliferation occurs, and reinvasion of the blood follows within a few days. The salmonellae are then widely dispersed to other organs, including the intestine, where an enteritis may result (Habermann and Williams, 1958). In chronic infections, organisms persist in the spleen and lymph nodes as well as in the liver and gallbladder. From these locations bacteria are discharged into the feces (Carlton and Hunt, 1978). In addition, in experimental studies with rats, Volkman and Collins (1976) have associated an immunologically mediated, chronic arthritis with *S. enteritidis* infection.

Gross and histologic changes seen in animals dying acutely include visceral hyperemia, congestion, pale liver, and catarrhal enteritis (Habermann and Williams, 1958). If the animal survives for 1 or 2 weeks, the more characteristic subacute lesions are observed in the intestine, liver, spleen, and lymph nodes. The intestine is distended, its walls congested, and the mucosa reddened. Dark red fluid may be present in the lumen. The liver is enlarged and has numerous subcapsular yellow foci. Because of congestion and hyperplasia of reticular and endothelial elements, the spleen may be enlarged three to four times normal and extend posteriorly to the pelvis. The capsule is taut and when cut, the parenchyma bulges. Numerous yellow-gray foci are scattered throughout the organ. Affected lymph nodes or nodules may be enlarged and red and contain yellow-gray foci. Occasionally these foci ulcerate through the intestinal mucosa (Loeffler, 1892; Seiffert *et al.*, 1928; Olitsky and Syverton, 1934; Bakken and Vogelsang, 1950; Habermann and Williams, 1958). Thrombosis, due to septic embolism, may occur in the venous system (Böhme *et al.*, 1959).

Histopathologic changes in affected tissue include hyperemia, hyperplasia, and focal necrosis. Necrotic foci are surrounded by polymorphonuclear and large mononuclear cells, but within 7 days postinfection these foci enlarge and transform into granulomata (Hsu and Nakoneczna, 1975). Degeneration of lymphoid follicles is seen in the spleen and lymph nodes. Clumps of bacilli are present in affected organs, and the reticuloendothelial cells may be enlarged and contain bacteria. Focal inflammation may occur in other organs, including the myocardium. Bone marrow activity may be depressed (Loeffler, 1892; Seiffert *et al.*, 1928; Habermann and Williams, 1958; Carlton and Hunt, 1978).

D. *Pseudomonas aeruginosa*

Although of low invasiveness and virulence in the healthy mouse, *Pseudomonas aeruginosa* can readily cause fatal disease in susceptible animals. Syndromes caused by this organism have, with few exceptions, been associated with the suppression of the host's immune system. Immunosuppressive agents such as irradiation (Miller *et al.*, 1950, 1951; Gordon *et al.*, 1955; Wensinck *et al.*, 1957; Flynn, 1963; Hammond, 1963), drugs, including cyclophosphamide (Pierson *et al.*, 1976; Brownstein, 1978; Urano and Maejima, 1978) and cortisone (Carlton and Hunt, 1978), and burn stress (Fox *et al.*, 1970; Pierson *et al.*, 1976) have all been related to fatal pseudomoniasis in mice. In diseased animals the organism may be isolated from many organs, particularly the blood, spleen, intestine, and lymph nodes (Hoag *et al.*, 1965; McDougall *et al.*, 1967).

Following immunosuppression of animals naturally harboring or exposed to the organism, severe disease develops within 3–14 days (Hammond, 1963; Flynn, 1963; Brownstein, 1978). Both the oropharyngeal and gastrointestinal routes of infection have been implicated (Wensinck *et al.*, 1957; Urano and Maejima, 1978). The bacteria penetrate the local lymphatic or vascular systems, and a bacteremia rapidly follows (Bartell *et al.*, 1968; Brownstein, 1978). Experimental studies of fatal *Pseudomonas* bacteremia in mice have indicated that organisms present in the intestinal flora may enter the circulatory system by lymphatic or vascular routes and initiate an exotoxin-induced hepatic necrosis (Pavlovskis *et al.*, 1976). Gordon *et al.* (1955) also found that small numbers of bacteria may pass the intestinal barrier to reach the regional lymph nodes but not the spleen or liver.

Clinical diagnosis is often difficult due to the rapid septicemic course of the infection (Hammond, 1963), and clinical signs are usually confined to listlessness, anorexia, and death, although circling from inner ear disease (Olson and Ediger, 1972), conjunctivitis, nasal discharge, weight loss, and head edema have also been reported (Brownstein, 1978). Since animals showing signs of disease have frequently been immunosuppressed, the lymphatic organs are often atrophic and nonreactive (Urano and Maejima, 1978). Multiple areas of focal hemorrhage and necrosis have been observed in spleens and other organs, including bone, in mice with *Pseudomonas* infection (Brownstein, 1978).

E. *Bacillus piliformis*

Spontaneous outbreaks of Tyzzer's disease have been reported in mouse colonies on several occasions (Tyzzer, 1917; Hagedoorn-La Brand and Hagedoorn, 1920; Tuffery, 1956; Saunders, 1958; Fujiwara *et al.*, 1963). The disease, an epizootic or enzootic diarrheal condition probably caused by the filamentous, intracellular organism *Bacillus piliformis*, was first described by Ernest Tyzzer in Japanese waltzing mice. Since then the still unclassified microbe has been associated with hepatic and enteric disease in many species. In mice, under conditions that reduce immunocompetence, such as cortisone treatment (Fujiwara *et al.*, 1963, 1964; Takagaki *et al.*, 1966), dietary manipulation (Cumming and Cumming, 1967), tumor transplantation (Tyzzer, 1917), and overcrowding (Rights *et al.*, 1947), a rapidly developing diarrhea, anorexia, weight loss, and death occur. Genetic background will also affect susceptibility (Hagedoorn-La Brand and Hagedoorn, 1920; Gowen and Schott, 1933; Fries, 1979). In spontaneous outbreaks, clinical signs and a rapidly fatal outcome occur more frequently in the recently weaned and the young (Fujiwara, 1967; Ganaway *et al.*, 1971).

The principal gross pathologic lesions are yellow-gray, circular, hepatic foci and, less often, intestinal inflammation (Tyzzer, 1917; Rights *et al.*, 1947; Fujiwara *et al.*, 1963; Fries, 1979). In addition, edema, hyperemia, and necrosis of mesenteric lymph nodes have been reported in mice (Saunders, 1958). Passage of the organism in cortisone-treated mice results in a bacteremia with organisms demonstrable in the spleen, adrenal gland, mesenteric lymph nodes, and myocardium (Ganaway *et al.*, 1971). Takagaki *et al.* (1966) observed necrotic heart lesions in mice infected experimentally and either X-irradiated or given cortisone.

F. *Mycobacterium lepraemurium*

The chronic, granulomatous disease caused by the gram-positive, acid-fast, obligate, intracellular bacterium *Mycobacterium lepraemurium* is usually associated with infection in wild rats (Krakower and González, 1937; Carlton and Hunt, 1978), although there are reports of spontaneous leprosy or the subclinical carrier state in mice (Hemmert-Halswick, 1934; Krakower and González, 1937; Sushida and Hirano, 1966; Pattyn and Verdoolaege-Van Loo, 1969). Hemmert-Halswick (1934) described a spontaneous epidemic in mice resembling human leprosy and associated with an acid-fast bacillus. Similar clinical and pathologic observations were made on a single wild mouse by Krakower and González (1937).

Nishimura *et al.* (1964) reported the isolation of *M. lepraemurium* from apparently healthy laboratory mice involved in studies of the human leprosy organism, *M. leprae*. Subsequent isolation studies from mice of several unrelated colonies indicated that the murine organism was latently present in some animals. Sushida and Hirano (1966) also noted *M. lepraemurium* in a colony of laboratory mice.

Clinical signs described in Nishimura *et al.* (1964) and Sushida and Hirano (1966) included alopecia, thickening of the skin, and subcutaneous swellings with reddening and ulceration of the overlying epidermis. Subclinical cases, spontaneous recoveries, and fatalities occurred. Nodules (lepromas) were observed in the subcutaneous tissues and in the reticuloendothelial organs such as the lungs, spleen, bone marrow, thymus, and lymph nodes. Lesions may also be found in skeletal muscle, myocardium, kidneys, nerve sheaths, and adrenal glands (Krakower and González, 1940; Closs and Haugen, 1973). Spleen and lymph nodes are often enlarged, and the thymus is involuted (Krakower and González, 1937; Ptak *et al.*, 1970; Chang *et al.*, 1972). The dissemination, replication, and pathologic effects of the bacterium require several months (Kawaguchi, 1959; Sushida and Hirano, 1966).

The usually perivascular granulomas caused by *M. lepraemurium* are characterized by a prominent accumulation of "lepra cells," large, foamy, epithelioid macrophages usually packed with acid-fast bacilli, which are slender, beaded or granular rods 3–5 μm long. Lymphocytes and neutrophils may be associated with the macrophage accumulations (Figs. 1 and 2) (Krakower and González, 1937, 1940; Chang *et al.*, 1972; Closs and Haugen, 1973; Stookey and Moe, 1978; Lagrange and Hurtel, 1979). Intravenous inoculation of mice and rats with *M. lepraemurium* results in a chronic granulomatous disease primarily involving the lymphatic organs. The paracortical areas of the lymph nodes are filled with bacteria-laden histiocytes, as are the periarteriolar lymphocytic sheaths in the splenic white pulp (Bullock, 1976).

Interest in experimental murine leprosy relates to the pathologic similarities to human leprosy (Bullock, 1976). The spontaneous and experimental disease in man and mouse is characterized by a chronic inflammatory response with granuloma formation. Mice infected with *M. lepraemurium*

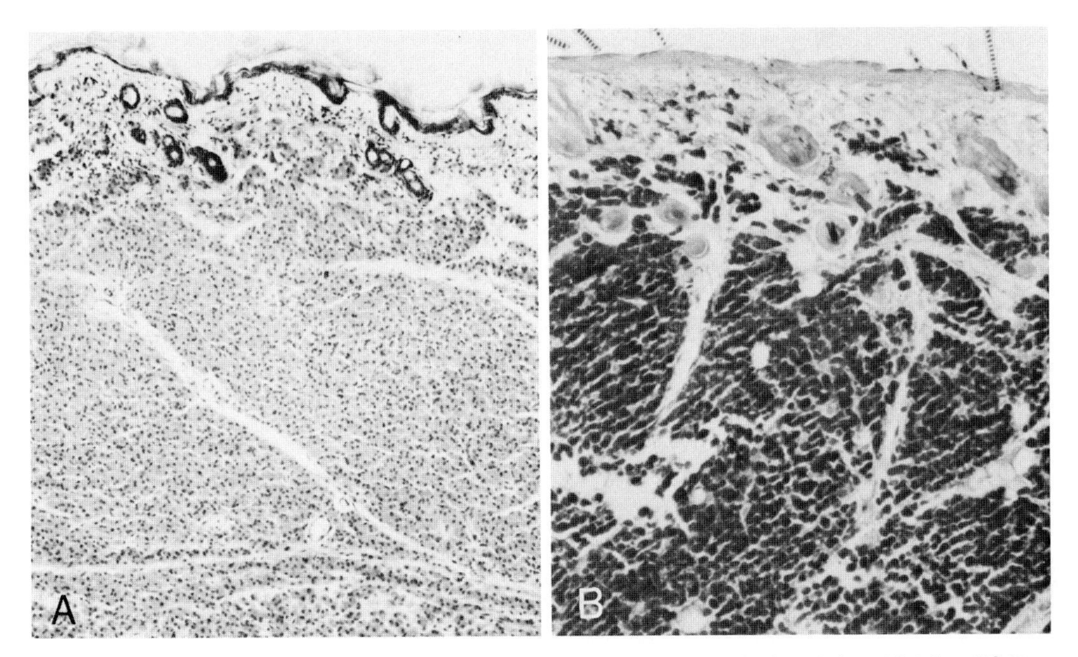

Fig. 1. (A) Macrophage infiltrate in the subcutis of a C3H/A mouse 30 weeks after inoculation with 1.5×10^8 *Mycobacterium lepraemurium* organisms. H&E. ×90. (B) Same lesion, showing large amounts of acid-fast (dark) material within macrophages. Modified Ziehl-Neelsen stain. ×190. (Photographs courtesy of Dr. Otto Closs, Oslo.)

have marked alterations in immune function, with significant impairment of cell-mediated immunity (Ptak *et al.*, 1970; Bullock *et al.*, 1977); however, the relationship of this impairment to the pathogenesis of the granuloma is unknown. Certain strains of mice, such as the C57BL/6 and BALB/c strains, appear to be more susceptible than other strains, such as DBA/2, to the generalized infection, but all mouse strains tested eventually succumbed (Lefford *et al.*, 1977). This variation in response, which may be noted in individuals within outbred or inbred populations, was attributed by Closs and Haugen (1974) to differences in cell-mediated immune responses. Experimentally infected C57BL/6 mice developed hard, well-circumscribed, benign nodules, whereas C3H mice exhibited the diffuse, usually fatal malignant form (Kawaguchi, 1959). It is important to note that even the more susceptible mice are able to mount a cell-mediated immune response. Comparisons of lesions among inbred mouse strains and the lepromatous to tuberculoid spectrum encountered in human leprosy (Ridley and Jopling, 1966), as well as the lymphocyte populations, recirculation patterns, and related immune responses associated with those lesions, require further investigation.

G. *Streptococcus* spp.

Mice may harbor streptococci in their upper respiratory tracts without signs of illness (Wensinck and Renaud, 1957b; Hook *et al.*, 1960; Stewart *et al.*, 1975); however, occasionally streptococci cause epizootics with considerable mortality. Streptococci of Lancefield groups A, B, D, and G have been associated with spontaneous disease outbreaks in mice (Loewenthal, 1932; Gledhill and Rees, 1952; Nelson, 1954; Hook *et al.*, 1960; Stewart *et al.*, 1975). Kutschera (1908) reported an outbreak among his laboratory mice that killed one to three mice daily over a period of several weeks. Clinical signs included depression, conjunctivitis, rough hair coat, hyperpnea, and emaciation of the young. Spleens were enlarged and contained pinhead-sized, yellow, multifocal abscesses. A hemolytic streptococcus was isolated from internal organs.

A similar condition destroyed 90% of Loewenthal's (1932) colony within 4 days of the onset of clinical signs. A group B streptococcus was isolated from the enlarged, softened spleens and from most other organs, urine, and feces. However, this highly pathogenic organism could be isolated from organs only 1–2 days before death. Mice not dying acutely developed a lymphadenitis.

Hook *et al.* (1960) reported 50% mortality among mice with group A streptococcal disease. Other reports (Glaser *et al.*, 1953; Nelson, 1954) have related group A streptococci to pharyngitis, cervical lymphadenitis, and fatal bacteremias. An experimental group A streptococcus produced and was recovered from cases of pneumonia and otitis media (Nelson, 1954), and Glaser *et al.* (1953) mentioned that group A streptococci were more likely to be localized in regional lymph nodes,

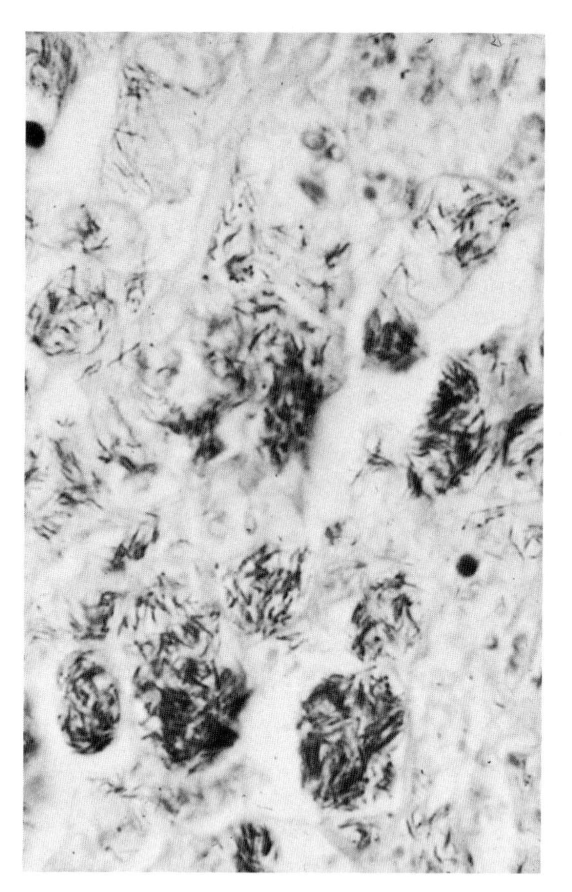
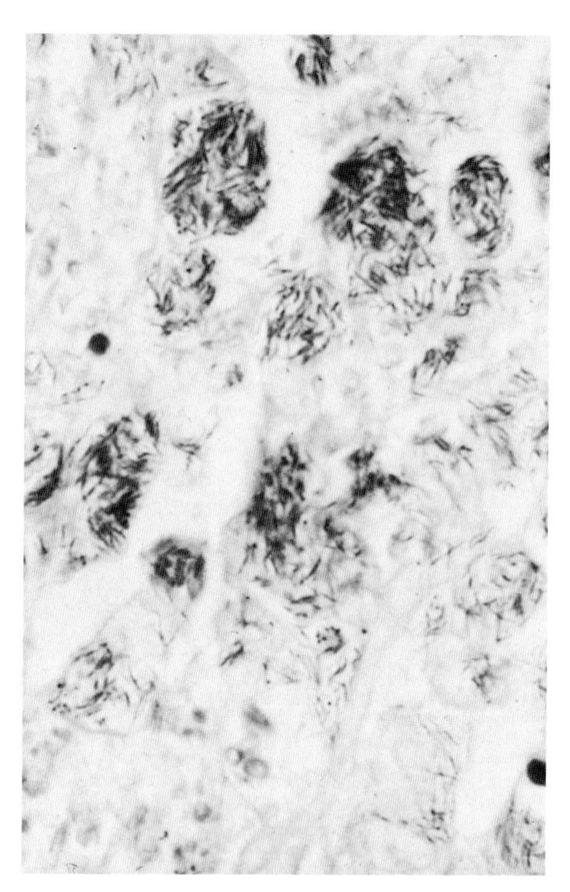

Fig. 2. Photomicrograph of acid-fast bacilli (*Mycobacterium lepraemurium*) in macrophages 4 weeks after subcutaneous inoculation of C57BL/6J. Ziehl-Neelsen stain. ×1000. (From Closs and Haugen, *Acta Pathol. Microbiol. Scand., Sect. A* **83,** 54, 1975; used by permission of Dr. Otto Closs and publisher.)

whereas group C organisms caused cellulitis in tissues adjacent to primary foci of infection (Sonkin, 1949).

Group G streptococci have been recovered from a case of ulcerative, necrotic dermatitis in male Caw:CFW(SW)SPF mice. Bacteria were cultured from visceral as well as cutaneous sites. Adjacent muscle was occasionally involved (Stewart *et al.,* 1975). Wensinck and Renaud (1957a,b) found large numbers of group G streptococci in CBA mice dying of a "bone marrow syndrome" following X irradiation.

Onset of streptococcal disease may be indicated by sudden high mortality, with few clinical or gross necropsy signs. Animals with more chronic forms of the disease developed roughened hair coats and, in group A infection, cervical lymphadenopathy. The enlarged nodes may ulcerate and discharge purulent content.

Gross pathologic change in streptococcal disease is related to the location of the lesions. In animals with lymphadenitis, a common manifestation, the nodes are enlarged and abscessed and may ulcerate through the skin or into a hollow organ. Microscopic lesions range from foci of inflammation to extensive abscessation with mixed inflammatory cell populations and necrosis. When the disease is systemic, other lymphatic organs may become involved.

H. *Staphylococcus aureus*

Spontaneous staphylococcal infection has been infrequently reported in laboratory mice (Shults *et al.,* 1973; Needham and Cooper, 1976; Clarke *et al.,* 1978). Staphylococci, however, are usually considered part of the resident microflora and are occasionally found on routine culture of the nasal passages of healthy mice (Markham and Markham, 1966; Blackmore and Francis, 1970). The susceptibility of animals to clinical infection with *S. aureus* is a subject of continuing research interest, particularly in regard to characteristics associated with high or low virulence of a particular staphylococcal strain (Gorrill and McNeil, 1963; Nutini and Berberich, 1965; Bonventre *et al.*, 1975; Iwata *et al.*, 1978).

After observing one mouse with a swollen jaw, Clarke *et al.* (1978) diagnosed staphylococcal infection in 57 similar cases in a large colony of 4- to 5-month-old mice. The swellings usually were located in the head and neck and were nodular with an intact, overlying epithelium. Some of the nodules were as large as 10 mm in diameter and distorted adjacent tissue, including facial muscles and bone. The lesions were firm and fibrous, and cut surfaces were yellow-white with grossly visible pale foci of necrosis. Regional lymph nodes were enlarged and edematous.

Peridontal abscesses frequently have been associated with cervical staphylococcal lesions, and this relationship suggests that foreign body injury to the oral cavity may provide a primary infection route. This may explain why most staphylococcal lesions in mice occur on or about the head (Shults *et al.*, 1973; Clarke *et al.*, 1978). Histologically, the nodules were suppurative and, as the disease progressed, granulomatous. Granulomas consisted of bacterial colonies, amorphous eosinophilic material surrounded by neutrophils, macrophages, and fibrous connective tissue. Mandibular rami had lesions detected histologically as foci of bone erosion with associated inflammatory cells and necrotic tissue.

I. *Pasteurella pneumotropica*

Pasteurella pneumotropica, a gram-negative coccobacillus first described by Jawetz (1948), has frequently been associated with latent and clinical infection of the respiratory tract of laboratory animals (Jawetz, 1950; Jawetz and Baker, 1950; Gray and Campbell, 1953; Hoag *et al.*, 1962; Brennan *et al.*, 1965b, 1969). Studies have shown that the organism can be cultured not only from the lung but also from other organs, including the spleen of apparently healthy mice.

The bacterium has been identified as a primary pathogen in clinical disease of mice (Sebesteny, 1973; Needham and Cooper, 1975). In one report (Wilson, 1976), the organism was isolated from a swelling on the lateral aspect of the jaw. The lesion, an abscess in the masseter muscle, contained green pus. A cervical lymphadenitis was noted in an outbreak of panophthalmitis among breeding mice (Fig. 3) (Weisbroth *et al.*, 1969), and *P. pneumotropica* was cultured from spleens of mice aborting with the same organism (Ward *et al.*, 1978). In an experimental study of *P. pneumotropica* (Jawetz and Baker, 1950), splenic and hilar lymph node enlargement with follicular hyperplasia was reported.

Disease expression is usually stress-related, and attempts to reproduce the disease in healthy animals have been unsuccessful (Jawetz, 1950; Urano and Maejima, 1978). The status of the host's immune system is important because immunologically incompetent mice, such as the athymic nude mice, are more susceptible than conventional mice to overt disease (Moore and Aldred, 1978).

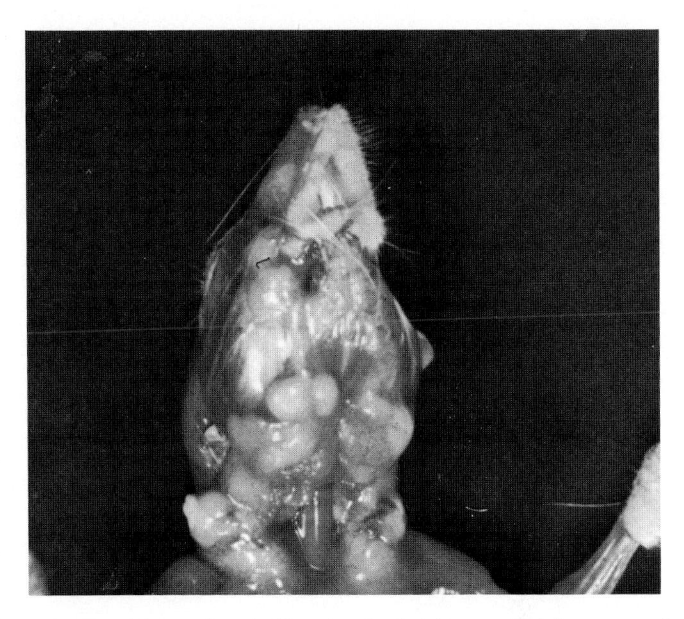

Fig. 3. Abscessed cervical lymph nodes draining an orbital abscess caused by *Pasteurella pneumotropica*. (Photograph used with permission of Steven H. Weisbroth and the American Veterinary Medical Association.)

J. *Klebsiella pneumoniae*

Although klebsiellae are common organisms found in the internal and external environments of man and animals, few reports exist of spontaneous clinical infection in laboratory mice. Flamm (1957) reported approximately 30 mice, mostly adult females, in a colony of 1200 with "cherry-size," movable, subcutaneous masses on the neck and shoulder. These abscesses, associated with cervical and supraclavicular lymph nodes, contained thick green pus and, microscopically, masses of encapsulated gram-negative bacilli. Abscesses often involved adjacent lymphatic vessels, glands, and muscle. Affected animals usually died before the abscesses ruptured, but in one case an abscess ruptured into the esophagus and filled the stomach with pus. *Klebsiella pneumoniae* was recovered in pure culture from the abscesses.

Additional signs encountered among the 30 mice were edema of the area normally drained by affected lymphatics, splenomegaly, hepatomegaly, leukocytic infiltration and thrombosis in the ventricular endocardium and myocardium, and purulent pleuropneumonia. Transmission and entry were

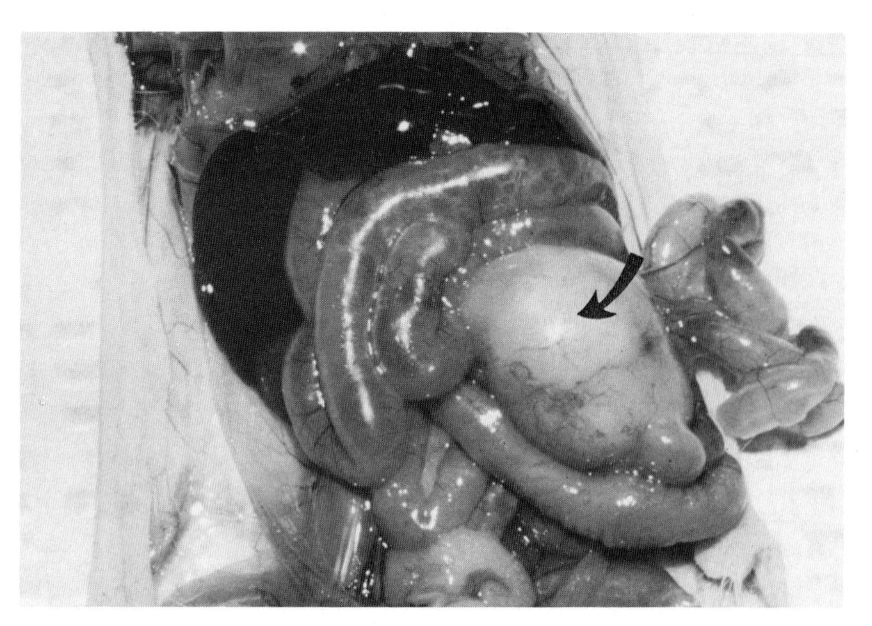

Fig. 4. An abscess caused by *Klebsiella pneumoniae* of mesenteric lymphatic tissue in an inbred Swiss white mouse. (Photograph courtesy of Helen D. Schneemilch, Institute of Medical and Veterinary Science, Adelaide, Australia.)

ascribed to oral trauma from sharp oat hulls in the feed, but trials with experimentally contaminated hulls did not confirm this assumption. Contact among mice did not spread the disease.

In a second report (Schneemilch, 1976), 73 of 226 inbred Swiss mice died from infection with *K. pneumoniae* capsule type 6, a variety not previously thought pathogenic in mice. Cervical lymph node enlargement, sneezing, and rapid respiration were prominent clinical signs. Other signs were nonspecific and included lethargy, rough hair coat, and hunched posture. Gross lesions were enlarged lymph nodes, abscesses in the kidneys, liver, and mesenteric lymph nodes (Fig. 4), and purulent pneumonia.

K. Miscellaneous Infections

The central role played by the lymphoreticular and cardiovascular systems in the processing and distribution of infecting bacteria virtually insures that some change, gross or microscopic, will occur in a vessel, lymph node, or the spleen during a local or systemic disease. Likewise, a septicemia or toxemia may affect a diversity of target organs, including those of the cardiovascular, musculoskeletal, and endocrine systems. The spontaneous diseases briefly discussed in the following paragraphs, therefore, represent rare or uncommon diseases of mice that in some way involve the systems discussed in this chapter.

Erysipelothrix insidiosa has been implicated as a primary pathogen by Wayson (1927) and Balfour-Jones (1935). Wayson (1927), studying a large natural migration of wild California mice (including *Mus musculus*), noted animals with humped backs, rough hair coats, labored breathing, and conjunctivitis. Necropsy revealed subcutaneous and lymphatic congestion and necrotic foci in lymph nodes, lungs, livers, and spleens. The intestine had subserosal, petechial hemorrhages. Balfour-Jones (1935) found focal hepatic necrosis, splenomegaly, and purulent conjunctivitis in dying mice of his laboratory stock. Hepatic foci were small pits seen microscopically as necrotic areas with peripheral leukocytic infiltration.

Sangiorgi (1911) reported a spontaneous epidemic that killed 20% of his stock of white mice. Affected animals had a rough hair coat, shallow respiration, and profuse diarrhea. Necrotic foci were present on the liver, the spleens were hypertrophied and congested, the kidneys pale, and the intestine gas- and fluid-filled. The etiologic agent or agents were not clearly defined or described, but a coliform-like bacillus was designated as the primary pathogen.

Proteus mirabilis is widely distributed in nature and may reside benignly in the respiratory and intestinal tracts of animals. In the role of a latent opportunist causing a postirradiation bacteremia in CBA and C57BL mice (Wensinck, 1961) or as a primary pathogen (Wilson, 1926; Jones *et al.,* 1972), the organism has been associated with disease in laboratory mice. Clinical signs included unthriftiness and humped posture. Necropsy signs were emaciation, splenomegaly, dehydration, watery blood, and swollen, pale hearts and kidneys. Although the disease is usually septicemic, focal suppurative lesions may also occur in the liver, kidney, and spleen. Focal lesions are accumulations of neutrophils and necrotic tissue. The splenomegaly is due primarily to extramedullary hematopoiesis and increased germinal center activity.

Baudet (1924) investigated an epidemic in which five to six mice died daily. On necropsy, splenomegaly and an enlarged intestine with red-brown content and congested mucosa were noted. *Pasteurella multocida* was recovered from the intestine. Greenwood and Topley (1925) reported deaths among laboratory mice apparently caused by *P. multocida,* although other enteric pathogens were present. Splenic enlargement and necrotic hepatic foci were seen on necropsy.

III. MYCOPLASMAL INFECTIONS

Three species of mycoplasma have been reported to cause disease in mice: *Mycoplasma pulmonis, M. arthritidis,* and *M. neurolyticum*. The last causes conjunctivitis in young mice and, experimentally, an exotoxin-induced rolling syndrome. *Mycoplasma arthritidis* causes spontaneous polyarthritis in rats but has not been reported to cause spontaneous arthritis in mice, although chronic joint disease may be experimentally induced (Cole *et al.,* 1973; Cole and Cassell, 1979). The third organism, *M. pulmonis,* is a significant pathogen in laboratory mice, in which it causes upper respiratory tract and pulmonary disease with related changes in peribronchiolar lymphoid tissue (Smith, 1973; Lindsey *et al.,* 1978).

In mice *M. pulmonis* infection is characterized by sudden onset with high mortality or survival with chronic wasting or uneventful recovery (Smith, 1973). Clinical signs may include respiratory "chattering," hunched posture, weight loss, and rough hair coat (Lindsey *et al.,* 1978).

Pathologic lesions in mice with mycoplasmosis vary with the duration and severity of the disease and include rhinitis, otitis, laryngotracheitis, bronchopneumonia, bronchiolectasis, and pulmonary abscessation. Histologically, the lungs of infected mice contain peribronchiolar aggregations of lymphocytes and plasma cells.

Inflammatory lesions outside the lungs and upper respiratory tract passages are rarely described in naturally occurring murine mycoplasmosis. Barden and Tully (1969) recovered *M. pulmonis* strain JB from arthritic ankle joints of mice injected intramuscularly with methylcholanthrene-induced tumor passage material. Affected joints, described as swollen and erythematous, occurred in injected limbs. The possibility of *Mycoplasma*-contaminated inocula was not pursued. Cole and Cassell (1979) mentioned that naturally occurring, chronic arthritis caused by *M. pulmonis* occurs in mice, but no details were given. Experimental arthropathies in mice following the injection of *M. pulmonis* are well documented (Harwick *et al.,* 1973; Keystone *et al.,* 1980).

Collier (1948) reported an outbreak of pneumonia, rolling disease, and polyarthritis in a mouse breeding colony in Indonesia. Using affected lung tissue as inoculum, Collier reproduced the disease and tentatively identified the causative organism as a pleuropneumonia-like bacterium.

In experimental studies with *M. pulmonis* (Lindsey and Cassell, 1973), histopathologic changes were seen in the spleen but not in peripheral lymph nodes. In the acute disease resulting from intranasal inoculation of high organism doses, mice developed mild lymphoid and granulocytic splenic hyperplasia and severe lung lesions. In chronic cases, splenic lymphoid and granulocytic hyperplasia occurred, the latter as late as 21 days postinoculation. In addition, the thymus glands in mice dying of pneumonia were depleted of cortical lymphocytes.

IV. MYCOTIC INFECTIONS

Mycotic infections are of low contagion and, as a result, only sporadic cases, often superimposed on other disease states, occur. Generally, the systemic mycoses, such as histoplasmosis, are characterized pathologically by the granulomatous nature of the lesion.

Reports of spontaneous, systemic mycotoxicoses in *Mus musculus* are uncommon, especially in the topic systems of this chapter (Smith and Austwick, 1967; Migaki *et al.,* 1978). Human deaths from histoplasmosis in Loudon County, Virginia, between 1922 and 1945 stimulated a thorough search of the local wildlife for a natural reservoir of the dimorphic fungus *Histoplasma capsulatum*. Eventually 988 mice, among hundreds of other animals, were examined; only one mouse was found with *Histoplasma,* and that mouse lived with an infected dog. No gross lesions were described, but microscopic, epithelioid granulomata were seen in the lungs, liver, and spleen (Emmons *et al.,* 1947; Olson *et al.,* 1947; Emmons, 1949).

Sacquet *et al.* (1959) reported cryptococcosis in mice. A female Af/A strain mouse had a splenic abscess approximately 1 mm in diameter as well as confluent, necrotic foci in the kidneys. *Cryptococcus neoformans* was isolated in pure culture from the spleen and in mixed culture from the kidney. The source of the infection was thought to be litter or caretakers' hands.

Mullink (1968) observed a lesion of suspected actinomycosis in an adult NZW mouse. The animal had a retroperitoneal yellow mass in the left abdominal wall. The mass contained multiple abscesses and fungal mycelia. The liver and spleen were enlarged and pale brown; no other lesions were observed. As cultures were negative, the diagnosis of actinomycosis was based on histopathologic observation.

Goetz and Taylor (1967) and Austwick (1974) reported spontaneous *Candida tropicalis* infections in laboratory mice. Lesions were confined to the kidneys, except for a young Swiss-Webster mouse with splenomegaly.

Stachybotryotoxicosis is a mycotoxicosis caused by *Stachybotrys chartarum,* a fungal contaminant of cellulosic grains. A test group of mice fed grain intentionally contaminated with *Stachybotrys* died; histopathologic examination revealed necrotic foci in the liver, kidney, and heart (Korpinen and Ylimäki, 1972).

REFERENCES

Aldred, P., Hill, A. C., and Young, C. (1974). The isolation of *Streptobacillus moniliformis* from cervical abscesses of guinea pigs. *Lab. Anim.* **8,** 275–277.

Alspaugh, M. A., and Van Hoosier, G. L., Jr. (1973). Naturally-occurring and experimentally-induced arthritides in rodents: A review of the literature. *Lab. Anim. Sci.* **23,** 724–742.

Amoss, H. L. (1922). Effect of the addition of healthy mice to a population suffering from mouse typhoid. *J. Exp. Med.* **36,** 45–69.

Antopol, W. (1950). Anatomic changes in mice treated with excessive doses of cortisone. *Proc. Soc. Exp. Biol. Med.* **73,** 262–265.

Antopol, W. (1951). Experimental observations with massive doses of cortisone. *Am. J. Pathol.* **27,** 705–706.

Austwick, P. K. C. (1974). Apparently spontaneous *Candida tropicalis* infection of a mouse. *Lab. Anim.* **8,** 133–136.

Bakken, K., and Vogelsang, T. M. (1950). Pathogenesis of *Salmonella typhimurium* infections in mice. *Acta Pathol. Microbiol. Scand.* **27,** 41–50.

Balfour-Jones, E. E. B. (1935). A bacillus resembling *Erysipelothrix muriseptica* isolated from necrotic lesions in the livers of mice. *Br. J. Exp. Pathol.* **16,** 236–243.

Barden, J. A., and Tully, J. G. (1939). Experimental arthritis in mice with *Mycoplasma pulmonis. J. Bacteriol.* **100,** 5–10.

Bartell, P. F., Orr, T. E., and Garcia, M. (1968). The lethal events in experimental *Pseudomonas aeruginosa* infection of mice. *J. Infect. Dis.* **118,** 165–172.

Baudet, E. A. R. F. (1924). Een bacil uit de groep der haemorrhagische septicaemie als oorzaak van een sterte onder muizen. *Tijdschr. Diergeneeskd.* **51,** 662–665.

Bicks, V. A. (1957). Infection of laboratory mice with *Corynebacterium murium. Aust. J. Sci.* **20,** 20–22.

Blackmore, D. K., and Francis, R. A. (1970). The apparent transmission of staphylococci of human origin to laboratory animals. *J. Comp. Pathol.* **80,** 645–651.

Böhme, P. H., Schneider, H. A., and Lee, J. M. (1959). Some physiopathological parameters of natural resistance to infection in murine salmonellosis. *J. Exp. Med.* **110,** 9–25.

Bongert, R. (1901). *Corynethrix pseudotuberculosis murium,* ein neuer pathogener Bacillus für Mäuse. *Z. Hyg. Infektionskr.* **37,** 449–475.

Bonventre, A., Nickol, R., Baughn, R., and Allison, A. (1975). Anomalous resistance of athymic mice to bacterial infection. *J. Reticuloendothel. Soc.* **18,** 43b.

Boyer, C. I., Bruner, D. W., and Brown, J. A. (1958). A streptobacillus. The cause of tendon sheath infection in turkeys. *Avian Dis.* **2,** 418–427.

Brennan, P. C., Flynn, R. J., and Fritz, T. E. (1965a). The isolation of Salmonella cerro from a closed colony of CF#1 mice. *Lab. Anim. Care* **15,** 260–261.

Brennan, P. C., Fritz, T. E., and Flynn, R. J. (1965b). *Pasteurella pneumotropica:* cultural and biochemical characteristics, and its association with disease in laboratory animals. *Lab. Anim. Care* **15,** 307–311.

Brennan, P. C., Fritz, T. E., and Flynn, R. J. (1969). Role of *Pasteurella pneumotropica* and *Mycoplasma pulmonis* in murine pneumonia. *J. Bacteriol.* **97,** 337–349.

Briody, B. A. (1966). The natural history of mousepox. *Natl. Cancer Inst. Monogr.* No. 20, 105–116.

Brown, T. McP., and Nunemaker, J. C. (1942). Rat-bite fever. A review of the American cases with reevaluation of etiology; report of cases. *Bull. Johns Hopkins Hosp.* **70,** 201–327.

Brownstein, D. G. (1978). Pathogenesis of bacteremia due to *Pseudomonas aeruginosa* in cyclophosphamide-treated mice and potentiation of virulence of endogenous streptococci. *J. Infect. Dis.* **137,** 795–801.

Bullock, W. E., Jr. (1976). Perturbation of lymphocyte circulation in experimental murine leprosy. I. Description of the defect. *J. Immunol.* **117,** 1164–1170.

Bullock, W. E., Jr., Evans, P. E., and Filomeno, A. R. (1977). Impairment of cell-mediated immune responses by infection with *Mycobacterium lepraemurium. Infect. Immun.* **18,** 157–164.

Carlton, W. W., and Hunt, R. D. (1978). Bacterial diseases. *In* "Pathology of Laboratory Animals" (K. Benirschke, F. M. Garner, and T. C. Jones, eds.), pp. 1367–1480. Springer-Verlag, Berlin and New York.

Carter, G. R. (1973). "Diagnostic Procedures in Veterinary Microbiology," 2nd ed. Thomas, Springfield, Illinois.

Carter, P. B., and Collins, F. M. (1974). The route of enteric infection in normal mice. *J. Exp. Med.* **139,** 1189–1203.

Carter, P. B., Woolcock, J. B., and Collins, F. M. (1975). Involvement of the upper respiratory tract in orally induced salmonellosis in mice. *J. Infect. Dis.* **131,** 570–574.

Chang, I. K., Yang, I. T., and Lew, J. (1972). A study of the pathogenesis of murine leprosy infection of mice. (In Korean.) *Yonsee Vidae Nonmunjip* **5,** 209–230.

Clarke, M. C., Taylor, R. J., Hall, G. A., and Jones, P. W. (1978). The occurrence in mice of facial and mandibular abscesses associated with *Staphylococcus aureus. Lab. Anim.* **12,** 121–123.

Closs, O., and Haugen, O. A. (1973). Experimental murine leprosy. 1. Clinical and histological evidence for varying susceptibility of mice to infection with *Mycobacterium lepraemurium. Acta Pathol. Microbiol. Scand., Sect. A* **81,** 401–410.

Closs, O., and Haugen, O. A. (1974). Experimental murine leprosy. 2. Further evidence for varying susceptibility of outbred mice and evaluation of the response of 5 inbred strains to infection with *Mycobacterium lepraemurium. Acta Pathol. Microbiol. Scand., Sect. A* **82,** 459–474.

Cole, B. C., and Cassell, G. H. (1979). Mycoplasma infections as models of chronic joint inflammation. *Arthritis Rheum.* **22,** 1375–1381.

Cole, B. C., Ward, J. R., and Golightly-Rowland, L. (1973). Factors influencing the susceptibility of mice to *Mycoplasma arthritidis. Infect. Immun.* **7,** 218–225.

Collier, W. A. (1948). Über eine pneumonie- und Arthritis-Epizootie bei weissen Mäusen. *Pathol. Microbiol.* **11,** 133–145.

Collins, F. M., and Carter, P. B. (1978). Growth of salmonellae in orally infected germfree mice. *Infect. Immun.* **21,** 41–47.

Cumming, E. L. W., and Cumming, C. N. W. (1967). News briefs. The laboratory mouse: Diseases. Tyzzer's disease. *Carworth Q. Lett.* No. 74.

Dienes, L. (1939). L organisms of Klieneberger and *Streptobacillus moniliformis*. *J. Infect. Dis.* **65,** 24–42.

Emmons, C. W. (1949). Histoplasmosis in animals. *Trans. N.Y. Acad. Sci.* **11,** 248–254.

Emmons, C. W., Bell, J. A., and Olson, B. J. (1947). Naturally occurring histoplasmosis in *Mus musculus* and *Rattus norvegicus*. *Public Health Rep.* **62,** 1642–1646.

Erling, H. G., and Rokey, N. W. (1960). *Salmonella dublin* in Arizona. *J. Am. Vet. Med. Assoc.* **136,** 381–387.

Fauve, R. M., Pierce-Chase, C. H., and Dubos, R. (1964). Corynebacterial pseudotuberculosis in mice. II. Activation of natural and experimental latent infections. *J. Exp. Med.* **120,** 283–304.

Fischl, V., Koech, M., and Kussat, E. (1931). Infekarthritis bei Muriden. *Z. Hyg. Infektionskr.* **112,** 421–425.

Flamm, H. (1957). *Klebsiella*-enzootic in einer Mäusezucht. *Schweiz. Z. Pathol. Bakteriol.* **20,** 23–27.

Flynn, R. J. (1963). *Pseudomonas aeruginosa* infection and radiobiological research at Argonne National Laboratory: effects, diagnosis, epizootiology, control. *Lab. Anim. Care* **13,** 25–35.

Fox, C. L., Jr., Sampath, A. C., and Stanford, J. W. (1970). Virulence of *Pseudomonas* infection in burned rats and mice. Comparative efficacy of silver sulfadiazine and mafenide. *Arch. Surg. (Chicago)* **101,** 508–512.

Freundt, E. A. (1956a). *Streptobacillus moniliformis* infection in mice. *Acta Pathol. Microbiol. Scand.* **38,** 231–245.

Freundt, E. A. (1956b). Experimental investigations into the pathogenicity of the L-phase variant of *Streptobacillus moniliformis*. *Acta Pathol. Microbiol. Scand.* **38,** 246–258.

Freundt, E. A. (1959). Arthritis caused by *Streptobacillus moniliformis* and pleuropneumonia-like organisms in small rodents. *Lab. Invest.* **8,** 1358–1375.

Fries, A. S. (1979). Studies on Tyzzer's disease: acquired immunity against infection and activation of infection by immunosuppressive treatment. *Lab. Anim.* **13,** 143–147.

Fujiwara, K. (1967). Complement fixation reaction and agar gel double diffusion test in Tyzzer's disease of mice. *Jpn. J. Microbiol.* **11,** 103–117.

Fujiwara, K., Takagaki, Y., Maejima, K., Kato, K., Naiki, M., and Tajima, Y. (1963). Tyzzer's disease in mice: Pathologic studies on experimentally infected animals. *Jpn. J. Exp. Med.* **33,** 183–202.

Fujiwara, K., Takagaki, Y., and Naiki, M. (1964). Tyzzer's disease in mice: Effects of corticosteroids on the formation of liver lesions and the level of blood transaminases in experimentally infected animals. *Jpn. J. Exp. Med.* **34,** 59–75.

Ganaway, J. R., Allen, A. M., and Moore, T. D. (1971). Tyzzer's disease. *Am. J. Pathol.* **64,** 717–730.

Glaser, R. J., Berry, J. W., and Loeb, L. H. (1953). Production of group A streptococcal cervical lymphadenitis in mice. *Proc. Soc. Exp. Biol. Med.* **82,** 87–92.

Gledhill, A. W., and Rees, R. J. W. (1952). A spontaneous enterococcal disease of mice and its enhancement by cortisone. *Br. J. Exp. Pathol.* **33,** 183–189.

Goetz, M. E., and Taylor, D. O. N. (1967). A naturally occurring outbreak of *Candida tropicalis* infection in a laboratory mouse colony. *Am. J. Pathol.* **50,** 361–369.

Gordon, L. E., Ruml, D., Hahne, H. J., and Miller, C. P. (1955). Studies on susceptibility to infection following ionizing radiation. IV. The pathogenesis of the endogenous bacteremias in mice. *J. Exp. Med.* **102,** 413–424.

Gorrill, R. H., and McNeil, E. (1963). Staphylococcal infection in the mouse. I. The effect of route of infection. *Br. J. Exp. Pathol.* **44,** 414–415.

Gowen, J. W., and Schott, R. G. (1933). Genetic predisposition to *Bacillus piliformis* infection among mice. *J. Hyg.* **33,** 370–378.

Gray, D. F., and Campbell, A. L. (1953). The use of chloramphenicol and foster mothers in the control of natural pasteurellosis in experimental mice. *Aust. J. Exp. Biol. Med. Sci.* **31,** 161–166.

Greenwood, M., and Topley, W. W. C. (1925). A further contribution to the experimental study of epidemiology. *J. Hyg.* **24,** 45–110.

Habermann, R. T., and Williams, F. P. (1958). Salmonellosis in laboratory animals. *J. Natl. Cancer Inst.* **20,** 933–947.

Hagedoorn-La Brand, A. C., and Hagedoorn, A. L. (1920). Inherited predisposition for a bacterial disease. *Am. Nat.* **54,** 368–375.

Hammond, C. (1963). *Pseudomonas aeruginosa* infection and its effects on radiobiological research. *Lab. Anim. Care* **13,** 6–10.

Harwick, H. J., Kalmanson, G. M., Fox, M. A., and Guze, L. B. (1973). Arthritis in mice due to infection with *Mycoplasma pulmonis*. I. Clinical and microbiological features. *J. Infect. Dis.* **128,** 533–534.

Heilman, F. R. (1940). Chemotherapy in experimental infections caused by *Streptobacillus moniliformis*. *Science* **91,** 366–367.

Hemmert-Halswick, A. (1934). Über eine lepraähnliche Erkrankung bei Mäusen. *Arch. Wiss. Prakt. Tierheilkd.* **67,** 534–539.

Hoag, W. G., Wetmore, P. W., Rogers, J., and Meier, H. (1962). A study of latent pasteurella infection in a mouse colony. *J. Infect. Dis.* **111,** 135–140.

Hoag, W. G., Strout, J., and Meier, H. (1964). Isolation of *Salmonella* spp. from laboratory mice and from diet supplements. *J. Bacteriol.* **88,** 534–536.

Hoag, W. G., Strout, J., and Meier, H. (1965). Epidemiological aspects of the control of *Pseudomonas* infection in mouse colonies. *Lab. Anim. Care* **15,** 217–225.

Hojo, E. (1939). On the bacillus Pseudotuberculosis murium prevalent in mouse. *Jpn. J. Exp. Med.* **10,** 113–114. (Abstr.)

Hook, E. W., Wagner, R. R., and Lancefield, R. C. (1960). An epizootic in Swiss mice caused by a Group A streptococcus, newly designated type 50. *Am. J. Hyg.* **72,** 111–119.

Hsu, H. S., and Nakoneczna, I. (1975). Antimicrobial function of R. E. cells—Part B. *J. Reticuloendothel. Soc.* **18,** 42b.

Iwata, K., Kanda, Y., and Yamaguchi, H. (1978). Electron microscope study on phagocytosis of staphylococci by mouse peritoneal macrophages. *Infect. Immun.* **19,** 649–658.

Jawetz, E. (1948). A latent pneumotropic pasteurella of laboratory animals. *Proc. Soc. Exp. Biol. Med.* **68,** 46–48.

Jawetz, E. (1950). A pneumotropic pasteurella of laboratory animals. I. Bacteriological and serological characteristics of the organism. *J. Infect. Dis.* **86,** 172–183.

Jawetz, E., and Baker, W. H. (1950). A pneumotropic pasteurella of laboratory animals. II. Pathological and immunological studies with the organism. *J. Infect. Dis.* **86,** 184–196.

Jones, J. B., Estes, P. C., and Jordan, A. E. (1972). *Proteus mirabilis* infection in a mouse colony. *J. Am. Vet. Med. Assoc.* **161,** 661–664.

Jones, P. W. (1976). Salmonellosis in wild mammals. *J. Hyg.* **77,** 51–54.

Jones, T. C. (1967). Pathology of the liver of rats and mice. *In* "Pathology of Laboratory Rats and Mice" (E. Cotchin and F. J. C. Roe, eds.), pp. 11–12. Davis, Philadelphia, Pennsylvania.

Juhr, N.-C., and Horn, J. (1975). Modellinfektion mit *Corynebacterium kutscheri* bei der Maus. *Z. Versuchstierkd.* **17,** 129–141.

Kawaguchi, Y. (1959). Classification of mouse leprosy. *Jpn. J. Exp. Med.* **29,** 651–663.

Keystone, E. C., Taylor-Robinson, D., Osborn, M. F., Ling, L., Pope, C., and Fornasier, V. (1980). Effect of T-cell deficiency on the chronicity of arthritis induced in mice by *Mycoplasma pulmonis*. *Infect. Immun.* **27,** 192–196.

Klieneberger, E. (1935). The natural occurrence of pleuropneumonia-like organisms in apparent symbiosis with *Streptobacillus moniliformis* and other bacteria. *J. Pathol. Bacteriol.* **40,** 93–105.

Klieneberger, E. (1936). Further studies on *Streptobacillus moniliformis* and its symbiont. *J. Pathol. Bacteriol.* **42,** 587–598.

Korpinen, E.-L., and Ylimäki, A. (1972). Discovery of toxicogenic *Stachybotrys chartarum* strains in Finland. *Experientia* **28,** 108–109.

Krakower, C., and González, L. M. (1937). Spontaneous leprosy in a mouse. *Science* **86,** 617–618.

Krakower, C., and González, L. M. (1940). Mouse leprosy. *Arch. Pathol.* **30,** 308–329.

Kutscher (1894). Ein Beitrag zur Kenntniss der bacillären Pseudotuberculose der Nagethiere. *Z. Hyg. Infektionskr.* **18,** 327–342.

Kutschera, F. (1908). Eine spontane Streptokokkenepidemie unter weissen Mäusen. *Zentralbl. Bakteriol., Parasitenkd. Infektionskr., Abt. 1: Orig.* **46,** 671–673.

Lagrange, P. H., and Hurtel, B. (1979). Local immune response to *Mycobacterium lepraemurium* in C3H and C57BL/6 mice. *Clin. Exp. Immunol.* **38,** 461–474.

Lawrence, J. J. (1957). Infection of laboratory mice with *Corynebacterium murium. Aust. J. Sci.* **20,** 147.

Lefford, M. J., Patel, P. J., Poulter, L. W., and Mackaness, G. B. (1977). Induction of cell-mediated immunity to *Mycobacterium lepraemurium* in susceptible mice. *Infect. Immun.* **18,** 654–659.

Lerner, E. M., and Sokoloff, L. (1959). The pathogenesis of bone and joint infection produced in rats by *Streptobacillus moniliformis. AMA Arch. Pathol.* **67,** 364–372.

Levaditi, C., Selbie, R.-F., and Schoen, R. (1932). La rhumatisme infectieux spontané de la souris provoqué par le *Streptobacillus moniliformis. Ann. Inst. Pasteur, Paris* **48,** 308–343.

Lindsey, J. R., and Cassell, G. H. (1973). Experimental *Mycoplasma pulmonis* infection in pathogen-free mice. *Am. J. Pathol.* **72,** 63–83.

Lindsey, J. R., Cassell, G. H., and Baker, H. J. (1978). Diseases due to mycoplasmas and rickettsias. *In* "Pathology of Laboratory Animals" (K. Benirschke, F. M. Garner, and T. C. Jones, eds.), pp. 1481–1550. Springer-Verlag, Berlin and New York.

Litchfield, J. H., and Margard, W. L. (1963). Occurrence of unusual salmonellae in laboratory mice. *J. Bacteriol.* **85,** 1451–1452.

Loeffler, F. (1892). Ueber epidemieen unter den im hygienischen Institute zu Greifswald gehattenen Mäusen und über die Bekämpfung der Feldmaus plage. *Zentralbl. Bakteriol. Parasitenkd.* **11,** 129–141.

Loewenthal, H. (1932). Ein Streptokokkenstamm von ungewöhnlicher Pathogenität. *Z. Hyg. Infektionskr.* **113,** 445–456.

Lynch, C. J. (1922). An outbreak of mouse typhoid and its attempted control by vaccination. *J. Exp. Med.* **36,** 15–23.

McDougall, P. T., Wolf, N. S., Stenback, W. A., and Trentin, J. J. (1967). Control of *Pseudomonas aeruginosa* in an experimental mouse colony. *Lab. Anim. Care* **17,** 204–214.

Mackie, T. J., Van Rooyen, C. E., and Gilroy, E. (1933). An epizootic disease occurring in a breeding stock of mice: bacteriological and experimental observations. *Br. J. Exp. Pathol.* **14,** 132–136.

Markham, N. P., and Markham, J. G. (1966). Staphylococci in man and animals. Distribution and characteristics of strains. *J. Comp. Pathol.* **76,** 49–56.

Migaki, G., Voelker, F. A., and Sagartz, J. W. (1978). Fungal diseases. *In* "Pathology of Laboratory Animals" (K. Benirschke, F. M. Garner, and T. C. Jones, eds.), pp. 1551–1586. Springer-Verlag, Berlin and New York.

Miller, C. P., Hammond, C. W., and Tompkins, M. (1950). The incidence of bacteremia in mice subjected to total body X-radiation. *Science* **111,** 540–541.

Miller, C. P., Hammond, C. W., and Tompkins, M. (1951). The role of infection in radiation injury. *J. Lab. Clin. Med.* **38,** 331–343.

Moore, G. J., and Aldred, P. (1978). Treatment of *Pasteurella pneumotropica* abscesses in nude mice (nu/nu). *Lab. Anim.* **12,** 227–228.

Mullink, J. W. (1968). A case of actinomycosis in a male NZW mouse. *Z. Versuchtierkd.* **10,** 225–227.

Needham, J. R., and Cooper, J. E. (1975). An eye infection in laboratory mice associated with *Pasteurella pneumotropica. Lab. Anim.* **9,** 197–200.

Needham, J. R., and Cooper, J. E. (1976). Bulbourethral gland infections in mice associated with *Staphylococcus aureus. Lab. Anim.* **10,** 311–315.

Nelson, J. B. (1954). Association of group A streptococci with an outbreak of cervical lymphadenitis in mice. *Proc. Soc. Exp. Biol. Med.* **86,** 542–545.

Nishimura, S., Kawaguchi, Y., Kohsaka, K., and Mori, T. (1964). Contamination of healthy mice with murine leprosy-like acid-fast bacillus. *Repura* **33,** 245–256.

Nutini, L. G., and Berberich, N. J., Jr. (1965). Effect of diet and strain difference on the virulence of *Staphylococcus aureus* for mice. *Appl. Microbiol.* **13,** 614–617.

Olitsky, P. K., and Syverton, J. T. (1934). Bacteriological studies on an epizootic of intestinal disease in suckling and newly weaned mice. *J. Exp. Med.* **60,** 385–394.

Olson, B. J., Bell, J. A., and Emmons, C. W. (1947). Studies on histoplasmosis in a rural community. *Am. J. Public Health* **37,** 441–449.

Olson, L. D., and Ediger, R. D. (1972). Histopathologic study of the heads of circling mice infected with *Pseudomonas aeruginosa. Lab. Anim. Sci.* **22,** 522–527.

Pattyn, S. R., and Verdoolaege-Van Loo, G. (1969). Isolation of a strain of *Mycobacterium lepraemurium* from normal laboratory mice. *Ann. Soc. Belge Med. Trop.* **49,** 465–468.

Pavlovskis, O. R., Voelker, F. A., and Shackelford, A. H. (1976). *Pseudomonas aeruginosa* exotoxin in mice: Histopathology and serum enzyme changes. *J. Infect.Dis.* **133,** 253–259.

Pierson, C. L., Johnson, A. G., and Feller, I. (1976). Effect of cyclophosphamide on the immune response to *Pseudomonas aeruginosa* in mice. *Infect. Immun.* **14,** 168–177.

Ptak, W., Gaugas, J. M., Rees, R. J. W., and Allison, A. C. (1970). Immune responses in mice with murine leprosy. *Clin. Exp. Immunol.* **6,** 117–124.

Read, W. J., and Savage, W. G. (1913). Gaertner group bacilli in rats and mice. *J. Hyg.* **13,** 343–352.

Reed, D. M. (1902). The bacillus *Pseudo-tuberculosis murium;* its streptothrix forms and pathogenic action. *Johns Hopkins Hosp. Rep.* **9,** 515–541.

Ridley, D. D., and Jopling, W. H. (1966). Classification of leprosy according to immunity. A five-group system. *Int. J. Lepr.* **34,** 255–273.

Rights, F. L., Jackson, E. B., and Smadel, J. E. (1947). Observations on Tyzzer's disease in mice. *Am. J. Pathol.* **23,** 627–633.

Robbins, S. L. (1974). "Pathologic Basis of Disease." Saunders, Philadelphia, Pennsylvania.

Sacquet, E., Drouhet, E., and Vallee, A. (1959). Un cas spontané de cryptococcose (*Cryptococcus neoformans*) chez la souris. *Ann. Inst. Pasteur, Paris* **97,** 252–253.

Sangiorgi, G. (1911). Ueber einen coliähnlichen Bacillus einer spontanen Epizootie der weissen Mäuse. *Zentralbl. Bakteriol., Parasitenkd. Infektionskr., Abt. 1: Orig.* **57,** 57–79.

Saunders, L. Z. (1958). Tyzzer's disease. *J. Natl. Cancer Inst.* **20,** 893–897.

Savage, N. L. (1972). Host-parasite relationships in experimental *Streptobacillus moniliformis* arthritis in mice. *Infect. Immun.* **5,** 183–190.

Sawicki, L., Bruce, H. M., and Andrewes, C. H. (1962). *Streptobacillus moniliformis* as a probable cause of arrested pregnancy and abortion in laboratory mice. *Br. J. Exp. Pathol.* **43,** 194–197.

Schneemilch, H. D. (1976). A naturally acquired infection of laboratory mice with *Klebsiella* capsule type 6. *Lab. Anim.* **10,** 305–310.

Sebesteny, A. (1973). Abscesses of the bulbourethral glands of mice due to *Pasteurella pneumotropica*. *Lab. Anim.* **7,** 315-317.

Seiffert, G., Jahncke, A., and Arnold, A. (1928). Zeitliche Untersuchungen über den Ablauf übertragbarer Krankheiten I.—III. (Mäusetyphus). *Zentralbl. Bakteriol., Parasitenkd. Infektionskr., Abt. 1: Orig.* **109,** 193-225.

Shechmeister, I. L., and Adler, F. L. (1953). Activation of pseudotuberculosis in mice exposed to sublethal total body radiation. *J. Infect. Dis.* **92,** 228-239.

Shimi, A., Keyhani, M., and Hedayati, K. (1979). Studies on salmonellosis in the house mouse, *Mus musculus*. *Lab. Anim.* **13,** 33-34.

Shults, F. S., Estes, P. C., Franklin, J. A., and Richter, C. B. (1973). Staphylococcal botryomycosis in a specific-pathogen-free mouse colony. *Lab. Anim. Sci.* **23,** 36-42.

Smith, A. W. (1973). "Aeromedical Review. Selected Topics in Laboratory Animal Medicine. Vol. XIX: The Mouse." USAF Sch. Aerosp. Med., Aerosp. Med. Div., Brooks Air Force Base, Texas.

Smith, J. M. B., and Austwick, P. K. C. (1967). Fungal diseases of rats and mice. *In* "Pathology of Laboratory Rats and Mice" (E. Cotchin and F. J. C. Roe, eds.), pp. 681-732. Davis, Philadelphia, Pennsylvania.

Smith, W. (1941). Cervical abscesses in guinea pigs. *J. Pathol. Bacteriol.* **53,** 29-37.

Sonkin, L. S. (1949). Infections induced in mice by local application of streptococci and pneumococci to the nasal mucosa and by intrapulmonary instillation. *J. Infect. Dis.* **84,** 290-305.

Speirs, R. S. (1956). Effect of oxytetracycline upon cortisone-induced pseudotuberculosis in mice. *Antibiot. Chemother.* **6,** 395-399.

Stewart, D. D., Buck, G. E., McConnell, E. E., and Amster, R. L. (1975). An epizootic of necrotic dermatitis in laboratory mice caused by Lancefield group G streptococci. *Lab. Anim. Sci.* **25,** 296-302.

Stookey, J. L., and Moe, J. V. (1978). The respiratory system. *In* "Pathology of Laboratory Animals" (K. Benirschke, F. M. Garner, and T. C. Jones, eds.), pp. 71-113. Springer-Verlag, Berlin and New York.

Strangeways, W. F. (1933). Rats as carriers of *Streptobacillus moniliformis*. *J. Pathol. Bacteriol.* **37,** 45-51.

Sushida, K., and Hirano, N. (1966). Investigation for acid-fast bacilli in normal mice with special reference to normal pregnant mice and their fetus. *Repura* **35,** 179-186.

Takagaki, Y., Ito, M., and Naiki, M. (1966). Experimental Tyzzer's disease in different species of laboratory animals. *Jpn. J. Exp. Med.* **36,** 519-534.

Tuffery, A. A. (1956). The laboratory mouse in Great Britain.—IV. Intercurrent infection (Tyzzer's disease). *Vet. Rec.* **68,** 511-515.

Tyzzer, E. E. (1917). A fatal disease of the Japanese waltzing mouse caused by a spore-bearing bacillus (*Bacillus piliformis* N. SP.) *J. Med. Res.* **37,** 307-338.

Urano, T., and Maejima, K. (1978). Provocation of pseudomoniasis with cyclophosphamide in mice. *Lab. Anim.* **12,** 159-161.

Van Rooyen, C. E. (1936). The biology, pathogenesis and classification of *Streptobacillus moniliformis*. *J. Pathol. Bacteriol.* **43,** 455-472.

Volkman, A., and Collins, F. M. (1976). Role of host factors in the pathogenesis of *Salmonella*-associated arthritis in rats. *Infect. Immun.* **13,** 1154-1160.

Ward, G. E., Moffatt, R., and Olfert, E. (1978). Abortion in mice associated with *Pasteurella pneumotropica*. *J. Clin. Microbiol.* **8,** 177-180.

Wayson, N. E. (1927). An epizootic among meadow mice in California, caused by the bacillus of murine septicemia or of swine erysipelas. *Public Health Rep.* **22,** 1489-1493.

Weisbroth, S. H., and Scher, S. (1968). *Corynebacterium kutscheri* infection in the mouse. I. Report of an outbreak, bacteriology, and pathology of spontaneous infections. *Lab. Anim. Care* **18,** 451-458.

Weisbroth, S. H., Scher, S., and Boman, I. (1969). *Pasteurella pneumotropica* abscess syndrome in a mouse colony. *J. Am. Vet. Med. Assoc.* **155,** 1206-1210.

Wensinck, F. (1961). The origin of endogenous *Proteus mirabilis* bacteremia in irradiated mice. *J. Pathol. Bacteriol.* **81,** 395-400.

Wensinck, F., and Renaud, H. (1957a). Bacteraemia in irradiated mice. *Br. J. Exp. Pathol.* **38,** 483-488.

Wensinck, F., and Renaud, H. (1957b). Murine group G streptococci. *Br. J. Exp. Pathol.* **38,** 489-492.

Wensinck, F., Van Bekkum, D. W., and Renaud, H. (1957). The prevention of *Pseudomonas aeruginosa* infections in irradiated mice and rats. *Radiat. Res.* **7,** 491-499.

Williams, S. (1941). An outbreak of infection due to *Streptobacillus moniliformis* among wild mice. *Med. J. Aust.* **28,** 357-359.

Wilson, G. S. (1926). A spontaneous epidemic in mice associated with Morgan's bacillus, and its bearing on the aetiology of summer diarrhoea. *J. Hyg.* **26,** 170-186.

Wilson, G. S., and Miles, A. (1975). "Topley's and Wilson's Principles of Bacteriology, Virology, and Immunity," 6th ed. Williams & Wilkins, Baltimore, Maryland.

Wilson, P. (1976). *Pasteurella pneumotropica* as the causal organism of abscesses in the masseter muscle of mice. *Lab. Anim.* **10,** 171-172.

Wittler, R. G., and Cary, S. G. (1974). Genus *Streptobacillus* Levaditi, Nicolau and Poincloux 1925, 1188. *In* "Bergey's Manual of Determinative Bacteriology" (R. E. Buchanan and N. E. Gibbons, eds.), 8th ed., pp. 378-381. Williams & Wilkins, Baltimore, Maryland.

Wolff, H. L. (1950). On some spontaneous infections observed in mice. *Antonie van Leeuwenhoek* **16,** 11-18.

Yamamoto, R., and Clark, G. T. (1966). *Streptobacillus moniliformis* infection in turkeys. *Vet. Rec.* **79,** 95-100.

Zucker, T. F. (1957). Pantothenate deficiency in rats. *Proc. Anim. Care Panel* **7,** 193-202.

Chapter 7

Rickettsial and Chlamydial Diseases

Paul K. Hildebrandt

I. Introduction 99
II. Rickettsial Infections 99
A. General 99
B. *Rickettsia akari* 100
C. *Eperythrozoon* and *Hemobartonella* 100
D. Immunity and Latency 104
E. Experimental Infection in Mice 104
F. Diagnosis and Control 105
III. Chlamydial Infections 105
A. General 105
B. *Chlamydia trachomatis* 105
C. *Chlamydia psittaci* (Latent Infection) 106
References 106

I. INTRODUCTION

This chapter provides a review of rickettsial and chlamydial agents which cause natural infection in mice. A general review of taxonomic position, morphology and other general characteristics of these agents, and the diseases they produce in mice is presented. The mouse as an experimental animal in rickettsial research is also discussed.

II. RICKETTSIAL INFECTIONS

A. General

Rickettsial diseases are caused by microorganisms of the order Rickettsiales. According to Moulder (1974) in "Bergey's Manual of Determinative Bacteriology," eighth edition, three families are accepted in this order: *Rickettsiaceae, Bartonellaceae,* and *Anaplasmataceae*. The classification of Rickettsiales and Clamydiales follows.

This group of fastidious microorganisms is intermediate between the bacteria and the viruses in the classification scheme. In general, and certainly most notably, the rickettsial diseases occurring in man and some animals belong to the family *Rickettsiaceae*. The generic designation *Rickettsia* of many of the microorganisms in this family pays honor to one of the early pioneers, Dr. H. T. Ricketts, who became infected and died of a rickettsial infection while studying Rocky Mountain Spotted Fever and typhus fever. Although members of the family are well known in humans, most are maintained in nature in a life cycle between biting and bloodsucking arthropods, i.e., ticks, mites, fleas, lice, and a large variety of small animals, includ-

ISBN 0-12-262502-1

Order II Rickettsiales	
Family I	Rickettsiaceae
Genus I	*Rickettsia*
Genus II	*Rochalimaea*
Genus III	*Coxiella*
Genus IV	*Cowdria*
Genus V	*Ehrlichia*
Genus VI	*Neorickettsia*
Family II	Bartonellaceae
Genus I	*Bartonella*
Genus II	*Grahamella*
Family III	Anaplasmataceae
Genus I	*Anaplasma*
Genus II	*Paranaplasma*
Genus III	*Aegyptianella*
Genus IV	*Hemobartonella*
Genus V	*Eperythrozoon*
Order II Clamydiales	
Family	Chlamydiaceae
Genus	*Chlamydia*
Species	*trachomatis*
Species	*psittaci*

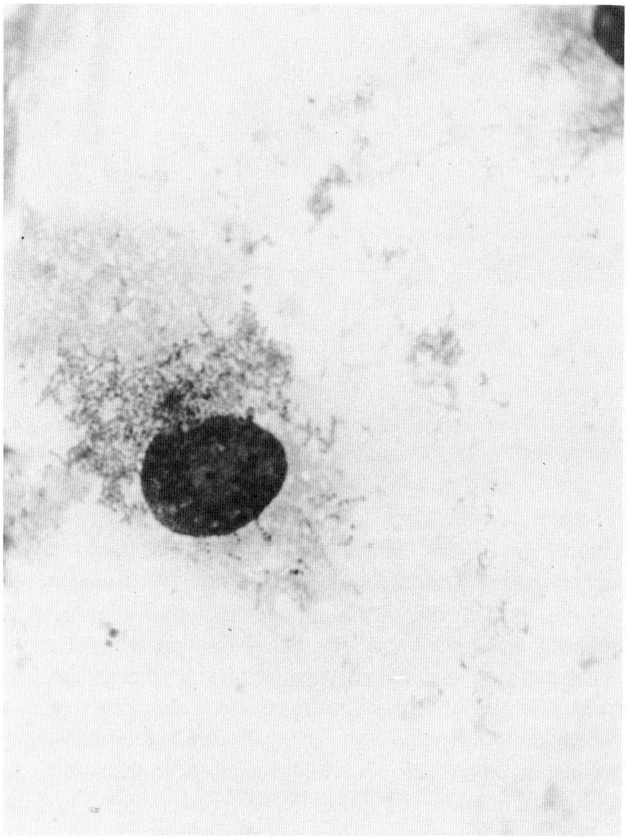

Fig. 1. Chick embryo cell infected with *Rickettsia tsutsugamuchi* which illustrates numerous small, rod-shaped organisms within the cytoplasm. Giemsa stain. ×1000. (Courtesy of Dr. E. H. Stevenson.)

ing rats and mice. In recognition of the historical background accompanying members of the family *Rickettsiaceae,* a brief description will follow.

Rickettsia in the family *Rickettsiaceae* generally are small and rod-shaped, coccoid, occasionally pleomorphic, having typical bacterial cell walls and absence of flagella. They are gram-negative and multiply by binary fission within host cells (Fig. 1). Most are insect transmitted, and those that parasitize vertebrates usually affect (1) reticuloendothelial cells, particularly endothelium, causing vasculitis or (2) erythrocytes causing hemolytic anemia with corresponding pathological changes. It should be noted that *Coxiella burnetii,* the causative agent of Q fever in man, is insect transmitted but that other means of transmission, such as inhalation or ingestion of infected particles, have been reported in natural outbreaks.

Few of the Rickettsiaceae cause natural disease in mice; however, laboratory mice are used in conjunction with other diagnostic tests, particularly in the diagnosis of *Rickettsia tsutsugamushi* (scrub typhus) and *R. akari* (rickettsialpox). The guinea pig is very susceptible to many organisms in this family and has been used extensively as a test animal for isolation and identification of rickettsia in suspect material (Burrows, 1973).

B. *Rickettsia akari*

Rickettsialpox of man is caused by *R. akari.* The house mouse, *Mus musculus,* is the natural mammalian reservoir for the agent, and a bloodsucking mite, *Allodermanyssus sanguineus,* transmits the organism among mice and occasionally to man (Huebner *et al.,* 1946; Moulder, 1973). Laboratory-infected mice usually become ill between the ninth and thirteenth day, showing clinical signs of lethargy, ruffled fur, labored breathing, and anorexia. Animals that die show inguinal and axillary lymphadenopathy, increased amounts of peritoneal fluid, and enlarged, congested spleens (Rose, 1949).

C. *Eperythrozoon* and *Hemobartonella*

1. General

There are two genera, *Eperythrozoon* and *Hemobartonella,* in the family Anaplasmaceae, that infect mice and rats as well as a wide variety of other animals. *Eperythrozoon coccoides* is the *Eperythrozoon* species of major concern in mice, and *Hemobartonella muris,* although infectious for mice, is of major concern in rats. These two organisms have several common characteristics: Both are transmitted by insects; they parasitize and have a characteristic position on erythrocytes; both can cause primary acute anemia with fever in mice; they can be latent infections for long periods of time; and they may complicate research when mice, unknowingly infected with these agents, are used in experimental study. Although there are similarities, differences such as morphology and occurrence to include the presence or absence of organisms in the plasma warrant the separation of these two genera. Even though the nature of these animal pathogens is still in doubt, it is hoped that this chapter will provide a better understanding of the infections in mice. Much of the information known about these two organisms comes from work done in mice (Gothe and Kreier, 1977; Baker *et al.*, 1971).

2. *Eperythrozoon*

The genus *Eperythrozoon* was created by Schilling (1928) for parasites he observed in mouse blood. They stain bluish- or pinkish-violet with Giemsa and Romanowsky-type stains. They appear as rings or coccoids, 0.4–1.5 μm in diameter, usually round with numerous annular or disk-shaped elements (Moulder, 1974). Rod-shaped forms are rare but may occur partly or entirely circling an erythrocyte. (Figs. 2 and 3). Organisms occur on erythrocytes and free in the plasma. In fresh blood preparations observed by contrast microscopy, only cocci 0.5–1.0 μm in diameter occur and ring forms are not observed (Kreier and Ristic, 1963). The ring form is believed to be an artifact resulting from a coccus having its contents dispersed around the margin during drying (Gothe and Kreier, 1977). Electron microscopic examination has demonstrated that *E. coccoides* has no cell wall and is surrounded by a single limiting membrane. No membrane-bound nucleus or other membrane-bound organelles are present.

In nature, the mouse louse, *Polyplax serratus,* is a known vector for transmission of *E. coccoides* (Eliot, 1936). The mouse, once infected, may remain clinically normal even with a high parasitemia (Ansari *et al.*, 1963) or develop anemia and splenomegaly, become febrile, and occasionally die (Baker *et*

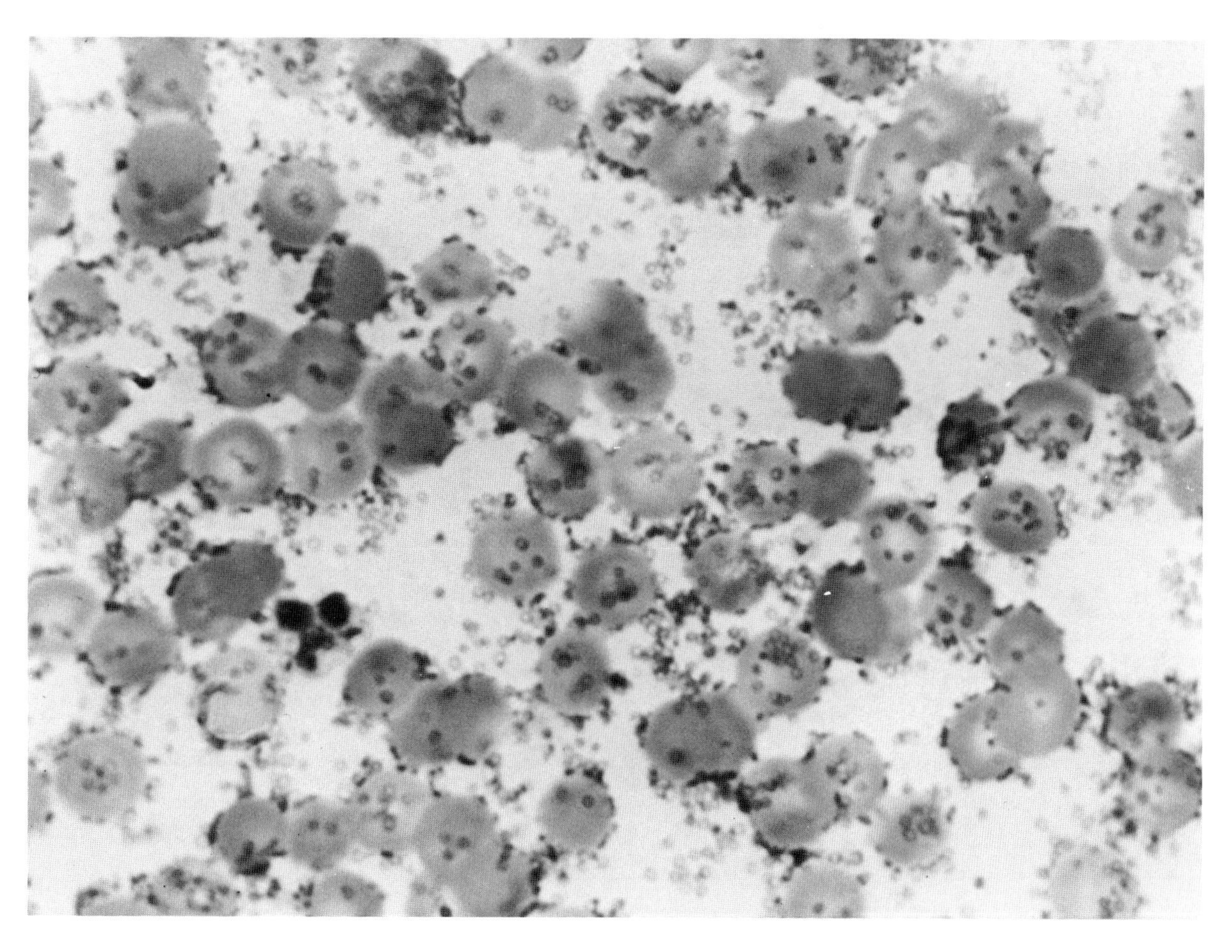

Fig. 2. Mouse blood infected with *Eperythrozoon coccoides*. Note the massive number of ring-shaped organisms attached to the erythrocytes and free in plasma. Giemsa stain. ×1500. (By permission of the *American Journal of Pathology*.)

Fig. 3. Scanning electron micrograph of mouse erythrocytes infected with *Eperythrozoon coccoides*. Note the extreme distortion of the erythrocytes and the massive number of organisms (fixed in glutaraldehyde and coated with gold). ×5000. (Courtesy of Dr. H. J. Baker.)

al., 1971). Baker and colleagues (1971) observed the temporal sequence of the splenic enlargement in pathogen-free mice infected with *E. coccoides*. By the fourth or fifth day postinfection, the splenic corpuscles had increased in size and contained many blast and stem cells. These were accompanied by scattered, large, phagocytic cells containing cellular debris. This cellular proliferation predominated for several days to the extent that the differentiation of red pulp and white pulp was less apparent. Erythroid elements then appeared and were followed by the occurrence of many plasma cells after the ninth day. The reticuloendothelial cell proliferation may also occur in lymph nodes and involve Kupffer cells of the liver. Hepatocellular degeneration and multifocal necrosis have been reported in acute infections with *E. coccoides* and *H. muris* (McCluskie and Niven, 1934).

3. *Hemobartonella*

The paucity of reports of *H. muris* infection in mice suggests that bartonellosis is not a major problem in mouse colonies. The agent is primarily a pathogen of rats; however, because it is reported to infect mice and most other rodents (Griesemer, 1958) and because of its similarity to *E. coccoides*, a discussion of this organism is in order.

Hemobartonella muris was observed for the first time in 1921 (Mayer, 1921). The organism was incorrectly placed in the genus *Bartonella*, which was subsequently split into two genera: *Bartonella*, which contains the single species *Bartonella bacilliformis*, a pathogen for man; and *Hemobartonella*, into which all species infecting animals other than man were placed (Weinman, 1944). With Giemsa-stained smears, the deep-purplish organisms appear as slender rods with rounded ends 0.7–3.0 μm long and 0.1–0.2 μm in diameter, or as cocci 0.3–0.7 μm in diameter. The organisms lie over the surface of the erythrocyte or are attached to the margin (Fig. 4). Ring forms and parasites free in the plasma are rare (Gothe and Kreier, 1977). Electron microscopically, *Hemobartonella*, like *Eperythrozoon*, has a single limiting membrane and no cell wall. No membrane-bound nucleus or other membrane-bound organelles are observed. The erythrocyte membranes

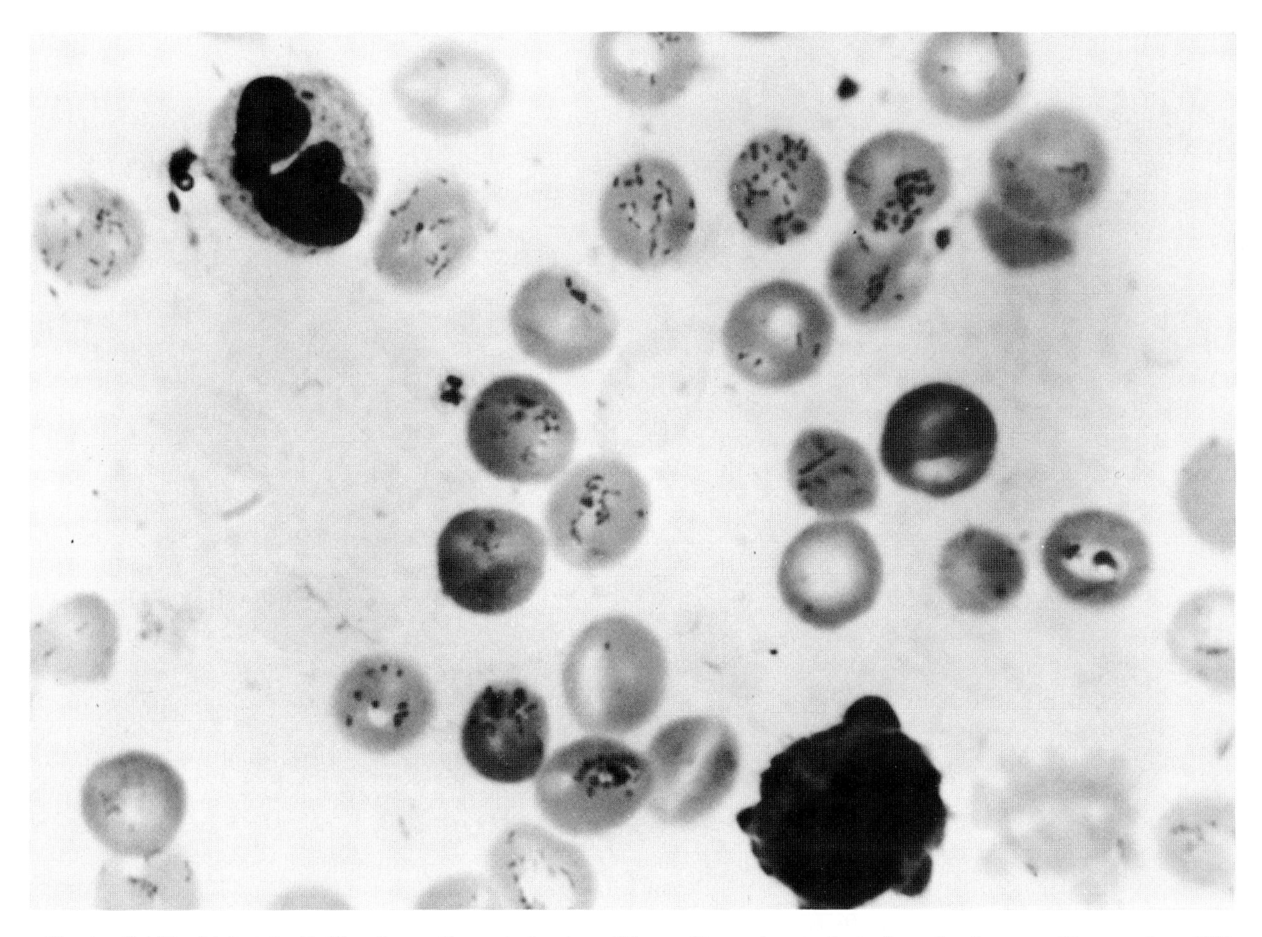

Fig. 4. Rat blood infected with *Hemobartonella muris* showing solid coccoid organisms on the surface of erythrocytes. Giemsa stain. ×2000. (Courtesy of Dr. H. G. Baker.)

are intact at the site of parasite adhesion, and a space exists between the parasite and the host membrane (Gothe and Kreier, 1977).

Transmission of *Hemobartonella* in nature has been shown to be spread by the rat louse, *Polyplax spinulosa* (Crystal, 1958). Transplacental transmission has not been proved. The essential elimination of this disease in recent years in cesarean-derived rat colonies would support nontransplacental transmission. Lindsey *et al.* (1978) and Baker *et al.* (1971) emphasize inadvertent transmission through contaminated biological material as being the major means of transmission in today's rat colonies, where modern ectoparasite control measures are adequate. The infectivity of this organism is very high; one organism can establish infection (Wigand, 1958, cited in Baker *et al.,* 1971).

Although the natural disease is usually inapparent, severe disease can occur by splenectomizing *H. muris*-infected rats or by experimentally infecting splenectomized rats. A severe hemolytic anemia with marked reduction in erythrocytes and accompanying reticulocytosis, hemoglobinuria, and death have been reported (Finch and Jonas, 1973). In contrast, rats with intact spleens rarely develop clinical disease; mild transient parasitemia and reticulocytosis have been reported (Rudnick and Hollingsworth, 1959). Histologically, the reticuloendothelial system, especially the spleen, displays the greatest pathophysiologic change and resembles the events described for *E. coccoides* in mice.

D. Immunity and Latency

The basic concept is that infection is long-lived and exhibits two modes of clinical manifestations: acute febrile anemia and latent or asymptomatic infection. The latter seems to be an actual expression of immunity toward the acute form and can be interrupted by spleen removal, hence "splenic immunity." Latent infections are not limited to postcontrol of the acute form, for mice with no history of the disease can be carriers and develop the acute disease when the spleen is removed. The incidence of carriers is not known; however, it is apparent that the carrier state may be for the life of the mouse, even when infection is contracted at an early age. Other procedures may have an effect similar to that of spleen removal, i.e., irradiation, immunosuppressive drugs, anti-lymphocytic serum, and certain associated diseases. The mouse organism, *E. coccoides,* has for a long time attracted attention because it is a striking example of a heterologous latent infection which may convert an otherwise benign hepatitis virus infection into a fatal disease, and at the same time, both parasitemia and virus titer may increase many-fold (Niven *et al.,* 1952; Gledhill, 1956). Other viral agents of mice cited to be potentiated by *E. coccoides* are the lymphocytic choriomeningitis virus and the lactic dehydrogenase virus (Seamer *et al.,* 1961; Riley, 1964; Riley *et al.,* 1964; Baker *et al.,* 1971; Lindsey *et al.,* 1978). Instead of potentiating certain *plasmodial* infections, as is the case with certain viral agents, *E. coccoides* appears to reduce the pathogenicity. Ott and Stauber (1967) reported that mixed *Plasmodium chabaudi*-*E. coccoides* infections were mild but, after elimination of *E. coccoides* by drug therapy, the *P. chabaudi* alone consistently killed mice. *Plasmodium berghei* is reported to be a milder infection in mice also infected with *E. coccoides* (Finerty *et al.,* 1973). There is concern that some stock inocula of *P. berghei* and other rodent malarial species may be contaminated with *E. coccoides* or *H. muris* (Baker *et al.,* 1971).

Increases in serum IgG_2 and IgM have been shown in murine eperythrozoones (Baker *et al.,* 1971). Other functional changes included hyperphagocytosis and a decreased production of interferon, which was most marked during the acute infection (Suntharasamai and Rytel, 1973). Cox and Calaf-Iturri (1976) demonstrated that serum factors (cold-active hemagglutinin and serum antigen) similar to those described for anaplasmosis, malaria, babesiosis, and viral equine infectious anemia were also present in sera of mice infected with *E. coccoides* and rats infected with *H. muris.* It has been presumed that these factors act as opsonin and cause erythrocytes to be sequestered and phagocytized by the spleen. It is postulated that these serum factors are kept in abeyance by the spleen. In splenectomized animals, these serum factors can reach sufficient titers to activate complement and trigger a hemolytic crisis.

E. Experimental Infection in Mice

Although mice are generally considered to be rather resistant to infection with rickettsial organisms, there are some strains of mice which are susceptible and are utilized in experimental research. Mouse strain susceptibility is important when attempting to determine the virulence of a rickettsial agent if utilizing the mouse as an experimental animal. Sammons *et al.* (1977) attempted to provide a better animal model for the study of rickettsia of the spotted fever group. The following strains of mice were studied: Mai;(S) BALB/cj, DBA/1J, DBA/2J, C3H/Mai, AKR/J and C57BL/6J. Mai;(S) mice were moderately susceptible to *R. siberica* and the BALB/cj mice were moderately susceptible to *R. australia.* With these exceptions, mice were relatively nonsusceptible to the spotted fever group of rickettsia.

The differences in susceptibility of certain strains of mice infected with *R. tsutsugamushi* have been shown to be genetic (Groves and Osterman, 1978). The natural resistance of mice to *R. tsutsugamushi* is apparently controlled by a single autosomal dominant gene, and susceptibility is related to the displacement or other aberration of this gene (Groves *et al.,* 1980).

Defective mouse macrophage function, of probable genetic

origin, is suspected to be the basis for the fact that some strains of mice seem to be more susceptible to *R. akari* infection (Meltzer and Nacy, 1980).

The mouse peritoneal mesothelial cell obtained from BALB/c mice has been utilized as a tissue system to study host–parasite relationships with the agent *R. tsutsugamushi*. An elegant study by Ewing *et al.* (1978) provided insight into the means by which the organism enters peritoneal mesothelial cells, replicates, and exits.

F. Diagnosis and Control

The intensified ectoparasite control and cesarean-derived animals in modern laboratory facilities have reduced the incidence of infections to such a low level that it has been unusual to encounter these diseases in recent years. The inadvertent introduction of *E. coccoides* in biological specimens from conventional colonies and specimens obtained from wild animals submitted for diagnostic purposes must be appreciated by investigators and diagnosticians.

The parenteral inoculation of test material into uninfected splenectomized mice is the most sensitive means of detecting infective material. Splenectomy of mice suspected of latent infection is also the method of choice. Usually, a high parasitemia commences in 48 hr and may be short in duration. Lindsey *et al.* (1978) recommend that blood smears be prepared every 6 hours beginning at 48 hr postsplenectomy to assure that the transient parasitemia is not missed. An indirect fluorescent antibody test has been utilized by Baker *et al.* (1971), and acridine orange staining for *H. felis* has been advocated by Small and Ristic (1967). Scanning and transmission electron microscopy have also been proposed for identification of these organisms (McKee *et al.*, 1973). Treatment of *E. coccoides*-infected mice is in most cases not considered practical; however, chlortetracycline and other tetracyclines, Neosalvarsan, and neoansphenamine have been reported to eliminate the organism (Gothe and Kreier, 1977).

III. CHLAMYDIAL INFECTIONS

A. General

Chlamydia are intracellular parasites which cause disease in a wide variety of animals and man. These intracellular organisms multiply in the cytoplasm of host cells and form membrane-bound cytoplasmic inclusions. The developmental cycle is unique and complicated, consisting of an elementary body which is specialized for extracellular survival and the reticulate body which multiplies within the host cell cytoplasm (Moulder, 1973). The elementary body is a coccus 300 μm in diameter, and the reticulate body ranges up to 1000 μm. *Chlamydia* stain readily with any Romanowsky stain. Storz and Spears (1977), in an excellent in-depth review of chlamydia, have divided the developmental cycle into four major phases:

1. Adsorption of infectious elementary bodies to cells and their uptake into a cytoplasmic site.
2. Primary reorganization of the infecting elementary body into a larger dispersing form and further differentiation until this form is ready to divide as a reticulate body.
3. Multiplication and division of the larger noninfectious but metabolically active reticulate bodies.
4. Secondary reorganization of the reticulate bodies into infectious elementary bodies through a transition stage which they call *condensing forms*.

There are two species of *Chlamydia* which are differentiated by stable metabolic characteristics (Storz and Spears, 1977; Storz, 1971). The generally accepted scheme is as follows: (1) *Chlamydia trachomatis* is characterized by compact, glycogen-positive microcolonies that are inhibited by sulfadiazine. (2) *Chlamydia psittaci* is characterized by diffuse histocytoplasmic microcolonies that do not produce glycogen and are resistant to sulfadiazine.

B. *Chlamydia trachomatis*

Man is the primary host of *C. trachomatis* strains which cause trachoma, inclusion conjunctivitis, lymphogranuloma venereum, and other urogenital tract infections. The only strain of *C. trachomatis* which causes disease other than in man is an agent responsible for causing a pneumonitis in mice. Nigg (1942) isolated a filterable agent from lungs of albino Swiss mice which would produce pneumonitis when passed intranasally into mice. Impression smears from cut sections of infected lungs and stained with Giemsa and other Romanowsky stains demonstrated bodies that were similar to those described for psittacosis and lymphogranuloma venereum. The "Nigg agent" produces an inclusion matrix which is iodine-positive, indicating glycogenlike material in the cytoplasmic inclusions, and its multiplication is inhibited by sulfadiazine (Gordon and Quan, 1965) and (Page, 1968). These characteristics established the agent as a strain of *C. trachomatis,* which is the only strain occurring naturally in a host other than man.

Clinical signs reported by Karr (1943) and Nigg and Eaton (1944) consisted of ruffled fur, hunched posture, indisposition to move about even when prodded, and increasingly labored respiration, with death often occurring in 24 hr. Mice inoculated with smaller doses developed the same signs more slowly and, in addition, showed progressive emaciation and cyanosis of the ears and tail. Gross lesions consisted of scattered focal, pinpoint, slightly elevated, grayish lesions in one or more lobes. The individual lesions increased in size with progression of the infection until they coalesced to produce complete pulmonary consolidation.

Microscopically, the lesion is principally an interstitial pneumonia (Gogolak, 1953). There is initially an alveolar septal cell proliferation, which is the predominant cell type until the first developmental cycle of the agent is complete. The organism, having completed its first cycle in these macrophages and bronchial epithelium, causes these cells to rupture, and an influx of neutrophils occurs. The severity of this pneumonitis depends on the size of the inoculum, and the lesion continues to either consolidation or resolution. Lindsey *et al.* (1978) emphasized that the pneumonitis is produced by passaging latently infected tissues, particularly lung, in mice. Clinically, the ensuing disease is indistinguishable from other respiratory diseases of mice, including respiratory mycoplasmosis.

C. *Chlamydia psittaci* (Latent Infection)

Chlamydia psittaci infects a wide range of avian and mammalian species, including man, and causes pneumonia, enteritis, abortion, conjunctivitis, encephalitis, and polyarthritis. Although there have been no outbreaks of disease in mice associated with *C. psittaci,* there have been two strains of chlamydia isolated from mouse colonies in recent years. In both instances, these were considered to be latent infections. Gerloff and Watson (1970) reported the isolation of a chlamydia from the peritoneal cavity of mice in their colony. The isolation was accomplished by a long series of intraperitoneal passages of liver and spleen. The agent produces ascites, spleen enlargement, and serofibrinous exudate on the liver and spleen. Initial intranasal inoculation of the agent in mice caused little disease; however, subsequent intranasal passages of diseased lungs resulted in consolidation of lungs in all animals inoculated and mortality of 50%. The agent was shown to be glycogen negative and sulfadiazine resistant, and thus, presumably, it is a strain of *C. psittaci*.

Ata *et al.* (1971) isolated three chlamydial agents from three strains of mice which had been inbred for over 25 years and were being utilized in genetic research. The agents were isolated by intranasal inoculation of lung tissue obtained from mice of the respective strain. After the second and third passages, small areas of consolidation or hyperemia were noted on the lungs. Chlamydial bodies were observed in Gimenez-stained impression smears. The agents were cultivated in embryonated eggs. All three agents were glycogen negative and sulfadiazine resistant, thus being compatible with a strain of *C. psittaci*.

Although chlamydia produce no serious spontaneous disease of mice, the fact that latent infections have been demonstrated and can be potentiated should serve as a warning to investigators utilizing the mouse in biomedical research.

REFERENCES

Ansari, K. A., Neilson, C. F., and Stansly, P. G. (1963). Pathogenesis of infectious splenic enlargement in mice. *Exp. Mol. Pathol.* **2,** 61-68.

Ata, F. A., Stephenson, E. H., and Storz, J. (1971). Inapparent respiratory infection of inbred swiss mice with sulfadiazine-resistant, iodine-negative chlamydia. *Infect. Immun.* **4,** 506-507.

Baker, H. J., Cassell, G. H., and Lindsey, J. R. (1971). Research complications due to *Hemobartonella* and *Eperythrozoon* infections in experimental animals. *Am. J. Pathol.* **64,** 625-656.

Burrows, W. (1973). "Textbook of Microbiology," pp. 831-852. Saunders, Philadelphia, Pennsylvania.

Cox, H. W., and Calaf-Iturri, G. (1976). Autoimmune factors associated with anaemia in acute *Haemobartonella* and *Eperythrozoon* infections of rodents. *Ann. Trop. Med. Parasitol.* **70,** 73-79.

Crystal, M. M. (1958). The mechanism of transmission of *Haemobartonella muris* (mayle) of rats by the spined rat louse, *Polyplax spinulosa* (Burmeister). *J. Parasitol.* **44,** 603-606.

Eliot, P. (1936). The insect vector for the natural transmission of eperthyrozoon. *Science* **84,** 397.

Ewing, E. P., Takeuchi, A., Shirai, A., and Osterman, J. V. (1978). Experimental infection of mouse peritoneal mesothelium with scrub typhus Rickettsial: An ultrastructural study. *Infect. Immun.* **19,** 1068-1075.

Finch, S. C., and Jonas, A. M. (1973). Ethyl palmitate-induced bartonellosis as an index of functional splenic ablation. *J. Reticuloendothel. Soc.* **13,** 20-26.

Finerty, J. F., Evans, C. B., and Hyde, C. L. (1973). *Plasmodium berghei* and *Eperythrozoon coccoides:* Antibody and immunoglobulin synthesis in germ-free and conventional mice simultaneously infected. *Exp. Parasitol.* **34,** 76-84.

Gerloff, R. K., and Watson, R. O. (1970). A *Chlamydia* from the peritoneal cavity of mice. *Infect. Immun.* **1,** 64-68.

Gledhill, A. W. (1956). Quantitative aspects of the enhancing of eperythrozoon on the pathogenicity of mouse hepatitis virus. *J. Gen. Microbiol.* **15,** 292-304.

Gogolak, F. M. (1953). The histopathology of murine pneumonitis infection and the growth of the virus in mouse lung. *J. Infect. Dis.* **92,** 254-272.

Gordon, F. B., and Quan, A. L. (1965). Occurance of glycogen in inclusions of the psittacosis-lymphogranuloma venereum-trachoma agents. *J. Infect. Dis.* **115,** 186-196.

Gothe, R., and Kreier, J. P. (1977). Aegyptionella, eperythrozoon, and haemobartonella. *In* "Parasitic Protozoa" (J. P. Kreier, ed.), Vol. 4, pp. 251-294. Academic Press, New York.

Griesemer, R. A. (1958). Bartonellosis. *J. Natl. Cancer Inst.* **20,** 949-954.

Groves, M. G., and Osterman, J. V. (1978). Host defenses in experimental scrub typhus: Genetics of natural resistance to infection. *Infect. Immun.* **19,** 583-588.

Groves, M. G., Rosenstreich, D. L., Taylor, B. A., and Osterman, J. V. (1980). Host defenses in experimental scrub typhus: Mapping the gene that controls natural resistance in mice. *J. Immunol.* **125,** 1395-1399.

Huebner, R. J., Jellison, W. L., and Pomerantz, C. (1946). Rickettsialpox. A newly recognized rickettsial disease. IV. Isolation of a rickettsia apparently identical with the causative agent of rickettsialpox from *Allodermanyssus sanguineus,* a rodent mite. *Public Health Rep.* **61,** 1677-1682.

Karr, H. V. (1943). Study of a latent pneumotropic virus of mice. *J. Infect. Dis.* **72,** 108-116.

Kreier, J. P., and Ristic, M. (1963). Morphologic, antigenic, and pathogenic characteristics of *Eperythrozoon ovis* and *Eperthrozoon wenzoni*. *Am. J. Vet. Res.* **24,** 488-500.

Lindsey, J. R., Cassell, G. H., and Baker, H. J. (1978). Diseases due to mycoplasmas and rickettsias. *In "Pathology of Laboratory Animals"* (K. Berinschke, F. M. Garner, and T. C. Jones, eds.), pp. 1482-1550. Springer-Verlag, Berlin and New York.

McCluskie, J. A. W., and Niven, S. F. (1934). The blood changes in rats and mice after splenectomy with observations on *Bartonella muris* and *Eperythrozoon coccoides*. *J. Pathol. Bacteriol.* **39,** 185-196.

McKee, A., Ziegler, R., and Giles, R. C. (1973). Scanning and transmission electron microscopy of *Haemobartonella canis* and *Eperythrozoon ovis*. *Am. J. Vet. Res.* **34,** 1196-1201.

Mayer, M. (1921). Über einige bakterienahnliche Parasiten der Erythrozyten bei Menshen and Tieren. *Arch. Schiffs- Trop.-Hyg.* **25,** 150-152.

Meltzer, M. S., and Nacy, C. A. (1980). Macrophages in resistance to rickettsial infection: Susceptibility to lethal effects of *Rickettsia akari* infection in mouse strains with defective macrophage function. *Cell. Immunol.* **54,** 487-490.

Moulder, J. W. (1973). Rickettsia. *In* "Microbiology" (B. D. Davis, R. Dulbecco, H. N. Eisen, and H. S. Ginsberg, eds.), pp. 898-913. Harper, New York.

Moulder, J. W. (1974). The rickittsias. *In* "Bergey's Manual of Determinative Bacteriology" (R. E. Buchanan and N. E. Gibbons, eds.), 8th ed., pp. 882-925. Williams & Wilkins, Baltimore, Maryland.

Nigg, C. (1942). Unidentified virus which produces pneumonia and systemic infection in mice. *Science* **95,** 49-50.

Nigg, C., and Eaton, M. D. (1944). Isolation from normal mice of a pneumatropic virus which forms elementary bodies. *J. Exp. Med.* **79,** 496-510.

Niven, J. F. S., Dick, G. W. A., Gledhill, A. W., and Andrews, C. H. (1952). Further light on mouse hepatitis. *Lancet* No. 263, p. 1061.

Ott, K. J., and Stauber, L. A. (1967). *Eperythrozoon coccoides*: Influence on cause of infection of *Plasmodium chabaudi* in mice. *Science* **155,** 1546-1548.

Page, L. A. (1968). Proposal for the recognition of two species in the genus *Chlamydia,* Jones, Rake, and Stearns, 1945. *Int. J. Syst. Bacteriol.* **18,** 51-66.

Riley, V. (1964). Synergism between a lactate dehydrogenase elevating virus and *Eperythrozoon coccoides*. *Science* **146,** 921-922.

Riley, V., Loveless, J. D., and Fitzmaurice, M. A. (1964). Comparison of a lactate dehydrogenase elevating virus-like agent and *Eperythrozoon cocciodes*. *Proc. Soc. Exp. Biol. Med.* **116,** 486-490.

Rose, H. M. (1949). The clinical manifestations and laboratory diagnosis of rickettsialpox. *Ann. Intern. Med.* **31,** 871-883.

Rudnick, P., and Hollingsworth, J. W. (1959). Lifespan of rat erythrocytes parasitized by *Bartonella muris*. *J. Infect. Dis.* **104,** 24-27.

Sammons, L. S., Kenyon, R. L., Hickman, R. L., and Pedersen, C. E., Jr. (1977). Susceptibility of laboratory animals to infection by spotted fever group rickettsiae. *Lab. Anim. Sci.* **27,** 229.

Schilling, V. (1928). *Eperythrozoon coccoides,* eine neue durch Splenektomie aktivierbare Dauerinfektion der weisse Maus. *Klin. Wochenschr.* **7,** 1853.

Seamer, J., Gledhill, A. W., Barlow, J. L., and Hotchin, J. (1961). Effect of *Eperythrozoon coccoides* upon lympocytic choriomeningitis in mice. *J. Immunol.* **86,** 512-515.

Small, E., and Ristic, M. (1967). Morphologic features of *Hemobartonella felis*. *Am. J. Vet. Res.* **28,** 845-851.

Storz, J. (1971). "Chlamydia and Chlamydia-Induced Diseases." Thomas, Springfield, Illinois.

Storz, J., and Spears, P. (1977). Chlamydiales properties cycle of development and effect on eukaryotic host cells. *Curr. Top. Microbiol. Immunol.* **76,** 167-214.

Suntharasamai, P., and Rytel, M. W. (1973). Interferon response and interferon action in *Eperythrozoon coccoides* infection of mice. *Proc. Soc. Exp. Biol. Med.* **142,** 811-816.

Weinman, D. (1944). Infectious anaemia due to bartonella and related red cell parasites. *Philos. Trans. Soc.* **33,** 243-350.

Wigand, R. (1958). "Morphologische Biologische and Serologische Eigenschaften der Bartonellen." Thieme, Stuttgart.

Chapter 8

Viral Diseases of the Respiratory System

John C. Parker and Conrad B. Richter

I. Introduction ... 110
II. Sendai Virus ... 110
A. Historical Background ... 110
B. Properties of the Virus ... 111
C. Epizootiology ... 116
D. Pathogenesis ... 123
E. Diagnosis ... 131
F. Control and Prevention ... 133
III. Pneumonia Virus of Mice (PVM) ... 134
A. Historical Background ... 134
B. Properties of the Virus ... 134
C. Epizootiology ... 136
D. Pathogenesis ... 138
E. Diagnosis ... 140
F. Control and Prevention ... 141
IV. K Virus ... 142
A. Historical Background ... 142
B. Properties of the Virus ... 142
C. Epizootiology ... 143
D. Pathogenesis ... 144
E. Diagnosis ... 148
F. Control and Prevention ... 150
Addendum ... 150
References ... 150
Sendai Virus ... 150
Pneumonia Virus of Mice (PVM) ... 155
K Virus ... 157

ISBN 0-12-262502-1

I. INTRODUCTION

Respiratory virus infections of mice are common and play a major role in the etiology of respiratory diseases in mice. They may occur as acute clinical or subclinical infections in all ages and strains of mice and can produce a wide variety of signs ranging from mild rhinitis to fatal pneumonia and devastating epizootics. The impact of these infections on the overall clinical health of laboratory animals provides one of the greatest potential variables in experiments with which scientists must contend.

Historically, three viruses have been primarily responsible for respiratory infections in mice: Sendai virus, pneumonia virus of mice (PVM), and K virus. Each virus is distinct in its epizootiology, serology, and pathogenesis. Only Sendai regularly produces clinical disease, PVM is latent but produces widespread subclinical infection, and K virus, also latent, is infrequently encountered.

The epizootiology of Sendai and PVM is complicated by the fact that these viruses cross-infect different laboratory animal species, notably the mouse, hamster, rat, and guinea pig. The likelihood of epizootics is increased by mixing animals from different sources. Traditionally, this problem has been addressed by attempting to select carefully animals from colonies without infections, or as a compromise, with similar infections, and to mix these animals carefully into the laboratory colony through a variety of quarantine procedures. Efforts by breeders to eliminate these virus infections from their colonies have been only partially successful.

II. SENDAI VIRUS

A. Historical Background

Sendai virus was first isolated by Kuroya *et al.* (1953) in 1952 and was described clinically by Sano *et al.* (1953) and pathologically by Noda (1953) during an outbreak of pneumonitis in 17 newborn human infants at the Tohoku University Hospital in Sendai, Japan (Table I). Lung suspensions from five fatal cases were inoculated intranasally into mice; three specimens produced death and two produced consolidation in the lungs of the mice within 7 days. One strain (Fushimi strain) was studied in detail and shown to grow in embryonated eggs and to agglutinate a variety of red blood cells. A virus (BM_2 strain) possessing similar properties was also isolated from normal laboratory mice (Kuroya *et al.*, 1953), and sera from 10 normal laboratory mice had hemagglutination inhibition (HAI) antibody to a similar strain, the Akitsugu strain (Misao *et al.*, 1954). Because the serum inhibitors could be removed by treatment with PR8 influenza A virus, they were believed to be nonspecific. Antibody titer rises were also demonstrated in patients, and isolates (Akitsugu and Yamao strain) were reported to produce disease in human volunteers (Kuroya *et al.*, 1953; Yamada, 1956). After attempts to show identity with other known viruses were unsuccessful, Kuroya *et al.* (1953) concluded that the virus they had isolated was a new virus of human origin and named it *Newborn virus penumonitis* (*type Sendai*). Later, during the period from 1953 to 1956, other investigators reported isolations of Sendai virus following inoculation of human specimens into mice or embryonated hens' eggs, and although seroepidemiologic surveys sometimes gave conflicting results, it was generally believed that Sendai virus was an etiologic agent of respiratory infection in humans (reviewed in Ishida and Homma, 1978).

Table I

Sendai Virus Synonyms

Year	Synonym	Reference
1952	Newborn pneumonitis virus (type Sendai)	Kuroya *et al.* (1953)
1953	Newborn virus pneumonitis, type Sendai	Noda (1953)
1953	Hemagglutinating virus of the mouse (HVM)	Fukai and Suzuki (1955)
1955	Hemagglutinating virus of Japan (HVJ) (Third Annual Meeting of the Society of Japanese Virologists)	Ishida and Homma (1978)
1955	Type D influenza virus	Jensen *et al.* (1955)
1957	Japanese haemagglutinating virus (JHV)	Gorbunova *et al.* (1957)
1958	Sendai virus	Chanock *et al.* (1958)

Fukumi *et al.* (1954) first raised doubts as to the human origin of Sendai virus when they showed that laboratory mice in Japan were frequently infected with the same virus that had been isolated on numerous occasions from mice inoculated with human materials. During this same period (1952), Matsumoto *et al.* (1954) isolated Sendai virus from hamsters, and in 1955, Sasahara (1955) implicated rats and guinea pigs as additional sources of virus.

An influenza-like illness occurred nationwide in Japan in swine between 1953 and 1956 that resulted in stillbirths and high death rates in young pigs (Sasahara *et al.*, 1954). Sendai virus isolates were made in mice and embryonated eggs, and antibody was demonstrated in pigs. The infection had subsided by 1956, and the frequency of seropositives likewise had decreased. No infections or seropositives have been seen since 1961, and this swine infection is no longer of any concern in Japan. Further confirmation that Sendai virus was the etiologic agent in this outbreak or in other swine infections elsewhere has not been made.

An epidemic of human influenza was reported from Vladivostock in the Soviet Union in 1956 in which five virus strains were isolated that shared properties with Sendai virus and where seroepidemiologic studies demonstrated four-fold rises in Sendai virus HAI antibody in some paired sera (Gerngross, 1957). Zhdanoff *et al.* (1957) studied an influenza epidemic in children at the Institute of Virology in Moscow and isolated three strains of hemagglutinating viruses from nasopharyngeal smears. These strains had the same characteristics as the Vladivostock virus and the Japanese virus, Fushimi (Sendai) strain (Gorbunova *et al.*, 1957; Zhdanoff *et al.*, 1957).

Since Sendai virus was originally recovered from mice inoculated with human autopsy materials, and since mice throughout the world are now known to be commonly infected with Sendai virus, there remains the question of the validity and true origin of these isolates. One must recall the state of the art in virology during the early 1950s and the lack of knowledge of the parainfluenza viruses in order to understand why the confusion developed and why to some degree it has persisted even until today. The lack of knowledge of the widespread prevalence of Sendai infection in laboratory animals and the ease with which Sendai can contaminate a laboratory which uses infected animals were but two of the problems. It was not until 1958 (Chanock *et al.*, 1958), when type 2 hemadsorption virus (HA-2) was isolated from humans, that the close relationship between Sendai virus and human HA-2 virus, both parainfluenza type 1 viruses, was shown (Fukumi and Nishikawa, 1961; Chanock *et al.*, 1963; Stark and Heath, 1967) and when the serologic cross-relationship between the parainfluenza viruses and mumps virus was demonstrated (DeMeio and Walker, 1957; Numazaki, 1960; Lennette *et al.*, 1963) that the actual relationships of Sendai virus and other viruses began to emerge. It was then possible to explain some of the earlier conflicting and confusing data. Serologic evidence of Sendai virus associated with human respiratory illnesses had also been reported from the United States (Jensen *et al.*, 1955; DeMeio and Walker, 1958), England (Gardner, 1957; White *et al.*, 1957) Germany (Sandow *et al.*, 1969), and Scotland (Sommerville and Carson, 1957). However, no virus isolations had been made, and because of the close antigenic sharing of Sendai virus with the human parainfluenza viruses, the likelihood of Sendai virus as the etiologic agent of the respiratory illnesses is questionable.

Given the history, it is not surprising that many investigators have questioned the authenticity of the original reports of human infection with Sendai virus. The consensus today is that humans are not natural hosts of Sendai virus (Fukumi *et al.*, 1954, 1959; Fukumi and Nishikawa, 1961; Chanock *et al.*, 1963; Ishida and Homma, 1978); rather, the natural hosts of Sendai virus are exclusively laboratory rodents.

B. Properties of the Virus

1. Classification

Sendai virus belongs in the family Paramyxoviridae, genus *Paramyxovirus* (Fenner, 1976), and is classified as the species *parainfluenza 1 (Sendai)* together with the human virus, hemadsorption type 2 (HA-2), with which it has a close antigenic relationship (Andrewes *et al.*, 1959; Chanock *et al.*, 1958, 1963; Cook *et al.*, 1959; Zhdanov and Bukrinskaya, 1959; Fukumi and Nishikawa, 1961; Chanock and Parrott, 1965; Kingsbury *et al.*, 1978). In the classification of Kingsbury *et al.* (1978), the distinction between the human and murine parainfluenza 1 members was not clarified. Unfortunately, this may perpetuate the host range confusion between these two viruses.

2. Strains of Virus

In spite of the early historical confusion between reported Sendai virus isolations from humans, swine, and laboratory rodents in Japan and Russia and possible, but unknown, differences of these original virus strains, there do not appear to be any variances in the strains of Sendai virus isolated from rodents other than minor differences in cell culture virulence (Sugamura *et al.*, 1974), virus yields from eggs or cell culture, and HAI or complement fixation (CF) titers. All strains are considered antigenically homologous and of a single serotype (Chanock and Parrott, 1965). Some newly isolated strains do not produce illness when first inoculated into the lungs of mice, although replication occurs; however, after passage, lung consolidation, illness, and death frequently occur (Robinson *et al.*, 1968). Adaptation may be partly dependent on selection of virus resistance to β inhibitor (Briody *et al.*, 1955). Egg-passaged strains may be 10 to 20 times less virulent for mice than mouse-passaged strains (van Nunen and van der Veen, 1967). Observed differences of pathogenicity in mice may be due to different properties of the virus strains (Andrewes and Pereira, 1972) but are more likely the result of

Table II

Commonly Used Laboratory Strains of Sendai Virus

Virus strain	Origin	Reference
Fushimi	Human/mouse	Kuroya *et al.* (1953)
Akitsugu	Human/mouse	Misao *et al.* (1954)
MN	Mouse	Fukumi *et al.* (1954)
Z	Mouse	Fukai and Suzuki (1955)
52	Human/mouse	ATCC[a] VR-105

[a] ATCC: American Type Culture Collection, Rockville, MD. Isolate of Kuroya *et al.* (1953).

a differential susceptibility of the host mouse strains (Parker *et al.*, 1978). Several manipulated and modified virus strains which have abnormal biologic characteristics (Nishiyama *et al.*, 1976; Matsumoto and Maeno, 1962; Homma, 1971; Silver *et al.*, 1978), such as temperature-sensitive strains which grow at 32°C but not at 38°C, have been produced in the laboratory (Kimura *et al.*, 1975; Nagata *et al.*, 1972). Some of the more common virus strains reported in the literature and frequently used in laboratory investigations are shown in Table II.

3. Physical and Chemical Properties

The properties of Sendai virus have been summarized by Kingsbury *et al.* (1978) in their review of the family Paramyxoviridae (Table III). Virus particles are pleomorphic but roughly spherical, measuring 150–250 nm in diameter (Nishikawa and Fukumi, 1954) and containing a helical nucleocapsid with a continuous single-stranded RNA genome. Hemagglutinating, neuraminidase, hemolytic, and cell fusion activities of Sendai virus are associated with the virion envelope (Hosaka, 1970). Treatment with 20% diethyl ether for 18 hr at 4°C destroys infectivity (Chanock *et al.*, 1963). Hemagglutinins are destroyed at 45°–50°C in 10–20 min and at pH values of less than 5.3 or greater than 9.8 (Fukai and Suzuki, 1955; Fukumi and Nishikawa, 1961). Sendai virus agglutinates red blood cells from a wide variety of animal species, including human type 0, guinea pig, chicken, rat, mouse, hamster, rabbit, cow, sheep, pigeon, monkey, and dog at room temperature (RT) and at 5°C (Fukumi *et al.*, 1954; J. C. Parker, 1979 unpublished observations).

A soluble, nonhemagglutinating, CF antigen can be prepared from egg fluid antigens by ether extraction and absorption with red blood cells (Cook *et al.*, 1959; Rott *et al.*, 1963). The soluble antigen is associated with the inner helical component of the virion (Rott *et al.*, 1963). The presence of soluble antigens in whole virion HA antigen preparations does not alter the sensitivity of the HA antigens in detecting antibody (Cook *et al.*, 1959).

Table III

Properties of the Mouse Respiratory Viruses

	Sendai	PVM	K
Classification			
Family	Paramyxoviridae	Paramyxoviridae	Papovaviridae
Genus	*Paramyxovirus*	*Pneumovirus*	*Polyomavirus*
Size (diameter)	150–250 nm	80–200 nm	35–50 nm
Morphology	Pleomorphic, roughly spherical, enveloped	Pleomorphic, roughly spherical, filamentous, enveloped	Spherical
Nucleic acid type	RNA	RNA	DNA
pH Stability	Stable (pH 5.3–9.8)	Acid sensitive (pH ≤3) Alkaline sensitive (pH ≥11)	Stable
Heat stability	Sensitive (48°C, 15 min)	Sensitive (56°C, 30 min)	Stable (60°C, 4 hr)
Lipid solvent	Sensitive	Sensitive	Stable
Neuraminidase	Present	Lacks	Lacks
Hemolysin	Present	Lacks	Lacks
Hemadsorption	Present	Present	Lacks
Hemagglutination	Human O, guinea pig, rat, chicken, others (RT, 4°C)	Mouse, rat, hamster (RT, 4°C)	Sheep, dog (4°C)
Replication site	Cytoplasm, nucleoli	Cytoplasm	Nucleus
Host range			
In vivo	Mice, rats, guinea pigs, hamsters	Mice, rats, guinea pigs, hamsters	Mice
In vitro	Primary monkey kidney cells, embryonate eggs (serial cell lines with protease treatment)	Primary hamster kidney cells, BHK-21, Vero	Not known

Parainfluenza virus types 1, 2, and 3 are antigenically distinct, but within the type 1 group, Sendai virus possesses a broader antigenic structure than other group members (Cook *et al.*, 1959; Bukrinskaya *et al.*, 1962; Lief *et al.*, 1975).

Guinea pigs immunized with Sendai virus develop high levels of CF antibody for soluble antigens of Sendai and HA-2 viruses; however, sera from guinea pigs immunized with HA-2 virus react with the hemologous soluble antigen to a considerably higher titer than with Sendai virus (Cook *et al.*, 1959). CF tests with viral antigens are similar; however, antibody titers are higher, particularly the homologous titers, resulting in a more specific test (Cook *et al.*, 1959). The HAI test is more specific than the CF test for differentiating Sendai from HA-2 virus because of lower heterologous titers (Cook *et al.*, 1959).

The Sendai virion contains either six (Homma and Ohuchi, 1973) or seven (Homma *et al.*, 1975) major proteins depending on the interpretation of the F glycoprotein as one glycoprotein or as two glycoproteins, F_1 and F_2, which are derived from the single F glycoprotein. Of the two major glycoproteins, F and HN (or HANA), the larger HN glycoprotein (VP-2) carries hemagglutination (HA) and neuraminidase activity (Homma and Ohuchi, 1973; Scheid and Choppin, 1974; Shimizu *et al.*, 1974; Homma *et al.*, 1975; Orvell and Norrby, 1977), and the smaller F glycoprotein (VP-4) is involved in hemolysis, infectivity, and cell fusion (Homma and Ohuchi, 1973; Scheid and Choppin, 1974; Homma *et al.*, 1975; Orvell and Norrby, 1977). Cell culture virions deficient in hemolytic, cell fusion, and cell culture infectivity contain the F glycoprotein which is not present in virions grown in eggs. However, cleavage of the cell culture F glycoprotein into two smaller glycoproteins, F_1 and F_2, restores these activities in cell culture (Homma *et al.*, 1975). Hemolysis is a laboratory artifact occurring only when the virion envelope is damaged, resulting in the loss of cell fusion capability (Homma *et al.*, 1975).

4. Growth in Mice, Tissue Culture, and Eggs

a. Growth in Eggs and Cell Culture

i. Eggs. The most common and reliable method for growth of Sendai virus is in embryonated hens' eggs. Normally, fertile eggs 8–13 days old are inoculated in the amniotic or allantoic cavity and incubated at 35°–38°C for 48–72 hr and the appropriate fluids harvested (Fukai and Suzuki, 1955; Jensen *et al.*, 1955; Fukumi and Nishikawa, 1961; Fukumi *et al.*, 1961; Shibuta *et al.*, 1971). Virus yields from eggs are higher than from cell culture and normally range from 10^8 to 10^{10} EID_{50}/0.1 ml, with HA titers of 1:100 to over 1:1000. Serial passage of undiluted Sendai virus in eggs leads to the formation of incomplete virus with decreased HA titer (Sokol *et al.*, 1964).

ii. Cell Culture. Primary or secondary cultures of rhesus (Fukumi *et al.*, 1959; Fukumi and Nishikawa, 1961; Chanock and Parrott, 1965) or cynomolgus (Shibuta *et al.*, 1971; Shibuta, 1972) monkey kidney cells (MK) show cytopathic effects (CPE) of virus replication and are the most widely used cells for detection and isolation of Sendai virus, although serial lines, with trypsin added to the culture medium, are sensitive and useful in plaquing (Nagata *et al.*, 1965; Shibuta *et al.*, 1971; Shibuta, 1972; Sugita *et al.*, 1974). Serially passaged rhesus monkey kidney (RMK) cells which support multiple-cycle replication and plaque formation in primary cultures lose their ability to do so after subpassage due to a loss of a trypsin-like proteolytic activity (Silver *et al.*, 1978). CPE in primary MK are characteristic: The cytoplasm of the infected cells elongates, with the ends becoming very long and slender (Fig. 1); the nucleus is also elongated (Fukumi and Nishikawa, 1961); cell fusion or syncytia formation is seen in the early stages of CPE (Okada and Tadokoro, 1963); and CPE or hemadsorption titration endpoints can generally be reached in 5–7 days (Chanock and Parrott, 1965). Sendai virus also grows to high titer ($10^{6.5}$ TCD_{50}/0.1 ml) and produces CPE and hemagglutinins within 3–4 days in primary cell cultures of the lung and kidney of the tortoise (*Testudo graeca*) (Shindarov *et al.*, 1969). Mouse organ cultures of nasal mucosa and trachea support optimal growth of Sendai virus to high titers at 33°–35°C (pH 7.1–7.2), whereas by comparison, lung cultures support growth poorly (Willems and van der Veen, 1968).

Reed *et al.*, (1975) examined the serial cell lines L929, BHK-21, and WI-38, BSC-1, and Vero and primary cell cultures of Swiss mouse embryo, Swiss mouse kidney, rat embryo, hamster kidney, and hamster embryo, for Sendai growth. Only BSC-1 and Vero were hemadsorption positive. Serial cell lines (BHK, L), under conditions of proper temperature, medium, and cell state, and with certain strains of virus, can become persistently infected (Nagata *et al.*, 1972; Ito *et al.*, 1974; Sugamura *et al.*, 1974; Kimura *et al.*, 1975; Collins and Flanagan, 1977, 1978).

Sugita *et al.* (1974), Silver *et al.* (1978), and Frank *et al.* (1979) showed that the serial cell line LLC-MK2 with trypsin added to the culture medium was similar to RMK for isolation or plaquing of Sendai virus, and Ito (1976) developed a plaque assay using a clonal line of porcine kidney cells (PS-Y15) which was as sensitive as egg titration.

Infection of susceptible cells occurs as the result of adsorption of the virus to the cell membrane, mediated through the affinity of glycosylated proteins associated with the surface projections of the virion to neuraminic acid-containing mucoprotein receptors (gangliosides) of the cell membrane (Kelen and McLeod, 1977). Penetration of the nucleocapsid into the cytoplasm results from fusion between the viral envelope and the cell membrane at the site of contact (Morgan and Howe, 1968; Scheid and Choppin, 1974). Synthesis of RNA com-

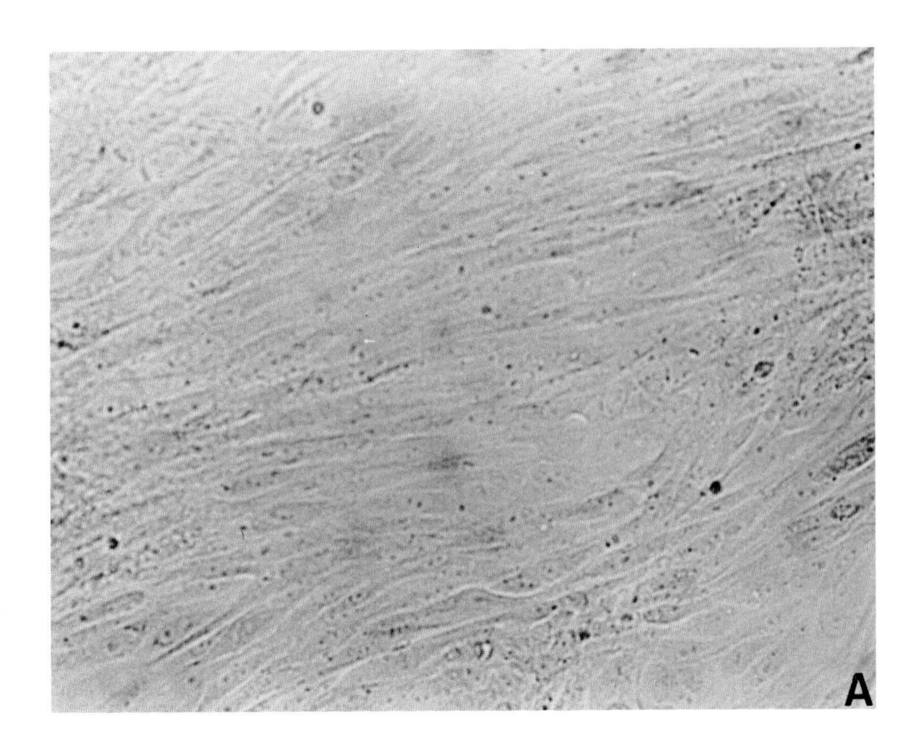

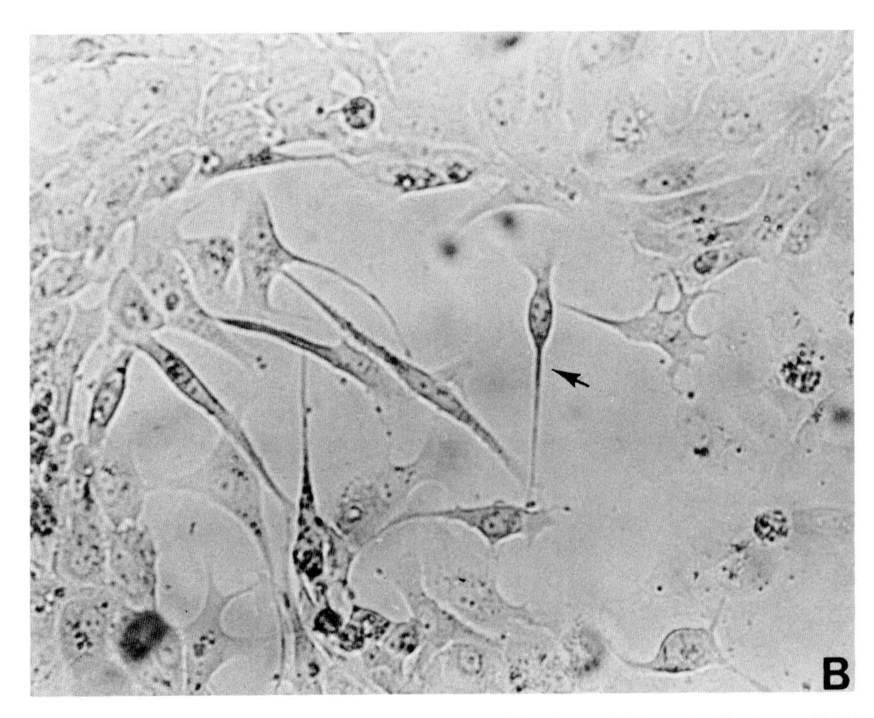

Fig. 1. Cytopathic effects (CPE) induced by Sendai virus in rhesus monkey kidney cell culture. (A) Normal cell culture. (B) Advanced cytopathology 8 days after infection with $10^{5.5}$ $TCID_{50}$ showing characteristic cells with elongated nuclei and cytoplasmic extensions. Approximately ×100. (Photographs courtesy of Dr. Michael J. Collins, Jr.)

plementary to virion RNA, structural protein synthesis, and assembly and maturation of progeny virions are intracytoplasmic events (Kelen and McLeod, 1977). The maturation of the virus which protrudes from the cell surface as a bud is completed by fusion of the base of the bud, and the complete virion is released from the cell.

Although Sendai virus grows to high titer in embryonated eggs (Jensen *et al.*, 1955), it does not grow well in most cultured mammalian cells (Blair and Robinson, 1970). In embryonated hens' eggs, Sendai virus replicates in multiple steps and the progeny viruses have all of their active biologic properties, including hemagglutinins, hemolysins, cell-fusing ability, neuraminidase, and infectivity (Ishida and Homma, 1978). When Sendai virus is inoculated onto a nonpermissive cell line (L, HeLa, FL, LLC-MK2, or RK-13), however, the progeny viruses are not infectious for the homologous cell line, and the hemolyzing and cell-fusing abilities are also lost (Matsumoto and Maeno, 1962; Ishida and Homma, 1978). This host-induced modified virus remains fully infectious for eggs, however (Ishida and Homma, 1960, 1961). Treatment of defective host-modified progeny virus with trypsin restores the phenotypically masked properties through the direct action of trypsin on the virion-coded glycoprotein (Homma, 1971; Homma and Ohuchi, 1973; Scheid and Choppin, 1974; Ohuchi and Homma, 1976; Ishida and Homma, 1978). Sendai virus penetrates cells by fusing the virus envelope with the host cell plasma membrane (Morgan and Howe, 1968; Apostolov and Almeida, 1972), and it is the lack of the cell-fusing ability in defective progeny virus which is restored by extracellular treatment with proteolytic enzymes or normal chorioallantoic fluid. Some Sendai virus mutants are activated by different proteoses (Scheid and Choppin, 1974), and it has been suggested that some cultured cells release an activating enzyme into the culture fluid. This would account for the permissiveness of some primary cells, such as monkey kidney cells (Ishida and Homma, 1978). This activator-dependent replication system, which results in the production of normal infectious virus, may account for such phenomena as organ specificity and host age differences in susceptibility (Ishida and Homma, 1978).

Permissive cell cultures such as primary chick embryo lung cells (CEL) produce 100 times more infectious virus than chick embryo fibroblast cells (CEF), and although both cell types contain abundant cytoplasmic virus nucleocapsids, budding particles and CPE are more abundant in CEL (Darlington *et al.*, 1970). This indicates a block in the virus maturation process in converting the viral nucleocapsid to complete enveloped virus (Blair and Robinson, 1970). Addition of trypsin to Sendai virus-infected cultures of CEF allows completion of the virus maturation, with production of viral plaques (Shibuta *et al.*, 1971; Shibuta, 1972).

b. Growth in Mice. Virus can be isolated from the nasal washings or saliva 24–48 hr after infection for approximately 9 days postinfection (PI) and from lungs from 24 hr after infection for approximately 9–14 days PI (Fig. 2). Lungs usually contain detectable virus slightly longer than nasal washings, with the overall length of infection being somewhat dose dependent (Parker and Reynolds, 1968; Sawicki, 1962; Robinson *et al.*, 1968; van Nunen and van der Veen, 1967; Appell *et al.*, 1971; Parker *et al.*, 1978). In intranasally infected mice, several different organs are occasionally found to contain virus for brief periods PI, including the kidney, liver, and spleen, but not the urine, blood or intestine; however, it is probable that virus replication does not normally occur outside of the respiratory tract (Parker and Reynolds, 1968; van Nunen and van der Veen, 1967; Appell *et al.*, 1971; Blandford and Heath, 1972). Tucker and Stewart (1976) and Stewart and Tucker (1978) isolated virus from a variety of organs from adult inbred mice inoculated by either the intravenous or intranasal route for as long as 21 days PI and suggested that the virus may have been disseminated by macrophages. The immune response is highly efficient and of long duration, as exemplified by the fact that low-titer chronic infections have not been demonstrated in recovered mice (van der Veen *et al.*, 1974). Infections in nude mice are an anomaly in that virus infection persists as a chronic infection in the respiratory mucosa for 5 weeks or longer (Ward and Young, 1976; Ward *et al.*, 1976b). The same may also be true to a lesser extent of the highly susceptible 129/J mouse, which demonstrates acute-phase pathology for as long as 5 weeks after experimental infection (Parker *et al.*, 1978).

Natural infections in mice usually cannot be established by inoculation routes other than the intranasal (Jensen *et al.*, 1955), although infection of organs not normally infected does occur under artificial conditions, leading to some confusion concerning where the virus replicates under natural conditions. Intrauterine transmission of Sendai virus can be demonstrated when virus is inoculated by the intravenous route, for example (Tucker and Stewart, 1976). Under these conditions, virus can be isolated from placentas and from different tissues of embryos and newborns. It may be noteworthy, however, that in studies of rats in which abnormalities such as weight loss, increased neonatal mortality, and longer gestation periods were seen, the cause was attributed to the constitutional upset of the pregnant animal caused by the respiratory disease and not a direct action by the virus on the fetus since the fetuses were free of infection (Coid and Wardman, 1971, 1972). Sendai virus will infect zona-free but not zona pellucide-intact mouse cleavage stage embryos *in vitro* (Bowen *et al.*, 1978). Tuffrey *et al.* (1972) found Sendai virus-specific immunofluorescent antigen on the surface of fertilized mouse eggs taken from mice from a Sendai virus-infected colony and on eggs infected *in vitro;* however, virus isolations were not attempted, and therefore the true significance of these observations is not yet clear. In addition, peritoneal macrophages may become infected after intraperitoneal inoculation, Kupffer cells of the liver and endothelial cells in large veins and auricles after intravenous inoculation, and the ependyma and choroid plexus epithelium after intracerebral inoculation (Mims and Murphy, 1973). These organ tropisms are not seen in natural Sendai virus infections and therefore should be considered laboratory-manipulated artifacts even though they may be useful in research as models for human diseases. Although tissues other than respiratory ones are not normally infected, the experimenter should be aware that manipulation of acutely infected mice might release virus into the circulation, resulting in visceral organ infection. Further discussion of this topic can be found in Section II,D,2,e.

C. Epizootiology

1. Host Range

a. Rodents. Naturally occurring infections of Sendai virus have been reported in laboratory-reared mice (Fukumi *et al.*, 1954, 1962; Fukai and Suzuki, 1955; Parker *et al.*, 1964, 1978; Grunert, 1967; Poiley, 1970; Fujiwara and Iida, 1974; Descoteaux *et al.*, 1977; Zurcher *et al.*, 1977; Fujiwara *et al.*, 1979), rats (Sasahara, 1955; Schels and Hartl, 1971; Makino *et al.*, 1973; Fujiwara and Iida, 1974; Burek *et al.*, 1977; Parket *et al.*, 1978; Fujiwara *et al.*, 1979), hamsters (Fukumi *et al.*, 1954; Matsumoto *et al.*, 1954; Soret and Buthala, 1967; Profeta *et al.*, 1969; Reed *et al.*, 1974; Parker *et al.*, 1978) and guinea pigs (Sasahara, 1955; Parker and Reynolds, 1968; Kunz and Hutton, 1971; Parker *et al.*, 1978). Infections in guinea pigs have not been substantially documented because of a lack of adequate confirmatory virus isolations. Most information pertaining to guinea pig infection is derived from serologic data, and since there is evidence that other parainfluenza viruses naturally infect the guinea pig (Sasahara, 1955; Iwai *et al.*, 1976; Van Hoosier and Robinette, 1976; Welch *et al.*, 1977), potential antigenic sharing between these viruses must be studied and evaluated. Infection in feral mice was not detected in a survey of 400 mice trapped in many locations from a four-state area in the United States (Parker and Reynolds, 1968). This is not surprising since the perpetuation of Sendai virus infection is dependent on a constant supply of susceptible mice, normally provided in laboratory breeder colonies but not in feral mouse niches. Infections in gerbils have not been detected (Parker *et al.*, 1978).

b. Age and Sex. The sequela of infection in suckling and aging mice is more severe than in adult mice. Suckling mice have extended infections and develop higher lung virus titers associated with lower lung interferon titers (Sawicki, 1961, 1962). They experience a severe infection of the trachea, bronchi, and bronchioles which is identical to that in older mice but more extensive (Mims and Murphy, 1973). Mortality can be high (Fukumi *et al.*, 1962), with many mice dying within 3 days after infection (Mims and Murphy, 1973), compared to 6–14 days for lethal infection in adults (Parker *et al.*, 1978). No difference in susceptibility has been observed in mice between 4 and 12 weeks of age (van Nunen and van der Veen, 1967) or between male and female mice (van Nunen and van der Veen, 1967; Parker *et al.*, 1978). Generally, older mice also show a heightened susceptibility compared to young adults (Parker *et al.*, 1978; Kay, 1978a,b; Kay *et al.*, 1979). In contrast, during an epizootic in a mouse and rat colony, Zurcher *et al.* (1977) noted that the highest death rates occurred in young mice (less than 10 weeks of age) compared to young adults (more than 10 weeks of age) but did not observe increased death rates in the oldest mice (Zurcher *et al.*, 1977; Burek *et al.*, 1977).

c. Human. The literature is replete with claims and counterclaims of Sendai virus infection in humans (see Section II, A). Some of the controversy stems from the fact that heterologous antibodies are often produced in response to seemingly unrelated antigenic stimuli caused by antigenic sharing with other viruses. In spite of the countless number of attempts to isolate viruses from patients with respiratory disease, claims of Sendai virus isolation from humans occurred only in the early and mid-1950s. Significantly, these reports preceded Chanock's (Chanock, 1956; Chanock *et al.*, 1958) discovery of the parainfluenza viruses. Rises in human serum antibody level, reactive with Sendai virus, are often encountered following infection with HA-1 and HA-2 viruses (Chanock *et al.*, 1958; Start and Heath, 1967), CA virus (Chanock, 1956), and mumps virus (DeMeio and Walker, 1957; Pentitinen and Cantell, 1967). Stark and Heath (1967) suggested that the increase in Sendai virus antibody level in human sera was produced by a series of these heterotypic infections. Both human and murine parainfluenza 1 subtypes exhibit a nonreciprocal serologic relationship in which Sendai virus antiserum exhibits broad reactivity, whereas HA-2 antiserum reacts more specifically (Cook *et al.*, 1959). Thus, the human parainfluenza 1 virus strain (HA-2) can be readily distinguished from Sendai virus by conventional HAI, CF, or hemadsorption neutralization tests (Cook *et al.*, 1959; Lief *et al.*, 1975), but Sendai virus cannot be as easily distinguished from HA-2 virus (see Section II,B,3).

A parainfluenza type 1 virus, 6/94 virus, isolated from the human brain tissues of two multiple sclerosis patients was shown to be closely related to Sendai virus by RNA–RNA hybridization and patterns of electrophoretic mobilities of viral polypeptides. It was, however, distinct antigenically (Lief *et al.*, 1975) and phenotypically, as indicated by temperature growth requirements, hemolytic activity, and cell-fusion capabilities (Koprowski and ter Meulen, 1975). There is no entirely satisfactory serologic assay which will differentiate human from rodent virus isolates in order to establish irrevocably the true host range of Sendai virus (Koprowski and ter Meulen, 1975).

d. Other Species. Ferrets infected by the intranasal route sustain high virus titers in the lungs and nasal turbinates and develop a strong immune response. The infection is severe in initial passages, causing fever, lung consolidation, and death (Jensen *et al.*, 1955). Asymptomatic infections with antibody rises occur in rhesus and cynomolgus monkeys after intranasal inoculation. Intracerebral inoculation results in temperature elevation, weakness, tremors, and ataxia (Jensen *et al.*, 1955).

e. Strains of Mice. Strains of mice differ remarkably in their susceptibility or sensitivity to Sendai virus infection, and this variance is important for the understanding and correct interpretation of epizootics in which resistant and sensitive mouse strains coexist (Tucker and Stewart, 1976; Iwai, 1978; Parker *et al.*, 1978; Stewart and Tucker, 1978) (Table IV). Patterns of mortality may vary extensively in spontaneous outbreaks from nearly 0% in the very resistant strains of mice to nearly 100% in highly susceptible strains. In a natural outbreak among naive adults, one of us (Richter) observed high mortality among 129/J mice, whereas SJL/J mice housed in the same room suffered less than 1% mortality. Parker *et al.* (1978) observed 66% mortality among 129/J, Snell, and C3H/Bi mice in a natural epizootic, whereas only 29% mortality was observed among C57L and C57BL/6 in the same outbreak. Zurcher *et al.* (1977) observed excessive mortality among young RFM (less than 10 weeks of age), old male BALB/c, and old females in general when compared to other strains, such as C57BL and NZB. High mortality was also observed in weanling DBA/2J mice suffering from concurrent mycoplasma infection (Richter, 1970). Both Parker *et al.* (1978) and Stewart and Tucker (1978) observed that individual strains of mice, regardless of their colony origin, always showed the same sensitivity to lethal infection and that this range in sensitivity between different strains was significant. Of 24 strains tested by Parker *et al.* (1978), the 129/J mouse was 25,000 times more sensitive to lethal infection than the most resistant (SJL/J) mouse; however, both strains were similarly permissive in support of viral replication in their lung tissues. The athymic nude mouse strain is an anomaly, reacting in a manner analogous to that of immunosuppressed mice (Blandford, 1975). Infected nude mice develop signs of illness later than infected immunocompetent mice and undergo progressive emaciation. Death occurs sporadically over a long period, and some mice survive for more than 10 weeks. Chronic virus infection is characteristic, and isolation of virus from lungs is readily achieved throughout a period of 10 weeks or longer (Ward and Young, 1976; Ward *et al.*, 1976a,b; Ueda *et al.*, 1977; Iwai *et al.*, 1979) (Fig. 2).

Table IV

Susceptibility of Inbred and Outbred Strains of Mice to Sendai Virus Infection

Mouse strain	No. of replicate titrations	$LD_{50} \pm SE^a$ (log 10)
129/ReJ	1	0.5
129/J	4	0.6 ± 0.4
Nude (Swiss)	1	0.7
DBA/1J	3	1.3 ± 0.4
C3H/Bi	1	1.4
DBA/2J	3	1.6 ± 0.3
DBA/2	1	2.0
A/HeJ	3	2.5 ± 0.1
A/J	2	2.5 ± 1.0
SWR/J	1	2.7
Swiss[b]	1	2.7
C57L/J	2	2.7 ± 0.5
C57BL/10Sn	2	2.8 ± 0.6
C3HeB/FeJ	2	2.8 ± 0.1
BALB/cJ	1	3.0
C57BL/6	1	3.0
Swiss[c]	1	3.1
C58/J	1	3.2
AKR/J	3	3.4 ± 0.2
Swiss[d]	1	3.4
Swiss[e]	4	4.4 ± 0.0
C57BL/6J	1	4.4
RF/J	2	5.0 ± 0.5
SJL/J	3	5.0 ± 0.4

[a] $TCID_{50}/LD_{50}$. SE, standard error.
[b] National Institutes of Health.
[c] Life Sciences.
[d] National Laboratory Animal Co.
[e] Microbiological Associates.

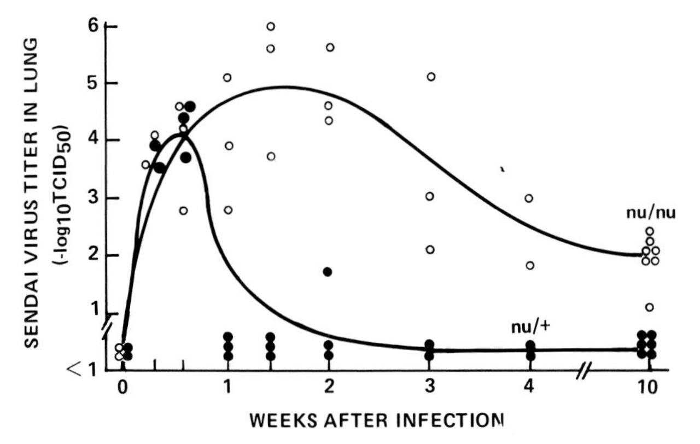

Fig. 2. Sendai virus titers in the lungs of nude and *nu*/+ mice after intranasal infection. (From Iwai *et al.*, 1979.)

Table V

Incidence of Respiratory Virus Antibody in Mouse Breeder Colonies

Colonies	Mouse colonies			Rat colonies		Hamster colonies	
	Sendai	PVM	K	Sendai	PVM	Sendai	PVM
Number positive	*47*	*44*	*4*	*14*	*17*	*9*	*17*
Number tested	73	73	73	25	25	18	18
Percent positive	64	63	5	56	68	50	94
Antibody prevalence in infected colonies (%)	42	20	21	64	62	64	45
Range of antibody prevalence in colonies (%)	2–100	1–86	9–23	8–100	7–94	4–100	4–92

2. Prevalence and Geographic Distribution

Sendai virus infection of laboratory mice is one of the most common and most serious problems in infectious disease control management of colonies and is the leading cause of viral respiratory disease in mice. Sendai virus has been found to be a frequent contaminant in colonies of mice throughout the world: Canada (Descoteaux *et al.*, 1977), China (Chanock *et al.*, 1963; Yun-De, 1962), England (Cooper *et al.*, 1977), Germany (Schels and Hartl, 1971), Japan (Fukumi *et al.*, 1954; Fujiwara and Iida, 1974; Fujiwara *et al.*, 1976, 1979; Iida *et al.*, 1973; Saito *et al.*, 1978), Netherlands (Zurcher *et al.*, 1977; van Nunen *et al.*, 1978), Russia (Gerngross, 1957), United States (Parker *et al.*, 1964; Poiley, 1970), and Australia, Denmark, Israel, France, and Switzerland (J. C. Parker, 1979 unpublished observations). The virus became apparent in Japan between 1951 and 1952, whereas in the United States the infection was not recognized before the early 1960s (Ishida and Homma, 1978; Parker *et al.*, 1964; Profeta *et al.*, 1969). In an extensive multiyear study, Parker (1973; Parker *et al.*, 1978) observed widespread infection of mouse, rat, and hamster colonies and also high rates of infection, generally in more than 50% of the animals within the infected colonies (Tables V and VI). No evidence of infection in feral mice has been reported (Parker *et al.*, 1966). In the United States the number of mouse colonies infected with Sendai virus increased from 44% in 1968 (Parker and Reynolds, 1968) to 64% in 1973 (Parker, 1973), 66% in 1977 (Parker *et al.*, 1978), and 85% in 1979 (J. C. Parker, 1979 unpublished observations). In general, specific pathogen-free colonies maintained within good barriers have lower incidences of infection and consequently better quality mice (Fujiwara, 1971; Poiley, 1970; Van Hoosier *et al.*, 1966). Gnotobiotic colonies maintained in germ-free isolators are normally free of infection (Parker *et al.*, 1965).

Table VI

Incidence of Sendai Virus Antibody within Infected Mouse Breeder Colonies

Mouse colony	No. tested	Age (months)	Hemagglutination inhibition		
			No. positive (%)	Mean titer of positives[a] (reciprocal)	Titer range of positives (reciprocal)
1	69	9	94	35	10–160
2	25	5	92	61	20–160
3	48	10	79	22	10–≥80
4	74	3–10	88	28	10–640
5	50	9	88	18	10–80
6	61	8–11	20	12	10–20
Total	327		76	27	

[a] Geometric mean titer.

3. Natural History of Infection

Sendai virus produces two basic patterns of infection in mouse colonies: (1) a chronic, usually clinically inapparent infection (enzootic) that is self-perpetuating in the colony; or (2) an acute, usually clinically apparent infection (epizootic) that is of a relatively short duration and either disappears (self-cure) or evolves into the enzootic type. Environmental and physiologic factors are important in determining the type and outcome of the infection (strain susceptibility, age, colony size, previous history and immunologic status, and perhaps the season). The authors' experience indicates that more than one-half of infected colonies show some clinical manifestations of infection.

In chronic enzootic infections (Fig. 3), virus infection occurs in a high percentage of mice from a period shortly after weaning to approximately 6 weeks of age. Normally the infection is subclinical, with virus persisting for approximately 2 weeks (Fig. 4), accompanied by seroconversion and followed by an uneventful recovery, with circulating antibody persisting for a year or longer (Parker *et al.*, 1964; Parker and Reynolds, 1968; Fujiwara *et al.*, 1976). There is no evidence that recovered immunocompetent mice harbor latent virus or have chronic infections which later can be activated (Sawicki, 1962;

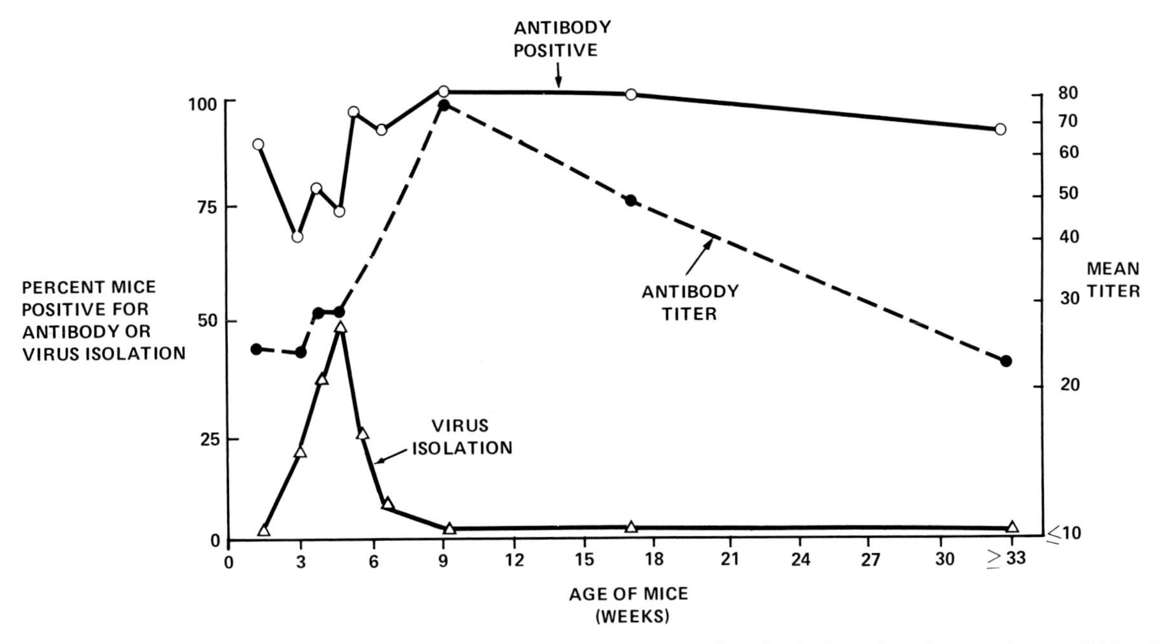

Fig. 3. Sendai virus infection pattern in an enzootically infected mouse breeder colony. Mean titer is the reciprocal geometric mean HAI antibody titer of positive mice. (From Parker and Reynolds, 1968.)

van der Veen *et al.,* 1974; Fujiwara *et al.,* 1976; J. C. Parker, 1979 unpublished observations). Grunert (1967) claimed that Sendai virus was present as a truly latent infection in Swiss Webster mice used in influenza virus–mouse adaptation studies; however, more recent studies indicate that he was probably the victim of infected supplier colonies and that his test mice were incubative when acquired. A limited age group of uninfected susceptible mice, approximately 4–6 weeks old, which are continuously added to a breeding colony, are responsible for the persistence of enzootic infections (Iida *et al.,* 1973; Fujiwara *et al.,* 1976; Goto and Shimizu, 1978).

Acute epizootic-type infections vary in their character but usually last for approximately 2–7 months and either disappear or become enzootic (Parker and Reynolds, 1968; Parker *et al.,* 1978; Fukumi *et al.,* 1962; Itoh *et al.,* 1978). Nonbreeding and occasionally breeding colonies which undergo epizootic infection with self-cure do not experience infection again unless the virus is reintroduced (Fukumi *et al.,* 1962; Parker and

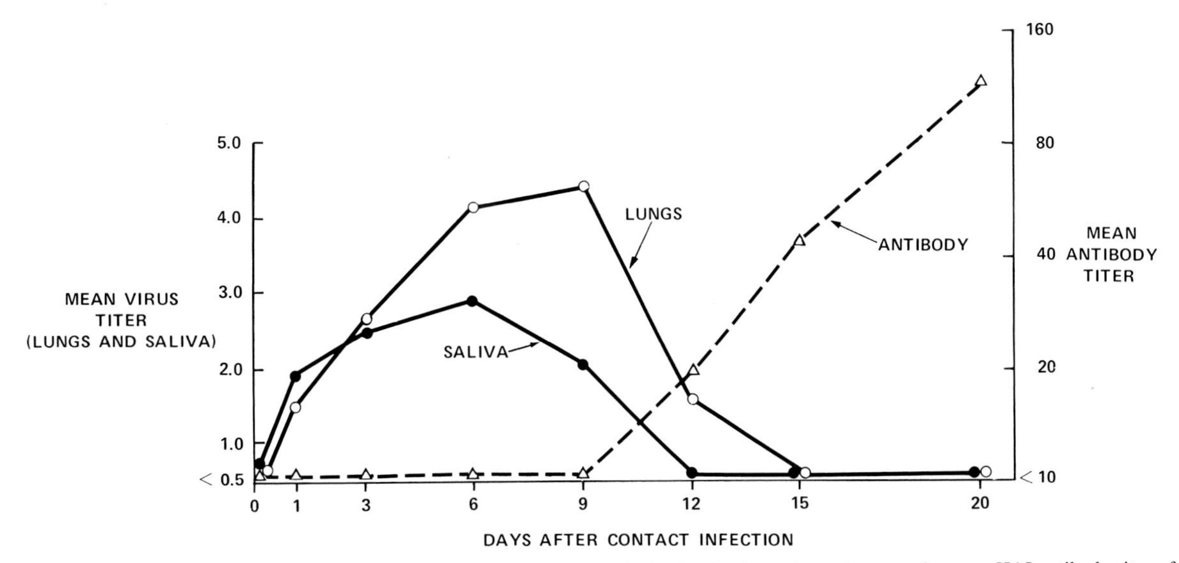

Fig. 4. Sendai virus titers in the lungs and saliva of infected mice. Mean antibody titer is the reciprocal geometric mean HAI antibody titer of positive mice. Virus titers are mean titers of positive specimens expressed as negative $\log_{10}$ $TCID_{50}$ per 0.05 ml. (From Parker and Reynolds, 1968.)

Reynolds, 1968; van der Veen *et al.*, 1974; Itoh *et al.*, 1978; J. C. Parker, 1979 unpublished observations). Epizootics which occur in previously uninfected mouse colonies are characterized by respiratory distress, mortality of newborns (Fig. 5), retarded growth in young mice and prolonged gestation in adults (Iwai *et al.*, 1977; Bhatt and Jonas, 1974). Weaning rates also decrease dramatically for a short period, after which normal or near-normal levels are resumed regardless of whether the infection disappears (Fig. 6) or becomes enzootic (Bhatt and Jonas, 1974; Iwai *et al.*, 1977; Zurcher *et al.*, 1977; Itoh *et al.*, 1978).

Although there is some evidence that enzootic Sendai infection in breeding colonies might increase weanling mortality (Van Hoosier *et al.*, 1966), this appears to be uncommon. Epizootics in breeder colonies usually cause high mortality in preweaning litters during the early weeks of the infection, as described by Bhatt and Jonas (1974), Zurcher *et al.* (1977), and Fujiwara *et al.* (1976). In the epizootic reported by Bhatt and Jonas, 30% of 184 affected litters suffered 100% mortality, with the greatest losses among the 8- to 14-day-olds. Litter mortality peaked on the tenth day of the epizootic and declined thereafter. In spite of the high morbidity and mortality, reproduction continued; and by the third week, the epizootic was brought under control. Mortality levels among preweanlings returned to preepizootic levels within 3 weeks and remained thus for an extended period even though the infection remained enzootic in the colony (Bhatt and Jonas, 1974). Fujiwara *et al.* (1976) reported a 50% reduction in the production index of a breeding colony of ICR mice when Sendai and mouse hepatitis virus (MHV) invaded a production barrier. Production returned to normal after 2 months, but both infections persisted in enzootic form even though 100% of breeders had seroconverted positive to Sendai and 83% to MHV. The epizootic in aging mice reported by Zurcher *et al.* (1977) provides one of the few good examples of the chronology of natural Sendai infection in a closed colony. From onset to termination, the episode took approximately 1½ months and involved 12 strains and hybrids. The outbreak started in a breeding colony and was manifested by increased mortality among preweanlings. Eventually, all strains were involved and the highest mortality was observed in preweanling ND2, C3H, and AKR strains (sometimes approaching 100%). The lowest mortality occurred among CBA and BALB/c strains, whereas intermediate levels were seen among C57BL, C57BL/Ka, RFM, and $BCBAF_1$ strains. From the breeding colony, the epizootic then spread to the aging colony. Mortality increased in the aging colony, but clinical disease was seldom seen. Necropsies showed, however, that approximately 80% of dead mice from the aging colony had foci of acute pneumonia.

In addition to strain and age influences on mortality patterns

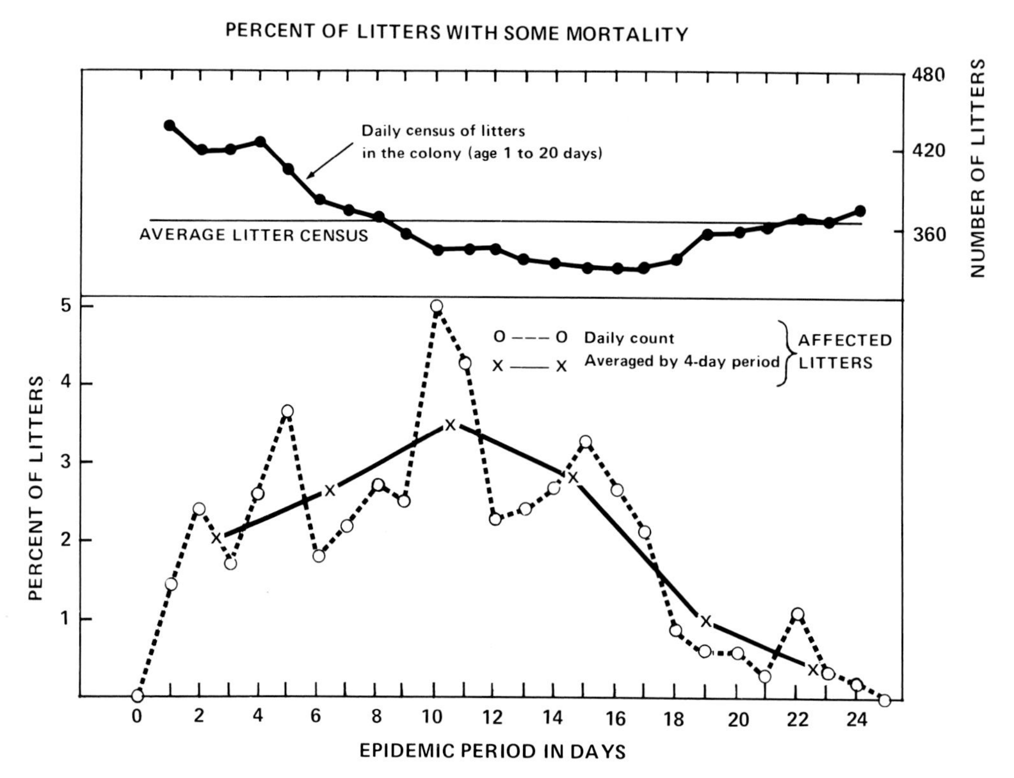

Fig. 5. Total population of litters (age 1–20 days) in a mouse colony and the percentage of litters with some mortality during an epizootic. (From Bhatt and Jonas, 1974.)

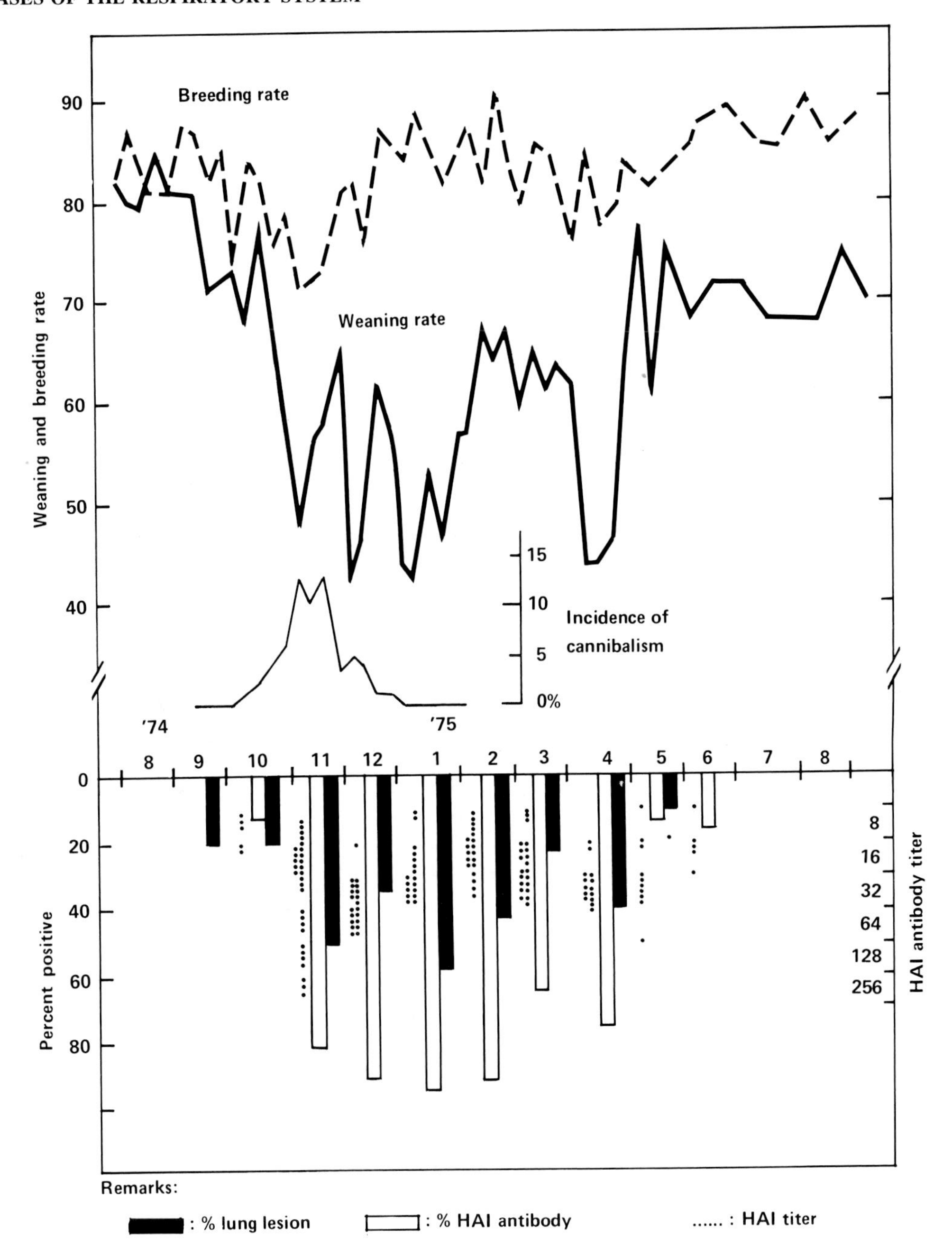

Fig. 6. Sendai virus infection and productivity in a mouse breeding colony. (From Itoh *et al.*, 1978.)

in Sendai infection, the gnotobiotic status is also likely to influence the outcome of spontaneous infections. Degré and Midtvedt (1971) have shown that axenic and conventional mice differ significantly in their survivability when challenged with identical doses of Sendai virus. Furthermore, conventional mice given penicillin-treated drinking water survived longer than nontreated conventionals when challenged with Sendai. Studies by Jakab and Dick (1973), Jakab (1974), and Degré and Glasgow (1968) on the role of intercurrent bacterial infection with Sendai virus have shown that unless mice are preimmunized against the specific challenge bacteria, Sendai infection suppresses the normal antibacterial activity of lungs, thus predisposing the mice to secondary bacterial penumonia. Prior immunization against Sendai virus also prevents bacterial pneumonia when the mice are challenged (Degré and Glasgow, 1968; Jakab and Green, 1973a). Degré and Glasgow (1968) showed that combined Sendai-*Hemophilus influenzae* infections killed more than twice as many mice as the sum of the mortality observed from the two agents given separately. Sendai-infected mice failed to clear bacteria, and it was shown that the effect was dependent upon the dose of both agents, except that very high Sendai doses did not increase lethality. Histopathologic changes were more pronounced in doubly infected mice. Degré and Solberg (1971) concluded that impaired ciliary activity following Sendai infection reduced the mucociliary flow from the lungs, thus inhibiting bacterial removal and promoting secondary infection. However, there is strong evidence that this effect is primarily caused by defective intracellular processing of phagocytized bacteria rather than by defective phagocytosis or impaired ciliary action (Jakab, 1975; Jakab and Green, 1972, 1973a,b, 1974, 1976).

Jakab (1977) showed that the suppression of pulmonary antibacterial activity in mice which follows aerosol inhalation of the irritant acrolein, formed in cigarette smoke, is significantly exacerbated by a concomitant Sendai virus infection. In addition, Nettesheim (1974) has shown that the interaction of respiratory infections, such as caused by Sendai virus, and carcinogens has an effect on urethane-induced pulmonary adenoma in the lung, although the mechanisms are unknown and the outcome, either enhancement or suppression, is dependent on several factors. Thus, although the role of Sendai virus respiratory infection in mice may provide models for lung cancer studies, the infection might also unknowingly interfere with lung carcinogen studies if the investigator is unaware of the presence of the infection.

Inhibitory effects of Sendai virus on other virus infections and tumor growth have also been demonstrated. Wheelock (1966, 1967) observed that inoculation (intraperitoneal route) of DBA/2 mice with Sendai virus up to 6 weeks before Friend leukemia virus (FV) inoculation inhibited the FV replication and subsequent pathologic response of FV infection. Matsuya *et al.* (1978) found that Ehrlich ascites tumor (EAT) cells treated with Sendai virus produced variants which possessed a greater degree of chromosome heterogeneity than untreated EAT cells and that cells with increased chromosome numbers were less tumorigenic. Persistent, noncytopathic infection of L1210 leukemia cells by Sendai led to lower transplantability in normal syngeneic mice, which was believed to be the result of a modification of the cell surface membrane of the L1210 cell by Sendai virus (Takeyama *et al.*, 1979).

Contact and airborne routes have been demonstrated as methods of transmission or spread of the virus (van der Veen *et al.*, 1970, 1972; Iida, 1972; Parker and Reynolds, 1968). Van der Veen *et al.* (1970) and Parker and Reynolds (1968) felt that direct contact was at least twice as effective as airborne transmission in spreading the virus; however, van der Veen *et al.* (1970) also showed that with increasing numbers of transmitters, the risk to susceptibles from airborne transmission climbed toward 100%, even when the susceptibles were exposed for only a short period. Thus, van der Veen *et al.* (1970) felt that the airborne route probably plays a highly significant role in disease. Mice that were experimentally exposed via the airborne route had detectable virus in their lungs by the fifth day, peak titers occurring between days 6 and 9 of the infection. Experimentally infected mice most readily transmitted the disease to secondaries between days 4 and 9 of the experimental infection and rarely before or after that period (van der Veen *et al.*, 1970). Iida (1972) showed that mice placed in contact with experimentally infected mice on the same day of the infection did not have virus in their lungs, trachea, or nasal washings for 5 days and concluded that infected contacts required an incubation period of 2–3 days before they became infective. On the other hand, Appell *et al.* (1971) have shown that aerosol-infected mice have demonstrable virus in the nasal washings regularly by PI day 2, and van Nunen and van der Veen (1967) reported that virus was recoverable from SPF Swiss albinos as early as 24 hr PI. Van der Veen *et al.* (1972) demonstrated that high relative humidity (RH) (60–70% versus 30–45%) significantly affected the transmission rate of experimentally induced Sendai infection, especially airborne infection over a longer distance. Thus 32% of airborne contacts exposed at a distance of 5–6 feet became infected when RH levels were maintained at 60–70% even though exposure was only for 24 hr. Their experiments also showed that true airborne transmission, e.g., via droplet nuclei, did occur and that higher transmission rates were achieved over short distances when 5 to 10 air exchanges per hour were used versus 20 air changes per hour.

Although chronic or latent infections have long been suspected to play a role in the natural history of Sendai infection (Grunert, 1967), the weight of evidence is against this (Fujiwara *et al.*, 1976; Parker and Reynolds, 1968), and most commercial breeders and experimenters feel that enzootic infections can be squelched by a temporary halt in breeding. Parker and Reynolds (1968) and Iida *et al.* (1973) concluded that the maintenance of enzootic infections is dependent upon a continuous supply of new susceptibles, conveniently supplied by the waning of protective maternal antibody in mice between 1 and 2 months of age. This is further substantiated by Fujiwara *et al.* (1976), who concluded from their study of a breeder colony epizootic → enzootic that mice between 5 and 7 weeks of age are the most important in maintaining virus dissemination in a breeding colony. Younger mice are passively immune, whereas older mice are actively immune and do not carry transmissible virus.

Although seasonal periodicity has not been adequately studied, there is a general consensus that whereas minor fluctuations in infection may occur, seasonal variation is not a significant factor in the epizootiology of Sendai virus infection. G. J. Eaton (1979 personal communication), however, has observed a definite pattern of disease for several years in a closed colony of Swiss ICR mice, in which the incidence of disease in litters increases each spring and fall, with little or no evidence of infection in winter and summer.

D. Pathogenesis

1. Clinical Disease

The absence of overt clinical signs is common and many infections, notably enzootics, go clinically undetected (Appell *et al.*, 1971; Parker and Reynolds, 1968; Robinson *et al.*, 1969; Bhatt and Jonas, 1974; Parker *et al.*, 1978; Zurcher *et al.*, 1977; Kay, 1978a). Furthermore, most observers agree that mortality is low in most strains when healthy adult animals become infected (Bhatt and Jonas, 1974; Parker and Reynolds, 1968). Although most infected adult mice show few if any clinical signs, sick mice typically sit in a hunched position with hair coat erect; eyes may be pasted shut and sunken; weight loss is precipitous; and respiration is labored. Ward (1974) also noted chattering in naturally infected weanling C57BL/6 Cr mice. Few severely affected animals survive.

2. Pathology

a. Gross Lesions. Lung lobes of mice infected with Sendai virus are plum-colored and exude a frothy, blood-stained fluid when cut. Whole lobes or parts of lobes may be involved, and the demarcation between normal and penumonic tissue is distinct (Iwai *et al.*, 1979; Jensen *et al.*, 1955; Ward, 1974; Zurcher *et al.*, 1977). Although Appell *et al.* (1971) suggested that apical, diaphragmatic, and hilar lobes and regions were more frequently infected, any or all lobes may be involved (Jakab and Green, 1973a; Richter, 1973). During recovery, red foci are replaced by gray-white spots (Ward, 1974; Zurcher *et al.*, 1977). Pulmonary consolidation may be severe enough to cause lung tissue to sink when placed in fixative. Robinson *et al.* (1968) pointed out that although none of their experimentally infected Swiss mice were clinically ill, serially sacrificed cohorts had an average increase in lung weight of 50% by the fifth day PI and 100% by the fourteenth day PI. Parker *et al.* (1978) also showed that lung weights increased with infection; furthermore, highly susceptible strains had heavier PI lungs than did resistant strains. Increased quantities of serous fluid are sometimes present in the pericardial and pleural cavities, and pleural adhesions and lung abscesses are occasionally seen (Appell *et al.*, 1971).

b. Histopathology. The histopathology of Sendai infection can be divided into three rather distinct phases: the *acute phase,* characterized by an inflammatory response to the virus and lysis of target cells; the *reparative phase,* characterized by proliferation of regenerating target epithelium; and the *recovery phase,* characterized by parenchymal scars and strictures that may persist for the life of the animal. Since it is possible to observe the same lesions in both experimentally and naturally infected mice, the following description is based on numerous experimental pathology studies (Appell *et al.*, 1971; Blandford and Heath, 1972; Degré and Glasgow, 1968; Degré and Solberg, 1971; Fukumi *et al.*, 1954; Iwai *et al.*, 1979; Jakab and Green, 1973a; Parker *et al.*, 1978; Richter, 1970, 1973; Robinson *et al.*, 1968). Detailed descriptions of the natural pathology of Sendai infection in euthymic mice may be found in Ueda *et al.* (1977), Ward (1974), and Zurcher *et al.* (1977).

The earliest acute-phase pathology includes edema and modest polymorphonuclear infiltration of the lamina propria of infected bronchial segments, alveoli, and alveolar ducts. By 48 hr, noticeable eosinophilia can be observed in the cytoplasm of the infected bronchial epithelium, and this progresses to marked cytoplasmic distention. Nuclear location in infected cells is often atypical, with a tendency for nuclei to move away from the normal basilar position (Fig. 7a). Eosinophilic intranuclear and cytoplasmic inclusion bodies surrounded by clear halo zones have been reported in experimental and natural infections (Appell *et al.*, 1971; Ward *et al.*, 1976a,b; Zurcher *et al.*, 1977), but such findings are uncommon. Congestion of septal capillaries and edema of the peribronchial and perivascular connective tissue are observed in the lungs. Small hemorrhages are also common, and erythrocytes, necrotic polymorphs, and desquamated epithelial cells may be found in the alveoli and alveolar ducts. Desquamation of the bronchial epithelium may begin as early as the third day of experimental infection (Blandford and Heath, 1972) but becomes more intense during the next few days. Denudation may be complete, and in those bronchial segments where exfoliated and exudative cells have been washed out, bronchi may be confused with small venules (Fig. 7b). In most strains of mice, bronchial exfoliation is complete by the ninth or tenth day of experimental infection; however, time sequences are modified by host resistance, and highly susceptible strains may continue to demonstrate epithelial cell hyperplasia and necrosis as late as the end of the second week (Parker *et al.*, 1978). Infection tends to be segmental within bronchi; hence a spectrum of pathologic changes ranging from hyperplastic epithelium to necrosis and exfoliation may be present on the same histologic slide. The probable mechanisms of bronchial epithelial cell necrosis and exfoliation are discussed below.

Type I alveolar epithelium is also desquamated, and by the fifth day of experimental infection, accumulations of necrotic cells and phagocytes plug the terminal airways. Infiltration of alveolar septa and alveoli by inflammatory leukocytes and the exfoliation of the alveolar lining epithelium obscure parenchymal architecture, producing the typical appearance of interstitial pneumonitis. Lymphocytes and small mononuclear cells appear in the peribronchial and perivascular spaces by the third

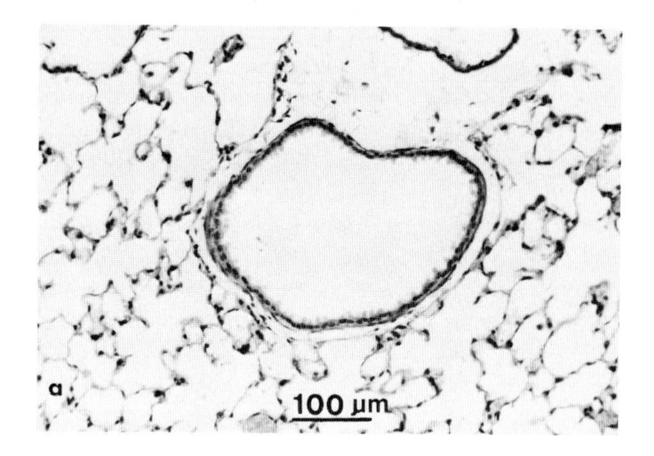

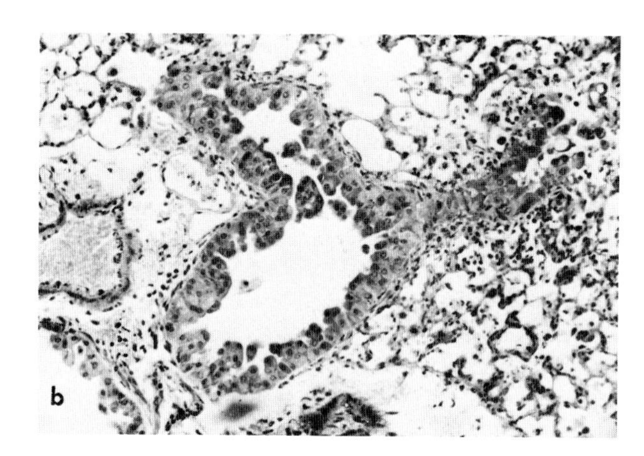

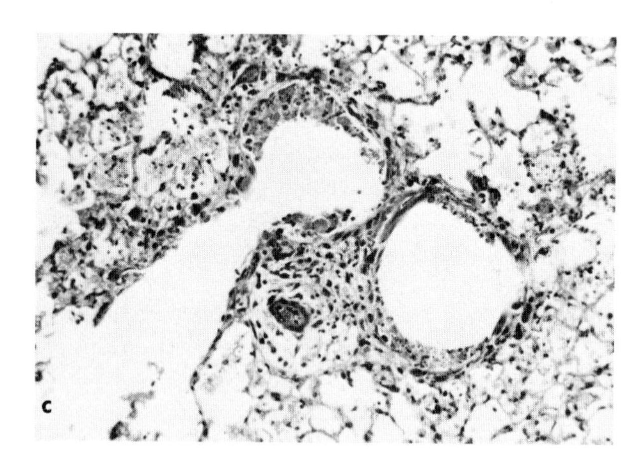

Fig. 7. (a) Normal bronchial epithelium and surrounding alveolar tissue. The sawtoothed appearance of the epithelium is caused by Clara cells bulging above ciliated cells. (b) Experimentally infected C3H mouse, PI day 5. The bronchial epithelium is markedly hypertrophic and atypic. An acute inflammatory reaction is present in the lung parenchyma. (c) Necrosis and desquamation of epithelium in small bronchioles. Experimental infection, BALB mouse, PI day 10. Parts (a) through (c) are the same magnification. H&E stain. (From Richter, 1973.)

day of experimental infection and increase in number over the next several days. These cells are highly pyroninophilic at this stage (Robinson *et al.*, 1968) and rapidly mature to plasma cells (Richter, 1973).

Regeneration and repair quickly follow the lytic phase of the infection. Evidence of regeneration can be found as early as the third day PI, but it reaches maximum levels about the tenth day. Histologically, this phase is characterized by marked proliferation of low cuboidal, or squamoid, highly basophilic cells in the bronchi and alveoli. These cells exhibit a high mitotic index and rapidly repopulate the damaged bronchi and alveoli (Richter, 1973) (Fig. 8). They may form hyperplastic and/or squamous metaplastic foci that can be extensive(Fig.9a). Squamous metaplasia may also extend from damaged terminal bronchioles into the alveoli as sheets of squamous cells (Fig. 9b) complete with rudimentary pearl formation. This sometimes gives the appearance of invasive carcinoma (R. P. Custer, personal communication; Iwai *et al.*, 1979; Richter, 1970). In addition, small polypoid outgrowths with myxomatous stalks are frequently seen in the small bronchi during this period (Fig. 9c); however, these disappear as repair is completed.

A second form of aberrant repair occurs in the alveoli and is

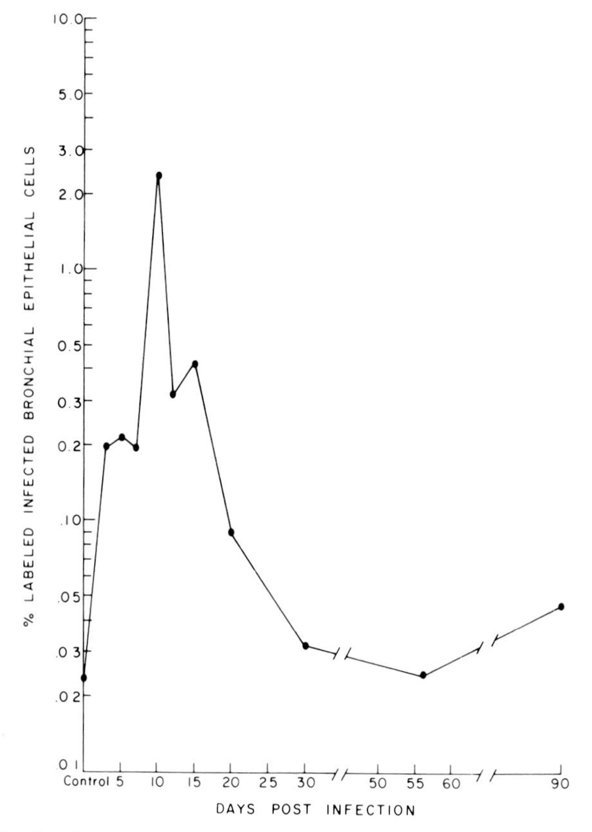

Fig. 8. Percent of bronchial epithelial cells showing [^{3}H]thymidine uptake following a single pulse label of experimentally infected DBA/2 mice. $\geq 2 \times 10^3$ cells per time point. (From Richter, 1973.)

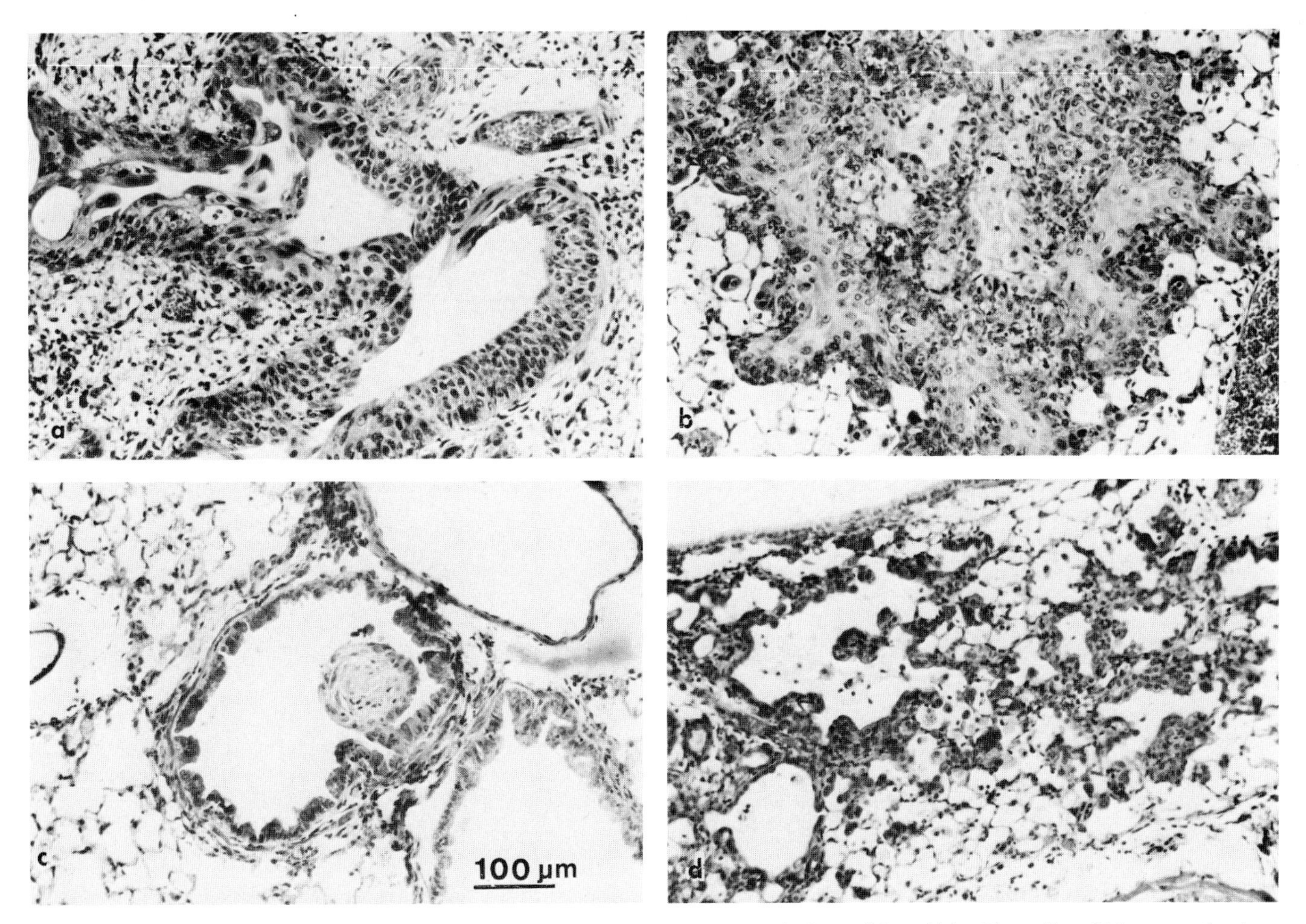

Fig. 9. Experimental infection in DBA/2 mice, PI day 15. (a) Severe metaplastic change in the small bronchioles. The peribronchiolar connective tissue has a myxomatous appearance. (b) Extension of squamous metaplasia into alveoli surrounding a terminal bronchiole. (c) Polypoid outgrowth in a small bronchus. These growths are common during the early repair phase but do not persist. (d) Adenomatoid repair of alveoli. Most lining cells are low cuboidal. Large macrophages and cell debris persist in the alveoli and alveolar duct. Parts (a) through (d) are the same magnification. H&E stain. (From Richter, 1973.)

characterized by proliferation of cuboidal epithelium along the septa. This produces a focal acinar or "adeomatoid" arrangement around terminal bronchioles and is sometimes referred to as *alveolar bronchiolization* (Fig. 9d). It is possible to demonstrate cilia on some of these cuboidal cells; however, most are nonciliated and of uncertain origin. During the repair phase, large numbers of mature plasma cells are present in perivascular and peribronchial connective tissue.

Repair of damaged lung parenchyma in surviving animals is generally complete, but there are some notable exceptions. Appell *et al.* (1971) concluded that most pathologic changes in Swiss albinos had subsided by day 21 except for mild focal hyperplasia of the bronchial epithelium, persistent collars of lymphoid tissue, and occasional collections of foamy macrophages. Richter (1973) found that reparative lesions persisted somewhat longer in DBA/2J mice, and Parker *et al.* (1978) showed that late acute and early reparative lesions were still present at 37 days PI in the highly susceptible 129/ReJ strain. Robinson *et al.* (1968) noted that the bronchial mucosa had returned to a normal appearance by day 33 but that peribronchial mononuclear infiltration persisted for longer periods. Appell *et al.* (1971) and Robinson *et al.* (1968) noted focal alveolar accumulations of foamy macrophages in the late stages of the disease.

Few studies have been conducted which show the residual pathology of Sendai infection in mice. Richter (1973) examined experimentally infected DBA/2 mice at PI days 30, 56, 90, 200, and 365. Eleven of 12 mice killed at 365 days had easily recognizable gross and/or microscopic lesions, and the twelfth mouse had subtle microscopic lesions. Most frequently observed microscopic lesions included dilated terminal air spaces containing inspissated secretions, cholesterol crystals, foreign body macrophages, and thickened septal walls; alveoli lined by cuboidal epithelium; focal alveolar collections of large

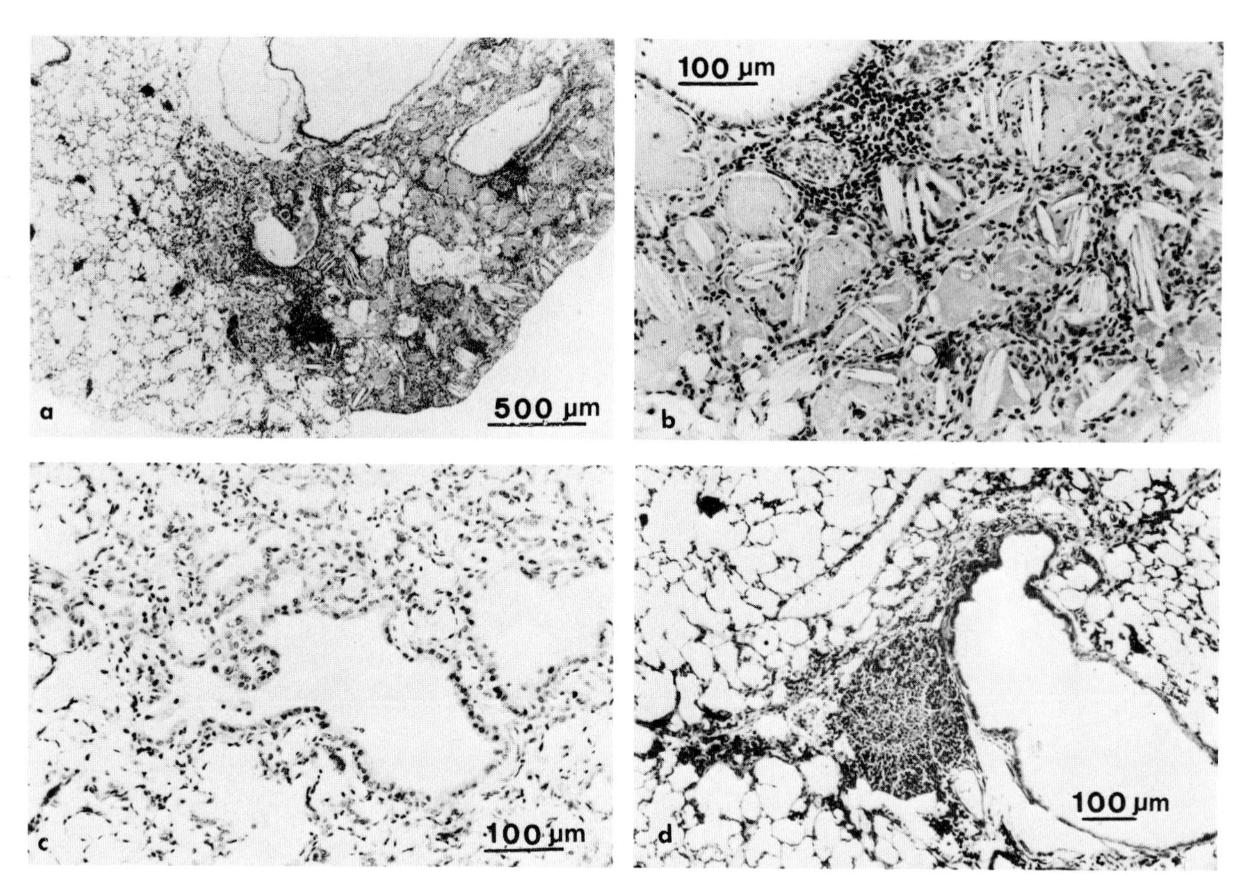

Fig. 10. Experimental infection in DBA/2 mice, PI day 365. (a) Focal residual lesion. Alveoli are distended with an inspissated secretion, cholesterol clefts, and macrophages. Focal lymphoid aggregates are present. (b) Higher magnification of a segment of (a). (c) There is residual focal bronchiolization. The lesion is fairly "clean." (d) Solitary lymphoid aggregate in peribronchial connective tissue. The surrounding septal walls contain scattered lymphocytes. H&E stain. (From Richter, 1973.)

foamy macrophages, sometimes multinucleate; bronchioles plugged by mucus and cell debris; peribronchial and perivascular lymphoid aggregates; and occasionally highly refractile intracytoplasmic crystals in macrophages and atypical bronchial-lining epithelium. These lesions were usually discretely focal; however, occasionally more diffuse thickening of septal walls with scattered excessive alveolar macrophages was seen. Examples of most of these lesions are shown in Fig. 10. Permanently wrinkled lung lobes were the most common macroscopic lesion seen. Zurcher *et al.* (1977) followed surviving mice from the outbreak they described, thus providing us with the best documentation of residual lesions resulting from a natural outbreak. Gross white glassy areas and microscopic lesions similar to those observed by Richter were found in 10–40% of mice examined, some as late as 2 years postepizootic.

Vasculitis is sometimes seen in the lung (Ward, 1974), and Ward also observed acute and chronic rhinitis as well as otitis media during the recovery phase. Kay *et al.* (1979) concluded that natural Sendai infection in their colony was responsible for linear deposits of immunoglobulins along the glomerular basement membrane and granular deposits in the mesangium of older mice that had been infected at 1 month of age. Mice infected at older ages also had granular deposits. In spite of the fact that Sendai is a potent syncytial virus, there are apparently no reports of polykaryotic or giant cells forming naturally in mouse lungs as a direct result of the virus.

c. Ultrastructure Pathology. The ultrastructure pathology of experimental Sendai infection has been studied by Richter (1970, 1973) and to a limited extent by Bruch *et al.* (1973) and

Fig. 11. (A) Bronchial epithelium replicating massive quantities of Sendai virus in the cytoplasma (v). Both ciliated cells (1) and Clara cells (2) replicate virus. Several large budded particles are seen in the lumen (→). PI day 5. Final magnification, X8950. The inset shows a modified host cell membrane over a linearly arranged tubule (→). Transversely arranged nucleocapsids are seen in the budding particle (⇉). Small budded particles are also present (P). PI day 4. Final magnification, X80,625. (B) Sendai virus replicating in the cytoplasm(C) and nucleus (N) (→) of bronchial Clara cell. PI day 7. Final magnification ×16,750.

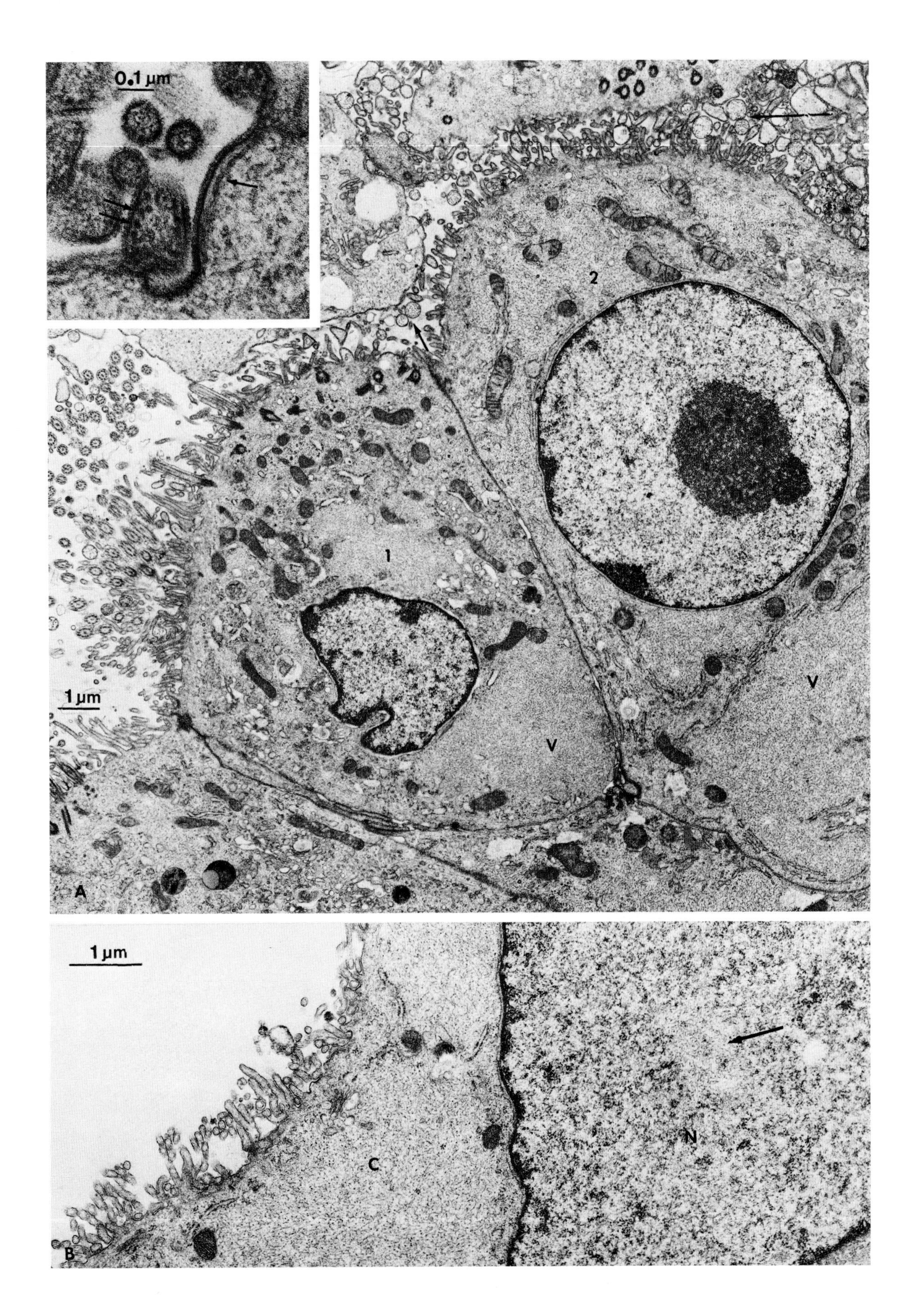

0.1 µm
2
1
1 µm
V
V
A
1 µm
C
N
B

Zurcher *et al.* (1977). Ward *et al.* (1976b) studied the ultrastructure pathology of spontaneously infected athymic nude mice. The ultrastructure of the replication cycle for parainfluenza viruses has been described by Compans and Choppin (1971), and most of these phenomena, including viral penetration, assembly of completed virions, alignment of virions along cell membranes, and budding from altered cell membranes can be shown ultrastructurally in experimentally infected mice. Ultrastructure studies demonstrate that Sendai virus replicates in both the nucleus and the cytoplasm of infected cells, but to a much greater extent in the cytoplasm (Fig. 11). The primary site of virus replication is the bronchial epithelium; however, virus can also be found replicating in the type 1 and type 2 alveolar epithelium (Richter, 1970; Ward *et al.*, 1976b; Zurcher *et al.*, 1977). Richter also found the virus replicating in septal endothelium but was unable to observe budding from this cell. Although virus budding can be seen from bronchial epithelium as early as 72 hr after infection, this process reaches its peak about the fifth day of experimental infection. This time also correlates with peak infectivity of infected animals (Appell *et al.*, 1971). Parainfluenza viruses typically produce irregular patterns of nucleocapsids in the host cell cytoplasm; however, Ward *et al.* (1976b) saw regular crystalline aggregates of viral nucleocapsids in the nucleus and cytoplasm of spontaneously infected athymic nude mice (Fig. 12). This observation correlates with the finding of inclusion bodies that they reported in these mice. Kay (1978a) reported particles "closely resembling Sendai virus" in the blood, spleen, and bone marrow of T cell-deficient and old mice during a natural outbreak.

Sequential studies of reparative phase DBA/2J mice showed that although squamous metaplastic cells contained numerous tonofilaments and desmosomes, in none of these cells were keratohyaline granules seen (Richter, 1970). Nevertheless, the use of the term *squamous metaplasia* to describe the epithelial changes seen at the light microscopic level seems to be justified. Richter (1970) and Bruch *et al.* (1973) observed degenerative changes in the basement membranes of denuded bronchi and alveoli. Ultrastructure and [^{3}H]thymidine uptake studies also showed that division of type 2 epithelium takes place, and during the reparative phase of the infection these cells can be found in clusters of three or more where only one would normally be seen.

d. Nude Mouse. Infections in athymic nude mice housed in proximity to conventional mice are common and inevitably disastrous. Mortality is extremely high, and lesions appear histologically stalled in the early acute phase. Ward *et al.* (1976b) and Ueda *et al.* (1977) have reported the pathology of natural infections, and additional observations can be found in Eaton *et al.* (1975) and Sharkey (1978). Ueda *et al.* (1977), Iwai *et al.* (1979), and Iwasaki (1978) have reported on the experimental pathology of nude mouse Sendai infection. The reports by Ward (1974) and Iwai *et al.* (1979) offer excellent descriptions of clinical and pathologic events and should be consulted for details. Although the basic aspects of Sendai

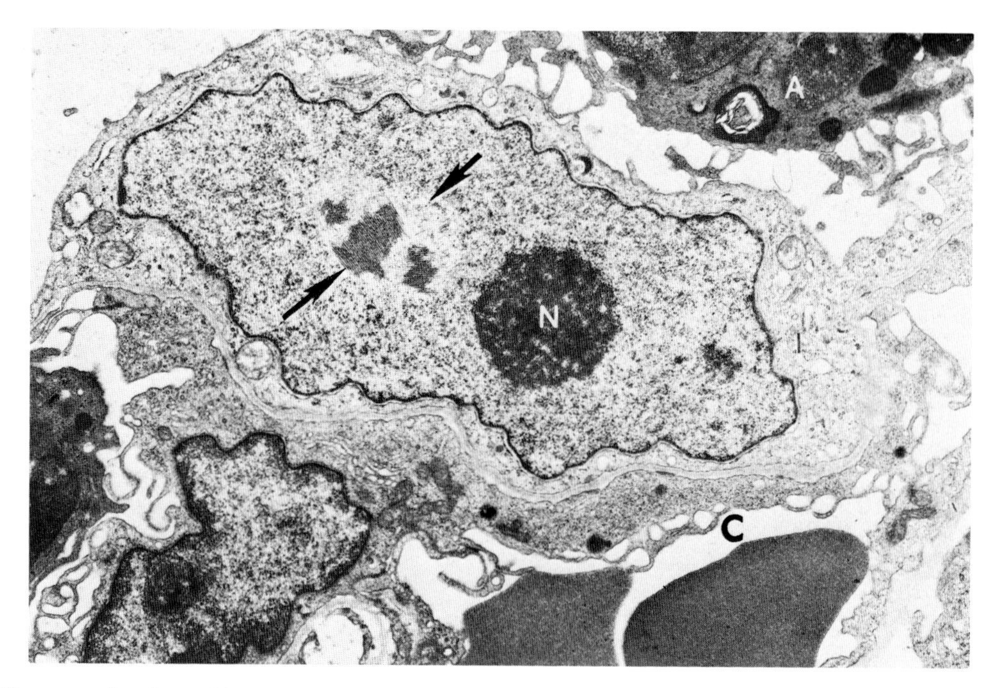

Fig. 12. Crystalline array of nucleocapsids in the nucleus of a type I alveolar cell in a naturally infected nude mouse (→). N, nucleus. Magnification not given. (From Ward *et al.*, 1976b; used by permission of the publisher.)

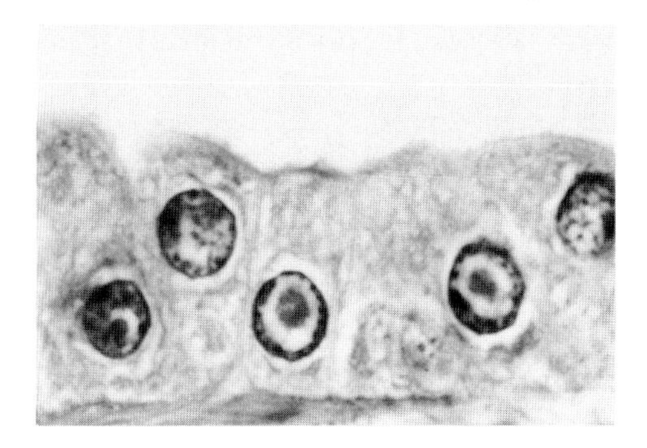

Fig. 13. Intranuclear inclusion bodies in the bronchial epithelium of naturally infected nude mice. Magnification not given. Phloxine stain. (From Ward *et al.*, 1976b; used by permission of the publisher.)

infection in nude mice are similar to those of infections in euthymic mice, several important distinctions are seen.

The clinical course of the infection is protracted, and infected nudes waste and become cyanotic. Lung consolidation is extensive, and histologic lesions are more diffuse. Bronchial epithelial necrosis is not as extensive as in euthymics; rather, these cells remain in a hypertrophied state and actively replicate virus for at least 70 days if the mice survive that long (Iwai *et al.*, 1979). Ward (1974) observed intranuclear and intracytoplasmic inclusion bodies surrounded by clear halos (Fig. 13), but Iwai *et al.* (1979) and Ueda *et al.* (1977) could make no similar finding. In late stages of the infection, aberrant epithelial proliferation similar to that seen in normal mice begins (Ward, 1974; Iwai *et al.*, 1979; Ueda *et al.*, 1977), and subsequently a mixture of hyperplastic and reparative lesions, including typical metaplasias, are seen. Ward observed secondary bronchopneumonia, vascular endothelial hypertrophy, and inflammation of the vessel walls and perivascular connective tissue. None of these workers observed the typical perivascular and peribronchial lymphoid aggregates seen in euthymic mice. Iwai *et al.* (1979) observed 70% mortality by PI day 51 in their experimental study, and survivors continued to exhibit signs of illness. They were unable to isolate virus from the brain, liver, spleen, or kidney of *nu*/+ or *nu*/*nu* mice and only occasionally from the heart of *nu*/*nu*. By using immunofluorescence (IFA) they were able to demonstrate viral antigen in hyperplastic bronchial epithelium, but not in neighboring metaplastic epithelium at 70 days PI.

e. Special Pathology. Mice have been experimentally infected by a number of unnatural routes, including intravenous, intraperitoneal, intracranial, and intrapleural (Charlton and Blandford, 1977; Jensen *et al.*, 1955; Mims and Murphy, 1973; Ohba, 1958; Shibuta *et al.*, *1978; Shimokata et al.*, 1976, 1977; Stewart and Tucker, 1978; Tanaka *et al.*, 1975; Tucker and Stewart, 1976). Viral antigens and replicating virus have thus been demonstrated experimentally in most body organs and tissues; however, the evidence for similar natural events is limited, and viremias apparently occur under unusual circumstances. It is of more than passing interest to note that virus inoculated by parenteral routes does not cause pneumonia. Intravenous inoculation of pregnant females resulted in virus in the placenta as well as the brain, choroid plexus, and various visceral organs as long as 21 days after infection (Tucker and Steward, 1976), and even though runting of the litter occurred, resorption of fetuses did not. Tuffrey *et al.* (1972) have reported that preimplanted fertilized eggs from naturally Sendai-infected CBA/H/T6T6 mice are frequently degenerate. When exposed to specific anti-Sendai serum and indirect IFA, these eggs stain brightly. The effect can be blocked by prior staining with unlabeled conjugate or elimination of the anti-Sendai serum, indicating that staining is specific. The route of infection was uncertain, but Tuffrey *et al.* (1972) felt that Sendai could play a natural role in fetal malformation or death. Ohba (1958) reported embryonic abnormalities following per nasal infection of pregnant females on gestation days 5–8 and examination on day 13, but no subsequent substantiating reports could be found.

A number of studies have employed Sendai virus inoculated intracranially into suckling mice. Jensen *et al.* (1955) observed ''neurological symptoms'' following intracerebral inoculation of infectious allantoic fluid and lung suspensions, but Shimokata *et al.* (1976) observed no clinical signs following intracerebral inoculation even though viral antigen was present in hippocampal neurons as long as 4 months afterward. Shimokata *et al.* (1976, 1977) and Shibuta (1978) observed late-developing hydrocephalus in intracranially inoculated sucklings, and Shimokata *et al.* (1977) were able to demonstrate viral envelope and nucleocapsid antigens in the perilymphatic and endolymphatic structures of the inner ear 2 days after infection. Antigen was gone from these structures by PI day 5, and there was never any evidence of inflammatory change.

3. Immune Response

a. Circulating Antibody. Positive evidence of seroconversion is probably the most sensitive means of detecting Sendai infection in a mouse colony. Blandford and Heath (1972) have demonstrated antibody as early as 3 days after infection; however, this early antibody is fixed to infected cells in the lungs and is not detectable by standard serodiagnostic means. Most workers agree that HAI antibody is not demonstrable in the blood before the sixth or eighth day of infection (van der Veen *et al.*, 1970; Appell *et al.*, 1971; Robinson *et al.*, 1968; Blandford and Heath, 1972; Charlton and Blandford, 1977), but this antibody may persist for at least 1 year (Richter, 1973; Zurcher *et al.*, 1977). Tucker and Stewart (1976) reported circulating HAI antibody as early as PI days

4–6 and also reported virus isolation as long as PI day 21 in the presence of circulating antibody.

Van der Veen *et al.* (1970) detected CF antibody at 12 days, and Iida (1972) detected it at 14 days but not 7 days after experimental infections. More recently, Parker *et al.* (1979) have shown that circulating CF antibody is detectable earlier and rises higher than HAI antibody when Swiss mice were challenged with 10 $TCID_{50}$ Sendai virus. In the same experiment, these workers showed not only that the ELISA test is more sensitive but also that ELISA antibodies are detectable as early as CF antibodies. Because of its high sensitivity and single dilution assay capability, the ELISA test seems destined for wide usage in Sendai serodiagnostic procedures.

Ward *et al.* (1976b) found low-titer CF antibody (1:10–1:20) in 9 of 18 *nu/nu* mice during a natural epizootic, but these positive results seemed to have little relationship to the time of infection and their meaning is unclear. With but one exception, Ueda *et al.* (1977) failed to detect CF antibody in naturally or experimentally infected *nu/nu* mice, and Iwai *et al.* (1977) failed to detect HAI, neutralizing, or CF antibodies in experimentally infected *nu/nu* mice. Sawicki (1961) and Robinson *et al.* (1968) presented evidence that interferon plays a role in recovery from Sendai infection. Interferon may appear as early as the third day of infection and peak by the fifth day. In the absence of a competent immune system, interferon may be an important mechanism of survival in nonimmunized very young mice (Sawicki, 1961).

In a study of an enzootic infection in a commercial colony, Iida *et al.* (1973) showed that 90% of pregnant females had significant CF titers to Sendai. Sixty-six percent of their 1-day-old offspring had titers matching those of their dams, whereas 84% of 7-day-olds had matching titers. These observations suggest that transplacental and gastrointestinal routes are both important in passive immunization of young animals. Immunoelectrophoresis showed that passive maternal antibody was IgG_1 and IgG_2 but not IgA. Titers peaked at 2 weeks and gradually declined thereafter. Mice challenged during the period of passive immunity were resistant to infection but did not actively synthesize antibody themselves. Iida *et al.* (1973) thus accurately concluded that in a breeding colony Sendai would exist as a continuous smoldering fire rather than as periodic epizootics, and Goto and Shimizu (1978) added the observation that maternally acquired antibody will protect 4- to 5-week-old mice against gross pulmonary lesions even though viral replication in the lungs does occur.

b. Cellular Response. The marked lymphocytic response to Sendai infection has already been described; however, the exact role that this local cellular response plays in recovery is not entirely clear. Mims and Murphy (1973) showed that mice infected with Sendai virus and treated with antithymocyte serum had mortality patterns similar to those of untreated mice, and they concluded that cell-mediated immunity was not important in the pathogenesis of the disease. On the other hand, Ward *et al.* (1976b) and Iwai *et al.* (1979) concluded from their studies of the athymic nude mouse that the absence of T cells in the nude mouse was responsible for the failure of the lytic phase of the infection and hence the protraction of the acute virus replication phase.

Studies by Robinson *et al.* (1969), Charlton and Blandford (1977), and Blandford and co-workers (1971; Blandford and Heath, 1972, 1974; Blandford, 1975; Blandford and Charlton, 1977) indicate that locally produced immunoglobulins play a significant role in the local and systemic response to Sendai virus. Specific subpopulations of immunoglobulin-containing cells (IgA, IgG, IgM) all increase in the infected respiratory submucosa as early as 2 days after infection. Over the course of the infection, IgG cells increase proportionately more than IgM or IgA cells, even though the latter are three to four times more populous before infection. This high ratio of IgG cells persists for at least 70 days after infection, but all three types of Ig cells remain three to four times more numerous during the 70-day period. The course of appearance of specific antibodies in bronchoalveolar washings parallels the increase in specific immunoglobulin-producing cells, e.g., IgG first, IgM second, and IgA third (Charlton and Blandford, 1977). In the absence of a splenic response to the antigen until PI day 24, the researchers concluded that the lungs are the major source of all three classes of serum antibody to Sendai virus. Blandford (1975) had previously reached this conclusion and showed that if the local immune response was suppressed, the lytic phase of the infection could be delayed. He concluded that complement and locally produced IgG and IgA were responsible for cell lysis.

The importance of cytotoxic thymus-derived lymphocytes (CTL) in target cell lysis in mouse Sendai infection is largely unknown; however, it is clear from *in vitro* studies that CTL capable of lysing Sendai-infected cells occur naturally and can be induced in mice (Anderson *et al.*, 1977; Anderson, 1978; Palmer *et al.*, 1977). CTL lysis is dependent upon H-2 histocompatibility receptors as well as virus-determinant receptors. Gething *et al.* (1978) have shown that virus fusion, and not just virus attachment, with target cell membrane is essential to CTL activity. Two distinct functions are involved in attachment and fusion that result in insertion of viral glycoprotein into the cell membrane and the viral capsid into the cell (see Section II,B,4,*a* for more detailed discussion). This apparently produces a recognizable change in the cell, and CTL lysis may follow. Since induced CTL do not appear prior to the fourth day of infection (Anderson *et al.*, 1977), it is possible to speculate that strain variability in the number of innate CTL may play some role in strain susceptibility variation (Parket *et al.*, 1978).

In an important series of studies on aging and autoimmune disease in a mouse colony following a natural outbreak of Sendai infection, Kay (1978a,b; Kay *et al.,* 1979) demonstrated a decline in T-cell population in long-lived CBA/T6T6 mice that preceded autoimmune disease and showed that this decline was the result of Sendai infection in immunologically immature young and immunodepressed old mice. She observed that of 63 weight, cellular, activity, and autoimmune indices, 55 remained abnormal 8 months after disappearance of the infection. Important examples of these indices that were reduced included body weight; thymus weight in old mice; cells per milligram wet weight of thymus, lymph node, and spleen; cells, per thymus and spleen; frequency of B-cells in lymph nodes, spleen, and bone marrow; phytohemagglutinin response; and hematocrits of mice 24 months old and older. Those indices that were increased included lymph node weight; bone marrow cellularity; percentage of dead thymus, lymph node, spleen, and bone marrow cells; IgM Coombs titer; and percentage of dead cells at 37°C in the presence of complement. *It is difficult to understate the significance of these observations to the experimenter who might choose to use previously Sendai-infected mice in sensitive studies.*

c. Immunofluorescence Studies. Numerous investigators have used IFA techniques to study Sendai infections (Charlton and Blandford, 1977; Blandford and Heath, 1972; Iwai *et al.,* 1979; Mims and Murphy, 1973; Richter, 1973; Shimokata *et al.,* 1977; Tucker and Stewart, 1976; Ueda *et al.,* 1977). Mims and Murphy (1973) demonstrated specific IFA in tracheal and bronchial epithelium 18 hr after experimental infection, and Blandford and Heath (1972) demonstrated superficial antigen in infected cells as early as 24 hr postinoculation. Most workers agree that by 48 hr PI, strong cytoplasmic staining is extensive. Specific fluorescence has been demonstrated in the ducts of the tracheal submucous glands, trachea, bronchi, type 2 alveolar epithelium, alveolar macrophages, and scattered foci in the nasal turbinate epithelium. Mims and Murphy (1973) observed that staining was more intense at the distal end of the trachea than at the pharyngeal end and felt that the main source of infectious particles was the lower rather than the upper respiratory tract. Specific IFA is no longer detectable after the ninth or tenth day of experimental infection. Because of the ease and rapidity with which indirect IFA techniques can be conducted in many laboratories, the diagnostician in larger facilities should consider this technique for rapid preliminary diagnosis.

E. Diagnosis

Sendai virus is antigenically distinct from all other viruses known to infect laboratory rodents; and since there is a variety of technically uncomplicated assays which have high sensitivity and accuracy, diagnosis of infection is not difficult.

1. Serology

Since only one serotype for Sendai virus is known, the problems of serodiagnosis are considerably simplified. Serologic

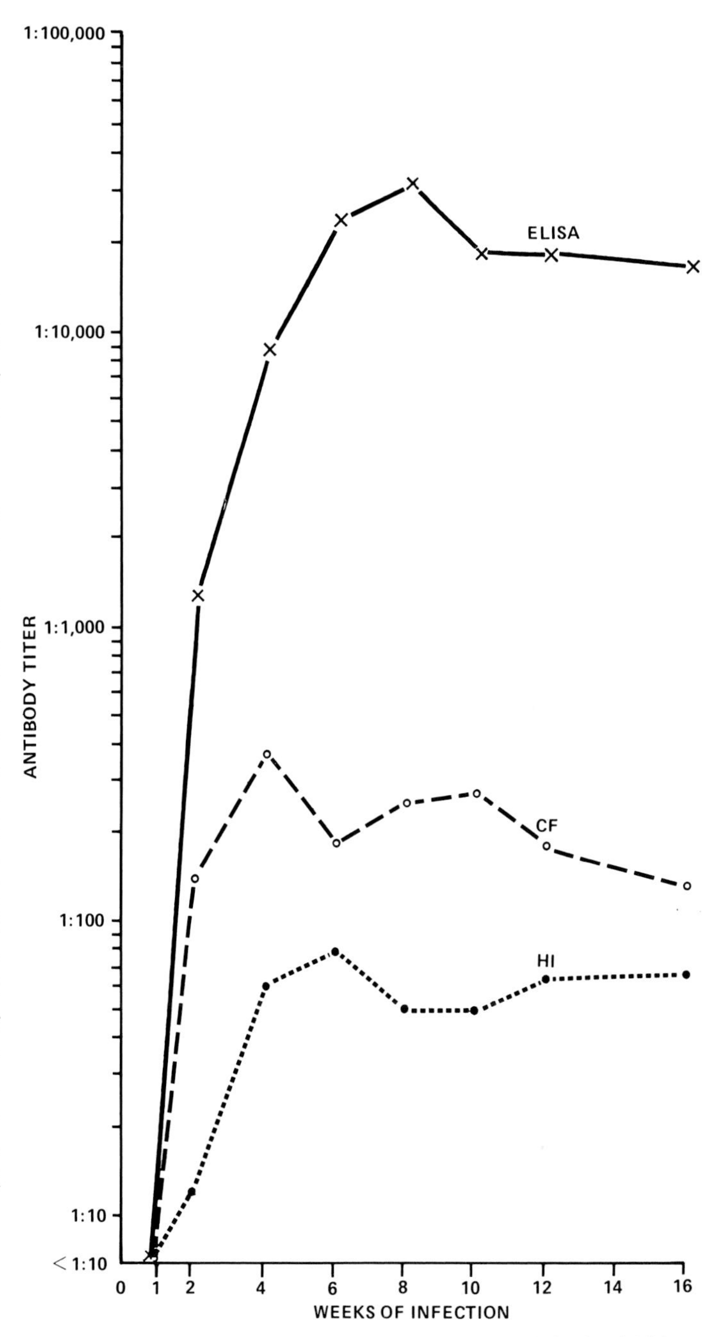

Fig. 14. ELISA, complement fixation (CF), and hemagglutination inhibition (HAI) antibody responses in Swiss mice inoculated with ten 50% tissue culture-infective doses of Sendai virus. (From Parker *et al.,* 1978.)

Table VII

Serodiagnostic Tests

Virus	Assay	Sensitivity	Antibody titers: Low positive	Antibody titers: High positive	Occurrence of false +'s or −'s	Approximate range of mean antibody titers in mice from infected colonies
Sendai	HAI	Fair	20[a]	80	Low	10–26
	CF	Good	10	160	Low	25–58
	ELISA	High	20 (0.15[b])	256 (0.80[b])	Low	72–657
PVM	HAI	High	20	160	Moderate	42–117
	CF	Fair	20	40	Low	
K	HAI	Fair–good	10	40	Low	20–40
	CF	Fair–good	10	320	Low	40–160

[a] Reciprocal antibody titer.
[b] Optical density at 405 nm.

tests developed for Sendai virus include HAI (Fukumi *et al.*, 1954; Heath *et al.*, 1962; Parker *et al.*, 1966, 1978, 1979), CF (Parker *et al.*, 1978, 1979), hemadsorption inhibition (HAdI) (Kashiwazaki *et al.*, 1965; Fernando and Heath, 1971), IFA (Kashiwazaki *et al.*, 1965; Stewart and Tucker, 1978), tissue culture neutralization (TC neut.) (Heath *et al.*, 1962), agar gel precipitation (Yachida *et al.*, 1975), radioimmunoassay (RIA) (Yung *et al.*, 1977; Charlton and Blandford, 1975), and enzyme immunoassay (EIA) (Parker *et al.*, 1979; Ertl *et al.*, 1979). The technical aspects of some of these assays were reviewed by Parker *et al.* (1965) and by Ohder (1971). The most frequently used and useful tests are summarized in Table VII.

One of the most important aspects of serologic diagnosis is the collection of the serum to be tested. In surveillance programs the sample size must be appropriate for the experimental design and the confidence level one wishes to achieve. An estimate of the sample size required can be made (ILAR, 1976). Finally, the blood samples must be collected at random from the colony at regular time intervals, usually quarterly or semiannually.

Serum for HAI and CF tests is normally heated at 56°C for 30 min prior to testing to remove heat-labile, nonspecific inhibitors of agglutination and to inactivate complement, but further serum treatment is not normally necessary. Fernando and Heath (1971) cautioned, however, that HAI and neutralization titers, particularly in early infections, may be reduced two- to eightfold by heating the serum. This heat-labile serum factor, which has not been identified but is not believed to be complement, can be restored by adding back unheated normal mouse serum.

The HAI and CF tests are the most often used assays in serodiagnosis. The HAI test uses four HA antigen units and human type O erythrocytes (or other sensitive erythrocytes; see Section II, B, 3) incubated at room temperature. The CF test uses the optimal number of CF antigen units as determined by checkerboard titration and five $C'H_{50}$ units of complement (USPHS, 1965). The CF test is the more sensitive test, although it requires 2 days to complete, whereas the HAI test can be performed in only a few hours (Parker *et al.*, 1978). Detectable CF antibody appears sooner in infected mice, and CF titers are higher than HAI titers (Fig. 14). The EIA and ELISA tests are more sensitive than either the HAI or CF test and with titers approximately 100- and 300-fold higher than those detected by the CF or HAI tests, respectively (Table VII; Fig. 14). Quantitative measurements of ELISA antibody amounts can be determined from a single dilution rather than from the serial titration required by the CF and HAI tests (Parker *et al.*, 1979).

2. Virus Isolation

Several techniques are available for isolation of virus from infected mice. Two of the most sensitive and popular are primary or secondary monkey (rhesus or cynomolgus) kidney cell cultures (RMK or CMK) (Parker *et al.*, 1964; Parker and Reynolds, 1968) and embryonated hens eggs (Fukumi *et al.*, 1962) (Table VIII). The sensitivity of the serial cell line LLC-MK2, with trypsin added to the medium, is equivalent to MK and has advantages over MK because of its lower cost, freedom from foamy-virus contamination, and high sensitivity (Sugita *et al.*, 1974; Silver *et al.*, 1978; Frank *et al.*, 1979).

Virus may be isolated easily from infected mice by taking mouth swabs (Parker and Reynolds, 1968); nasopharyngeal washings (turbinates), or lung tissue extracts. The latter two are the most common specimens used in virus isolations; however, they require killing the mice. Normally, virus is present in infected mice 12–14 days PI, with peak titers occurring

Table VIII

Virus Isolation Techniques[a]

Virus	Host	Route	Incubation (days)	Isolate identification
Sendai	MKCC	Cell monolayer	5–12	CPE, HAdI, HAI, CF, IFA
	Chick eggs	Allantoic	3–5	HAI
	Weanling mice	Intranasal	5–12	HAI, clinical disease, histopathology, MAP
PVM	HKCC, BHK-21	Cell monolayer	7–10	CPE, HAdI, HAI, CF, IFA
	Weanling mice	Intranasal	7–14	HA and HAI on lung extract, MAP, histopathology
K	Newborn mice	Intracranial	6–15	HA and HAI on liver extracts, MAP, histopathology

[a] MKCC, monkey kidney cell culture—primary rhesus or cynomologus; HKCC, hamster kidney cell culture (primary); BHK-21, baby hamster kidney (serial); MAP, mouse antibody production test.

around the third to sixth day of infection after intranasal instillation. When mice are infected via the airborne route, peak titers are reached between PI days 6 and 9 (van der Veen *et al.*, 1970). Titers in the saliva or nasal washings reach $10^{2.5}$–$10^{4.5}$ TCD_{50}/0.05 ml and in the lungs 10^4–10^6 TCD_{50}/0.05 ml (van Nunen and van der Veen, 1967; Parker and Reynolds, 1968; Parker *et al.*, 1978) (Fig. 4). Lung suspensions are prepared as 10–20% homogenated extracts and inoculated onto monolayers of cell cultures that are examined for CPE from the fifth to the fourteenth day PI. CPE-positive cultures may be tested either by HA and HAI, CF, HAd and HAdI, or IFA to identify the isolate. A blind passage is sometimes necessary (Chanock *et al.*, 1958; van Nunen and van der Veen, 1967; Parker and Reynolds, 1968; Bhatt and Jonas, 1974; Tucker and Stewart, 1976). For isolation in eggs, 8- to 10-day-old fertile embryonated hens' eggs are inoculated into the amniotic or allantoic cavity (amniotic preferred for new isolates); after 2–4 days' incubation at 35°–37°C, the appropriate fluids are tested for HA. Two blind passages may be required (Fukai and Suzuki, 1955; Fukumi and Nishikawa, 1961; Fukumi *et al.*, 1962).

F. Control and Prevention

Sendai virus infections in mouse colonies have proved to be one of the most difficult virus infections to control because the virus is highly infectious and easily disseminated (Parker and Reynolds, 1968). High titers of virus present in the nasopharyngeal passages and saliva of infected mice cause large amounts of virus to be shed into the environment (van Nunen and van der Veen, 1967; Parker and Reynolds, 1968). Efforts to protect uninfected colonies or to eliminate virus from infected colonies have been largely unsuccessful; however, there are some important exceptions, and these are worth consideration if we are to gain by experience. The best record on a large scale in this regard is that of the Jackson Laboratory, Bar Harbor, Maine, United States. Known throughout the world as an important repository and supplier of inbred strains, congenics, and mutants, Jackson Laboratory's success is doubly significant because of its need to repopulate most strains and stocks in 1948 following the disastrous fire that destroyed much of the laboratory. It is probably fortuitous that Sendai infection was not introduced at that time. In the late 1950s the laboratory adopted a 6-week quarantine period for all newly received animals, and presently the operating scheme requires that all new breeding stock be delivered in germ-free containers and housed in germ-free isolators in a separate facility. These animals are serum-tested for antibodies to murine viruses and contact-housed with known negative mice that are further serum tested. Finally, new stock is hysterectomy-derived into the breeding facility after a total time lapse of approximately 8 weeks. It is also important to point out that since 1967, all production breeding cages are filter-capped. Contributing factors might be strict visitor rules and work force stability (D. D. Myers, Jackson Laboratory, 1980, personal communication).

A second example of successful Sendai-free mouse production and maintenance can be found in the production and experimental barriers of the Biology Division, Oak Ridge National Laboratory, Oak Ridge, Tennessee, United States. These facilities have remained free of Sendai virus since the early 1960s in spite of the occurrence of this disease in the conventional animal facilities on the same premises (Richter, 1970; E. A. Bingham, Oak Ridge National Laboratory, 1980, personal communication). Contributing factors to this success include closed colonies, with new introductions only through the germ-free route, strict barrier operating discipline and protocols, stabilized work force, and the standard use of cage filter devices.

Large-scale contract production of mice for research has provided contract monitors at the National Cancer Institute, U.S. Department of Health and Human Services, with insight into the problem of Sendai-free mouse production. Only a few commercial colonies have remained free of Sendai infection for extended periods, and the overriding consideration appears to be failure to eliminate infected stock from an entire premise or improper quarantine of incoming stock. Failure to eliminate all infected stock inevitably results in barrier failure, usually within 1 year (J. G. Mayo, National Cancer Institute, Frederick Cancer Research Center, Frederick, Maryland, 1980, personal communication).

Once the infection is diagnosed, only two options are available: (1) destroy the colony and restock by rederivation, using germ-free technology, or restock from a disease-free colony, or (2) remove all newborns, weanlings, and pregnant mice from the colony, allowing only adult mice to remain. This nonbreeding colony can then be held in a static condition until the infection is extinguished (approximately 2 months) and then breeding and other normal activities are resumed. This second method has been shown to be effective when it was possible to maintain the control criteria (Iwai *et al.*, 1977, J. C. Parker, 1979, unpublished observations). Cesarean rederivation is also an effective method for eliminating nearly all adventitious virus infections, its only drawback being the long period of time required to rear sufficient quantities of animals for restocking—perhaps 1–2 years—plus the necessity to operate two facilities during the rederivation program (van Hoosier *et al.*, 1966). The development of formalin-killed Sendai virus vaccines (Fukumi and Takeuchi, 1975; J. C. Parker, 1979 unpublished observations) and temperature-sensitive mutant strains of Sendai virus (Kimura *et al.*, 1979) give the investigator an additional method by which to protect naive mice or to eliminate infection from colonies under certain conditions. In addition, the intraperitoneal or aerosol administration of ribavirin to experimentally infected mice has been shown to delay lung pathology and increase survivability (Larson *et al.*, 1976; Sidwell *et al.*, 1975). Lethal infection can be significantly inhibited by intraperitoneal ribavirin treatment up to 24 hr after virus inoculation and moderately inhibited up to 96 hr PI in experimental situations. There are no reports of the clinical application of these observations in naturally infected mice.

Following elimination of infection, a barrier system must be developed and used to ensure the isolation of the colony from extraneous pathogens (Flynn, 1967; ILAR, 1976; Flynn *et al.*, 1967; Cooper *et al.*, 1977). A health surveillance program (ILAR, 1976) should be established, but it must be borne in mind that such programs are more meaningful as a preventive measure than as a curative measure since most Sendai breaks seem to run an inevitable course to natural completion. Surveillance programs should be used to monitor source suppliers of mice, quarantine stations, biologic materials such as transplantable tumors that are intended for inoculation into mice (Collins and Parker, 1972), and sentinel mice that are strategically distributed throughout the facility. Most importantly, it may be well to emulate closely the physical plants and operating schemes of those institutions with long histories of success in Sendai prevention.

III. PNEUMONIA VIRUS OF MICE (PVM)

A. Historical Background

During attempts to isolate human viruses from the nasopharyngeal washings of patients with respiratory infections, Horsfall and Hahn (1939, 1940) observed that the lungs of both inoculated and uninoculated mice often became consolidated during serial passage. Subsequently, a pneumotropic virus (*pneumonia virus of mice,* or PVM, Horsfall and Hahn, 1939), which was avirulent until serially passaged was isolated from the lungs of normal mice and shown to be a prevalent latent infection in mouse colonies. Mills and Dochez (1944, 1945), under circumstances similar to Horsfall and Hahn (1940), also isolated a strain of PVM from mice, which they termed *pneumonitis virus*. In addition, other agents causing pneumonia in mice were also isolated during the late 1930s to mid-1940s and the literature is replete with reports on the isolation of "pneumonitis viruses" [*sic*] (reviewed in Tennant, 1963; Tennant *et al.*, 1966) which, with one possible exception, were not viruses. Nelson (1937) described the agent of enzootic bronchiectasis, which appeared to be a virus, and the agent of infectious catarrh, which was a mycoplasma species (Joshi *et al.*, 1961; Nelson, 1963). The "Gray lung virus" (Andrewes and Glover, 1946) is also probably a mycoplasma, and the agents described by Eaton *et al.* (1941), Gönnert (1941), Nigg and Eaton (1944), Gordon *et al.* (1938), and Dochez *et al.* (1937) likely belong to the family Chlamydiaceae (Hilleman and Gordon, 1943; Nigg and Eaton, 1944). In 1940, Pearson and Eaton (1940) isolated PVM from Syrian hamsters, and in 1944 Eaton and van Herick (1944) isolated PVM from cotton rats.

B. Properties of the Virus

1. Classification

PVM belongs to the family Paramyxoviridae, genus *Pneumovirus* (Fenner, 1976), which also contains human and bovine respiratory syncytial virus species (Fenner, 1976; Kingsbury *et al.*, 1978). PVM is antigenically distinct from all other members of the family Paramyxoviridae; however, it shares other properties with respiratory syncytial virus (Berthiaume *et al.*, 1974) and mumps virus (Ginsberg *et al.*, 1947, 1948; Ginsberg and Horsfall, 1948, 1951c).

2. Strains of Virus

There is no evidence that different isolates or strains of PVM, whether from mice or hamsters, show any unique properties not shared by all other strains, and therefore, all PVM strains can be considered antigenically homologous (Horsfall and Hahn, 1939, 1940; Pearson and Eaton, 1940). Virulence of mouse lung isolates increases on serial mouse lung passage, and therefore, some strains may show variations in their virulence for mice (Horsfall and Hahn, 1940). PVM grown in BHK-21 cells is antigenically similar to mouse lung virus but is somewhat less virulent in mice (Harter and Choppin, 1967).

3. Physical and Chemical Properties

PVM particles occur in pleomorphic forms, most commonly as filaments 100 nm in diameter and up to 3 μ in length and less frequently as spheres 80–200 nm in diameter, and contain a helical nucleocapsid of single-stranded RNA (Tennant *et al.*, 1966; Compans *et al.*, 1967; Berthiaume *et al.*, 1974) (Table III). Virus replication occurs in the cytoplasm, and complete virus buds from the cell membrane mainly in the filamentous form (Berthiaume *et al.*, 1974). The virus envelope is studded with projections 12 nm in length and 6 nm apart (Berthiaume *et al.*, 1974). Infective virus sediments at a density of 1.15 and noninfective viral hemagglutinin at a density of 1.13. Sedimentation by centrifugation or ammonium sulfate precipitation results in a loss of infectivity and an increase in HA titer. Stock virus preparations normally contain a large amount of noninfective hemagglutinin (Harter and Choppin, 1967).

Agglutination of mouse, rat, and hamster erythrocytes occurs at RT and 5°C (Mills and Dochez, 1944, 1945; Curnen and Horsfall, 1946; J. C. Parker, 1979 unpublished observations). HA will not occur in nonelectrolyte solutions (Davenport and Horsfall, 1948). Whereas mouse erythrocytes will hemadsorb onto infected cell cultures (Tennant and Ward, 1962) and virus grown in cell cultures freely agglutinates erythrocytes (Tennant and Ward, 1962; Harter and Choppin, 1967), virus obtained from mouse, rat, or hamster lung suspensions is cell-associated with lung tissue components and must be treated to release the hemagglutinins (Ginsberg, 1951; Mills and Dochez, 1944, 1945; Curnen *et al.*, 1947; Volkert and Horsfall, 1947; Curnen and Horsfall, 1947). PVM from lung tissue can be treated to yield virus in various states (Curnen and Horsfall, 1947; Ginsberg, 1951): (1) tissue-combined noninfectious virus; homogenized lung tissue suspensions are heated at 70°C for 30 min or at 80°C for 10 min to release hemagglutinins (Curnen and Horsfall, 1946); (2) tissue-combined infectious virus; unheated lung tissue suspensions; (3) free noninfectious virus; extracted from lung tissues without grinding (clarified fluids from perfused, minced lung tissue) and the fluids heated; and (4) free infectious virus; as in (3) but not heated. Within the infected lung, virus exists mainly in the free infectious state. However, when tissue suspensions are prepared, the virus combines with tissue components, which then must be either heated or treated with strong alkaline solutions (pH 11) to release the stable viral hemagglutinin which inactivates the virus (Curnen and Horsfall, 1947).

PVM can also be dissociated from erythrocytes or lung tissue particles at low electrolyte concentrations (0.01 *M* phosphate). Optimal dissociation from erythrocytes is a function of both the pH and electrolyte concentrations and will occur without hemolysis in 0.25 *M* dextrose or sucrose buffered at pH 7.2 with 0.01 *M* phosphate (Davenport and Horsfall, 1948).

PVM is labile at room temperature, losing 99% of its infectivity in 1 hr, and is completely inactivated by heating at 56°C for 30 min (Horsfall and Hahn, 1939) or by treatment with ether (20% for 18 hr at 4°C) (Franklin and Gematos, 1961). Trypsin destroys the hemagglutinin (Girardi and Hilleman, 1960; Curnen and Horsfall, 1946). No neuraminidase or other enzyme has been described (Kingsbury *et al.*, 1978), and virus may repeatedly be eluted from and agglutinated by erythrocytes (Davenport and Horsfall, 1948).

Replication of PVM in the mouse lung following intranasal instillation is inhibited by certain polysaccharides of bacterial or nonbacterial origin (Horsfall and McCarty, 1947). These polysaccharides do not act directly on or combine with the virus, but they appear to combine or compete with and deplete the target cells of an essential material necessary for virus maturation (Ginsberg and Horsfall, 1948, 1949, 1951b; Ginsberg *et al.*, 1948).

4. Growth of Virus in Mice and Cell Culture

a. Mice. PVM is strictly pneumotropic (Horsfall and Hahn, 1940), and multiplication of the virus in the lung is relatively constant, occurring in discrete cycles, averaging a 7.9-fold increase in titer per day until reaching a plateau around the sixth day of infection (Horsfall and Ginsberg, 1951). After intranasal inoculation, a large proportion (90%) of the inoculum disappears, and a 15 hr latent period occurs that is followed by increasing amounts of virus in the lung over the next 10–15 hr for a total cycle interval of 24–30 hr (Ginsberg and Horsfall, 1951a). The yield per cycle is approximately 16-fold and maximum virus titers in lung tissues are approximately $10^{2.5}$ mouse LD_{50}/ml and $10^{5.5}$ mouse ID_{50}/ml (Tennant, 1963; Horsfall and Hahn, 1940). There is a direct relationship between the time required to achieve maximal viral titers in the lung and the size of the inoculum; however, the maximum titer is not affected by the size of the infecting dose (Horsfall and Ginsberg, 1951). The highest titers of virus are usually observed 6–8 days after infection and normally decrease after the ninth day. Virus cannot be detected after 2 weeks. Maximum HA titers in lung tissue extracts range from 1:256 to 1:1024 (Horsfall and Ginsberg, 1951; Ginsberg and Horsfall, 1951a; Tennant and Ward, 1962; Tennant, 1963).

b. Cell Culture. PVM replicates in primary hamster kidney cell culture (HKCC) (Tennant and Ward, 1962), BHK-21 (Harter and Choppin, 1967), Vero (Berthiaume *et al.*, 1974), and hamster embryo (Reed *et al.*, 1975). In Vero cells, CPE is slowly progressive without syncytia and is still incomplete at 11 days. Inclusion bodies are seen in the cytoplasm 24 hr after infection. Cytoplasmic accumulation of viral antigens is detectable by IFA 16–24 hr after infection. Maturation of the virus occurs at the cytoplasmic membrane by budding at or near the time of appearance of the inclusions. Filamentous virus forms are more common than spherical particles (Berthiaume *et al.*, 1974). Mature particles are seen only extracellularly (Compans *et al.*, 1967). An increased virus yield over input occurs after 18 hr, with a first and second peak of 10^4 $TCID_{50}$/ml and 10^7 $TCID_{50}$/ml at 48 and 96 hr, respectively (Berthiaume *et al.*, 1974). Infectious virus and hemagglutinins are produced more rapidly and at higher titers in BHK-21 than in HKCC; HA of 1:4096 and 1:32, respectively, 10 days after infection. HKCC produces only low titers of virus ($10^{3.5}$ mouse ID_{50}) which decrease further on serial passage (Tennant and Ward, 1962), whereas 6.3 10^6 $TCID_{50}$/ml are produced from BHK-21 cells. Generally, CPE in HKCC starts on the seventh day after infection and is characterized by focal cell destruction with regrowth of cells; therefore, HKCC is not suitable for quantitative titration (Tennant and Ward, 1962). CPE in BHK-21 consists of a scattered rounding and detachment of cells from the surface beginning about 48 hr after infection but, like HKCC, it does not progress to complete destruction of the cell monolayer (Harter and Choppin, 1967). After multiple passages in BHK-21, clones of persistently infected cells can be isolated which contain PVM viral antigens and hemadsorb murine erythrocytes but do not yield infectious virus (Gallaspy *et al.*, 1978). A plaque assay for PVM using BHK-21 with proteolytic enzymes in the overlay medium has been developed by Shimonaski and Came (1970).

c. Eggs. Horsfall and Hahn (1939, 1940) claimed to have maintained PVM infectivity for mice through 10 serial transfers in embryonated chicken eggs, but there was no evidence that viral replication occurred. Volkert and Horsfall (1947) subsequently showed that both chick and duck embryos were not susceptible to infection.

C. Epizootiology

1. Host Range

a. Species. The natural host range of PVM is restricted to laboratory rodents: mice, hamsters, rats [cotton rats (*Sigmodon*) and albino rats (*Rattus*)] (Horsfall and Hahn, 1939, 1940; Pearson and Eaton, 1940; Horsfall and Curnen, 1946b; Eaton and van Herick, 1944; Parker, 1973), and perhaps guinea pigs (Iwai *et al.*, 1976). Only serologic evidence of PVM virus infection has been reported in guinea pigs (Van Hoosier and Robinette, 1976; Horsfall and Curnen, 1946b; J. C. Parker, 1979, unpublished observations). In a survey of 94 feral mice trapped in the state of Maryland during the 1960s, no PVM HAI antibody was detected (J. C. Parker, 1979, unpublished observations). Horsfall and Curnen (1946b) reported the presence of neutralizing antibody in rabbits, monkeys, chimpanzees, and humans, in addition to mice, hamsters, cotton rats, and guinea pigs. Natural infections in rabbits, monkeys, chimpanzees, and humans have not been confirmed, and it is not generally accepted that they are natural hosts; however, the nature and avidity of the serum inhibitors for PVM are not fully understood (Tennant *et al.*, 1966). Although Horsfall and Hahn (1940) showed that 6 of 30 (20%) human sera would neutralize 100 mouse LD_{50}'s of PVM, Tennant *et al.* (1966) could not detect HAI antibody in 70 human sera obtained from persons with prolonged laboratory contact with PVM in infected rodents. Inoculated ferrets do not develop symptoms, and virus does not become adapted but can be recovered from their lungs (Horsfall and Hahn, 1939).

Although young mice appear to be more susceptible than older mice (Mirick *et al.*, 1952 and unpublished observations), a more important influence on susceptibility is diet (see Section III, D). No differences in susceptibility influenced by host sex have been reported.

b. Strains of Mice. Variation in the susceptibility of different mouse strains was first reported by Horsfall and Hahn (1939, 1940). Tennant *et al.* (1966) confirmed this observation when they observed that mice from different breeders showed various degrees of susceptibility to infection. The most susceptible strains were those from colonies free of indigenous PVM infection. Therefore, the strain variation seen by Horsfall and Hahn (1939, 1940) was probably correlated with prior contact and immunologic resistance of some of the mice to PVM and thus had no genetic basis (Tennant, 1963).

2. Prevalence and Geographic Distribution

On the basis of neutralizing antibody, PVM appeared to be prevalent in mice and numerous other mammalian species, including humans (Horsfall and Hahn, 1939, 1940; Horsfall and Curnen, 1946a,b); however, on the basis of HAI antibody, the prevalence appears to be more restricted (Tennant *et al.*, 1966). PVM infection occurs primarily and probably exclusively in rodent breeder colonies throughout the world and is one of the most frequently encountered virus infections in mice (Table V).

Serological surveys conducted in the United States (Parker *et al.*, 1966; Van Hoosier *et al.*, 1966; Trentin *et al.*, 1966; Parker, 1973; Poiley, 1970; Reed *et al.*, 1974; Tennant *et al.*, 1964, 1966), Canada (Descoteaux *et al.*, 1977), Japan (Iwai, 1978), Germany (Jakubik and Shoda, 1972), and the Netherlands (Zurcher *et al.*, 1977; van Nunen *et al.*, 1978) have shown evidence of prevalent and geographically worldwide infections in rodent colonies. The incidence of mouse colony infections is approximately 63% in the United States (Table V). The number of mice in infected colonies with HAI antibody is highly variable, ranging from 1% to 100%, and is probably due in part to the focal nature of the infection and consequent sampling errors. However, it is noteworthy that the number of rats and hamsters with HAI antibody in infected colonies is significantly higher than in mouse colonies: 62% in rats and 45% in hamsters, compared to 20% in mice (Table V). HAI antibody titers likewise are two- to fourfold higher than observed in mice, and the number of infected rat and hamster colonies is also higher (68% in rats and 94% in hamsters), all of which may be indicative of a difference in the epizootiology of PVM virus infection in rats and hamsters.

Infection is less common in cesarean-derived mice that are subsequently reared under specific pathogen-free (SPF) or germ-free conditions (Parker *et al.*, 1965; Trentin *et al.*, 1966; Van Hoosier *et al.*, 1966; Poiley, 1970; van Nunen *et al.*, 1978).

3. Natural History and Transmission

Within mouse colonies, PVM probably induces an acute enzootic infection that may be focal in the colony, reflecting the low contagiousness of the virus (Tennant *et al.*, 1966; Parker *et al.*, 1966; Horsfall and Hahn, 1940). Low transmissibility is also evident by the modal distribution of antibody in mice in infected colonies (Fig. 15), where, in the majority of infected colonies, less than 25% of the mice have demonstrable HAI antibody (Tennant *et al.*, 1966). As a result, sampling errors may be common in epizootiologic serologic surveys. In two infected colonies studied by Parker *et al.* (1966), antibody was detected in 2-month-old mice, and the incidence increased slowly in groups of progressively older mice. In the second colony, seropositive mice were only rarely detected until the seventh month of age. In other infected colonies, which were tested repeatedly over 1½ years, it was not uncommon to take samples from the colonies which contained no positive sera. This is believed to reflect sampling errors as a consequence of the focal nature of infection rather than seasonal or other influences. Clearly, there is a need for further study in order to understand fully the natural history of PVM infection in rodent species.

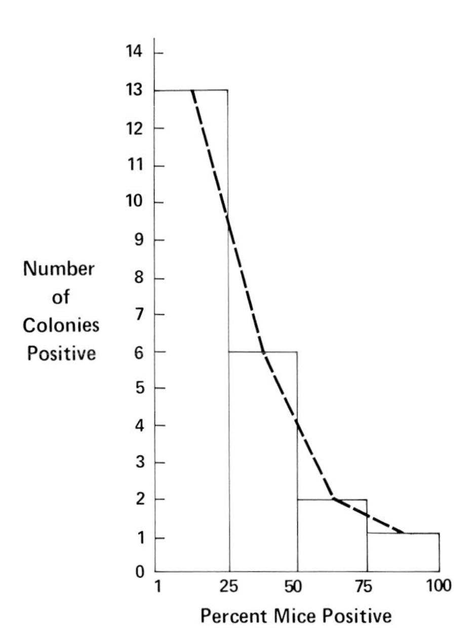

Fig. 15. Frequency distribution of HAI antibody to pneumonia virus of mice in positive mouse colonies. (From Tennant *et al.*, 1966).

Although there have been numerous claims of latent PVM infection in rodents, various definitions of what constitutes a latent infection have led to interpretive difficulties. The term *latent* has been used interchangeably with *chronic-* and *acute-inapparent* infection as described for the individual mouse, together with the inapparent or latent infection as described for an entire colony of mice (Horsfall and Hahn, 1939, 1940; Horsfall and Curnen, 1946b; Verlinde and Winsser, 1949). The term *latency* may be defined as "a state of seeming inactivity" or as "not manifest" or "concealed," any of which in defining a biologic infection might have quite different meanings and interpretation. Therefore, the term *latency* must be further defined when describing virus infections. Horsfall and Curnen (1946b) found that intranasal inoculation of normal chick embryo suspensions into cotton rats and hamsters resulted in the development of PVM-neutralizing antibody and, together with isolation of virus from asymptomatic animals, provided evidence for latent infection. However, during periods of low antibody incidence, evidence for virus persistence in the cotton rats could not be found. Horsfall and Hahn (1939, 1940) also reported that PVM could be isolated with relative ease from mice when lungs were blind-passaged serially in mice at 7- to 9-day intervals. Thus, the term *latent infection* became associated with PVM infection, and it was not until the work of Tennant (Tennant *et al.*, 1964, 1966; Tennant, 1963) that the true natural history of PVM infection in mice became more fully understood. A series of experiments, including serial lung passage in mice, mouse antibody production (MAP) tests on lung suspensions, tissue culture isolation, and challenge of mice with normal chick embryo extract, were unable to detect and activate any virus which might have been harbored as a chronic-inapparent infection. This led Tennant (Tennant, 1963; Tennant *et al.*, 1964, 1966) to propose that Horsfall and Curnen (1946b) had isolated or demonstrated virus in acute but inapparently infected mice and that chronic-inapparent infections of mice do not occur. Thus, the normal pattern of infection in a colony would be one of acute-inapparent infections with virus present in the mice for only short periods of time. This infection pattern would then account for its apparent epizootiology and focal occurrence in enzootically infected colonies. Because of the rapidity with which PVM is inactivated at RT, it seems likely that intimate contact between infected and susceptible individuals is essential to sustain colony enzootics. Horsfall and Hahn (1940) attempted to transmit PVM by direct contact for average periods of 48 hr during the period 12 hr to 4 days following experimental infection of the infectors. Uninoculated contacts were negative for symptoms of PVM infection and negative for pulmonary consolidation at PI day 12. Unfortunately, no serologic examination of uninoculated contacts was done, and because of the large dose of virus required by direct intranasal inoculation to produce lesions and morbidity, neither clinical signs nor gross pulmonary consolidation may be sensitive enough indicators of contact transmission. This type of experiment has apparently not been repeated.

A seasonal variation in the incidence of neutralizing antibody was observed over a 3-year period in hamsters and cotton rats, with higher incidences in the late winter and spring than during the summer and fall (Horsfall and Curnen, 1946b); however, the evidence is weak and needs confirmation.

D. Pathogenesis

1. Clinical Disease

There are no published reports of natural clinical disease caused by PVM and only brief descriptions of the clinical effects of experimental infection. Horsfall and Hahn (1940) observed loss of activity and appetite 5–7 days after experimental intranasal inoculation. Subsequently, weight gains ceased and mice actually lost weight. Sick mice were ruffled, dirty, and hunched, whereas moribund mice appeared emaciated. Respiration became slow and deep and sometimes labored. Cyanosis of the ears and tail was sometimes seen, and deaths occurred up to the twelfth or thirteenth day, depending on dose administered.

2. Pathology

a. Gross. Horsfall and Hahn (1940) described spontaneous pulmonary consolidation, usually limited to a portion of one lobe, in 1–2% of mice from eight separate commercial breeders and retrospectively attributed this to PVM. Mills and Dochez (1944) also reported areas of consolidation in the lungs of mice used for experimental studies of human respiratory agents and succeeded in producing fatal pneumonias by the fifth serial passage. The causative agent was neutralized by anti-PVM serum and shared other characteristics with PVM, such as the ability to agglutinate erythrocytes. Whether or not the consolidations described in these reports were entirely due to PVM is uncertain. There are no recent reports of proved PVM-induced lesions in natural enzootics.

Experimental infection with PVM was first reported in 1939 by Horsfall and Hahn when they serially blind-passaged lung suspensions at 7-day intervals in 21 groups of Swiss mice. By the third passage 43% of the inoculated groups had areas of pulmonary consolidation, and by the sixth passage 24% of the inoculated groups had deaths. These investigators then demonstrated that the causative agent was filterable. Experimental infections cause extensive plum-colored consolidation affecting one-half to three-fourths of the lungs. Lesions are more pronounced in the hilar region during the early phase of infection but later radiate outward in a striated manner (Horsfall and Hahn, 1940). Horsfall and Ginsberg (1951) determined that the extent of pneumonia that developed was dependent upon the amount of virus inoculated and that when viral replication ceased, the pneumonia underwent gradual resolution. Since naturally infected mice are not likely to be infected by large amounts of virus, naturally occurring gross lesions are likely to be limited or not observed. Experimental infection is apparently possible only by the intranasal route, and the virus appears to be strictly pneumotropic (Horsfall and Hahn, 1939, 1940). Germ-free and conventional mice appear to be equally susceptible to experimental PVM infection (Tennant *et al.*, 1966).

The severity of experimental PVM infection can be altered by a number of procedures. Horsfall and McCarty (1947) discovered that the intranasal administration of streptococcal broth culture, 2 days after the administration of a normally fatal dose of PVM, reduced the severity of lung lesions by 50% and the virus titer in infected lungs by 1.25 log units. Reversal of the order of administration did not alter these results. This effect with streptococcus could be extended from 3 days pre- to 4 days post-PVM infection. The researchers demonstrated that this effect was caused by polysaccharides that were present in a variety of bacterial and nonbacterial sources. As little as 0.1

mg purified capsular polysaccharide from Friedländer type B bacillus was capable of abruptly interrupting virus replication. Ginsberg and Horsfall (1951b) used Friedländer type B bacillus capsular polysaccharide 2 days post-PVM infection to reduce PI day 6 lung consolidation from 100% to 43% and mortality from 100 to 0%. When administered 3 days post-PVM, the PI day 6 scores were reduced to 57% consolidation and 66% survival. Survivors treated with polysaccharide showed evidence of lung lesion resolution by PI day 10, and only small gray areas of consolidation remained at PI day 16. This effect was produced by as little as 0.02 mg polysaccharide, although on the other hand, high-titer-specific PVM antiserum given at 2 and 3 days PI had no demonstrable effect on the pneumonic process. Leftwich and Mirick (1949) and Mirick and Leftwich (1949) showed that mice fed pyridoxine-deficient diets the first 5 days after experimental infection were less susceptible to PVM than mice on an adequate pyridoxine diet. They concluded that pyridoxine was essential for virus replication. On the other hand, when pyridoxine was restricted for longer periods prior to infection, mice were more susceptible to PVM even though they did not respond differently serologically to PVM than did adequately pyridoxine-nourished mice. Mirick *et al.* (1952) also showed that urethane administered by the oral, intranasal, and intraperitoneal routes had an enhancing effect on PVM infection but did not alter the antibody response. Tennant (1963) and Tennant *et al.* (1966) have shown that ether anesthesia potentiates experimental PVM infection.

b. Histopathology. There are few published micrographs of natural or experimental PVM lesions, and the details of the histopathology are somewhat limited. The following account

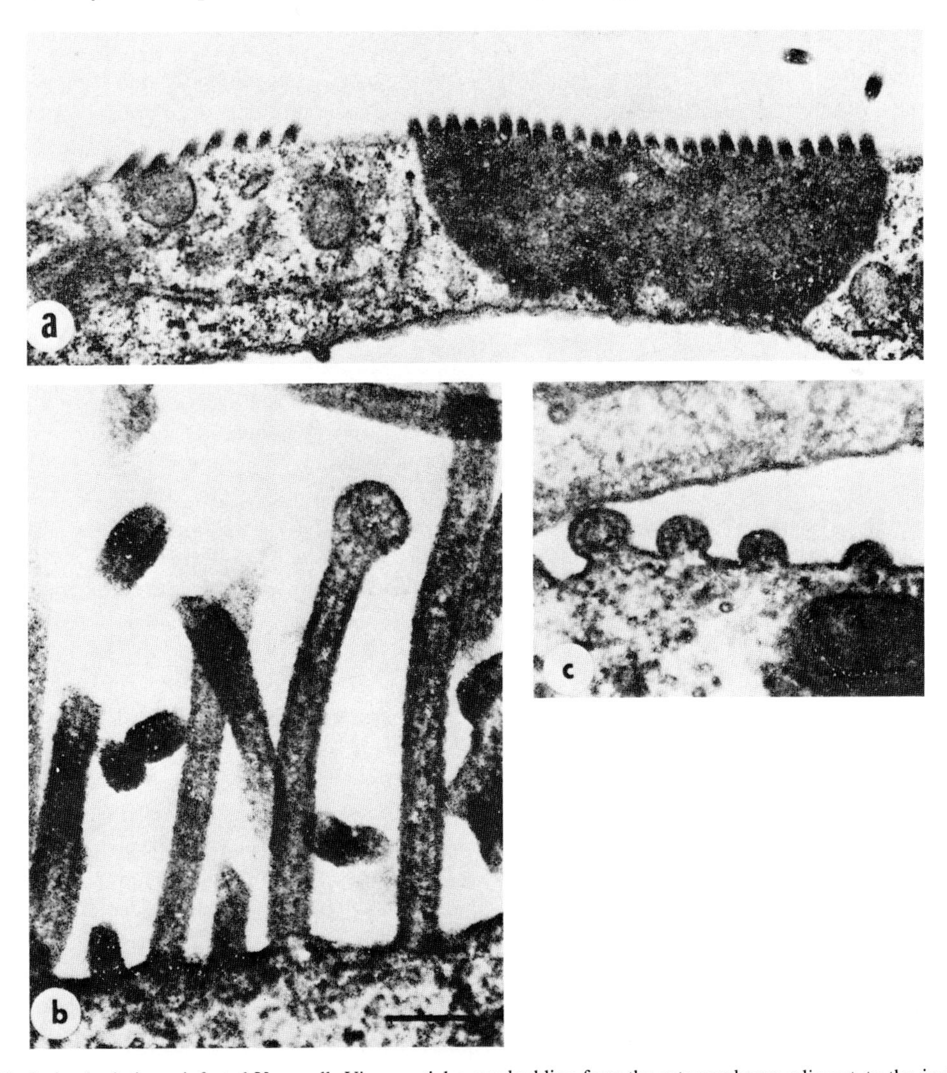

Fig. 16. (a) PVM inclusion body in an infected Vero cell. Virus particles are budding from the cytomembrane adjacent to the inclusion body. (b) PVM filamentous forms budding from the cytomembrane. One form has a terminal swelling. (c) Budding PVM round forms. Bar in each micrograph is 200 nm. (Reproduced by permission of Springer-Verlag, New York.)

is taken from Horsfall and Hahn (1939, 1940). There is a dense perivascular and peribronchial accumulation of mononuclear cells, alveolar septa are thickened and contain cellular exudate, and alveolar spaces contain a mixture of fluid, mononuclear cells, and erythrocytes. Polymorphs are common in the bronchi but not the alveoli. Bronchial epithelium appears intact but is hyperplastic. Unfortunately, these experiments by Horsfall and Hahn were complicated by mycoplasma, which they readily isolated from control and experimental groups. Carthew and Sparrow (1980) studied the experimental pathology of PVM infection in ex-germ-free mice using immunoperoxidase and histopathology (see Addendum). Franklin and Gomatos (1961) described the lesions of experimental PVM infection in mice as "typical viral pneumonitis." They observed scattered foci of edema of the walls of bronchi and nearby alveoli, mononuclear cell exudate, and few polymorphs and necrotic epithelial cells. Eosinophilic exudate lining alveolar walls was also present, but hemorrhage was not observed.

Compans *et al.* (1967) and Harter and Choppin (1967) observed eosinophilic cytoplasmic inclusion bodies in PVM-infected BHK-21 cultured cells. These inclusions stained bright red with acridine orange (single-stranded RNA), but there was no evidence of viral nucleic acid in infected cell nuclei. Specific IFA was also limited to the cytoplasm and was most intense 72 hr PI. Berthiaume *et al.* (1974) made similar observations with acridine orange and IFA using Vero cell cultures. These cells also exhibited cytoplasmic inclusions. No detailed studies of this type have been carried out on whole mouse lungs.

c. Ultrastructure. Ultrastructure studies of PVM-infected cells have also apparently been limited to tissue culture. The physical properties of PVM have been described above, but brief mention of the ultrastructure studies of cell cultures by Compans *et al.* (1967), Berthiaume *et al.* (1974), and Gallaspy *et al.* (1978) is made here. Compans *et al.* (1967) and Berthiaume *et al.* (1974) observed that the microscopically visualized inclusion bodies consisted of strandlike elements (nucleocapsids) that are 12 nm in diameter and sometimes contiguous with budding particles at the cytomembrane. Budding forms are usually filamentous, about 100 nm in diameter and up to 3μm in length. Budding particles are covered by projections approximately 12 nm long (Fig. 16). The fascinating study by Gallaspy *et al.* (1978) demonstrated that BHK-21 tissue culture cells could be persistently infected by PVM antigens. Infected cells, trypsinized after 72 hr and subsequently serially passaged more than 100 times, remained capable of adsorbing mouse erythrocytes. They retained cell membrane affinity for ferritin-labeled specific antibody and IFA evidence of cytoplasmic antigen. Electron microscopic examination of these persistently infected cells did not disclose morphologic evidence of viral nucleocapsids. No attempt was made to see if these cells were infectious to mice. There is no evidence that similar persistent infection occurs in mice.

E. Diagnosis

Diagnostic techniques have been described and reviewed by Parker *et al.* (1965) and Ohder (1971). The available assays are sensitive and not technically difficult, and interpretation is uncomplicated since PVM is antigenically distinct from all other viruses known to infect laboratory rodents.

1. Serology

The most useful serologic test for diagnosis is the HAI test (Mills and Dochez, 1944, 1945) (Table VII). Other tests available are the neutralization test in mice (Horsfall and Hahn, 1940) or cell culture (Harter and Choppin, 1967; Tennant and Ward, 1962; Shimonaski and Came, 1970), the CF test (Tennant *et al.*, 1966), and the HAdI test (Tennant and Ward, 1962). All strains of PVM are antigenically homologous in diagnostic tests, and suitable antigen may be prepared from mouse lung extracts, heated or otherwise treated (Mills and Dochez, 1944, 1945; Parker *et al.*, 1965; Davenport and Horsfall, 1948; Curnen and Horsfall, 1947), or prepared from infected cell cultures (Tennant and Ward, 1962; Harter and Choppin, 1967).

The HAI test employs eight HA antigen units and mouse erythrocytes incubated at room temperature, and the degree of agglutination or inhibition is evaluated by allowing the settled erythrocytes to slide after tilting the test plate or tube at an angle (Parker *et al.*, 1965). Sera to be tested are heated at 56°C for 30 min, and pooling of several sera is not recommended because of the low antibody titers normally encountered. Titers of HAI antibody in mice range widely from the lowest titer considered to be significant of 1:20 to a typical high titer of 1:16, with an average range of 1:20–1:40. In addition, there appears to be a low-titer (1:10–1:20), nonspecific inhibitor of HA that is occasionally present in the sera of normal mice. Therefore, if low incidences (<5%) of low-titer positives (1:20) are observed, diagnosis should be tentative until further confirmatory assays can be performed.

The CF test is not recommended for general antibody assay or surveys but is useful for determining recent infections (Fig. 17). Infected mice usually seroconvert by HAI and CF 9–10 days after infection; however, whereas HAI antibody persists in recovered mice at high titers for at least 4 months, CF antibody persists for a shorter period, declining after 2 weeks and reaching minimal detectable levels at 70 days (Tennant *et al.*, 1966).

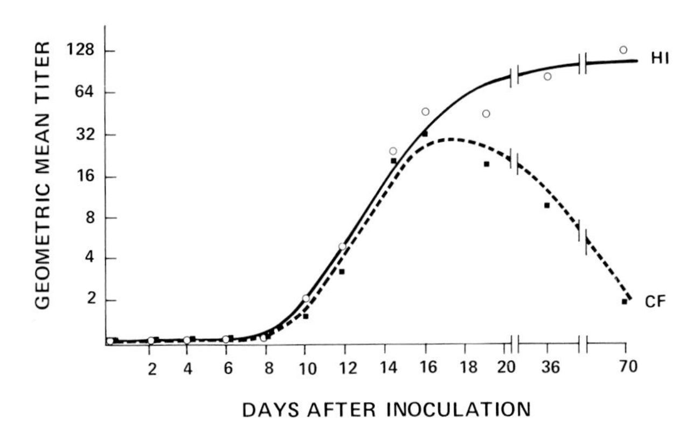

Fig. 17. Hemagglutination inhibition (HAI) and complement fixation (CF) antibody responses of mice to infection with PVM. Each point represents the mean of four to ten mice. (From Tennant *et al.*, 1966.)

Neutralization and HAdI tests, either in mice (Horsfall and Hahn, 1940) or in cell culture (Tennant, 1962), are less frequently used but offer alternative tests for confirmation. An enzyme immunoassay (ELISA) that is eight times more sensitive than the HAI test has been developed for quantifying PVM antibody in rat sera (Payment and Descoteaux, 1978).

2. Virus Isolation

The recommended procedure for isolation of PVM is inoculation of cell culture and subsequent demonstration of CPE, hemagglutinins, IFA, or hemadsorption (Table VIII). BHK-21 cells (Harter and Choppin, 1967) are more sensitive than HKCC cells (Tennant and Ward, 1962). Vero cells have also been shown to be sensitive (Berthiaume *et al.*, 1974). A plaque assay using BHK-21 cells with proteolytic enzyme in the overlay medium was developed by Shimonaski and Came (1970) and is useful in cloning virus. Virus multiplication can usually be detected 5–10 days after inoculation by the production of hemagglutinins and may be accompanied by incomplete destruction of the cell sheet. Serial blind passages may be necessary but normally are not required for isolation.

Isolation in mice, the oldest and previously the most often used technique, is accomplished by intranasal instillation of the specimen (using ether anesthesia) and serial subpassage of lung extracts at 7- to 9-day intervals until consolidation of the lung is observed or the mice develop clinical signs of pneumonia (Horsfall and Hahn, 1940). Virus identification can be made by demonstration and inhibition of hemagglutinins specific for PVM virus. Virus isolation by serial blind passage in the lungs of mice is not the method of choice, however, since it is expensive, time- and facilities-dependent, and subject to a high probability of both false-positive and false-negative results. Obtaining a virus-free source of mice and maintaining them free of accidental infection in an experimental animal colony is difficult and can easily result in false-positives; furthermore, improper handling of the specimens or too quick a serial passage in mice (4–5 days rather than the required 7–9) may cause the virus to be lost, resulting in false-negatives. An average of four serial passages in antibody-free mice may be required to isolate virus (Horsfall and Hahn, 1940).

Mice can be used more effectively, however, in the MAP test (Rowe *et al.,* 1962) to demonstrate virus indirectly by seroconversion after inoculation with a test specimen. The limitation of this procedure is a 3- to 4-week test period, the necessity of a quarantine animal facility, and the possibility that the MAP test may be less sensitive than other assays.

F. Control and Prevention

Mice that are hysterectomy derived and maintained in isolation from all other rodents infected with PVM may be reared free of PVM virus indefinitely (Van Hoosier *et al.*, 1966; Trentin *et al.*, 1966; Parker *et al.*, 1965), suggesting that the virus is not transmitted by the intrauterine route and that other mammalian species (including man) are not natural hosts for the virus (Tennant *et al.*, 1966). On the other hand, practical control of PVM infection in conventional rodent colonies has not been very successful without the employment of these extraordinary procedures. Good animal husbandry procedures and rearing under SPF conditions such as those described in the publication "Long-Term Holding of Laboratory Rodents" (ILAR, 1976) are excellent but have been useful only at best in suppressing, not eliminating, infection.

Vaccination and chemotherapy are unproved alternatives. Live or heat-inactivated virus given intraperitoneally or sublethal doses given intranasally are effective in immunization of mice (Andrewes and Pereira, 1972; Horsfall and Curnen, 1946b; Horsfall and Hahn, 1940); however, an effective vaccine has not been developed. The replication of PVM, however, has been shown to be inhibited by several types of polysaccharides (Horsfall and McCarty, 1947; Ginsberg *et al.*, 1948), and the first example of successful chemotherapy of a virus infection in animals was demonstrated by Ginsberg and Horsfall in 1951 (1951b) when they treated and cured otherwise fatal infections in mice with the polysaccharide of Friedländer bacillus type B. The polysaccharide blocks an essential step in virus maturation, with restriction in the development of pneumonia in the mouse lung. Therapy was effective with a single intranasal inoculation of only 0.02 mg of the polysaccharide given up to 10 hr after infection. Other polysaccharides derived from both bacteria and similar materials not of bacterial origin are effective in lessening the severity of infection. The practical aspects of broadcast use of these methods in an infected mouse colony are untried and appear to be of theoretical rather than practical value.

IV. K VIRUS

A. Historical Background

The mouse K papovavirus is not to be confused with "Kilham rat virus," a parvovirus of rats and completely unrelated to mouse K virus (Kilham and Oliver, 1959; May *et al.*, 1967; Brailowsky, 1966) but sometimes erroneously referred to as *rat K virus*. Mouse K virus was isolated by Kilham in 1952 (Kilham, 1952). During a study of the "Bittner milk agent" (mammary tumor virus) in female C3H mice, 1-day-old Swiss mice were inoculated by the intracerebral route with an extract of liver, spleen, and mammary adenocarcinoma from adult C3H mice. Twelve days later, the lungs of one mouse with respiratory distress and lung consolidation were harvested and passaged serially, brain to brain, in 1-day-old Swiss mice. K virus was isolated from this material (Kilham, 1952). Subsequently, additional isolations (total of five) were made from C3H mice, and the biology and pathology were described (Kilham and Murphy, 1953; Fisher and Kilham, 1953).

B. Properties of the Virus

1. Classification

K virus belongs to the family Papovaviridae and the genus *Polyomavirus,* which also contains one other murine species, polyoma virus (Fenner, 1976). Polyoma virus has a host range similar to that of K virus but is a distinct virus without significant serologic or antigenic sharing (Dalton *et al.*, 1963; Bond *et al.*, 1978). Differences between strains of K virus isolated from mice have not been reported. Synonyms for K virus that appear in the literature include: *K papovavirus* (Takemoto and Fabisch, 1970; Jordon and Doughty, 1969), *Kilham virus* (Kraus *et al.*, 1968), and *mouse pneumonitis virus* (Parsons, 1963).

2. Physical Properties

Ultrathin sections of tissues infected with K virus contain intranuclear spherical virus particles approximately 35–45 nm in diameter (Dalton *et al.*, 1963), and negatively stained particles are approximately 50 nm in diameter with icosahedral morphology and probably 72 capsomers (Mattern *et al.*, 1963; Parsons, 1963; Fenner, 1976). The virion contains double-stranded DNA (Mattern *et al.*, 1963; Bond *et al.*, 1978); is stable at room temperature for 16 days (Kilham, 1952) and at 60°C for 30 min (Kilham, 1961); resists 20% ether for 30 min at room temperature (Kilham and Murphy, 1953) and for 18 hr at low temperatures (Kilham, 1961); and is acid stable (Fenner, 1976) (Table III). K virus agglutinates sheep erythrocytes at 37° or 4°C (erythrocyte patterns are best at 4°C) (Kilham, 1961) and dog erythrocytes at 4°C (J. C. parker, 1979, unpublished observations). Specific antibody inhibits virus agglutination of erythrocytes of both species. Elution from erythrocytes does not occur (Kilham, 1961). Hemagglutinins may be obtained from 10% suspensions of infected livers or lungs; however, liver suspensions provide the richest source of hemagglutinins (Kilham, 1961). Heating the suspensions at 60°C for 30 min unmasks the hemagglutinins in lung tissue but is unnecessary for liver (Kilham, 1961). Treatment of liver and lung tissue suspensions with 2.5% trypsin and heat (at 56°C for 30 min) increases the hemagglutinin titer (Parker *et al.*, 1965).

Because of the generic similarities and taxonomic rank of K and polyoma viruses, several investigators studied the differential properties of the two viruses (Parsons, 1963; Mattern *et al.*, 1963; Dalton *et al.*, 1963; Bond *et al.*, 1978). Both have a similar host range, are of similar size, contain double-stranded DNA, and produce primary intranuclear inclusions and secondary cytoplasmic inclusions. They resist heat, lipid solvents, and acid treatments. On the other hand, contrary to polyoma, K virus is not tumorigenic in mice, hamsters, or rats (Parsons, 1963), has a different pathogenesis in mice (Fisher and Kilham, 1953), hemagglutinates different species of erythrocytes, and is clearly different immunologically (Bond *et al.*, 1978; Dalton *et al.*, 1963). Although K virus has not been shown to replicate in any cell culture, it will transform mouse lung cells *in vitro* (Takemoto and Fabisch, 1970). No cross-reactivity is seen associated with the viral antigens by IFA, HAI, or neutralization, or with tumor antigens by IFA. An antigenic relationship among the structural proteins (VP1) of K and polyoma and other members of the papovavirus group after disruption with detergent has been shown (Bond *et al.*, 1978). No similarity is seen between the DNAs of polyoma and K viruses when analyzed by six different restriction endonucleases or by hybridization studies (Bond *et al.*, 1978). Thus, despite the fact that both viruses have similar hosts and physical properties, they are distinct both immunologically and biochemically.

3. Growth in Mice and Cell Culture

a. Cell Culture. No cell culture system has been found which will support the replication of K virus (Reed *et al.*, 1975; Bond *et al.*, 1978), although possible growth or persistence of virus has been observed in mouse embryo cell culture (W. P. Rowe, J. W. Hartley, and J. C. Parker, 1970, unpublished observations), opossum kidney cell culture, and HK cell culture for a few passages (Tennant *et al.*, 1966). Cell cultures prepared from the lung tissues of infected C57 mice have been reported to transform after subpassage during a 6-week culture period (Takemoto and Fabisch, 1970). The transformed cell line had all of the properties of such cell lines, including altered morphology, immortality, growth in soft agar, and synthesis of a K virus-specific, nonvirion intranuclear antigen detectable by CF or IFA. A C57, K virus-transformed cell line did not produce tumors in syngeneic mice; however, a similar BALB/c-transformed cell line did produce tumors in newborn and X-irradiated mice (Takemoto and Fabisch, 1970).

b. Mice. Following infection, virus titers in organs increase rapidly, reaching their maximum shortly before death. The intracranial route is often used for preparation of infective virus and serologic antigens. Resultant mouse infectivity titers of liver and lung pools contain approximately 10^8 mouse LD_{50}/0.03 ml after 8 days of infection (Kilham and Murphy, 1953). Brain, intestine, blood, spleen, kidney, and saliva contain less virus than liver and lungs, but significant amounts ranging from 10^1 to 10^7 mouse LD_{50}/0.03 ml are found (Kilham and Murphy, 1953; Rowe *et al.,* 1962; Greenlee, 1979). Smaller amounts of virus are present in urine, feces, and mammary gland tissue (Kilham and Murphy, 1953; Holt, 1959; Rowe *et al.,* 1962).

Using IFA, Greenlee (1979) confirmed the earlier electron microscopic work of Dalton *et al.* (1963) and Jordan and Doughty (1969). In addition, Greenlee (1979) detected intranuclear virus in cells lining the hepatic sinusoids in numbers paralleling those observed in pulmonary endothelium, and he suggested that the liver was a major extrapulmonary site of viral replication. With the exception of experimental oral inoculation, when viral antigen first appears in intestinal capillaries by PI day 2–3, antigen can first be detected in experimental infections in the lung and liver sinusoidal lining cells. Following intraperitoneal or intracranial infection, viral antigen first appears in the cytoplasm of infected cells at PI day 3 and subsequently in the nucleus at PI day 5. The endothelium of larger pulmonary arteries and veins is not infected until PI day 7. In the case of oral inoculation, viral antigen is not detectable in the lungs or liver until 2–3 days after its presence is observed in the jejunal endothelium. Endothelium in other organs such as skin, skeletal muscle, lymph nodes, and renal glomeruli is slightly and irregularly involved only later. Intranasal inoculation delays the appearance of viral antigen in alveolar endothelium and liver sinusoidal lining cells until PI day 7, but moribund mice that had been inoculated by this route demonstrate a greater quantity of viral antigen in brain tissue than mice inoculated by any other route. Viral antigen is present not only in brain endothelium but also in unidentified larger cells not related to blood vessels and is seen most commonly in cerebral white matter.

C. Epizootiology

1. Host Range

K virus infects laboratory-reared and feral mice (*Mus musculus*) exclusively. Young suckling mice of either sex and all strains tested, including C3H, C57BL, A, C, and Swiss, have been susceptible to lethal infection (Kilham, 1952; Holt, 1959). Attempts to infect embryonated eggs, suckling rabbits, suckling and adult hamsters, adult guinea pigs, suckling albino rats, meadow mice (*Microtus*), and deer mice (*Peromyscus*) have been unsuccessful (Kilham, 1952; Holt, 1959). All ages of mice are susceptible to infection; however, lethal infection occurs only in suckling mice up to approximately 8 days of age. Between 8 and 18 days of age, resistance to lethal infection becomes complete (Kilham and Murphy, 1953; Holt, 1959).

2. Prevalence and Geographic Distribution

K virus infection in mice has been reported in the United States (Kilham, 1952; Rowe *et al.,* 1962; Parker, 1973; Parker *et al.,* 1966), Canada (Descoteaux *et al.,* 1977), Germany (Kraus *et al.,* 1968), and Australia (Derrick and Pope, 1960; Holt, 1959), and therefore probably occurs worldwide.

Prevalence of infection within colonies from different geographic areas is not fully understood, perhaps due partly to insensitive diagnostic techniques and partly to a lack of knowledge of the biology of the normal chronic latent infection. The results of serologic surveys by Rowe *et al.* (1962) using the CF test showed that a small proportion of the mice (feral and laboratory) were positive for antibody in the majority of the colonies tested, and they speculated that few, if any, conventional mouse colonies might be entirely free of the virus; on the other hand, SPF and germ-free colonies were free of the infection. Parker *et al.* (1966) used the HAI test to conduct an extensive serologic survey of laboratory mouse colonies and found that only 1 of 34 colonies tested had antibody, and in this colony approximately 35% of the mice were seropositive. In a controlled study, HAI and CF seroconverters showed good correlation 1 month after inoculation; however, CF antibody titers were severalfold higher than HAI titers (J. C. Parker, 1979 unpublished observations). Comparative HAI and CF tests on 19 mouse sera from three infected colonies showed that the CF test was more sensitive, titers usually being severalfold higher than HAI titers, and it is noteworthy that 7 of the 18 CF-positive sera were HAI-negative (J. C. Parker, 1979, unpublished observations). Thus, undetected latent infections may be accounted for by the use of the less sensitive HAI test (Kraus, *et al.,* 1968; Gleiser and Heck, 1972).

In other serologic surveys using the HAI test, Poiley (1970) did not detect K virus HAI antibody during a 1½-year study which tested 2400 mice produced in six NIH Genetic Production Centers. In Canada, however, Descoteaux *et al.* (1977) detected K virus HAI antibody at prevalence rates of 15–60% with mean antibody titers ranging from 1:13 to 1:21 in four of five colonies examined. The findings of Descoteaux *et al.* (1977) in Canada are in marked contrast to those of comparable studies in the United States, where the results of serologic surveys using the HAI tests have shown that K virus infection is either rare or absent in colony-reared mice (Parker *et al.*, 1965, 1966; Tennant *et al.*, 1966; Van Hoosier *et al.*, 1966; Poiley, 1970; J. C. Parker, 1979 unpublished observations). On the contrary, the prevalence data obtained by Rowe *et al.* (1962), using the CF test, would support and be consistent with an epizootiologic hypothesis that natural K virus infections in mice are prevalent and geographically widespread as latent enzootic infections in only a small proportion of mice in a given population. It is also possible that the CF test may detect some nonspecific or cross-reactive antibody and that the HAI test is the more accurate measure of infection (Tennant *et al.*, 1966); however, current data would not seem to support this theory.

3. Natural History and Transmission

Beyond the observations on experimental transmission, little is known about this aspect of the natural biology of K virus. Virus has been demonstrated in lactating mammary glands from C3H mice (Kilham, 1952), lung, liver, spleen, kidney and, to a lesser extent, in brain, saliva, intestinal content, blood, and urine (Kilham and Murphy, 1953; Rowe *et al.*, 1962; Greenlee, 1979). Blackmore *et al.* (undated) have stated that the virus is spread via the nasal or oral routes but cite no specific reference. On the basis of the known presence of the virus in the kidneys and urine, urine aerosols would seem to be one of the possible methods of spread, as would feces. Since the virus is stable outside the host, contaminated food, bedding, soil, or water would remain infectious for long periods, enabling the virus to remain viable and thus persist at low infectious levels in a mouse population. The possibility that insect vectors and ectoparasites could play a role should not be overlooked since the virus is present in the blood. Experimental infection of older resistant mice results in carriers (Kilham, 1952), and indeed, the first isolations of K virus were made from inapparent carriers. Thus, even though mortality is limited to very young animals, transmission among older animals might be more important in natural maintenance of colony infections since the urine and feces may remain infective for 4 weeks or more when adults are infected (Holt, 1959; Kilham, 1952). It is noteworthy that the natural pattern of infection is enzootic. Epizootics of K virus have not been reported.

Greenlee (1979) states that the natural routes of papovavirus infections in animals remain unknown, and he concluded from his study that since the liver and lung invariably demonstrated viral antigen simultaneously, these organs might well be involved secondarily by the hematogenous route. Furthermore, K virus does not appear to infect the upper or lower respiratory tract epithelium; hence these cells can hardly be the primary site. Since oral infection is more lethal than intranasal infection and is accompanied by the early appearance of viral antigen in the intestinal villi, the oral route is a strong possibility for natural transmission. It is equally possible that the jejunal endothelium is the target organ in natural infection. It should also be mentioned that given the crude state of intranasal inoculations in mice, it can be assumed that a reasonable portion of such inoculae will either be swallowed directly or coughed up and swallowed. Thus, the intranasal route would merely serve as a means of diluting an oral inoculation where the respiratory tract epithelium is not the target organ. Even less well understood at present is the observation by Greenlee (1979) that K virus replicates in the brains of neonates without apparent pathologic or behavioral consequences. It is not yet known whether replication also occurs in the brain of asymptomatically infected older mice, or what the possible late effects of brain infection might be.

Within naturally infected mouse colonies, recovered mothers confer passive immunity on their litters via maternal antibody, and because of the early development of age resistance to lethal infection, newborn mortality is rare. After weaning, any mice becoming infected would experience a subclinical and probably prolonged chronic infection in which small amounts of virus would be voided via the urine, feces, and perhaps saliva over several weeks (Rowe *et al.*, 1962; Holt, 1959; Kilham and Murphy, 1953). The cycle would be complete when immune female virus carriers give birth to passively immunized newborn mice and simultaneously infect them via their virus-containing milk (Kilham, 1952), saliva, urine, or feces. This infection model might suggest that infection would therefore occur in mice of various ages within an infected colony. This pattern of infection was observed by Kilham and Murphy (1953) when they isolated K virus from three nursing C3H mice 6 months of age, by Kraus *et al.* (1968) when virus was isolated from adult mice, and by Parker *et al.* (1966) in a seroepizootiologic study of K virus infection in a mouse breeder colony. Holt (1959) and Derrick and Pope (1960) made similar observations. Parker *et al.* (1966) monitored a colony for HAI antibody over a period of 1½ years. Mice 4–5 months of age seldom had HAI antibody, whereas mice 7 months of age and older were positive. Virus was not isolated from 1- to 2-month-old mice but was isolated from 4- to 5-month-old and 16-month-old mice, suggesting that the mice became infected at 4–5 months of age and subsequently seroconverted. Isolation from 16-month-old mice suggests a delayed focal infection or possibly a chronic infection lasting up to 1 year (Parker *et al.*, 1966). It is noteworthy that the three wild mice from which Holt (1959) isolated K virus were positive for neutralizing antibody.

D. Pathogenesis

1. Clinical Disease

On the basis of current knowledge, this disease is entirely inapparent in mouse colonies under natural circumstances (Kilham and Murphy, 1953), even though serologic evidence of infection is widespread (Rowe *et al.,* 1962). Clinical disease becomes significant where manipulatory experiments are conducted that involve the injection of mouse tissues or excreta into newborn mice in infected colonies. A lengthy discussion is presented by Rowe *et al.* (1962) that clearly describes the problems that latent infections may cause by producing extreme data aberration in animal experiments.

The virus causes fatal disease when inoculated into 1- to 3-day-old neonatal mice by almost any route; however, most workers agree that the intracerebral route is the most sensitive (Kilham and Murphy, 1953; Kilham, 1952; Fisher and Kilham, 1953; Kraus *et al.,* 1968; Jordan and Doughty, 1969; Gleiser and Heck, 1972; Greenlee, 1979; Margolis *et al.,* 1976; Dalton *et al.,* 1963; Holt, 1959). Greenlee (1979) also demonstrated that the intraperitoneal route was more lethal than the oral route and that the intranasal route was the least fatal. Furthermore, in mice less than 48 hr old, the time of death after inoculation began as early as 8 days PI for intracranial inoculation and as late as 14 days PI for intranasal inoculation. Kilham and Murphy (1953) demonstrated that susceptibility to the clinical manifestations of experimental infection began to decline by 12 days of age, and resistance was complete by 18 days of age. Hence, neither weanlings or adults exhibited signs of disease when they were inoculated. In Holt's (1959) experiments, mortality from intracerebral inoculation began to decline after 8 days of age, and no mice died when they were inoculated at 11 days or older. Regardless of the exact age at which resistance to experimental infection becomes complete, several important aspects of this infection are noteworthy: (1) sometime during the preweaning period, complete resistance to overt evidence of infection develops and is carried over to adulthood; (2) recovered individuals and nonsick inoculated adults without signs may carry virus for extended periods; (3) most inbred strains and feral *M. musculus* appear to be susceptible (Kilham, 1952).

Experimental infection produces no apparent signs of disease until 6–15 days PI, when sudden onset of labored, ''pumping'' respiration begins and leads rapidly to death in less that 24 hr (Holt, 1959; Kilham and Murphy, 1953; Kraus *et al.,* 1968; Jordan and Doughty, 1969). Incubative animals grow normally until clinical signs develop and show no clinical evidence of CNS involvement even when inoculated by the intracerebral route (Greenlee, 1979; Kilham, 1952). Studies by Margolis *et al.* (1976) suggest that hypoxia (10% O_2) promotes survivability of experimentally infected mice because of the high O_2 requirements of K virus for replication. This appears to explain the marked preference of this virus for endothelium, particularly pulmonary endothelium.

2. Pathology

a. Gross Lesions. Serially killed, experimentally infected mice show that grossly detectable pathology is present as early as 3 days PI. These changes become progressively more severe with time and include petechial hemorrhage, congestion, edema, atelectasis, and excessive pleural fluids (Margolis *et al.,* 1976; Holt, 1959). In mice left to die, the lungs are a deep plum color and appear consolidated. Reportedly, gross pathologic changes are limited entirely to the lungs.

b. Histopathology. The most obvious histologic change occurs in the endothelial cells of the small blood vessels of the lungs. Characteristically, these cells exhibit swollen nuclei and Feulgen-positive intranuclear inclusion bodies (Fig. 18). These inclusions may be multiple and are amphophilic or basophilic when stained with H&E stain (Kraus *et al.,* 1968). Kraus *et al.* (1968) attributed K virus fatalities to massive interference with the pulmonary microcirculation by swollen nuclei. Margolis *et al.* (1976) stated that K virus replicates exclusively in the pulmonary endothelium; however, Greenlee's study (1979) using IFA has established that viral antigen is widely distributed in the body endothelium, reticuloendothelium, and even the brain. Nevertheless, pulmonary endothelium and hepatic

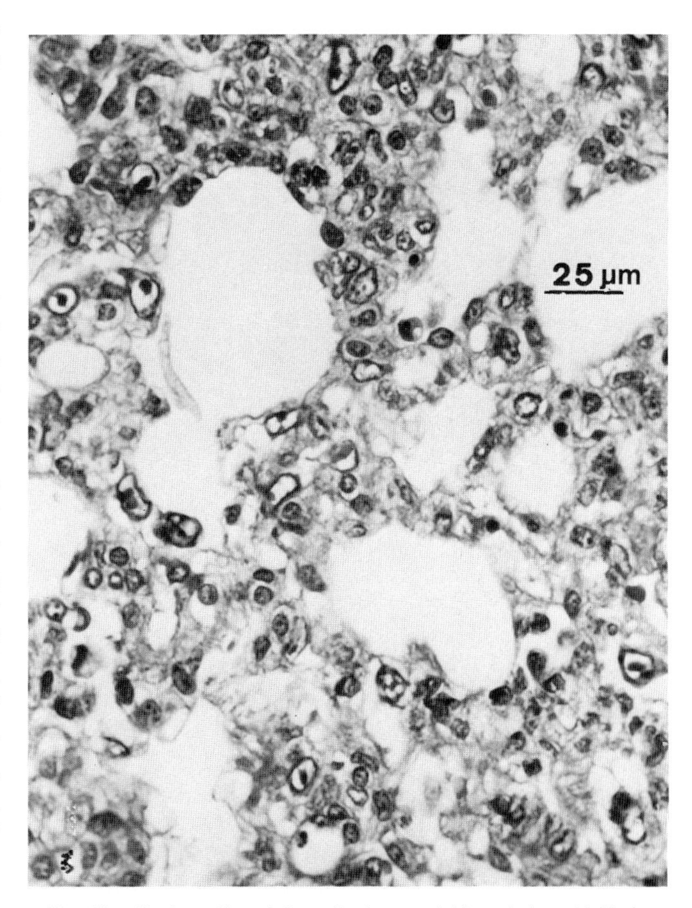

Fig. 18. Section of lung 8 days after intracranial inoculation with K virus. The alveolar walls are thickened, with a sparse inflammatory infiltrate. Intranuclear inclusions are present in numerous cells within the alveolar walls. H&E. ×455. (Courtesy of the American Society for Microbiology and Dr. J. E. Greenlee.)

sinusoidal lining cells are the principal sites of antigen localization.

Possibly the most outstanding pathologic change in infected endothelium is the occurrence of intranuclear inclusion bodies, and Margolis *et al.* (1976) showed that inclusions could be seen as early as PI day 3 or 4 after experimental infection. By this time the cytoplasm also showed increased granularity, basophilia, and swelling. By PI days 6 and 7 affected endothelium began to desquamate, possibly causing vascular occlusions that led to the gross lesions observed by Margolis and his co-workers. The most conspicuous involvement was found in the capillaries, but venous involvement was also notable (Fig. 19), especially peripherally, whereas arterial endothelium was least involved. Jordan and Doughty (1969) did not see histologic changes until PI day 6, but they also reported enlarged nuclei as the first noticeable change. Gleiser and Heck (1972) could not detect inclusion bodies before PI day 8, and they felt that the earliest significant changes were thickened alveolar septa and lymphocyte infiltration. The most detailed study on the development and distribution of intranu-

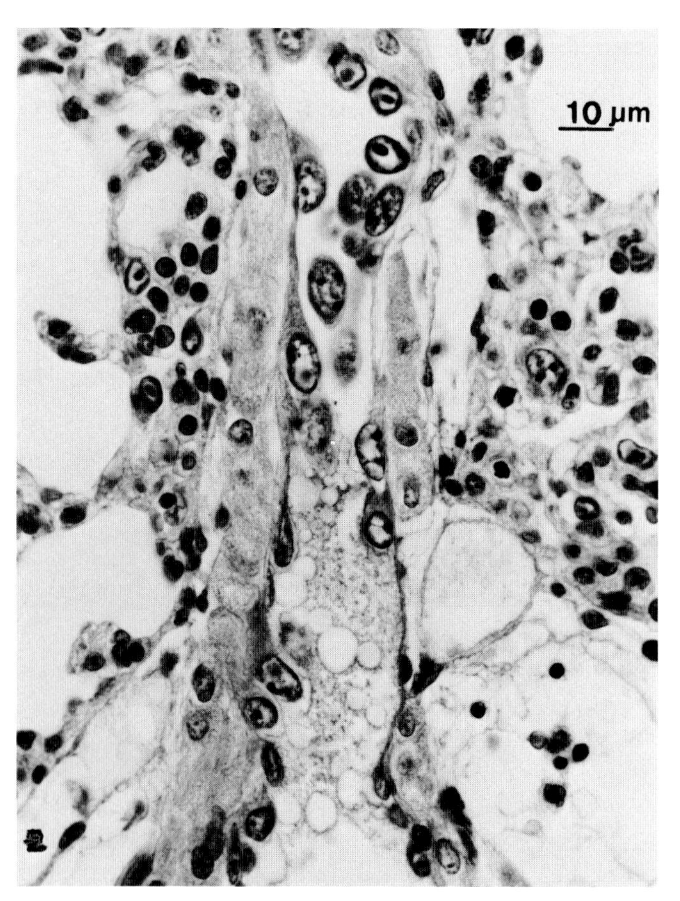

Fig. 19. Pulmonary vein with walls composed of cardiac muscle. Note both the attached and loose endothelial cells parasitized with K virus. The latter bear little resemblance to endothelium. Modified Lendrum stain. ×750. (Courtesy of the American Lung Association and Dr. G. Margolis.)

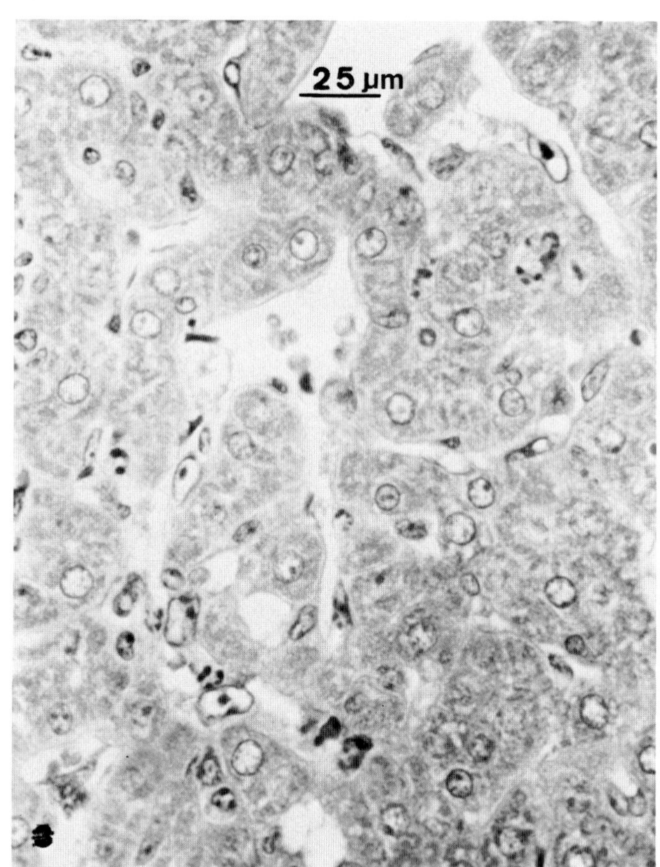

Fig. 20. Section of liver 10 days after oral inoculation with K virus. Intranuclear inclusions are present in several cells lining the hepatic sinusoids. H&E ×455. (Courtesy of the American Society for Microbiology and Dr. J. E. Greenlee.)

clear inclusion bodies is that of Greenlee (1979). Employing 3-μm sections, Greenlee (1979) showed that inclusions could be found most frequently, and in increasing numbers as illness progressed, in the lungs. The next most common site was the sinusoidal lining cells of the liver (Fig. 20). Less frequently, they were found in splenic sinusoidal lining cells, renal glomerular endothelium, and jejunal villous endothelium (Fig. 21). Greenlee also reported occasional intranuclear inclusions in alveolar epithelium but did not state whether it was type I or type II epithelium. This observation will require further examination, since most authorities, including Greenlee, feel that respiratory tract epithelium is not a target organ.

In addition to the pathologic changes in pulmonary vascular endothelium, small numbers of lymphocytes, monocytes, and polymorphs infiltrate the alveolar septa, but there is no cellular exudation into the air spaces, and pleura and bonchi remain free of abnormalities (Fisher and Kilham, 1953). Greenlee (1979), however, reported that many alveoli in experimentally infected mice were filled with proteinaceous material. Although Gleiser and Heck (1972) observed marked perivascular

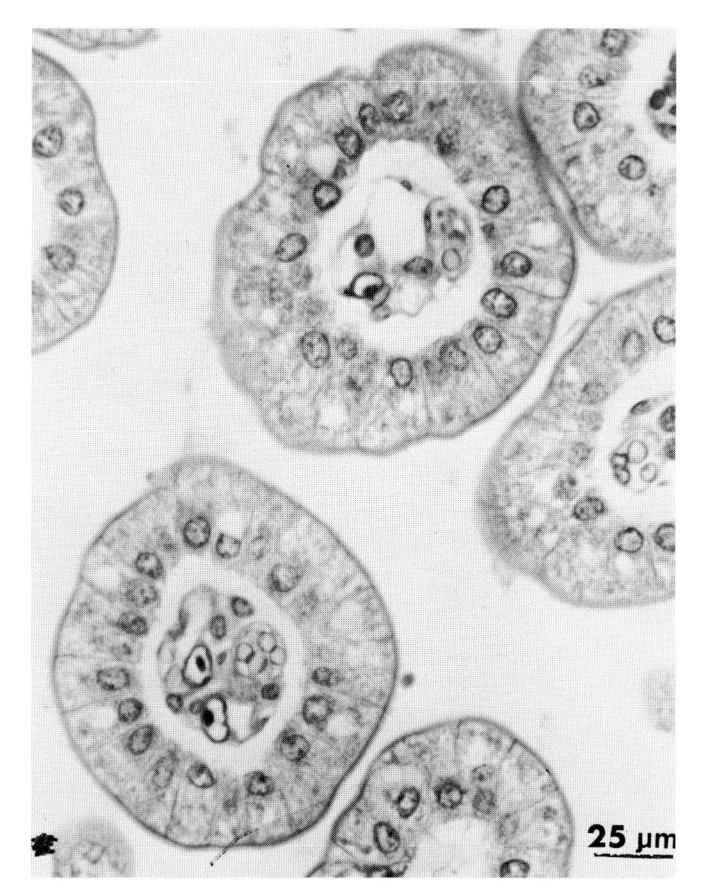

Fig. 21. Cross section of jejunal villi 10 days after oral inoculation with K virus showing intranuclear inclusions within the capillary endothelial cells. H&E. ×420. (Courtesy of the American Society for Microbiology and Dr. J. E. Greenlee.)

infiltration of lymphocytes and plasma cells in one animal surviving 15 days, it is worth noting that, unlike Sendai virus and PVM infections, K virus infection does not regularly result in the formation of perivascular and peribronchiolar lymphoid nodules.

Kraus *et al.* (1968) studied spontaneous lung disease in 88 neonatal fatalities in a closed colony known to be infected with K virus. They observed endothelial nuclear swelling, hyperchromasia, hyperemia, and desquamation in postcapillary venules in 26 lungs of 2- to 3-week old animals, and ascribed this pathology to natural K virus infection. They also observed changes in alveolar capillaries but felt that these were too difficult to assess. Affected veins exhibited perivascular edema, modest extravasation of erythrocytes, and a thin mononuclear cell infiltrate. Similar changes were present in experimentally induced K virus infection, but to a lesser degree. So far as we know, this is the only published description of lesions which are claimed to be caused by natural K virus infection. Unfortunately, the source colony for these observations showed serologic evidence of Sendai virus and PVM, and 25 of the 88 mice studied had fully developed lobar pneumonia; hence, the separation of purely K virus components of the pathologic changes would probably depend on methods other than morphologic ones.

Histopathologic changes have been described in the liver of experimentally infected mice (Rowe *et al.*, 1962; Holt, 1959; Fisher and Kilham, 1953; Greenlee, 1979; Gleiser and Heck, 1972; Tisdale, 1963). Rowe *et al.* (1962) described these changes in the liver as a "Swiss cheese effect" resulting from the formation of round, empty spaces lined by a delicate membrane and speculated that this represented a severe form of fatty degeneration. Kraus *et al.* (1968) interpreted the Swiss cheese effect in the liver as focal expansion of the sinusoids, to which swollen nuclei may contribute but are not solely responsible for; however, the consistently round appearance of these spaces (never longitudinal or oblong), the absence of erythrocytes within, and the observations of Gleiser and Heck (1972) argue strongly against this interpretation. Fisher and Kilham (1953) described changes in the sinusoidal lining cells as ballooned nuclei accompanied by the appearance of intranuclear inclusion bodies. Inclusion bodies have not been reported in Kupffer cell nuclei. Gleiser and Heck (1972) made a specific study of the experimental hepatic pathology of K virus infection in crossbred, barrier-sustained, cesarean-derived mice and concluded that the membrane lined spaces reported by Rowe *et al.* (1962) were degenerating hepatocytes. The spaces were first seen at PI day 6 but were more prevalent by day 10. Between days 7 and 10, hepatocyte vacuolation was common, but this quickly disappeared. In one 15-day survivor, the sinusoids remained dilated, cytomegaly and binucleate hepatocytes were common, and portal connective tissue was infiltrated by polymorphs, plasma cells, and lymphocytes. Greenlee (1979) observed increased numbers of lymphocytes and polymorphs throughout the hepatic parenchyma in moribund experimentally infected mice. Tisdale (1963) saw scattered hepatocellular necrosis in experimentally infected weanling mice, but when MHV-S infection was superimposed, necrosis extended throughout the lobule and frequently became confluent. Nevertheless, mortality did not differ between the MHV-S-only and MHV-S-plus K virus groups, and by PI day 14 the severe necrosis seen earlier had largely resolved.

c. Ultrastructure. Dalton *et al.* (1963) have shown that virus particles are consistently seen in the nucleus and less frequently the cytoplasm of infected endothelium. Intranuclear virus particles are associated directly with intact nuclear chromatin rather than with inclusion bodies; hence the composition of the light microscopic inclusion bodies remains uncertain. Occasionally, cytoplasmic inclusions with closely packed virus particles are seen, but the available evidence indicates that cytoplasmic virus is secondary to replication in the nucleus (Jordan and Doughty, 1969; Dalton *et al.*, 1963) (Fig. 22).

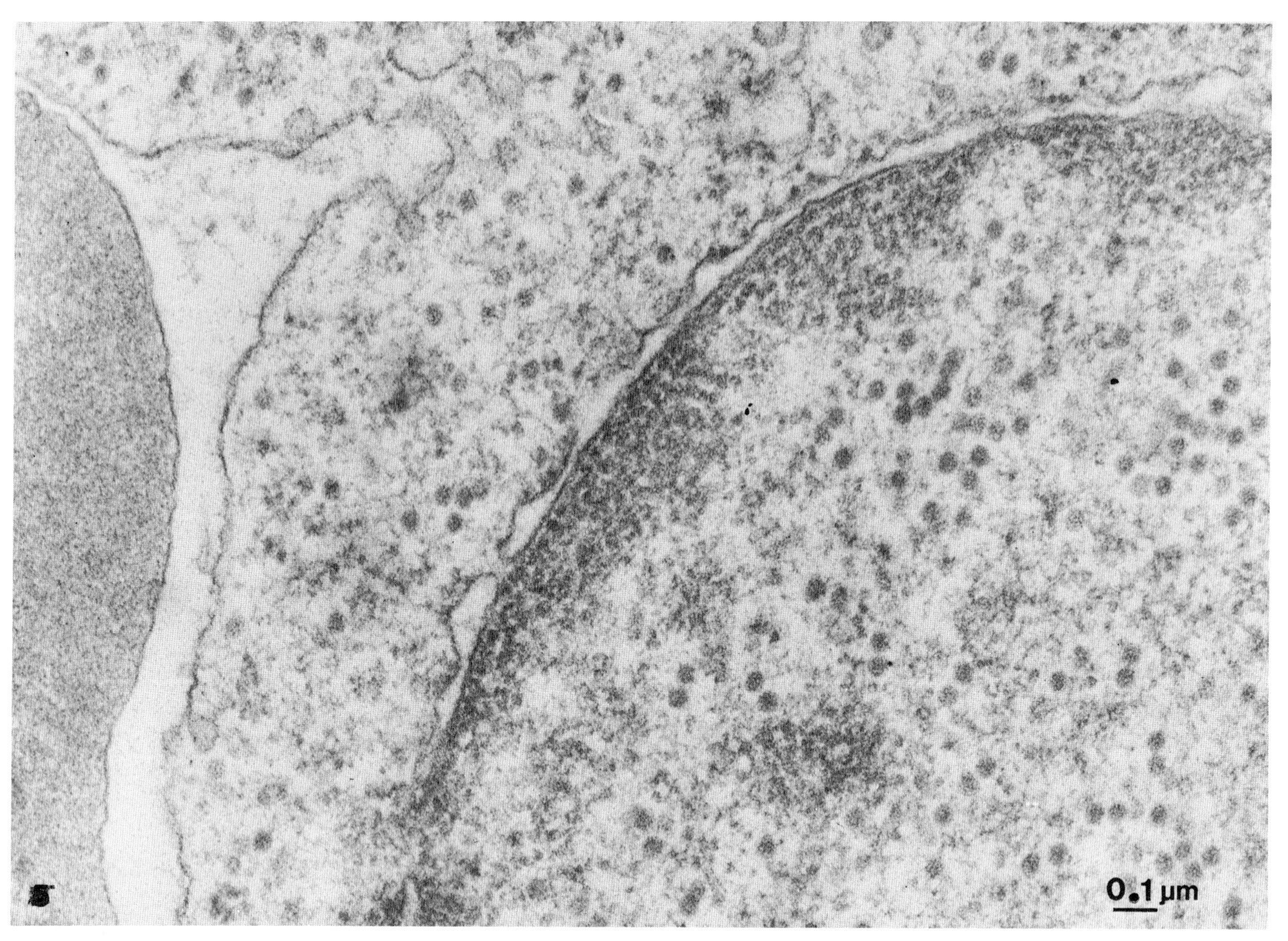

Fig. 22. Pulmonary endothelial cell showing nuclear and cytoplasmic K virus particles, PI day 8. An erythrocyte is present at the extreme left. ×62,400. (Courtesy of Academic Press and Dr. S. W. Jordan.)

Some infected nuclei consist almost entirely of virus particles, with only the nucleolus and remnants of unmodified chromatin persisting. Dalton *et al.* (1959) speculated that infected nuclei degenerate, thus allowing virus to escape to the cytoplasm. Mitochondrial swelling can be observed in conjunction with the appearance of virus in the cytoplasm.

Virus can readily be found in the cytoplasm of pulmonary and splenic macrophages, as well as polymorphs, and in large numbers in Kupffer cells. Virus particles in these cells are usually present in membrane-bound organelles (Fig. 23), possibly phagosomes or lysosomes; however, sometimes they are packed in a dense crystallinelike arrangement or scattered loosely in edematous cytoplasm (Jordan and Doughty, 1969; Dalton *et al.*, 1959). Jordan and Doughty (1969) concluded, as did Greenlee (1979), that the reticuloendothelial cells of the liver, spleen, and lymph nodes were probably not sites of viral replication, but rather sites of viral clearance (phagocytosis). Gleiser and Heck (1972) have observed degenerative changes and virus in hepatocytes as well as sinusoidal lining cells in experimentally infected mice, and speculated that hepatocyte infection was secondary to sinusoidal lining cell infection. Since it seems likely that hepatocytes would phagocytize large numbers of virus particles, their presence in these cells may be related to the hepatocellular changes first described by Rowe *et al.* (1962) and further elucidated by Gleiser and Heck (1972). Greenlee (1979) did not report viral antigen in hepatic parenchyma using IFA, so the extent and importance of hepatocellular involvement appear to be another aspect of K virus infection that requires further study.

E. Diagnosis

Although diagnosis of clinical disease using virus isolation or histopathology is not difficult, interpretation of serologic data in seroepizootiological surveys requires special attention. K virus is antigenically homologous, and cross-reactions with other viruses have not been reported. A second papovavirus, polyoma virus, frequently infects mice but is serologically and biologically distinct and presents no problems in differential

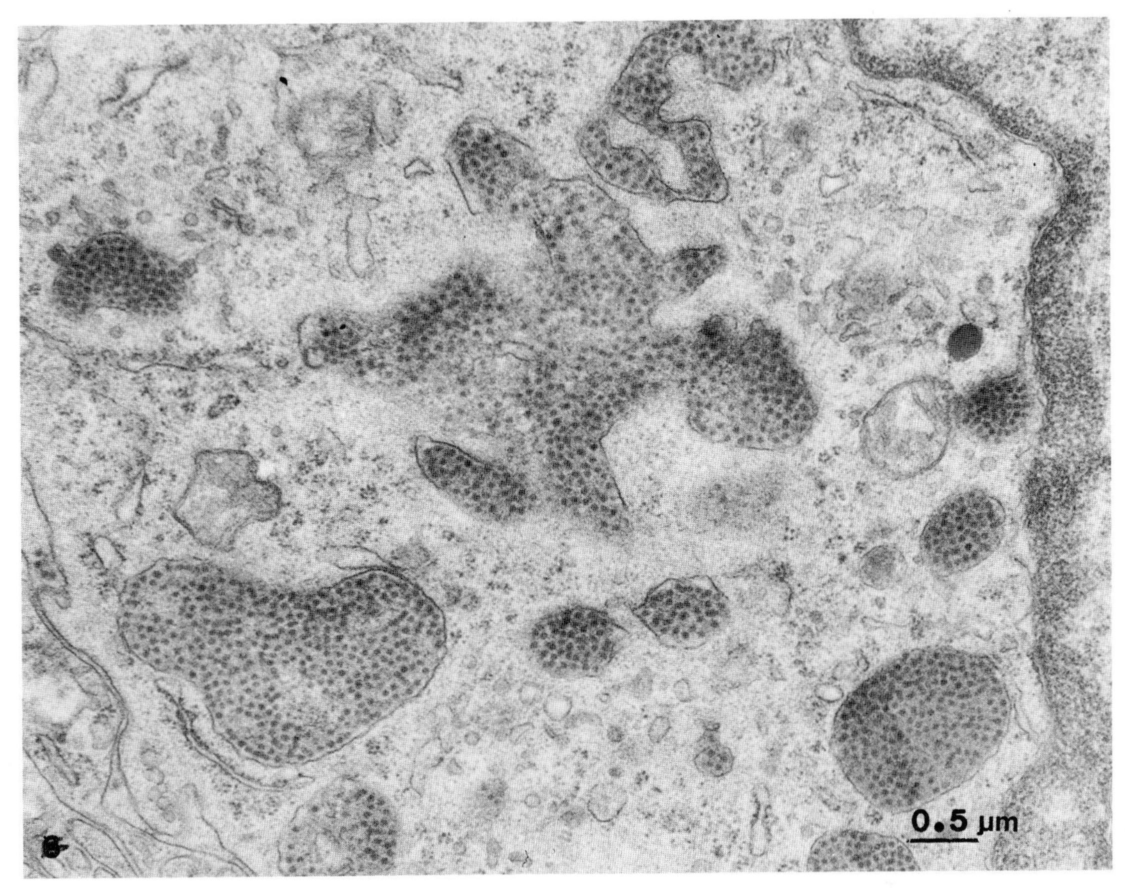

Fig. 23. Kupffer cell showing numerous membrane-bound cytoplasmic K virus particles. ×20,000. (Courtesy of Academic Press and Dr. S. W. Jordan.)

diagnosis (Bond *et al.*, 1978; Dalton *et al.*, 1963). Preparations of antigens and diagnostic procedures have been described by Parker *et al.* (1965) and Ohder (1971).

1. Serology

The CF test is more sensitive than the HAI test; however, there is a question of which test is more appropriate for use in seroepizootiologic surveys (see Section IV,C,2). Kraus *et al.* (1968) were unable to demonstrate K virus antibody using the HAI test in a colony known to be infected and doubted the usefulness of this test. The serum neutralization test and IFA test are also useful. Suitable CF and HA antigens are prepared from livers of moribund suckling mice (Rowe *et al.*, 1962; Parker *et al.*, 1965). HA and HAI tests are improved by treatment of the antigen with receptor-destroying enzyme or trypsin and heat (Parker *et al.*, 1965; Tennant *et al.*, 1966). Sera are heated to remove possible nonspecific HA inhibitors, or autohemagglutinins. Eight HA units (using sheep erythrocytes) are used in the HAI test, and antibody titers of 1:10 or greater are considered significant in either HAI or CF tests. CF antibody titers are normally severalfold higher than HAI titers (J. C. Parker, 1979 unpublished observations) (Table IV), and CF antibody correlates closely with neutralization antibody (Rowe *et al.*, 1962). Neutralizing antibody is assayed using the constant serum (heated at 56°–60°C for 30 min), varying virus method with intracerebral inoculation into 3-day-old mice (Kilham and Murphy, 1953; Holt, 1959).

Indirect IFA tests for the detection of viral antigens in experimentally infected mice have been developed (Bond *et al.*, 1978; Greenlee, 1979) and might prove useful if adapted for use in serodiagnosis. Infected newborn mouse lung cells, air-dried onto coverslips, or cryostat sections of infected lungs provide the proper antigens and staining with a suitable fluorescein-conjugated antimouse globulin is uncomplicated.

In conducting serologic surveys of mouse colonies for possible infection, it is important that careful consideration be given to the number of mice sampled (ILAR, 1976), the age of the mice in the sample, and randomization of the sample population, as well as the serologic test to be used. The predominant characteristics of K virus infection are latency, chronicity, low incidence, low antibody titers in recovered mice, and infection in older mice. Thus, testing large numbers of mice at frequent intervals and testing mice of all ages, especially those 7 months and older, may be required to certify a population free of infection.

2. Virus Isolation

Procedures for virus isolations are sensitive and uncomplicated (Table VIII). Inoculation of suckling mice is generally used; however, Greenlee (1979) developed an IFA technique which may have certain advantages in some instances over isolation in mice.

Isolation in mice is made directly, by inoculation of sucklings (Kilham and Murphy, 1953; Kilham, 1952; Holt, 1959) or indirectly in weanlings by the MAP procedure (Rowe *et al.*, 1962; Parker *et al.*, 1966; Collins and Parker, 1972). Suckling mice, preferably 1 to 3 but less than 8 days old, can be inoculated by any route; however, the intracranial route is the most sensitive (Holt, 1959). Isolations of virus are usually made from homogenated, clarified suspensions of pooled mouse viscera; however, feces, urine, and saliva are also likely to contain virus in chronic, latent infected mice (Kilham and Murphy, 1953; Holt, 1959; Rowe *et al.*, 1962). Blind subpassages are not required for isolation of virus. Diagnosis is made by demonstration of sheep erythrocyte agglutinins in liver-lung suspensions and specific inhibition of HA with a K virus reference serum.

The MAP test is based on the appearance of CF or HAI antibody in the serum of inoculated weanling mice, and although quite specific, it is 10- to 100-fold less sensitive than intracerebral inoculation of suckling mice (Rowe *et al.*, 1962).

F. Control and Prevention

Control or elimination of K virus infection from laboratory mouse colonies is complicated by possible difficulties in detection of infection due to insensitivity of some diagnostic tests and also the chronic, latent, enzootic nature of the infection. Mice with latent infections may (Holt, 1959) or may not (Kilham and Murphy, 1953) have detectable antibody, making diagnosis, and thus control, more difficult. Cesarean derivation and subsequent rearing of mice under SPF or germ-free conditions is an effective although expensive and time-consuming method of virus elimination (Parker *et al.*, 1965; Van Hoosier *et al.*, 1966; Trentin *et al.*, 1966; ILAR, 1976). Since wild mice are natural hosts of K virus, colonies must be reared in rodent-proof facilities with adequate wild rodent control programs. Rodents other than mice, reared in the same facility, do not pose a potential threat to mice, since the mouse is the only recognized host for K virus. Holt (1959) has shown that vaccination of their mothers (dams) with suspensions is an effective method of providing sucklings with passive immunity to experimental infection when the dams are vaccinated prior to pregnancy; however, nothing is known of the duration of this immunity.

Serially passaged tumors, mouse leukemias, or other mouse-derived materials may rarely become contaminated during passage in infected mice and thus pose a potential risk to uninfected mice; however, they should be screened periodically using the MAP or other tests to detect possible virus contamination (Rowe *et al.*, 1962). For example, Rowe *et al.* (1962) observed that when Gross leukemia virus began to produce a recognizable disease in suckling mice, it was due not to a change in the leukemia virus but rather to a K virus contaminant acquired during mouse passage of the Gross virus.

ADDENDUM

Carthew and Sparrow (1980) mention a natural outbreak of PVM but do not provide details of the clinical or epizootiologic aspects of the outbreak. They state that the pathology in lung sections taken from mice in the outbreak closely resembled the pathology they found in experimentally infected, ex-germ-free NMRI mice. Earliest changes were seen on PI day 1 and appeared as granular hyperchromasia of the bronchial epithelium and some debris in the bronchial lumen. Desquamation of bronchial epithelium was prominent by PI day 3 with the formation of bronchial plugs in heavily infected mice. By PI day 4, hyperemia and edema of alveolar walls and septal and alveolar infiltration by polymorphs and macrophages were evident. By PI day 5, there was marked cellular infiltration of interstitial tissue, and by PI day 7, normal architecture was obscured in infected areas. In contrast to Sendai infection, bronchial hyperplasia was mild. Surviving mice had marked interstitial inflammation, but by PI day 12 bronchial epithelium appeared normal. No mice survived high doses of PVM (10^5 $TCID_{50}$) beyond PI day 11. Using immunoperoxidase methods, Carthew and Sparrow (1980) showed that Sendai virus was detectable sometimes as early as 2 days PI, and sometimes as late as 9 days PI. On the other hand, PVM was always present by PI day 2 and never beyond PI day 7. Virus was present in bronchial epithelium and alveolar macrophages and possibly other alveolar epithelium (type not determined). These authors encourage the use of conjugated antisera and tissue sections as a means of rapid diagnosis, a technique espoused earlier in this chapter.

REFERENCES

Sendai Virus

Anderson, M. J. (1978). Innate cytotoxicity of CBA mouse spleen cells to Sendai virus-infected L cells. *Infect. Immun.* **20,** 608–612.

Anderson, M. J., Bainbridge, D. R., Pattison, J. R., and Heath, R. B. (1977). Cell-mediated immunity to Sendai virus infection in mice. *Infect. Immun.* **15,** 239–244.

Andrewes, C. H., and Pereira, H. G. (1972). "Viruses of Vertebrates," pp. 228–230, 249–250, 301. Williams & Wilkins, Baltimore, Maryland.

Andrewes, C. H., Bang, F. B., Chanock, R. M., and Zhdanov (sic), V. M.

(1959). Parainfluenza viruses 1, 2, and 3: suggested names for recently described myxoviruses. *Virology* **8,** 129–130.

Apostolov, K., and Almeida, J. D. (1972). Interaction of Sendai (HVJ) virus with human erythrocytes: a morphological study of haemolysis cell fusion. *J. Gen. Virol.* **15,** 227–234.

Appell, L. H., Kovatch, R. M., Reddecliff, J. M., and Gerone, P. J. (1971). Pathogenesis of Sendai virus infection in mice. *Am. J. Vet. Res.* **32,** 1835–1841.

Bhatt, P. N., and Jonas, A. M. (1974). An epizootic of Sendai infection with mortality in a barrier-maintained mouse colony. *Am. J. Epidemiol. 100,* 222–229.

Blair, C. D., and Robinson, W. S. (1970). Replication of Sendai virus. II. Steps in virus assembly. *J. Virol.* **5,** 639–650.

Blandford, G. (1975). Studies on the immune response and pathogenesis of Sendai virus infection of mice. III. The effects of cyclophosphamide. *Immunology* **28,** 871–883.

Blandford, G., and Charlton, D. (1977). Studies of pulmonary and renal immunopathology after nonlethal primary Sendai virus infection in normal and cyclophosphamide-treated hamsters. *Am. Rev. Respir. Dis.* **115,** 305–314.

Blandford, G., and Heath, R. B. (1972). Studies on the immune response and pathogenesis of Sendai virus infection of mice. I. The fate of viral antigens. *Immunology* **22,** 637–649.

Blandford, G., and Heath, R. B. (1974). Studies on the immune response and pathogenesis of Sendai virus infection of mice. II. The immunoglobulin class of plasma cells in the bronchial submucosa. *Immunology* **26,** 667–671.

Blandford, G., Cureton, R. J. R., and Heath, R. B. (1971). Studies of the immune response in Sendai virus infection of mice. *J. Med. Microbiol.* **4,** 351–356.

Bowen, R. A., Storz, J., and Leary, J. (1978). Interaction of viral pathogens with preimplantation embryos. *Theriogenology* **9,** 88.

Briody, B. A., Cassell, W. A., and Medill, M. A. (1955). Adaptation of influenza virus to mice. III. Development of resistance to β-inhibitor. *J. Immunol.* **74,** 41–45.

Bruch, J., Muntefering, H., and Koch, I. (1973). Submicroscopic changes in the alveolar region of the lungs after infection with Sendai virus. *Verh. Dtsch. Ges. Pathol.* **57,** 439.

Bukrinskaya, A. G., Yun-De, H., and Gorbunova, A. S. (1962). Further investigations on the antigenic relationships between Type 2 Haemadsorption (Ha-2) and Sendai virus. *Acta Virol. (Prague)* **6,** 352–356.

Burek, J. D., Zurcher, C., van Nunen, M. C. J., and Hollander, C. F. (1977). A naturally occurring epizootic caused by Sendai virus in breeding and aging rodent colonies. II. Infection in the rat. *Lab. Anim. Sci.* **27,** 963–971.

Chanock, R. M. (1956). Association of a new type of cytopathogenic myxovirus with infantile croup. *J. Exp. Med.* **104,** 555–576.

Chanock, R. M., and Parrott, R. H. (1965). Para-influenza viruses. *In* "Viral and Rickettsial Infections of Man" (F. L. Horsfall and L. Tamm, eds.), pp. 741–754. Lippincott, Philadelphia, Pennsylvania.

Chanock, R. M., Parrott, R. H., Cook, K., Andrews, B. E., Bell, J. A., Reichelderfer, T., Kapikian, A. Z., Mastrota, F. M., and Huebner, R. J. (1958). Newly recognized myxoviruses from children with respiratory disease. *N. Engl. J. Med.* **258,** 207–213.

Chanock, R. M., Parrott, R. H., Johnson, K. M., Kapikian, A. Z., and Bell, J. A. (1963). Myxoviruses: Parainfluenza. *Am. Rev. Respir. Dis.* **88,** Suppl., 152–166.

Charlton, D., and Blandford, G. (1975). A solid phase microradioimmunoassay to detect minute amounts of Ig class specific antiviral antibody in a mouse model system. *J. Immunol. Methods* **8,** 319–330.

Charlton, D., and Blandford, G. (1977). Immunoglobulin class-specific antibody response in serum, spleen, lungs, and bronchoalveolar washings after primary and secondary Sendai virus infection of germfree mice. *Infect. Immun.* **17,** 521–527.

Coid, C. R., and Wardman, G. (1971). The effect of parainfluenza type 1 (Sendai) virus infection on early pregnancy in the rat. *J. Reprod. Fertil.* **24,** 39–43.

Coid, C. R., and Wardman, G. (1972). The effect of maternal respiratory disease induced by parainfluenza type 1 (Sendai) virus on foetal development and neonatal mortality in the rat. *Med. Microbiol. Immunol.* **157,** 181–185.

Collins, A. R., and Flanagan, T. D. (1977). Interferon production and response to exogenous interferon in two cell lines of mouse brain origin persistently infected with Sendai virus. *Arch. Virol.* **53,** 313–321.

Collins, A. R., and Flanagan, T. D. (1978). Recovery of a Sendai virus variant with temperature sensitive hemolytic activity from persistently infected cells from mouse brain. *Arch. Virol.* **58,** 81–93.

Collins, M. J., Jr., and Parker, J. C. (1972). Murine virus contaminants of leukemia viruses and transplantable tumors. *J. Natl. Cancer Inst.* **49,** 1139–1143.

Compans, R. W., and Choppin, P. W. (1971). The structure and assembly of influenza and parainfluenza virus. *In* "Comparative Virology" (K. Maramorosch and F. Kurstak, eds.), pp. 407–432. Academic Press, New York.

Cook, M. K., Andrews, B. E., Fox, H. H., Turner, H. C., James, W. D., and Chanock, R. M. (1959). Antigenic relationships among the newer myxoviruses (parainfluenza). *Am. J. Hyg.* **69,** 250–264.

Cooper, J. E., Needham, J. R., and Hetherington, C. M. (1977). The use of a simple barrier system to exclude murine pathogens. *Lab. Anim.* **11,** 47–48.

Darlington, R. W., Portner, A., and Kingsbury, D. W. (1970). Sendai virus replication: An ultrastructural comparison of productive and abortive infections in avian cells. *J. Gen. Virol.* **9,** 169–177.

Degré, M., and Glasgow, L. A. (1968). Synergistic effect in viral–bacterial infection. I. Combined infection of the respiratory tract in mice with parainfluenza virus and hemophilus influenza. *J. Infect. Dis.* **118,** 449–462.

Degré, M., and Midtvedt, T. (1971). Respiratory infection with parainfluenza 1, Sendai virus in gnotobiotic and conventional mice. *Acta Pathol. Microbiol. Scand., Sect. B* **79,** 123–124.

Degré, M., and Solberg, L. A. (1971). Synergistic effect in viral bacterial infection. III. Histopathologic changes in the trachea of mice following viral and bacterial infection. *Acta Pathol. Microbiol. Scand., Sect. B* **79,** 129–136.

DeMeio, J. L., and Walker, D. L. (1957). Demonstration of antigenic relationship between mumps virus and hemagglutinating virus of Japan. *J. Immunol.* **78,** 465–471.

DeMeio, J. L., and Walker, D. L. (1958). Sendai virus antibody in acute respiratory infections and infectious mononucleosis. *Proc. Soc. Exp. Biol. Med.* **98,** 453–458.

Descoteaux, J. P., Grignon-Archambault, D., and Lussier, G. (1977). Serologic study on the prevalence of murine viruses in five Canadian mouse colonies. *Lab. Anim. Sci.* **27,** 621–626.

Eaton, G. J., Outzen, H. C., Custer, R. P., and Johnson, F. N. (1975). Husbandry of the "nude" mouse in conventional and germfree environment. *Lab. Anim. Sci.* **25,** 309–314.

Ertl, H. C. J., Gerlich, W., and Koszinowski, U. H. (1979). Detection of antibodies to Sendai virus by enzyme-linked immunosorbent assay (ELISA). *J. Immunol. Methods* **28,** 163–176.

Fenner, F. (1976). Classification and nomenclature of viruses. *Intervirology* **7,** 1–116.

Fernando, N. I., and Heath, R. B. (1971). Studies of the heat lability of mouse anti-Sendai virus immune sera. *Arch. Gesamte Virusforsch.* **34,** 295–300.

Flynn, R. J. (1967). The control of disease in laboratory animals. *In* Husbandry of Laboratory Animals'' (M. L. Conalty, ed.), pp. 361-372. Academic Press, New York.

Flynn, R. J., Fritz, T. E., Poole, C. M., Camden, R. W., Brennan, P. C., and Tolle, D. V. (1967). Disease control studies. ANL-7409, Argonne Natl. Lab., Biological and Medical Research Division, Argonne, Illinois. 216-217.

Frank, A. L., Couch, R. B., Griffis, C. A., and Baxter, R. D. (1979). Comparison of different tissue culture for isolation and quantitation of influenza and parainfluenza viruses. *J. Clin. Microbiol.* **10,** 32-36.

Fujiwara, K. (1971). Problems in checking inapparent infections in laboratory mouse colonies: an attempt at serological checking by anamnestic response. *Defining Lab. Anim. Symp., 4th, Washington, D.C., 1969* pp. 77-92.

Fujiwara, K., and Iida, T. (1974). Sendai virus infections in mouse-rat colonies. (In Jpn.) *Exp. Anim.* **23,** 265-270.

Fujiwara, K., Takenaka, S., and Shumiya, S. (1976). Carrier state of antibody and viruses in a mouse breeding colony persistently infected with Sendai and mouse hepatitis viruses. *Lab. Anim. Sci.* **26,** 153-159.

Fujiwara, K., Tanishima, Y., and Tanaka, M. (1979). Seromonitoring of laboratory mouse and rat colonies for common murine pathogens. (In Jpn.) *Exp. Anim.* **28,** 297-306.

Fukai, K., and Suzuki, T. (1955). On the characteristics of a newly-isolated hemagglutinating virus from mice. *Med. J. Osaka Univ.* **6,** 1-15.

Fukumi, H., and Nishikawa, F. (1961). Comparative studies of Sendai and HA-2 viruses. *Jpn. J. Med. Sci. Biol.* **14,** 109-120.

Fukumi, H., and Takeuchi, Y. (1975). Vaccination against parainfluenza 1 virus (*typus muris*) infection in order to eradicate this virus in colonies of laboratory animals. *Dev. Biol. Stand.* **28,** 477-481.

Fukumi, H., Nishikawa, F., and Kitayama, T. (1954). A pneumotropic virus from mice causing hemagglutination. *Jpn. J. Med. Sci. Biol.* **7,** 345-363.

Fukumi, H., Nishikawa, F., Sugiyama, T., Yamaguchi, Y., Nanba, J., Matsuura, T., and Oikawa, R. (1959). An epidemic due to HA-2 virus in an elementary school in Tokyo. *Jpn. J. Med. Sci. Biol.* **12,** 307-317.

Fukumi, H., Mizutani, H., Takeuchi, Y., Tajima, Y., Imaizumi, K., Tanaka, T., and Kaneko, J. I. (1962). Studies on Sendai virus infection in laboratory mice. *Jpn. J. Med. Sci. Biol.* **15,** 153-163.

Gardner, P. S. (1957). Serological evidence of infection with Sendai virus in England. *Br. Med. J.* **I,** 1143-1145.

Gerngross, O. G. (1957). Peculiarities of the 1956 influenza outbreak in Vladivostock due to D virus. *Vopr. Virusol.* **2,** 71-75.

Gething, M. J., Koszinowski, U., and Waterfield, M. (1978). Fusion of Sendai virus with the target cell membrane is required for T cell cytotoxicity. *Nature (London)* **274,** 689-691.

Gorbunova, A. S., Gerngross, O. G., Gnorizova, V. M., and Bukrinskaya, A. G. (1957). Strains of influenza virus of type D isolated in Vladivostock and their role in the aetiology of the 1956 outbreak. *Vopr. Virusol.* **2,** 76-84.

Goto, H., and Shimizu, K. (1978). The role of maternal antibody in contact infection of mice with Sendai virus. *Exp. Anim.* **27,** 423-426.

Grunert, R. R. (1967). Isolation of Sendai virus as a latent respiratory virus in mice. *Lab. Anim. Care* **17,** 164-171.

Heath, R. B., Tyrrell, D. A. J., and Peto, S. (1962). Serological studies with Sendai virus. *Br. J. Exp. Pathol.* **43,** 444-450.

Homma, M. (1971). Trypsin action on the growth of Sendai virus in tissue culture cells. I. Restoration of the infectivity for L cells by direct action of trypsin on L cell-borne Sendai virus. *J. Virol.* **8,** 619-629.

Homma, M., and Ohuchi, M. (1973). Trypsin action on the growth of Sendai virus in tissue culture cells. III. Structural difference of Sendai viruses grown in eggs and tissue culture cells. *J. Virol.* **12,** 1457-1465.

Homma, M., Tozawa, H., Shimizu, K., and Ishida, N. (1975). A proposal for designations of Sendai virus proteins. *Jpn. J. Microbiol.* **19,** 467-470.

Hosaka, Y., (1970). Biological activities of sonically treated Sendai virus. *J. Gen. Virol.* **8,** 43-54.

Iida, T. (1972). Experimental study on the transmission of Sendai virus in specific pathogen-free mice. *J. Gen. Virol.* **14,** 69-75.

Iida, T., Tajima, M., and Murata, Y. (1973). Transmission of maternal antibodies to Sendai virus in mice and its significance in enzootic infection. *J. Gen. Virol.* **18,** 247-254.

Institute of Laboratory Animal Resources (ILAR), Committee on Long-Term Holding of Laboratory Rodents, National Research Council (1976). Long-Term Holding of Laboratory Rodents. *ILAR News* **16,** 28.

Ishida, N., and Homma, M. (1960). A variant Sendai virus, infectious to egg embryos but not to L cells. *Tohoku J. Exp. Med.* **73,** 56-59.

Ishida, N., and Homma, M. (1961). Host controlled variation observed with Sendai grown in mouse fibroblast (L) cells. *Virology* **14,** 486-488.

Ishida, N., and Homma, M. (1978). Sendai virus. *Adv. Virus Res.* **23,** 349-383.

Ito, H. (1976). Plaque assay of Sendai virus in monolayers of a clonal line of porcine kidney cells. *J. Clin. Microbiol.* **3,** 91-95.

Ito, Y., Kimura, Y., Nagata, I., and Kunii, A. (1974). Effects of L-glutamine deprivation on growth of HVJ (Sendai virus) in BHK cells. *J. Virol.* **13,** 557-566.

Itoh, T., Kagiyama, N., Iwai, H., Okada, O., Takashina, S., and Goto, H. (1978). Sendai virus infection in a small mouse breeding colony. *Jpn. J. Vet. Sci.* **40,** 615-618.

Iwai, H. (1978). Serological examinations on natural infections with mouse pathogens in inbred mouse strains: Difference in antibody detection among the strains. (In Jpn.) *Exp. Anim.* **27,** 17-20.

Iwai, H., Parker, J. C., Quist, K. D., and Van Hoosier, G. L. (1976). Serologic reactivity to paramyxoviruses in guinea pig sera. Unpublished data.

Iwai, H., Itoh, T., and Shumiya, S. (1977). Persistence of Sendai virus in a mouse breeder colony and possibility to re-establish the virus free colonies. *Exp. Anim.* **26,** 205-212.

Iwai, H., Goto, Y., and Ueda, K. (1979). Response of athymic nude mice to Sendai virus. *Jpn. J. Exp. Med.* **49,** 123-130.

Iwasaki, Y. (1978). Experimental virus infection in nude mice. *In* ''The Nude Mouse in Experimental and Clinical Research'' (J. Fogh and B. C. Giovanella, eds.), pp. 457-475. Academic Press, New York.

Jakab, G. J. (1974). Effect of sequential inoculations of Sendai virus and *Pasteurella pneumotropica* in mice. *J. Am. Vet. Med. Assoc.* **164,** 723-728.

Jakab, G. J. (1975). Suppression of pulmonary antibacterial activity following Sendai virus infection in mice: Dependence on virus dose. *Arch. Virol.* **48,** 385-390.

Jakab, G. J. (1977). Adverse effect of a cigarette smoke component, acrolein, on pulmonary antibacterial defenses and on viral-bacterial interactions in the lung. *Am. Rev. Respir. Dis.* **115,** 33-38.

Jakab, G. J., and Dick, E. C. (1973). Synergistic effect of viral-bacterial infection: Combined infection of the murine respiratory tract with Sendai virus and *Pasteurella pneumotropica*. *Infect. Immun.* **8,** 762-768.

Jakab, G. J., and Green, G. M. (1972). The effect of Sendai virus infection on bactericidal and transport mechanisms of the murine lung. *J. Clin. Invest.* **51,** 1989-1998.

Jakab, G. J., and Green, G. M. (1973a). Immune enhancement of pulmonary bactericidal activity in murine virus pneumonia. *J. Clin. Invest.* **52,** 2878-2884.

Jakab, G. J., and Green, G. M. (1973b). Effects of pneumonia on intrapulmonary distribution of inhaled particles. *Am. Rev. Respir. Dis.* **107,** 675-678.

Jakab, G. J., and Green, G. M. (1974). Pulmonary defense mechanisms in consolidated and nonconsolidated regions of lungs infected with Sendai virus. *J. Infect. Dis.* **129,** 263-270.

Jakab, G. J., and Green, G. M. (1976). Defect in intracellular killing of ***Staphylococcus aureus*** within alveolar macrophages in Sendai-virus-infected murine lungs. *J. Clin. Invest.* **57,** 1533-1539.

Jensen, K. E., Minuse, E., and Ackermann, W. W. (1955). Serologic evidence of American experience with newborn pneumonitis virus (type Sendai). *J. Immunol.* **75,** 71-77.

Kashiwazaki, H., Homma, M., and Ishida, N. (1965). Assay of Sendai virus by immunofluorescence and hemadsorbed cell-counting procedures. *Proc. Soc. Exp. Biol. Med.* **120,** 134-138.

Kay, M. M. B. (1978a). Immunologic aging patterns: Effect of parainfluenza type 1 virus infection on aging mice of eight strains and hybrids. *Birth Defects, Orig. Artic. Ser.* **14,** 213-240.

Kay, M. M. B. (1978b). Long term subclinical effects of parainfluenza (Sendai) infection on immune cells of aging mice. *Proc. Soc. Exp. Biol. Med.* **158,** 326-331.

Kay, M. M. B., Mendoza, J., Hausman, S., and Dorsey, B. (1979). Age-related changes in the immune system of mice of eight medium and long-lived strains and hybrids. II. Short- and long-term effects of natural infection with parainfluenza type 1 virus (Sendai). *Mech. Ageing Dev.* **11,** 347-362.

Kelen, A. E., and McLeod, D. A. (1977). Paramyxoviruses: comparative diagnosis of parainfluenza, mumps, measles, and respiratory syncytial virus infections. *In* "Comparative Diagnosis of Viral Diseases, Vol. I, Human and Related Viruses" Part A (E. Kurstak and C. Kurstak, eds.), pp. 503-607. Academic Press, New York.

Kimura, Y., Ito, Y., Shimokata, K., Nishiyama, Y., Nagata, I., and Kitoh, J. (1975). Temperature-sensitive virus derived from BHK cells persistently infected with HVJ (Sendai virus). *J. Virol.* **15,** 55-63.

Kimura, Y., Aoki, H., Shimokata, K., Ito, Y., Takano, M., Kirabayashi, N., and Norrby, E. (1979). Protection of mice against virulent virus infection by a temperature-sensitive mutant derived from an HVJ (Sendai virus) carrier culture. *Arch. Virol.* **61,** 297-304.

Kingsbury, D. W., Bratt, M. A., Choppin, P. W., Hanson, R. P., Hosaka, Y., ter Meulen, V., Norrby, E., Plowright, W., Rott, R., and Wunner, W. H. (1978). Paramyxoviridae. *Intervirology* **10,** 137-152.

Koprowski, H., and ter Meulen, V. (1975). Multiple sclerosis and parainfluenza 1 virus. *J. Neurol.* **208,** 175-190.

Kunz, L. L., and Hutton, G. M. (1971). Diseases of laboratory guinea pigs. *Vet. Scope* **16,** 12-20.

Kuroya, M., Ishida, N., and Shiratori, T. (1953). Newborn virus pneumonitis (type Sendai). II. Report: The isolation of a new virus possessing hemagglutinin activity. *Yokohama Med. Bull.* **4,** 217-233.

Larson, E. W., Stephen, E. L., and Walker, J. S. (1976). Therapeutic effects of small-particle aerosols of ribavirin on parainfluenza (Sendai) virus infection of mice. *Antimicrob. Agents Chemother.* **10,** 770-772.

Lennette, E. H., Jensen, F. W., Guenther, R. W., and Magoffin, R. L. (1963). Serologic responses to parainfluenza viruses in patients with mumps virus infection. *J. Lab. Clin. Med.* **61,** 780-788.

Lief, F. S., Loh, W., ter Meulen, V., and Koprowski, H. (1975). Antigenic variation among parainfluenza type 1 (Sendai) viruses: Analysis of 6/94 virus. *Intervirology* **5,** 1-9.

Makino, S., Seko, S., Nakao, H., and Mikazuki, K. (1973). An epizootic of Sendai virus infection in a rat colony. (In Jpn.) *Exp. Anim.* **22,** 275-280.

Matsumoto, T., and Maeno, K. (1962). A host-induced modification of hemagglutinating virus of Japan (HVJ, Sendai virus) in its hemolytic and cytopathic activity. *Virology* **17,** 563-570.

Matsumoto, T. Nagata, I., Kariya, Y., and Ohashi, K. (1954). Studies on a strain of pneumotropic virus of hamster. *Nagoya J. Med. Sci.* **17,** 93-97.

Matsuya, Y., Kusano, T., Endo, S., Takahashi, N., and Yamane, I. (1978). Reduced tumorigenicity by addition *in vitro* of Sendai virus. *Eur. J. Cancer* **14,** 837-850.

Mims, C. A., and Murphy, F. A. (1973). Parainfluenza virus Sendai infection in macrophages, ependyma, choroid plexus, vascular endothelium, and respiratory tract of mice. *Am. J. Pathol.* **70,** 315-328.

Misao, T., Kanehisa, T., Kaji, M., Yamada, H., Shinohara Y., and Tajima, R. (1954). *Ann. Meet. Jpn. Assoc. Infect. Dis., 28th, Nagoya,* p. 3.

Morgan, C., and Howe, C. (1968). Structure and development of viruses as observed in the electron microscope. IX. Entry of parainfluenza I (Sendai) virus. *J. Virol.* **2,** 1122-1132.

Nagata, I., Kimura, Y., Ito, Y., and Tanaka, T. (1972). Temperature-sensitive phenomenon of viral maturation observed in BHK cells persistently infected with HVJ. *Virology* **49,** 453-461.

Nettesheim, P. (1974). Review and introductory remarks: Multifactorial respiratory carcinogenesis. *In* "Experimental Lung Cancer, Carcinogenesis and Bioassays" (Eberhard, K., and Park, J. F., eds.), Springer-Verlag, New York.

Nishikawa, F., and Fukumi, H. (1954). Shape and size of hemagglutinating virus of mice (HVM). *Jpan. J. Med. Sc. Biol.* **7,** 513-522.

Nishiyama, Y., Ito, Y., Shimokata, K., Kimura, Y., and Nagata, I. (1976). Relationship between establishment of persistent infection of haemagglutinating virus of Japan and the properties of the virus. *J. Gen. Virol.* **32,** 73-83.

Noda, K. (1953). Newborn virus penumonitis, type Sendai. III. Report pathological studies on the 9 autopsy cases and the mice inoculated with the newfound virus. *Yokohama Med. Bull.* **4,** 281-287.

Numazaki, Y. (1960). Common antigens shared by Sendai and mumps viruses. First Report: Serological response of mumps patients when examined with various antigens derived from Sendai virus, with particular emphasis on the nonreactivity against soluble antigen of Sendai virus. (In Jpn.) *Virus (Osaka)* **10,** 359-363.

Ohba, N. (1958). Formation of embryonic abnormalities of the mouse by a viral infection of mother animals. *Acta Pathologica Japonica* **8,** 127-138.

Ohder, H. von (1971). Serologischer nachweis von virusinfektionen in mäuse- und rattenzuchten. *Z. Versuchstierk Bd.* **138,** 137-150.

Ohuchi, M., and Homma, M. (1976). Trypsin action on the growth of Sendai virus in tissue culture cells. IV. Evidence for activation of Sendai virus by cleavage of a glycoprotein. *J. Virol.* **18,** 1147-1150.

Okada, Y., and Tadokoro, J. (1963). The distribution of cell fusion capacity among several cell strains or cells caused by HVJ. *Exper. Cell Res.* **32,** 417-430.

Orvell, C. and Norrby, E. (1977). Immunologic properties of purified Sendai virus glycoproteins. *J. Immunol.* **119,** 1882-1887.

Palmer, J. C., Lewandowski, L. J., and Waters, D. (1977). Non-infectious virus induces cytotoxic T lymphocytes and binds to target cells to permit their lysis. *Nature (London)* **269,** 595-597.

Parker, J. C. (1973). Discussion of indigenous murine virus infections and epidemiology of an LCM epizootic. *In* "Biohazards in Biological Research" (A. Hellman, M. N. Oxman, and R. Pollack, eds.), pp. 65-69. Cold Spring Harbor Lab., Cold Spring Harbor, New York.

Parker, J. C., and Reynolds, R. K. (1968). Natural history of Sendai virus infection in mice. *Am. J. Epidemiol.* **88,** 112-125.

Parker, J. C., Tennant, R. W., Ward, T. G., and Rowe, W. P. (1964). Enzootic Sendai virus infections in mouse breeder colonies within the United States. *Science* **146,** 936-938.

Parker, J. C., Tennant, R. W., and Ward, T. G. (1965). Virus studies with germfree mice. I. Preparation of serologic diagnostic reagents and survey of germfree and monocontaminated mice for indigenous murine viruses. *J. Natl. Cancer Inst.* **34,** 371-380.

Parker, J. C., Tennant, R. W., and Ward, T. G. (1966). Prevalence of viruses in mouse colonies. *Natl. Cancer Inst. Monogr.* No. 20, 25-36.

Parker, J. C., Whiteman, M. D., and Richter, C. B. (1978). Susceptibility of inbred and outbred mouse strains to Sendai virus and prevalence of infection in laboratory rodents. *Infect. Immun.* **19,** 123–130.

Parker, J. C., O'Beirne, A. J., and Collins, M. J., Jr. (1979). Sensitivity of the enzyme-linked immunosorbent assay, complement fixation, and hemagglutination-inhibition serologic tests for detection of Sendai virus antibody in laboratory mice. *J. Clin. Microbiol.* **9,** 444–447.

Pentitinen, K., and Cantell, K. (1967). Parainfluenza 1 (Sendai) antibodies and mumps vaccination. *Ann. Med. Exp. Biol. Fenn.* **45,** 202–205.

Poiley, S. M. (1970). A survey of indigenous murine viruses in a variety of production and research animal facilities. *Lab. Anim. Care* **20,** 643–650.

Profeta, M. L., Lief, F. S., and Plotkin, S. A. (1969). Enzootic Sendai infection in laboratory hamsters. *Am. J. Epidemiol.* **89,** 316–324.

Reed, J. M., Schiff, L. J., Shefner, A. M., and Henry, M. C. (1974). Antibody levels to murine viruses in Syrian hamsters. *Lab. Anim. Sci.* **24,** 33–38.

Reed, J. M., Schiff, L. J., Shefner, A. M., and Poiley, S. M. (1975). Murine virus susceptibility of cell cultures of mouse, rat, hamster, monkey, and human origin. *Lab. Anim. Sci.* **25,** 420–424.

Richter, C. B. (1970). Application of infectious agents to the study of lung cancer: Studies on the etiology and morphogenesis of metaplastic lung lesions in mice. *In* "Morphology of Experimental Respiratory Carcinogenesis" (P. Nettesheim, M. G. Hanna, Jr., and J. W. Deatherage, Jr., eds.), AEC Symposium Series, No. 21, pp. 365–382. USAEC, Oak Ridge, Tennessee.

Richter, C. B. (1973). Experimental pathology of Sendai virus infection in mice. *J. Am. Vet. Med. Assoc.* **163,** 1204.

Robinson, T. W. E., Cureton, R. J. R., and Heath, R. B. (1968). The pathogenesis of Sendai virus infection in the mouse lung. *J. Med. Microbiol.* **1,** 89–95.

Robinson, T. W. E., Cureton, R. J. R., and Heath, R. B. (1969). The effect of cyclophosphamide on Sendai virus infection of mice. *J. Med. Microbiol.* **2,** 137–145.

Rott, R., Waterson, A. P., and Reda, I. M. (1963). Characterization of "soluble" antigens derived from cells infected with Sendai and Newcastle disease viruses. *Virology* **21,** 663–665.

Saito, M., Nakagawa, M., Kinoshita, K., and Imaizumi, K. (1978). Etiological studies on natural outbreaks of pneumonia in mice. *Jpn. J. Vet. Sci.* **40,** 283–290.

Sandow, D., Spencker, F. B., and Schmidt, J. (1969). Usefulness of the hemagglutination inhibition test and the complement fixation test for detection of Sendai virus antibodies. *Z. Gesamte Hyg. Ihre Grenzgeb.* **15,** 598–602.

Sano, T., Niitsu, I., Nakagawa, I., and Ando, T. (1953). Newborn virus pneumonitis (type Sendai). I. Report: Clinical observation of a new virus pneumonitis of the newborn. *Yokohama Med. Bull.* **4,** 199–216.

Sasahara, J. (1955). Studies on the HVJ (Hemagglutinating virus of Japan) newly isolated from the swine. (In Jpn.) *Bull. Natl. Inst. Anim. Health* **30,** 13–38.

Sasahara, J., Hayashi, S., Kumagai, T., Yamamoto, Y., Hirasawa, S., Munakata, K., Okaniwa A., and Kato, K. (1954). A swine disease newly discovered in Japan. Its characteristic traits of pneumonia, 1. Isolation of the virus, 2. Some properties of the virus. (In Jpn.) *Virus* **4,** 131–139, 297–301, 302–308.

Sawicki, L. (1961). Influence of age of mice on the recovery from experimental Sendai virus infection. *Nature (London)* **192,** 1258–1259.

Sawicki, L. (1962). Studies on experimental Sendai virus infection in laboratory mice. *Acta Virol. (Eng. Ed.)* **6,** 347–351.

Scheid, A., and Choppin, P. W. (1974). Identification of biological activities of paramyxovirus glycoproteins. Activation of cell fusion, hemolysis, and infectivity by proteolytic cleavage of an inactive precursor protein of Sendai virus. *Virology* **57,** 475–490.

Schels, H., and Hartl, G. (1971). Occurrence of parainfluenza 1 virus (Sendai virus) infection in an experimental rat colony. *Zentralbl. Veterinaermed., Reihe* **18,** 396–399.

Sharkey, F. E. (1978). Histological observations on a nude mouse colony. *In* "The Nude Mouse in Experimental and Clinical Research" (J. Fogh and B. C. Giovanella, eds.), pp. 75–93. Academic Press, New York.

Shibuta, H. (1972). Effect of trypsin on the infectivity of Sendai virus grown in several host cells. *Jpn. J. Microbiol.* **16,** 193–198.

Shibuta, H., Akami, M., and Matumoto, M. (1971). Plaque formation by Sendai virus of parainfluenza virus group, type 1 on monkey, calf kidney, and chick embryo cell monolayers. *Jpn. J. Microbiol.* **15,** 175–183.

Shibuta, H., Adachi, A., Kanda, T., and Shimada, H. (1978). Experimental parainfluenza virus infection. I. Hydrocephalus of mice due to infection with parainfluenza virus type 1 and type 3. *Microbiol. Immunol.* **22,** 505–508.

Shimizu, K., Shimizu, Y. K., Kohama, T., and Ishida, N. (1974). Isolation and characterization of two distinct types of HVJ (Sendai virus) spikes. *Virology* **62,** 90–101.

Shimokata, K., Nishiyama, Y., Ito, Y., Kimura, Y., Nagata, I., Iida, M., and Sobue, I. (1976). Pathogenesis of Sendai virus infection in the central nervous system of mice. *Infect. Immun.* **13,** 1497–1502.

Shimokata, K., Nishiyama, Y., Ito, Y., Kimura, Y., and Nagata, I. (1977). Affinity of Sendai virus for the inner ear of mice. *Infect. Immun.* **16,** 706–708.

Shindarov, L., Galabov., A., Vassileva, V., and Runeuski, N. (1969). Multiplication of myxovirus parainfluenza 1 (Sendai) in lung tissue of tortoise (*Testudo graeca*). *Zentralbl. Veterinaermed. B Reihe* **16,** 832–839.

Sidwell, R. W., Khare, G. P., Allen, L. B., Huffman, J. H., Witkowski, J. T., Simon, L. N., and Robins, R. K. (1975). *In vitro* and *in vivo* effect of 1-β-D-Ribofuranosyl-1, 2, 4-Triazole-3-Carboxamide (Ribavirin) on types 1 and 3 parainfluenza virus infections. *Chemotherapy* **21,** 205–220.

Silver, S. M., Scheid, A., and Choppin, P. W. (1978). Loss on serial passage of Rhesus monkey kidney cells of proteolytic activity required for Sendai virus activation. *Infect. Immun.* **20,** 235–241.

Sokol, F., Neurath, A. R., and Vilcek, J. (1964). Formation of incomplete Sendai virus in embryonated eggs. *Acta Virol. (Engl. Ed.)* **8,** 59–67.

Sommerville, R. G., and Carson, H. G. (1957). Newborn pneumonitis virus (type Sendai). Evidence of infection in South-West Scotland. *Br. Med. J.* **1,** 1145–1148.

Soret, M. G., and Buthala, D. A. (1967). Enzootic Sendai virus infection in hamsters. *Fed. Proc., Fed. Am. Soc. Exp. Biol.* Abstr., p. 163.

Stark, J. E., and Heath, R. B. (1967). The development of antibodies against Sendai virus in childhood. *Arch. Gesamte Virusforsch.* **20,** 438–444.

Stewart, R. B., and Tucker, M. J. (1978). Infection of inbred strains of mice with Sendai virus. *Can. J. Microbiol.* **24,** 9–13.

Sugamura, K., Tozawa, H., Homma, M., and Ishida, N. (1974). Factors influencing the establishment of persistent infection of HVJ (Sendai virus) in L cells. *Jpn. J. Microbiol.* **18,** 349–355.

Sugita, K., Maru, M., and Sato, K. (1974). A sensitive plaque assay for Sendai virus in an established line of monkey kidney cells. *Jpn. J. Microbiol.* **18,** 262–264.

Takeyama, H., Kawashima, K., Yamada, K., and Ito, Y. (1979). Induction of tumor resistance in mice by L1210 leukemia cells persistently infected with HVJ (Sendai virus). *Gann* **70,** 493–501.

Tanaka, R., Iwasaki, Y., and Koprowski, H. (1975). Experimental parainfluenza type-1 virus-induced encephalopathy in the adult mouse. *Am. J. Pathol.* **79,** 335–346.

Tucker, M. J., and Stewart, R. B. (1976). Intrauterine transmission of Sendai virus in inbred mouse strains. *Infect. Immun.* **14,** 1191–1195.

Tuffrey, M., Zisman, B., and Barnes, R. D. (1972). Sendai (parainfluenza 1) infection of mouse eggs. *Br. J. Exp. Pathol.* **53,** 638–640.

Ueda, K., Tamura, T., Machii, K., and Fujiwara, K. (1977). An outbreak of

Sendai virus infection in a nude mouse colony. *Proc. Int. Workshop Nude Mice,* **2,** 61-69.

U.S. Public Health Service (1965). "Standardized Diagnostic Complement Fixation Method and Adaptation to Microtest," PHS Publ. No. 1228 (Public Health Monogr. No. 74). U.S. Gov. Print. Off., Washington, D.C.

van der Veen, J., Poort, Y., and Birchfield, D. J. (1970). Experimental transmission of Sendai virus infection in mice. *Arch. Gesamte Virusforsch.* **31,** 237-246.

van der Veen, J., Poort, Y., and Birchfield, D. J. (1972). Effect of relative humidity on experimental transmission of Sendai virus in mice. *Proc. Soc. Exp. Biol. Med.* **140,** 1437-1440.

van der Veen, J., Poort, Y., and Birchfield, D. J. (1974). Study of the possible persistence of Sendai virus in mice. *Lab. Anim. Sci.* **24,** 48-50.

Van Hoosier, G. L., and Robinette, L. R. (1976). Viral and chlamydial diseases. *In* "The Biology of the Guinea Pig" (J. E. Wagner and P. Manning, eds.), pp. 137-150. Academic Press, New York.

Van Hoosier, G. L., Trentin, J. J., Shields, J., Stephens, K., Stenback, W. A., and Parker, J. C. (1966). Effect of caesarean-derivation, gnotobiotic foster nursing and barrier maintenance of an inbred mouse colony on enzootic virus status. *Lab. Anim. Care* **16,** 119-128.

van Nunen, M. C. J., and van der Veen, J. (1967). Experimental infection with Sendai virus in mice. *Arch. Gesamte Virusforsch.* **22,** 388-397.

van Nunen, M. C. J., Koopman, J. P., Mullink, J. W. M. A., and van der Veen, J. (1978). Prevalence of viruses in colonies of laboratory rodents. *Z. Vershuchstierkd.* **208,** 201-208.

Ward, J. M. (1974). Naturally occurring Sendai virus disease of mice. *Lab. Anim. Sci.* **24,** 938-942.

Ward, J. M., and Young, D. M. (1976). Persistent parainfluenza (Sendai) virus infection of athymic nude mice. *Lab. Invest.* **34,** 336.

Ward, J. M., Collins, M. J., and Parker, J. C. (1976a). Hepatitis and Sendai virus pneumonia in nude mice: Unusual responses to virus infection. *Ann. Sess. Am. Assoc. Lab. Anim. Sci., 27th* Abstr., No. 42.

Ward, J. M., Houchens, D. P., Collins, M. J., Young, D. M., and Reagan, R. L. (1976b). Naturally-occurring Sendai virus infection of athymic nude mice. *Vet. Pathol.* **13,** 36-46.

Welch, B. G., Snow, E. J., Hegner, J. R., Adams, S. R., and Quist, K. D. (1977). Development of a guinea pig colony free of complement-fixing antibodies to parainfluenza viruses. *Lab. Anim. Sci.* **27,** 976-979.

Wheelock, E. F. (1966). The effects of nontumor viruses on virus-induced leukemia in mice: Reciprocal interference between Sendai virus and Friend leukemia virus in DBA/2 mice. *Proc. Natl. Acad. Sci. U.S.A.* **55,** 774-780.

Wheelock, E. F. (1967). Inhibitory effects of Sendai virus on Friend leukemia in mice. *J. Natl. Cancer Inst.* **38,** 771-778.

White, G. B. B., Gardner, P. S., and Simpson, R. E. H. (1957). Infection with Sendai virus in an outbreak of respiratory illness. *Br. Med. J.* **1,** 381-383.

Willems, F. T. C., and van der Veen, J. (1968). Growth of Sendai virus in organ cultures of mouse tissue. *Arch. Gesamte Virusforsch.* **23,** 148-156.

Yachida, S., Iritani, Y., Makino, S., and Seko, S. (1975). Application of agargel precipitin test for the detection of Sendai virus antibody in rat sera. *Lab. Anim. Sci.* **25,** 434-436.

Yamada, H. (1956). Studies on HVJ (Hemagglutinating Virus of Japan). On the biologic character of HVJ Akitsugu strain and its pathogenecity in human beings as revealed after experimental inoculation of it in volunteers. *Igaku Kenkyu* **26,** 2653-2667.

Yun-De, H. (1962). Study of the antigenic variants of the para-influenza Sendai virus. *J. Hyg., Epidemiol., Microbiol., Immunol.* **6,** 154-157.

Yung, L. L. L., Loh, W., and ter Meulen, V. (1977). Solid phase indirect radioimmunoassay: Standardization and applications in viral serology. *Med. Microbiol. Immunol.* **163,** 111-123.

Zhdanov, V., and Bukrinskaya, A. (1959). Further considerations on nomenclature of parainfluenza viruses. *Virology* **10,** 146-149.

Zhdanoff (sic), V. M., Ritova, V. V., and Golygina, L. A. (1957). Influenza D in early infancy. *Acta Virol.* (*Engl. Ed.*) **1,** 216-219.

Zurcher, C., Burek, J. D., van Nunen, M. C. J., and Meihuizen, S. P. (1977). A naturally occurring epizootic caused by Sendai virus in breeding and aging rodent colonies. I. Infection in the mouse. *Lab. Anim. Sci.* **27,** 955-962.

Pneumonia Virus of Mice (PVM)

Andrewes, C. H., and Glover, R. E. (1946). Grey lung virus: An agent pathogenic for mice and other rodents. *Br. J. Exp. Pathol.* **26,** 379.

Andrewes, C. H., and Pereira, H. G. (1972). "Viruses of Vertebrates," pp. 249-250, 301. Williams & Williams, Baltimore, Maryland.

Berthiaume, L., Joncas, J., and Pavilanis, V. (1974). Comparative structure, morphogenesis, and biological characteristics of the respiratory syncytial (RS) virus and the pneumonia virus of mice (PVM). *Arch. Gesamte Virusforsch.* **45,** 39-51.

Carthew, P., and Sparrow, S. (1980). A comparison in germ-free mice of the pathogenesis of Sendai virus and mouse pneumonia virus infections. *J. Pathol.* **130,** 153-158.

Compans, R. W., Harter, D. H., and Choppin, P. W. (1967). Studies on pneumonia virus of mice (PVM) in cell culture. II. Structure and morphogenesis of the virus particle. *J. Exp. Med.* **126,** 267-276.

Curnen, E. C., and Horsfall, F. L. (1946). Studies on pneumonia virus of mice (PVM). III. Hemagglutination by the virus; the occurrence of combination between the virus and a tissue substance. *J. Exp. Med.* **83,** 105-132.

Curnen, E. C., and Horsfall, F. L. (1947). Properties of pneumonia virus of mice (PVM) in relation to its state. *J. Exp. Med.* **85,** 39-53.

Curnen, E. C., Pickels, E. G., and Horsfall, F. L. (1947). Centrifugation studies on pneumonia virus of mice (PVM). The relative sizes of free and combined virus. *J. Exp. Med.* **85,** 23-38.

Davenport, F. M., and Horsfall, F. L. (1948). The associative reactions of pneumonia virus of mice (PVM) and influenza viruses: The effect of pH and electrolytes upon virus-host cell combination. *J. Exp. Med.* **88,** 621-644.

Descoteaux, J. P., Grignon-Archambault, D., and Lussier, G. (1977). Serologic study on the prevelance of murine viruses in five Canadian mouse colonies. *Lab. Anim. Sci.* **27,** 621-626.

Dochez, A. R., Milles, K. C., and Mulliken, B. (1937). A virus disease of Swiss mice transmissible by intranasal inoculation. *Proc. Soc. Exp. Biol. Med.* **36,** 683-686.

Eaton, M. D., and van Herick, W. (1944). Demonstration in Cotton rats and rabbits of a latent virus related to pneumonia virus of mice. *Proc. Soc. Exp. Biol. Med.* **57,** 89-92.

Eaton, M. D., Beck, M. D., and Pearson, H. E. (1941). A virus from cases of atypical pneumonia; relation to viruses of meningopneumonitis and psittacosis. *J. Exp. Med.* **73,** 641-654.

Fenner, F. (1972). Genetic aspects of viral diseases of animals. *Prog. Med. Genet.* **8,** 1-60.

Fenner, F. (1976). Classification and nomenclature of viruses. *Intervirology* **7,** 1-116.

Franklin, R. M., and Gomatos, P. J. (1961). A comparison of some properties of polyoma virus and pneumonia virus of mice. *Proc. Soc. Exp. Biol. Med.* **108,** 651-653.

Gallaspy, S. E., Coward, J. E., and Howe, C. (1978). Persistent infection of BHK-21 cells with pneumonia virus of mice. *J. Virol.* **26,** 110-114.

Ginsberg, H. S. (1951). *In vitro* reactions of pneumonia virus of mice (PVM). *Fed. Proc., Fed. Am. Soc. Exp. Biol.* **10,** 570-572.

Ginsberg, H. S., and Horsfall, F. L. (1948). Studies on the mechanism of polysaccharide inhibition of virus multiplication. *Bull. N.Y. Acad. Med.* **24,** 541–542.

Ginsberg, H. S., and Horsfall, F. L. (1949). Concurrent infection with influenza virus and mumps virus or pneumonia virus of mice (PVM) as bearing on the inhibition of virus multiplication by bacterial polysaccharides. *J. Exp. Med.* **89,** 37–52.

Ginsberg, H. S., and Horsfall, F. L. (1951a). Characteristics of the multiplication cycle of pneumonia virus of mice (PVM). *J. Exp. Med.* **93,** 151–160.

Ginsberg, H. S., and Horsfall, F. L. (1951b). Therapy of infection with pneumonia virus of mice (PVM). Effect of a polysaccharide on the multiplication cycles of the virus and on the course of the virus pneumonia. *J. Exp. Med.* **93,** 161–171.

Ginsberg, H. S., and Horsfall, F. L. (1951c). Interference between mumps virus and pneumonia virus of mice (PVM). Fate of mumps virus in the mouse lung. *J. Immunol.* **67,** 369.

Ginsberg, H. S., Goebel, W. F., and Horsfall, F. L. (1947). Inhibition of mumps virus multiplication by a polysaccharide. *Proc. Soc. Exp. Biol. Med.* **66,** 99–100.

Ginsberg, H. S., Goebel, W. F., and Horsfall, F. L. (1948). The inhibitory effect of polysaccharide on mumps virus multiplication. *J. Exp. Med.* **87,** 385–410.

Girardi, A. J., and Hilleman, M. R. (1960). Lack of identity of polyoma virus and pneumonia virus of mice (PVM). *Virology* **11,** 648–649.

Gönnert, R. (1941). Die bronchopneumonie, eine neue viruskrankheit der mäus. Zentralbl. Bakteriol., Parasitenkd, Infektionskr., Abt. 1: Orig. **147,** 161–174.

Gordon, F. B., Freeman, G., and Clampit, J. M. (1938). A pneumonia producing filterable agent from stock mice. *Proc. Soc. Exp. Biol. Med.* **39,** 450–453.

Harter, D. H., and Choppin, P. W. (1967). Studies on pneumonia virus of mice (PVM) in cell culture. I. Replication in baby hamster kidney cells and properties of the virus. *J. Exp. Med.* **126,** 251–266.

Hilleman, M. R., and Gordon, F. B. (1943). A protective antiserum against mouse pneumonitis virus. *Science* **98,** 347–348.

Horsfall, F. L., and Curnen, E. C. (1946a). Studies on pneumonia virus of mice (PVM). I. The precision of measurements *in vivo* of the virus and antibodies against it. *J. Exp. Med.* **83,** 25–42.

Horsfall, F. L., and Curnen, E. C. (1946b). Studies on pneumonia virus of mice (PVM). II. Immunological evidence of latent infection with the virus in numerous mammalian species. *J. Exp. Med.* **83,** 43–64.

Horsfall, F. L., and Ginsberg, H. S. (1951). The dependence of the pathological lesion upon the multiplication of pneumonia virus of mice (PVM). Kinetic relation between the degree of viral multiplication and the extent of pneumonia. *J. Exp. Med.* **93,** 139–150.

Horsfall, F. L., and Hahn, R. G. (1939). A pneumonia virus of Swiss mice. *Proc. Soc. Exp. Biol. Med.* **40,** 684–686.

Horsfall, F. L., and Hahn, R. G. (1940). A latent virus in normal mice capable of producing pneumonia in its natural host. *J. Exp. Med.* **71,** 391–408.

Horsfall, F. L., and McCarty, M. (1947). The modifying effects of certain substances of bacterial origin on the course of infection with pneumonia virus of mice (PVM). *J. Exp. Med.* **85,** 623–646.

Institute of Laboratory Animal Resources (ILAR), Committee on Long-Term Holding of Laboratory Rodents, National Research Council (1976). Long-Term Holding of Laboratory Rodents. *ILAR News* **16,** 28.

Iwai, H. (1978). Serological examinations on natural infections with mouse pathogens in inbred mouse strains: Difference in antibody detection among the strains. (In Jpn.) *Exp. Anim.* **27,** 17–20.

Iwai, H., Parker, J. C., Quist, K. D., and van Hoosier, G. L. (1976). Serologic reactivity to paramyxoviruses in guinea pig sera. Unpublished data.

Jakubik, J. von, and Shoda, R. (1972). Untersuchen über das vorkommen von latenen virusinfektionen in einigen europäischen zuchten von knoventionellen SPF und keinfreinen mäusen. *Z. Versuchstierkd.* **14,** 281.

Joshi, N. N., Blackwood, A. C., and Dale, D. G. (1961). Chronic murine pneumonia: A review. *Can. J. Comp. Med. Vet. Sci.* **25,** 267–273.

Kingsbury, D. W., Bratt, M. A., Choppin, P. W., Hanson, R. P., Hosaka, Y., ter Meulen, V., Norrby, E., Plowright, W., Rott, R., and Wunner, W. H. (1978). Paramyxoviridae. *Intervirology* **10,** 137–152.

Leftwich, W. B., and Mirick, G. S. (1949). The effect of diet on the susceptibility of the mouse to pneumonia virus of mice (PVM). I. Influence of pyridoxine in the period after the inoculation of virus. *J. Exp. Med.* **89,** 155–173.

Mills, K. D., and Dochez, A. R. (1944). Specific agglutination of murine erythrocytes by a pneumonitis virus in mice. *Proc. Soc. Exp. Biol. Med.* **57,** 140–143.

Mills, K. D., and Dochez, A. R. (1945). Further observation of red cell agglutinating agent present in lungs of virus-infected mice. *Proc. Soc. Exp. Biol. Med.* **60,** 141–143.

Mirick, G. S., and Leftwich, W. B. (1949). The effect of diet on the susceptibility of the mouse to pneumonia virus of mice (PVM). *J. Exp. Med.* **89,** 175–184.

Mirick, G. S., Smith, J. M., Leftwich, C. I., Jr., and Leftwich, W. B. (1952). The enhancing effect of urethane on the severity of infection with pneumonia virus of mice (PVM). *J. Exp. Med.* **95,** 147–160.

Nelson, J. B. (1937). Infectious catarrh of mice I–III. *J. Exp. Med.* **65,** 833–841, 843–849, 851–860.

Nelson, J. B. (1963). Chronic respiratory disease in mice and rats. *Lab. Anim. Sci.* **13,** Suppl., 137–143.

Nigg, C., and Eaton, M. D. (1944). Isolation from normal mice of a pneumotropic virus which forms elementary bodies. *J. Exp. Med.* **79,** 497–510.

Ohder, V. H. (1971). Serologischer Nachweis von Virus infektionen in mäuse- und rattenzuchten. *Z. Versuchstierkd.* **13,** 137–150.

Parker, J. C. (1973). Discussion of indigenous murine virus infections and epidemiology of a LCM epizootic. *In* "Biohazards in Biological Research" (A. Hellman, M. N. Oxman, and R. Pollack, Eds.), pp. 65–69. Cold Spring Harbor Lab., Cold Spring Harbor, New York.

Parker, J. C., Tennant, R. W., and Ward, T. G. (1965). Virus studies with germfree mice. I. Preparation of serologic diagnostic reagents and survey of germfree and monocontaminated mice for indigenous murine viruses. *J. Natl. Cancer Inst.* **34,** 371–380.

Parker, J. C., Tennant, R. W., and Ward, T. G. (1966). Prevalence of viruses in mouse colonies. *Natl. Cancer Inst. Monogr.* No. 20, 25–36.

Payment, P., and Descoteaux, J. P. (1978). Enzyme-linked immunosorbent assay for the detection of antibodies to pneumonia virus of mice in rat sera. *Lab. Anim. Sci.* **28,** 676–679.

Pearson, H. E., and Eaton, M. D. (1940). A virus pneumonia of Syrian hamsters. *Proc. Soc. Exp. Biol. Med.* **45,** 677–679.

Poiley, S. M. (1970). A survey of indigenous murine viruses in a variety of production and research animal facilities. *Lab. Anim. Care* **20,** 643–650.

Reed, J. M., Schiff, L. J., Shefner, A. M., and Henry, M. C. (1974). Antibody levels to murine viruses in Syrian hamsters. *Lab. Anim. Sci.* **24,** 33–38.

Reed, J. M., Schiff, L. J., Shefner, A. M., and Poiley, S. M. (1975). Murine virus susceptibility of cell cultures of mouse, rat, hamster, monkey, and human origin. *Lab. Anim. Sci.* **25,** 420–424.

Rowe, W. P., Hartley, J. W., and Huebner, R. J. (1962). Polyoma and other indigenous mouse viruses. *In* "The Problems of Laboratory Animal Disease" (R. J. C. Harris, ed.), pp. 131–142. Academic Press, New York.

Shimonaski, G., and Came, P. E. (1970). Plaque assay for pneumonia virus of mice. *Appl. Microbiol.* **20,** 775–777.

Tennant, R. W. (1963). Studies on pneumonia virus of mice (PVM) in an *in vitro* cell system and in its natural host. Ph.D. Thesis, Georgetown Univ., Washington, D.C.

Tennant, R. W. (1966). Taxonomy of murine viruses. *Natl. Cancer Inst. Monogr.* No. 20, 47–53.

Tennant, R. W., and Ward, T. G. (1962). Pneumonia virus of mice (PVM) in cell culture. *Proc. Soc. Exp. Biol. Med.* **111,** 395–398.

Tennant, R. W., Parker, J. C., and Ward, T. G. (1964). Studies on the natural history of pneumonia virus of mice. *Bacteriol. Proc.* Abstr., 125–126.

Tennant, R. W., Parker, J. C., and Ward, T. G. (1966). Respiratory virus infections of mice. *Natl. Cancer Inst. Monogr.* No. 20, 93–104.

Trentin, J. J., Van Hoosier, G. L., Shields, J., Stephens, K., Stenback, W. A., and Parker, J. C. (1966). Limiting the viral spectrum of the laboratory mouse. *Natl. Cancer Inst. Monogr.* No. 20, 147–160.

Van Hoosier, G. L., and Robinette, L. R. (1976). Viral and chlamydial diseases. *In* "The Biology of the Guinea Pig" (J. E. Wagner and P. Manning, eds.), pp. 137–150. Academic Press, New York.

Van Hoosier, G. L., Trentin, J. J., Shields, J., Stephens, K., Stenback, W. A., and Parker, J. C. (1966). Effect of caesarean-derivation, gnotobiotic foster nursing and barrier maintenance of an inbred mouse colony on enzootic virus status. *Lab. Anim. Care* **16,** 119–128.

van Nunen, M. C. J., Koopman, J. P., Mullink, J. W. M. A., and van der Veen, J. (1978). Prevalence of viruses in colonies of laboratory rodents. *Z. Versuchstierkd.* **208,** 201–208.

Verlinde, J. D., and Winsser, J. (1949). Activation of inapparent infection with ectromelia and pneumonia virus in mice. *Antonie van Leewenhoek* **15,** 40–48.

Volkert, M., and Horsfall, F. L. (1947). Studies on a lung tissue component which combines with pneumonia virus of mice (PVM). *J. Exp. Med.* **86,** 393–407.

Zurcher, C., Burek, J. D., van Nunen, M. C. J., and Meihuizen, S. P. (1977). A naturally occurring epizootic caused by Sendai virus in breeding and aging rodent colonies. I. Infection in the mouse. *Lab. Anim. Sci.* **27,** 955–962.

K Virus

Blackmore, D. K., Guillon, J. C., and Schwanzer, V. (undated). "The Viruses of Laboratory Rodents and Lagomorphs," ICLA Virus Reference Laboratory Sub-Committee. Adlard, Dorking, Surrey, England.

Bond, S. B., Howley, P. M., and Takemoto, K. K. (1978). Characterization of K virus and its comparison with polyoma virus. *J. Virol.* **28,** 337–343.

Brailowsky, C. (1966). Recherches sur le virus K du rat (parvovirus ratti) 1.-une methode de titrage par plages et son application a l'etude du cycle de multiplication du virus. *Ann. Inst. Pasteur, Paris* **110,** 49.

Collins, M. J., Jr., and Parker, J. C. (1972). Murine virus contaminants of leukemia viruses and transplantable tumors. *J. Natl. Cancer Inst.* **49,** 1139–1143.

Dalton, A. J., Moore, A. E., and Mottram, F. C. (1959). Electron microscopic observations of the K virus of mice. *J. Appl. Physiol.* **30,** 2025.

Dalton, A. J., Kilham, L., and Zeigel, R. F. (1963). A comparison of polyoma, K, and Kilham rat viruses with the electron microscope. *Virology* **20,** 391–398.

Derrick, E. H., and Pope, J. H. (1960). Murine typhus mice, rats, and fleas on the Darling Downs. *Med. J. Aust.* **2,** 924–928.

Descoteaux, J. P., Grignon-Archambault, D., and Lussier, G. (1977). Serologic study on the prevalence of murine viruses in five Canadian mouse colonies. *Lab. Anim. Sci.* **27,** 621–626.

Fenner, F. (1976). Classification and nomenclature of viruses. *Intervirology* **7,** 1–116.

Fisher, E. R., and Kilham, L. (1953). Pathology of a pneumotropic virus recovered from C3H mice carrying the Bittner milk agent. *AMA Arch. Pathol.* **55,** 14–19.

Gleiser, C. A., and Heck, F. C. (1972). The pathology of experimental K-virus infection in suckling mice. *Lab. Anim. Sci.* **22,** 865–869.

Greenlee, J. E. (1979). Pathogenesis of K virus infection in newborn mice. *Infect. Immun.* **25,** 705–713.

Holt, D. (1959). Presence of K virus in wild mice in Australia. *Aust. J. Exp. Biol. Med. Sci.* **37,** 183–191.

Institute of Laboratory Animal Resources (ILAR), Committee on Long-Term Holding of Laboratory Rodents, National Research Council (1976). Long-Term Holding of Laboratory Rodents. *ILAR News* **16,** 28.

Jordon, W. D., and Doughty, W. E. (1969). Ultrastructural pathology of murine pneumonitis caused by K-papovavirus. *Exp. Mol. Pathol.* **11,** 1–7.

Kilham, L. (1952). Isolation in suckling mice of a virus from C3H mice harboring Bittner milk agent. *Science* **116,** 391–392.

Kilham, L. (1961). Hemagglutination by K-virus. *Virology* **15,** 384–385.

Kilham, L., and Murphy, H. W. (1953). A pneumotropic virus isolated from C3H mice carrying the Bittner milk agent. *Proc. Soc. Exp. Biol. Med.* **82,** 133–137.

Kilham, L., and Oliver, L. J. (1959). A latent virus of rats isolated in culture. *Virology* **7,** 428–437.

Kraus, G. E., Soule, H., and Gruber, J. (1968). Uber Vorkommen, Diagnostik and Pathologie latenter Infektionen mit Kilham-virus in laboratoriumsmäusen. *Arch. Exp. Veterinaermed.* **22,** 1203–1210.

Margolis, G., Jacobs, L. R., and Kilham, L. (1976). Oxygen tension and the selective tropism of K virus for mouse pulmonary endothelium. *Am. Rev. Respir. Dis.* **114,** 45–51.

Mattern, C. F. T., Allison, A. C., and Rowe, W. P. (1963). Structure and composition of K virus, and its relation to the "papovavirus" group. *Virology* **20,** 413–419.

May, P., Niveleau, A., Berger, G., and Brailovsky, C. (1967). Research on the DNA of K rat virus. *J. Mol. Biol.* **27,** 603–614.

Ohder, H. von (1971). Serologischer nachweis von virusinfektionen in mäuse- und rattenzuchten. *Z. Versuchstierkd* **138,** 137–150.

Parker, J. C. (1973). Discussion of indigenous murine virus infections and epidemiology of a LCM epizootic- *In* "Biohazards in Biological Research" (A. Hellman, M. N. Oxman, and R. Pollack, eds.), pp. 65–69. Cold Spring Harbor Lab. Cold Spring Harbor, New York.

Parker, J. C., Tennant, R. W., and Ward, T. G. (1965). Virus studies with germfree mice. I. Preparation of serologic diagnostic reagents and survey of germfree and monocontaminated mice for indigenous murine viruses. *J. Natl. Cancer Inst.* **34,** 371–380.

Parker, J. C., Tennant, R. W., and Ward, T. G. (1966). Prevalence of viruses in mouse colonies. *Natl. Cancer Inst. Monogr.* No. 20, 25–36.

Parsons, D. G. (1963). Morphology of K virus and its relation to the Papova group of viruses. *Virology* **20,** 385–387.

Poiley, S. M. (1970). A survey of indigenous murine viruses in a variety of production and research animal facilities. *Lab. Anim. Care* **20,** 943–950.

Reed, J. M., Schiff, L. J., Shefner, A. M., and Poiley, S. M. (1975). Murine virus susceptibility of cell cultures of mouse, rat, hamster, monkey, and human origin. *Lab. Anim. Sci.* **25,** 420–424.

Rowe, W. P., Hartley, J. W., and Huebner, R. J. (1962). Polyoma and other indigenous mouse viruses. *In* "The Problems of Laboratory Animal Disease" (R. J. C. Harris, ed.), pp. 131–142. Academic Press, New York.

Takemoto, K. K., and Fabisch, P. (1970). Transformation of mouse cells by K-papovavirus. *Virology* **40,** 135–143.

Tennant, R. W., Parker, J. C., and Ward, T. G. (1966). Respiratory virus infections of mice. *Natl. Cancer Inst. Monogr.* No. 20, 93–103.

Tisdale, W. A. (1963). Potentiating effect of K virus on mouse hepatitis virus (MHV-5) in weanling mice. *Proc. Soc. Exp. Biol. Med.* **114,** 774–777.

Trentin, J. J., Van Hoosier, G. L., Shields, J., Stephens, K., Stenback, W. A., and Parker, J. C. (1966). Limiting the viral spectrum of the laboratory mouse. *Natl. Cancer Inst. Monogr.* No. 20, 147–160.

Van Hoosier, G. L., Trentin, J. J., Shields, J., Stephens, K., Stenback, W. A., and Parker, J. C. (1966). Effect of caesarean-derivation, gnotobiotic foster nursing and barrier maintenance of an inbred mouse colony on enzootic virus status. *Lab. Anim. Care* **16,** 119–128.

Chapter 9

Viral Diseases of the Digestive System

Lisbeth M. Kraft

I. Introduction 159
II. Epizootic (Epidemic) Diarrhea of Infant Mice (EDIM), Mouse Rotavirus Enteritis 160
A. Historical Background 160
B. Properties of the Virus 160
C. Pathogenesis 163
D. Epizootiology 165
E. Diagnosis 166
F. Control and Prevention 166
III. Reovirus 3 Infection (Hepatoencephalomyelitis, ECHO 10 Virus Infection) 167
A. Historical Background 167
B. Properties of the Virus 167
C. Pathogenesis 170
D. Epizootiology 172
E. Diagnosis 173
F. Control and Prevention 173
IV. Murine (Mouse) Hepatitis Virus Infection (MHV) 173
A. Historical Background 173
B. Properties of the Virus 174
C. Pathogenesis 177
D. Epizootiology 182
E. Diagnosis 182
F. Control and Prevention 183
References 183

I. INTRODUCTION

At a symposium on the viruses of laboratory rodents held in 1965, Parker *et al.* (1966) described the results of antibody tests for a number of virus infections in 34 colonies of mice throughout the United States. In these, reovirus 3 infection occurred in 28 (82%) and murine hepatitis in 25 (74%). Although there was at that time no routine serologic test for

ISBN 0-12-262502-1

epidemic diarrhea of infant mice, it is common knowledge that virtually every colony of conventional mice suffered that infection to a greater or lesser extent.

In recent years, with the advent of measures such as cesarean derivation, barrier-sustained breeding and maintenance procedures, routine serologic surveillance, laminar flow hoods or rooms, and filter-top cages, the problems associated with these diseases have become somewhat less critical. Nevertheless, they are at times and under certain circumstances still troublesome.

All three infections may cause serious, debilitating, and sometimes fatal diarrheal disease in nursling and weanling mice. Thus, the economic impact on commercial mouse colonies can be severe. Furthermore, newborn mice, pooled from many dams and then redistributed to them at random, are used for the isolation and identification of certain viruses in diagnostic and epidemiological studies, for example arboviruses (Shope, 1980), coronaviruses (McIntosh *et al.*, 1967), and reoviruses (Stanley, 1977). Clearly, indigenous infection with related or identical agents in only a few such infants could confound and compromise the validity of observations following the inoculation of test materials. It should be pointed out, too, that Collins and Parker (1972) demonstrated both reovirus 3 and murine hepatitis virus as contaminants in some murine leukemia and transplantable tumor specimens.

Whereas it is mainly from these standpoints that the diseases in question derive importance, they should not be dismissed without considering their intrinsic value as models for elucidating the pathogenesis and control of related infections in man and other animals as well as their utility for the molecular biologist.

II. EPIZOOTIC (EPIDEMIC) DIARRHEA OF INFANT MICE (EDIM), MOUSE ROTAVIRUS ENTERITIS

A. Historical Background

Cheever and Mueller (1947, 1948), Pappenheimer and Enders (1947), and Pappenheimer and Cheever (1948) were the first to describe the pathological changes and epidemiology of EDIM. Runner and Palm (1953) and Cheever (1956) also contributed to knowledge concerning the epidemiology and etiology of the disease. Thereafter Kraft (1957, 1958, 1961, 1962b, 1966) reported on studies regarding the etiology, mode of transmission, carrier state, immune response, pathogenesis, and control of the disease. Serologic studies were also undertaken by Blackwell *et al.* (1966).

Adams and Kraft (1963, 1967) first demonstrated the agent in electron micrographs of infected infant mouse intestinal epithelium. Banfield *et al.* (1968) enlarged on those findings, comparing the virions to those of the reoviruses. Particles of similar morphology were subsequently observed in the diarrheal feces of many species, primarily in the young: cattle (Mebus *et al.*, 1969), man (Flewett *et al.*, 1974), horse (Flewett *et al.*, 1975), pig (Rodger *et al.*, 1975), sheep (Snodgrass *et al.*, 1976), rabbit (Bryden *et al.*, 1976), deer (Tzipori *et al.*, 1976), goat (Scott *et al.*, 1978), and dog (England and Poston, 1980). In addition, one virus, SA-11, was isolated from a nondiarrheal monkey, and another, OA (offal agent), was recovered from intestinal washings of sheep in an abattoir (Els and Lecatsas, 1972).

Much and Zajac (1972) purified EDIM virus from diarrheal infant mice and further characterized it.

Based on its morphology and other known characters, EDIM virus has been placed into the genus *Rotavirus* in the family Reoviridae (Matthews, 1979). During the past decade, knowledge of the rotaviruses as agents of diarrheal disease of young mammals has burgeoned, especially with regard to those affecting children, calves, and piglets. Reviews concerning the genus have been published by Wyatt *et al.* (1978), McNulty (1978, 1979), Flewett and Woode (1978), Andrewes *et al.* (1978), and Holmes (1979). Additional comments may be found in the American Veterinary Medical Association *Panel Report of the Colloquium on Selected Diarrheal Diseases of the Young* (Anonymous, 1978).

B. Properties of the Virus

1. Classification

The classification of EDIM virus in the genus *Rotavirus,* family Roeviridae, derives from its morphology and mode of replication as seen in electron micrographs (Adams and Kraft, 1967; Banfield *et al.*, 1968), which were later shown to resemble those observed in ultrathin sections in human intestinal material containing "reovirus-like" particles (Bishop *et al.*, 1973). The term *rotavirus* (L. *rota,* wheel) was proposed by Flewett *et al.* (1974) because of its morphology as viewed in negative-contrast electron micrographs. It has become widely used in preference to *duovirus,* which is synonymous (Davidson *et al.*, 1975).

2. Strains of EDIM Virus

There is no evidence that antigenic or pathogenetic variants of EDIM virus exist. Analysis of RNA segment and structural polypeptide variation, however, as has been accomplished for other rotaviruses (Kalica *et al.*, 1978; Derbyshire and Woode, 1978; Rodger and Holmes, 1979), may reveal differences among strains. Further, several serotypes have been described among the human rotaviruses by means of the serum neutralization test (Beards *et al.*, 1980). This technique may also prove useful in future studies of EDIM virus strains.

3. Physical Properties

a. Morphology, Size, and Composition. Kraft (1962b), using Millipore filters, determined the size of the infective EDIM virion to be between 33 and 100 nm. In electron micrographs, Adams and Kraft (1967) recognized two principal particle types: a spherical one, 75–80 nm in diameter, with two double membranes surrounding a nucleoid or core, and another, likewise spherical but less frequently seen, 65 nm in diameter with only one double membrane. Tubular structures as well as coreless (electron-lucent) spherical particles were also occasionally encountered. Banfield *et al.* (1968) described the tubular structures more extensively and demonstrated them within nuclei as well as in the cytoplasm.

Holmes *et al.* (1975) studied EDIM virus by negative-contrast electron microscopy as well as in ultrathin sections of mouse intestines and reached similar conclusions as to the size and morphology of the virions. Much and Zajac (1972) determined the mean diameter of 100 virions of purified EDIM virus to be 54.4 ± 2 nm. On the other hand, Melnick (1979) gives the diameter of the *Rotavirus* genus as 70 nm. Possibly Much and Zajac (1972) measured only incomplete particles, those lacking the outer capsid layer.

Specific details for EDIM virus with regard to RNA molecular weight, number of capsid polypeptides, number and shape of capsomers, etc., are lacking in the literature. However, all rotaviruses are regarded as possessing essentially spherical capsids, 60–80 nm in diameter, with icosahedral symmetry (Fig. 1). The genome consists of 10–12 molecules of double-stranded RNA with a total molecular weight of about 15×10^6.

b. Effect of Heat. As a filtrate of intestinal suspension, infectivity of EDIM virus is retained at 4°C for 1 hr, whereas

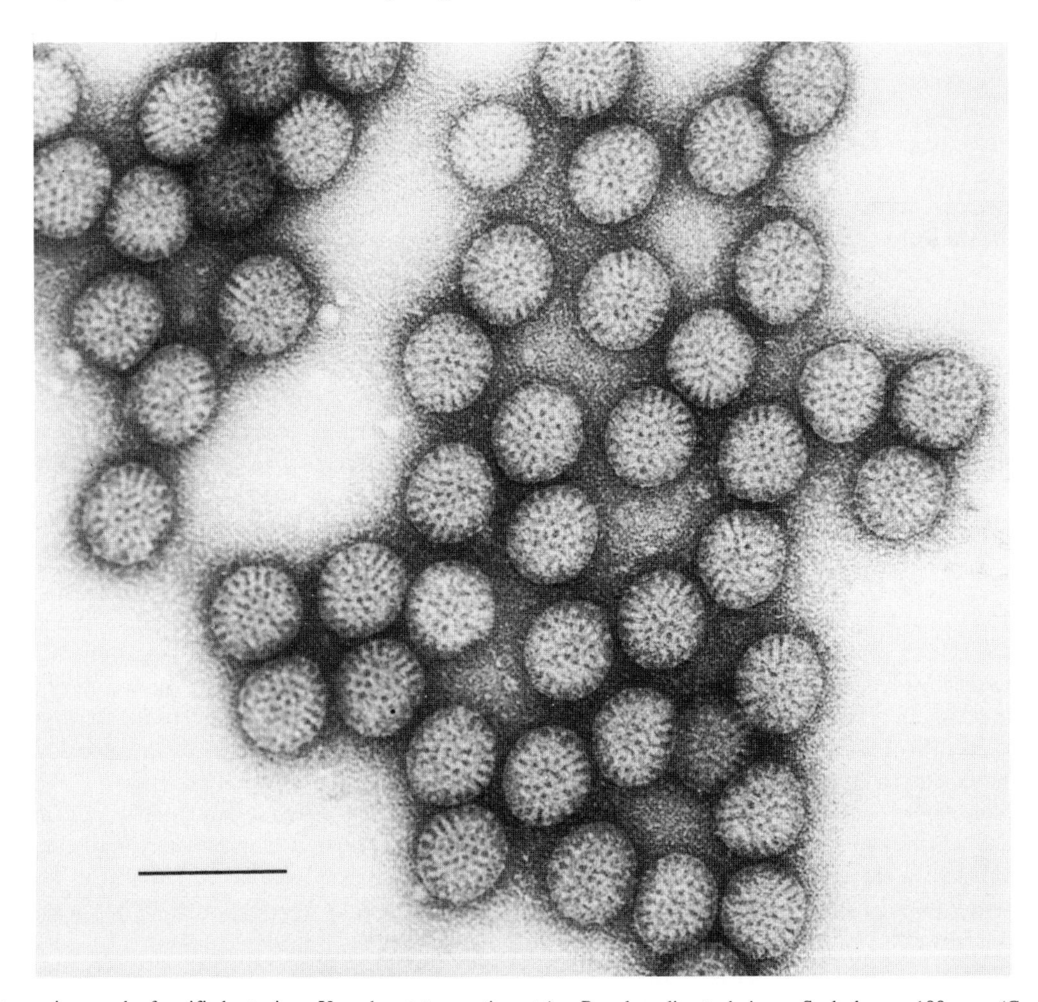

Fig. 1. Electron micrograph of purified rotavirus. Uranyl acetate negative stain. Pseudoreplica technique. Scale bar = 100 nm. (Courtesy of Erskine Palmer, Center for Disease Control, Atlanta.)

about 50% is lost at 4°C for 24 hr or at 37°C for 1 hr. Although some infectivity (>0.05%) remains at either 56°C or 60°C for 30 min, it is abolished at 70°C for 15 min (Cheever and Mueller, 1947; Kraft, 1957, 1962b). Much and Zajac (1972) found that the purified virus is unstable at both 4° and −24°C for 2 weeks but found that at −70°C infectivity is retained for at least 4 weeks.

c. Effect of Chemicals. In an intestinal filtrate, EDIM virus titer is not significantly reduced when held in ether or 0.1% sodium deoxycholate at 4°C for 24 hr (Kraft, 1962b). According to Much and Zajac (1972), the purified virus is stable in 20% ether, 5% chloroform, or 0.1% sodium deoxycholate at 4°C for 1 hr and, as is true for other rotaviruses, is resistant to pancreatin.

Ward and Ashley (1980a,b) examined the effects of the anionic detergent sodium dodecyl sulfate and of the chelating agent ethylenediaminetetraacetate on purified simian (SA-11) virus and determined that low concentrations and mild temperature conditions readily inactivated the agent. Both chemicals modified the viral capsid to prevent adsorption of the inactivated virions to cells. Indeed, this study was the outgrowth of a need to determine the survival of enteric viruses in wastewater. It had been determined that wastewater sludge reduced the heat necessary for simian rotavirus inactivation. Ionic detergents in the sludge were identified as the active components. Nonionic detergents did not destabilize the virus; further, these compounds protected the virus from the destabilizing effect of sodium dodecyl sulfate. Destabilization by both cationic and anionic detergents was found to be dependent on the pH of the medium.

d. Effect of pH. A systematic study of the stability of EDIM virus to extremes of pH has not been reported. Much and Zajac (1972) treated virus preparations during purification procedures at pH 9.0 without apparent loss of titer.

e. Antigenic Determinants. EDIM virus shares a common antigen with other rotaviruses that is demonstrable by a variety of serologic tests. This determinant is associated with the inner capsid of the virion (Mathan *et al.*, 1977; Woode *et al.*, 1976). Based on the neutralization test, on the other hand, EDIM virus appears to be type specific, neutralization of infectivity being associated with the intact (complete) virion (Thouless *et al.*, 1977b).

Kraft (1958) was unable to demonstrate an EDIM virus hemagglutinin using guinea pig, adult and embryonic chicken, adult and embryonic mouse, rabbit, hamster, human 0, and sheep erythrocytes. However, Fauvel *et al.* (1978), studying calf and human rotaviruses by means of osmotic shock to release cell-bound virus from BS-C-1 tissue culture cells, prepared a hemagglutinin for human 0, guinea pig, and bovine red cells that is apparently associated with the intact virion. They also succeeded in this regard with the simian SA-11 rotavirus.

4. Relationship to Other Viruses

Kraft (1966) determined that EDIM virus is antigenically distinct from reovirus 3. Hyperimmune sera prepared in mice against EDIM virus by Blackwell *et al.* (1966) did not react with pneumonia virus of mice, mouse encephalomyelitis virus, reovirus 3, newborn mouse pneumonitis (K) virus, polyomavirus, or murine (mouse) hepatitis virus.

Using gel electrophoresis, Smith and Tzipori (1979) examined the RNA of rotaviruses from six animal species. Reproducible differences in RNA migration patterns were found between all isolates (from calf, pig, mouse, deer, foal, and dog-adapted human isolates). Pig, mouse, and foal viruses each yielded 11 bands; calf, 9; deer, 10; and dog-adapted human isolates 12 bands.

5. Growth and Distribution in Mice

Replication of EDIM virus occurs in epithelial cells of the villi of the small intestine. In electron micrographs, Adams and Kraft (1967), Banfield *et al.* (1968), and Holmes *et al.* (1975) demonstrated replication by budding in distended cisternae of the endoplasmic reticulum. The significance of intracytoplasmic and intranuclear tubular structures encountered is not clear.

Using direct immunofluorescence, Wilsnack *et al.* (1969) found that EDIM virus antigen in infant mice is limited to the cytoplasm of the villous epithelium from the duodenum to the colon and is detectable within 48 hr after peroral inoculation. Stainable virus was also seen in the intestines of contact infant mice without necessarily causing clinical signs to appear. The virus could not be stained in the stomach or liver of mice infected either naturally or experimentally.

Kraft (1958) studied the distribution of infectious EDIM virus in 3-day-old mice following peroral inoculation. At 3 hr, virus was detected in the stomach and small and large intestines (the cecum was not tested) and could not be recovered from the lungs, liver, spleen, kidney, or blood. At 22 hr, all those tissues were positive except for the kidney. The bladder and urine, as well as the brain, were also devoid of the agent, but at 30 hr, these too were positive.

At 72 hr, the liver, spleen, kidney, and intestines yielded virus, but the brain was negative. The blood, stomach, lungs, bladder, and urine were not tested at that interval. At 6 days, blood, liver, and intestines, the only tissues examined, still contained virus.

Ingested virus from a diarrheal litter brings about an active infection in the nursing dam that previously had been nondiarrheal herself and had had only nondiarrheal litters (Kraft,

1958). Blood, liver, spleen, and feces all contain infectious virus in the dam 1 week after her litter is given virus perorally as an intestinal filtrate. Later, Kraft (1961) determined that adult male mice can also be intestinal carriers for at least 17 days after a single peroral exposure.

With regard to the mechanism of cell penetration, Holmes *et al.* (1976) proposed that lactase of the villous brush border of the intestine may be the receptor that uncoats the rotavirus virion by attacking the glycolysated polypeptides of the outer capsid. In addition, pancreatic enzymes within the lumen may also be instrumental in the infectious process (see Section II,B,6).

The question of cofactors that might enhance pathogenicity or of interfering substances, in addition to antibodies, that might inhibit EDIM virus replication is open. Kraft (1958) found increased numbers of *Clostridium tertium* in the intestine of diarrheal animals, and Pappenheimer and Enders (1947) remarked on the persistent presence of "coccoid bodies" in the intestines of diarrheal animals as seen by light microscopy. In both instances, these may be considered opportunistic organisms. Their effect on the severity of the clinical disease is unknown, however.

One report (LaBonnardiere and deVaurieux, 1979) concerns the question of interferon production by bovine rotavirus and EDIM virus. Orally instilled bovine rotavirus, which is nonpathogenic for infant mice, was shown to delay the onset of EDIM virus-induced diarrhea when the latter agent was given orally 48 hr after the bovine agent. Only intact infectious bovine virus was capable of eliciting this effect. Interferon could not be found in the mouse small intestine at various times after instillation of the bovine agent or after EDIM virus infection; therefore the nature of the interference is not clear. It may be of significance for possible immunization of mice with active virus that some bovine rotavirus replication could be visualized in mouse intestinal cells by means of immunofluorescence, although detectable antigen was not evident from utilization of the ELISA test.

6. Growth in Tissue and Organ Culture

The use of standard methods to grow EDIM virus in tissue culture has resulted for the most part in failure.

Habermann (1959) made 10 serial passages of EDIM virus in Chang liver cells but reported results only after the second passage, when supernatant fluid produced diarrhea in infant mice. Kraft (1958) was unable to grow the agent in monkey kidney epithelial cells, mouse embryo, two strains of mouse fibroblasts, HeLa cells, human intestine (Henle), and human conjunctiva. Thouless *et al.* (1977b) cultivated EDIM virus as well as other rotaviruses in LC MK2 cells but could not serially subculture them. Viral multiplication was detectable only by means of immunofluorescence. On the other hand, Rubenstein *et al.* (1971) grew the agent for three passages in organ cultures of embryonic mouse ileum, cecum, and colon but not in esophagus or duodenum cultures. Subculture in mouse embryo fibroblasts was unsuccessful.

It may be of some importance for successful cultivation of EDIM virus that treatment of tissue culture passage material with pancreatin enhances the virus yield of porcine rotavirus (Thiel *et al.*, 1978). Further, Babiuk *et al.* (1977) and Almeida *et al.* (1978) state that calf rotavirus production is markedly enhanced when passaged in the presence of trypsin. Clark *et al.* (1979) produced high-titer calf rotavirus in a variety of continuous cell lines, also by utilizing trypsin, and Matsuno *et al.* (1977) were successful in producing plaques with bovine rotavirus in monkey kidney cell monolayers when trypsin was incorporated in the overlay medium. Estes and Graham (1979) found that simian virus titers were enhanced by both trypsin and elastase in Vero as well as MA104 cells. Ramia and Sattar (1980) studied additional proteolytic enzymes utilizing plaque formation by simian virus SA-11 in MA104 cells as the endpoint. Elastase, α-chymotrypsin, subtilase, pronase, and pancreatin were as effective as trypsin, whereas pepsin, papain, and thermolysin were ineffective.

Another approach with human rotavirus has been to incorporate the agent into cells by means of low-speed centrifugation of the two together (Banatvala *et al.*, 1975; Bryden *et al.*, 1977; Schoub *et al.*, 1979; Wyatt *et al.*, 1980).

It remains to be seen if similar treatments will lead to successful cultivation of EDIM virus in tissue culture.

7. Growth in Fertile Hens' Eggs

Chick embryos appeared unaffected when the chorioallantois was inoculated. The agent did survive up to 5 days, but an increase in infectious titer could not be demonstrated (Kraft, 1958).

C. Pathogenesis

Moon (1978) has discoursed on pathogenetic mechanisms in diarrheal diseases, comparing EDIM, calf rotavirus infection, transmissible gastroenteritis coronavirus infection of pigs (TGE), and feline panleukopenia as to the predilection of each virus for particular regions of enteric epithelium. The character of each disease appears to be reflected not only in the particular group of cells but also in the numbers of host cells involved.

1. The Clinical Disease

Lucid descriptions of the natural clinical disease are given by Cheever and Mueller(1947), Seamer (1967), and McClure *et al.* (1978). The experimental disease does not differ significantly (Kraft, 1957, 1962b). Overt illness is confined to pre-

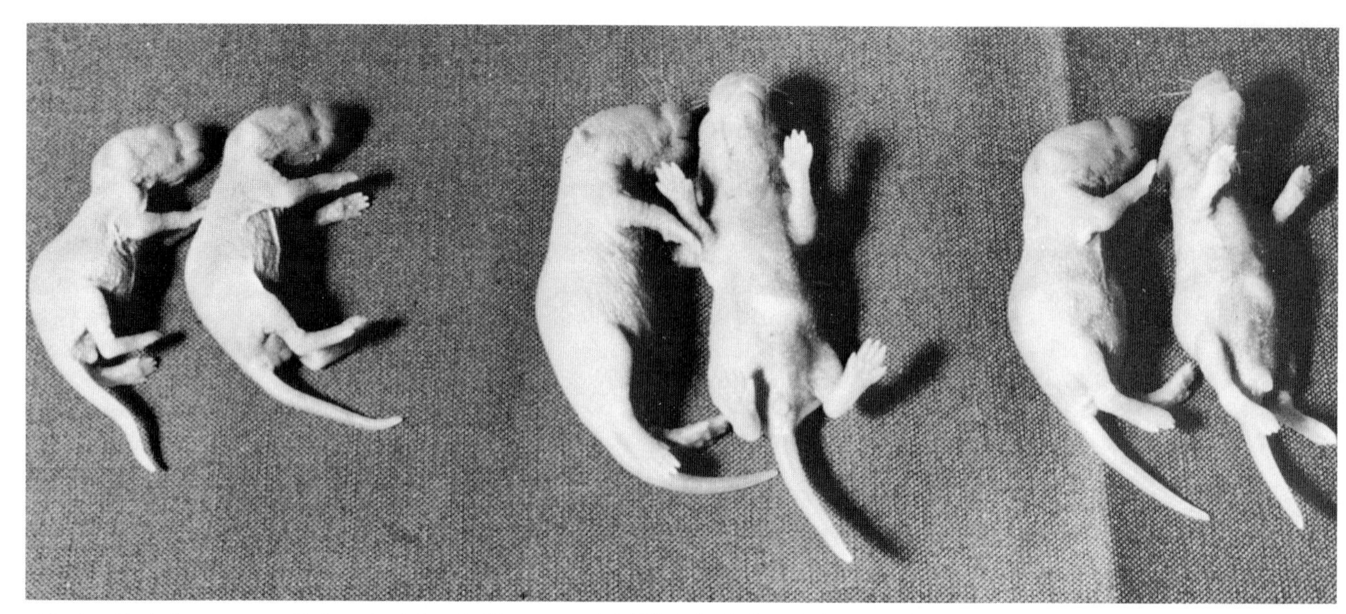

Fig. 2. Typical appearance of infant mice infected with enterotropic MHV (pair of mice at left) or rotavirus (pair at right) compared with normal mice of the same age (center pair). Note the evidence of inanition and dehydration in the MHV-infected animals, whereas those infected with rotavirus show mainly a somewhat prominent abdomen and only a slightly smaller size than the normal control mice.

weanling mice. The first signs usually appear at 7–8 days of age. In mild cases diarrhea is manifested by a minimal amount of pasty fecal material about the perineum. In severe cases the amount is copious, the entire infant becoming soiled. Rectal impaction may occur at about 12–16 days of age, and death can ensue if the impacted mass is not removed spontaneously or deliberately. Death does not seem to be the result of the virus infection per se, but rather as a consequence of protracted obstipation.

In pure EDIM virus infections, mice continue to nurse throughout their illness (in the absence of impaction). They may be slightly stunted (Fig. 2), but those with mild cases soon attain the weight of their nondiarrheal peers.

Especially when diarrhea is mild, morbidity in the natural disease may be difficult to ascertain, but one could quite correctly assume that, if diarrheal signs are observed in a few litters, it is likely that all infants in a colony will sooner or later be affected. Concerning mortality, Cheever and Mueller (1947) state that there was both 100% recovery and 100% lethality in their outbreaks. Based on present knowledge, the first is suggestive of EDIM virus infection and the latter of another disease, perhaps murine (mouse) hepatitis.

At no time do adult mice exhibit signs of illness ascribable to EDIM virus infection.

2. Pathology

a. Gross Pathology. In the experimentally induced disease, the absence of external clinical signs cannot be relied upon for a negative diagnosis (Kraft, 1957). Necropsy of each animal is essential. In this way, the appearance of the colonic contents can be observed, which in normal infant mice are burnt orange in color and semisolid but formed in consistency. The colon is not distended.

In diarrheal mice, however, even in those with no external soiling, the colonic contents are fluid or mucoid, bright lemon yellow to amber, or gray-green, with no formed feces in evidence. Gas is often seen in the colon and cecum, which are distended. The stomach, too, is distended with curdled milk (except in terminal cases with anal impaction). All other organs appear normal.

Gross change is not seen in the intestines of adults exposed perorally. It is noteworthy that in germ-free infant mice, experimental EDIM virus infection appears in the gross as it does in conventional mice (Kraft, 1966).

Findings in the spontaneous disease are identical.

b. Microscopic Pathology. Pappenheimer and Enders (1947) noted that the only changes were in the small intestine, where there was a slight increase in the cellularity of the stroma of some of the villi. Frank inflammation was lacking. They, together with Pappenheimer and Cheever (1948), described fuchsinophilic intracytoplasmic inclusions that were always related to clinical EDIM and were found only in the villous epithelial cells of the small intestine. They described degenerating, inclusion-bearing cells as well. Inclusions were seen only in the early stages of the disease. Correspondence between these inclusions and any of the ultrastructures later seen in electron micrographs, such as immature viroplasm, has not been reported.

The histologic picture has been confirmed by others (Adams and Kraft, 1967; McClure *et al.,* 1978). There is no inflammatory reaction in the intestines. Inclusions may indeed be found in enterocytes near the tips of villi, especially in the jejunum, and infrequently in the sloughed cells seen in the lumen. Epithelial cells near and at the tips of villi are frequently vacuolated.

Acres and Babiuk (1978) have pointed out that in all species studied, the rotaviruses cause diarrhea by attacking and destroying the columnar epithelium of the small intestine. Middleton (1978) considers that cell migration from the crypts is speeded in response to diarrhea and that the infected cells are relatively immature. Enzyme levels, high thymidine kinase and low sucrase, are similar in such cells to those of normal crypt cells and thus support this view.

The microscopic appearance belies the severity of the gross and clinical findings of EDIM. In this regard, D-xylose absorption has been studied in calves in which 60–90% reduction occurred during the acute illness with the calf rotavirus (Woode *et al.,* 1978). Aberrations in sodium transport have also been investigated in other animals and in humans (Middleton, 1978), but such studies have not been carried out in mice using the mouse agent.

3. Transmission

Cheever and Mueller (1947) and Kraft (1957) demonstrated that the agent is transmissible perorally. Kraft (1957) showed further that dissemination in a mouse colony is mediated principally by the airborne route.

Vertical transmission of the agent has not been demonstrated. On the other hand, the fact that viremia can occur in the adult suggests that transplacental transmission should be investigated.

Fomites, arthropods, or humans acting as passive vectors cannot be excluded as initiators of an epizootic, but they are probably not as important as the mice themselves. Since adult mice can become carriers for some days after a single exposure to the virus, these are likely to be instrumental in bringing about an epizootic.

4. Immune Response

Kraft (1961) examined the neutralization test in EDIM and found that antibodies were not always formed following infection, and when they were present, they tended to be low in titer. Sera from adult mice having had lifelong contact with the agent (in the form of consistently diarrheal litters) were more apt to neutralize the virus than those of mice exposed for the first time as adults. On the other hand, hyperimmunization of mice has resulted in the production of significant serum titers of both complement-fixing and neutralizing antibodies (Blackwell *et al.,* 1966).

Antibody response measured by other serologic tests has not been investigated for EDIM, nor have immune substances in lacteal secretions been measured directly. Suggestive of their presence, however, are data showing that infants of primiparae who themselves had been diarrheal in infancy resisted about 10 ID_{50} more of EDIM virus than did offspring of previously nondiarrheal dams (Kraft, 1961). Antirotaviral immunoglobulins have been found in both colostrum and milk in man (Yolken *et al.,* 1978c; Simhon and Mata, 1978; Thouless *et al.,* 1977a), bovines (Acres and Babiuk, 1978), and lambs (Snodgrass and Wells, 1976). Indeed, local immunity afforded by lacteal immune substances is regarded as crucial for protection against any intestinal infection in the young (Welliver and Ogra, 1978; Snodgrass and Wells, 1976).

D. Epizootiology

1. Host Range

a. Species Affected. There is no evidence that species other than the mouse (*Mus musculus*) are susceptible to EDIM virus infection. On the other hand, other rotaviruses cross species barriers. In the future, this may also prove true for the mouse agent, EDIM.

b. Age and Sex Susceptibility. As indicated, mice of all ages can be infected, but overt disease is restricted to animals up to about 12–13 days of age at the time of first exposure.

There seems to be no predilection for a particular sex.

c. Influence of Parity (Birth Order). The observation that first litters are more apt to show copious and protracted soiling than those of later parity has been reported by Cheever and Mueller (1947, 1948) and by Runner and Palm (1953). Kraft (1962b) reported similar findings in a mixed infection in a conventional colony. It is not clear if the cause for this resides in the immune response of the dam, cofactors such as opportunistic microbiota that can influence the disease picture as secondary factors, or genetics of the host and/or viral agent.

d. Strains of Mice. Cheever and Mueller (1948) studied the disease in four strains of mice: Harvard, Schwentker, CFW, and C (National Cancer Institute). Using percentage of animals weaned as the endpoint, the CFW strain was the most susceptible.

2. Prevalence and Distribution

EDIM Virus is probably worldwide in distribution. Whether it occurs in wild mouse populations is unknown. Its prevalence is difficult to estimate, since serologic tests have not been readily available, and it is not customary to sacrifice animals for the purpose of examining the appearance of their intestinal tract or for electron microscopic visualization of fecal contents.

3. Latent and Chronic Infection

As has been shown experimentally, latent carriers can exist. The carrier rate in a diarrheal colony is not known, nor is the frequency of viral shedding under colony conditions.

4. Seasonal Periodicity

Cheever and Mueller (1948) examined seasonal variations in the weaning percentage in their mouse strains and found that there was a significant effect in only the CFW mice. They experienced the lowest weaning rate in the late fall and winter. Runner and Palm (1953), studying C3H mice, indicated that there was a higher incidence of diarrhea in December/January than in September/November, although their figures are not vastly different: 59% and 44%, respectively.

E. Diagnosis

1. Serology

A number of serologic tests have been developed for the rotaviruses. The references which follow are cited as examples and are not intended to be all-inclusive. Furthermore, the techniques have been applied mainly to rotaviruses other than EDIM virus: neutralization (Kraft, 1961; Blackwell *et al.*, 1966), complement fixation (Wilsnack *et al.*, 1969; Kapikian *et al.*, 1976; Thouless *et al.*, 1977b), direct immunofluorescent staining or precipitin (Wilsnack *et al.*, 1969; Spence *et al.*, 1975; Foster *et al.*, 1975; Peterson *et al.*, 1976), immune electron microscopy (Kapikian *et al.*, 1974; Bridger and Woode, 1975), immunoelectroosmophoresis (Tufvesson and Johnsson, 1976; Middleton *et al.*, 1976), enzyme-linked immunosorbent assay (ELISA) (Scherrer and Bernard, 1977; Ellens *et al.*, 1978; Yolken *et al.*, 1978a,b,c), radioimmunoassay (Acres and Babiuk, 1978; Kalica *et al.*, 1977; Middleton *et al.*, 1977), immunodiffusion (Woode *et al.*, 1976), hemagglutination inhibition (Fauvel *et al.*, 1978), enzyme-linked fluorescence assay (ELISA) (Yolken and Stopa, 1979), an unlabeled soluble enzyme peroxidase–antiperoxidase method (Graham and Estes, 1979), plaque reduction test (Estes and Graham, 1980), serologic trapping on antibody-coated electron microscope grids (Nicolaieff *et al.*, 1980), a solid phase system (SPACE, solid phase aggregation of coupled erythrocytes) for detection of rotaviruses in feces (Bradburne *et al.*, 1979), and immune electron microscopy with serum in agar diffusion (Lamontagne *et al.*, 1980).

Large-scale serologic surveys by any of these techniques for detection of EDIM virus infection have not appeared in the literature. Impetus for developing a routine serologic test for EDIM may be afforded by the report of Ghose *et al.* (1978) that the ELISA technique has proved more useful than the complement fixation test for extensive epidemiologic studies of human rotavirus infection.

More recently, Sheridan and Aurelian (1981) have described an ELISA test for EDIM virus which should prove beneficial for both practical (serologic) purposes and for investigations of the antigenic structure of the virion.

2. Virus Isolation and Visualization

In the absence of a reliable tissue culture system, EDIM virus isolation is generally impractical. Bacteria-free filtrates of intestinal suspensions can, of course, be given to diarrhea-free animals (gnotobiotes or axenics), but this is expensive and inefficient. On the other hand, such filtrates may be concentrated by ultracentrifugation and examined in the electron microscope for characteristic virus particles (Bryden and Davies, 1975; Flewett, 1978). The particles may also be identified by immune electron microscopy.

3. Necropsy of Mice; Sentinel Animals

A practical approach to a presumptive diagnosis would be to kill selected animals in order to examine the appearance of the colonic contents. The inclusion of sentinel dams with litters in a breeding colony should be considered. These might be sacrificed at intervals to determine the presence of EDIM in the colony. Together with the clinical history of the mouse colony, this practice may provide a fairly reliable, although not pathognomonic, indication of the presence of EDIM. Histopathologic examination could also be of value.

F. Control and Prevention

Based on experimental results, Kraft *et al.* (1964) proposed the use of air-filter devices, essentially dust caps for each cage, for the practical control of airborne transmission of EDIM in a commercial mouse colony. It was subsequently shown (Kraft, 1966) that 43% of first litters and 79% of all other litter parities were weaned from cages without filters, whereas in cages provided with filters, weaning percentages were 96 and 99%, respectively, during the same observation period. It should be pointed out that both EDIM and mouse hepatitis virus (LIVIM) were present simultaneously in that colony. Although the filter devices did not eliminate the disease(s), they probably reduced the pathogenic microbial load in the immediate environment of the susceptible animals, but the precise mechanism by which this control method succeeds to the extent that it does is obscure.

Since that time (1964), various devices based on a filter cage or filter-top design have been utilized. Some of these are described by Simmons and Brick (1970). Woods *et al.* (1974) evaluated them from the viewpoint of environmental factors within the cages. They found that dry bulb temperature differentials, comparing environments inside and outside the cage, were not significantly different (about 2°C) between filtered and unfiltered cages, but that dew point differentials were significantly greater in the filtered cages (about 5°C) than in unfiltered cages (about 3°C). However, they concluded that a suitable cage size for a particular species and number of animals could compensate for the higher wet bulb readings under filters to maintain acceptable conditions for the animals. Currently, several types of filter covers or bonnets are available commercially.

Vaccination as a method of control has not been attempted. Judging from the lack of reports to the contrary, caesarian derivation together with barrier maintenance apparently eliminates and controls the infection.

Kunstýř (1962) attempted to control EDIM by decontamination of air by the use of triethylene glycol but was unable to do so. The use of antibiotics, too, is without value, although there seems to be amelioration of signs for a short period as secondary organisms are temporarily reduced in number.

As indicated by Kraft (1966), it may be relatively easy to establish a colony of mice free from EDIM virus infection. This may be accomplished by eliminating those breeding pairs whose first litter is diarrheal. Filter devices are required for this, and they and the animals must be handled with the aid of a transfer or laminar flow hood using sterile techniques during observation and handling of the animals. The method is suitable for small colonies, but it is impractical for commerce where caesarian derivation and barrier maintenance are the methods of choice.

III. REOVIRUS 3 INFECTION (HEPATOENCEPHALOMYELITIS, ECHO 10 VIRUS INFECTION)

A. Historical Background

Reovirus 3 was first isolated from the feces of an Australian child manifesting a cough, fever, vomiting, hypertrophic tonsils, and bilateral bronchopneumonia. It was named *hepatoencephalomyelitis virus* by Stanley *et al.* (1953), who originally recovered the agent. Sabin (1959) proposed the name *reovirus* for a group of agents associated with the respiratory and enteric tracts of humans. They were found to be ether resistant, about 70 nm in size by membrane filtration but of unknown shape, and caused distinctive cytopathic effects in monkey kidney tissue cultures. One of the ECHO group of viruses, ECHO 10, now synonymous with reovirus, became a member of the new group on that basis. (ECHO is the acronym for *e*nteric *c*ytopathic *h*uman *o*rphan, agents isolated in tissue culture from asymptomatic humans—so-called viruses in search of disease.) Stanley (1961) then demonstrated that the hepatoencephalomyelitis virus was serologically identical to reovirus 3.

Additional strains have since been recovered from humans, other mammals, marsupials, birds, insects, and reptiles (for review, see Stanley, 1974), and from mollusks (Meyers, 1979; Meyers and Hirai, 1980). Reoviruses have been divided into three serotypes on the basis of hemagglutination inhibition and neutralization tests (Sabin, 1959; Rosen, 1960).

Reovirus 3 was established as an indigenous murine virus by Hartley *et al.* (1961) and by Cook (1963). Reviews concerning the biological and clinical aspects of disease caused by this agent have been prepared (Stanley, 1974, 1977).

When Gomatos *et al.* (1962) and Gomatos and Tamm (1962) discovered that reoviruses possess double-stranded RNA, a unique characteristic among viruses, molecular biologists were inspired to study them in minute detail. Currently, knowledge of reovirus replication on biochemical and biophysical planes is as extensive as that available for any other virus and is, for the most part, outside the scope of this chapter. Interested readers are therefore referred to reviews by Shatkin (1969) and Joklik (1974) for extended literature coverage and detailed discussion and to Andrewes *et al.* (1978) for a condensed overview.

B. Properties of the Virus

1. Classification

Reovirus 3 belongs to the genus *Orthoreovirus,* or *Reovirus,* in the family Reoviridae (Melnick, 1979).

2. Strains of Virus

Wild-type strains of reovirus 3 include: Dearing, the prototype strain (Sabin, 1959), isolated from a child with diarrhea; Abney (Rosen, 1960), isolated from a child with a febrile upper respiratory infection; CAN 230, from a case of Burkitt's lymphoma (Bell *et al.,* 1964); and several strains obtained from naturally infected cattle (Rosen, 1960). Mutant, temperature-sensitive (*ts*) strains have been developed in the laboratory (Fields and Joklik, 1969) and have been used for studying the synthesis of viral RNA and peptides (Cross and Fields, 1972; Fields *et al.,* 1972) as well as for examining problems of pathogenesis. A neurotropic strain was also derived from the more hepatotropic original isolate (Stanley *et al.*, 1954).

3. Physical Properties of the Virus

a. Morphology, Size, and Composition. The reovirus 3 virion (Fig. 3) has a mean diameter of about 60–76 nm and is icosahedral in shape, with 5:3:2 symmetry. Particles have a core, an inner layer or shell containing a number of capsomers 50 Å in diameter, and an outer capsid composed of 92 capsomers 90 Å in diameter and 40 Å apart (Vasquez and Tournier, 1962, 1964; Jordan and Mayor, 1962; Luftig *et al.*, 1972).

As reviewed by Joklik (1970), the genome consists of 10 discrete segments of double-stranded RNA that can be grouped into three classes on the basis of their molecular weight. The polypeptides encoded by each of the RNA segments have been characterized (McCrae and Joklik, 1978).

Genetic maps of reovirus 1, 2, and 3 have been constructed by means of polyacrylamide gel electrophoresis of genome RNAs of recombinants between the three types (Sharpe *et al.*, 1978). Further, the genome segments that contain the *ts* lesion of several mutants of reovirus 3 have been determined by similar genetic mapping procedures (Ramig *et al.*, 1978).

b. Effect of Heat. Stanley *et al.* (1953) and Cook (1963) stated that the virus survives heating for 2 hr at 56°C or 30 min at 60°C. A variant recovered from L cells by Gomatos *et al.* (1962) was found to have a half-life at 37°C of 157 min, at 45°C of 33 min, and at 56°C of 1.6 min.

Temperature-sensitive mutants have been alluded to above (Sections B,2 and B,3,a).

c. Effect of Chemicals. Stanley *et al.* (1953) found the virus to be relatively resistant to hydrogen peroxide, 1% phenol, 3% formalin, and 20% lysol but sensitive to 70% ethanol for 1 hr at room temperature. It was sensitive to 3%

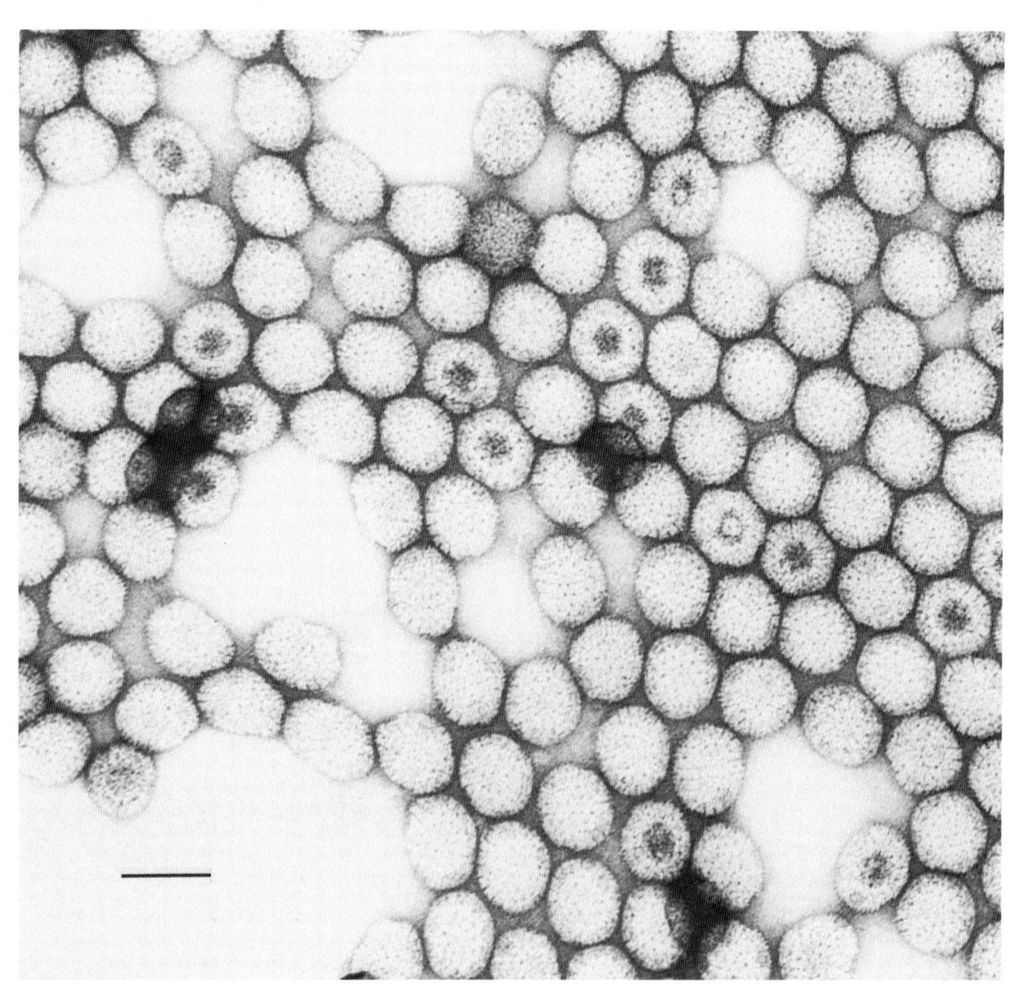

Fig. 3. Electron micrograph of purified reovirus 3. Uranyl acetate negative stain. Pseudoreplica technique. Scale bar = 100 nm. (Courtesy of Erskine Palmer, Center for Disease Control, Atlanta.)

formalin at 56°C. Ether treatment was ineffective. Rozee and Leers (1967) determined that although cholorform does not affect infectivity, it does destroy the hemagglutinin. Wallis *et al.* (1964) found that Mg^{2+} enhances the titer of reovirus at 50°C. The infective titer increased four to eight times by heating for 5–15 min in 2 *M* $MgCl_2$. Other divalent cations and NaCl were ineffective. Hemagglutinin was not affected. It is thought that the high temperature and Mg^{2+} caused activation of reovirus particles that were inactive in the original preparations.

As with rotavirus, Ward and Ashley (1978) found that reovirus was sensitive to anionic detergents in wastewater sludge, i.e., these chemicals decreased the temperature needed to inactivate the virus. Cationic detergents were more active than anionic, and nonionic detergents were inactive in decreasing reovirus thermal stability.

Mutagens (nitrous acid, nitrosoguanidine, and proflavine) have been applied to reovirus 3 (Fields and Joklik, 1969). The resulting mutants are of interest not only to the molecular biologist but to the clinical virologist as well, since some of them produce altered disease pictures (Fields and Raine, 1972). For example, when inoculated into rats, wild-type virus produced a necrotizing encephalitis, whereas a mutant gave rise to a slowly progressive communicating hydrocephalus.

The effects of enzymes are described below (see Section III,B,5).

d. Effect of pH. In phosphate–citrate buffers, Stanley *et al.* (1953) ascertained that the virus is stable between pH 2.2 and 8.0.

e. Antigenic Determinants. Sabin (1959) demonstrated that the mammalian reoviruses known at that time could be divided into three serologic groups by neutralization tests. They could also be differentiated by hemagglutination inhibition (Rosen, 1960), Hull *et al.* (1956) having discovered that ECHO 10 virus possessed a hemagglutinin for human O erythrocytes. Later, Gomatos and Tamm (1962) reported that reovirus 3, but not reovirus 1 or 2, agglutinated ox erythrocytes. The reovirus 3 hemagglutinin was inhibited by nonspecific substances such as normal mouse, rabbit, or rat serum, and by *Vibrio cholerae* filtrate.

Weiner *et al.* (1978) were able to show that the *S1* RNA segment, which is associated with type specificity, encodes the polypeptide that determines the hemagglutinating properties of the virion.

Complement-fixing antigens were prepared by Stanley *et al.* (1953, 1954) and by J. C. Parker *et al.* (1965, 1966). These are group specific.

Leers *et al.* (1968) determined that reovirions display at least one type-specific and one to two group-specific antigens when studied by immunodiffusion. With the more sensitive immunoelectrophoresis technique, however, these authors encountered two type-specific and four group-specific precipitin lines.

4. Growth in Mice

The agent is regarded as pantropic in mice. In neonates, Stanley *et al.* (1953) observed that within 48 hr after intraperitoneal inoculation, virus could be recovered from the liver. Subsequently, it was found in the central nervous system in very high titer. Both cellular and plasma viremia are present. Kundin *et al.* (1966) used immunofluorescence to locate reovirus 3 antigen in suckling mice dying after subcutaneous inoculation. Antigen was seen in the brain, spinal cord, meninges, liver, pancreas, spleen, lymph nodes, and blood vessels.

By electron microscopy, Papadimitriou (1967) traced the evolution of the virus in the central nervous system of neonatal mice inoculated intracerebrally. The earliest appearance of virions was in lymphocytes within vessels. They were then found in capillary endothelial cells, and finally, about 1 week after inoculation and preceded by the appearance of perinuclear "viroplasm," both complete and coreless virions could be seen in neurons.

More recently, Weiner *et al.* (1980c) determined that the *S1* gene segment, which encodes the viral hemagglutinin, is responsible for binding the virion to lymphocytes as well as to neuronal cells (Weiner *et al.*, 1977).

Following ingestion, reovirus 1 has been demonstrated to enter intestinal M cells, those specialized epithelial cells covering Peyer's patches, but not other intestinal epithelial cells of suckling mice. It would be of interest to know if the same holds true for reovirus 3 infection (Wolf *et al.*, 1981).

Papadimitriou (1965, 1966, 1968) also studied viral replication by electron microscopy in the mucosa of the common bile duct and in the liver.

Reovirus 3 has been reported to be oncolytic for a mouse ascites tumor by Bennette (1960), Bennette *et al.* (1967,a,b), and Nelson and Tarnowski (1960).

5. Growth in Tissue Culture

Most of the studies dealing with replication of the reoviruses have been accomplished in tissue culture, frequently in plaque assays in L-cell monolayers. Other cells that have been successfully employed are primary kidney monolayers of rhesus, patas, and capuchin monkeys, as well as those of pigs, cats, and dogs. Continuous lines, such as FL human amnion, BS-C-1, and KB, have also been used (Hsiung, 1958; Cook, 1963; McClain *et al.*, 1967; Rhim and Melnick, 1961; Harford *et al.*, 1962). Harford *et al.* (1962) described large masses of viruslike particles, often in crystalline array, in the cytoplasm of KB cells. Extracellular virus was seen in position, as if entering cells by pinocytosis or phagocytosis.

The effects of enzymes are of interest on theoretical as well as practical grounds. Pancreatin treatment was studied by Wallis *et al.* (1966), who ascertained that virus yield was greatly enhanced, perhaps because the enzyme permits rapid transmission of virus from cell to cell. Other proteolytic enzymes that enhance reovirus infectivity are ficin, papain, pepsin, pronase, and trypsin. Enhancement was also achieved with 0.1 *N* HCl. Carboxypeptidase, collagenase, and leucine aminopeptidase were ineffective.

Spendlove *et al.* (1970), Nonoyama *et al.* (1970), and Joklik (1972) studied the enhancing effect of chymotrypsin on reovirus 3. The resulting product, termed *a subviral particle* (SVP), depends upon the virus concentration, enzyme concentration, and ionic strength of the suspending medium. It may consist of virions lacking a portion of the capsid proteins, or of cores only, which are noninfectious but do possess transcriptase activity.

Borsa *et al.* (1979) have detected two modes of entry of reovirus particles into cultures of L cells. The complete particle enters almost exclusively by viropexis involving phagocytic vacuoles, whereas the intermediate subviral particles apparently gain direct access without the aid of phagocytosis.

Early steps in virus replication in L cells were described by Silverstein *et al.* (1972). Soon after entering the cell, the virions are sequestered in lysosomes. The outer layer of the viral capsomers is digested by hydrolases, but the core proteins remain untouched. The resulting structure is similar in protein composition to SVP produced *in vitro* by chymotrypsin digestion of intact virions. Within the cells, 10 hr after infection, synthesis of the progeny progresses with conservation of the parental RNA, which retains its macromolecular state throughout the replicative cycle. Rhim *et al.* (1962) reported that in monkey kidney cells, virus protein antigen was first detectable 12 hr after infection, and RNA, as determined by acridine orange staining, was increased in the cytoplasm at 16 hr, with maximum staining at 54 hr after infection. The nucleus did not seem to play a role in virus synthesis. For additional details, see Borsa *et al.* (1973a,b,c, 1974, 1979).

6. Growth in Fertile Hens' Eggs

Stanley *et al.* (1953, 1954) reported the development of pocks on the chorioallantois of 12-day-old chick embryos inoculated with infectious brain and liver. The embryos appeared unaffected, and with succeeding passages, the pocks could no longer be observed, although oral inoculation of suckling mice with chorioallantoic suspensions resulted in active disease. Essentially the same results were found following amniotic inoculation.

C. Pathogenesis

1. Clinical Disease, Morbidity, and Mortality

The experimental and natural diseases appear identical except for variations in intensity of signs, perhaps due to differences in infectious dose. Up to 16 days after intraperitoneal inoculation, the mice appear emaciated and uncoordinated. The hair is oily and matted—the so-called oily hair effect (OHE)—an effect that can be demonstrated in contact animals as well. However, in these mice it disappears as soon as the diseased animal is removed from the healthy ones. The effect was ultimately traced to a high proportion of fat in the intestinal contents (steatorrhea): 12.9% in infected animals as compared with 4.6% in normal mice. Feces may contain as much as 29% fat in infected mice (Stanley *et al.,* 1953, 1954).

Branski *et al.* (1980b) have reported that reovirus 3 inoculated intraperitoneally into adult and suckling mice resulted in no histologic or pancreatic zymogen changes in the former, whereas in the latter, amylase and lipase activities were significantly decreased and trypsin and chymotrypsin were increased. Peptidase A and B remained unchanged in infected animals. Histologic change in the suckling mice was observed 6 days after infection and was confined to a mild mononuclear infiltrate. Branski *et al.* (1980a) extended their studies further with an examination of the brush border of intestinal epithelium in suckling mice, observing that no significant changes were found in intestinal morphology or activity of enzymes tested at 3 days after infection. By the sixth day, villi were shortened and there was a mild mononuclear infiltrate in the lamina propria. Lactase and enterokinase activities were significantly decreased and alkaline phosphatase remained unchanged, whereas maltase and isoleucine activities were increased in infected mice.

Jaundice may be apparent in the ears, feet, nose, and tail. Neurologic signs are observed in many mice: incoordination, with tremors and paralysis occurring just before death. Central nervous system tropism has been linked to the viral hemagglutinin by Weiner *et al.* (1980a).

One of the earliest signs in some mice is marked abdominal distention. In mice that are recovering and in which OHE and neurologic signs had been seen, alopecia may occur. The virus has been implicated in the "runting syndrome" as well (Stanley and Leak, 1963).

Stanley *et al.* (1964) studied the disease in survivors of the acute illness between 28 and 700 days after peroral inoculation with 1 ID_{50} of virus. Ascites, rather prevalent in parenterally infected animals, was not frequent among them, but all had previously shown OHE and stunting. Neurologic signs had been most common 2–20 days after infection. In 7 of 25 animals, jaundice had lasted for 3 weeks, and alopecia over the lumbar and occipital regions occurred in the same number.

Infectious virus could not be recovered from any tissue during this "chronic" stage. Of the 1528 mice exposed in that study, 639 (42%) developed clinical signs in the acute phase. Of these, 25 survived as runts.

Onodera *et al.* (1978) studied reovirus 3 infection in α and β pancreatic cells by viral and insulin immunofluorescent techniques using SJL/J mice and the Abney strain of reovirus that had been passaged at least seven times in cultures of pancreatic B cells from the same strain of mice. They found viral antigen only in insulin-containing β cells and none in glucagon-producing α cells. The virus destroyed β cells, resulting in decreased insulin production and an altered response to the glucose tolerance test. The implication was that this could serve as an animal model for juvenile onset diabetes in man.

2. Pathology

a. Gross Pathology. The liver is enlarged, dark in color, with focal circular yellow regions of varying diameter up to 3 mm. The intestine may appear reddened and distended. In some cases, the heart shows small circular epicardial foci, and hemorrhagic areas are occasionally found in the lungs. The brain may be swollen and congested. In infants, the intestinal contents are often lemon yellow in color. Other viscera appear normal (Stanley *et al.*, 1953; Walters *et al.*, 1963). As noted above (Section III,B,1), Stanley (1974) regards the acute phase of the disease as limited to 28 days.

In the chronic phase, the mice are wasted, sometimes jaundiced, and may display OHE or areas of alopecia. The slightly enlarged liver is dark, and small subcapsular yellowish foci are seen. The peritoneum is congested; exudate is sometimes observed. Splenomegaly may be moderate late in the course of the disease (Stanley *et al.*, 1964).

b. Microscopic Pathology. Walters *et al.* (1963) described the histopathology of the acute disease following oral inoculation of large or small doses of virus in suckling mice. The same picture was encountered in both instances, except that the lesions were less intense following small doses. The liver showed minimal change at 4 days. Beneath the capsule and near the centrilobular veins were aggregates of mononuclear cells and a few segmented neutrophils. Eosinophilic degeneration of hepatocytes could be seen at this time. Throughout the next few days, the foci of hepatocyte necrosis underwent cystic change due to lysis of the eosinophilic material. Dense eosinophilic structures, similar to Councilman bodies, could be seen lying free or within macrophages. Foci of proliferating mesenchymal cells occurred in all lobular zones. They varied from small ones, the result of recent necrosis, to very large regions that coalesced with other similar foci. At the same time, about the seventh day after infection, hepatocytes enlarged to three to four times the normal size. The nuclei were not affected, although they seemed slightly less basophilic than usual. Hyperplasia of Kupffer cells and sinusoidal collections of macrophages, lymphocytes, polymorphonuclear leukocytes, and debris were evident. By the fourteenth day, many necrotic cells had lysed, and organization of cells peripheral to the necrotic foci began. Nevertheless, new necrotic regions were still developing.

In the pancreas, changes progressed from cytoplasmic vacuolization and eosinophilic degeneration beginning on the third day following inoculation. Duct cells were prominent. By 10–14 days, total necrosis of the pancreas occurred. The islets remained normal throughout.

In the salivary glands, changes similar to those seen in the pancreas took place, whereas in the heart, necrotic lesions appeared during the second week after inoculation in the papillary muscles of the left ventricle. The degeneration was eosinophilic and associated with edema and lymphocytic and macrophagic infiltrates. Repair began on the twelfth day, and by 2 weeks, all hearts appeared normal.

Some degenerative changes also took place in the skeletal musculature, and in the lungs scattered hemorrhages and pulmonary edema with a leukocytic reaction were noted. The gastrointestinal tract showed only dilatation of the central villous lymphatics and submucosal lymph channels. In some animals the spleen manifested follicular hyperplasia.

The thymus and other organs appeared normal.

In the central nervous system, neuronal degeneration began about the ninth day and was most prominent in the brain stem and cerebral hemispheres. By the tenth day, perivascular cuffing as well as neuronal satellitosis was evident. The meninges were infiltrated with round cells and netrophilic leukocytes. By the fourteenth day encephalitis was severe and widespread, with small hemorrhages occurring in necrotic regions.

In suckling rats inoculated intracerebrally with reovirus types 1, 2, or 3, viral cytoplasmic inclusions and intranuclear bodies corresponding to Cowdry type B inclusions have been observed (Margolis *et al.*, 1975). The latter were seen in cells that were free from intracytoplasmic inclusions; they were readily found in weanlings, were unassociated with inflammatory changes, and persisted for long periods without cytolysis. Electron microscopy of the choroid plexus showed no secondary virus particles present and demonstrated that the nuclear inclusions were composed of granular elements sometimes enclosed in lamellar membranes similar to "nuclear bodies." The authors believe that the Cowdry type B inclusion is identical to the nuclear body. In the chronic disease, Stanley *et al.* (1964) found that at 4 weeks after inoculation the liver still displayed small necrotic foci and areas of recent necrosis undergoing resolution. Hyaline bodies were numerous, and in the unaffected parenchyma the cells were large and often binucleate. Mitoses were rare. Recurrence of the acute picture was seen at 5 weeks: Large regions of necrosis with central lysis and suppuration were found. Thereafter the picture of bile duct occlusion developed, and the remaining parenchyma showed many mitoses. In some animals, it was evident that there were continuous cycles of necrosis and resolution throughout life.

Changes leading to acinar necrosis took place in the pancreas and salivary glands, whereas in the central nervous system lesions were variable and absent after 12 weeks. Lesions in the heart and skeletal muscles occurred about 5 weeks after infection. Interstitial pneumonia was seen from the fourth to the tenth week. The thymus remained normal, and the adrenals appeared unaffected. Some animals showed atrophy of the epidermis with loss of hair, subcutaneous edema, and infiltration with polymorphonuclear leukocytes.

In experimentally induced disease using large inocula intraperitoneally, chronic biliary obstruction may be brought about in weanlings (Phillips *et al.*, 1969).

Ultrastructurally, Papadimitriou (1965) ascertained that the virus reaches the liver in membrane-bound inclusions in the cytoplasm of leukocytes, which then undergo degeneration. The resulting debris is engulfed by Kupffer cells, and by phagocytosis the virus then enters the hepatocytes, where it replicates. In the biliary tract, Papadimitriou (1968) determined that virus invades many of the lining cells of the common bile duct, which eventually die. Bile canaliculi are dilated and their microvilli swollen. The common bile duct is dilated, and the ampullary region becomes obstructed with debris. It is this obstruction that leads to both hepatic and pancreatic dysfunction, for after studying the pancreatic lesions further, Papadimitriou and Walters (1967) concluded that, even though virions were seen in pancreatic acinar cells, the principal cause of acinar degeneration is ductal obstruction.

3. Transmission

It is clear that the agent can be transmitted by the oral route as well as by parenteral inoculation. Further, L. Parker *et al.* (1965) isolated a strain of reovirus 3 from a pool of *Aedes vigilax* and another from a pool of *Culex fatigans* in Western Australia. A third strain was isolated from members of a litter of sentinel mice that had been exposed overnight in the same region. The authors believe that mosquito transmission may account for the ubiquity of reovirus antibodies in vertebrates throughout the world. Subsequently, McCrea *et al.* (1968) reported the maintenance of reovirus 3 through a number of the developmental stages of *Culex pipiens fatigans*.

4. Immune Response

Mice respond to the natural infection with neutralizing, hemagglutination-inhibiting, and complement-fixing antibodies. As indicated earlier (Section III,B,3,*e*), precipitating antibodies can also be demonstrated.

Based on the neutralization test, Weiner and Fields (1977) determined that the *S1* genome segment is linked to type specificity, and Finberg *et al.* (1979) found that the same genome segment is also responsible for the production of cytolytic T lymphocytes after reovirus infection.

Tytell *et al.* (1967), working with reovirus 3 RNA, found that it was highly active in inducing interferon in rabbits and tissue culture, and Lai and Joklik (1973) showed that coreless virions as well as those lacking the outer capsid shell induce no interferon. The question of the role of interferon in protection of mice from either the acute or chronic infection remains an intriguing problem.

In an effort to permit a precise definition of the host cellular immune response to viral antigens, Weiner *et al.* (1980b) and Greene and Weiner (1980) have examined delayed type hypersensitivity (DTH) in mice infected with reovirus 3 following foot pad inoculation with active virus. They found that DTH could be transferred by lymph node cells, that it was mediated by T cells and was type specific. By means of recombinant viral clones of types 1, 2, and 3, they showed that the *S1* gene determined serotype specificity, whereas from adaptive transfer experiments, it was clear that certain structures of the major histocompatibility complex were needed for transfer of reactivity. Route of inoculation and viral infectivity (intact as opposed to uv-irradiated virus) determined the ability of mice to produce immunocompetent T cells that mediated the DTH reaction. DTH was not conferred by UV-irradiated virus. This tolerance was due to active suppression by means of generation of suppressor T cells. Serotype-specific tolerance by suppressor T cells was regarded as a property of the viral hemagglutinin, which also determined serotype-specific humoral (neutralizing) and cytolytic T-cell responses.

D. Epizootiology

1. Host Range

a. Species and Strains of Mice Affected. There is no evidence that any mouse strain is more or less susceptible to reovirus 3 infection than any other, provided the animals come from a colony that is free of the infection.

The host range is broad. Stanley (1974) cited at least 60 species that may be infected with reoviruses, and, as mentioned above, it is thought that the prevalence of antibodies in otherwise normal mammals is related to this fact and that mosquitoes or other insects may be operational in the spread of the infection.

b. Age and Sex Susceptibility. The acute disease affects mainly sucklings and weanlings, whereas the chronic disease is encountered in animals over 28 days of age. There is no indication that either sex is more or less susceptible than the other.

c. Influence of Parity (Birth Order). In an epizootic described by Cook (1963), 130 of 800 first litters were affected. Later-parity litters were involved hardly at all.

2. Prevalence and Distribution

In view of the absent or low complement-fixing antibody titers in the presence of significant hemagglutination-inhibiting titers that follow natural infection, prevalence estimation by immunologic means may be difficult to assess. The data cited by Parker *et al.* (1966), 82% of 34 colonies positive, and by Descoteaux *et al.* (1977), 100% of five colonies positive, may be typical incidences for conventional mouse colonies.

As already indicated, reovirus 3 infection is regarded as worldwide in distribution.

3. Latent and Chronic Infection

This aspect of reovirus 3 infection has been covered above (Section III,B,1,2*a,b*).

4. Seasonal Periodicity

There is no evidence that epizootics occur preferentially at a particular time of the year.

E. Diagnosis

1. Serology

J. C. Parker *et al.* (1965, 1966) discussed the serologic diagnosis of reovirus 3 infection, concluding that for these purposes the hemagglutination inhibition test was the most reliable. In preparing type-specific antisera for standardization and controls, Behbehani *et al.* (1966) found that the ubiquity of inhibitory substances (antibodies included) in most mammalian species precluded accurate work. They therefore used domestic geese, in which virtually no hemagglutination inhibition or neutralizing antibodies could be detected prior to immunization. In any event, for routine surveillance, the hemagglutination inhibition test is currently utilized.

2. Virus Isolation and Visualization

Comments similar to those expressed for EDIM virus recovery and visualization apply. Although it is possible to perform these procedures, they are inefficient for routine diagnostic purposes. Stanley (1977) has outlined methods for this purpose. In brief, infectious material can be inoculated into tissue cultures (primary rhesus monkey or human kidney) or newborn mice (from reovirus 3 free colonies!), or the material may be subjected to immunofluorescent methods. An immunoperoxidase method (enzyme-labeled antibody) has been employed for reovirus 1 (Ubertini *et al.,* 1971) and may find application for reovirus 3 diagnosis as well.

3. Observation of Clinical Signs; Necropsy of Animals

Although OHE may not be absolutely pathognomonic for reovirus 3 infection, it seems distinctive enough so that a provisional diagnosis may be made when it is seen. Stronger evidence is afforded if the animals are also jaundiced and wasted. Necropsy coupled with histopathologic examination is never a mistake and is to be encouraged.

4. Sentinel Mice

The placement of sentinel animals at strategic locations in an animal colony should be considered. Such animals may be regarded as expendable for sacrific, necropsy, and virus isolation as well as for the acquisition of serum for antibody determinations.

F. Control and Prevention

Cesarean derivation and barrier maintenance are believed to be suitable techniques for control and prevention of reovirus 3 infection. Although no experimental evidence has been found, it is possible that the use of filter devices in conventional colonies might also be helpful in preventing the spread of infection. In the absence of information on the vertical transmission of the agent, it is impossible to evaluate the influence of that route on successful control of the endemic disease.

Although therapy is impractical from the standpoint of controlling epizootics of reovirus infection, it is nonetheless of considerable interest that Willey and Ushijima (1980) found that thymosin given intraperitoneally to 7-day-old mice that had been neonatally infected with reovirus 2 (2 LD_{50}) significantly increased their mean survival time, provided it was administered at 2200 hr. When given at 0800 hr, significantly increased survival time was not observed, but when inoculated at 1200 hr, there was an apparent decrease in mean survival time.

IV. MURINE (MOUSE) HEPATITIS VIRUS INFECTION (MHV)

A. Historical Background

Knowledge of the murine hepatitis viruses dates from the isolation of a spontaneously occurring neurotropic murine virus, designated JHM, during the course of experiments on the epidemiology of EDIM. The most distinctive feature of JHM virus infection was widespread destruction of myelin, regardless of the parenteral route of inoculation. Giant cells were noted in a number of tissues, and the liver manifested focal regions of necrosis. Young rats, cotton rats, and hamsters were also susceptible when inoculated intracerebrally (Cheever *et al.,* 1949; Bailey *et al.,* 1949).

Following those reports, a number of agents, differing from JHM virus and from each other mainly in pathogenicity, were isolated by others under a variety of circumstances. Even today, new strains may appear, resembling or differing from those already described with respect to virulence, tissue tropism, host range, and immunogenicity. The recovery of two coronavirus strains from cases of multiple sclerosis in man, one of which was isolated in suckling mice (Burks *et al.,* 1980), draws attention to the great care that must be exercised in ascertaining that such agents do not originate in the mice, especially since, as is true in the present case, antigenic similarity with known murine coronaviruses may be demonstrated.

Because of the characteristic disease picture seen with each strain of MHV, several investigators are utilizing variants or mutants to elucidate the mechanisms of virus replication, viral nucleic acid and capsid protein structure, and virus-host interactions that result in different expressions of disease by very closely related agents (see, e.g., Haspel *et al.,* 1978; Robb and Bond, 1979; Robb *et al.,* 1979; Bond *et al.,* 1979). Bang (1978) has addressed the genetics of resistance of mice to murine coronavirus infection.

Reviews on the subject include those by Piazza (1969), McIntosh (1977), Kapikian (1975), Andrewes *et al.* (1978), Holmes (1979), Garwes (1979), and Robb and Bond (1980).

B. Properties of the Virus

1. Classification

Based on their own electron microscopic studies and on those of David-Ferreira and Manaker (1965), Becker *et al.* (1967) considered that MHV might be an "IBV-like" (infectious bronchitis viruslike) agent. Virions of similar morphology seen in electron micrographs and also ether sensitive, as is IBV, were being isolated at that time from human cases of colds (Almeida and Tyrell, 1967). The following year, the term *coronavirus* was proposed for the group (Tyrell *et al.,* 1968), and in 1975 the Coronaviridae became an official family with a single genus, *Coronavirus* (Tyrell *et al.,* 1975). The name refers to coronalike surface projections (peplomers) which, when seen in negatively stained electron micrographs, resemble the sun's corona.

Members of the group, their natural hosts, and the associated diseases are: human coronavirus (HCV) (common cold, perhaps pneumonia); infectious bronchitis virus of chickens (IBV) (infectious bronchitis, nephrosis, and uremia); transmissible gastroenteritis virus of pigs (TGEV) (gastroenteritis); hemagglutinating encephalomyelitis of pigs (HEV)* (encephalitis, vomiting, wasting); bluecomb disease virus of turkeys (TBDV) (infectious diarrhea, bluecomb disease); neonatal calf diarrhea coronavirus (NCDCV) (diarrhea); murine hepatitis virus (MHV) (hepatitis, encephalitis, wasting); rat coronavirus (RCV) (pneumonia of newborn rats); and rat sialodacryoadenitis virus (SDAV) (sialodacryoadenitis). Additional agents may be added to this group, e.g., "runde" virus (Traavik *et al.,* 1977), a coronavirus causing cardiomyopathy in rabbits (Small *et al.,* 1979).

2. Strains of Virus

As noted, many strains of MHV have been recovered from mice under various circumstances. In addition to JHM, these include: MHV1, from "white mice" (P or Parkes strain), during attempts to adapt human hepatitis virus to animals (the strain was originally isolated as a dual agent consisting of the virus and a protozoon parasite, *Eperythrozoon coccoides* (Gledhill and Andrewes, 1951); MHV2, from mice used to propagate murine leukemia virus (Nelson, 1952a,b); MHV3, also found during studies on adaptation of human hepatitis virus to mice (Dick *et al.,* 1956); MHV-B (EHF-120), from mice used for human epidemic hemorrhagic fever (HEHF) adaptation attempts (Buescher, 1952); an unnamed strain from mice undergoing murine leukemia chemotherapy trials (Braunsteiner and Friend, 1954); H747, following intracerebral inoculation of suckling mice with HEHF materials (Morris, 1959); MHV-A59, during transfer of Moloney leukemia virus in mice (Manaker *et al.,* 1961); MHV-S, from cesarean-derived mice that had been barrier-maintained before exposure to conventional mice (Rowe *et al.,* 1963); MHV-C (MHV-BALB/c), isolated during passage of spleens from leukemic mice (Nelson, 1955); four additional strains from spleens of leukemic mice or from natural outbreaks (MHV-SR1, -SR2, -SR3, -SR4) (Nelson, 1963, 1965); lethal intestinal virus of infant mice (LIVIM), from infant mice dying of a spontaneous infection (Kraft, 1962a), identified as an MHV strain by Broderson *et al.* (1976) and Hierholzer *et al.* (1979); and NuU, NuA, Nu66, from nude mice with hepatitis and wasting syndrome (Hirano *et al.,* 1975). Other isolates have also been derived from nude mice (Sebesteny and Hill, 1974; Ward *et al.,* 1977), and Fox *et al.* (1977) have described a strain that appeared during passage of an ascites myeloma cell line in BALB/c mice.

*HEV may also denote the *hepatoencephalomyelitis virus* of Stanley *et al.* (1954), which is identical to reovirus 3.

3. Physical Properties of the Virus

a. Morphology, Size, and Composition. As reviewed by McIntosh (1974), coronaviruses are pleomorphic, enveloped, and variable in size, measuring about 80–150 nm in diameter. Peplomers are 12–24 nm in length (Fig. 4).

Using various techniques, a number of workers have confirmed the diameter of MHV virions to fall within the range of the coronaviruses (Gledhill *et al.*, 1955; Miyazaki *et al.*, 1957; Kraft, 1962a; Starr *et al.*, 1960). Svoboda *et al.* (1962) further described the virion as consisting of a nucleoid separated from an outer membrane by an electron-lucent space. David-Ferreira and Manaker (1965), working with MHV-A59 in tissue culture, found that the virions had a mean diameter of 75 nm with an electron-dense inner shell, 55 nm in diameter, separated from the outer double membrane by an electron-lucent space 8 nm wide.

Hirano *et al.* (1978) studied MHV2 and determined that the virus had a bouyant density of 1.183 gm/cm^3 in sucrose, and that that fraction contained coronaviruslike particles measuring 70–130 nm in diameter. In negative contrast preparations, Davies and Macnaughton (1979) ascertained that the envelope diameter of MHV3 ranged from 76 to 121 nm (mean, 100 nm). Peplomers were 16.6–23.4 nm long. Not only was the virion significantly smaller than that of IBV, but the peplomers differed from those of both IBV and HCV 229E, being cone- rather than club-shaped.

Mallucci (1965) established that MHV3 is an RNA-containing virus with no DNA-dependent replicative phase. Structural proteins have been investigated by means of polyacrylamide gel electrophoresis. In purified preparations of JHM virus, Wege *et al.* (1979) identified six polypeptides, whereas Sturman (1977) found only four in MHV-A59. Also utilizing polyacrylamide gels, Macnaughton (1980) compared the polypeptide composition of two coronaviruses, HCV 229E and MHV3, and found similar patterns for both agents. He felt that the importance of this finding rests in the fact that not only were different hosts involved here, but different diseases as well, implying that factors other than those under study are instrumental in the outcome of a disease picture.

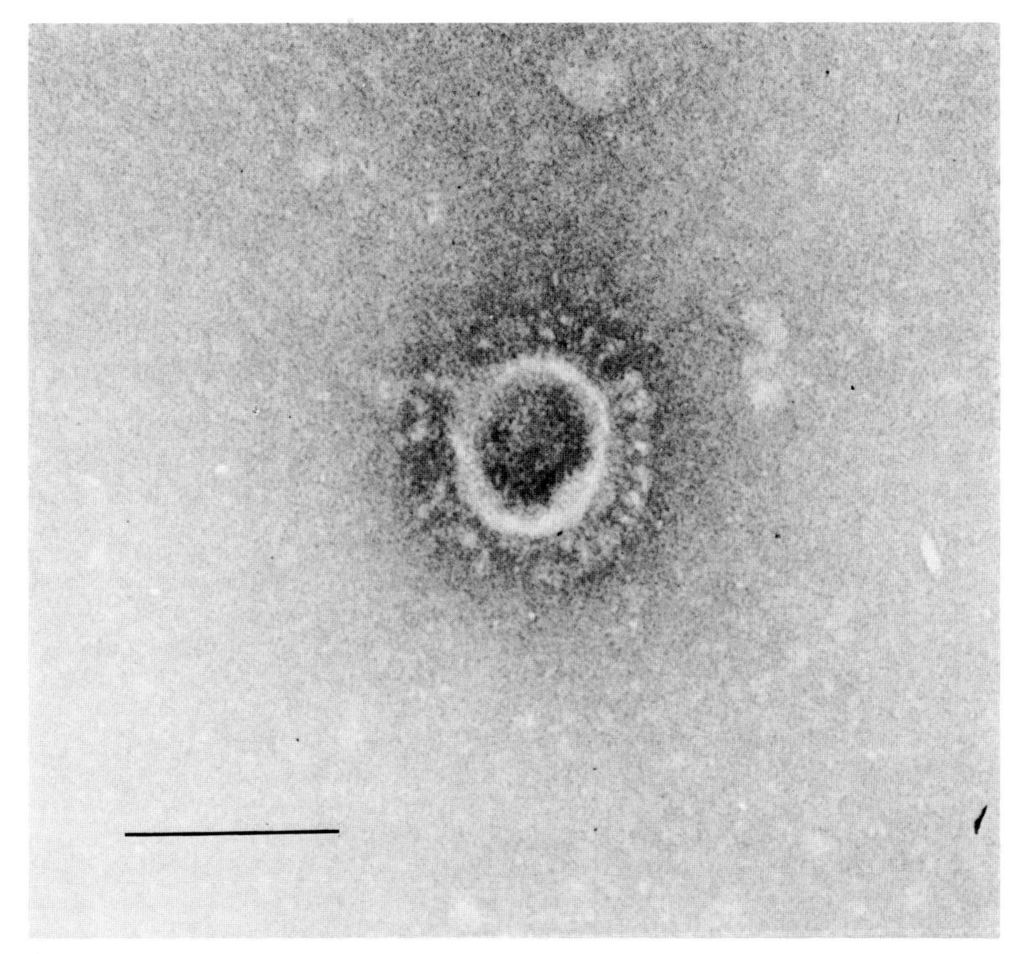

Fig. 4. Electron micrograph of an MHV particle prepared from the brain of a mouse inoculated intracerebrally. Pseudoreplica technique. Scale bar = 100 nm. (Courtesy of Harold S. Kaye, Center for Disease Control, Atlanta.)

b. Effect of Heat. In general, coronaviruses are inactivated at 56°C in 10–15 min, at 37°C in several days, and at 4°C in several months (Tyrell *et al.*, 1968; Kapikian, 1975). Wild-type MHV isolates also fall into this range of sensitivity (Cheever *et al.*, 1949; Gledhill and Andrewes, 1951; Gledhill *et al.*, 1955; Kraft, 1962a). Hirano *et al.* (1978), however, indicated that MHV2 is not completely destroyed at 56°C for 30 min and is stable at 50°C for 15 min in 1 *M* $MgCl_2$ or $MgSO_4$ but not in water. Freezing and thawing or sonication at 20 kc for 3 min does not affect the virus titer.

c. Effect of Chemicals. All coronaviruses are sensitive to ether when exposed overnight at 2°–4°C. Chloroform also destroys or reduces infectivity (McIntosh, 1974). Fifty percent glycerol inactivates MHV1 after 6 weeks at 2°C (Gledhill and Andrewes (1951). Sodium deoxycholate reduces the titer of LIVIM significantly (Kraft, 1962a), but Calisher and Rowe (1966) regard MHV virus as moderately resistant. According to Hirano *et al.* (1978), MHV2 is completely inactivated by ether, chloroform, sodium deoxycholate, and β-propiolactone, but it is completely resistant to trypsin.

Mutagenesis has been reported by means of *N*-methyl-*N′*-nitroguanidine or 5-fluorouridine (Robb *et al.*, 1979) and by 5-azacytidine or 5-fluorouracil (Haspel *et al.*, 1978).

d. Effect of pH. The pH stability of all known murine hepatitis viruses has not been reported. For coronaviruses in general, acid sensitivity is regarded as variable (McIntosh, 1974). Hirano *et al.* (1978) found that MHV2 is stable between pH 3 and 9 at 37°C for 30 min.

e. Antigenic Determinants. Calisher and Rowe (1966) developed a soluble complement-fixing antigen for MHV. All strains tested, JHM, MHV1, MHV3, A59, and H747, as well as a number of field isolates, shared both a common complement-fixing and a neutralizing antigen in their tests. No two strains from different sources appeared to be identical, however.

Childs *et al.* (1980) examined the serologic relationships among five MHV strains by a plaque reduction test and confirmed some serologic relatedness among all of them, although two distinct groups emerged: one, MHV4 (JHM), and two, MHV1, MHV-S, MHV-A59, and MHV2.

Although they have been sought, hemagglutinins have not been demonstrated for MHV strains (see, e.g., Hirano *et al.*, 1978; Kraft, 1962a; Bradburne, 1970; Miyazaki *et al.*, 1957), but two human coronavirus strains do agglutinate human 0 erythrocytes (Kaye and Dowdle, 1969).

Precipitating antigens have been described. Bradburne (1970) found two precipitin arcs when MHV3 reacted with homologous hyperimmune serum.

4. Relationship to Other Viruses

Cheever *et al.* (1949) found JHM virus to be unrelated to other neurotropic viruses, including GD VII, pseudorabies, Lansing poliomyelitis, and Mengo virus. Kraft (1962b) found no relationship between LIVIM, EDIM, and reovirus 3.

With regard to other coronaviruses, the picture is somewhat different, for as a group, coronaviruses display complex serologic variability (Bradburne, 1970; McIntosh *et al.*, 1969). MHV is serologically closely related to RCV and SDAV in complement fixation tests and distantly related to RCV in cross-neutralization tests (Parker *et al.*, 1970; Bhatt *et al.*, 1972). Several strains of MHV are closely related to human coronaviruses OC 38 and OC 43 (McIntosh *et al.*, 1967), and MHV3 is related to HCV-229E (Bradburne, 1970). Antibody to MHV strains commonly found in human sera is probably present because of endemic human infection with related coronaviruses (Hartley *et al.*, 1964).

5. Growth in Mice

Electron micrographs and studies using fluorochrome stains indicate that coronaviruses develop exclusively in the cytoplasm of infected cells, that the virions collect in cytoplasmic vesicles of diverse size, that particles may also be seen in the matrix outside of the endoplasmic reticulum as well as in the Golgi apparatus, and that they are not observed in the nucleus. In the main, replication involves budding into cytoplasmic cisternae, but tubular structures have also been seen within the cytoplasm during virus formation (Ruebner *et al.*, 1967; Starr *et al.*, 1960; Watanabe, 1969a,b).

Wilsnack (1971), confirming the work of Boss and Jones (1963), elicited immunofluorescent antigen staining in sinusoidal lining cells in necrotic liver foci of weanling mice within 24 hr after intraperitoneal inoculation of the A59 strain. Stainable antigen in intestinal impression smears of mice infected by cage contact was also demonstrated.

Piazza *et al.* (1967) examined the fate of MHV3 after intravenous inoculation. The agent was not demonstrable between 40 min and 3.5 hr, when it appeared first in the spleen, then in the liver (4 hr) and blood (4.5 hr). At 5 hr it was recoverable from brain and kidney. High titers were then reached in all organs. Barinsky and Dementiev (1968) studied bone marrow involvement in MHV3 infection and found that the marrow yielded high virus titers long before the appearance of clinical signs, persisting almost until death. Chromosomal aberrations were frequently noted in the infected marrow.

Watanabe (1969a,b) described MHV2 replication in the liver by means of electron micrographs. Budding occurred in hepatocytes but never in Kupffer cells, although virions could be seen in these. In hepatocytes, particles were undetectable 64 hr after intraperitoneal inoculation, but they were subsequently observed 8 hr later. They were most numerous within perisinusoidal spaces. Viral multiplication was demonstrated in the pancreas by Fujiwara *et al.* (1975). Virions were present in both the matrix and the endoplasmic cisternae of secretory pancreatic cells.

Concerning the neurotropic variants of MHV, Lampert *et al.* (1973) saw particles of the JHM strain in cells identified as oligodendrocytes in the brain stem and spinal cord in weanling mice, and Weiner (1973) demonstrated immunofluorescent staining in cells of the white matter (glia) of the spinal cord.

Taguchi *et al.* (1979b) have shown that the low-virulence strain, MHV-S, multiplies first in the nasal mucosa after intranasal inoculation, antigen being found in both the neurosensory olfactory cells and supporting (sustentacular) cells. Thereafter, the cells of the olfactory bulb and other brain regions are affected.

6. Growth in Tissue Culture

A number of cell systems have been successfully employed for *in vitro* growth of mouse hepatitis viruses: MHV-C in mouse embryo explants (Mosley, 1961); MHV1 in newborn mouse kidney explants (Starr and Pollard, 1959); MHV-S in mouse embryo explants (Gompels, 1953) and in liver (Gallily *et al.,* 1964); MHV-B in liver cell monolayers (Paradisi and Piccinino, 1968); MHV3 in liver explants (Vainio, 1961); MHV-B in liver cells (Miyazaki *et al.,* 1957); MHV2 and MHV3 in DBT cells (Hirano *et al.,* 1978; Takayama and Kirn, 1978); and various strains in NCTC 1469 cells (David-Ferreira and Manaker, 1965; Wilsnack *et al.,* 1971; Hartley and Rowe, 1963).

Mallucci (1965), Seamer (1965), and Lewis and Starr (1972) described syncytium formation by MHV in mouse macrophage cultures, and a plaque assay for MHV2 in primary peritoneal macrophage cultures was described by Shif and Bang (1966). Laufs (1967) also described multinucleated giant cells with as many as 200 nuclei per cell in macrophage cultures infected with MHV3. Using autoradiography, he ascertained that there was no DNA synthesis in them and that they originated from cell fusion.

Macrophages derived from either liver or peritoneal washings are of enormous interest for the question of host cell–virus interactions. Bang and Warwick (1959, 1960) showed that virulent MHV2 had a selective destructive effect on macrophages cultured from the liver of newborn mice without affecting fibroblasts or epithelial cells. They concluded that such tissue susceptibility is a property of the reticuloendothelial system. Macrophage cultures from resistant mouse strains were not destroyed, whereas those from susceptible strains were killed. Those obtained from F_1 animals of susceptible × resistant crosses demonstrated that *in vitro* as well as *in vivo* susceptibility is inherited and that genetic segregation of susceptibility and resistance occurs in the F_2 generation and in backcrosses. They proposed that susceptibility was a dominant trait and unifactorial. Shif and Bang (1970) enlarged on those findings, showing that adsorption of virus to cells was not impaired.

Gallily *et al.* (1967) determined that susceptibility of macrophages in culture reflects alterations in the host. For example, C3H infant mice are susceptible to and die from infection with MHV2, whereas weanlings and adults are more resistant. Macrophages from the mice mirror these changes in resistance as the animals age. Kantoch *et al.* (1963) determined that temporary susceptibility could be induced in resistant cells in culture if they were exposed to homogenates of susceptible cells, and Gallily *et al.* (1964) showed that macrophages from genetically resistant mice treated with cortisone to enhance susceptibility behave in culture as if they were from susceptible animals.

Macrophages from mice susceptible to MHV2 virus can be converted to resistance by the intraperitoneal inoculation of concanavalin A in the donor mice. This enhanced resistance is also expressed *in vivo* (Weiser and Bang, 1977).

7. Growth in Fertile Hens' Eggs

Cheever *et al.* (1949), Nelson (1952b), and Kraft (1962a) all reported failure to propagate JHM, MHV2, and LIVIM, respectively, in embryonated hens' eggs. Evidence of positive results concerning other strains has not been found in the literature.

C. Pathogenesis

1. Clinical Signs, Morbidity, and Mortality

The acute and chronic clinical signs of MHV infection in sucklings, weanlings, and adults vary with the virus strain and its tissue tropism, the mouse strain, the age of the mice, and the presence or absence of enhancing or inhibitory factors in the mice or their environment. Signs of the acute disease are distinct but not pathognomonic: ruffled hair, depression or lassitude, inanition, dehydration, weight loss, huddling with cagemates, muscle tremors, deeply colored urine, and reluctance to move when prodded. These signs are seen in various combinations. In older animals, ascites and wasting tend to occur. In infections with neurotropic variants, such as JHM, the principal sign in weanlings and adults is a flaccid paralysis of the hindlimbs that may be preceded by ruffled hair and depression. Conjunctivitis may occur. Convulsions and hyperirritability are infrequent, as are tremors. The righting ability may be compromised, and circling is occasionally seen.

In suckling mice, rapid wasting with or without neurologic

signs may take place, accompanied in some cases by diarrhea, inanition, and dehydration (Fig. 2).

Mortality and morbidity are variable, ranging between almost 0 and 100% depending on factors like those affecting clinical signs.

Of importance to users and breeders of mice alike is the fact that a number of agents and procedures are known to modify the reactivity (and therefore the clinical signs) of mice to both spontaneous and experimental infection. Examples of these, together with pertinent references, are presented in Table I.

LePrévost *et al.* (1975a,b) have taken the view that there are three types of sensitivity to MHV3 infection in mice: resistance, full susceptibility, and semisusceptibility. These are reflected in the susceptibility of their macrophages (Virelizier and Allison, 1976). Further, they consider two distinct phases of the disease, acute and chronic, as virtually distinct entities. C3H mice, regarded by others as resistant, are considered by them to be semisusceptible, since approximately 50% of adult infected animals resist the acute disease, and young animals remain susceptible up to 8 weeks of age. In most of the survivors of the acute disease, chronic illness with wasting and occasional paralysis results. In this period, virus can regularly be recovered from brain, liver, spleen, and lymph nodes.

Table I

Some Modifying Factors in MHV Infection

Factor	Reference
MHV enhanced by:	
Eperythrozoon coccoides	Gledhill and Andrewes (1951)
Eperythrozoon coccoides	Nelson (1952b)
Eperythrozoon coccoides	Niven *et al.* (1952)
Eperythrozoon coccoides	Lavelle and Bang (1973)
Moloney and Friend leukemia	Gledhill (1961)
Urethane, methylformamide	Braunsteiner and Friend (1954)
X irradiation	Vella and Starr (1965)
X irradiation	Dupuy *et al.* (1975)
Cortisone	Lavelle and Starr (1969)
Cortisone	Ruebner *et al.* (1967)
Cortisone	Datta and Isselbacher (1969)
Cyclophosphamide	Willenborg *et al.* (1973)
Cyclophosphamide	Weiner (1973)
Neonatal thymectomy	East *et al.* (1963)
Neonatal thymectomy	Sheets *et al.* (1978)
Neonatal thymectomy	Dupuy *et al.* (1975)
Antilymphocyte serum	Dupuy *et al.*, (1975)
K virus (papovavirus)	Tisdale (1963)
Splenectomy	Stone *et al.* (1967)
MHV suppressed by:	
Triolein (triglyceryl oleate)	Lavelle and Starr (1969)
S. typhosa endotoxin	Vella and Starr (1965)
MHV infection enhances:	
Response of *nu/nu* mice to sheep erythrocytes	Tamura and Fujiwara (1979)

DBA/2 mice, on the other hand, are regarded as fully susceptible since deaths begin 4–6 days after infection, even when the mice are 90 days old, whereas the A/J strain is resistant, being susceptible to the acute disease only up to 3–4 weeks of age. Although C3H mice are partially susceptible to the acute phase of the disease, they are fully susceptible to the chronic stage.

Virelizier *et al.* (1975) have described the neuropathologic effects of chronic MHV3 infection in C3H mice. Chronic illness lasted up to 12 months, with eventual paralysis of the limbs. In contrast to the susceptible strains, C57BL and DBA2, which die within 1 week after infection, most of the C3H mice appeared normal until 2–12 weeks after inoculation. Ruffled hair and loss of condition and activity then appeared, followed by neurologic signs: incoordination and paresis of one or more limbs, especially the hindlimbs. Circling when suspended by the tail was also seen. The signs were somewhat variable from one animal to the next and tended to be progressively more severe up to the time of death, 2–12 months after infection.

In neonatally thymectomized mice, East *et al.* (1963) described a wasting syndrome: gradual deterioration in physical condition, progressive loss of weight, and a ''curious, high-stepping but coordinated gait'' [p. 1069]. Depletion of lymphocytes was found in the lymph nodes, spleen, and peripheral blood.

In nude (*nu/nu*) mice, MHV takes on special significance. Indeed, perhaps the best description of the clinical signs may be found in the original report describing this mutant mouse (Flanagan, 1966), published before *runt disease,* as the wasting syndrome was called, came to be recognized as something other than a genetic effect.

MHV-infected nude mice lose weight slowly or rapidly. They move stiffly with a stilted gait, and their faces assume a pointed, anxious appearance. Partial paralysis may develop first in the hindlimbs and then in the forelimbs, resulting in almost total immobility. *Nu/+* heterozygotes are not affected in this way. Flanagan (1966) found that at weaning, the nude animals were much smaller than controls (heterozygotes), that 55% died within 2 weeks of birth, and that 100% were dead by 25 weeks of age, whereas only 6% of controls died in the same period.

It was not until Sebesteny and Hill (1974) described a spontaneous MHV infection in nude mice that the agent was considered as a cause of runt disease. Hirano *et al.* (1975) confirmed these findings when he isolated the NuU strain from nude mice, and Tamura *et al.* (1977) then established persistent infection with wasting in *nu/nu* mice after intraperitoneal inoculation of MHV-NuU (Tamura *et al.,* 1976) as further proof.

Tamura *et al.* (1980) demonstrated increased phagocytic activity of macrophages in MHV-NuU infected *nu/nu* mice. UV-inactivated virus did not elicit the effect. The authors believe that fixed macrophages in athymic mice may be activated, as was indicated by Nickol and Bonventre (1977) and by Cheers and Waller (1975) in certain bacterial infections in nude mice. Tamura *et al.* (1979) further found that inoculation of *nu/nu* mice with silica, which is toxic for macrophages, simultaneously with MHV-NuU caused death within 2 weeks, whereas those mice receiving no silica survived longer than 3 weeks. Silica given 4 days after the virus had no effect, indicating the importance of the intact macrophage in early resistance to MHV in these athymic mice.

2. Pathology

a. Gross Pathology. In weanling and adult mice, the most striking and constant feature, regardless of virus strain, appears in the liver, which shows sparse to extensive mottling during the acute disease. The incubation period in susceptible mice is generally short. For example, liver lesions occur as early as 41 hr after infection with MHV3 (Jones and Cohen, 1962).

Other gross findings include occasional ascites and exudate on the surface of the liver in MHV3-infected weanlings, with ascites a more constant finding in adults (Dick *et al.*, 1956). Dark brown material suggesting decomposed blood can be seen in the intestine in JHM infection (Bailey *et al.*, 1949) and in MHV2 infection (Nelson, 1952b). Kidney pallor, deep yellow urine, occasional jaundice, and hemorrhagic peritoneal exudate are sometimes evident.

In suckling mice, the liver may not be grossly affected, but Rowe *et al.* (1963) indicated that the only gross findings in MHV-S infection were "small yellow or whitish spots on the liver" [p. 162]. Kraft (1962a) and Biggers *et al.* (1964) found no gross abnormalities in the liver even after repeated oral passage of LIVIM virus, nor did Broderson *et al.* (1976) mention them in their report. Dick *et al.* (1956), on the other hand, found that sucklings showed extensive liver change within 72 hr of inoculation with MHV3, occasionally accompanied by jaundice and peritoneal hemorrhagic exudate. Ishida *et al.* (1978), too, saw numerous necrotic foci in the liver of nurslings dying of MHV infection during an epizootic.

It is probably indicative of the protean nature of MHV infection that neither Rowe *et al.* (1963) nor Dick *et al.* (1956) mentioned the gross appearance of the gastrointestinal tract in infected infant mice, whereas Kraft (1962a), Ishida *et al.* (1978), Broderson *et al.* (1976), and Hierholzer *et al.* (1979) encountered shrunken, empty stomachs, intestines filled with watery to mucoid yellowish, sometimes gaseous, contents, and an intestinal wall so thin that it sometimes ruptured during life.

b. Microscopic Pathology. In the liver, focal necrosis is found in a high proportion of mice during the acute phase of the disease. It may be assumed that lesions are present in all animals but that sectioning techniques cause some of them to be missed. The periphery of the lesions displays a sharp boundary between normal hepatocytes and cells undergoing hyaline degeneration. The hyaline material is intensely eosinophilic and has been compared to Councilman bodies in human hepatitis (Svoboda *et al.*, 1962) (Fig. 5). Early in the course of infection, basophilic cytoplasmic inclusions may be seen. Hepatocyte nuclei degenerate after the cytoplasm. In the center of the necrotic focus, hepatocytes have disappeared, leaving only collapsed reticulum and fat-laden phagocytes. Polymorphonuclear leukocytes appear early but then die and are phagocytosed along with the necrotic debris. Multinucleated giant cells are seen, and in animals surviving for some time, necrotic cells may become calcified. Aside from the focal lesions, the remaining liver parenchyma appears unaffected.

Slight variations in liver pathology may be seen with different routes of infection, virus strains, age of animals, etc. (see, e.g., Ruebner and Bramhall, 1960; Gledhill *et al.*, 1952; Nelson, 1953).

In sucklings infected with the LIVIM strain, the liver is minimally affected. Necrotic foci are very small and inconspicuous (D. C. Biggers, personal communication), whereas in sucklings infected with other strains, necrotic foci are obvious and distinct and not difficult to locate in histologic preparations.

Liver regeneration may take place as early as 10–14 days after infection (Ruebner and Bramhall, 1960), ranging from complete healing to chronic scarring with intermediate gradations.

Kupffer cells were examined by Ruebner and Miyai (1962) and Ruebner *et al.* (1967). They may undergo nuclear pyknosis and karyorrhexis 24 hr after intravenous inoculation of MHV3.

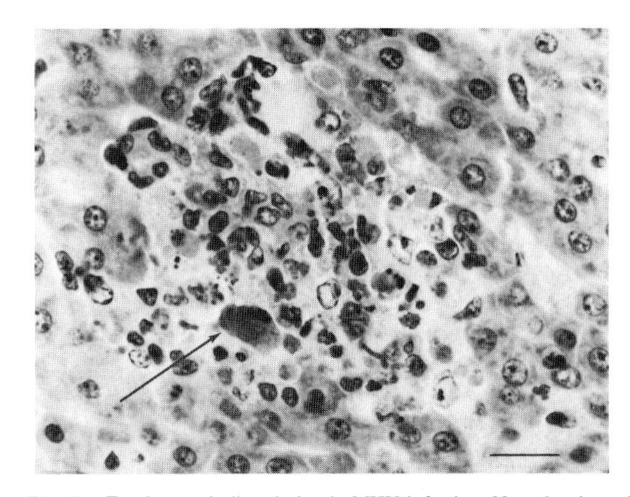

Fig. 5. Focal necrotic liver lesion in MHV infection. Note the sharp demarcation between normal hepatocytes and the periphery of the lesion. Intensely eosinophilic material (Councilman body) can be seen within a hepatocyte (arrow). Scale bar = 20 μm.

On the ultrastructural level, Svoboda *et al.* (1962) found virions of MHV2 or MHV3 to be more numerous in the space of Disse (between hepatocytes and endothelial lining cells) than in the neighboring cells themselves.

In infection with neurotropic variants, such as JHM, the principal lesions appear in the central nervous system (Bailey *et al.*, 1949). Meningitis is present but varies in degree and location. In the brain, lesions may be found in all regions, but the hippocampus and its connections, the olfactory lobes, the periependymal tissues, and the brain stem seem to be affected most often. Necrotizing lesions predominate in the olfactory lobes and hippocampal regions, whereas demyelination is the major change in the brain stem. Some exudate may be found around blood vessels associated with lesions, and at about 5 days after infection, proliferating pericytes and scant lymphocytic cuffing can be seen. Peripheral nerves show no change.

In sucklings, JHM virus produces extensive lesions in the brain and cord at 6–8 days. Meningitis is present, and large regions of necrosis with many giant cells occur throughout the brain. In the cord, the lesions consist of spongy necrosis of the central gray matter. Ganglion cells appear unaffected.

Powell and Lampert (1975) described the ultrastructural changes taking place in the oligodendroglia. Infected cells undergo hypertrophy before degenerating. The hypertrophic cells show abundant microtubules, filaments, and mitochondria, aggregates of electron-dense particles, and numerous unusual plasma membrane connections to myelin lamellae. Vacuolar and hydropic changes are prominent in the degenerating cells, and postinfection recurrence of the demyelination develops subsequent to the acute disease. The question of recurrent demyelination has been addressed by Herndon *et al.* (1975), who found evidence of renewed demyelination in mice 16 months after infection, at a time when the acute lesions had resolved into small foci of fibrillary gliosis with an increased size and number of astrocytic processes. The importance of this finding for investigations into the cause of multiple sclerosis and other demyelinating diseases in man is obvious and has been addressed by, among others, Lucas *et al.* (1977) and Lampert (1978).

In sucklings infected with enterotropic strains (Kraft, 1962a; Broderson *et al.*, 1976; Ishida *et al.*, 1978; Ishida and Fujiwara, 1979; Rowe *et al.*, 1963; Hierholzer *et al.*, 1979), the entire intestinal tract may be involved. Multinucleated giant cells, regarded as altered or fused enterocytes, may be numerous. Some are pinched off and appear free in the lumen. Villi are stunted or disappear altogether (Fig. 6). There is no inflammatory reaction early in the infection. Basophilic intracytoplasmic material may also be seen at this time. Eosinophilic hyaline degeneration of enterocytes and giant cells is noted.

Necrosis was not seen in the spleen in JHM infection (Bailey *et al.*, 1949), but others have observed it with hepatotropic strains (Dick *et al.*, 1956; Ruebner and Bramhall, 1960; Hirano and Ruebner, 1966), and Biggart and Ruebner (1970) attribute the change to virus replication in lymphocytes.

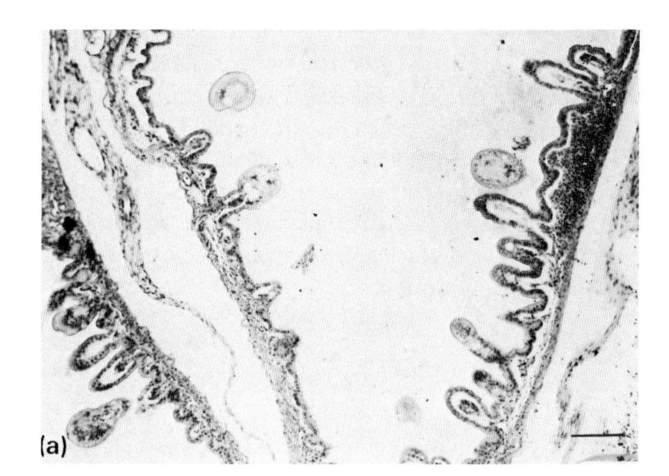

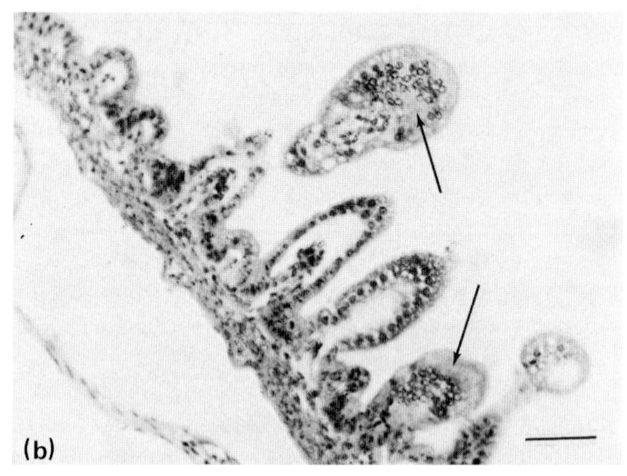

Fig. 6. Small intestine in enterotropic MHV infection in an infant mouse. (a) The remaining villi are stunted. Balloon-like structures are multinucleated enterocytes that develop at the tips of some villi. Scale bar = 100 μm. (b) Higher magnification showing multinucleated enterocytes (arrows). Scale bar = 50 μm.

In other organs, minute superficial necrotic foci may be found in the stomach. No changes are seen in the heart, lungs, pancreas, kidney, adrenals, voluntary muscle, femoral or vertebral bone marrow, or pituitary gland, although virus may be isolated from some of those organs. Occasional giant cells are found in peripancreatic lymph nodes and in Peyer's patches.

Flanagan (1966) described the histopathology of the hepatitis seen in the original nude mice. With time, progressively larger regions of liver tissue became necrotic until the foci joined, sometimes consuming an entire lobe. Giant cells, called *balloon cells*† by Flanagan, contained basophilic bodies. Necrotic foci were infiltrated with neutrophilic leukocytes and phagocytes. Even where necrosis reached an advanced stage, a few

†Kraft (1962b) independently had applied the same name to giant cells in the intestinal epithelium of LIVIM-infected infant mice.

islands of hepatocytes, markedly hypertrophied, remained. Changes in other organs were not described.

The liver lesions observed in other nude mice infected with MHV were similar. Sebesteny and Hill (1974) noted central nervous system lesions in their nude mice. Ward *et al.* (1977) also encountered central nervous system lesions in addition to vascular changes, giant cell peritonitis, ascites, and giant cells in the villous epithelium of the intestines.

3. Transmission

Since virus can be transmitted perorally and intranasally, it may be assumed that these are the principal routes of natural infection. Transmission may be mediated by both the airborne and contact modes. Feces, nasopharyngeal exudates, and perhaps urine would be sources of infection.

Evidence of vertical transmission is at hand, but reports are conflicting. Piccinino *et al.* (1966) stated that, although virus (MHV3) antigen can be stained in the placenta by immunofluorescence, infectious virus could not be isolated from fetuses or newborn animals. On the other hand, Katami *et al.* (1978), using JHM and slightly different techniques and timing of virus inoculation in pregnant females, demonstrated virus-specific antigen in the placentas, we well as the visceral yolk sac and fetal livers. Furthermore, they visualized characteristic coronavirus particles in these locations by electron microscopy. Histopathologically, degenerative and necrotic changes were seen in all those tissues and also in the fetal bone marrow. In commenting on these findings, it is probably important to consider that different virus strains were utilized, and in addition, Piccinino *et al.* (1966) inoculated the females late in pregnancy, whereas Katami *et al.* (1978) took advantage of the early and midgestational periods. It will be of great importance for the epizootiology of MHV to ascertain if strains other than JHM are capable of being transmitted vertically.

Mosquitoes (*A. aegypti* and *C. pipiens fatigans*) have been incriminated as potential mechanical carriers of MHV (Ishii *et al.*, 1974).

4. Immune Response

The dynamics of the rise in neutralizing and complement-fixing antibodies following natural infection are not readily apparent from the literature. In general, however, titers appear to be low or absent in infected colonies. Thus, Fujiwara *et al.* (1976b), for example, described an epizootic that suddenly appeared in a colony in which retired breeders had been monitored for complement-fixing antibodies with consistently negative results. After the epizootic was underway, only mice 6 weeks of age or older demonstrated antibodies, but the precent positive was low.

In order to increase the sensitivity of the complement fixation test, Fujiwara (1971) had devised a method utilizing a booster dose of inactivated MHV polyvalent antigen to elicit an anamnestic response. Without the booster, for example, only 2.4% of 40 retired breeder females had demonstrable antibodies, whereas 32.4% of 41 were positive after the booster dose was given. On the other hand, 4-week-old male weanlings, with or without booster, showed no antibody in the same colony. In a conventional colony of DDD mice, furthermore, seropositivity increased from 0 to 12% and in ddY mice from 6.3 to 45% as a result of the booster technique. In barrier-maintained animals, the booster did not elicit antibodies where there had been none before. Presumably, these animals were free from MHV infection.

From the foregoing, it is evident that the classic humoral immune response to MHV seems to be weak, and that a colony in which serologic tests are consistently negative is not necessarily free from MHV infection. Perhaps of even greater importance to an understanding of resistance mechanisms in this infectious disease are studies concerning immunopathology. Using MHV3 as a model, LePrévost *et al.* (1975b) examined both humoral and cell-mediated immunity and the effect of immunosuppression by antilymphocyte serum. Levy-LeBlond and Dupuy (1977) extended these studies to conclude that at least two types of mature cells are required for transfer of MHV resistance to newborn mice of the A/J (resistant) strain: T lymphocytes and an adherent spleen cell population. Transfer of isologous cells consisting of purified lymphocytes and adherent spleen cells did not, however, confer resistance on newborn mice (A/J strain) (Tardieu *et al.*, 1980). Addition of nonadherent spleen cells (or bone marrow or peritoneal exudate cells) to T lymphocytes and adherent spleen cells resulted in 88% protection. Clearly, cell-mediated immune functions play a significant role in resistance of mice to MHV3 infection.

In contrast to the findings of Bang and Warwick (1960), who concluded that one gene (or factor) was responsible for host susceptibility in MHV2 infection, Stohlmann and Frelinger (1978) showed that two genes are required for resistance of the central nervous system in SJL mice to fatal disease due to MHV-JHM. Further, Stohlmann *et al.* (1980) reported that there is an age-related change in resistant mice that protects them from acute central nervous system disease. They identified this change as due to a maturing adherent spleen or peritoneal exudate cell population.

In extensive genetic studies, Lévy-LeBlond *et al.* (1979) found no correlation between the H-2 locus and either the acute or chronic disease in C57BL (susceptible) or A/J (resistant) animals. Using congenic C3H lines, however, they were able to show that the H-2^{f} allele enables both heterozygous and homozygous animals to resist the development of the chronic disease. They believe, therefore, that MHV sensitivity appears to be influenced by at least two major genes: one for the acute disease, and the other, linked to the H-2 gene complex, for the chronic disease.

Mice also produce interferon as a result of MHV infection

(Virelizier *et al.*, 1976; Virelizier and Gresser, 1978), and Taguchi *et al.* (1979a) ascribe the greater susceptibility of suckling C3H/HeJms mice to serum levels of interferon that are considerably lower than in weanling and adult mice, as well as to greater macrophage sensitivity in the neonates. MHV-induced interferon may be regarded as the principal mechanism by which the virus modifies the immune responsiveness of mice to sheep red blood cells, for example. (See also Tamura and Fujiwara, 1979.) Further, of practical importance to users and breeders of mice, MHV3 infection interferes with the secretion of lymphocyte interferon that is normally induced by Sendai virus infection, thus perhaps operating to increase the susceptibility of the MHV-infected mice to Sendai virus infection.

D. Epizootiology

1. Host Range

a. Species Affected. A number of attempts have been made to infect other species. Cheever *et al.* (1949) infected cotton rats, rats, and hamsters intracerebrally, but rabbits and guinea pigs failed to respond. Sebesteny and Hill (1974) attempted to infect infant Wistar rats and hamsters using virus recovered from nude mice. All survived at least 21 days without signs of illness.

Of considerable importance is a report by Taguchi *et al.* (1979c) concerning asymptomatic MHV-S infection in suckling rats following intranasal inoculation. The agent multiplied mainly in the nasal epithelium without any clinical signs. Neutralizing antibodies were produced, however, and could also be demonstrated in adult rats following infection. Necrotic changes took place in the nasal mucosa, and cytoplasmic immunofluorescence was demonstrated in the nasal epithelium 2 days after intranasal inoculation of 10-day-old rats.

b. Age and Sex Susceptibility. Age susceptibility has been amply addressed above in the consideration of virus growth in mice and tissue culture and in the discussion of the clinical picture. There appears to be no difference in susceptibility between the sexes (Taguchi *et al.*, 1976). Parker *et al.* (1966) found that the incidence of complement-fixing antibodies was greater in females than in males under colony conditions, ascribing this difference to continual exposure to virus-infected litters.

c. Influence of Parity (Birth Order). There appears to be no evidence that first litters are significantly more susceptible than later ones.

d. Strains of Mice. As noted above (Section IV,B, 2 and C,1), the susceptibility of various mouse strains depends on the age of the host at the time of infection, the virus strain, and the host genotype. A completely resistant mouse strain has not been reported.

2. Prevalence and Distribution

Prevalence of MHV infection in an animal colony is often difficult to ascertain. In addition to examples given earlier, Descoteaux *et al.* (1977) found a low prevalence of hepatitis antibodies in three of five colonies studied in Canada. Complement-fixing antibodies ranged in titer from 8 to 32, 8 being considered positive. Fewer than 20% of the animals, which were 6–9 months old at the time of testing, were positive.

In distribution, MHV is regarded as occurring worldwide.

3. Latent and Chronic Infections

These have been defined as occurring more than 2 months after intracerebral, intraperitoneal, intranasal, or intravenous inoculation (as reviewed in Robb and Bond, 1980). The chronic clinical manifestations range from none to porencephaly, paralysis, hepatitis, immunodeficiency manifestations, encephalitis, lymph node adenopathy, and vasculitis. Virus may or may not be isolated. Cells stainable by immunofluorescence may be found. Demyelination with or without remyelination may occur. Occasional scattered mononuclear cell infiltrates may be evident.

4. Seasonal Periodicity

There is no evidence that seasonal changes influence the occurrence of epizootics of MHV infection.

E. Diagnosis

1. Serology

As indicated above, serologic testing is routinely carried out by means of the complement fixation test. Cross-reactions with other coronaviruses must be taken into account when interpreting the results, however. Neutralization tests performed in tissue culture systems are also possible.

Employing virus strain A59 as antigen in the ELISA, Peters *et al.* (1979) found a high prevalence of MHV antibodies in colonies with a low incidence of both complement-fixing and neutralizing antibodies.

Hierholzer and Tannock (1977) have used the single radial hemolysis test for human coronavirus serodiagnosis. They then applied it to some MHV strains (Hierholzer *et al.*, 1979).

2. Virus Isolation and Visualization

These techniques are helpful under certain circumstances, e.g., experimental investigations, but they are inefficient for field conditions.

3. Necropsy; Sentinel Animals

Necropsy of dead or sick animals is always useful, although a definitive diagnosis in the absence of serologic evidence cannot be made. Nevertheless, wherever possible, gross and microscopic pathologic examination should be undertaken. Sentinel animals, especially gnotobiotes, could be incorporated into a program of colony health surveillance. These animals can then be checked at predetermined intervals for clinical, serologic, and histopathologic evidence of endemic disease in the colony.

F. Control and Prevention

Control of MHV is difficult in mouse colonies unless caesarian derivation coupled with barrier maintenance is undertaken. With the finding that vertical transmission is possible, however, barrier maintenance alone may not be adequate if, for example, such transmission is frequent.

Using small number of animals for experimental purposes, Kraft (1962a), Fox *et al.* (1977), and Rowe *et al.*, 1963) found that filters on the animal cages were efficacious in separating healthy from diseased animals.

In nude mice Fujiwara *et al.* (1976a) attempted vaccination. Antibody levels could be enhanced by transfer of spleen cells from vaccinated heterozygotes (*nu*/+) to the homozygotes. Tamura *et al.* (1976) found that *nu*/*nu* mice could resist MHV infection when they previously received thymocytes from weanling *nu*/+ littermates. They were then not only able to produce antibody but survived a challenge infection as well.

REFERENCES

Acres, S. D., and Babiuk, S. A. (1978). Studies on rotaviral antibody in bovine serum and lacteal secretions, using radioimmunoassay. *J. Am. Vet. Med. Assoc.* **173,** 555–559.

Adams, W. R., and Kraft, L. M. (1963). Epidemic diarrhea of infant mice. Identification of the etiologic agent. *Science* **141,** 359–360.

Adams, W. R., and Kraft, L. M. (1967). Electron microscopic study of the intestinal epithelium of mice infected with the agent of epizootic diarrhea of infant mice (EDIM virus). *Am. J. Pathol.* **51,** 39–60.

Almeida, J. D., and Tyrell, D. A. J. (1967). The morphology of three previously uncharacterized human respiratory viruses that grow in organ culture. *J. Gen. Virol.* **1,** 175–178.

Almeida, J. D., Hall, T., Bantvala, J. E., Totterdell, B. M., and Chrystie, I. L. (1978). The effect of trypsin on the growth of rotavirus. *J. Gen. Virol.* **40,** 213–218.

Andrewes, C. H., Pereira, H. G., and Wildy, P. (1978). "Viruses of Vertebrates," 4th ed. Baillière, London.

Anonymous (1978). Panel Report of the Colloquium on Selected Diarrheal Diseases of the Young. *J. Am. Vet. Med. Assoc.* **173,** 315–318.

Babiuk, L. A., Mohammed, K., Spence, L., Fauvel, M., and Petro, R. (1977). Rotavirus isolation and cultivation in the presence of trypsin. *J. Clin. Microbiol.* **6,** 610–617.

Bailey, O. T., Pappenheimer, A. W., Cheever, F. S., and Daniels, J. B. (1949). A murine virus (JHM) causing disseminated encephalomyelitis with extensive destruction of myelin. II. Pathology. *J. Exp. Med.* **90,** 195–212.

Banatvala, J. E., Totterdell, B., Chrystie, I. L., and Woode, G. N. (1975). *In vitro* detection of human rotaviruses. *Lancet* **ii,** 821.

Banfield, W. G., Kasnic, G., and Blackwell, J. H. (1968). Further observations on the virus of epizootic diarrhea of infant mice. An electron microscopic study. *Virology* **36,** 411–421.

Bang, F. B. (1978). Genetics of resistance of animals to viruses: I. Introduction and studies in mice. *Adv. Virus Res.* **23,** 270–348.

Bang, F. B., and Warwick, A. (1959). Macrophages and mouse hepatitis. *Virology* **9,** 715–717.

Bang, F. B., and Warwick, A. (1960). Mouse macrophages as host cells for the mouse hepatitis virus and the genetic basis of their susceptibility. *Proc. Natl. Acad. Sci. U.S.A.* **46,** 1065–1075.

Barinsky, I. F., and Dementiev, I. V. (1968). Virological and cytogenetic studies on the involvement of bone marrow of mice in some hepatoencephalotropic viral infections. *Acta Virol.* (*Engl. Ed.*) **12,** 464–467.

Beards, G. M., Pilford, J. N., Thouless, M. E., and Flewett, T. H. (1980). Rotavirus serotypes by serum neutralisation. *J. Med. Virol.* **5,** 231–237.

Becker, W. B., McIntosh, K., Dees, J. H., and Chanock, R. M. (1967). Morphogenesis of avian infectious bronchitis virus and a related human virus (Strain 229E). *J. Virol.* **1,** 1019–1027.

Behbehani, A. M., Foster, L. C., and Weiner, H. A. (1966). Preparation of type-specific antisera to reoviruses. *Appl. Microbiol.* **14,** 1051–1053.

Bell, T. M., Massie, A., Ross, M. G. R., and Williams, M. C. (1964). Isolation of a reovirus from a case of Burkitt's lymphoma. *Br. Med. J.* **1,** 1212–1213.

Bennette, J. G. (1960). Isolation of a non-pathogenic tumour-destroying virus from mouse ascites. *Nature* (*London*) **187,** 72–73.

Bennette, J. G., Bush, P. V., and Steele, R. D. (1967a). Characteristics of a newborn runt disease induced by neonatal infection with an oncolytic strain of reovirus type 3 (reo3MH). II. Immunological aspects of the disease in mice. *Br. J. Exp. Pathol.* **48,** 267–284.

Bennette, J. G., Bush, P. V., and Steele, R. D. (1967b). Characteristics of a newborn runt disease induced by neonatal infection with an oncolytic strain of reovirus type 3 (reo3MH). I. Pathological investigations in rats and mice. *Br. J. Exp. Pathol.* **48,** 251–266.

Bhatt, P. N., Percy, D. H., and Jonas, A. M. (1972). Characterization of the virus of sialodacryoadenitis of rats: A member of the coronavirus group. *J. Infect. Dis.* **126,** 123–130.

Biggart, J. D., and Ruebner, B. H. (1970). Lymphoid necrosis in the mouse spleen produced by mouse hepatitis virus (MHV3): An electron-microscopic study. *J. Med. Microbiol.* **3,** 627–632.

Biggers, D. C., Kraft, L. M., and Sprinz, H. (1964). Lethal intestinal virus infection of mice (LIVIM). An important new model for study of the response of the intestinal mucosa to injury. *Am. J. Pathol.* **45,** 413–418.

Bishop, R. F., Davidson, G. P., Holmes, I. H., and Ruck, B. J. (1973). Virus particles in epithelial cells of duodenal mucosa from children with acute non-bacterial gastroenteritis. *Lancet* **ii,** 1281–1283.

Blackwell, J. H., Tennant, R. W., and Ward, T. G. (1966). Serological studies with an agent of epizootic diarrhea of infant mice. *Natl. Cancer Inst. Monogr.* No. 20, 63–66.

Bond, C. W., Leibowitz, J. L., and Robb, J. A. (1979). Pathogenic murine coronaviruses. II. Characterization of virus-specific proteins of murine coronaviruses JHMV and A59V. *Virology* 94, 371–384.

Borsa, J., Sargent, M. D., Copps, T. P., Long, D. G., and Chapman, J. D. (1973a). Specific monovalent cation effects on modification of reovirus infectivity by chymotrypsin digestion *in vitro*. *J. Virol.* **11,** 1017–1019.

Borsa, J., Sargent, M. D., Long, D. G., and Chapman, J. D. (1973b). Extraordinary effects of specific monovalent cations on activation of reovirus transcriptase by chymotrypsin *in vitro*. *J. Virol.* **11,** 207–217.

Borsa, J., Copps, T. P., Sargent, M. D., Long, D. G., and Chapman, J. D. (1973c). New intermediate subviral particles in the *in vitro* uncoating of reovirus virions by chymotrypsin. *J. Virol.* **11,** 552–564.

Borsa, J., Long, D. G., Sargent, M. D., Copps, T. P., and Chapman, J. D. (1974). Reovirus transcriptase activation *in vitro:* involvement of an endogenous uncoating activity in the second stage of the process. *Intervirology* **4,** 171–188.

Borsa, J., Morash, B. D., Sargent, M. D., Copps, T. P., Lievaart, P. A., and Szekely, J. G. (1979). Two modes of entry of reovirus particles into L cells. *J. Gen. Virol.* **45,** 161–170.

Boss, J. H., and Jones, W. A. (1963). Hepatic localization of infectious agent in murine viral hepatitis. *Arch. Pathol.* **76,** 4–8.

Bradburne, A. F. (1970). Antigenic relationships amongst coronaviruses. *Arch. Gesamte Virusforsch.* 31, 352–364.

Bradburne, A. F., Almeida, J. D., Gardner, P. S., Moosai, R. B., Nash, A. A., and Coombs, R. R. A. (1979). A solid-phase system (SPACE) for the detection and quantification of rotavirus in faeces. *J. Gen. Virol.* **44,** 615–623.

Branski, D., Lebenthal, E., Faden, H. S., Hatch, T. F., and Krasner, J. (1980a). Small intestinal epithelial brush border enzymatic changes in suckling mice infected with reovirus type 3. *Pediatr. Res.* **14,** 803–805.

Branksi, D., Lebenthal, E., Faden, H. S., Hatch, T. F., and Krasner, J. (1980b). Reovirus type 3 infection in a suckling mouse: the effects on pancreatic structure and enzyme content. *Pediatr. Res.* **14,** 8–11.

Braunsteiner, H., and Friend, C. (1954). Viral hepatitis associated with transplantable mouse leukemia. I. Acute hepatic manifestations following treatment with urethane or methylformamide. *J. Exp. Med.* **100,** 665–677.

Bridger, J. C., and Woode, G. N. (1975). Neonatal calf diarrhoea: identification of reovirus-like (rotavirus) agent in faeces by immunofluorescence and immune electron microscopy. *Br. Vet. J.* **131,** 528–535.

Broderson, J. R., Murphy, F. A., and Hierholzer, J. C. (1976). Lethal enteritis in infant mice caused by mouse hepatitis virus. *Lab. Anim. Sci.* **26,** 824.

Bryden, A. S., and Davies, H. A. (1975). The laboratory diagnosis of epizootic diarrhoea of infant mice. *J. Inst. Anim. Tech.* **26,** 63–67.

Bryden, A. S., Thouless, M. E., and Flewett, T. H. (1976). A rabbit rotavirus. *Vet. Rec.* **99,** 323.

Bryden, A. S., Davies, H. A., Thouless, M. E., and Flewett, T. H. (1977). Diagnosis of rotavirus infection by cell culture. *J. Med. Microbiol.* 10, 121–125.

Buescher, E. L. (1952). "A Hepatitis Virus of Mice," Prof. Rep., pp. 46–51. U.S. Army 406th Med. Gen. Lab., Tokyo.

Burks, J. S., DeVald, B. L., Jankovsky, L. D., and Gerdes, J. C. (1980). Two coronaviruses isolated from central nervous system tissue of two multiple sclerosis patients. *Science* **209,** 933–934.

Calisher, C. H., and Rowe, W. P. (1966). Mouse hepatitis, reo-3, and the Theiler viruses. *Natl. Cancer Inst. Monogr.* No. 20, 67–75.

Cheers, C., and Waller, R. (1975). Activated macrophages in congenitally athymic "nude" mice and in lethally irradiated mice. *J. Immunol.* **115,** 844–847.

Cheever, F. S. (1956). Epidemic diarrheal disease of suckling mice. *Ann. N.Y. Acad. Sci.* **66,** 196–203.

Cheever, F. S., and Mueller, J. H. (1947). Epidemic diarrheal disease of suckling mice. I. Manifestations, epidemiology, and attempts to transmit the disease. *J. Exp. Med.* **85,** 405–416.

Cheever, F. S., and Mueller, J. H. (1948). Epidemic diarrheal disease of suckling mice. III. The effect of strain, litter, and season upon the incidence of the disease. *J. Exp. Med.* **88,** 309–316.

Cheever, F. S., Daniels, J. B., Pappenheimer, A. M., and Bailey, O. T. (1949). A murine virus (JHM) causing disseminated encephalomyelitis with extensive destruction of myelin. I. Isolation and biological properties of the virus. *J. Exp. Med.* **90,** 181–194.

Childs, J., Stohlmann, S., and Russell, R. (1980). Serological interrelationships of murine hepatitis viruses. *Ann. Meet. Am. Soc. Microbiol., 80th* Abstr. T 99, p. 252.

Clark, S. M., Barnett, B. B., and Spendlove, R. S. (1979). Production of high-titer bovine rotavirus with trypsin. *J. Clin. Microbiol.* **9,** 413–417.

Collins, M. J., and Parker, J. C. (1972). Murine virus contaminants of leukemia viruses and transplantable tumors. *J. National Cancer Inst.* **49,** 1139–1143.

Cook, I. (1963). Reovirus type 3 infection in laboratory mice. *Aust. J. Exp. Biol.* **41,** 651–660.

Cross, R. K., and Fields, B. N. (1972). Temperature-sensitive mutants of reovirus type 3: Studies on the synthesis of viral RNA. *Virology* 50, 799–809.

Datta, D. V., and Isselbacher, K. J. (1969). Effect of corticosteroids on mouse hepatitis virus infection. *Gut* **10,** 522–529.

David-Ferreira, J. F., and Manaker, R. A. (1965). An electron microscope study of the development of a mouse hepatitis virus in tissue culture cells. *J. Cell Biol.* **24,** 57–78.

Davidson, G. P., Bishop, R. F., Townley, R. R. W., Holmes, K. H., and Ruck, B. J. (1975). Importance of a new virus in acute sporadic enteritis in children. *Lancet* **1,** 242–246.

Davies, H. A., and Macnaughton, M. R. (1979). Comparison of the morphology of three coronaviruses. *Arch. Virol.* **59,** 25–33.

Derbyshire, J. B., and Woode, G. N. (1978). Classification of rotaviruses: Report from the World Health Organization/Food and Agriculture Organization Comparative Virology Program. *J. Am. Vet. Med. Assoc.* **173,** 519–521.

Descoteaux, J.-P., Grignon-Archambault, D., and Lussier, G. (1977). Serologic study on the prevalence of murine viruses in five Canadian mouse colonies. *Lab. Anim. Sci.* **27,** 621–626.

Dick, G. W. A., Niven, J. S. F., and Gledhill, A. W. (1956). A virus related to that causing hepatitis in mice (MHV). *Br. J. Exp. Pathol.* **37,** 90–98.

Dupuy, J. M., Levey-LeBlond, E., and LeProvost, C. (1975). Immunopathology of mouse hepatitis virus type 3. II. Effect of immunosuppression in resistant mice. *J. Immunol.* **114,** 226–230.

East, J., Parrott, D. M. V., Chesterman, F. C., and Pomerace, A. (1963). The appearance of a hepatotrophic virus in mice thymectomized at birth. *J. Exp. Med.* **118,** 1069–1082.

Ellens, D. J., deLeeuw, P. W., Straver, P. J., and van Balken, J. A. M. (1978). Comparison of five diagnostic methods for the detection of rotavirus antigens in calf faeces. *Med. Microbiol. Immunol.* **166,** 157–163.

Els, H. J., and Lecatsas, G. (1972). Morphological studies on simian virus SA 11 and the "related" 0 agent. *J. Gen. Virol.* **17,** 129–132.

England, J. J., and Poston, R. P. (1980). Electron microscopic identification and subsequent isolation of a rotavirus from a dog with fatal neonatal diarrhea. *Am. J. Vet Res.* **41,** 782–783.

Estes, M. K., and Graham, D. Y. (1979). Enhancement of rotavirus infectivity by trypsin and elastase. *In* "Viral Enteritis in Humans and Animals" (F. Bricout and R. Scherrer, eds.), Vol. 90, pp. 83–86. Inst. Natl. Santé Rech. Méd., Paris.

Estes, M. K., and Graham, D. Y. (1980). Identification of rotaviruses of different origins by the plaque-reduction test. *Am. J. Vet. Res.* **41,** 151–152.

Fauvel, M., Spence, K., Babiuk, L. A., Petro, R., and Bloch, S. (1978). Hemagglutination and hemagglutination-inhibition studies with a strain of Nebraska calf diarrhea virus (bovine rotavirus). *Intervirology* **9,** 95–105.

Fields, B. N., and Joklik, W. K. (1969). Isolation and preliminary genetic and biochemical characterization of temperature-sensitive mutants of reovirus. *Virology* **37,** 335–342.

Fields, B. N., and Raine, C. S. (1972). Altered disease in rats due to mutants of reovirus type 3. *J. Clin. Invest.* **51,** 30a.

Fields, B. N., Laskov, R., and Scharff, M. D. (1972). Temperature-sensitive mutants of reovirus type 3: Studies on the synthesis of viral peptides. *Virology* **50,** 209–215.

Finberg, R., Weiner, H. L., Fields, B. N., Benacerraf, B., and Burakoff, S. J. (1979). Generation of cytolytic T lymphocytes after reovirus infection: role of the S1 gene. *Proc. Natl. Acad. Sci. U.S.A.* **76,** 442–446.

Flanagan, S. P. (1966). "Nude", a new hairless gene with pleiotropic effects in the mouse. *Genet. Res.* **8,** 295–309.

Flewett, T. H. (1978). Electron microscopy in the diagnosis of infectious diarrhea. *J. Am. Vet. Med. Assoc.* **173,** 538–543.

Flewett, T. H., and Woode, G. N. (1978). The rotaviruses. *Arch. Virol. (Engl. Ed.)* **57,** 1–23.

Flewett, T. H., Bryden, A. S., Davies, H., Woode, G. N., Bridger, J. C., and Derrick, J. M. (1974). Relation between viruses from acute gastroenteritis of children and newborn calves. *Lancet* **ii,** 61–63.

Flewett, T. H., Bryden, A. S., and Davies, H. (1975). Virus diarrhoea in foals and other animals. *Vet. Rec.* **96,** 477.

Foster, L. G., Peterson, M. W., and Spendlove, R. S. (1975). Fluorescent virus precipitin test. *Proc. Soc. Exp. Biol. Med.* **150,** 155–160.

Fox, J. G., Murphy, J. C., and Igras, V. E. (1977). Adverse effects of mouse hepatitis virus on ascites myeloma passage in the Balb/cJ mouse. *Lab. Anim. Sci.* **27,** 173–179.

Fujiwara, K. (1971). Problems in checking inapparent infections in laboratory mouse colonies: An attempt at serological checking by anamnestic response. *Defin. Lab. Anim., Symp., 4th, Washington, D.C., 1969* pp. 77–92.

Fujiwara, K., Tamura, T., Hirano, N., and Takenaka, S. (1975). Implication pancréatique chez la souris infectée avec le virus de l'hépatite murine. *C. R. Seances Soc. Biol. Ses Fil.* **169,** 477–480.

Fujiwara, K., Tamura, T., Taguchi, F., Machii, K., and Suzuki, K. (1976a). Immunisation de la souris "nude" contre le virus de l'hépatite murine par transfert de lymphocytes sensibilisés. *C. R. Seances Soc. Biol. Ses Fil.* **170,** 509–513.

Fujiwara, K., Takenaka, S., and Shumiya, S. (1976b). Carrier state of antibody and viruses in a mouse breeding colony persistently infected with Sendai and mouse hepatitis viruses. *Lab. Anim. Sci.* **26,** 153–159.

Gallily, R., Warwick, A., and Bang, F. B. (1964). Effect of cortisone on genetic resistance to mouse hepatitis virus *in vivo* and *in vitro*. *Proc. Natl. Acad. Sci. U.S.A.* **51,** 1158–1164.

Gallily, R., Warwick, A., and Bang, F. B. (1967). Ontogeny of macrophage resistance to mouse hepatitis *in vivo* and *in vitro*. *J. Exp. Med.* **125,** 537–548.

Garwes, D. J. (1979). Structure and physicochemical properties of coronaviruses. *In* "Viral Enteritis in Humans and Animals" (F. Bricout and R. Scherrer, eds.), Vol. 90, pp. 141–162. Inst. Natl. Santé Rech. Méd., Paris.

Ghose, L. H., Schnagl, R. D., and Holmes, I. H. (1978). Comparison of an enzyme-linked immunosorbent assay for quantitation of rotavirus antibodies with complement fixation in an epidemiological survey. *J. Clin. Microbiol.* **8,** 268–276.

Gledhill, A. W. (1961). Enhancement of the pathogenicity of mouse hepatitis virus (MHV1) by prior infection of mice with certain leukaemia agents. *Br. J. Cancer* **15,** 531–538.

Gledhill, A. W., and Andrewes, C. H. (1951). A hepatitis virus of mice. *Br. J. Exp. Pathol.* **32,** 559–568.

Gledhill, A. W., Dick, G. W. A., and Andrewes, C. H. (1952). Production of hepatitis in mice by the combined action of two filterable agents. *Lancet* **ii,** 509–511.

Gledhill, A. W., Dick, G. W. A., and Niven, J. S. F. (1955). Mouse hepatitis virus and its pathogenic action. *J. Pathol. Bacteriol.* **69,** 299–309.

Gomatos, P. J., and Tamm, I. (1962). Reactive sites of reovirus type 3 and their interaction with receptor substances. *Virology* 17, 455–461.

Gomatos, P. J., Tamm, I., Dales, S., and Franklin, R. M. (1962). Reovirus type 3: Physical characteristics and interaction with L cells. *Virology* **17,** 444–454.

Gompels, A. E. H. (1953). The propagation of S virus of mouse hepatitis in tissue culture. *J. Pathol. Bacteriol.* **66,** 567–569.

Graham, D. Y., and Estes, M. K. (1979). Comparison of methods for immunocytochemical detection of rotavirus infections. *Infect. Immun.* **26,** 686–689.

Greene, M. I., and Weiner, H. L. (1980). Delayed hypersensitivity in mice infected with reovirus. II. Induction of tolerance and suppressor T cells to viral specific gene products. *J. Immunol.* **125,** 283–287.

Habermann, R. T. (1959). Spontaneous diseases and their control in laboratory animals. *Public Health Rep.* **74,** 165–169.

Harford, C. J., Hamlin, A., Middelkamp, J. N., and Briggs, D. D., Jr. (1962). Electron microscopic examination of cells infected with reovirus. *J. Lab. Clin. Med.* **60,** 179–193.

Hartley, J. W., and Rowe, W. P. (1963). Tissue culture cytopathic and plaque assays for mouse hepatitis viruses. *Proc. Soc. Exp. Biol. Med.* **113,** 403–406.

Hartley, J. W., Rowe, W. P., and Huebner, R. J. (1961). Recovery of reoviruses from wild and laboratory mice. *Proc. Soc. Exp. Biol. Med.* **108,** 390–395.

Hartley, J. W., Rowe, W. P., Bloom, H. H., and Turner, H. C. (1964). Antibodies to mouse hepatitis viruses in human sera. *Proc. Soc. Exp. Biol. Med.* **115,** 414–418.

Haspel, M. V., Lampert, P. W., and Oldstone, M. B. A. (1978). Temperature-sensitive mutants of mouse hepatitis virus produce a high incidence of demyelination. *Proc. Natl. Acad. Sci. U.S.A.* **75,** 4033–4036.

Herndon, R. M., Griffin, D. E., McCormick, U., and Weiner, L. P. (1975). Mouse hepatitis virus-induced recurrent demyelination. *Arch. Neurol.* **32,** 32–35.

Hierholzer, J. C., and Tannock, G. A. (1977). Quantitation of antibody to non-hemagglutinating viruses by single radial hemolysis: Serological test for human coronaviruses. *J. Clin. Microbiol.* **5,** 613–620.

Hierholzer, J. C., Broderson, J. R., and Murphy, F. A. (1979). New strain of mouse hepatitis virus as the cause of lethal enteritis in infant mice. *Infect. Immun.* **24,** 508–522.

Hirano, N., Tamura, T., Taguchi, F., Ueda, K., and Fujiwara, K. (1975). Isolation of low-virulent mouse hepatitis virus from nude mice with wasting syndrome and hepatitis. *Jpn. J. Exp. Med.* **45,** 429–432.

Hirano, T., and Ruebner, B. H. (1966). Studies on the mechanism of destruction of lymphoid tissue in murine hepatitis virus (MHV3) infection. I. Selective prevention of lymphoid necrosis by cortisone and puromycin. *Lab. Invest.* **15,** 270–282.

Hirano, T., Hino, S., and Fujiwara, K. (1978). Physico-chemical properties of mouse hepatitis virus (MHV-2) grown on DBT cell culture. *Microbiol. Immunol.* **22,** 377–390.

Holmes, I. H. (1979). Viral gastroenteritis. *Prog. Med. Virol.* **25,** 1–36.

Holmes, I H., Ruck, B. J., Bishop, R. F., and Davison, G. P. (1975).

Infantile enteritis viruses: morphogenesis and morphology. *J. Virol.* **16,** 937–943.

Holmes, I. H., Rodger, S. M., Schnagl, R. D., Ruck, B. J., Gust, I. D., Bishop, R. F., and Barnes, G. L. (1976). Is lactase the receptor and uncoating enzyme for infantile enteritis (rota) viruses? *Lancet* **1,** 1387–1389.

Hsiung, G. D. (1958). Some distinctive biological characteristics of ECHO-10 virus. *Proc. Soc. Exp. Biol. Med.* **99,** 387–390.

Hull, R. N., Minner, J. R., and Smith, J. W. (1956). New viral agents recovered from tissue cultures of monkey kidney cells. I. Origin and properties of cytopathogenic agents SV_1, SV_2, SV_4, SV_5, SV_6, SV_{11}, SV_{12}, and SV_{15}. *Am. J. Hyg.* 63, 204–215.

Ishida, T., and Fujiwara, K. (1979). Pathology of diarrhea due to mouse hepatitis virus in the infant mouse. *Jpn. J. Exp. Med.* **49,** 33–41.

Ishida, T., Taguchi, F., Lee, Y.-S., Yamada, A., Tamura, T., and Fujiwara, K. (1978). Isolation of mouse hepatitis virus from infant mice with fatal diarrhea. *Lab. Anim. Sci.* **28,** 269–276.

Ishii, A., Yago, A., Nariuchi, H., Shirasaka, A., Wada, Y., and Matubashi, T. (1974). Some aspects on the transmission of hepatitis B antigen; model experiments by mosquitoes with murine hepatitis virus. *Jpn. J. Exp. Med.* **44,** 495–501.

Joklik, W. K. (1970). The molecular biology of reovirus. *J. Cell Physiol.* **76,** 289–301.

Joklik, W. K. (1972). Studies on the effect of chymotrypsin on reovirions. *Virology* **49,** 700–715.

Joklik, W. K. (1974). Reproduction of reoviridae. *Compr. Virol.* **2,** 231–344.

Jones, W. A., and Cohen, R. B. (1962). The effect of a murine hepatitis virus on the liver. *Am. J. Pathol.* **41,** 329–347.

Jordan, L. E., and Mayor, H. D. (1962). The fine structure of reovirus, a new member of the icosahedral series. *Virology* **17,** 597–599.

Kalica, A. R., Purcell, R. H., Sereno, M. M., Wyatt, R. G., Kim, H. W., Chanock, R. M., and Zapikian, A. Z. (1977). A microtiter solid phase radioimmunoassay for detection of the human reovirus-like agent in stools. *J. Immunol.* **188,** 1275–1279.

Kalica, A. R., Wyatt, R. G., and Kapikian, A. Z. (1978). Detection of differences among human and animal rotaviruses using analysis of viral RNA. *J. Am. Vet. Med. Assoc.* **173,** 531–537.

Kantoch, M., Warwick, A., and Bang, F. B. (1963). The cellular nature of genetic susceptibility to a virus. *J. Exp. Med.* **117,** 781–797.

Kapikian, A. Z. (1975). The coronaviruses. *Dev. Biol. Stand.* **28,** 42–64.

Kapikian, A. Z., Kim, H. W., Wyatt, R. G., Rodriquez, W. J., Ross, S., Cline, W. L., Parrott, R. H., and Chanock, R. M. (1974). Reoviruslike agent in stools: Association with infantile diarrhea and development of serologic tests. *Science* **185,** 1049–1053.

Kapikian, A. Z., Cline, W. L., Kim, H. W., Kalica, A. R., Wyatt, R. G., van Kirk, D. H., Chanock, R. M., James, H. D., Jr., and Vaughn, A. L. (1976). Antigentic relationships among five reovirus-like agents by complement fixation. *Proc. Soc. Exp. Biol. Med.* **152,** 535–539.

Katami, K., Taguchi, F., Nakayama, M., Goto, N., and Fujiwara, K. (1978). Vertical transmission of mouse hepatitis virus infection in mice. *Jpn. J. Exp. Med.* **48,** 481–490.

Kaye, H. S., and Dowdle, W. R. (1969). Some characteristics of hemagglutination of certain strains of "IBV-like" virus. *J. Infect. Dis.* **120,** 576–581.

Kraft, L. M. (1957). Studies on the etiology and transmission of epidemic diarrhea of infant mice. *J. Exp. Med.* **106,** 743–755.

Kraft, L. M. (1958). Observations on the control and natural history of epidemic diarrhea of infant mice (EDIM). *Yale J. Biol. Med.* **31,** 121–137.

Kraft, L. M. (1961). Responses of the mouse to the virus of epidemic diarrhea of infant mice. Neutralizing antibodies and carrier state. *Proc. Anim. Care Panel* **11,** 125–136.

Kraft, L. M. (1962a). Two viruses causing diarrhea in infant mice. *In* "The Problems of Laboratory Animal Disease" (R. J. C. Harris, ed.), pp. 115–127. Academic Press, New York.

Kraft, L. M. (1962b). An apparently new lethal virus disease of infant mice. *Science* **137,** 282–283.

Kraft, L. M. (1966). Epizootic diarrhea of infant mice and lethal intestinal virus infection of infant mice. *Natl. Cancer Inst. Monogr.* No. 20, 55–61.

Kraft, L. M., Pardy, R. F., Pardy, D. A., and Zwickel, H. (1964). Practical control of diarrheal disease in a commercial mouse colony. *Lab. Anim. Care* **14,** 16–19.

Kundin, W. D., Liu, C., and Gigstad, J. (1966). Reovirus infection in suckling mice: Immunofluorescent and infectivity studies. *J. Immunol.* **97,** 393–401.

Kunstýř, I. (1962). Discussion of Kraft, L. M. Two viruses causing diarrhea in infant mice. *In* "The Problems of Laboratory Animal Disease" (R. J. C. Harris, ed.), pp. 127–130. Academic Press, New York.

LaBonnardiere, C., and deVaurieux, C. (1979). *In vivo* interference between heterologous rotaviruses. *In* "Viral Enteritis in Humans and Animals" (F. Bricout and R. Scherrer, eds.), Vol. 90, pp. 95–98. Inst. Natl. Santé Rech. Méd., Paris.

Lai, M.-H., and Joklik, W. K. (1973). The induction of interferon by temperature-sensitive mutants of reovirus, UV-irradiated reovirus, and subviral reovirus particles. *Virology* **51,** 191–204.

Lamontagne, L., Marsolais, G., Marois, P., and Assaf, R. (1980). Diagnosis of rotavirus, adenovirus, and herpes virus infections by immune electron microscopy using a serum-in-agar diffusion method. *Can. J. Microbiol.* **26,** 261–264.

Lampert, P. W. (1978). Autoimmune and virus-induced demyelinating diseases. *Am. J. Pathol.* **91,** 176–208.

Lampert, P. W., Sims, J. K., and Kniazeff, A. J. (1973). Mechanism of demyelination of JHM virus encephalomyelitis. *Acta Neuropathol.* **24,** 76–85.

Laufs, R. (1967). Untersuchungen über die Entstehung von Riesenzellen in Mäusemakrophagenkulturen nach Infektion mit dem Mäusehepatitis-virus (MHV-3). *Virchows Arch. Pathol. Anat. Physiol* **342,** 169–183.

Lavelle, G. C., and Bang, F. B. (1973). Differential growth of MHV(PRI) and MHV(C_3H) in genetically resistant C_3H rendered susceptible by *Eperythrozoon coccoides. Arch. Gesamte Virusforsch.* **41,** 175–184.

Lavelle, G. C., and Starr, T. J. (1969). Relationship of phagocytic activity to pathogenicitiy of mouse hepatitis virus as affected by triolein and cortisone. *Br. J. Exp. Pathol.* **50,** 475–480.

Leers, W. D., Rozee, K. R., and Wardlaw, A. C. (1968). Immunodiffusion and immunoelectrophoretic studies of reovirus antigens. *Can. J. Microbiol.* **14,** 161–164.

LePrévost, C., Virelizier, J. L., and Dupuy, J. M. (1975a). Immunopathology of mouse hepatitis virus type 3 infection. III. Clinical and virologic observation of a persistent viral infection. *J. Immunol.* **115,** 640–643.

LePrévost, C., Lévy-LeBlond, E., Virelizier, J. L., and Dupuy, J. M. (1975b). Immunopathology of mouse hepatitis virus type 3 infection. I. Role of humoral and cell-mediated immunity in resistance mechanisms. *J. Immunol.* **114,** 221–225.

Lévy-LeBlond, E., and Dupuy, J. M. (1977). Neonatal susceptibility to MHV3 infection in mice. I. Transfer of resistance. *J. Immunol.* **118,** 1219–1222.

Lévy-LeBlond, E., Oth, D., and Dupuy, J. M. (1979). Genetic study of mouse sensitivity to MHV3 infection: Influence of the H-2 complex. *J. Immunol.* **122,** 1359–1362.

Lewis, L., and Starr, T. J. (1972). Polykaryocytosis and replication of mouse hepatitis virus in peritoneal macrophages. *Br. J. Exp. Pathol.* **53,** 202–205.

Lucas, A., Flintoff, W., Anderson, R., Percy, D., Coulter, M., and Dales, S. (1977). *In vivo* and *in vitro* models of demyelinating diseases: tropism of the JHM strain of murine hepatitis virus for cells of glial origin. *Cell* **12,** 553–560.

Luftig, R. B., Kilham, S. S., Hay, A. J., Zweerink, H. J., and Joklik, W. K. (1972). An ultrastructural study of virions and cores of reovirus type 3. *Virology* **48,** 170–181.

McClain, M. E., Spendlove, R. S., and Lennette, E. H. (1967). Infectivity assay of reoviruses: Comparison of immunofluorescent cell count and plaque methods. *J. Immunol.* **98,** 1301–1308.

McClure, H. M., Chapman, W. L., Hooper, B. E., Smith, F. G., and Fletcher, O. J. (1978). The digestive system. *In* "Pathology of Laboratory Animals" (K. Benirschke, F. M. Garner, and T. C. Jones, eds.), p. 214. Springer-Verlag, Berlin and New York.

McCrae, M. A., and Joklik, W. K. (1978). The nature of the polypeptide encoded by each of the 10 double-stranded segments of reovirus type 3. *Virology* **89,** 578–593.

McCrea, A. W. R., Bell, T. M., Henderson, B. E., Munube, G. M. R., and Mukwaya, L. K. (1968). Trans-stadial maintenance of reovirus type 3 in the mosquito *Culex pipiens fatigans* Weidmann and its implications. *East Afr. Med. J.* **45,** 677–686.

McIntosh, K. (1974). Coronaviruses: A comparative review. *Curr. Top. Microbiol. Immunol.* **63,** 85–129.

McIntosh, K. (1977). Coronaviruses as causes of diseases: Clinical observations and diagnosis. *In* "Comparative Diagnosis of Viral Diseases, Vol. I, Human and Related Viruses" Part A (E. Kurstak, and C. Kurstak, eds.), pp. 609–619. Academic Press, New York.

McIntosh, K., Becker, W. B., and Chanock, R. M. (1967). Growth in suckling-mouse brain of "IBV-like" viruses from patients with upper respiratory tract disease. *Proc. Natl. Acad. Sci. U.S.A.* **58,** 2268–2273.

McIntosh, K., Kapikian, A. Z., Hardison, K. A., Hartley, J. W., and Chanock, R. M. (1969). Antigenic relationships among the coronaviruses of man and between human and animal coronaviruses. *J. Immunol.* **102,** 1109–1118.

Macnaughton, M. R. (1980). The polypeptides of human and mouse coronaviruses. *Arch. Virol.* **63,** 75–80.

McNulty, M. S. (1978). Rotaviruses—a review. *J. Gen. Virol.* **40,** 1–18.

McNulty, M. S. (1979). Morphology and chemical composition of rotaviruses. *In* "Viral Enteritis in Humans and Animals" (F. Bricout and R. Scherrer, eds.), Vol. 90, pp. 111–140. Inst. Natl. Santé Rech. Méd., Paris.

Mallucci, L. (1965). Observations on the growth of mouse hepatitis virus (MHV-3) in mouse macrophages. *Virology* **25,** 30–37.

Manaker, R. A., Piczak, C. V., Miller, A. A., and Stanton, M. F. (1961). A. hepatitis virus complicating studies with mouse leukemia. *J. Natl. Cancer Inst.* **27,** 29–51.

Margolis, G., Kilham, L., and Baringer, R. (1975). Identity of Cowdry type B inclusions and nuclear bodies: observations in reovirus encephalitis. *Exp. Mol. Pathol.* **84,** 63–74.

Mathan, M., Almeida, J. D., and Cole, J. (1977). An antigenic subunit present in rotavirus infected faeces. *J. Gen. Virol.* **34,** 325–329.

Matsuno, S., Inouye, S., and Kono, R. (1977). Plaque assay of neonatal calf diarrhoea and the neutralising antibody in human sera. *J. Clin. Microbiol.* **5,** 1–4.

Matthews, R. E. F. (1979). The classification and nomenclature of viruses. *Intervirology* **11,** 133–135.

Mebus, C. A., Underdahl, M. R., Rhodes, M. B., and Twiehaus, M. J. (1969). Calf diarrhea (scours) reproduced with a virus from a field outbreak. *Stn. Bull.—Nebr., Agric. Exp. Stn.* **233,** 1–16.

Melnick, J. L. (1979). Taxonomy of viruses, 1979. *Prog. Med. Virol.* **25,** 160–166.

Meyers, T. R. (1979). A reo-like virus isolated from juvenile American oysters (*Crassostrea virginica*). *J. Gen. Virol.* **43,** 203–212.

Meyers, T. R., and Hirai, K. (1980). Morphology of a reo-like virus isolated from juvenile American oysters (*Crassostrea virginica*). *J. Gen. Virol.* **46,** 249–253.

Middleton, P. J. (1978). Pathogenesis of rotaviral infection. *J. Am. Vet. Med. Assoc.* **173,** 544–545.

Middleton, P. J., Petric, M., Hewitt, C. M., Szymanski, M. T., and Tam, J. S. (1976). Counter-immunoelectro-osmophoresis for the detection of infantile gastroenteritis virus (orbi-group) antigen and antibody. *J. Clin. Pathol.* **29,** 191–197.

Middleton, P. J., Holdaway, M. D., Petric, M., Szymanski, M. T., and Tam, J. S. (1977). Solid phase radioimmunoassay for the detection of rotavirus. *Infect. Immun.* **16,** 439–444.

Miyazaki, Y., Katsuta, H., Aoyama, Y., Kawai, K., and Takaoka, T. (1957). Experimental studies on hepatitis virus of mice in tissue culture. *Jpn. J. Exp. Med.* **27,** 381–399.

Moon, H. W. (1978). Mechanisms in the pathogenesis of diarrhea: a review. *J. Am. Vet. Med. Assoc.* **172,** 443–448.

Morris, J. A. (1959). A new member of hepatoencephalitis group of murine viruses. *Proc. Soc. Exp. Biol. Med.* **100,** 875–877.

Mosley, J. W. (1961). Multiplication and cytopathogenicity of mouse hepatitis virus in mouse cell cultures. *Proc. Soc. Exp. Biol. Med.* **108,** 524–529.

Much, D. H., and Zajac, I. (1972). Purification and characterization of epizootic diarrhea of infant mice virus. *Infect. Immun.* **6,** 1019–1024.

Nelson, J. B. (1952a). Acute hepatitis associated with mouse leukemia. I. Etiology and host range of the causal agent in mice. *J. Exp. Med.* **96,** 303–312.

Nelson, J. B.(1952b). Acute hepatitis associated with mouse leukemia. I. Pathological features and transmission of the disease. *J. Exp. Med.* **96,** 293–300.

Nelson, J. B. (1953). Acute hepatitis associated with mouse leukemia. IV. The relationship of *Eperythrozoon coccoides* to the hepatitis virus of Princeton mice. *J. Exp. Med.* **98,** 441–449.

Nelson, J. B. (1955). Acute hepatitis associated with mouse leukemia. V. The neurotropic properties of the causal virus. *J. Exp. Med.* **102,** 581–594.

Nelson, J. B. (1963). Recovery and behavior of hepatitis virus from Swiss mice injected with ascites tumor. *Proc. Soc. Exp. Biol. Med.* **113,** 909–912.

Nelson, J. B. (1965). Pathogenicity of murine hepatitis virus recovered from infant Swiss mice. *Proc. Soc. Exp. Biol. Med.* **120,** 41–44.

Nelson, J. B., and Tarnowski, G. S. (1960). An oncolytic virus recovered from Swiss mice during passage of an ascites tumour. *Nature (London)* **185,** 866–867.

Nickol, A. D., and Bonventre, P. F. (1977). Anomalous high native resistance of athymic mice to bacterial pathogens. *Infect. Immun.* **18,** 636–645.

Nicolaieff, A., Obert, G., and Van Regenmortel, M. H. V. (1980). Detection of rotavirus by serological trapping on antibody-coated electron microscope grids. *J. Clin. Microbiol.* **12,** 101–104.

Niven, J. S. F., Gledhill, A. W., Dick, G. W. A., and Andrewes, C. H. (1952). Further light on mouse hepatitis. *Lancet* **ii,** 1061.

Nonoyama, M., Watanabe, Y., and Graham, A. F. (1970). Defective virions of reovirus. *J. Virol.* **6,** 226–236.

Onodera, T., Jenson, A. B., Yoon, J.-W., and Notkins, A. L. (1978). Virus-induced diabetes mellitus: reovirus infection of pancreatic β-cells in mice. *Science* **201,** 529–531.

Papadimitriou, J. M. (1965). Electron micrographic features of acute murine reovirus hepatitis. *Am. J. Pathol.* **47,** 565–585.

Papadimitriou, J. M. (1966). Ultrastructural features of chronic murine hepatitis after reovirus type 3 infection. *Br. J. Exp. Pathol.* **47,** 624–631.

Papadimitriou, J. M. (1967). An electron microscopic study of murine reovirus-3 encephalitis. *Am. J. Pathol.* **50,** 59–75.

Papadimitriou, J. M. (1968). The biliary tract in acute murine reovirus 3 infection. *Am. J. Pathol.* **52,** 595–611.

Papadimitriou, J. M. and Walters, M. N.-I. (1967). Studies on the exocrine pancreas. II. ultrastructural investigation of reovirus pancreatitis. *Am. J. Pathol.* **51,** 387–403.

Pappenheimer, A. W., and Cheever, F. S. (1948). Epidemic diarrheal disease of suckling mice. IV. Cytoplasmic inclusion bodies in intestinal epithelium in relation to the disease. *J. Exp. Med.* **88,** 317–324.

Pappenheimer, A. W., and Enders, J. F. (1947). An epidemic diarrheal disease of suckling mice. II. Inclusions in the intestinal epithelial cells. *J. Exp. Med.* **85,** 417–422.

Paradisi, F., and Piccinino, F. (1968). Propagation of mouse hepatitis virus (MHV-3) in monolayer cell cultures from liver of newborn mice. *Experientia* **24,** 373–374.

Parker, J. C., Tennant, R. W., Ward, T. G., and Rowe, W. P. (1965). Virus studies with germfree mice. I. Preparation of serologic diagnostic reagents and survey of germfree and monocontaminated mice for indigenous murine viruses. *J. Natl. Cancer Inst.* **34,** 371–380.

Parker, J. C., Tennant, R. W., and Ward, T. G. (1966). Prevalence of viruses in mouse colonies. *Natl. Cancer Inst. Monogr.* No. 20, 25–36.

Parker, J. C., Cross, S. S., and Rowe, W. P. (1970). Rat coronavirus (RCV): A prevalent, naturally occurring pneumotropic virus of rats. *Arch. Gesamte Virusforsch.* **31,** 293–302.

Parker, L., Baker, E., and Stanley, N. F. (1965). The isolation of reovirus type 3 from mosquitoes and a sentinel infant mouse. *Aust. J. Exp. Biol. Med. Sci.* **43,** 167–170.

Peters, R. L., Collins, M. J., O'Beirne, A. J., Howton, P. A., Hourihan, S. L., and Thomas, S. F. (1979). Enzyme-linked immunosorbent assay for detection of antibodies to murine hepatitis virus. *J. Clin. Microbiol.* **10,** 595–597.

Peterson, M. W., Spendlove, R. S., and Smart, R. A. (1976). Detection of neonatal calf diarrhea virus, infant reovirus-like diarrhea virus, and a coronavirus using the fluorescent virus precipitin test. *J. Clin. Microbiol.* **3,** 376–377.

Phillips, P. A., Keast, D., Papadimitrious, J. M., Walters, M. N.-I., and Stanley, N. F. (1969). Chronic obstructive jaundice induced by reovirus type 3 in weanling mice. *Pathology* **1,** 193–203.

Piazza, M. (1969). "Experimental Viral Hepatitis." Thomas, Springfield, Illinois.

Piazza, M., Pane, G., and DeRitis, F. (1967). The fate of murine hepatitis virus (MHV-3) after intravenous injection into susceptible mice. *Arch. Gesamte Virusforsch.* **22,** 472–474.

Piccinino, F., Galanti, B., and Giusti, G. (1966). Lack of transplacental transmissibility of MHV-3 virus. *Arch. Gesamte Virusforsch.* **18,** 327–332.

Powell, H. C., and Lampert, P. W. (1975). Oligodendrocytes and their myelin-plasma membrane connections in JHM mouse hepatitis virus encephalomyelitis. *Lab. Invest.* **33,** 440–445.

Ramia, S., and Sattar, S. A. (1980). Proteolytic enzymes and rotavirus SA-11 plaque formation. *Can. J. Comp. Med.* **44,** 232–236.

Ramig, R. F., Mustoe, T. A., Sharpe, A. H., and Fields, B. N. (1978). A genetic map of reovirus. II. Assignment of the double-stranded RNA-negative mutant groups C, D, and E to genome segments. *Virology* **85,** 531–544.

Rhim, J. S., and Melnick, J. L. (1961). Plaque formation by reoviruses. *Virology* **15,** 80–81.

Rhim, J. S., Jordan, L. E., and Mayor, H. D. (1962). Cytochemical, fluorescent-antibody and electron microscopic studies on the growth of reovirus (ECHO 10) in tissue culture. *Virology* **17,** 342–355.

Robb, J. A., and Bond, C. W. (1979). Pathogenic murine coronaviruses. I. Characterization of biological behavior *in vitro* and virus-specific intracellular RNA of strongly neurotropic JHMV and weakly neurotropic A59V viruses. *Virology* **94,** 352–370.

Robb, J. A., and Bond, C. W. (1980). Coronaviridae. *Compr. Virol.* **14,** 193–248.

Robb, J. A., Bond, C. W., and Leibowitz, J. L. (1979). Pathogenic murine coronaviruses. III. Biological and biochemical characterization of temperature-sensitive mutants of JHMV. *Virology* **94,** 385–399.

Rodger, S. M., Craven, J. A., and Williams, I. (1975). Demonstration of reovirus-like particles in intestinal contents of piglets with diarrhoea. *Aust. Vet. J.* **51,** 536.

Rodger, S. M., and Holmes, I. A. (1979). Comparison of the genomes of simian, bovine, and human rotaviruses by gel electrophoresis and detection of genomic variation among bovine isolates. *J. Virol.* **30,** 839–846.

Rosen, L. (1960). Serologic grouping of reoviruses by hemagglutination-inhibition. *Am. J. Hyg.* **71,** 242–249.

Rowe, W. P., Hartley, J. W., and Capps, W. I. (1963). Mouse hepatitis virus infection as a highly contagious, prevalent, enteric infection of mice. *Proc. Soc. Exp. Biol. Med.* **112,** 161–165.

Rozee, K. R., and Leers, W. D. (1967). Chloroform inactivation of reovirus hemagglutinins. *Can. Med. Assoc. J.* **96,** 597–599.

Rubenstein, D., Milne, R. G., Buckland, R., and Tyrrell, D. A. J. (1971). The growth of the virus of epidemic diarrhea of infant mice (EDIM) in organ cultures of intestinal epithelium. *Br. J. Exp. Pathol.* **52,** 442–445.

Ruebner, B. H., and Bramhall, J. L. (1960). Pathology of experimental hepatitis in mice. *Arch. Pathol.* **69,** 190–198.

Ruebner, B. H., and Miyai, K. (1962). The Kupffer cell reaction in murine and human viral hepatitis with particular reference to the origin of acidophilic bodies. *Am. J. Pathol.* **40,** 425–435.

Ruebner, B. H., Hirano, T., and Slusser, R. J. (1967). Electron microscopy of the hepatocellular and Kupffer-cell lesions of mouse hepatitis, with particular reference to the effect of cortisone. *Am. J. Pathol.* **51,** 163–189.

Runner, M. N., and Palm, J. (1953). Factors associated with the incidence of infantile diarrhea in mice. *Proc. Soc. Exp. Biol. Med.* **82,** 147–150.

Sabin, A. B. (1959). Reoviruses. A new group of respiratory and enteric viruses formerly classified as ECHO type 10 is described. *Science* **130,** 1387–1389.

Scherrer, R., and Bernard, S. (1977). Application d'une technique immunoenzymologique (ELISA) à la détection du rotavirus bovin et des anticorps dirigés contre lui. *Ann. Microbiol. (Paris)* **128A,** 499–510.

Schoub, B. D., Kalica, A. R., Greenberg, H. B., Bertran, D. M., Sereno, M. M., Wyatt, R. G., Chanock, R. M., and Kapikian, A. Z. (1979). Enhancement of antigen incorporation and infectivity of cell cultures by human rotavirus. *J. Clin. Microbiol.* **9,** 488–492.

Scott, A. C., Luddington, J., and Lucas, M. (1978). Rotavirus in goats. *Vet. Rec.* **103,** 145.

Seamer, J. (1965). Mouse macrophages as host cells for murine viruses. *Arch. Gesamte Virusforsch.* **17,** 654–663.

Seamer, J. (1967). Some virus infections of mice. *In* "Pathology of Laboratory Rats and Mice" (E. Cotchin, and F. J. C. Roe, eds.), pp. 537–567. Blackwell, Oxford.

Sebesteny, A., and Hill, A. C. (1974). Hepatitis and brain lesions due to mouse hepatitis virus accompanied by wasting in nude mice. *Lab. Anim.* **8,** 317–326.

Sharpe, A. H., Ramig, R. F., Mustoe, T. A., and Fields, N. B. (1978). A genetic map of reovirus. I. Correlation of genome RNAs between serotypes 1, 2, and 3. *Virology* **84,** 63–74.

Shatkin, A. J. (1969). Replication of reovirus. *Adv. Virus Res.* **14,** 63–87.

Sheets, P., Shah, K. V., and Bang, F. B. (1978). Mouse hepatitis virus (MHV) infection in thymectomized C_3H mice. *Proc. Soc. Exp. Biol. Med.* **159,** 34–38.

Sheridan, J. F., and Aurelian, L. (1981). Rotavirus infection of neonatal mice: characterization of the humoral immune response. *Annu. Meet. Am. Soc. Microbiol. 81st, Dallas, Tex* Abstr. No. E109, p. 73.

Shif, I., and Bang, F. B. (1966). Plaque assay for mouse hepatitis virus (MHV-2) on primary macrophage cell cultures. *Proc. Soc. Exp. Biol. Med.* **121,** 829–931.

Shif, I., and Bang, F. B. (1970). *In vitro* interaction of mouse hepatitis virus and macrophages from genetically resistant mice. I. Adsorption of virus and growth curves. *J. Exp. Med.* **131,** 843–850.

Shope, R. E. (1980). Arboviruses. *In* "Manual of Clinical Microbiology" (E. H. Lennette, A. Balows, W. J. Hausler, Jr., and J. P. Truant, eds.), pp. 869–874. Am. Soc. Microbiol., Washington, D.C.

Silverstein, S. C., Astell, C., Levin, D. H., Schonberg, M., and Acs, G. (1972). The mechanisms of reoviris uncoating and gene activation *in vivo*. *Virology* **47,** 797–806.

Simhon, A., and Mata, L. (1978). Anti-rotavirus antibody in human colostrum. *Lancet* **1,** 39–40.

Simmons, M. L., and Brick, J. O. (1970). "The Laboratory Mouse. Selection and Management," pp. 153–156. Prentice-Hall, Englewood Cliffs, New Jersey.

Small, J. D., Aurelian, L., Squire, R. A., Strandberg, J. D., Melby, E. C., Jr., Turner, T. B., and Newman, B. (1979). Rabbit cardiomyopathy, associated with a virus antigenically related to human coronavirus strain 229E. *Am. J. Pathol.* **95,** 709–724.

Smith, M., and Tzipori, S. (1979). Gel electrophoresis of rotavirus RNA derived from six different animal species. *Aust. J. Exp. Biol. Med. Sci.* **57,** 583–585.

Snodgrass, D. R., and Wells, P. W. (1976). Rotavirus infection in lambs: Studies on passive protection. *Arch. Virol.* (*Engl. Ed.*) **52,** 201–205.

Snodgrass, D. R., Smith, W., Gray, E. W., and Herring, J. A. (1976). A rotavirus in lambs with diarrhoea. *Res. Vet. Sci.* **20,** 113–114.

Spence, L., Fauvel, M., Bouchard, S., Babiuk, L., and Saunders, J. R. (1975). Test for reovirus-like agent. *Lancet* **ii,** 322.

Spendlove, R. S., McClain, M. E., and Lennette, E. H. (1970). Enhancement of reovirus infectivity by extracellular removal or alteration of the virus capsid by proteolyitic enzymes. *J. Gen. Virol.* **8,** 83–94.

Stanley, N. F. (1961). Relationship of hepatoencephalomyelitis virus and reoviruses. *Nature* (*London*) **189,** 687.

Stanley, N. F. (1974). The reovirus murine models. *Prog. Med. Virol.* **18,** 257–272.

Stanley, N. F. (1977). Diagnosis of reovirus infection: Comparative aspects. *In* "Comparative Diagnosis of Viral Diseases, Vol. I, Human and Related Viruses," Part A (E. Kurstak and C. Kurstak, eds.), pp. 385–421. Academic Press, New York.

Stanley, N. F., and Leak, P. J. (1963). Murine infection with reovirus type 3 and the runting syndrome. *Nature* (*London*) **199,** 1309–1310.

Stanley, N. F., Dorman, D. C., and Ponsford, J. (1953). Studies on the pathogenesis of a hitherto undescribed virus (Hepatoencephalomyelitis) producing unusual symptoms in suckling mice. *Aust. J. Exp. Biol.* **31,** 147–160.

Stanley, N. F., Dorman, D. C., and Ponsford, J. (1954). Studies on the hepato-encephalomyelitis virus (HEV). *Aust. J. Exp. Biol.* **32,** 543–562.

Stanley, N. F., Leak, P. J., Walters, M. N.-I., and Joske, R. A. (1964). Murine infection with reovirus. II. The chronic disease following reovirus type 3 infection. *Br. J. Exp. Pathol.* **45,** 142–149.

Starr, T. J., and Pollard, M. (1959). Propagation of mouse hepatitis virus (Gledhill) in tissue culture. *Proc. Soc. Exp. Biol. Med.* **100,** 97–100.

Starr, T. J., Pollard, M., Duncan, D., and Dunaway, M. R. (1960). Electron and fluorescence microscopy of mouse hepatitis virus. *Proc. Soc. Exp. Biol. Med.* **104,** 767–769.

Stohlmann, S. A., and Frelinger, J. A. (1978). Resistance to fatal central nervous system disease by mouse hepatitis virus, strain JHM. I. Genetic analysis. *Immunogenetics* **6,** 277–281.

Stohlmann, S. A., Frelinger, J. A., and Weiner, L. P. (1980). Resistance to fatal central nervous system disease by mouse hepatitis virus, strain JHM. II. Adherent cell-mediated protection. *J. Immunol.* **124,** 1733–1739.

Stone, H. H., Stanley, D. G., and DeJarnette, R. H. (1967). Postsplenectomy viral hepatitis. *J. Am. Med. Assoc.* **199,** 851–853.

Sturman, L. S. (1977). Characterization of a coronavirus. I. Structural proteins: Effects of preparative conditions on the migration of protein in polyacrylamide gels. *Virology* **77,** 637–649.

Svoboda, D., Nielson, A., Werder, A., and Higginson, J. (1962). An electron microscopic study of viral hepatitis in mice. *Am. J. Pathol.* **41,** 204–224.

Taguchi, F., Hirano, N., Kiuchi, Y., and Fujiwara, K. (1976). Difference in response to mouse hepatitis virus among susceptible mouse strains. *Jpn. J. Microbiol.* **20,** 293–302.

Taguchi, F., Yamada, A., and Fujiwara, K. (1979a). Factors involved in the age-dependent resistance of mice infected with low virulence mouse hepatitis virus. *Arch. Virol.* (*Engl. Ed.*) **62,** 333–340.

Taguchi, F., Goto, Y., Aiuchi, M., Hayashi, T., and Fujiwara, K. (1979b). Pathogenesis of mouse hepatitis infection. The role of nasal epithelial cells as a primary target of low virulence virus, MHV-S. *Microbiol. Immunol.* **23,** 249–262.

Taguchi, F., Yamada, A., and Fujiwara, K. (1979c). Asymptomatic infection of mouse hepatitis virus in the rat. *Arch. Virol.* (*Engl. Ed.*) **59,** 275–279.

Takayama, N., and Kirn, A. (1978). *In vitro* growth characteristics and heterogeneity of mouse hepatitis virus type 3. *Arch. Virol.* (*Engl. Ed.*) **58,** 29–34.

Tamura, T., and Fujiwara, K. (1979). IgM and IgG response to sheep red blood cells in mouse hepatitis virus-infected nude mice. *Microbiol. Immunol.* **23,** 177–183.

Tamura, T., Ueda, K., Hirano, N., and Fujiwara, K. (1976). Response of nude mice to a mouse hepatitis virus isolated from a wasting nude mouse. *Jpn. J. Exp. Med.* **46,** 19–30.

Tamura, T., Taguchi, F., Ueda, K., and Fujiwara, K. (1977). Persistent infection with mouse hepatitis virus of low virulence in nude mice. *Microbiol. Immunol.* **21,** 683–691.

Tamura, T., Kai, C., Sakaguchi, A., Ishida, T., and Fujiwara, K. (1979). The role of macrophages in the early resistance to mouse hepatitis virus infection in nude mice. *Microbiol. Immunol.* **23,** 965–974.

Tamura, T., Sakaguchi, A., Kai, C., and Fujiwara, K. (1980). Enhanced phagocytic activity of macrophages in mouse hepatitis virus-infected nude mice. *Microbiol. Immunol.* **24,** 243–247.

Tardiue, M., Hery, C., and Dupuy, J. M. (1980). Neonatal susceptibility to MHV3 infection in mice. II. Role of natural effector marrow cells in transfer of resistance. *J. Immunol.* **124,** 418–423.

Thiel, K. W., Bohl, E. H., and Saif, L. J. (1978). Techniques for rotaviral propagation. *J. Am. Vet. Med. Assoc.* **173,** 548–551.

Thouless, M. E., Bryden, A. S., and Flewett, T. H. (1977a). Rotavirus neutralization by human milk. *Br. Med. J.* **ii,** 1390.

Thouless, M. E., Bryden, A. S., Flewett, T. H., Woode, G. N., Bridger. J. C., Snodgrass, D. R., and Herring, J. A. (1977b). Serological relationships between rotaviruses from different species as studied by complement fixation and neutralization. *Arch. Virol.* (*Engl. Ed.*) **53,** 287–294.

Tisdale, W. A. (1963). Potentiating effect of K-virus on mouse hepatitis virus (MHV-S) in weanling mice. *Proc. Soc. Exp. Biol. Med.* **114,** 774–777.

Traavik, T., Mehl, R., and Kjeldsberg, E. (1977). "Runde" virus, a coronavirus-like agent associated with seabirds and ticks. *Arch. Virol.* (*Engl. Ed.*) **55,** 25–38.

Tufvesson, B., and Johnsson, T. (1976). Immunoelectroosmophoresis for detection of reo-like virus: Methodology and comparision with electron microscopy. *Acta Pathol. Microbiol. Scand.* **84,** 225–228.

Tyrell, D. A. J., Almeida, J. D., Berry, D. M., Cunningham, C. H., Hamre, D., Hofstad, M. S., Mallucci, L., and McIntosh, K. (1968). Coronaviruses. *Nature* (*London*) **220,** 650.

Tyrell, D. A. J., Almeida, J. D., Cunningham, C. H., Dowdle, W. R., Hofstad, M. S., McIntosh, K., Tajima, M., Zakstelskaya, L. Y., Easterday, B. C., Kapikian, A., and Bingham, R. W. (1975). Coronaviridae. *Intervirology* **76,** 76–82.

Tytell, A. A., Lampson, G. P., Field, A. K., and Hilleman, M. R. (1967). Inducers of interferon and host resistance. III. Double-stranded RNA from reovirus type 3 virions (Reo 3-RNA). *Proc. Natl. Acad. Sci. U.S.A.* **58,** 1719–1722.

Tzipori, W., Caple, I. W., and Butler, R. (1976). Isolation of rotavirus from deer. *Vet. Rec.* **99,** 398.

Ubertini, T., Wilkie, B. N., and Noronha, F. (1971). Use of horseradish peroxidase labelled antibody for light and electron microscopic localization of reovirus antigen. *Appl. Microbiol.* **21,** 534–538.

Vainio, T. (1961). Studies on murine hepatitis virus (MHV3) *in vitro*. *Proc. Soc. Exp. Biol. Med.* **107,** 326–331.

Vasquez, C., and Tournier, P. (1962). The morphology of reovirus. *Virology* **17,** 503–510.

Vasquez, C., and Tournier, P. (1964). New interpretation of reovirus structure. *Virology* **24,** 128–130.

Vella, P. P., and Starr, T. J. (1965). Effect of X radiation and cortisone on mouse hepatitis virus infection in germfree mice. *J. Infect. Dis.* **115,** 271–177.

Virelizier, J.-L., and Allison, A. C. (1976). Correlation of persistent mouse hepatitis virus (MHV-3) infection with its effect on mouse macrophage cultures. *Arch. Virol. (Engl. Med.)* **50,** 279–285.

Virelizier, J.-L., and Gresser, I. (1978). Role of interferon in the pathogenesis of viral diseases of mice as demonstrated by the use of anti-interferon serum. V. Protective role in mouse hepatitis virus type 3 infection of susceptible and resistant strains of mice. *J. Immunol.* **120,** 1616–1619.

Virelizier, J.-L., Dayan, A. D., and Allison, A. C. (1975). Neuropathological effects of persistent infection of mice by mouse hepatitis virus. *Infect. Immun.* **12,** 1127–1140.

Virelizier, J.-L., Virelizier, A.-M., and Allison, A. C. (1976). The role of circulating interferon in the modifications of immune responsiveness by mouse hepatitis virus (MHV-3). *J. Immunol.* **117,** 748–753.

Wallis, C., Smith, K. O., and Melnick, J. L. (1964). Reovirus activation by heating and inactivation by cooling in $MgCl_2$ solutions. *Virology* **22,** 608–619.

Wallis, C., Melnick, J. L., and Rapp, F. (1966). Effects of pancreatin on the growth of reovirus. *J. Bacteriol.* **92,** 155–160.

Walters, M. N.-I., Joske, R. A., Leak, P. J., and Stanley, N. F. (1963). Murine infection with reovirus: I. Pathology of the acute phase. *Br. J. Exp. Pathol.* **44,** 427–436.

Ward, J. M., Collins, M. J., Jr., and Parker, J. C. (1977). Naturally occurring mouse hepatitis virus infection in the nude mouse. *Lab. Anim. Sci.* **27,** 372–376.

Ward, R. L., and Ashley, C. S. (1978). Identification of detergents as components of wastewater sludge that modify the thermal stability of reovirus and enteroviruses. *Appl. Environ. Microbiol.* **36,** 889–897.

Ward, R. L., and Ashley, C. S. (1980a). Effects of wastewater sludge and its detergents on the stability of rotavirus. *Appl. Environ. Microbiol.* **39,** 1154–1158.

Ward, R. L., and Ashley, C. S. (1980b). Comparative study on the mechanisms of rotavirus inactivation by sodium dodecyl sulfate and ethylenediaminetetracetate. *Appl. Environ. Microbiol.* **39,** 1148–1153.

Watanabe, K. (1969a). Electron microscopic studies of experimental viral hepatitis in mice. II. Ultrastructural changes of hepatocytes associated with virus multiplication. *J. Electron Microsc.* **18,** 173–188.

Watanabe, K. (1969b). Electron microscopic studies of experimental viral hepatitis in mice. I. Virus particles and their relationship to hepatocytes and Kupffer cells. *J. Electron Microsc.* **18,** 158–172.

Wege, H., Wege, H., Nagashima, K., and ter Meulen, V. (1979). Structural polypeptides of the murine coronavirus, JHM. *J. Gen. Virol.* **42,** 37–47.

Weiner, H. L., and Fields, B. N. (1977). Neutralization of reovirus: the gene responsible for the neutralization antigen. *J. Exp. Med.* **146,** 1305–1310.

Weiner, H. L., Drayna, D., Averill, D. R., Jr., and Fields, B. N. (1977). Molecular basis of reovirus virulence: role of the S1 gene. *Proc. Natl. Acad. Sci. U.S.A.* **74,** 5744–5748.

Weiner, H. L., Ramig, R. F., Mustoe, T. A., and Fields, B. N. (1978). Identification of the gene coding for the hemagglutinin of reovirus. *Virology* **86,** 581–584.

Weiner, H. L., Powers, M. L., and Fields, B. N. (1980a). Absolute linkage of virulence and central nervous system cell tropism of reoviruses to viral hemagglutinin. *J. Infect. Dis.* **141,** 609–616.

Weiner, H. L., Greene, M. I., and Fields, B. N. (1980b). Delayed hypersensitivity in mice infected with reovirus. I. Identification of host and viral gene products responsible for the immune response. *J. Immunol.* **125,** 278–282.

Weiner, H. L., Ault, K. A., and Fields, B. N. (1980c). Interaction of reovirus with cell surface receptors. I. Murine and human lymphocytes have a receptor for the hemagglutinin of reovirus type 3. *J. Immunol.* **124,** 2143–2148.

Weiner, L. P. (1973). Pathogenesis of demyelination induced by a mouse hepatitis virus (JHM virus). *Arch. Neurol.* **28,** 298–303.

Weiser, W. Y., and Bang, F. B. (1977). Blocking of *in vitro* and *in vivo* susceptibility to mouse hepatitis virus. *J. Exp. Med.* **146,** 1467–1472.

Welliver, R. C., and Ogra, P. L. (1978). Importance of local immunity in enteric infection. *J. Am. Vet. Med. Assoc.* **173,** 560–564.

Willenborg, D. O., Shah, K. V., and Bang, F. B. (1973). Effect of cyclophosphamide on the genetic resistance of C_3H mice to mouse hepatitis virus. *Proc. Soc. Exp. Biol. Med.* **142,** 762–766.

Willey, D. E., and Ushijima, R. N. (1980). The circadian rhythm of thymosin therapy during acute reovirus type 3 infection of neonatal mice. *Clin. Immunol. Immunopathol.* **16,** 72–74.

Wilsnack, R. E. (1971). Immunofluorescent detection of murine virus antigens. *In* "Defining the Laboratory Animal" (H. A. Schneider, ed.), pp. 93–109. Natl. Acad. Sci., Washington, D.C.

Wilsnack, R. E., Blackwell, J. H., and Parker, J. C. (1969). Identification of an agent of epizootic diarrhea of infant mice by immunofluorescent and complement-fixation tests. *Am. J. Vet. Res.* **30,** 1195–1204.

Wolf, J. L., Rubin, D. H., Kauffman, R. S., Sharpe, A. H., Trier, J. S., and Fields, B. N. (1981). Intestinal M cells: a pathway for entry of reovirus into the host. *Science* **212,** 471–472.

Woode, G. N., Bridger, J. C., Jones, J. M., Flewett, T. H., Bryden, A. S., Davies, H. A., and White, G. B. B. (1976). Morphological and antigenic relationships between viruses (rotaviruses) from acute gastroenteritis of children, calves, piglets, mice, and foals. *Infect. Immun.* **14,** 804–810.

Woode, G. N., Smith, C., and Dennis, M. J. (1978). Intestinal damage in rotavirus infected calves assessed by D-xylose malabsorption. *Vet. Rec.* **102,** 340–341.

Woods, J. E., Besch, E. L., and Nevins, R. G. (1974). Heat and moisture transfer in filter-top rodent cages. Publ. 74-6, Abstr. No. 77. Am. Assoc. Lab. Anim. Sci., Jolet, Illinois.

Wyatt, R. G., Kalica, A. R., Mebus, C. A., Kim, H. W., London, W. T., Chanock, R. M., and Kapikian, A. Z. (1978). Reovirus-like agents (Rotaviruses) associated with diarrheal illness in animals and man. *Perspect. Virol.* **10,** 121–145.

Wyatt, R. G., James, W. D., Bohl, E. H., Theil, K. W., Saif, L. J., Kalica, A. R., Greenberg, H. B., Kapikian, A. Z., and Chanock, R. M. (1980). Human rotavirus type 2: cultivation *in vitro*. *Science* **207,** 189–191.

Yolken, R. H., and Stopa, P. J. (1979). Enzyme-linked fluorescence assay: ultrasensitive solid-phase assay for detection of human rotavirus. *J. Clin. Microbiol.* **10,** 317–321.

Yolken, R. H., Barbour, B., Wyatt, R. G., Kalica, A. R., Kapikian, A. Z., and Chanock, R. M. (1978a). Enzyme-linked immunosorbent assay for identification of rotaviruses from different animal species. *Science* **201,** 259–261.

Yolken, R. H., Wyatt, R. G., Barbour, B. A., Kim, H. W., Kapikian, A. Z., and Chanock, R. M. (1978b). Measurement of rotavirus antibody by an enzyme-linked immunosorbent assay blocking assay. *J. Clin. Microbiol.* **8,** 283–287.

Yolken, R. H., Wyatt, R. G., Mata, L., Urrutia, J. J., Garcia, B., Chanock, R. M., and Kapikian, A. Z. (1978c). Secretory antibody directed against rotavirus in human milk—measurement by means of enzyme-linked immunosorbent assay. *J. Pediatr.* **93,** 916–921.

Chapter 10

Lactate Dehydrogenase-Elevating Virus

Margo A. Brinton

I. Introduction ... 194
A. Host Specificity ... 194
B. History of Isolation ... 194
C. Classification ... 194
II. Characteristics of Infection ... 194
A. Transmission ... 194
B. Site of Virus Replication *in Vivo* ... 195
C. Kinetics of Replication ... 195
D. Persistence ... 196
E. Antibody Response and Immune Complexes ... 196
F. Effects on Immune System Components ... 196
G. Splenomegaly ... 197
H. Impaired Serum Enzyme Clearance ... 197
I. Pathology ... 199
J. Association with Tumors ... 199
III. Methods for Detecting the Presence of LDV ... 200
A. LDH Assay ... 200
B. Method of Freeing Transplantable Tumor Suspensions and Virus Pools of LDV ... 200
IV. Replication in Tissue Culture ... 200
A. Characteristics of Target Cells ... 200
B. Kinetics of Replication ... 202
V. Properties of the Virus ... 203
A. Morphology ... 203
B. Susceptibility to Various Chemical Agents and Temperature ... 203
C. Physicochemical Properties ... 203
D. Intracellular Viral RNA Synthesis ... 204
E. Interferon ... 204
F. Antigenic Nature ... 205
VI. Conclusion ... 205
References ... 205

ISBN 0-12-262502-1

I. INTRODUCTION

A. Host Specificity (Only Mice)

Lactate dehydrogenase-elevating virus (LDV) is unusual in its extreme host specificity. To date, LDV is known to infect only mice. The virus replicates rapidly in all strains of mice so far tested, producing a persistent infection. Even though a high titer of infectivity is continually present in the blood, no clinical disease is normally observed in mice infected with LDV. The infection is characterized by elevated levels of certain serum enzymes. This characteristic was actually responsible for the original discovery of the virus and is utilized currently for the assay of its infectivity.

The intent of this chapter is to discuss briefly the parameters of an LDV infection. A more detailed review of previous experimental work has been prepared by Rowson and Mahy (1975).

B. History of Isolation

During a search for methods that could be used in the early diagnosis of tumors, Riley and Wroblewski (1960) found that, following the inoculation of mice with Ehrlich carcinoma cells, a 5- to 10-fold increase in lactate dehydrogenase (LDH) levels in the serum occurred before detectable tumor growth. A further increase in LDH activity was observed as the tumor mass began to enlarge. When mice were subsequently treated with *o*-phenylenediamine, an antitumor drug, a decrease in LDH levels accompanied tumor regression, but LDH levels remained above normal. The early rise in LDH levels observed in the tumor-bearing mice was duplicated in normal mice by the injection of serum from the mice with tumors (Riley *et al.*, 1960); serum from these infected mice subsequently raised levels of LDH in other normal mice. These results suggested the presence of an infectious agent. That this agent is a virus was indicated by its ability to pass through a bacteria-retaining filter and by its susceptibility to inactivation at 70°C. Riley (1968) subsequently found that this virus (LDV) was a contaminant of 26 murine tumors which had been maintained by transplantation in mice. However, that LDV is not a necessary component of tumor lines was indicated by the finding that spontaneous or induced tumors were not contaminated with LDV (Riley, 1961, 1962; Crispens, 1963; Mundy and Williams, 1961; Notkins *et al.*, 1962; Goergii *et al.*, 1963). Subsequently, several transplantable tumors were also shown to be free of LDV (Mundy and Williams, 1961; Notkins *et al.*, 1962).

C. Classification

Although LDV has been referred to in the literature by a number of names, including *Riley virus* (Adams *et al.*, 1961) and *enzyme-elevating factor* (Riley *et al.*, 1960), it is now commonly referred to as either *lactate* or *lactic dehydrogenase-elevating virus*.

Currently, LDV is classified as a togavirus (Fenner, 1977), but its characteristics distinguish it from members of the four named groups of togaviruses, the alphavirus, flavivirus, rubivirus, and pestivirus groups. It seems most similar to a few ungrouped togaviruses such as hog cholera virus, bovine diarrhea virus, and equine arteritis virus (Brinton, 1979; Horzinek *et al.*, 1971).

II. CHARACTERISTICS OF INFECTION

A. Transmission

Although LDV replicates rapidly once it enters its host, it is not readily transferred from one mouse to another by natural means. No insect vector for this virus has been identified, nor is anything known about its incidence in wild mouse populations. However, viruses with biological properties identical to those of LDV have been isolated from small groups of wild mice in Australia (Pope, 1961), Germany (Georgii and Kirschenhofer, 1965), the United States (Pope and Rowe, 1964), and England (Rowson, 1963; Field and Adams, 1968).

Mice infected with LDV have been found to excrete the virus in their feces, urine, milk, and probably saliva (Plagemann *et al.*, 1963; Notkins, 1965a; Crispens, 1964a,b).

In the laboratory, LDV is rarely transferred between mice, even when they are housed in the same cage. Infected females infrequently transmit the virus to their young. However, it has been reported that, when fighting males were housed in the same cage, virus was apparently transferred from saliva to an open wound (Notkins, 1963). In addition, when normal males were housed with infected males whose incisors had been removed, the virus was again transmitted at a high rate, suggesting that the ingestion of blood and/or tissue of infected animals can lead to infection.

The major mode of transmission of LDV among laboratory mice has most likely been by means of experimental procedures. Since the virus remains at high titers in the blood throughout the lifetime of an infected mouse, the transfer of serum or tissue from one mouse to another could result in the transfer of virus. The use of the same needle for the injection of several mice could also inadvertently spread an infection. In the past, an occasional mouse or sometimes a whole shipment of mice obtained from a commercial supplier has been infected with LDV, but the incidence of this is now relatively rare, as indicated by the results of spot checks carried out by the author over the last 7 years.

Although LDV infection usually produces no clinical disease in infected mice, it does induce certain alterations in the host which can affect the host's response to other infectious agents, tumor cells, or experimental procedures. Therefore, it is important for investigators working with mice and with primary murine cell cultures to be aware of the possibility of LDV infection among their experimental animals. Fairly simple methods, to be described in this chapter, exist for the detection and eradication of LDV.

B. Site of Virus Replication *in Vivo*

That LDV replicates in macrophages has been suggested by a number of different investigations. The replication of LDV is normal in mice subjected to whole-body irradiation (DuBuy and Johnson, 1966; M. A. Brinton, unpublished observations). After irradiation, the number of lymphocytes in the blood is reduced by 99%, whereas the number of macrophages in the peritoneal fluid is only slightly reduced (DuBuy and Johnson, 1966). Using an indirect immunofluorescent technique with mouse anti-LDV serum, two laboratories have demonstrated the presence of cells containing LDV-specific antigen in sections of the spleen and liver (Porter *et al.*, 1969; Rowson and Michaels, 1973). Antigen-containing cells were not observed in sections of kidney, lung, thymus, or salivary gland. The maximum number of stained cells (1500–5000 per 6–8 nm^2) were observed in spleen sections between 18 and 24 hr after infection. These stained cells were nucleated and were located in the red pulp, which suggests that they were monocytes or macrophages. A reduction in the number of nucleated cells in the spleen was observed 36 hr after infection, and only about 200 cells showed positive staining in a comparable section. In the liver sections, staining was confined to Kupffer cells (Porter *et al.*, 1969).

Beginning at day 3 after infection, an increase in the number of immunoglobulin-producing cells was noted. This increase in antibody-producing cells caused a high background fluorescence in the indirect immunofluorescence assay. However, Porter and his co-workers observed only faint staining with the direct immunofluorescence technique, and studies of the number of cells showing LDV-specific antigen after 36 hr have therefore not been carried out. Electron microscopic examination of the spleens and lymph nodes of infected mice revealed increasing numbers of virus particles 12 hr after infection, in close association with the plasmalemma of reticular cells located in the marginal zone of lymphoid nodules in the spleen and medulla of the mesenteric lymph nodes (Snodgrass *et al.*, 1972).

In attempts to locate a target organ, a large number of investigators have removed various tissues from infected animals and assessed the titer of virus each contained (Bailey and Monroe, 1972; DuBuy and Johnson, 1966; Plagemann *et al.*, 1963). Such experiments are complicated by the fact that LDV is found in very high titers in the blood. Titers in the spleen, lymph nodes, liver, and thymus are similar to those in the serum, whereas such tissues as kidney, lung, small intestine, pancreas, and brain contain less virus than the serum. K. E. K. Rowson (1964 unpublished observations) looked at tissue titers soon after infection and was unable to demonstrate the appearance of virus in any organ prior to its appearance in the serum. However, perfusion of the spleen or lymph nodes with saline before titration did not reduce the virus titer significantly (DuBuy and Johnson, 1966). These experiments are consistent with the hypothesis that LDV replicates in macrophage-like cells and that such cells are present in many tissues as well as in the bloodstream.

C. Kinetics of Replication

LDV replicates rapidly in mice after infection, reaching an unusually high titer in the serum of 10^{10}–10^{11} ID_{50}/ml 12–14 hr after infection (Riley, 1974; Notkins, 1965a). Few other known mammalian viruses replicate as efficiently. The titer subsequently drops to about 10^7 ID_{50}/ml 72–96 hr after infection; thereafter, there is a further gradual decrease until a stable level of 10^5 ID_{50}/ml is reached approximately 2 weeks after infection (Fig. 1). The cause of the first rapid decrease in titer is presently unknown. It has been suggested that this decrease might be due to the death of infected target cells and the resulting scarcity of infectable cells. However, no histologic evi-

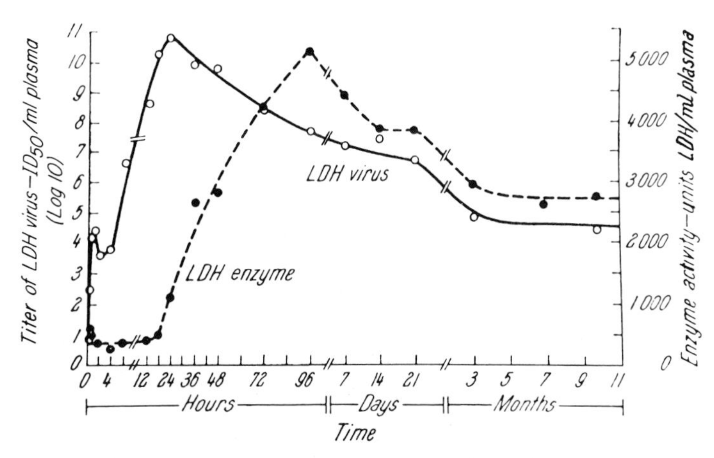

Fig. 1. Titer of LDV and activity of LDH in the plasma of mice at various times after infection with LDV. (Used with permission from Notkins and Schochat, 1963.)

dence of target cell damage has been found (Rowson and Mahy, 1975). Interferon is produced during the first 48 hr after infection, and its action may cause the early decrease in virus titer (Evans and Riley, 1968). Interferon production will be discussed later (Section V,E). Riley (1968) reported the presence of an additional viral inhibitor in plasma from LDV-infected mice 72 hr after infection. Although its identity is not known, this inhibitor did not seem to have the properties of either interferon or antibody.

The efficiency of LDV replication in mice is demonstrated by the fact that the injection of as little as 10 ID_{50} LDV yields serum titers of 10^{10} ID_{50}/ml 24 hr after infection (M. A. Brinton, unpublished observations; Rowson and Mahy, 1975).

D. Persistence

Although LDV infection results in a lifelong viremia, during which the average titer of infectivity in the serum is 10^5 ID_{50}/ml, infected mice usually show no sign of illness and live a normal life span (Notkins, 1965a). As shown in Fig. 1, LDV infection is also characterized by a permanent elevation in the plasma level of LDH (Riley *et al.*, 1960) and of several other enzymes (Plagemann *et al.*, 1962). Since LDV infection usually causes no obvious clinical disease and LDV does not agglutinate red blood cells (Rowson and Mahy, 1975; Brinton-Darnell and Plagemann, 1975), this increase in plasma LDH level has been utilized for the titration of LDV infectivity (Riley *et al.*, 1960; Plagemann *et al.*, 1963).

E. Antibody Response and Immune Complexes

The viremic stage of most virus infections ends with the appearance of neutralizing antibodies in the plasma. Initial attempts to demonstrate the presence of antiviral antibody in the plasma of LDV-infected mice were unsuccessful, and it was suggested that LDV infection might induce a state of immunologic tolerance. It was subsequently shown that anti-LDV antibodies are indeed synthesized, but are present in the plasma in a complex with virus. When sera from chronically infected mice were first heated at 58°C for 1 hr (Rowson *et al.*, 1966), treated with uv irradiation, or extracted with ether (Notkins *et al.*, 1966a,b) to destroy virus within the antibody–virus complexes, a neutralizing activity against virus obtained from mice 24 hr after infection was demonstrated. However, even under conditions of antibody excess, only 99% of the infectivity was neutralized. In initial experiments, neutralizing activity measured in animals was first detectable 34 days after infection (Notkins *et al.*, 1966a,b). However, the methods used to inactivate virus in the complex may well have also destroyed some antiviral antibody. By means of an indirect immunofluorescence technique, anti-LDV antibodies were subsequently detected as early as 6 days after infection (Porter *et al.*, 1969).

Although 99% of the infectivity in sera collected 24 hr after infection could be neutralized by anti-LDV antiserum, no neutralization of infectivity in sera obtained from mice 48 days after infection could be demonstrated. The titer of infectivity in 48-day serum samples could be reduced by incubation with antiserum to mouse γ-globulin (Notkins *et al.*, 1966a,b); however, significant residual infectivity was still observed. Sufficient data have not yet been collected to provide an adequate explanation for the resistance of this LDV population to neutralization. This phenomenon may be due to the types of antibody molecules elicited by LDV infection, to unique properties of the viral membrane-binding sites, to the presence of infectious subviral particles, or to properties of the macrophage target cells.

F. Effects on Immune System Components

An increased level of γ-globulin is observed in the serum of mice between 6 and 10 days after LDV infection. Data obtained by Notkins (Notkins *et al.*, 1966a,b; Notkins, 1966b) show that this increase is due to the synthesis of anti-LDV antibody and not to a decrease in the rate of clearance of γ-globulins. LDV infection was also found to stimulate the formation of germinal centers in gnotobiotic mice (Notkins *et al.*, 1966a,b).

During the lifelong viremia, LDV is present in the bloodstream in the form of infectious complexes with antiviral antibody (Notkins, 1966a,b; Notkins *et al.*, 1968). Although a significant number of immune complex deposits have been observed in the glomeruli of chronically infected mice, only mild subclinical lesions develop (Porter and Porter, 1971) and no symptoms of an immune complex disease are displayed.

Increased levels of antibody to human γ-globulin are produced after immunization of LDV-infected mice. However, uninfected and LDV-infected mice synthesize similar quantities of antibody after immunization with sheep or goat red blood cells, hemocyanin, or keyhole limpet (Crispens, 1968; Notkins *et al.*, 1966a,b; Salaman and Wedderburn, 1966; Oldstone *et al.*, 1974).

In contrast, LDV infection in New Zealand (NZ) mice leads to a decrease in the production of antibodies directed against nuclear antigen and red blood cells. New Zealand black (NZB) and New Zealand white (NZW) mice produce these antibodies and develop immune complex glomerulonephritis and autoimmune hemolytic anemia. In NZW mice, which show an increasing incidence of autoantibodies with age, LDV infection reduces the number of animals which begin to produce autoantibodies. In (NZB × NZW)F_1 hybrid mice, which display

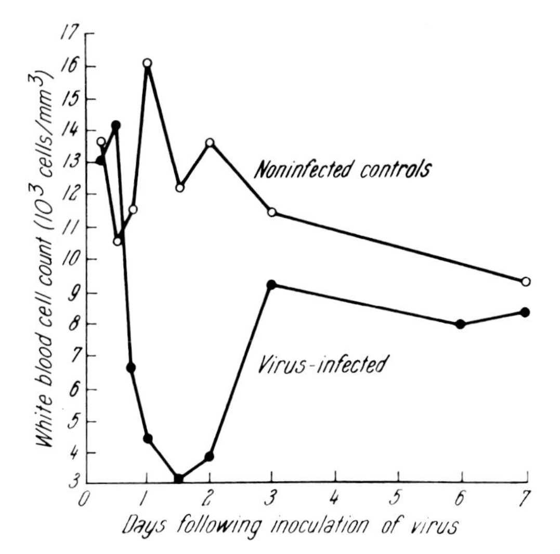

Fig. 2. A transient decrease in the total white blood cell count is observed after LDV infection in mice. (Used with permission from Riley, 1968.)

a higher incidence of autoantibodies and a more severe disease than NZW mice, LDV infection decreases the amount of antibody synthesized by four- to fivefold and significantly reduces the incidence of mortality (Oldstone and Dixon, 1972). The mechanism by which LDV infection reduces the production of autoantibodies is not known. Its effect may be on either the antigen which elicits the abnormal response or the production of the autoantibodies.

Several investigators have observed a marked necrosis of lymphocytes from thymic-dependent areas during the first 4 days after LDV infection (Hanna *et al.*, 1970a,b; Proffitt *et al.*, 1972). A transient fall in the total white blood cell count was observed by Riley (1968) beginning at about 24 hr after LDV infection (Fig. 2). Adrenalectomy prior to infection with LDV prevents this involution of the thymus (Santisteban *et al.*, 1970, 1972; Speckman *et al.*, 1974). These data suggest that LDV infection causes an increase in the levels of circulating adrenal cortical hormones, which in turn are responsible for necrosis in the thymic-dependent areas.

A depression of the cellular immune response was noted when mice infected with LDV were tested for skin allograft rejection or for a graft-versus-host reaction. Skin allografts survived several days longer on LDV-infected recipients than on controls, and the intensity of the graft-versus-host response was lower (Howard *et al.*, 1969).

That the depression of the cellular immune response by LDV is transient was indicated by the growth of MOPC-315 tumor cells injected in mice at various intervals after LDV infection (Michaelides and Schlesinger, 1974). Tumor growth was significantly enhanced when small numbers of tumor cells (4×10^5) were injected into mice within the first week after infection with LDV. However, tumor growth was less in mice infected 2–3 weeks with LDV than in control mice, which were not infected. Tumor growth in mice infected with LDV for 5 months was identical to that in control mice.

The transient decrease in the cellular immune response after LDV infection was coincident with a transitory decrease in thymus weight. Thymus weight began to decrease 24 hr after infection, and by 3–4 days after infection, had been reduced by about 40%. Thymus weight increased again, and by the seventh day exceeded that of the controls (Santisteban *et al.*, 1972; Proffitt and Congdon, 1970).

G. Splenomegaly

An increase in spleen weight has been reported to occur after LDV infection. Splenomegaly is slight but significant, as shown by Notkins (1965a), who noted that only 21 of 55 mice showed a more than 30% increase in spleen weight by 24 hr after infection which persisted up to 1 month after infection. Substantial increases in spleen weight observed after LDV infection have been attributed to infection of the mice with another pathogen, such as *Eperythrozoon coccoides* (Riley, 1964).

H. Impaired Serum Enzyme Clearance

As can be seen in Table I, the levels of a number of serum enzymes are elevated after LDV infection. Enzyme levels begin to rise by 24 hr after infection, and the maximum increase is observed after 72–96 hr. Although enzyme levels fall somewhat during the next 2 weeks after infection, elevated levels persist in infected animals (Fig. 1) (Rowson *et al.*,

Table I

Plasma Enzymes Elevated in Plasma of Mice Infected with LDV

Enzyme	Degree of elevation
Lactate dehydrogenase	8- to 11-fold
Isocitrate dehydrogenase	5- to 8-fold
Malate dehydrogenase	2- to 3-fold
Phosphoglucose isomerase	2- to 3-fold
Glutathione reductase	2- to 3-fold
Aspartate transaminase	2- to 3-fold
Glutamate-oxalacetate transaminase	2- to 3-fold
Alanine transaminase	Slight
Acid phosphatase	None
Alkaline phosphatase	None
Aldolase	None
α-Glycerophosphate	None
Glucose-6-phosphate dehydrogenase	None
Glutamate pyruvate transaminase	None
Leucine aminopeptidase	None

1963; Georgii, 1962; Notkins, 1965a). Furthermore, although the initial rate at which enzyme levels increase is somewhat dependent on the dose of virus (rising faster with increasing amounts of virus), the final elevated levels are similar, regardless of the amount of virus inoculum (Notkins and Shochat, 1963).

There is as yet no good explanation for the increase of only particular plasma enzymes during LDV infection. Some serum enzymes are increased 5- to 10-fold; others 2- to 4-fold; still others are unaffected by LDV infection (Table I). Those enzymes which are elevated show no obvious similarities in function or tissue origin, except that they are all cytoplasmic (Plagemann *et al.*, 1962). A possible explanation for which there is currently no supporting evidence might be that LDV infection damages a subpopulation of cells responsible for the clearance of only certain enzymes.

Observed plasma enzyme levels represent a balance between influx and clearance of enzymes. Normally, an increase in the plasma levels of tissue enzymes is the result of cell damage; however, with LDV infection, enzyme levels are permanently raised without evidence of significant tissue damage. The increase in LDH levels after LDV infection appears to result mainly from a decreased rate of enzyme clearance. Mahy *et al.* (1964) first postulated this theory to explain their observation that animals with necrosing tumors infected with LDV exhibit a synergistic increase in the LDH level in the plasma. This may result from the increased release of LDH from cells combined with a decreased rate of enzyme clearance. Subsequent studies of the rates of elimination of exogenous LDH from the plasma of LDV-infected and normal mice indicated that the rate of clearance is indeed significantly lower in infected than in normal animals (Mahy, 1964; Notkins and Scheele, 1964; Bailey and Wright, 1965; Notkins, 1965b; Riley *et al.*, 1965; Mahy *et al.*, 1965a,b). Mahy and Wachsmuth (1973) have demonstrated that the enzyme protein is actually removed from the blood, not merely inactivated, after the intravenous injection of exogenous LDH.

Although there are five naturally occurring LDH isozymes in mouse plasma, only isozyme LDH V has been found in increased amounts in mice infected with LDV (Plagemann *et al.*, 1963; Warnock, 1964). When clearance rates of isozymes LDH I and LDH V were compared in LDV-infected and normal mice, it was found that in normal mice, the clearance of LDH I is slower than that of LDH V. In contrast, in LDV-infected mice, the clearance rate of LDH V decreases to a level equal to that of LDH I, whereas the clearance of LDH I is unaffected (Mahy and Rowson, 1965; Rowson and Mahy, 1975).

Studies by Riley (1968) suggest that there may also be an increase in the influx of enzyme into the bloodstream of LDV-infected mice and that this may account for a portion of the observed enzyme increase. However, the tissue source of this released enzyme is not known.

The SJL/J mouse strain displays a unique elevation of LDH after LDV infection. Whereas a 5- to 10-fold increase in the plasma LDH levels is normally observed, SJL/J mice show a 15- to 20-fold increase after LDV infection (Crispens, 1971). It has been found that this trait is under the control of a recessive Mendelian gene (Crispens, 1972). The plasma levels of uninfected SJL and Swiss mice are similar. Moreover, the temporal course of virus replication and the peak titer of virus are the same in the two types of mice, as well as in peritoneal macrophage cultures prepared from them. The rate of clearance of exogenously injected LDH V as well as exogenous LDH released from damaged tissue after intraperitoneal injection of 0.001 ml carbon tetrachloride is similar in LDV-infected SJL/J and Swiss mice (Brinton and Plagemann, 1977). Since the higher LDH levels observed in infected SJL/J mice do not seem to result either from an increase in virus synthesis or from an increase in the impairment of enzyme clearance, an alternative explanation might be that there is a greater influx of LDH into the bloodstream in infected SJL/J mice. SJL/J mice have a number of unusual characteristics, at least two of which appear to be T cell-mediated phenomena (Mozes *et al.*, 1975; Ben-Yaakov and Haran-Ghera, 1975). However, whether or not any of these properties are related to the inheritable trait which controls the plasma levels of LDH after LDV infection remains to be investigated.

No increase in LDH levels is observed in LDV-infected cultures of peritoneal exudate cells. Peritoneal exudate cultures do not reduce levels of exogenous LDH added to culture fluid (M. A. Brinton, unpublished data). It is possible that only a few cells (6–20%) in peritoneal exudate cultures are infected by LDV (Tong *et al.*, 1977) and that the released LDH from this small number of cells, if they are damaged by LDV infection, would not be detectable in the culture fluid. *In vivo*, thymocytes, in which LDV does not replicate, are rapidly depleted during the first 48 hr after LDV infection (Snodgrass *et al.*, 1972) and destruction of these could add to the increased serum levels of LDH.

The reticuloendothelial system (RES) is thought to be responsible for plasma enzyme clearance (Wakim and Fleisher, 1963a,b). Assay of the disappearance of intravenously injected carbon particles has been used to evaluate the functioning of the RES system (Halpern *et al.*, 1957). Carbon clearance is suppressed in LDV-infected mice only during the first 3 days after infection; thereafter it returns to normal (Mahy, 1964; Notkins and Scheele, 1964). It is still not clear how elevated enzyme levels persist in infected mice, since the major effect on the RES is detectable only during the first 3 days after infection. It is also not known why the clearance of only certain enzymes is affected. It is possible that a population of RES cells that is unaffected by LDV infection is responsible for the clearance of the enzymes whose levels remain normal after infection.

I. Pathology

Until relatively recently, LDV was not known to cause any overt disease symptoms in infected mice. However, it has now been demonstrated that LDV can induce a polioencephalomyelitis in C58 mice when these mice are immunosuppressed during the initial stage of the infection (Martinez *et al.*, 1980).

It was found in studies of the immune response of C58 mice to leukemia cells of the syngeneic Ib line that a polioencephalomyelitis develops in old mice injected with tumor cells (Murphy *et al.*, 1970). C58 mice become spontaneously immunosuppressed between 6 and 12 months. It was subsequently shown that cell-free Ib cell extracts can also induce paralysis and that young (3-month-old) C58 mice become susceptible to the induction of paralysis by such extracts only after immunosuppression (Duffey *et al.*, 1976). The etiologic agent present in these extracts has been identified as LDV (Martinez *et al.*, 1980). Considering the restricted conditions that are required for the induction of paralysis by LDV, it is not surprising that the paralytogenic activity of this virus was not previously detected. The LDV which was isolated from C58 tumor-bearing mice replicated efficiently in 13 strains of mice tested but induced paralysis only in two, AKR and C58 (Duffey *et al.*, 1976; Martinez, 1979). A state of immunosuppression is necessary for the induction of paralysis by LDV. The incidence of LDV-induced paralysis is lower in AKR than in C58 mice, and it appears that other isolates of LDV are less efficient at inducing paralysis than the one isolated from C58 mice (Table II). This C58 LDV strain may have been selected during prolonged inadvertent passage in C58 mice as a contaminant of the serially transplanted leukemia cells. The time at which the leukemic cell line was contaminated with LDV is not known, since the tumor cells were initially passaged in young C58 mice which had not yet become immunosuppressed by the aging process.

Table II

Test for Paralytogenicity of LDV Isolates in C58 Mice[a]

	Incidence of paralysis			
	6-Month-old mice		12-Month-old mice	
Virus injected[b]	Proportion	Mean day ± SD	Proportion	Mean day ± SD
LDV-1[c]	0/10		9/11	16.9 ± 5.1
LDV-2	1/10	18	6/10	15.2 ± 5.2
LDV-3	0/10		8/9	17.1 ± 5.2
LDV-4	0/18		6/19	19.0 ± 5.0
ADPE agent	10/10	12.1 ± 2.3	10/10	10.0 ± 0.5

[a] Used with permission from Martinez *et al.* (1980).

[b] Mice were given cyclophosphamide 1 day before challenge with 10^7 ID_{50} of the indicated virus (as determined by enzyme elevation assay).

[c] The four LDV isolates were obtained as follows: LDV-1 from M. A. Brinton, LDV-2 from A. L. Notkins, LDV-3 from S. Schlesinger, and LDV-4 from V. Riley.

Preliminary studies indicate that control of susceptibility to LDV-induced paralysis may be multigenic and that a gene(s) of the *H-2* complex may be involved (Martinez, 1979). The two susceptible mouse strains, C58 and AKR, show a high incidence of spontaneous leukemia, and this may in some as yet unknown way be connected to the induction of paralysis in these mice by LDV (Pease and Murphy, 1980).

J. Association with Tumors

LDV has been found in association with more than 50 different transplanted murine tumors (Riley, 1968). It is widely accepted that this association is due to chance contamination of a tumor suspension by the blood of an LDV-infected host and that this then leads to the infection of each sequential tumor host with LDV. When mice receive an injection of tumor cells or tumor virus during the first week after LDV infection, an enhancement of tumor growth and oncogenicity is observed. This phenomenon appears to be related to the transient depression in cellular immunity that occurs during the first few days after LDV infection (Howard *et al.*, 1969; Michaelides and Schlesinger, 1974).

It has been reported that LDV infection delays foreign body (FB) tumorigenesis (Brinton-Darnell and Brand, 1977). Mice that received subcutaneous implants of unplasticized vinyl chloride–vinyl acetate copolymer films (0.2 × 15 × 22 mm) 2 weeks after LDV infection developed FB tumors at the same 100% incidence as uninfected mice, but at a rate that was slower by 2 months. Since the development of this type of tumor is not affected by immunosuppression (Pick, 1974; Michelich *et al.*, 1977), it seems unlikely that such a delay in FB tumorigenesis is mediated by the effect that LDV infection has on the cellular immune response. The effect of LDV on FB tumorigenesis may well be mediated through LDV-induced alterations in the monocyte population or in macrophage functions (Mahy *et al.*, 1965a,b; Steigbigel *et al.*, 1974; Riley, 1974). Although macrophages are not the originator cells of FB-induced sarcomas (Johnson *et al.*, 1973), their presence seems to be intricately involved in the multistage process of FB tumorigenesis (Brand *et al.*, 1975).

LDV synergistically increases plasma enzyme levels

produced by tumor growth (Georgii *et al.*, 1966; Adams *et al.*, 1961; Mahy *et al.*, 1964; Notkins, 1965a; Ebert *et al.*, 1967). The very high levels of certain plasma enzymes in tumor-bearing, LDV-infected mice are most likely due to the impairment of the plasma clearance system by LDV.

III. METHODS FOR DETECTING THE PRESENCE OF LDV

A. LDH Assay

LDH plasma levels increase in all mice infected with LDV. Maximum levels are observed 4 days after infection, but elevated levels persist (Fig. 1). Small samples of plasma are obtained from the retro-orbital sinus of mice with a heparinized Nathelson capillary tube. These tubes are sealed at the bottom with clay and centrifuged at 500× *g* for 20 min in a CRU-5000 Damon/IEC centrifuge. This method yields plasma which is usually free of hemolysis. If a plasma sample has a reddish color after centrifugation, we do not use it; red blood cells contain high levels of cytoplasmic LDH, and lysis of red blood cells gives a false-positive result.

LDH activity is assayed in a coupled reaction with nicotinamide-adenine dinucleotide (NAD). The activity of LDH can be assessed by measuring either the increase or the decrease in NADH with time:

$$\text{Lactate} + \text{NAD}^+ \underset{}{\overset{\text{LDH}}{\rightleftharpoons}} \text{pyruvate} + \text{NADH} \qquad (1)$$

We normally measure the disappearance of NADH. Our reaction mixture contains 2.6 ml 0.5 *M* phosphate buffer, pH 7.4, 0.2 ml plasma, and 0.1 ml NADH (2.5 mg/ml). This is mixed thoroughly in a cuvette, and then 0.1 ml sodium pyruvate (2.5 mg/ml) is added. The decrease in optical density is measured during a 1- to 2-min period at 340 nm. Information on other methods used to measure LDH activity can be found in the review by Rowson and Mahy (1975). One conventional unit of LDH activity produces a decrease in optical density of 0.001/min/ml plasma. One conventional unit/ml is equal to 0.5 IU. Normal plasma levels of LDH for mice are 400–800 conventional units/ml, whereas levels in LDV-infected mice can range from 1,800 to 16,000 units/ml.

If a sample of mouse plasma by the above assay is found to contain an elevated level of LDH, proof that this increase in LDH was caused by LDV infection can be obtained by the intraperitoneal injection of serial dilutions of this plasma into normal mice. Two or more mice are used per dilution, and a sample of their plasma is then obtained on the fourth day after injection and is in turn assayed for LDH activity. A titer of about 10^5 ID_{50}/ml is normal for plasma obtained from a mouse chronically infected with LDV.

If further analysis is required, small amounts (0.2–0.4 ml) of plasma from a mouse suspected of being infected with LDV could be used to infect 1-day-old cultures of primary mouse peritoneal exudate cells in 25-cm^2 flasks. Culture fluid is supplemented with [5-^{3}H]uridine 1–2 hr after infection, harvested 24 hr later, and layered on a sucrose density gradient. If LDV has replicated in the cultures, a peak of acid-insoluble radioactivity is observed in gradient fractions in the region of the gradient corresponding to a density of 1.13–1.14 gm/ml (Brinton-Darnell and Plagemann, 1975).

LDV does not agglutinate red blood cells and so cannot be assayed by hemagglutination. We have developed a radioimmunoassay which can be used to quantitate LDV antigens (M. A. Brinton and T. G. Tachovsky, unpublished observations), and Lagwinska and co-workers (1975) have reported on an assay method using interferon induction in macrophage cultures. At present, these alternative methods are not in general use for the titration of LDV.

B. Method of Freeing Transplantable Tumor Suspensions and Virus Pools of LDV

LDV does not replicate in any animal species other than the mouse, nor does it replicate in tumor cells in culture. If a tumor cell line is inadvertently injected into a mouse with a chronic LDV infection at some point during serial transplantation, LDV will be transferred with the tumor tissue at each subsequent passage (Riley, 1968). LDV can be eliminated either by passage of the tumor cells in another rodent species or by maintenance of the tumor cells in tissue culture (Plagemann and Swim, 1963, 1966b). The tumor cells must be maintained in culture for several passages in order to eliminate any macrophages or macrophage precursor cells which may be present. Thereafter, the tumor cells can again be maintained in mice. It would be wise to check the LDH levels of recipient mice before injection of tumor cells. Plasma LDH levels should also be assayed 4 days after injection of the cultured tumor cells to check that all LDV virions had indeed been eliminated. Likewise, pools of other viruses prepared in mice can be freed of LDV by passage in cultures of continuous cell lines derived from other animal species.

IV. REPLICATION IN TISSUE CULTURE

A. Characteristics of Target Cells

In vitro, LDV replicates only in primary cultures of normal mouse tissue (DuBuy and Johnson, 1966; Evans, 1967; Plagemann and Swim, 1966a,b; Yaffe, 1962a). Primary cultures of murine spleen, bone marrow, and embryofibroblasts propagate LDV (Evans and Salaman, 1965; Anderson *et al.*, 1965, 1966; Brinton-Darnell and Plagemann, 1975). It seems likely that monocytes or macrophages in these cultures may actually be the cells in which LDV replicates, since cultures of peritoneal exudate cells, which are rich in macrophages, seem to yield the highest titers of virus. Virus replication usually occurs in primary murine cultures only during the first week in culture; virus replicates little in cells infected later than 1 week after planting (see Fig. 4B) (Evans and Salaman, 1965; Brinton-Darnell *et al.*, 1975). Although primary peritoneal exudate cultures support LDV replication, cultures of SV40-transformed macrophages do not (Brinton-Darnell *et al.*, 1975). Neither primary cultures of rat (Evans, 1964) nor human peritoneal macrophages support LDV replication. A number of other cells, such as suckling hamster kidney cells (Tennant and Ward, 1962), murine tumor cell lines (Yaffe, 1962a,b; Plagemann and Swim, 1966a) HeLa cells (Plagemann and Swim, 1966a), and rhesus monkey kidney cells (Evans and Salaman, 1975), have been found not to replicate LDV.

Even in primary peritoneal exudate cell cultures, which produce the highest yields of LDV in culture, only a small proportion of cells seems to be infected. Although 95% of the cells in peritoneal exudate cultures prepared from starch-stimulated mice were capable of phagocytosis of latex particles, only 6–20% of these cells, when analyzed autoradiographically, showed evidence of a productive LDV infection (Tong *et al.*, 1977). The highest proportion of positive cells was observed when labeling was carried out 6–8 hr after infection. The number of positive cells decreased progressively when labeling was performed at later times after infection. Electron microscopic examination of thin sections of infected peritoneal exudate cultures revealed virions in about 3–8% of the cells (J. Stueckemann and P. G. W. Plagemann, personal communication).

Lagwinska and co-workers (1975) were able to detect pseudotypes in culture fluid after the coinfection of peritoneal exudate cells with sindbis and LDV. Pseudotypes composed of the genome of sindbis and the envelope of LDV were able to infect chick embryo cells and L cells and to produce sindbis progeny, the reverse gave no virus. This result suggests the restricted host range of LDV may not be due to an inability of the virus to adsorb to cells other than murine macrophages.

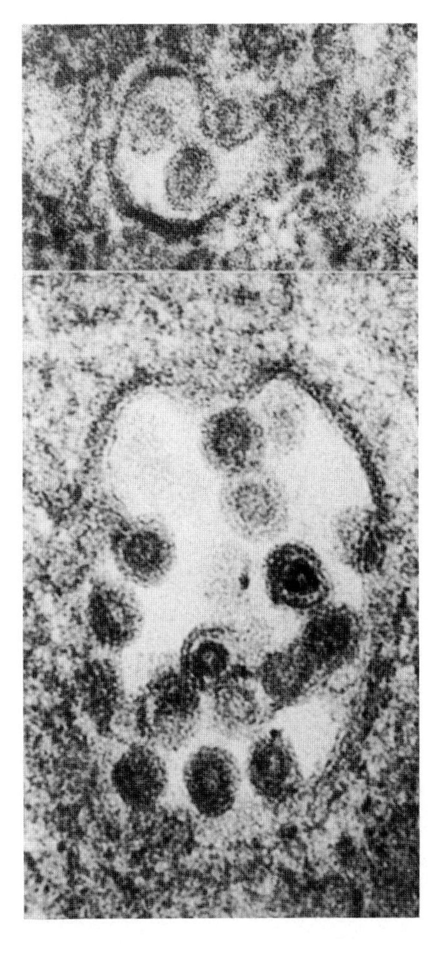

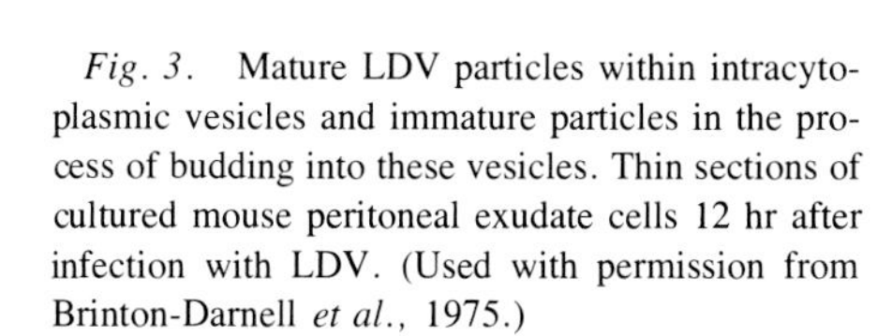
Fig. 3. Mature LDV particles within intracytoplasmic vesicles and immature particles in the process of budding into these vesicles. Thin sections of cultured mouse peritoneal exudate cells 12 hr after infection with LDV. (Used with permission from Brinton-Darnell *et al.*, 1975.)

It has been observed that LDV virions mature by budding from the cytoplasm into intracytoplasmic vesicles (Fig. 3) (Brinton-Darnell *et al.*, 1975). Large numbers of free viral cores are often observed around the membrane of cytoplasmic vesicles in cells infected with alpha or flavi togaviruses, but no accumulation of cores has been observed in LDV-infected cells. The core structure of LDV is visible only after it has begun to bud into the center of the vesicle.

Whether or not LDV has an adverse effect on the cells in which it replicates is presently not known. It was observed that, within 24 hr after infection, peritoneal exudate cells with virion-containing vesicles also contain a perinuclear accumulation of empty double-membrane vesicles. These vesicles were not observed in replicate uninfected cells and may represent a cytotoxicity induced by LDV infection (J. Stueckemann and P. G. W. Plagemann, personal communication). The gradual loss of 3–8% of the cells from an already sparse culture, due to an LDV-induced cytopathic effect, would be difficult to detect technically.

B. Kinetics of Replication

A representative growth curve for LDV in primary cultures of murine peritoneal exudate cells can be seen in Fig. 4A. The initial lag period lasts about 5 hr, and a maximum titer of virus of $10^{8.5}$ ID_{50}/ml culture fluid is observed about 16 hr after infection. Thereafter, the titer of infectious virus in the culture fluid progressively decreases. If the culture fluid is removed daily and replaced with fresh medium, the synthesis of virus during each subsequent 24-hr period progressively decreases. When cultures are infected on successive days after planting, the virus yield decreases progressively with increasing age of the culture at the time of infection (Fig. 4B). Seven-day cultures produce almost no virus.

There are several reports of prolonged production of low levels of LDV in primary murine cell cultures (Evans and Salaman, 1965; Anderson *et al.*, 1965, 1966; Plagemann and Swim, 1966a). However, prolonged shedding of virus from such cultures has not been consistently observed (Georgii and Lenz, 1964; DuBuy and Johnson, 1966). Lagwinska and coworkers (1975) reported that murine peritoneal exudate cells cultured in the presence of macrophage colony-stimulating factor, a substance present in L cell-conditioned medium (Virolainen and Defendi, 1967; Stewart *et al.*, 1975), retained their susceptibility to infection with LDV for several weeks. Although replication of peritoneal exudate cells is usually minimal in culture, cultivation in the presence of the colony-stimulating factor greatly increases the mitotic index of these cells. Stueckemann and Plagemann (1978) found that cells in

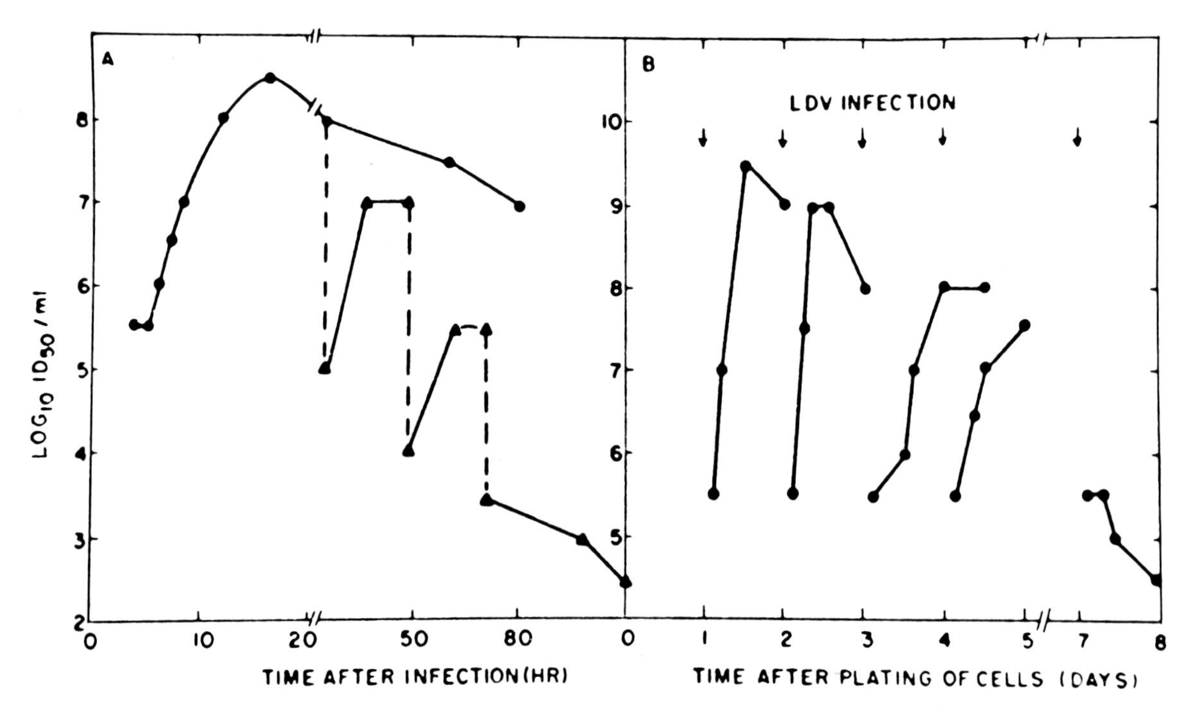

Fig. 4. Replication of LDV in cultures of mouse peritoneal exudate cells. (A) Duplicate 24-hr cultures were infected with LDV (20 ID_{50}/cell). From one culture (●- - - - -●) 0.2-ml samples were removed and infectivity assayed at the indicated times. The entire volume of medium of the second culture (▲- - - - -▲) was removed and replaced with fresh medium at 24, 48, and 72 hr. (B) Replicate cultures were infected with LDV (50 ID_{50}/cell) at 1-day intervals after planting of the cells. Samples of culture fluid were removed at the times indicated and assayed for infectivity. (Used with permission from Brinton-Darnell *et al.*, 1975.)

such cultures are still infectible by LDV months after their placement in culture. A possible explanation of this prolongation of infectibility may be that the mitogens in the media allow differentiation of stem cells present in the cultures; thus, susceptible cells are constantly being synthesized. Cultivation of peritoneal exudate cells in L cell-conditioned medium seems to promote resistance to the nonspecific cytotoxic effects of actinomycin D (Brinton and Plagemann, 1979). The growth factors in this medium may provide a culture environment which more nearly approximates the *in vivo* environment.

The production of LDV decreases by about 99% 1–2 days after infection regardless of the age of the culture at the time of infection (Stueckemann and Plagemann, 1978). A small number of infected cultures continued to shed virus at a low level (10^3–10^5 ID_{50}/ml); however, in the rest of the cultures tested, the replication of infectious LDV ceased completely. All cultures were resistant to superinfection with LDV. A possible explanation of this phenomenon may be that once a cell is infected, it synthesizes LDV for a period of only a few days Therefore, if new susceptible cells are not continuously synthesized from stem cells in the culture, the culture stops producing virus and ceases to be infectible by a superinfecting inoculum.

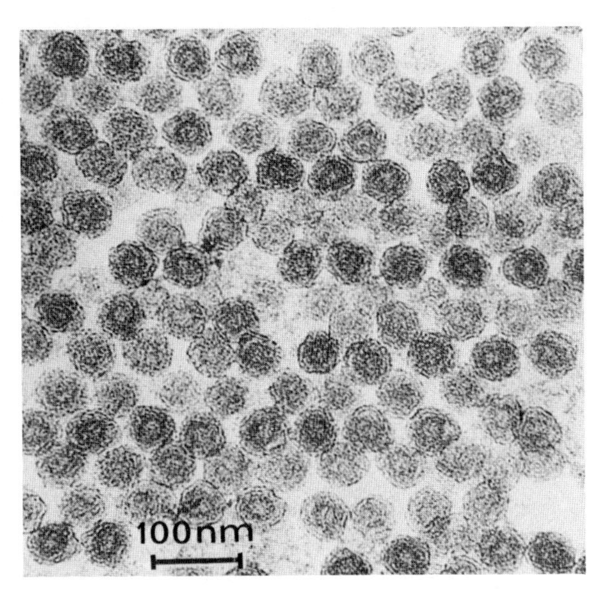

Fig. 5. A positively stained thin section of LDV pelleted from mouse plasma which was collected 24 hr after infection. (Used with permission from Brinton-Darnell and Plagemann, 1975.)

V. PROPERTIES OF THE VIRUS

A. Morphology

LDV is an enveloped virus. Its surface appears in electron micrographs to be free of spike projections (Horzinek *et al.*, 1975); however, the carbohydrate moiety of the LDV glycoprotein is most likely located on the outer surface of the viral envelope, since LDV can be precipitated by concanavalin A (Brinton-Darnell and Plagemann, 1975).

The diameter of LDV has been reported to range from 40 to 70 nm (Rowson and Mahy, 1975). Unfixed LDV is very susceptible to distortion by the usual procedures of negative staining. Several reports agree that the average diameter of LDV is 50–55 nm, with some preparations containing a few virions as large as 80 nm; such large particles may contain two nucleocapsids (Fig. 5) (Horzinek *et al.*, 1975; Brinton-Darnell and Plagemann, 1975). The diameter of the nucleocapsid of LDV is approximately 30–35 nm.

B. Susceptibility to Various Chemical Agents and Temperature

LDV is most stable in undiluted mouse plasma. However, upon purification or dilution, the virions become more sensitive to inactivation by heating, variation in the salt content of the suspending medium, and treatment with various chemicals.

LDV in mouse plasma can be stored indefinitely at −70°C without loss of titer (Notkins and Shochat, 1963). However, storage at 4°C results in a loss of infectivity with a decrease in titer by about 3.5 logs in 32 days (Riley, 1968). At room temperature, virus-containing plasma or feces retain their infectivity undiminished for about 24 hr. Infectivity is lost more rapidly at higher temperatures. It has been reported that LDV infectivity can be completely inactivated by heating at 58°C for 1 hr (Bailey *et al.*, 1963; Rowson *et al.*, 1966). Virus suspended in tissue culture fluid is more susceptible to heating at 37°C than is virus in mouse plasma (Evans and Salaman, 1965; Riley, 1968).

LDV is fairly stable between pH 6 and pH 8 but is inactivated at pH 3 (Riley, 1968; Crispens, 1965a). The virus is inactivated by lipid solvents such as ether, butanol, and chloroform (Andrewes and Horstmann, 1949; Notkins and Shochat, 1963; Crispens, 1965b; Mahy *et al.*, 1966) and is extremely sensitive to detergent treatment. A brief incubation with a nonionic detergent, such as 0.01% nonidet P-40 or Triton X-100, is sufficient for the disruption of LDV; in contrast, the same concentration of detergent has no effect on the alpha togavirus, sindbis (Brinton-Darnell and Plagemann, 1975). LDV is resistant to digestion with trypsin or papain (Crispens, 1965a,b).

C. Physicochemical Properties

It has been reported that LDV in glycerol gradients has a sedimentation coefficient of about 200 S; in contrast, its nucleocapsid sediments at 176 S (Horzinek *et al.*, 1975). The most recent studies indicate that LDV has an unusually low density in sucrose gradients (1.13–1.14 gm/ml) (Michaelides and Schlesinger, 1973; Horzinek *et al.*, 1975; Brinton-Darnell and Plagemann, 1975).

The genome of LDV is a 48 S RNA molecule with a molecular weight of 5–6 × 10^6. That the RNA is sensitive to ribonuclease (RNase) digestion shows that it is single-stranded. Purified viral RNA is infectious when injected into mice intracerebrally, a route which minimizes the chance of RNA digestion by ribonucleases (Notkins and Scheele, 1963; Rowson *et al.*, 1968a,b; Darnell and Plagemann, 1972; Niwa *et al.*, 1973).

LDV virions are composed of at least three structural proteins. The largest of these, VP3, is a glycoprotein. VP2 is a non-glycosylated protein which can be removed with the viral envelope by detergent treatment. VP1 is the capsid protein. The molecular weights of the LDV proteins have been estimated by two laboratories. Michaelides and Schlesinger (1973) reported molecular weights of 13,000, 17,000, and 28,000 for VP1, VP2, and VP3, respectively, whereas Brinton-Darnell and Plagemann (1975) estimated molecular weights of 15,000, 18,000 and 24,000–44,000. The molecular weight of VP3 could not be estimated with accuracy because it always appeared as a wide heterogeneous band. The molecular ratios of VP1, VP2, and VP3 within LDV virions have been estimated at 3–5:1:1 by Michaelides and Schlesinger (1973) and 2–3:1:1 by Brinton-Darnell and Plagemann (1975). Recent analysis on 5–20% gradient acrylamide slab gels indicate that as many as three bands may be separable within the VP3 band (M. A. Brinton, unpublished observations; S. Schlesinger, personal communication).

D. Intracellular Viral RNA Synthesis

The synthesis of LDV RNA can be followed in cultures of peritoneal exudate cell cultures in the presence of actinomycin D. Heterogeneous RNA sedimenting at 10–40 S can be detected in infected cells within 3 hr after infection. It has been found that some of the heterogeneous RNA species are partially resistant to RNase treatment; a predominant species which sediments at 29 S remains after RNase treatment. Although this heterogeneous RNA may represent degradation products of viral RNA, the partial RNase resistance indicates that viral replication complexes may be present in this region of the gradient. The 48 S viral RNA is first observed 4–5 hr after infection (Brinton-Darnell *et al.*, 1975). The time period required for a completed genome to appear extracellularly within a mature virion is about 1.5 hr at 37°C.

The synthesis of intracellular viral RNA is completely inhibited by the addition of cycloheximide to the culture medium 0.5–1 hr after infection, which indicates that protein synthesis is necessary for the initiation of viral replication. Furthermore, no virion-associated RNA polymerase activity could be detected in preparations of partially purified virions.

Yamazaki and Notkins (1973) reported that the 24-hr yield of infectious LDV from primary mouse embryo cultures was reduced by 75–90% when actinomycin D was added to the culture medium during the first 9 hr after infection. However, since actinomycin D has a rapid toxic effect on macrophages in culture, the effect observed by Yamazaki and Notkins may reflect a general cytotoxic effect rather than a specific dependence of LDV replication on DNA-dependent RNA synthesis (Brinton and Plagemann, 1979). Cultures grown in L cell-conditioned medium appear to be more resistant to the cytotoxic effect of actinomycin D.

Because of technical difficulties which have so far not allowed viral proteins to be distinguished from the background of cellular proteins, the synthesis of LDV-specific proteins has not yet been studied intracellularly.

E. Interferon

LDV in tissue culture appears to be a poor inducer of interferon. Notkins (1971a) and Yamazaki and Notkins (1973) were unable to detect any interferon activity in primary mouse embryo cultures infected with LDV, and DuBuy and co-workers (1973) could not detect interferon in mouse peritoneal exudate cultures after LDV infection.

In vivo, interferon may be involved in bringing about the decline in plasma viral titers that begins about 24 hr after infection (Fig. 1). High levels of interferon have been found in the plasma at this time (Baron *et al.*, 1964, 1966; DuBuy and Johnson, 1965; Falke and Rowe, 1965; Evans and Riley, 1968; DuBuy *et al.*, 1973). Results of experiments using the interferon inducer statolon support the theory that interferon plays a role in halting the initial rapid replication of LDV. The plasma titer of LDV is significantly reduced 24 hr after infection when an intraperitoneal injection of 4.5 mg statolon is given 6–12 hr before the virus is inoculated (Crispens, 1970).

LDV replication is sensitive to the action of interferon. Primary mouse embryo cultures incubated for 16 hr with mouse interferon before infection with LDV produced 99% less LDV than untreated cultures (Notkins, 1971; Yamazaki and Notkins, 1973). LDV replication in peritoneal exudate cultures is not very sensitive either to exogenous mouse interferon or to interferon induced in these cultures with Newcastle disease virus or with defective-interfering particles of vesicular stomatitis virus containing a double-stranded RNA (Stueckmann and Plagemann, 1979).

During the first few days after planting, peritoneal exudate cultures do not produce interferon upon infection with LDV

(Evans, 1970; Yamazaki and Notkins, 1973; Lagwinska *et al.*, 1975). However, Lagawinska and co-workers (1975) have shown that if exudate cells are cultivated in L cell-conditioned medium, the ability of these cells to produce interferon after LDV infection increases substantially and seems to correlate with the initiation of cell replication. It is possible that only proliferating cells may be induced to produce interferon.

F. Antigenic Nature

Serologic studies of LDV have not been carried out because conventional neutralization tests cannot be performed. Virus-antibody complexes retain some degree of infectivity in mice (DuBuy and Johnson, 1965; Notkins *et al.*, 1966a,b; Rowson *et al.*, 1966), and no plaque assay exists for LDV. Two further complicating factors are (1) that only virus harvested from infected mice during the first 5 days after infection is free of antiviral antibody, and (2) that because of the persistent viremia, mouse serum which contains antiviral antibody also contains infectious LDV (Notkins, 1971a,b). Since immune complexes retain infectivity, complete neutralization of LDV infectivity could not be achieved by addition of excess antiviral antibody or by addition of antimouse γ-globulin antibody (Notkins *et al.*, 1966a,b).

It is not clear whether or not anti-LDV antibody can be produced in animal species other than mice. A few attempts have been made to obtain antibody from rats and hamsters (Bailey *et al.*, 1964; Riley, 1968); however, whether or not anti-LDV antibody was actually produced was not rigorously assessed.

We have developed a radioimmunoassay for LDV (M. A. Brinton and T. G. Tachovsky, unpublished observations). This assay is currently being employed in the assessment of the antigenic relationship of LDV to alpha and flavi togaviruses and in the analysis of possible antigenic differences between various isolates of LDV. It had not been possible to study these aspects of LDV previously.

VI. CONCLUSION

Although much has been learned about the structure and replication of LDV since its discovery in 1960 by Riley *et al.* (1960), many questions about this interesting virus remain unanswered. The inability of LDV to replicate in transformed cell lines and to cause a detectable cytopathic effect in cultures in which it does replicate, such as mouse peritoneal exudate cells, represent annoying technical difficulties to further analysis of LDV.

Additional viruses able to produce chronic infections which are accompanied by no obvious clinical symptoms in their hosts may well exist in nature. There is no reason to suppose that such viruses would induce an increase in the host's serum enzyme levels, the indicator which led to the fortuitous discovery of LDV. The detection of these viruses might well prove difficult. However, such viruses may indeed be responsible for certain effects, which are currently attributed to known infectious agents or to tumors. This situation certainly occurred during the course of murine tumor research, in which many transplantable tumors were inadvertently contaminated with LDV. The unknown presence of LDV in mouse tumor and oncogenic virus preparations led investigators incorrectly to ascribe certain observed biological effects to the presence of the tumor or the oncogenic virus which were, in actuality, caused by the contaminating LDV. LDV infection has been demonstrated to alter host immune competence as well as the rate of clearance of therapeutic drugs (Riley, 1974).

The further study of the interaction of relatively "silent" viruses such as LDV with their host species should lead to new insights into the mechanisms by which viruses can establish and maintain relatively harmless, persistent infections in their hosts.

ACKNOWLEDGMENT

Dr. Brinton is supported in part by grant NS-11036 from the National Institutes of Health.

REFERENCES

Adams, D. H., Rowson, K. E. K., and Salaman, M. H. (1961). The effect of tumours, of leukemia, and of some viruses associated with them, on the plasma lactic dehydrogenase activity of mice. *Br. J. Cancer* **15**, 860–867.

Anderson, H. C. V., Riley, V., Wade, P., and Moore, A. E. (1965). Quantitative evidence for propagation of the lactate dehydrogenase (LDH) elevating virus in mouse embryo cell cultures. *Proc. Am. Assoc. Cancer Res.* **6**, 2.

Anderson, H. C. V., Riley, V., Fitzmaurice, M. A., Loveless, J. D., Wade, P., and Moore, A. E. (1966). Quantitative study of the lactate dehydrogenase-elevating virus in mouse embryo cultures. *J. Natl. Cancer Inst.* **36**, 89–95.

Andrewes, C. H., and Horstmann, D. M. (1949). The susceptibility of viruses to ethyl ether. *J. Gen. Microbiol.* **3**, 290–297.

Bailey, J. M., and Monroe, M. (1972). Studies on replication of LDH virus. *Fed. Proc., Fed. Am. Soc. Exp. Biol.* **31**, 836.

Bailey, J. M., and Wright, D. A. (1965). Plasma enzyme elevations with LDH viruses from different tumors. *Proc. Soc. Exp. Biol. Med.* **120**, 346–350.

Bailey, J. M., Stearman, M., and Clough, J. (1963). LDH levels in blood and tissues of mice infected with LDH agent. *Proc. Soc. Exp. Biol. Med.* **114**, 148–153.

Bailey, J. M., Clough, J., and Stearman, M. (1964). Clearance of plasma enzymes in normal and LDH-agent infected mice. *Proc. Soc. Exp. Biol. Med.* **117**, 350–354.

Baron, S., Buckler, C. E., Friedman, R. M., and McCloskey, R. V. (1964). Role of interferon during viraemia in mice. *Bacteriol. Proc.* p. 116.

Baron, S., Buckler, C. E., McCloskey, R. V., and Kirschstein, R. L. (1966). Role of interferon during viraemia. 1. Production of circulating interferon. *J. Immunol.* **96,** 12–16.

Ben-Yaakov, M., and Haran-Ghera, N. (1975). T and B lymphocytes in thymus of SJL/J mice. *Nature (London)* **255,** 64–66.

Brand, K. G., Buoen, L. C., and Johnson, K. H. (1975). Etiological factors, stages, and the role of the foreign body in foreign body tumorigenesis: A review. *Cancer Res.* **35,** 279–286.

Brinton, M. A. (1980). Non-arbo togaviruses. *In* "Togaviruses" (W. Schlesinger, ed.), pp. 623–666. Academic Press, New York.

Brinton, M. A., and Brand, I. (1977). Delayed foreign-body tumorigenesis in mice infected with lactate dehydrogenase-elevating virus: Brief communication. *J. Natl. Cancer Inst.* **59,** 1027–1029.

Brinton, M. A., and Plagemann, P. G. W. (1977). Unique response of SJL/J mice to infection with lactate dehydrogenase-elevating virus (LDV). *Annu. Meet. Am. Soc. Microbiol.* Abstr., p. 327.

Brinton, M. A., and Plagemann, P. G. W. (1979). Actinomycin D cytotoxicity for mouse peritoneal macrophages and effect on lactate dehydrogenase-elevating virus replication. *Intervirology* **12,** 349–356.

Brinton-Darnell, M., and Plagemann, P. G. W. (1975). Structure and chemical-physical characteristics of lactate dehydrogenase-elevating virus and its RNA. *J. Virol.* **16,** 420–433.

Brinton-Darnell, M., Collins, J. K., and Plagemann, P. G. W. (1975). Lactate dehydrogenase-elevating virus replication, maturation and viral RNA synthesis in primary mouse macrophage cultures. *Virology* **65,** 187–195.

Clough, J. D., and Bailey, J. M. (1965). Mechanism of plasma enzyme elevation by tumor LDH agent. *Tex. Rep. Biol. Med.* **23,** 644–645.

Crispens, C. G. (1963). Serum lactic dehydrogenase levels in mice during the development of autochthonous and chemically induced tumors. *J. Natl. Cancer Inst.* **30,** 361–366.

Crispens, C. G. (1964a). The lactic dehydrogenase agent; its possible implications for the virologist and the oncologist. *Bull., Univ. Md. Sch. Med.* **49,** vii.

Crispens, C. G. (1964b). On the epizootiology of the lactic dehydrogenase agent. *J. Natl. Cancer Inst.* **32,** 497–505.

Crispens, C. G. (1965a). On the properties of the lactic dehydrogenase agent. *J. Natl. Cancer Inst.* **35,** 975–979.

Crispens, C. G. (1965b). Properties of lactic dehydrogenase-elevating agents. *Anat. Rec.* **151,** 448–449.

Crispens, C. G. (1968). Antibody response in normal and neonatally thymectomized mice infected with the lactate dehydrogenase virus. *Anat. Rec.* **160,** 466.

Crispens, C. G. (1970). Effect of statolon on lactate dehydrogenase virus infection in mice. *Arch. Gesamte Virusforsch.* **31,** 191–195.

Crispens, C. G. (1971). Studies on the response of SJL/J mice to infection with the lactate dehydrogenase virus. *Arch. Gesamte Virusforsch.* **35,** 177–182.

Crispens, C. G. (1972). Genetic control of the response of SJL/J mice to LDH virus infection. *Arch. Gesamte Virusforsch.* **38,** 225–227.

Darnell, M. B., and Plagemann, P. G. W. (1972). Physical properties of lactic dehydrogenase-elevating virus and its ribonucleic acid. *J. Virol.* **10,** 1082–1085.

DuBuy, H. G., and Johnson, M. L. (1965). Some properties of the lactic dehydrogenase agent of mice. *J. Exp. Med.* **122,** 587–600.

DuBuy, H. G., and Johnson, M. L. (1966). Studies on the *in vitro* and *in vivo* multiplication of the LDH virus of mice. *J. Exp. Med.* **123,** 985–998.

DuBuy, H., Baron, S., Uhlendorf, C., and Johnson, M. L. (1973). Role of interferon in murine lactic dehydrogenase virus infection, *in vivo* and *in vitro*. *Infect. Immun.* **8,** 977–984.

Duffey, P. S., Martinez, D., Abrams, G. D., and Murphy, W. H. (1976). Pathogenetic mechanisms in immune polioencephalomyelitis: induction of disease in immunosuppressed mice. *J. Immunol.* **116,** 475–481.

Ebert, P. S., Chirigos, M. A., Fields, L. A., and Ellsworth, P. A. (1967). Plasma lactate dehydrogenase and spleen heme biosynthetic activity following Friend and Rauscher leukemia virus infections. *Life Sci.* **6,** 1963–1971.

Evans, R. (1964). Replication of Riley's plasma enzyme-elevating virus *in vitro*. *J. Gen. Microbiol.* **37,** vii.

Evans, R. (1967). Replication of Riley's plasma enzymes-elevating virus in tissue culture: the importance of the cellular composition. *J. Gen. Virol.* **1,** 363–374.

Evans, R. (1970). Further studies on the replication of the lactate dehydrogenase-elevating virus (LDH virus) in peritoneal macrophage. *J. Gen. Microbiol.* **57,** XXI.

Evans, R., and Riley, V. (1968). Circulating interferon in mice infected with the lactate dehydrogenase-elevating virus (LDH virus). *J. Gen. Virol.* **3,** 449–452.

Evans, R., and Salaman, M. H. (1965). Studies on the mechanism of action of Riley virus. III. Replication of Riley's plasma enzyme-elevating virus *in vitro*. *J. Exp. Med.* **122,** 993–1002.

Falke, D., and Rowe, W. P. (1965). Die interferenz zwischen dem polyomavirus und dem stomatitis-vesicularis-virus in der mäus. *Arch. Gesamte Virusforsch.* **15,** 210–219.

Fenner, F. (1977). Classification and nomenclature of viruses. *Intervirology* **7,** 44–47.

Field, E. J., and Adams, D. H. (1968). Riley virus in wild mice. *Lancet* **i,** 868.

Georgii, A. (1962). Die aktivitätsänderung der lactatdehydrogenase im serum nach infektion mit geschwulstvirus der mäus als regulationsstörung *in vivo*. *Verh. Dtsch. Pathol. Ges.* **46,** 357–358.

Georgii, A., and Kirschenhofer, I. (1965). Über die isolierung von lactatdehydrogenase-erhöhendem virus aus wilden mäusen. *Z. Naturforsch., Teil B* **20,** 1310.

Georgii, A., and Lenz, I. (1964). Failure to propagate a lactic dehydrogenase-elevating agent from mice tumours in mice embryo cultures. *Nature (London)* **202,** 1228–1229.

Georgii, A., Bayerle, H., Brdiczka, D., and Zobl, H. (1963). Über ein die lactatdehydrogenase im serum aktivierendes virus aus geschwülsten der mäus. *Z. Krebsforsch.* **65,** 334–341.

Georgii, A., Thorn, L., and Wrba, H. (1966). Action of Riley's enzyme-elevating virus on tumour-bearing mice. *Nature (London)* **209,** 929–930.

Halpern, B. N., Biozzi, G., Nicol, T., and Bilbey, D. L. J. (1957). Effect of experimental biliary obstruction on the phagocytic activity of the reticuloendothelial system. *Nature (London)* **180,** 503–504.

Hanna, M. G., Szakal, A. K., and Tyndall, R. L. (1970a). Histoproliferative effect of Rauscher leukemia virus on lymphatic tissue: histological and ultrastructural studies of germinal centers and their relation to leukemogenesis. *Cancer Res.* **30,** 1748–1763.

Hanna, M. G., Walburg, H. E., Tyndall, R. C., and Snodgrass, M. J. (1970b). Histoproliferative effect of Rauscher leukemia virus on lymphatic tissue. II. Antigen-stimulated germfree and conventional BALB/c mice. *Proc. Soc. Exp. Biol. Med.* **134,** 1132–1141.

Horzinek, M., Maess, J., and Laufs, R. (1971). Studies on the substructure of togaviruses. *Arch. Gesamte Virusforsch.* **33,** 306–318.

Horzinek, M. C., van Wielink, P. S., and Ellens, D. J. (1975). Purification and electron microscopy of lactic dehydrogenase virus in mice. *J. Gen. Virol.* **26,** 217–226.

Howard, R. J., Notkins, A. L., and Mergenhagen, S. E. (1969). Inhibition of cellular immune reactions in mice infected with lactic dehydrogenase virus. *Nature (London)* **221,** 873–874.

Johnson, K. H., Ghobrial, H. K., Buoen, L. C., Brand, I., and Brand, K. G. (1973). Intracisternal type A particles occurring in foreign body-induced sarcomas. *Cancer Res.* **33,** 1165–1168.

Lagwinska, E., Stewart, C. C., Adles, C., and Schlesinger, S. (1975). Replication of lactic dehydrogenase virus and sindbis virus in mouse peritoneal macrophages. Induction of interferon and phenotypic mixing. *Virology* **65,** 204–214.

Mahy, B. W. J. (1964). Action of Riley's plasma enzyme-elevating virus in mice. *Virology* **24,** 481–483.

Mahy, B. W. J., and Rowson, K. E. K. (1965). Isoenzymic specificity of impaired clearance in mice infected with Riley virus. *Science* **149,** 756–757.

Mahy, B. W. J., and Wachsmuth, E. D. (1973). Studies on the clearance of lactic dehydrogenase (LDH) isoenzymes from plasma of normal mice and mice infected with lactic dehydrogenase virus (LDV). *J. Med. Microbiol.* **6,** Px.

Mahy, B. W. J., Rowson, K. E. K., Salaman, M. H., and Parr, C. W. (1964). Plasma enzyme levels in virus infected mice. *Virology* **23,** 528–541.

Mahy, B. W. J., Rowson, K. E. K., and Parr, C. W. (1965a). Studies on the mechanism of action of Riley virus. IV. The reticuloendothelial system and impaired plasma enzyme clearance in infected mice. *J. Exp. Med.* **125,** 277–288.

Mahy, B. W. J., Rowson, K. E. K., Parr, C. W., and Salaman, M. H. (1965b). Studies on the mechanism of action of Riley virus. I. Action of substances effecting the reticuloendothelial system on plasma enzyme levels in mice. *J. Exp. med.* **122,** 967–981.

Mahy, B. W. J., Harvey, J. J., and Rowson, K. E. K. (1966). Some physical properties of a murine sarcoma virus (Harvey). *Tex. Rep. Biol. Med.* **24,** 620–628.

Martinez, D. (1979). Histocompatibility-linked genetic control of susceptibility to age-dependent polioencephalomyelitis in mice. *Infect. Immun.* **23,** 133–139.

Martinez, D., Brinton, M. A., Tachovsky, T. G., and Phelps, A. H. (1980). Identification of lactate dehydrogenase-elevating virus as the etiologic agent of the genetically restricted age-dependent polioencephalomyelitis of mice. *Infect. Immun.* **27,** 979–987.

Michaelides, M. C., and Schlesinger, S. (1973). Structural proteins of lactic dehydrogenase virus. *Virology* **55,** 211–217.

Michaelides, M. C., and Schlesinger, S. (1974). Effect of acute or chronic infection with lactic dehydrogenase virus (LDV) on the susceptibility of mice to plasmacytoma MOPC-315. *J. Immunol.* **112,** 1560–1564.

Michelich, V. J., Buoen, L. C., and Brand, K. G. (1977). Immunosuppression studies in foreign body tumorigenesis: No evidence for tumor-specific antigenicity. *J. Natl. Cancer Inst.* **58,** 757–761.

Mozes, E., Isac, R., and Taussig, M. J. (1975). Antigen-specific T-cell factors in the genetic control of the immune response to poly (TYR, GLU)-poly dl ALA-polylys. Evidence for T-cell and B-cell defects in SJL mice. *J. Exp. Med.* **141,** 703–707.

Mundy, J., and Williams, P. C. (1961). Transmissible agent associated with some mouse neoplasms. *Science* **134,** 834–835.

Murphy, W. H., Tam, M. R., Lanzi, R. L., Abell, M. R., and Kauffman, C. (1970). Age dependence of immunologically induced central nervous system disease in C58 mice. *Cancer Res.* **30,** 1612–1622.

Niwa, A., Yamazaki, S., Bader, J., and Notkins, A. L. (1973). Incorporation of labeled precursors into the RNA and proteins of lactic dehydrogenase virus. *J. Virol.* **12,** 401–404.

Notkins, A. L. (1963). Studies of the properties and transmission of the lactic dehydrogenase agent. *Proc. Am. Assoc. Cancer Res.* **4,** 48.

Notkins, A. L. (1965a). Lactic dehydrogenase virus. *Bacteriol. Rev.* **29,** 143–160.

Notkins, A. L. (1965b). Studies on the mechanism of enzyme elevation in mice infected with the lactic dehydrogenase virus. *Fed. Proc., Fed. Am. Soc. Exp. Biol.* **24,** 378.

Notkins, A. L. (1966a). Infectious virus-antibody complexes during chronic viremia. *Proc. Int. Cancer Congr., 9th, Tokyo.* p. 310.

Notkins, A. L. (1966b). Catabolism of γ-globulin and increased antibody production in mice infected with lactic dehydrogenase virus. *Proc. Int. Congr. Microbiol., 9th, Moscow* p. 628.

Notkins, A. L. (1966c). Infectious virus-antibody complex. *Fed. Proc., Fed. Am. Soc. Exp. Biol.* **25,** 615.

Notkins, A. L. (1971a). Enzymatic and immunologic alterations in mice infected with lactic dehydrogenase virus. *Am. J. Pathol.* **64,** 733–746.

Notkins, A. L. (1971b). Infectious virus-antibody complexes-interaction with anti-immunoglobulins, complement, and rheumatoid factor. *J. Exp. Med.* **134,** 41s–51s.

Notkins, A. L., and Scheele, C. (1963). An infectious nucleic acid from the lactic dehydrogenase agent. *Virology* **20,** 640–642.

Notkins, A. L., and Scheele, C. (1964). Impaired clearance of enzymes in mice infected with the lactic dehydrogenase agent. *J. Natl. Cancer Inst.* **33,** 741–749.

Notkins, A. L., and Shochat, S. J. (1963). Studies on the multiplication and the properties of the lactic dehydrogenase agent. *J. Exp. Med.* **117,** 735–747.

Notkins, A. L., Berry, R. J., Moloney, J. B., and Greenfield, R. E. (1962). Relationship of the lactic dehydrogenase factor to certain murine tumors. *Nature (London)* **193,** 79–80.

Notkins, A. L., Mahar, S., Scheele, C., and Goffman, J. (1966a). Infectious virus-antibody complex in the blood of chronically infected mice. *J. Exp. Med.* **124,** 81–97.

Notkins, A. L., Mergenhagen, S. E., Rizzo, A. A., Scheele, C., and Waldmann, T. A. (1966b). Elevated γ-globulin and increased antibody production in mice infected with lactic dehydrogenase virus. *J. Exp. Med.* **123,** 347–364.

Notkins, A. L., Mage, M., Ashe, W. K., and Mahar, S. (1968). Neutralization of sensitized lactic dehydrogenase virus by anti-γ-globulin. *J. Immunol.* **100,** 314–320.

Oldstone, M. B. A., and Dixon, F. J. (1972). Inhibition of antibodies to nuclear antigen and to DNA in New Zealand mice infected with lactate dehydrogenase virus. *Science* **175,** 784–786.

Oldstone, M. B. A., Tishon, A., and Chiller, J. M. (1974). Chronic virus infection and immune responsiveness. II. Lactic dehydrogenase virus infection and immune response to non-viral antigens. *J. Immunol.* **112,** 1260–1263.

Pease, L. R., and Murphy. W. H. (1980). Co-infection by lactic dehydrogenase virus and C-type retrovirus elicits neurological disease. *Nature (London)* **286,** 398–400.

Pick, C. R. (1974). Effects of long-term immunosuppression on spontaneous and polymer-induced tumours in rats. *Imp. Cancer Res. Fund Sci. Rep., London* pp. 128–129.

Plagemann, P. G. W., and Swim, H. E. (1963). Studies of the plasma lactic dehydrogenase-elevating virus (PLDEV) of mice. *Proc. Am. Assoc. Cancer Res.* **4,** 53.

Plagemann, P. G. W., and Swim, H. E. (1966a). Propagation of lactic dehydrogenase-elevating virus in cell culture. *Proc. Soc. Exp. Biol. Med.* **121,** 1147–1152.

Plagemann, P. G. W., and Swim, H. E. (1966b). Relationship between the lactic dehydrogenase-elevating virus and transplantable murine tumors. *Proc. Soc. Exp. Biol. Med.* **121,** 1142–1146.

Plagemann, P. G. W., Watanabe, M., and Swim, H. E. (1962). Plasma lactic dehydrogenase-elevating agent of mice: effect on levels of additional enzymes. *Proc. Soc. Exp. Biol. Med.* **111,** 749–754.

Plagemann, P. G. W., Gregory, K. F., Swim, H. E., and Chan, K. K. W. (1963). Plasma lactic dehydrogenase-elevating agent of mice: distribution in tissues and effect on lactic dehydrogenase isozyme patterns. *Can. J. Microbiol.* **9,** 75–86.

Pope, J. H. (1961). Studies of a virus isolated from a wild house mouse, *mus musculus,* and producing splenomegaly and lymph node enlargement in mice. *Aust. J. Exp. Biol. Med. Sci.* **39,** 521–536.

Pope, J. H., and Rowe, W. P. (1964). Identification of WMI as LDH virus, and its recovery from wild mice in Maryland. *Proc. Soc. Exp. Biol. Med.* **116,** 1015–1019.

Porter, D. D., and Porter, H. G. (1971). Deposition of immune complexes in the kidneys of mice infected with lactic dehydrogenase virus. *J. Immunol.* **106,** 1264–1266.

Porter, D. D., Porter, H. G., and Deerhake, B. B. (1969). Immunofluorescence assay for antigen and antibody in lactic dehydrogenase virus infection of mice. *J. Immunol.* **102,** 431–436.

Proffitt, M. R., and Congdon, C. C. (1970). The effect of a large dose of LDH virus on mouse lymphatic tissue. *Fed. Proc., Fed. Am. Soc. Exp. Biol.* **29,** 559.

Proffitt, M. R., Congdon, C. C., and Tyndall, R. L. (1972). The combined action of Rauscher leukemia virus and lactic dehydrogenase virus on mouse lymphatic tissue. *Int. J. Cancer* **9,** 193–211.

Riley, V. (1961). Virus-tumor synergism. *Science* **134,** 666–668.

Riley, V. (1962). Role of viruses in glycolysis of tumors and hosts. *Fed. Proc., Fed. Am. Soc. Exp. Biol.* **21,** 21–87.

Riley, V. (1964). Synergism between lactate dehydrogenase-elevating virus and *Eperythrozoon coccoides. Science* **146,** 921–923.

Riley, V. (1968). Lactate dehydrogenase in the normal and malignant state in mice and the influence of a benign enzyme-elevating virus. *Methods Cancer Res.* **4,** 493–618.

Riley, V. (1974). Persistence and other characteristics of the lactate-dehydrogenase-elevating virus (LDH-virus). *Prog. Med. Virol.* **18,** 198–213.

Riley, V., and Wroblewski, F. (1960). Serial lactic dehydrogenase activity in plasma of mice with growing or regressing tumors. *Science* **132,** 151–152.

Riley, V., Lilly, F., Huerto, E., and Bardell, D. (1960). Transmissible agent associated with 26 types of experimental mouse neoplasms. *Science* **132,** 545–547.

Riley, V., Loveless, J. D., Fitzmaurice, M. A., and Siler, W. M. (1965). Mechanism of lactate dehydrogenase (LDH) elevation in virus infected hosts. *Life Sci.* **4,** 487–507.

Rowson, K. E. K. (1963). Riley virus in wild mice, effect of drugs on replication of Riley viruses. *Br. Emp. Cancer Campaign Annu. Rep.* **41,** 222–223.

Rowson, K. E. K., and Mahy, B. W. J. (1975). Lactic dehydrogenase virus. *Virol. Monogr.* **13,** 1–121.

Rowson, K. E. K., and Michaels, L. (1973). Lactic dehydrogenase (LDH) virus and its localization by immunofluorescence. *J. Med. Microbiol.* **6,** Pxi.

Rowson, K. E. K., Adams, D. H., and Salaman, M. H. (1963). Riley's enzyme-elevating virus; a study of the infection in mice and its relation to virus-induced leukemia. *Acta Unio Int. Cancrum* **19,** 404–406.

Rowson, K. E. K., Mahy, B. W. J., and Bendinelli, M. (1966). Riley virus neutralizing activity in the plasma of infected mice with persistent viraemia. *Virology* **28,** 775–778.

Rowson, K. E. K., Parr, I. B., and Alper, T. (1968a). The radiation target size of Riley virus infectivity. *J. Gen. Microbiol.* **50,** v.

Rowson, K. E. K., Parr, I. B., and Alper, T. (1968b). Radiation target size of Riley virus. *Virology* **36,** 157–159.

Salaman, M. H., and Wedderburn, N. (1966). The immunodepressive effect of Friend virus. *Immunology* **10,** 445–458.

Santisteban, G. A., Riley, V., and Willhight, K. (1970). Studies in virus-tumor relationships: responses of the adrenocortical-thymolymphatic system to the LDH-elevating virus. *Int. Cancer Congr., Abstr., 10th, Houston, Tex.* p. 302.

Santisteban, G. A., Riley, V., and Fitzmaurice, M. A. (1972). Thymolytic and adrenal cortical responses to the LDH-elevating virus. *Proc. Soc. Exp. Biol. Med.* **139,** 202–206.

Snodgrass, M. J., Lowrey, D. S., and Hanna, M. G. (1972). Changes induced by lactic dehydrogenase virus in thymus and thymus-dependent areas of lymphatic tissue. *J. Immunol.* **108,** 877–892.

Speckman, D., Riley, V., Santisteban, G. A., Kirk, W., and Bredberg, L. (1974). The role of stress in producing elevated corticosterone levels and thymus involution in mice. *Int. Cancer Congr., Abstr., 11th* p. 382.

Steigbigel, R. T., Oldstone, M. B., and Remington, J. S. (1974). Induction of non-specific resistance by viral infections. *Annu. Meet. ICAAC, 14th, San Francisco, Calif.* Abstr. No. 11.

Stewart, C. C., Lin, H. S., and Adles, C. (1975). Proliferation and colony-forming ability of peritoneal exudate cells in liquid culture. *J. Exp. Med.* **141,** 1114–1132.

Stueckemann, J., and Plagemann, P. G. W. (1978). Persistent infection of mouse peritoneal exudate cells by lactate dehydrogenase-elevating virus (LDV) *in vitro.* ICN-UCLA Symposia on Molecular and Cellular Biology, p. 247. Academic Press, New York.

Tennant, R. W., and Ward, T. G. (1962). Pneumonia virus of mice (PVM) in cell culture. *Proc. Soc. Exp. Biol. Med.* **111,** 395–398.

Tong, S. L., Stueckemann, J., and Plagemann, P. G. W. (1977). An autoradiographic method for detection of lactate dehydrogenase-elevating virus (LDV)-infected cells in primary mouse macrophage cultures. *J. Virol.* **22,** 219–227.

Virolainen, M., and Defendi, V. (1967). Dependence of macrophage growth *in vitro* upon interaction with other cells. *In* "Growth Regulating Substances for Animal Cells in Culture" (V. Defendi and M. Stoker, eds.), pp. 67–83. Wistar Inst. Press, Philadelphia, Pennsylvania.

Wakim, K. G., and Fleisher, G. A. (1963a). The fate of enzymes in body fluids—an experimental study. II. Disappearance rates of glutamic-oxalacetic transaminase I under various conditions. *J. Lab. Clin. Med.* **61,** 86–97.

Wakim, K. G., and Fleisher, G. A. (1963b). The fate of enzymes in body fluids—an experimental study. IV. Relationship of the reticuloendothelial system to activities and disappearance rates of various enzymes. *J. Lab. Clin. Med.* **61,** 107–119.

Warnock, M. L. (1964). Isozymic patterns in organs of mice infected with LDH agent. *Proc. Soc. Exp. Biol. Med.* **115,** 448–452.

Yaffe, D. (1962a). The distribution and *in vitro* propagation of an agent causing high plasma lactic dehydrogenase activity. *Cancer Res.* **22,** 573–580.

Yaffe, D. (1962b). Studies on an agent associated with high plasma lactic dehydrogenase activity. *Acta Unio Int. Cancrum* **19,** 407–409.

Yamazaki, S., and Notkins, A. L. (1973). Inhibition of replication of lactic dehydrogenase virus by actinomycin. *J. Virol.* **11,** 473–478.

Chapter 11

Mousepox

Frank Fenner

I. Introduction ... 209
II. Viral Agent ... 210
A. Historical Background ... 210
B. Properties of the Virus ... 210
C. Pathogenesis ... 211
D. Epizootiology ... 220
E. Diagnosis ... 224
F. Control and Prevention ... 226
III. Summary: The Practical Problems ... 227
A. Problems of the Mouse Breeder ... 227
B. Problems of the Research Worker ... 227
References ... 228

I. INTRODUCTION

First recognized in 1930 (Marchal, 1930), when the use of mice as experimental animals in virology was just beginning (see Burnet, 1960, Table 1), infectious ectromelia, or mousepox, has had a rather different history in the four continents where laboratory mice have been extensively used: Europe, North America, Australia, and Japan. In Europe and Japan it was soon found to be present, usually as an unrecognized enzootic infection, in many breeding colonies. As well as threatening potentially valuable mouse stocks, this infection complicated, and continues to complicate, much virological research involving serial passages of viruses or tumors in mice, and the presence of enzootic mousepox has also rendered suspect several studies of the nature of the disease itself. By chance, most mouse colonies established in North America were free from the disease. When the virus was inadvertently imported into the United States from Europe with mouse stocks or mouse tissues (cell lines or tumors), there were sometimes disastrous outbreaks in breeding colonies; hence quarantine precautions were instituted, and research with the virus in the United States was forbidden. Enzootic mousepox does not occur in Australia. Here, following the recognition of the relation of the virus to vaccinia virus by Burnet and Boake (1946), mousepox has been used as a model for research in several fields, notably the pathogenesis of generalized infections, experimental epidemiology, and the cellular immune response.

ISBN 0-12-262502-1

II. VIRAL AGENT

A. Historical Background

The disease was first recognized by Marchal (1930) as an epizootic disease of laboratory mice in England, following the investigation of an unusually high mortality in mice received from commercial breeders by the National Institute of Medical Research at Hampstead. She named it *infectious ectromelia* because of the frequent amputation of the extremities that occurred in the outbreak that she investigated. Soon after, Barnard and Elford (1931) demonstrated by ultraviolet microscopy that the virion ("elementary particle") was similar in size and shape to that of vaccinia virus. However, no serological comparisons with other poxviruses were made until 1946, when Burnet and Boake (1946) demonstrated by the hemagglutination inhibition (HI) test, newly developed for vaccinia virus by Nagler (1944), that ectromelia and vaccinia viruses were closely related.

During the 1930s, Greenwood and collaborators (1936) used ectromalia virus as one of the model pathogens in their experimental epidemics in mice, and in 1946 Fenner (for review, see Fenner, 1949a) revived this work with the added knowledge of the taxonomic status of the virus. He soon found (Fenner, 1948b) that the disease did in fact produce a rash, and he developed a picture of the pathogenesis of mousepox (Fenner, 1948a), which remains a useful model for generalized viral exanthemata (Fenner, 1948c). He suggested that the disease should be called *mousepox* but that the virus should be called *ectromelia virus* (cf. smallpox and variola virus).

During the 1950s and early 1960s, Mims and colleagues took up Fenner's work on pathogenesis using fluorescent antibody staining to probe more deeply into events at the cellular level (for reviews, see Mims, 1964, 1966), and Blanden and colleagues (for reviews, see Blanden, 1974; Cole and Blanden, 1981) have exploited the ectromelia–mouse system in exploring the role of cell-mediated immunity in poxvirus infections and demonstrating the role of major histocompatibility complex (MHC) antigens in viral infections in general.

Virtually all fundamental studies with mousepox have been carried out in Australia, first at the Walter and Eliza Hall Institute in Melbourne and then at the John Curtin School of Medical Research in Canberra. Elsewhere, it has been known only as a troublesome disease that in Europe often interfered with experiments involving mice and that has been periodically imported into the United States, sometimes with disastrous consequences. This chapter will examine the natural history of mousepox in laboratory colonies of mice and the measures that can be used for its control. It will also explore, in a less detailed manner, some of the results which have emerged from the use of the disease as a model infection by Australian virologists and immunologists.

B. Properties of the Virus

1. Classification

Among the 16 named families of viruses that infect vertebrate animals (Fenner, 1976), viral species which occur as natural infections of mice are found in all except five (Iridoviridae, Bunyaviridae, Togaviridae, Rhabdoviridae, and Orthomyxoviridae), and mice can be artificially infected with viruses from each of these families as well. Two distinct species of poxvirus produce natural infections of rodents: infectious ectromelia, or mousepox virus, and the Turkmenia rodent poxvirus (Marennikova *et al.*, 1978), which is a close relative of cowpox virus. Both of these are orthopoxviruses and show strong serological cross-reactivity with other members of the genus *Orthopoxvirus* (Table I).

Comparative studies of ectromelia virus DNA and the DNAs of other orthopoxviruses have shown that they are closely related, but ectromelia DNA differs more from vaccinia, rabbitpox, and cowpox DNAs than they differ from each other (Bellett and Fenner, 1968; Muller *et al.*, 1978). Ectromelia virus is clearly a distinct species of *Orthopoxvirus*. Serological comparisions with other orthopoxviruses have been made mainly by neutralization and HI tests. Cross-neutralization by vaccinia-immune sera is readily demonstrable in pock-reduction tests on the chorioallantoic membrane (McCarthy and Downie, 1948), and in plaque-reduction assays in cultured cells, in which ectromelia virus could be clearly distinguished from vaccinia and cowpox viruses (McNeill, 1968). Although they cross-react, the HI titers of ectromelia and vaccinia hemagglutinins are substantially higher with homologous combinations of serum and antigen (Fenner, 1947a).

Table I

Taxonomic Position of Infectious Ectromelia Virus[a]

Species	Animals found naturally infected	Host range in laboratory animals
Camelpox	Camel	narrow
Cowpox	Cow, man, rodents, carnivores, elephant, okapi	broad
Ectromelia	Mice, ?voles	narrow
Monkeypox	African monkeys, anteater, great apes, man	broad
Raccoonpox	Raccoon	? broad
Taterapox	*Tatera kempi* (an African gerbil)	narrow
Vaccinia	Man, cow, buffalo, rabbit	broad
Variola	Man	narrow

[a] Family: Poxviridae; subfamily: Chordovirinae; genus: *Orthopoxvirus*.

2. Strains of Virus

Isolates from Manchester, England (McGaughey and Whitehead, 1933), Paris (Schoen, 1938), Germany (Kikuth and Gönnert, 1940), and Japan (Ichihashi and Matsumoto, 1966) were found to be serologically indistinguishable from the original Hampstead strain of Marchal (1930). Two strains, Hampstead and Moscow, have been extensively used for experimental studies by Australian workers. Their detailed histories are given below.

a. Hampstead. The original isolation was made by Marchal (1930). This virus was used in Greenwood's epidemiological experiments (Greenwood *et al.*, 1936) and in most of the early work with ectromelia virus. A mouse-passaged Hampstead strain retained its high virulence, but egg passage led to a substantial reduction in its virulence for mice (Fenner, 1949b), although it was then more readily adapted to growth on the rabbit cornea (Paschen, 1936).

b. Moscow. The Moscow strain of virus was isolated in the laboratory of Professor V. D. Soloviev in Moscow and is highly virulent and highly infectious. It was used extensively in experiments by a succession of Australian workers (Fenner, Mims, Roberts, Blanden) and by Andrewes and Elford (1947).

Japanese workers have made considerable use of strains recovered in Japan, as well as the Hampstead strain. Of these, the Ishibashi strain differed in several properties from the Hampstead strain (Ichihashi and Matsumoto, 1966).

3. Physical Properties

Ectromelia virus is a typical orthopoxvirus, morphologically indistinguishable from the prototype species of that genus, vaccinia virus. Like most other orthopoxviruses, it produces a hemagglutinin which, by analogy with vaccinia virus (Payne and Norrby, 1976), is part of the viral envelope, the presence of which is not necessary for infectivity. Just as several strains of vaccinia virus fail to produce a hemagglutinin (Cassel, 1957; Fenner, 1958), it is said (Guillon, 1975) that in Europe, where the disease is enzootic in many mouse stocks, "numerous strains are only slightly hemagglutinogenic on isolation" (p. 20).

Like other orthopoxviruses, ectromelia virus is relatively resistant to heat and to many disinfectants (see Fenner, 1949a, Table III).

4. Growth in Tissue Culture and Eggs

a. Tissue Culture. Ectromelia virus multiplies in HeLa cells and human amnion cells, L cells, mouse fibroblasts, and chick embryo fibroblasts. Ichihashi and Matsumoto (1966) found that a Japanese strain of ectromelia virus (Ishibashi) produced much larger plaques in chick embryo fibroblasts than the Hampstead strain. Plaque production in mouse fibroblasts is improved by the inclusion of DEAE-dextran in the overlay medium and assay in L cells or mouse fibroblasts is about as sensitive as assay on the chorioallantoic membrane, but more reproducible, and has been adopted by the Canberra immunologists as the standard assay method (Blanden, 1974).

b. Chick Embryo. Infection of the chick embryo was described simultaneously by Paschen (1936) and Burnet and Lush (1936a). Both workers grew the virus on the chorioallantoic membrane, and Burnet and Lush showed that if dilute suspensions of virus were inoculated, separate foci developed, which could be counted. This pock-counting method of titration of the virus was exploited by Fenner (1948a) for quantitative studies of mousepox infection. Chorioallantoic inoculation with large doses of virus was usually followed by death of the embryo 4 or 5 days later, and in the livers and spleens of these embryos there were often scattered areas of necrosis.

Serial passage of ectromelia virus on the chorioallantois sometimes modified its character. Paschen (1936) found that egg-passaged virus was more suitable for infection of the rabbit and guinea pig cornea and skin than was mouse liver virus. Serial chorioallantoic passage of the Hampstead strain of ectromelia virus (50–60 passages intermittently over a period of 10 years) resulted in greatly reduced virulence of the strain for mice (Fenner, 1949b). No change in the high virulence of the Moscow strain of virus occurred after 20 consecutive passages on the chorioallantois.

C. Pathogenesis

1. Clinical Disease

Early workers (Marchal, 1930; McGaughey and Whitehead, 1933; Schoen, 1938) described two forms of the disease: a rapidly fatal form in which apparently healthy mice died within a few hours of the first signs of illness and showed extensive necrosis of the liver and spleen at autopsy, and a chronic form characterized by ulcerating lesions of the feet, tail, and snout. Fenner (for review, see 1949a) showed that in every case there was a stage in which virus multiplied to a high titer in the liver and spleen. Some mice died at this stage, but if they survived they almost invariably developed a rash which occurred over the whole body, not only on the hairless extremities. Subsequently it was shown that the symptomatology of the disease was greatly affected by the mouse genotype (see Section II,D,1,*b*).

Fenner (1947b, 1949b) believed that the usual portal of entry of the virus in natural infections was through small abrasions of the skin. Using fluorescent antibody staining to follow the process, Roberts (1962b) showed that after scarification, the first cells infected were macrophages in the dermis. Spread in the dermis initiated "island foci" of epidermal infection in advance of the main, more slowly spreading dermal focus, and spread into the lymphatics, often via infected macrophages, initiated infection of local lymph nodes.

After an incubation period of 7 or 8 days, during which the virus multiplied locally and also invaded the internal organs, a primary lesion developed at the site of entry of the virus (Fig. 1). The subsequent course depended on the degree of multiplication of the virus in the liver and spleen. If this reached a high level, death occurred within 0–4 days of the appearance of the primary lesion. If the virus failed to reach a lethal concentration in the internal organs, virus which had been deposited locally in the epidermal cells throughout the body multiplied and eventually caused necrosis of these epidermal cells, producing a generalized rash (Fig. 2). Healing of the skin lesions usually occurred within a week or so, often leaving hairless scars. The severity of the rash depended upon the degree of viremia, which appeared to be directly related to the viral content of the liver and spleen. In acutely fatal cases, in which no macroscopic skin lesions could be seen, the virus content of the skin was often very high, but not enough time had elapsed for skin lesions to develop.

It is important to note that all these observations refer to mousepox in young adult mice of a genetically susceptible strain; the symptomatology may be very different in genetically resistant mouse strains (Section II,D,1,*b*).

Age affects the response of genetically susceptible mice (Fenner, 1949d). Both the Moscow and Hampstead (egg) strains produced higher mortalities in suckling mice and in mice about a year old than in the 8-week-old mice, the differences being more pronounced with the less virulent Hampstead strain. The increased severity was evident after footpad inoculation and in the naturally spreading disease. In suckling mice there was a very short delay between peripheral inoculation of the virus and its appearance in the liver and spleen. In the old animals this interval, and the survival time of the mice, were the same as in 8-week-old animals. However, lethal titers of virus in the liver and spleen were attained only in occasional 8-week-old mice but did occur in most of the older animals. The causes of these differences were not elucidated.

2. Pathology

a. Intracytoplasmic Inclusion Bodies. Like all other poxviruses, ectromelia virus multiplies in the cytoplasm. Marchal (1930) demonstrated that infected epithelial cells, but not infected cells of the liver and spleen, contain characteristic prominent acidophilic cytoplasmic inclusion bodies (Fig. 3). Subsequently Kato *et al.* (1955) distinguished two types of cytoplasmic inclusion body in poxvirus-infected cells: what he called *B type,* in which viral replication occurred [the "viral factories" of Cairns (1960)] and *A type,* the prominent acidophilic inclusion bodies found in epithelial cells infected with ectromelia (Marchal bodies) and cowpox (Downie bodies), but not in cells infected with variola virus or some strains of vaccinia (Kato *et al.,* 1959). Kato *et al.* (1963) showed that all ectromelia-infected cells contained B-type inclusions, whereas A-type inclusions were rare in infected liver cells but prominent in infected epithelial cells.

Electron microscopic examination of cultured L cells infected with two strains of ectromelia virus showed that with

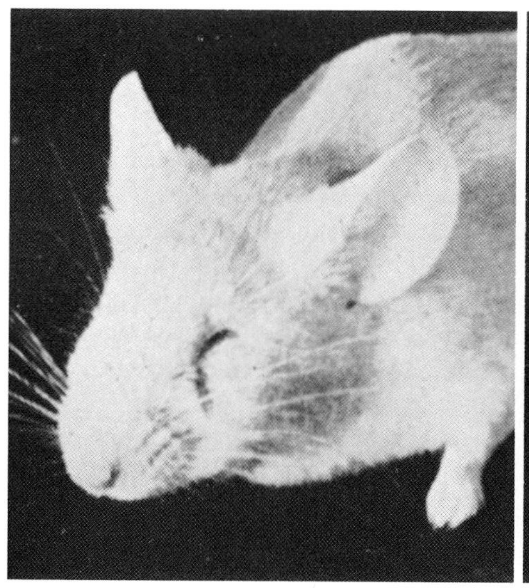
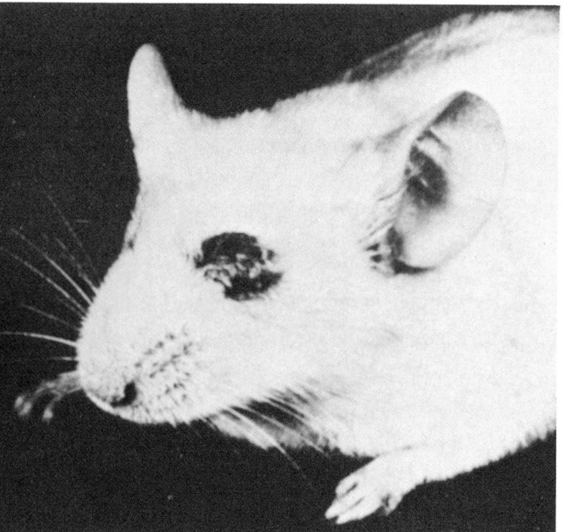

Fig. 1. The primary lesion of mousepox on the left eyebrow of a naturally infected mouse, 8 days (left) and 14 days (right) after infection.

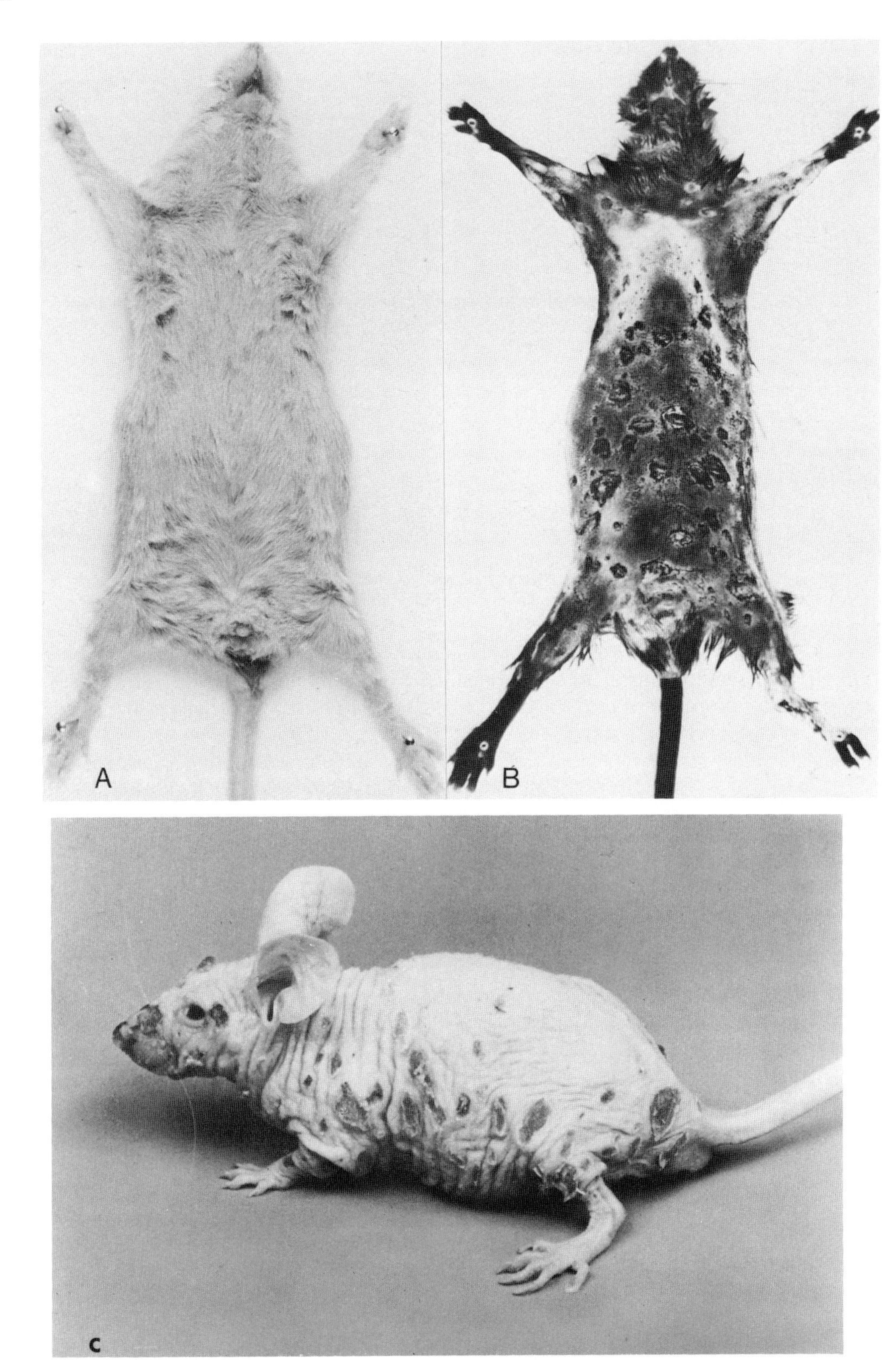

Fig. 2. The rash of mousepox as it appears 14 days after infection. (A) Normal mouse of a susceptible strain. (B) Normal mouse after depilation to reveal the rash. (C) In a naturally infected hairless mutant mouse (not athymic). [(C) Courtesy of the Zentralinstitut für Versuchstiere, Hannover, Federal Republic of Germany.]

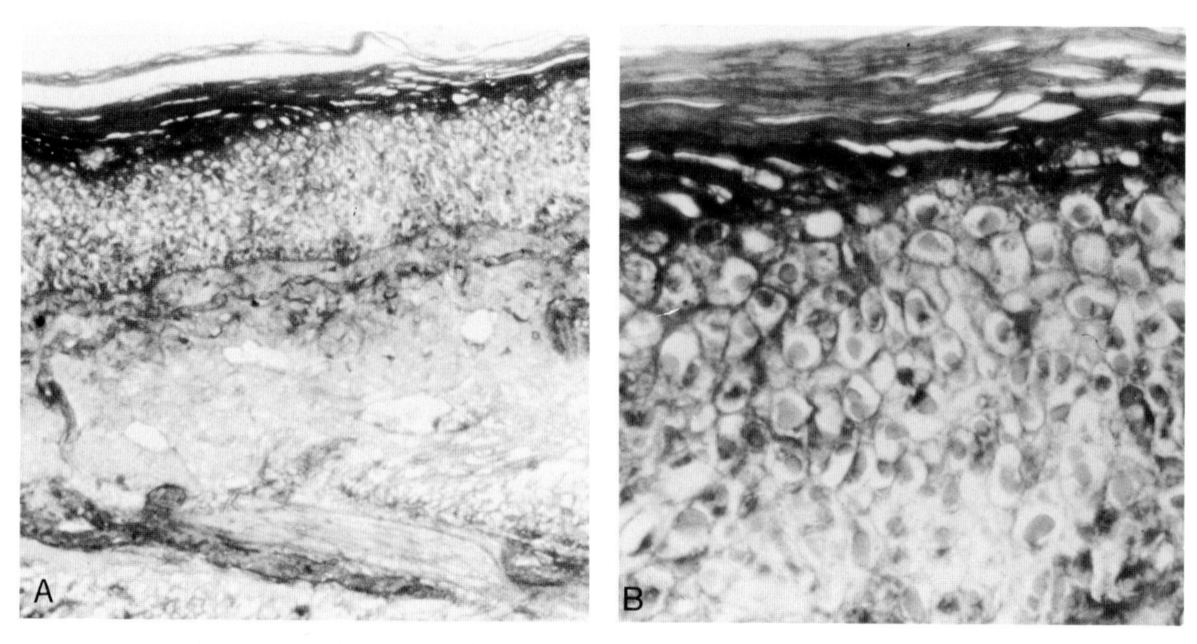

Fig. 3. Section of the skin of the foot of a mouse injected with ectromelia virus in the footpad 6 days earlier. (A) Low power. (B) High power. Mann's strain. Almost every epithelial cell contains an eosinophilic inclusion body.

one strain (Hampstead), all the mature viral particles that had developed in the B-type inclusion were finally included within the Marchal body, whereas with another strain (Ishibashi), the inclusion bodies were devoid of virions (Matsumoto, 1958; Ichihashi and Matsumoto, 1966).

b. Skin Lesions. The primary lesion appeared first as a localized swelling usually surmounted by a minute breach of the surface (Fig. 1). It rapidly increased in size, with pronounced edema of the surrounding tissues. Later, a hard, adherent scab formed and fell off after a week or two, leaving the site of the primary lesion marked by a deep, hairless scar which often persisted for life (Fig. 4).

The earliest primary lesions that could be recognized were the seat of advanced histological changes, for viral multiplication had then been in progress for a week. There was no macroscopic breach of the skin surface, but the dermis and subcutaneous tissue were edematous and there was widespread

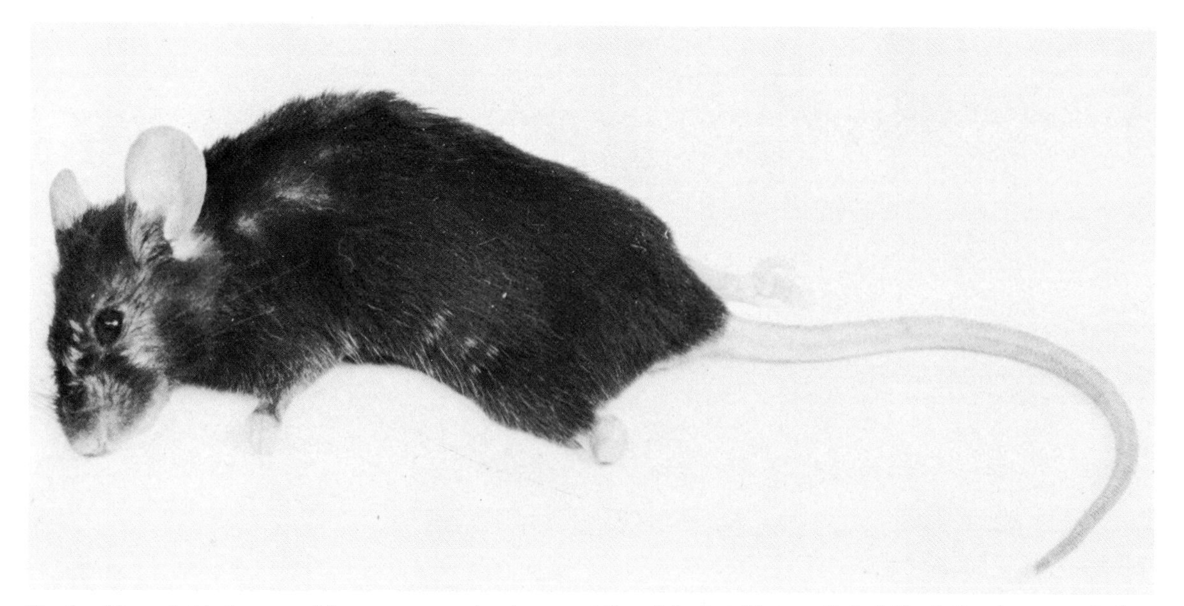

Fig. 4. Mouse that had recovered from mousepox, showing amputation of the foot (''ectromelia'' of Marchal) and scars on the face.

lymphocytic infiltration of the dermis. Inclusion bodies could be seen in the epidermal cells at the summit of the lesion (Fig. 3). Necrosis of these epidermal cells was followed by ulceration of the surface and widespread necrosis through the dermis. The exudate dried and formed a scab beneath which healing occurred.

Individual lesions of the secondary rash appeared 2 or 3 days after the development of the primary lesion, and when first seen on the shaved skin, they were slightly raised, pale areas 2 or 3 mm in diameter. They increased in number and size, ulcerated, and in animals that survived, they healed by the end of the third week after infection. Conjunctivitis and blepharitis occurred frequently during the stage of the secondary rash, and in severe cases ulcers could be found on the tongue and buccal mucous membrane.

Histologically, the first changes of the secondary rash were seen on the same day the primary lesion was detected, when a few localized areas of the epidermis appeared hyperplastic, with dark-staining nuclei surrounded by vacuoles; occasionally these cells contained intracytoplasmic inclusion bodies. The areas of proliferation and edema increased in size until they became macroscopically visible as pale, slightly raised macules. Numerous inclusion bodies were then present in the epidermal cells. Fresh foci appeared in the intervening regions of previously normal skin, and by the next day, necrosis of the superficial cells of the early lesions had commenced. Massive necrosis followed quickly, with accompanying widespread edema and lymphocytic infiltration of the dermis, and the papules were converted to ulcers with closely adherent scabs. The lesions progressed in size and number for a couple of days, and then healing commenced and was complete in a few days.

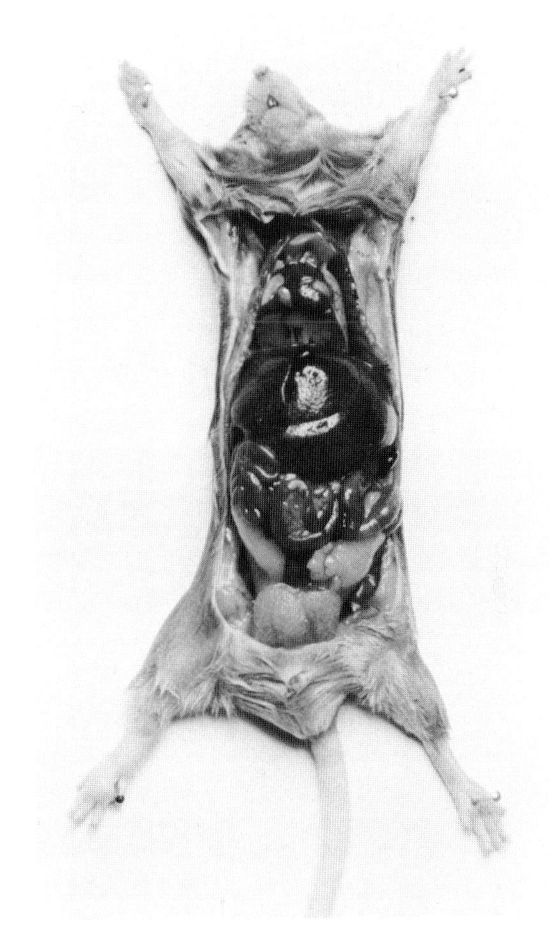

Fig. 5. Acute hepatic necrosis in mousepox. A fatal case 6 days after subcutaneous inoculation of a large dose of the Moscow strain ectromelia virus.

c. *Lesions of the Liver.* The liver and spleen were invariable invaded during the incubation period, and considerable multiplication of virus occurred there. The liver remained macroscopically normal, even in cases which would prove fatal, until within 24 hr of death, when it appeared enlarged and studded with many minute white foci (Fig. 5). The necrotic process extended rapidly, and at the time of death the liver was enlarged with many large, semiconfluent necrotic foci. The fat content of such livers was about 13% of the liver weight, compared with a normal figure of 3–4%. In animals which survived, the liver usually returned to its normal macroscopic appearance, but occasionally numerous white necrotic foci occurred, especially along the anterior border of the median lobe

Histologically, little change was apparent until macroscopic changes had also appeared, that is, within a day or so of death in rapidly fatal cases, although with fluorescent antibody staining, it could be shown that infection always occurred first in the littoral cells of the hepatic ducts, from which it spread to contiguous parenchymal cells (Mims, 1959). Numerous scattered foci of necrosis then appeared throughout the liver parenchyma, and in fatal cases these foci rapidly extended until they became semiconfluent (Fig. 6). At the margins of the necrotic areas the hepatic polygonal cells usually showed active regeneration with many multinucleate cells, even in fatal cases. The portal tracts showed slight infiltration with lymphoid cells, but Marchal bodies were rarely found in infected hepatic parenchymal cells.

The necrotic foci occasionally seen after recovery from infection consisted of hyaline necrotic tissue. In the liver of most animals which survived and showed no macroscopic lesions at autopsy there were small accumulations of lymphocytes, usually around branches of the portal tract but occasionally in the parenchyma of the liver.

The necrotic process in the liver, which was the dominant histological finding, was always focal and random in distribution (Fig. 6) and showed no regular relationship to the normal liver architecture. Liver regeneration commenced early and was active, especially in nonfatal cases, and fibrosis did not occur.

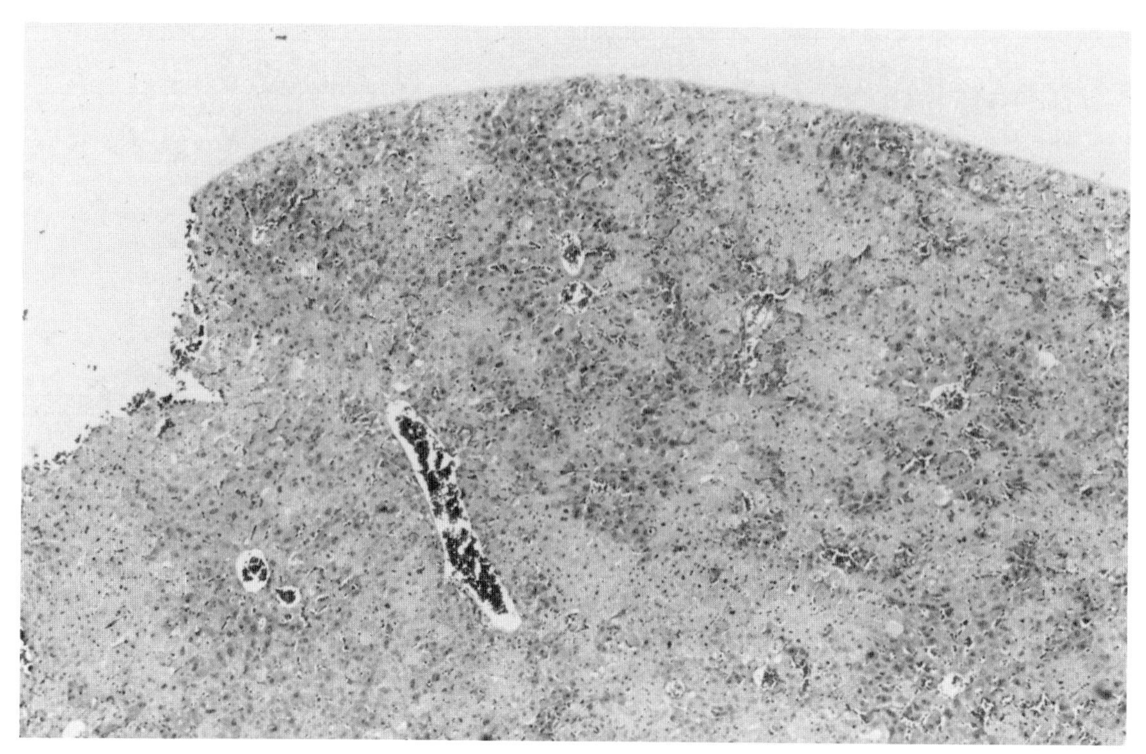

Fig. 6. Section of the liver of a mouse sacrificed when moribund, 7 days after the injection of a large dose of ectromelia virus in the footpad. There is extensive irregular necrosis of the liver, with little inflammatory reaction.

d. Lesions of the Spleen. The spleen showed macroscopic changes at least a day earlier than the liver, and higher titers of virus were found in the spleen each day until death, when the titers of the spleen and liver were approximately the same. Fluorescent antibody studies (Mims, 1964) revealed that virus probably reached the spleen in infected lymphocytes, which initiated infection in the substance of the follicles. Whereas infected follicles were destroyed by the spreading infection, neighboring follicles showed the proliferative response characteristic of antibody production in the spleen.

The spleen was at first engorged, and later pale, slightly depressed areas of necrosis appeared either in isolated patches or were semiconfluent. In surviving mice, obvious lesions of the spleen were much more common than were lesions of the liver, and varied in severity from small raised plaques about 1 mm in diameter to strands of fibrous tissue which after severe attacks almost completely replaced the normal splenic tissue. These changes (Fig. 7) constitute the most frequent and reliable autopsy evidence that a mouse has recovered from an attack of mousepox.

Histologically, the early changes consisted of lymphoblastic hyperplasia of the follicles and congestion of the sinuses of the red pulp, and focal necrosis with fragmentation of the lymph follicles. The necrosis rapidly extended, and the red and white pulp was characterized by transformation to endothelial-type cells. Very extensive necrosis of the spleen occurred in some mice which ultimately recovered. The plaques seen on the spleen in mice which had recovered from mild attacks of mousepox consisted of localized areas of hyperplasia of the serosal cells, and the scarred areas consisted of fibrous tissue.

e. Lesions of Other Organs. The regional lymph nodes draining the site of the primary lesion were enlarged from the time that the primary lesions could be detected, and they usually showed localized areas of necrosis, with pyknotic nuclear debris in a featureless background. Sometimes these necrotic areas were almost confluent, and often numerous inclusion bodies were present. In the later stages, the majority of lymph nodes showed lymphoblastic hyperplasia and small foci of necrosis.

The intestine was often engorged, and the lymphoid follicles were enlarged in fatal cases of mousepox. Greenwood *et al.* (1936) reported that a careful histological survey of the intestines showed that small necrotic foci with typical inclusion bodies occurred in about 65% of acutely fatal cases of mousepox in which characteristic lesions of the liver and spleen occurred. Briody (1955, 1959) has commented upon the frequency with which a hyperemic or blood-filled small intestine was observed during epizootics of mousepox in the United States, especially in genetically highly susceptible strains of mice.

No other organs were regularly affected in natural mousepox, but occasionally, especially in very young mice, there were hemorrhagic foci in the kidneys and occasionally in

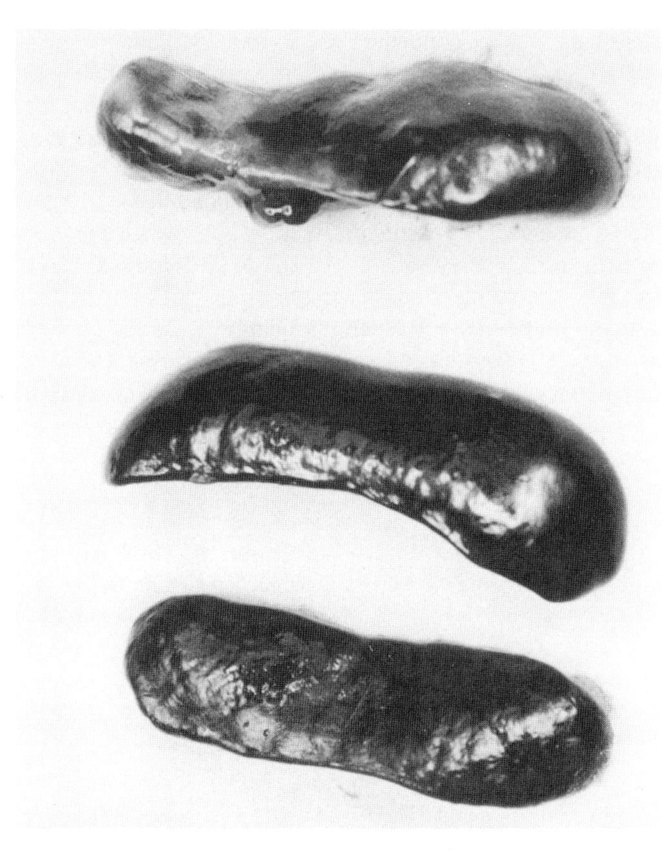

Fig. 7. Scarring of the spleen after recovery from mousepox.

the bladder. Here also the picture in the acute stage was one of widespread focal necrosis, especially in the region of the convoluted tubules. Inclusion bodies could be seen in the epithelial cells, and sometimes they appeared to occur free in the lumina of the tubules. Later, in the few animals which survived, the necrotic areas were replaced by areas of fibrosis and cellular infiltration.

3. Transmission

a. Intradermal Inoculation and Scarification. Infection by scarification, by pad inoculation, or by instillation of virus into the cornea was followed by a disease indistinguishable from naturally acquired mousepox except for the localization of the primary lesion.

Other methods of inoculation were followed by pathological changes which differed considerably from those found with the natural disease. These changes are discussed in greater detail below.

b. Feeding. Fenner (1947b) found that infection by feeding was possible only if very large doses of virus were used, the disease which followed resembling natural infection. However Gledhill (1962a,b) and subsequently Horzinek and Höpken (1965) found that occasionally mice could be infected orally with much smaller doses of virus. More important, such infections were usually inapparent. Gledhill showed that chronic infection of Peyer's patches occurred, and that small amounts of virus could be excreted for prolonged periods of time in the feces, sometimes associated with chronic tail lesions that also released small amounts of virus. In Gledhill's experience, such carrier mice did not spread mousepox to uninfected mice by contact, nor was he able to "activate" acute mousepox in the carriers. Nevertheless they clearly constitute a reservoir from which virus could be transferred to other mice by the inoculation of tissue suspensions.

c. Arthropod Vectors. Although this mode of transfer is important in several other poxvirus diseases, e.g., fowlpox and myxomatosis, Fenner (unpublished experiments) was unable to demonstrate mechanical transmission of mousepox by mosquitoes. Guillon (1970) has suggested another possible mode of transfer by arthropods. He found that the rat mite, *Ornithonyssus bacoti,* became infectious after a blood meal and suggested that it [and perhaps also cockroaches (Guillon, 1975)] might act as a passive vector.

d. Intraperitoneal Inoculation. This has been a commonly used way of passing ectromelia virus. The lesions differ considerably from those observed in the natural disease, and it is obvious that some accounts of the pathological changes in mousepox are based upon the results of intraperitoneal inoculation. There is, of course, no primary skin lesion, but in acutely fatal cases the necrosis of the liver and spleen resembles that found after natural infection. In addition, there is usually some increase in intraperitoneal fluid and a considerable amount of pleural fluid, and the pancreas is often grossly edematous. In animals which survive the acute infection, the signs of general peritonitis are much more pronounced (Fig. 8). There is a great excess of peritoneal and pleural fluid, the peritoneal surfaces of the liver and spleen are covered with a white exudate, the walls of the intestine are thickened and rigid, and there is often fat necrosis in the intraperitoneal fat. Extensive adhesions develop later between the abdominal viscera. Animals infected by the intraperitoneal route which survive long enough develop a characteristic rash.

The survival times in fatal cases inoculated intraperitoneally are usually 2 or 3 days shorter than when the same dose of virus is inoculated into the foot, for reasons which will become apparent when the pathogenesis of mousepox is considered.

e. Intranasal Inoculation. One common way in which ectromelia used to be encountered in laboratory practice was during the mouse lung passage of influenza virus. Some workers (see, e.g., Kikuth and Gönnert, 1940) have suggested that ectromelia virus acquires marked pneumotropic properties

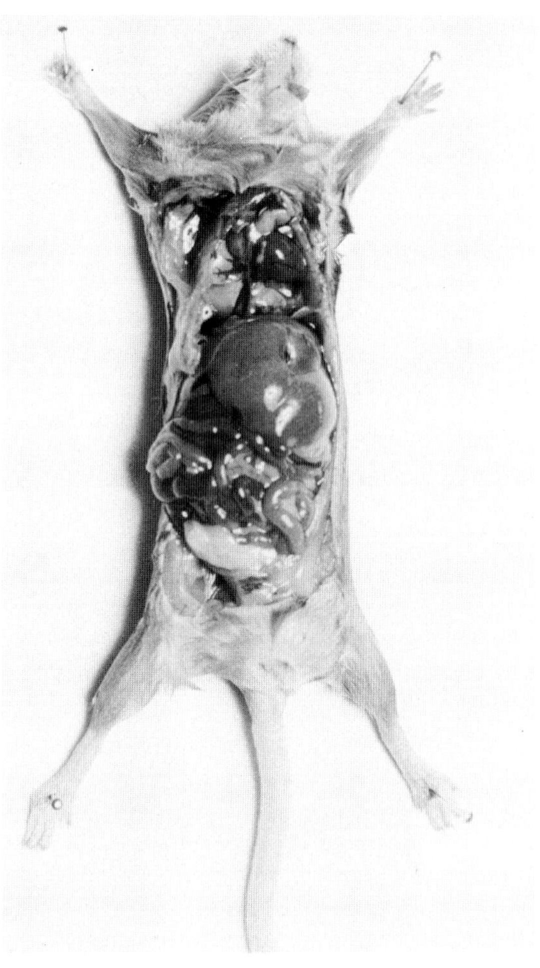

Fig. 8. Postmortem appearances after intraperitoneal injection of ectromelia virus. Necrosis of the liver, hyperemia of the intestine, and peritoneal and pleural effusions are seen.

after serial lung passage, because deaths from pneumonia then occur with little macroscopic evidence of involvement of the liver and spleen. However, when small doses of virus are inoculated intranasally, there is usually little change in the lungs except for the development of patchy congestion. The survival time of fatal cases is approximately the same as in fatal cases inoculated in the foot pad with the same dose of virus, and the changes in the liver and spleen are those characteristic of acute, naturally acquired mousepox. With larger doses of virus, congestion of the lungs becomes more pronounced and consolidation may occur. When very large doses are given, death occurs, with patchy or complete consolidation of the lungs and little change in the liver and spleen. It is this picture which led investigators to speak of the pneumotropism of the virus, and the detailed histological description given by Kikuth and Gönnert (1940) is of such lungs. There was early exudation into the alveoli and bronchi, and small foci of necrosis of the bronchial epithelium occurred. Inclusion bodies were seen in the bronchial epithelial cells, in histiocytes, and in cells of the pleural epithelium and eventually of the alveolar wall. When the lungs, liver, and spleen of such fatal cases were examined for their virus content, it was found that the viral titers of the apparently normal liver and spleen were very high, just below the threshold at which demonstrable necrosis occurred (F. Fenner, unpublished observations; Ipsen, 1945). Indeed, such necrosis was found in a few mice which survived for 6 instead of the usual 4 or 5 days after the intranasal inoculation of a large dose of virus.

Using fluorescent antibody staining, Roberts (1962a) showed that either macrophages or alveolar mucosal cells were initially infected, but it was the macrophages that carried virus to the pulmonary lymph nodes and thus to the bloodstream, from which it was taken up by the liver and spleen, in which multiplication then proceeded. The apparent pneumotropism is due to the fact that the local reaction which occurs after the intranasal inoculation of very large doses of virus kills the animal before there is time for the characteristic changes in the liver and spleen to occur.

f. Intracerebral Inoculation. Kanazawa (1937), Jahn (1939), and Schoen (1938) described intracerebral passage of ectromelia virus, and Kikuth and Gönnert (1940) confirmed their results. Jahn (1939) could not find inclusion bodies in the brain, but Kikuth and Gönnert (1940) described them in neural cells and also in macrophages. It is apparent that after intracerebral inoculation, the virus rapidly enters the systemic circulation, for the virus titers of the liver and spleen of fatal cases are always high (F. Fenner, unpublished observations), and except after large doses of virus, characteristic changes are present in both liver and spleen (Kanazawa, 1937). Jahn (1939) found no evidence of increased neurotropism after 20 brain-to-brain passages in mice.

g. Intrauterine Infection. Mims (1969) showed that many pregnant mice infected intradermally with the attenuated Hampstead egg strain of ectromelia virus survived, but there was extensive growth of virus in the placenta and infection of the fetuses. Infected fetuses died either *in utero* or soon after birth, and fluorescent antibody staining showed that there was widespread growth of virus throughout their bodies. Schwanzer *et al.* (1975) obtained similar results.

In resistant strains of mice from colonies enzootically infected with mousepox, intrauterine infection might pose problems in raising "clean" stock by caesarian section delivery, but it is unlikely that infected fetuses would survive. However, problems could arise from the use of mouse embryo cell cultures derived from infected embryos (Germer *et al.,* 1960).

4. The Mechanism of Infection

The close resemblance between the disease initiated by the inoculation of a small dose of virus in the footpad and the natural disease suggested that this mode of inoculation should be used in experiments upon the pathogenesis of the rash in mousepox. Accordingly, large groups of mice were inoculated in the hindfoot with small dosese of virus, both the Hampstead and Moscow strains being used (Fenner, 1948a). The incubation period with the doses used resembled that found in natural infections. The course of events was followed by killing mice at daily intervals and titrating the viral content of certain organs: the inoculated foot, the spleen, the skin, and the blood. In subsidiary experiments, the regional lymph node was examined at 2 hr intervals during the first 24 hr after infection.

The results obtained with both strains of virus, and in several subsequent experiments with immunized mice (Fenner, 1949c) indicated that the sequence of events in mousepox followed a constant pattern. If the appearance of the primary lesion was taken as the end of the incubation period, it was evident that the latter consisted of a complex series of events in which the virus passed in a stepwise fashion: infection, multiplication, and liberation, usually accompanied by cell necrosis, first through the skin, then the regional lymph node, and then presumably the deeper lymph nodes, until it reached the bloodstream. It seems, from the work of Mims (1964), that the phagocytic cells of the liver and spleen then ingested the viral particles, and the same cycle proceeded in these two organs. After an interval of a day or so, larger amounts of virus were liberated into the circulation, and during the secondary viremia, focal infection of the skin and sometimes of the kidneys, lungs, intestines, etc. occurred. There was again an interval during which the virus multiplied to reach a high titer before visible changes were produced, so that an interval of about 2 or 3 days usually elapsed between the appearance of the primary lesion and the sores of the secondary rash (Fig. 9).

Growth curves of the virus in the different organs were constructed, and during the first 3 or 4 days, viral multiplication in each organ followed a logarithmic course. Then, coincidentally with the appearance of circulating antibody, the growth curves flattened and eventually fell steeply. In acute fatal cases, and only in these cases, the viral titers in the spleen and liver reached levels of 10^9 or 10^{10} infective particles per gram.

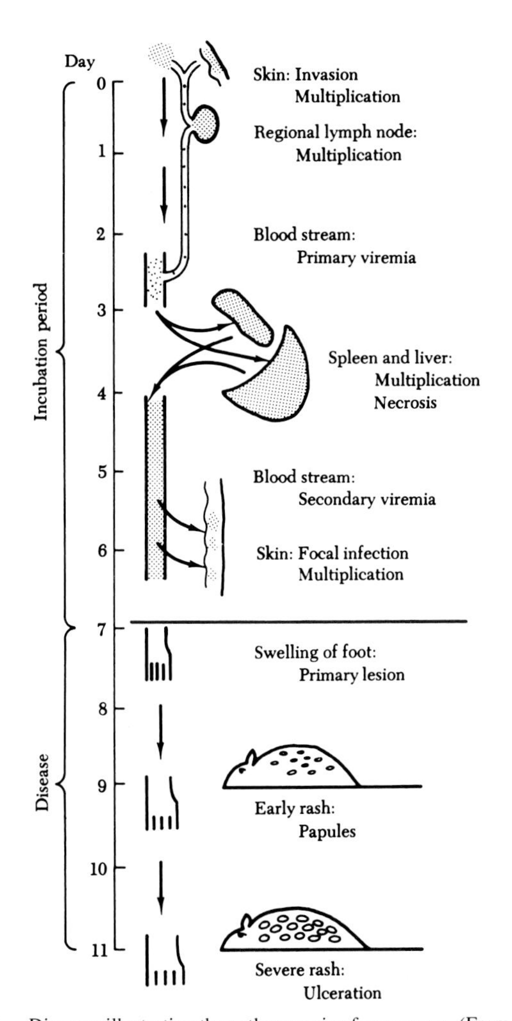

Fig. 9. Diagram illustrating the pathogenesis of mousepox. (From Fenner, 1948c.)

5. Immune Response

Early investigations showed that mice that had recovered from mousepox were immune to reinfection (Marchal, 1930; Greenwood *et al.*, 1936). Fenner (1949c) showed that 2 weeks after infection, mice were solidly immune to reinfection by footpad inoculation of the virus. This immunity declined slowly, but even after a year, multiplication of the virus was confined to the local skin lesion, and only in occasional animals could virus be isolated from the spleen. When multiplication of the virus in the foot occurred, with consequent swelling, the HI titer usually rose significantly, but if no multiplication occurred, there was no antibody rise. Long-continued epidemics showed the epidemiological importance of this durable immunity, for in only 3 mice out of 168 which had recovered from mousepox did any sign of reinfection or recurrence occur, and in none of these did it proceed beyond a local lesion in the foot (Fenner, 1948d).

Serum from recovered mice contains neutralizing (Burnet and Lush, 1936a), HI (Burnet and Boake, 1946), and precipitating (Horzinek, 1965) antibodies, and mice that have recovered from mousepox also exhibit delayed type hypersensitivity (Fenner, 1948a; Owen *et al.*, 1975).

During the last decade, Blanden and colleagues (for reviews, see Blanden, 1974; Cole and Blanden, 1981) have used mousepox as a model disease to elucidate the role of humoral and cell-mediated immunity in recovery from poxvirus infection. The full story is complicated and involves different classes of T cells and the H-2 antigens. Briefly, mechanisms controlling viral growth in the major visceral target organs (liver and spleen) are operating 4–6 days after primary infection by the natural route (simulated by subcutaneous inoculation into the foot). Cell-mediated immune responses occur soon after infection, e.g., virus-specific cytotoxic T cells are detectable 4 days after infection and reach peak levels in the spleen 1–2 days later, whereas delayed type hypersensitivity is detectable by the footpad test 5–6 days after infection. In contrast, significant neutralizing antibody is not detectable in the circulation until the eighth day.

Mice pretreated with anti-thymocyte serum die from otherwise sublethal doses of virus due to uncontrolled viral growth in target organs. These mice have impaired cell-mediated responses but normal neutralizing antibody responses, elevated interferon levels in the spleen, and unchanged innate resistance in target organs.

Very large doses of interferon or immune serum transferred to preinfected recipients are relatively ineffective against the established infection in target organs, though high antibody titers can be demonstrated in the sera of the recipients of immune serum. On the other hand, immune spleen cells harvested 6 days after donor immunization transfer specific and highly efficient antiviral mechanisms which rapidly eliminate infection from the target organs of the recipients, although neither antibody nor interferon is detectable in the recipients. The active cells in the immune population are θ-positive and immunoglobulin-negative, i.e., they are T cells. The kinetics of their generation and the requirement for sharing of *H-2K* or *H-2D* genes between donor and recipient identify them as cytotoxic T cells. Other T cells of the helper class, which recognize antigens dependent on the *I* region of the *H-2* complex, may be important in the generation of cytotoxic T cells but are not important at effector level.

Mononuclear phagocytes of immune T-cell recipients, labeled with tritiated thymidine before T-cell transfer, appear in foci of infection in the liver after T-cell transfer, and prior irradiation of immune T-cell recipients in a regimen designed to reduce blood monocyte levels significantly reduces the antiviral efficiency of the transferred cells. Further, subpopulations of immune spleen cells labeled with tritiated thymidine show immunologically specific localization in liver lesions of cell recipients.

These findings support the idea that blood-borne cytotoxic T cells specific for virus-induced antigenic changes in infected cell surface membranes enter infectious foci and retard viral spread by lysing infected cells before the maturation and assembly of progeny virions. This T-cell activity attracts blood monocytes, which contribute to the elimination of infection by phagocytosis and intracellular destruction of virus or by becoming unproductively infected. Macrophage activation and locally produced interferon may increase the efficiency of virus control and elimination but are less important than T cells.

From the point of view of the natural history of mousepox, two features of the immunological response are important. First, the effectiveness of the cell-mediated immune response provides at least part of the basis for the absence of clinical symptoms evident in some mouse genotypes (see Section II,D,1,*b*). Second, maternal antibody, transmitted in the milk for at least the first week after birth, is very effective for a few weeks in saving young mice of a highly susceptible stock from death due to virulent ectromelia virus infection (Fenner, 1948b). Limited investigations on the spread of the disease in two breeding colonies suggested that maternal antibody may play an important role in the persistence of the virus in laboratory colonies (see Section II,D,4).

D. Epizootiology

1. Host Range

a. Species. The mouse can be infected by all routes of inoculation; the pathology and pathogenesis of the disease and genetic differences in the response of mice to infection are elaborated elsewhere in this chapter. Several other laboratory animals have been tested for susceptibility, with the results given below.

i. Other Species of Mouse. Except for Gröppel's (1962) observations on wild mice of several species captured in a rural location in Germany (a finding which should be reexamined), mousepox is known only as a disease of laboratory colonies of *Mus musculus* (which is sometimes associated with infections in wild house mice). No records exist of the inoculation of other species of mouse with ectromelia, although several such colonies exist. However, Dr. V. M. Chapman (personal communication to Dr. J. D. Small, 1979) relates that he inoculated mice of a laboratory colony of *Mus caroli* with vaccinia virus to protect them from mousepox, to which they might have been inadvertently exposed because of the importation of mice into the colony from other parts of the United States and from Europe. In contrast to laboratory *M. musculus,* which were inoculated at the same time, almost all the *M. caroli* animals died of vaccinia infection within 2 weeks.

ii. Rat. Burnet and Lush (1936b) showed that when large doses of virus were inoculated intranasally into the rat, inapparent infection and multiplication of the virus in the cells of the olfactory mucosa occurred. Circulating antibody was detected in the serum of such rats by neutralization tests in mice and on the chorioallantois. This was confirmed by Reames (1940).

Mooser (1943) found that when peritoneal fluid from moribund mice which were infected with both ectromelia virus and *Rickettsia prowaseki* was inoculated intraperitoneally into rats, there was a severe rickettsial peritonitis, but characteristic ectromelia inclusion bodies could also be found in smears of the peritoneal fluid.

Intradermal inoculation of large doses of ectromelia virus produced no lesions, or at the most a tiny papule in the skin of the rat, but the HI titer of the sera of such animals had risen to a high level 14 days later (F. Fenner, unpublished observations). Intraperitoneal inoculation was also without effect except for the production of antibody. Repeated inoculation of the virus into such animals produced no further increase in antibody titer, suggesting that inapparent infection must have occurred.

Reports that rats can be naturally infected with ectromelia virus and may indeed be a reservoir of the virus in nature (see, e.g., Iftimovici *et al.*, 1976) are unconvincing.

iii. Rabbit. Most early workers found that the rabbit was quite resistant to ectromelia infection, although Paschen (1936) reported that virus which had been passed several times on the chorioallantois produced infection of the rabbit's cornea, with the production of inclusion bodies, and could then be passed to the rabbit skin. Burnet and Boake (1946) found that two rabbits inoculated intravenously with a large dose of virus died 6 and 10 days later, and hemagglutinin was present in suspensions of the liver and spleen of both animals. They found that intradermal inoculation of the virus resulted in the appearance of indurated papules. Subsequent experiments (F. Fenner, unpublished observations) showed that such papules regularly appeared about 4 days after the intradermal inoculation of large doses of virus and sometimes ulcerated a few days later. In rabbits which had been immunized by an earlier infection, an accelerated reaction occurred, the papules reaching their maximum size by the second day and fading by the sixth day. Such intradermal infections were followed by the appearance of ectromelia HI antibodies in the rabbit serum. Christensen *et al.* (1966) showed that ectromelia could be used to immunize rabbits against rabbitpox, although because of the danger to mouse stocks, it was preferable to use vaccinia virus for this purpose.

iv. Guinea Pig. Paschen (1936) found that the plantar surface of the guinea pig foot and the cornea were both susceptible to infection with egg-passaged ectromelia virus but not with mouse liver virus, typical inclusion bodies being produced in both situations. Intradermal inoculation of egg membrane preparations of the virus regularly produced local indurated lesions (F. Fenner, unpublished observations), and HI antibodies could be found in the sera 14 days later. Intraperitoneal inoculation of large doses caused no symptoms but was followed by the appearance of circulating antibody.

v. Hamster. According to Flynn and Briody (1962), Syrian hamsters are not susceptible to infection with ectromelia virus.

vi. Man. The only conscious attempt at infection of humans with ectromelia virus (F. Fenner, unpublished observations) was scarification by the multiple-pressure technique of the arms of two men who had been vaccinated with vaccinia virus several times previously. In both a small papule appeared on the second day, became slightly vesicular in one on the fourth day, and disappeared by the eighth day. No change occurred in the ectromelia HI antibody titer of the serum, a not unexpected result in view of Nagler's (1944) observation that after revaccination an increase in vaccinia HI antibody occurred only if there had been a good "take" with definite vesiculation.

Packalén (1947) has shown that the Laigret-Durand strain of "mouse-pathogenic murine typhus rickettsia," like the epidemic typhus strain studied by Mooser (1943), owed its mouse pathogenicity to its ectromelia virus content. He considered that the strain received by his laboratory in 1940 was probably pure ectromelia virus at that time, so that subsequent experiments on "mouse pathogenic murine typhus rickettsiae" (Ipsen, 1944, 1945; Kling and Packalén, 1947) were really experiments with ectromelia virus. This means that active ectromelia virus, either alone or mixed with rickettsiae, has been inoculated subcutaneously into hundreds of thousands of humans in doses of up to 1000 "mouse units" (Laigret and Durand, 1939, 1941; Packalén, 1945). No local or general reaction of any significance was reported (Laigret and Durand, 1941).

b. Strains of Mice. Analysis of epizootics of mousepox in the United States (Trentin, 1953; Briody, 1955, 1959; Briody *et al.*, 1956) revealed that the ability of ectromelia virus to produce acute death with visceral lesions and no rash, a generalized rash, or a chronic or even subclinical infection was critically dependent on the genotype of the mouse. The data which most adequately and accurately reflect the role of the genotype in epizootic mousepox are summarized in Table II. Mortality was greatest in DBA/1, A, and C3H mice. On the other hand, MA/Nd, C56BL/6, BALB/c, AKR, and MA mice experienced little or no mortality. The C57BL/6 mice are particularly interesting because they were shown to include asymptomatic carriers of ectromelia virus and served as the local source of the epizootic. The same C57BL/6 mice had initiated an epizootic in another laboratory 1 month earlier, in which they had remained healthy whereas their contacts died.

Briody (1966) assembled information on the response of different mouse genotypes in all the epizootics in the United States on which he was able to collect data (Table III) (Briody, 1955, 1959; Briody *et al.*, 1956; Trentin, 1953). He concluded from this survey, and from experiments with individual mice, that as a rule, mice of the strains most susceptible to the lethal effects of the virus (A, BC, DBA/1, DBA/2, and CBA) usually did not survive long enough to permit the primary lesion or the generalized rash to develop to the stage of virus release, so that they had a weak potential for disseminating the virus into the environment. Development of an epizootic in these mice was

Table II

Influence of the Genotype on the Case Fatality Rate in an Epizootic of Mousepox[a]

Genotype	Case fatality rate	
	Stock mice (%)	Breeding colony (%)
DBA/1	84	61
A	84	71
C3H	71	28
MA/Nd	2	<1
C57BL/6	<1	<1
BALB/c	<1	15
AKR	<1	<1
MA	<1	2
Sy N	<1	—
Sy D	<1	—

[a] The figures given refer to the 3-week period after onset, when no effort was made to control the disease in the 4000 stock mice but during which all ill mice were immediately removed from the breeding colony of 2,000 mice. From Briody *et al.* (1956).

critically dependent on other mice (C3H, A, SWR, and CF1) which had extensive skin lesions. Enzootic infection but not obvious disease would be expected if the population consisted solely of C57BL/6 mice, which (like BDF1, BALB/c and CAF1) Briody regarded as ideal for maintaining the enzootic infection over long periods at a subclinical level, but able to initiate explosive epizootics when other suitable strains (C3H, A, SWR, and CF1) were present in the same environment. He suggested that neither infection nor disease would result if the population groups were MA/Nd, AKR, MA, or C58 mice.

Briody may have overemphasized the importance of genotype on the epizootic behavior of mousepox, for other factors undoubtedly play a role, notably passive and active immunity (see Section II,D,4). For example, in both the Yale outbreak (Trentin, 1953; Trentin and Ferrigno, 1957) and European laboratory mouse colonies (Guillon, 1975), the first outbreaks were characterized by acute deaths with predominantly visceral lesions (even among C57BL mice). Subsequently, if enzootic disease was established in the colony, few acute deaths were observed (although they may have occurred in young mice that were then eaten), and the situation appeared to be dominated by subacute, chronic or inapparent infections.

Table III

The Response of Mice in Epizootics of Mousepox in the United States[a]

Mice	Skin lesions	Usual result
MA/Nd, AKR, MA, Sy N, Sy D	Absent	No infection
C57BL6, BDF_1, BALB/c, CAF_1	Minimal	Uneventful recovery
A, BC, DBA/1, DBA/2, CBA	Minimal	Death
C3H, A	Extensive	Death
SWR, CF1	Extensive	Death or recovery

[a] From Briody (1966).

All Fenner's studies on the clinical picture, pathology, and pathogenesis (Sections II,C and D,3) were carried out with a highly susceptible noninbred mouse stock. The clinical picture and pathological findings might be rather different if experimental inoculations had been made in genetically resistant mice. Some studies reported by Schell (1960a,b) in C57BL mice, a strain found by Trentin (1953) and Briody (1959) to be resistant in laboratory outbreaks of mousepox, are pertinent. Schell found no difference in the infectivity endpoint of a viral suspension titrated in susceptible stock mice and C57BL, but the titer of virus in the footpad of C57BL mice ceased to rise 6 days after infection, and the highest titers in blood, liver, and spleen were 2 to 3 logs lower than in noninbred susceptible mice. Cultured cells from resistant and susceptible strains were equally susceptible to ectromelia infection, but neutralizing antibody, delayed type hypersensitivity, and active immunity were demonstrable 1 or 2 days earlier in C57BL than in stock mice. Breeding experiments by Schell (1960b) and Ermolaeva *et al.* (1974) demonstrated that resistance was dominant over susceptibility and was largely determined by a single gene which was not linked to sex, certain color genes, or the *H-2* genotype (R. V. Blanden, 1978, personal communication).

Schell showed that strain differences between mice were best demonstrated after footpad inoculation or, as Briody found, in natural epizootics. C57BL mice were relatively susceptible by intranasal, intracerebral, or intraperitoneal infection, being even more susceptible than stock mice after large doses inoculated intranasally. Since recovered stock mice also die after receiving large doses intranasally, this response may be due to the more vigorous immune response of the C57BL mice.

By comparing results obtained in several experiments, Roberts (1964) showed that the rate of growth of ectromelia virus in the littoral (reticuloendothelial) cells of the livers of mice depended upon the virulence of the strain of virus used, whereas the subsequent growth of virus in the parenchymal cells of the liver depended upon the mouse strain.

2. Prevalence and Distribution

There are two reports suggesting that mousepox might occur as a disease of wild rodents, other than in wild house mice associated with an outbreak in laboratory mice (see, e.g., McGaughey and Whitehead, 1933). Gröppel (1962) examined wild mice belonging to three genera (*Microtus, Apodemus,* and *Cletrionomys*) which were captured in several rural localities in Germany, distant from possible contamination from laboratory mice. His observations are difficult to interpret, but they suggest that several *Apodemus* individuals were infected with an infectious agent that was probably ectromelia virus. More recently, Kaplan *et al.* (1980) found that sera from several voles and woodmice captured in the wild in the United Kingdom contained complement-fixing antibodies for ectromelia virus. Because of the cross-reactivity of orthopoxviruses, this result suggests recent infection with an orthopoxvirus, not necessarily ectromelia virus.

Since 1930 mousepox has been recognized as a relatively common infection of laboratory mouse colonies in Europe, and in Japan and China. It has been so common in parts of Europe that few of the many outbreaks were reported, but recovery of the virus from laboratory mice (usually associated with a laboratory outbreak) has been described in England (Marchal, 1930; Fairbrother and Hoyle, 1937; McGaughey and Whitehead, 1933; Gledhill, 1962a), France (Hornus and Thibault, 1939; Schoen, 1938), Germany (Kikuth and Gönnert, 1940; Schell, 1964), Russia (cited in Andrewes and Elford, 1947), Switzerland (Mooser, 1943), and Sweden, Czechoslovakia, and Israel (Briody, 1959). The disease is still common in laboratory colonies of mice in the People's Republic of China (Dr. Jiang Yutu, July 1979 personal communication) but is now much less prevalent than formerly in Japan (Dr. M. Nakagawa, October 1979 personal communication). This is attributed to the introduction of better management and the extensive use of specific pathogen-free (SPF) and "clean" conventional colonies of mice. Vaccination was never used in Japan for the control of mousepox.

The situation is different in North America. Apart from two vague early reports (Hon, 1918; Thompson, 1934) which Trentin and Briody (1953) did not regard as reliable, mousepox does not seem to have been enzootic in mouse colonies in the United States. For example, Briody (1966) found no evidence of enzootic mousepox in serological surveys involving over 100,000 serum examinations from various colonies of mice in different parts of the United States. However, the virus has been unwittingly imported into the United States from Europe several times with mice or mouse tissues, sometimes causing devastating outbreaks (Melnick and Gaylord, 1953; Trentin, 1953; Dalldorf and Gifford, 1955; Briody, 1955; Briody *et al.*, 1956; Whitney, 1974).

With the development over the last decade or so of SPF mouse colonies for the large-scale production of healthy mice, ectromelia-free stocks of widely used mouse strains have been established. However, the simultaneous large-scale development of many strains of mice with special genetic features, which are usually not produced as SPF animals, and the extensive international exchange that occurs in these animals or in tumor material derived from them, have increased the risk of importation of the virus into previously uninfected mouse stocks. A rather serious situation appears to have developed in the United States since 1979, with evidence of ectromelia in mouse breeding colonies and among animals under experiment in several parts of the United States (Wallace, 1981; New, 1981). There have been periodic efforts (Shope, 1954; Report, 1973; Anslow *et al.*, 1975; Wallace, 1981) to limit the spread of the virus between stocks of mice in different laboratories.

Several examples have been reported, and others have undoubtedly occurred, in which ectromelia virus has been confused with other viruses potentially present in human or other material inoculated into mice (see, e.g., Mooser, 1943; Fairbrother and Hoyle, 1937; Dalldorf and Gifford, 1955; MacCallum *et al.*, 1957; Palmer *et al.*, 1968). The most bizarre such incident was the production of a "live vaccine against murine typhus" by serial intranasal passage of *Rickettsia mooseri* (Laigret and Durand, 1941) and its administration to hundreds of thousands of persons in north Africa. Packalén (1947) demonstrated that this "typhus vaccine" owed its mouse pathogenicity to its high content of ectromelia virus.

3. Epizootic Mousepox

Outbreaks of mousepox have occurred in laboratory mice when ectromelia virus has been inadvertently introduced into colonies of susceptible animals ("natural epizootics"), and deliberately planned epizootics have been used as a tool for the experimental study of epidemiology.

a. Natural Epizootics. Mousepox is notorious for the damage it can cause if unwittingly introduced into large colonies of laboratory mice. As noted above, mousepox has been and remains a recurrent problem in European laboratories, and outbreaks are not held to justify detailed investigation and reporting. The position is quite different in the United States because of the rarity of outbreaks and the large size of the laboratory mouse industry there, and several reports of outbreaks have been published (Trentin, 1953; Melnick and Gaylord, 1953; Poel, 1954; Briody, 1955, 1959; Briody *et al.*, 1956; Whitney, 1974; New, 1981). The most interesting feature of these epizootics was the striking differences in the symptomatology and mortality in different strains of mice (see, e.g., Briody *et al.*, 1956; Briody, 1959; see also Section II,D,1,*b*).

This variability in symptomatic response emphasizes the need for some method of diagnosis other than the clinical picture. Different procedures are recommended for the confirmation of a suspicious case and the screening of mouse stocks (see Section II,E).

b. Experimental Epidemiology. In the early 1920s, the distinguished British bacteriologist W. W. C. Topley embarked upon a long-term study of experimental epidemics in mice housed in specially designed cages (Topley, 1923). Initially he used a variety of bacteria, but with Marchal's discovery of ectromelia virus he undertook studies with this agent, which have been summarized in Greenwood *et al.* (1936). Their studies of long-continued epizootics (1.75 and 3.25 years) in herds of mice maintained by the regular addition of normal mice suffered from the fact that the only indication of infection was death. Conducting similar experiments after the demonstration of the nature of the virus and the nature of the disease, Fenner (1948d) constructed life tables for mice exposed to a virulent and an attenuated strain of virus. The characteristic virulence of each strain was maintained throughout the 190 and 290 days of the two experiments.

Anderson and May (1979) have recently used the results of the experiments by Greenwood and Fenner in an analysis of the population biology of infectious diseases.

Greenwood *et al.* (1936) also used closed epidemics to test the effects of vaccination with small doses of active ectromelia virus. Fenner considerably extended and refined this approach by regularly examining mice for primary lesions and rash, as well as observing mortality, and demonstrated the protective effects of prior vaccination with vaccinia virus (Fenner and Fenner, 1949) and the differing virulence and infectivity of three strains of ectromelia virus (Fenner, 1949b).

4. Enzootic Mousepox

Mousepox appears to be enzootic in many mouse-breeding establishments in Europe. A variety of mechanisms probably operate to maintain the virus without so disrupting the mouse-breeding program as to make control mandatory. One important factor is probably the high level of genetic resistance and trivial symptomatology exhibited by many mouse genotypes (Section II,D,1,*b*). Another may be maternal antibody, which Fenner (1948b) found was transmitted in the milk for at least the first week after birth and which was very effective for a few weeks in saving young mice of a highly susceptible stock from death due to virulent ectromelia virus infection. However, it did not prevent the development of infectious lesions in many of these young animals. A combined serological and clinical study of a large mouse-breeding colony in England (F. Fenner and E. M. B. Fenner, unpublished observations) suggested that only a few breeding cages were actively infected at any time and that infection was maintained partly by handling by the animal attendants. Infection of young mice which had suckled immune mothers resulted in nonfatal disease, but the lesions on such animals were infectious. When young mice from several litters were assembled in large cages after weaning, prior to their despatch to the laboratory, these animals acted as a source of infection. Since the protective effect of maternal antibody is lost within 4 weeks (Fenner, 1948b), mice not infected as weanlings were highly susceptible to the disease, and deaths were frequent. A self-contained epidemic could thus be established in mice awaiting shipment to the laboratory, and the large number of severe cases which then occurred acted as a further source from which virus might be spread to "clean" cages.

A third possibility, clearly involving mouse genotype, is chronic, clinically inapparent infection. The most convincing evidence of this is the report by Gledhill (1962a,b) of infection of Peyer's patches and lesions in tail skin, referred to in Section II,C,3,*b*. Some authors (see, e.g., Kikuth and Gönnert, 1940) believe that latent infection occurs that can be activated by various kinds of stress, including the inoculation of tissue homogenates. Kikuth and Gönnert (1940) described two "latently" infected stocks of mice. Strain "Do," which they thought contained more unhealthy-looking mice than their other four strains, never showed external signs which they regarded as typical for ectromelia. Some animals had ruffled hair, occasionally "pyodermia" occurred on different parts of the body and extremities, and a few mice had blepharitis, but no other significant signs were apparent. The other infected strain, "Vo," showed even less. It is likely that the pyodermia and blepharitis were lesions of the secondary rash of mousepox, and these animals probably constituted the source of infection for other members of the stock. No mention is made of undue mortality in the infected strains. Fenner (1948d) was unable to demonstrate latency in the mouse strains that he used, although he found that the lungs and spleen of two mice that had recovered from mousepox 28 and 75 days earlier yielded virus.

Both clinically inapparent and latent infections may be dependent upon the host genotype. If they do occur in a mouse stock, they are important both as potential sources of virus for natural mouse-to-mouse spread and as a source of virus that might be unwittingly transferred by subinoculation.

E. Diagnosis

1. Diagnosis of a Suspicious Case

Mousepox can be diagnosed by the microscopic (Marchal, 1930) or electron microscopic (Leduc and Bernhard, 1962) examination of the tissues of suspected cases (Fig. 10), the diagnostic features being the distinctive eosinophilic cytoplasmic inclusion bodies (see Fig. 3) and the poxvirus particles, respectively. Passage of concentrated viral suspensions on the chorioallantoic membrane should produce confluent lesions which when ground up exhibit characteristic hemagglutination (Burnet and Boake, 1946); suitably diluted suspensions will produce small pocks on the chorioallantoic membrane. Inoculation of genetically susceptible, nonimmune mice will pro-

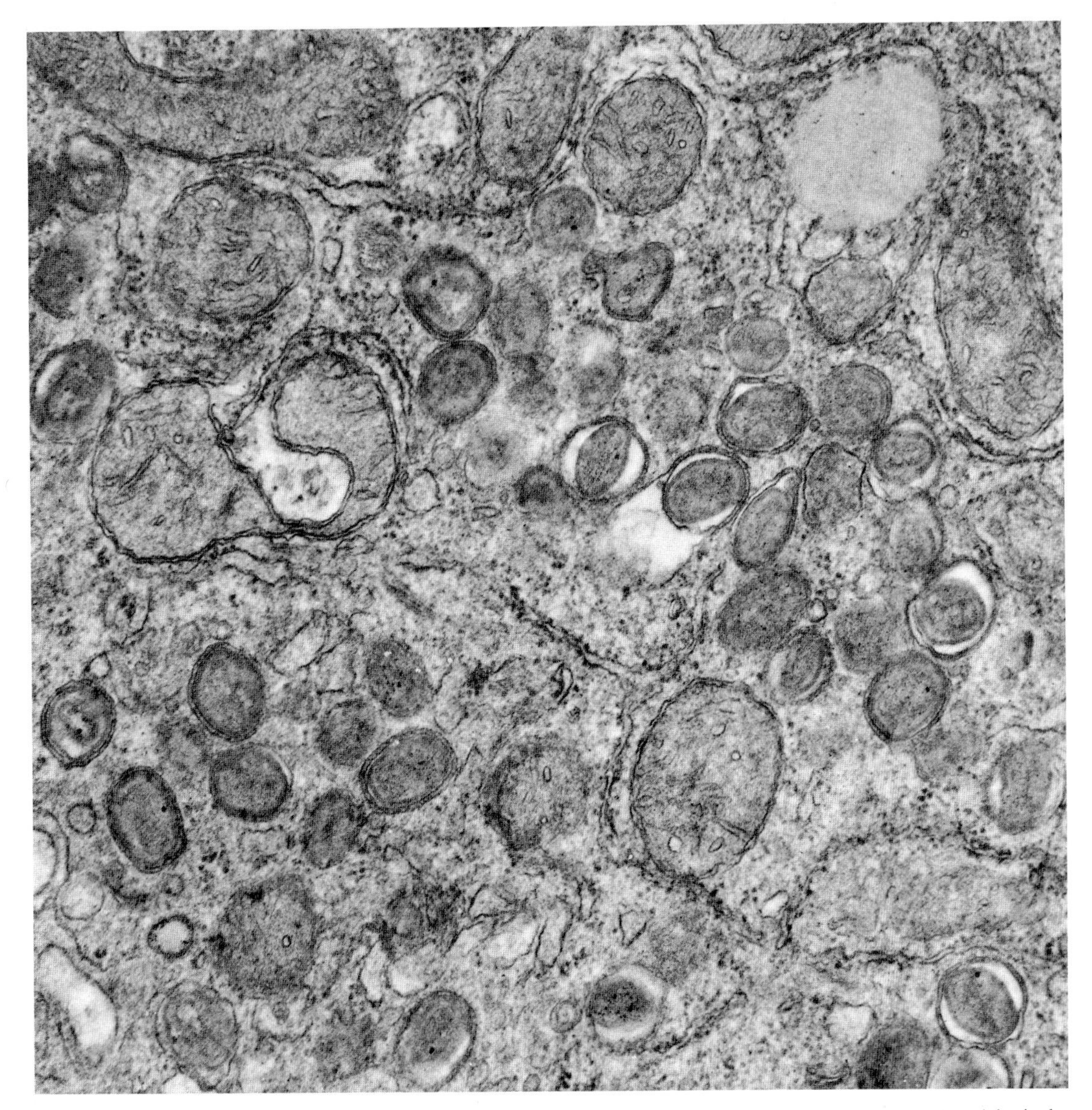

Fig. 10. Electron micrograph of the liver of a mouse that had died with acute mousepox. There are many viral particles in the liver. ×45,000. (Courtesy of D. O. Irving, John Curtin School of Medical Research.)

duce the symptoms and signs described earlier, whereas vaccinated mice of the same strains should prove resistant (see Section II,F,2) Gel diffusion tests with scab material have also been used for rapid diagnosis (Carthew *et al.*, 1977).

Following the pattern established for the diagnosis of smallpox in the Smallpox Eradication Program of the World Health Organization, rapid diagnosis could be most readily effected by searching for poxvirions in the homogenized tissues (or scabs) from suspected cases. No other poxvirus is likely to be found in laboratory mice, and the specific diagnosis could be readily confirmed by inoculation in mice or eggs.

2. Screening Tests for Mouse Stocks

Briody, who developed a Mouse Pox Service Laboratory in 1956 under contract to the National Cancer Institute, National Institutes of Health, United States, has demonstrated the value of the HI test, using vaccinia virus to provide the hemagglutinin (Briody, 1959). Positive results were obtained in the epizootics described earlier and in several enzootically infected European stocks. On the basis of the serological examination of over 100,000 sera from various colonies of mice throughout the United States, Briody (1966) concluded that enzootic mousepox probably did not occur in that country. But international transfer of mice, mouse cells, or mouse tumor material poses the constant threat that it might be introduced, as was evident as recently as 1974 in laboratories of the National Institutes of Health in Bethesda following the importation of a tumor from London, to which it had earlier (1972) been sent from Prague, from a laboratory where mousepox was enzootic (J. D. Small, 1978, personal communication).

Some doubt has been cast on the wisdom of relying only on the HI test for excluding mousepox in serological surveys. It is well known that mutants of vaccinia virus can occur which fail to produce hemagglutinin (Cassel, 1957; Fenner, 1958), and similar variants might occur with ectromelia virus. Indeed, Christensen *et al.* (1966) and more recently S. P. Pakes (1978 personal communication to Dr. J. D. Small) noted that HI antibodies were not found in screening tests of recovered animals in several outbreaks in the United States, and Guillon (1975) commented that several strains of ectromelia virus were only slightly hemagglutinogenic upon isolation. Although in many years of experience in the United States demonstrated the practical value of the HI test in mousepox surveillance, recent experience with false positive and false negative results has not been reassuring (see New, 1981). It is now considered desirable that the HI test be supplemented by another serological test, e.g., immunofluorescence, used as a test for the presence of either virus or antibody (Christensen *et al.*, 1966), or gel precipitation with either sera (Horzinek, 1965) or scab material from suspected cases (Carthew *et al.*, 1977), or the ELISA test (P. Bhatt, personal communication, 1981).

F. Control and Prevention

Quarantine and regulation of the importation and distribution, of ectromelia virus or materials infected with it are mandatory in the United States in order to protect the large, uninfected mouse colonies of that country (Shope, 1954; Anslow *et al.*, 1975). However, there is some doubt as to how well such regulations apply (Briody, 1959), and in any case, they offer no protection against unsuspected sources of infection. Briody (1959) suggested that all mice imported into the United States from other countries should be effectively immunized with vaccinia virus. Only those mice which develop the typical lesion in 6–7 days following intradermal scarification of the tail with vaccinia virus should be admitted to the colony, since failure to react might indicate prior infection with ectromelia virus. Upon arrival, the mice should be isolated from the main colony, kept under rigid quarantine, and carefully observed. This period of isolation and quarantine should persist until the first-generation offspring have been effectively vaccinated. Healthy, genetically susceptible mice should be placed as sentinel animals in pens with the imported mice for a few weeks to reveal the possible presence of ectromelia virus. Where the imported mice are to be used as breeding stock, this test necessitates segregating the males and females during this period of contact, or alternatively, introducing only ovariectomized females as susceptible contact mice. Tumors derived from mice obtained from other countries should be especially suspect in view of the predilection of ectromelia virus and other viruses for tumors. Such tumors should be passed for at least two transplant generations in vaccinated mice.

The establishment by Briody of the Mouse Pox Service Laboratory in 1956 provided scientists in the United States with a valuable diagnostic facility. Since his death in 1975, this service has been provided by Microbiological Associates Inc. under contract to the Mammalian Genetics and Animal Production Section, National Cancer Institute, National Institutes of Health, Bethesda.

If mousepox is present in a colony, two procedures may be followed: eradication by slaughter and disinfection, or control by vaccination.

1. Slaughter and Disinfection

The National Institutes of Health Committee of which Shope was chairman recommended (Shope, 1954) that infected colonies should be destroyed and materials (viral stocks, tumors, etc.) derived from such animals incinerated. Colonies of mice suspected of infection should be isolated and quarantined. All materials harvested from mice received from foreign sources, and mouse strains themselves, should be regarded as potentially infected with ectromelia virus, and any suspected or proved outbreaks should be promptly reported and their sources investigated.

2. Vaccination

Vaccination with vaccinia virus, by any route of inoculation, will protect mice against ectromelia, although the protection is not absolute, especially against virulent strains of virus, and it wanes with time (Fenner, 1947a, 1949c). Closed epidemics (Fenner and Fenner, 1949) showed that vaccination conveyed substantial but not complete protection against the naturally spreading disease.

The practical use of vaccination for the control of mousepox in infected colonies has been elaborated by Trentin and Ferrigno (1957) and Briody (1959) and by several European scientists (Salaman and Tomlinson, 1957; Zeller and Reckzeh, 1965a,b; Munz *et al.*, 1974, 1976). Although the differential titer of HI antibodies to ectromelia and vaccinia can be used to determine whether mice have been infected with vaccinia, ectromelia, or both (Fenner, 1949c), Briody's (1959) procedure of vaccinating with a strain of vaccinia virus (IHD-T; Cassel, 1957) that does not produce hemagglutinin simplifies serological studies.

During the epizootics in the United States in 1957 and 1958, a number of different control procedures were used by breeders (Briody, 1959). In some laboratories, the newly established mouse colony was immunized with vaccinia virus for one or two generations; in others vaccination was established on a continuing basis after cleanup of the epizootic situation. As a result of routine vaccination of all mice in their colony over a period of about 2 years, Trentin and Ferrigno (1957) were able to eradicate ectromelia virus from the colony without the slaughter of infected mice. Within any given mouse colony, Briody suggested that universal vaccination represented the most effective weapon for the prevention of mousepox and should guarantee freedom from the epizootic disease. Universal vaccination should be practiced as soon as exposure to ectromelia virus is known or suspected.

Intradermal scarification of the dorsal surface of the base of the tail is rapid and effective (Salaman and Tomlinson, 1957; Briody, 1959; Flynn, 1963), but some European workers have suggested oral vaccination (Munz *et al.*, 1974), vaccination by aerosol (Munz *et al.*, 1976), or intraperitoneal inoculation (Zeller and Reckzeh, 1965a,b). Oral vaccination proved quite unreliable. Aerosol dispersion of vaccinia virus presents problems in achieving safety for man and effective dosage for the mice. Zeller and Rechzeh (1965b) conclude that their experiences over a 5-year period show that vaccination of all mice received by an institute will eliminate the occurrence of mousepox. They believe that the expenditure of time and money on vaccination and revaccination by intraperitoneal injection is low in comparison with other methods of vaccination and with the losses which may arise from uncontrolled intercurrent mousepox infections.

The strain of vaccinia virus, the age and genotype of the mouse (Owen *et al.*, 1975), the concentration of virus, and the route of inoculation all affect the response, which may vary from lethal infection at one extreme to failure to produce an antibody response at the other. In general, strains used for human vaccination seem to be satisfactory for mice. There are some advantages in following Briody's (1959) suggestion that IHD-T strain should be accepted as the standard strain for vaccination.

III. SUMMARY: THE PRACTICAL PROBLEMS

Although Australian virologists and immunologists have found mousepox to be a very useful model disease for the study of pathogenesis, immunology, and epidemiology, for most laboratory workers it is essentially a hazard which can ruin their experiments and decimate their stocks of mice. This concluding section is devoted to the problems of the mouse breeder and the research worker who uses mice or their cells and tissues.

A. Problems of the Mouse Breeder

Although there may be some difficulty in establishing breeding stocks free from ectromelia because transplacental transmission can occur (although it is likely that infected fetuses will always die), the development and maintenance of SPF stocks of mice constitutes the most effective way of controlling this and other enzootic infections of mice. However, this procedure has not been widely followed in many countries because of its expense, and it cannot be adopted for each of the many genetically marked mouse strains that constitute such a valuable resource for biomedical research.

It is important, therefore, that the mouse breeder be aware of the potential existence of mousepox in such stocks, of the great differences in genetic susceptibility and symptomatology found among different strains of mice, and of the appearance of the lesions of mousepox in acute and chronic cases of infection. He must be able to confirm the diagnosis by histological and serological means and, if necessary, to survey his breeding stocks by a serological screening test.

The procedure to be followed upon discovery of enzootic mousepox depends upon a number of factors: the presence of mousepox in other stocks in the country, the value of the mouse strain found to be infected, and the extent of infection in the mouse-breeding establishment concerned.

Two principal procedures can be followed. If early diagnosis has been made and there is a reasonable probability that the virus has not been widely disseminated (for example, with mouse stocks or tissues sent to other laboratories), the most effective procedure is the slaughter of the infected stock of mice, followed by thorough sterilization of mouse rooms, cages, equipment, etc. The other procedure is vaccination with vaccinia virus. As in human smallpox, this does not give absolute protection, and protection wanes with time, but it has been conclusively shown that in most mouse stocks mousepox can be eradicated by systematic, and if necessary repeated, vaccination of all mice. To do this effectively, the breeder must have available, or be able to obtain, a strain of vaccinia virus that will be potent but safe in all strains of mice that he wishes to vaccinate. Some mouse breeders in Europe, where mousepox is much more common than in the United States and risks of reintroduction are ever-present, routinely vaccinate all their breeding stock on a permanent basis. Others vaccinate for perhaps two generations to eradicate ectromelia virus. Two routes of vaccination are used: intraperitoneal and scarification of the base of the tail. The latter method has the advantage that mice that fail to react (and thus presumably have had mousepox) can be recognized.

B. Problems of the Research Worker

With the increasing replacement of intact animals by cultured cells for viral isolation and diagnosis, some of the problems due to enzootic mousepox that face the laboratory worker are lessening. But intact animals are still required for many purposes. Any research worker who uses mice other than those from reliable SPF colonies needs to be aware of the existence of mousepox, especially if he imports mice, or mouse tissues or tumors, from countries where infection is rife, as in Europe. Examples of the unintentional "pickup" of ectromelia virus during serial passage of material in mice are legion, and in such cases ectromelia virus usually replaces whatever agent was first isolated.

As well as knowing how to diagnose classic acute or chronic mousepox, the research worker should be aware of the different signs to be found in animals subjected to serial intraperitoneal or intranasal (Sections II,C,3,*c* and *d*) passage of material that contains ectromelia virus, and of the fact that different mouse genotypes may exhibit widely differing symptomatology when infected with ectromelia virus.

Finally, any research worker who discovers ectromalia virus in material that might have been disseminated more widely should ensure that his colleagues are informed immediately and should seek to ascertain the source of the infection.

REFERENCES

Anderson, R. M., and May, R. M. (1979). Population biology of infectious diseases: Part 1. *Nature (London)* **280,** 361–367.

Andrewes, C. H., and Elford, W. J. (1947). Infectious ectromelia: Experiments on interference and immunization. *Br. J. Exp. Pathol.* **28,** 278–285.

Anslow, R. A., Ewald, B. H., Pakes, S. P., Small, J. D., and Whitney, R. A. (1975). Letter: Institutional outbreak of ectromelia. *Lab. Anim. Sci.* **25,** 532–533.

Barnard, J. E., and Elford, W. J. (1931). The causative organism in infectious ectromelia. *Proc. R. Soc. London, Ser. B* **109,** 360–374.

Bellett, A. J. D., and Fenner, F. (1968). Study of base-sequence homology among some deoxyriboviruses of vertebrate and invertebrate animals. *J. Virol.* **2,** 1374–1380.

Blanden, R. V. (1974). T cell response to viral and bacterial infections. *Transplant. Rev.* **19,** 56–88.

Briody, B. A. (1955). Mouse pox (ectromelia) in the United States. *Lab. Anim. Care* **6,** 1–8.

Briody, B. A. (1959). Response of mice to ectromelia and vaccinia viruses. *Bacteriol. Rev.* **23,** 61–95.

Briody, B. A. (1966). The natural history of mousepox. *Natl. Cancer Inst. Monogr.* No. 20, pp. 105–116.

Briody, B. A., Hauschka, T. S., and Mirand, E. A. (1956). The role of genotype in resistance to an epizootic of mouse pox (ectromelia). *Am. J. Hyg.* **63,** 59–68.

Burnet, F. M. (1960). "Principles of Animal Virology," 2nd ed. Academic Press, New York.

Burnet, F. M., and Boake, W. D. (1946). The relationship between the virus of infectious ectromelia of mice and vaccinia virus. *J. Immunol.* **53,** 1–13.

Burnet, F. M., and Lush, D. (1936a). The propagation of the virus of infectious ectromelia of mice in the developing egg. *J. Pathol. Bacteriol.* **43,** 105–120.

Burnet, F. M., and Lush, D. (1936b). Inapparent (subclinical) infection of the rat with the virus of infectious ectromelia of mice. *J. Pathol. Bacteriol.* **42,** 469–476.

Cairns, J. (1960). The initiation of vaccinia infection. *Virology* **11,** 603–623.

Carthew, P., Hill, A. C., and Verstraete, A. P. (1977). Some observations on the diagnosis of an outbreak of ectromelia in 1976. *Vet. Rec.* **100,** 293.

Cassel, W. A. (1957). Multiplication of vaccinia virus in the Ehrlich ascites carcinoma. *Virology* **3,** 514–526.

Christensen, L. R., Weisbroth, S., and Matanic, B. (1966). Detection of ectromelia virus and ectromelia antibodies by immunofluorescence. *Lab. Anim. Care* **16,** 129–141.

Cole, G. A., and Blanden, R. V. (1981). Immunology of poxvirus infection. *In* "Immunology of Human Infections." Vol. 9, Part II, Viruses and Parasites; Immunodiagnosis and Prevention of Infectious Diseases (A. J. Nahmias and R. J. O'Reilly, eds.). Plenum Press, New York.

Dalldorf, G., and Gifford, R. (1955). Recognition of mouse ectromelia. *Proc. Soc. Exp. Biol. Med.* **88,** 290–292.

Deerberg, F., Kästner, W., Pittermann, W., and Schwanzer, V. (1973). Nachweis einer Ektromelie-Enzootie bein haarlosen Mäusen. *Dtsch. Tieraerztl. Wochenschr.* **80,** 78–81.

Ermolaeva, S. N., Blandova, Z. K., and Dushkin, V. A. (1974). Genetic study on susceptibility of different mouse lines to ectromelia virus. *Sov. Genet.* **8,** 681–783.

Fairbrother, R. W., and Hoyle, L. (1937). Observations on the aetiology of influenza. *J. Pathol. Bacteriol.* **44,** 213–223.

Fenner, F. (1947a). Studies in infectious ectromelia of mice. I. Immunization of mice against ectromelia with living vaccinia virus. *Aust. J. Exp. Biol. Med. Sci.* **25,** 257–274.

Fenner, F. (1947b). Studies in infectious ectromelia of mice. II. Natural transmission: the portal of entry of the virus. *Aust. J. Exp. Biol. Med. Sci.* **25,** 275–282.

Fenner, F. (1948a). The clinical features of mouse-pox (infectious ectromelia of mice) and the pathogenesis of the disease. *J. Pathol. Bacteriol.* **60,** 529–552.

Fenner, F. (1948b). The epizootic behaviour of mouse-pox (infectious ectromelia of mice). *Br. J. Exp. Pathol.* **29,** 29–91.

Fenner, F. (1948c). The pathogenesis of the acute exanthems. An interpretation based upon experimental investigations with mouse-pox (infectious ectromelia of mice). *Lancet* **ii,** 915–920.

Fenner, F. (1948d). The epizootic behaviour of mouse-pox (infectious ectromelia of mice). II. The course of events in long-continued epidemics. *J. Hyg.* **46,** 383–393.

Fenner, F. (1949a). Mouse pox (infectious ectromelia of mice): A review. *J. Immunol.* **63,** 341–373.

Fenner, F. (1949b). Studies in mouse-pox (infectious ectromelia of mice). VI. A comparison of the virulence and infectivity of three strains of ectromelia virus. *Aust. J. Exp. Biol. Med. Sci.* **27,** 31–43.

Fenner, F. (1949c). Studies in mouse-pox (infectious ectromelia of mice). IV. Quantitative investigations on the spread of virus through the host in actively and passively immunized animals. *Aust. J. Exp. Biol. Med. Sci.* **27,** 1–18.

Fenner, F. (1949d). Studies in mousepox (infectious ectromelia of mice). VII. The effect of age of the host upon the response to infection. *Aust. J. Exp. Biol. Med. Sci.* **27,** 45–53.

Fenner, F. (1958). The biological characters of several strains of vaccinia, cowpox, and rabbitpox viruses. *Virology* **5,** 502–529.

Fenner, F. (1976). Classification and Nomenclature of Viruses: Second Report of the International Committee on Taxonomy of Viruses. *Intervirology* **7,** 1–115.

Fenner, F., and Fenner, E. M. B. (1949). Studies in mouse-pox (infectious ectromelia of mice). V. Closed epidemics in herds of normal and vaccinated mice. *Aust. J. Exp. Biol. Med. Sci.* **27,** 19–30.

Flynn, R. J. (1963). The diagnosis and control of ectromelia infection of mice. *Lab. Anim. Care* **13,** 130–136.

Flynn, R. J., and Briody, B. A. (1962). Relative susceptibilities of vaccinated and non-vaccinated Syrian hamsters and mice to ectromelia virus. *Proc. Anim. Care Panel* **12,** 263–266.

Germer, W.-D., Diefenthal, W., and Habermehl, K.-O. (1961). Latente Infection von Mäuse-embryonen mit Ektromelievirus (Mäusepocken) infolge diaplacentarer Übertragung. *Z. Hyg.* **147,** 269–276.

Gledhill, A. W. (1962a). Latent ectromelia. *Nature (London)* **196,** 298.

Gledhill, A. W. (1962b). Viral diseases in laboratory animals. *In* "The Problems of Laboratory Animal Disease" (R. J. C. Harris, ed.), pp. 99–112. Academic Press, New York.

Greenwood, M., Hill, A. B., Topley, W. W. C., and Wilson, J. (1936). Experimental epidemiology. *Med. Res. Counc. (G.B.), Spec. Rep. Ser.* No. 209.

Gröppel, K.-H. (1962). Über das Vorkommen von Ektromelie (Mäusepocken) unter Wildmäusen. *Arch. Exp. Veterinaermed.* **16,** 243–278.

Guillon, J.-C. (1970). Recherches sur la transmission du virus de l'ectromélie de la souris par un acarien hématophage: "Ornithonyssus bacoti." *Exp. Anim.* **3,** 177–182.

Guillon, J.-C. (1975). Prophylaxie sanitaire et médicale de l'ectromélie de la souris. *Rec. Med. Vet.* **151,** 19–24.

Hon, C. (1918). How to breed mice. *J. Am. Med. Assoc.* **70,** 1225.

Hornus, G., and Thibault, P. (1939). Ectromélie de la souris après inoculation de sang humain. *C. R. Seances Soc. Biol. Ses Fil.* **130,** 640–641.

Horzinek, M. (1965). Untersuchungen zur Diagnose der Mäusepocken (Ektromelie) mittels der Agargel-Präzipitation. *Arch. Gesamte Virusforsch.* **17,** 264–272.

Horzinek, M., and Höpken, W. (1965). Untersuchungen über die inapparente Infecktion mit dem Ektromelievirus. *Arch. Gesamte Virusforsch.* **17,** 125–138.

Ichihashi, Y., and Matsumoto, S. (1966). Studies on the nature of Marchal bodies (A-type inclusion) during ectromelia virus infection. *Virology* **29,** 264–275.

Iftimovici, R., Iacobeson, V., Mutiu, A., and Pucă, D. (1976). Enzootic with ectromelia symptomatology in Sprague–Dowley rats. *Rev. Roum. Med. Virol.* **27,** 65–66.

Ipsen, J. (1944). Quantitative Studien über mäusepathogene Fleckfieberrickettsien. I. Methode zur Messung der Virusmenge in infizierten Organen. *Schweiz. Z. Pathol. Bakteriol.* **7,** 129–151.

Ipsen, J. (1945). Quantitative Studien über mäusepathogene Fleckfieberrickettsien. II. Das Wachstum der Infektion und ihre Ausbreitung im Mäuseorganismus. *Schweiz. Z. Pathol. Bakteriol.* **8,** 57–70.

Jahn, H. (1939). Experimentelle Untersuchungen über das Virus der Ecktromelia Infectiosa. *Arch. Gesamte Virusforsch.* **1,** 91–103.

Kanazawa, K. I. (1937). Expériences sur l'infection cérébrale de la souris avec le virus de l'ectromélie infectieuse. *Trans. Soc. Pathol. Jpn.* **27,** 585–591.

Kaplan, C., Healing, T. D., Evans, N., Healing, L., and Prin, A. (1980). Evidence of infection by viruses in small British field rodents. *J. Hyg.* **84,** 285–294.

Kato, S., Hagiwara, K., and Kamahora, J. (1955). The mechanism of growth of ectromelia virus propagated in the ascites tumor cells. I. Study of the inclusion bodies of ectromelia virus. *Med. J. Osaka Univ.* **6,** 39–50.

Kato, S., Takahashi, M., Kameyama, S., and Kamahora, J. (1959). A study of the morphological and cyto-immunological relationship between the inclusions of variola, cow pox, rabbit pox, vaccinia (variola origin) and vaccinia IHD and a consideration of the term "Guarnieri body". *Biken J.* **2,** 353–363.

Kato, S., Aoyama, Y., and Kamahora, J. (1963). Autoradiography of the tissues of mice infected with ectromelia virus using ^{3}H-thymidine. *Biken J.* **6,** 9–16.

Kikuth, W., and Gönnert, R. (1940). Erzeugung von Ektromelie durch Provokation. *Arch. Gesamte Virusforsch.* **1,** 295.

Kling, C., and Packalén, T. (1947). An experimental study on a derivative of Laigret–Durand's strain of murine rickettsiae for the preparation of living typhus vaccine. I. Serological and immunological analysis. *Acta Pathol. Microbiol. Scand.* **24,** 371–374.

Laigret, J., and Durand, R. (1939). La vaccination contre le typhus exanthématique. Nouvelle technique de préparation du vaccin: emploi des cervaux de souris. *Bull. Soc. Pathol. Exot.* **32,** 735–751.

Laigret, J., and Durand, R. (1941). Précisions techniques sur le vaccin vivant en enrobe contre le typhus exanthématique. *Bull. Soc. Pathol. Exot.* **34,** 193–198.

Leduc, E. H., and Bernhard, W. (1962). Electron microscope study of mouse liver infected with ectromelia virus. *J. Ultrastruct. Res.* **6,** 466–488.

MacCallum, F. O., Scott, T. F. M., Dalldorf, G., and Gifford, R. (1957). "Pseudo-lymphocytic choriomeningitis": a correction. *Br. J. Exp. Pathol.* **38,** 120–121.

McCarthy, K., and Downie, A. W. (1948). An investigation of immunological relationships between the viruses of variola, vaccinia, cowpox, and ectromelia by neutralization tests on the chorioallantois of chick embryos. *Br. J. Exp. Pathol.* **29,** 501–510.

McGaughey, C. A., and Whitehead, R. (1933). Outbreaks of infectious ectromelia in laboratory and wild mice. *J. Pathol. Bacteriol.* **37,** 253–256.

McNeill, T. A.(1968). The neutralization of pox viruses. II. Relations between vaccinia, rabbitpox, cowpox and ectromelia. *J. Hyg.* **66,** 549–555.

Marchal, J. (1930). Infectious ectromelia. A hitherto undescribed virus disease of mice. *J. Pathol. Bacteriol.* **33,** 713–718.

Marennikova, S. A., Ladnyj, I. D., Ogorodnikova, Z. I., Shelukhima, E. M., and Maltseva, N. N. (1978). Identification and study of a poxvirus isolated from wild rodents in Turkmenia. *Arch. Virol.* **56,** 7–14.

Matsumoto, S. (1958). Electron microscope studies of ectromelia virus multiplication. *Annu. Rep. Inst. Virus Res., Kyoto Univ.* **1,** 151–184.

Melnick, J. E., and Gaylord, W. H., Jr. (1953). Problems with spontaneous ectromelia (mouse pox) in a virus laboratory. *Proc. Soc. Exp. Biol. Med.* **83,** 315–318.

Mims, C. A. (1959). The response of mice to large intravenous injections of ectromelia virus. II. The growth of virus in the liver. *Br. J. Exp. Pathol.* **40,** 543–550.

Mims, C. A. (1964). Aspects of the pathogenesis of virus diseases. *Bacteriol. Rev.* **28,** 30–71.

Mims, C. A. (1966). Pathogenesis of rashes in virus diseases. *Bacteriol. Rev.* **30,** 739–760.

Mims, C. A. (1969). Effect on the fetus of maternal infection with lymphocytic choriomeningitis (LCM) virus. *J. Infect. Dis.* **120,** 582–597.

Mooser, H. (1943). Über die Mischinfektion der weissen Maus mit einem Stamm Klassischen Fleckfiebers und dem Virus der infectiösen Ektromelie. *Schweiz. Z. Pathol. Bakteriol.* **6,** 463–472.

Müller, H. K., Wittek, R., Schaffner, W., Schümperli, D., Menna, A., and Wyler, R. (1978). Comparison of five poxvirus genomes by analysis with restriction endonucleases *Hin* d III, *Bam* I and *Eco* RI. *J. Gen. Virol.* **38,** 135–147.

Munz, E., Reimann, M., Hoffman, R., and Gobel, E. (1974). Perorale Immunisierungsversuche mit Vaccinia-Virus gegen Mäusepocken. *Zentralbl. Bakteriol., Parasitenkd., Infektionskr, Hyg., Abt. 1: Orig., Reihe B* **159,** 10–30.

Munz, E., Reimann, M., and Zschekel, W. D. (1976). Die Aerosol-Impfung gegen Mäusepocken (Infektiose Ektromelie) in Vergleich zu anderer Vakzinationsverfahren unter Verwerdung von Vaccinia-und Mäusepockenvirushaltigen Impfstoffen. *Zentralbl. Veterinaermed., Reihe B* **23,** 431–446.

Nagler, F. P. O. (1944). Red cell agglutination by vaccinia virus. Application to a comparative study of vaccination with egg vaccine and standard calf lymph. *Aust. J. Exp. Biol. Med. Sci.* **22,** 29-35.

New, A. E. (1981). Ectromelia (mousepox) in the United States. *Lab. Anim. Sci.* **31,** No. 5, Part II.

Owen, D., Hill, A., and Argent, S. (1975). Reaction of mouse strains to skin test for ectromelia using an allied virus as inoculum. *Nature (London)* **254,** 598-599.

Packalén, T. (1945). Rickettsial agglutination and complement fixation studies in epidemic typhus fever. *Acta Pathol. Microbiol. Scand.* **22,** 573-592.

Packalén, T. (1947). An experimental study on a derivative of Laigret-Durand's strain of murine rickettsiae for the preparation of living typhus vaccine II. Isolation of ectromelia virus from a vaccine strain. *Acta Pathol. Microbiol. Scand.* **24,** 375-378.

Palmer, E. L., Ziegler, D. W., Kissling, R. E., Hutchinson, H. D., and Murphy, F. A. (1968). Poxvirus nature of the Motol virus. *J. Infect. Dis.* **118,** 500-509.

Paschen, E. (1936). Zuchtung der Ektromelievirus auf der Chorionallantois Membran von Huhnerembryonen. *Zentralbl. Bakteriol., Parasitenkd., Infektionskr. Hyg., Abt. 1: Orig.* **135,** 445-452.

Payne, L. G., and Norrby, E. (1976). Presence of haemaggluttinim in the envelope of extracellular vaccinia virus particles. *J. Gen. Virol.* **32,** 63-720.

Poel, W. E. (1954). "Report of a Mouse-Pox Epizootic (Ectromelia) in the Environmental Carcinogenesis Laboratory of the Graduate School of Public Health, University of Pittsburgh," Supplement to Annual Report. F.I.D.B. Natl. Cancer Inst., Natl. Inst. Health, Bethesda, Maryland.

Reames, H. R. (1940). Pathogenesis and immunity in ectromelia virus infection in the nasal mucosa of the rat. *J. Infect. Dis.* **66,** 254-262.

Report (1973). "Institute of Laboratory Animal Resources, Subcommittee on Procurement Standards for Defined Laboratory Rodents and Rabbits. Procurement Specification (Contract Clause) IX. Defined Laboratory Rodents and Rabbits."

Roberts, J. A. (1962a). Histopathogenesis of ectromelia. I. Respiratory infection. *Br. J. Exp. Pathol.* **43,** 451-461.

Roberts, J. A. (1962b). Histopathogenesis of ectromelia. II. Cutaneous infection. *Br. J. Exp. Pathol.* **43,** 462-468.

Roberts, J. A. (1964). Growth of ectromelia virus in the liver parenchymal cells of different strains of mice. *Nature (London)* **202,** 1140-1141.

Salaman, M. H., and Tomlinson, A. J. H. (1957). Vaccination against mousepox: with a note on the suitability of vaccinated mice for work on chemical induction of tumours. *J. Pathol. Bacteriol.* **74,** 17-24.

Schell, K. (1960a). Studies on the innate resistance of mice to infection with mouse-pox. I. Resistance and antibody production. *Aust. J. Exp. Biol. Med. Sci.* **38,** 271-288.

Schell, K. (1960b). Studies on the innate resistance of mice to infection with mouse-pox. II. Route of inoculation and resistance; and some observations on the inheritance of resistance. *Aust. J. Exp. Biol. Med. Sci.* **38,** 289-299.

Schell, K. (1964). On the isolation of ectromelia virus from the brains of mice from a "normal" mouse colony. *Lab. Anim. Care* **14,** 506-513.

Schoen, R. (1938). Recherches sur une maladie spontanée mortelle des souris blanches. Transmission expérimentelle. *C. R. Seances Soc. Biol. Ses Fil.* **128,** 695-698.

Schwanzer, V., Deerberg, F., Frost, J., Liess, B., Schwanzerova, I., and Pittermann, W. (1975). Zur intrauterinen Infektion der Maus mit Ektromelie-virus. *Z. Versuchstierkd.* **17,** 110-120.

Shope, R. E. (1954). Report of a committee on infectious ectromelia of mice (mousepox). *J. Natl. Cancer Inst.* **15,** 405-408.

Thompson, J. (1934). Inclusion bodies in the salivary glands and liver of mice and rats. *Am. J. Pathol.* **10,** 676.

Topley, W. W. C. (1923). The spread of bacterial infection; some general considerations. *J. Hyg.* **21,** 226-236.

Trentin, J. J. (1953). An outbreak of mouse-pox (infectious ectromelia) in the United States. I: Presumptive diagnosis. *Science* **117,** 226-227.

Trentin, J. J., and Briody, B. A. (1953). An outbreak of mouse-poc (infectious ectromelia) in the United States. II: Definitive diagnosis. *Science* **117,** 227-228.

Trentin, J. J., and Ferrigno, M. A. (1957). Control of mouse pox (infectious ectromelia) by immunization with vaccinia virus. *J. Natl. Cancer Inst.* **18,** 757-767.

Wallace, G. D. (1981). Mouse pox threat. *Science* **211,** 438.

Whitney, R. A. (1974). Ectromelia in mouse colonies. *Science* **184,** 609.

Zeller, H., and Reckzeh, G. (1965a). Zur Immunisierung der weissen Maus gegen Ektromelie mit aktiven Vaccine Virus I. Mitteilung: Methoden und Grundlagen. *Zentralbl. Bakteriol. Parasitenkd., Infektionskr. Hyg., Abt. 1: Orig., Reihe B* **195,** 282-295.

Zeller, H., and Reckzeh, G. (1965b). Zur Immunisierung der weissen Maus gegen Ektromelie mit aktiven Vaccine Virus II. Mitterlung: Praktische Erfahrungen. *Zentralbl. Bakteriol., Parasitenkd., Infektionskr. Hyg., Abt. 1: Orig., Reihe B* **197,** 34-47.

Chapter 12

Lymphocytic Choriomeningitis Virus

Fritz Lehmann-Grube

I. Introduction 231
II. Definitions 233
III. Properties of the Agent 233
A. Classification 233
B. Physical and Chemical Properties 234
C. Virus Strains and Growth of Virus in Laboratory Hosts 235
D. Virus Carrier Cultures 236
IV. Infection of Mice 237
A. Clinical Disease 237
B. Pathology 239
C. Immunity 242
D. Persistent Infection (Carrier State) 249
V. Epizootiology 254
A. Host Range 254
B. Geographic Distribution 254
C. Transmission 255
VI. Diagnosis 256
VII. Control and Prevention 256
References 258

I. INTRODUCTION

Lymphocytic choriomeningitis (LCM) virus was discovered at about the same time but independently in three different laboratories in the United States. Armstrong and Lillie (1934), working at the U.S. Public Health Service's Hygienic Laboratory in Bethesda, Maryland (the forerunner of the National Institutes of Health), encountered it when they passaged intracerebrally in monkeys "infectious materials" (no details) from patient C.G., who had died in the 1933 epidemic of St. Louis encephalitis. On the basis of the pathologic picture the new agent caused in intracerebrally inoculated monkeys and mice, it was designated *virus of experimental lymphocytic choriomeningitis*. Its true source was not verified.

At the Rockefeller Institute for Medical Research in Princeton, New Jersey, Traub (1935) recovered an infectious agent

ISBN 0-12-262502-1

from white mice which produced an illness in mice closely resembling the one described by Armstrong and Lillie. Its origin remained unknown, but wild house mice were suspected. Two further agents with similar properties were isolated by Rivers and Scott (1935) from the cerebrospinal fluid of two men, W.E. and R.E.S., who were treated for virus meningitis at the hospital of the Rockefeller Institute, New York City. R.E.S. had worked with mice from that institute's colony (shown by Traub to be infested); however, W.E. was unlikely to have had contact with infected animals (Rivers and Scott, 1936). The close similarity of these isolates was soon established (Armstrong and Dickens, 1935), and the name *lymphocytic choriomeningitis virus* was adopted.

The salient feature of the relationship between LCM virus and the mouse is an apparent paradox. Infection of the adult animal results in a characteristic illness which either terminates in death or leads to recovery with elimination of the virus. By contrast, introduction of the agent early in life, i.e., before or soon after birth, results in persistent infection which remains clinically inapparent although the virus is present throughout life in high concentrations in all organs (Fig. 1).

The biological relevance of the persistent infection of the

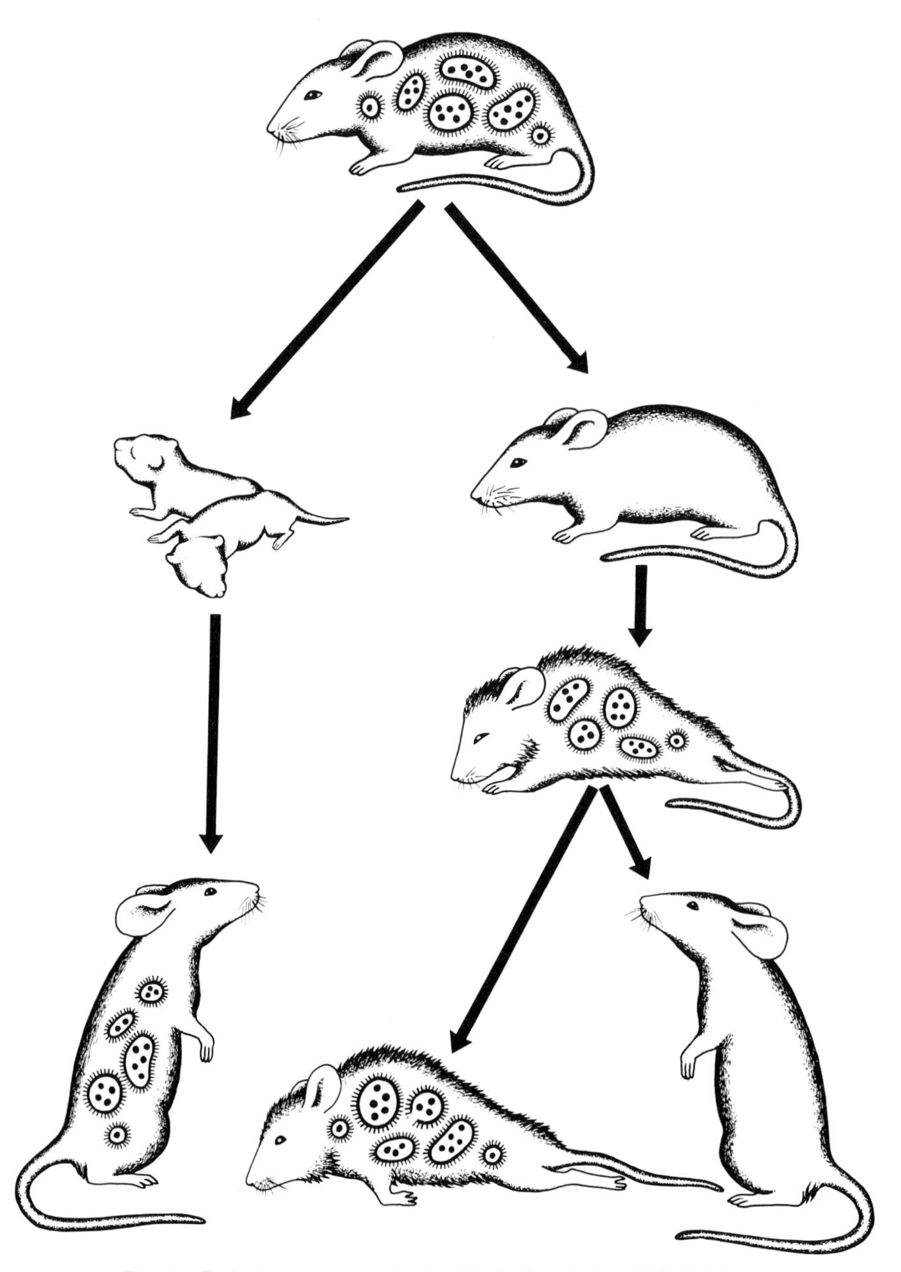

Fig. 1. Basic phenomena associated with infection of mice with LCM virus.

mouse with LCM virus was recognized by Burnet and Fenner (1949). Together with the phenomenon of erythrocyte chimerism in cattle twins (Owen, 1945), it formed the basis for their theory of immunologic recognition of self which—especially after its merger with the concept of "actively acquired tolerance" of Medawar and his colleagues (Billingham *et al.*, 1953)—was to become one of the most fruitful ideas in modern immunology. It was also Burnet (1955) who pointed out that the virus had to be essentially harmless for the host in order to make prolonged persistence possible, and Hotchin (1962a) advanced the notion that illness and death following infection of adult mice are pathologic immune phenomena. Today, the LCM virus-infected mouse is considered by many to be an excellent system in which to study such diverse phenomena as persistent viral infections, virus-specific immunologic tolerance, and pathologic immune reactions in virus diseases. The LCM virus-infected mouse is the theme of this chapter; that is, this agent's interaction with the murine host will be predominantly considered. The work on LCM virus published up to 1969 has been discussed by Lehmann-Grube (1971); further information is contained in a monograph on persistent and slow virus infections written by Hotchin (1971). In the present report, an attempt has been made to consider all relevant publications irrespective of the date of their appearance; yet for the sake of economy, the older work has been cited sparingly. For a complete bibliography of the LCM virus up to 1969, the reader should consult the accounts mentioned. For information on further investigations of this agent that are outside the scope of this chapter and extend it, the reader is referred to meeting reports and review articles (Meeting Report, 1973, 1975, 1977; Pfau, 1974; Hotchin, 1974; Cole and Nathanson, 1974; Doherty and Zinkernagel, 1974; Casals, 1975; Doherty *et al.*, 1976a; Murphy, 1977; Zinkernagel, 1978; Zinkernagel and Doherty, 1977, 1979; Oldstone, 1975a, 1979; Oldstone and Peters, 1978; Bro-Jørgensen, 1978; Pedersen, 1979; Rawls and Leung, 1979; Buchmeier *et al.*, 1980; Lehmann-Grube, 1972, 1975, 1980a,b).

II. DEFINITIONS

Throughout this chapter, the following definitions will be used. *Lymphocytic choriomeningitis* is any illness of man or animals caused by the LCM virus. A strict distinction will be made between *infection* and *(infectious) disease,* the former signifying uptake by an organism or a cell of an agent, followed by its multiplication, the latter referring to all pathologic consequences of this event.

Carrier mouse and *persistently infected mouse* will be used synonymously, both meaning lifelong coexistence of the LCM virus with its murine host, whether disease signs develop or not. A mouse that becomes persistently infected due to contact with the virus soon after birth is known as a *neonatal carrier,* and a mouse that acquires its persistent infection through transfer of the virus *in utero* is a *congenital carrier.* A congenital carrier mouse born to a neonatal carrier mother is also referred to as a *first-generation carrier mouse,* and a mouse born to a first-generation carrier mother is also referred to as a *second-generation carrier mouse.*

Immunity is the state of an organism after it has made contact—once or repeatedly—with an antigen (immunogen). It is revealed by immune reactions which are accompanied or followed by immune phenomena: for instance, protective immunity or footpad swelling. *Immunologic tolerance* is the antigen (tolerogen)-induced state of an organism whose ability to respond to an immunogen of the same specificity is depressed or abolished. Both terms, *immunity* and *immunologic tolerance,* will be used operationally, meaning irrespective of the underlying mechanisms.

For virological terms, the rules of Caspar *et al.* (1962) will be followed. In particular, the word *virus* refers to all forms of the viral life cycle (including virion), whereas *virion* is reserved for the final stage. The term *virion* does not necessarily imply a capacity to infect, but an infectious virus particle is always a virion.

III. PROPERTIES OF THE AGENT

A. Classification

For many years, the LCM virus was thought to be unique among animal viruses until biological similarities to Machupo virus, a member of the Tacaribe complex, were noted by Johnson (1965) and Webb (1965). Efforts to find further links continued and were successful when Murphy and colleagues (1969) examined by electron microscopy cultivated cells infected with either LCM virus or Machupo virus and observed strikingly similar alterations. Rowe *et al.* (1970b) then found that antibodies to all Tacaribe complex viruses (including Machupo virus) bound to cultivated LCM virus-infected cells, as demonstrated by indirect immunofluorescence microscopy. Irregular results were obtained with anti-LCM virus antiserum and cells infected with certain members of the Tacaribe complex, and also when serological connections between LCM virus and Tacaribe complex viruses were sought by means of the complement fixation test. Nevertheless, the relatedness had been established, and this, together with the biological and morphological similarities, led to the proposition (Rowe *et al.*, 1970a) that a new virus group be formed comprising LCM virus, Lassa virus, and the eight agents belonging to the Tacaribe complex (Junin, Tacaribe, Machupo, Amapari,

Tamiami, Parana, Pichinde, and Latino viruses). The suggested name was *arenoviruses*—later changed to *arenaviruses* (Pfau *et al.*, 1974)—which was to reflect the characteristic granules seen by electron microscopy of infected cells (from the Latin word *arenosus*, meaning sandy). By approval of the International Committee on Taxonomy of Viruses, this group has been given the status of a genus named *Arenavirus* belonging to the family Arenaviridae; type species is the LCM virus (Fenner, 1976).

B. Physical and Chemical Properties

The LCM virus had been known and experimentally employed for over 20 years. Nevertheless, when the next of the viruses which were later to be assembled as the genus *Arenavirus* was discovered, with regard to physical and chemical properties more is known of other members of the group than of LCM virus. However, where the data allow comparison, these agents are so similar that in this chapter the arenaviruses will be described collectively and the LCM virus referred to only when its individuality is concerned.

Electron microscopic visualization of infected cells (Fig. 2) reveals particles which are formed by a budding process. The outer cellular membrane is transformed into the viral envelope by an increase in the density of both layers, enlargement of the width of the intermediary zone, and insertion of club-shaped surface projections. The particles thus released are round or pleomorphic, varying in diameter from 50 to more than 300

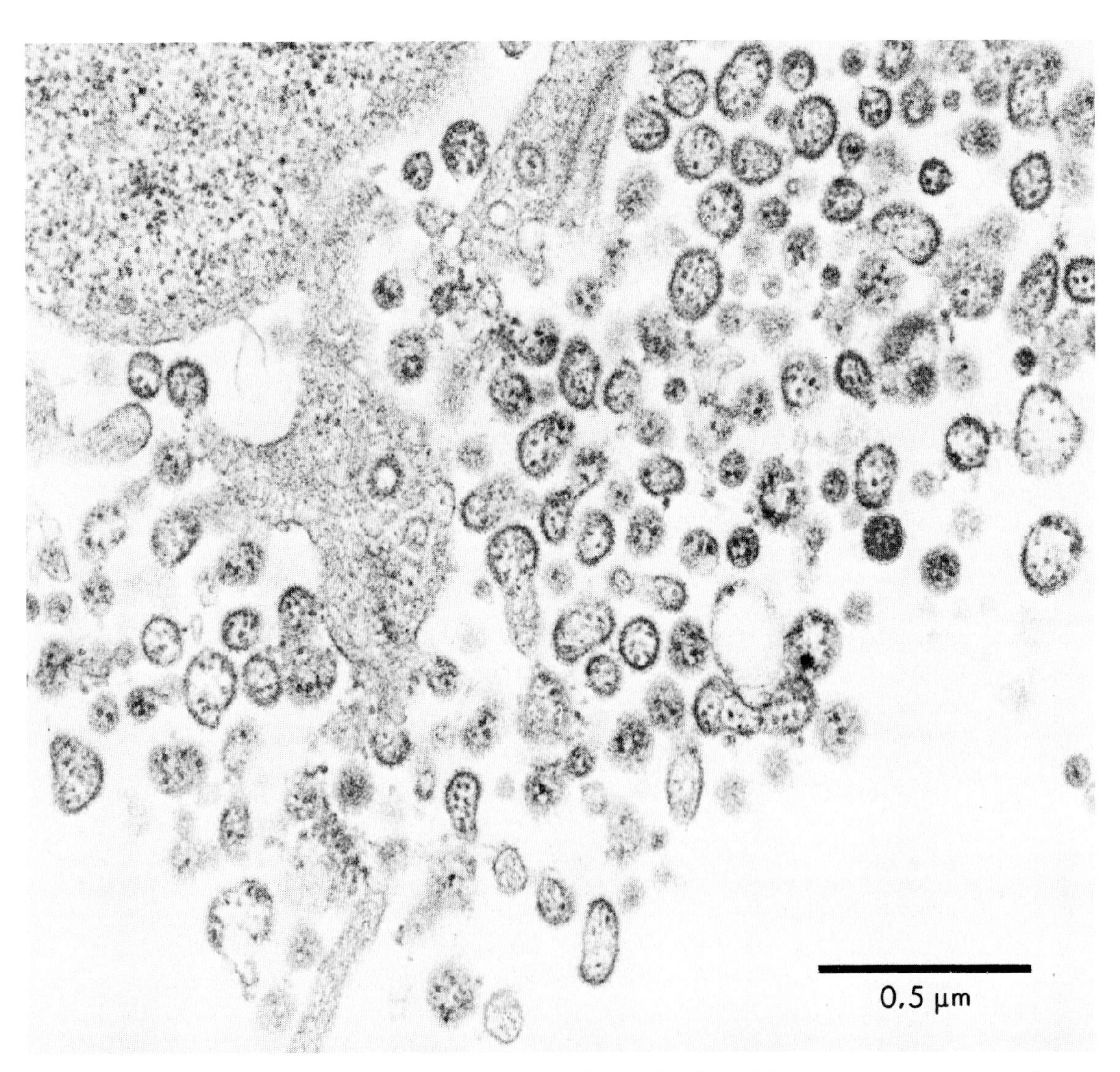

Fig. 2. Thin section electron micrograph of an LCM virus-infected cultivated mouse L cell. Multiple arenavirus particles are seen either released from the plasma membrane or still budding. Cells, grown as monolayer on glass, were infected with strain WE virus at a multiplicity of infection of 0.01. After incubation for 44 hr at 37°C they were fixed with glutaraldehyde, removed from the surface, postfixed with osmium tetroxide, and embedded in ERL. Sections were contrasted with uranyl acetate and lead citrate. (Photograph kindly supplied by Dr. G. Müller, Hamburg.)

nm, and contain one or several internal granules (Dalton *et al.*, 1968; Murphy *et al.*, 1969; Murphy and Whitfield, 1975). They are viral (Abelson *et al.*, 1969; Mannweiler and Lehmann-Grube, 1973), but for budding virions they have unusual morphological properties. Neither a spatial association with viral nucleoprotein nor cores have ever been seen, although several of the arenaviruses were shown to release nucleocapsids if treated with detergents.

The RNA of arenaviruses is single-stranded and consists of four components with 31–33 S (*L*), 28 S, 22–25 S (*S*), and 18 S, respectively. Small molecules which sediment with 4–6 S have also been found, and in the case of Pichinde virus, a 15 S component has been identified (Pedersen, 1979; Rawls and Leung, 1979).

Arenaviruses are not as uniform with respect to their protein composition. Whereas two of them (Tacaribe and Tamiami viruses) contain two major structural polypeptides, in other members of this group three were found, of which two are glycosylated (Pedersen, 1979; Rawls and Leung, 1979). In this latter category falls the LCM virus, whose major structural proteins, termed *GP-1, GP-2,* and *NP,* have molecular weights of 44,000, 35,000, and 63,000, respectively. A further glycopeptide (*GP-C*) of molecular weight 74,000–75,000 was demonstrated in extracts of infected cells. Probably GP-C is a proteolytic cleavage precursor of the structural glycopeptides GP-1 and GP-2 (Buchmeier and Oldstone, 1978a, 1979; Buchmeier *et al.*, 1978).

The buoyant density of LCM virus was found to be 1.22 if determined in cesium chloride, 1.18 if determined in sucrose, and 1.14 if determined in a gradient formed of iodinated organic compounds (Gschwender *et al.*, 1975).

Since arenaviruses are sensitive to ether and detergents, they probably contain host cell-derived lipids which may be assumed to maintain the viral structural integrity. The glycoproteins have been shown to be located on the surface of the virion, whereas the non-glycosylated nucleoprotein was found to be an internal component intimately associated with the viral RNA. Knowledge of the function of the individual components may be summarized as follows. The 31–33 S and the 22–25 S RNA classes are probably virus-coded and carry the viral genetic information. Data obtained with Pichinde virus strongly suggest that its RNA—and by analogy, probably the RNA of all arenaviruses—is negative-stranded (Rawls and Leung, 1979). Attempts to prepare infectious RNA from virions of LCM virus were unsuccessful, pointing in the same direction (Welsh *et al.*, 1975).

In all probability, the 28 and 18 S RNA components originate from host cell ribosomes. However, ribosomes do not play a role in infectivity of these agents, and it may be assumed that the same is true for the ribosome-like granules—provided they are indeed host cell-derived ribosomes (Rawls and Leung, 1979). Glycogen granules, which have occasionally been observed inside arenavirus particles (Kajima and Majde, 1970) are presumably accidentally incorporated during the budding process.

Replication of infectious LCM virus both *in vitro* and *in vivo* is accompanied by the production of a class of particles which lack infectivity but have the ability to prevent cytopathic effects and to abolish the infectious yield. These interfering particles are much more resistant to ultraviolet light (Welsh and Pfau, 1972; Popescu *et al.*, 1976; Popescu and Lehmann-Grube, 1977), which is characteristic of defective interfering (*DI*) particles. Other properties, however, are in striking contrast. DI particles are defined as deletion mutants which require help by standard virus for their own replication, interfere with the replication of standard virus, and increase their proportion in the yield from cells coinfected with standard virus ("enrichment") (Huang and Baltimore, 1977). By definition, LCM virus-interfering particles have the ability to interfere. However, they do not require help, nor can the phenomenon of enrichment be demonstrated. Consequently, passaging LCM virus repeatedly at low dilution does not result in a von Magnus phenomenon (Lehmann-Grube, unpublished observations). LCM virus-interfering particles seem to multiply whenever infectious virus does, and we may conclude that they are by-products of virus replication rather than the progeny of deletion mutants (Lehmann-Grube *et al.*, 1975).

As to the biochemical composition of LCM virus-interfering particles, not much is known. Their density has been determined by use of Urografin and sucrose (Gschwender and Popescu, 1976; Pedersen, 1979). The results vary, but obviously the density of standard virions is only slightly different from that of interfering particles, which has made separation of the latter for their characterization an as yet unresolved task. (For the same reason, analysis of standard virus is always complicated by the fact that every preparation is contaminated with a certain quantity of interfering particles.)

It is often said that LCM standard virus and interfering particles share antigenic determinants because both infectivity and interfering potential may be blocked by one and the same anti-LCM virus antiserum. This opinion overlooks the fact that in cell cultures as well as in mice (and probably in all other hosts), replication of standard virus is inevitably associated with replication of interfering particles. Hence, it may be assumed that antibodies are produced against both entities irrespective of whether these are antigenically related or not.

C. Virus Strains and Growth of Virus in Laboratory Hosts

Strictly speaking, there exist as many virus strains as there are carrier house mice throughout the world, and to these we must add all the isolates which are kept in numerous laboratories. The strains most widely used for experimental purposes are WE (Rivers and Scott, 1936), E-350 (isolated by the late Dr. C. Armstrong under unknown circumstances), W or Traub (Traub, 1975a), and CA 1371 (isolated by Dr. C. Armstrong and possibly identical to E-350). Minor antigenic differences between strains are likely to exist but have—to my knowledge—not been defined. Nor has a systematic study of the biological properties of different virus strains been conducted. However, from published and unpublished observations made in this laboratory and elsewhere, it is concluded that biological variations are considerable. Parameters which vary widely among virus strains are pathogenicity for adult mice following extracerebral inoculation, magnitude of humoral and cellular immune responses they elicit in mice, distribution and extent of replication in various tissues of acutely infected adult mice, ability to kill guinea pigs, and kinetics of replication and concentration attained in cell cultures, to name but a few.

As is true for most viruses, the passage history greatly influences biological properties. Traub (1937) observed attenuation for guinea pigs following transfer of the virus through mice. Conversely, Shwartzman (1946) noted changes of pathogenicity for mice by maintenance of the virus in guinea pigs. Hotchin and Sikora (1973) recovered LCM virus of low pathogenicity for mice from persistently infected L cells, and Lehmann-Grube *et al.* (1969) described a line of carrier L cells which was essentially free of infectious virus, even though more than 95% of the cells were infected, as shown by the immunofluorescence procedure (see Section III,D). Prolonged replication of LCM virus in certain organs of the mouse may also affect its character (Hotchin *et al.*, 1971; Popescu and Lehmann-Grube, 1976). Of particular interest is the observation that virus produced in different organs of one and the same mouse may have different properties (Hotchin, 1972; Popescu and Lehmann-Grube, 1976), which again illustrates the great diversity of this agent. In addition, one virus preparation may have quite dissimilar effects if inoculated into different mouse strains. It is mainly for these reasons that results often vary between laboratories, and this emphasizes the necessity to define the experimental conditions as rigidly as possible and to keep them constant.

The LCM virus multiplies in a wide variety of laboratory hosts, a statement which applies to the intact animal irrespective of its age as well as to cells propagated *in vitro;* even certain cultivated insect cells support its multiplication (Řeháček, 1968). As for the mouse, persistent infection is a natural phenomenon (although this state can be established in the laboratory), and this animal may be considered a laboratory host only when the virus is introduced by artificial means; natural infection by contact may occur but is rare (see Section V,C). A detailed account of replication of the virus in the organs of adult or newborn mice following introduction of the virus would require consideration of all possible variables, such as dose and strain of virus, route of inoculation, and strain and age of mice. Two examples will be given for illustration. (1) the WE strain of LCM virus multiplies to high titers in the spleen and other internal organs of adult mice after intracerebral, intravenous, or intraperitoneal inoculation. In contrast, the E-350 strain virus attains low concentrations, although both strains multiply with similar kinetics to similar titers in the brain after intracerebral inoculation (Lehmann-Grube, 1964a and unpublished observations). (2) Intraperitoneal infection of adult NMRI and AKR mice with the WE strain of LCM virus leads to rapid multiplication in the spleens, and titers of approximately 10^9 mouse infectious doses per gram of tissue are reached after 2 days in both mouse strains. In the liver, the virus multiplies more slowly, but in AKR mice a similar concentration is attained by day 5. However, in the livers of NMRI mice the virus concentration remains more than 10-fold lower throughout the period of observation (Lehmann-Grube and Löhler, 1981).

D. Virus Carrier Cultures

With the appropriate experimental conditions, contact between cultivated cells and LCM virus regularly results in persistent infection (Lehmann-Grube, 1967; Lehmann-Grube *et al.*, 1969; Pfau, 1977). Such cultures, in whom the virus multiplies with little or no harmful consequences for the cells, have frequently been experimentally analyzed. With respect to the aim, namely, deriving an understanding of the mechanism which allows the lifelong coexistence of the infectious agent and the host in an LCM virus carrier mouse, the results have not been as rewarding as expected, one reason being that the courses of such infections vary. During a few passages, the cells produce infectious virus, usually with great fluctuation in quantity. Later changes may occur which, as a rule, arise quite abruptly. Whereas most of the cells continue to synthesize viral material, as demonstrated by the immunofluorescence procedure, a virus is released into the medium, whose infectious potential for cultivated cells and mice is greatly depressed. Sometimes a few cells remain fully infectious (Stanwick and Kirk, 1976), and sometimes fully infectious virus transiently reappears either spontaneously or after the cells have been kept frozen for some time (Staneck *et al.*, 1972). Often, however, infectivity in the usual sense cannot be detected in either medium or cells (Lehmann-Grube *et al.*, 1969; Hotchin *et al.*, 1975; Andzhaparidze *et al.*, 1977; Oldstone *et al.*, 1977; Cherednichenko *et al.*, 1977; Welsh and Buchmeier, 1979). The virus produced in such carrier cultures interferes with the replication of fully infectious LCM virus both *in vitro* and *in vivo,* and some investigators believe that it represents defective interfering particles. Others (including the author) maintain that this virus should be considered of a different nature (see Section III,B). Whatever the outcome of this discussion, the interfering virus from this type of carrier culture has been characterized to some extent, and more is known of its biological and biochemical properties than of the properties of interfering particles as they are generated during acute infection (Lehmann-Grube *et al.*, 1969; Welsh *et al.*, 1972, 1975, 1977; Welsh, 1975; Welsh and Oldstone, 1977; Dutko and Pfau, 1978; Welsh and Buchmeier, 1979). Where comparisons are possible, both these entities appear to be similar; whether they are indeed identical remains to be established.

Infection of cultivated cells may take an entirely different course: the virus first multiplies but then completely disappears from the culture. When this has happened, not a trace of viral activity remains. The cells are free of antigen and can readily be reinfected with LCM virus (Hotchin, 1973; Jacobson *et al.*, 1979). In still other cases, infectious virus keeps multiplying for prolonged periods of time (Lehmann-Grube, 1967; Hotchin *et al.*, 1975). Whether this outcome is different from the other two or whether complete or partial shutdown of virus synthesis eventually always ensues is not known. However, in one case replication of fully infectious LCM virus in L cells has been observed for more than 160 days (Lehmann-Grube, 1967). Other courses of persistent LCM virus infections of cultivated cells are still more complex (Hotchin *et al.*, 1975).

Why the fate of LCM virus-infected cultures varies is not known with certainty. Virus strain, type of cells, and initial multiplicity of infection have been named as determining factors. As to the carrier state, several mechanisms have been proposed (Hotchin, 1974; Welsh, 1975; Popescu and Lehmann-Grube, 1975; Pfau, 1977; Pedersen, 1979). Gaidamovich *et al.* (1978) have reported the transmission of persistent infection from an LCM virus carrier culture with intracellular viral antigens but without infectious virus (see above) to another cell line by transfection. They concluded that integration of a DNA transcript of the viral genome into cell DNA was the basic mechanism. Similar attempts by Holland *et al.* (1976) and Andzhaparidze *et al.* (1977) had been unsuccessful. Jacobson *et al.* (1979) isolated a virus variant (SP) from the same kind of carrier culture. SP was found to resist interference, and it was hypothesized that this entity provided help for the continued production of defective interfering virus, which was considered necessary to maintain persistence.

IV. INFECTION OF MICE

A. Clinical Disease

Over the years, the clinical response to LCM virus has been experimentally determined for numerous animals; both mammals and birds were tested. Early reports on the pathogenic potential of this agent frequently contained contradictory findings. We now know that there are great differences between virus strains and, given a certain strain of virus, between strains of experimental animals as well. Nonetheless, the conclusion may be drawn that the virus often multiplies but is of low pathogenicity, causing clinical signs in only a few species. One of these is man.

In mice, depending on age, route of infection, and other factors, four entirely different types of illness are observed. These are (1) the cerebral form of lymphocytic choriomeningitis in adult mice infected by the intracerebral route, (2) the visceral form of lymphocytic choriomeningitis in adult mice infected by the peripheral route, (3) late onset disease in adult neonatal or congenital carriers, and (4) runting and early death in mice infected immediately after birth.

1. The Cerebral Form of Murine Lymphocytic Choriomeningitis

Intracerebral inoculation of the virus into adult mice results in a highly characteristic disease which is influenced minimally by virus strain, virus dose, or mouse strain:

> During the first 5 days after inoculation the mice appear well. Occasionally on the 5th, but more commonly on the 6th day, symptoms appear, at which time some of the mice may be found dead although none of them were obviously sick on the preceding day, while others with dirty, ruffled fur, half-closed eyes, and hunched backs remain motionless. When disturbed, they occasionally leap up and down in the jar and fall over backwards; but the characteristic reaction, especially when the animals are suspended by the tail, is for them to exhibit coarse tremors of the head and extremities frequently going on to a series of clonic convulsions terminating in a tonic extension of the hind legs. In male mice an erection sometimes occurs during the convulsions. The convulsions, often the cause of death, may also occur spontaneously either in sick mice or even in those that appear to be normal. As a rule, the animals either die within 1 to 3 days after the onset of symptoms or quickly recover in 5 or 6 days. Paralyses have never been observed. (Rivers and Scott, 1936, p. 420.)

As far as a comparison is permissible, the convulsive attacks characterizing the cerebral form of murine lymphocytic choriomeningitis resemble generalized grand mal occurring in human epilepsy. Electroencephalograms recorded during the spastic seizure of a mouse suffering from lymphocytic choriomeningitis disclosed high-amplitude bursts of epileptiform potentials (Chastel *et al.*, 1978). The severity of the murine illness can be ameliorated by anticonvulsant therapy (Camenga *et al.*, 1977; Walker *et al.*, 1977).

The shape of the dose-response curve relating quantity of LCM virus and rate of lethality of mice infected intracerebrally is peculiar; more animals die after inoculation of low doses than after inoculation of high doses (Fig. 3). Differences exist between virus strains, but experiments of the author (unpublished) indicate that this phenomenon is always observed, provided high enough virus doses are employed. The mechanism is not known with certainty; different explanations have been

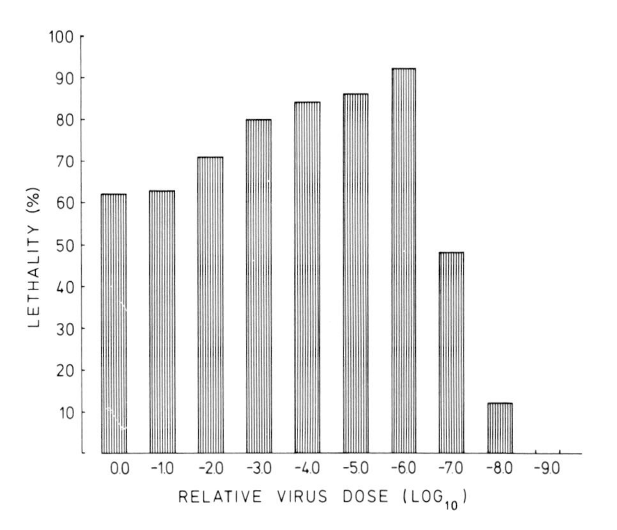

Fig. 3. Dose-response relationship in mice infected intracerebrally with LCM virus. Groups of 50 C57BL/6J mice were inoculated with increasing dilutions of strain WE virus, each mouse receiving 0.03 ml, and mortality was recorded from the fifth to the twenty-first day. (Lehmann-Grube, unpublished observations.)

given (Hotchin and Benson, 1963; Suzuki and Hotchin, 1971; Dunlop and Blanden, 1977a; Zinkernagel and Doherty, 1979).

2. The Visceral Form of Murine Lymphocytic Choriomeningitis

After extraneural infection of adult mice, the clinical response is quite different and greatly influenced by route of inoculation, strain and dose of virus, and strain of mice. Subcutaneous infection with any virus strain usually remains inapparent, although more subtle pathologic changes may be revealed by demonstrating loss of weight (Hotchin and Benson, 1963). The same is true when the E-350 strain of LCM virus is inoculated intraperitoneally or intravenously, irrespective of the mouse strain employed. In contrast, intraperitoneal infection with the WE strain of virus results in severe illness in mice of certain strains, though not in others (Tosolini and Mims, 1971; Löhler *et al.*, 1974; Löhler and Lehmann-Grube, 1981; Lehmann-Grube and Löhler, 1981). Signs consisting of ruffled fur and conjunctivitis start appearing 6–7 days after infection. The mice become inactive, may be found hunched in a corner of the cage, and often die. If they survive, complete recovery may take weeks. A few animals develop ascites, but signs of neurological involvement are rarely seen after extraneural inoculation of the virus.

Intraperitoneal infection of adult C3H mice with the Traub strain of LCM virus greatly enhanced their sensitivity to X irradiation and, at the same time, impaired the capacity of their hemopoietic system (Bro-Jørgensen and Volkert, 1972a,b; Bro-Jørgensen and Knudtzon, 1977). This effect was also observed, though less marked, in nude C3H/*nu*/*nu* mice (Christoffersen and Bro-Jørgensen, 1977). The mechanism is not known with certainty, but Bro-Jørgensen (1978) has presented evidence suggesting that these effects are caused by interferon. Silberman *et al.* (1978), who performed similar experiments also using C3H mice but a different virus strain, came to the conclusion that of several possible mechanisms, a direct inhibitory effect of the LCM virus on a subpopulation of radioresistant hemopoietic stem cells was the most probable cause. Both groups of investigators agreed that the transient depression of immunological reactivity of mice following acute LCM virus infection (see Section IV,C,3) is a separate phenomenon.

Infection of mice with LCM virus increased markedly their sensitivity to endotoxins (lipopolysaccharides) from gram-negative bacteria (Barlow, 1964). Surprisingly, mice dying after the administration of endotoxin exhibited signs characteristic of the cerebral form of lymphocytic choriomeningitis, even when the time interval between intracerebral infection and death was as short as 2 days and the virus was inoculated subcutaneously.

3. Late Onset Disease

A third type of illness is observed in neonatally infected LCM virus carrier mice. Long after birth, these animals seem to be healthy, although the virus multiplies continuously in all their organs. However, when they are 9–12 months old, signs of a chronic illness may appear which consist of ruffled fur, a hunched posture, and loss of weight. These signs are sometimes accompanied by degenerative changes of the eyes and the skin; often there is protein in the urine, and ascites may be prominent. It should be stressed that this severe type of late onset disease (Hotchin and Collins, 1964) is not always seen. In many virus strain–mouse strain combinations, only a thorough inspection reveals abnormalities such as behavioral alterations (Hotchin and Seegal, 1977), slower weight gain, smaller litter size, and shorter life span (as compared with noninfected controls), and even these may be absent (see Section V,B).

Late onset disease is also observed in congenital carrier mice, i.e., mice infected *in utero* from their persistently infected mothers. As in neonatal carrier mice, the incidence varies between mouse strains, and Traub (1975a,b) found many more cases in colony-bred NMRI than in inbred CBA/J mice. However, a systematic study of the influence that mouse strain and virus strain have on late onset disease in congenitally infected mice has, to the author's knowledge, not been conducted.

4. Early Death in Neonatally Infected Mice

The above statement, that mice infected soon after birth develop quite normally during the earlier part of their life, requires qualification. Under certain poorly defined conditions, neonatal infection is lethal (Hotchin *et al.*, 1962; Lehmann-Grube, 1971; Oldstone and Dixon, 1973; von Boehmer *et al.*, 1974; Gaidamovich and Kocherovskaya, 1976; Doyle *et al.*, 1980). Again, initially these mice appear outwardly unaffected. During the second week, however, they begin to show signs of illness which progress to death. Alternatively, the mice slowly recover, although in many instances they remain smaller and look less healthy than noninfected controls.

Both types of acute illness following infection of adult mice as well as the late onset disease in carrier mice are predominantly, if not exclusively, immunopathologic in nature and hence will be further dealt with in the section on immune responses. Whether the same mechanism is in effect with early lymphocytic choriomeningitis disease in mice infected with LCM virus soon after birth is uncertain. This is not solely a function of the strain of virus or its passage history, nor is the mouse strain alone decisive (Lehmann-Grube, 1971). In C3H mice less than 18 hours old infected with the Traub strain of LCM virus, development of the immune system, both morphologically and functionally, was found to be markedly retarded (Bro-Jørgensen and Volkert, 1974). Other factors have to be considered, such as neglect by the mother undergoing infection contracted from her offspring and stress of the young (Hotchin *et al.*, 1970). Also, in some unknown way, interferon seems to participate in this illness (Rivière *et al.*, 1977). As a rule, lymphocytic choriomeningitis disease in suckling mice is not seen among the progeny of congenital carriers.

B. Pathology

Alterations of tissues following the infection with LCM virus have been described for many species, including man. In the mouse, the LCM virus causes four types of illness and, consequently, four patterns of pathologic changes have to be considered.

1. Central Nervous System

Just as the neurological disease following infection by intracerebral inoculation of the LCM virus is rather uniform and not much influenced by mouse strain or virus strain, so is the pathologic picture presented by the central nervous system after intracerebral infection. This has repeatedly been described in the past, but the following account is largely based on unpublished work performed by my colleague, Jürgen Löhler. By the third day, slight inflammatory infiltrates in the leptomeninx and plexus choroidei consisting of mononuclear cells, mostly lymphocytes in various stages of blast transformation, occur. The infiltrates increase rapidly, reaching maximal numbers between the sixth and ninth day (Fig. 4); they are most intense in the cisternae of the arachnoidea, and the choroid plexuses of all ventricles are heavily involved. Electron microscopically, characteristic intracytoplasmic inclusions and necroses of the cells of the plexus epithelium and, less frequently, of the leptomeninx and ependyma can be observed. Lymphocytic infiltrations often extend into the Virchow-Robin spaces, but signs of encephalitis are always absent (Lillie and Armstrong, 1945; Walker *et al.*, 1975). Edema of the brain, which has been incriminated as the cause of death (Doherty and Zinkernagel, 1974), cannot be demonstrated by electron microscopy and other methods (Walker *et al.*, 1975; Camenga *et al.*, 1977; G. Schwendemann and J. Löhler, unpublished observations). The pathophysiological mechanism leading to convulsions and death of mice infected with LCM virus by intracerebral inoculation has yet to be determined.

2. Organs Other Than the Central Nervous System

Since inoculation by means of syringe and needle of fluid material directly into the brain opens the blood–brain barrier, the greater part of the inoculum immediately enters the circulation; an intracerebral inoculation therefore is always also an intravascular one (Mims, 1960). Consequently, if the inoculated material affects extraneural organs after intravenous inoculation, it will do so almost as efficiently after intracerebral inoculation. However, inoculation of infectious material into the venous system (or the peritoneal cavity) of a mouse does not usually alter the blood–brain barrier. Hence, after intravenous inoculation the brain tissue is not necessarily involved with LCM virus infection of the mouse. This consideration explains why organs other than the central nervous system are affected to approximately the same extent whether the virus is administered by the neural or the extraneural route, whereas the brain remains undamaged, as a rule, if the virus is inoculated peripherally (J. Löhler, personal communication).

In sharp contrast to the rather uniform response of the brain of intracerebrally infected mice, alterations of other organs vary greatly and are determined by differences in mouse and virus strain. Indeed, the multitude of reactions is rather confusing, and only a few illustrative examples will be given (J. Löhler, personal communication). After intracerebral, intravenous, or intraperitoneal infection with LCM virus strain E-350, pathologic alterations in tissues other than the central nervous system are absent except for proliferation of lymphoid cells and activation of macrophages in spleens and lymph nodes, probably indicating immunological reactions to the viral antigens. With the WE strain of LCM virus, the pathologic responses vary greatly, and whether and how a mouse reacts depend on the strain employed. Alterations are prominent in NMRI mice, with all organs showing inflammation; these are severe and most characteristic in liver and lymphoid organs.

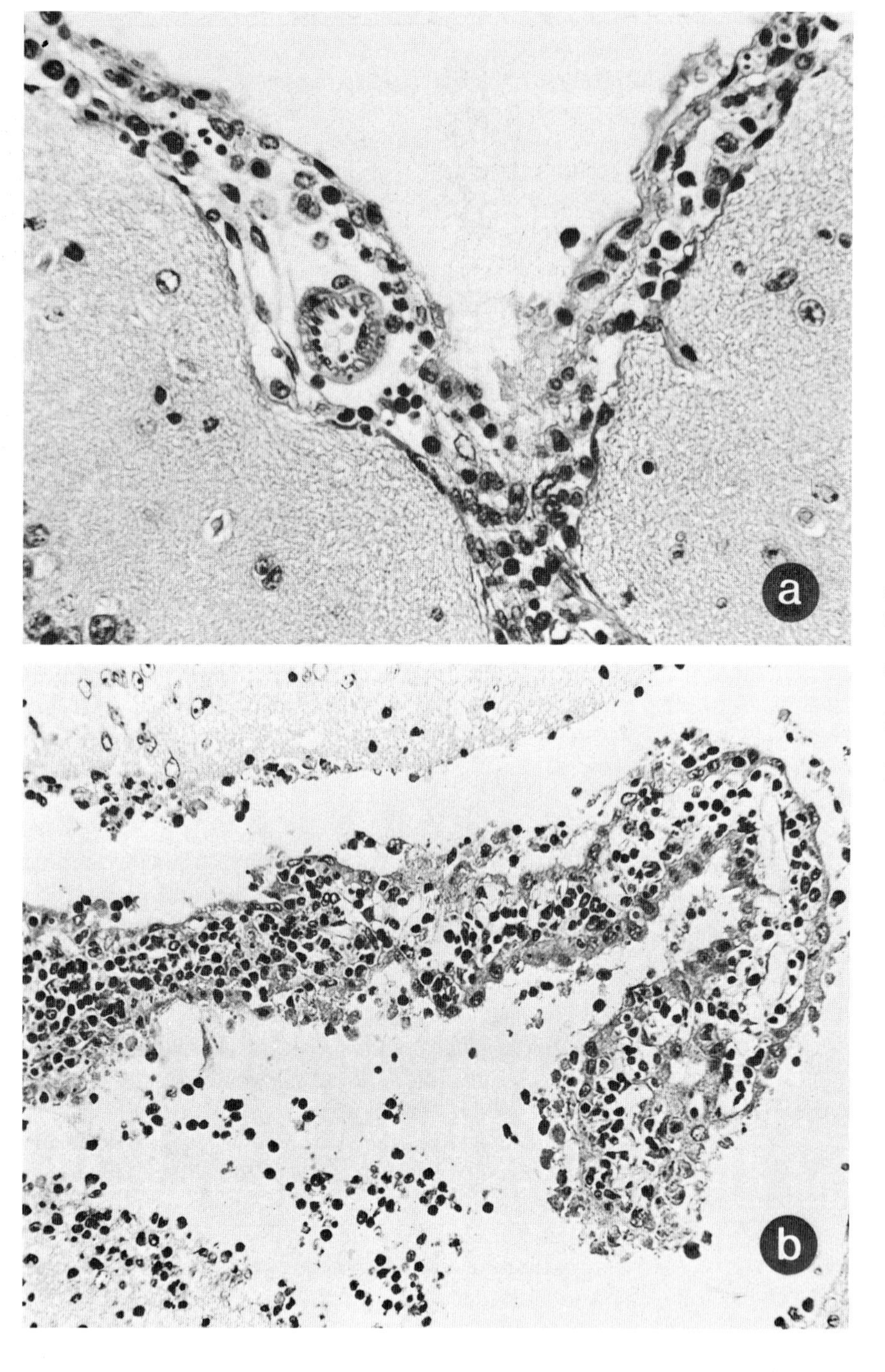

Fig. 4. Histopathologic changes of (a) leptomeninx and (b) plexus choroideus of a lateral ventricle of a mouse 6 days after intracerebral infection with LCM virus. These are the alterations which, together with similar ones in infected monkeys, have led to the name of the agent and the illness it causes. Hematoxylin and eosin. Magnification ×420 (a) and ×250 (b). (Photographs kindly supplied by Dr. J. Löhler, Hamburg.)

In the liver, 5–6 days after infection, a lobular hepatitis is found with disseminated eosinophilic cell necroses. These are usually associated with nodular infiltrates, which consist of transformed lymphocytes (lymphoblasts), monocytes, activated Kupffer cells, activated endothelial cells derived from the sinusoid liver capillaries, and, rarely, polymorphonuclear cells and megakaryocytes. Similar infiltrates are found in the periportal portions of the liver. On day 7 after infection, glycogen has disappeared from hepatocytes and fatty degeneration becomes increasingly prominent.

Because this type of illness is caused by an aberrant immune response just like the convulsive disease following intracerebral infection (see Section IV,C,2), the lymphoid organs are of special interest (Mims and Tosolini, 1969; Hanaoka *et al.*, 1969). The following description will be based on observations made in this laboratory (Löhler and Lehmann-Grube, 1981).

After the intraperitoneal infection of NMRI mice with strain WE of LCM virus, three types of alterations can be distinguished which partly evolve in parallel. These are (1) destruction of lymphocytes and cells of the mononuclear phagocytic system, (2) proliferation of lymphocytes, and (3) fibrinoid necroses of reticular cells and macrophages. In spleen and lymph nodes, lymphocytolysis is observed on day 3 after infection. Initially, T cells are more affected than B cells, but beginning with day 4 and thereafter, the opposite is true. In parallel with cytolysis, other lymphoid cells proliferate. On days 4–5, macrophages are activated. At the same time, small foci of fibrinoid necrosis begin to appear. Both lymphocytolysis and lymphoblastoid proliferation are maximal 4–5 days after infection, whereas fibrinoid necrosis progresses until day 6 (Fig. 5). In the thymus, changes are first observed on day 6 and consist of massive necroses of cortical thymocytes leading to extensive

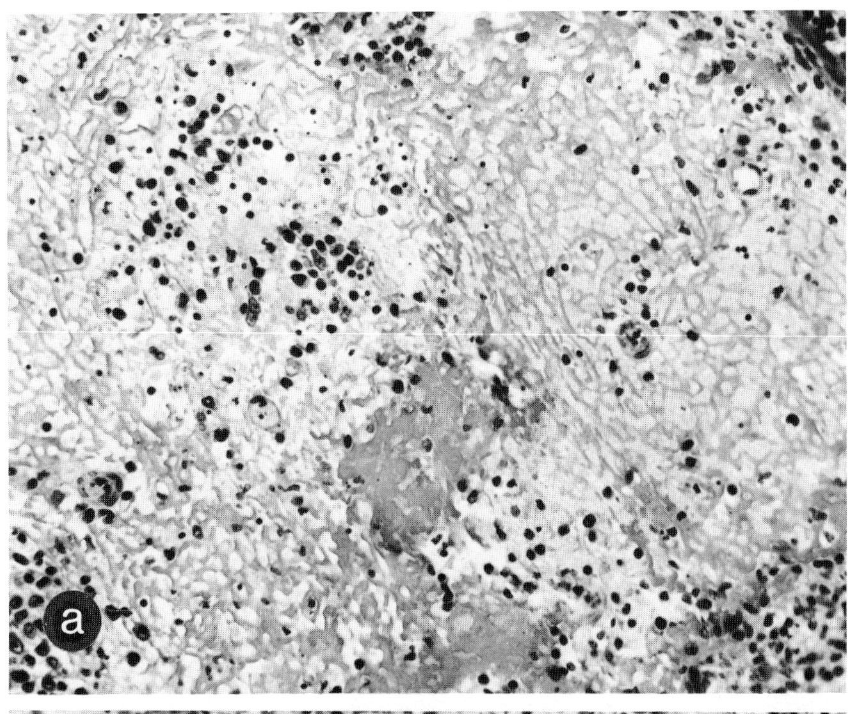

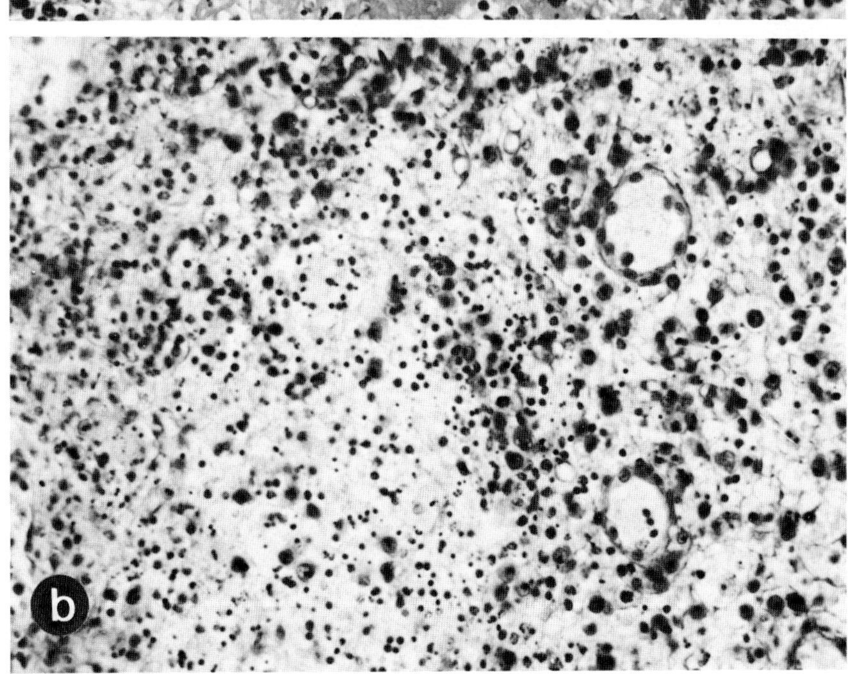

Fig. 5. Histopathologic alterations of (a) spleen and (b) lymph node of an NMRI mouse 5 days after intraperitoneal infection with strain WE of LCM virus. Hematoxylin and eosin. Magnification ×250 (a) and ×200 (b). (Photographs kindly supplied by Dr. J. Löhler, Hamburg.)

involution of the organ. As in spleen and lymph nodes, many phagocytic cells of the thymus exhibit increased activity of hydrolytic enzymes. Unless the mice die, recovery commences about the tenth day and is usually complete 3 weeks after infection.

In addition to NMRI mice, 20 other mouse strains have been investigated (Lehmann-Grube and Löhler, 1981). Of these, 19 were inbred (including five hybrids) and one was the gray house mouse. When infected by intraperitoneal inoculation with WE strain virus, the agent in the spleens multiplied essentially to identical concentrations. Yet, the histological alterations varied between strains from predominantly proliferative to predominantly cytolytic. Since a correlation was found to exist between the severity of destructive lesions, on the one hand, and parameters of LCM virus-specific cell-mediated immunity, on the other, we concluded that the lesions in the lymphoid organs of mice infected with LCM virus by extraneural inoculation of the agent are pathologic immune phenomena (see Section IV,C,2). Wallnerova and Mims (1971) observed a marked reduction of the number of small lymphocytes in the thoracic duct lymph of Walter and Eliza Hall Institute (WEHI) mice but not in C57BL mice after intravenous infection with the WE strain of LCM virus. This finding mirrored the pathologic alterations in the lymphoid organs, which were severe and affected essentially all cell types in WEHI mice but were virtually absent in C57BL mice.

3. Late Onset Disease

The so-called late onset disease of neonatal or congenital carrier mice should be designated more accurately *chronic immune complex disease*. Glomeruli of the kidneys (Fig. 6) and the choroid plexus of the brain are most severely affected (Baker and Hotchin, 1967; Oldstone and Dixon, 1969; Kajima and Pollard, 1970; Lampert and Oldstone, 1974). Synovial membranes, ciliar processes, the walls of blood vessels, and the dermo-epidermal junctions of the skin are also involved, though less extensively (Löhler *et al.*, 1980). The high filtration rate of blood plasma which is associated with these structures, favors trapping and deposition of circulating immune complexes. The immune complex disease of LCM virus carrier mice is accompanied by the formation, in various organs, of extensive diffuse or nodular infiltrates developing from perivascular spaces. These infiltrates, resembling lymphocytic follicles, consist of lymphocytes, numerous plasma cells (which may contain Russell's bodies), and monocytes; in close proximity, tubular lesions are frequently found (Pollard and Sharon, 1969; Sharon and Pollard, 1971; Accinni *et al.*, 1978; J. Löhler, personal communication).

In the central nervous system of persistently infected aging NMRI mice, many neurons have been found to undergo vacuolar degeneration, causing in severe cases a status spongiosus (Löhler *et al.*, 1980). Whether this applies to all carrier mice or is confined to this particular mouse strain has yet to be determined.

4. Early Death in Neonatally Infected Mice

To the author's knowledge, the histopathologic picture in suckling mice suffering from early lymphocytic choriomeningitis disease after infection soon after birth has not been described in any detail. According to Rivière *et al.* (1977), the alterations are almost identical to those described in newborn mice undergoing an illness induced by repeated injections of mouse interferon. In these animals, extensive liver cell degeneration and necroses without any inflammatory reaction are the principal findings (Gresser *et al.*, 1975). In contrast, Doyle *et al.* (1980) found neither inflammation nor tissue necroses in the liver, brain, spleen, thymus, kidney, and heart of mice undergoing early lymphocytic choriomeningitis after neonatal infection.

C. Immunity

In this section, only the functioning of the immune system during acute LCM virus infection of the adult mouse will be explained; its interaction with the LCM virus during persistent infection will be discussed later (see Section IV,D).

1. Participation of the Immune System in the Infectious Process

Immune responses determine the pathologic consequences of infection in the adult mouse (see Section IV,C,2). To aid in understanding, an account will be presented of current knowledge concerning participation of the lymphatic system in acute infection. After intracerebral, intraperitoneal, or intravenous inoculation, the LCM virus readily multiplies in the spleen, lymph nodes, and thymus (Traub, 1964; Mims and Tosolini, 1969; Hanaoka *et al.*, 1969; Löhler *et al.*, 1974; Löhler and Lehmann-Grube, 1981). White blood cells can also be infected by this agent. According to Mims, lymphocytes in newborn and adult mice undergo infection (Mims and Wainwright, 1968). Wilsnack and Rowe (1964) inoculated the WCP strain intraperitoneally into HA/ICR mice and used an immunofluorescence procedure for locating cells containing viral antigen. In the white pulp of the spleen, a few positive lymphocytes were found. With the same method, Wallnerova and Mims (1971) determined that infected cells were present in the thoracic duct lymph of WEHI mice infected intravenously with the WE strain of LCM virus. On days 6 and 7 after inoculation of the virus, 1.0–1.4% of the cells were positive; further identification was not attempted.

Cells of the mononuclear phagocyte system also undergo infection (Seamer, 1965b; Mims and Subrahmanyan, 1966; Tosolini, 1970; Zinkernagel and Doherty, 1975a; Schwartz *et*

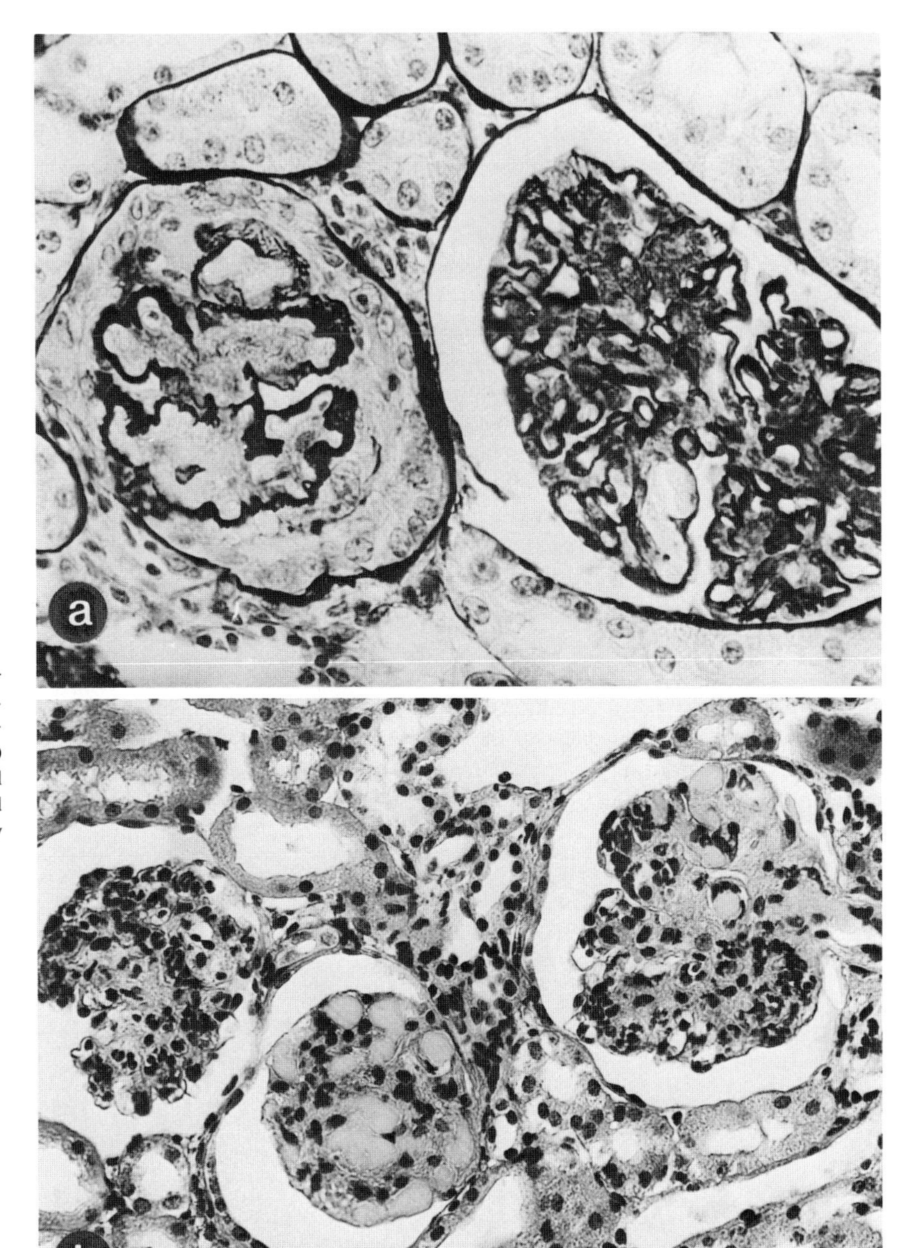

Fig. 6. Glomerular alterations in late onset (immune complex) disease of LCM virus carrier mouse. (a) Thickening of the glomerular basement membrane. Methenamine silver impregnation. ×570. (b) Final stage with marked proliferation of mesangial cells and hyaline degeneration. Hematoxylin and eosin. Magnification ×360. (Photographs kindly supplied by Dr. J. Löhler, Hamburg.)

al., 1978; Löhler and Lehmann-Grube, 1981). According to Doyle and Oldstone (1978), the same is true for B cells. The findings differ for T lymphocytes—the only cells whose relevance for the LCM virus-specific immune reactions of the mouse has been proved (see Section IV,C,2). Doyle and Oldstone reported that a proportion of them scored as infectious centers if taken from acutely infected adult mice. Numerous attempts by the author using several methods failed to demonstrate infection of T lymphocytes from adult mice both *in vivo* and *in vitro* (Lehmann-Grube, unpublished observations). It was found, however, that a proportion of these cells could be infected in newborn mice (Tijerina *et al.,* 1980).

2. LCM Virus-Specific Immunogens, Immune Reactions, and Immune Phenomena

Only one of the various immunogens inducing LCM viral immune reactions during LCM virus infection has been defined to some extent. This is the complement-fixing antigen, which is produced in abundance in infected cells and is also an internal component of the virion. It is not, however, represented on the surface of either the virion or the infected cell (Gschwender *et al.*, 1976). By analogy, with members of the Tacaribe complex arenaviruses, the major complement-fixing antigen of the virion of the LCM virus is probably the nucleoprotein (Buchmeier and Oldstone, 1978b).

The immunogen which induces neutralizing antibody is of different specificity. It is situated on the surface of the virion, but its chemical nature has not been determined; probably it is one of the structural glycoproteins (see Section III,B). The virion surface also carries an immunogen which induces the formation of complement-fixing antibody. Its specificity is possibly identical to that of the immunogen leading to the formation of neutralizing antibody, but it is different from the specificity of the internal component inducing complement-fixing antibody (Lehmann-Grube *et al.*, 1975).

LCM virus-specific, cell-mediated immunity is induced by some immunogen which acts in conjunction with the surface of the infected cell. No data are available which would allow its identification for LCM virus. In other virus–host systems, the viral part of the cell surface immunogen inducing cellular immunity is a structural component of the virion (Hapel *et al.*, 1978), and observations on the effector phase of cell-mediated immunity against Sendai virus and vesicular stomatitis virus suggest that it is a viral glycoprotein (Koszinowski *et al.*, 1977; Zinkernagel *et al.*, 1978b). Perhaps with LCM virus, it is GP-1, found by Buchmeier and Oldstone (1979) to be situated on the outer surface of virus-infected culture cells. The part of the cell surface immunogen complex contributed by the cell is a gene product of the major histocompatibility complex and is also present on noninfected cells. In the mouse, it is coded by the *D* and/or *K* regions, the same regions which code for the major transplantation antigens (Zinkernagel and Doherty, 1977, 1979; Wagner *et al.*, 1980); its exact biochemical nature has not yet been determined.

The spatial relationship of the two components of the immunogenic complex on virus-infected cells has not been resolved in any virus–host system. Two hypotheses are being discussed. These have been given various names but are best known as the *altered self model* and the *dual recognition model* (Zinkernagel and Doherty, 1977).

In members of most species, though not in man, the LCM virus-specific immune reaction most regularly detected is the development of complement-fixing antibody (Traub and Schäfer, 1939). In mice this immunoglobulin may attain detectable levels as early as 1 week after infection. It then reaches moderate titers of up to 1000 and apparently persists throughout life (Fig. 7). By immunofluorescence procedures, immunoglobulins with at least two different specificities can be

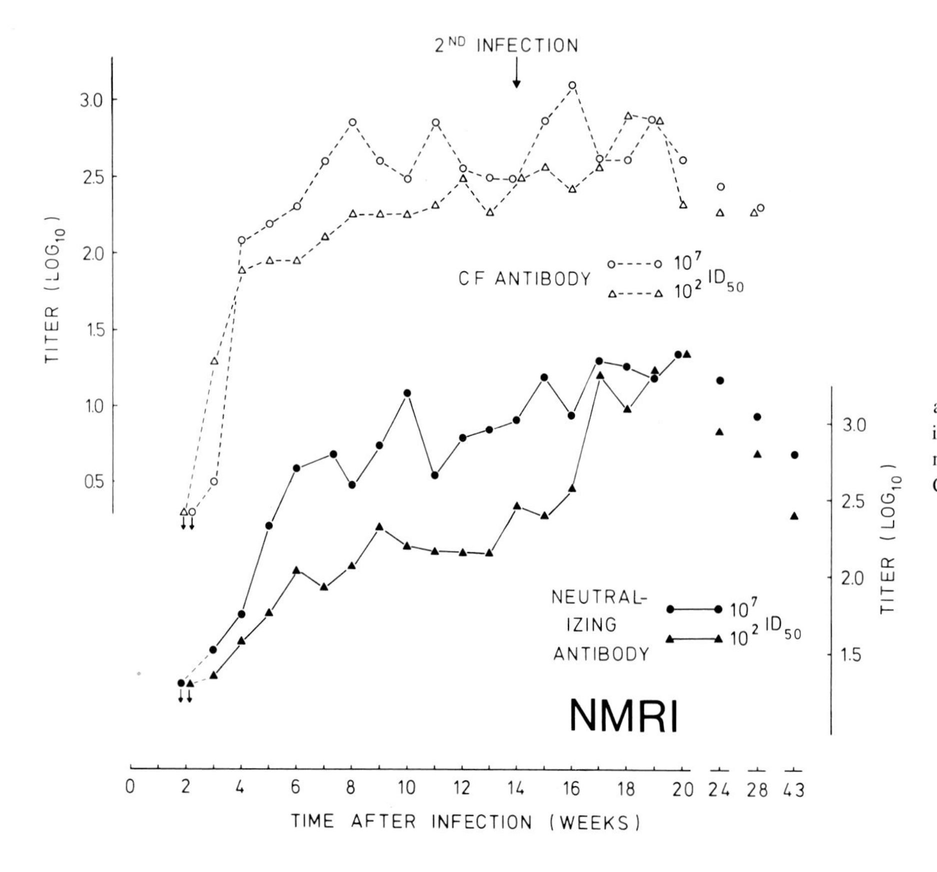

Fig. 7. Development of LCM viral neutralizing and complement-fixing antibodies in NMRI mice infected with strain WE of LCM virus. (From Kimmig and Lehmann-Grube, 1979, with permission of Cambridge University Press, Cambridge, England.)

detected. One is identical to the previously mentioned complement-fixing antibody. Of the other, it is only known that it is directed at material inside and on virus-infected cells (Gschwender *et al.*, 1976). Two serologic specificities are also detected by immunodiffusion tests; one corresponds to the complement-fixing antigen, whereas the other is complementary with a heat-labile, protease-susceptible substance present in virus-infected cells (Bro-Jørgensen, 1971; Gschwender *et al.*, 1976).

Until recently, it was uncertain whether the mouse was capable of forming LCM virus-neutralizing antibody, and there were about as many positive as negative reports. Employing a plaque reduction assay, Kimmig and Lehmann-Grube (1979) showed that mice readily responded to LCM virus infection with the production of neutralizing antibody (Fig. 7), but there were marked differences between virus strains.

Immunoglobulin binding to the virion surface can also be revealed by its ability to bring about precipitation of virus or reduction of its infectivity by antimouse immunoglobulin antiserum (Oldstone and Dixon, 1971; Kimmig and Lehmann-Grube, 1979). LCM virus-infected mice also produce an antibody which, in the presence of complement, lyses infected cells (Cole *et al.*, 1973).

LCM virus-specific, cell-mediated immune phenomena have attracted much attention. In adult mice, a nonlethal infection by the subcutaneous route rapidly leads to protection against an otherwise lethal intracerebral challenge (Table I). As shown by Volkert *et al.* (1974) and Johnson and Cole (1975), protective immunity against LCM virus is mediated by T cells.

After acute infection of adult mice the virus multiplies in many organs, the extent being dependent on virus strain and route of inoculation. If the animals survive, the agent is gradually eliminated, although traces may remain for many months (Haas, 1954; Rowe, 1954; Lehmann-Grube, 1964a; Volkert and Lundstedt, 1968). As shown by Mims and Blanden (1972) and using an improved method by Zinkernagel and Welsh (1976), virus elimination from acutely infected adult mice also requires T cells. Related is elimination of virus from carrier mice following transfer of lymphoid cells from actively immunized mice, which is also T cell-dependent (Volkert *et al.*, 1974).

A measure of LCM virus-specific, delayed type hypersensitivity in the mouse is swelling of the foot following inoculation of virus into the footpad. Then, 6–7 days later, the foot begins to increase in size, reaching up to twice its usual thickness (Hotchin, 1962b; Roger, 1963; Tosolini and Mims, 1971; Lehmann-Grube and Löhler, 1981) (Fig. 8). This footpad response is also mediated by T cells (Zinkernagel, 1976b).

It is well established that the characteristic neurological disease of mice following intracerebral infection is an immunopathologic phenomenon (see below), and all evidence indicates that the same is true of illness and death following extraneural inoculation of the virus (Mims and Tosolini, 1969; Löhler *et al.*, 1974; Ticho *et al.*, 1974a,b; Lehmann-Grube and Löhler, 1981). Indeed, present knowledge permits stating as a rule that all signs of disease in adult mice following LCM virus infection of any strain and irrespective of the route of inoculation are immunopathologically mediated.

The first indications of the immunological nature of lymphocytic choriomeningitis of the mouse were derived from the work of Rowe (1954, 1956), who observed that mice could be protected against the lethal consequences of intracerebral inoculation of the virus by their irradiation. However, it was Hotchin and Cinits (1958; Hotchin, 1962a) who explicitly in-

Table I

Development of Protective Immunity in Mice by Subcutaneous LCM Virus Infection against Death from Intracerebral LCM Virus Infection[a]

Time (hours) between immunizing infection and intracerebral challenge[b]	First experiment		Second experiment		Third experiment	
	NMRI	CBA/J	NMRI	C57BL/6J	NMRI	DBA/1
60	20/20[c]	20/20	n.d.[d]	n.d.	n.d.	n.d.
72	20/20	20/20	25/25	24/24	20/20	28/30
84	20/20	19/19	23/24	24/25	24/24	22/24
96	13/20	19/20	8/25	18/25	19/25	16/25
108	1/20	20/20	1/23	2/25	5/25	8/25
120	0/20	8/19	0/22	0/24	1/23	1/24
132	0/20	1/20	0/23	1/25	1/25	1/24
144	0/20	1/19	0/25	0/22	1/23	1/25
156	0/19	0/17	0/24	0/24	1/24	0/20

[a] Data taken from Lehmann-Grube and Löhler (1981) and Lehmann-Grube (unpublished observations).
[b] Mice were immunized and challenged with 10^4 mouse infectious units.
[c] Number of mice dead/number inoculated.
[d] Not done.

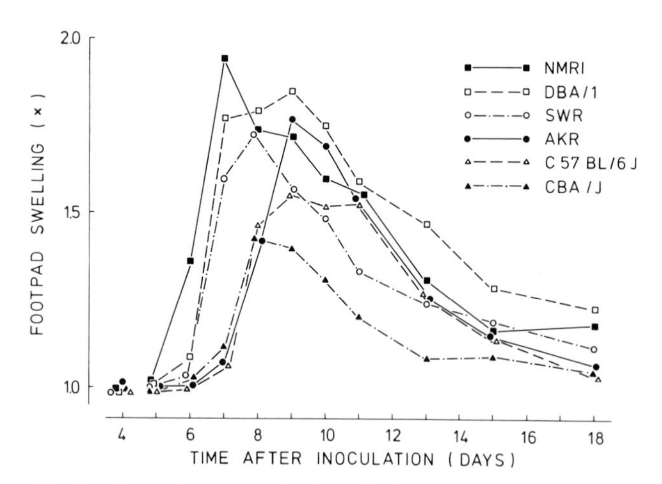

Fig. 8. Footpad response of mice to LCM virus. Inbred mice or NMRI colony-bred mice were inoculated subcutaneously into the left hindfeet, each foot receiving 10^4 mouse infectious doses of WE strain virus in 0.03 ml. At intervals the degree of swelling was determined as the factor by which the thickness of the left foot exceeded the thickness of the right foot. Data points are means obtained from 10 mice. (Lehmann-Grube, unpublished observations.)

terpreted illness and death of the LCM virus-infected mouse as consequences of responses against viral or virus-induced antigens. Numerous confirmatory observations have since been made, and today the validity of this conclusion is unquestioned (reviewed in Hotchin, 1971; Lehmann-Grube, 1971; Cole and Nathanson, 1974; Doherty and Zinkernagel, 1974; Nathanson *et al.,* 1975; Buchmeier *et al.,* 1980). Over the years, circumstantial evidence has suggested that the pathologic immune response of the mouse leading to lymphocytic choriomeningitis is cell mediated. Direct proof was presented by Cole *et al.* (1971; Gilden *et al.,* 1972b), who succeeded in provoking fatal lymphocytic choriomeningitis in cyclophosphamide-induced virus carriers by transfer of LCM virus-immune spleen cells. Again, this effect was shown to be dependent on the presence of T cells in the inoculum (Cole *et al.,* 1972).

It may seem surprising that two mutually exclusive consequences of LCM virus infection of the mouse, namely illness and protection against illness, should be immunologically mediated by one cell type, the T lymphocyte. However, an explanation can be given by assuming that it is the relationship between the number of virus-infected target cells and the degree of cell-mediated immunity which determines the outcome (Cole and Johnson, 1975; Nathanson *et al.,* 1975; Zinkernagel, 1978). When an early and effective cell-mediated immune response curtails virus replication, the number of target cells remains below a critical level. If immunity develops more slowly, allowing virus replication to proceed initially unchecked, the number of target cells is high when cell-mediated immunity is at its maximum and the interaction has (immuno)pathologic consequences. Data of Marker *et al.* (1976) and Thomsen *et al.* (1979) were interpreted to support this view.

Although LCM viral protective immunity, footpad response, and illness of the mouse, as well as virus elimination, have all been shown to be cell-mediated immune phenomena, do antibodies play subsidiary roles? Numerous observations support the assertion that they do not, but direct evidence exists only in the case of disease: in mice depleted of B cells but not of T cells by treatment with anti-μ-chain immunoglobulin, the disease of the central nervous system following intracerebral inoculation of the virus was clinically and histologically indistinguishable from that of control mice (Johnson *et al.,* 1978).

A further immunological observation made with LCM virus-infected mice is an apparent histoincompatibility between carriers and their uninfected syngeneic counterparts. Holtermann and Majde (1971) transplanted skin from congenital or neonatal SWR/J carrier mice onto syngeneic noninfected recipients and observed rejection which commenced on the eighth day. When the recipients had been preimmunized either by a skin transplant from a carrier mouse or by subcutaneous inoculation of the LCM virus, the rejection was accelerated. A similar response was seen when cells from an adenocarcinoma of SWR/J origin, which had been infected by passages through SWR/J carrier mice, were inoculated into syngeneic recipients.

The alteration of tissues by virus infection in such a way as to elicit rejection by a nominally syngeneic recipient—also called *skin heterogenization* (Svet-Moldavsky *et al.,* 1968)—is probably the consequence of the exposition of new cell surface structures. Though of great theoretical and possibly practical interest, the further analysis of the phenomenon of skin heterogenization in LCM virus carrier mice may turn out to be a difficult task. Employing different mouse strains persistently infected with different virus strains, and normal as well as LCM-immune recipients, the author has transplanted hundreds of skin grafts, of which the majority was permanently accepted. In a number of cases, however, there was partial or complete rejection, which confirms, in principle, the findings of Holtermann and Majde (Lehmann-Grube, 1971, and unpublished observations).

The method most often employed to quantify LCM virus-specific cellular immunity *in vitro* is determining the activity of cytotoxic lymphocytes. Tests based on viability of target cells yield reliable information (Lundstedt, 1969a; Holtermann and Majde, 1971; Cihak and Lehmann-Grube, 1974), but a modification of the Brunner technique is usually preferred: spleen, lymph node, or peritoneal exudate cells are incubated together with virus-infected and radioactively labeled target cells, the counts released into the medium being a measure of cytotoxic cell activity (Marker and Volkert, 1973). The pattern thus revealed is quite characteristic (Fig. 9). After primary infection, the threshold of detection is attained on days 3–4 and

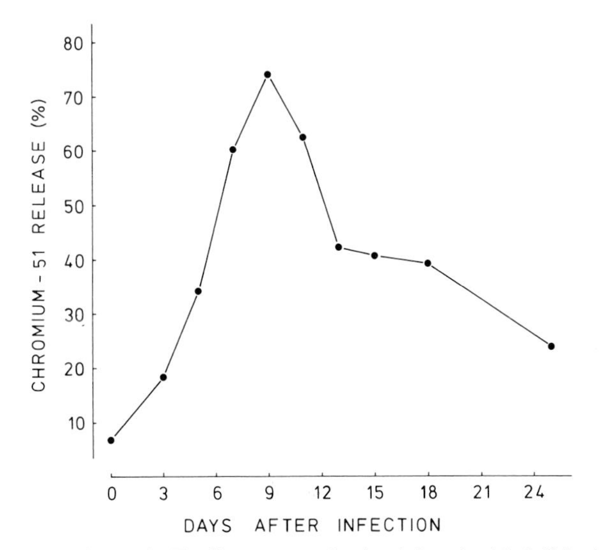

Fig. 9. Cytotoxic T-cell response of mice infected with LCM virus. C3H/HeJ mice were infected by intravenous inoculation of 10^2 mouse infectious units of strain WE virus, and their cytotoxic T-cell responses were determined at intervals as percentages of ^{51}Cr release from LCM virus-infected L cells by spleen cells. Each data point depicts the mean from two mice. (D. Moskophidis and F. Lehmann-Grube, unpublished observations.)

the maximum on days 8–10. Cytotoxicity then declines and reaches low or nondetectable levels about the twentieth day after infection. Upon challenge infection, an anamnestic response is observed. As has been shown for numerous virus–host systems, LCM virus-specific cytotoxic activity *in vitro* is mediated by thymus-derived T cells (Doherty *et al.*, 1974; Zinkernagel and Doherty, 1974d). Cytotoxic T lymphocytes specifically directed at syngeneic LCM virus-infected target cells can also be induced *in vitro,* and both primary and secondary responses have been observed (Dunlop *et al.*, 1976, 1977; Dunlop and Blanden, 1976, 1977b; Blanden *et al.* 1977; Marker *et al.*, 1977). Immunologically active T cells generated *in vitro* by specific stimulation of memory cells have been shown to fulfill at least two functions in the mouse, namely, virus elimination (Dunlop, 1978) and protection from illness due to intracerebral challenge (Marker and Andersen, 1979). Interestingly, immune protection was not conferred by such cells if these had been obtained through stimulation of memory cells with concanavalin A rather than with infected cells, although the cytotoxic potential for cultivated targets of both types of immune cells was indistinguishable.

In addition to cytotoxic T cells, in LCM virus-infected mice cytotoxic cells are generated (or preexisting cells boosted) which lyse infected as well as uninfected syngeneic, allogeneic, and xenogeneic target cells; they are neither B nor T cells and have been identified as natural killer cells (Herberman *et al.*, 1977; Nunn *et al.*, 1977; Welsh and Zinkernagel, 1977; Welsh, 1978; Welsh *et al.*, 1979; Welsh and Doe, 1980). Their role with respect to LCM virus infection of the mouse is uncertain (Welsh and Kiessling, 1980). The relationship between LCM virus-induced natural killer cells and θ antigen-bearing cells which appear during murine lymphocytic choriomeningitis and have the ability to destroy noninfected syngeneic but not allogenic target cells (Pfizenmaier *et al.*, 1975, 1976b) has not been determined.

An interesting feature of anti-viral, cell-mediated immunity is the phenomenon of *H-2* restriction. Its first discovery was made with LCM virus and mice by Zinkernagel and Doherty (1974a; Doherty and Zinkernagel, 1975b). It has already been pointed out that the immunogen inducing LCM viral cell-mediated immunity (including cytotoxic T cells) is converted into the active form by combination with gene products of the *H-2* major histocompatibility complex so that it consists of two components, one being of viral and the other of cellular origin. For its recognition as foreign by immunologically competent cells of the host, the latter have to "know" the *H-2K* or *H-2D* region antigens on the virus-infected cells; this is, of course, always the case during the development of immunity to a virus following either natural or experimental infection. By sophisticated experimentation, it has been shown that the corresponding antigens on cells of a different haplotype serve the same purpose, provided the responding host had previously been made to accept them as self. Similar rules apply during the effector phase of anti-viral cell-mediated immunity.

Although most work aimed at understanding the mechanism and biological relevance of *H-2* restriction has been and is being done by using cytotoxicity assays, cytotoxic T cell activity is by no means the only immune phenomenon associated with *H-2* restriction. On the contrary, the latter is generally considered to be a general principle applicable not only to viral immunity but to all cases in which cells interact with cells during either the induction or the effector phase of cell-mediated immunity (Zinkernagel and Doherty, 1977, 1979; Wagner *et al.*, 1980). In the LCM virus–mouse system, in addition to activity of cytotoxic T cells (Zinkernagel and Doherty, 1974a,c, 1975b; Doherty and Zinkernagel, 1975b,c; Blanden *et al.*, 1975, 1976; Zinkernagel, 1976a,c,d; Zinkernagel *et al.*, 1976, 1977, 1978a; Pfizenmaier *et al.*, 1976a,c; Zinkernagel and Oldstone, 1976), *H-2* restriction has been demonstrated to govern acute lymphocytic choriomeningitis disease (Doherty and Zinkernagel, 1975a; Doherty *et al.*, 1976b), virus elimination (Zinkernagel and Welsh, 1976), and delayed type hypersensitivity as measured by the footpad response (Zinkernagel, 1976b). For practical purposes, it is important to know that *H-2* identity is not confined to members of inbred mouse strains. Zinkernagel *et al.* (1975) observed marked LCM virus-specific cytotoxic T-cell activity and succeeded in transferring fatal lymphocytic choriomeningitis between outbred mice, provided these had been reared in closed colonies of long standing; essentially the same was found with colony-bred NMRI mice (Lehmann-Grube, unpublished observations).

Data of Marker and Andersen (1976) indicate that in certain effector cell–target cell combinations, matching of only the *K* regions (and not the *D* regions) of *H-2* resulted in low cytotoxicity, and when the effector cells were from LCM virus-immune hybrid mice, (haploid) identity at both *K* and *D* regions with the LCM virus-infected target cells (homozygous for *H-2*) sometimes led to unexpectedly low chromium release. On the other hand, immunologic interaction between effector and target cells is not always absent if they differ at *H-2*. Thus, Cole *et al.* (1973) recorded lysis when LCM virus-immune BALB/c effector cells ($H\text{-}2^d$) interacted with LCM virus-infected L target cells ($H\text{-}2^k$), and Cihak and Lehmann-Grube (1974), who employed a target reduction assay, observed considerable cytotoxic activity of spleen cells from CBA/Ca mice for virus-infected Vero (monkey) targets. A one-way cross-reactivity between LCM virus-induced murine cytotoxic T lymphocytes and trinitrophenyl-modified syngeneic target cells (Starzinski-Powitz *et al.,* 1976) was explained by sharing of antigenic determinants between LCM virus-infected and trinitrophenyl-conjugated target cells.

It is often said that cytotoxic T-cell activity is an *in vitro* correlate of LCM virus-induced immunity in the mouse. Although this statement is undoubtedly true if it implies a general correlation, it cannot be accepted if it signifies a causal relationship. Murine T cells are not uniform; rather, they comprise several functional subclasses which can be differentiated by a variety of markers (McKenzie and Potter, 1979). Currently, the most effective technique for separating subsets of T cells uses alloantisera directed at certain cell surface components known as *Lyt* (*Ly*) antigens. T cells lysing allogeneic target cells against which they are sensitized, as well as suppressor T cells, carry Lyt-2,3. T cells involved in delayed type hypersensitivity reactions (e.g., footpad response) and helper T cells are characterized by Lyt-1 (Cantor and Boyse, 1977). Probably in the LCM virus-infected mouse, too, different immune responses are mediated by different T-cell subsets. Whether the cytotoxic T cells as determined *in vitro* are identical to the T cells involved in any of the LCM virus-specific immune phenomena observed in the mouse is not known. For influenza virus-infected mice, it has been shown that virus elimination depends on Lyt-2,3 cells (Yap *et al.,* 1978) and virus-specific delayed type hypersensitivity on Lyt-1 cells (Leung *et al.,* 1980). There is some evidence supporting the generally held view that cytotoxic T cells participate directly in the immune-mediated pathologic processes underlying murine lymphocytic choriomeningitis. Zinkernagel and Doherty (1973) demonstrated the presence of cytotoxic T lymphocytes in the cerebrospinal fluid of mice suffering from the cerebral form of the illness. Also, cortisone resistance of cells that mediates LCM-viral immunopathology in the mouse and the conclusion derived from this finding that for its development macrophages are not required have been interpreted to mean that lymphocytic choriomeningitis results from direct T cell-mediated damage to virus-infected cells (Zinkernagel and Doherty, 1975c).

The assumption that elimination of LCM virus from its murine host is the result of a direct cytolytic interaction between T lymphocytes and virus-infected tissue cells occurring *in vivo* just as it does *in vitro* (Zinkernagel and Doherty, 1979) is less substantiated. In fact, several observations militate against such a mechanism; two will be mentioned. The tissues of neonatal, congenital, and drug-induced carrier mice contain high concentrations of infectious virus; numerous cells in essentially all organs harbor virus antigen, as demonstrated by the immunofluorescence procedure. Transfer of spleen and lymph node cells from LCM virus-immune syngeneic donors always results in the rapid reduction of infectious virus in blood and organs (Volkert, 1963; Volkert and Hannover Larsen, 1965; Gilden *et al.,* 1972b; Hoffsten *et al.,* 1977). The effect of this procedure on cell-associated viral antigen has been found to vary among organs; whereas it persists in the brain and renal tubules, it completely disappears from the liver (Hoffsten *et al.,* 1977). Adoptive immunization often leads to illness and death in carrier mice (Hotchin, 1959; Lundstedt, 1969b; Gilden *et al.,* 1972b; Nathanson *et al.,* 1975; Lehmann-Grube, unpublished observations). However, there are well-documented cases in which overt illness does not develop and histopathologic changes are minimal or absent, although the virus disappears just as rapidly (Volkert and Hannover Larsen, 1965; Hoffsten *et al.,* 1977; M. Volkert, personal communication; Lehmann-Grube, unpublished observations). This is difficult to reconcile with a mechanism of virus elimination based on cytolytic destruction of infected cells. A more general argument is the following. To be recognized as target cells, the expression of *H-2* coded antigens on the surface of cultivated virus-infected cells is *conditio sine qua non* (Zinkernagel and Oldstone, 1976; Doherty *et al.,* 1977). If the rules of *H-2* restriction are to be valid for all tissue cells *in situ,* these must express *H-2* antigens, which does not appear to be always the case. Thus, on certain types of murine epithelial as well as pancreatic cells, none were found (Parr and Kirby, 1979; Parr, 1979a) and hepatocytes, too, expressed few or no *H-2* antigens (Parr, 1979b). Reservations about accepting cytolysis of virus-infected cells in the tissues of the intact animal by T lymphocytes as one mechanism which helps the host to get rid of the infectious agent are not meant to deny the possibility that cytotoxic T cells play a role, but effects other than destruction of target cells may be relevant. Whether cytotoxic factors and factors inhibiting the migration of macrophages reported by Oldstone and Dixon (1970) and Tubergen and Oldstone (1971) to be released from LCM virus-immune lymphoid cells upon their contact *in vitro* with infectious or ultraviolet light-inactivated LCM virus are active in the infected mouse is not known.

Whereas a great deal of information is available on the

genetic control of humoral immune responses, corresponding knowledge about cell-mediated immunity is limited (Gasser and Silvers, 1974). Because of the ease with which anti-LCM viral cellular immune phenomena in the mouse can be detected, attempts were made to analyze these phenomena genetically, so far with little success. The initial observation of Oldstone and Dixon (1968) that the quantity of infectious virus required to cause the (immunopathologic) death of a mouse varied markedly among members of different inbred strains, and the later finding that such differences in susceptibility were controlled in part by a dominant gene which is closely linked to the *H-2* locus (Oldstone *et al.*, 1973a), have not been confirmed by others (Lehmann-Grube, 1969, 1975; Neustadt *et al.*, 1978). Oldstone (1975b) has had difficulty in reproducing his original findings. Differences in LCM viral cell-mediated immune reactions among mouse strains have also been observed by measuring swelling of feet after intraplantar inoculation of the virus (Tosolini and Mims, 1971), but these were found not to be linked to *H-2* (Lehmann-Grube and Löhler, 1981). Other measures said to be useful in this respect are the time intervals between intracerebral inoculation of the virus and death, and footpad inoculation and onset of footpad swelling (Skinner and Knight, 1973b). Also, the speed with which protective immunity develops following subcutaneous infection varies among mouse strains (Table I), but none of these parameters has as yet been adequately analyzed.

3. General Immunosuppression by LCM Virus

Numerous viruses have been shown to influence the ability of the host to respond immunologically to other antigens (Notkins *et al.*, 1970; Woodruff and Woodruff, 1975). In the case of the mouse acutely infected with LCM virus, depression of immune functions is the rule (Mims and Wainwright, 1968; Lehmann-Grube *et al.*, 1972; Bro-Jørgensen and Volkert, 1974; Bro-Jørgensen *et al.*, 1975; Güttler *et al.*, 1975), and both humoral and cellular responses are affected. The mechanism is not known. Observations of Jacobs and Cole (1976) point to a virus-induced macrophage defect—possibly caused by cytotoxic T lymphocytes (Silberman *et al.*, 1978)—but Bro-Jørgensen (1978) considers other explanations such as action of interferon, viral damage to lymphoid precursor cells, and antigenic competition more likely. In view of the enhancing effect an LCM virus infection has on the ability of the cells of the mononuclear phagocyte system of the mouse to eliminate *Listeria monocytogenes* (Blanden and Mims, 1973), it is difficult to understand why these cells should be deficient in another function, namely, participation in immune responses. Immune depression by LCM virus is probably not a consequence of the severe destruction of lymphoid tissues of certain mouse strains infected with certain virus strains because depression is not less marked in mouse strain–virus strain combinations in which alterations of this kind are absent (Mims and Wainwright, 1968; Lehmann-Grube *et al.*, 1972).

D. Persistent Infection (Carrier State)

The unique relationship between LCM virus and mouse has been extensively investigated by Traub (1939, 1960). His findings, confirmed and extended by Rowe (1954), Haas (1954), Skinner and Knight (1974), and others, have resulted in a clear picture of persistence of the virus in its murine host.

How the first LCM virus carrier mouse came into existence no one knows, but carrier mice are easily produced experimentally by infection of newborn mice. Once the carrier state is established, intrauterine transmission occurs with 100% efficiency. Only young born of experimentally produced neonatal (0-generation) carrier mothers are sometimes free of virus, even if the dams had been shown to be viremic before and after parturition. Infection *in utero* occurs very early in pregnancy, and Traub (1960) concluded that the ova are infected perhaps before implantation. Mims (1966) demonstrated by the immunofluorescence method that reproductive cells in female carrier mice may be infected, but we do not know whether these develop normally.

The role of the father has been investigated by Traub (1960) and Skinner and Knight (1969, 1973a). When a normal female mouse is mated with a male carrier, the young are sometimes born with virus. All findings suggest that their carrier state is not the result of fertilization of normal ova by infected spermatozoa. Rather, these animals acquire the virus either from their mother, who undergoes an acute infection by contact with her mate, or after birth from the father. Detection of viral antigen by the immunofluorescence procedure in a proportion of the spermatozoa in carrier mice (J. Löhler, personal communication) proves that these cells may undergo infection, but again it is not known whether they are capable of fertilizing. There is no evidence indicating that the viral genome is integrated into the DNA of either reproductive or other cells of the mouse.

If a virus is to persist for long periods of time in a mammalian host, three conditions have to be met. These are: (1) tissues are not damaged, either directly (e.g., by cytopathic destruction of cells in which the virus multiplies) or indirectly (e.g., by inflammation), to such a degree as to impair vital functions; (2) the immune apparatus is prevented from accomplishing its task; and (3) replication of virus is regulated so that its concentration remains limited, yet there is compensation from loss due to excretion and natural decay (Lehmann-Grube, 1977). Proof that the replicating LCM virus does not damage vital tissues of the mouse to any extent is obtained from the general observation that carrier mice may develop normally, even though essentially all of their tissues continuously rep-

Table II

Concentration of Infectious LCM Virus of the WE Strain in Blood and Organs of Carrier Mice 10 ± 2 Weeks after Neonatal Infection[a]

Organ	Strain of mice				
	NMRI (colony bred)	C57BL/6J	CBA/J	C3H/HeJ	DBA/2
Blood	5.76[b] ± 0.11 (5)	5.86 ± 0.09 (3)	5.52 ± 0.15 (5)	5.87 ± 0.22 (5)	6.27 ± 0.15 (6)
Brain	7.29 ± 0.13 (5)	7.05 ± 0.18 (6)	7.33 ± 0.08 (6)	7.57 ± 0.14 (5)	7.66 ± 0.16 (4)
Liver	6.88 ± 0.17 (4)	7.52 ± 0.22 (7)	7.53 ± 0.34 (6)	7.17 ± 0.19 (5)	7.78 ± 0.18 (7)
Kidney	8.50 ± 0.19 (5)	8.38 ± 0.11 (7)	8.38 ± 0.08 (6)	8.68 ± 0.05 (4)	8.47 ± 0.10 (7)

[a] From von Boehmer *et al.* (1974), with permission of Cambridge University Press, Cambridge, England.
[b] Mean $\log_{10}$ mouse infectious units per 1.0 ml blood or 1.0 g tissue ± single standard error; number of mice in parentheses.

licate the virus to high concentrations (Table II). More specifically, mice infected with LCM virus when less than 24 hr old gained weight for at least 5 days as rapidly as uninfected controls, although the virus multiplied extensively, reaching concentrations of close to 10^{10} infectious units per gram of tissue (Lehmann-Grube, 1971).

If, as has already been pointed out, the disease of mice infected with LCM virus as adults is an immunopathologic phenomenon and not the result of direct viral effects on infected cells, immunosuppression should be of therapeutic value. Over the years, many measures have been shown to protect mice from LCM virus disease, the common denominator being immunosuppression (Table III). Mice thus treated may remain essentially free of disease signs, although the virus multiplies to high concentrations.

One observation which seems to contradict the assertion that in carrier mice tissues are not damaged is the late onset disease seen in older neonatal and also congenital carrier mice. True, in mice persistently infected with LCM virus, late onset disease is a frequent occurrence, but the extent varies considerably among mouse strains and virus strains, and sometimes this illness is completely absent (see Section V,B), proving, in principle, that the virus does not damage directly the cells in which it multiplies, although indirect effects may have harmful consequences. Nonetheless, this agent has cytolytic potential *in vitro,* which raises the question of the difference between these cells and cells left in their natural environment. Destruction by LCM virus of cultivated cells occurs only when they undergo rapid multiplication. Cytopathic effects are absent when the monolayer is complete, i.e., when the cells rest or multiply slowly (M. Popescu and F. Lehmann-Grube, unpublished), which applies to the vast majority of cells in the mouse. Interfering particles have been shown to protect cultivated cells against cytopathic destruction by LCM standard virus (Welsh and Pfau, 1972; Popescu *et al.*, 1976; Dutko and Pfau, 1978), and it is sometimes assumed that cells in a persistently infected mouse are shielded by these entities. LCM virus interference by interfering particles has been shown to be an all-or-none phenomenon (Popescu *et al.*, 1976), and whereas it is easy to understand that a cell which is prevented from producing infectious virus is thereby spared cytolysis, it is difficult to imagine how protection could occur in a cell which does produce infectious virus. Furthermore, the concentration of interfering particles in the persistently infected mouse (Popescu and Lehmann-Grube, 1977) is hardly sufficient to explain the observed degree of protection.

Most investigators agree that it is immunologic tolerance which prevents the immune system of the mouse from fulfilling its physiological task (Volkert and Hannover Larsen, 1965; Hotchin, 1971; Lehmann-Grube, 1971; Zinkernagel and Doherty, 1974b; Volkert *et al.*, 1975a). Since immune elimination of the virus is dependent on T cell function, this should

Table III

Protection of Adult Mice from Disease due to Infection with the LCM Virus[a]

X Irradiation
Cyclophosphamide
Methotrexate (= amethopterin) (reversed by citrovorum factor)
Folic acid-deficient diet
8-Azaguanine (= guanazolo)
5-Fluorouracil
Chlorambucil (= leukeran)
1,3-Bis(2-chloroethyl)-1-nitrosourea
Azaserine
6-Diazo-5-oxo-nor-1-leucine (DON)
Surgical or genetic depletion of T lymphocytes
Anti-lymphocytic serum
Graft-*versus*-host disease
High virus dosis

[a] Modified from a review by Lehmann-Grube (1971), where most references may be found. More recent observations are protection by treatment with cyclophosphamide (Gilden *et al.*, 1972a) and genetic absence (Christoffersen *et al.*, 1976; Mori *et al.*, 1976) or surgical removal (G. A. Cole and P. C. Doherty, cited in Doherty and Zinkernagel, 1974) of T lymphocytes from adult mice.

be absent from carrier mice. Cihak and Lehmann-Grube (1974) searched for cytotoxic T cells in such animals and found none. Also, spleen cells did not take up tritiated thymidine *in vitro* when incubated with syngeneic LCM virus-infected and mitomycin-treated cells, and more recent experiments (Lehmann-Grube, unpublished observations) showed that the LCM virus did not induce footpad swelling—known to depend on T cells—in carriers. Other authors, too, failed to detect virus-specific T cell activity (Marker and Vokert, 1973; Zinkernagel and Doherty, 1974b), and it is concluded that LCM virus-specific, cell-mediated immunity is absent from LCM virus carrier mice. As would be expected, the response of carrier mice to other immunogens is, as a rule, normal (Holtermann and Majde, 1971; Lehmann-Grube, 1971; Oldstone *et al.*, 1973b; Güttler *et al.,* 1975), and the same is true of other parameters of integrity of the immune system (Schwenk *et al.*, 1971; Oldstone, 1973). There are, however, exceptions. Pincus *et al.* (1970) could not induce tolerance to bovine γ-globulin in neonatal carrier mice. Oldstone *et al.* (1973b) observed depressed antibody responses to human, bovine, and avian immunoglobulins and prevention of the development of immunological hyporesponsiveness to human immunoglobulin. Also, LCM virus carrier mice are unusually sensitive to X-rays, relatively low doses causing death and associated marked reduction of numbers of red and white blood cells (Bro-Jørgensen and Volkert, 1972a,b; Sharon and Pollard, 1972). These effects have been attributed to the low contents of hemopoietic precursor cells in carrier mice (Bro-Jørgensen, 1978).

Specific immunological unresponsiveness had been explained by Burnet (1959) by elimination of cell clones, and in many instances of tolerance this is probably the mechanism (Howard and Mitchison, 1975). In others, blockade of potentially active cells (Hellström and Hellström, 1974) or active suppression (Gershon and Kondo, 1971) are more likely explanations. In experiments aimed at understanding the mechanism of immunologic tolerance in LCM virus carrier mice, Cihak and Lehmann-Grube (1978) could not find in such animals either blocking factors or suppressor cells. This indicates that immunologic tolerance in carrier mice follows deletion or irreversible inactivation of cells destined to develop into T cells mediating LCM virus-specific immunity.

The mechanism by which clones are eliminated cannot as yet be definitely explained. On the basis of experiments performed with acutely infected adult mice, Dunlop and Blanden (1977a) suggested that cells programmed to respond immunologically to LCM virus are eliminated by becoming targets for cytotoxic T cells. The observation of Popescu *et al.* (1977, 1979) and Doyle and Oldstone (1978) that in LCM virus carrier mice a small proportion of lymphocytes is infected has led to the hypothesis that these are potentially LCM virus-specific immune cells which have made contact with infectious virus via their antigen receptors, resulting in antigen-triggered clonal expansion followed by infection and functional inactivation (Popescu *et al.,* 1979). However, more recent experiments of Tijerina *et al.* (1980) show clearly that during acute infection of adult mice, T lymphocytes resist infection whereas in newborn mice a proportion of these cells are susceptible. Thus, if the above hypothesis is to be upheld, it has to be modified by assuming that infection and inactivation of immunologically LCM virus-programmed T lymphocytes occur only in newborn (and presumably unborn) mice. This tallies with the well-established fact that the carrier status is most easily induced—i.e., without further experimental manipulation—by infection of immature mice.

The belief that the LCM virus persists because of immunologic tolerance was challenged (Oldstone and Dixon, 1967, 1969) because of virus-specific antibodies in carrier mice which are often present (Benson and Hotchin, 1969; Oldstone and Dixon, 1969). There can be no doubt that these animals are capable of producing LCM virus-specific antibodies. Buchmeier, Oldstone, and their colleagues have demonstrated the presence of immunoglobulin with specificities for all the major structural virus proteins (Buchmeier and Oldstone, 1978a; Oldstone *et al.*, 1980). In carrier mice of certain strains, even neutralizing antibody has been detected which, in the case of the NMRI mouse, may attain titers as high as those reached after acute immunizing infection (M. Popescu and F. Lehmann-Grube, unpublished observations). It is nonetheless questionable whether the presence of LCM virus-specific antibodies in carrier mice is a relevant finding when considering these animals' inability to remove the virus by immunological means. As has already been pointed out, virus elimination is mediated by cells, and antibodies probably play no role. Indeed, both these immune phenomena appear to develop independently of each other. Hannover Larsen (1969) showed that elimination of the virus and production of an antibody can be uncoupled simply by changing the time interval between birth and infection. Also, the concentration of infectious virus in carrier mice of different strains is essentially identical (von Boehmer *et al.*, 1974); yet there is great variation in their ability to make antibodies, and traces as well as high concentrations may be found (Oldstone *et al.,* 1980).

As to their T cell compartment, mice persistently infected with the LCM virus fulfill all criteria of LCM virus-specific immunologic tolerance; the same cannot be said of B cells. The relationship among the various viral immunogens inducing antibodies and the immunogen(s) inducing cell-mediated immunity is not known. But even if the plausible assumption should turn out to be correct—namely, that the component of the virion which induces neutralizing antibody is identical to the viral moiety of the cell membrane-associated complex which activates T cells—the fact that cell-mediated immune phenomena are *H-2* restricted, whereas neutralization is not, proves that the attachment sites of antibody molecules and the attachment sites of T cells are different. One explanation for the frequent finding of LCM virus-specific antibodies in mice lacking LCM viral T cell activities is immunity with respect to B lymphocytes and tolerance with respect to T lymphocytes. The author has seen no data which would allow one to determine whether B cells in carrier mice producing anti-LCM viral antibodies receive T-cell help or whether their activity is independent of T cells. There is at least one observation suggesting that LCM viral immunologic tolerance may also extend to the B cell compartment. Volkert *et al.* (1975b) showed convincingly that in C3H LCM virus carrier mice, neither T helper cells sensitized against LCM virus complement-fixing antigen nor the corresponding antibody-producing B cells were present. This proves that an antibody which regularly develops in mice infected as adults may be entirely absent in a true carrier mouse. In this case, however, it is almost certain that the tolerogen inducing unresponsiveness with respect to antibody and the tolerogen inducing unresponsiveness with respect to cellular immunity have different specificities.

Whereas the finding of LCM virus-specific antibodies in carrier mice is of doubtful relevance when considering virus persistence, the same cannot be said when such antibodies are regarded in connection with the late onset disease in aging carrier mice. In diseased kidneys and other organs, immunoglobulin of host origin is deposited in a pattern virtually identical to the one seen in animals with experimentally induced immune complex diseases (Oldstone and Dixon, 1969), and by the double labeling method, immunoglobulin and virus antigen have been localized in identical sites (Fig. 10) (Löhler *et al.*, 1980). Also, a correlation was found to exist between glomerulonephritis in carrier mice and quantity of accumulated immunoglobulin. The conclusion that the clinical signs in older carrier mice are the pathologic consequence of deposition in various tissues of antigen–antibody complexes (Oldstone, 1975a) is entirely convincing. In contrast, the question about their immunological specificity cannot as yet be answered to everybody's satisfaction. Whereas the detection of immunoglobulin, binding to all three major structural polypeptides of the virion, in acid eluates from diseased kidneys of carrier mice (Buchmeier and Oldstone, 1978a; Oldstone *et al.*, 1980) makes the dispute as to whether such eluates do (Oldstone and Dixon, 1969) or do not (Klein *et al.*, 1977) contain LCM virus-specific, complement-fixing antibody irrelevant, this finding suggests, but does not prove, that the LCM virus–immunoglobulin complexes are indeed the ones which cause glomerulonephritis.

The difficulties in sorting out the various factors contributing to an immune complex disease when agents are involved that are both infectious and antigenic have been reviewed by Dixon *et al.* (1971). That more than one element may be relevant in the renal disease of persistently infected mice is illustrated by the study of Tonietti *et al.* (1970), who induced the LCM virus carrier state in New Zealand mice and observed augmentation of glomerulonephritis, which occurs spontaneously in these animals. Also, the persistent LCM virus infection increased the incidence and concentration of anti-nuclear antibody but not that of anti-erythrocyte antibody in New Zealand mice.

The obstacles encountered when interpreting data from elution experiments with aging LCM virus carrier mice are increased by the fact that in most of these animals' organs, and especially in the kidneys, plasma cells are present which may be so numerous as to form aggregates resembling plasma cell tumors (Pollard and Sharon, 1969; Sharon and Pollard, 1971; Accinni *et al.*, 1978; J. Löhler, personal communication). Probably they are not removed from the minced tissues prior to incubation under eluting conditions. We do not know whether these cells produce LCM virus-specific immunoglobulin, but if they do—and this is a plausible assumption—they are likely to cause contamination of eluates with antibody not derived from complexes. One observation indicates that other factors are of pathogenic relevance in this condition. Gresser *et al.* (1978) treated neonatally LCM virus-infected mice with anti-interferon globulin and observed a retardation of the evolution of glomerulonephritis, which they interpreted to mean that interferon contributed in some unknown way to the development of this disease.

The virus concentrations in the tissues of LCM virus carrier mice are remarkably constant. This leads to the question of how this lifelong equilibrium between elimination and spontaneous decay on the one hand, and new production, on the other, is maintained. In theory, interfering particles could function as part of a feedback mechanism controlling LCM virus replication in a carrier mouse. With a quantitative and highly sensitive procedure (Popescu *et al.*, 1976), Popescu and Lehmann-Grube (1977) detected interfering particles in the tissues of LCM virus carrier mice. However, their number was found to be very low, hardly sufficient for effectively curtailing the replication of infectious virus. That interfering particles play a decisive role in keeping infectious virus in a carrier mouse in check becomes even more doubtful given a report of Jacobson and Pfau (1980), who described an LCM virus variant (A) which they had isolated—together with another variant (D)—from a 10-month-old neonatal carrier mouse. Variant A behaved serologically and biologically as a characteristic LCM virus but had the peculiar property of replicating normally (in cultivated cells) in the presence of LCM virus-interfering particles, whether these had been derived from either the variants (A and D) or the original virus (T). It is difficult to understand how in these mice replication of A should have been influenced by interfering particles. Probably the cell itself is of greater importance. Hotchin (1973) had described the phenomenon of cyclical transient infection, which he explained on the basis of a natural defense mechanism of the cell against the LCM virus. Experiments of M. Popescu (personal communication)

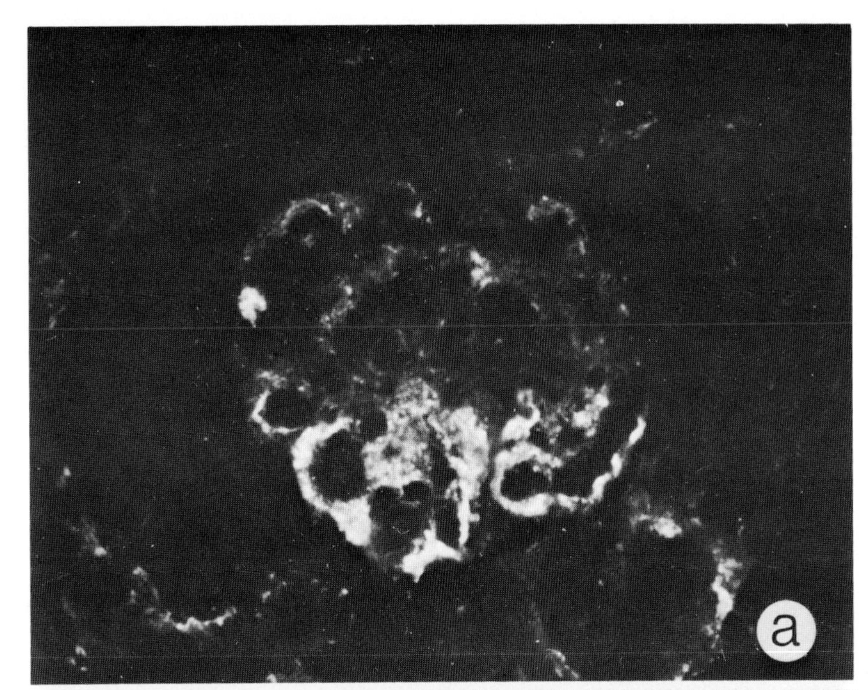

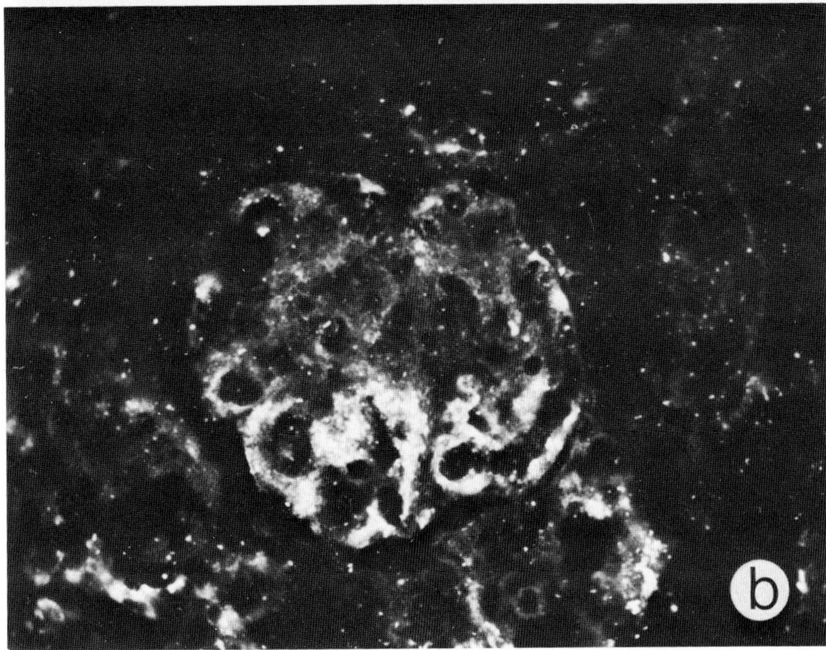

Fig. 10. Immune complexes in the mesangium and subendothelial space of a glomerulus and also of tubuli of the kidney of an LCM virus carrier mouse with late onset disease demonstrated by the immunofluorescence procedure. Double labeling applied to one section. Visualization of host immunoglobulin by rhodamine-conjugated anti-mouse immunoglobulin antibody (a) and of virus antigen by fluorescein isothiocyanate-conjugated anti-LCM virus antibody (b). Magnification ×400. (Photographs kindly supplied by Dr. J. Löhler, Hamburg.)

have shown that the rate of virus multiplication depends on the state of the cell. As soon as replication of the latter ceases due to completeness of the monolayer, virus replication is greatly reduced. In *Mutatis mutandis,* the same was found by Pfau *et al.* (1973) who worked with suspension cultures. Probably virus replication in cells *in situ* is similarly restricted.

Since immunosuppressive treatment with either X-rays or cyclophosphamide did not alter the levels of viremia in carrier mice (Bro-Jørgensen and Volkert, 1972a; Sharon and Pollard, 1972; Hoffsten and Dixon, 1973), the immune apparatus of the host is probably not involved. Interferon, which theoretically could have a regulating function, has been shown to be transiently produced in newborn and adult mice following infection with LCM virus (Padnos *et al.,* 1971; Rivière and Bandu, 1977; Merigan *et al.,* 1977; Bro-Jørgensen and Knudtzon, 1977) but has, to the author's knowledge, never been detected in established carrier mice (Wagner and Snyder, 1962; Mims and Subrahmanyan, 1966). When congenital carrier mice were challenged with Newcastle disease virus, they produced less interferon than their uninfected counterparts (Holtermann and Havell, 1970). In tissues and serum of adult mice acutely infected with the E-350 (Armstrong) strain—but not with the WE strain—of LCM virus, Veltri and Kirk (1971) found an antiviral substance with properties which distinguished it from interferon. Whether it affected replication of LCM virus was not determined.

V. EPIZOOTIOLOGY

A. Host Range

The first indication of an association of the LCM virus with *Mus musculus* came from Traub's encounter with the virus in a colony of laboratory mice (Traub, 1935, 1936). His report, together with a similar one from France (Lépine and Sautter, 1936) and one from Great Britain (Findlay *et al.,* 1936), prompted Armstrong and Sweet (1939) to investigate house mice. They isolated the virus from mice trapped in the homes of persons with lymphocytic choriomeningitis, and their conclusion "that gray mice constitute a reservoir of choriomeningitis infection" (p. 683) has since been confirmed numerous times.

As will be more fully described in the section on transmission (Section V,C), the LCM virus carrier mouse serves as the principal source of the virus in nature. In view of the prevalence of the commensal house mouse, transmission of virus from persistently infected mice to members of other species is likely to occur, an assumption which is borne out by virological findings in many cases. However, only the golden hamster and man are of importance, the golden hamster because it is the one animal (besides the mouse) known to transmit the virus to other individuals of the same or a different species and man because infection may have severe consequences. From published reports, the impression is gained that humans become infected more often than individuals from other species, but this may reflect the direction of virological efforts rather than true differences in rates of infection.

Since the LCM virus is infectious for man, in whom it may cause a variety of illnesses (Armstrong, 1941; Scheid, 1957; Lehmann-Grube, 1971, 1981a), research work with this agent carries a certain risk which may be reduced by taking adequate precautions. Sometimes, however, the presence of the virus is not suspected, in which case the consequences may be severe. A variety of laboratory materials such as cell lines, vaccines, and infectious agents have been found harboring the virus. In relation to the many years during which this virus has been known to exist, such encounters must be considered rare events. The same cannot be said of inadvertently contaminated neoplasms transplanted experimentally in mice and hamsters (Stewart and Haas, 1956; Biggar *et al.,* 1976), and several outbreaks of lymphocytic choriomeningitis among laboratory personnel could be traced to this source (Lewis *et al.,* 1965; Baum *et al.,* 1966; Hinman *et al.,* 1975; Biggar *et al.,* 1977). Secondary infection from person to person did not occur under these circumstances (Biggar *et al.,* 1975a).

As has already been pointed out (see Section III,C), the experimental host range of the LCM virus is wide.

B. Geographic Distribution

The often expressed opinion that LCM virus exists throughout the world is not well substantiated. In fact, the evidence indicates that the virus is present in certain regions but absent in others. Thus, reliable positive information is available for Europe (including Western Russia) and North and South America, but in large areas such as Australia and South Africa, attempts at demonstrating the virus either in mice or as a cause of human infections have remained unsuccessful (F. Fenner, personal communication; Bayer and Gear, 1955; O. W. Prozesky, personal communication). For all other parts of the world, either no information is obtainable or alleged isolations are not sufficiently controlled to allow their acceptance as evidence for the presence of the virus in some natural host.

Patchy occurrence of LCM virus carrier mice is also obvious if one considers smaller geographic areas. In an extensive survey conducted by Ackermann and colleagues (1964) in the Federal Republic of Germany, carriers were found among the mice in the north and west, but in the south they were conspicuously absent.

The uneven distribution of LCM virus carrier mice has been explained by slow spread of the virus among mice. However, considering the way these animals live in nature (Freye and Freye, 1960), infection by contact between carriers and virus-free newborn mice or pregnant dams—the only mode of transmission which could create new carrier mice (see Section V,C)—is unlikely to occur. Vectors which could aid in dissemination of the virus have never been identified. Hence, an alternative explanation must be sought. In doing so, other considerations have to be taken into account. Europe is the home of two different forms (usually considered subspecies) of *M. musculus: M. m. domesticus* in the west and *M. m. musculus* in the east. Except for a narrow contact zone in the center of the continent, these two forms are well separated. The mice which colonized North and South America and Australia are all pure *M. m. domesticus* (Schwarz and Schwarz, 1943; R. D. Sage, personal communication). To my knowledge, no wild carrier house mouse has ever been analyzed as to subspecies, but from the places in which they were trapped it is safe to assume that there are carriers among *M. m.*

domesticus as well as *M. m. musculus,* and it can be concluded that the common ancestor has already lived with the virus. From this, one would expect carrier mice to be evenly distributed among the subspecies mentioned, which, as we have seen, is not the case. One explanation is that an occasional mouse loses the virus and, being biologically superior, outgrows the carriers. Since mouse families do not intermingle (Freye and Freye, 1960), a patchy occurrence of carriers among wild house mice would be expected.

Attempts to confirm this hypothesis experimentally have met with little success. Years ago, the author observed two mice which had been infected after birth and which were viremic at age 50–80 days but had lost the virus at age 180–190 days (Lehmann-Grube, 1964b). More recently, Traub (1975a) saw one congenital carrier which cleared the virus between two pregnancies, "the first case of autosterilization observed by us in many years of experience with LCM" (p. 772). To establish more accurately the frequency with which spontaneous elimination of virus from carrier mice occurs, I have followed virus persistence in a total of 342 neonatal and congenital carrier mice and did not find one animal which harbored the virus when less than 4 months old but was virus-free at age 10–12 months (Table IV). As to the postulated biological inferiority of congenital carrier mice, the reports differ. Pollard *et al.* (1968) found neither growth rate nor litter size to be reduced and saw no overt disease signs for at least 8 months in white mice of a carrier colony of long standing. In a carefully controlled study, Skinner and Knight (1979a) observed normal outward appearances for up to 18 months and normal weights in laboratory mice from an established carrier colony into which the LCM virus had probably been introduced by a gray house mouse. Significantly, the litters born to carrier mice were consistently smaller when compared with litters born to control mice. Other investigators observed slight or severe defects of one kind or another in mice from established carrier colonies (Traub, 1938, 1941; Seamer, 1965a; Mims, 1968, 1970).

Table IV

Carrier Status with Increasing Age of NMRI Mice Infected either Neonatally or Congenitally with LCM Virus[a]

Strain of virus	Carrier[b] generation	Age of mice when tested (months) <4	4–6	7–9	10–12	>12
Traub (= W)	I	52/52[c]	n.t.	48/48	35/35	32/32[d]
WE	0	n.t.	53/53	51/51	40/40	16/16
WE	I	194/199[e]	n.t.[f]	n.t.	155/159[e]	121/125[e]
WE	II	148/148	n.t.	124/124	108/108	21/21

[a] Lehmann-Grube (unpublished observations).

[b] Neonatal (0) carrier mice infected when less than 24 hr old; first (I) and second (II) generation of carrier mice born of neonatal or first-generation carrier mothers, respectively.

[c] Number of mice viremic/number tested.

[d] Mice older than 1 year were often used for other experimental purposes. Hence the low numbers in this column do not necessarily indicate high mortality.

[e] The same mice were always nonviremic.

[f] Mice not tested during this time interval.

Carrier house mice are outwardly inconspicuous, but to my knowledge, their biological or histological status has not been investigated. Hence, it is not known for certain whether persistently infected gray house mice remain truly asymptomatic. That this may be the case is indicated by the results of a recent study by F. Lehmann-Grube and J. Löhler (unpublished observations), who searched for immune complex disease and found none in members of a carrier colony of gray house mice infected with an LCM virus strain isolated from a wild carrier mouse and never passaged in a laboratory host. Nonetheless, their reproductive capacity was reduced compared with that of uninfected house mice kept in parallel. From this observation and the evidence cited above, it can be concluded that LCM virus-free mice have selective advantages over persistently infected mice even when the latter may not be afflicted with immune complex disease.

C. Transmission

Although members of other species are occasionally infected, only the persistently infected house mouse (*M. musculus*) and the acutely infected golden hamster (*Mesocricetus auratus*) are known to transmit the virus.

Having high concentrations of virus in all its organs throughout life, including kidneys and salivary glands, a carrier mouse excretes much virus with saliva, nasal secretions, and urine; if other mice are nearby, they can be easily infected. In an adult mouse, a self-limited and immunizing infection may result, in which case the infectious chain usually ends. If, however, the virus enters a newborn mouse or, via a pregnant dam, an unborn mouse, a lifelong infection results. As previously discussed, mice infected shortly after birth usually reach maturity, and since essentially all mice born of a persistently infected mother are again persistently infected, it is easy to visualize how a colony of mice may become infested if only one LCM virus carrier mouse, either male or female, is inadvertently introduced (Traub, 1938; Haas, 1941; Skinner and Knight, 1973a, 1974). Indeed, after a few years, literally all mice in a closed colony may be persistently infected carriers (Traub, 1939; Skinner and Knight, 1971).

Spread of the virus as just described occurs only if the mice live in close proximity, as is the case in the laboratory or a breeding facility. In nature the *lebensraum* of every wild house mouse is said to be small and migration to other areas, leading to contact with other mice, is assumed to be rare (Freye and Freye, 1960). Thus, dissemination of virus by horizontal transmission from persistently infected to normal mice is not likely to occur often. The principal mode by which the LCM virus is maintained among house mice is *in utero* transmission (see Section IV,D). Of course, this statement is valid only as long as the mice live undisturbed. How the virus may be transmitted from wild carriers to laboratory mice when man tampers with the murine habitat has been vividly described by Skinner *et al.* (1977).

Infection of members of other species, including man, occurs also by contact with carrier mice. As a rule, no further transmission to other individuals of the same or different species occurs. This, together with the above considerations, implies that the infectious chain by which the LCM virus is horizontally transmitted usually consists of only two links. There is, however, one exception which, because of its importance for man, has attracted much attention. The animal which—besides the carrier mouse—is known to spread the virus readily to members of its own and other species is the golden hamster (*M. auratus*) (D. Armstrong *et al.*, 1969; Ackermann *et al.*, 1972, 1974; Ackermann, 1973; Förster and Wachendörfer, 1973; Anonymous, 1974; Hirsch *et al.*, 1974; Biggar *et al.*, 1975b, 1977; Maetz *et al.*, 1976; Skinner and Knight, 1979a; Lehmann-Grube *et al.*, 1979a). Hamster colonies are frequently also inhabited by mice. If the latter happen to be carriers, transmission of the virus is easily accomplished. Once individual hamsters in a colony are infected, the virus is distributed by a mechanism closely resembling the one described for a mouse colony. However, unlike carrier mice, which harbor the virus lifelong, the hamster remains infected (and infectious) for long periods of time, but eventually the agent is eliminated (Smadel and Wall, 1942; Skinner *et al.*, 1976). With a particular strain of virus, lifelong infections of hamsters have been observed, but these animals underwent a chronic, progressive illness leading to early death (Parker *et al.*, 1976).

The ease with which the LCM virus circulates in a mouse or hamster colony, and the infectious spread to members of other species if these are in close contact, are probably consequences of the high concentration of virus in excretions of carrier mice and infected hamsters. It is also not difficult to explain laboratory infections. However, numerous transmissions outside the laboratory, especially to humans, are on record in which association with infected mice or hamsters was only indirect; in such cases, the mode of spread is not well understood. Because of this virus' great lability, dry dust is not a likely vehicle and fomites can be assumed to play a role only if the material containing the virus has not dried out. In view of the efficiency with which monkeys and guinea pigs can be infected by aerosols (Daneš *et al.*, 1963; Benda *et al.*, 1964), perhaps these and droplets carry the virus from carrier mice to, for instance, people. Also, the portals of entry are not known for certain, but probably these are broken skin and broken mucous membranes (Skinner and Knight, 1979b). Vectors such as insects have sometimes been implicated, but the experimental evidence that these are of relevance is scant.

VI. DIAGNOSIS

Infections with LCM virus are diagnosed serologically and by detecting the agent. In man, transient viremia seems to be the rule, and the virus is most readily isolated by inoculating blood into the brains of mice. In cases with involvement of the central nervous system, the virus may also be found in the cerebrospinal fluid. Several methods are available to demonstrate the presence of LCM virus-specific antibodies in human sera, but a search for complement-fixing antibody cannot be recommended (Lehmann-Grube *et al.*, 1979b).

In contrast, for the serodiagnosis of LCM virus infections in individuals from other species (e.g., mouse, golden hamster, guinea pig), the complement fixation test gives reliable information, but only if the antigen is of adequate potency, which is often not the case. A preparation of sufficient quality for most purposes may be obtained by the method of Smadel *et al.* (1939). If the titer is to be higher or if some degree of purity is desired, other procedures may be employed (Schell *et al.*, 1966; Gschwender *et al.*, 1976).

Virus isolations are accomplished by inoculating body fluids or tissue homogenates into the brains of laboratory mice, provided these have been shown to be free of LCM virus (see Section VII). The easiest way to identify an isolate is by cross-protection test employing mice as assay hosts. Other methods in use are neutralization and complement fixation tests and immunofluorescence procedures (Armstrong and Dickens, 1935; Smadel and Wall, 1941; Ackermann *et al.*, 1964; Förster, 1973).

A variety of procedures have been recommended to quantify infectious LCM virus. Although the mouse is still widely employed as assay host, quantitative methods based on plaque formation (Pfau *et al.*, 1973; Lehmann-Grube and Ambrassat, 1977) are greatly superior and should be preferred wherever possible.

VII. CONTROL AND PREVENTION

The view, sometimes expressed, that laboratory mice are often infected with LCM virus is not correct. Since discovery of this virus 45 years ago, four colonies have been described as harboring it (Traub, 1935; Findlay *et al.*, 1936; Lépine and Sautter, 1936; Skinner and Knight, 1969). Even if additional reports on recovery of LCM virus from experimental or diagnostic materials are considered where the circumstances make one suspect that the agent had existed in laboratory mice, the number remains small. [The few mice found by Poiley (1970) in one of six genetic production centers and in one of 16 production colonies to have anti-LCM virus (complement-fixing?) antibody cannot be considered adequate evidence for the presence of the virus in these breeding units.] Of course, other pertinent observations might have remained unpublished and infested colonies may exist which have gone unnoticed, but the overall conclusion is that infections of laboratory mice are rare. They do occur, however, and since the virus is a hazard for man and, furthermore, may falsify experimental results, it is sometimes desirable to exclude its presence.

As a rule, LCM virus in a mouse colony is not suspected from the appearance or the behavior of its members. There are several procedures by which mice may be tested, but only two give reliable information even if employed by the inexperienced. One is based on the fact that adult mice are not naturally resistant to the virus and die if infected via the brain. The procedure consists of challenging mice by intracerebral inoculation of LCM virus. If, contrary to expectation, they survive, they are either persistently infected or were immunologically protected by previous contact with the virus. In either case, the agent must be present in the colony. The advantage of this approach is its simplicity. The disadvantages, however, are considerable. A large number of mice must be sacrificed in order to obtain the desired degree of probability that the whole colony is virus free. This may be regrettable with ordinary albino mice but may be intolerable in the case of valuable inbred mice. Furthermore, challenge virus of known potency must be available.

The second approach used to inspect a mouse colony for LCM virus is based on the fact that carrier mice are always viremic. The procedure is as follows. Blood from the mouse to be tested is added to balanced salt solution containing a trace of heparin, and 0.03 ml of this mixture is injected intracerebrally into each of three mice. If these mice remain free from clinical signs for at least 10 days, the donor mouse may be assumed not to be a carrier. Since blood of a persistently infected mouse contains 10^5 infectious units or more per millimeter, for screening purposes blood from up to 20 mice may be pooled before being inoculated into test mice. The advantages of this procedure are that mice may be examined without being killed and that LCM virus for challenge purposes is not required. The disadvantage is that carrier mice are detected but not mice which are immunologically protected as a consequence of previous infection. In a colony in which the virus is firmly established, this distinction is of no concern because most or even all of its members will be persistently infected. However, if the virus has been introduced more recently, only a small proportion of the mice is likely to carry it, whereas many more should be immune. These resist challenge, but their blood—unless taken shortly after infection—is free of virus; hence, their status would not become apparent using this procedure. In theory, the only way to make sure that a mouse colony is free of LCM virus is to test all its members for viremia. This is often impractical, but an acceptable compromise would be to test all breeders—male and female—and make certain that they subsequently have no contact with mice whose status as to infection with LCM virus has not been ascertained in parallel or very recently. A more elaborate account of how mouse stocks may be monitored for LCM virus has been presented by Skinner and Knight (1971).

The observation which induced Traub (1935) to search for a pathogenic agent in his colony of mice and led to recovery of LCM virus was the development of clinical signs in apparently healthy animals following intracerebral inoculation of sterile broth. Similar findings were subsequently reported by Findlay *et al.* (1936) and Lépine and Sautter (1936). Later, Traub (1936) himself could not provoke disease in carrier mice, and in the following years, workers in numerous laboratories have tried by various means to convert a persistent infection with LCM virus into illness. Only the special procedure of adoptive immunization led sometimes but not always (see Section IV,C) to clinical signs; all other attempts were unsuccessful. We do not know why in earlier reports injury to the brain resulted in overt disease, but it should be emphasized that the intracerebral inoculation of inert materials such as bacterial broth or starch is not a procedure to detect LCM virus carrier mice.

If a mouse colony is found to be infected, measures to be taken depend on the degree at the time the virus is detected. If a stage is encountered in which there are only carriers, all mice should be destroyed. If the colony is of great value—and provided syngeneic, noninfected mice are available—an attempt could be made with adoptive immunization (Volkert and Hannover Larsen, 1965). However, whether mice so treated remain fertile is not known.

The outlook is less bleak if LCM virus is detected soon after its introduction into the colony. In this case, nonviremic breeders should be selected and a new breeding colony established. It is only necessary to avoid true carriers and mice undergoing an acute infection, because previous contact with the virus leading to immunity would not matter. If the supply of mice is to continue, a more elaborate procedure may be adopted (Hoyland *et al.*, 1972).

To keep a mouse colony free from LCM virus is a simple task. It merely requires preventing wild house mice from entering the premises; no other animal or man is likely to act as a vector. In the author's opinion, special precautions, as sometimes proposed (Cooper, 1973), need not be taken. The same applies to hamster colonies.

ACKNOWLEDGMENTS

The excellent secretarial help of Miss Evelyne Danckers is greatly appreciated. Work done in the author's laboratory was financially supported by the Deutsche Forschungsgemeinschaft.

The Heinrich-Pette-Institut is financially supported by Freie und Hansestadt Hamburg and Bundesministerium für Jugend, Familie und Gesundheit, Bonn.

REFERENCES

Abelson, H. T., Smith, G. H., Hoffman, H. A., and Rowe, W. P. (1969). Use of enzyme-labeled antibody for electron microscope localization of lymphocytic choriomeningitis virus antigens in infected cell cultures. *J. Natl. Cancer Inst.* **42,** 497–515.

Accinni, L., Archetti, I., Branca, M., Hsu, K. C., and Andres, G. (1978). Tubulo-interstitial (TI) renal disease associated with chronic lymphocytic choriomeningitis viral infection in mice. *Clin. Immunol. Immunopathol.* **11,** 395–405.

Ackermann, R. (1973). Epidemiologic aspects of lymphocytic choriomeningitis in man. *In* "Lymphocytic Choriomeningitis Virus and Other Arenaviruses" (F. Lehmann-Grube, ed.), pp. 233–237. Springer-Verlag, Berlin and New York.

Ackermann, R., Bloedhorn, H., Küpper, B., Winkens, I., and Scheid, W. (1964). Über die Verbreitung des Virus der Lymphocytären Choriomeningitis unter den Mäusen in Westdeutschland. I. Untersuchungen überwiegend an Hausmäusen (Mus musculus). *Zentralbl. Bakteriol., Parasitenkd., Infektionskr. Hyg., Abt. 1: Orig.* **194,** 407–430.

Ackermann, R., Stille, W., Blumenthal, W., Helm, E. B., Keller, K., and Baldus, O. (1972). Syrische Goldhamster als Überträger von Lymphozytärer Choriomeningitis. *Dtsch. Med. Wochenschr.* **97,** 1725–1731.

Ackermann, R., Körver, G., Turss, R., Wönne, R., and Hochgesand, P. (1974). Pränatale Infektion mit dem Virus der Lymphozytären Choriomeningitis. Bericht über zwei Fälle. *Dtsch. Med. Wochenschr.* **99,** 629–632.

Andzhaparidze, O. G., Boriskin, Y. S., Bogomolova, N. N., Bektemirov, T. A., Sheinbergas, M. M., and Lotte, V. D. (1977). Persistencija antigena arenavirusa v perevivaemyh kletočnyh linijah. Popytki vydelenija infekcionnogo virusa. (Persistence of arenavirus antigen in continuous cell lines. Attempts at isolation of an infectious virus.) *Vopr. Virusol.* No. 5, 557–561.

Anonymous (1974). Follow-up on hamster-associated LCM infection—United States. *Morbid. Mortal. Wkly. Rep.* **23,** 131–132.

Armstrong, C. (1941). Studies on choriomeningitis and poliomyelitis. *Bull. N.Y. Acad. Med.* **17,** 295–318.

Armstrong, C., and Dickens, P. F. (1935). Benign lymphocytic choriomeningitis (acute aseptic meningitis). A new disease entity. *Public Health Rep.* **50,** 831–842.

Armstrong, C., and Lillie, R. D. (1934). Experimental lymphocytic choriomeningitis of monkeys and mice produced by a virus encountered in studies of the 1933 St. Louis encephalitis epidemic. *Public Health Rep.* **49,** 1019–1027.

Armstrong, C., and Sweet, L. K. (1939). Lymphocytic choriomeningitis. Report of two cases, with recovery of the virus from gray mice (*Mus musculus*) trapped in the two infected households. *Public Health Rep.* **54,** 673–684.

Armstrong, D., Fortner, J. G., Rowe, W. P., and Parker, J. C. (1969). Meningitis due to lymphocytic choriomeningitis virus endemic in a hamster colony. *J. Am. Med. Assoc.* **209,** 265–266.

Baker, F. D., and Hotchin, J. (1967). Slow virus kidney disease of mice. *Science* **158,** 502–504.

Barlow, J. L. (1964). Hyperreactivity to endotoxin in mice infected with lymphocytic choriomeningitis virus. *In* "Bacterial Endotoxins" (M. Landy and W. Braun, eds.), pp. 448–454. Quinn & Boden, Rahway, New Jersey.

Baum, S. G., Lewis, A. M., Rowe, W. P., and Huebner, R. J. (1966). Epidemic nonmeningitic lymphocytic-choriomeningitis-virus infection. An outbreak in a population of laboratory personnel. *N. Engl. J. Med.* **274,** 934–936.

Bayer, P., and Gear, J. (1955). Virus meningo-encephalitis in South Africa. A study of the cases admitted to the Johannesburg Fever Hospital. *S. Afr. J. Lab. Clin. Med.* **1,** 22–35.

Benda, R., Daneš, L., and Fuchsová, M. (1964). Experimental inhalation infection of guinea-pigs with the virus of lymphocytic choriomeningitis. *J. Hyg., Epidemiol., Microbiol., Immunol.* **8,** 87–99.

Benson, L., and Hotchin, J. (1969). Antibody formation in persistent tolerant infection with lymphocytic choriomeningitis virus. *Nature (London)* **222,** 1045–1047.

Biggar, R. J., Douglas, R. G., and Hotchin, J. (1975a). Lymphocytic choriomeningitis associated with hamsters. *Lancet* **1,** 856–857.

Biggar, R. J., Woodall, J. P., Walter, P. D., and Haughie, G. E. (1975b). Lymphocytic choriomeningitis outbreak associated with pet hamsters. Fifty-seven cases from New York State. *J. Am. Med. Assoc.* **232,** 494–500.

Biggar, R. J., Deibel, R., and Woodall, J. P. (1976). Implications, monitoring, and control of accidental transmission of lymphocytic choriomeningitis virus within hamster tumor cell lines. *Cancer Res.* **36,** 551–553.

Biggar, R. J., Schmidt, T. J., and Woodall, J. P. (1977). Lymphocytic choriomeningitis in laboratory personnel exposed to hamsters inadvertently infected with LCM virus. *J. Am. Vet. Med. Assoc.* **171,** 829–832.

Billingham, R. E., Brent, L., and Medawar, P. B. (1953). "Actively acquired tolerance" of foreign cells. *Nature (London)* **172,** 603–606.

Blanden, R. V., and Mims, C. A. (1973). Macrophage activation in mice infected with ectromelia or lymphocytic choriomeningitis viruses. *Aust. J. Exp. Biol. Med. Sci.* **51,** 393–398.

Blanden, R. V., Doherty, P. C., Dunlop, M. B. C., Gardner, I. D., Zinkernagel, R. M., and David, C. S. (1975). Genes required for cytotoxicity against virus-infected target cells in K and D regions of H-2 complex. *Nature (London)* **254,** 269–270.

Blanden, R. V., Dunlop, M. B. C., Doherty, P. C., Kohn, H. I., and McKenzie, I. F. C. (1976). Effects of four H-2K mutations on virus-induced antigens recognized by cytotoxic T cells. *Immunogenetics* **3,** 541–548.

Blanden, R. V., Kees, V., and Dunlop, M. B. C. (1977). *In vitro* primary induction of cytotoxic T cells against virus-infected syngeneic cells. *J. Immunol. Methods* **16,** 73–89.

Bro-Jørgensen, K. (1971). Characterization of virus-specific antigen in cell culture infected with lymphocytic choriomeningitis virus. *Acta Pathol. Microbiol. Scand., Sect. B* **79,** 466–474.

Bro-Jørgensen, K. (1978). The interplay between lymphocytic choriomeningitis virus, immune function, and hemopoiesis in mice. *Adv. Virus Res.* **22,** 327–369.

Bro-Jørgensen, K., and Knudtzon, S. (1977). Changes in hemopoiesis during the course of acute LCM virus infection in mice. *Blood* **49,** 47–57.

Bro-Jørgensen, K., and Volkert, M. (1972a). Haemopoietic defects in mice infected with lymphocytic choriomeningitis virus. 1. The enhanced X-ray sensitivity of virus infected mice. *Acta Pathol. Microbiol. Scand., Sect. B* **80,** 845–852.

Bro-Jørgensen, K., and Volkert, M. (1972b). Haemopoietic defects in mice infected with lymphocytic choriomeningitis virus. 2. The viral effect upon the function of colony-forming stem cells. *Acta Pathol. Microbiol. Scand., Sect. B* **80,** 853–862.

Bro-Jørgensen, K., and Volkert, M. (1974). Defects in the immune system of mice infected with lymphocytic choriomeningitis virus. *Infect. Immun.* **9,** 605–614.

Bro-Jørgensen, K., Güttler, F., Jørgensen, P. N., and Volkert, M. (1975). T lymphocyte function as the principal target of lymphocytic choriomeningitis virus-induced immunosuppression. *Infect. Immun.* **11,** 622–629.

Buchmeier, M. J., and Oldstone, M. B. A. (1978a). Virus-induced immune complex disease: identification of specific viral antigens and antibodies deposited in complexes during chronic lymphocytic choriomeningitis virus infection. *J. Immunol.* **120,** 1297–1304.

Buchmeier, M. J., and Oldstone, M. B. A. (1978b). Identity of the viral protein responsible for serologic cross reactivity among the Tacaribe complex arenaviruses. *In* "Negative Strand Viruses and the Host Cell" (B. W. J. Mahy and R. D. Barry, eds.), pp. 91–97. Academic Press, New York.

Buchmeier, M. J., and Oldstone, M. B. A. (1979). Protein structure of lymphocytic choriomeningitis virus: evidence for a cell-associated precursor of the virion glycopeptides. *Virology* **99,** 111–120.

Buchmeier, M. J., Elder, J. H., and Oldstone, M. B. A. (1978). Protein structure of lymphocytic choriomeningitis virus: identification of the virus structural and cell associated polypeptides. *Virology* **89,** 133–145.

Buchmeier, M. J., Welsh, R. M., Dutko, F. J., and Oldstone, M. B. A. (1980). The virology and immunobiology of lymphocytic choriomeningitis virus infection. *Adv. Immunol.* **30,** 275–331.

Burnet, F. M. (1955). "Principles of Animal Virology." Academic Press, New York.

Burnet, F. M. (1959). "The Clonal Selection Theory of Acquired Immunity." Cambridge Univ. Press, Cambridge, U.K.

Burnet, F. M., and Fenner, F. (1949). "The Production of Antibodies," 2nd ed. Macmillan, New York.

Camenga, D. L., Walker, D. H., and Murphy, F. A. (1977). Anticonvulsant prolongation of survival in adult murine lymphocytic choriomeningitis. I. Drug treatment and virologic studies. *J. Neuropathol. Exp. Neurol.* **36,** 9–20.

Cantor, H., and Boyse, E. A. (1977). Lymphocytes as models for the study of mammalian cellular differentiation. *Immunol. Rev.* **33,** 105–124.

Casals, J. (1975). Arenaviruses. *Yale J. Biol. Med.* **48,** 115–140.

Caspar, D. L. D., Dulbecco, R., Klug, A., Lwoff, A., Stoker, M. G. P., Tournier, P., and Wildy, P. (1962). Proposals. *Cold Spring Harbor Symp. Quant. Biol.* **27,** 49–50.

Chastel, C., Mabin, D., and Barthelemy, L. (1978). Chorioméningite lymphocytaire expérimentale de la souris. Corrélations anatomo-cliniques, virologiques et électriques. *Pathol. Biol.* **26,** 467–473.

Cherednichenko, Y. N., Mikhailova, G. R., Gaidamovich, S. Y., Monastyreva, L. A., and Demidova, S. A. (1977). Persistencija virusa limfocitarnogo horiomeningita v perevivaemoj linii kletok Detroit-6. (Persistence of lymphocytic choriomeningitis virus in continuous Detroit-6 cell line.) *Vopr. Virusol.* No. 6, 703–707.

Christoffersen, P. J., and Bro-Jørgensen, K. (1977). Erythropoietic activity and interferon production in LCM virus-infected nude mice. *Acta Pathol. Microbiol. Scand., Sect. B* **85,** 435–439.

Christoffersen, P. J., Volkert, M., and Rygaard, J. (1976). Immunological unresponsiveness of nude mice to LCM virus infection. *Acta Pathol. Microbiol. Scand., Sect. C* **84,** 520–523.

Cihak, J., and Lehmann-Grube, F. (1974). Persistent infection of mice with the virus of lymphocytic choriomeningitis: virus-specific immunological tolerance. *Infect. Immun.* **10,** 1072–1076.

Cihak, J., and Lehmann-Grube, F. (1978). Immunological tolerance to lymphocytic choriomeningitis virus in neonatally infected virus carrier mice: evidence supporting a clonal inactivation mechanism. *Immunology* **34,** 265–275.

Cole, G. A., and Johnson, E. D. (1975). Immune responses to LCM virus infection *in vivo* and *in vitro*. Mechanisms of immune-mediated disease. *Bull. W. H. O.* **52,** 465–469.

Cole, G. A., and Nathanson, N. (1974). Lymphocytic choriomeningitis. Pathogenesis. *Prog. Med. Virol.* **18,** 94–110.

Cole, G. A., Gilden, D. H., Monjan, A. A., and Nathanson, N. (1971). Lymphocytic choriomeningitis virus: pathogenesis of acute central nervous system disease. *Fed. Proc., Fed. Am. Soc. Exp. Biol.* **30,** 1831–1841.

Cole, G. A., Nathanson, N., and Prendergast, R. A. (1972). Requirement for θ-bearing cells in lymphocytic choriomeningitis virus-induced central nervous system disease. *Nature (London)* **238,** 335–337.

Cole, G. A., Prendergast, R. A., and Henney, C. S. (1973). *In vitro* correlates of LCM virus-induced immune response. *In* "Lymphocytic Choriomeningitis Virus and Other Arenaviruses" (F. Lehmann-Grube, ed.), pp. 61–71. Springer-Verlag, Berlin and New York.

Cooper, J. E. (1973). Maintenance of a lymphocytic choriomeningitis-free mouse unit in Kenya. *Trop. Anim. Health Prod.* **5,** 174–180.

Dalton, A. J., Rowe, W. P., Smith, G. H., Wilsnack, R. E., and Pugh, W. E. (1968). Morphological and cytochemical studies on lymphocytic choriomeningitis virus. *J. Virol.* **2,** 1465–1478.

Daneš, L., Benda, R., and Fuchsová, M. (1963). Experimentální inhalační nákaza opic druhů Macacus cynomolgus a Macacus rhesus virem lymfocitárni choriomeningitidy (kmenem WE). *Bratisl. Lek. Listy* **43,** 71–79.

Dixon, F. J., Oldstone, M. B. A., and Tonietti, G. (1971). Pathogenesis of immune complex glomerulonephritis of New Zealand mice. *J. Exp. Med.* **134,** 65s–71s.

Doherty, P. C., and Zinkernagel, R. M. (1974). T-cell-mediated immunopathology in viral infections. *Transplant. Rev.* **19,** 89–120.

Doherty, P. C., and Zinkernagel, R. M. (1975a). Capacity of sensitized thymus-derived lymphocytes to induce fatal lymphocytic choriomeningitis is restricted by the H-2 gene complex. *J. Immunol.* **114,** 30–33.

Doherty, P. C., and Zinkernagel, R. M. (1975b). H-2 compatibility is required for T-cell-mediated lysis of target cells infected with lymphocytic choriomeningitis virus. *J. Exp. Med.* **141,** 502–507.

Doherty, P. C., and Zinkernagel, R. M. (1975c). Enhanced immunological surveillance in mice heterozygous at the H-2 gene complex. *Nature (London)* **256,** 50–52.

Doherty, P. C., Zinkernagel, R. M., and Ramshaw, I. A. (1974). Specificity and development of cytotoxic thymus-derived lymphocytes in lymphocytic choriomeningitis. *J. Immunol.* **112,** 1548–1552.

Doherty, P. C., Blanden, R. V., and Zinkernagel, R. M. (1976a). Specificity of virus-immune effector T cells for H-2K or H-2D compatible interactions: implications for H-antigen diversity. *Transplant. Rev.* **29,** 89–124.

Doherty, P. C., Dunlop, M. B. C., Parish, C. R., and Zinkernagel, R. M. (1976b). Inflammatory process in murine lymphocytic choriomeningitis is maximal in H-2K or H-2D compatible interactions. *J. Immunol.* **117,** 187–190.

Doherty, P. C., Solter, D., and Knowles, B. B. (1977). H-2 gene expression is required for T cell-mediated lysis of virus-infected target cells. *Nature (London)* **266,** 361–362.

Doyle, L. B., Doyle, M. V., and Oldstone, M. B. A. (1980). Susceptibility of newborn mice with H-2^k backgrounds to lymphocytic choriomeningitis virus infection. *Immunology* **40,** 589–596.

Doyle, M. V., and Oldstone, M. B. A. (1978). Interactions between viruses and lymphocytes. I. *In vivo* replication of lymphocytic choriomeningitis virus in mononuclear cells during both chronic and acute viral infections. *J. Immunol.* **121,** 1262–1269.

Dunlop, M. B. C. (1978). Secondary cytotoxic cell response to lymphocytic choriomeningitis virus. III. *In vivo* protective activity of effector cells generated *in vitro*. *Immunology* **34,** 291–302.

Dunlop, M. B. C., and Blanden, R. V. (1976). Secondary cytotoxic cell response to lymphocytic choriomeningitis virus. I. Kinetics of induction *in vitro* and yields of effector cells. *Immunology* **31,** 171–180.

Dunlop, M. B. C., and Blanden, R. V. (1977a). Mechanisms of suppression of cytotoxic T-cell responses in murine lymphocytic choriomeningitis virus infection. *J. Exp. Med.* **145,** 1131–1143.

Dunlop, M. B. C., and Blanden, R. V. (1977b). Induction of a primary cytotoxic T cell response to lymphocytic choriomeningitis virus-infected cells *in vitro*. I. Kinetics of response and nature of effector cells. *Cell. Immunol.* **28,** 190–197.

Dunlop, M. B. C., Doherty, P. C. Zinkernagel, R. M., and Blanden, R. V. (1976). Secondary cytotoxic cell response to lymphocytic choriomeningitis virus. II. Nature and specificity of effector cells. *Immunology* **31,** 181–186.

Dunlop, M. B. C., Doherty, P. C., Zinkernagel, R. M., and Blanden, R. V. (1977). Cytotoxic T cell response to lymphocytic choriomeningitis virus. Properties of precursors of effector T cells, primary effector T cells and memory T cells *in vitro* and *in vivo*. *Immunology* **33,** 361–368.

Dutko, F. J., and Pfau, C. J. (1978). Arenavirus defective interfering particles mask the cell-killing potential of standard virus. *J. Gen. Virol.* **38,** 195–208.

Fenner, F. (1976). Classification and nomenclature of viruses. Second report of the International Committee on Taxonomy of Viruses. *Intervirology* **7,** 1–115.

Findlay, G. M., Alcock, N. S., and Stern, R. O. (1936). The virus aetiology of one form of lymphocytic meningitis. *Lancet* **1,** 650–654.

Förster, U. (1973). Zur Anwendung der Immunofluoreszenz in der Routinediagnostik der Lymphozytären Choriomeningitis. *Zentralbl. Veterinaermed., Reihe B* **20,** 622–628.

Förster, U., and Wachendörfer, G. (1973). Inapparent infection of Syrian hamsters with the virus of lymphocytic choriomeningitis. *In* "Lymphocytic Choriomeningitis Virus and Other Arenaviruses" (F. Lehmann-Grube, ed.), pp. 113–120. Springer-Verlag, Berlin and New York.

Freye, H.-A., and Freye, H. (1969). "Die Hausmaus." Ziemsen Verlag, Wittenberg Lutherstadt, German Democratic Republic.

Gaidamovich, S. Y., and Kocherovskaya, M. Y. (1976). Ostraja i hroničeskaja infekcija myšej, vyzvannaja virusom limfocitarnogo horiomeningita, vydelennym v moskve ot Mus musculus. (Acute and chronic infection of mice caused by lymphocytic choriomeningitis virus isolated from Mus musculus in Moscow.) *Vopr. Virusol.* No. 2, 223–226.

Gaidamovich, S. Y., Cherednichenko, Y. N., and Zhdanov, V. M. (1978). On the mechanism of the persistence of lymphocytic choriomeningitis virus in the continuous cell line Detroit-6. *Intervirology* **9,** 156–161.

Gasser, D. L., and Silvers, W. K. (1974). Genetic determinants of immunological responsiveness. *Adv. Immunol.* **18,** 1–66.

Gershon, R. K., and Kondo, K. (1971). Infectious immunological tolerance. *Immunology* **21,** 903–914.

Gilden, D. H., Cole, G. A., Monjan, A. A., and Nathanson, N. (1972a). Immunopathogenesis of acute central nervous system disease produced by lymphocytic choriomeningitis virus. I. Cyclophosphamide-mediated induction of the virus-carrier state in adult mice. *J. Exp. Med.* **135,** 860–873.

Gilden, D. H., Cole, G. A., and Nathanson, N. (1972b). Immunopathogenesis of acute central nervous system disease produced by lymphocytic choriomeningitis virus. II. Adoptive immunization of virus carriers. *J. Exp. Med.* **135,** 874–889.

Gresser, I., Tovey, M. G., Maury, C., and Chouroulinkov, I. (1975). Lethality of interferon preparations for newborn mice. *Nature (London)* **258,** 76–78.

Gresser, I., Morel-Maroger, L., Verroust, P., Rivière, I., and Guillon, J.-C. (1978). Anti-interferon globulin inhibits the development of glomerulonephritis in mice infected at birth with lymphocytic choriomeningitis virus. *Proc. Natl. Acad. Sci. U.S.A.* **75,** 3413–3416.

Gschwender, H. H., and Popescu, M. (1976). Equilibrium sedimentation of virus in density gradients of iodinated compounds. *In* "Biological Separations in Iodinated Density-Gradient Media" (D. Rickwood, ed.), pp. 145–158. Inf. Retr., London.

Gschwender, H. H., Rutter, G., and Popescu, M. (1975). Use of iodinated organic compounds for the density gradient centrifugation of viruses. *Arch. Virol.* **49,** 359–364.

Gschwender, H. H., Rutter, G., and Lehmann-Grube, F. (1976). Lymphocytic choriomeningitis virus. II. Characterization of extractable complement-fixing activity. *Med. Microbiol. Immunol.* **162,** 119–131.

Güttler, F., Bro-Jørgensen, K., and Jørgensen, P. N. (1975). Transient impaired cell-mediated tumor immunity after acute infection with lymphocytic choriomeningitis virus. *Scand. J. Immunol.* **4,** 327–336.

Haas, V. H. (1941). Studies on the natural history of the virus of lymphocytic choriomeningitis in mice. *Public Health Rep.* **56,** 285–292.

Haas, V. H. (1954). Some relationships between lymphocytic choriomeningitis (LCM) virus and mice. *J. Infect. Dis.* **94,** 187–198.

Hanaoka, M., Suzuki, S., and Hotchin, J. (1969). Thymus-dependent lymphocytes: destruction by lymphocytic choriomeningitis virus. *Science* **163,** 1216–1219.

Hannover Larsen, J. (1969). On the induction of immunological tolerance to a self-reproducing antigen. *Immunology* **16,** 15–23.

Hapel, A. J., Bablanian, R., and Cole, G. A. (1978). Inductive requirements for the generation of virus-specific T lymphocytes. I. The nature of the host cell-virus interaction that triggers secondary poxvirus-specific cytotoxic T lymphocyte induction. *J. Immunol.* **121,** 736–743.

Hellström, K. E., and Hellström, I. (1974). Lymphocyte-mediated cytotoxicity and blocking serum activity to tumor antigens. *Adv. Immunol.* **18,** 209–277.

Herberman, R. B., Nunn, M. E., Holden, H. T., Staal, S., and Djeu, J. Y. (1977). Augmentation of natural cytotoxic reactivity of mouse lymphoid cells against syngeneic and allogeneic target cells. *Int. J. Cancer* **19,** 555–564.

Hinman, A. R., Fraser, D. W., Douglas, R. G., Bowen, G. S., Kraus, A. L., Winkler, W. G., and Rhodes, W. W. (1975). Outbreak of lymphocytic choriomeningitis virus infections in medical center personnel. *Am. J. Epidemiol.* **101,** 103–110.

Hirsch, M. S., Moellering, R. C., Pope, H. G., and Poskanzer, D. C. (1974). Lymphocytic-choriomeningitis-virus infection traced to a pet hamster. *N. Engl. J. Med.* **291,** 610–612.

Hoffsten, P. E., and Dixon, F. J. (1973). Effect of immunosuppression on chronic LCM virus infection of mice. *J. Exp. Med.* **138,** 887–899.

Hoffsten, P. E., Oldstone, M. B. A., and Dixon, F. J. (1977). Immunopathology of adoptive immunization in mice chronically infected with lymphocytic choriomeningitis virus. *Clin. Immunol. Immunopathol.* **7,** 44–52.

Holland, J. J., Villarreal, L. P., Welsh, R. M., Oldstone, M. B. A., Kohne, D., Lazzarini, R., and Scolnick, E. (1976). Long-term persistent vesicular stomatitis virus and rabies virus infection of cells *in vitro*. *J. Gen. Virol.* **33,** 193–211.

Holtermann, O. A., and Havell, E. A. (1970). Reduced interferon response in mice congenitally infected with lymphocytic choriomeningitis virus. *J. Gen. Virol.* **9,** 101–103.

Hoitermann, O. A., and Majde, J. A. (1971). An apparent histoincompatibility between mice chronically infected with lymphocytic choriomeningitis virus and their uninfected syngeneic counterparts. *Transplantation* **11,** 20–29.

Hotchin, J. (1959). Discussion. *In* "'Allergic' Encephalomyelitis" (M. W. Kies and E. C. Alvord, eds.), pp. 511–513. Thomas, Springfield, Illinois.

Hotchin, J. (1962a). The biology of lymphocytic choriomeningitis infection: virus-induced immune disease. *Cold Spring Harbor Symp. Quant. Biol.* **27,** 479–499.

Hotchin, J. (1962b). The foot pad reaction of mice to lymphocytic choriomeningitis virus. *Virology* **17,** 214–216.

Hotchin, J. (1971). "Persistent and Slow Virus Infections," Monographs in Virology, Vol. 3. Karger, Basel.

Hotchin, J. (1972). Slow viruses and neurological damage. *Monogr. Hum. Genet.* **6,** 172–181.

Hotchin, J. (1973). Transient virus infection: spontaneous recovery mechanism of lymphocytic choriomeningitis virus-infected cells. *Nature (London), New Biol.* **241,** 270–272.

Hotchin, J. (1974). The role of transient infection in arenavirus persistence. *Prog. Med. Virol.* **18,** 81–93.

Hotchin, J., and Benson, L. (1963). The pathogenesis of lymphocytic choriomeningitis in mice: the effects of different inoculation routes and the footpad response. *J. Immunol.* **91,** 460–468.

Hotchin, J. E., and Cinits, M. (1958). Lymphocytic choriomeningitis infection of mice as a model for the study of latent virus infection. *Can. J. Microbiol.* **4,** 149–163.

Hotchin, J., and Collins, D. N. (1964). Glomerulonephritis and late onset disease of mice following neonatal virus infection. *Nature (London)* **203,** 1357–1359.

Hotchin, J., and Seegal, R. (1977). Virus-induced behavioral alteration of mice. *Science* **196,** 671–674.

Hotchin, J., and Sikora, E. (1973). Low-pathogenicity variant of lymphocytic choriomeningitis virus. *Infect. Immun.* **7,** 825–826.

Hotchin, J., Benson, L. M., and Seamer, J. (1962). Factors affecting the induction of persistent tolerant infection of newborn mice with lymphocytic choriomeningitis. *Virology* **18,** 71–78.

Hotchin, J., Benson, L., and Gardner, J. (1970). Mother–infant interaction in lymphocytic choriomeningitis virus infection of the newborn mouse: the effect of maternal health on mortality of offspring. *Pediatr. Res.* **4,** 194–200.

Hotchin, J., Kinch, W., and Benson, L. (1971). Lytic and turbid plaque-type mutants of lymphocytic choriomeningitis virus as a cause of neurological disease or persistent infection. *Infect. Immun.* **4,** 281–286.

Hotchin, J., Kinch, W., Benson, L., and Sikora, E. (1975). Role of substrains in persistent lymphocytic choriomeningitis virus infection. *Bull. W. H. O.* **52,** 457–462.

Howard, J. G., and Mitchison, N. A. (1975). Immunological tolerance. *Prog. Allergy* **18,** 43–96.

Hoyland, F., Knight, E. H., and Skinner, H. H. (1972). Eradication of lymphocytic choriomeningitis (LCM) virus—a human pathogen—from a mouse breeding colony. *J. Inst. Anim. Technol.* **23,** 6–23.

Huang, A. S., and Baltimore, D. (1977). Defective interfering animal viruses. *Compr. Virol.* **10,** 73–116.

Jacobs, R. P., and Cole, G. A. (1976). Lymphocytic choriomeningitis virus-induced immunosuppression: a virus-induced macrophage defect. *J. Immunol.* **117,** 1004–1009.

Jacobson, S., and Pfau, C. J. (1980). Viral pathogenesis and resistance to defective interfering particles. *Nature (London)* **283,** 311–313.

Jacobson, S., Dutko, F. J., and Pfau, C. J. (1979). Determinants of spontaneous recovery and persistence in MDCK cells infected with lymphocytic choriomeningitis virus. *J. Gen. Virol.* **44,** 113–122.

Johnson, E. D., and Cole, G. A. (1975). Functional heterogeneity of lymphocytic choriomeningitis virus-specific T lymphocytes. I. Identification of effector and memory subsets. *J. Exp. Med.* **141,** 866–881.

Johnson, E. D., Monjan, A. A., and Morse, H. C. (1978). Lack of B-cell participation in acute lymphocytic choriomeningitis disease of the central nervous system. *Cell. Immunol.* **36,** 143–150.

Johnson, K. M. (1965). Epidemiology of Machupo virus infection. III. Significance of virological observations in man and animals. *Am. J. Trop. Med. Hyg.* **14,** 816–818.

Kajima, M., and Majde, J. (1970). LCM virus as a carrier of non-viral cellular components. Electron microscopic evidence. *Naturwissenschaften* **57,** 93.

Kajima, M., and Pollard, M. (1970). Ultrastructural pathology of glomerular lesions in gnotobiotic mice with congenital lymphocytic choriomeningitis (LCM) virus infection. *Am. J. Pathol.* **61,** 117–140.

Kimmig, W., and Lehmann-Grube, F. (1979). The immune response of the mouse to lymphocytic choriomeningitis virus. I. Circulating antibodies. *J. Gen. Virol.* **45,** 703–710.

Klein, B., Hill, C., and Hoffsten, P. E. (1977). Studies of the immunoglobulin eluted from the glomeruli of mice chronically infected with lymphocytic choriomeningitis virus. *J. Immunol.* **119,** 707–713.

Koszinowski, U., Gething, M. J., and Waterfield, M. (1977). T-cell cytotoxicity in the absence of viral protein synthesis in target cells. *Nature (London)* **267,** 160–163.

Lampert, P. W., and Oldstone, M. B. A. (1974). Pathology of the choroid plexus in spontaneous immune complex disease and chronic viral infections. *Virchows Arch. A: Pathol. Anat. Histol.* **363,** 21–32.

Lehmann-Grube, F. (1964a). Lymphocytic choriomeningitis in the mouse. I. Growth in the brain. *Arch. Gesamte Virusforsch.* **14,** 344–350.

Lehmann-Grube, F. (1964b). Lymphocytic choriomeningitis in the mouse. II. Establishment of carrier colonies. *Arch. Gesamte Virusforsch.* **14,** 351–357.

Lehmann-Grube, F. (1967). A carrier state of lymphocytic choriomeningitis virus in L cell cultures. *Nature (London)* **213,** 770–773.

Lehmann-Grube, F. (1969). Lymphocytic choriomeningitis in the mouse. III. Comparative titrations of virus strains in inbred mice. *Arch. Gesamte Virusforsch.* **28,** 303–307.

Lehmann-Grube, F. (1971). "Lymphocytic Choriomeningitis Virus," Virology Monographs, Vol. 10. Springer-Verlag, Wien and New York.

Lehmann-Grube, F. (1972). Persistent infection of the mouse with the virus of lymphocytic choriomeningitis. *J. Clin. Pathol., Suppl. (R. Coll. Pathol.)* No. 6, 8–21.

Lehmann-Grube, F. (1975). Pathogenesis of disease associated with persistent infection of the mouse with the LCM virus. *In* "Proceedings of the First Intersectional Congress of IAMS," Tokyo, 1975 (T. Hasegawa, ed.), Vol. 3, pp. 125–142. Science Council of Japan.

Lehmann-Grube, F. (1977). Lymphocytic choriomeningitis virus carrier mice. Factors determining virus persistence. *Medicina (Buenos Aires)* **37,** Suppl. 3, 78–89.

Lehmann-Grube, F. (1981a). Lymphocytic choriomeningitis. *In* "International Textbook of Medicine, Vol. II, Medical Microbiology and Infectious Diseases" (A. I. Braude, ed.), pp. 1254–1258. Saunders Co., Philadelphia, Pennsylvania.

Lehmann-Grube, F. (1981b). Arenaviruses. *In* "International Textbook of Medicine, Vol. II, Medical Microbiology and Infectious Diseases" (A. I. Braude, ed.), pp. 602–610. Saunders Co., Philadelphia, Pennsylvania.

Lehmann-Grube, F., and Ambrassat, J. (1977). A new method to detect lymphocytic choriomeningitis virus-specific antibody in human sera. *J. Gen. Virol.* **37,** 85–92.

Lehmann-Grube, F., and Löhler, J. (1981). Immunopathologic alterations of lymphatic tissues of mice infected with lymphocytic choriomeningitis virus. II. Pathogenetic mechanism. *Lab. Invest.* **44,** 205–213.

Lehmann-Grube, F., Slenczka, W., and Tees, R. (1969). A persistent and inapparent infection of L cells with the virus of lymphocytic choriomeningitis. *J. Gen. Virol.* **5,** 63–81.

Lehmann-Grube, F., Niemeyer, I., and Löhler, J. (1972). Lymphocytic choriomeningitis of the mouse. IV. Depression of the allograft reaction. *Med. Microbiol. Immunol.* **158,** 16–25.

Lehmann-Grube, F., Popescu, M., Schaefer, H., and Gschwender, H. H. (1975). LCM virus infection of cells *in vitro*. *Bull. W. H. O.* **52,** 443–455.

Lehmann-Grube, F., Ibscher, B., Bugislaus, E., and Kallay, M. (1979a). Untersuchungen über die Rolle des Goldhamsters (*Mesocricetus auratus*) bei der Übertragung der Lymphozytären Choriomeningitis auf den Menschen. *Med. Microbiol. Immunol.* **167,** 205–210.

Lehmann-Grube, F., Kallay, M., Ibscher, B., and Schwartz, R. (1979b). Serologic diagnosis of human infections with lymphocytic choriomeningitis virus: comparative evaluation of seven methods. *J. Med. Virol.* **4,** 125–136.

Lépine, P., and Sautter, V. (1936). Existence en France du virus murin de la chorio-méningite lymphocytaire. *C. R. Hebd. Seances Acad. Sci.* **202,** 1624–1626.

Leung, K.-N., Ada, G. L., and McKenzie, I. F. C. (1980). Specificity, Ly phenotype, and H-2 compatibility requirements of effector cells in delayed-type hypersensitivity responses to murine influenza virus infection. *J. Exp. Med.* **151,** 815–826.

Lewis, A. M., Rowe, W. P., Turner, H. C., and Huebner, R. J. (1965). Lymphocytic-choriomeningitis virus in hamster tumor: spread to hamsters and humans. *Science* **150,** 363–364.

Lillie, R. D., and Armstrong, C. (1945). Pathology of lymphocytic choriomeningitis in mice. *Arch. Pathol.* **40,** 141–152.

Löhler, J., and Lehmann-Grube, F. (1981). Immunopathologic alterations of lymphatic tissues of mice infected with lymphocytic choriomeningitis virus. I. Histopathologic findings. *Lab. Invest.* **44,** 193–204.

Löhler, J., Ehlerding, I., and Lehmann-Grube, F. (1974). Pathologie des lymphatischen Gewebes bei der Lymphozytären Choriomeningitis der Maus. *Zentralbl. Bakteriol., Parasitenkd. Infektionskr. Hyg., Abt. 1: Orig., Reihe A* **227,** 458–468.

Löhler, J., Schwendemann, G., and Lehmann-Grube, F. (1980). Lymphocytic choriomeningitis of mice. Pathogenesis of the chronic disease. *In* "Search for the Cause of Multiple Sclerosis and other Chronic Diseases of the Central Nervous System" (A. Boese, ed.), pp. 202–213. Verlag Chemie, Weinheim.

Lundstedt, C. (1969a). Interaction between antigenically different cells. Virus-induced cytotoxicity by immune lymphoid cells *in vitro*. *Acta Pathol. Microbiol. Scand.* **75,** 139–152.

Lundstedt, C. (1969b). Effect of antilymphocyte serum on adoptive immunization of lymphocytic choriomeningitis virus carrier mice. *Acta Pathol. Microbiol. Scand.* **77,** 518–526.

McKenzie, I. F. C., and Potter, T. (1979). Murine lymphocyte surface antigens. *Adv. Immunol.* **27,** 179–338.

Maetz, H. M., Sellers, C. A., Bailey, W. C., and Hardy, G. E. (1976). Lymphocytic choriomeningitis from pet hamster exposure: a local public health experience. *Am. J. Public Health* **66,** 1082–1085.

Mannweiler, K., and Lehmann-Grube, F. (1973). Electron microscopy of LCM virus-infected L cells. *In* "Lymphocytic Choriomeningitis Virus and Other Arenaviruses" (F. Lehmann-Grube, ed.), pp. 37–48. Springer-Verlag, Berlin and New York.

Marker, O., and Andersen, G. T. (1976). The cytotoxicity of specifically sensitized lymphocytes from mouse strains of varying H-2 specificities on LCM virus-infected L cells. *Acta Pathol. Microbiol. Scand., Sect. C* **84,** 447–454.

Marker, O., and Andersen, G. T. (1979). A comparison between LCM virus-specific secondary cytotoxic T lymphocytes generated by Con A and by the homologous antigen. *Immunology* **38,** 235–244.

Marker, O., and Volkert, M. (1973). Studies on cell-mediated immunity to lymphocytic choriomeningitis virus in mice. *J. Exp. Med.* **137,** 1511–1525.

Marker, O., Thörner Andersen, G., and Volkert, M. (1976). The interplay between target organ concentrations of lymphocytic choriomeningitis virus and cell mediated immunity in baby mice. *Acta Pathol. Microbiol. Scand., Sect. C* **84,** 23–30.

Marker, O., Thomsen, A. R., and Andersen, G. T. (1977). Concanavalin A-induced activation of lymphocytic choriomeningitis virus memory lymphocytes into specifically cytotoxic T cells. *Acta Pathol. Microbiol. Scand., Sect. C* **85,** 483–486.

Meeting Report (1973). "Lymphocytic Choriomeningitis Virus and Other Arenaviruses," Hamburg, 1972 (F. Lehmann-Grube, ed.). Springer-Verlag, Berlin and New York.

Meeting Report (1975). International Symposium on Arenaviral Infections of Public Health Importance, Atlanta, Georgia, 1975. *Bull. W. H. O.* **52,** 381–765.

Meeting Report (1977). Hemorrhagic Fevers Produced by Arenavirus, Buenos Aires, 1976. *Medicina (Buenos Aires)* **37,** Suppl. 3.

Merigan, T. C., Oldstone, M. B. A., and Welsh, R. M. (1977). Interferon production during lymphocytic choriomeningitis virus infection of nude and normal mice. *Nature (London)* **268,** 67–68.

Mims, C. A. (1960). Intracerebral injections and the growth of viruses in the mouse brain. *Br. J. Exp. Pathol.* **41,** 52–59.

Mims, C. A. (1966). Immunofluorescence study of the carrier state and mechanism of vertical transmission in lymphocytic choriomeningitis virus infection in mice. *J. Pathol. Bacteriol.* **91,** 395–402.

Mims, C. A. (1968). Pathogenesis of viral infections of the fetus. *Prog. Med. Virol.* **10,** 194–237.

Mims, C. A. (1970). Observations on mice infected congenitally or neonatally with lymphocytic choriomeningitis (LCM) virus. *Arch. Gesamte Virusforsch.* **30,** 67–74.

Mims, C. A., and Blanden, R. V. (1972). Antiviral action of immune lymphocytes in mice infected with lymphocytic choriomeningitis virus. *Infect. Immun.* **6,** 695–698.

Mims, C. A., and Subrahmanyan, T. P. (1966). Immunofluorescence study of the mechanism of resistance to superinfection in mice carrying the lymphocytic choriomeningitis virus. *J. Pathol. Bacteriol.* **91,** 403–415.

Mims, C. A., and Tosolini, F. A. (1969). Pathogenesis of lesions in lymphoid tissue of mice infected with lymphocytic choriomeningitis (LCM) virus. *Br. J. Exp. Pathol.* **50,** 584–592.

Mims, C. A., and Wainwright, S. (1968). The immunodepressive action of lymphocytic choriomeningitis virus in mice. *J. Immunol.* **101,** 717–724.

Mori, R., Hino, Y., Taniguchi, T., Nomoto, K., and Takeya, K. (1976). The ability of athymic nude mice to survive intracerebral infection of lymphocytic choriomeningitis (LCM) virus. *Jpn. J. Microbiol.* **20,** 249–250.

Murphy, F. A. (1977). Arenaviruses: diagnosis of lymphocytic choriomeningitis, Lassa, and other arenaviral infections. *In* "Comparative Diagnosis of Viral Diseases, Vol. 1, Human and Related Viruses," Part A (E. Kurstak and C. Kurstak, eds.), pp. 759–791. Academic Press, New York.

Murphy, F. A., and Whitfield, S. G. (1975). Morphology and morphogenesis of arenaviruses. *Bull. W. H. O.* **52,** 409–419.

Murphy, F. A., Webb, P. A., Johnson, K. M., and Whitfield, S. G. (1969). Morphological comparison of Machupo with lymphocytic choriomeningitis virus: basis for a new taxonomic group. *J. Virol.* **4,** 535–541.

Nathanson, N., Monjan, A. A., Panitch, H. S., Johnson, E. D., Petursson, G., and Cole, G. A. (1975). Virus-induced cell-mediated immunopathological disease. *In* "Viral Immunology and Immunopathology" (A. L. Notkins, ed.), pp. 357–391. Academic Press, New York.

Neustadt, P. M., Cody, T. S., and Monjan, A. A. (1978). Failure to find H-2-associated susceptibility to LCM disease. *J. Immunogenet.* **5,** 397–400.

Notkins, A. L., Mergenhagen, S. E., and Howard, R. J. (1970). Effect of virus infections on the function of the immune system. *Annu. Rev. Microbiol.* **24,** 525–538.

Nunn, M. E., Herberman, R. B., and Holden, H. T. (1977). Natural cell-mediated cytotoxicity in mice against non-lymphoid tumor cells and some normal cells. *Int. J. Cancer* **20,** 381–387.

Oldstone, M. B. A. (1973). Thymus-dependent (T) cell competence in chronic LCM virus infection. *In* "Lymphocytic Choriomeningitis Virus and Other Arenaviruses" (F. Lehmann-Grube, ed.), pp. 185–193. Springer-Verlag, Berlin and New York.

Oldstone, M. B. A. (1975a). Virus neutralization and virus-induced immune complex disease. *Prog. Med. Virol.* **19,** 84–119.

Oldstone, M. B. A. (1975b). Discussion. *Bull. W. H. O.* **52,** 485.

Oldstone, M. B. A. (1979). Immune responses, immune tolerance, and viruses. *Compr. Virol.* **15,** 1–36.

Oldstone, M. B. A., and Dixon, F. J. (1967). Lymphocytic choriomeningitis: production of antibody by "tolerant" infected mice. *Science* **158,** 1193–1195.

Oldstone, M. B. A., and Dixon, F. J. (1968). Susceptibility of different mouse strains to lymphocytic choriomeningitis virus. *J. Immunol.* **100,** 355–357.

Oldstone, M. B. A., and Dixon, F. J. (1969). Pathogenesis of chronic disease associated with persistent lymphocytic choriomeningitis viral infection. I. Relationship of antibody production to disease in neonatally infected mice. *J. Exp. Med.* **129,** 483–505.

Oldstone, M. B. A., and Dixon, F. J. (1970). Tissue injury in lymphocytic choriomeningitis viral infection: virus-induced immunologically specific release of a cytotoxic factor from immune lymphoid cells. *Virology* **42,** 805–813.

Oldstone, M. B. A., and Dixon, F. J. (1971). The immune response in lymphocytic choriomeningitis viral infection. *Immunopathology* **6,** 391–397.

Oldstone, M. B. A., and Dixon, F. J. (1973). Change in susceptibility of C3H/HeJ mice to LCM virus infection. *J. Immunol.* **111,** 1613–1615.

Oldstone, M. B. A., and Peters, C. J. (1978). Arenavirus infections of the nervous system. *In* "Infections of the Nervous System," Part II (P. J. Vinken, G. W. Bruyn, and H. L. Klawans, eds.), Handbook of Clinical Neurology, Vol. 34, pp. 193–207. North-Holland Publ., Amsterdam.

Oldstone, M. B. A., Dixon, F. J., Mitchell, G. F., and McDevitt, H. O. (1973a). Histocompatibility-linked genetic control of disease susceptibility. Murine lymphocytic choriomeningitis virus infection. *J. Exp. Med.* **137,** 1201–1212.

Oldstone, M. B. A., Tishon, A., Chiller, J. M., Weigle, W. O., and Dixon, F. J. (1973b). Effect of chronic viral infection on the immune system. I. Comparison of the immune responsiveness of mice chronically infected with LCM virus with that of non-infected mice. *J. Immunol.* **110,** 1268–1278.

Oldstone, M. B. A., Holmstoen, J., and Welsh, R. M. (1977). Alterations of acetylcholine enzymes in neuroblastoma cells persistently infected with lymphocytic choriomeningitis virus. *J. Cell. Physiol.* **91,** 459–472.

Oldstone, M. B. A., Buchmeier, M. J., Doyle, M. V., and Tishon, A. (1980). Virus-induced immune complex disease: specific anti-viral antibody and C1q binding material in the circulation during persistent lymphocytic choriomeningitis virus infection. *J. Immunol.* **124,** 831–838.

Owen, R. D. (1945). Immunogenetic consequences of vascular anastomoses between bovine twins. *Science* **102,** 400–401.

Padnos, M., Shimonaski, G., and Came, P. E. (1971). Interferon in mice acutely infected with M-P virus. *J. Gen. Virol.* **13,** 163–165.

Parker, J. C., Igel, H. J., Reynolds, R. K., Lewis, A. M., and Rowe, W. P. (1976). Lymphocytic choriomeningitis virus infection in fetal, newborn, and young adult Syrian hamsters (*Mesocricetus auratus*). *Infect. Immun.* **13,** 967–981.

Parr, E. L. (1979a). The absence of H-2 antigens from mouse pancreatic β-cells demonstrated by immunoferritin labeling. *J. Exp. Med.* **150,** 1–9.

Parr, E. L. (1979b). Diversity of expression of H-2 antigens on mouse liver cells demonstrated by immunoferritin labeling. *Transplantation* **27,** 45–48.

Parr, E. L., and Kirby, W. N. (1979). An immunoferritin labeling study of H-2 antigens on dissociated epithelial cells. *J. Histochem. Cytochem.* **27,** 1327–1336.

Pedersen, I. R. (1979). Structural components and replication of arenaviruses. *Adv. Virus Res.* **24,** 277–330.

Pfau, C. J. (1974). Biochemical and biophysical properties of the arenaviruses. *Prog. Med. Virol.* **18,** 64–80.

Pfau, C. J. (1977). The role of defective interfering (DI) virus in arenavirus infections. *Medicina (Buenos Aires)* **37,** Suppl. 3, 32–38.

Pfau, C. J., Welsh, R. M., and Trowbridge, R. S. (1973). Plaque assays and current concepts of regulation in arenavirus infections. *In* "Lymphocytic Choriomeningitis Virus and Other Arenaviruses" (F. Lehmann-Grube, ed.), pp. 101–111. Springer-Verlag, Berlin and New York.

Pfau, C. J., Bergold, G. H., Casals, J., Johnson, K. M., Murphy, F. A., Pedersen, I. R., Rawls, W. E., Rowe, W. P., Webb, P. A., and Weissenbacher, M. C. (1974). Arenaviruses. *Intervirology* **4,** 207–213.

Pfizenmaier, K., Trostmann, H., Röllinghoff, M., and Wagner, H. (1975). Temporary presence of self-reactive cytotoxic T lymphocytes during murine lymphocytic choriomeningitis. *Nature (London)* **258,** 238–240.

Pfizenmaier, K., Starzinski-Powitz, A., Rodt, H., Röllinghoff, M., and Wagner, H. (1976a). Virus and trinitrophenol hapten-specific T-cell-mediated cytotoxicity against H-2 incompatible target cells. *J. Exp. Med.* **143,** 999–1004.

Pfizenmaier, K., Starzinski-Powitz, A., Röllinghoff, M., and Wagner, H. (1976b). Further evidence for the presence of selfreactive cytotoxic T cells during murine lymphocytic-chorio-meningitis. *Z. Immunitaetsforsch.* **152,** 107.

Pfizenmaier, K., Trostmann, H., Röllinghoff, M., and Wagner, H. (1976c). Cell-mediated immunity in lymphocytic choriomeningitis. I. The specificity of the cytotoxic T lymphocytes. *Z. Immunitaetsforsch.* **151,** 224–236.

Pincus, T., Rowe, W. P., Staples, P. J., and Talal, N. (1970). Inability to induce tolerance to bovine gamma globulin in mice infected at birth with lymphocytic choriomeningitis virus. *Proc. Soc. Exp. Biol. Med.* **133,** 986–988.

Poiley, S. M. (1970). A survey of indigenous murine viruses in a variety of production and research animal facilities. *Lab. Anim. Care* **20,** 643–650.

Pollard, M., and Sharon, N. (1969). Immunoproliferative effects of lymphocytic choriomeningitis virus in germfree mice. *Proc. Soc. Exp. Biol. Med.* **132,** 242–246.

Pollard, M., Sharon, N., and Teah, B. A. (1968). Congenital lymphocytic choriomeningitis virus infection in gnotobiotic mice. *Proc. Soc. Exp. Biol. Med.* **127,** 755–761.

Popescu, M., and Lehmann-Grube, F. (1975). LCM virus interference. *In* "International Virology 3, Abstracts" (H. S. Bedson, R. Nájera, L. Valenciano, and P. Wildy, eds.), p. 190. Hoechst AG, Frankfurt.

Popescu, M., and Lehmann-Grube, F. (1976). Diversity of lymphocytic choriomeningitis virus: variation due to replication of the virus in the mouse. *J. Gen. Virol.* **30,** 113–122.

Popescu, M., and Lehmann-Grube, F. (1977). Defective interfering particles in mice infected with lymphocytic choriomeningitis virus. *Virology* **77,** 78–83.

Popescu, M., Schaefer, H., and Lehmann-Grube, F. (1976). Homologous interference of lymphocytic choriomeningitis virus: detection and measurement of interference focus-forming units. *J. Virol.* **20,** 1–8.

Popescu, M., Löhler, J., and Lehmann-Grube, F. (1977). Infectious lymphocytes in mice persistently infected with lymphocytic choriomeningitis virus. *Z. Naturforsch., Teil C* **32,** 1026–1028.

Popescu, M., Löhler, J., and Lehmann-Grube, F. (1979). Infectious lymphocytes in lymphocytic choriomeningitis virus carrier mice. *J. Gen. Virol.* **42,** 481–492.

Rawls, W. E., and Leung, W.-C. (1979). Arenaviruses. *Compr. Virol.* **14,** 157–192.

Řeháček, J. (1968). The growth of arboviruses in mosquito cells in vitro. *Acta Virol. (Engl. Ed.)* **12,** 241–246.

Rivers, T. M., and Scott, T. F. M. (1935). Meningitis in man caused by a filterable virus. *Science* **81,** 439–440.

Rivers, T. M., and Scott, T. F. M. (1936). Meningitis in man caused by a filterable virus. II. Identification of the etiological agent. *J. Exp. Med.* **63,** 415–432.

Rivière, Y., and Bandu, M.-T. (1977). Induction d'interféron par le virus de la chorioméningite lymphocytaire chez la souris. *Ann. Microbiol. (Paris)* **128A,** 323–329.

Rivière, Y., Gresser, I., Guillon, J.-C., and Tovey, M. G. (1977). Inhibition by anti-interferon serum of lymphocytic choriomeningitis virus disease in suckling mice. *Proc. Natl. Acad. Sci. U.S.A.* **74,** 2135–2139.

Roger, F. (1963). Études sur le pouvoir pathogène expérimental du virus de la chorioméningite lymphocytaire. III.—Une réaction inflammatoire directement visible chez la souris: l'oedème viral du membre inférieur. *Ann. Inst. Pasteur, Paris* **104,** 347–360.

Rowe, W. P. (1954). Studies on pathogenesis and immunity in lymphocytic choriomeningitis infection of the mouse. *Nav. Med. Res. Inst., Res. Rep.* **12,** 167–219.

Rowe, W. P. (1956). Protective effect of pre-irradiation on lymphocytic choriomeningitis infection in mice. *Proc. Soc. Exp. Biol. Med.* **92,** 194–198.

Rowe, W. P., Murphy, F. A., Bergold, G. H., Casals, J., Hotchin, J., Johnson, K. M., Lehmann-Grube, F., Mims, C. A., Traub, E., and Webb, P. A. (1970a). Arenoviruses: proposed name for a newly defined virus group. *J. Virol.* **5,** 651–652.

Rowe, W. P., Pugh, W. E., Webb, P. A., and Peters, C. J. (1970b). Serological relationship of the Tacaribe complex of viruses to lymphocytic choriomeningitis virus. *J. Virol.* **5,** 289–292.

Scheid, W. (1957). Das Virus der lymphozytären Choriomeningitis und seine Bedeutung für die Neurologie. *Fortschr. Neurol. Psychiatr. ihrer Grenzgeb.* **25,** 73–99.

Schell, K., Huebner, R. J., and Turner, H. C. (1966). Concentration of complement fixing viral antigens. *Proc. Soc. Exp. Biol. Med.* **121,** 41–46.

Schwartz, R., Löhler, J., and Lehmann-Grube, F. (1978). Infection of cultivated mouse peritoneal macrophages with lymphocytic choriomeningitis virus. *J. Gen. Virol.* **39,** 565–570.

Schwarz, E., and Schwarz, H. K. (1943). The wild and commensal stocks of the house mouse, *Mus musculus* Linnaeus. *J. Mammal.* **24,** 59–72.

Schwenk, H.-U., Slenczka, W., and Lehmann-Grube, F. (1971). Phytohemagglutinin-induced DNA synthesis in blood lymphocytes from mice persistently infected with the virus of lymphocytic choriomeningitis. *Arch. Gesamte Virusforsch.* **33,** 197–199.

Seamer, J. (1965a). The growth, reproduction and mortality of mice made immunologically tolerant to lymphocytic choriomeningitis virus by congenital infection. *Arch. Gesamte Virusforsch.* **15,** 169–177.

Seamer, J. (1965b). Mouse macrophages as host cells for murine viruses. *Arch. Gesamte Virusforsch.* **17,** 654–663.

Sharon, N., and Pollard, M. (1971). Effects of cyclophosphamide on lesions induced by persistent LCM virus infection in gnotobiotic mice. *Arch. Gesamte Virusforsch.* **34,** 278–286.

Sharon, N., and Pollard, M. (1972). Responses of lymphocytic choriomeningitis carrier mice to X-ray-induced lymphoma. *Infect. Immun.* **6,** 255–257.

Shwartzman, G. (1946). Alterations in pathogenesis of experimental lymphocytic choriomeningitis caused by prepassage of the virus through heterologous host. *J. Immunol.* **54,** 293–304.

Silberman, S. L., Jacobs, R. P., and Cole, G. A. (1978). Mechanisms of hemopoietic and immunological dysfunction induced by lymphocytic choriomeningitis virus. *Infect. Immun.* **19,** 533–539.

Skinner, H. H., and Knight, E. H. (1969). Studies on murine lymphocytic choriomeningitis within a partially infected colony. *Lab. Anim.* **3,** 175–184.

Skinner, H. H., and Knight, E. H. (1971). Monitoring mouse stocks for lymphocytic choriomeningitis virus—a human pathogen. *Lab. Anim.* **5,** 73–87.

Skinner, H. H., and Knight, E. H. (1973a). Natural routes for post-natal transmission of murine lymphocytic choriomeningitis. *Lab. Anim.* **7,** 171–184.

Skinner, H. H., and Knight, E. H. (1973b). Mice with a slow response to lymphocytic choriomeningitis virus as a host of possible value for virus-induced immune disease studies. *Arch. Gesamte Virusforsch.* **41,** 185–190.

Skinner, H. H., and Knight, E. H. (1974). Factors influencing pre-natal infection of mice with lymphocytic choriomeningitis virus. *Arch. Gesamte Virusforsch.* **46,** 1–10.

Skinner, H. H., and Knight, E. H. (1979a). The potential role of Syrian hamsters and other small animals as reservoirs of lymphocytic choriomeningitis virus. *J. Small Anim. Pract.* **20,** 145–161.

Skinner, H. H., and Knight, E. H. (1979b). Epidermal tissue as a primary site of replication of lymphocytic choriomeningitis virus in small experimental hosts. *J. Hyg.* **82,** 21–30.

Skinner, H. H., Knight, E. H., and Buckley, L. S. (1976). The hamster as a secondary reservoir host of lymphocytic choriomeningitis virus. *J. Hyg.* **76,** 299–306.

Skinner, H. H., Knight, E. H., and Grove, R. (1977). Murine lymphocytic choriomeningitis: the history of a natural cross-infection from wild to laboratory mice. *Lab. Anim.* **11,** 219–222.

Smadel, J. E., and Wall, M. J. (1941). Identification of the virus of lymphocytic choriomeningitis. *J. Bacteriol.* **41,** 421–430.

Smadel, J. E., and Wall, M. J. (1942). Lymphocytic choriomeningitis in the Syrian hamster. *J. Exp. Med.* **75,** 581–591.

Smadel, J. E., Baird, R. D., and Wall, M. J. (1939). A soluble antigen of lymphocytic choriomeningitis. I. Separation of soluble antigen from virus. *J. Exp. Med.* **70,** 53–66.

Staneck, L. D., Trowbridge, R. S., Welsh, R. M., Wright, E. A., and Pfau, C. J. (1972). Arenaviruses: cellular response to long-term in vitro infection with Parana and lymphocytic choriomeningitis viruses. *Infect. Immun.* **6,** 444–450.

Stanwick, T. L., and Kirk, B. E. (1976). Analysis of baby hamster kidney cells persistently infected with lymphocytic choriomeningitis virus. *J. Gen. Virol.* **32,** 361–367.

Starzinski-Powitz, A., Pfizenmaier, K., Koszinowski, U., Röllinghoff, M., and Wagner, H. (1976). Shared determinants between virus-infected and trinitrophenyl-conjugated H-2-identical target cells detected in cell-mediated lympholysis. *Eur. J. Immunol.* **6,** 630–634.

Stewart, S. E., and Haas, V. H. (1956). Lymphocytic choriomeningitis virus in mouse neoplasms. *J. Natl. Cancer Inst.* **17,** 233–245.

Suzuki, S., and Hotchin, J. (1971). Initiation of persistent lymphocytic choriomeningitis infection in adult mice. *J. Infect. Dis.* **123,** 603–610.

Svet-Moldavsky, G. J., Mkheidze, D. M., Liozner, A. L., and Bykovsky, A. P. (1968). Skin heterogenizing virus. *Nature (London)* **217,** 102–104.

Thomsen, A. R., Volkert, M., and Marker, O. (1979). The timing of the immune response in relation to virus growth determines the outcome of the LCM infection. *Acta Pathol. Microbiol. Scand., Sect. C* **87,** 47–54.

Ticho, U., Cole, G. A., and Silverstein, A. M. (1974a). Immunopathologic uveitis in the mouse due to lymphocytic choriomeningitis virus. *Invest. Ophthalmol.* **13,** 33–38.

Ticho, U., Silverstein, A. M., and Cole, G. A. (1974b). Immunpathogenesis of LCM virus-induced uveitis: the role of T lymphocytes. *Invest. Ophthalmol.* **13,** 229–231.

Tijerina, R., Löhler, J., Chaturvedi, U. C., and Lehmann-Grube, F. (1980). Infection of murine T lymphocytes with lymphocytic choriomeningitis virus: effect of age of mice on susceptibility. *Z. Naturforch. Teil C* **35,** 1062–1065.

Tonietti, G., Oldstone, M. B. A., and Dixon, F. J. (1970). The effect of induced chronic viral infections on the immunologic diseases of New Zealand mice. *J. Exp. Med.* **132,** 89–109.

Tosolini, F. A. (1970). The response of mice to the intravenous injection of lymphocytic choriomeningitis virus. *Aust. J. Exp. Biol. Med. Sci.* **48,** 445–460.

Tosolini, F. A., and Mims, C. A. (1971). Effect of murine strain and viral strain on the pathogenesis of lymphocytic choriomeningitis infection and a study of footpad responses. *J. Infect. Dis.* **123,** 134–144.

Traub, E. (1935). A filterable virus recovered from white mice. *Science* **81,** 298–299.

Traub, E. (1936). The epidemiology of lymphocytic choriomeningitis in white mice. *J. Exp. Med.* **64,** 183–200.

Traub, E. (1937). Immunization of guinea pigs with a modified strain of lymphocytic choriomeningitis virus. *J. Exp. Med.* **66,** 317–324.

Traub, E. (1938). Factors influencing the persistence of choriomeningitis virus in the blood of mice after clinical recovery. *J. Exp. Med.* **68,** 229–250.

Traub, E. (1939). Epidemiology of lymphocytic choriomeningitis in a mouse stock observed for four years. *J. Exp. Med.* **69,** 801–817.

Traub, E. (1941). Ueber den Einfluss der latenten Choriomeningitis-Infektion auf die Entstehung der Lymphomatose bei weissen Mäusen. *Zentralbl. Bakteriol., Parasitenkd., Infektionskr. Hyg., Abt. 1:Orig.* **147,** 16–25.

Traub, E. (1960). Über die natürliche Übertragungsweise des Virus der lymphocytären Choriomeningitis (LCM) bei Mäusen und ihre Parallelen zum Übertragungsmodus gewisser muriner Krebsviren. *Zentralbl. Bakteriol., Parasitenkd., Infektionskr. Hyg., Abt. 1: Orig.* **177,** 453–471.

Traub, E. (1964). Studies on the mechanism of immunity in murine LCM. *Arch. Gesamte Virusforsch.* **14,** 65–86.

Traub, E. (1975a). Observations on "late onset disease" and tumor incidence in different strains of laboratory mice infected congenitally with LCM virus. I. Experiments with random-bred NMRI mice. *Zentralbl. Veterinaermed., Reihe B* **22,** 764–782.

Traub, E. (1975b). Observations on "late onset disease" and tumor incidence in different strains of laboratory mice infected congenitally with LCM virus. II. Experiments with inbred CBA/J mice. *Zentralbl. Veterinaermed., Reihe B* **22,** 783–792.

Traub, E., and Schäfer, W. (1939). Serologische Untersuchungen über die Immunität der Mäuse gegen die lymphozytische Choriomeningitis. *Zentralbl. Bakteriol., Parasitenkd., Infektionskr. Hyg., Abt. 1: Orig.* **144,** 331–345.

Tubergen, D. G., and Oldstone, M. B. A. (1971). Release of macrophage migration inhibitory factor by virus-infected spleen cells. *J. Immunol.* **107,** 1483–1485.

Veltri, R. W., and Kirk, B. E. (1971). An antiviral substance in the tissues of mice acutely infected with lymphocytic choriomeningitis virus. *J. Gen. Virol.* **10,** 17–27.

Volkert, M. (1963). Studies on immunological tolerance to LCM virus. 2. Treatment of virus carrier mice by adoptive immunization. *Acta Pathol. Microbiol. Scand.* **57,** 465–487.

Volkert, M., and Hannover Larsen, J. (1965). Immunological tolerance to viruses. *Prog. Med. Virol.* **7,** 160–207.

Volkert, M., and Lundstedt, C. (1968). The provocation of latent lymphocytic choriomeningitis virus infections in mice by treatment with antilymphocytic serum. *J. Exp. Med.* **127,** 327–339.

Volkert, M., Marker, O., and Bro-Jørgensen, K. (1974). Two populations of T lymphocytes immune to the lymphocytic choriomeningitis virus. *J. Exp. Med.* **139,** 1329–1343.

Volkert, M., Bro-Jørgensen, K., and Marker, O. (1975a). Persistent LCM virus infection in the mouse. Immunity and tolerance. *Bull. W. H. O.* **52,** 471–478.

Volkert, M., Bro-Jørgensen, K., Marker, O., Rubin, B., and Trier, L. (1975b). The activity of T and B lymphocytes in immunity and tolerance to the lymphocytic choriomeningitis virus in mice. *Immunology* **29,** 455–464.

von Boehmer, H., Lehmann-Grube, F., Flemer, R., and Heuwinkel, R. (1974). Multiplication of lymphocytic choriomeningitis virus in cultivated foetal inbred mouse cells and in neonatally infected inbred carrier mice. *J. Gen. Virol.* **25,** 219–228.

Wagner, H., Pfizenmaier, K., and Röllinghoff, M. (1980). The role of the major histocompatibility gene complex in murine cytotoxic T cell responses. *Adv. Cancer Res.* **31,** 77–124.

Wagner, R. R., and Snyder, R. M. (1962). Viral interference induced in mice by acute or persistent infection with the virus of lymphocytic choriomeningitis. *Nature (London)* **196,** 393–394.

Walker, D. H., Murphy, F. A., Whitfield, S. G., and Bauer, S. P. (1975). Lymphocytic choriomeningitis: ultrastructural pathology. *Exp. Mol. Pathol.* **23,** 245–265.

Walker, D. H., Camenga, D. L., Whitfield, S., and Murphy, F. A. (1977). Anticonvulsant prolongation of survival in adult murine lymphocytic choriomeningitis. II. Ultrastructural observations of pathogenetic events. *J. Neuropathol. Exp. Neurol.* **36,** 21–40.

Wallnerova, Z., and Mims, C. A. (1971). Thoracic lymph duct cannulation of mice infected with lymphocytic choriomeningitis (LCM) and ectromelia viruses. *Arch. Gesamte Virusforsch.* **35,** 152–160.

Webb, P. A. (1965). Properties of Machupo virus. *Am. J. Trop. Med. Hyg.* **14,** 799–802.

Welsh, R. M. (1975). The role of defective lymphocytic choriomeningitis virus in *in vitro* and *in vivo* infections. *In* "International Virology 3, Abstracts" (H. S. Bedson, R. Nájera, L. Valenciano, and P. Wildy, eds.), p. 43. Hoechst AG, Frankfurt.

Welsh, R. M. (1978). Cytotoxic cells induced during lymphocytic choriomeningitis virus infection of mice. I. Characterization of natural killer cell induction. *J. Exp. Med.* **148,** 163–181.

Welsh, R. M., and Buchmeier, M. J. (1979). Protein analysis of defective interfering lymphocytic choriomeningitis virus and persistently infected cells. *Virology* **96,** 503–515.

Welsh, R. M., and Doe, W. F. (1980). Cytotoxic cells induced during lymphocytic choriomeningitis virus infection of mice: natural killer cell activity in cultured spleen leukocytes concomitant with T-cell-dependent immune interferon production. *Infect. Immun.* **30,** 473–483.

Welsh, R. M., and Kiessling, R. W. (1980). Natural killer cell response to lymphocytic choriomeningitis virus in beige mice. *Scand. J. Immunol.* **11,** 363–367.

Welsh, R. M., and Oldstone, M. B. A. (1977). Inhibition of immunologic injury of cultured cells infected with lymphocytic choriomeningitis virus: role of defective interfering virus in regulating viral antigenic expression. *J. Exp. Med.* **145,** 1449–1468.

Welsh, R. M., and Pfau, C. J. (1972). Determinants of lymphocytic choriomeningitis interference. *J. Gen. Virol.* **14,** 177–187.

Welsh, R. M., and Zinkernagel, R. M. (1977). Heterospecific cytotoxic cell activity induced during the first three days of acute lymphocytic choriomeningitis virus infection in mice. *Nature (London)* **268,** 646–648.

Welsh, R. M., O'Connell, C. M., and Pfau, C. J. (1972). Properties of defective lymphocytic choriomeningitis virus. *J. Gen. Virol.* **17,** 355–359.

Welsh, R. M., Burner, P. A., Holland, J. J., Oldstone, M. B. A., Thompson, H. A., and Villarreal, L. P. (1975). A comparison of biochemical and biological properties of standard and defective lymphocytic choriomeningitis virus. *Bull. W. H. O.* **52,** 403–408.

Welsh, R. M., Lampert, P. W., and Oldstone, M. B. A. (1977). Prevention of virus-induced cerebellar disease by defective-interfering lymphocytic choriomeningitis virus. *J. Infect. Dis.* **136,** 391–399.

Welsh, R. M., Zinkernagel, R. M., and Hallenbeck, L. A. (1979). Cytotoxic cells induced during lymphocytic choriomeningitis virus infection of mice. II. "Specificities" of the natural killer cells. *J. Immunol.* **122,** 475–481.

Wilsnack, R. E., and Rowe, W. P. (1964). Immunofluorescent studies of the histopathogenesis of lymphocytic choriomeningitis virus infection. *J. Exp. Med.* **120,** 829–840.

Woodruff, J. F., and Woodruff, J. J. (1975). The effect of viral infections on the function of the immune system. *In* "Viral Immunology and Immunopathology" (A. L. Notkins, ed.), pp. 393–418. Academic Press, New York.

Yap, K. L., Ada, G. L., and McKenzie, I. F. C. (1978). Transfer of specific cytotoxic T lymphocytes protects mice inoculated with influenza virus. *Nature (London)* **273,** 238–239.

Zinkernagel, R. M. (1976a). H-2 compatibility requirement for virus-specific T-cell-mediated cytolysis. The H-2K structure involved is coded by a single cistron defined by H-2K^b mutant mice. *J. Exp. Med.* **143,** 437–443.

Zinkernagel, R. M. (1976b). H-2 restriction of virus-specific T-cell-mediated effector functions in vivo. II. Adoptive transfer of delayed-type hypersensitivity to murine lymphocytic choriomeningitis virus is restricted by the K and D region of H-2. *J. Exp. Med.* **144,** 776–787.

Zinkernagel, R. M. (1976c). Virus-specific T-cell-mediated cytotoxicity across the H-2 barrier to virus-altered alloantigen. *Nature (London)* **261,** 139–141.

Zinkernagel, R. M. (1976d). H-2 restriction of virus-specific cytotoxicity across the H-2 barrier. Separate effector T-cell specificities are associated with self-H-2 and with the tolerated allogeneic H-2 in chimeras. *J. Exp. Med.* **144,** 933–945.

Zinkernagel, R. M. (1978). Major transplantation antigens in T cell-mediated immunity: a comparison of the transplantation reaction with antiviral immunity. *Fed. Proc., Fed. Am. Soc. Exp. Biol.* **37,** 2379–2384.

Zinkernagel, R. M., and Doherty, P. C. (1973). Cytotoxic thymus-derived lymphocytes in cerebrospinal fluid of mice with lymphocytic choriomeningitis. *J. Exp. Med.* **138,** 1266–1269.

Zinkernagel, R. M., and Doherty, P. C. (1974a). Restriction of *in vitro* T cell-mediated cytotoxicity in lymphocytic choriomeningitis within a syngeneic or semiallogeneic system. *Nature (London)* **248,** 701–702.

Zinkernagel, R., and Doherty, P. (1974b). Indications of active suppression in mouse carriers of lymphocytic choriomeningitis virus. *In* "Immunological Tolerance. Mechanisms and Potential Therapeutic Applications" (D. H. Katz and B. Benacerraf, eds.), pp. 403–412. Academic Press, New York.

Zinkernagel, R. M., and Doherty, P. C. (1974c). Immunological surveillance against altered self components by sensitized T lymphocytes in lymphocytic choriomeningitis. *Nature (London)* **251,** 547–548.

Zinkernagel, R. M., and Doherty, P. C. (1974d). Characteristics of the interaction in vitro between cytotoxic thymus-derived lymphocytes and target monolayers infected with lymphocytic choriomeningitis virus. *Scand. J. Immunol.* **3,** 287–294.

Zinkernagel, R. M., and Doherty, P. C. (1975a). Peritoneal macrophages as target cells for measuring virus-specific T cell mediated cytotoxicity *in vitro*. *J. Immunol. Methods* **8,** 263–266.

Zinkernagel, R. M., and Doherty, P. C. (1975b). H-2 compatibility requirement for T-cell-mediated lysis of target cells infected with lymphocytic choriomeningitis virus. Different cytotoxic T-cell specificities are associated with structures coded for in H-2K or H-2D. *J. Exp. Med.* **141,** 1427–1436.

Zinkernagel, R. M., and Doherty, P. C. (1975c). Cortisone-resistant effector T cells in acute lymphocytic choriomeningitis and *Listeria monocytogenes* infection of mice. *Aust. J. Exp. Biol. Med. Sci.* **53,** 297–303.

Zinkernagel, R. M., and Doherty, P. C. (1977). Major transplantation antigens, viruses, and specificity of surveillance T cells. *Contemp. Top. Immunobiol.* **7,** 179–220.

Zinkernagel, R. M., and Doherty, P. C. (1979). MHC-restricted cytotoxic T cells: studies on the biological role of polymorphic major transplantation antigens determining T-cell restriction-specificity, function, and responsiveness. *Adv. Immunol.* **27,** 51–177.

Zinkernagel, R. M., and Oldstone, M. B. A. (1976). Cells that express viral antigens but lack H-2 determinants are not lysed by immune thymus-derived lymphocytes but are lysed by other antiviral immune attack mechanisms. *Proc. Natl. Acad. Sci. U.S.A.* **73,** 3666–3670.

Zinkernagel, R. M., and Welsh, R. M. (1976). H-2 compatibility requirement for virus-specific T cell-mediated effector functions *in vivo*. I. Specificity of T cells conferring antiviral protection against lymphocytic choriomeningitis virus is associated with H-2K and H-2D. *J. Immunol.* **117,** 1495–1502.

Zinkernagel, R. M., Dunlop, M. B. C., and Doherty, P. C. (1975). Cytotoxic T cell activity is strain-specific in outbred mice infected with lymphocytic choriomeningitis virus. *J. Immunol.* **115,** 1613–1616.

Zinkernagel, R. M., Dunlop, M. B. C., Blanden, R. V., Doherty, P. C., and Shreffler, D. C. (1976). H-2 compatibility requirement for virus-specific T-cell-mediated cytolysis. Evaluation of the role of H-2I region and non-H-2 genes in regulating immune response. *J. Exp. Med.* **144,** 519–532.

Zinkernagel, R. M., Adler, B., and Althage, A. (1977). The question of derepression of H-2 specificities in virus-infected cells: failure to detect specific alloreactive T cells after systemic virus infection or alloantigens detectable by alloreactive T cells on virus-infected target cells. *Immunogenetics* **5,** 367–378.

Zinkernagel, R. M., Althage, A., Cooper, S., Kreeb, G., Klein, P. A., Sefton, B., Flaherty, L., Stimpfling, J., Shreffler, D., and Klein, J. (1978a). Ir-genes in H-2 regulate generation of antiviral cytotoxic T cells. Mapping to K or D and dominance of unresponsiveness. *J. Exp. Med.* **148,** 592–606.

Zinkernagel, R. M., Althage, A., and Holland, J. (1978b). Target antigens for H-2-restricted vesicular stomatitis virus-specific cytotoxic T cells. *J. Immunol.* **121,** 744–748.

Chapter 13

Cytomegalovirus and Other Herpesviruses

June E. Osborn

I. Introduction 267
II. Murine Cytomegalovirus 268
A. Historical Background 268
B. Properties of the Virus 269
C. Pathogenesis 273
D. Epizootiology 285
E. Techniques for Identification and Handling of MCMV 285
F. Control and Prevention 285
III. Mouse Thymic Virus 288
A. Introduction 288
B. Properties of the Virus 288
C. Pathogenesis 288
D. Control and Prevention 289
References 289

I. INTRODUCTION

The recognition that cytomegaloviruses (CMVs) were frequent parasites of several species preceded awareness of the important etiologic role of human CMV in significant human disease (Smith, 1959). Mouse cytomegalovirus (MCMV) was first recognized in 1934 (McCordock and Smith, 1936) and successfully recovered and propagated in mouse embryo explants by Smith in 1954, just as human CMV was being recognized as the occasional cause of devastating fetal infection. Since all the CMVs display strict species specificity, at least *in vivo,* animal models were clearly necessary for study of CMV pathogenicity, and MCMV infection became the focus of increasingly intense and careful study.

The pace of progress with MCMV matched the emerging revelation that human CMV was a far more common cause of disease than had been appreciated, and in the past decade the mouse model of CMV infection and disease has attracted the attention of many able investigators. The status of MCMV as a member of the herpesvirus group has been affirmed; its

ISBN 0-12-262502-1

biophysical as well as biological similarity to other CMVs has been established; and its effect on its murine host is sufficiently subtle and intricate to have provided insight into a variety of basic immunopathologic problems of intrinsic biological interest.

As these findings were accumulating, early reports appeared of another herpeslike virus of mice—called *mouse thymic virus* (Rowe and Capps, 1961)—which has been little studied thus far except to establish firmly that it is quite distinct from MCMV.

Neither MCMV nor mouse thymic virus is known to have major pathologic effects during natural infection of wild mice, although this may represent lack of recognition rather than lack of effect. MCMV has the potential to establish a chronic, subclinical carrier state and as such can represent a source of unwanted infection in laboratory mice (Mannini and Medearis, 1961). However, the frequency of this problem seems to be low when contrasted to that of other chronic mouse virus infections (Rowe *et al.,* 1962). Thus the major biomedical importance of MCMV at present appears to lie in the insights its study can yield concerning analogous events in human CMV infection and disease.

MCMV has been well reviewed as recently as 1975 (Weller, 1971; Lussier, 1975b).* This chapter will focus, therefore, on the abundant literature that has appeared since then, with particular emphasis on the revelations of the model for issues of human disease. A separate section of the chapter will summarize the information currently available concerning the alternative agent, now referred to as *mouse thymic virus*.

II. MURINE CYTOMEGALOVIRUS

A. Historical Background

The first identification of most of the CMVs followed from the recognition by pathologists of characteristic enlarged cells bearing intranuclear inclusions, most commonly located in salivary gland acinar cells. These were not commonly associated with active disease, and for many years the term *salivary gland virus* was used to designate the presumed agents causing this cytopathology in man and several species of animals. In fact, it was only in 1960 that Weller proposed that the term *cytomegalovirus* replace the earlier designation, since many organs besides the salivary glands were significantly involved when acute or reactivated infection occurred.

The presence or absence of CMV infection was recognized in early studies by the appearance of distinctive histologic changes, and using this method, McCordock and Smith (1936) first established the probable viral nature of MCMV by passage of salivary gland extracts through several series of mice, each time causing evidence of systemic involvement as well as the cellular enlargement and intranuclear inclusions of salivary gland acinar cells. They noted inclusions in the liver, spleen, pancreas, and ovary, as well as in the salivary glands, and also reported variations in susceptibility of different strains of mice. It is of interest that this first study established that serial passage could not be maintained if liver and spleen extracts were used, even though these tissues were much involved by cytopathologic criteria.

The first isolation of MCMV was achieved by Smith in 1954 (reviewed in Smith, 1959) using mouse embryo explants as indicator cells which developed a cytomegalic cytopathic effect *in vitro* in the presence of infected tissue extracts. Within the next 2 years, human CMV was isolated in several different laboratories (reviewed in Weller, 1971), and in one of these, Rowe and colleagues initiated the systematic study of MCMV as a possible mouse analog of human CMV. In their first efforts, they established the fact that subclinical infection of weanling mice could result in chronic salivary excretion of infectious virus for at least 1 year (Brodsky and Rowe, 1958). They also noted that chronic urinary excretion of virus did not occur, thus establishing a difference between mouse and human chronic CMV infection.

Shortly after Brodsky and Rowe's observations were published, Medearis and his colleagues began a series of studies designed to establish the basic pathogenetic patterns of MCMV infection in mice (Mannini and Medearis, 1961). They found that, whereas natural infection of wild mice seemed to occur with a high frequency, it was rare to cultivate MCMV from the salivary glands of laboratory mice. In fact, it is of interest that their only three successes in MCMV isolation from nearly 300 animals were from salivary glands of pregnant mice.

Medearis and his group also studied the effect of route of inoculation, dose, preexisting maternal immunity, and viral effect on pregnancy and assessed viral distribution, titer, and humoral immune response (Medearis, 1964a and 1964b). Their detailed findings will be described below.

The next stage in the development of interest in MCMV as a model infection came with the recognition that MCMV significantly affected a variety of immunologic responses to other antigenic stimuli, at least during acute infection. The humoral response to sheep erythrocytes was depressed, as was the interferon response to Newcastle disease virus (Osborn and Medearis, 1966, 1967; Osborn *et al.,* 1968; Howard and Najarian, 1974). These immunosuppressive effects were of special interest since at the time of the initial studies, no such observations had been made concerning non-oncogenic viruses.

Efforts to reactivate or disseminate chronic MCMV had been generally unsuccessful (Henson *et al.,* 1967; Medearis,

*An additional thorough review of MCMV appeared as this manuscript was in final revision (Hudson, 1979).

1964b), but the question of immunologic perturbation raised by these studies and later observations of depressed cellular immunity (Howard *et al.*, 1974) and undue susceptibility to bacterial and fungal pathogens in acutely infected mice (Hamilton *et al.*, 1976; Hamilton and Overall, 1978) led to renewed interest in the usefulness of the MCMV model as a possible analog of human CMV. The latter was then being recognized as a major pathogen in immunosuppressed patients, and it was suggested that immunologic changes induced by human CMV might even be a trigger for rejection of transplanted tissues.

A final stimulus leading to the current high interest in MCMV was the demonstration that latency (as well as chronic, productive infection) might occur in MCMV-infected mice and that reactivation could indeed occur under certain restricted conditions (Cheung and Lang, 1977a,b; Dowling *et al.*, 1977; Jordan *et al.*, 1977; Mayo *et al.*, 1977; Olding *et al.*, 1975, 1976; Wu *et al.*, 1975). This has caused numerous investigators to begin using this model in their explorations of immunopathogenesis and cellular immunity.

B. Properties of the Virus

1. Classification

The initial tentativeness of classification of MCMV as a member of the herpesvirus group reflected a lack of hard data. In recent years this situation has been remedied, and it is now firmly established that MCMV, like other herpesviruses, is an enveloped virus with icosahedral symmetry of its capsid, which has 162 capsomers, containing a double-stranded DNA genome of sufficient size to give a coding capacity for several dozen gene products. The genome is significantly larger than that of most herpesviruses, which has led some to argue that MCMV may be unique (Misra *et al.*, 1978).

Its classification as a cytomegalovirus is somewhat more problematic, primarily because the definition of *cytomegaloviruses* is considerably more biological and less precise. The term usually conveys a rather strict species specificity, the ability to induce cytomegalia in infected cells, and a relatively strong degree of cell associatedness in viral replication in cell cultures (Weller, 1971; Wright, 1973). In addition, the establishment of chronic, productive infections in the natural host is generally considered a hallmark of cytomegaloviruses. MCMV fits these criteria relatively well, with the exception that infectious virus is readily released into the extracellular fluid in cell culture. This "exception" does not make it categorically different from other CMVs, however, since many isolates of human CMV can be adapted to greatly increased efficiency of release of infectious virus after sustained laboratory passage (Smith, 1959; Weller, 1971).

2. Strains of Virus

The predominant strain of virus used for laboratory study of MCMV is the Smith strain, deriving from the original isolate of Margaret G. Smith in 1954. This has been maintained primarily by mouse-to-mouse passage, with salivary gland extraction as the means of preparing inocula. However, notable biological differences exist among several strains, all of which bear the name *Smith strain,* resulting in some apparently conflicting findings in the literature emanating from different laboratories.

The most frequently used Smith strain is that which was employed by Rowe, Medearis and Osborn and is now represented in most of the laboratories actively studying immunologic reactivity of MCMV infection. The hallmarks of this variant are its ability to cause acute immunosuppression (Osborn *et al.*, 1968; Howard and Najarian, 1974; Howard *et al.*, 1974), its suppression of interferon response (Osborn and Medearis, 1966, 1967), its lifelong persistence in submaxillary glands (Brodsky and Rowe, 1958; Medearis, 1964b), and its failure to infect macrophages productively (Selgrade and Osborn, 1974).

Another agent, also called the *Smith strain,* has been used in a series of studies from Henson's laboratory. This variant produces a good interferon response, neutralizing antibody to MCMV appears in good titer by 10 days postinoculation, and submaxillary gland infections are self-limited except under conditions of immunosuppression (Henson and Neapolitan, 1970). Furthermore, a chronic hepatitis can be produced with this virus, whereas the other causes significant but self-limited hepatic involvement. The behavior of this agent in macrophages *in vitro* has not been described.

Yet another variation among Smith strains is seen in the virus used by Craighead and his group, which readily induces productive infection of macrophages *in vitro* (Tegtmeyer and Craighead, 1968) and can be used to establish an interstitial pneumonitis in immunosuppressed mice upon subcutaneous infection (both models have been difficult to establish with other strains) (Brody and Craighead, 1974).

Other isolates of MCMV have also been studied. Most of these reports were well reviewed by Lussier (1975b). Although an early report suggested differences in DNA characteristics among strains, subsequent studies have not revealed any major biophysical or antigenic differences (Kim *et al.*, 1974, 1975; Mosmann and Hudson, 1973). However, data on this point are incomplete. There has been some suggestion of substantial biological differences, particularly in a broadening of host range. Diosi's group studied an agent isolated by Raynaud from *Apodemus sylvaticus* and reported that after several tissue culture passages, the Raynaud isolate had the capacity to infect human diploid cells as well as cells from hamsters and monkeys (Diosi *et al.*, 1972). This finding was not confirmed, however, by Kim *et al.* (1974), who showed abortive but not productive infection of human diploid cells with the Raynaud strain. However, the conditions of the attempted confirmation were not identical, and thus the point is still unresolved.

Another isolate studied by Diosi's group (isolated from *Microtus arvalis*) was also of interest in that Swiss mice and cell cultures derived from them seemed to be insusceptible (see Lussier, 1975b).

It should be noted that standard techniques used in handling the various laboratory strains are apparently effective in isolating virus from wild mice with a high frequency (Mannini and Medearis, 1961; Gardner *et al.*, 1974). Thus, whereas differences among isolates have been reported, a common pattern of tissue tropism and cytopathogenicity seems to occur, and differences may be found to be minor. The same issue continues to present conundra in the study of HCMV (Weller, 1971), so its resolution in mice would be of substantial interest.

3. Physical Properties

The virions of MCMV are relatively unstable, and storage without stabilizers results in rapid loss of infectivity. This effect can be largely prevented by storage in 25% sorbitol (Medearis, 1964b) or 10% DMSO (Osborn and Walker, 1971). Like other herpesviruses, the MCMV virion is composed of a double-layered envelope derived from the nuclear membrane, enclosing a capsid composed of 162 capsomers arranged in icosahedral symmetry. A peculiar characteristic of MCMV morphology in cell culture is the occurrence of multicapsid virions, i.e., envelopes containing anywhere from 2 to more than 20 capsids (Hudson *et al.*, 1976b). This phenomenon has not yet been reported for other herpesviruses, but it has been suggested that it may be related to the occurrence of so-called dense bodies seen in human CMV-infected cultures (Hudson *et al.*, 1976b).

The MCMV virions have a buoyant density of 1.20 gm/cm^3 in sucrose gradients, and the viral DNA has a density of 1.718 gm/cm^3 in CsCl (Mosmann and Hudson, 1973; Lakeman and Osborn, 1979). It was earlier reported by Plummer that MCMV DNA banded at two different densities, 1.717 and 1.722, suggesting a striking heterogeneity of genomes; however, subsequent studies in Hudson's laboratory seemed to resolve this issue (Mosmann and Hudson, 1973, 1974). They demonstrated that shearing of MCMV genomes to one-fourth their size or less yielded the appearance of two DNA bands, whereas when the genome was handled to preserve full integrity—or broken to yield half-size fragments—a homogeneous DNA band with a density of 1.718 was found. They proposed as an explanation that the guanosine + cytosine (G + C) distribution within stretches of the genome was sufficiently variable so that fragmentation yielded two apparently different bands.

The size of the viral DNA was first reported by Mosmann and Hudson (1973) to be 132×10^6 daltons, using biophysical means of measurement. Lakeman and Osborn used infectious DNA as a biological measure of genome size and found a sharp peak of infectivity associated with DNA of molecular weight 136×10^6 (Lakeman and Osborn, 1979). In that same study, they found that human CMV DNA had a bimodal peak of infectivity, both peaks being associated with DNA larger than 125×10^6. Thus the suggestion of Hudson that MCMV may be unique with respect to its large genome (herpes simplex is approximately 100×10^6) may extend to cytomegaloviruses as a group, giving a firmer basis to their subgrouping within the herpesvirus category.

Relatively little has been learned as yet about the functions of the various proteins of the MCMV virion. Gonczol *et al.* (1978) have shown that heat-inactivated (but not ultraviolet-inactivated) virions stimulate cellular DNA synthesis, which they interpret to mean that a heat-sensitive structural protein of MCMV is responsible for the usual shutdown of host cell DNA synthesis. As will be discussed later, two groups have shown that MCMV neither codes for nor carries a thymidine kinase (Muller and Hudson, 1977a; Eizuru *et al.*, 1978). Chantler and Hudson (1978) identified at least 30 proteins labeled in virus-infected cells which could be precipitated with viral-specific antisera; 22 of these had electrophoretic mobilities similar to those of structural viral proteins, whereas an additional eight were not represented in purified virions. Moon *et al.* (1979) reported identifying at least 14 virus-induced proteins but stated that improved resolution might identify as many as 36 additional proteins.

Although these results are preliminary and additional studies need to be conducted, it is interesting that Chantler (1978) reported that the use of 150–160-mm NaCl selectively favored translation of viral rather than host proteins from mRNA. This should allow better definition of viral proteins, particularly those which appear early in MCMV replication, since MCMV shutoff of host cell protein synthesis is relatively slow and has hampered earlier studies.

4. Multiplication of MCMV in Mice and in Tissue Culture

a. Mice. The pattern of multiplication of MCMV in mice is much influenced by route of inoculation, strain of mice, and age at the time of inoculation. The details of these variations will be dealt with in the section on pathogenesis. An additional variable which significantly affects the behavior of MCMV in mice is tissue culture passage: The virus virulence is remarkably affected by even one passage in cell culture, as will be discussed in Section II,F (Osborn and Walker, 1971).

b. Tissue Culture

i. Kinds of Cultures Permissive for MCMV Replication. There is some conflict in the literature concerning the range of cells in culture which are permissive for MCMV replication. The various descendants of the Smith strain of MCMV all multiply relatively efficiently in fibroblastic cultures of murine cells. The most commonly used are either secondary or tertiary cultures of 13-16-day mouse embryos; in general, the variation in susceptibility of different genetic lines of mice to MCMV has not been found to extend to this *in vitro* environment, since mouse embryo cells derived from relatively resistant strains of mice appear to support the growth of MCMV as efficiently as do embryonic cells from susceptible strains (Selgrade and Osborn, 1974; Nedrud *et al.,* 1979).

Another common fibroblastic tissue culture system which has been used is the 3T3 or 3T6 cell from various strains. These cells appear to be at least as permissive as mouse embryo fibroblasts, and their use in Hudson's laboratory has facilitated studies of cell cycle dependence of MCMV replication (Muller and Hudson, 1977b).

The permissiveness of other murine cells has been less consistently demonstrable. The most clear-cut demonstration of a different productive system has been the use of mouse tracheal ring explant cultures (Mantyjarvi *et al.,* 1977; Nedrud *et al.,* 1979). The results are of particular interest in that the predominant permissive cells are epithelial rather than fibroblastic in origin. Using this system, the investigators have shown some inverse correlation between genetic resistance of mouse strains and the extent of multiplication of MCMV in tracheal ring explants, whereas no such relation is observed with fibroblast cultures from those strains. This has led them to suggest that innate epithelial cell properties may mediate the genetic susceptibility and resistance seen among mouse strains to MCMV.

Several other kinds of cell cultures have been reported to support the multiplication of MCMV. Kim and Carp (1971) reported success in cultivation of MCMV in a considerably more diverse species range, including monkey (BSC-1), hamster (BHK-21), and rabbit (primary kidney) cells. Other cell systems in their study (fetal sheep brain, L cells, and RK-13 rabbit kidney cells) also supported MCMV replication after one or more blind passages. The investigators did not demonstrate productive infection with human cells, but in a later report (Kim and Carp, 1972) they presented data suggesting that WI-38 cells supported a partial replicative cycle of MCMV and sustained a cytopathic effect which may have been mediated (based on metabolic inhibitor studies) by an early protein coded for by the parental genome.

Plummer and Goodheart (1974) pursued the primary rabbit kidney system, which they found to be moderately permissive for MCMV. They were able to enhance this permissiveness significantly by preparation of the cells in the presence of 100 μg/ml IUdR prior to infection.

Other specialized tissue culture observations with the Smith strain of MCMV have yielded positive evidence of MCMV replication in several elements of neuronal or ganglionic explants from mice, including Schwann and satellite cells and, less consistently, neuronal cells (Davis *et al.*, 1979).

Nonpermissive ME-D cells were developed by Misra and Hudson (1977) by passage of mouse embryo cells past a "crisis" at about the eighth passage level. These cells were then resistant to MCMV during passes 15–45 and showed no evidence of viral DNA synthesis, although by hybridization analysis some late viral RNA was represented and 70% of input genomes could be shown to have survived intact in the nucleus. After passage 45, progeny cells again began to regain permissiveness for MCMV.

The findings summarized above relate to the behavior of the Smith strain variants. The other two MCMV isolates studied have yielded somewhat different results. The isolate from *Mus muscularis* studied by Plavosin and Diosi (1974) would not grow either in Swiss mice or in mouse embryo fibroblasts prepared from Swiss mouse embryos. This is the only report of any mouse embryo fibroblastic system being nonpermissive for MCMV. As noted earlier, Raynaud's isolate from *Apodemus sylvaticus* was initially reported to have a rather broad host range which, after cell culture passage, included productive infection of human diploid cells. This finding was not confirmed by Kim *et al.* (1974), who found that both the Smith and Raynaud strains produced only abortive infection in human cells.

In summary, although some variation exists among MCMV strains and in reports from different laboratories, MCMV generally replicates most productively *in vitro* in mouse embryonic fibroblastic cells and has a relatively—but not absolutely—restricted host range.

ii. Centrifugal Enhancement. A nearly unique property of MCMV in cell culture is its marked enhancement by centrifugation of the inoculum against the cell monolayer during adsorption. This maneuver results in as much as an 80- to 100-fold increase in efficiency of identification of infectious units compared to more standard methods of viral adsorption (Osborn and Walker, 1968; Hudson *et al.,* 1976a). The mechanism of this centrifugal enhancement remains obscure despite Hudson's vigorous efforts to clarify it. The phenomenon has also been reported for the AD-169 strain of human CMV, but similar centrifugal inoculation with other herpesviruses (herpes simplex types 1 and 2 and pseudorabies) results in only a three- to fivefold enhancement, leading Hudson to suggest that it may be a unique property of cytomegaloviruses (Hudson *et al.,* 1976a).

iii. Multiplication Cycle in Tissue Culture. (a) Macromolecular synthesis during productive infection. Early studies of replication kinetics of MCMV in mouse embryonic fibroblast cultures yielded fairly consistent results, with a lag phase before the appearance of infectious progeny virions of approximately 18 hr and a log-linear increase in extracellular infectious virus of 18–36 hr (Osborn and Walker, 1968; Kim *et al.,* 1974; Kurimura *et al.,* 1977). As noted earlier, no appreciable fraction of progeny virus remains cell associated, in contrast to human CMV, and yields of 100 pfu/cell are possible in permissive systems.

Although the results of various investigators were fairly uniform, some variability existed. Moon *et al.* (1976) found that viral DNA synthesis began 10–12 hr postinfection. Later studies shed some light on sources of variability in the earlier investigations. By use of synchronous 3T3 cell cultures, a more precise description of the MCMV replicative cycle has been formulated by Hudson's group. Muller and Hudson (1977b, 1978) have found that MCMV replication is dependent on the host cells being in the S phase. Under these conditions, host cell DNA synthesis is turned off fairly quickly, new DNA synthesis begins at 6–8 hr, and by 22 hr the equivalent of 900 viral genomes per cell have been synthesized (Misra *et al.*, 1977). When 3T3 cells were infected in the G_1 phase, however, the latent period could be protracted from 12 to 24 hours (Muller and Hudson, 1977b). Pursuing this observation, these investigators studied the effect of MCMV infection on nuclear monolayers and found that only S phase—and not G_0 phase—cell nuclei were capable of synthesizing viral DNA (Muller and Hudson, 1978). They demonstrated the appearance of a new DNA polymerase which was thought to be coded for by the virus in that it was stimulated (rather than inhibited) by increased concentrations of ammonium sulfate.

The reason for the cell cycle dependency of MCMV is unclear, but it is not related to thymidine kinase. It has been demonstrated that MCMV neither codes for nor stimulates thymidine kinase activity and can multiply in 3T3(tk^-) cells (Muller and Hudson, 1977a; Eizuru *et al.*, 1978).

Misra *et al.* (1978) used reassociation kinetics with ^{125}I-labeled viral DNA to follow the sequence of transcription in productively infected mouse embryo cells. They found that approximately 25% of the genome was transcribed prior to DNA synthesis, and after viral DNA replication, approximately 38% of the genome was transcribed. Both early and late RNA comprised two abundance classes differing 8- to 10-fold in concentration. They concluded that MCMV displayed temporal, quantitative, and posttranscriptional controls over gene expression but that the pattern differed considerably from that of herpes simplex.

Two groups have studied the sequence of viral protein synthesis. Chantler and Hudson (1978) identified three "immediate-early" proteins appearing before 4 hr, after which no new proteins were identified until 10 hr, when a structural protein (VP2) was synthesized in association with DNA synthesis. After this, many additional structural (29) and nonstructural (8) proteins were identified. The investigators commented that the paucity of early proteins was unusual in view of the large viral genome, but that it correlated well with the cell cycle dependency of MCMV since one can infer a requirement for a number of host enzymes.

Similar findings were reported by Moon *et al.* (1979), who identified at least 14 virus-induced proteins, of which two major and three minor ones appeared by 4 hr. They commented that further resolution might permit the identification of as many as 36 additional proteins induced by MCMV. This group sought but found no evidence of posttranslational modification of major virion proteins.

(b) Macromolecular events during nonproductive infection. Several variations in the multiplication cycle have been described in special host–virus systems. Kurimura *et al.* (1977) reported that tissue culture-passaged MCMV had a shorter lag period in mouse embryo fibroblasts than did virus extracted from salivary glands of mice. They were unable to correlate this with differences in morphogenesis by electron microscopy, however.

Misra and Hudson (1977) studied the events during nonproductive infection of ME-D cells. They found no evidence of DNA synthesis, but some late viral RNA was identified by reassociation kinetic analysis. Their data suggested that as many as 70% of the viral input genomes remained intact in the nuclei of these cells.

This group later studied the events associated with nonproductive infection of 3T3 cells arrested in the G_0 phase (Muller *et al.*, 1978). Again, they found that the genomes remained intact (as identified by infectious centers) but that only 20% of the genome was transcribed, yielding at least five new polypeptides, none of which were structural proteins. They proposed that this system might be studied further as an *in vitro* model of MCMV latency (Hudson *et al.,* 1979).

(c) Cytopathic effects of MCMV. During productive infection, the cytopathic effect of MCMV on host cells is quite characteristic. The enlargement of nuclei and cytoplasm results in so-called cytomegalia and characteristically leads to cell death. This is apparently the case even in the abortive infection of WI-38 human diploid cells reported by Kim and Carp (1972), in which the cytopathic effect was seen at 12 hr and infected cells could not be induced to multiply further (in contrast to cells inoculated with ultraviolet-inactivated virus). They suggested that the observed cytopathic effect was due to synthesis of a protein coded for by the input genome without a requirement for new DNA synthesis.

Relatively few variations in cytopathic effect have been reported. An interesting exception was the study by Margolis and Kilham (1976) of MCMV encephalitis following intracranial inoculation of newborn mice in which frequent neuronal syncytium formation was reported. Characteristically, one inclusion-bearing and one normal nucleus were enveloped in a single membrane, suggesting recruitment of normal neurons and parasitism by infected neurons. With the exception of this report, however, MCMV has not been reported to induce cell fusion.

Before leaving the topic of cytopathic effect, however, it should be stressed that the suspicion has been raised repeatedly in the literature that MCMV infection may be present and productive in the absence of characteristic cytopathic effect, particularly *in vivo,* where the correlation between virus concentration and cytopathology is sometimes poor.

C. Pathogenesis

1. Overt Disease

a. Wild Mice. No overt disease has been observed in the few studies which have been done in wild mice, but it has been consistently found that a majority of adult mice have MCMV in their salivary glands or saliva (Gardner *et al.,* 1974; Mannini and Medearis, 1961; Rowe *et al.,* 1962). Newborn wild mice are not infected, and since it is known from laboratory studies that colostrum can be protective (Medearis, 1964a), it is reasonable to assume that in feral mice infection is acquired during early months of life and then persists throughout the life of the animal.

The presence of persistent MCMV infection in this context is closely analogous to the situation observed in human populations, and it is of considerable interest that Gardner *et al.* (1974) were able to convert subclinical to lethal MCMV infection by treating wild mice with antithymocyte serum. This result is strongly reminiscent of the clinical CMV problem in organ transplant patients and suggests that naturally infected wild mice might be an interesting model for study of that facet of human CMV disease.

b. Laboratory Mice. As implied above, the bulk of information concerning MCMV pathogenesis derives from manipulative studies of laboratory mice. The results of such studies are somewhat difficult to summarize since variables of virus strain and passage history, mouse strain, mouse age, virus dose, and route of inoculation all influence the results to a remarkable extent. Several examples can be presented to emphasize this point.

First, it has been repeatedly observed that between the ages of 21 and 28 days, the host resistance to MCMV increases dramatically (Booss and Wheelock, 1975; Selgrade and Osborn, 1974). Therefore, a shipment of weanlings which is received at an inconvenient time and inoculated a few days later may yield markedly atypical results.

Second, the dose effect is so striking that a difference of fourfold is the range between 0 and 100% mortality in susceptible animals (Selgrade and Osborn, 1974; Selgrade *et al.,* 1976).

Finally, even one passage of MCMV through mouse embryo fibroblast cell cultures significantly diminishes the virulence of the progeny virions for mice (Osborn and Walker, 1971). With this potential variability in mind, an effort will be made to describe the several patterns of pathogenesis reported from various laboratories.

2. Age, Dose, and Route of Inoculation

MCMV virulence for mice decreases with age, and newborn animals are exquisitely susceptible. Less than 1000 infectious units inoculated intraperitoneally usually results in nearly 100% mortality within a week (Mannini and Medearis, 1961; Osborn and Walker, 1971). This same age-related susceptibility can be shown by intracranial inoculation, in which case newborns (1 day old) and sucklings (6 days old) sustain a lethal necrotizing encephalitis with demonstrable infection of neuronal cells, whereas older (3-week-old) animals sustain a subclinical, self-limited meningoencephalitis without ganglionic involvement (Lussier, 1973).

The newborn animal infected by either the intraperitoneal or intracranial route supports abundant MCMV multiplication in a wide variety of organs and tissues including the spleen, liver, pancreas, lungs, kidneys, ovaries, and adrenals (Osborn and Walker, 1971; Mims and Gould, 1979). The animals characteristically die between the third and eighth day of infection, and survival is uncommon unless very low doses of inoculum are used. Even the differences in susceptibility among strains of mice noted in older animals are minimal in the newborn animals (Chalmer, 1979).

The cause of neonatal death is not yet clear. Schwartz *et al.* (1975) demonstrated thymic atrophy and runting in survivors after intraorbital inoculation and further showed bacteremia to occur in runted sucklings which subsequently succumbed. However, no other group has found the thymus to be affected by MCMV, and it is a common finding that MCMV pools may be contaminated with mouse thymic agent (Cross *et al.,* 1979), so that the observed thymic disease may have reflected a mixed inoculum.

An alternative, potentially lethal aspect of the acute newborn infection is the occurrence of myocarditis. Lussier (1974) has shown that focal necrosis of myocardial and striated muscle occurs in newborn infection and that residual focal myocarditis can be demonstrated as late as 56 days postinfection in survivors.

The abrupt lethality seen when newborn mice are inoculated with mouse-passaged virus is strikingly altered with one or more passages of virus through tissue culture. When the inoculum is thus modified, the equivalent number of plaque-forming units injected intraperitoneally will not result in any deaths of newborns during the first 10 days (in which interval virtually all deaths occur with a "virulent" inoculum) (Osborn and Walker, 1971). However, subsequently as many as half the inoculated animals sicken and die in the ensuing weeks. The cause of such deaths is not known.

Cruz and Waner (1978) facilitated the survival of 1-day-old mice by using an inoculum passaged twice in tissue culture. They observed runting of survivors for at least 4 weeks in all inoculees. This growth retardation was much potentiated by undernutrition, which they found to have a synergistic effect on both growth and immunosuppression.

In summary, the "attenuation effect" of tissue culture passage of MCMV is a very relative matter, as will be discussed further when we consider the model for studying cytomegalovirus vaccines.

The effect of inoculation of weanling mice with mouse-passaged MCMV is strongly dose dependent, with a fourfold variation making the difference between 0 and 100% mortality. (As noted earlier, the difference between 21, 24, and 28 days of age is incrementally significant in increasing resistance to a given dose; Booss and Wheelock, 1975.) When more than 10^5 pfu are inoculated intraperitoneally into 21-day-old weanlings of susceptible mouse strains, mortality of 50–100% is likely, again occurring primarily during the first week of infection. Virus concentrations are initially highest in the liver, spleen, and pancreas (Mannini and Medearis, 1961; Osborn and Walker, 1971). Viremia has been demonstrated frequently, although there is some variability regarding its duration. Cheung and Lang (1977b) reported isolation of MCMV from blood only on days 1–3; Booss and Wheelock (1977a) found evidence of viremia in the first 2 weeks of infection and reported that some of the circulating virus was cell-free. Quinnan *et al.* (1980) also reported viremia to be sustained from days 3 to 18, but in their study the virus was associated almost entirely with the gravity-sedimented buffy coat.

MCMV appears in the salivary glands only toward the end of the first week of infection, often after liver and spleen concentrations of virus have abruptly declined (Mannini and Medearis, 1961; Mims and Gould, 1979; Osborn and Walker, 1971). However, liver and spleen infections are generally self-limited, whereas the salivary glands and (variably) pancreas are the sites of chronic high-titer infection for a period of weeks or months, if not for the lifetime of the mouse (Medearis, 1964b; Brodsky and Rowe, 1958; Mims and Gould, 1979; Osborn and Walker, 1971).

There is a gradient of tissue susceptibility even among salivary glands, with the submaxillary gland by far the most susceptible to chronic infection, the sublingual gland next, and the parotid gland almost uninvolved (Mims and Gould, 1979). There may also be some effect of dosage on the duration of chronic infection of the submaxillary gland, and susceptibility of different mouse genotypes also plays a role. Some investigators have manipulated these variables successfully to yield apparent termination of submaxillary gland infections (Cheung and Lang, 1977a; Henson and Strano, 1972; Mayo *et al.*, 1977; Mims and Gould, 1979), which has allowed them to study latency and reactivation in this organ.

It is of interest that renal infection does occur, but in most studies it has been found to be much less prominent than in human CMV infection, with lower titers of virus and less durable persistence (Brodsky and Rowe, 1958; Medearis, 1964b). One exception was the study of Mims and Gould (1979), who found the kidney to be the site of chronic infection in a greater number of mice than was the submaxillary gland. The persistent renal infection seems to have little pathophysiologic significance, analogous to the human infection. However, Lussier (1976) found that treatment of chronically infected mice with anti-thymocyte serum resulted in the development of a striking acute glomerulonephritis.

When adult mice are inoculated intraperitoneally, only very high doses (more than 10^6 pfu) result in mortality in susceptible strains (Mannini and Medearis, 1961), whereas some resistant strains are refractory even to the highest attainable doses of inoculum (Chalmer *et al.*, 1977; Chalmer, 1979; Selgrade and Osborn, 1974). As noted earlier, the transition to this refractory state seems to occur between 3 and 4 weeks of age (Booss and Wheelock, 1975), and the specific mechanism of this maturation is not known.

Inoculation of adult animals by intravenous or intraperitoneal routes with sublethal doses generally does not produce any sign of overt illness. However such sublinical infections are not entirely innocuous, as was shown by Osborn *et al.* (1968) in their demonstration that humoral immunosuppression and decreased interferon responsiveness occurred during the first 2 weeks of infection of adult animals (as a property of "virulent" mouse-passaged virus, since it did not occur with tissue-culture-passaged inoculum) (Osborn and Walker, 1971).

Variation in the route of inoculation causes several changes in the pattern of pathogenesis. When intravenous rather than intraperitoneal inoculation is studied, lower doses yield a pathologic effect (Mannini and Medearis, 1961). When inoculation is by the intracranial route, as noted above, newborns sustain a lethal necrotizing encephalitis, whereas weanlings undergo subclinical meningoencephalitis (Lussier, 1974). However, antithymocyte serum treatment of such animals renders them much more susceptible to intracranial MCMV, suggesting that cellular immunity is protective in this context (in contrast, for instance, to lymphocytic choriomeningitis) (Lussier, 1975a).

Brody and Craighead (1974) found that subcutaneous inoculation of adult mice with a low dose (100 pfu) of MCMV resulted only in submaxillary gland infection, whereas treatment of such mice with anti-lymphocyte serum from 5 days before until 21 days after virus inoculation resulted in the development of a severe interstitial pneumonitis with features quite similar to those associated with human CMV infection.

Jordan (1978) used an intranasal route of inoculation of 4 to 6-week-old mice and found that a dose of 100 pfu resulted in subclinical systemic infection, whereas a higher dose (10^4 pfu) resulted in severe interstitial pneumonitis with a 20% mortality rate.

Quinnan *et al.* (1980) proposed that the intranasal route of infection was likely to mimic natural conditions most closely and chose that route for studies of the immune response to MCMV. They found that viral titers in the spleen were lower during acute infection than when intraperitoneal inoculation was used, but that the spleen cytotoxic lymphocyte response was the same with either route.

3. Genetic Factors in Susceptibility to MCMV

Variation in susceptibility to MCMV among strains of mice was noted in the first study of this agent by McCordock and Smith (1936). They reported that outbred Swiss mice were more susceptible to MCMV than were the "Buffalo" strain, and that inbred C57BL were least susceptible. Selgrade and Osborn (1974) found CBA mice to be more resistant than C57BL and reported that F_1 hybrids remained as resistant as the CBA parent.

Mims and Gould (1978b) observed that weanling or adult mice of some strains (CD1, LACA, NIH Swiss, BALB/c or A) underwent striking spleen necrosis following intravenous or intraperitoneal inoculation of large doses of MCMV whereas other strains (C57BL/6, C57BL/10, and CBA) did not. The latter group had less virus growth in their spleens. However, virus-induced immunosuppression was comparable in the two groups, and newborn CD1 mice sustained high titers of MCMV multiplication in their spleens in the absence of necrosis.

Chalmer has conducted the most detailed studies to date of the genetics of mouse resistance to MCMV (Chalmer *et al.*, 1977; Chalmer, 1979). They compared C3H (*H-2*^{k}), CBA (*H-2*^{k}), BALB/c (*H-2*^{d}), C57BL (*H-2*^{b}), DK-black, Simpson, and PHI mice. The C3H and CBA strains were relatively resistant to the lethal effect of MCMV (intraperitoneally), whereas the remaining strains were relatively susceptible. The F_1 hybrid between C3H and BALB/c mice had a susceptibility intermediate between that of the parents, and the backcross of the F_1 hybrid to the resistant C3H mice produced a segregation of resistant and susceptible progeny in the ratio of 1 : 1. This result was interpreted to mean that there was a single gene involved which was partially dominant.

Additional studies by Chalmer (1979) to determine the effect of *H-2* haplotypes using congenic strains of BALB/c identified the *H-2*^{k} haplotype as resistant. Use of recombinant strains suggested that there were two loci in the *H-2* region which determined resistance.

It is of interest that differences in susceptibility were not apparent in very young mice and were not found in tissue cultures derived from the various strains. Chalmer postulated that, since death due to MCMV was an early event, mobilization of natural killer or M cells—rather than variation in development of specific immune responses—might mediate the observed resistance.

As noted earlier, Nedrud *et al.* (1979) found that tracheal organ cultures reflected susceptibility characteristics of the mouse strain of origin, whereas fibroblast monolayers did not. They suggested that one mediator of resistance might be some inherent property of epithelial cells since these cells seemed to be the permissive element in the tracheal organ culture system.

A final point of interest concerning genetic factors controlling susceptibility to MCMV is the lack of correlation between resistance patterns to MCMV and those to herpes simplex in mice (Selgrade and Osborn, 1974).

4. Usefulness of MCMV as a Model of HCMV Infection and Disease

Much of the intense interest in MCMV reflects the inability of virologists to do manipulative biological studies of human CMV *in vivo* because of the strict species specificity. Thus it is useful to assess mouse infection with MCMV as a model in several contexts.

a. Intrauterine Infection and Congenital Disease. Since the earliest awareness of HCMV pathogenicity came from the recognition of rare, devastating intrauterine infection of human fetuses, efforts were directed early at the establishment of MCMV as a model of intrauterine infection. Medearis (1964b) attempted to produce fetal infection by both intravenous and direct intrauterine inoculation of pregnant mice. He found that extensive fetal wastage resulted from infection of mothers in the second half of gestation but that surviving young were uninfected and virus could not be recovered from fetal tissues. Thus, unlike the probable human CMV situation, MCMV effects on fetuses appeared to be indirect.

Johnson (1969) extended these studies and found that he could reproduce the extensive fetal wastage observed by Medearis, again without evidence of productive fetal infection. He showed that one of the three layers of placental trophoblasts was infected and proposed that physiologic malfunction of the placenta was the mediator of fetal wastage. He further pointed out that although human and mouse placentas were both hemichorionic, the human placenta had only one trophoblastic layer, whereas the mouse had three, only one of which could be shown to be infected. He therefore propopsed that this anatomic difference might help to explain the better protection of the mouse fetus.

Chantler *et al.* (1979) observed that the pregnancy outcomes of mice chronically infected with MCMV seemed to be adversely affected. During the year following inoculation as weanlings, female mice were followed through several pregnancies; litter sizes were observed to be decreased and, in three instances, embryos retrieved from survivors at 1 year yielded MCMV on serial passage, suggesting latent and persistent infection in the embryos. If such vertical infection occurred, its origin is not clear. There is support for this finding, however, in the studies of Olding *et al.* (1975), who initiated latent infection by inoculation of pregnant mice and showed that spleen cells from their offspring 6 weeks later could be induced to MCMV productivity by cocultivation with allogeneic cells.

Several other lines of evidence can be invoked to support the suggestion that vertical transmission of MCMV may occur. Dutko and Oldstone (1979) have shown that adult male mice, infected as newborns with 100 pfu MCMV, have 4–6 genome equivalents per 100 cells in spermatogonia several months later. Cheung and Lang (1977a) demonstrated apparently latent prostatic infection following low-dose inoculation of adults. Mims and Gould (1979) found MCMV in ovarian tissue although not in ova, as studied by immunofluorescence, confirming the original report of ovarian infection by McCordock and Smith (1936). It is of interest, parenthetically, that Mims did not find virus or viral antigens in mouse mammary glands or milk, in contrast to human CMV infection.

In an effort to circumvent the murine placenta as a barrier against fetal MCMV infection, several groups have done studies in which two- or more celled mouse embryos were bathed *in vitro* in medium containing MCMV. Heggie and Gaddis (1979) found that MCMV—and several other, but not all, viruses tested—caused failure of subsequent cleavage and blastocyst formation of two-cell mouse embryos after *in vitro* exposure to virus. However, there was no direct evidence of MCMV multiplication in this system. Neighbour (1978), in a similar study, had demonstrated that no productive MCMV infection occurred in the preimplantation mouse embryo even if the zona pellucida was removed. Using a somewhat different protocol, Young *et al.* (1977) mixed sperm and MCMV prior to artificial insemination of females and then prepared tissue cultures from 14-day embryos of resulting pregnancies. They recovered virus from blind-passaged, experimental embryonic fibroblasts in one of 28 instances and felt that this finding might support the hypothesis that prenatal CMV infection in the mouse might be induced in some instances at the time of fertilization.

It is evident that these kinds of study require extension to resolve apparent contradictions. It is less clear what pertinence this has to either MCMV or human CMV effect *in vivo*.

b. Interstitial Pneumonitis. The dominance of CMV-induced interstitial pneumonitis as a complication of organ transplantation is well recognized, although there is some debate as to its importance as the sole lethal factor in this clinical setting. The MCMV model has been tested for partial adaptation for study of MCMV-induced pneumonitis.

Brody and Craighead (1974) found that subcutaneous inoculation of 100 pfu MCMV into normal adult mice produced little effect beyond subclinical establishment of chronic submaxillary gland infection. However, immunosuppression with antilymphocyte serum (from −5 to +21 days) resulted in a severe interstitial pneumonitis, with very high viral titers peaking at 21–25 days postinfection. Murphy *et al.* (1975) and Brody *et al.* (1978) have studied additional pathologic features of this model pneumonitis, including an analysis of the role of CMV-infected monocytes in its pathogenesis. They demonstrated that infected cells migrated through vascular endothelium and basement membranes of alveoli and bronchioles into alveolar spaces and concluded that these cells ultimately became alveolar macrophages.

Jordan (1978) designed studies to assess the respiratory route as a possible route of infection in the initiation of interstitial pneumonitis. He administered an inoculum of 100 pfu intranasally and found that systemic but subclinical infection ensued. Higher doses (10^4 pfu) resulted in severe interstitial pneumonitis with 20% mortality.

Finally, Gardner *et al.* (1974) used antithymocyte serum to produce immunosuppression of persistently infected wild mice and found that disseminated, often lethal infection ensued within a month. Widespread pathologic changes were associated, including the appearance of inclusion-bearing cells in alveolar septa. This last study is appealing in its apparent relevance to human CMV, although there is some reason to suspect that the truly dire outcomes of HCMV infection in transplant patients represent primary rather than reactivation infection.

c. Subclinical CMV Infection. Since the vast majority of cases of primary human CMV infection are clinically silent or unrecognized, several studies with MCMV have been aimed at shedding light on possible host effects of subclinical CMV infection. Osborn and her co-workers (Osborn and Medearis, 1967; Osborn *et al.*, 1968; Osborn and Shahidi, 1973) found that immunosuppression, interferon response suppression, thrombocytopenia, and undue susceptibility to Newcastle disease virus (NDV) infection all occurred during the first 2 weeks of silent MCMV infection in adult mice. Howard *et al.* (1974, 1977) and Selgrade *et al.* (1976) extended those observations to include depressed cell-mediated immunity, and Glasgow's group has established that mice thus compromised may suffer synergistic ill effects of MCMV and various weakly pathogenic microorganisms (Hamilton *et al.*, 1976; Hamilton and Overall, 1978; Kelsey *et al.*, 1977; Stringfellow *et al.*, 1977). These findings have some possible relevance to human disease in that it has been suggested that HCMV may alter human immune responsiveness in a variety of ways.

d. Effect on the Developing Mouse Ear. One notable advance in our appreciation of the range of pathogenic effects of HCMV has been the recognition that late-onset deafness can be a major if not a sole sequel of congenital HCMV infection. Davis and Hawrisiak (1977) studied the MCMV model in this context, starting with the assumption that immaturity of the ear in the newborn mouse was equivalent to that of a 3–5-month-old human fetus. With that in mind, they inoculated newborn mice intracranially and then studied subsequent histopathologic events in the inner ear (other routes of inoculation did not affect the ear at all). They observed that only perilabyrinthitis occurred, whereas the few CMV-infected human infants studied pathologically have shown striking endolabyrinthitis. They concluded that MCMV bore a closer resemblance to other human herpesviruses—notably varicella zoster—than to HCMV with respect to its effect on the developing ear.

In more recent studies (Davis *et al.*, 1979), this group explanted trigeminal ganglia from mice which had been inoculated intracranially at different ages with MCMV. They found that Schwann cells, satellite cells, and neurons of mice inoculated on day 1 or day 6 supported MCMV replication, whereas 21-day-old weanlings did not show signs of MCMV infection of any ganglionic elements.

e. Chorioretinitis and Encephalitis. Schwartz *et al.* (1974) inoculated MCMV intraocularly into weanling mice to see if they could reproduce the striking necrotizing chorioretinitis often associated with congenital HCMV infection. They found that viral titers peaked at 4 days, that neuronal cells (including those in the ganglion cell and inner and outer nuclear layers) showed no evidence of viral replication, but that major involvement of the uvea and retinal pigment epithelium was similar to changes in human infection. The model did not, however, duplicate the extensive necrosis and disruption of the sensory retina often seen in HCMV chorioretinitis.

As noted earlier, Lussier (1973) studied the necrotizing encephalitis which followed intracranial inoculation of newborn but not weanling mice. He further noted that immunosuppression with antithymocyte serum increased the susceptibility of weanling mice to intracranial MCMV (Lussier, 1975a). Medearis and Prokay (1978) reported the sustained presence of infectious MCMV in the central nervous systems of mice inoculated by systemic routes. They regularly recovered virus from brain homogenates from days 5–60; however, this was not associated with major pathologic changes in the brain. Thus the analogy to congenital human infection is again somewhat tenuous, for marked necrotizing encephalitis with consequent intracranial calcifications and microcephaly are among the hallmarks of severe intrauterine HCMV infection.

f. Hematologic Effects. Thrombocytopenia and anemia can be striking manifestations of congenital HCMV infection, and it has been suggested that isolated thrombocytopenia might be caused by HCMV. Osborn and Shahidi (1973) studied 4-week-old mice inoculated with 10^5 pfu MCMV intraperitoneally and found that the animals underwent a significant thrombocytopenia coincident with viral replication in the spleen and liver. Morphologic and immunofluorescent studies of megakaryocytes suggested direct viral infection as at least a partial mechanism for decreasing circulating platelet numbers, although the rapidity of change in platelet count could not be fully explained by decreased production. They also documented a transient leukocytosis and mild anemia during the acute infection.

g. Latency and Reactivation. One of the more disappointing aspects of the MCMV–mouse model until recently was the failure to establish a convincing system of latency and reactivation. Whereas HCMV seemed clearly to enter a latent state beyond the interval of chronic excretion—from which it could be regularly reactivated under conditions of natural or iatrogenic immunosuppression—MCMV seemed not to do so. Medearis (1964a) established chronic infections of adult mice and then attempted dissemination by treatment with cortisone or methotrexate, without success. Henson was able to convert self-limited to chronic submaxillary gland infection and to establish a model of chronic hepatitis by treatment of mice with cortisone (Henson *et al.*, 1967); again, however, dissemination was not a feature.

Gardner *et al.* (1974) achieved a more striking success in their study of naturally infected wild mice. Antithymocyte serum treatment resulted in disseminated and often lethal infection in a high proportion of mice.

Since 1975, however, reports of latency and reactivation in laboratory mice have begun to appear with some frequency and the model has proved to yield some interesting biologic features, as will be discussed in Section V,B.

h. Vaccine. The eagerness of some investigators to create and use a live attenuated human CMV vaccine in human populations has kindled interest in MCMV as a possible model system. Osborn and Walker (1971) observed that MCMV lost its characteristic virulence for suckling mice after even one passage in tissue culture and that mice infected with such attenuated virus were in fact protected when challenged with a potentially lethal dose of virulent virus. However, they also noted that delayed mortality occurred in nearly half the vaccinated mice.

Howard and Balfour (1977) also showed the protective effect of tissue culture-passaged virus against exogenous challenge; and Medearis and Prokay (1978) used sublethal doses of wild virus to establish immunity in female mice which was protective for subsequent litters if immune females suckled them. Chantler *et al.* (1979) followed mice for a year after inoculation with either virulent or tissue culture-passaged MCMV and noted pregnancy wastage and possible vertical transmission of virus which occurred with equal frequency regardless of inoculum.

Thus, studies with attenuated MCMV as a vaccine have yielded mixed but somewhat worrisome results, upholding the theoretical concern of many virologists when contemplating live attenuated herpesvirus vaccines: Whereas acute virulence properties can be modified and inoculated animals can be protected against exogenous challenge, the attenuated virus itself appears to develop pathogenic properties in a delayed but nonetheless significant manner.

5. Pathology

The pathologic effects of MCMV infection have been alluded to in previous sections. However, several pathologic studies of interest may warrant additional emphasis.

a. Cell Types and Cytoplasmic Fate of MCMV. Ruebner and co-workers (1964, 1966) described electron microscopic findings of MCMV multiplication in various organs of mice. They noted that morphologic features of intranuclear multiplication appeared identical in the several organs studied, whereas the cytoplasmic appearance of assembled virions varied considerably depending on the cell type. Submaxillary gland acinar cells accumulated large numbers of intact virions in their cytoplasm; by contrast, in spleen and liver cells, viral particles were enmeshed in an electron-dense matrix in the cytoplasm and complete virions were rarely seen. It was proposed that this electron-dense material, which was acid phosphatase-positive by histochemical staining, might represent lysosomal activity. The authors hypothesized that the fate of virions budding from the nuclear membrane in those cell types was in effect to enter a "cytoplasmic graveyard," and that the chronicity of infection of submaxillary gland acinar cells reflected their ineffectiveness at cytoplasmic inactivation.

These differences in viral fate depending on cell type have been confirmed and emphasized by Lussier *et al.* (1974). Hudson *et al.* (1976b) also confirmed the morphologic differences between submaxillary glands and other organs but suggested that the dense bodies observed in the liver and spleen might represent the intracellular form of multicapsid virions. Since that group found multicapsid virions to retain their infectivity, such a hypothesis would be inconsistent with the idea of cytoplasmic inactivation.

It has often been asserted that, whereas productive MCMV infection *in vitro* almost always involves cells of fibroblastic origin, a major feature of MCMV infection *in vivo* is the involvement of epithelial cells. However, Mims and Gould (1978a) investigated the distribution of inclusion-bearing cells and of viral antigen by histopathology and immunofluorescence in a variety of epithelia and found that, outside of the salivary gland, epithelial cells were not prominently infected by MCMV.

A final cytological study of note was done by Tegtmeyer *et al.* (1969) to determine the effect of MCMV on chromosomes of fibroblasts in culture. They observed inhibition of mitosis and the induction of chromosomal abnormalities in this system, although the changes occurred relatively late in the infectious cycle when a gross cytopathic effect was well developed.

b. Lymphoid Tissues and Organs. The spleen is the site of abundant viral multiplication during acute MCMV infection initiated by the intravenous or intraperitoneal route. It is noteworthy that Quinnan *et al.* (1980) found lower titers of MCMV in the spleen after intranasal inoculation, although the generation of splenic cytotoxic lymphocytes was equivalent regardless of the route of inoculation. Several descriptions of the histopathology associated with splenic infection have appeared (Howard and Najarian, 1974; Osborn and Shahidi, 1973; Ruebner *et al.*, 1964) and Mims and Gould (1978b) reported a detailed study of changes in the spleen. They found that some mouse strains underwent acute splenic necrosis when doses of more than 10^4 pfu were inoculated systemically. This striking necrotic effect occurred only in weanling or adult mice, even though virus multiplied to higher titers in the spleens of newborn animals. Necrotic changes were first seen in perifollicular areas at 3 days and predominantly involved cells which had no surface markers, although occasional B cells were also seen to be infected. In some animals, residual gross evidence of necrosis was still visible 1–2 months postinfection. The pathogenetic mechanism of this striking effect was not determined. As noted above, its occurrence did not correlate directly with viral titer, and Mims and Gould also concluded that immunologic mechanisms were not directly responsible for the observed pathology, since immunosuppressive treatments did not alter the necrotic effect.

When low doses of MCMV are used to initiate infection, splenomegaly rather than splenic necrosis may result (Kelsey *et al.*, 1977).

In contrast to splenic changes during MCMV infection, lymph nodes are much less strikingly involved (Howard and Najarian, 1974). Occasional cytomegalic cells have been identified, but little necrosis or other histopathologic change occurs in the course of systemic infection.

Most investigators have reported that the thymus is not involved in MCMV infection, using virus titer, viral antigen localization, or histopathologic changes as indicators of infection (Howard and Najarian, 1974; Mims and Gould, 1978a; Selgrade and Osborn, 1974). Furthermore, Olding *et al.* (1976) found no evidence of viral DNA by nuclei acid hybridization in thymuses of latently infected mice. There have been two differing reports concerning thymic involvement (Minamishima *et al.*, 1978; Schwartz *et al.*, 1975), reporting extensive thymic necrosis as a feature of acute infection. Schwartz *et al.* (1975) described a rather dramatic thymic atrophy, runting of survivors, and undue susceptibility to bacteremia following neonatal infection. The pathologic changes described in this report, however, were essentially identical to those ascribed to mouse thymic agent (see below), which may explain these discrepant results, since both MCMV and mouse thymic agent infect submaxillary glands and it is not uncommon to have both agents present in mouse populations. In each study reporting thymic atrophy, the attributes of the inoculum were not rigorously described, so that mixed infection is a distinct possibility.

c. Liver. The hepatitis induced by MCMV varies with the strain of both virus and mouse. McCordock and Smith (1936) presented a detailed description of foci of inflammation surrounding cytomegalic cells, especially in periportal regions. In their report and in most subsequent studies, hepatic parenchymal cells are most frequently infected. Only occasionally are Kupffer cells involved, although Mims and Gould (1978a) have suggested that they may play a barrier role in genetically resistant mice.

The studies of Ruebner *et al.* (1964, 1966) presented electron microscopic and histochemical details of the acute hepatic pathology as contrasted to that in other organs, as noted earlier. Henson *et al.* (1967) found that they could produce a model of chronic hepatitis by treatment of acutely infected mice with cortisone.

Selgrade and Osborn (1974) found that the nature of inflammatory changes in the liver provided the sole histologic clue to observed genetic differences in susceptibility of CBA and C57BL mice to MCMV. The more susceptible C57BL mice regularly had inflammatory cells surrounding inclusion-bearing parenchymal cells, with foci scattered widely in both portal and parenchymal areas. The more resistant CBA mice, by contrast, had no inflammatory cell reaction surrounding inclusion-bearing cells, and inflammatory cell foci were confined to periportal areas.

d. Exocrine and Endocrine Glands. McCordock and Smith (1936), using histopathologic criteria of infection, noted in their initial study that the adrenal glands were regularly and extensively involved during acute MCMV infection. Mims and Gould (1979) have extended this observation and report that a sizable fraction of survivors of neonatal infection have characteristically small adrenals with pale, depleted cortices and a zone of hyaline, eosinophilic, acellular material between the cortex and the medulla. The functional importance of this effect is not known.

Pancreatic infection was also noted by McCordock and Smith (1936), and the pancreas was noted by Osborn and Walker (1971) to be the site of productive viral replication over a relatively long interval. However, histologic changes have not been noted to be extensive, and Mims and Gould (1979) reported that only occasional acinar cells contained viral antigen by immunofluorescence and no infection of islets was observed. Again, the functional consequences of pancreatic infection are not known.

Submaxillary gland pathology has been studied extensively (Brodsky and Rowe, 1958; Henson and Neapolitan, 1970; Henson and Strano, 1972; McCordock and Smith, 1936; Ruebner *et al.*, 1966). Acinar cells are the predominantly infected cell type. They do not appear to die during acute infection, but rather become progressively compressed by chronic inflammatory cells infiltrating the interstitium surrounding infected acini. Whereas scattered infected cells can be demonstrated during chronic infection, it has been noted repeatedly that the extent of viral multiplication does not correlate well with histologic change, and that virus isolation from saliva or salivary gland extracts is a much more sensitive measure of chronic infection than is the identification of cytomegalic cells.

e. Musculoskeletal Tissues. Lussier (1974) reported that newborn mice underwent distinctive pathologic changes of both striated and myocardial muscle and brown fat. Whereas striated muscle and brown fat underwent limited focal necrosis, active foci of myocardial inflammation could be noted as long as 56 days after neonatal infection. Mims and Gould (1979) confirmed the changes in muscle and brown fat and noted that viral antigen could be found in cartilaginous tissue.

f. Central Nervous System. No evidence of pathologic change in the central nervous system as a consequence of infection has been described when indirect routes of inoculation are used, although small amounts of virus can be detected in brains of suckling mice 5–60 days after intraperitoneal inoculation (Medearis and Prokay, 1978). Direct intracranial inoculation of newborn mice results in an extensive necrotizing meningoencephalitis (Lussier, 1973; Margolis and Kilham, 1976), but weanling animals undergo only a mild subclinical meningeal inflammatory response unless immunosuppressed with antithymocyte serum (Lussier, 1975b). Margolis and Kilham (1976) noted that the neuropathologic changes in newborn encephalitis included a striking element of cell fusion and neuronal parasitism, with apparent recruitment of normal by infected neurons.

g. Kidney. The kidney has generally been reported to be less prominently involved in the MCMV than in human CMV infection (Brodsky and Rowe, 1958; Medearis, 1964a). Mims and Gould (1979) reported that viral antigen could be identified primarily in glomeruli and interstitial tissues but not in tubular cells of the kidney, except in very young mice in which an occasional tubular cell also contained antigen.

Lussier (1976) reported that the relative lack of renal involvement in normal mice could be altered strikingly by immunosuppressive treatment. One of the major pathologic effects of MCMV in animals given antihymocyte serum was the development of acute glomerulonephritis, with intranuclear inclusions in glomerular epithelial cells followed by eosinophilic thickening of the glomerular capillary loops, proliferation of capsular epithelial cells, and prominent glomerular sclerosis.

Immunologically mediated glomerulonephritis has also been described by Olding *et al.* (1976), who demonstrated virus–antivirus immune complex deposition in renal glomeruli of all strains of mice they studied during both latent and chronic MCMV infection.

h. Lung. Several studies of interstitial pneumonitis have been reported. Brody and Craighead (1974) inoculated immunosuppressed mice subcutaneously. Beginning 12 days after inoculation, extensive interstitial pneumonitis developed, with maximum virus titers present at 4–6 weeks. The pulmonary disease was frequently fatal and was characterized by the presence of infected and uninfected monocytes in vessels and the accumulation of these cells and proteinaceous fluid in the interstitium and alveoli. Interstitial fibrogenesis developed later in the course of infection. Murphy *et al.* (1975) studied this model further to ascertain the mechanism of monocyte migration across pulmonary membranes. Brody *et al.* (1978) studied the formation of alveolar exudate and proposed that intraepithelial pinocytotic transport of macromolecules played an important role.

Jordan (1978) also studied interstitial pneumonia produced by MCMV, but his investigations were done using the intranasal route of inoculation. Whereas low-dose inocula resulted in subclinical and self-limited changes, doses of more than 10^4 pfu produced as much as 20% mortality from severe, diffuse interstitial pneumonitis which closely resembled that seen in immunocompromised human adults and in congenitally infected newborns. Jordan described the extensive histopathologic changes associated with this higher dose. Normal pulmonary architecture was obliterated by sheets of histiocytes, many of which bore intranuclear inclusions, and by the accumulation of proteinaceous exudate.

6. Transmission

a. Wild Mice. As described earlier, most adult wild mice studied have been found to be persistently infected with MCMV (Gardner *et al.*, 1974; Mannini and Medearis, 1961; Rowe *et al.*, 1962), whereas when newborn wild mice have been screened for virus, they seem to be uninfected. The route of transmission by which such prevalent adult infection is acquired is unknown, although the fact that colostrum is protective and that saliva is generally positive for infectious virus in adult mice suggests that close-contact acquisition after weaning could be expected to occur with reasonable frequency.

b. Laboratory Mice. Mannini and Medearis (1961) established the fact that close contact was important in lateral spread of MCMV. They placed inoculated and uninoculated mice in the same cages and found that uninfected cagemates acquired MCMV readily from infected animals, whereas sentinel mice in cages adjacent to those of infected animals remained virus-free.

Screenings of laboratory mice have generally found MCMV to be a lesser problem than other endemic mouse viral pathogens. However, it is of interest that Mannini and Medearis (1961) found three virus-positive mice among nearly 300 they screened, and all three positive animals were pregnant. As will be discussed below, certain studies have established clearly that MCMV infection may become latent and undetectable by ordinary means, so that the apparently negative results from most mouse colonies may be somewhat misleading.

7. Immune Response

a. Humoral Immunity to MCMV. Mannini and Medearis (1961) studied the appearance of neutralizing antibody during the course of MCMV infection. They concluded that the virus was a relatively weak antigen, since relatively low titers of serum-neutralizing antibody appeared during acute infection and antibody was often not detectable during chronic infection. In a subsequent study, Medearis (1964a) tested sera of chronically infected mice for both neutralizing and complement-fixing (CF) antibody. He found very little antibody during the first month of infection, with subsequent appearance of CF titers of 40–80 during the ensuing 6 months; neutralizing antibody titers remained low (≤ 4) throughout the study. He also found a positive correlation between initial dose of viral inoculum and subsequent CF antibody titer: High doses of inocula produced an antibody response, whereas low doses of MCMV initiated chronic, productive submaxillary gland infection without demonstrable antibody. Similar results of low neutralizing antibody titers were reported by Osborn and Medearis (1967) using a plaque-reduction assay, and it was suggested that this relative unresponsiveness might be a reflection of general humoral immunosuppression (Osborn *et al.*, 1968).

However, Kim and Carp (1973b) subsequently reported that the neutralization of MCMV by hyperimmune rabbit antisera was significantly enhanced by the addition of complement. Since complement was not included in previous studies, this observation could explain in part the apparently weak antigenicity earlier reported. This possibility was reinforced by the study of Araullo-Cruz *et al.* (1978), who identified complement-requiring, neutralizing antibody in MCMV-infected mice as early as 3 days after inoculation. Their results were rather curious in that the earliest antibody response to MCMV was found exclusively in the IgG fraction, whereas later in the infection they could identify neutralizing activity in both IgG and IgM fractions.

b. Secretory Immune Response. Little is known about secretory immunity to MCMV. The sole data that bear on this point come from Medearis' laboratory (Mannini and Medearis, 1961; Medearis, 1964b; Medearis and Prokay, 1978). These showed clearly that colostrum and breast milk conveyed protection to suckling young, although the protective substance was not fully characterized.

c. Cellular Immune Responses to MCMV. Since 1975, numerous studies have addressed various aspects of the cellular immune responses to MCMV antigens during acute MCMV infection. As will be summarized below, acute MCMV infection has been shown to have a suppressive effect on a variety of cellular immunologic functions, so the response to MCMV itself was of considerable interest. Booss and Wheelock (1975) observed that survival during MCMV infection correlated with maintenance of spleen cell responsiveness to the T cell mitogen, concanavalin A. Only in the context of lethal infection were T cells from infected mice unresponsive. They inferred from this correlation that T cells played an important role in the effective host response to MCMV. In a subsequent study, these investigators were able to demonstrate a consistent parallel in time between depression of T cell function and development of signs of clinical illness (Booss and Wheelock, 1977b).

Starr and Allison (1977) studied MCMV infection in congenitally athymic nude mice and found them to be far more susceptible than their heterozygous euthymic littermates. They noted that the livers of resistant mice had a striking cellular inflammatory response, whereas the *Nu/Nu* animals had little hepatic involvement. They were able to convey protection to *Nu/Nu* mice by passive transfer of immune spleen cells, but pretreatment of those cells with anti-θ serum plus complement significantly reduced their protective effect.

Howard *et al.* (1978) studied the kinetics of the appearance of cellular immunity to MCMV antigens by use of an *in vitro* lymphocyte proliferation assay. They found that spleen cells from infected mice proliferated in the presence of MCMV antigens as early as 1 week postinfection. The peak response occurred at 15 days, but spleen cells were still significantly responsive as late as 60 days postinfection.

Quinnan *et al.* (1978) first reported direct MCMV-specific cytotoxic T lymphocyte responses during acute infection initiated by the intraperitoneal route. They found that cytotoxic lymphocytes could be detected as early as 3 days after inoculation, with peak responses at 10 days. In addition to demonstrating that the effector cells were T cells, they showed that the response was virus-specific and *H-2* restricted. In a subsequent study (Quinnan *et al.*, 1980), this group found that intranasal inoculation resulted in an equally vigorous spleen cytotoxic lymphocyte response even though MCMV titers in spleen were significantly lower. They proposed that specific cytotoxic T lymphocytes were an important element in the control of acute MCMV infection.

Quinnan and Manischewitz (1979) have expanded their study of cellular responses in acute MCMV infection. In addition to the specific cytotoxic responses noted above, they reported the appearance of infection on days 3–6 of increased numbers of nonspecific natural killer (NK) and antibody-dependent killer (K) cells in MCMV-infected mouse spleens. These were not genetically restricted and began to increase 24 hr after the appearance in serum of significant levels of interferon. This nonspecific cytotoxicity preceded measurable MCMV-specific humoral or cellular responses. The stimulus for the appearance of these cells is unclear, but it is likely, as the authors suggest, that they constitute an important first line of defense against acute MCMV infection.

Kelsey *et al.* (1978) addressed the question of cellular immune response to a variety of antigens—including MCMV—during acute MCMV infection. They confirmed that spleen cell blastogenic responses to T and B cell mitogens were suppressed in their system but that nonetheless a specific blastogenic response to MCMV could be measured as early as 2 days postinfection and lasted as long as 75 days. The cells mediating this response were sensitive to anti-θ serum. They concluded that a subset of T cells was functionally responsive to MCMV antigens and that the other perturbations of immune and interferon responses caused by MCMV might represent alterations in populations of T cells having a regulatory function.

Finally, Howard *et al.* (1979) investigated the effect of immunosuppression on the immune response to MCMV. Using either prednisolone or antilymphocyte globulin treatments from -2 to $+7$ days (with day 0 as the time of virus inoculation), they found that cellular immune responses to MCMV failed, whereas humoral immune responses were not affected. Preexisting cellular (but not humoral) immunity to MCMV could also be abolished by immunosuppressive treatment.

d. Immune Responses to Other Stimuli during MCMV Infection. In 1968 Osborn *et al.* (1968) reported that adult mice undergoing acute, sublethal MCMV infection had markedly diminished hemolysin and hemagglutinin responses to sheep erythrocytes during the first 2 weeks of MCMV infection. This effect was shown to be a property of virulent virus, since tissue culture passage abolished the suppressive effect (Osborn and Walker, 1971). Howard and Najarian (1974) confirmed and extended the finding of humoral immunosuppression, showing that secondary responses to sheep erythrocytes were also depressed during acute MCMV infection.

It is interesting that Tinghitella and Booss (1979) showed that a significantly enhanced response to sheep erythrocytes could be elicited if the primary stimulus were given in the third or fourth week of MCMV infection. They suggested that a defect in the regulation of immune responses underlay both the suppressive and the hyperresponsive phenomena observed.

Selgrade *et al.* (1976) and Kelsey *et al.* (1977) demonstrated that splenic B and T cell responses to mitogens *in vitro* were normal early in infection but then were significantly depressed beginning between days 3 and 6. This decreased mitogen responsiveness lasted through the second week of infection before returning to normal. Several hypotheses were put forward to explain this suppression, including a possible direct viral effect on spleen cells (Kelsey *et al.*, 1977). In this context, it is noteworthy that Hudson *et al.* (1977) studied the effect of MCMV on mouse spleen cells *in vitro*. They showed that the frequency of productive infection, as demonstrated by infectious center assay, was very low (less than 1% of cells) but that MCMV nevertheless had an inhibitory effect on DNA synthesis in both untreated and mitogen-stimulated spleen cell cultures. Cell viability was not altered and inhibition occurred even when ultraviolet-inactivated MCMV was used.

A number of other studies have documented the decreased mitogen responsiveness of spleen cells taken from MCMV-infected mice (Booss and Wheelock, 1977a; Howard *et al.*, 1974, 1977; Kelsey *et al.*, 1978). Booss and Wheelock (1977a) also ascribed some of the decreased responsiveness to a direct viral effect and postulated that viremia *in vivo* might partially mediate the immunosuppression observed in the intact, infected mouse. However, in their study the immunosuppressive effect was sustained beyond the interval of demonstrable viremia, making it likely that other mechanisms also played an important role.

Loh and Hudson (1980) reported decreased mitogen responses of infected spleen cells *in vitro*. Their studies suggested that infected macrophages had an impaired capacity to mediate the response of T lymphocytes to concanavalin A.

As noted earlier, Kelsey *et al.* (1978) showed that, despite diminished blastogenic responses to a number of B and T cell mitogens, a subset of T cells from spleens of infected mice underwent vigorous blastogenesis when MCMV antigen was used as a stimulus.

Duration of skin graft survival has been used by several groups as a different measure of cellular immune capability during MCMV infection. Howard *et al.* (1974) reported that skin graft survival was significantly prolonged across the strong *H-2* and weak *H-Y* histocompatibility barriers in mice when the graft was performed during the first 5 days of MCMV infection of the recipient mouse. Lang *et al.* (1976) reported briefly that skin grafts applied even 3 weeks after sublethal MCMV infection still had a significantly increased survival time.

A somewhat different result was obtained by Hamilton *et al.* (1978), who studied the survival of Ehrlich ascites cells as a homograft during subclinical MCMV infection of adult mice. Their results varied from the reports of increased skin graft survival in that viral infection did not significantly affect the rate of homograft rejection, even though the recipient mice could be shown to have undergone inhibition of mitogen responsiveness during viral infection.

In summary, acute MCMV infection of mice has been shown by numerous groups to exert a significant but self-limited suppressive effect on a number of immunologic functions, including the humoral response to erythrocyte antigens, the blastogenic response to both B and T cell mitogens, and skin homograft rejection. However, some immune functions appear to remain intact, leading to the suggestion that the specific viral effect is on regulatory cells rather than on T cells as a whole. There are somewhat conflicting reports concerning murine responses to MCMV itself. Neutralizing antibody and CF responses have been delayed in several studies, but this may reflect technical rather than biological variations, since complement may be required for efficient MCMV neutralization. The cellular immune responses to MCMV seem to be brisk, and cytotoxic T lymphocytes have been proposed as an important mechanism of host control of acute MCMV infection.

A final point of interest in this context is the effect of MCMV immunosuppression on the host response to other pathogens. Osborn and Medearis (1967) demonstrated that mice acutely infected with MCMV were briefly permissive for NDV multiplication in the spleen (the mouse is not normally a permissive host for NDV). Hamilton *et al.* (1976) studied the possible synergistic effect of MCMV with concurrent *Pseudomonas aeruginosa* or *Staphylococcus aureus* infections. Using doses of virus (2×10^6 pfu) and concentrations of bacteria which were ordinarily less than 20% lethal for adult mice when given alone, they found that introduction of superinfecting bacteria 3 days after intraperitoneal inoculation of MCMV resulted in 90–100% mortality within 24–48 hr. When *Candida albicans* was used at day 3 as the challenge organism, synergism was again observed, with 80% mortality occurring over the next 5 days.

Hamilton *et al.* (1978) studied the MCMV–*P. aeruginosa* synergism in greater detail and showed that viral titers were not affected but that the increased mortality correlated with a marked increase in bacterial multiplication. Thus it appears that MCMV alters the host response of acutely infected mice sufficiently to facilitate undue susceptibility to other microbial pathogens.

e. Interferon

i. Effects on MCMV Infection. Osborn and Medearis (1966) first reported that MCMV was relatively refractory to the effects of mouse interferon (induced by NDV) and did not induce measurable titers of interferon in cell cultures or in mice. In a subsequent study, they showed that MCMV infection actually caused a profound suppression of the mouse serum interferon response to intravenous challenge with NDV, beginning on day 3 of MCMV infection and lasting through the second week of infection. As noted above, NDV challenge on day 4 actually resulted in NDV replication in spleens of MCMV-infected mice, an effect partially reversed by administration of exogenous mouse serum interferon. At the time of these studies, mouse interferon was known to be of two types, operationally defined as virus-induced and endotoxin-induced interferon. Osborn and Medearis (1967) found that only the former response was suppressed by MCMV infection. In view of both the insensitivity of MCMV to the interferon effect and the failure of its production during acute MCMV infection, they concluded that interferon played a minor role in MCMV infection.

Two studies have reexamined the MCMV–interferon system, with somewhat discrepant results. Oie *et al.* (1975) investigated the effect of interferon on MCMV infection in cell culture and reported that variation in MCMV dosage significantly affected the observed effect of interferon over a wide range. They proposed that an "interferon-resistant fraction" of MCMV might be represented in virus pools. This fraction dominated when high concentrations of virus were used but was diluted out at lower doses of inoculum. However, Araullo-Cruz *et al.* (1978) found no evidence that either exogenous or endogenous interferon influenced the course or viral titer of acute murine infection with MCMV.

ii. Interferon Response during MCMV Infection. Kelsey *et al.* (1977) measured serum interferon levels during acute MCMV infection. They reported that interferon was readily identifiable in the sera of mice 12–36 hours postinoculation with MCMV, but that no interferon was subsequently found and the response to other interferon stimuli was significantly depressed on days 5–9 of MCMV infection. Stringfellow *et al.* (1977) showed that a number of viruses, including MCMV, induced a depressed interferon response state during acute infection. They found that serum from such animals contained a "serum hyporesponsive factor" which might be a soluble mediator of this effect. Tarr *et al.* (1978) reported similar results, with early, brief appearance of interferon in the sera of MCMV-infected mice.

The report of Quinnan and Manischewitz (1979) correlates the early appearance of serum interferon with the increase in NK and K cell activity observed 24 hr later (day 3 of infection). This fits well with current suggestions that interferon may have an important role as a mediator of K cell activity.

Rytel and Hooks (1977) reported that lymphocytes obtained from mice during days 10–14 of MCMV infection produced immune interferon when exposed to MCMV antigens. Since immune interferon is considered to be one of many lymphokines reflecting specific cellular immunity, this finding supported the contention that some facets of cellular immunity were intact during MCMV infection.

f. Macrophages. There have been relatively few studies of the effect of MCMV on macrophages, and those that exist are somewhat contradictory. The first report was that of Tegtmeyer and Craighead (1968), who studied a Smith strain of MCMV, obtained from Rowe, in cultures of unstimulated peritoneal macrophages from CD-1 mice. Infection was initiated with a relatively low dose of virus (120 pfu/ml) After a 72-hr lag, they measured a progressive increase of extracellular viral titers to 2×10^7 pfu/ml by 15 days. Furthermore, they were able to recover virus from thioglycollate-stimulated (but not unstimulated) macrophages obtained from latently infected mice and proposed that the macrophage might play an important role in MCMV latency.

The finding that macrophages were readily permissive for MCMV replication was not confirmed, however, by Selgrade and Osborn (1974); reasons for the differences in these studies remain unclear. Selgrade and Osborn (1974) used either virulent or attenuated MCMV, also of the Smith strain, as well as a strain of MCMV supplied by the American Type Culture Collection. They infected cultures of unstimulated macrophages from either susceptible (C57BL/6) or resistant (CBA) mice but were unable to establish significant, productive infection *in vitro*. Over a 9-day interval, a gradual increase in infectious centers could be demonstrated but there was no difference among virus strains or between sources of mouse macrophages, and titers of infectious virus remained low in the extracellular fluid of the cultures. These investigators nevertheless proposed that macrophages played a protective role *in vivo* in the host response to MCMV, since passive transfer of adult macrophages to syngeneic newborns conveyed significant protection, and administration of silica to adult mice at a dose reported to be selectively toxic to macrophages increased the lethality of MCMV. Selgrade and Osborn proposed that macrophages might exert their protective effect during an early, inductive phase of the cellular immune response.

Aminzadeh *et al.* (1976) provided additional evidence for a protective role of macrophages in MCMV infection. Using an antimacrophage serum (AMS) prepared in rabbits—which had some cross-reactivity with neutrophils and lymphocytes—they showed that AMS treatment of immature mice (age 16 days) rendered them significantly more susceptible. A dose of MCMV which caused 35% mortality in animals treated with normal rabbit serum caused 100% mortality in AMS-treated animals. When 21-day-old mice were studied, mortality went from 0 to 7% with AMS treatment. Furthermore, AMS-treated mice showed more widespread histopathologic evidence of viral infection, with prominent infection of adrenals, lungs, and kidneys (which were not infected in mice that did not receive AMS).

Mims and Gould (1978a) provided additional data on the interaction of MCMV with macrophages *in vitro* and *in vivo*. Using an MCMV strain originally obtained from Osborn, their findings in tissue culture were similar to those reported by Selgrade and Osborn (1974). The efficiency of infection of cultured peritoneal macrophages was low (less than 5%), although eight passages of MCMV through tissue culture seemed partially to adapt the virus to growth in macrophages, with an increase in efficiency of as much as 15-fold. However, efficiency of infection *in vitro* was decreased when so-called activated macrophages were used and was further decreased if macrophages were obtained from mice on day 6 of MCMV infection. There was a slight effect of age in that macrophages from 3-day-old mice were two- to threefold more permissive than those from older animals.

Loh and Hudson (1979) reported the variable interactions of MCMV with splenic B and T cells and macrophages separated and infected *in vitro*. B lymphocyte subfractions were permissive as tested by infectious center assays, whereas T cells were not. Macrophages took up the bulk of radiolabeled input virus but were only briefly permissive.

In their studies *in vivo*, Mims and Gould (1978a) described the initial infection of Kupffer cells in the liver, which appeared to occur at approximately the same rate on day 1 whether virulent or attenuated virus was used. However, virulent virus also caused direct infection of hepatic parenchymal cells, whereas attenuated virus did not spread. They further reported observing little or no evidence of infection in spleen or lymph node macrophages. From these several findings, they concluded that macrophages were relatively insusceptible to MCMV, that they might play a role in MCMV infection in restricting viral replication, but that observed major differences in susceptibility to MCMV among mouse strains could not be explained by variation in macrophage function.

A report by Brautigam *et al.* (1979) completes the cycle of conflicting results in that it is most similar to the original report of Tegtmeyer and Craighead (1968). Using thioglycollate-stimulated macrophages, these investigators found that cells from any of six strains were permissive for MCMV infection. They inoculated macrophages with a dose of 5 pfu per cell. A lag of 5 days occurred, after which viral titers rose rapidly to 10^7 pfu/ml by day 9. When unstimulated macrophages were used, less than 20% were productively infected as measured by infectious center assays, but *in situ* hybridization showed the MCMV genome to be present in 82% of cultured cells. They proposed that the macrophage was therefore an important permissive cell for replication and a potential vehicle for the establishment of latency.

At present, therefore, there is general agreement among investigators that mouse macrophages may play a role in the host response to MCMV infection. However, there are inexplicably wide differences in results from different laboratories as to the permissiveness of macrophages for MCMV. One possible explanation is that MCMV strains may vary widely in some aspects of their biologic behavior, even when they bear the common Smith strain designation.

8. Chronicity and Latency

The subject of chronic infection has been discussed at length in Section C. In summary, it is possible to demonstrate with most strains of MCMV that susceptible mice, once infected, may continue to undergo productive infection of submaxillary glands for many months, if not for the lifetime of the mouse. Studies of wild mouse populations have documented regularly that more than half of a cross-sectional sample will have MCMV in their saliva, suggesting that this lifelong infection may occur in a natural setting as a rule.

The kidney has been found by most investigators to be less productively infected with MCMV and for a shorter interval than is the case with humans infected with CMV (Brodsky and Rowe, 1958; Medearis, 1964a). However, Mims and Gould (1979) found that the kidney in some animals was a more important site of chronic productive infection than the salivary gland.

The issue of MCMV latency has been subjected to intense study because of its possible relevance to human disease caused by human CMV. The first report of successful reactivation of MCMV from latency was that of Tegtmeyer and Craighead (1968), who found that thioglycollate-stimulated (but not unstimulated) macrophages from presumed latently infected mice could be induced to produce infectious MCMV. This report has been reinforced by Brautigam *et al.* (1979), who showed that even apparently uninfected macrophages could be shown to contain the MCMV genome by *in situ* hybridization techniques. Thus the macrophage has been proposed as one site of MCMV latency.

Several investigators have provided evidence that lymphoid cell populations can harbor latent MCMV. Henson *et al.* (1972) reported successful reactivation of MCMV from both the spleen and lymph nodes by cocultivation. Subsequently, the hypothesis gained credence that immunologic events involved in homograft rejection reactions mediated the reactivation of MCMV. Wu *et al.* (1975) reported that infected mice given incompatible skin grafts showed increased MCMV titers in the spleen and submaxillary glands during the ensuing host-vs.-graft reaction. Olding *et al.* (1975) studied latently infected mice of several strains. They established latent infection either by infecting pregnant mice 3–4 days prior to delivery with 100 $TCID_{50}$ or, alternatively, by inoculating newborn mice intraperitoneally with 100 $TCID_{50}$. They then studied surviving, virus-negative progeny or inoculated mice as adults. When spleen cells from such mice were cocultivated with syngeneic mouse embryonic fibroblasts, no MCMV was produced, whereas cocultivation with allogeneic cells yielded infectious MCMV after a 2–3-week lag. Sonication of spleen cells before cocultivation prevented reactivation of the virus even when allogeneic feeder layers were used. They further provided evidence that the latently infected cells were B rather than T cells and that relatively few cells contained the MCMV genome in a latent state.

Conflicting data have come from other laboratories concerning the requirement for a homograft reaction to reactivate MCMV. Mayo *et al.* (1977) reported that transfer of spleen cells from previously MCMV-infected mice converted susceptible syngeneic recipients to virus positivity, but that there was a significantly greater likelihood of reactivation when allogeneic recipients were used. This group also studied the effect of graft-vs.-host reactions on MCMV infection (Dowling *et al.*, 1977). Using infected F_1 hybrid mice and multiple injections of parental spleen cells, they found that the resultant graft-vs.-host reaction provoked dissemination of MCMV to the spleen, liver, lung, and kidney. They proposed that the cell lytic effect of the graft-vs.-host reaction might have facilitated dissemination to a variety of organs, in contrast to the host-vs.-graft reaction, which resulted only in increased viral titers in the spleen and kidney.

Cheung and Lang (1977b) reported that blood transfusions from latently infected mice commonly resulted in infection of susceptible allogeneic mice, whereas syngeneic recipients rarely became infected. Mayo *et al.* (1978) presented data suggesting that either allogeneic or syngeneic tissues could induce spleen cells to reactivate latent MCMV infection, and Wu and Ho (1979) subsequently reported that T as well as B cells were latently infected. They demonstrated that T cells produced MCMV under conditions of cocultivation even when kept in chambers to avoid direct contact with fibroblasts; B cells, however, required direct contact with feeder layers in order to reactivate MCMV. These reactions could be demonstrated whether the fibroblasts were syngeneic or allogeneic. The investigators proposed that B cells might be a more important vehicle of latency than T cells in that the latter were more readily reactivable and therefore might be more susceptible to elimination by host immune surveillance.

Other facets of latent lymphoid cell infection have been reported. Olding *et al.* (1976) reaffirmed their finding that, in their system, allogeneic interactions were necessary for reactivation of latent MCMV. They also used DNA–DNA reassociation techniques to search for MCMV genomes in various tissues of latently infected adult mice. They reported finding three to four MCMV genome copies per 100 spleen cells and one to two copies per 100 salivary gland cells. By contrast, no viral DNA sequences were detected in the thymus, liver, kidney or brain.

The latent infection of spleen cells has been demonstrated by Wise *et al.* (1979) using an explant technique which they found to be more sensitive than either cocultivation or nucleic acid hybridization methods. They were able to activate and recover MCMV regularly from spleen explants even though DNA–DNA reassociation was negative at the level of sensitivity of five genomes per 100 cells and cocultivation did not yield virus.

Latent infections of other tissues have also been reported. Submaxillary glands which were initially virus-negative could be provoked to virus production by cyclophosphamide (Mayo *et al.*, 1977) or explant culture (Cheung and Lang, 1977a). Jordan *et al.* (1977) and Shanley *et al.* (1979) demonstrated reactivation of MCMV in both submaxillary glands and spleens after treatment of latently infected mice with antilymphocyte serum alone. When cortisone as well as antilymphocyte serum was used, widespread viral dissemination resulted.

Latent infection of reproductive tract tissues has also been documented. Young *et al.* (1977) mixed mouse sperm with MCMV prior to artificial insemination. After retrieval of the resultant embryos, 1 of 28 pools of mouse embryonic fibroblasts yielded kMCMV but only after two blind passages in culture. Cheung and Lang (1977a) also reported MCMV replication after cocultivation of prostatic tissue. Dutko and Oldstone (1979) reported finding four to six viral genome equivalents per 100 cells in mouse testes and localized the latent viral genomes to spermatogonia.

In summary, evidence is strong that MCMV can produce latent infection of B cells, probably T cells, submaxillary glands, and possible prostatic and testicular cells as well. Although allogeneic tissue reactions have been reported by some to be a requirement for reactivation of MCMV from latency, recent evidence suggests that this is not a necessary condition.

D. Epizootiology

The topics of MCMV occurrence in wild mice and spread among laboratory mice have been covered in detail in Section II,C,3.

E. Techniques for Identification and Handling of MCMV

1. Isolation and Identification

a. Cytopathic Effect in Cell Culture. The use of fibroblastic murine cells in culture permits ready presumptive identification of MCMV from test specimens. Either secondary mouse embryo fibroblasts or mouse 3T3 cells are readily permissive for MCMV infection. Usually 24–48 hr elapse after inoculation before the appearance of foci of rounded cells with increased refractility. These foci expand to form lytic plaques with gaps in the cell monolayer surrounded by characteristic enlarged cells which have large intranuclear inclusions and margination of chromatin. The cytopathic effect is sufficiently distinctive to allow a presumptive identification of MCMV, but this can be affirmed by demonstrating the limitation of the permissive host cell range and by staining with specific hyperimmune immunofluorescent sera.

Additionally, when putative virus is recovered from submaxillary glands, virulence for sucklings and infection of the liver and spleen with sparing of the thymus may be helpful to ensure that the mouse thymic agent is not a contaminant (see Section III).

b. Quantitation of MCMV. Since MCMV is readily released into the extracellular fluid, measurement of virus concentration in test specimens must be done either by tube dilution technique or, preferably, by plaque enumeration after incubation with a semisolid gel to prevent secondary plaque formation. Several variations of the latter method have been used. The chief potential problem with MCMV plaque enumeration is the inhibitory effect of agar on viral multiplication. This problem can be circumvented by the use of agarose or starch gels with subsequent neutral red staining. Alternatively, methocel or tragacanth overlays can be used, with the advantage that they can be removed after incubation, and monolayers can then be fixed and stained to facilitate plaque enumeration. By any of these means, microplaques can be counted under a dissecting microscope by 4 days postinoculation or by direct visualization at 6 days. Since variability in plaque size occurs even after cloning (Osborn and Walker, 1971), it may be preferable to use a dissecting microscope even when most plaques are visible to the naked eye.

As discussed in Section II,B,4, the efficiency of adsorption of MCMV can be increased 10- to 100-fold by centrifugation of inoculum against monolayers grown either on coverslips or in 1-oz bottles in a basket rotor (Osborn and Walker, 1968; Hudson *et al.*, 1976a).

c. Tissue Extraction, Cell Association, and Cocultivation. MCMV is readily released from tissue culture cells, and the yield is not significantly enhanced by harvesting infected cells or cell sonicates. Similarly, acutely infected organs and tissues to be assayed for MCMV can be ground manually and efficiently extracted without the need for sonication. However, reports reviewed in Section II,C,5 suggest that latency may occur later in MCMV infection after chronic viral excretion has ceased. Under these conditions, cocultivation of intact cells is necessary for rescue of infectious virus, usually with a lag time of 2 or more weeks. As discussed above, there is some disagreement in the literature as to whether there is a need for allogeneic feeder layers to stimulate expression of the latent MCMV genome. The most recent reports suggest that syngeneic and allogeneic cells work equally well.

Finally, Wise *et al.* (1979) report that explant techniques are even more sensitive than cocultivation for detection of latent MCMV.

2. Preparation and Maintenance of Virus Pools

Optimally, MCMV stock virus is prepared by intraperitoneal inoculation of 10^5 pfu of MCMV into 3-week-old mice. When a new stock virus is received from another laboratory in limited quantity, it is prudent to bracket this estimated dose by fourfold in either direction to allow for variation in mouse strains and virus storage conditions. An ideal dose produces 10–20% mortality in recipients, with deaths occurring at 4–8 days postinoculation. Submaxillary glands of surviving mice are best harvested at 14–21 days postinfection, and titers of more than 10^7 pfu/ml of 10% submaxillary gland extract are usual. Tissue suspensions, clarified by low-speed centrifugation, are optimally stored with a stabilizer (10% DMSO or 25% sorbitol) at −70°C.

It has been the practice of this laboratory to repeat mouse-to-mouse passage at approximately 6-month intervals since even with osmotic stabilization, freezer-held virus on occasion drops in titer with time for undetermined reasons.

When preparing pools of tissue culture-passaged virus, Muller and Hudson (1977b) report that pools of virus are best prepared by inoculation of 3T3 cells which have been synchronized by serum activation of quiescent G_1 arrested cells. They report yields of greater than 10^8 pfu/ml following high-multiplicity inoculation (20–30 pfu per cell).

3. Serologic Techniques

Complement-fixing antibody tests have been found to be relatively insensitive for identification of humoral immunity to MCMV. Neutralizing antibody detection was similarly unsatisfactory in early studies, but the report of Araullo-Cruz *et al.* (1978) that the addition of complement facilitates neutralization suggests that plaque reduction neutralization may be the method of choice for detection of specific humoral immunity.

4. Cellular Immunity Techniques

As reviewed in Section II,C,4, several laboratories have described techniques for studying manifestations of cellular immune responses to MCMV. These include blastogenic transformation of lymphocytes in the presence of MCMV antigen (Kelsey *et al.*, 1978), lymphocyte proliferation assay (Howard *et al.,* 1978), lymphocyte-mediated cellular cytotoxicity (Quinnan *et al.,* 1978), and production of immune interferon (Rytel and Hooks, 1977). These studies are as yet sufficiently preliminary and diverse that precise recommendations concerning optimal measurement of cellular immunity to MCMV would be premature.

5. Hyperimmune Sera

Rabbits can be induced to respond with antibody to repeated injections of MCMV. The resultant antibody neutralizes MCMV in the presence of complement (Kim and Carp, 1973b) and provides one kind of useful serologic tool for study. However, such sera are difficult to rid of nonspecific staining properties when used in direct immunofluorescent studies, even with extensive adsorption against various mouse tissues. A satisfactory alternative for immunofluorescent staining is the hyperimmunization of adult mice (Osborn and Shahidi, 1973), which yields potent sera which are virtually free of nonspecific staining properties upon direct conjugation with fluorescein isothiocyanate.

6. Purity of Mouse MCMV Pools

Because of the requirement for mouse-to-mouse passage to maintain virulence of MCMV pools, contamination with other endogenous mouse viruses is a potential problem. Mouse thymic agent is of particular concern because of its ability to establish chronic salivary gland infection without other demonstrable effects on adult mice (see Section III). Contamination with mouse thymic agent can be detected by the occurrence of thymic atrophy in newborn inoculees. Lactic dehydrogenase-elevating virus is another easily acquired noncytopathic agent which can contaminate pools of MCMV. Both of these agents can be biologically eliminated by passage through secondary mouse embryonic fibroblast cultures, which are nonpermissive for the contaminants, followed by two or more passages in virus-free mice to reestablish virulence (Osborn and Walker, 1971). Other known mouse viruses should be screened for periodically because of the steady possibility of their introduction into MCMV stocks during mouse-to-mouse passage.

F. Control and Prevention

1. Control

Techniques for control of MCMV in laboratory mouse colonies have not been developed, since spontaneous infection seems to be infrequent in this setting. It may be most efficient, in screening for prevalence of infection in colonies, to test the saliva of pregnant mice selectively (Mannini and Medearis, 1961).

Since cage-to-cage transmission has not been demonstrated, the use of filter tops for mouse cages probably represents excessive prudence, but it can serve as a useful protective barrier against superinfection of experimental mice with other unwanted mouse viruses.

2. Prevention

No pragmatic need exists for active MCMV prevention at present. However, several studies have employed vaccines derived from MCMV as possible model systems for the assessment of vaccine approaches to human CMV.

Osborn and Walker (1971) first reported that as little as one tissue culture passage of MCMV resulted in a marked decrease in acute lethality for suckling mice. Furthermore, the suppressive effect on humoral immunity and interferon response were also lost in tissue culture passage, along with the ability to multiply in the livers and spleens of weanling mice. However, the suckling mice that received vaccine virus began to die during the second week of infection with a delayed, cumulative mortality of as much as 50%. Surviving animals were fully protected against otherwise lethal challenge doses of exogenous virus.

Selgrade *et al.* (1976) also found that attenuated virus was not entirely innocuous. They reported that, although suppression of mitogen responses was a regular property of virulent virus, high doses of attenuated MCMV could also suppress the blastogenic response to phytohemagglutinin (but not to concanavalin A).

Hamilton *et al.* (1976) reported the use of attenuated MCMV in their studies of mixed infections. They found that vaccine virus ablated the synergistic effect of MCMV infection and superinfection with *P. aeruginosa, S. aureus,* or *C. albicans.* Howard and Balfour (1977) reported that attenuated virus given at 9–12 days of age conferred protection against subsequent challenge 2 weeks later with virulent virus and prevented the immunosuppressive effect of MCMV as demonstrated by normal skin graft survival times.

Minamishima *et al.* (1978) studied both inactivated and attenuated MCMV as possible approaches to active immunization. They found that inactivated MCMV had no protective effect but that attenuated virus protected 4–5-week-old ICR mice against challenge doses of 10^6 pfu of virulent virus. Using hyperimmune rabbit sera raised against the attenuated virus, they found only a one-way cross-reactivity between virulent and attenuated viruses; virulent virus was not neutralized by such serum despite the observed protection. This differs from the findings of Osborn and Walker (1971), who did not detect an antigenic change in association with the development of attenuation.

In summary, a number of laboratories have reported that live, attenuated MCMV prepared by brief passage through cell culture will protect against challenge with virulent virus. However, the observation that such protected animals suffer delayed effects of their endogenous vaccine virus should lend caution to the efforts to establish live, attenuated human CMV vaccines.

3. Chemotherapy and Chemical Inactivation

Very little has been reported concerning chemical treatment of MCMV. Kim and Carp (1973a) showed that MCMV was rapidly inactivated by proteolytic enzymes such as trypsin or chymotrypsin. Kim *et al.* (1978) reported that murine and human CMV strains were inactivated by butylated hydroxytoluene, an antioxidant that is presumed to act by degradation of lipid-containing structures.

Several studies with candidate chemotherapeutic agents have led to mixed success, in that efficacy against MCMV *in vitro* is fairly uniform, whereas protection of infected animals has generally not been found. Dowling *et al.* (1976) reported that ribavirin was effective in decreasing both the cytopathic effect and the virus yield in cell cultures infected with MCMV; however, it was completely ineffective in treating mice either acutely or chronically infected with MCMV. Similarly, Kelsey *et al.* (1976) demonstrated that both cytosine arabinoside and 5-iodo-2′-deoxyuridine were significantly inhibitory for MCMV in tissue culture but were ineffective in mice. In a subsequent study, this group found that adenine arabinoside was also effective *in vitro* but not *in vivo,* but their results with phosphonoacetic acid were more promising (Overall *et al.*, 1976). With the latter agent, they were able to provide significant protection to mice as late as 24 hr after a potentially lethal dose of MCMV. When lower doses of inoculum were used, protection (as measured by decreased titers of virus in organs) could be conferred as late as 48 hr after virus inoculation. Huang *et al.* (1976) also demonstrated the efficacy of phosphonoacetic acid in inhibition of MCMV DNA synthesis in tissue culture.

Thus phosphonoacetic acid stands as the sole exception to date to the generalization that *in vitro* viral inhibition does not correlate with efficacy in the intact mouse in the chemotherapy of MCMV.

III. MOUSE THYMIC VIRUS

A. Introduction

In 1961 Rowe and Capps published data reporting the existence of a second mouse herpesvirus which they called *mouse thymic virus*. The agent was isolated during study of mouse mammary tumor virus and was noted to cause isolated necrosis of the thymus when inoculated into mice less than 10 days of age (Rowe and Capps, 1961). The technical difficulty which hampered their initial studies—notably, lack of a cell culture assay system—has persisted, and subsequent published studies have been limited. However, several additional reports have since appeared characterizing the virus morphologically and measuring the immunologic consequences of infection of neonatal mice. The mouse thymic virus has intrinsically interesting biologic properties, but in addition, it represents a source of potential confusion in studies of MCMV. This section will be devoted to a summary of available information about this second herpesvirus of mice.

B. Properties of the Virus

The morphology of mouse thymic virus is entirely consistent with that of other herpesviruses. Parker *et al.* (1973) described its electron microscopic appearance in infected thymocytes; intranuclear particles measured approximately 100 nm in diameter, and cytoplasmic and extracellular particles measured 135 nm. Furthermore, the virus was shown to have properties of heat and ether lability compatible with membership in the herpesvirus group. In addition to virion physical properties, the intranuclear inclusions seen in infected thymic lymphocytes were similar to those caused by other herpesviruses, although Parker *et al.* (1973) reported that margination of chromatin was less striking than that caused by MCMV.

In an effort to establish a permissive tissue culture system, Rowe and Capps (1961) attempted to grow the virus in primary mouse embryo and mouse kidney monolayers, in explant cultures of mouse thymus and spleen, and in several continuous lines of mouse tumor cells. None of these efforts yielded either a cytopathic effect or cell-free virus. Although several reports have appeared concerning biologic effects of the virus in mice, no permissive system has yet been found for study of the agent *in vitro,* and therefore no information is available concerning the replicative cycle, viral genome, or gene products.

C. Pathogenesis

The mouse thymic virus has not caused lethal disease in any mouse strain tested. When inoculated intraperitoneally, it causes macroscopic disease solely in the thymus, and that occurs only if mice are less than 10 days of age at inoculation (Rowe and Capps, 1961). For optimal demonstration of the pathologic effect, 24-hr-old mice are inoculated intraperitoneally. Using the effect on suckling mice as an assay system, it was possible to demonstrate a viremia between 3 and 7 days postinfection. Virus was present in fairly high concentrations in the thymus, with peak titers on day 7 and a rapid decline thereafter (Rowe and Capps, 1961; Cross, 1973). Viral assay of long-term survivors of neonatal infection yielded occasional positive results from blood, suggesting a recurrent, cyclic viremia; and frequent positive isolations from mouth swabs and/or salivary glands were made over a period of several months (Cross, 1973). When mice were inoculated as adults, only salivary glands yielded positive viral isolations. When the intracerebral route of inoculation was used in newborn mice, results were similar to those following intraperitoneal inoculation.

The pathologic changes caused by the mouse thymic virus were restricted to the thymus, with the exception that an occasional lymph node cell was seen to have immunofluorescence-positive intranuclear inclusions and there was a fairly diffuse lymphadenopathy which was not associated with viral antigen by immunofluorescence (Cross, 1973).

In the thymus, progressive necrotic changes were observed, first evident on day 5 and lagging significantly behind peak viral titers. Initial necrosis was seen in the medulla but soon extended to include cortical areas. Virtually all thymocytes contained antigen-positive intranuclear inclusions as the necrosis progressed, but subcapsular collections of large, blastlike thymocytes appeared to be spared (Cohen *et al.*, 1975). Macroscopically, the thymuses of infected animals shrank to 15% or less of control weight. The peak necrotic effect occurred at days 12–14, but full recovery did not occur until 6–8 weeks postinfection. It was interesting that granulomatous changes were frequently seen during the recovery phase (Cross, 1973). Careful immunofluorescent search has yielded only occasional antigen-positive cells in a variety of other organs and tissues, and no gross pathology is found outside the thymus.

Several alterations in immunologic responses accompany the neonatal infection with mouse thymic virus. During acute infection, spleen cells were found to have markedly diminished reactivity to T cell mitogens and to allogeneic cells. However, normal numbers of immunoglobulin-bearing cells were present in spleens, and thymic cell reactivity to mitogens was unimpaired (Cohen *et al.*, 1975). Four and 8 weeks after neonatal infection, there was a profound reduction (70–80%) in direct graft-vs.-host reactivity of thymocytes (Cross *et al.*, 1976). Lesser effects were noted in other cell populations. Direct graft-vs.-host reactivity of spleen cells was markedly depressed 4 weeks after infection but returned to normal at 8 weeks. Graft-vs.-host reactivity of lymph node cells was unaffected, whereas amplifier T cell activity in lymph node cells was markedly depressed at 8 weeks. B cell activity matured rapidly despite thymic virus infection as measured by plaque-forming responses to type III pneumococcal polysaccharide (Morse *et al.*, 1976), and suppressor T cell activity was similarly unaffected. In summary, the functional effects of mouse thymic virus were highly selective for specific subpopulations of T cells.

The mode of transmission of mouse thymic virus is not known, but the frequent occurrence of the agent in salivary glands and isolation from mouth swabs of infected adults suggest that horizontal spread could occur readily. Cross (1973) found that 4 of 15 breeder colonies had adult mice positive for mouse thymic virus both by antibody screening and by isolation of virus from saliva. Similarly, 15 wild mice trapped in the area of Frederick, Maryland, were studied and four were found to be positive for the mouse thymic virus. Cross also found MCMV in both those settings. No antigenic similarities were found between the two mouse herpesviruses by immunofluorescence.

Immune response to mouse thymic virus was found to be a direct function of age at the time of infection. Cross (1973) and Cohen *et al.* (1975) were unable to detect any antibody response in animals inoculated as newborns. However adult mice developed high titers of neutralizing and complement-fixing antibodies by day 7 postinfection, and their sera reacted with viral antigen by immunofluorescence. The neutralizing antibody test was noted to be relatively unsatisfactory in that endpoints (prevention of thymic necrosis in newborn mice) were indistinct (Cross, 1973). No information is available about cellular immunity or the interferon response to mouse thymic virus.

Rowe and Capps (1961) documented considerable variability among mouse strains with respect to susceptibility to the mouse thymic virus. Newborns of some strains uniformly underwent extensive thymic necrosis, whereas mice of other strains had less regular necrosis, sometimes limited to a focal area of the thymus.

D. Control and Prevention

Even with the preliminary data available, it is possible to conclude that the mouse thymic agent is fairly widespread in nature and in breeder colonies of mice. It is likely to represent a significant source of contamination of MCMV pools since both viruses cause chronic, productive salivary gland infection and can readily coinfect the same host. Additional detailed information about mouse thymic virus, including means of prevention and control, awaits improved techniques for its study.

REFERENCES

Aminzadeh, M., Lussier, G., and Boudreault, A. (1976). Effect of anti-mouse macrophage serum on murine cytomegalovirus infection. *Rev. Can. Biol.* **35**, 11–16.

Araullo-Cruz, T. P., Ho, M., and Armstrong, J. A. (1978). Protective effect of early serum from mice after cytomegalovirus infection. *Infect. Immun.* **21**, 840–842.

Booss, J., and Wheelock, E. F. (1975). Correlation of survival from murine cytomegalovirus infection with spleen cell responsiveness to concanavalin A. *Proc. Soc. Exp. Biol. Med.* **149**, 443–446.

Booss, J., and Wheelock, E. F. (1977a). Role of viremia in the suppression of T-cell function during murine cytomegalovirus infection. *Infect. Immun.* **17**, 378–381.

Booss, J., and Wheelock, E. F. (1977b). Progressive inhibition of T-cell function preceding clinical signs of cytomegalovirus infection in mice. *J. Infect. Dis.* **135,** 478–481.

Brautigam, A. R., Dutko, F. J., Olding, L. B., and Oldstone, M. B. A. (1979). Pathogenesis of murine cytomegalovirus infection: the macrophage as a permissive cell for cytomegalovirus infection, replication and latency. *J. Gen. Virol.* **44,** 349–360.

Brodsky, I., and Rowe, W. P. (1958). Chronic subclinical infection with mouse salivary gland virus. *Proc. Soc. Exp. Biol. Med.* **99,** 654–655.

Brody, A. R., and Craighead, J. E. (1974). Pathogenesis of pulmonary cytomegalovirus infection in immunosuppressed mice. *J. Infect. Dis.* **129,** 677–689.

Brody, A. R., Kelleher, P. C., and Craighead, J. E. (1978). A mechanism of exudation through intact alveolar epithelial cells in the lungs of cytomegalovirus-infected mice. *Lab. Invest.* **39,** 281–288.

Chalmer, J. E. (1979). Genetics of infection by murine cytomegalovirus. Ph.D. Thesis, Univ. of Western Australia, Perth.

Chalmer, J. E., Mackenzie, J. S., and Stanley, N. F. (1977). Resistance to murine cytomegalovirus linked to the major histocompatibility complex of the mouse. *J. Gen. Virol.* **37,** 107–114.

Chantler, J. K. (1978). The use of hypertonicity to selectively inhibit host translation in murine cytomegalovirus-infected cells. *Virology* **90,** 166–169.

Chantler, J. K., and Hudson, J. B. (1978). Proteins of murine cytomegalovirus: identification of structural and nonstructural antigens in infected cells. *Virology* **86,** 22–36.

Chantler, J. K., Misra, V., and Hudson, J. B. (1979). Vertical transmission of murine cytomegalovirus. *J. Gen. Virol.* **42,** 621–625.

Cheung, K.-S., and Lang, D. J. (1977a). Detection of latent cytomegalovirus in murine salivary and prostate explant cultures and cells. *Infect. Immun.* **15,** 568–575.

Cheung, K.-S., and Lang, D. J. (1977b). Transmission and activation of cytomegalovirus with blood transfusion: a mouse model. *J. Infect. Dis.* **135,** 841–845.

Cohen, P. L., Cross, S. S., and Mosier, D. C. (1975). Immunologic effects of neonatal infection with mouse thymic virus. *J. Immunol.* **115,** 706–710.

Cross, S. S. (1973). Development of bioassays and studies on the biology of mouse thymic virus. Ph.D. Thesis, George Washington Univ., Washington, D.C.

Cross, S. S., Morse, H. C., III, and Asofsky, R. (1976). Neonatal infection with mouse thymic virus. Differential effects on T cells mediating the graft-*versus*-host reaction. *J. Immunol.* **117,** 635–638.

Cross, S. S., Parker, J. C., Rowe, W. P., and Robbins, M. L. (1979). Biology of mouse thymic virus, a herpesvirus of mice, and the antigenic relationship to mouse cytomegalovirus. *Infect. Immun.* **26,** 1186–1195.

Cruz, J. R., and Waner, J. L. (1978). Effect of concurrent cytomegaloviral infection and undernutrition on the growth and immune response of mice. *Infect. Immun.* **21,** 436–441.

Davis, G. L., and Hawrisiak, M. M. (1977). Experimental cytomegalovirus infection and the developing mouse inner ear. *Lab. Invest.* **37,** 20–29.

Davis, G. L., Krawczyk, K. W., and Hawrisiak, M. M. (1979). Age-related neurocytotropism of mouse cytomegalovirus in explanted trigeminal ganglia. *Am. J. Pathol.* **97,** 261–275.

Diosi, P., Arcan, P., Babusceac, B., and Moldovan, E. (1972). Identification of a cytomegalovirus recovered from field mice. *Pathol. Microbiol.* **38,** 178–183.

Dowling, J. N., Postic, B., and Guevarra, L. O. (1976). Effect of ribavirin on murine cytomegalovirus infection. *Antimicrob. Agents Chemother.* **10,** 809–813.

Dowling, J. N., Wu, B. C., Armstrong, J. A., and Ho, M. (1977). Enhancement of murine cytomegalovirus infection during graft-*versus*-host reaction. *J. Infect. Dis.* **135,** 990–994.

Dutko, F. J., and Oldstone, M. B. A. (1979). Murine cytomegalovirus infects spermatogenic cells. *Proc. Natl. Acad. Sci. U.S.A.* **76,** 2988–2991.

Eizuru, Y., Minamishima, Y., Hirano, A., and Kurimura, T. (1978). Replication of mouse cytomegalovirus in thymidine kinase-deficient mouse cells. *Microbiol. Immunol.* **22,** 755–764.

Gardner, M. B., Officer, J. E., Parker, J., Estes, J. D., and Rongey, R. W. (1974). Induction of disseminated virulent cytomegalovirus infection by immunosuppression of naturally chronically infected wild mice. *Infect. Immun.* **10,** 966–969.

Gonczol, E., Stone, J., and Melero, J. M. (1978). The effect of heat-inactivated murine cytomegalovirus on host DNA synthesis of different cells. *J. Gen. Virol.* **39,** 415–426.

Hamilton, J., and Overall, J. C., Jr. (1978). Synergistic infection with murine cytomegalovirus and *Pseudomonas aeruginosa* in mice. *J. Infect. Dis.* **137,** 775–782.

Hamilton, J., Overall, J. C., Jr., and Glasgow, L. A. (1976). Synergistic effect on mortality in mice with murine cytomegalovirus and *Pseudomonas aeruginosa, Staphylococcus aureus,* or *Candida albicans* infections. *Infect. Immun.* **14,** 982–989.

Hamilton, J., Fitzwilliam, J. F., Cheung, K. S., Shelburne, J., Lang, D. J., and Amos, D. B. (1978). Viral infection-homograft interactions in a murine model. *J. Clin. Invest.* **62,** 1303–1312.

Heggie, A. D., and Gaddis, L. (1979). Effects of viral exposure of the two-cell mouse embryo on cleavage and blastocyst formation *in vitro*. *Pediatr. Res.* **13,** 937–941.

Henson, D., and Neapolitan, C. (1970). Pathogenesis of chronic mouse cytomegalovirus infection in submaxillary glands of C3H mice. *Am. J. Pathol.* **58,** 255–267.

Henson, D., and Strano, A. J. (1972). Mouse cytomegalovirus. Necrosis of infected and morphologically normal submaxillary gland acinar cells during termination of chronic infection. *Am. J. Pathol.* **68,** 183–195.

Henson, D., Smith, R. D., Gehrke, J., and Neapolitan, C. (1967). Effect of cortisone on nonfatal mouse cytomegalovirus infection. *Am. J. Pathol.* **51,** 1001–1011.

Henson, D., Strano, A. J., Slotnik, M., and Goodheart, C. (1972). Mouse cytomegalovirus: isolation from spleen and lymph nodes of chronically infected mice. *Proc. Soc. Exp. Biol. Med.* **140,** 802–806.

Howard, R. J., and Balfour, H. H., Jr. (1977). Prevention of morbidity and mortality of wild murine cytomegalovirus by vaccination with attenuated cytomegalovirus. *Proc. Soc. Exp. Biol. Med.* **156,** 365–368.

Howard, R. J., and Najarian, J. S. (1974). Cytomegalovirus-induced immune suppression. I. Humoral immunity. *Clin. Exp. Immunol.* **18,** 109–118.

Howard, R. J., Miller, J., and Najarian, J. S. (1974). Cytomegalovirus-induced immune suppression. II. Cell-mediated immunity. *Clin. Exp. Immunol.* **18,** 119–126.

Howard, R. J., Balfour, H. H., Jr., Seidel, M. V., Simmons, R. L., and Najarian, J. S. (1977). Effect of murine cytomegalovirus on cell-mediated immunity. *Transplant. Proc.* **9,** 355–358.

Howard, R. J., Mattsson, D. M., Seidel, M. V., and Balfour, H. H., Jr. (1978). Cell-mediated immunity to murine cytomegalovirus. *J. Infect. Dis.* **138,** 597–604.

Howard, R. J., Mattsson, D. M., and Balfour, H. H., Jr. (1979). Effect of immunosuppression on humoral and cell-mediated immunity to murine cytomegalovirus. *Proc. Soc. Exp. Biol. Med.* **161,** 341–346.

Huang, E.-S., Huang, C.-H., Huong, S.-M., and Selgrade, M. K. (1976). Preferential inhibition of herpes-group viruses by phosphonoacetic acid: effect on virus DNA synthesis and virus-induced DNA polymerase activity. *Yale J. Biol. Med.* **49,** 93–98.

Hudson, J. B. (1979). The murine cytomegalovirus as a model for the study of viral pathogenesis and persistent infections. *Arch. Virol.* **62,** 1–29.

Hudson, J. B., Misra, V., and Mosmann, T. R. (1976a). Cytomegalovirus infectivity: analysis of the phenomenon of centrifugal enhancement of infectivity. *Virology* **72,** 235–243.

Hudson, J. B., Misra, V., and Mosmann, T. R. (1976b). Properties of the multicapsid virions of murine cytomegalovirus. *Virology* **72,** 224–234.

Hudson, J. B., Loh, L., Misra, V., Judd, B., and Suzuki, J. (1977). Multiple interactions between murine cytomegalovirus and lymphoid cells *in vitro*. *J. Gen. Virol.* **38,** 149–159.

Hudson, J. B., Chantler, J. K., Loh, L., Misra, V., and Muller, M. T. (1979). Model systems for analysis of latent cytomegalovirus infections. *Can. J. Microbiol.* **25,** 245–253.

Johnson, K. P. (1969). Mouse cytomegalovirus: placental infection. *J. Infect. Dis.* **120,** 445–450.

Jordan, M. C. (1978). Interstitial pneumonia and subclinical infection after intranasal inoculation of murine cytomegalovirus. *Infect. Immun.* **21,** 275–280.

Jordan, M. C., Shanley, J. D., and Stevens, J. G. (1977). Immunosuppression reactivates and disseminates latent murine cytomegalovirus. *J. Gen. Virol.* **37,** 419–423.

Kelsey, D. K., Kern, E. R., Overall, J. C., Jr., and Glasgow, L. A. (1976). Effect of cytosine arabinoside and 5-iodo-2′-deoxyuridine on a cytomegalovirus infection in newborn mice. *Antimicrob. Agents Chemother.* **9,** 458–464.

Kelsey, D. K., Olsen, G. A., Overall, J. C., Jr., and Glasgow, L. A. (1977). Alteration of host defense mechanisms by murine cytomegalovirus infection. *Infect. Immun.* **18,** 754–760.

Kelsey, D. K., Overall, J. C., Jr., and Glasgow, L. A. (1978). Correlation of the suppression of mitogen responsiveness and the mixed lymphocyte reaction with the proliferative response to viral antigen of splenic lymphocytes from cytomegalovirus-infected mice. *J. Immunol.* **121,** 464–470.

Kim, K. S., and Carp, R. I. (1971). Growth of murine cytomegalovirus in various cell lines. *J. Virol.* **7,** 720–725.

Kim, K. S., and Carp, R. I. (1972). Abortive infection of human diploid cells by murine cytomegalovirus. *Infec. Immun.* **6,** 793–797.

Kim, K. S., and Carp, R. I. (1973a). Effect of proteolytic enzymes on the infectivity of a number of herpesviruses. *J. Infect. Dis.* **128,** 788–790.

Kim, K. S., and Carp, R. I. (1973b). Influence of complement on the neutralization of murine cytomegalovirus by rabbit antibody. *J. Virol.* **12,** 1620–1621.

Kim, K. S., Sapienza, V., and Carp, R. I. (1974). Comparative studies of the Smith and Raynaud strains of murine cytomegalovirus. *Infect. Immun.* **10,** 672–674.

Kim, K. S., Sapienza, V., and Carp, R. I. (1975). Comparative characteristics of three cytomegaloviruses from rodents. *Am. J. Vet. Res.* **36,** 1495–1500.

Kim, K. S., Moon, H. M., Sapienza, V., Carp, R. I., and Pullarkat, R. (1978). Inactivation of cytomegalovirus and Semliki Forest virus by butylated hydroxytoluene. *J. Infect. Dis.* **138,** 91–94.

Kurimura, T., Kimura, M., Eizuru, Y., and Minamishima, Y. (1977). Electron microscopic studies on the replication of mouse cytomegalovirus in mouse embryo cells. *Acta Virol. (Engl. Ed.)* **21,** 491–494.

Lakeman, A. D., and Osborn, J. E. (1979). Size of infectious DNA from human and murine cytomegaloviruses. *J. Virol.* **30,** 414–416.

Lang, D. J., Cheung, K. S., Schwartz, J. N., Daniels, C. A., and Harwood, S. E. (1976). Cytomegalovirus replication and the host immune response. *Yale J. Biol. Med.* **49,** 45–58.

Loh, L., and Hudson, J. B. (1979). Interaction of murine cytomegalovirus with separated populations of spleen cells. *Infect. Immun.* **26,** 853–860.

Loh, L., and Hudson, J. B. (1980). Immunosuppressive effect of murine cytomegalovirus. *Infect. Immun.* **27,** 54–60.

Lussier, G. (1973). Encephalitis caused by murine cytomegalovirus in newborn and weanling mice. *Vet. Pathol.* **10,** 366–374.

Lussier, G. (1974). Pathology of murine cytomegalovirus infection in newborn mice. Muscle, heart and brown fat lesions. *Can. J. Comp. Med.* **38,** 179–184.

Lussier, G. (1975a). Effect of anti-thymocyte serum on murine cytomegalovirus encephalitis in weanling mice. *Rev. Can. Biol.* **34,** 1–10.

Lussier, G. (1975b). Murine cytomegalovirus (MCMV). *Adv. Vet. Sci. Comp. Med.* **19,** 223–247.

Lussier, G. (1976). Glomerulonephritis caused by murine cytomegalovirus. *Vet. Pathol.* **13,** 123–130.

Lussier, G., Berthiaume, L., and Payment, P. (1974). Electron microscopy of murine cytomegalovirus: development of the virus *in vivo* and *in vitro*. *Arch. Gesamte Virusforsch.* **46,** 269–280.

McCordock, H. A., and Smith, M. G. (1936). The visceral lesions produced in mice by the salivary gland virus of mice. *J. Exp. Med.* **63,** 303–310.

Mannini, A., and Medearis, D. N., Jr. (1961). Mouse salivary gland virus infections. *Am. J. Hyg.* **73,** 329–343.

Mantyjarvi, R. A., Selgrade, M. K., Collier, A. M., Hu, S.-C., and Pagano, J. S. (1977). Murine cytomegalovirus infection of epithelial cells in mouse tracheal ring organ culture. *J. Infect. Dis.* **136,** 444–448.

Margolis, G., and Kilham, L. (1976). Neuronal parasitism and cell fusion in mouse cytomegalovirus encephalitis. *Exp. Mol. Pathol.* **25,** 20–30.

Mayo, D. R., Armstrong, J. A., and Ho, M. (1977). Reactivation of murine cytomegalovirus by cyclophosphamide. *Nature (London)* **267,** 721–723.

Mayo, D. R., Armstrong, J. A., and Ho, M. (1978). Activation of latent murine cytomegalovirus infection: cocultivation, cell transfer, and the effect of immunosuppression. *J. Infect. Dis.* **138,** 890–896.

Medearis, D. N., Jr. (1964a). Mouse cytomegalovirus infection. II. Observations during prolonged infections. *Am. J. Hyg.* **80,** 103–112.

Medearis, D. N., Jr. (1964b). Mouse cytomegalovirus infection. III. Attempts to produce intrauterine infections. *Am. J. Hyg.* **80,** 113–120.

Medearis, D. N., Jr., and Prokay, S. L. (1978). Effect of immunization of mothers on cytomegalovirus infection in suckling mice. *Proc. Soc. Exp. Biol. Med.* **157,** 523–527.

Mims, C. A., and Gould, J. (1978a). The role of macrophages in mice infected with murine cytomegalovirus. *J. Gen. Virol.* **41,** 143–153.

Mims, C. A., and Gould, J. (1978b). Splenic necrosis in mice infected with cytomegalovirus. *J. Infect. Dis.* **137,** 587–591.

Mims, C. A., and Gould, J. (1979). Infection of salivary glands, kidneys, adrenals, ovaries and epithelia by murine cytomegalovirus. *J. Med. Microbiol* **12,** 113–122.

Minamishima, Y., Eizuru, Y., Yoshida, A., and Fukunishi, R. (1978). Murine model for immunoprophylaxis of cytomegalovirus infection. I. Efficacy of immunization. *Microbiol. Immunol.* **22,** 693–700.

Misra, V., and Hudson, J. B. (1977). Murine cytomegalovirus infection in a non-permissive line of mouse fibroblasts. *Arch. Virol.* **55,** 305–313.

Misra, V., Muller, M. T., and Hudson, J. B. (1977). The enumeration of viral genomes in murine cytomegalovirus-infected cells. *Virology* **83,** 458–461.

Misra, V., Muller, M. T., Chantler, J. K., and Hudson, J. B. (1978). Regulation of murine cytomegalovirus gene expression. I. Transcription during productive infection. *J. Virol.* **27,** 263–268.

Moon, H. M., Sapienza, V. J., and Carp, R. I. (1976). DNA synthesis in mouse embryo fibroblast cells infected with murine cytomegalovirus. *Virology* **75,** 376–383.

Moon, H. M., Sapienza, V. J., Carp, R. I., and Kim, K. S. (1979). Murine cytomegalovirus-induced protein synthesis. *J. Gen. Virol.* **42,** 159–169.

Morse, H. C., III, Cross, S. S., and Baker, P. J. (1976). Neonatal infection with mouse thymic virus: effects on cells regulating the antibody response to type III pneumococcal polysaccharide. *J. Immunol.* **116,** 1613–1617.

Mosmann, T. R., and Hudson, J. B. (1973). Some properties of the genome of murine cytomegalovirus (MCMV). *Virology* **54,** 135–149.

Mosmann, T. R., and Hudson, J. B. (1974). Structural and functional heterogeneity of the murine cytomegalovirus genome. *Virology* **62,** 175–183.

Muller, M. T., and Hudson, J. B. (1977a). Thymidine kinase activity in mouse 3T3 cells infected by murine cytomegalovirus (MCV). *Virology* **80,** 430–433.

Muller, M. T., and Hudson, J. B. (1977b). Cell cycle dependency of murine cytomegalovirus replication in synchronized 3T3 cells. *J. Virol.* **22,** 267–272.

Muller, M. T., and Hudson, J. B. (1978). Murine cytomegalovirus DNA synthesis in nuclear monolayers. *Virology* **88,** 371–378.

Muller, M. T., Misra, V., Chantler, J. K., and Hudson, J. B. (1978). Murine cytomegalovirus gene expression during nonproductive infection in G_0-phase 3T3 cells. *Virology* **90,** 279–287.

Murphy, G. F., Brody, A. R., and Craighead, J. E. (1975). Monocyte migration across pulmonary membranes in mice with cytomegalovirus. *Exp. Mol. Pathol.* **22,** 35–44.

Nedrud, J. G., Collier, A. M., and Pagano, J. S. (1979). Cellular basis for susceptibility to mouse cytomegalovirus: evidence from tracheal organ culture. *J. Gen. Virol.* **45,** 737–744.

Neighbour, P. A. (1978). Studies on the susceptibility of the mouse preimplantation embryo to infection with cytomegalovirus. *J. Reprod. Fertil.* **54,** 15–20.

Oie, H. K., Easton, J. M., Ablashi, D. V., and Baron, S. (1975). Murine cytomegalovirus: induction of and sensitivity to interferon *in vitro*. *Infect. Immun.* **12,** 1012–1017.

Olding, L. B., Jensen, F. C., and Oldstone, M. B. A. (1975). Pathogenesis of cytomegalovirus infection. I. Activation of virus from bone marrow-derived lymphocytes by *in vitro* allogeneic reaction. *J. Exp. Med.* **141,** 561–572.

Olding, L. B., Kingsbury, D. T., and Oldstone, M. B. A. (1976). Pathogenesis of cytomegalovirus infection. Distribution of viral products, immune complexes and autoimmunity during latent murine infection. *J. Gen. Virol.* **33,** 267–280.

Osborn, J. E., and Medearis, D. N., Jr. (1966). Studies of relationship between mouse cytomegalovirus and interferon. *Proc. Soc. Exp. Biol. Med.* **121,** 819–824.

Osborn, J. E., and Medearis, D. N., Jr. (1967). Suppression of interferon and antibody and multiplication of Newcastle disease virus in cytomegalovirus infected mice. *Proc. Soc. Exp. Biol. Med.* **124,** 347–353.

Osborn, J. E., and Shahidi, N. T. (1973). Thrombocytopenia in murine cytomegalovirus infection. *J. Lab. Clin. Med.* **81,** 53–56.

Osborn, J. E., and Walker, D. L. (1968). Enhancement of infectivity of murine cytomegalovirus *in vitro* by centrifugal inoculation. *J. Virol.* **2,** 853–858.

Osborn, J. E., and Walker, D. L. (1971). Virulence and attenuation of murine cytomegalovirus. *Infect. Immun.* **3,** 228–236.

Osborn, J. E., Blazkovec, A. A., and Walker, D. L. (1968). Immunosuppression during acute murine cytomegalovirus infection. *J. Immunol.* **100,** 835–844.

Overall, J. C., Jr., Kern, E. R., and Glasgow, L. A. (1976). Effective antiviral chemotherapy in cytomegalovirus infection of mice. *J. Infect. Dis.* **133S,** A237–A244.

Parker, J. C., Vernon, M. L., and Cross, S. S. (1973). Classification of mouse thymic virus as a herpesvirus. *Infect. Immun.* **7,** 305–308.

Plavosin, L., and Diosi, P. (1974). Non-productive infection of genetically resistant mouse cells by murine cytomegalovirus. *Arch. Gesamte Virusforsch.* **45,** 165–168.

Plummer, G., and Goodheart, C. R. (1974). Growth of murine cytomegalovirus in a heterologous cell system and its enhancement by 5-iodo-2′-deoxyuridine. *Infect. Immun.* **10,** 251–256.

Quinnan, G. V., and Manischewitz, J. E. (1979). The role of natural killer cells and antibody-dependent cell-mediated cytotoxicity during murine cytomegalovirus infection. *J. Exp. Med.* **150,** 1549–1554.

Quinnan, G. V., Manischewitz, J. E., and Ennis, F. A. (1978). Cytotoxic T-lymphocyte response to murine cytomegalovirus infection. *Nature (London)* **273,** 541–543.

Quinnan, G. V., Manischewitz, J. E., and Ennis, F. A. (1980). Role of cytotoxic T lymphocytes in murine cytomegalovirus infection. *J. Gen. Virol.* **47,** 503–508.

Rowe, W. P., and Capps, W. I. (1961). A new mouse virus causing necrosis of the thymus in newborn mice. *J. Exp. Med.* **113,** 831–844.

Rowe, W. P., Hartley, J. W., and Huebner, R. J. (1962). Polyoma and other indigenous mouse viruses. *In* "The Problems of Laboratory Animal Disease" (R. C. J. Harris, ed.), pp. 131–142. Academic Press, New York.

Ruebner, B. H., Miyai, K., Slusser, R. J., Wedemeyer, P., and Medearis, D. N., Jr. (1964). Mouse cytomegalovirus infection. An electron microscopic study of hepatic parenchymal cells. *Am. J. Pathol.* **44,** 799–821.

Reubner, B. H., Hirano, T., Slusser, R., Osborn, J., and Medearis, D. N., Jr. (1966). Cytomegalovirus infection. Viral ultrastructure with particular reference to the relationship of lysosomes to cytoplasmic inclusions. *Am. J. Pathol.* **48,** 971–989.

Rytel, M. W., and Hooks, J. J. (1977). Induction of immune interferon by murine cytomegalovirus. *Proc. Soc. Exp. Biol. Med.* **155,** 611–614.

Schwartz, J. N., Daniels, C. A., Shivers, J. C., and Klintworth, G. K. (1974). Experimental cytomegalovirus ophthalmitis. *Am. J. Pathol.* **77,** 477–486.

Schwartz, J. N., Daniels, C. A., and Klintworth, G. K. (1975). Lymphoid cell necrosis, thymic atrophy and growth retardation in newborn mice inoculated with murine cytomegalovirus. *Am. J. Pathol.* **79,** 509–518.

Selgrade, M. K., and Osborn, J. E. (1974). Role of macrophages in resistance to murine cytomegalovirus. *Infect. Immun.* **10,** 1383–1390.

Selgrade, M. K., Ahmed, A., Sell, K. W., Gershwin, M. E., and Steinberg, A. D. (1976). Effects of murine cytomegalovirus on the *in vitro* responses of T and B cells to mitogens. *J. Immunol.* **116,** 1459–1465.

Shanley, J. D., Jordan, M. C., Cook, M. L., and Stevens, J. G. (1979). Pathogenesis of reactivated latent murine cytomegalovirus infection. *Am. J. Pathol.* **95,** 67–78

Smith, M. G. (1959). The salivary gland viruses of man and animals (cytomegalic inclusion disease). *Prog. Med. Virol.* **2,** 171–202.

Starr, S. E., and Allison, A. C. (1977). Role of T lymphocytes in recovery from murine cytomegalovirus infection. *Infect. Immun.* **17,** 458–462.

Stringfellow, D. A., Kern, E. R., Kelsey, D. K., and Glasgow, L. A. (1977). Suppressed response to interferon induction in mice infected with encephalomyocarditis virus, Semliki Forest virus, influenza A2 virus, *Herpesvirus hominis* type 2 or murine cytomegalovirus. *J. Infect. Dis.* **135,** 540–551.

Tarr, G. C., Armstrong, J. A., and Ho, M. (1978). Production of interferon and serum hyporeactivity factor in mice infected with murine cytomegalovirus. *Infect. Immun.* **19,** 903–907.

Tegtmeyer, P. J., and Craighead, J. E. (1968). Infection of adult mouse macrophages with cytomegalovirus. *Proc. Soc. Exp. Biol. Med.* **129,** 690–694.

Tegtmeyer, P. J., Ming, P. M. L., and Craighead, J. E. (1969). Cytological studies on cytomegalovirus infected mouse cells. *Arch. Gesamte Virusforsch.* **26,** 334–340.

Tinghitella, T. J., and Booss, J. (1979). Enhanced immune response late in primary cytomegalovirus infection of mice. *J. Immunol.* **122,** 2442–2446.

Weller, T. H. (1971). The cytomegaloviruses: ubiquitous agents with protean clinical manifestations. *N. Engl. J. Med.* **285,** 203–214, 267–274.

Wise, T. G., Manischewitz, J. E., Quinnan, G. V., Aulakh, G. S., and Ennis, F. A. (1979). Latent cytomegalovirus infection of BALB/c mouse spleens detected by an explant culture technique. *J. Gen. Virol.* **44,** 551–556.

Wright, H. T., Jr. (1973). Cytomegaloviruses. *In* "The Herpesviruses" (A. S. Kaplan, ed.), pp. 353–388. Academic Press, New York.

Wu, B. C., and Ho, M. (1979). Characteristics of infection of B and T lymphocytes from mice after inoculation with cytomegalovirus. *Infect. Immun.* **24,** 856–864.

Wu, B., Dowling, J. N., Armstrong, J. A., and Ho, M. (1975). Enhancement of mouse cytomegalovirus infection during host-*versus*-graft reaction. *Science* **190,** 56–58.

Young, J. A., Cheung, K. S., and Lang, D. J. (1977). Infection and fertilization of mice after artificial insemination with a mixture of sperm and murine cytomegalovirus. *J. Infect. Dis.* **135,** 837–840.

Chapter 14

Polyomavirus

Bernice E. Eddy

I.	Introduction	293
II.	History	293
III.	Related Viruses	295
IV.	The Virus Particle	295
V.	Resistance of the Virus or Its DNA to Physical and Chemical Agents	297
VI.	Cultivation	298
VII.	Infection of Mice and Other Animals	300
VIII.	Spread of Polyomavirus and Methods for Its Prevention	304
IX.	Conclusions	305
	References	305

I. INTRODUCTION

Polyomavirus, one of the smallest oncogenic viruses known, is a highly stable deoxyribonucleic acid (DNA) virus existing in certain laboratory mouse colonies and in some wild mice, both in urban and rural areas, as a silent infection. The addition of polyomavirus extracts of organs of mice carrying the virus, or of polyomavirus-induced tumors, to susceptible cell cultures with incubation at 37°C can significantly increase virus concentration. Fluids from such cultures injected into newborn mice cause a variety of histologically different neoplasms. The virus can induce tumors in other animal species if they are infected when very young, and it is antigenic when injected into mature animals. The concentration of virus or antibody can be determined *in vitro* or *in vivo* (Eddy, 1969).

II. HISTORY

When Gross (1953a,b) was carrying out his important work on leukemia by inoculating newborn mice with filtered extracts of mouse leukemic tissue, he noted that some mice developed progressively growing neck or parotid tumors which were not characteristic of leukemia. Leukemia did not occur in mice bearing these neck tumors. Gross postulated that the filtered leukemic extract contained two agents, one heavier than the other. When newborn mice were injected with the supernatant fluid of extracts centrifuged at 144,000 *g* or filtered through a fine-porosity filter, more mice developed neck tumors than leukemia. No mice infected with leukemic extract heated to 68°C for 30 min developed leukemia, but one mouse developed a tumor in the neck area. The agent responsible for the

ISBN 0-12-262502-1

neck tumors was viable after storage in a sealed glass ampoule at 4°C for 13 months and lyophilization for a year.

Stewart (1953) reported that newborn (C3H × AKR) F_1 hybrid mice inoculated with filtrate prepared from tissues of AKR leukemic mice developed unilateral or bilateral sarcomas of the parotid gland, whereas only a small percentage had leukemia. The results of these experiments, in which two different neoplasms were induced, were ascribed to differences in the strains of mice.

Tumors involving the parotid glands unexpectedly occurred in C58 mice or (C58 × AKR) F_1 hybrid mice that were treated subcutaneously with 1 mg cortisone on three successive days each month throughout life. These tumors were similar to those described by Gross following injections of newborn mice with filtered leukemic extracts (Woolley, 1954).

These observations stimulated others to extend the results (Law *et al.*, 1955; Stewart, 1955; Schmidt, 1956; Dulaney, 1956). For further details, see Gross (1970).

The second advance was the discovery that the agent would increase in rhesus monkey kidney cell cultures. Cytologic changes were not seen, but fluid from cultures incubated for at least 2 weeks and injected subcutaneously into newborn mice induced parotid, submaxillary, and sublingual tumors, renal cortical lesions, thymic tumors, and mammary and other tumors after 2.5–10 months. Fluid from cultures incubated for 2 weeks, passed to fresh monkey kidney cell cultures, and incubated for an additional 2 weeks induced similar tumors in newborn mice (Stewart *et al.*, 1957).

When the agent was transferred to Swiss mouse embryo cell cultures, cytopathic changes in the cells were observed. Fluids from these cultures passed to freshly prepared mouse embryo cultures caused similar cytopathic changes (Eddy *et al.*, 1958b).

This same fluid injected into mature guinea pigs or rabbits induced antiviral antibodies. Antibody was demonstrated by incubating the serum with virus and adding it to susceptible cell cultures (Eddy *et al.*, 1958b), by injecting serum into newborn mice (Stewart and Eddy, 1959), or by testing for inhibition of hemagglutination with guinea pig or human O erythrocytes (Eddy *et al.*, 1958a).

Polyomavirus plaques from susceptible cells were used to determine if the histologically different neoplasms were caused by one virus. Twice-plaqued polyomavirus induced tumors in newborn hamsters that were indistinguishable from those induced by virus that had not been plaqued (Eddy and Stewart, 1959a). Similar results were obtained in mice (Stewart *et al.*, 1959b). Plaque assays for determining the amount of virus in a preparation were described by Winocour and Sachs (1959), Dulbecco and Freeman (1959), and Sheinin (1961). Plaque-induced tumors in hamsters were carried out and reported a year before those in mice. (See the report of the Third Canadian Cancer Conference held June 17–21, 1958.)

Hamsters infected as newborns develop neoplasms more quickly than infected mice (Eddy *et al.*, 1958d). The virus multiplies to some extent in hamster embryo cells *in vitro* and can be recovered with ease from extracts of hamster tumors cultured in mouse embryo monolayers (Eddy, 1963; Negroni and Chesterman, 1960). Malignant tumors can be induced in newborn rats (Eddy *et al.*, 1959b), guinea pigs (Eddy *et al.*, (1960), mastomys (Rabson *et al.*, 1960b), and ferrets (Harris *et al.*, 1961). Infected newborn rabbits develop benign tumors which eventually regress (Eddy *et al.*, 1959a).

The viral DNA, which can be separated from other viral constituents, is capable of initiating new viral particles (Di-Mayorca *et al.*, 1959). Weil (1961) found that infectivity could be titrated by the plaque method if susceptible cells were pretreated with hypertonic sodium chloride, thus enabling the DNA to enter the cells.

Transformation of permanent inheritable traits acquired by cells in culture was noted by investigators before polyomavirus was known (Gey, 1941; Gey *et al.*, 1949). Abercombie *et al.* (1957), studying confronted tissue culture outgrowths of mouse muscle, chick heart muscle, and a mouse sarcoma (S37 Crocker), observed that normal fibroblasts obstruct each other's movement, since rarely one cell moved over another (contact inhibition). Some influence appears to be exerted from a distance, for some cells travel farther toward an opposing explant. No obstruction of sarcoma cell movement by fibroblasts is detectable, but some obstruction of fibroblast movement by sarcoma outgrowths occurs. It was hypothesized that contact inhibition occurs between fibroblasts of different origins, and malignancy involves a loss or diminution of the response to contact.

Vogt and Dulbecco (1960, 1962) and Dulbecco and Vogt (1960) studied the phenomenon of transformed cells in relation to polyomavirus release in both mouse and hamster cells. Degenerative changes occur in infected mouse cells beginning on the third or fourth day, with a simultaneous increase in virus titer. By 4 weeks the cells release less virus and change morphologically. They are more elongated and tend to form interwoven netlike structures, unlike contact-inhibited cells in noninfected cultures. By 2 weeks, cultures are made up of the new cell types. Virus-infected hamster cells at 1 week look much like uninfected hamster cells, but by 2 weeks infected cultures contain approximately three times more cells than uninfected cultures. If infected cells are dispersed with trypsin and recultured, within 2 weeks these cells show whirls of heavy strands.

Many transformed mouse and, occasionally, transformed hamster cells continue to release virus indefinitely. When clonal isolates of neoplastic mouse embryo cells were maintained under conditions that minimized reinfection by carryover virus, about half the clones did not release virus, whereas the other half released virus from the beginning. The conclusion was that neoplastic cells in either mouse or hamster embryo cultures are intrinsically nonvirus releasers. A factor in the phenomenon of virus release appears to be the resistance of the neoplastic cells to superinfection with polyomavirus. In transformed hamster cells resistance to reinfection is high, whereas in transformed mouse cells resistance is less.

The successful transplantation of a solid tumor or suspension of tumor cells to nontumorous animals is one of the crucial tests for determining malignancy (Stewart, H. L., *et al.*, 1959; Gross, 1970). Transplantation, to be successful, must be done in genetically similar animals. Athymic animals are an exception. Newborn animals are more prone to accept a transplant than older animals, certain sites are more sensitive to tumor "takes" than others, and treatment of the recipient animal with cortisone or X irradiation favors successful transplantation. Some malignant tumors cannot be transplanted; others that grow with difficulty can be more easily transplanted after continued growth.

Many tumors induced with polyomavirus can be transplanted. Law *et al.* (1955) successfully transplanted tumors induced in C3H mice to C3H mice in 7 of 15 attempts, but none to AKR or C58 mice. Negroni *et al.* (1959) serially transplanted a virus-induced mammary tumor in mice and a renal sarcoma in hamsters.

III. RELATED VIRUSES

Antigenically, the polyomavirus (meaning "many tumors") is distinct but shares some characteristics with other viruses. Melnick (1962) pointed out similarities of the rabbit papillomavirus, polyomavirus, the human wart virus, and simian virus 40 (SV40), also known as the vacuolating virus, and gave the name *papova* to the group, using the first two letters from the names of three of the viruses. Later, the International Committee on Nomenclature of Viruses (Melnick *et al.*, 1974; Fenner, 1976) proposed *Papoviridae* for the family name and divided the family into two genera: *papovavirus A group,* comprising the human papillomavirus (wart), Shope rabbit papillomavirus, canine papillomavirus, and hamster papillomavirus, and *papovavirus B group,* made up of polyomavirus, SV40, K papovavirus, rabbit kidney vacuolating virus (RKV), the BK virus recovered from urine of a human patient who had received a renal transplant (Gardner *et al.,* 1971), and the JC virus recovered from the brain of a patient with progressive multifocal leukoencephalopathy (Padgett *et al.,* 1971). An extensive review of the papovaviruses was published by Weil (1978). In addition, a papovavirus B has been recovered from a stump-tailed macaque (*Macaca arctoides*), STM virus (Reissig *et al.,* 1976). The size and morphology of the virus particle are the same as those of SV40 and polyomavirus, but many of the particles appear to have an additional outer envelope. It differs from the other type B papovaviruses, but, like the BK and JC viruses, its T antigen may be related to SV40.

All of the papovaviruses are small DNA viruses, icosahedral in shape, covered by 72 projecting knobs or capsomers (hexons and pentons), with an inner core of double-stranded, circular DNA which carries the genetic information. Although the viruses resemble each other in structure, type A viruses are larger, about 55 nm in diameter, and contain approximately 5 $\times 10^6$ daltons of DNA; type B viruses are about 45 nm in diameter and contain 3×10^6 daltons of DNA. Differences in base sequences have been reported by Crawford and Crawford (1963), Crawford (1965), and Ferguson and Davis (1975).

The papillomaviruses found in common skin warts, plantar warts, oral and genital condylomata, and oral and laryngeal papillomas were assumed to be the same, but evidence is now available, supported by molecular hybridization, that DNA from congenital condylomata and laryngeal papillomas show no homologous sequences, and skin warts and condylomata are antigenically distinct and have different polypeptid patterns (Orth *et al.,* 1977). Law *et al.* (1979) failed to demonstrate nucleotide sequence homology between human papillomavirus and cottontail papilloma (Shope), but under less stringent conditions they found some regions of homology between the two viruses.

Among the type B group, each virus is distinct, with SV40 showing minor crosses with the BK and the JC viruses and, apparently, with the STM virus as well.

IV. THE VIRUS PARTICLE

Infectious polyomavirus, shadowed with chromium, appeared spherical with an average diameter of 44 nm (Kahler *et al.,* 1959). In reality, the polyomavirus particle, like other papoviridae, is an icosahedron with 72 morphological units (hexons and pentons) arranged over the surface in an irregular fashion (Klug and Finch, 1965; Klug, 1965).

In tissue culture the virus particle either multiplies and destroys susceptible cells or transforms them so that they increase in number and overgrow each other in a bizarre fashion (Vogt and Dulbecco, 1960; Dulbecco, 1969). Usually a lytic phase occurs first; then a few remaining cells become transformed and repopulate the culture. Transformed cells may cease to shed virus, yet carry the tumor (T) antigen and become transplantable to animals of the same species. Polyomavirus-infected cells, whether lysed or transformed, go through an early phase that lasts until the start of viral DNA replication, during which early messenger RNA (mRNA) is primarily furnished by the host (Tooze, 1973; Acheson and Miéville, 1978).

A cell may be transformed by two viruses, for example,

SV40 and polyomavirus (Todaro *et al.*, 1964; Todaro and Green, 1965; Takemoto and Habel, 1966), or by polyomavirus and an unrelated ribonucleic acid (RNA) virus, e.g., Maloney leukemia virus (Sjögren and Hellström, 1965).

The polyomavirus is highly antigenic, most of the antigens being associated with the capsid structures. The virus causes hemagglutination of guinea pig or human O erythrocytes at 4°C, and elution from the cells occurs at 37°C.

Viruses injected into animals, particularly adult animals such as rabbits or guinea pigs, induce serum antibodies which can inhibit cytopathogenicity due to the virus in susceptible cell cultures (Eddy *et al.*, 1958b). These antibodies, when mixed with the virus and injected into newborn susceptible animals, can prevent tumor development (Stewart *et al.*, 1959a). The antiserum can also inhibit hemagglutination (Eddy *et al.*, 1958a).

The cellular tumor (T) antibody which develops is virus specific (Habel, 1961; Sjögren, 1961). Animals immunized with polyomavirus are resistant to transplantation with virus-free, polyoma-induced tumor cells but not to tumor cells induced by SV40. Conversely, animals immunized with SV40 are resistant to transplantation when challenged with SV40 tumor cells but not to tumor cells induced by the polyomavirus (Habel and Eddy, 1963; Koch and Sabin, 1963; Defendi, 1963). The T antigen can also be demonstrated by complement fixation (Habel, 1965) and immunofluorescence (Takemoto *et al.*, 1966). It appears to be produced in the early phase of polyomavirus infection just prior to viral DNA synthesis (Wintersburger and Wintersburger, 1976).

Two nonspecific antigens have been reported to be induced by the polyomavirus: Forssman antigen and interferon. Forssman antigen was observed to increase in hamsters with polyomavirus-induced kidney tumors and in *in vitro* cultures of infected hamster and mouse cells (Fogel and Sachs, 1962).

Polyomavirus-infected mouse cells acquired resistance to vesicular stomatitis virus just before, or at the time, hemagglutinins became detectable and before cytopathic changes occurred. Propagation of the challenge virus can be completely inhibited by prior infection with polyomavirus, but polyomavirus growth is not affected (Deinhardt and Henle, 1960). The growth of other viruses can be inhibited by polyomavirus infection: encephalomyocarditis and vaccinia (Allison, 1961, 1963; Inglot and Niedźwiedzka, 1965), herpesvirus (Glasgow and Habel, 1963), and pseudorabies (Coto *et al.*, 1965). Less oncogenic strains of polyomavirus induce more interferon production (Friedman *et al.*, 1963). Interferon production was increased when cultures were incubated at 39°C rather than at 35°C or when the medium contained a low concentration of bicarbonate (Allison, 1963; Inglot and Niedźwiedzka, 1965). The cytopathic effect of polyomavirus can be delayed by prior infection of cells with the PR8 strain of influenza virus or by treatment of the cells with interferon prepared with the PR8 strain of influenza virus (Allison, 1961).

Not all virions in a culture are identical. Those containing *DNA I* have double-stranded, circular molecules with a supercoiled helix and a molecular weight of 3×10^6. One strand of the DNA, the early (E) strand, codes for the events that occur prior to replication of viral DNA, and the late (L) strand codes for late viral functions. A single break in one strand of the DNA I converts it to *DNA II*, and when this strand is broken into short linear fragments, it is known as *DNA III* (Weil, 1978). The strands of DNA I have the unique property of separating on boiling but spontaneously renaturing on cooling. The strands of DNA II and DNA III, which account for 10–20% of the total DNA, also separate on boiling but do not renature. Empty viral capsids or capsids filled with linear fragments of host DNA known as *pseudovirions* also occur (Michael *et al.*, 1967; Winocour, 1968; Basilico and Burstin, 1971; Yelton and Aposhian, 1972; Türler, 1975). The proportion of pseudovirions in a virus preparation depends on the type of host cell and varies from undetectable amounts to 90%. It is not known whether the pseudovirions are produced by loss of their nucleic acid or whether they represent an early development of the infective particle. In the course of pseudorabies infection, there is evidence that they are produced without a DNA core (Reissig and Kaplan, 1962).

A permissive cell must have receptor sites to which virus particles can attach. Virus adsorption on a permissive cell takes place at 4° or at 37°C, but the virion does not enter the cytoplasm at 4°C. The optimum pH for adsorption is 7.4 and the time required is 3 hr, although 50% of the virions are taken up in the first 30 min. Nitrogen ions are also required for adsorption (McKay and Consigli, 1976).

Penetration across the cell membrane differs for particles with electron-dense DNA cores and transparent cores without DNA. Virions with dense cores are engulfed by the cell membrane to form monopinocytotic vesicles which migrate toward the nuclear membrane. Pseudovirions enter the cytoplasm in large, membrane-enclosed groups, much like phagocytic vesicles, which do not appear to migrate toward the nuclear membrane or to enter the nucleus (McKay and Consigli, 1976).

The next event, uncoating of the virus, has been reported to take place in the nucleus or at the nuclear membrane, since no virus particles are seen in the nucleus (Khare and Consigli, 1965; Mattern *et al.*, 1966; McKay and Consigli, 1976). Calcium ions appear to have a role in maintaining virus integrity, and the nuclear membrane may have the ability to remove calcium from the virus particle, thus causing its uncoating. With chelation of calcium ions, 98% of viral hemagglutinin is destroyed within 30 min and the virus dissociates into capsomers and a DNA-protein complex which is predominantly DNA I.

Once the outer protective structural proteins are removed, transcription and translation begin and viral genes are expressed (Winston *et al.*, 1980). T antigen-positive nuclei increase in number, and the nucleoli increase in size and become darker (Weil and Kára, 1970). This event occurs with stimulation of host cellular mRNA, the earliest known step in the virus-induced mitogenic reaction of the host cell (Weil, 1978).

On precipitation with serum prepared against polyomavirus tumors (anti-T serum), three proteins with molecular weights of 105,000, 63,000, and 20,000 are isolated (Simmons *et al.*, 1979). The large and small proteins appear to be identical to at least two genes, denoted *hr-t* and *ts-a*. The product of the *hr-t* gene does not seem to be required for lytic growth of the virus in permissive cells (Goldman and Benjamin, 1975; Staneloni *et al.*, 1977), but a functional *ts-a* gene is required to initiate viral DNA in productively infected cells (Francke and Eckhart, 1973; Paulin and Cuzin, 1975; Eckhart, 1974).

Soon after T antigen synthesis, the late phase of polyomavirus infection commences. In the lytic cycle, densely staining material accumulates in the nucleus in about 20 hr, and by 24 hr, bundles of filaments which have the same diameters as the virus particles are seen. In a few more hours, spherical particles but not filamentous forms are present in the cytoplasm, and some nuclei contain large inclusions (Henle *et al.*, 1959; Williams and Scheinin, 1961; Mattern, 1962; Khare and Consigli, 1965).

During this late phase viral DNA is synthesized, structural proteins or capsid proteins are made, and the mature virus particle emerges from the cytoplasm. Though capsid proteins can usually be demonstrated only in the nucleus, Tachovsky and Hare (1975) described a strain in which the protein was present in both the nucleus and cytoplasm. "Giant" molecules (several times larger than the viral genome) of polyoma-specific RNA appearing in the nucleus are subsequently cleaved into smaller RNAs of specific sizes (Acheson *et al.*, 1971; Lev and Manor, 1977).

The shell of the polyomavirus is composed of one major capsid protein (VP1), which comprises 75% of the total protein and has a molecular weight of 45,000 daltons, and two minor proteins, VP2 and VP3, with molecular weights of approximately 35,000 and 23,000, respectively, as well as four cellular histones, VP4–VP7. Plaque formation and hemagglutination appear to be properties of the VP1 protein (Brady and Consigli, 1978). Tryptic peptide analysis of the VP2 and VP3 revealed that VP2 had the same tryptic peptides as VP3 but had other peptides not present in VP3. VP2 and VP3 had few or no peptides in common with VP1 (Sidell and Smith, 1978). All three proteins, VP1–VP3, are synthesized independently in response to discrete mRNAs and are found in polyomavirus-infected cells but not in noninfected cells (Hunter and Gibson, 1978).

The histones (VP4–VP7) are acid-soluble nuclear proteins containing lysine or arginine but not tryptophan (Shimino and Kaplan, 1969). They are closely associated with internal viral DNA and are believed to be external on the capsid structures. The main structural protein which makes up the 60 hexons on the faces of the icosahedron is VP1, whereas the 12 penton capsomers are made up of VP2 and VP3. The external histones are in close association with the proteins VP1, VP2, and VP3. This arrangement is believed to enable the cells to distinguish between complete and incomplete virions, whereas calcium ions in the penton capsomer subunits are important in the maintenance of viral structure (Brady *et al.*, 1978).

A number of enzymes have been found to show increased activity during the period of DNA and protein synthesis: thymidine kinase, deoxycytidine deaminase, thymidine synthetase, and DNA polymerase (Hartwell *et al.*, 1965; Kit *et al.*, 1966a,b; Pettijohn and Kamiya, 1967; Winters and Consigli, 1967; Wintersberger and Wintersberger, 1976; Närkhammar and Magnusson, 1976). Polynucleotide lipases are believed to be enzymes which seal breaks in double-stranded DNA and are involved in DNA repair and replication (Sambrook and Shotkin, 1969). Because of the small size of the viral DNA, Weil *et al.* (1967) excluded the possibility that the enzymes which increase during polyomavirus infection are all coded by the viral DNA. More likely, they are coded by cellular DNA.

Although polyomaviruses are essentially alike, they express a large number of functions. Certain strains, for example, have a predilection for inducing tumors or lesions in specific sites or produce an excess of capsid proteins. Others produce large or small plaques or vary in their ability to induce interferon production. Crawford *et al.* (1974) examined different viral strains by cleavage with endonucleases and noted that there was variation in the DNA fragment patterns. They suggested that it may be possible to correlate these biological properties with those of the DNA fragment patterns.

V. RESISTANCE OF THE VIRUS OR ITS DNA TO PHYSICAL AND CHEMICAL AGENTS

Polyomavirus, like other viruses in the papova type B group, is resistant to the action of many physical and chemical agents. Its spread in nature and/or in rooms where polyomavirus-infected animals are being held is dependent on its survival in bedding and its capacity to be airborne (McGarrity *et al.*, 1976). The virus is highly resistant to heating, and tumor induction is not affected when the virus is heated to 60°C for 30 min. Heating to 60°C for 1 hr delays tumor production, and heating to 70°C does not completely abolish its capacity to induce tumors in hamsters (Eddy *et al.*, 1958c; Eddy and Stewart, 1959b). The virus also retains its ability to induce tumors in animals after storage for long periods in 50 or 33% glycerol or after lyophilization (Gross, 1953b; Eddy *et al.*, 1958c).

Hemagglutination is a relatively stable characteristic. Heating the virus to 23°C or to 60°C for 30 min caused a fourfold decrease in hemagglutination titer and a 3-log loss in infectivity; heating to 70°C destroyed the hemagglutinating capacity and reduced infectivity to $<10^{-1}$. The virus kept at $-70°$, $-60°$, $-20°$ and 4°C for 8 weeks showed no significant changes in hemagglutinating titer, but the same virus preparation kept at 37°C for 8 weeks showed a drop in hemagglutinating titer of 2.5 logs (Brodsky *et al.*, 1959).

A mixture of equal volumes of polyomavirus and disinfectants, held at room temperature, was tested for hemagglutination after 30 min and 24 hr, and for the ability to induce antibodies after 24 hr. Two percent phenol reduced hemagglutination by one-half in 30 min and to one-fourth after 24 hr. Fifty percent ethyl alcohol reduced the hemagglutinin titers to one-fourth in 30 min and to one-eighth in 24 hr. Neither 2% phenol nor 50% ethyl alcohol significantly altered antibody production. Both 1 : 1000 Zephiran (benzalkonium chloride) and 95% ethyl alcohol containing iodine destroyed the capacity to hemagglutinate erythrocytes or to induce antibody (Brodsky *et al.*, 1959).

A freshly prepared polyomavirus preparation to which formalin had been added, to a concentration of 1 : 4000, retained its ability to induce tumors when given to newborn hamsters; however, tumors were not produced with a preparation containing the same formalin concentration and stored for several weeks at $-20°C$ (Eddy, 1960).

Ethyl ether does not inactivate the virus. Gross (1956) kept the virus in contact with ether for 23 hr at 4°C, and Eddy *et al.* (1958c) for 14 days at 22°C. This property distinguishes polyomavirus from the ether-sensitive leukemia virus (Gross, 1956) and can be used for freeing polyomavirus preparations from bacterial contamination (Eddy, 1960).

Polyomavirus is resistant to the action of trypsin, ribonuclease (RNase), and deoxyribonuclease (DNase). Since polyoma DNA is quickly destroyed by DNase, it appears that the capsid proteins of the infectious virion protect the DNA from DNase.

When the virus is subjected to any one of a number of inactivating procedures, the reproductive (plaque-forming) ability is destroyed much more rapidly than the transforming properties and often before the hemagglutinating or complement-fixation abilities. Exposure of polyomavirus to ultraviolet (uv) radiation, gamma radiation, nitrous acid, or ^{32}P destroyed its ability to give rise to progeny virus while retaining its capacity to cause transformation of susceptible cells (Latarjet *et al.*, 1967; Benjamin, 1972). Certain concentrations of β-propiolactone had much the same effect. Virus particles inactivated with 0.125% β-propiolactone, though relatively noninfectious, were capable of eliciting cell transformation and T antigens (Brown *et al.*, 1974).

A number of compounds are known to affect polyomavirus by inhibiting DNA and protein synthesis: puromycin (Bourgaux and Bourgaux-Ramsey, 1972), fluordeoxyuridine (Wintersberger and Wintersberger, 1975), 1-β-arabinofuranosyl (Hunter and Francke, 1975), actinomycin D and cyclohexamide (Wintersberger and Wintersberger, 1976; Yu and Cheevers, 1976), and 2′-dioxy-2-azidocylidine (Bjursell *et al.*, 1977). Some of these compounds in lower concentrations did not entirely block DNA synthesis but allowed DNA to form in long, relaxed strands or other abnormal forms.

Brady and Consigli (1978) found that the addition of 0.1 *M* β-mercaptoethanol and 6 *M* guanidine to purified polyomavirus resulted in the immediate loss of both hemagglutinating and plaque-forming activities of the virus and the separation of the VPI capsid protein.

Polyoma DNA can be separated from the virion by 80% phenol (DiMayorca *et al.*, 1959) and is infectious when added to permissive cells pretreated with a hypertonic salt solution to stimulate cell penetration.

VI. CULTIVATION

Polyomavirus was first propagated in rhesus monkey kidney monolayers because such cultures were available, but no specific cytocidal changes in the cells were observed. After 1 week's incubation, fluids from these cultures failed to induce tumors when injected into newborn mice, but after 2–3 weeks' incubation they induced a variety of tumors. Fluids from seven of the positive monkey kidney cell cultures were used to infect freshly prepared monkey kidney cell cultures, and after 2 weeks fluids from these cultures also induced tumors in newborn mice (Stewart *et al.*, 1957). Some years later, Cordray *et al.* (1962) added polyomavirus to cultures made from tissues of a number of animal species, including monkey kidney. They failed to demonstrate the virus in the monkey kidney cell cultures by hemagglutination or by freezing and thawing the cells and injecting them into mice. They did state that their methods did not exclude a lower order of proliferation, but there was no quantitative likeness to the growth of the virus in tissues of the mouse. Adzhigitov (1963, 1964) succeeded in growing the polyomavirus in monkey kidney tissue cultures from 6-week-old and 3-month-old monkeys but not from monkeys 5 months old or more.

When Swiss mouse embryo cell cultures are infected with polyomavirus, small, dark, pyknotic cells appear on the surface of the cell sheet and after 4–7 days; many cells fall off the glass, whereas some seem to be attached to long, threadlike cells (Eddy *et al.*, 1958b).

This lytic phase of the polyomavirus was shown to occur not

only in mouse embryo cultures but in cell cultures made from the kidney, heart, thymus, and spleen of adult mice (Negroni *et al.,* 1959; Porwit-Bób *et al.,* 1962).

Strains of mice vary in their susceptibility to tumor induction, but cultures made from such mice fail to show differences in their capacity to support growth of the polyomavirus *in vitro* (Jahkola and Vainio, 1964). Polyomavirus has also been grown *in vitro* in cells of mouse tumors such as lymphoma (Rabson and Legallais, 1959) and ascites (Nordenskjöld and Nordenskjöld, 1966) in established mouse cell lines 3T3 (Mulder and Vogt, 1973) and 3T6 (Winnacher *et al.,* 1972) and in cells from other rodents.

Attempts to grow polyomavirus in suspension cultures have not resulted in high yields of virus. One method involves the infection of confluent monolayers of mouse cells with virus, removing the fluid 3 hr later, dispersing the cells with trypsin, and transferring them to a spinner culture with an automatic stirrer. At daily intervals one-fourth of the culture is removed, and the cells are separated by centrifugation, resuspended in fresh medium, and returned to the spinner flask. After 5 days, cellular multiplication stops (Jonsen and Eng, 1962).

Feeder cultures have also been employed in investigations of the polyomavirus. Cell cultures prevented from further multiplication by exposure to 3000 or to 1500–1700 R of X irradiation are found to retain their capacity to support the growth of the virus (Levine, 1963; Stoker and Sussman, 1965).

Polyomavirus, except for elaboration of T antigen, is similar to other nononcogenic viruses in the lytic phase of development. It can either enter a cell, multiply and destroy it, or stimulate abnormal cell growth (transformation) with little or no increase in infectious virus. This last characteristic makes it important in attempting to understand malignancies.

Transformation is not confined to virus-infected mouse cells since cells from other animals often transform with greater ease. Hamster embryo and hamster heart (Stoker and Macpherson, 1961; Porwit-Bóbr *et al.,* 1963), rat cells (Vandeputte and DeSomer, 1962), and fetal bovine skin and lungs (Thomas and LeBouvier, 1967; Diderholm, 1967) have been transformed by the polyomavirus. The polyomavirus does not multiply in or transform cells from the dog, swine, human embryonic kidney, or HeLa line of human carcinoma (Cordray *et al.,* 1962; Diderholm and Wesslén, 1963).

Transformation is not a one-step phenomenon, and normal cells have been recovered from transformed cells (Marin and Littlefield, 1968). When early transformed cells were cloned, there were variant clones, some dense, others thin, with dense clones more persistent than thin ones. The thin clones, on further cultivation, acquired a high degree of transplantability to an isologous host (Vogt and Dulbecco, 1963).

The use of semisolid agar suspension cultures has proved a useful method for identifying early transformation, which begins in a few cells. Colonies showing altered morphology are removed from the agar with a capillary pipette and grown on glass (Macpherson and Montagnier, 1964; Montagnier and Macpherson, 1964).

Virus-free transformed cells have variable susceptibility to reinfection with polyomavirus (Vogt and Dulbecco, 1962; Winocour and Sachs, 1962). This has been interpreted to be due to differences in interferon production or to virus adsorption (Hellström and Hellström, 1963).

Plaques, the growth of single virus particles on a solid nutrient, are useful not only for determining the amount of virus in a preparation but for selecting variant particles in a mixed culture, since all plaques are not alike. Large (3–4 mm in diameter) and small (1–1.5 mm in diameter) plaques were seen; the large plaques appeared diffuse and did not destroy the cells in them, whereas the small plaques had necrotic centers. The large plaques had low cytopathic titers, $10^{3.5}$–$10^{5.5}$, and high hemagglutinating titers, 1 : 960 ±-1 : 3800 and were strongly oncogenic for mice and hamsters. The small plaques had high cytocidal titers, 10^{6}–$10^{6.5}$, and low hemagglutinating titers ranging from 1 to 1240, and the capacity to induce tumors was less than that of the large plaque or the parent strain (Gotlieb-Stematsky and Leventon, 1960). One large-plaque strain was found to induce production of 4–12 times more interferon than a small-plaque strain (Gotlieb-Stematsky *et al.,* 1966b,c).

Dulbecco and Vogt (1960) noted that the mouse or hamster cells that were transformed with a large-plaque strain and persisted in releasing virus released small-plaque and not large-plaque virus mutants. Mutants with different-sized plaques have also been observed by Sachs and Medina (1960).

Prior to the use of endonucleases which can cleave polyoma DNA in precise locations, there was extensive use of lethal mutants induced by treatment of the virus with nitrous acid or other chemicals or radiation and cultivation at temperaturs of 31.5°–32°C and 38.5°–39°C. In general, mutants fell into two classes. The early mutants grew at the low permissive temperature but not at the restrictive (higher) temperature and were defective in transforming cells at the restrictive temperature. Late mutants synthesized viral DNA and transformed virus at the restrictive temperature but not infectious virus (DiMayorca *et al.,* 1969; Fried, 1970; Eckhart, 1974; Paulin and Cuzin, 1975).

Since young animals are more susceptible to polyomavirus than older animals, the effect on cell differentiation has been extensively studied in mouse embryos. The effect of polyomavirus on two-celled embryos and morulae removed from the uteri of ICR mice was studied. The two-celled embryos developed into blastocytes within 3 days of culture, morulae within 1 day, and 8–16-cell embryos within 2 days. Blastocytes were transferred to cloning rings after 3 days, and a layer of trophoblastic cells with a clump of inner cells grew out. The inner cell masses were isolated by immunosurgery and similarly cultured. Only the late genes of polyomavirus were expressed. Indirect immunofluorescence showed the presence of antigen in the nucleus of infected trophoblastic cells; virus-specific antigen was not found in the inner cell mass (Abramczuk *et al.*, 1978).

The organ cultures studied for polyomavirus-induced tumors have been rudiments of the salivary gland, the kidney, and tooth buds. All three types of virus-infected organ cultures revealed the importance of the interaction of mesenchymal and epithelial components and were in accord with experiments reported by Grobstein (1955, 1956, 1962). There is a developmental dependency between epithelial and mesenchymal components, and neither alone can continue morphogenesis. Tubule production from metanephrogenic mesenchyma can be induced by embryonic submandibular salivary epithelium or by chick embryonic central nervous tissue, but not by a number of other living embryonic and adult tissues. Pancreatic epithelium can also be stimulated to form tubules and acini by mesenchyma. The tubule-inducing effect of embryonic mouse spinal cord on metanephrogenic mesenchyma can be transmitted across membrane filters 20–30 μm thick, which completely excludes exchange between the interacting cells.

The salivary glands are the most prevalent sites of polyomavirus-induced tumors in most strains of mice but not in other susceptible animal species. In polyomavirus-infected mice, the release of esterase but not amylase in the salivary gland was significantly increased. Whether this altered enzyme has a bearing on the site of tumors in the mouse is unknown. Infected salivary gland rudiments have been studied by Dawe and Law (1959), Dawe *et al.* (1962), Morgan and Dawe (1963), and Dawe *et al.* (1964). The rudiments of the salivary gland at all stages of development, including stage 1, show morphological transformation *in vitro*. Rudiments in stages 2 or 3 can be successfully transplanted in some mice, and the incidence of transplants which produce tumors rises to 60–78% when stage 5 rudiments are used. When stages 2–5 rudiments are separated into epithelial and mesenchymal components by trypsinization, neither component gives rise to tumors, but recombination of the components before transplantation restores the ability to develop tumors.

The kidney is a common site of polyomavirus-induced tumors in mice and in other animal species. Experiments carried out on polyomavirus-infected kidney rudiments (Vainio *et al.*, 1962, 1963a,b; Saxén *et al.*, 1962, 1963) were followed by immunofluorescence, since no proliferation occurred in mesenchyma or in epithelium derived from the ureter bud or from tubules developing from the metanephrogenic mesenchyma. The mesenchyma was highly susceptible to the cytopathic effects of the virus, but only minor changes were seen in epithelial elements. When the ureter bud (epithelial) and mesenchyma were separated, only the mesenchyma showed the presence of antigen. When an artificial inducer (spinal cord) was added to the mesenchyma, it was not induced to form tubules or to synthesize antigen.

Tumors of the tooth germ can be more frequently induced by certain strains of polyomavirus. Fetal ddO mice were removed from their mother 15 days after gestation, and the molar tooth germs were removed with the aid of a dissecting microscope. Of 144 tooth germs, 82 were infected with polyomavirus, transplanted into other mice, and the transplants removed after 5, 10, 15, 20, and 25 days. On histological examination, epithelial cells showed greater changes than mesenchymal cells, with growth of inner enamel epithelium into the dental pulp and alterations in the enamel organs. In tooth germs transplated for 20 or 25 days, osteoid-like tissue, resembling bone, was seen in place of dentine (Nagai *et al.*, 1963).

Main and Dawe (1966) also removed tooth germs from mice, infected them with polyomavirus, and transplanted them to other mice; 15 of 39 transplants developed ameloblastoma. Ameloblastoma developed in some infected transplants, and no tooth development could be found in serial sections. Ameloblastomas occurred in all the infected host animals that survived for 20 days, with tumors appearing to originate from the outer enamel epithelium. Coexisting in three mice with ameloblastomas were sarcomas, two of which arose from dental pulp and one from the periodontal membrane.

Main and Waheed (1971) separated the odontogenic epithelium and mesenchyma from 14-day-old mouse tooth germs. The epithelial component survived no longer than 10 days, whether infected with polyomavirus or not, whereas the odontogenic mesenchyma grew in culture for 28 days. At that point, fresh odontogenic epithelium and polyomavirus were added, and in 10 more days epithelial proliferation occurred.

When virus-infected and uninfected organ cultures of tooth germs taken from 13- or 14-day-old mice were examined at 18 days and at 38 days or more, virus was seen in epithelial cells in all infected cultures, whether transformed or not. In the nucleus, both spherical and filamentous forms were seen, whereas intracytoplasmic virus appeared only as spheres. No virus was seen in cells undergoing mitosis (Radden *et al.*, 1973).

VII. INFECTION OF MICE AND OTHER ANIMALS

The appearance of polyomavirus-induced tumors in infected mice under natural conditions seldom occurs. Tumor induction is a laboratory phenomenon, and the greatest success is achieved when newborn animals from polyomavirus-free mothers are infected with a high-titer virus preparation. The virus is usually administered by the subcutaneous or intramuscular route but may also be given intraperitoneally, intracerebrally, intraspinally, intratracheally, or injected into the uterus, heart, or salivary glands.

The oral route is not an effective means of inducing tumors; in nature, the nasal route may be important in transmitting the virus to young mice. To ensure that the virus given to the animals is responsible for tumor development, the animals should be kept in a clean area, away from other polyomavirus-infected animals. Tumors in mice may develop as early as 2½–3 months, with most appearing at 4–7 months, and some developing as late as 11–12 months or even longer after infection.

Since mouse cells at a very early stage of development are permissive to certain functions of the polyomavirus, it is understandable that the virus might induce histologically different tumors in a variety of sites (Abramczuk *et al.,* 1978).

Stanton *et al.* (1959) infected mice *in utero* on the seventeenth to twenty-first days of gestation and noted that, although no new histological types of tumors were induced, a greater proportion of these mice developed tumors. Some of these tumors regressed, and others were limited in their capacity to invade adjacent tissues or metastasize.

Stanley *et al.* (1964) have observed that nearly all the tissues in which epithelial tumors are induced are appendages of either the skin or the upper portion of the alimentary tract. These include the parotid, submaxillary, sublingual and minor salivary glands, the lacrimal glands, thymus, dental organ, hair follicles, mammary glands, thyroid, and harderian glands.

In most mouse strains, the parotid and other salivary glands are the most prevalent sites of tumors induced by the polyomavirus. Every tumor-bearing mouse reported by Stewart *et al.* (1957, 1958) developed salivary gland tumors, with tumors in other sites as well (Fig. 1). In a study by Law *et al.* (1955) using extracts of mouse leukemic tissue which contained polyomavirus, most of the mice had bilateral parotid gland tumors, but of two mice with subcutaneous tumors, one had no parotid tumor and the other had only a microscopic tumor in the parotid gland. In A/SN mice, Sjögren and Ringertz (1962) noted a high incidence of bone tumors and a low incidence of salivary gland tumors.

Gross (1953b, 1970) submitted sections of mouse "neck" tumors to four pathologists. One diagnosed the tumors as anaplastic carcinoma and three as salivary gland carcinomas. Stewart (1953) referred to the tumors as sarcomas of the parotid gland, then later as pleomorphic, since both fusiform and epithelial elements were present and frequently there were cysts filled with a mucinous substance (Stewart *et al.,* 1957). Earlier, Law *et al.* (1955) had also called them pleomorphic because, histologically, epithelial and mesenchymal or fibroblastic components were present. Rudiments of salivary glands grown *in vitro* show that both epithelial and mesenchymal cells are necessary for tumor development (see Section VI).

It is evident that some of the polyomavirus-induced tumors are pleomorphic. Dawe (1960, 1963, 1967) has discussed the tumors, particularly in regard to the origin of the cells. Tumors in the thymus occurring in 26% of Swiss mice (Stewart, 1960) have been described as epithelioid thymomas by Stewart *et al.* (1958) and by Gross (1970), as reticulum cell sarcomas by Buffett *et al.* (1958), and as thymic lymphomas by McCulloch *et al.* (1961). Dawe considered them closely related to parotid gland tumors.

Fig. 1. Mouse bearing polyomavirus-induced salivary gland and thymic tumors.

About 50% of the Swiss mice and 100% of the hybrid mice developed lesions in the convoluted tubules of the kidney

(Stewart *et al.*, 1957; Stewart, 1960). In a study of kidney lesions, Leuchtenberger *et al.* (1961) found that epithelial cells of tubules displayed no mitosis and no striking increase in DNA content but showed viral activity as denoted by DNA inclusion bodies resulting from destruction of cells. The stroma cells displayed no viral activity but did show mitosis, proliferation leading to sarcoma formation, and an increase in DNA. The significance of the interaction of epithelial cells was shown when kidney rudiments were cultured *in vitro:* The metanephrogenic mesenchyma, until tubules developed from it, was highly susceptible to the cytotoxic action of polyomavirus. After that, it was not susceptible (see Section VI).

Mammary adenocarcinomas occurred in 25% of both female and male mice at an early age (Stewart *et al.*, 1958; Stewart, 1960), appearing earlier than parotid gland tumors in some mice. About two-thirds of the mice with mammary tumors also had tumors of the hair follicles which were pale yellow and occurred in groups on the ventral surface, sometimes extending to the legs but seldom to the back. These often coalesced but rarely ulcerated. Histologically, they ranged from severe degeneration to atrophy, through hyperplasia, to tumors with cytological characteristics of epidermoid carcinoma. Metastases to the lungs occurred from one of these tumors (Stewart, 1960). Because he did not see metastasis, Dawe (1960) disagreed that the hair follicle tumors were epidermoid carcinoma. Dawe (1960) also questioned whether two primary lung tumors listed by Stewart *et al.* (1958) and primary lung tumors reported by Latarjet and DeJaco (1958) were established. Epidermoid carcinomas of the forestomach and oral mucosa, the latter invading the mandible, were described by Stewart *et al.* (1958). Dawe considered these tumors to be spontaneous carcinomas that occasionally arise.

Bone tumors (frequently multiple) involving the flat bones of the skull in the mandibles, scapula, ribs, vertebrae, sacrum, and long bones of the femur and humerus were seen in 22% of the polyomavirus-infected Swiss mice. The tumors ranged from those that appeared benign to those appearing highly malignant. In some mice, metastases to the liver and lung occurred (Stewart, 1960).

Mesotheliomas occurred in 22% of virus-inoculated Swiss mice. These usually developed in the serous lining of the lungs and occasionally in the epicardium, pericardium, and peritoneum. Some of the mesotheliomas were isolated in the pleura and had diameters of 2–5 mm. The generalized lesions of the serous membranes showed fingerlike projections of flattened cuboidal cells. These were continuous with those of the normal mesothelial lining of the organs and were supported by a fine network of connective tissue. The lesions were often discrete, but in a few instances, they were confluent and produced masses of acidophilic cells.

Hemangioendotheliomas occurred in a few virus-infected mice at an early age. They were similar to those that are known to arise spontaneously or to be induced by certain carcinogens.

Subcutaneous sarcomas occurred in infected mice kept for 6 months or longer. A few were associated with salivary gland tumors; others were independent of salivary gland tumors or any other tumors.

Medullary adrenal carcinomas occurred more frequently in polyomavirus-infected C3H mice (36%) but only in 4% of the Swiss mice. These tumors were often anaplastic, and the centers of the large masses were necrotic. Sometimes invasion to the kidneys occurred, and metastasis to the liver, peripheral lymph nodes, and ovaries was noted with large tumors.

Thyroid lesions were observed in 6% of the Swiss mice. These consisted of certain enlarged acinar cells containing a small amount of colloid and lined by small cells and irregular cords of small cuboidal cells. A few carcinomas also developed (Stewart, 1960).

Selected tumors noted in a few virus-infected mice were sweat gland adenocarcinomas, liver hemangiomas (Stewart *et al.*, 1958), and epidermoid carcinomas of the buccal membranes (Stewart, 1960). Other cases of anemia and glomerulonephritis developed in virus-infected mice whose littermates developed tumors.

Some mice had retarded growth, remained small throughout their life span, and had abnormal hair development; in some, acute inflammatory reactions of the eyes with resultant opacity occurred (Stewart *et al.*, 1957, 1958; Stewart, 1960). Vandeputte and DeSomer (1965) attributed runting to a depletion of lymphocytes, since Peyer's patches were atrophied and there was a depletion of lymphocytes in the cortex of the lymph nodes and spleen. In severe runting, thymic lesions were also seen.

The sites of tumors or lesions induced by various viral strains are known to vary. Many investigators have made no mention of tooth germ abnormalities in polyomavirus-infected mice. Yet, at least two strains of the virus, one from the Laboratory of Pathology at the National Cancer Institute and another from the Laboratory of Infectious diseases at the National Institute of Health, have induced bizarre changes in the oral tissues which have been of great interest to students of dental and oral pathology. Dawe *et al.* (1959) noted the similarity of the lesions in mice to ameloblastomas in man, and further studies bore out this observation (Fleming, 1963; Fleming and Soni, 1964; Stanley *et al.*, 1964, 1965). The viral inoculation route was not the same in different studies; some mice were injected subcutaneously in the nape of the neck or the base of the tail, whereas others were injected in the peritoneal cavity. Organ cultures of tooth bud rudiments were also cultured *in vitro* and transplanted to other mice; occasionally these transplants developed ameloblastosis (see Section VI).

Polyomavirus-induced hydrocephalus is another infrequently observed manifestation. Polyomavirus propagated in sarcoma 180 and passed intracerebrally in DBA mice caused hydrocephalus when injected into Swiss mice but not when injected into hamsters (Li and Jahnes, 1959; Holtz *et al.*, 1964, 1966). Vandeputte and Brucher (1962) produced hydrocephalus in rats with the SE polyomavirus strain.

Hamsters infected with polyomavirus at 1–3 days of age develop tumors in many sites: kidneys, liver, heart, lungs, thymus, gastrointestinal tract, subcutaneous tissue, and brain. Tumors in hamsters arise in less time than tumors in mice (one hamster died 12 days postinfection with a renal carcinoma), with most of the tumors occurring 1–2 months after infection. Histologically, hamster tumors are not as varied as those occurring in mice, the majority being spindle cell sarcomas and angiomatous tumors. Hemangiomas and hemangioendotheliomas also occur (Eddy *et al.*, 1958d; McCulloch *et al.*, 1959; Romanul *et al.*, 1961; Stoker, 1960; Defendi and Lehman, 1965). Virus-induced hepatic hemangiomas have a low incidence of transplantability, growing as cystic hemangiomas after transplantation. One line of this transplantable tumor acquired the ability to metastasize rapidly and consistently (Stanton, 1965). Lung tumors were induced in hamsters by intratracheal inoculation of virus (Rabson *et al.*, 1960a).

Newborn rats infected with polyomavirus usually develop angiosarcomas in the subcutaneous tissues or kidneys after approximately 2–3 months. Tumors showing osteoid formation and calcification, cavernous hemangiomas (Eddy *et al.*, 1959b), and hydrocephalus (Vandeputte and Brucher, 1962) have also been shown in some polyomavirus-infected rats.

Newborn rabbits, after 2 weeks–2 months postinfection, respond to the polyomavirus by developing multiple, small, fibromatous skin nodules, some of which are filled with a yellowish fluid. The nodules are well circumscribed but not encapsulated and regress after 3–4½ months. After regression of the nodules, rabbits have developed humoral antibodies to polyomavirus. Virus can be recovered by incubating a finely minced nodule from a rabbit infected 51 days earlier with mouse embryo monolayers. These growths in the infected rabbit are of interest because rabbits are the only hosts known in which the tumors regress (Eddy *et al.*, 1959a). No attempt was made to grow the virus in rabbit tissue.

Newborn guinea pigs infected with polyomavirus develop large tumors at the site of inoculation after a few weeks to over a year. These tumors are firm and encapsulated, invade adjacent muscle and connective tissue in some areas, grow progressively, and eventually kill the animals. In addition to subcutaneous tumors, small spindle cell tumors or pleomorphic, undifferentiated sarcomas are present in the kidneys, liver, spleen, and lungs of some animals (Eddy *et al.*, 1960).

Virus-infected mastomys, an African rodent, develop renal sarcomas, liver angiomatous tumors or adenosarcomas, or sarcomas of the heart muscle similar to the heart muscle tumors induced in hamsters (Rabson *et al.*, 1960b).

Ferrets are also susceptible to tumor induction with either polyomavirus or its DNA. Most of the neoplasms are fibrosarcomas, but one osteosarcoma in the diaphragm with metastasis to the lungs and liver was reported (Harris *et al.*, 1961; Pomerance and Chesterman, 1965).

The concentration of the infecting virus dose, the susceptibility of the animal, and the age at infection are important determinants of whether tumors are induced. When the concentration of the infecting virus is decreased or the age of the animal at the time of infection is increased, tumors may occur only at the site of inoculation and the latent period for tumor development may be prolonged (Platz *et al.*, 1963; Flocks *et al.*, 1965).

All polyomavirus-free strains of mice are susceptible to infection with the polyomavirus, but there are minor genetic differences in strain susceptibility. The most resistant mice are the C57 Black and C57 Brown (Mirand *et al.*, 1960; Gross, 1970), while an A strain of mice reacted with a high incidence of bone tumors (Sjögren and Ringertz, 1962).

Animal mothers infected with live virus or immunized with killed virus prior to delivery gave birth to offspring that were less susceptible to tumor induction with polyomavirus (Eddy *et al.*, 1959c; Stewart and Eddy, 1959; Stewart *et al.*, 1960).

Law (1957) noted that transfer of thymic tissue to thymectomized mice restored their ability to develop lymphocytic neoplasms. Thymectomy was later shown to increase oncogenesis in polyomavirus-infected animals and to decrease resistance in older animals (Vandeputte *et al.*, 1963; Malmgren *et al.*, 1964; Miller *et al.*, 1964; Mori *et al.*, 1964; Law, 1965; Defendi and Roosa, 1965; Diderholm *et al.*, 1966). The effect of the thymus on tumor induction was further demonstrated when antilymphocytic sera were found to stimulate oncogenicity by polyomavirus (Allison and Law, 1968; Vandeputte, 1969).

Nude or athymic mice are more sensitive to tumor development than their heterozygous counterparts. The types of tumors induced in the nude mice do not differe significantly from those observed in normal mice, but the latent period of tumor development is shortened and there is a high incidence of tumors in mice infected some weeks after birth. The mice do become partially resistant to tumor induction if infected at 120 days of age, but this late resistance appears to be mediated by B cells and is thymus independent (Vandeputte *et al.*, 1974; Stutman, 1975).

Radiation, the application of certain chemicals, and high-intensity sound stress have also been used to increase tumor induction in polyomavirus-infected mice. Mice infected when 2 months old and exposed to 300 R X irradiation developed tumors, whereas the control nonirradiated group did not (Law and Dawe, 1960). When 9,10-dimethyl-1,2-benzanthracene, 3,4-benzopyrene, or croton oil was painted on the skins of virus-infected mice starting at 6 weeks of age, a synergetic effect was demonstrated, the chemically treated mice developing more tumors than infected but untreated mice (Rowson *et al.*, 1961). Prolonged exposure to high-intensity sound stress increased tumor development. In two groups of mice 14–34 days old infected intracardially with concentrated polyomavirus, the group exposed to sound stress developed more tumors than the unstressed group (Chang and Rasmussen, 1964).

Infection with another microorganism may influence either polyomavirus infection or the growth of the other microorganism. Some polyomavirus strains, particularly the large-plaque variants, are more oncogenic and are known to induce interferon production, which tends to inhibit certain other viruses (see Section VI).

Tumor induction in hamsters was reduced and longevity increased if the hamsters were inoculated with BCG tubercle bacilli, Calmette-Guerin strain, when 1 week old and later infected with the polyomavirus. (Disease due to the tubercle bacilli was prevented by streptomycin.) Tumor incidence was not affected when the virus was given to newborn hamsters, and the bacilli were given later (Lemonde and Clode, 1962; Lemonde and Clode-Hyde, 1966).

Tumor incidence was enhanced when newborn mice were infected with a mixture of polyomavirus and either type 1 or 5 (both nononcogenic) adenovirus (Gotlieb-Stematsky *et al.*, 1966a). An adenovirus is known to be able to form a genetic hybrid with SV40 (Rowe and Baum, 1964), but a hybrid with polyomavirus and an adenovirus has not been reported.

In nature, the polyomavirus is elusive, since it can exist as a silent infection, showing no gross signs of its presence. Studies of wild mice have been carried out to detect infection using hemagglutination inhibition or a sensitive mouse antibody production (MAP) test. The MAP test consists of injecting weanling mice intraperitoneally with material suspected of containing small amounts of virus. The mice are then bled after 21–35 days and their sera tested for hemagglutination inhibition (Rowe *et al.*, 1959a).

In New York City, polyomavirus infection was localized in mice caught in certain apartment houses or on certain floors, with 40% of the mice serologically positive in some areas. Sera from humans, cats, and dogs in the area were antibody free, and virus was not detected in mites and roaches in the area (Rowe *et al.*, 1961). In rural areas in Maryland, polyomavirus was found in mice on 5 of 16 farms and in three grain mills. During August, November, and January, 38 of 179 captured mice had antibody for polyomavirus, and virus was recovered from four of eight hay specimens. At one mill, 21 of 155 mice had humoral antibodies, and virus was isolated from 5 of 10 grain specimens (Huebner *et al.*, 1962).

A number of colonies of inbred laboratory mice had antibodies to polyomavirus when tested by the hemagglutination inhibition or complement fixation tests (Rowe *et al.*, 1959b; Yabe *et al.*, 1961; Jahkola and Vainio, 1962; Kondo and Matsuura, 1963; Parker *et al.*, 1964). Infection ranged from 1 to 86% in some colonies, whereas others were free from infection. Evidence was found for the presence of the virus in mice in Europe and Japan; these stocks had originated from mice obtained in the United States (Rowe *et al.*, 1961). No evidence of infection in mice in another laboratory was found (Mori and Amako, 1963). Sera from wild mice (*Peromyscus, Microtus,* and *Perognathus*), rats (*Neotoma* and *Dipodomys*), chipmunks, squirrels, nutria, rabbits, muskrats, woodchucks, porcupines, monkeys, mule deer, moose, sheep, seals, cats, dogs, foxes, raccoons, and skunks from Dugway, Utah, were all negative when tested for polyomavirus antibodies. Seven opossum sera showed positive hemagglutination inhibition antibody titers but were negative in virus neutralization tests and for complement fixation (Rowe *et al.*, 1961).

VIII. SPREAD OF POLYOMAVIRUS AND METHODS FOR ITS PREVENTION

Soon after the tumor-inducing agent was identified as a virus, it was noted that uninoculated mice kept in the same room as infected animals also became infected (Buffett *et al.*, 1958; Rowe *et al.*, 1958; Sachs and Heller, 1959; Neriishi *et al.*, 1961). Infected mice shed virus in urine (Rowe *et al.*, 1958; Sachs and Heller, 1959; Salaman and Rowson, 1960; Yabe *et al.*, 1961), in saliva (Rowe *et al.*, 1958; Sachs and Heller, 1959), and in feces (Sachs and Heller, 1959; Sato *et al.*, 1961).

Newborn mice can be infected intranasally (Rowe *et al.*, 1961) and intratracheally (Rabson *et al.*, 1960a). In nature, the intranasal route of infection is believed to account for the virus spread. Ingestion of the virus in food and drink is not an efficient method of transmission. The virus is known to persist in contaminated bedding, hay, or wherever mice live. Because it is highly stable, the virus can also exist in aerosols (Rowe *et al.*, 1961).

Young mice are important in virus dissemination, since the virus often multiplies to a high titer in their tissues before they subsequently develop antiviral antibodies (Eddy and Stewart, 1959a). [The viral antibodies secreted in the milk of infected mothers may account for the failure of young mice to develop tumors (Stewart *et al.*, 1960).]

Polyomavirus-free mouse colonies can be maintained by the initial use of virus-free breeders housed in clean quarters entirely separate from areas where other experimental work with microorganisms is being performed, having facilities for frequent cage sterilization, caretakers who are not working with other polyomavirus-infected animals, and where animals of unknown origins are never introduced into the colony until, or unless, they are found to be free of polyomavirus. Monitoring of the mice at intervals by the MAP test (Rowe *et al.*, 1959a) is a useful procedure, but the testing of sera must be done away from the mouse colony. Direct testing of mouse sera by hemagglutination inhibition or complement fixation can also be used, but these methods are generally less sensitive. *In vitro* cultivation of suspected positive tissue or biological material in susceptible cell cultures can also be used.

The potential of airborne virus spread is especially important in rooms where polyomavirus-infected animals are being housed or where cell culture work with the virus is being performed. Twenty-four of 32 mice held for a month in a room with virus-infected mice developed hemagglutination inhibiting antibodies, and 3 of 15 control mice held in an air mass flow cabinet showed seroconversion. The handling of contaminated bedding and cages increases polyomavirus aerosols, and a 1- and 3-hr exposure resulted in seroconversion of 40% (6/15) and 72% (23/32) of exposed animals, respectively. Ventilation is an important factor. In one study, a conversion rate of 50% (15/30) occurred in a conventionally ventilated room during a 12-week period, only 10% (3/30) in an air flow room with a vertical air velocity of 26 ft/min, and 0/30 with a vertical air velocity of 30 ft/min (McGarrity *et al.*, 1976).

A novel and useful method for detecting airborne polyoma involves the use of a Litton high-volume air sampler that concentrates a large volume of air in a relatively small amount of culture medium that can then be further concentrated by centrifugation. Four of six such concentrated air samples given the MAP test were positive (McGarrity and Dixon, 1978).

Cell cultures to be infected with polyomavirus are best handled in small cubicles or cabinets equipped with uv light and washed down with a disinfectant before and after use. Monitoring for the presence of the virus is important, but there is no substitute for constant attention to all known methods for preventing contamination.

IX. CONCLUSIONS

Polyomavirus, one of the smallest known viruses, causes a natural infection in mice. Because of the small size of the viral DNA which carries the entire genome, it can code for only a few proteins and is dependent on host cells for most of its early development. It can multiply and destroy cells, or it may transform a few cells into malignant cells which can grow into lethal tumors. It is therefore not surprising that nonmalignant lesions are seen in tissues of polyomavirus-infected animals, or that cancers in animals or man are generally defined in histopathological and clinical terms (Weil, 1978).

The lytic phase of the virus is similar to that of many nononcogenic viruses with one exception, which is not evident unless specific tests are carried out; most polyomavirus strains induce a T, or tumor, antigen which can inhibit transplantation of polyomavirus in isologous animals.

Transformation usually occurs in other animal species, i.e., the hamster or rat, with greater ease than in mice. Transformation or tumor induction is favored by a high concentration of infective virus, the young age of the animal when infected, and use of cells or animals from virus-free mothers. The transformed cells' mitotic ability is altered by the virus, and in many transformed cells no virus can be detected, yet the cells resist transplantation to an isologous host that has received prior polyomavirus immunization.

A method for reversing the transformation still eludes investigators. As long as mice naturally infected with the polyomavirus are not subjected to stress due to irradiation, certain chemical compounds, or other microbial infections, the silent infections can remain unnoticed. However, such a silent infection may affect experiments with other microorganisms either by stimulating them, by repressing them, or by the appearance of tumors. With constant vigilance, laboratory mouse colonies can be kept free of polyomavirus infection.

REFERENCES

Abercombie, M., Heaysman, J. E. M., and Karthauser, H. M. (1957). Social behavior of cells in tissue culture. III. Mutual influence of sarcoma cells and fibroblasts. *Exp. Cell Res.* **13,** 276–291.

Abramczuk, J., Vorbrodt, A., Sotter, D., and Koprowski, H. (1978). Infection of mouse preimplantation embryos with simian virus 40 and polyoma virus. *Proc. Natl. Acad. Sci. U.S.A.* **75,** 999–1003.

Acheson, N. H., and Miéville, F. (1978). Extent of transcription of the E strain of polyomavirus DNA during the early phase of productive infection. *J. Virol.* **28,** 885–894.

Acheson, N. H., Buetti, E., Sherrer, K., and Weil, R. (1971). Transcription of the polyomavirus genome. Synthesis and cleavage of giant late polyoma specific RNA. *Proc. Natl. Acad. Sci. U.S.A.* **68,** 2231–2235.

Adzhigitov, F. I. (1963). On relationship of polyomavirus and tissue culture of monkeys. *Acta Unio Int. Cancrum* **19,** 332–333.

Adzhigitov, F. I. (1964). [Some forms of interaction between polyomavirus and cells from tissue cultures from different animals.] *Vopr. Virusol.* **9,** 667–670.

Allison, A. C. (1961). Interference with, and interferon production by, polyomavirus. *Virology* **15,** 47–51.

Allison, A. C. (1963). Interference and interferon in relation to tumor viruses and tumor cells. *In* "Viruses, Nucleic Acid and Cancer," University of Texas, M. D. Anderson Hospital and Tumor Institute, No. 17, pp. 462–484. Williams & Wilkins, Baltimore, Maryland.

Allison, A. C., and Law, L. W. (1968). Effects of anti-lymphocytic serum on virus oncongenesis. *Proc. Soc. Exp. Biol. Med.* **127,** 207–212.

Basilico, C., and Burstin, S. J. (1971). Multiplication of polyomavirus in mouse-hamster somatic hybrids: A hybrid cell line which produces viral particles containing predominantly host deoxyribonucleic acid. *J. Virol.* **7,** 802-812.

Benjamin, T. L. (1972). Physiological and genetic studies of polyomavirus. *Curr. Top. Microbiol. Immunol.* **59,** 107-133.

Bjursell, G. L., Skoog, L., Thelander, L., and Södermann, G. (1977). 2'Deoxy-2'-azidocytidine inhibits the initiation of polyoma DNA synthesis. *Proc. Natl. Acad. Sci. U.S.A.* **74,** 5310-5313.

Bourgaux, P., and Bourgaux-Ramoisy, D. (1972). Is a specific protein responsible for the supercoiling of polyoma DNA? *Nature (London)* **235,** 105-107.

Brady, J. N., and Consigli, R. A. (1978). Chromatographic separation of the polyomavirus proteins and renaturation of the isolated VPI major capsid protein. *J. Virol.* **27,** 436-442.

Brady, J. N., Winston, V. D., and Consigli, R. A. (1978). Characterization of a DNA protein complex and capsomere subunits derived from polyoma virus by treatment with ethylene-glycol bis-n.n'-tetra acetic acid and ditheiothreitol. *J. Virol.* **27,** 193-204.

Brodsky, I., Rowe, W. P., Hartley, J. W., and Lane, W. T. (1959). Studies of mouse polyomavirus infection. II. Virus stability. *J. Exp. Med.* **109,** 439-447.

Brown, A., Consigli, R. A., Zabielski, J., and Weil, R. (1974). Effect of β-propiolactone inactivation of polyomavirus on viral functions. *J. Virol.* **14,** 840-845.

Buffett, R. F., Commerford, S. L., Furth, J., and Hunter, M. J. (1958). Agent in AK leukemia tissues, not sedimented at 105000 g causing neoplastic and non-neoplastic lesions. *Proc. Soc. Exp. Biol. Med.* **99,** 401-407.

Chang, S. S., and Rasmussen, A. F. (1964). Effect of stress on susceptibility of mice to polyomavirus infection. *Bacteriol. Proc.* **64,** 134. (Abstr.)

Cordray, D. R., Rogers, H. C., and Mogabgab, W. J. (1962). Growth of polyomavirus in the lungs of C3H mice and susceptibility of other host systems to infection. *J. Infect. Dis.* **111,** 146-154.

Coto, V., Galeota, C. A., Fantoni, V., Lavegas, E., and Coraggio, F. (1965). Sulla produzione di interferon da cellule embrionali di topo coltivate *in vitro* ed infettate col virus del polioma. *Riv. Ist. Sieroter. Ital.* **40,** 104-108.

Crawford, L. V. (1965). A study of human papilloma virus DNA. *J. Mol. Biol.* **13,** 362-372.

Crawford, L. V., and Crawford, E. M. (1963). A comparative study of polyoma and papilloma viruses. *Virology* **21,** 258-263.

Crawford, L. V., Robbins, A. K., Nicklin, P. M., and Osborn, K. (1974). Polyoma DNA replication. Location of the origin of different virus strains. *Cold Spring Harbor Symp. Quant. Biol.* **39,** 219-225.

Dawe, C. J. (1960). Cell sensitivity and specificity of response to polyoma virus. *Natl. Cancer Inst. Monogr.* No. 4, 67-125.

Dawe, C. J. (1963). Skin-appendage tumors induced by polyomavirus in mice. *Natl. Cancer Inst. Monogr.* No. 10, 459-488.

Dawe, C. J. (1967). Neoplasms induced by polyomavirus in the upper respiratory tracts of mice. *In* "Cancer of the Nasopharynx" (C. S. Muir and K. Shanmugaratnam, eds.), Union Int. Cancer Monogr. Ser. No. 1, pp. 179-196. Munksgaard, Copenhagen.

Dawe, C. J., and Law, L. W. (1959). Morphologic changes in salivary-gland tissue of the newborn mouse exposed to parotid tumor agent in vitro. *J. Natl. Cancer Inst.* **23,** 1157-1177.

Dawe, C. J., Law, L. W., and Dunn, T. B. (1959). Studies of parotid tumor agent in cultures of leukemic tissues of mice. *J. Natl. Cancer Inst.* **23,** 717-797.

Dawe, C. J., Law, L. W., Morgan, W. D., and Shaw, M. G. (1962). Morphologic response to tumor viruses. *Fed. Proc., Fed. Am. Soc. Exp. Biol.* **21,** 5-13. (Abstr.)

Dawe, C. J., Slatick, M. S., Morgan, W. D., and Hensley, E. (1964). Induction of neoplasms in salivary gland rudiments recovered from frozen storage. *Proc. Soc. Exp. Biol. Med.* **116,** 149-153.

Defendi, V. (1963). Effect of SV40 virus immunization on growth of transplantable SV40 and polyoma virus tumors in hamsters. *Proc. Soc. Exp. Biol. Med.* **113,** 12-16.

Defendi, V., and Lehman, J. M. (1965). Transformation of hamster embryo cells *in vitro* by polyomavirus; morphological, karological, immunological and transplantation characteristics. *J. Cell. Comp. Physiol.* **66,** 351-410.

Defendi, V., and Roosa, R. A. (1965). Effect of thymectomy on induction of tumors and on the transplantability of polyoma-induced tumors. *Cancer Res.* **25,** 300-306.

Deinhardt, F., and Henle, G. (1960). Inteference between polyoma and vesicular stomatitis viruses in tissue culture. *J. Immunol.* **84,** 608-614.

Diderholm, H. (1967). Transformation of bovine cells *in vitro* by polyoma virus, and the properties of the transformed cells. *Proc. Soc. Exp. Biol. Med.* **124,** 1197-1201.

Diderholm, H., and Wesslén, T. (1963). Interaction between polyomavirus and different tissues of mouse and hamster in vitro. *Arch. Gesamte Virusforsch.* **14,** 45-54.

Diderholm, H., Estola, T., and Wesslén, T. (1966). Effect of thymectomy on tumour production by polyomavirus in rabbits. *Acta Pathol. Microbiol. Scand.* **66,** 396-400.

DiMayorca, G. A., Eddy, B. E., Stewart, S. E., Hunter, W. S., Friend, C., and Bendich, A. (1959). Isolation of infectious deoxyribonucleic acid from SE polyoma-infected tissue cultures. *Proc. Natl. Acad. Sci. U.S.A.* **45,** 1805-1808.

DiMayorca, G., Callender, J., Marin, G., and Giordano, R. (1969). Temperature-sensitive mutants of polyoma virus. *Virology* **38,** 126-133.

Dulaney, A. D. (1956). Parotid gland tumor in AKR mice inoculated when newborn with cell-free AK leukemic extracts. *Cancer Res.* **16,** 877-879.

Dulbecco, R. (1969). Cell transformation by viruses. Two minute viruses are powerful tools for analyzing the mechanism of cancer. *Science* **166,** 962-968.

Dulbecco, R., and Freeman, G. (1959). Plaque production by the polyomavirus. *Virology* **8,** 396-397.

Dulbecco, R., and Vogt, M. (1960). Significance of continued virus production in tissue cultures rendered neoplastic by polyomavirus. *Proc. Natl. Acad. Sci. U.S.A.* **46,** 1617-1623.

Eckhart, W. (1974). Properties of temperature-sensitive mutants of polyomavirus. *Cold Spring Harbor Symp. Quant. Biol.* **39,** 37-40.

Eddy, B. E. (1960). The polyomavirus. *Adv. Virus Res.* **7,** 91-102.

Eddy, B. E. (1963). Comparison of properties of two viruses. SV40 and polyomavirus, oncogenic for hamsters. *Perspect. Virol.* **3,** 138-158.

Eddy, B. E. (1969). Polyoma virus. *Virol. Monogr.* **7,** 1-114.

Eddy, B. E., and Stewart, S. E. (1959a). Physical properties and hemagglutinating and cytopathic effects of the SE polyomavirus. *Proc. Can. Cancer Res. Conf.* **3,** 307-324.

Eddy, B. E., and Stewart, S. E. (1959b). Characteristics of the SE polyomavirus. *Am. J. Public Health* **49,** 1486-1492.

Eddy, B. E., Rowe, W. P., Hartley, J. W., Stewart, S. E., and Huebner, R. J. (1958a). Hemagglutination with the SE polyomavirus. *Virology* **6,** 290-291.

Eddy, B. E., Stewart, S. E., and Berkeley, W. (1958b). Cytopathogenicity in tissue cultures by a tumor virus from mice. *Proc. Soc. Exp. Biol. Med.* **98,** 848-851.

Eddy, B. E., Stewart, S. E., and Grubbs, G. E. (1958c). Influence of tissue culture passage, storage, temperature and drying on the viability of SE polyomavirus. *Proc. Soc. Exp. Biol. Med.* **99,** 289-292.

Eddy, B. E., Stewart, S. E., Young, R., and Mider, G. B. (1958d). Neoplasms in hamsters induced by a mouse tumor agent passed in tissue culture. *J. Natl. Cancer Inst.* **29,** 747–761.

Eddy, B. E., Stewart, S. E., Kirschstein, R. L., and Young, R. D. (1959a). Induction of subcutaneous nodules in rabbits with the SE polyomavirus. *Nature (London)* **183,** 766–767.

Eddy, B. E., Stewart, S. E., Stanton, M. F., and Marcotte, J. M. (1959b). Induction of tumors in rats by tissue culture preparation of SE polyomavirus. *J. Natl. Cancer Inst.* **22,** 161–171.

Eddy, B. E., Stewart, S. E., and Touchette, R. H. (1959c). Effect of immunization of adult female hamsters on the latency of infection in offspring inoculated with the SE polyomavirus. *Fed. Proc., Fed. Am. Soc. Exp. Biol.* **18,** 565. (Abstr.)

Eddy, B. E., Borman, G. S., Kirschstein, R. L., and Touchette, R. H. (1960). Neoplasms in guinea pigs infected with SE polyomavirus. *J. Infect. Dis.* **107,** 361–368.

Fenner, F. (1976). Classification and nomenclature of viruses. *Intervirology* **3,** 106–120.

Ferguson, J., and Davis, R. W. (1975). An electron microscopic method for studying and mapping the region of weak sequence homology between SV40 and polyomavirus. *J. Mol. Biol.* **94,** 135–149.

Fleming, H. S. (1963). SE polyomavirus—tumor and teeth. *J. Dent. Res.* **42,** 1405–1415.

Fleming, H. S., and Soni, N. N. (1964). SE polyomavirus and periodontum. *Periodontics* **2,** 115–118.

Flocks, J. S., Weiss, T. P., Kleinman, D. C., and Kirsten, W. H. (1965). Dose-response studies to polyomavirus in rats. *J. Natl. Cancer Inst.* **35,** 259–284.

Fogel, M., and Sachs, L. (1962). Studies on the antigenic composition of hamster tumors induced by polyomavirus and of normal hamster tissues *in vivo* and *in vitro*. *J. Natl. Cancer Inst.* **29,** 239–252.

Francke, B., and Eckhart, W. (1973). Polyoma gene function required for viral DNA synthesis. *Virology* **55,** 127–135.

Fried, M. A. (1970). Characterization of a temperature-sensitive mutant of polyomavirus. *Virology* **40,** 605–617.

Friedman, R. M., Rabson, A. S., and Kirkham, W. R. (1963). Variation in interferon production by polyomavirus strains of different oncogenicity. *Proc. Soc. Exp. Biol. Med.* **112,** 347–349.

Gardner, S. D., Field, A. M., Coleman, D. V., and Hume, B. (1971). New human papova virus (BK) isolated from urine after renal transplantation. *Lancet* **1,** 1253–1257.

Gey, G. O. (1941). Cytological and cultural observations on transplantable rat sarcomata produced by the inoculation of altered normal cells maintained in continuous culture. *Proc. Am. Assoc. Cancer Res.* **1,** 737. (Abstr.)

Gey, G. O., Gey, M. K., Firor, W. M., and Self, W. O. (1949). Cultural and cytological studies on autologous normal and malignant cells. *Acta Unio Int. Cancrum* **6,** 706–712.

Glasgow, L. A., and Habel, K. (1963). Role of polyomavirus and interferon in herpes simplex infection *in vitro*. *Virology* **19,** 328–339.

Goldman, E., and Benjamin, T. L. (1975). Analysis of host range of nontransforming polyomavirus mutants. *Virology* **66,** 372–384.

Gotlieb-Stematsky, T., and Leventon, S. (1960). Studies on the biological properties of two plaque variants isolated from the SE polyomavirus. *Br. J. Exp. Pathol.* **41,** 507–519.

Gotlieb-Stematsky, T., Karby, S., and Allison, A. C. (1966a). Increased tumour formation by polyomavirus in the presence of nononcogenic viruses. *Nature (London)* **212,** 421–422.

Gotlieb-Stematsky, T., Karby, S., and Eylan, E. (1966b). Antibody response and antigenic difference between high and low oncogenic variants of SE polyomavirus. *Arch. Gesamte Virusforsch.* **19,** 161–175.

Gotlieb-Stematsky, T., Rotem, Z., and Karby, S. (1966c). Production of susceptibility to interferon of polyomavirus variants of high and low oncogenic properties. *J. Natl. Cancer Inst.* **37,** 99–103.

Grobstein, C. (1955). Inductive interaction in the development of the mouse metanephros. *J. Exp. Zool.* **130,** 319–340.

Grobstein, C. (1956). Trans-filter induction of tubules in mouse metanephrogenic mesenchyme. *Exp. Cell Res.* **10,** 424–440.

Grobstein, C. (1962). Interactive processes in cytodifferentiation. *J. Cell. Comp. Physiol.* **60,** 35–48.

Gross, L. (1953a). A filterable agent recovered from AK leukemic extracts, causing salivary gland carcinomas in C3H mice. *Proc. Soc. Exp. Biol. Med.* **83,** 414–421.

Gross, L. (1953b). Neck tumors, or leukemia, developing in adult C3H mice following inoculation, in early infancy, with filtered (Berkefeld, N.), or centrifuged (144,000 × g.) AK leukemic extracts. *Cancer (Philadelphia)* **6,** 948–957.

Gross, L. (1956). Influence of ether, *in vitro,* on pathogenic properties of mouse leukemia extracts. *Acta Haematol.* **15,** 273–277.

Gross, L. (1970). The parotid tumor (polyoma)virus. *In* "Oncogenic Viruses" (L. Gross, ed.), 2nd ed., pp. 651–750. Pergamon, London.

Habel, K. (1961). Resistance of polyomavirus immune animals to transplanted polyoma tumors. *Proc. Soc. Exp. Biol. Med.* **106,** 722–725.

Habel, K. (1965). Specific complement-fixing antigens in polyoma tumors and transformed cells. *Virology* **25,** 55–61.

Habel, K., and Eddy, B. E. (1963). Specificity of resistance to tumor challenge of polyoma and SV40 virus-immune hamsters. *Proc. Soc. Exp. Biol. Med.* **113,** 1–4.

Harris, R. J. C., Chesterman, F. C., and Negroni, G. (1961). Induction of tumours in newborn ferrets with Mill Hill polyomavirus. *Lancet* **1,** 788–791.

Hartwell, L. H., Vogt, M., and Dulbecco, R. (1965). Induction of cellular DNA synthesis by polyomavirus. II. Increase in the rate of enzyme synthesis after infection with polyomavirus in mouse kidney cultures. *Virology* **27,** 262–272.

Hellström, I., and Hellström, K. E. (1963). Correlation between the sensitivity of polyoma-induced mouse tumors to polyoma and vaccinia virus *in vitro*. *J. Natl. Cancer Inst.* **31,** 1525–1532.

Henle, G., Deinhardt, F., and Rodriguez, J. (1959). The development of polyomavirus in mouse embryo cells as revealed by fluorescent antibody staining. *Virology* **8,** 388–391.

Holtz, A., Borman, G., and Li, C. P. (1964). Further studies on hydrocephalus in mice inoculated with SE polyomavirus (abstract). *Fed. Proc., Fed. Am. Soc. Exp. Biol.* **73,** 193.

Holtz, A., Borman, G., and Li, C. P. (1966). Hydrocephalus in mice infected with polyomavirus. *Proc. Soc. Exp. Biol. Med.* **121,** 1196–2000.

Huebner, R. J., Rowe, W. P., Hartley, J. W., and Lane, W. T. (1962). Mouse polyomavirus in rural ecology. *Tumor Viruses Murine Origin, Ciba Found. Symp.* No. 11, pp. 314–328.

Hunter, T., and Francke, B. (1975). *In vitro* polyoma synthesis. Inhibition by 1-β-D-arabinofuranosyl CTP. *J. Virol.* **15,** 759–775.

Hunter, T., and Gibson, W. (1978). Characterization of the mRNAs for the polyomavirus capsid proteins, VP1, VP2 and VP3. *J. Virol.* **28,** 240–253.

Inglot, A. D., and Niedźwiedzka, E. (1965). Observations on the production of interferon by polyomavirus. *Arch. Immunol. Ther. Exp.* **13,** 506–515.

Jahkola, M., and Vainio, T. (1962). Occurrence of polyomavirus H1 and antibodies in the rodent colonies of the State Serum Institute, Helsinki. *Ann. Med. Exp. Fenn.* **40,** 97–102.

Jahkola, M., and Vainio, T. (1964). Polyomavirus and mouse strain susceptibility. *Acta Pathol. Microbiol. Scand.* **61,** 60–66.

Jonsen, J., and Eng, J. (1962). Production of SE polyomavirus in suspension cultures. *Acta Pathol. Microbiol. Scand., Suppl. No. 154,* pp. 139–140.

Kahler, H., Rowe, W. P., Lloyd, B. J., and Hartley, J. W. (1959). Electron microscopy of mouse parotid tumor (polyoma)virus. *J. Natl. Cancer Inst.* **22,** 647–657.

Khare, G. P., and Consigli, R. A. (1965). Multiplication of polyomavirus. I. Use of selectively labeled (H_3) virus to follow the course of infection. *J. Bacteriol.* **90,** 819–821.

Kit, S., Dubbs, D. R., DeTorres, R. A., and Frearson, P. M. (1966a). Enzymology of cell cultures infected with oncogenic viruses. *Int. Cancer Congr., Abstr., 9th, Tokyo,* p. 228.

Kit, S., Dubbs, D. R., and Frearson, P. M. (1966b). Enzymes of nucleic acid metabolism in cells infected with polyomavirus. *Cancer Res.* **26,** 638–646.

Klug, A. (1965). Structure of viruses of the papilloma-polyoma type II. Comments on other works. *J. Mol. Biol.* **11,** 424–431.

Klug, A., and Finch, J. F. (1965). Structure of viruses of papilloma polyoma Type 1. Human wart virus. *J. Mol. Biol.* **11,** 403–423.

Koch, M. A., and Sabin, A. B. (1963). Specificity of virus-induced resistance to transplantation of polyoma and SV40 tumors in adult hamsters. *Proc. Soc. Exp. Biol. Med.* **113,** 4–12.

Kondo, T., and Matsuura, H. (1963). Distribution of virus antibodies in mice in Japan. *Proc. Soc. Exp. Biol. Med.* **112,** 170–173.

Latarjet, R., and DeJaco, M. (1958). Production de cancers multiples chez des souris AkR ayant reçu un extract leucémique a-cellulaire isologue. *C. R. Hebd. Seances Acad. Sci.* **246,** 499–501.

Latarjet, R., Craemer, R., and Montagnier, L. (1967). Inactivation of UV, X, and λ radiations on the infecting and transforming capacities of polyomavirus. *Virology* **33,** 104–111.

Law, L. W. (1957). Present status of nonviral factors in the etiology of reticular neoplasms of the mouse. *Ann. N.Y. Acad. Sci.* **68,** 616–625.

Law, L. W. (1965). Neoplasms in thymectomized mice following room infection with polyomavirus. *Nature (London)* **205,** 672–673.

Law, L. W., and Dawe, C. J. (1960). Influence of total body X-irradiation on tumor induction by parotid tumor agent in adult mice. *Proc. Soc. Exp. Biol. Med.* **105,** 414–419.

Law, L. W., Dunn, T. B., and Boyle, P. G. (1955). Neoplasms in C3H strain and in F_1 hybrid mice of two crosses following introduction of extracts and filtrates of leukemic tissues. *J. Natl. Cancer Inst.* **16,** 495–539.

Law, M.-F., Lancaster, W. D., and Howley, P. M. (1979). Conserved polynucleotide sequences among the genomes of papilloma viruses. *J. Virol.* **32,** 199–207.

Lemonde, P., and Clode, M. (1962). Effect of BCG infection on leukemia and polyoma in mice and hamsters. *Proc. Soc. Exp. Biol. Med.* **111,** 739–742.

Lemonde, P., and Clode-Hyde, M. (1966). Influence of bacilli Calmette–Guérin infection on polyoma in hamsters and mice. *Cancer Res.* **26,** 585–589.

Leuchtenberger, R., Leuchtenberger, C., Stewart, S. E., and Eddy, B. E. (1961). Difference in host cell-virus relationship between tubular epithelium and stroma in kidneys of mice infected with SE polyomavirus. A correlated cytological, histological and cytochemical study. *Cancer (Philadelphia)* **14,** 567–576.

Lev, Z., and Manor, H. (1977). Amount and distribution of virus-specific sequences in giant RNA molecules isolated from polyoma-infected mouse kidney cells. *J. Virol.* **21,** 831–842.

Levine, S. (1963). Effect of X-irradiation on the response of animal cells to virus. *Prog. Med. Virol.* **5,** 127–168.

Li, C. P., and Jahnes, W. G. (1959). Hydrocephalus in suckling mice inoculated with the SE polyomavirus. *Virology* **9,** 489–492.

McCulloch, E. A., Howatson, A. F., Siminovitch, L., Axelrad, A. A., and Ham, A. W. (1959). A cytopathogenic agent from a mammary tumor in a C3H mouse that produces tumours in Swiss mice and hamsters. *Nature (London)* **183,** 1535–1536.

McCulloch, E. A., Siminovitch, L., Ham, A. W., Axelrad, A. A., and Howatson, A. F. (1961). Carcinogenesis in vitro by polyomavirus. *Proc. Can. Cancer Res. Conf.* **4,** 253–270.

McGarrity, G. J., and Dixon, A. S. (1978). Detection of airborne polyomavirus. *J. Hyg.* **81,** 9–13.

McGarrity, G. J., Coriell, L. L., and Ammen, V. (1976). Airborne transmission of polyomavirus. *J. Natl. Cancer Inst.* **56,** 159–162.

McKay, R. L., and Consigli, R. A. (1976). Early events in polyomavirus infection, attachment, penetration and nuclear entry. *J. Virol.* **19,** 620–636.

Macpherson, I., and Montagnier, L. (1964). Agar suspension culture for the selective assay of cells transformed by polyomavirus. *Virology* **23,** 291–294.

Main, J. H. P., and Dawe, C. J. (1966). Tumor induction in transplanted tooth buds infected with polyomavirus. *J. Natl. Cancer Inst.* **36,** 1121–1136.

Main, J. H. P., and Waheed, M. A. (1971). Epitheliomesenchymal interaction in the proliferative response evoked by polyomavirus in odontogenic epithelium *in vitro. J. Natl. Cancer Inst.* **47,** 711–726.

Malmgren, R. A., Rabson, A. S., and Carney, P. G. (1964). Immunity and viral carcinogenesis. Effect of thymectomy on polyoma carcinogenesis in mice. *J. Natl. Cancer Inst.* **33,** 101–104.

Marin, G., and Littlefield, J. W. (1968). Selection of morphologically normal cell lines from polyoma-transformed BHK 21/13 hamster fibroblasts. *J. Virol.* **2,** 69–77.

Mattern, C. F. T. (1962). Polyoma and papilloma viruses. Do they have 42 or 92 subunits? *Science* **137,** 612–613.

Mattern, C. F. T., Takemoto, K. K., and Daniel, W. A. (1966). Replication of polyomavirus in mouse embryo cells. Electron microscopic observations. *Virology* **30,** 242–256.

Melnick, J. L. (1962). Papova virus group. *Science* **135,** 1128–1130.

Melnick, J. L., Allison, A. C., Butel, J. S., Eckhart, W., Eddy, B. E., Kit, S., Levine, A. J., Miles, J. A. R., Pagano, J. S., Sachs, L., and Vonka, V. (1974). Papoviridae. *Intervirology* **3,** 106–120.

Michael, M. R., Hirt, B., and Weil, R. (1967). Mouse cellular DNA enclosed in polyomaviral capsids (pseudovirions). *Proc. Natl. Acad. Sci. U.S.A.* **58,** 1381–1388.

Miller, J. F., Ting, R. C., and Law, L. W. (1964). Influence of thymectomy on tumor induction by polyomavirus in C57BL mice. *Proc. Soc. Exp. Biol. Med.* **116,** 323–327.

Mirand, E., Grace, J. T., Moore, G. E., and Mount, D. (1960). Relationship of viruses to malignant disease. I. Tumor induction by SE polyomavirus. *Arch. Intern. Med.* **105,** 469–481.

Montagnier, L., and Macpherson, I. (1964). Croissance sélective in gélose de cellules de hamster transformées par le virus du polyome. *C. R. Hebd. Seances Acad. Sci.* **258,** 4171–4173.

Morgan, W. D., and Dawe, C. J. (1963). Long term microcinematographic studies of action of polyomavirus in organ cultures of mouse salivary-gland rudiments. *Acta Unio Int. Cancrum* **19,** 309–319.

Mori, R., and Amako, K. (1963). Polyomavirus in experimental animal colonies in the Faculty of Medicine, Kyushu University. *Fukuoka Igaku Zasshi* **54,** 303–306.

Mori, R., Namoto, K., and Takeya, K. (1964). Tumor formation by polyomavirus in neonatally thymectomized mice. *Proc. Jpn. Acad.* **40,** 445–447.

Mulder, C., and Vogt, M. (1973). Production of nondefective and defective oligomers of viral DNA in mouse 3T3 cells transformed by a thermosensitive mutant of polyomavirus. *J. Mol. Biol.* **75,** 601–608.

Närkhammer, M., and Magnusson, G. (1976). DNA polymerase activities induced by polyomavirus infection of 3T3 mouse fibroblasts. *J. Virol.* **18,** 1–6.

Nagai, I., Yoshioka, W., Kumegawa, M., Arita, J., and Ikeda, J. (1963). Early influence of polyomavirus on transplanted tooth germs. *J. Dent. Res.* **42,** 1131–1139.

Negroni, G., and Chesterman, F. C. (1960). Virus cell relationship in kidney tumours induced in golden hamsters by the Mill Hill polyomavirus. *Br. J. Cancer* **14,** 672–678.

Negroni, G., Dourmashkin, R., and Chesterman, F. C. (1959). A "polyoma" virus derived from a mouse leukemia. *Br. Med. J.* **ii,** 1359–1360.

Neriishi, S., Yabe, Y., Oda, N., Ida, N., Gonzales, F., Sutlow, W., Kirschbaum, A., Taylor, H. G., and Trentin, J. J. (1961). Effects in newborn mice of tumor-producing agents carried in tissue culture. *J. Natl. Cancer Inst.* **26,** 611–619.

Nordenskjöld, B. O., and Nordenskjöld, K. (1966). Polyomavirus replication in tumor cells. *Int. Cancer Congr., Abstr., 9th, Tokyo*, p. 226.

Orth, G., Favre, M., and Croissant, O. (1977). Characterization of a new type of human papilloma virus that causes skin warts. *J. Virol.* **24,** 108–120.

Padgett, B. L., Walker, D. L., ZuRhein, G. M., Eckroade, R. J., and Dessel, B. H. (1971). Cultivation of papovalike virus from human brain with progressive multifocal leucoencephalopathy. *Lancet* **1,** 1257–1260.

Parker, J. C., Tennant, R. W., and Ward, T. G. (1964). Prevalence of murine viruses in mouse breeder colonies. *Fed. Proc., Fed. Am. Soc. Exp. Biol.* **23,** 580. (Abstr.)

Paulin, D., and Cuzin, F. (1975). Polyomavirus T antigen. I. Synthesis of modified heat-labile T antigen in cells transformed with a ts-a mutant. *J. Virol.* **15,** 393–397.

Pettijohn, D., and Kamiya, J. (1967). Interaction of RNA polymerase with polyoma DNA. *J. Mol. Biol.* **29,** 275–295.

Platz, C. E., Flocks, J. S., and Kirsten, W. H. (1963). Dose-response studies to polyoma virus in rats. *Fed. Proc., Fed. Am. Soc. Exp. Biol.* **22,** 606. (Abstr.)

Pomerance, A., and Chesterman, F. C. (1965). The pathology of polyoma-induced tumours in ferrets. *Br. J. Cancer* **19,** 211–215.

Porwit-Bóbr, Z., Przybylkiewicz, Z., and Chlap, Z. (1962). Studies on the oncogenic activity of polyomavirus *in vitro* and *in vivo*. I. Course of polyomavirus infection of mouse embryo cells in tissue culture *in vitro*. *Folia Biol. (Krakow)* **10,** 187–198.

Porwit-Bóbr, Z., Chlap, Z., Rokitowa, Z., and Jaszcz, W. (1963). Studies on the oncogenic activity of polyomavirus *in vitro* and *in vivo*. III. Malignant transformation of hamster heart cells as result of polyomavirus infection. *Acta Med. Pol.* **4,** 209–220.

Rabson, A. S., and Legallais, F. Y. (1959). Cytopathogenic effect produced by polyomavirus in cultures of milk-adapted murine lymphoma cells (Strain P388D). *Proc. Soc. Exp. Biol. Med.* **100,** 229–233.

Rabson, A. S., Branigan, W. J., and Legallais, F. Y. (1960a). Lung tumors produced by intratracheal inoculation of polyomavirus in Syrian hamsters. *J. Natl. Cancer Inst.* **25,** 937–965.

Rabson, A. S., Branigan, W. J., and Legallais, F. Y. (1960b). Production of tumors in *Rattus* (Mastomys) *natalensis* by polyomavirus. *Nature (London)* **187,** 423–425.

Radden, B. G., Main, T. H. P., and TenCote, A. R. (1973). Electron microscopic studies of organ cultures of tooth germs infected with polyomavirus. *J. Oral Pathol.* **2,** 272–279.

Reissig, M., and Kaplan, A. S. (1962). The morphology of noninfective pseudorabies virus produced by cells treated with 5-fluorouracil. *Virology* **16,** 1–8.

Reissig, M., Kelly, T. J., Jr., Daniel, R. W., Rangan, S. R. S., and Shah, K. V. (1976). Identification of the stumptailed macaque virus as a new papova virus. *Infect. Immun.* **14,** 225–231.

Romanul, F. C. A., Roizman, B., and Luttrell, C. N. (1961). Studies of polyomavirus: Pathology and distribution of tumors in Syrian hamsters following intracranial or subcutaneous inoculation. *Bull. Johns Hopkins Hosp.* **108,** 1–15.

Rowe, W. P., and Baum, S. G. (1964). Evidence for a possible genetic hybrid between adenovirus type 7 and SV40 viruses. *Proc. Natl. Acad. Sci. U.S.A.* **52,** 1340–1347.

Rowe, W. P., Hartley, J. W., Brodsky, I., Huebner, R. J., and Law, L. W. (1958). Observations on the spread of mouse polyomavirus infection. *Nature (London)* **182,** 1617.

Rowe, W. P., Hartley, J. W., Estes, J. D., and Huebner, R. J. (1959a). Studies of mouse polyomavirus infection. I. Procedures for quantitation and detection of virus. *J. Exp. Med.* **109,** 379–391.

Rowe, W. P., Hartley, J. W., Law, L. W., and Huebner, R. J. (1959b). Studies of mouse polyomavirus infection. III. Distribution of antibody in laboratory mouse colonies. *J. Exp. Med.* **109,** 449–462.

Rowe, W. P., Huebner, R. J., and Hartley, G. W. (1961). The ecology of mouse tumor virus. *Perspect. Virol.* **2,** 177–194.

Rowson, K. E. K., Roe, F. J. C., Ball, J. K., and Salaman, M. H. (1961). Induction of tumours by polyomavirus. Enhancement by chemical agents. *Nature (London)* **191,** 893–895.

Sachs, L., and Heller, E. (1959). The *in vitro* and *in vivo* analysis of mammalian tumour viruses. III. Experiments on the epidemiology of the polyomavirus. *Br. J. Cancer* **13,** 452–460.

Sachs, L., and Medina, D. (1960). Polyomavirus mutant with a reduction in tumour formation. *Nature (London)* **187,** 715–716.

Salaman, M. H., and Rowson, K. E. K. (1960). The epidemiology of polyomavirus in a mouse colony. *Lab. Anim. Cent. Symp., R. Soc. Med., London* **9,** 61–66.

Sambrook, J., and Shotkin, A. J. (1969). Polynucleotide ligase activity in cells infected with simian virus 40, polyomavirus or vaccinia virus. *J. Virol.* **4,** 719–726.

Sato, Y., Taylor, H. G., and Trentin, J. J. (1961). Excretion of polyoma virus from mouse bearing polyoma-infected tumor transplants (abstract). *Proc. Am. Assoc. Cancer Res.* **3,** 280.

Saxén, L., Vainio, T., and Toivonen, S. (1962). Effect of polyomavirus on mouse kidney rudiments *in vitro*. *J. Natl. Cancer Inst.* **29,** 597–631.

Saxén, L., Vainio, T., and Toivonen, S. (1963). Viral Susceptibility and embryonic differentiation. I. The histopathology of mouse kidney rudiments infected with polyomavirus and vesicular stomatitis virus in vitro. *Acta Pathol. Microbiol. Scand.* **58,** 191–204.

Schmidt, F. (1956). Über das Auftreten verschiedenartiger Tumoren bei Mäusen nach Injektion einer Zytoplasmafrakion aus Ehrlich-Ca-Zellen. *Z. Gesamte Inn. Med. Ihre Grenzgeb.* **11,** 640–644.

Sheinin, R. (1961). A rapid plaque assay for polyomavirus. *Virology* **15,** 85–87.

Shimino, H., and Kaplan, A. S. (1969). Correlation between the synthesis of DNA and histones in polyomavirus-infected mouse embryo cells. *Virology* **37,** 690–694.

Sidell, S. G., and Smith, A. E. (1978). Polyomavirus has three late mRNAs, one for each virion protein. *J. Virol.* **27,** 427–431.

Simmons, D. T., Chang, C., and Martin, M. A. (1979). Multiple forms of polyomavirus tumor antigens from infected and transformed cells. *J. Virol.* **29,** 881–887.

Sjögren, H. O. (1961). Further studies on the induced resistance against isotransplantation of polyoma tumors. *Virology* **15,** 214–219.

Sjögren, H. O., and Hellström, I. (1965). Induction of polyoma specific transplantation antigenicity in Moloney leukemic cells. *Exp. Cell Res.* **40,** 208–212.

Sjögren, H. O., and Ringertz, N. (1962). Histopathology and transplantability of polyoma induced tumors in strain A/SN and three coisogenic resistant (IR) substrains. *J. Natl. Cancer Inst.* **28,** 859–895.

Staneloni, R. J., Fluck, M. M., and Benjamin, T. L. (1977). Host range selection of transformation-defective hr.t. mutants of polyomavirus. *Virology* **77,** 598–609.

Stanley, H. R., Dawe, C. J., and Law, L. W. (1964). Oral tumors induced by polyomavirus in mice. *Oral Surg., Oral Med. Oral Pathol.* **17,** 547–558.

Stanley, H. R., Baer, P. N., and Kilham, L. (1965). Oral tissue alterations in mice inoculated with the Rowe substrain of polyomavirus. *Periodontics* **3,** 178–183.

Stanton, M. F. (1965). Transplantability, morphology and behavior of polyomavirus-induced hepatic hemangiomas of hamsters. *J. Natl. Cancer Inst.* **35,** 201–213.

Stanton, M. F., Stewart, S. E., Eddy, B. E., and Blackwell, R. H. (1959). The oncogenic effect of tissue culture preparations of polyomavirus on fetal mice. *J. Natl. Cancer Inst.* **23,** 1441–1475.

Stewart, H. L., Snell, K. C., Dunham, L. J., and Schylen, S. M. (1959). Transplantable and transmissible tumors of animals. *In* "Atlas of Pathology," Sect. 12, Fasc. 40, pp. 1–378. Armed Forces Inst. Pathol., Washington, D.C.

Stewart, S. E. (1953). Leukemia in mice produced by a filterable agent present in AKR leukemic tissue with notes on a sarcoma produced by the same agent. *Anat. Rec.* **117,** 532. (Abstr.)

Stewart, S. E. (1955). Neoplasms in mice inoculated with cell-free extracts or filtrates of leukemic mouse tissues. I. Neoplasms of the parotid and adrenal glands. *J. Natl. Cancer Inst.* **15,** 1391–1415.

Stewart, S. E. (1960). The polyomavirus, Section A. *Adv. Virus Res.* **1,** 61–90.

Stewart, S. E., and Eddy, B. E. (1959). Tumor induction by SE polyomavirus and the inhibition of tumors by specific neutralizing antibodies. *Am. J. Public Health* **49,** 1493–1497.

Stewart, S. E., Eddy, B. E., Gochenour, A. M., Borgese, N. G., and Grubbs, G. E. (1957). The induction of neoplasms with a substance released from mouse tumors by tissue culture. *Virology* **3,** 380–400.

Stewart, S. E., Eddy, B. E., and Borgese, N. G. (1958). Neoplasms in mice inoculated with a tumor agent carried in tissue culture. *J. Natl. Cancer Inst.* **20,** 1223–1243.

Stewart, S. E., Eddy, B. E., and Stanton, M. F. (1959a). Induction of neoplasms in mice and other mammals by a tumor agent carried in tissue culture. *Proc. Can. Cancer Res. Conf.* **3,** 287–305.

Stewart, S. E., Eddy, B. E., Stanton, M. F., and Lee, S. L. (1959b). Tissue culture plaques of SE polyomavirus. *Proc. Am. Assoc. Cancer Res.* **3,** 67. (Abstr.)

Stewart, S. E., Eddy, B. E., Irvin, M., and Lee, S. (1960). Development of resistance in mice to tumor induction by the SE polyomavirus. *Nature (London)* **186,** 615–617.

Stoker, M. (1960). Studies on the oncogenic activity of the Toronto strain of polyomavirus. *Br. J. Cancer* **14,** 679–689.

Stoker, M., and Macpherson, I. (1961). Studies on transformation of hamster cells by polyomavirus *in vitro. Virology* **14,** 359–370.

Stoker, M. G. P., and Sussman, M. (1965). Studies on the action of feeder layers in cell culture. *Exp. Cell Res.* **38,** 645–653.

Stutman, O. (1975). Tumor development after polyoma infection in athymic mice. *J. Immunol.* **114,** 1213–1217.

Tachovsky, T. G., and Hare, J. D. (1975). Polyomavirus strain with enhanced synthesis of capsid protein. *J. Virol.* **16,** 116–122.

Takemoto, K. K., and Habel, K. (1966). Hamster tumor cells doubly transformed by SV40 and polyomaviruses. *Virology* **30,** 20–28.

Takemoto, K. K., Malmgren, R. A., and Habel, K. (1966). Immunofluorescent demonstration of polyoma tumor antigen in lytic infection of mouse embryo cells. *Virology* **28,** 485–488.

Thomas, M., and LeBouvier, G. (1967). Transformation of bovine cell cultures by preparations of polyomavirus. *J. Gen. Virol.* **1,** 125–130.

Todaro, G. J., and Green, H. (1965). Successive transformations of an established cell line by polyomavirus and SV40. *Science* **147,** 513–514.

Todaro, G. J., Green, H., and Goldberg, B. D. (1964). Transformation of properties of an established cell line by SV40 and polyomavirus. *Proc. Natl. Acad. Sci. U.S.A.* **51,** 66–73.

Tooze, J. (1973). Polyoma and SV40 lytic cycle. *In* "The Molecular Biology of Tumour Viruses," Cold Spring Harbor Monograph Series, pp. 305–349. Cold Spring Harbor, New York.

Türler, H. (1975). Interaction of polyoma and mouse DNAs. III. Mechanisms of polyoma pseudovirion formation. *J. Virol.* **15,** 1158–1167.

Vainio, T., Saxén, L., and Toivonen, S. (1962). The effect of polyomavirus upon the development of mouse kidney rudiments *in vitro. Proc. Am. Assoc. Cancer Res.* **3,** 369. (Abstr.)

Vainio, T., Saxén, L., and Toivonen, S. (1963a). The acquisition of cellular resistance to polyomavirus during embryonic differentiation. *Virology* **20,** 350–385.

Vainio, T., Saxén, L., Toivonen, S., and Rapola, G. (1963b). Polyomavirus and mouse kidney organogenesis *in vitro. Acta Unio Int. Cancrum* **19,** 306–308.

Vandeputte, M. (1969). Antilymphocytic serum and polyoma oncogenesis in rats. *Transplant. Proc.* **1,** 100–105.

Vandeputte, M., and Brucher, J. M. (1962). Sarcomatose méningés expérimentale provoquée chez la ration par le virus polyome. *Acta Neuropathol.* **1,** 397–405.

Vandeputte, M., and DeSomer, P. (1962). Virus cell relation in rat polyoma tumors. *J. Gen. Microbiol.* **29,** 105–111.

Vandeputte, M., and DeSomer, P. (1965). Runting syndrome in mice inoculated with polyoma virus. *J. Natl. Cancer Inst.* **35,** 237–250.

Vandeputte, M., Denys, P., Jr., Leyten, R., and DeSomer, P. (1963). The oncogenic activity of the polyoma in thymectomized rats. *Life Sci.* **7,** 475–478.

Vandeputte, M., Eyssen, H., Sobis, H., and DeSomer, P. (1974). Induction of polyoma tumors in athymic nude mice. *Int. J. Cancer* **14,** 445–450.

Vogt, M., and Dulbecco, R. (1960). Virus cell interaction with a tumor producing virus. *Proc. Natl. Acad. Sci. U.S.A.* **46,** 365–370.

Vogt, M., and Dulbecco, R. (1962). Studies on cells rendered neoplastic by polyomavirus. The problem of the presence of virus-related materials. *Virology* **16,** 41–51.

Vogt, M., and Dulbecco, R. (1963). Steps in the neoplastic transformation of hamster embryo cells by polyomavirus. *Proc. Natl. Acad. Sci. U.S.A.* **49,** 171–179.

Weil, R. (1961). A quantitative assay for a subviral infective agent related to polyomavirus. *Virology* **14,** 46–53.

Weil, R. (1978). Viral "tumor antigens." A novel type of mammalian regulator proteins. *Biochim. Biophys. Acta* **516,** 301–388.

Weil, R., and Kára, J. (1970). Polyoma "tumor antigen" an activator of chromosome replication? *Proc. Natl. Acad. Sci. U.S.A.* **67,** 1011–1017.

Weil, R., Pétursson, G., Kára, J., and Diggelman, H. (1967). On the interaction of polyomavirus with the genetic apparatus of host cells. *In* "Molecular Biology of Viruses" (J. S. Colter and W. Paranchych, eds.), pp. 593–626. Academic Press, New York.

Williams, M. G., and Scheinin, R. (1961). Cytological studies of mouse embryo cells infected with polyomavirus using acridine orange and fluorescent antibody. *Virology* **13,** 368–370.

Winnacher, E. L., Magnusson, G., and Reichard, A. (1972). Replication of polyoma DNA in isolated nuclei. I. Characterization of the system from mouse fibroblast 3T6 cells. *J. Mol. Biol.* **72,** 523–537.

Winocour, E. (1968). Further studies on the incorporation of cell DNA into polyoma related particles. *Virology* **34,** 571–582.

Winocour, E., and Sachs, L. (1959). A plaque assay for the polyomavirus. *Virology* **8,** 397–400.

Winocour, E., and Sachs, L. (1962). Challenge infection of polyoma parotid tumor cells. *Virology* **16,** 496–498.

Winston, V. D., Bolen, J. B., and Consigli, R. A. (1980). Isolation and characterization of polyoma uncoating intermediates from the nuclei of infected mouse cells. *J. Virol.* **33,** 1173–1181.

Winters, A. L., and Consigli, R. A. (1967). Ornithine transcarbamylase activity associated with polyomavirus-infected mouse embryo culture. *Proc. Am. Soc. Microbiol.* **67,** 142. (Abstr.)

Wintersberger, E., and Wintersberger, U. (1976). Induction of DNA polymerase in polyomavirus-infected mouse cells requires transcription and translation. *J. Virol.* **19,** 291–295.

Wintersberger, U., and Wintersberger, E. (1975). DNA polymerase in polyomavirus-infected mouse kidney cells. *J. Virol.* **16,** 1095–1100.

Woolley, G. W. (1954). Occurrence of "neck tumors" in cortisone-treated leukemic strain mice. *Proc. Am. Assoc. Cancer Res.* **I,** 52. (Abstr.)

Yabe, Y., Neriishi, S., Sato, Y., Liebelt, A., Taylor, H. G., and Trentin, J. J. (1961). Distribution of hemagglutination-inhibiting antibodies against polyomavirus in laboratory mice. *J. Natl. Cancer Inst.* **26,** 621–628.

Yelton, D. B., and Aposhian, H. V. (1972). Polyoma pseudovirions. I. Sequence of events in primary mouse embryo cells leading to pseudovirus production. *J. Virol.* **10,** 340–346.

Yu, K., and Cheever, W. P. (1976). DNA synthesis in polyomavirus infection. Kinetic evidence for two requirements for protein synthesis during viral DNA replication. *J. Virol.* **17,** 415–421.

Chapter 15

Minute Virus of Mice

David C. Ward and Peter J. Tattersall

I.	Introduction	313
II.	Biology of MVM	314
	A. Original and Subsequent Isolations	314
	B. Methods of Detection and Quantitation	314
	C. Antigenic Relationships	316
	D. Epidemiology and Prevalence	316
	E. Pathogenesis of Experimental Infection	317
	F. Virus–Cell Interactions *in Vitro*	318
III.	Structure and Replication of MVM	322
	A. Types of MVM Virions	322
	B. Virion Proteins: Sequence Homology and Posttranslational Modification	323
	C. Properties of MVM DNA	324
	D. Replicative Life Cycle	326
	References	333

I. INTRODUCTION

Minute virus of mice (MVM) is a highly contagious agent which is extremely widespread among mouse populations both in the wild and in the laboratory. It is endemic in the majority of conventional and specific pathogen-free (SPF) mouse-breeding colonies and is a very common contaminant of serially transplanted mouse tumors and leukemia virus stocks. The prevalence of the virus is such that investigators using mice as laboratory tools should be aware that, unless they are certain to the contrary, they are probably introducing MVM into their experimental system. The inadvertent and uncontrolled addition of MVM to many biological assays in common use may grossly affect their outcome. For instance, under some circumstances the virus may interfere with the development of neoplastic disease, whereas under other conditions it will suppress *in vitro* immunological reactions. Current understanding of the structure and replication of MVM, and of its interaction with mouse cells in the whole animal and in culture, suggests reasons for these properties and allows a preliminary description of viral pathogenesis in molecular terms.

Familiarity with MVM and its biology is obviously impor-

ISBN 0-12-262502-1

tant to the experimenter and breeder, as the virus is a ubiquitous, although generally asymptomatic, infection of laboratory mice. However, the interaction of MVM with its host is becoming an important system in its own right for the study of specific tissue tropism, virulence, persistence, immunosuppression and, indeed, the mechanisms of mammalian cell differentiation. In addition, MVM is playing an increasingly significant role in attempts to understand replication and gene expression in higher eukaryotes.

MVM is a member of the Parvoviridae family, which contains a large number of physically and chemically similar viruses infecting species as diverse as man and moth (Rose, 1974; Tattersall and Ward, 1978). These agents are small, nonenveloped, isometric virions approximately 200 Å in diameter, containing a single-stranded DNA genome. The vertebrate parvoviruses are divided into two genera or subgroups on the basis of their requirement for helper viruses. All members of the *adeno-associated virus* subgroup are defective for replication and are, as their name implies, entirely dependent upon adenovirus coinfection for successful multiplication (Berns and Hauswirth, 1979). Viruses of the larger subgroup are usually referred to as the *autonomous parvoviruses*. These are capable of productive replication in the vast majority of host cells studied to date. Despite its original discovery as a contaminant of a mouse adenovirus stock, MVM was shown to be independent of the adenovirus for multiplication (Crawford, 1966) and to infect mouse cells with single-hit kinetics (Tattersall, 1972); it thus belongs in the autonomous subgroup.

In the past decade, much has been learned of the biology of virus–host interaction, both *in vivo* and *in vitro,* for the parvoviruses in general and MVM in particular. In the past few years, advances in biochemical technology have led to an increased understanding of the replication of these viruses at the molecular level. It is our purpose here to review the current state of knowledge in both of these general areas, with specific reference to what is now known about MVM.

II. BIOLOGY OF MVM

A. Original and Subsequent Isolations

In 1966 Crawford reported the isolation of a new virus from a stock of mouse adenovirus. This small agent formed a dense band in cesium chloride, separated distinctly from the adenovirus and from the polyomavirus which also contaminated the stock (Crawford, 1966). The virus, given the name *minute virus of mice,* also proved to be separable from the other viruses biologically, in that it appeared to be capable of infecting mouse embryo cultures, multiplying, and causing cytopathic changes in the absence of the other viruses. When grown on its own, it was shown to give rise to two major density classes of particle, both of which agglutinated guinea pig red blood cells. The denser of the two appeared to be the infectious virus, and the other looked like empty capsids in the electron microscope. Further physical characterization of the virus led Crawford to propose that MVM belonged to the same group as Kilham's rat virus, Toolan's H viruses, and the X14 virus (Kilham and Olivier, 1959; Toolan, 1961; Payne *et al.*, 1964). Chemical analysis of the infectious virion revealed that it contained a single-stranded DNA genome (Crawford, 1966; Crawford *et al.,* 1969), now accepted as the major common feature of this group.

Since the original isolation of MVM-CR by Crawford, MVM has been isolated from many different sources, including naturally infected mice. For instance, MVM strain 890 was isolated in rat embryo culture from a kidney suspension of a 40-day-old Swiss mouse (Parker, *et al.,* 1970a). This stock was subsequently passed two or three times in rat embryo culture and therefore is very close to being a field strain of MVM. MVM-CR, on the other hand, was grown in mouse embryo culture for many passages before being plaque purified (Tattersall, 1972), as described below. The resulting strain, MVM-T, has since been distributed to many of the laboratories working on the molecular biology of the virus and so has become the prototype MVM strain for study *in vitro*. In addition to these strains, many isolations have been made from contaminated leukemia virus stocks and transplantable tumors (Parker *et al.,* 1970b; Collins and Parker, 1972). An immunosuppressive virus carried by a transplantable murine lymphoma has been identified as MVM by serotyping (Bonnard *et al.,* 1976), and this has now been cloned by terminal dilution assay and designated strain MVM-i (Tattersall, unpublished observation). The properties of this isolate will be compared later to those of MVM-T.

B. Methods of Detection and Quantitation

MVM has been shown to agglutinate erythrocytes of many mammalian species, and although this is a general property of the autonomous parvoviruses, each serotype displays a considerable species specificity in the red cells it will agglutinate. Crawford originally used the inability of MVM to agglutinate sheep red blood cells and its hemagglutinating activity (HA) with guinea pig erythrocytes at pH 8.5 and 37°C to distinguish it from polyomavirus hemagglutinin (Crawford, 1966). Although Crawford reported that MVM would not agglutinate human red blood cells, subsequently it has been reported that human type O erythrocytes are efficiently agglutinated (Hallauer *et al.,* 1972; Siegl, 1976). The ability to agglutinate human but not simian erythrocytes distinguishes MVM from all other parvovirus isolates reported to date (Toolan, 1967;

Hallauer *et al.*, 1972; Siegl, 1976). The standard laboratory hemagglutination assay for MVM is usually performed using guinea pig erythrocytes, and it is a quick and effective measure of total virion concentration, allowing location and quantitation of virus during purification and serving as a preliminary estimate for infectivity titrations. HA with guinea pig red blood cells is unaffected by a pH between 6.6 and 8.5 or by temperatures between 4° and 37°C (Crawford, 1966; Hallauer *et al.*, 1972). The virus does not spontaneously elute from erythrocytes but will do so reversibly when the pH is raised to 9. The *Vibrio cholera* receptor-destroying enzyme, and the neuraminidases of influenza virus and Newcastle disease virus, remove the agglutinin receptor from guinea pig erythrocytes, suggesting that the receptor contains an essential sialic acid residue (Hallauer *et al.*, 1972).

Several methods have been described for the assay of virus infectivity. First, the cytopathic effect (CPE) of the virus in monolayers of primary or secondary rat or mouse embryo cells has been used to determine the mean tissue culture infectious dose ($TCID_{50}$) titer of virus preparations (Parker *et al.*, 1970a). The CPE in such cultures is associated with the development of large intranuclear inclusion bodies as the monolayer degenerates. The monolayer, however, is not always completely destroyed (Parker *et al.*, 1970a), and islands of apparently resistant cells can be distinguished in infected mouse embryo cultures maintained for several days after the maximal CPE has occurred (Tattersall, unpublished observations). The development of the CPE depends upon high mitotic activity of cells in the monolayer, and consequently the CPE develops slowly, if at all, in confluent monolayers. In addition, batches of serum, especially calf serum, may contain a potent inhibitor of virus growth. The nature of this inhibitor has not been well established. It is known to interfere with the adsorption of virus to the host cell and thus reduces spread of the virus within the infected monolayer (Tattersall, unpublished observations). Thus serum batches used in the isolation or assay of MVM should be pretested for possible inhibitory activity by comparison with a standard virus stock infecting the same test cells in a noninhibitory serum. The inhibitor has almost always been absent from commercially available fetal calf serum batches we have tested (Tattersall, unpublished results).

In addition to the development of CPE and HA in infected cultures, the infectivity titers of virus preparations have been estimated by the detection of infected cells by fluorescent antibody (FA) staining. The number of infected cells elicited by infection with a sample of virus is compared with that produced by a standard preparation with a known $TCID_{50}$ (Parker *et al.*, 1970a). This technique has the advantage of taking only 2 days, rather than the 9 days needed for the development of detectable endpoint CPE. In a standard infection at a multiplicity of 20 $TCID_{50}$ per cell, fluorescent nuclei can be detected by 12 hr postinfection (8%), rising to a maximum (41%) at 48 hr and followed by a sharp decline in the next 24 hr, with a concomitant rise in the appearance of cells giving cytoplasmic staining. Development of both nuclear and cytoplasmic viral antigen was abolished by the DNA synthesis inhibitor cytosine arabinoside (Parker *et al.*, 1970a).

Parker and his colleagues have developed a further sensitive test for the presence of MVM and other viruses as contaminants of transplantable tumors and leukemia virus stocks (Parker *et al.*, 1970b). In this procedure, mice inoculated with an aliquot of the specimen under test are assayed for the production of virus-specific antibodies by hemagglutination inhibition (HAI) with serum obtained 28 days or so postinoculation. The major advantage of this mouse antibody production (MAP) test is that the serum of one animal can be screened for positive seroconversion for as many viruses as one cares to test, thus allowing rapid and sensitive screening of many specimens. A comparison of these techniques, that is, HA, CPE, FA, and MAP, for their sensitivity of detection for MVM revealed some interesting strain differences. The MAP test was some 10^3- to 10^4-fold lower in sensitivity compared to CPE for the detection of the CR strain, whereas for the 890 strain, MAP was some 10-fold more sensitive than CPE. Since CR is an MVM strain which has been passaged for several years in cell culture and the 890 strain is essentially a field strain recently isolated in rat embryo culture, this suggests that the virus might become adapted for *in vitro* passage at the expense of losing virulence when tested *in vivo*. Such attenuation in cell culture will be discussed later. Both virus strains showed more efficient infection of mice when introduced by the intraperitoneal rather than the intranasal route.

The development of a plaque assay for MVM (Tattersall, 1972) allowed more convenient and accurate estimations of infectivity titers of the MVM-CR strain and led to the establishment of virus–host systems *in vitro* which have been studied intensively over the last few years at the molecular level. The assay uses the ability of the virus to cause distinct plaques in monolayers of A9 cells, an $HGPRT^-$ mutant of the L cell line, growing under agar. Again, the monolayer must be growing rapidly for plaques to develop, and the average plaque size is determined by the extent of cell growth in the monolayer. Assay plates seeded at confluence failed to show detectable plaques. Using this assay, it was possible to show directly that the particle banding at a density of 1.43 gm/cm^3 in cesium chloride was the infectious form and that approximately 1 in 300–400 of these particles would cause a plaque. The single-hit dose response shown by these purified infectious virions also demonstrated not only that MVM is helper independent but also that a single particle could give rise to a plaque. As mentioned previously, hemagglutination is a property of the viral capsid and therefore is not necessarily an accurate reflection of the number of infectious particles in a virus preparation. A suspension of infectious virions from the 1.43 density peak in cesium chloride contains approximately 10^6 plaque-forming units (pfu) per milliliter at endpoint dilution in the HA assay, that is to say, one HA unit per milliliter is equivalent to 10^6 pfu per milliliter. Since most unpurified stocks of MVM, such as infected cell lysates, contain about 10% infectious virions, one HA unit of such a suspension is usually equivalent to 10^5 pfu per milliliter. The plaque assay technique also allowed the genetic purification of the virus through five consecutive single plaque isolations. This isolate, designated strain MVM-T, has been the predominant virus strain used since that time for studies on the molecular biology of virus replication.

C. Antigenic Relationships

The MVM-CR strain was initially compared with Kilham's rat virus (KRV) and found to be antigenically distinct in reciprocal HAI tests using the two viruses and their corresponding rabbit antisera (Crawford, 1966). Subsequently MVM was compared with 11 other parvovirus isolates and antisera raised in rabbits against each of them (Hallauer *et al.*, 1971). The 12 isolates studied by HAI fell into seven distinct groups within which there was high reciprocal cross-reaction. MVM-CR did not cross-react reciprocally with any other isolate. Low levels of one-way cross-reaction were detected, however. Anti-MVM antiserum showed HAI with H-1 virus (about 1% of the homologous reaction), but the reciprocal reaction was undetectable (less than 0.1%). Of the other antisera, only anti-H3 and anti-KRV cross-reacted with MVM hemagglutinin (about 3% and 25% of the homologous reaction), but again, the reciprocal reactions were undetectable. These one-way cross-reactions may be specific since all the sera were kaolin adsorbed before use, but possible cross-reactivity of preimmune sera from the individual rabbits used was not reported (Hallauer *et al.*, 1971). Possible antigenic relationships among MVM-CR, KRV, and H-1 have also been explored using antisera raised to each in rats (Cross and Parker, 1972). In this study no cross-reaction at all between these viruses was demonstrable by reciprocal HAI, complement fixation (CF), or CPE neutralization tests. However, the anti-KRV and anti-H-1 antisera efficiently stained MVM-infected rat cells in a manner indistinguishable from staining by homologous anti-MVM antiserum, although anti-MVM antiserum did not stain KRV or H-1 infected cells. This suggests that during infection of rat cells *in vitro* all three viruses elicit a common antigen, which is not part of the capsid involved in hemagglutination, or interaction with complement-fixing or neutralizing antibody. Presumably in the rat *in vivo*, this common antigen is expressed in immunogenic amounts during KRV and H-1 infections, but not when the infecting virus is MVM.

D. Epidemiology and Prevalence

The single aspect of the biology of MVM of most importance to all experimenters using mice is the ubiquity of the virus. MVM is widespread among mouse colonies used as supply sources for biomedical research. A survey of mouse breeder colonies in the United States showed that 38 out of 44 conventional colonies and three out of eight SPF colonies contained mice which had or were experiencing MVM infection, as monitored by the presence of HAI antibodies (Parker *et al.*, 1970b). None of the five germ-free colonies tested yielded positive mice. The virus was found to be endemic within affected colonies, with an average of 74% of mice positive for HAI antibodies. The immunoglobulin nature of this HAI was, in many cases, confirmed by parallel FA, CF, and CPE neutralization studies. Individual mice showed considerable antibody titers, often exceeding 1 : 100, with titers of 1 : 1000 not being unusual. In addition to the many laboratory mice included in this very extensive study, the sera from wild mice trapped in four different states were examined. About 20% of sera from such mice were positive for MVM HAI antibodies, with titers ranging from 1 : 20 to 1 : 160. Seropositive animals were found in all four states, showing the virus to be widespread outside laboratory breeding centers as well.

The natural history of the disease within an endemically infected colony was studied by examining the presence of HAI antibody as a function of the animal's age (Parker *et al.*, 1970b). From 2 through 6–8 weeks of age, the number of seropositive animals declined, then rose sharply to a rate of 90–100% positive by 12 weeks. This implies that the young mice in an enzootically infected colony are protected in the first few weeks of life by maternal antibody. When this passive protection decreases during the second and third months of life, they become susceptible. These animals are then infected by contact with other animals in the infectious phase of disease or by contaminated fomites. Since recently infected mice excrete virus in feces and urine and the agent is very stable to desiccation, this would ensure that animals becoming susceptible are exposed to infectious virus, although the main reservoir of virus in the environment is not known. Transmission between infected and susceptible animals is affected by direct contact or by limited nasal–oral contact as well as by urine or fecal contamination. However, transmission did not appear to occur by airborne dissemination at all, even across a space as little as 8 in. Susceptible mice in contact with infected animals or their fomites seroconverted by 3 weeks of exposure, but limited contact took a week or more longer for comparable antibody responses to occur (Parker *et al.*, 1970b).

When animals with low or absent titers at 8 weeks of age became infected in an endemically infected colony, their individual HAI titers increased to a maximum by the sixteenth week of life and remained high for at least an additional 24 weeks and perhaps for the life of the animal. The long-term maintenance of high circulating antibody is indicative of persistent, low-level infection and continual representation of viral antigen to the immune system. In keeping with this possibility is the demonstration that infectious virus could be isolated from the blood and kidneys of mice having a coincidentally high circulating HAI antibody titer. Such virus isolation studies showed that during the phase of maternal antibody protection no virus could be isolated from young mice in infected colonies. However, at 6–7 weeks, virus could be isolated from blood and kidneys. This increase in isolable virus was concomitant with the production of HAI antibodies.

Given the widespread distribution of MVM, its success in establishing enzootic infections in mouse colonies, and its predilection for dividing cells, it is perhaps not surprising that the virus is a frequent contaminant of tumor cell lines and leukemia virus stocks passaged in mice. A survey of transplantable tumors by Parker and colleagues (1970b) using the MAP test showed that 44% of virus-induced leukemias, 18% of chemical- or radiation-induced tumors, and 47% of spontaneous tumors regularly transplanted in mice were contaminated with MVM. These tumors were obtained from various sources, including the Cancer Chemotherapy National Service Center at the National Institutes of Health. The presence of MVM was not correlated with any particular biological type of tumor. A further study which screened transplantable tumors and leukemias from similar sources showed that MVM contamination, with an incidence of 32%, was second only to lactic dehydrogenase virus in frequency of isolation out of the 11 agents studied (Collins and Parker, 1972). This incidence is startlingly high, and one wonders how often experiments in cancer chemotherapy and immunology have been influenced by the unsuspected presence of MVM. The isolation of strain MVM-i underlines the profound effect the presence of MVM might have on the interpretation of observations *in vivo* or *in vitro*. In 1975 Bonnard and Herberman reported that a factor produced by the EL-4 (G-) lymphoma markedly suppressed the generation of cytotoxic lymphoblasts in allogeneic mixed lymphocyte cultures (Bonnard and Herberman, 1975). On further investigation, this factor was shown to be an infectious agent specifically neutralized by reference anti-MVM antiserum. However, MVM-CR had a slight enhancing effect on cell-mediated cytotoxicity in equivalent tests, indicating that the immunosuppressive virus was a variant of MVM with a virus–cell interaction different from that of the prototype laboratory strain (Bonnard *et al.*, 1976). We have subsequently isolated this virus in cell culture and found it to be remarkably similar to MVM-T in its genetic structure at the nucleotide sequence level, but with distinguishable biological properties *in vitro* (Tattersall, unpublished results). Under the appropriate conditions, this variant, strain MVM-i, is highly cytopathic *in vitro* for the original tumor cell line EL-4 (G-) from which it was isolated, suggesting that during passage *in vivo*, the developing transplanted tumor was probably protected by the immune response of the mouse to the contaminating MVM. Alternatively, the colony of mice from which animals were drawn for passage of the tumor might have been endemically infected, and thus most of the mice were already immune. At what point the virus was picked up by the tumor in its passage history is not clear, but its isolation in this way exemplifies the caution with which *in vitro* results obtained with *in vivo*-derived material must be treated. One should also bear in mind that not only may the passage of a tumor in an MVM-infected animal introduce the virus into the tumor stock, but also that the passage of an MVM-carrying tumor into mice in a hitherto MVM-free colony will almost certainly lead to the establishment of an endemic MVM infection in that colony.

E. Pathogenesis of Experimental Infection

Studies on the epidemiology and prevalence of MVM indicate that it is a virus of both laboratory and wild mice. The virus is, however, only mildly pathogenic for mice compared with its disease potential in other rodent species, especially hamsters. Infections of neonatal hamsters with MVM-CR are generally fatal, with survivors showing the classical signs of parvorvirus teratogenesis, that is, mongoloid-like craniofacial lesions, runting, and periodontal disease (Kilham and Margolis, 1970; Baer and Kilham, 1974). Inoculation of neonatal mice by intracranial and intraperitoneal routes, on the other hand, did not usually produce fatal disease but resulted in generalized growth retardation, with proliferation of virus to high titers in the brain, gut and urine but only to low levels in the blood. On histological examination, these mice showed signs of another classical parvovirus-induced syndrome, that of cerebellar hypoplasia due to a selective viral attack on the rapidly proliferating tissue of the external germinal layer of the cerebellum. The mildness of the cerebellar lesions observed in mice was consistent with the absence of clinical ataxia. This selective attack on the cerebellar cortex did not appear to be dependent upon intracranial inoculation because uninfected littermates of infected neonates developed cerebellar lesions, apparently as a result of direct contact-mediated natural infection, within 9 days of the inoculation of the original animals (Kilham and Margolis, 1970). These authors also noted that MVM-CR passaged through hamsters increased in virulence. Although virus carried through nine consecutive passages in neonatal mice showed no increase in virulence, virus carried for three passages in hamsters appeared to be more virulent for infant mice, causing mortality in several cases and a 100%

incidence of dwarfism and cerebellar lesions (Kilham and Margolis, 1970). This hamster-passaged virus with increased virulence was still specifically neutralized by anti-MVM but not by anti-KRV or anti H-1 antisera, indicating that a change in the MVM itself was responsible. However, its association with an unknown synergistic agent derived from the hamster was not ruled out.

The infection of susceptible adult mice leads to a viremia, peaking at 5–6 days postinoculation, in which the circulating virus is intimately associated with the erythrocyte fraction of the blood (Harris *et al.,* 1974). Perhaps a tight erythrocyte association is responsible for the sparing of virus from coexisting high levels of circulating neutralizing antibody, an anomalous situation reported for infected adult mice (Parker *et al.,* 1970b; Kilham and Margolis, 1971). Interferon induction is minimal during infection of adults and is detectable only in the plasma of female mice. The levels of induction are one to two orders of magnitude lower than titers induced by encephalomyocarditis virus. However, when MVM was tested for sensitivity to interferon *in vitro* by a dilution vs. CPE assay, its growth was shown to be about five times more sensitive to inhibition than that of vesicular stomatitis virus (Harris *et al.,* 1974).

Infection of pregnant mice with MVM is also a very mild process compared to similar infections of pregnant hamsters or rats. The virus can replicate to high titers in maternal tissues, placentas, and fetuses, all without any histological evidence of infection (Kilham and Margolis, 1971). Not all placentas and fetuses were necessarily infected, but a high titer of virus in a particular fetus corresponded with a high titer in its associated placenta.

Thus there appears to be a brief period during early neonatal development when the mouse becomes susceptible to MVM-induced disease, mild though it may be, as opposed to infection. In endemically infected populations, this disease-susceptible period usually occurs when the infant mouse is protected by maternal antibody. The persistence of the virus in the population, and the passage of each animal through an infection-susceptible phase some weeks after the disease-susceptible period, ensure that the number of neonates simultaneously nonimmune and susceptible to disease is low. This would allow high levels of virus within the affected colony in the absence of obvious symptoms. This enzootic balance is of particular significance to researchers using mice as experimental tools, since it is seldom apparent that a particular colony harbors the virus, setting MVM apart from most cytolytic viruses affecting mice. The consequences of this highly evolved parasite–host relationship to the experimenter are manyfold. For instance, endemic infection with MVM may lead to a consistent failure to establish long-term cell cultures *in vitro* from such mice. Alternatively, endemic infection may be manifest by generally low tumor ''takes'' in experiments using transplantable tumors.

F. Virus–Cell Interactions *in Vitro*

1. The Host Cell Component

The development, over the last decade, of *in vitro* cell culture systems for the growth and assay of many of the parvoviruses has made possible rapid progress in the understanding of the viruses, both in terms of structure and function at the molecular level and in terms of the biochemical events of productive infection. MVM was initially isolated and grown in whole mouse embryo cultures and was subsequently plaque-purified in monolayers of A-9 cells, a derivative of the long-established L cell line. The strains MVM-CR and MVM-T have been shown to grow productively in primary and secondary cultures derived from mouse or rat embryos and in cell lines established from a number of rodent species, namely, the rat, Syrian hamster, mouse, and Chinese hamster. Indeed, MVM is the only parvovirus known to infect cells of both the mouse and the Chinese hamster (Crawford, 1966; Parker *et al.*, 1970a; Tattersall, 1972; Astell, 1977). Although MVM and H-1 both grow in BHK21 and HAK cells, Syrian hamster cell lines, H-1 does not grow lytically in any mouse cell line tested (Hallauer *et al.*, 1972; Siegl, 1976; S. L. Rhode, personal communication; P. Tattersall, unpublished observations).

Early in the study of parvovirus replication in cell culture, it was discovered that the viruses grow better in rapidly dividing cultures than in confluent or resting monolayers, and they appear to require a cellular function expressed only transiently during the S phase of the host cell cycle (Tennant *et al.,* 1969; Tennant and Hand, 1970; Hampton, 1970; Tattersall, 1972). In addition, experiments using cell monolayers rendered quiescent by serum starvation showed that, unlike tumor viruses, MVM is not capable of stimulating resting cells to enter the S phase (Tattersall, 1972). The exact nature of the S phase host function is not yet known, but the point at which it may act in the virus growth cycle and the biochemical event it may catalyze will be discussed in the next section. The requirement for gene expression under the temporal regulation of the host and the virus's inability to induce its expression neatly explain the affinity for dividing cells that the experimental pathologists had previously described for members of the parvovirus group (Kilham and Margolis, 1975; Margolis and Kilham, 1975).

While extending this *in vitro* study of MVM replication, it became clear that not all mouse cell lines were susceptible to lytic infection by the virus, even when growing rapidly in culture. At present, such cell lines can be divided into two classes: those which are resistant because they do not have the virus-specific receptor on their surface, and those which survive MVM infection because of some intracellular defect in virus replication. This latter virus–host interaction was first observed when an attempt was made to use Lewis lung carcinoma cells cultured *in vitro* as a host for MVM-T (P. Tattersall and P. Cawte, unpublished results). Although infection of clonally derived tumor cells always resulted in the production of a small amount of virus with some concomitant cell death, the remaining population was extremely resistant to cell killing by virus both *in vitro* and *in vivo*. This limited infection, which has been termed *restrictive interaction*, has been extensively compared with the normal productive interaction between MVM-T and A9 cells. The results of this study are summarized in Table I. The number of cells which become infected in a restrictive host culture is dependent upon the multiplicity of infection, and infection appears to be an all-or-none phenomenon at the level of the individual cell. Infected cultures seldom produce more virus than the input required to elicit the infection, and the surviving culture appears to be depleted of cells capable of supporting infection at the subsequent lower multiplicity in the next round of infection. These two factors cause the infection to dilute out very rapidly as the surviving culture is passaged. The level of resistance at a given multiplicity varied from subclone to subclone, suggesting that the sensitive cells were separate subpopulations which arose spontaneously within the culture during clonal growth *in vitro*. The high frequency with which these sensitive cells were found suggested that they did not arise by spontaneous mutation but were due to some epigenetic event. The host restriction on virus growth did not correlate with the mouse strain, since whole mouse embryo cultures from the same strain as the restrictive cell line are susceptible to productive infection by MVM-T. The converse of this restrictive interaction has been found to exist in populations of productive host cells, where surviving subpopulations can be selected at high frequency, and this refractory trait is maintained quite stably over generations in culture in the absence of selective pressure. Again, even recently cloned cell cultures appear to comprise subpopulations with different inheritable degrees of susceptibility to virus takeover (P. Tattersall and M. Van Wiles, unpublished results). Thus the outcome of infection is determined, at the level of the individual cell, by at least one variable factor other than mitotic activity.

Although it appears that proliferation is a prerequisite for being a target organ for the parvoviruses, it is clear that not all tissues which turn over rapidly are necessarily subject to virus-induced damage. Although most adult tissues are mitotically quiescent compared to those of the fetus and neonate, many, such as gut epithelium and the lymphopoietic system, contain large numbers of dividing cells at all times throughout the animal's life. These cells are essential to the animal's well-being and one might expect them to be targets for attack by such a "mitolytic" agent as MVM. That they are not affected points to the existence of further constraints upon virus multiplication in otherwise apparently susceptible host cells. In addition, the specific tissue tropism that many members of the parvovirus group exhibit during teratogenesis suggested that susceptibility to infection might be dependent upon the differentiated state of the host cell in question, and that the restrictive interaction was

Table I

Comparison of Productive and Restrictive Host Responses to MVM-T Infection

Multiplicity of infection (pfu/cell)	Response of productive host culture	Response of restrictive host culture
10^4–10^6	More than 99.9% of cells die	Up to 50% of cells die
5–10	Most cells infected Yield of 100–1000 pfu per cell More than 97% of cells die	0.1–5.0% of cells are productively infected and die Yield per infected cell is normal Yield from culture is always less than the multiplicity required to obtain it Survivors (95.0–99.9%) continue to grow and are not infectable at the original multiplicity therefore Infections dilute out rapidly
0.01–0.0001	Two or three rounds of infection expanding to give final values similar to those above	No infected cells are detectable in the culture

an accurate reflection of this requirement, rather than an *in vitro* artifact. The first study to approach this aspect of MVM replication was the attempt to infect rapidly dividing cells of the early mouse embryo in culture. Mohanty and Bachmann (1974) reported that MVM did not lytically infect two-cell-stage fertilized mouse eggs. Although viral antigen could be demonstrated within the cells, the embryos appeared to develop normally to the morula stage and beyond. Since MVM grows lytically in cells derived from mid- to late-gestation mouse embryos, these results suggested that virus replication is affected by a developmental component operating in the host at the cellular level, and that cell cycling, though necessary, is not sufficient for productive MVM infection. Subsequently, it proved possible to study this aspect of MVM replication by infecting teratocarcinoma stem cells, which can be induced to differentiate *in vitro* (Martin and Evans, 1975; Nicolas *et al.*, 1975). Several clonal lines of these cells, and their differentiated derivatives, have been used to study developmentally regulated susceptibility to MVM in culture (Miller *et al.*, 1977; Tattersall, 1978).

Undifferentiated stem cells, either nullipotent or pluripotent, were found to be restrictive hosts for MVM-T. By using the sensitive infectious center assay, it was possible to show that, at multiplicities of 5–10 pfu per cell, as little as 0.07% of cells in some stem cell populations responded productively compared to 39% for a fibroblast culture from the same strain of mice. The yield from the productively infected cells in the stem cell population, however, was normal or often higher than normal, compared to the fibroblast culture. Comparison of cultures induced to differentiate in the presence and absence of infectious virus showed that one predominant cell type, with a fibroblast morphology, was missing from infected cultures, whereas the appearance of a wide range of differentiated cell types was unaffected (Tattersall, 1978). When stem cells were induced to differentiate and primary cultures were prepared from them, the fraction of infectable cells rose dramatically by over 100-fold. Such primary cultures are enriched for a fibroblast-like cell which appears to be the primary target cell for MVM-T *in vitro*. These cells have been characterized as fibroblasts because they are attached, well spread but mobile, and most carry the surface alloantigen Thy-1 (P. Tattersall and S. Cotmore, unpublished results). Thus it seemed quite possible that most cell lines which were not derived from fibroblasts might be resistant to MVM-T because of a developmentally controlled block to replication. In order to learn more about this block, the interaction of MVM-T and undifferentiated stem cells has been studied in more detail. Stem cells can be killed to only about 50% by an input multiplicity of 10^5 pfu per cell, in contrast to the extremely low survival (<0.02%) of L cells under the same conditions. All cells in the population carry the virus-specific receptor on their surfaces. The virus enters the cell, since virus capsid antigen synthesis has been shown to be rescued by the fusion of uninfected permissive cells to infected stem cells which had been treated with antiviral antiserum and receptor-destroying enzyme. The growth of virus in such heterokaryons provides strong evidence that virus replication is not blocked by a diffusible inhibitor in the stem cell, but rather is due to the lack of a host function in the stem cell which is required for virus replication (Tattersall, 1978). Indeed, stable fusion hybrids between stem cells and sensitive fibroblasts are permissive for virus replication. This appears to be at variance with the results of Miller *et al.* (1977), who found that stem cell × Friend erythroleukemia cells are not permissive for MVM-T. The latter result perhaps is understandable in the light of recent results which show that Friend cells themselves are a restrictive host for MVM-T (Tattersall, unpublished results). Further studies on the virus-producing cells in stem cell cultures showed that the virus produced by stem cells was no better at infecting stem cells than was the standard virus, and that the lytically infected cells comprised a relatively stable subpopulation susceptible in an inheritable fashion to the original input multiplicity of infection. This situation closely parallels the restrictive interaction previously described for Lewis lung carcinoma cells, and it has subsequently been demonstrated for the interaction of MVM-T with several differentiated cell lines of non-fibroblast origin.

Studies on the infection of teratocarcinoma stem cells and their differentiated derivatives strongly suggest that MVM, and perhaps other parvoviruses, require for their replication host gene expression which is developmentally controlled. This host gene (or genes) is (or are) suppressed in stem cells and most differentiated cells but expressed as a consequence of differentiation down a small number of distinct pathways. It appears that visceral yolk sac endoderm cells, in addition to fibroblasts, normally express the full complement of such genes required by MVM-T (Miller *et al.*, 1977; Tattersall, 1978). Clonal lines of fibroblasts, resistant to high multiplicities of MVM-T, have been selected from sensitive populations, as previously described, and compared to teratoma stem cells and differentiated cell lines by the technique of somatic cell fusion. These studies have shown that such fibroblast clones have become resistant to virus infection by switching down, or off, the expression of one or more of the developmentally regulated genes which are suppressed in stem cells (M. Van Wiles and P. Tattersall, unpublished results). Presumably, the products of these genes are normal components of particular differentiated cells and are competed for by the incoming virus to establish infection in an all-or-none fashion. The level of expression of these products in individual cells determines the success or failure of a given input "dose" of viral genomes in taking over the cell.

The other major class of MVM-resistant cells are those which do not have the virus-specific receptor on their surfaces.

Such cells can be selected as stable mutants from a virus-susceptible cell line (Linser *et al.*, 1977; Linser and Armentrout, 1978) or may occur naturally as transplantable tumors or established cell lines. For instance, the receptor appears to be absent from the surface of the teratocarcinoma-derived parietal yolk sac (PYS) endoderm cell line (M. Van Wiles and P. Tattersall, unpublished results) as well as the murine leukemia lines L1210 and L5578Y (Ward, unpublished results). That this may be a result of cell differentiation rather than a cell line artifact is suggested by recent results, again using the induction of differentiation of teratocarcinoma stem cells *in vitro*. When stem cells, which are restrictive for infection by MVM-T, are induced to differentiate to an endoderm-type cell by treatment with retinoic acid (Strickland and Mahdavi, 1978), the cells become entirely refractory to MVM-T infection, although they continue to divide. Concomitant with the morphological differentiation and onset of complete resistance to infection, the virus-specific receptor disappears from the cell surface (M. Van Wiles and P. Tattersall, unpublished results).

2. The Virus Component

The discovery of a serologically related variant of MVM, designated strain i, which suppresses mixed lymphocyte cytotoxicity assays *in vitro* (Bonnard *et al.*, 1976), a property not shared by MVM-T, suggests that the fibroblast specificity of MVM-T might not be an invariant property of MVM. As described earlier, MVM-i was discovered as a contaminant of the transplantable T cell lymphoma EL-4, which had been passaged in mice. The immunosuppressive activity of the virus appears to be due to its ability to kill cytotoxic T lymphocyte precursors which are responding to allogeneic cells by entering the S phase as the first step in clonal expansion (H. Engers and B. Hirt, personal communication). These responding cells do not appear to be susceptible to MVM-T infection.

Strain MVM-i has been cloned by terminal dilution in EL-4 cultures, its antigenic relatedness to MVM-T confirmed by serum neutralization, and its properties compared with those of MVM-T as summarized below (Tattersall, unpublished results). MVM-i grows productively in a number of lymphoid tumor cell lines, of T but not of B cell origin, which are restrictive for MVM-T. The normal host cells of MVM-T, such as A9, are restrictive for MVM-i. The defective step in MVM-i replication in fibroblasts lies between adsorption and viral antigen synthesis and behaves as a recessive trait in somatic cell hybrids between fibroblast and lymphocyte. Indeed, hybrids between A9 and EL-4 are permissive for both viruses and give plaques with approximately equal particle : infectivity ratios for both viruses. MVM-T is over 10^6-fold more efficient than MVM-i in giving plaques on A9 monolayers. A simple interpretation of these results is that these two dissimilar differentiated cell types express different developmentally regulated helper functions which are required and exploited by the respective virus variant. The relationship between the two virus strains, which have been called *allotropic variants,* has been examined at both the biological and the structural level. Complementation tests have been performed in which lymphocytes were coinfected with MVM-i and MVM-T, and the yield was assayed on both A9 and A9 × EL-4 hybrid monolayers (which distinguishes between MVM-T and MVM-T + MVM-i). The results of such complementation tests demonstrate that ongoing MVM-i replication does not help MVM-T to replicate in the absence of the fibroblast-specific helper function. This implies that the differentiated end cell-specific helper function interacts in cis with a determinant within the viral genome. The genomes of the two variants have been compared by a number of biochemical techniques, including length measurement on alkaline gels, restriction endonuclease mapping, and heteroduplex formation. The genome of MVM-i is some 50–100 nucleotides (single-stranded length) shorter than that of MVM-T. So far, over 20 sites for five different restriction enzymes have been mapped and are coincident between MVM-T and MVM-i, reading from the left-hand (molecular 3′) end. This analysis has placed the length difference between the two genomes extremely close to the right-hand end, perhaps overlapping the stem of the terminal hairpin structure, as described in the following section.

The existence of these two allotropic variants of MVM obviously poses the question of their origin. Do they exist in nature as separate field strains or as complex field strains comprising several distinct allotropic variants? Perhaps allotropic variants arise by mutation and selection *in vitro* either by passage in differentiated tumors in the whole animal or by direct isolation in cultures of differentiated cells. Research directed toward answering these questions may also shed light on some of the other unexplained strain differences shown by MVM. For instance, Parker *et al.* (1970b) have shown that MVM-CR is not transmitted from infected animals to susceptible animals, even by direct contact, whereas strain 890 is extremely infectious. Perhaps an alteration in tissue-tropic specificity, or the proportion of different allotropic variants in the virus population, is responsible for this difference as well as for the increased virulence for infant mice of MVM passaged through neonatal hamsters (Kilham and Margolis, 1970).

Three factors, therefore, appear to act at the cellular level to confer competence as a host for MVM. First, the cell must be of the correct species; second, the cell must be traversing the S phase of the cell cycle; and third, the cell must be of the right differentiated phenotype. Only the last factor appears to be variable, and in this case the outcome of the interaction depends upon a genetic locus within the viral chromosome for which there exist, at present, two alleles.

III. STRUCTURE AND REPLICATION OF MVM

A. Types of MVM Virions

Three classes of virus particles have been extracted from cells lytically infected with MVM-T, and procedures for the large-scale growth and purification of these virion forms have been described (Clinton and Hayashi, 1975; Tattersall *et al.*, 1976; Richards *et al.*, 1977; Faust and Ward, 1979). Some of the salient features of these virions and their components are summarized in Table II and Fig. 1. The empty particles (e), which are usually the most abundant species, contain two polypeptides, A (MW = 83,300) and B (MW = 64,300), that comprise 15–18% and 82–85% of the virion mass, respectively (Tattersall *et al.*, 1976). Infectious, full particles (f) contain the single-stranded viral genome (~ 5100 nucleotides) and a third protein, C (MW = 61,400), in addition to polypeptides A and B. Again, the A polypeptide comprises 15–18% of the protein mass, but the amounts of B and C polypeptides vary inversely from preparation to preparation (Tattersall *et al.*, 1976). Neither virion type nor their component proteins contain detectable carbohydrate or RNA. The e particles band as a sharp peak in CsCl with a buoyant density of 1.32 gm/cm^3,

Table II

Characteristics of MVM Virions

Property	Virion type		
	Infectious	Empty	Defective
Particle diameter (Å)	200–250	200–250	200–250
Symmetry	Isometric	Isometric	Isometric
Molecular weight	5.8×10^6	4.2×10^6	Variable
Buoyant density in CsCl	1.46–1.41	1.32	1.33–1.39
s_{20W}	110 S	70 S	Variable
Particles per hemagglutin in unit	$1\text{-}2 \times 10^8$	2×10^8	—
Particle-infectivity ratio	200–400:1	—	—
Absorption E_{260}/E_{280}	1.38	0.67	—
Extinction coefficient ($E^{1\%}_{280}$)	71.2	17.8	—
Components			
Structural proteins			
A	83,300	83,300	83,300
B	64,300	64,300	64,300
C	61,400		61,400
DNA			
% mass	26.3	—	Variable
Molecular weight	1.6×10^6	—	Variable
Buoyant density in CsCl	1.722	—	Variable
s_{20W} (neutral pH)	20 S	—	Variable
(alkaline pH)	15 S	—	Variable

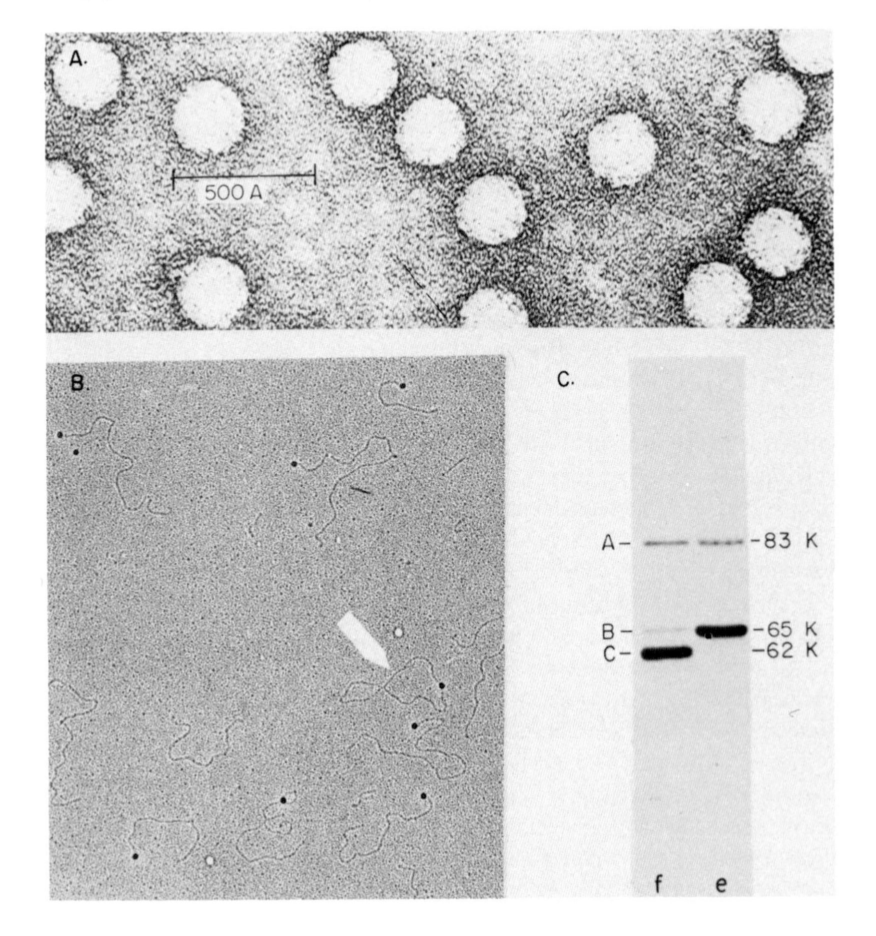

Fig. 1. Minute virus of mice (MVM) and its components. (A) Electron micrograph of purified infectious virus particles, negatively stained with uranyl acetate. (B) Electron micrograph of virions after treatment with 30% formamide-0.1 *M* Tris-HCl, pH 8.7, for 10 min at room temperature. The sample was spread by the formamide modification of the Kleinschmidt technique, stained with uranyl acetate, and rotary shadowed with platinum-paladium as per Bourguignon *et al.* (1976). Under these conditions, the majority of the capsids remain associated with the released DNA. The arrow indicates one molecule of MVM DNA which has both its termini associated with the capsid. (C) A fluorogram of an SDS-polyacrylamide gel displaying the ^{14}C amino acid-labeled polypeptides present in purified full (f) and empty (e) MVM virions. The resolving gel contained 7.5% acrylamide and 0.193% bisacrylamide as per Tattersall *et al.* (1976).

whereas the f particles band as a broader peak with buoyant densities ranging from 1.46 to 1.41 gm/cm^3. This peak, with a median density of 1.42–1.43 gm/cm^3, can, on occasion, be resolved as a doublet (see below). The third class is a heterogeneous population of virions with densities in CsCl between 1.33 and 1.39 gm/cm^3. Although these particles normally constitute only a few percent of the total obtained after a single high-multiplicity passage of MVM, their relative abundance can be increased significantly by serial undiluted passage. These particles contain all three structural proteins in approximately the same molar ratios as seen in the f virions. However, they possess viral DNA that varies in size from 10 to 70% of the complete genome and thus are designated *defective (D)* virions. The size of the DNA within the virion varies with particle density; particles which band at a greater density in CsCl contain larger DNA molecules than those which band at a lesser density. The structure and sequence organization of these incomplete viral genomes will be discussed later.

B. Virion Proteins: Sequence Homology and Posttranslational Modification

The three capsid proteins are synthesized only after virus infection, and their molecular weights and molar ratios are independent of the species of host cell infected, suggesting that all three proteins are virus coded (Tattersall *et al.*, 1976). However, since the combined molecular weights of these structural proteins exceeds the theoretical coding capacity of the MVM genome by some 30% (assuming the use of a single, nonoverlapping reading frame), it was important to establish that these three proteins are indeed of viral origin. Evidence suggesting that the B and C polypeptides are structurally related initially came from an analysis of infectious virions that could be resolved into two subpopulations on the basis of their density in CsCl (Clinton and Hayashi, 1975, 1976). Both virion subpopulations, which banded at 1.46 and 1.41 gm/cm^3, respectively, contained a complete viral genome. Pulse-chase experiments demonstrated that the denser particles, which contain the B polypeptide as their major protein, are precursors to the lighter particles, which have polypeptide C as their major component (Clinton and Hayashi, 1976; Richards *et al.*, 1977). The reason for such a large density shift is not understood, but it appears to be the result of the cleavage of approximately 30 amino acid residues from the B protein late in the infective cycle, presumably as a maturation step (Clinton and Hayashi, 1976; Tattersall *et al.*, 1976). An analysis of the tryptic and chymotryptic fingerprints of the three structural proteins after radio-iodination of their tyrosyl residues *in vitro* subsequently confirmed the suspected precursor–product relationship between the B and C proteins (Tattersall *et al.*, 1977). Although the proteolytic fingerprints of polypeptides B and C are almost identical, the fact that a single peptide fragment was observed in the digests of polypeptide C which was not detected in the fingerprints of polypeptide B would suggest that the proteolytic cleavage that takes place in the cell does not occur at a tryptic or chymotryptic site. In addition to the relationship between B and C proteins, it was apparent that all of the iodopeptides of the B protein were present in the largest A polypeptide. Indeed, only about 20% of the peptides obtained from the A protein were unique to that polypeptide. The extensive sequence homology between the structural proteins clearly resolves the potential coding capacity problem. Although it has not yet been established definitively whether the A and B proteins are coded by separate transcription units, certain studies (see below) would suggest that both are translated from RNA derived predominately from the 5′ half of the viral genome.

Unlike polyoma (Frearson and Crawford, 1972) and SV40 virions (Lake *et al.*, 1973), MVM particles are not assembled with cellular histones (Tattersall *et al.*, 1976). The A polypeptide of MVM, however, has an isoelectric point (PI = 8.2–8.5) which is considerably higher than that of the B and C polypeptides (PI = 7.0–7.2) (Peterson *et al.*, 1978). The observations that several of the A-specific iodopeptides seen in the chymotryptic fingerprints are quite basic (Tattersall *et al.*, 1977) and that the A protein has a much larger arginine content than the B protein (P. Paradiso, personal communication) provide additional evidence to suggest that the peptide portion unique to the A protein has a histone-like character. Although at present there is little evidence to indicate specific biochemical or enzymatic functions for any of the capsid proteins, it is tempting to speculate that the A protein may at least play a role in the charge neutralization and packaging of the viral DNA. By having one or more basic regions of the A protein located internally in the assembled virion, the A polypeptide could mimic the histone–DNA interactions of the papovaviruses. Indeed, the difference between the circular dichroic spectrum of the dense form of f particles and the e particles is almost identical to the difference observed between the gene 32 DNA-binding protein of bacteriophage T4 in the presence and absence of DNA (D. Ward, unpublished results).

All three MVM capsid proteins are phosphoproteins (Peterson *et al.*, 1978). Each protein can also be resolved by two-dimensional gel electrophoresis into two to four distinct species which differ in their apparent isoelectric point by 0.05 pH unit. This microheterogeneity appears to reflect subpopulations of the viral protein which possess different levels of phosphorylation. The nature of the protein–phosphate linkage is still unclear. However, the ^{32}P label was not found to be in phosphoserine, phosphothreonine, phosphohistidine, or phosphoarginine, nor was it in a form sensitive to bacterial alkaline phosphatase (Peterson *et al.*, 1978). Although viral polypeptides could also be radiolabeled with [^{3}H]adenosine, the resistance of both [^{32}P]phosphate and [^{3}H]adenosine labels to mild acid or pyrophosphatase argues against the presence of an ADP–ribose substituent. The observation that the T antigens of polyomavirus (Schaffhausen and Benjamin, 1979) and the transforming protein of Rous sarcoma virus (Hunter and Sefton, 1980) contain phosphorylated tyrosine residues should prompt further investigation of the MVM proteins for a similar type of modification.

The biological significance of the different isoelectric forms of viral proteins is unclear. The microheterogeneity may simply reflect viral proteins at different stages of posttranslational modification or, alternatively, the type or extent of derivatization may regulate or modulate virion maturation events, such as DNA encapsidation or capsid–protein proteolysis. Proteolytic digestion of intact full particles *in vivo* has shown that the cleavage of the B protein seen *in vivo* can be closely mimicked by trypsin and to a lesser extent by chymotrypsin (Tattersall *et al.*, 1977). However, the B protein in the empty virions is totally resistant to cleavage by either enzyme *in vitro*. This correlates well with the observation that e particles *in vivo* contain only the A and B proteins. Interestingly, the A proteins in either particle are completely resistant to cleavage by both proteases, although they contain the amino acid sequence that constitutes the proteolytic cleavage site. Although this differential susceptibility to proteolysis could be determined by the type or extent of posttranslational modification or by differences in the conformation that the proteins adopt within the two types of virions, the precise mechanism is unknown.

MVM virions appear as isometric particles with a diameter of approximately 200 Å when examined by electron microscopy (see Fig. 1A). The total number of protein molecules in each virion (8–9 of A and 53–57 of B or C), is close to 60, the largest number of equivalent units that can be arranged on a sphere in such a way that each is identically situated (Caspar and Klug, 1962). Because the majority of the sequence of protein A is the same as that of protein B, these two types of molecules could interact equivalently to construct such a spherical virion. However, several investigators have reported that other members of the autonomous parvovirus subgroup possess an icosahedral symmetry and contain 32 morphological capsomers (Vasquez and Brailovsky, 1965; Karasaki, 1966; Mayor and Jordan, 1966). If the capsomers of the MVM virions are arranged in an icosahedral fashion, the morphological subunits must be dimers of the viral polypeptides. However, since it is not clear how the eight to nine molecules of protein A might be arranged symmetrically in such a capsid structure, a definitive statement on the number of morphological capsomers and their organization within the MVM virion will require further investigation.

C. Properties of MVM DNA

Initial studies on the genome of MVM were performed on the MVM-CR strain by Crawford and associates (Crawford, 1966; Crawford *et al.*, 1969; McGeoch *et al.*, 1970). They concluded that MVM contained a linear, single-stranded DNA genome on the basis of its reactivity with formaldehyde, its base composition, its nearest-neighbor nucleotide frequency, and its appearance in the electron microscope. The molecular weight of the DNA was determined to be 1.5×10^6 from both velocity sedimentation and contour length measurements. Although MVM DNA was relatively rich in thymidine (with a base composition of 32.7% thymine, 26.5% adenine, 19.5% guanine, and 21.4% cytosine) and, like duplex cellular DNA, had a low frequency of the dinucleoside phosphate CpG, no other distinguishing features of the DNA were detected. Subsequent studies on the genome of MVM-T, the more widely studied strain, yielded similar results but, in addition, revealed that only 93% of the viral DNA is truly single-stranded. Both the 3′- and 5′-termini of MVM DNA contain palindromic nucleotide sequences which exist in the form of stable hairpin duplexes (Bourguignon *et al.*, 1976; Chow and Ward, 1978). The 3′-terminal hairpin structure is composed of 115 nucleotides (Astell *et al.*, 1979a), whereas the 5′-terminal hairpin contains approximately 260 nucleotides (Bourguignon *et al.*, 1976; Chow and Ward, 1978). The genomic termini, however, do not possess an inverted terminal repetition such as that seen in the DNA from the defective adeno-associated viruses (AAV) (Koczot *et al.*, 1973). Thus, MVM DNA does not form the circular "panhandle" structures that are observed when single-stranded AAV DNA is self-annealed (Bourguignon *et al.*, 1976). In addition, whereas AAV encapsidate equal amounts of both plus and minus strands of DNA in separate virions (Rose *et al.*, 1969; Berns and Adler, 1972), MVM, like other autonomous parvoviruses, packages almost exclusively (≧99%) the minus strand (Bourguignon *et al.*, 1976), which appears to be the only strand of DNA used as a transcription template (D. Dadachanji and D. Ward, unpublished results). Virion DNA encapsidation is a very selective process which is tightly coupled with strand-displacement DNA synthesis *in vivo* (Richards *et al.*, 1977) and likely is mediated or promoted by specific protein–DNA interactions between one or more capsid proteins and nucleotide sequences at or near the 5′-terminus of the viral genome (Faust and Ward, 1979). Indeed, when MVM virions are disrupted by treatment at room temperature with 30% formamide (in 0.1 *M* Tris-HCl buffer, pH 8.7) and then examined directly by electron microscopy, a significant number of capsids are seen to remain preferentially bound to terminal regions of the DNA molecules (see Fig. 1B). Because of the inverted terminal repetition at the ends of AAV DNA, which by direct sequence analysis involves 145 nucleotides (Berns *et al.*, 1979), the 5′-terminal nucleotide sequences of both plus and minus strands of AAV DNA are identical. A DNA encapsidation process involving a specific interaction with 5′-terminal nucleotide sequences would thus be expected to package either strand of AAV DNA with equal efficiency.

The observation that the genome of MVM-T was partially duplex DNA prompted a reevaluation of its molecular weight by electrophoresis in agarose gels under denaturing conditions (D. Dadachanji and D. Ward, unpublished observations). When compared with reference DNAs of known sequence (PBR-322, SV40, polyoma, and ϕX174 RF) that had been linearized by treatment with single-cut restriction endonuclease, the MVM-T genome was found to be 5150 ± 25 nucleotides long, which is slightly larger than the value reported previously for MVM-CR DNA. When analyzed under similar conditions, the genome of MVM-i was found to be about 50 nucleotides shorter than that of MVM-T (Tattersall, unpublished observations). This length difference appears to reside close to the 5′-terminus of the genome and may be intimately involved in determining the specific cell tropism of each strain (see Section II,F).

Terminal palindromic sequences in the form of hairpin duplexes are a feature common to all eight parvovirus genomes so far examined (Berns *et al.*, 1979; Astell *et al.*, 1979a). As will be discussed later, both the initiation and termination of DNA replication occur within the terminal regions of these viral genomes. Studies on the sequence organization of the DNA isolated from spontaneously arising deletion mutants of MVM-T have also provided evidence that all of the critical cis-acting sites for MVM DNA replication are entirely within the 200 to 300 nucleotides which encompass the self-complementary sequences at both ends of the viral genome (Faust and Ward, 1979). These deletion mutants, which possess 0.1–0.7 genome equivalents of DNA, contain two distinct types of viral DNA, designated *type I D-DNA* and *type II D-DNA*. Type I D-DNAs are predominantly single-stranded, "recombinant" molecules which have selectively conserved the palindromic sequences from both genomic termini. Type II D-DNAs are double-stranded hairpin molecules which contain viral sequences that map entirely at the 5′-end of the genome between map coordinates 85 and 100. Virtually all of the MVM genome sequences are found in the total heterogeneous population of type I D-DNAs isolated after a single high-multiplicity passage, although the extent and position of the deletions in individual molecules vary significantly. The shortest molecules in the population lack ~90% of the genome sequence and consist almost exclusively of sequences derived from within 5 map units (255 nucleotides) at both ends of the DNA. Since these miniature recombinant molecules are selectively amplified during serial undiluted passage in the presence of wild-type MVM, they must contain all of the critical recognition sites necessary for MVM DNA replication. In contrast, the type II DNA species are gradually lost from the total D-DNA population during serial undiluted passage (from ~50% at passage 1 to <5% at passage 10). This suggests that the type II D-DNA molecules are not competent for DNA replication and that they may arise as the result of fatal replication errors. If an appropriate method for cloning the type I D-DNA-containing deletion mutants can be developed, they could become extremely useful aids to future studies on MVM DNA replication and transcription.

Analysis of the MVM-T genome at the nucleotide level has been progressing rapidly, and to date approximately half of the DNA sequence has been determined (Astell *et al.*, 1979a; C. Astell, M. Smith, M. Chow, and D. Ward, unpublished observations). This has led to the identification of potential transcriptional promoters and translational start sites that will be discussed below. In addition, a comparison of the nucleotide sequences at the 3′-termini of the DNA from four autonomous rodent parvoviruses (MVM, KRV, H-1, and H-3) has revealed several interesting features. Although there are some very minor differences, the sequences of the first 150 nucleotides of each genome are essentially identical. This suggests either a strong pressure to conserve this region of the genome or a close evolutionary relationship between these viruses. Of the first 115 nucleotides, 102 can be base-paired to create a stable, Y-shaped hairpin structure (see Fig. 2). The topology or conformation of this hairpin structure rather than the absolute sequence, however, may be of major significance for parvovirus DNA replication in general. For example, the hairpin termini of the genome from AAV-2, which replicates in human and simian cells, exhibit no sequence homology with that of the rodent virus genomes, yet they are very similar topologically (Berns *et al.*, 1979). In addition, although the 3′-termini of the DNA from autonomous parvoviruses of bovine (BPV) and

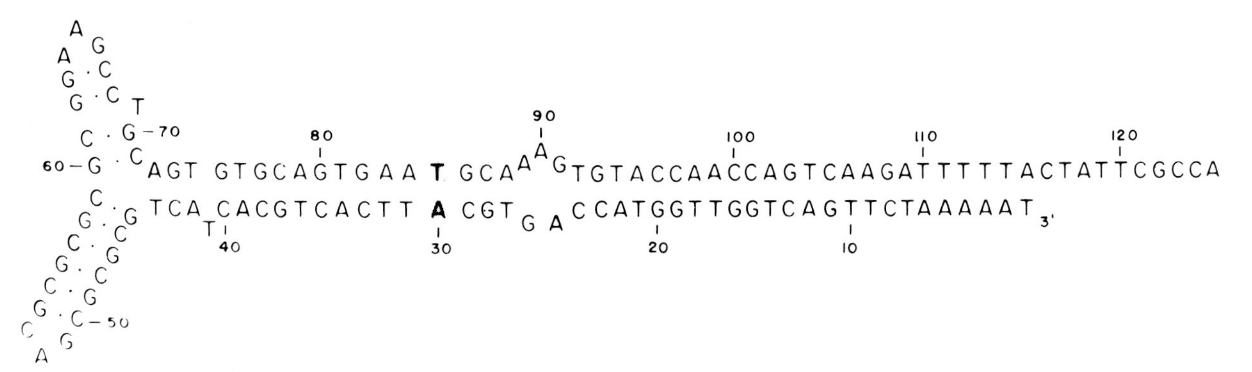

Fig. 2. The nucleotide sequence of the 3′-terminal hairpin structure of MVM DNA as determined by Astell *et al.* (1979a).

human (LuIII) hosts possess Y-shaped hairpin structures, the nucleotide sequences are again quite different from those of the rodent viruses or AAV (C. Astell and M. Chow, unpublished observations). One feature common to the hairpin structure of all these viral genomes is that the sequences at the tip of the Y (for example, between nucleotides 45 and 71 in Fig. 2) are extremely G-C rich. The structure of an oligonucleotide containing an alternating sequence of G-C base pairs, such as that seen in the MVM 3′ hairpin, has been solved at atomic resolution (Wang *et al.*, 1979) and shown to be a left-handed double helix with novel conformational properties that differ significantly from those of the familiar right-handed helical B form of DNA. This conformation of DNA, termed *Z-DNA,* can be generated from, or converted to, the B-DNA conformation in solution by manipulation of the ionic environment. The possibility that proteins are capable of interacting with specific conformers of unique DNA sequences or of inducing these conformational changes suggests that such interactions may play a major role in regulating biological processes. An analysis of specific interactions between cellular or viral proteins and the 3′-termini of parvovirus DNAs could provide direct evidence for this hypothesis.

Although the 5′-hairpin structure of MVM DNA appears to have characteristics similar to those of the 3′-hairpin structure, determination of its nucleotide sequence has proven to be much more difficult. The 5′-terminus of the genome as extracted from purified virions is often blocked with a peptide fragment, presumably a residual of the protein that is found *in vivo* on the 5′-ends of replicative forms of viral DNA (M. Chow and D. Ward, unpublished observations). That fraction of DNA which has an unblocked 5′-terminus, and thus is accessible for [^{32}P]phosphate labeling by ATP and polynucleotide kinase, exhibits sufficient sequence heterogeneity so that direct sequence analysis by the method of Maxam and Gilbert (1977) has yielded ambiguous results (Chow and Ward, 1978, and unpublished observations). The possibility that the target cell specificity of MVM-T and MVM-i may be determined, at least in part, by the nucleotide composition at or near the 5′-terminus of their genomes provides a strong incentive to characterize further the nucleotide sequences in this region of their DNAs.

D. Replicative Life Cycle

In the previous sections, we have concentrated on the biological consequences of the interaction of MVM with host cells both *in vivo* and *in vitro* and have presented a description of the virion and its structural components. The discussion to follow will outline some of the biochemical events that occur during the productive infection of fibroblast cells by MVM-T, the general features of which are illustrated schematically in Fig. 3.

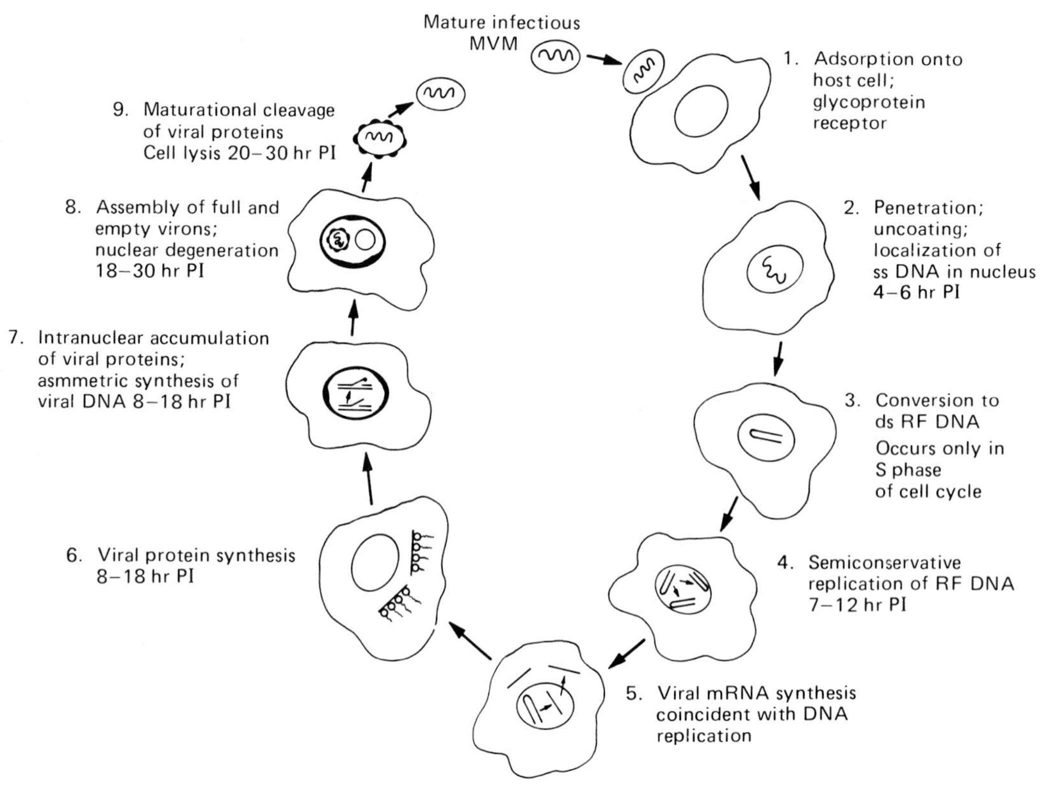

Fig. 3. A schematic representation of the MVM replicative life cycle.

1. Adsorption, Penetration, and Uncoating

Infection of cultured cells is initiated via an interaction with a specific receptor protein on the cell surface (Linser and Armentrout, 1978; Linser *et al.*, 1979). The binding of virus to A-9 or RT-7 cells, a rat brain tumor line, exhibits a pH optimum of 7.2 and is markedly independent of temperature over the range from 4° to 37°C. Although studies on the kinetics of virus adsorption are normally performed at 4°C to minimize uptake into the cell, the results obtained at either 4° or 37°C were qualitatively similar. An analysis of the amount of virus bound as a function of input multiplicity shows a distinctly biphasic profile, the initial component of which occurs rapidly and linearly until about 5–7 $\times$ 10^5 virus particles are bound per cell. The second component of the binding curve is detected at high input virus concentrations, is relatively slow, and is not saturable. The possibility that the latter component represents a form of low-affinity, nonspecific binding to the cell surface was confirmed by analyzing the kinetics of MVM adsorption to virus-resistant A-9 cell derivatives that were cloned from cells surviving long-term infections. These clonal isolates bound only low levels of virus with monophasic and nonsaturable kinetics which paralleled the second component of the binding curve seen with wild-type A-9 cells. These results suggest that only an interaction with the specific high-affinity receptor will lead to a productive infection of the cell. Although both density classes of full virus as well as empty virions appear to bind to the same cell surface receptor, the relative affinity of the different virion forms for the receptor has not been determined quantitatively. Whereas Clinton and Hayashi (1976) found that the 1.46 gm/cm^3 density class of f particles adhered less tightly to A-9 cells than f particles from the 1.42 gm/cm^3 peak, Linser and Armentrout (1978) observed that both types of virus particles bound equally well to cellular receptors at 4°C. It is therefore still unclear whether proteolytic cleavage of the major capsid protein results in an increased affinity for the receptor.

The virus receptor is present in varying amounts on different cultured cell lines, but virtually all permissive host cells examined bind in excess of 10^5 virus particles per cell. Although the number of receptors on the membrane does not change significantly during the growth cycle of individual cell lines (Linser *et al.*, 1979; R. Evans, personal communication), the number of, or accessibility to, receptors does vary with the type or extent of cellular differentiation (Section II,F,1). A differential expression of the receptor protein on the cell surface can, at least in part, account for the spectrum of cell tropisms exhibited by the virus *in vivo* (Section II,E). The distribution of receptor-bound virions on the surface of A-9 cells has been examined by electron microscopy using thin sectioning (Linser *et al.*, 1977) and by scanning electron microscopy of whole cells (P. Male, personal communication). Particles were observed singly or in clusters on filopodia, on membrane clefts with electron-dense, inner-surface glycocalyces which resemble clatherin-rich coated pits, as well as on the smooth, structurally undefined regions of the plasma membrane. Thus, there appears to be no unique distribution of receptors on the cell surface.

Cell lines treated with or grown in the presence of purified neuraminidase or trypsin are refractory to MVM infection and do not bind virus, even though the cell growth characteristics are unaltered (Tattersall, unpublished observations; R. Evans, personal communication). In addition, a ricin-resistant line of the normally virus-sensitive baby hamster kidney cell line BHK21, which has an altered pattern of surface glycosylation (Meager *et al.*, 1976), is completely resistant to cell killing by MVM-T (Tattersall, unpublished observations). Furthermore, addition of α-methylmannoside to the culture medium inhibits adsorption of MVM to both A-9 cells and guinea pig erythrocytes (R. Evans, personal communication). Taken together, these results suggest that the cellular receptor is a glycoprotein, although it has not been rigorously excluded that conformation or topological changes induced by these perturbations abolish virus binding in a nonspecific fashion.

The uptake of MVM into cultured rodent cell lines has been examined biochemically and by electron microscopy (Linser *et al.*, 1979). Cell–virus complexes were formed at 4°C, and the entry of virus into the cell was monitored after the cells were shifted to 37°C. Whereas virus bound at 4°C can be eluted rapidly from the cell surface upon the addition of EDTA even as long as 2 hr postadsorption, virtually all of the virus becomes resistant to EDTA elution within 30 min after the shift to 37°C. However, a significant part of the infectivity of the adsorbed virus remains sensitive to anti-MVM antiserum for up to 2 hr after the temperature shift, suggesting that the entry process involves at least two distinguishable steps.

Studies on the uptake of bound virus using electron microscopy demonstrated that virus penetration can occur both by general pinocytosis into smooth vesicles and by coated vesicles formed by the invagination of the endocytotic clefts. It has not been established, however, which uptake process is biologically relevant in facilitating subsequent virus replication. An analysis of the subcellular distribution of radiolabeled virus particles as a function of time after infected cells were placed at 37°C revealed that virus–nucleus interactions could be detected within 1 hr. However, maximal nuclear uptake required 4–6 hr, by which time approximately 40% of the input virus was found in the nuclear fraction. Although a high multiplicity of virus was necessary in these experiments to permit efficient quantification, parallel studies indicated that infectious particles enter the cell at essentially the same rate as the bulk of virus particles which do not initiate infection (Linser *et al.*, 1979). Since virus binding and uptake occur during all phases of the cell cycle, it is clear that these early events are not the cause of the S phase dependency of MVM replication.

Virtually nothing is known at the moment concerning the process by which the genome is released from the virion. However, the majority of the virus that becomes associated with the nucleus remains intact for many hours beyond the point at which viral DNA replication is initiated (J. Rommelaere, personal communication). Although the uncoating step may be an S phase-specific event, the available evidence (see below) suggests that conversion of the input viral genome into a double-stranded form is most likely the rate-limiting step of virus replication which requires the expression of S phase-specific host cell functions.

2. DNA Replication

The structure of MVM replicative forms of DNA and the kinetics of their appearance have been studied in some detail after selective fractionation from cellular DNA, mainly using the procedure of Hirt (1967). Collectively, these studies (Tattersall *et al.*, 1973; Tattersall and Ward, 1976; Ward and Dadachanji, 1978; Wolter *et al.*, 1980) indicate that MVM DNA replication occurs in three stages. The first is the conversion of the parental DNA into a monomer-length DNA duplex, termed *parental RF DNA*. This DNA then undergoes a second type of replication (RF DNA replication), a process in which both monomeric and concatemeric RF DNA species are generated. The final stage of DNA replication involves the asymmetric strand-displacement synthesis of single-stranded progeny DNA from the pool of RF DNA.

The initial studies of Tattersall *et al.* (1973) showed that viral DNA synthesis can be detected about 8 hr postinfection in nonsynchronized A-9 cells and that the amount of viral DNA rises rapidly 8–20 hr postinfection, by which time it represents approximately 20% of the total DNA synthesized. Analysis of this DNA by hydroxyapatite and benzyoylated DEAE-cellulose chromatography, by velocity sedimentation on neutral and akaline sucrose gradients, and by its sensitivity to specific nucleases revealed several interesting and unexpected features. First, the ratio of single-stranded to double-stranded forms of viral DNA remains essentially constant throughout the infection. Second, some of the duplex molecules appeared to be concatemers containing up to eight genome equivalents of DNA. These molecules, however, contained no single strands longer than two genome equivalents, and on denaturation and reannealing they formed monomer-length duplexes. The majority of these concatemers were also reduced to monomer-length duplexes when heated to submelting temperatures, suggesting that they contained staggered, closely spaced, single-stranded nicks. Third, a large fraction of monomer-length duplex DNA reannealed in a monomolecular fashion after denaturation, further suggesting that viral (V) and complementary (C) strands were linked covalently. Similar spontaneous renaturation behavior has been reported subsequently for the replicative DNA forms of many parvoviruses, both defective and autonomous (for review, see Berns and Hauswirth, 1978). It was initially suggested (Tattersall *et al.*, 1973) that duplex DNA with covalently linked V and C strands could be generated by using the 3′-end of the V strand as a primer for the synthesis of C strand DNA. The demonstration that MVM DNA contained terminal palindromic sequences at both the 3′- and 5′-termini which formed stable hairpin duplexes (Section III,C) provided evidence that the genome had structural features compatible with such a self-priming function. Furthermore, since virtually all the viral genome between these hairpin duplexes can be deleted in spontaneously arising mutants without impairing their ability to undergo DNA replication, these terminal segments of the genome must contain all of the critical recognition sites necessary for DNA replication (see Section III,C).

The structure of the viral DNA, the properties of *in vivo* replicative-form DNA, and the efficient utilization of MVM DNA as a template-primer *in vitro* by a variety of DNA polymerases, including reverse transcriptases (Bourguignon *et al.*, 1976), led Tattersall and Ward (1976) to propose a model for parvovirus DNA replication. Although this scheme, designated the *rolling hairpin model,* will not be discussed here in detail, some of its salient features should be mentioned since subsequent studies on MVM DNA replication have been in general agreement with the original hypothesis. It was suggested that synthesis of the C strand of parental RF DNA was primed by the 3′-hydroxyl terminus of the input viral DNA and proceeded, by a gap-filling mechanism, to the 5′-terminal hairpin structure. After the 5′-hairpin sequence was copied by strand-displacement synthesis, it was proposed that this terminal duplex palindrome was rearranged to form a "rabbit-eared" structure

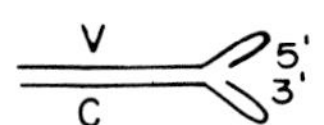

Such a rearrangement process would re-create the hairpin originally present at the 5′-end of the parental genome and generate a copy of this hairpin, on the 3′-end of the complementary strand, which could serve as a primer for the synthesis of a new viral strand. Dimer-length DNA duplexes would be produced by strand-displacement synthesis from this C strand-primed reaction in the absence of nucleolytic processing. A site-specific endonucleolytic cleavage of the C strand hairpin primer would subsequently generate the 5′-hairpin structure of the future progeny genome. This process of "hairpin transfer" following self-primed initiation is the major essential feature of the replication model. Intact dimer-length DNA duplexes would be fully capable of continued synthesis by the same hairpin transfer mechanism, yielding even larger oligomeric DNA. Concatemers resulting from such a replication process would have viral genomes situated alternately on opposite strands of the duplex with their palindromic ends overlapping. For example, dimer RF DNA molecules would have the structural organization

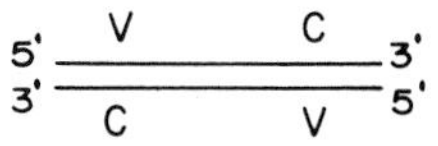

$$\begin{array}{l} 5' \overset{V \qquad\qquad C}{\rule{3cm}{0.4pt}} 3' \\ 3' \underset{C \qquad\qquad V}{\rule{3cm}{0.4pt}} 5' \end{array}$$

In a later study (Ward and Dadachanji, 1978), parasynchronized A-9 cells were infected at the beginning of the S phase (at an MOI of 15) with MVM containing ^{32}P-labeled DNA of high specific activity. Samples were withdrawn at various times postinfection and viral DNA, extracted by a modification of the Hirt (1967) procedure, was fractionated into single- and double-stranded forms by chromatography on hydroxyapatite. Synthesis of parental RF DNA was first detected 6 hr postinfection, and by 16 hr approximately 10% of the input DNA was coverted into duplex DNA. The majority (60–80%) of the parental RF DNA isolated at 6 or 8 hr postinfection exhibited reassociation kinetics of $C_0t = 0$; however, the percentage of these molecules which renatured in a monomolecular fashion declined sharply as the infection progressed. This observation suggests strongly that an initial event in the replication of viral DNA yields parental RF DNA molecules in which V and C strands are covalently attached; however, it does not provide definite proof. Since MVM-T has a particle : infectivity ratio of 300–400 : 1 (Tattersall, 1972), one cannot be absolutely certain that studies on the fate of the total input DNA are truly representative of viral DNA which is actually initiating infection.

Additional studies on the synthesis of parental RF DNA have been described by Wolter *et al.* (1980). Highly synchronized populations of RT cells, obtained by mitotic detachment, were infected in early G_1 with MVM, labeled in the DNA with [^{3}H]thymidine. Following an experimental protocol similar in design to that outlined above but utilizing gel electrophoresis and Southern (1975) transfer methods for analysis, significant amounts of parental RF DNA could be detected only when the infected cells entered the early S phase; by the mid-S phase (14 hr postinfection), between 50 and 100 copies of RF DNA per cell were synthesized. In contrast, little if any parental RF DNA was observed during the initial 10 hr of infection when the cells were in G_1. These results, in combination with those reported previously (Linser *et al.*, 1977, 1979; Tattersall, 1972), suggest that the cell cycle dependence of MVM replication is due to some DNA synthesis function controlled by the host cell which is necessary for the conversion of the parental genome into its double-stranded form. Richards and Armentrout (1979) explored the possibility that integration of viral DNA into the chromosome of the host cell may be an obligatory step to initiate viral DNA replication. If such a hypothesis were correct, one would expect integration to occur prior to, or concomitant with, the onset of detectable, nonintegrated viral DNA replication intermediates. However, during a single-step infection of RT cells, no viral sequences could be detected in association with cellular DNA during the first 12 hr of infection, by which time more than 100 genome equivalents of viral DNA per cell had been synthesized.

Wolter *et al.* (1980) also observed that 5-bromodeoxyuridine (BUdR) has a pronounced inhibitory effect on MVM DNA replication (and protein synthesis) under conditions in which it did not inhibit host cell DNA synthesis or alter the cell cycle kinetics. The inhibitory effect of 5 μM BUdR on viral RF DNA synthesis was dependent on the time that the drug was added during the viral replication cycle and on the multiplicity of virus infection. When BUdR is present from the onset of the infection in early G_1 (1 hr postmitosis), the yield of RF DNA that accumulates by the late S phase (22 hr postmitosis) is reduced by over 90%. However, RF DNA synthesis becomes increasingly resistant to BUdR inhibition if the drug is added after the infection had proceeded 11 or 12 hr, when the cells are entering the early S phase. In synchronized RT cells, the period between 12 and 18 hours postmitosis is the time when parental RF DNA undergoes an "activation event" which renders it capable of undergoing further replication as double-stranded DNA (i.e., RF DNA replication). Thus the presence of BUdR during the initial stages of virus DNA replication appears to block selectively, by a mechanism as yet undefined, the synthesis of parental RF DNA or its ability to undergo further rounds of replication. Wolter *et al.* (1980) suggest that the inhibitory effect of BUdR may result from the incorporation of the analog into parental RF DNA. If the latter is normally used as an early transcription template, BUdR substitution may render it transcriptionally inactive, thereby inhibiting the production of a gene product(s) essential for further RF DNA replication. It is also intriguing to consider the possibility that BUdR acts as a trigger to lower the expression of epigenetically controlled host genes (see Section II,F) that are necessary for virus replication. BUdR has previously been shown to alter markedly the expression of genes that influence the developmental or differentiated status of the cell (Bick, 1977; Cullen and Bick, 1978).

In addition to analyzing the structural properties of parental RF DNA molecules, the experiments of Ward and Dadachanji (1978) further demonstrated that the concatemeric forms of viral DNA observed in infected cells were true, metabolically active replication intermediates and that they possessed the genomic sequence organization predicted by the rolling hairpin replication model. Infected cells were pulse-labeled for 20 min with [^{3}H]thymidine 14 hr postinfection, and viral DNA was purified from samples taken immediately after the pulse label and after being chased with an excess of cold thymidine for periods of up to 2.5 hr. The various species of viral DNA were resolved by gel electrophoresis, and the radioactivity associated with each was quantitated. These results clearly showed that the radiolabel in both monomeric and oligomeric RF DNA molecules can be chased into progeny single-stranded

DNA. Monomer, dimer, and tetramer RF DNA species, labeled with [^{32}P]phosphate, were also excised from preparative agarose gels and subjected to structural studies. Analysis of the fragments obtained after digestion with restriction endonucleases demonstrated that only those fragments containing the 3′- or 5′-terminal portions of the viral genome were capable of spontaneous renaturation and that the oligomeric forms of RF DNA did indeed contain an alternating arrangement of V and C strands. In addition, these studies showed that many of the oligomeric RF DNA molecules possessed single-stranded nicks spaced one genome length apart at both ends of each genome copy, further implying that the maturation of progeny DNA from RF DNA requires the participation of one or more site-specific endonucleases.

In all of the studies described above, it was observed that little, if any, free, single-stranded viral DNA is found in infected cells. Significant amounts of virion DNA are observed only when the cell extracts have been exhaustively treated with pronase or proteinase K in the presence of sodium dodecylsulfate, conditions which disrupt fully or partially assembled particles. The strand-displacement synthesis of progeny DNA in synchronized RT cells has been shown by Richards *et al.* (1978) to be tightly coupled with the assembly of full virions, suggesting that packaging of progeny strands may drive the strand-displacement synthesis.

Although the structural features of MVM replicative intermediates now appear to be fairly well defined, considerable additional data will be required before a definitive picture of the replication process emerges. For example, the observation that the 3′-hairpin of the viral DNA had a unique nucleotide sequence (Astell *et al.*, 1979a) (Section III,C) raises some questions as to how the 3′-termini of progeny DNA are matured. If a hairpin transfer process was used to complete the maturation of viral DNA from monomer-length RF DNA molecules, then two sequence orientations of this hairpin would be predicted from the model unless the sequence of the hairpin was a perfect palindrome, which it is not. Since the sequencing data were derived from DNA that had been packaged into mature virions, the possibility still remains that viral strands containing only a single sequence orientation were selectively encapsidated. If this hypothesis is correct, then viral DNA strands with both sequence orientations should exist in the intracellular pool of RF DNA. The sequence data of Astell *et al.* (1979a) predict that 5′-end-labeled fragments derived from the 3′-terminal region of each RF DNA subpopulation would differ in size. When monomer RF DNA was extracted from infected cells, 5′-end-labeled with [^{32}P]phosphate and then digested with the restriction enzyme *Fnu*DII, only fragments of a size consistent with a single sequence orientation were observed (M. Chow and D. Ward, unpublished observations). This result could be misleading, however, if only one subpopulation of RF DNA molecule was end-labeled in the kinase reaction. Since replicative forms of MVM DNA contain a protein that is covalently bound to the 5′-terminus of both V and C strands (M. Chow and D. Ward, unpublished observations), this is a likely possibility which requires further investigation.

If all monomer RF DNA molecules do contain a unique terminal sequence in the viral DNA strand, then the replication and maturation of the 3′-hairpin sequence by a rolling hairpin mechanism cannot occur at the level of monomer RF DNA. Conservation of a unique 3′ sequence could occur, however, if the dimer RF DNA molecule were used as an obligatory precursor to mature monomer RF DNA. A site-specific double-stranded cleavage of the dimer could generate a metabolically active monomer RF DNA molecule which retains the same sequence orientation as the parental genome and an RF DNA species, presumably metabolically inactive, which lacks the 3′-hairpin sequence entirely. Studies on the fate of labeled parental DNA in synchronized A-9 cells demonstrate that parental RF DNA is indeed converted to a dimer molecule, which is subsequently converted back to monomer RF DNA, with kinetics consistent with such a proposal (J. Rommelaere and D. Ward, unpublished observations). Another possible pathway, involving circularization of dimer RF DNA and progeny DNA generation by a rolling circle mechanism, seems unlikely since circular DNA replication intermediates have never been observed in any parvovirus-infected cell (Berns and Hauswirth, 1978). It is interesting to note, however, that relaxed circular forms of DNA, generated from unusual "lasso-like" structures, do exist in MVM-infected cells (Bratosin *et al.*, 1979). These molecular forms are not seen in cell lysate supernatants prepared by the method of Hirt (1967), which utilizes ionic detergents, but they can be extracted if cell lysates are prepared using a combination of ionic and nonionic detergents. This suggests that these circular DNA species may be associated preferentially with nuclear substructures that are disrupted by the nonionic detergent Triton X-100 but are precipitated by the Hirt fractionation method. The reason some of the double-stranded DNA undergoes circularization is not known at the present time. However, it may be a means of providing a suitable template for transcription or a mechanism for separating the intracellular pools of RF DNA active in transcription and replication.

The genetic origin (viral or cellular) and biological function of the 60,000-dalton protein attached to the 5′-termini of some RF DNA molecules have yet to be established. Although it is conceivable that this protein may play a role in the initiation of DNA replication, by a mechanism similar to that proposed for the terminal protein on adenovirus DNA (Rekosh *et al.*, 1977), this seems unlikely in view of the studies already described. Alternatively, it may form part of an initial maturation complex which is used to nucleate the biogenesis of full virus particles during strand displacement of progeny DNA, or it may represent a covalent enzyme-substrate intermediate, similar to those seen with nicking-closing enzymes and topoisomerases (Cozzarelli, 1980) formed during the nucleolytic processing of the terminal DNA sequences.

Attempts to develop cell-free or reconstituted *in vitro* DNA replication systems have had limited success until recently. Whereas neither DNA polymerases α, β, nor γ, alone or in combination, are capable of utilizing MVM DNA as a template-primer, HeLa cell or A-9 cell DNA polymerase α "holoenzyme" complexes which contain three additional protein components (E. Baril, personal communication) are capable of synthesizing a complete monomer RF DNA species *in vitro* (H. Allaudeen and D. Ward, unpublished observations). This observation is of potential significance since DNA polymerase α is the only cellular polymerase that has been directly implicated in DNA replication of autonomous parvoviruses (Bates *et al.*, 1978, and personal communication). Of further interest is the observation that when purified ancillary proteins used in T4 bacteriophage DNA replication (products of genes 32, 44, 45, and 62) are incubated with MVM DNA and T4 DNA polymerase, both monomer and dimer RF DNA are observed (P. Tattersall and N. Nossal, unpublished observations). The bacteriophage proteins thus appear to be capable of copying the 5′-hairpin sequence, promoting the hairpin rearrangement process, and catalyzing the subsequent strand-displacement synthesis which leads to dimer RF DNA. The development of cell-free systems (R. Bates, personal communication; M. Goulian, personal communication), which faithfully produce all of the replicative forms of viral DNA seen *in vivo* during the replication of related parvoviruses (BPV, KRV, and H-1), should greatly facilitate further studies on the mechanism of DNA replication and the roles that viral and host proteins play in this process.

3. Transcription and Translation

Although this facet of the MVM replicative life cycle has received little attention until recently, some interesting features are beginning to emerge concerning the transcription products made late in infection (20 hr or more postinfection). Tal *et al.* (1979) examined, by electron microscopy, the hybrids formed between nuclear or cytoplasmic poly(A)-containing RNA and purified virion DNA. Over 95% of the hybrids formed with nuclear RNA were double-stranded over virtually the entire length of DNA. Thus the major species of nuclear RNA is a complete, or nearly complete, transcript of the genome. In contrast, the major species of cytoplasmic RNA observed in the electron microscope yielded hybrids which possessed a large loop of single-stranded DNA containing 30% of the genome. This intervening sequence had map coordinates between 8.8 ± 1.2 and 38.1 ± 2.0, where one map unit corresponds to 1% of the length of the whole genome and the 3′- and 5′-ends of the DNA are 0 and 100 map units, respectively. This transcript had a leader that mapped from the left-end splice junction (8.8) to near the extreme 3′-end of the chromosome and a main body that spanned virtually all of the remaining DNA beyond map coordinate 38. In addition to suggesting a precursor–product relationship between the nuclear and cytoplasmic transcripts, these results indicate that a major promoter for viral transcription is located near the 3′-terminal hairpin of the genome. Three additional pieces of evidence indicate that this is indeed the case. First, the sequence 5′-TATATTCG-3′ occurs between 178 and 185 nucleotides inboard of the 3′-terminus of the viral DNA (Astell *et al.*, 1979b). Grosschedl and Birnsteil (1980) have summarized evidence which clearly implies that 5′-TATAAATA-3′ and closely related sequences function in the initiation of transcription by RNA polymerase II. Second, the 5′-ends of these RNAs, designated the *4.9* and *3.4-Kb* transcripts (see below), map between nucleotides 200 and 210. Third, a translational initiation followed by an extensive open reading frame begins at nucleotide 262. Furthermore, the A polypeptide of MVM possesses N-terminal peptide sequences which are consistent with this initiator codon assignment (P. Paridiso, personal communication).

Subsequent studies (D. Dadachanji and D. Ward, unpublished observations), although in general agreement with the results of Tal *et al.* (1979), indicate that the transcription process is somewhat more complex than that revealed by electron microscopy. Four polyadenylated viral transcripts, with estimated sizes of 4.9, 3.4, 3.1, and 1.8 Kb, were detected by hybridization with V-strand DNA following the "Northern" transfer technique developed by Alwine *et al.* (1977). The 3.1-Kb transcript was the most abundant (~70%), with the relative molar amounts of the 4.9-, 3.4-, and 1.8-Kb species being 12, 15, and 3%, respectively. Since no transcripts were found to hybridize with C-strand sequences, either in solution or on Northern blots, it is apparent that, at least during the late stages of infection, only one of the two strands of DNA is transcriptionally active.

The regions of the genome which encode the three largest transcripts were mapped using the method of Berk and Sharp (1978). The 4.9-Kb transcript, which was found predominantly in the nucleus of infected cells, maps between coordinates 4.5 and 95/100. The uncertainty in mapping the precise termination point for this and other transcripts is a technical one related to the spontaneous renaturation of the 5′-hairpin sequence in the DNA. Although transcription certainly does not terminate before the hairpin, it is still unclear whether part or all of the terminal palindromic sequence is transcribed. Although some 4.9-Kb transcripts are unspliced, the majority have approximately 20–40 nucleotides, mapping between 45 and 45.5 on the DNA, spliced out. The 3.4-Kb transcript is composed of two "leader" sequences, mapping between positions 4.5 and 10.7 and between 39 and 45.0, respectively, and a main body mapping between 45.5 and 95/100. (The term *leader* is used here merely for convenience and is not intended to imply that these regions are not translated.) The sizes and map coordinates of the 4.9- and 3.4-Kb transcripts agree well with the data obtained by Tal *et al.* (1979), since one would not have expected to detect the small intervening sequence at position 45–45.5 by electron microscopy. However, Tal *et al.* (1979) suggest that the 3.4-Kb transcript is the major species of cytoplasmic RNA, whereas the data of Dadachanji and Ward (unpublished observation) shows that a 3.1-Kb transcript with a single leader, mapping between 39 and 45, and a main body with map coordinates between 45.5 and 95/100 is the most abundant. This apparent discrepancy could be rationalized if only hybrid molecules with obvious intervening sequence loops were analyzed in the electron microscopic study. Attempts to detect a small leader sequence in the 3.1-Kb transcript derived from the 3′-end of the viral genome have failed (D. Dadachanji, unpublished observations), indicating that if such a leader does exist, it may be too small to be detected by standard hybridization techniques. Should the 5′-end of the 3.1-Kb transcript actually start at or near map coordinate 39, this would suggest that the MVM genome contains at least two transcriptional promoters. The translation products of these viral RNAs have not been determined to date. It is tempting, however, to speculate that the 3.1-Kb transcript, because of its relatively high abundance, represents the messenger for the major capsid protein. Similarly, the low amounts of the unmapped 1.8-Kb transcript make it a potential candidate for the mRNA of the 60,000-dalton protein found covalently bound to the 5′-termini of replicative forms of MVM DNA (see Section III,C). These speculations should be resolved quickly, since restriction fragments of MVM RF DNA cloned in the plasmid pBR322 are now available (M. Merchlinsky, unpublished observations) for use as hybridization probes to select specific RNA species for *in vitro* translation.

4. Maturation and Lysis

Although viral peptide synthesis can be detected as early as 8–10 hr postinfection, virion morphogenesis occurs much later. For example, in highly synchronized RT cells infected 3 hr postmitosis, assembled virions first become evident in the nucleoplasm some 18–20 hr later (Richards *et al.*, 1977, 1978). The rate of virion assembly increases dramatically between 20 and 30 hr postinfection, being maximal about 8 hr after the end of the S phase. Since the extent of cellular degeneration increases in parallel with the accumulation of assembled virions, it is not surprising that these infected cells do not undergo subsequent mitosis (Richards *et al.*, 1977).

The normal ultrastructure of the cell appears unaltered during the first 18 hr of the infection, and the synthesis of cellular macromolecules continues unabated. However, with the onset of virion assembly, cytopathic and biochemical changes become apparent. The fibrillar and granular elements of the nucleolus begin to segregate and undergo condensation, and bands of heterochromatin at the periphery of the nucleus become denser and begin to thicken. At this time, no apparent changes in the cytoplasm of the cell can be detected. By 24 hr postinfection, the nucleolus has hypertrophied to the extent that the granular portion has condensed into compact spherical masses, and only remnants of condensed fibrillar material remain. The margination of the heterochromatin also becomes much more pronounced, and condensation of the euchromatin fraction becomes evident. Whereas large numbers of virion particles are seen in association with the euchromatin fibers at this stage, virtually no virions are found in the heterochromatin. Between 24 and 30 hr postinfection, the period of peak virion assembly, the nuclear elements undergo further degeneration, the perinuclear space distends and fills with a light, amorphous matrix, and the cytoplasm shows the first signs of CPE, an increasing concentration of free ribosomes appearing in the cytosol with a concomitant decrease in the amount of rough endoplasmic reticulum. Between 30 and 38 hr postinfection, the intracytoplasmic concentration of virions rises dramatically, and some of the cells begin to lyse and release virus into the medium. By 38 hr, general degradation of the cytoplasm has occurred and only a few identifiable cytoplasmic constituents remain. The lytic cycle is now complete.

The rates of accumulation of full and empty particles within infected cells are similar, although the synthesis of empty particles appears to begin slightly before that of the full virions (Richards *et al.*, 1977). The first forms of full virus synthesized are of the 1.46 gm/cm^3 density class, and these are rapidly chased into the mature 1.42 gm/cm^3 density virion while still within the nucleus (Richards *et al.*, 1978), a maturation process which appears to be mediated by a cellular enzyme (Clinton and Hayashi, 1976). The evidence demonstrating that all structural polypeptides are modified posttranslationally, and that a proteolytic cleavage of the B polypeptide within the 1.46 gm/cm^3 full virion is the maturation event that leads to the production of mature virus, has been presented in Section III,B and will not be discussed further. However, one final point on virion maturation that should be mentioned is that the assembly of the 1.46 gm/cm^3 class of "precursor" full virion is immediately halted when DNA synthesis is inhibited (Richards *et al.*, 1977). Since large amounts of empty capsids are present in infected cells when the pool of free, single-stranded DNA is minute or nonexistent, it is apparent that ongoing viral DNA synthesis is necessary for the assembly of infectious virus.

In summary, although a substantial body of information concerning the biological and biochemical characteristics of MVM has been obtained during the past decade, it is clear that much additional data are required before a detailed understanding of the virus–host cell interaction in molecular terms is at hand.

REFERENCES

Alwine, J. C., Kemp, D. J., and Stark, G. R. (1977). Method for detection of specific RNAs in agarose gels by transfer to diazobenzyloxymethyl-paper and hybridization with DNA probes. *Proc. Natl. Acad. Sci. U.S.A.* **74,** 5350–5354.

Astell, C. R. (1977). Replication of minute virus of mice in chinese hamster ovary fibroblasts. *J. Gen. Virol.* **35,** 587–591.

Astell, C. R., Smith, M., Chow, M. B., and Ward, D. C. (1979a). Structure of the 3′ hairpin termini of four rodent parvovirus genomes: nucleotide sequence homology at origins of DNA replication. *Cell* **17,** 691–703.

Astell, C. R., Smith, M., Chow, M. B., and Ward, D. C. (1979b). Sequence of the 3′-terminus of the genome from Kilham rat virus, a nondefective parvovirus. *Virology* **96,** 669–674.

Baer, P. N., and Kilham, L. (1974). Dental defects in hamsters infected with minute virus of mice. *Oral Surg.* **37,** 385–389.

Bates, R. C., Kuchenbuch, C. P., Patton, J. T., and Stout, E. R. (1978). DNA-polymerase activity in parvovirus-infected cells. *In* "Replication of Mammalian Parvoviruses" (D. C. Ward and P. Tattersall, eds.), pp. 367–382. Cold Spring Harbor Lab., Cold Spring Harbor, New York.

Berk, A. J., and Sharp, P. (1978). Spliced early mRNAs of simian virus 40. *Proc. Natl. Acad. Sci. U.S.A.* **75,** 1274–1278.

Berns, K. I., and Adler, S. (1972). Separation of two types of adeno-associated virus particles containing complementary polynucleotide chains. *J. Virol.* **9,** 394–396.

Berns, K. I., and Hauswirth, W. W. (1978). Parvovirus DNA structure and replication. *In* "Replication of Mammalian Parvovirus" (D. C. Ward and P. Tattersall, eds.), pp. 13–32. Cold Spring Harbor Lab., Cold Spring Harbor, New York.

Berns, K. I., and Hauswirth, W. W. (1979). Adeno-associated viruses. *Adv. Virus Res.* **25,** 407–449.

Berns, K. I., Hauswirth, W. W., Fife, K. H., and Lusby, E. (1979). Adeno-associated virus DNA replication. *Cold Spring Harbor Symp. Quant. Biol.* **47,** 781–788.

Bick, M. D. (1977). Bromodeoxyuridine inhibition of Friend leukemia cell induction. Mechanism of reversal by deoxycytidine. *Biochim. Biophys. Acta* **476,** 279–286.

Bonnard, G. D., and Herberman, R. B. (1975). Suppression of the generation of cytotoxic lymphoblasts by murine lymphoma cells. *Fed. Proc., Fed. Am. Soc. Exp. Biol.* **34,** 1002.

Bonnard, G. D., Manders, E. K., Campbell, D. A., Jr., Herberman, R. B., and Collins, M. J., Jr. (1976). Immunosuppressive activity of a subline of the mouse EL-4 lymphoma. Evidence for minute virus of mice causing the inhibition. *J. Exp. Med.* **143,** 187–205.

Bourguignon, G. J., Tattersall, P. J., and Ward, D. C. (1976). DNA of minute virus of mice: self-priming, non-permuted, single-stranded genome with a 5′-terminal hairpin duplex. *J. Virol.* **20,** 290–306.

Bratosin, S., Laub, O., Tal, J., and Aloni, Y. (1979). A novel mechanism for circulation of linear DNAs: circular parvovirus MVM DNA is formed by a "noose" sliding in a "lasso" like DNA structure. *Proc. Natl. Acad. Sci. U.S.A.* **76,** 4289–4293.

Caspar, D. L. D., and Klug, A. (1962). Physical principals in the construction of regular viruses. *Cold Spring Harbor Symp. Quant. Biol.* **27,** 1–24.

Chow, M. B., and Ward, D. C. (1978). Comparison of the terminal nucleotide structures in the DNA of nondefective parvoviruses. *In* "Replication of Mammalian Parvoviruses" (D. C. Ward and P. Tattersall, eds.), pp. 205–217. Cold Spring Harbor Lab., Cold Spring Harbor, New York.

Clinton, G. M., and Hayashi, M. (1975). The parvovirus MVM: particles with altered structural proteins. *Virology* **66,** 261–267.

Clinton, G. M., and Hayashi, M. (1976). The parvovirus MVM: a comparison of heavy and light particle infectivity and their density conversion *in vitro*. *Virology* **74,** 57–63.

Collins, M. J., Jr., and Parker, J. C. (1972). Murine virus contaminants of leukemia viruses and transplantable tumors. *J. Natl. Cancer Inst.* **49,** 1139–1143.

Cozzarelli, N. R. (1980). DNA gyrase and the supercoiling of DNA. *Science* **207,** 953–960.

Crawford, L. V. (1966). A minute virus of mice. *Virology* **29,** 605–612.

Crawford, L. V., Follett, E. A. C., Burdon, M. G., and McGeoch, D. J. (1969). The DNA of a minute virus of mice. *J. Gen. Virol.* **4,** 37–46.

Cross, S. S., and Parker, J. C. (1972). Some antigenic relationships of the murine parvoviruses: minute virus of mice, rat virus, and H-1 virus. *Proc. Soc. Exp. Biol. Med.* **139,** 105–108.

Cullen, B. R., and Bick, M. D. (1978). Bromodeoxyuridine induction of deoxycytidine deaminase activity in a hamster cell line. *Biochim. Biophys. Acta* **517,** 158–168.

Faust, E. A., and Ward, D. C. (1979). Incomplete genomes of the parvovirus minute virus of mice: selective conservation of genome termini, including the origin for DNA replication. *J. Virol.* **32,** 276–292.

Frearson, P. M., and Crawford, L. V. (1972). Polyoma virus basic proteins. *J. Gen. Virol.* **14,** 141–155.

Grosschedl, R., and Birnsteil, M. L. (1980). Identification of regulatory sequences in the prelude sequences of an H2A histone gene by the study of specific deletion mutants *in vivo*. *Proc. Natl. Acad. Sci. U.S.A.* **77,** 1432–1436.

Hallauer, C., Kronauer, G., and Siegl, G. (1971). Parvoviruses as contaminants of permanent human cell lines. I. Virus isolations from 1960–1970. *Arch. Gesamte Virusforsch.* **35,** 80–90.

Hallauer, C., Siegl, G., and Kronauer, G. (1972). Parvoviruses as contaminants of permanent human cell lines. III. Biological properties of the isolated viruses. *Arch. Gesamte Virusforsch.* **38,** 366–382.

Hampton, E. G. (1970). H-1 virus growth in synchronized rat embryo cells. *Can. J. Microbiol.* **16,** 266–268.

Harris, R. E., Coleman, P. H., and Morahan, P. S. (1974). Erythrocyte association and interferon production by minute virus of mice. *Proc. Soc. Exp. Biol. Med.* **145,** 1288–1292.

Hirt, B. (1967). Selective extraction of polyoma DNA from infected mouse cell cultures. *J. Mol. Biol.* **26,** 365–369.

Hunter, T., and Sefton, B. M. (1980). Transforming gene product of Rous Sarcoma Virus phosphorylates tyrosine. *Proc. Natl. Acad. Sci. U.S.A.* **77,** 1311–1315.

Karasaki, S. (1966). Size and ultrastructure of the H-viruses as determined with the use of specific antibodies. *J. Ultrastruct. Res.* **16,** 109–122.

Kilham, L., and Margolis, G. (1970). Pathogenicity of minute virus of mice (MVM) for rats, mice and hamsters. *Proc. Soc. Exp. Biol. Med.* **133,** 1447–1452.

Kilham, L., and Margolis, G. (1971). Fetal infections of hamsters, rats, and mice induced with the minute virus of mice (MVM). *Teratology* **4,** 43–62.

Kilham, L., and Margolis, G. (1975). Problems of human concern arising from animal models of intrauterine and neonatal infections due to viruses: A review. I. Introduction and virologic studies. *Prog. Med. Virol.* **20,** 113–143.

Kilham, L., and Olivier, L. J. (1959). A latent virus of rats isolated in tissue culture. *Virology* **7,** 428–437.

Koczot, F. J., Carter, B. J., Garon, C. F., and Rose, J. A. (1973). Self-complementarity of terminal sequences within plus or minus strands of adeno-associated virus DNA. *Proc. Natl. Acad. Sci. U.S.A.* **70,** 215–219.

Lake, R. S., Barban, S., and Salzman, N. P. (1973). Resolution and identification of the core deoxynucleoproteins of the simian virus 40. *Biochem. Biophys. Res. Commun.* **54,** 640–647.

Linser, P., and Armentrout, R. W. (1978). Binding of minute virus of mice to cells in culture. *In* "Replication of Mammalian Parvoviruses" (D. C. Ward and P. Tattersall, eds.), pp. 151–160. Cold Spring Harbor Lab., Cold Spring Harbor, New York.

Linser, P., Bruning, H., and Armentrout, R. W. (1977). Specific binding sites for a parvovirus, minute virus of mice, on cultured mouse cells. *J. Virol.* **24,** 211–221.

Linser, P., Bruning, H., and Armentrout, R. W. (1979). Uptake of minute virus of mice into cultured rodent cells. *J. Virol.* **31,** 537–545.

McGeoch, D. J., Crawford, L. V., and Follett, E. A. C. (1970). The DNAs of three parvoviruses. *J. Gen. Virol.* **6,** 33–40.

Margolis, G., and Kilham, L. (1975). Problems of human concern arising from animals models of intrauterine and neonatal infections due to viruses: A review. II. Pathologic studies. *Prog. Med. Virol.* **20,** 144–179.

Martin, G. R., and Evans, M. J. (1975). Multiple differentiation of clonal teratocarcinoma stem cells following embryoid body formation *in vitro*. *Cell* **6,** 467–474.

Maxam, A. M., and Gilbert, W. (1977). A new method for sequencing DNA. *Proc. Natl. Acad. Sci. U.S.A.* **74,** 560–564.

Mayor, H. D., and Jordan, E. L. (1966). Electron microscopic study of the rodent "picodnavirus" X14. *Exp. Mol. Pathol.* **5,** 580–589.

Meager, A., Ungkitchanukit, A., and Hughes, R. C. (1976). Variants of hamster fibroblasts resistant to *Ricinus Communis* toxin (Ricin). *Biochem. J.* **154,** 113–124.

Miller, R. A., Ward, D. C., and Ruddle, F. H. (1977). Embryonal carcinoma cells (and their somatic cell hybrids) are resistant to infection by the murine parvovirus MVM, which does infect other teratocarcinoma-derived cell lines. *J. Cell. Physiol.* **91,** 393–402.

Mohanty, S. B., and Bachmann, P. A. (1974). Susceptibility of fertilized mouse eggs to minute virus of mice. *Infect. Immun.* **9,** 762–763.

Nicolas, J. F., Dubois, P., Jakob, H., Gaillard, J., and Jacob, F. (1975). Tératocarcinome de la souris: Différenciation en culture d'une lignée de cellules primitives á potentialités multiples. *Ann. Microbiol. (Paris)* **126A,** 3.

Parker, J. C., Cross, S. S. Collins, M. J., Jr., and Rowe, W. P. (1970a). Minute virus of mice. I. Procedures for quantitation and detection. *J. Natl. Cancer Inst.* **45,** 297–303.

Parker, J. C., Collins, M. J., Jr., Cross, S. S., and Rowe, W. P. (1970b). Minute virus of mice. II. Prevalence, Epidemiology, and occurrence as a contaminant of transplanted tumors. *J. Natl. Cancer Inst.* **45,** 305–310.

Payne, F. E., Beals, T. F., and Preston, R. E. (1964). Morphology of a small DNA virus. *Virology* **23,** 109–133.

Peterson, J. L., Dale, R. M. K., Karess, R., Leonard, D., and Ward, D. C. (1978). Comparison of parvovirus structural proteins: evidence for post translational modification. *In* "Replication of Mammalian Parvoviruses" (D. C. Ward and P. Tattersall, eds.), pp. 431–445. Cold Spring Harbor Lab., Cold Spring Harbor, New York.

Rekosh, D. M. K., Russell, W. C., Bellett, A. J. D., and Robinson, A. J. (1977). Identification of a protein linked to the ends of adenovirus DNA. *Cell* **11,** 283–295.

Richards, R. G., and Armentrout, R. W. (1979). Early events in parvovirus replication: lack of integration by minute virus of mice into host cell DNA. *J. Virol.* **30,** 397–399.

Richards, R., Linser, P., and Armentrout, R. W. (1977). Kinetics of assembly of a parvovirus, minute virus of mice in synchronized rat brain cells. *J. Virol.* **22,** 778–793.

Richards, R., Linser, P., and Armentrout, R. W. (1978). Maturation of minute virus of mice particles in synchronized rat brain cells. *In* "Replication of Mammalian Parvoviruses" (D. C. Ward and P. Tattersall, eds.), pp. 447–458. Cold Spring Harbor Lab., Cold Spring Harbor, New York.

Rose, J. A. (1974). Parvovirus reproduction. *Compr. Virol.* **3,** 1–61.

Rose, J. A., Berns, K. I., Hoggan, M. D., and Koczot, F. J. (1969). Evidence for a single-stranded adeno-associated virus genome: formation of a DNA-density hybrid on release of viral DNA. *Proc. Natl. Sci. U.S.A.* **64,** 863–869.

Schaffhausen, B. S., and Benjamin, T. L. (1979). Phosphorylation of polyoma T-antigens. *Cell* **18,** 935–946.

Siegl, G. (1976). "The Parvoviruses," Virology Monographs, No. 15. Springer-Verlag, Berlin and New York.

Southern, E. M. (1975). Detection of specific sequences among DNA fragments separated by gel electrophoresis. *J. Mol. Biol.* **98,** 503–517.

Strickland, S., and Mahdavi, V. (1978). The induction of differentiation in teratocarcinoma stem cells by retinoic acid. *Cell* **15,** 393–403.

Tal, J., Ron, D., Tattersall, P., Bratosin, S., and Aloni, Y. (1979). About 30% of minute virus of mice RNA is spiced out following polyadenylation. *Nature (London)* **279,** 649–651.

Tattersall, P. (1972). Replication of parvovirus MVM. I. Dependence of virus multiplication and plaque formation on cell growth. *J. Virol.* **10,** 586–590.

Tattersall, P. (1978). Susceptibility to minute virus of mice as a function of host-cell differentiation. *In* "Replication of Mammalian Parvoviruses" (D. C. Ward and P. Tattersall, eds.), pp. 131–149. Cold Spring Harbor Lab., Cold Spring Harbor, New York.

Tattersall, P., and Ward, D. C. (1976). Rolling hairpin model for replication of parvovirus and linear chromosomal DNA. *Nature (London)* **263,** 106–109.

Tattersall, P., and Ward, D. (1978). The parvoviruses—an introduction. *In* "Replication of Mammalian Parvoviruses" (D. C. Ward and P. Tattersall, eds.), pp. 3–12. Cold Spring Harbor Lab., Cold Spring Harbor, New York.

Tattersall, P., Crawford, L. V., and Shatkin, A. J. (1973). Replication of the parvovirus MVM. II. Isolation and characterization of intermediates in the replication of the viral deoxyribonucleic acid. *J. Virol.* **12,** 1446–1456.

Tattersall, P., Cawte, P. J., Shatkin, A. J., and Ward, D. C. (1976). Three structural polypeptides coded for by minute virus of mice, a parvovirus. *J. Virol.* **20,** 273–289.

Tattersall, P., Shatkin, A. J., and Ward, D. C. (1977). Sequence homology between the structural polypeptides of minute virus of mice. *J. Mol. Biol.* **111,** 375–394.

Tennant, R. W., and Hand, R. E., Jr. (1970). Requirement of cellular synthesis for Kilham rat virus replication. *Virology* **42,** 1054–1063.

Tennant, R. W., Layman, K. R., and Hand, R. E., Jr. (1969). Effect of cell physiological state on infection by rat virus. *J. Virol.* **4,** 872–878.

Toolan, H. W. (1961). A virus associated with transplantable human tumors. *Bull. N.Y. Acad. Med.* **37,** 305–310.

Toolan, H. W. (1967). Agglutination of the H-viruses with various types of red blood cells. *Proc. Soc. Exp. Biol. Med.* **124,** 144–146.

Vasquez, C., and Brailovsky, C. (1965). Purification and fine structure of Kilham's rat virus. *Exp. Mol. Pathol.* **4,** 130–140.

Wang, A. H. J., Quigley, G. J., Kolpak, F. J., Crawford, J. L., van Boom, J. H., van der Marel, G., and Rich, A. (1979). Molecular structure of a lefthanded double helical DNA fragment at atomic resolution. *Nature (London)* **282,** 680–686.

Ward, D. C., and Dadachanji, D. K. (1978). Replication of minute virus of mice DNA. *In* "Replication of Mammalian Parvovirus" (D. C. Ward and P. Tattersall, eds.), pp. 297–313. Cold Spring Harbor Lab., Cold Spring Harbor, New York.

Wolter, S., Richards, R., and Armentrout, R. W. (1980). Cell cycle-dependent replication of the DNA of minute virus of mice, a parvovirus. *Biochim. Biophys. Acta* **607,** 420–431.

Chapter 16

Mouse Adenovirus

James A. Otten and Raymond W. Tennant

I. Introduction ... 335
II. Isolations of Mouse Adenoviruses and Pathological Effects in Mice ... 336
III. Physical Characteristics ... 337
IV. Serological Reactions ... 338
V. Epidemiology and Prevalence ... 338
References ... 339

I. INTRODUCTION

The name *adenovirus* was first applied by Enders *et al.* (1956) to a group of viruses that Rowe *et al.* (1953) and Hilleman and Werner (1954) independently isolated from the respiratory tracts of man. Over 80 different adenovirus serotypes have since been isolated from a variety of animal species (Wadell, 1970). The adenoviruses have become increasingly important in providing model systems for studies of DNA replication, gene expression, and the synthesis of macromolecules in eukaryotic cells (Flint, 1977; Wold *et al.*, 1978). Taxonomically, the adenoviruses are in the family Adenoviridae. Two genera have been established (Norrby *et al.*, 1976): *Mastadenovirus,* which includes all mammalian adenoviruses, and *Aviadenovirus,* into which the avian adenoviruses have been placed. A primary characteristic separating the two genera is the absence of any immunological cross-reactivity.

Adenoviruses are classified and subclassified primarily on the basis of molecular-physical and biological criteria. Some of these criteria include chemical composition (types of nucleic acid, proteins, lipids, and carbohydrates), physiochemical properties (density, sedimentation coefficient, stability to pH, heat, solvents, and radiation), structure, and morphology. In addition, the site of replication, host range, pathogenicity, transmission, and antigenic specificity are used. An excellent discussion of these criteria can be found in a report by Norrby *et al.* (1976). The present classification criteria of adenoviruses are similar to those described by Rowe and Hartley (1962). These include (1) cytopathogenicity, primarily for homologous cells in culture; (2) lack of pathogenicity for distantly related species of animals; (3) ether resistance (adenoviruses have no envelope) and heat lability (adenoviruses are inactivated when heated at 56°C for 30 min); and (4) group- and type-specific antigens. In addition, they contain DNA and fit other structural and biophysical properties of adenoviruses, and they produce

ISBN 0-12-262502-1

intranuclear inclusions. Although hemagglutination is a property of many adenoviruses, it is not a property of the two murine adenoviruses to be described. All adenoviruses have a classic cubic structure with a diameter of 65–80 nm. Excellent reviews on the adenoviruses have been written (Pettersson, 1973; Philipson and Pettersson, 1973; Philipson *et al.*, 1975; Wold *et al.*, 1978) and should be consulted for detailed information on the biophysical and chemical properties of this virus class.

Most adenoviruses cause only mild respiratory diseases in humans, but cases of conjunctivitis, enteritis, myocarditis, and diseases of lymph nodes have also been described (Sohier *et al.*, 1965). One property of adenoviruses that has received considerable attention is the oncogenicity of some strains in hamsters (Trentin *et al.*, 1962; Huebner *et al.*, 1962; Girardi *et al.*, 1964; Sarma *et al.*, 1965; Green, 1969, 1970) and the transformation of cultured cells (McBride and Wiener, 1964). However, none of the mouse adenovirus isolates have demonstrated any oncogenicity *in vivo* or the ability to transform cells *in vitro*.

II. ISOLATIONS OF MOUSE ADENOVIRUSES AND PATHOLOGICAL EFFECTS IN MICE

There have been many isolates of adenovirus from mice (*Mus musculus*), but only two distinct strains have been identified. The first, discovered by Hartley and Rowe (1960), involved a cytopathic agent isolated during attempts to establish Friend mouse leukemia virus in tissue culture; it was called *strain FL*. These investigators subsequently isolated the same type of agent from Swiss mice obtained from the National Institutes of Health animal production colony. Cytopathic changes were produced in mouse embryo and kidney cultures, but not in cultured cells of other species, and when strain FL adenovirus was inoculated into suckling mice by the intraperitoneal, intracerebral, or intranasal route, a rapidly fatal disease occurred that involved the brown fat, myocardium, adrenal cortex, salivary glands, and kidneys. When adult or weanling mice were inoculated with strain FL by the same route, no deaths occurred, and the animals responded by producing high levels of complement-fixing and neutralizing antibodies (Hartley and Rowe, 1960).

Hashimoto *et al.* (1966) isolated another distinct strain of mouse adenovirus from the feces of an inbred strain of DK1 mice while searching for viruses from apparently healthy mice. Their adenovirus isolate was given the designation *strain K87*. Strain K87 was destructive to mouse kidney tissue culture but not to cultured cells of other species. Hashimoto *et al.* (1966) demonstrated that strain K87, when given to suckling mice by the intracerebral route, produces no deaths or recognizable tissue changes. When strain K87 is inoculated into adult mice by the intracerebral, intraperitoneal, intramuscular, subcutaneous, or oral route, the virus localizes in the intestinal area (Hashimoto *et al.*, 1966, 1970, 1973; Sugiyama *et al.*, 1967). No deaths result, and the only measurements of virus infection are the continued excretion of the virus in the feces and the high levels of neutralizing antibody in the serum. Cyclophosphamide has an inhibitory effect upon the establishment and duration of infection in the intestines of experimentally infected mice (Hashimoto *et al.*, 1973). Strain K87 has not been isolated from the urine or nasal tissue of infected mice (Sugiyama *et al.*, 1967).

Strain FL adenovirus has been used in a large number of experimental studies because of its varied tissue affinity. In their original report describing the isolation of strain FL, Hartley and Rowe (1960) reported that fatal disease was produced in suckling mice inoculated intraperitoneally, intracerebrally, or intranasally, with pathologic changes occurring in the brown fat, myocardium, adrenal cortex, salivary glands, and kidneys. Weanling and adult mice did not develop overt disease, but they did produce complement-fixing and neutralizing antibodies by either artificial or natural infection. Since viruses can damage cardiovascular tissues in man (Rabin and Melnick, 1965), and since chronic heart valve disease develops in numerous patients who have had no history of rheumatic fever (Clawson *et al.*, 1926), it has been suggested that viruses might be the cause of these heart lesions in man (Burch and DePasquale, 1964). Burch *et al.* (1966) have reported that endocarditis can be experimentally induced in mice with coxsackievirus B4. Blailock *et al.* (1967) reported that a significant incidence of endocarditis developed in mice infected with strain FL adenovirus, with invasion and replication of the virus in cells of the cardiac valve tissue. Of 50 mice inoculated as newborns with Strain FL adenovirus, 12 had necrotic foci in the heart valves. Adenovirions in crystalline arrays were observed in the nuclei of the endothelial cells and fibroblasts of the valves. It is assumed that these acute valve lesions heal, with scar formation and dystrophic calcification occurring at the site of the lesion (Blailock *et al.*, 1967).

In another study, Blailock *et al.* (1968) demonstrated that myocarditis developed in mice following infection with strain FL adenovirus. Characteristic adenovirus lesions were observed in the myocardium, mural endocardium, and heart valves, and in the endothelium of the ascending aorta. This is an interesting observation since electrocardiographic changes have been observed following adenovirus pneumonia in children (Chaney *et al.*, 1958). Heck *et al.* (1972) used strain FL adenovirus infection as a model for a pathogenesis study. Suckling mice were inoculated intraperitoneally, and brain, heart, spleen, and kidney tissues were collected and analyzed for evidence of virus infection. Virus was detected in the blood at 1 day postinfection, and by day 3 virus was isolated from the spleen, kidney, brain, and heart. All suckling mice died by day 10 postinfection, but all adult mice survived and were positive serologically but without clinical illness. The cells of affected tissues developed type A intranuclear inclusion bodies and areas of necrosis. Two points of interest were observed during this study: (1) blood-associated virus is of interest because there are only a few reports of a viremic state associated with adenovirus infections (Davis *et al.*, 1973; Sinha *et al.*, 1960); (2) prior to this report, lesions in the central nervous system had not been described following infection with mouse adenovirus.

A study that used adult mice (4 weeks of age) was designed by Van Der Veen and Mes (1973) to investigate persistent adenovirus infection. Mice were inoculated intraperitoneally with strain FL adenovirus and observed for 2 years. Adenovirus was detected in the thymus and spleen by day 3 postinoculation, and other tissues and white blood cells became positive by days 6–9. The highest titers of virus were detected in the spleen on day 10, and by 2 weeks postinfection all tissues became negative. On day 14, the urine became positive for adenovirus, and the viuria persisted for the duration of the 2-year study. Ten of the 250 animals that were inoculated with adenovirus died between days 12 and 16 postinoculation. The only signs of illness in these mice were hunched posture, ruffled coat, and a decrease in food consumption. No autopsy data were presented. This is the only report of deaths in adult mice following infection by mouse adenovirus. A similar picture was found after intranasal inoculation except that no deaths occurred in mice inoculated by this route. All mice responded with high levels of neutralizing antibody, and this antibody response was probably the reason that no adenovirus was detected in tissue after 2 weeks.

Since adrenal changes in Addison's disease, consisting of cortical atrophy and focal lymphocytic infiltration (Petri and Nerup, 1971), have an unknown etiology, the possibility of viral involvement has been hypothesized and studied by Margolis *et al.* (1974) in mice using strain FL adenovirus. The virus had a highly selective affinity for the adrenal gland, and when suckling and adult mice were inoculated by the subcutaneous or intracranial route, adrenal changes were seen by day 3, with the absence of significant disease in other tissues and organs. The adenovirus damage centered in the cortical epithelium, with all adrenal cortical zones affected. As many as 80% of the epithelial cells contained inclusions. Although other tissues became involved, including the brain, only the adrenal glands showed cellular changes severe enough to suggest functional alteration. A complete description of the progression of the nuclear and cellular changes following infection of the mouse adrenal gland with strain FL adenovirus is discussed in a report by Hoenig *et al.* (1974).

There has been one report of spontaneous adenovirus infection in nude mice (Cohen and de Groot, 1976). The strain of adenovirus was not identified, but amphophilic inclusion bodies were found in the mucosal cells of the duodenum, jejunum, or ileum of the diseased mice, but not in any other organs. Experimentally, athymic (nude) mice developed a hemorrhagic duodenal lesion and lethal wasting disease following inoculation with type 1 mouse adenovirus (strain FL). Intranuclear inclusions were observed in endothelial and mucosal cells of the villi and crypts of the duodenum (Winters and Brown, 1980).

III. PHYSICAL CHARACTERISTICS

Two studies considered the biological and physical properties of strain FL adenovirus (Wigand *et al.*, 1977; Larsen and Nathans, 1977). Wigand *et al.* (1977) reported that strain FL produced significantly higher titers of virus in primary as opposed to secondary cells in culture, with an average yield of about 1000 infectious particles per cell. A typical one-step growth curve involves an eclipse period of about 18 hr. The cytopathic effect begins at 24 hr and involves the whole culture by 72 hr postinfection with strain FL. The released adenovirus particles had a buoyant density of 1.34 gm/ml, which is identical to that of human adenovirus types 9–15 (Wigand and Klein, 1974). In addition, Wigand *et al.* (1977) showed that strain FL particle assembly occurred in close proximity to large, electron-dense nuclear inclusions, as did Takeuchi and Hashimoto (1976) using strain K87 mouse adenovirus. Wigand *et al.* (1977) also found that inhibitors of DNA synthesis such as hydroxyurea, cytosine arabinoside (AraC), and 5-iodo-3′-deoxyuridine inhibited strain FL multiplication. These findings were similar to those of earlier studies done with human adenoviruses (Wigand and Schmieder, 1973). Hashimoto *et al.* (1966) demonstrated that bromodeoxyuridine inhibited Strain K87 adenovirus.

Larsen and Nathans (1977) also found that strain FL adenovirus reached maximum titers on BALB/3T3 cell monolayer cultures; the released particles had a buoyant density of 1.34 gm/ml. In addition, when the purified virus was examined by electron microscopy and length measurements were taken relative to form II SV40 DNA, it was found that the molecular weight of strain FL adenovirus DNA was $19.5 \pm 1.1 \times 10^6$. Larsen and Nathans (1977) also reported that strain FL adenovirus DNA, like human adenovirus DNA, yields circular duplex forms and that the DNAs also share terminal inverted repeat sequences. Restriction endonuclease maps of strain FL differed markedly from cleavage maps constructed from human Ad2 or Ad5. Analysis of the nucleotide sequence homology between strain FL DNA and the DNA of three human adenoviruses (Ad2, Ad7, and Ad12) showed that 85–90% of the human adenovirus DNAs hybridized, whereas less than 10% of the strain FL DNA was converted to the duplex form. This means that strain FL DNA has little or no nucleotide sequence homology with the DNAs of human Ad2, Ad7, or Ad12. Another significant difference between strain FL DNA and human adenovirus DNAs is in the base composition. In CsCl, strain FL DNA had a low G + C content (44%, as compared to 57% for human Ad2 DNA).

IV. SEROLOGICAL REACTIONS

Hartley and Rowe (1960) reported that a serological relationship existed between strain FL adenovirus and the adenovirus group as indicated by cross-complement-fixing antibody responses in guinea pigs infected with strain FL or human Ad1 and Ad5. In contrast, when antisera were prepared in mice against strain FL adenovirus and other adenovirus serotypes, the antisera reacted only with the homologous antigens. Hashimoto *et al.* (1966) reported similar results in that strain K87 reacted by complement fixation with serum from guinea pigs immunized with human Ad3. Antisera prepared in mice against strain K87 reacted only with strain K87 antigen but not with human adenovirus antigen. Neither of the two murine adenovirus strains has demonstrated hemagglutinating activity for various erythrocytes.

In studies done to determine the serological relationship between strains FL and K87, Van Der Veen and Mes (1974) reported that the two murine adenoviruses were antigenically distinct. They used antisera that had been prepared in mice or guinea pigs immunized to either strain FL or strain K87 adenoviruses. Cross-reactions were not observed in complement fixation and neutralization tests between the two strains of adenoviruses irregardless of the source of antisera. Antisera to the mouse adenoviruses also failed to react with complement-fixing antigen from human Ad2, but cross-reactions were observed between human and mouse adenoviruses in complement fixation tests using convalescent sera from humans with adenovirus infections. In contrast to this study, Wigand *et al.* (1977) reported that serological cross-reactions did exist between the two mouse adenoviruses. Cross-neutralization tests revealed a one-sided relationship between strains FL and K87. Antiserum to strain K87 neutralized both virus strains with equal titers, but antiserum to strain FL only weakly neutralized strain K87. A partial crossing by complement fixation of mouse adenovirus with human adenovirus antigen was observed for the FL strain as well as for the K87 strain. Larsen and Nathans (1977) reported that antisera against human Ad-2 reacted with strain FL antigen, but antisera prepared against the T antigens of human adenoviruses (Ad2-SV40 hybrid, Ad7, Ad12, Ad21) showed no reactivity with strain FL, indicating the absence of T antigens.

In a study by Hamada and Uetake (1975), strain FL adenovirus was used to immunize mice; immune spleen cells from these mice were then used to protect mice from infection with strain FL virus by adoptive transfer. An explanation for their results was that virus-specific antigen(s) appeared on the surface of strain FL-infected cells that acted as target sites for cell-mediated immunity. Inada and Uetake (1977) subsequently demonstrated that a virus-specific, cell-surface antigen was induced on strain FL-infected cells and could be detected using an immunofluorescent antibody technique. The nature and specificity of the effector cells in cell-mediated cytolysis of strain FL-infected cells were characterized by Inada and Uetake (1978). They reported that immune spleen cells from mice immunized with strain FL adenovirus lost their cytolytic activity when pretreated with antimouse thymocyte or anti-Thy 12 serum, but not when treated with antimouse immunoglobulin. From these data, it appeared that the effector cells in the immune spleen cell population were T cells.

V. EPIDEMIOLOGY AND PREVALENCE

Rowe *et al.* (1961) and Stansly (1965) called attention to problems that face researchers because of the prevalence of viruses that are indigenous to mice and that can be covertly transmitted in tumors, mice, and other experimental material. Such covert agents can alter the outcome of the experimental transmission of another disease and also affect tissue changes, biochemistry, and other experimental parameters. Mouse adenovirus is indigenous to mice and therefore should be considered as a potential covert contaminant.

As indicated earlier, Hartley and Rowe (1960) isolated strain FL adenovirus while attempting to establish Friend mouse leukemia virus in tissue culture. In this and another report (Rowe *et al.,* 1961), they also discussed testing the serum from older breeding mice of a "general purpose" colony; the results revealed a 45% incidence of complement-fixing antibody to mouse adenovirus, whereas mice from separately maintained colonies were free of antibody to mouse adenovirus. In natural epizootics, antibody to adenovirus was not found in mice less than 4–5 months of age (Hartley and Rowe, 1960; Parker *et al.,* 1966). Of seven other colonies tested by Hartley and Rowe (1960), evidence of natural infection was found in two. In addition, infection could be transmitted by artificially infected mice to their uninoculated cage contacts but not to mice held in the same animal room but in separate cages. In one report (Jacoby *et al.,* 1979), complement-fixing antibodies to mouse adenovirus in rats were detected, but there were no clinical disease symptoms or lesions attributable to infection of the rats, and no virus has been isolated from rats infected with mouse adenovirus.

In another study (Parker *et al.,* 1966), mice were obtained from either commerical or other mouse breeder colonies that were not associated with animals inoculated with tumors or other test materials. Colonies of animals tested were from germ-free colonies; caesarian-derived, specific pathogen-free colonies; and conventional colonies from several geographic areas of the United States. The mice were usually more than 9 months of age and were examined over a 2-year period. Of 34 colonies tested, mouse adenovirus was found in only four conventional colonies. Trentin *et al.* (1966) described another study in which a conventional colony and a caesarian-derived colony that, in turn, had been derived from the conventional colony were sampled for antibody to mouse adenovirus. Mice from both colonies were tested during a period of about 2 years. In the conventional colony, 15% of 741 mice tested were found to be positive for adenovirus antibody, whereas none of 583 mice from the caesarian-derived colony were found to be positive. This indicates again that when mice are reared under good standards of husbandry, adenovirus can be eliminated. A serological study (Descoteaux *et al.*, 1977) conducted in Canada using mice obtained from research and commerical colonies (utilizing a total of 139 mice that were retired breeders of both sexes, 6-9 months of age) also demonstrated the absence of antibody to adenovirus. In addition, thousands of tests performed more recently on numerous colonies in the United States and elsewhere by the National Cancer Institute and Microbiological Associates, Inc., have failed to detect antibody to mouse adenovirus. These results indicate that the health status of mouse colonies has been improved and that mouse adenovirus has been essentially eliminated as an ambient indigenous agent (M. Collins, 1980 personal communication).

ACKNOWLEDGMENT

Operated by Union Carbide Corporation, under contract W-7405-eng-26 from the U.S. Department of Energy.

REFERENCES

Blailock, Z. R., Rabin, E. R., and Melnick, J. L. (1967). Adenovirus endocarditis in mice. *Science* **157**, 69-70.

Blailock, Z. R., Rabin, E. R., and Melnick, J. L. (1968). Adenovirus myocarditis in mice. An electron microscopic study. *Exp. Mol. Pathol.* **9**, 84-96.

Burch, G. E., and DePasquale, N. P. (1964). Viral endocarditis. *Am. Heart J.* **67**, 721-723.

Burch, G. E., DePasquale, N. P., Sun, S. C., Hale, A. R., and Mogabgab, W. J. (1966). Experimental coxsackie virus endocarditis. *J. Am. Med. Assoc.* **196**, 349-352.

Chaney, C., Lepine, P., Lelong, M., Le-Tan-Vinh, S. P., and Virat, J. (1958). Severe and fatal pneumonia in infants and younger children associated with adenovirus infection. *Am. J. Hyg.* **67**, 367-378.

Clawson, B. J., Bell, E. T., and Hartzell, T. B. (1926). Valvular diseases of the heart with special reference to the pathogenesis of old valvular defects. *Am. J. Pathol.* **2**, 193-234.

Cohen, B. J., and de Groot, F. G. (1976). Adenovirus infection in athymic (nude) mice. *Lab. Anim. Sci.* **26**, 955-956.

Davis, B. D., Dulbecco, R., Eisen, H. N., Ginsberg, H. S., Wood, W. B., and McCarty, M. (1973). Adenoviruses. *In* "Microbiology," pp. 1222-1236. Harper (Hoeber), New York.

Descoteaux, J. P., Grignon-Archambault, D., and Lussier, G. (1977). Serologic study on the prevalence of murine viruses in five Canadian mouse colonies. *Lab. Anim. Sci.* **27**, 621-626.

Enders, J. R., Bell, J. A., Dingle, J. H., Francis, T., Jr., Hilleman, M. R., Huebner, R. J., and Payne, A. (1956). Adenoviruses: Group name proposed for new respiratory tract viruses. *Science* **124**, 119-120.

Flint, J. (1977). The topography and transcription of the adenovirus genome. *Cell* **10**, 153-166.

Girardi, A. J., Hilleman, M. R., and Zwickey, R. E. (1964). Tests in hamsters for oncogenic quality of ordinary viruses including adenovirus type 7. *Proc. Soc. Exp. Biol. Med.* **115**, 1141-1150.

Green, M. (1969). Nucleic acid homology as applied to investigations on the relationship of viruses to neoplastic diseases. *In* "Recent Results in Cancer Research" (M. Mizell, ed.), pp. 445-454. Springer-Verlag, New York.

Green, M. (1970). Search for adenovirus messenger RNA in cancers of man. *In* "Oncology, 1970" (R. L. Clark, R. W. Cumley, J. E. McCay, and M. M. Copeland, eds.), pp. 156-165. Year Book Publ., Chicago, Illinois.

Hamada, C., and Uetake, H. (1975). Mechanism of induction of cell-mediated immunity to virus infections: *In vitro* inhibition of intracellular multiplication of mouse adenovirus by immune spleen cells. *Infect. Immun.* **11**, 937-943.

Hartley, J. W., and Rowe, W. P. (1960). A new mouse virus apparently related to the adenovirus group. *Virology* **11**, 645-647.

Hashimoto, K., Sugiyama, T., and Sasaki, S. (1966). An adenovirus isolated from the feces of mice. I. Isolation and identification. *Jpn. J. Microbiol.* **10**, 115-125.

Hashimoto, K., Sugiyama, T., Yoshikawa, M., and Sasaki, S. (1970). Intestinal resistance in the experimental enteric infection of mice with a mouse adenovirus. I. Growth of the virus and appearance of a neutralizing substance in the intestinal tract. *Jpn. J. Microbiol.* **14**, 381-395.

Hashimoto, K., Okada, Y., Tajiri, T., Amano, H., Aoki, N., and Sasaki, S. (1973). Intestinal resistance in the experimental enteric infection of mice with a mouse adenovirus. III. Suppressive effect of cyclophosphamide on the establishment and duration of the intestinal resistance. *Jpn. J. Microbiol.* **17**, 503-511.

Heck, F. C., Jr., Sheldon, W. G., and Gleiser, C. A. (1972). Pathogenesis of experimentally produced mouse adenovirus infection in mice. *Am. J. Vet. Res.* **33**, 841-846.

Hilleman, M. R., and Werner, J. H. (1954). Recovery of new agent from patients with acute respiratory illness. *Proc. Soc. Exp. Biol. Med.* **85**, 183-188.

Hoenig, E. M., Margolis, G., and Kilham, L. (1974). Experimental adenovirus infection of the mouse adrenal gland. II. Electron microscopic observations. *Am. J. Pathol.* **75**, 372-386.

Huebner, R. J., Rowe, W. P., and Lane, W. T. (1962). Oncogenic effects in hamsters of human adenovirus type 12 and 18. *Proc. Natl. Acad. Sci. U.S.A.* **48**, 2051-2058.

Inada, T., and Uetake, H. (1977). Virus-induced specific cell surface antigen(s) on mouse adenovirus-infected cells. *Infect. Immun.* **18**, 41-45.

Inada, T., and Uetake, H. (1978). Nature and specificity of effector cells in cell-mediated cytolysis of mouse adenovirus-infected cells. *Infect. Immun.* **22**, 119-124.

Jacoby, P. O., Bhatt, P. N., and Jonas, A. M. (1979). Viral diseases. *In* "The Laboratory Rat" (H. J. Baker, J. R. Lindsey, and S. H. Weisbroth, eds.), Vol. 1, p. 284. Academic Press, New York.

Larsen, S. H., and Nathans, D. (1977). Mouse adenovirus: Growth of plaque-purified FL virus in cell lines and characterization of viral DNA. *Virology* **82,** 182–195.

McBride, W. D., and Wiener, A. (1964). *In vitro* transformation of hamster kidney cells by human adenovirus type 12. *Proc. Soc. Exp. Biol. Med.* **115,** 870–874.

Margolis, G., Kilham, L., and Hoenig, E. M. (1974). Experimental adenovirus infection of the mouse adrenal gland. I. Light microscopic observations. *Am. J. Pathol.* **75,** 363–372.

Norrby, E., Bartha, A., Boulanger, P., Oreizin, R. S., Ginsberg, H. S., Kalter, S. S., Kawamura, H., Rowe, W. P., Russell, W. C., Schlesinger, R. S., and Wigand, R. (1976). Adenoviridae. *Intervirology* **7,** 117–125.

Parker, J. C., Tennant, R. W., and Ward, T. G. (1966). Prevalence of viruses in mouse colonies. *Natl. Cancer Inst. Monogr.* No. 20, 25–36.

Petri, M., and Nerup, J. (1971). Addison's adrenalitis. *Acta Pathol. Microbiol. Scand.* **79,** 381–388.

Pettersson, U. (1973). The adenoviruses. *In* "Molecular Biology of Tumor Viruses" (J. Tooze, ed.), pp. 420–469. Cold Spring Harbor Lab., Cold Spring Harbor, New York.

Philipson, L., and Pettersson, U. (1973). Structure and function of virion proteins of adenoviruses. *Prog. Exp. Tumor Res.* **18,** 1–55.

Philipson, L., Pettersson, U., and Lindberg, U. (1975). Molecular biology of adenoviruses. *In* "Virology Monographs" (H. V. S. Gard and C. Hallauer, eds.), pp. 1–115. Springer-Verlag, New York.

Rabin, E. R., and Melnick, J. L. (1965). Viral myocarditis. *Cardiovasc. Res. Cent. Bull.* **4,** 2–4.

Rowe, W. P., and Hartley, J. W. (1962). A general review of the adenoviruses. *Ann. N.Y. Acad. Sci.* **101,** 466–474.

Rowe, W. P., Huebner, R. J., Gilmore, L. K., Parrott, R. H., and Ward, T. G. (1953). Isolation of a cytopathogenic agent from human adenoids undergoing spontaneous degeneration in tissue culture. *Proc. Soc. Exp. Biol. Med.* **84,** 570–573.

Rowe, W. P., Hartley, J. W., and Huebner, R. J. (1961). Polyoma and other indigenous mouse viruses. *In* "The problems of Laboratory Animal Disease" (R. J. C. Harris, ed.), pp. 131–142. Academic Press, New York.

Sarma, P. S., Huebner, R. J., and Lane, W. T. (1965). Induction of tumors in hamsters with an avian adenovirus (CELO). *Science* **149,** 1108.

Sinha, S. K., Fleming, L. W., and Scholes, S. (1960). Current considerations in public health of the role of animals in relation to human viral diseases. *J. Am. Vet. Med. Assoc.* **136,** 481–485.

Sohier, R., Chardonnet, Y., and Prumieras, M. (1965). Adenoviruses, status of current knowledge. *Prog. Med. Virol.* **7,** 253–325.

Stansly, P. G. (1965). Non-oncogenic infectious agents associated with experimental tumors. *Prog. Exp. Tumor Res.* **7,** 224–258.

Sugiyama, T., Hashimoto, K., and Saski, S. (1967). An adenovirus isolated from the feces of mice. II. Experimental infection. *Jpn. J. Microbiol.* **11,** 33–42.

Takeuchi, A., and Hashimoto, K. (1976). Electron microscope study of experimental enteric adenovirus infection in mice. *Infect. Immun.* **13,** 569–580.

Trentin, J. J., Yabe, Y., and Taylor G. (1962). The quest for human cancer viruses. *Science* **137,** 835–841.

Trentin, J. J., Van Hoosier, G. L., Jr., Shields, J., Stephens, K., Stenback, W. A., and Parker, J. C. (1966). Limiting the viral spectrum of the laboratory mouse. *Natl. Cancer Inst. Monogr.* No. 20, 147–160.

Van Der Veen, J., and Mes, A. (1973). Experimental infection with mouse adenovirus in adult mice. *Arch. Gesamte Virusforsch.* **42,** 235–241.

Van Der Veen, J., and Mes, A. (1974). Serological classification of two mouse adenoviruses. *Arch. Gesamte Virusforsch.* **45,** 386–387.

Wadell, G. (1970). Structural and biological properties of capsid components of human adenoviruses. Ph.D. Thesis, Karolinska Inst., Stockholm.

Wigand, R., and Klein, W. (1974). Properties of adenovirus substituted with iododeoxyuridine. *Arch. Gesamte Virusforsch.* **45,** 298–300.

Wigand, R., and Schmieder, J. (1973). Inhibition of adenovirus multiplication by metabolic inhibitors. *Arch. Gesamte Virusforsch.* **42,** 324–338.

Wigand, R., Gelderblom, H., and Ozel, M. (1977). Biological and biophysical characteristics of mouse adenovirus, strain FL. *Arch. Virol.* **54,** 131–142.

Winters, A. L., and Brown, H. K. (1980). Duodenal lesions associated with adenovirus infection in athymic "nude" mice. *Proc. Soc. Exp. Biol. Med.* **164,** 280–286.

Wold, W. S. M., Green, M., and Buttner, W. (1978). Adenoviruses. *In* "The Molecular Biology of Animal Viruses" (D. P. Nayak, ed.), pp. 673–768. Dekker, New York.

Chapter 17

Mouse Encephalomyelitis Virus

Wilbur G. Downs

I. Introduction 341
II. Viral Agent 341
A. Historical Background 341
B. Properties of the Virus 342
C. Pathogenesis 345
D. Epizootiology 349
E. Diagnosis 350
F. Control and Prevention 350
References 350

I. INTRODUCTION

Theiler's mouse encephalomyelitis viruses (TMEVs) are found in many mouse colonies in the United States, Europe, Japan, and undoubtedly elsewhere. They have been, and continue to be, unwelcome contaminants of such colonies, capable of seriously interfering with studies on isolation or actions of many viruses in mice. Apart from their nuisance value, TMEVs have provided models for studies of epidemiology, pathogenesis, virus multiplication and inhibition by naturally occurring or contrived inhibitory agents and for recent immunopathological studies.

The virus epidemiology provided a landmark model for studies on the fecal–oral transmission cycle of the polioviruses (Theiler, 1941; Gard, 1943).

II. VIRAL AGENT

A. Historical Background

TMEVs were first encountered in 1933 during the course of yellow fever virus studies involving the intracerebral inoculation of white mice (Theiler, 1934, 1937). The first strains studied were designated *TO* (*Theiler's Original*). In 1940 additional strains of a more virulent type, GD VII (Theiler and Gard, 1940a), were described. GDs I–VII were named for *George's Disease,* George Martine being a laboratory technician skilled in singling out affected mice. Early studies of host range, pathogenicity, and epidemiology in mice (Gard, 1940; Theiler and Gard, 1940b) established bases for measurement of virus activity in neutralization tests. The discovery of an easily

ISBN 0-12-262502-1

prepared hemagglutinin (Lahelle and Horsfall, 1949) greatly amplified the usefulness of the virus in model systems, and the growth of the virus in minced mouse brain tissue cultures (Pearson and Winzler, 1949), and in due course in cell cultures of several types (Falke, 1957; Calisher, 1964), further increased its usefulness. Radioisotope studies (Rafelson *et al.*, 1949, 1951a,b) have been employed in investigations of cell metabolism as affected by virus infection, and immunofluorescent studies (Lipton, 1975) have given insight into pathophysiological events. More recently, immunologic techniques have been used to identify cells involved in cell-mediated immunity and to understand the processes involved in the demyelination often seen as a sequel of the encephalitis (Lipton, 1975; Lipton and Dal Canto, 1979a,b).

B. Properties of the Virus

1. Classification

TMEVs are RNA viruses (Franklin *et al.*, 1959) close in size and other physical and chemical characteristics to the polioviruses and enteroviruses. They are very distinct from the latter two in terms of serology and host range, and yet are similar to some of them in epidemiology and pathophysiology. The TMEVs were considered to be picornaviruses (Nakamura, 1961; Calisher, 1964). Lipton and Friedmann (1980) investigated their physical properties as well as their structural polypeptide composition and established definitively that they fulfill the criteria necessary for their inclusion in the genus *Enterovirus* of the family Picornaviridae.

2. Isolates of the Virus

The original isolates of the virus have been designated *TO strains* (Theiler, 1934, 1937). Later isolates were related to the TO strains but with greater virulence for mice (Theiler and Gard, 1940a). One isolate was designated *GD VII* and a closely related isolate *FA*. A number of other isolates have been recovered and characterized, and all fall in the broad groupings of TO and GD VII. Isolates cited in references include: DA (Daniels *et al.*, 1952) ~ TO; UIF, UIIF (Gard, 1944) ~ TO; UVN (Gard, 1944) ~ GD VII or FA; FA (Theiler and Gard, 1940a) related to GD VII; MHG (Hemelt *et al.*, 1974) (no HA); TO (Theiler, 1934); and GD VII (Theiler and Gard, 1940a,b).

Table I presents a simplified picture of the similarities and differences among the isolates. More detailed data will be found in the sections on virus properties and pathophysiology. Serological cross-reactivity in complement fixation, hemagglutination inhibition, virus neutralization (mouse assay), and virus neutralization (cell culture plaque reduction assay) is shown for all isolates (Tables II–IV; see Hemelt *et al.*, 1974).

3. Morphological, Biochemical, and Biophysical Properties

Purified and negatively stained TMEV particles appear to have icosahedral symmetry and measure 28 nm in diameter. A buoyant density of 1.34 gm/ml was demonstrated by isopyknic centrifugation (Lipton and Friedmann, 1980). The nucleic acid composition was shown to be that of RNA (Franklin *et al.*, 1959).

The viruses are separable into two groups based on biological behavior: highly virulent isolates unable to cause persistent infection, and less virulent isolates which regularly produce persistent central nervous system (CNS) infection in mice. Mature virions of both groups possess three major structural polypeptides—VP1, VP2, and VP3—in the molecular weight range of 25,000–35,000, plus a fourth major polypeptide, VP4, weighing 6000. Measurements were made employing a sodium dodecyl sulfate-polyacrylamide gel electrophoresis

Table I

Comparison of TMEV Isolates

Virus isolate	Mortality in mice inoc. ic		Mouse brain hemagglutinin at 4°C[a]		CPE in BHK-21 cells	Plaque size on L 929 cell sheets	Late demyelination in mice	Disease
	Weanling	Infant	Weanling	Infant				
FA	Low	High	No	Yes	− After passage	—	—	Encephalitis
GDVII	High	High	Yes	Yes	+ On initial inoculation	2–4 mm	Survival time too short	Encephalitis
MHG	Low to none	High	No	Yes	+ After first passage	0.5–1.0 mm	—	Poliomyelitis
TO	Low	Low	No	Yes	+ After first passage	1–2 mm	Yes	Poliomyelitis Chronic demyelination

[a] All isolates after adaptation to cell culture hemagglutinate human erythrocytes.

Table II

Serologic Comparison of GDVII, TO, MHG, and FA Viruses by the Cross-Neutralization Test[a]

Virus	Reciprocal of antibody titer with indicated serum[b]				
	GDVII	TO	MHG	FA	Normal
GDVII	*1024 (1024)*	256 (1024)	64 (16)	256 (1024)	0 (0)
TO	256 (1024)	*1024 (1024)*	64 (16)	256 (1024)	0 (0)
MHG	256 (1024)	1024 (256)	*64 (16)*	256 (256)	0 (0)

[a] From Hemelt *et al.* (1974).
[b] Highest dilution at which 80% plaque reduction occurred; nonbracketed values = mouse sera; bracketed values = rat sera; 0 = < 1:10.

technique. A precursor of VP2 and VP4, VP0, a minor polypeptide of mature picornavirus particles, was also identified (Lipton and Friedmann, 1980).

These workers have shown a slight but consistent difference in several of the capsid polypeptides between the highly virulent and less virulent TMEVs, VP1 being slightly heavier in the highly virulent strains and VP2 and VP0 slightly lighter. Also, trypsin preferentially cleaves a 2000-dalton fragment or fragments from VP1 of only the less virulent isolates.

The TMEVs are rapidly destroyed at temperatures of over 50°C, and the inactivation rate at 37°C is 40–80 times greater than at 4°C. The viruses do not withstand lyophilization well but can be stored for long periods of time at −60°C. There are two optima of stability, one in the vicinity of pH 8 and the other at about pH 3.3. The virus is very slowly oxidized in air but is rapidly oxidized by H_2O_2 at 37°C (and very little at 4°C). Ether does not inactivate the virus, but at a 50% acetone or alcohol concentration, it is inactivated (Theiler and Gard, 1940a). The virus is not inactivated in sodium deoxycholate (Theiler, 1957). These inactivation characteristics have been very useful in field studies on arbovirus isolation, the arboviruses in general being very susceptible (togaviruses) to sodium deoxycholate action, or on partially susceptible (orbiviruses), whereas the TMEVs, encountered under field conditions in open mouse colonies, are fully resistant.

Table III

Serologic Comparison of GDVII, TO, MHG, and FA Viruses by the Complement Fixation Test[a]

Virus	Reciprocal of antibody titer with indicated serum[b]				
	GDVII	TO	MHG	FA	Normal
GDVII	*160 (160)*	80 (160)	40 (160)	80 (160)	0 (0)
TO	640 (160)	*320 (160)*	160 (80)	320 (160)	0 (0)
MHG	80 (80)	80 (80)	*640 (640)*	40 (80)	0 (0)
FA	160 (160)	80 (160)	40 (160)	*160 (320)*	0 (0)

[a] From Hemelt *et al.* (1974).
[b] Nonbracketed values = mouse sera; bracketed values = rat sera; 0 = < 1:10.

Table IV

Serologic Comparison of GDVII, TO, MHG, and FA Viruses by the Hemagglutination Inhibition Test[a]

Virus	Reciprocal of antibody titer with indicated serum[b]				
	GDVII	TO	MHG	FA	Normal
GDVII	*320 (320)*	80 (80)	0 (0)	80 (40)	0 (0)
TO	80 (16)	*640 (320)*	40 (0)	20 (40)	0 (0)

[a] From Hemelt *et al.* (1974).
[b] Nonbracketed values = mouse sera; bracketed values = rat sera; 0 = < 1:20.

Detailed properties of the hemagglutinin, the range of cells agglutinated, and the phenomena of elution and stability are described in a series of papers (Lahelle and Horsfall, 1949; Lahelle and Ward, 1951; Fastier, 1950, 1951a,b; Morris, 1952, 1953; Calisher and Rowe, 1966).

An inhibitor of hemagglutination and infectivity has been recovered from intestinal tissues of adult mice and guinea pigs (Mandel and Racker, 1953a,b), and there is a fecal enzyme of mice (also of human and rat feces) which destroys the inhibitor (Mandel and Racker, 1957).

The union between the mucopolysaccharide inhibitor of GD VII hemagglutinin and the virus is probably determined by weak electrostatic forces and can be broken by reducing the concentration of electrolytes, allowing the recovery of both components, the inhibitor and the virus, in active form.

A nonspecific inhibition can be demonstrated in lipid-free extracts of normal mouse brain and to a lesser degree of other organs (Fastier, 1950, 1951a), and also in normal serum of mice, rats, guinea pigs, and rabbits. It can be partially inactivated by trypsin (Morris, 1952). A fairly pure inhibitor can be isolated from human urine and chick allantoic fluid, which is heat stable, combines with virus at 4°C, and elutes at 22°C (Tamm and Tyrrell, 1954). It is electrophoretically distinct from the mucoid inhibitor of the influenza virus.

4. Growth in Mice, Tissue Culture, and Eggs

The various TMEVs, TO, and GD VII isolates (see Section II,B,2), occur naturally only in colonies of laboratory mice (Thompson *et al.*, 1951; Theiler, 1934; Theiler and Gard, 1940a; Gahagan and Stevenson, 1941; Gildemeister and Ahlfeld, 1938; Iguchi, 1939; Melnick and Riordan, 1947; Parker *et al.*, 1966), with the exception of the MGH strain, isolated from adult laboratory rats (McConnell *et al.*, 1964) (see Table I). There are no reports of isolation of TMEVs from the wild mouse, *Mus musculus*.

The viruses are found in the intestinal contents and intestinal mucosa of infected mice in low titer. When TO virus is inoculated intracerebrally into mice, there is a grading of susceptibility, infant mice often dying without showing signs of disease, young mice sickening with paralysis after an incubation period of 7–30 days, and adult mice often showing no signs of infection, although indeed infected (Theiler, 1937). Intranasal instillation of virus produces paralysis in only a small percentage of the mice. The GD VII and FA strains of virus, by comparison, are highly virulent (Theiler and Gard, 1940a). Six-week-old mice were infected by both the intranasal and intraperitoneal routes with resulting high morbidity and mortality. The average time of death following intracerebral inoculation was 4–5 days, 9–10 days following intraperitoneal inoculation with the FA strain, and 14–15 days following intranasal instillation of virus.

Virus titers of 10^{-3} were attained with the TO virus (Theiler, 1937) in adult mouse brain and 10^{-6} with the FA strain in adult mouse brain (Theiler and Gard, 1940a) following intracerebral inoculation.

By inoculation, the infant or adult laboratory mouse is the animal most commonly used in experimental studies. Infant or young cotton rats, hamsters, and laboratory rats are, however, susceptible to the intracerebrally inoculated virus.

The intranasal route of inoculation is almost as effective as the intracerebral one, but other routes—intraperitoneal, subcutaneous, gastric—are distinctly less so (see Fig. 1). Differences in pathogenicity by various inoculation routes can serve as a means of differentiating virus strains, such as GD VII and TO.

It is emphasized that these agents produce chronic CNS infection in mice. Although this is seen primarily in experimental situations (after intracerebral inoculation), it probably also takes place after natural infections in those rare instances in which viremia develops and virus is seeded in the CNS. In this connection, chronic CNS infection may be totally asymptomatic (as it can be after intracerebral inoculation), and these animals may then be a source of virus contaminating other researchers' experiments.

Early studies (Pearson, 1950; Shaw, 1953; Falke, 1957) reported the growth of TMEVs in minced mouse brain, minced duck embryo, minced mouse kidney, and chick embryo fibroblast tissue cultures. Growth of virus in several continuous cell culture lines has since been reported (Sturman and Tamm, 1966; Hemelt *et al.*, 1974; Lipton, 1978a,b). Sturman and Tamm (1966) give a list of permissive and nonpermissive cell lines for GD VII virus.

The BHK 21 cell line is now used instead of *in vivo* systems for TMEV multiplication, titration, and neutralization (Lipton and Dal Canto, 1979b).

Pearson *et al.* (1952, 1955, 1956; Pearson and Winzler, 1949, 1950; Pearson, 1950; Pearson and Lagerborg, 1955) tested the effects of many substances on virus propagation in cell culture, including nucleic acids, hormones, and bacteriostatic agents, azides and cyanides, salts, amino acids, imides and amines, and CNS depressant drugs. Inhibitory effects, when seen, were presumed to be related to effects on cell substrates rather than direct effects on the virus. Studies of effects of virus on uptake of $^{32}PO_4$ by infected cells (Rafelson *et al.*, 1949, 1950a,b, 1951a,b) indicated an increased uptake of $^{32}PO_4$ by cells at the time of maximum virus proliferation, whereas tyrosine, histidine, tryptophan, and 5-chlorouridine

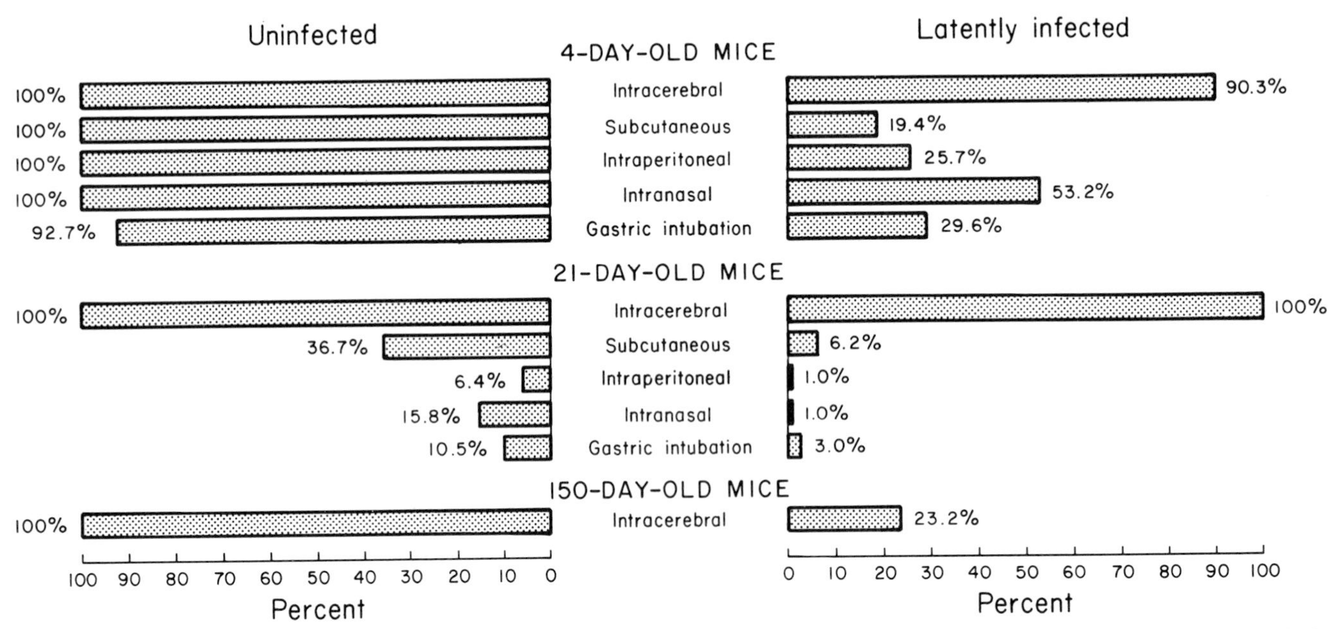

Fig. 1. Response of mice following intracerebral challenge with a TO virus isolate. The response is judged by the induction of paralysis or death. (After Dean, 1951.)

resulted in inhibition of virus multiplication and reduction of $^{32}PO_4$ incorporation in the cells. Studies using ^{14}C indicated that the virus inhibited the incorporation of glucose fragments into lysine and histidine.

C. Pathogenesis

1. Clinical Disease: Morbidity and Mortality

Later descriptions of the disease as observed in colonies of laboratory mice have added little to Theiler's (1934, 1937) original descriptions of the TO and Theiler's and Gard's (1940a) descriptions of the GD VII and FA infections.

The cardinal symptom observed in naturally infected mice is flaccid paralysis of the hindleg. Apart from this, the mice appear healthy. There is no mortality. Many later reports have confirmed the observations (Theiler, 1937) that the virus is widespread in open mouse colonies and that only unusually susceptible mice or possibly mice exposed to massive dosages of virus develop clinical signs of infection (see Section II, D,2).

If the TO virus is intracerebrally inoculated (Theiler, 1937), the clinical course of disease is much more pronounced in observable morbidity and mortality, both being dependent on the age of the mice. With suckling mice, almost all die in 2–3 days with no paralytic symptoms having been observed. With increasing age of the mice, the time to onset of paralysis increases and the mortality decreases. Almost all mice inoculated intracerebrally or intranasally develop paralysis.

In weanling (or older) mice infected intracerebrally after an incubation period ranging from 7 to over 30 days, the first sign is a weakness in one of the limbs, often a forelimb. This weakness progresses rapidly to paralysis, often first of a forelimb and later of both hindlimbs. At times, however, the initial paralysis is observed in the hindlimb. The extent of paralysis is usually much more marked in the hind- than in the forelimbs, frequently progressing to the extent that the animal is left entirely without the use of its hindlimbs. In these cases, mobility is then possible only by the use of the forelimbs. In some mice the progress of the disease is arrested spontaneously at this stage and the animal gradually recovers from the paralysis; in most instances, paralysis, if continued, leads to death. Complete recovery is seldom observed, however, and then only in animals which show only a mild degree of paralysis. Mice that live after severe paralysis show emaciation of the hindlegs and other deformities of their extremities. Throughout the entire course of the disease, the mice appear normal except for the obvious paralysis. In those markedly paralyzed in the hindquarters, a constant dribbling of urine is often observed. The tail seems never to become paralyzed.

The GD VII and FA strains of virus (Theiler and Gard, 1940a) differed from the TO in having a shorter incubation period (after repeated passage, as short as 2–3 days). GD VII-infected mice (intracerebral inoculation) died within 24–48 hr following the onset of symptoms. Hyperexcitability was a frequent sign. Infected mice, as a rule, looked perfectly well until shortly before death. With the FA strains, extreme hyperexcitability was common. Mice sat huddled up with fur ruffled and looked obviously sick. Convulsions involving fore- and hindlimb spasticity were common.

Circulating virus was not detected in early studies (Theiler, 1937). However, Daniels *et al.* (1952) demonstrated virus circulation early in the course of a DA (~ TO) infection.

Virus is detectable in the spinal cord and brain of paralyzed mice (Theiler, 1937). Following the poliomyelitic type of disease, it is present for at least a year, particularly in spinal cord tissue. Lipton and Dal Canto (1979a) have demonstrated the presence of virus late in the course of infection, particularly in the spinal cord tissue, in mice with spinal cord lesions of the late demyelinating type. It is possible that both the above reports refer to the same phenomenon. Daniels *et al.* (1952)

Figs. 2–5. Epon-embedded thoracic spinal cord sections (1 μm thick) stained with toluidine blue from inbred mice sacrificed 3 months after intracerebral inoculation with one LD_{50} of brain-derived DA virus.

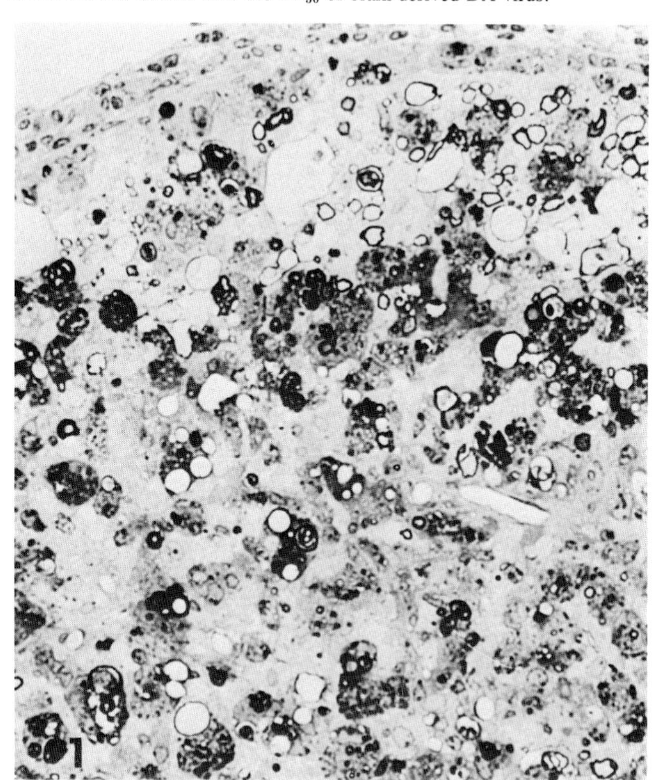

Fig. 2. Anterior column from an SJL mouse, showing mononuclear cells in the leptomeninges and an extensive area of demyelination in the underlying white matter. Numerous naked axons and debris-laden macrophages are present. ×260.

were the first to describe a demyelinating lesion in infected mice, these having been multiply inoculated with material containing mouse nervous tissue. Dal Canto and Lipton (1975) describe demyelination in mice which received a single infecting inoculum, visible by light microscopy and well demonstrated by electron microscopy, particularly in the regions showing perivascular infiltration. Figs. 2, 3, 4, and 5 [from Lipton and Dal Canto (1979b)] illustrate the demyelinating process in the anterior and lateral columns of the spinal cords of inbred mice. Theiler mentions flaccid paralysis of the extremities (see Fig. 6) but does not mention the late spastic paralysis described by Lipton (1975). Although Theiler describes perivascular infiltrating lesions in the cord in the late disease, he does not note a demyelinating process.

After intracerebral inoculation, TO virus is only occasionally recovered early and in low titer from tissues other than nervous tissue, including the lymphatic glands and liver and not at all from the lungs, spleen, kidney, and adrenals (Iguchi, 1939).

Virus is found frequently in low titer in the intestinal mucosa and fecal contents of mice in infected colonies, demonstrable at the age of 20 days or older (Olitsky, 1940). After intracerebral inoculation, virus is readily found in intestinal mucosa and feces, appearing when the titer in the brain reaches its maximum (Theiler and Gard, 1940a). It may be surmised that

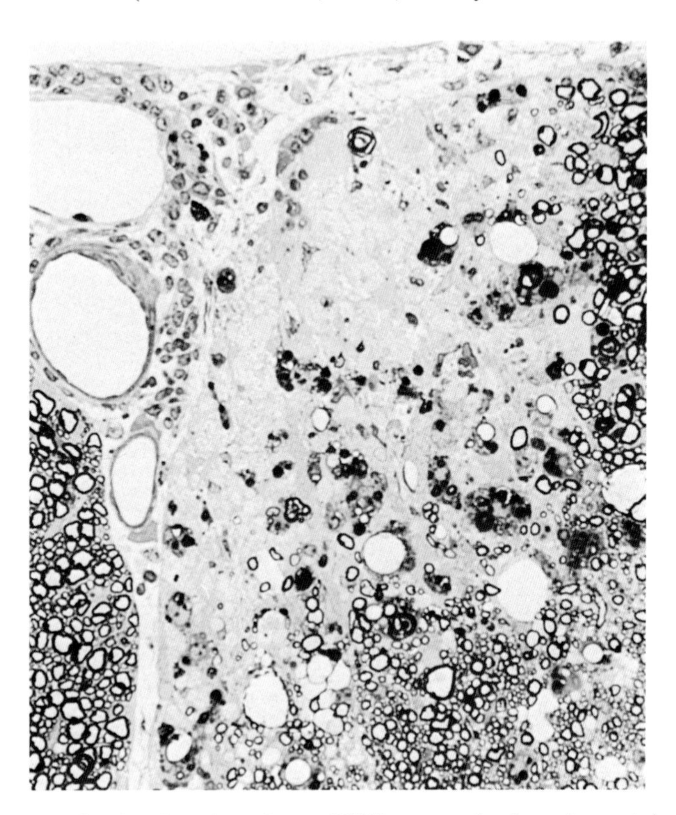

Fig. 3. Anterior column from a C3H/He mouse, showing a demarcated plaque of demyelination in the white matter. Leptomeningeal inflammation, naked axons, and macrophages are present. ×300.

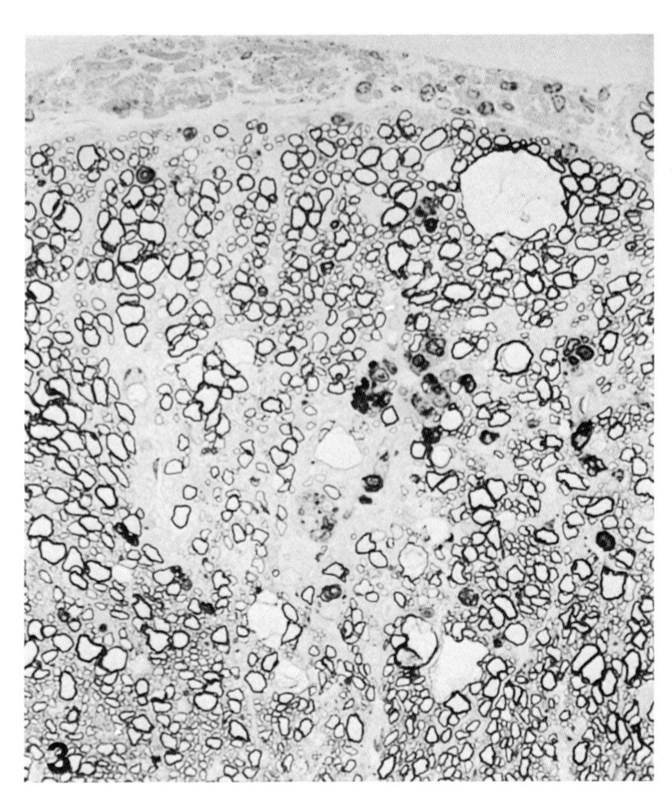

Fig. 4. Lateral column from a CBA mouse, showing some mononuclear cells in the leptomeninges and a small area of demyelination. ×255.

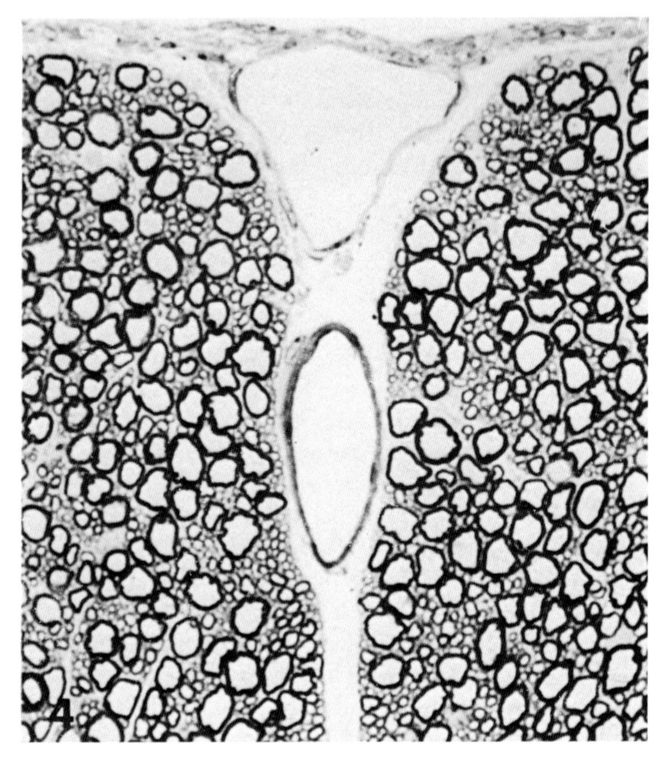

Fig. 5. Anterior columns of a C57BL/6 mouse, showing no abnormality. ×370.

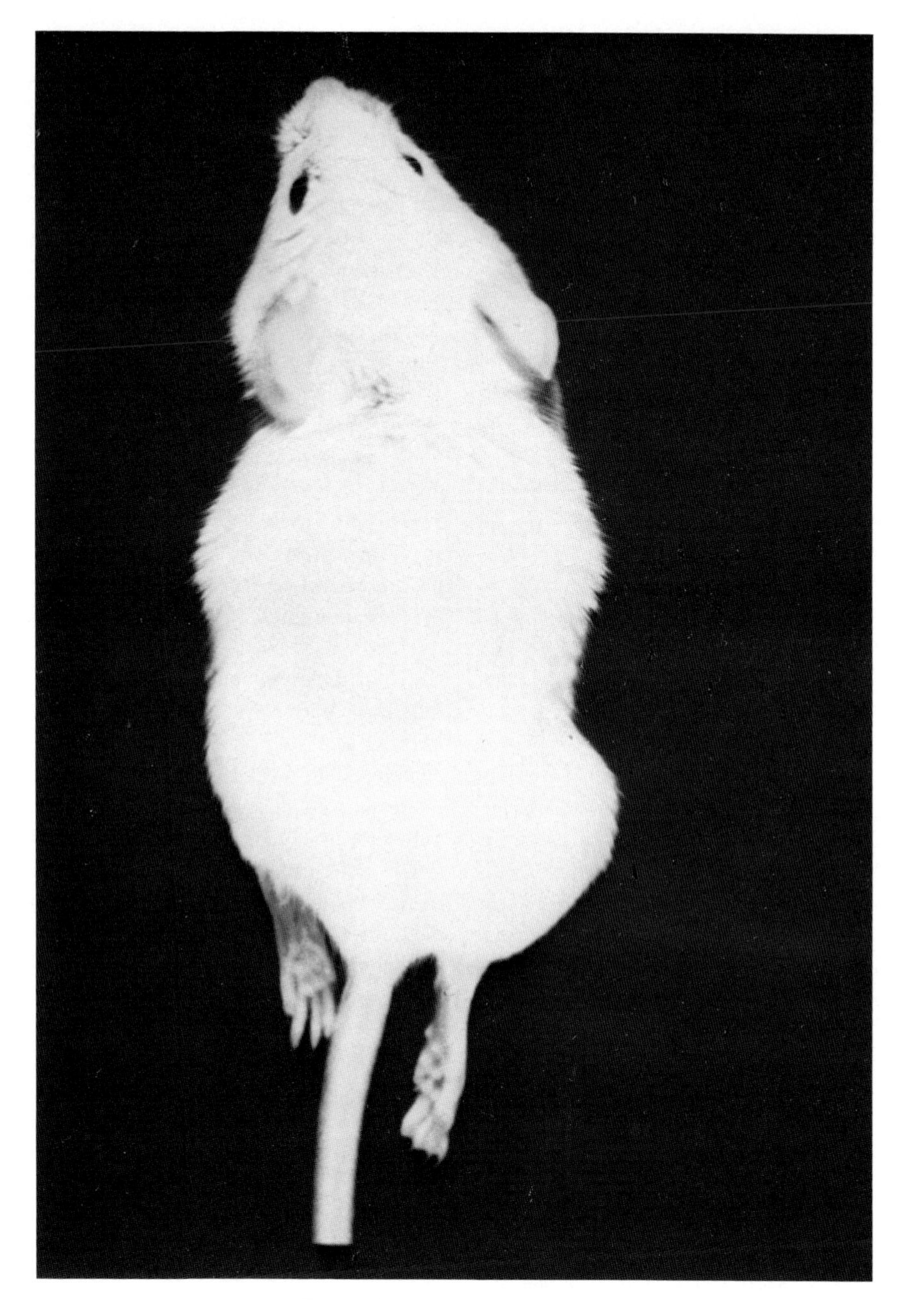

Fig. 6. Paralyzed white mouse.

in mice which recover, intestinal infection could persist, making such mice potential sources of infection for clean mice if permitted to mix with them. Direct observations in support of this conjecture are lacking, which is surprising considering the importance of this matter for animal colony managers.

There are three different clinical responses to TMEV infection, which are largely dependent on the isolate used, the route of inoculation, and the strain of mouse used. (The asymptomatic response is proportionally by far the most common one to natural infection.) (1) Encephalitic (following intracerebral or intranasal inoculation); (2) poliomyelitis-like (in naturally occurring disease and following inoculation); (3) chronic demyelinating disease (following inoculation). Further discussion of this topic can be found in the next section.

The pathogenesis of the TMEV infection in young adult mice following intracerebral inoculation of brain-derived DA virus is shown in Fig. 7. Temporal relationships of disease/pathology, virus titer in CNS tissue, antibody titer, and cell-mediated immunity are shown. Note (Lipton and Dal Canto, 1979a,b) that the poliomyelitic stage is bypassed when cell

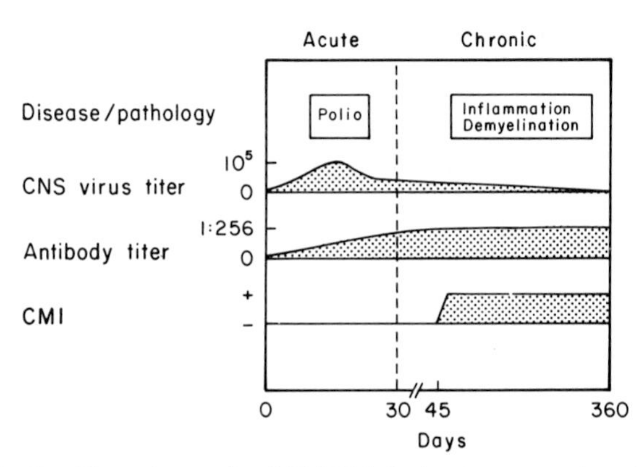

Fig. 7. The pathogenesis of TMEV infection in young mice after intracerebral inoculation of brain-derived DA (~ TO) virus. The figure shows the temporal relationships of virus replication and host immunity in the acute (early disease) and chronic (late disease) phases of the infection. CMI, cell-mediated immunity. (After Lipton, 1978a.)

culture-derived virus is used as inoculum. The relationship of mouse age and immune status of the mouse to challenge with a TO strain of virus is shown in Fig. 1.

2. Pathology

No abnormal pathological changes can be seen macroscopically. In the encephalitic type of disease, microscopically there is a perivascular round cell infiltration throughout the CNS, which is particularly marked in the spinal cord. When FA virus is inoculated either intracerebrally or intranasally, the cardinal symptoms and pathological changes produced are referable to the brain and not the cord.

In the poliomyelitis-like type of disease, an acute necrosis of ganglion cells, especially marked in the anterior horn cell, is a striking feature. Such necrotic changes can be observed before the onset of paralysis. Necrosis is soon followed by neuronophagia. The cells of the posterior root ganglia appear to remain unaffected. This type of disease is seen in spontaneous infections and after intraperitoneal inoculation, when the disease feature simulates that seen in mice with a spontaneous infection of the CNS. That is, apart from a flaccid paralysis, usually of the hindlimb, the mice appear well, with an almost complete absence of cerebral symptoms (Theiler and Gard, 1940a).

Mice paralyzed for several months have a decreased number of anterior horn cells and perivascular infiltration is still present, although less marked. No intranuclear inclusions in neuronal cells have been reported. Virus can still be demonstrated in the late paralytic stage in the spinal cord (Theiler and Gard, 1940a).

In the demyelinating disease, an intense mononuclear inflammatory cell lesion evolves in the spinal cord leptomeninges and white matter in mice which survive paralysis following an induced DA (~ TO) infection. Patchy demyelination is seen in areas of inflammation. The only clinical sign of this late phase of a biphasic infection is a mild gait disturbance in some mice (Lipton, 1975). Virus can be demonstrated by immunofluorescence in the cytoplasm of spinal cord neurons in the early paralytic phase, but less often in cells of the dorsal horn and only occasionally in cells of the neocortex. After day 30, virus antigen can be seen only in isolated cells in the white matter following a search through serial sections of the cord. In the areas of demyelination, lipid-laden macrophages can be found.

Ultrastructural studies (Dal Canto and Lipton, 1975) show extensive spinal cord lesions consisting of leptomeningeal and white matter mononuclear cell infiltrates with concomitant primary demyelination by 15 days after intracerebral inoculation of DA virus. The patterns of myelin breakdown consist of stripping of myelin lamellae by invading mononuclear cell processes and vesicular disruption of the myelin. Oligodendrocytes in the region of demyelinating lesions show no degenerative changes. No viral inclusions can be found in any cells of the CNS. Demyelinated axons can be detected by 21 days after infection, and active demyelination can be observed a year after infection. The SJL/J inbred strain of mice show more changes than Swiss mice. The morphogenesis of the white matter lesions in TMEV infection is noted to be quite similar to that seen in experimental allergic encephalomyelitis.

The demyelinating process can be halted by treatment of mice with cytoxan or antilymphocyte serum, suggesting that the demyelinating process induced by the TO strains is immune mediated.

A comparison of large-plaque variants (GD VII) and small-plaque variants (TO) with respect to virulence following intracerebral inoculation of 4- to 6-week-old outbred mice showed the large-plaque variants to be at least 1000-fold more virulent and indicated that infection with such variants did not lead to long persistence of virus in surviving animals, as is seen following small-plaque (TO) infection (Lipton, 1978b).

Adaptation of TMEV strains to cell culture and subsequent inoculation into mice revealed that with the adapted strains the earlier clinical phase of poliomyelitis was bypassed and the first observed clinical response was associated with demyelinating disease. The similarity of this progression to that seen in a slow virus infection has been noted by Lipton and Dal Canto (1979a).

3. Transmission

It was early shown (Theiler, 1937) that normal mice in contact with obviously infected animals, either those found paralytic or those artificially infected by intranasal or intracerebral inoculation, do not acquire an infection of the CNS or develop a greater degree of immunity than control mice of the same strain and age kept under the same conditions but not exposed to mice having an infection of the CNS.

After demonstration of virus in the intestinal mucosa and feces of a high proportion of mice held under open-colony conditions (Olitsky, 1940), it was determined that young mice acquire intestinal infection shortly after weaning and that almost all are infected by 30 days of age. Virus can be recovered only irregularly in mice over 6 months of age, and mice develop an increasing immunity with age.

Recovery of TMEVs, usually of the TO type, has been described from mice not inoculated with any materials (Theiler, 1937; Olitsky, 1940, 1945; Gildemeister and Ahlfeld, 1938; Gahagan and Stevenson, 1941; Melnick and Riordan, 1947; Thompson *et al.*, 1951). The GD VII and FA strains have usually come from mice inoculated with unrelated viruses (Theiler and Gard, 1940a; Thompson *et al.*, 1951). The TMEVs are present in many open mouse colonies (Parker *et al.*, 1966) and are considered the most ubiquitous mouse colony viruses.

The viruses cause no paralysis or disease in rabbits, guinea pigs, puppies (Gildemeister and Ahlfeld, 1938), rhesus monkeys, and hamsters (Melnick and Riordan, 1947). Paralysis has been produced in the cotton rat (Melnick and Riordan, 1947). The GD VII, MHG, and FA strains kill suckling rats upon intracerebral and intraperitoneal inoculation; the TO strain does not. None of the strains kill weanling rats by any route.

4. Immune Response

It was early noted that paralytic mice are immune to a subsequent intracerebral infection of virus (Theiler, 1937). It was also shown (Theiler and Gard, 1940a) that at such a time, virus was still present in the CNS. Also, a large proportion of mice which remained well after an intracerebral injection of virus were immune to a subsequent intracerebral inoculation of the virus.

Gard (1940) developed a technique for estimating virus titers of FA virus based on the (harmonic) mean of the reciprocals of the length of the incubation period in intracerebrally inoculated mice. This technique permitted calculation of virus titers replicable within an acceptable limit of variation. He also demonstrated an inhibitory substance in infected mouse brain. Virus relationships between TMEV strains were outlined principally by mouse neutralization tests and by intracerebral challenge of mice infected and recovered from the infection. When challenged by homologous virus, they showed solid protection; when challenged by heterologous TMEVs, they showed high-grade protection as well.

Not until virus-free mouse colonies were developed (von Magnus and von Magnus 1948; Dean, 1951) could further nuances of the immune process be explored satisfactorily.

An intraperitoneal inoculation route in baby mice from a virus-free colony (von Magnus, 1951) provides clear cut *in vivo* neutralization test results. Cell culture systems (Lipton and Dal Canto, 1979b) have almost entirely supplanted *in vivo* systems in virus growth and neutralization studies.

Fig. 7 illustrates the response of uninfected and infected mice to a TO virus strain challenge, showing the gradual development of immunity to inoculation by routes other than intracerebral with increase in the age of mice.

In connection with studies on the demyelinating process in the CNS of chronically infected mice (Lipton, 1975), Rabinowitz and Lipton (1976) showed that immunologically specific spleen cell reactivity develops, as judged by the *in vitro* incorporation of [^{3}H]TdR, into DNA in response to inactivated TMEV antigen. Spleen cell reactivity correlates with the temporal development of serum-neutralizing antibody. Antiviral antibody inhibits virus-induced spleen cell reactivity. The antigen-reactive lymphocyte subpopulation within the spleen responsible for the proliferation to TMEV antigen consists of T and not B cells.

D. Epizootiology

1. Host Range

a. Species. The laboratory mouse is the only animal involved in the endemic epizootic cycle. A single reported exception was an epidemic in adult laboratory rats (McConnell *et al.*, 1964), which was exceptional in that the MHG strain of virus isolated is not pathogenic for adult rats inoculated intracerebrally (Hemelt *et al.*, 1974). All TMEV strains are pathogenic for infant rats inoculated intracerebrally.

b. Strain of Mice. There are no reports of strain differences in mice with respect to susceptibility to infection under natural colony conditions or following inoculation. However, the SJL/J strain of mouse has been shown to be more apt to develop severe late disease than are several other mouse strains, with a disproportionately large number of macrophages in demyelinating lesions compared with similar lesions in CBA and C3H/He mice (Dal Canto and Lipton, 1975; Lipton and Dal Canto, 1979a,b). There are no reports of TMEVs in native *M. musculus*.

2. Prevalence and Distribution

TMEVs are considered to be the most prevalent and ubiquitous of the mouse pathogens. Many stock lines are infected even though the clinical disease has been reported only rarely. Infection may be presumed to be worldwide even though the number of countries in which the disease has been reported is not large (see Section II,C,3).

3. Seasonal Periodicity

There are no data on seasonal periodicity.

E. Diagnosis

1. Serology

The first line in diagnosis is the recognition of a paralyzed mouse among colony animals, an event which, even in nonbarrier (infected) colonies occurs no more frequently than once in 10,000 animals.

The presence of disease in infected colonies or breeding stocks is usually made by hemagglutination inhibition testing of serum specimens from individual mice or of serum pools of several mice. Inhibition at a 1:64 serum dilution is enough to arouse suspicion. Higher levels are an assurance of virus endemicity. Other tests such as complement fixation, virus neutralization in a cell culture system, radioimmunoassay can be used. If the exercise is of sufficient importance, confirmation of virus endemicity can be made by virus isolation. Many users of laboratory mice in the United States insist on TMEV-free mouse strains from commercial breeders.

2. Virus Isolation

Virus isolation can be an unwelcome accompaniment of experiments involving intracerebral inoculation of mice with other viruses. Many of the recorded TMEV isolations have been made in this fashion. Diagnosis of a TMEV infection can often be made by observing the clinical manifestations in the originally inoculated mouse or mice or in the passage mice (see Section II,C,1).

In searching for virus in a suspected endemically infected breeding stock, fecal material from one or more mice can be made into a suspension with antibiotics added, centrifuged at 10,000 rpm for 10 min, and the supernatant inoculated intracerebrally into mice or into a cell culture system.

Virus will usually not be encountered in brain material from colony mice unless there is the presence of, or history of, paralysis in the individual mouse assayed.

Brains of mice which become paralyzed following inoculation of test material can be assayed for hemagglutinin by making a 10% suspension, centrifuging at 10,000 rpm for 10 min, and setting up the supernatant in twofold dilutions against a 2.5% suspension of human O red blood cells. The specificity of the reaction can be checked with appropriate immune sera.

F. Control and Prevention

Disease-free stocks were originally developed by removing infant mice at birth from the mother and putting them to suckle on a lactating laboratory rat which had had its litter removed. There may have been a considerable loss of infant mice, but perseverance led to success (von Magnus and von Magnus, 1948). Dean (1951) established a disease-free colony by selecting a mouse from a mother which had had an infection 8 months before and was determined to be virus free (in feces). The strain was continued by brother–sister matings of the litter.

With modern laboratory animal management, one would merely need to use barrier colony mice to establish a TMEV-free colony from gnotobiotic breeders.

Unless a disease-free stock is shielded from infected stocks, reinfection of the stock can occur (von Magnus and von Magnus, 1948).

When a disease-free stock is supplying mice for experimental work under open conditions, the possibility of introduction of virus from contaminated inoculum into disease-free mice must be considered, and steps taken to protect the incoming mice in isolation rooms, until they are ready to be used in the (potentially) contaminated environment. When mice under open conditions are found to be infected, the recommended policy is to destroy them, thoroughly clean the room, and start again with reintroduction of clean mice.

REFERENCES

Calisher, C. H. (1964). Replication of GDVII virus in cell culture. *Fed. Proc., Fed. Am. Soc. Exp. Biol.* **23,** 400.

Calisher, C. H., and Rowe, W. P. (1966). Mouse hepatitis, REO-3 and Theiler virus. *Natl. Cancer Inst. Monogr.* No. 20, 67–75.

Dal Canto, M. C., and Lipton, H. L. (1975). Primary demyelination in Theiler's virus infection: An ultrastructural study. *Lab. Invest.* **33,** 626–637.

Daniels, J. B., Pappenheimer, A., and Richardson, S. (1952). Observation on encephalomyelitis of mice (DA strain). *J. Exp. Med.* **96,** 517–530.

Dean, D. J. (1951). Mouse encephalomyelitis: Immunologic studies of a non-infected colony. *J. Immunol.* **66,** 347–359.

Falke, D. (1957). Über die zuchtung des Theiler-TO-Virus in der Gewebekultur. *Z. Hyg. Infektionskr.* **143,** 645–655.

Fastier, L. B. (1950). Studies on hemagglutination with the GDVII strain of murine encephalomyelitis. *J. Immunol.* **65,** 323–330.

Fastier, L. B. (1951a). The inhibition of GDVII virus hemagglutination by normal tissue extracts. *J. Immunol.* **66,** 87–97.

Fastier, L. B. (1951b). Factors involved in hemagglutination by the GDVII strain of murine encephalomyelitis virus. *J. Immunol.* **66,** 365–378.

Franklin, R., Wecker, E., and Henry, C. (1959). Some properties of an infectious ribonculeic acid from mouse encephalomyelitis virus. *Virology* **7,** 220–235.

Gahagan, L., and Stevenson, L. D. (1941). A strain of virus producing meningoencephalomyelitis in mice, with special reference to pathogenesis. *J. Infect. Dis.* **69,** 232–237.

Gard, S. (1940). Encephalomyelitis of mice. II. A Method for the measurement of virus activity. *J. Exp. Med.* **72,** 69–77.

Gard, S. (1943). Purification of poliomyelitis viruses. *Acta Med. Scand., Suppl.* No. 143, 1–173.

Gard, S. (1944). Tissue immunity in mouse poliomyelitis. *Acta Med. Scand.* **119,** 27–46.

Gildemeister, E., and Ahlfeld, L. (1938). Über eine bei der weissen maus spontan aufgetretene meningo-enzephalomyelitis. *Zentralbl. Bakteriol., Parasitenkd. Infektionskr., Abt. 1: Orig.* **142,** 144–148.

Hemelt, I. E., Huxsoll, D. L., and Warner, A. J., Jr. (1974). Comparison of MHG virus with mouse encephalomyelitis viruses. *Lab. Anim. Sci.* **24,** 523–529.

Iguchi, M. (1939). On the spontaneous encephalomyelitis of mice and its virus. *Kitasato Arch. Exp. Med.* **16,** 56–79.

Lahelle, O., and Horsfall, F. L., Jr. (1949). Hemagglutination with the GDVII strain of mouse encephalomyelitis virus. *Proc. Soc. Exp. Biol. Med.* **71,** 713–718.

Lahelle, O., and Ward, T. J. (1951). Purification and concentration of mouse encephalomyelitis virus by hemagglutination. *J. Immunol.* **67,** 75–81.

Lipton, L. (1975). Theiler's virus infection in mice: An unusual biphasic disease process leading to demyelination. *Infect. Immun.* **11,** 1147–1155.

Lipton, H. L. (1978a). The relationship of Theiler's mouse encephalomyelitis virus plaque size with persistent infection. *In* "Persistent Viruses" (J. G. Stevens, G. J. Todaro, and C. F. Fox, eds.), pp. 679–689. Academic Press, New York.

Lipton, H. L. (1978b). Characterization of the TO strains of Theiler's mouse encephalomyelitis virus. *Infect. Immun.* **20,** 869–872.

Lipton, H. L., and Dal Canto, M. C. (1979a). The TO strains of Theiler's viruses cause "Slow-Virus-Like" infections in mice. *Ann. Neurol.* **6,** 25–28.

Lipton, H. L., and Dal Canto, M. C. (1979b). Susceptibility of inbred mice to chronic central nervous system infection by Theiler's murine encephalomyelitis virus. *Infect. Immun.* **26,** 369–374.

Lipton, H. L., and Friedmann, A. (1980). Purification of Theiler's murine encephalomyelitis virus and analysis of the structural virion polypeptides. Correlation of the polypeptide profile with virulence. *J. Virol.* **33,** 1165–1172.

McConnell, S. J., Huxsall, D. L., Garner, F. M., Spertzel, R. O., Warner, A. R., Jr., and Yager, R. H. (1964). Isolation and characterization of a neurotropic agent (MHG virus) from adult rats. *Proc. Soc. Exp. Biol. Med.* **115,** 362–367.

Mandel, B., and Racker, E. (1953a). Inhibition of Theiler's encephalomyelitis virus (GDVII strain) of mice by an intestinal mucopolysaccharide. I. Biological properties and mechanism of action. *J. Exp. Med.* **98,** 399–415.

Mandel, B., and Racker, E. (1953b). Inhibition of Theiler's encephalomyelitis virus (GDVII strain) of mice by an intestinal mucopolysaccharide. II. Purification and properties of the mucopolysaccharide. *J. Exp. Med.* **98,** 417–426.

Mandel, B., and Racker, E. (1957). Inhibition of Theiler's encephalomyelitis virus (GDVII strain) of mice by an intestinal mucopolysaccharide. III. Studies on factors that influence the virus-inhibitor reaction. *Virology* **3,** 444–463.

Melnick, J. L., and Riordan, J. T. (1947). Latent mouse encephalomyelitis. *J. Immunol.* **57,** 331–342.

Morris, M. C. (1952). The effect of trypsin on hemagglutination by mouse encephalomyelitis virus (GDVII). *J. Immunol.* **68,** 97–108.

Morris, M. C. (1953). The relation between the infectivity and the hemagglutinin of murine encephalomyelitis virus (GDVII). *J. Immunol.* **70,** 39–49.

Nakamura, M. (1961). A comparison of the yields of infectious ribonucleic acid from heated and ultraviolet irradiated mouse encephalomyelitis virus (GDVII strain). *J. Immunol.* **87,** 530–535.

Olitsky, P. K. (1940). Further studies of the agent in intestines of normal mice which induces encephalomyelitis. *Proc. Soc. Exp. Biol. Med.* **43,** 296–300.

Olitsky, P. K. (1945). Certain properties of Theiler's virus, especially in relation to its use as a model for poliomyelitis. *Proc. Soc. Exp. Biol. Med.* **58,** 77–81.

Parker, J. C., Tennant, R. W., and Ward, T. G. (1966). Prevalence of viruses in mouse colonies. *Natl. Cancer Inst. Monogr.* No. 20, 25–36.

Pearson, H. E. (1950). Factors affecting the propagation of Theiler's GDVII mouse encephalomyelitis virus in tissue cultures. *J. Immunol.* **64,** 447–454.

Pearson, H. E., and Lagerborg, D. L. (1955). Effects on CNS depressant drugs on mouse encephalomyelitis virus. *J. Bacteriol.* **69,** 193–194.

Pearson, H. E., and Winzler, R. J. (1949). Amino acids, analogues and propagation of Theiler's GDVII virus in mouse brain tissue culture. *Fed. Proc., Fed. Am. Soc. Exp. Biol.* **8,** 409.

Pearson, H. E., and Winzler, R. J. (1950). Amino acid inhibition of Theiler's GDVII virus in mouse brain mince. *Fed. Proc., Fed. Am. Soc. Exp. Biol.* **9,** 389.

Pearson, H. E., Lagerborg, D. L., and Winzler, R. J. (1952). Effects of certain amino acids and related compounds on propagation of mouse encephalomyelitis virus. *Proc. Soc. Exp. Biol. Med.* **79,** 409–411.

Pearson, H. E., Lagerborg, D. L., Winzler, R. J., and Visser, D. W. (1955). Methionine compounds and growth of mouse encephalomyelitis virus in tissue culture. *J. Bacteriol.* **69,** 225.

Pearson, H. E., Lagerborg, D. L., and Visser, D. W. (1956). Chemical inhibitors of Theiler's virus. *Proc. Soc. Exp. Biol. Med.* **93,** 61–63.

Rabinowitz, S. G., and Lipton, H. L. (1976). Cellular immunity in chronic Theiler's virus central nervous system infection. *J. Immunol.* **117,** 357–363.

Rafelson, M. E., Jr., Winzler, R. J., and Pearson, H. E. (1949). Effects of Theiler's GDVII virus on P^{32} uptake by minced one day-old mouse brain. *J. Biol. Chem.* **181,** 583–593.

Rafelson, M. E., Jr., Pearson, H. E., and Winzler, R. J. (1950a). Oxygen consumption and radiophosphate uptake by minced brain from mice of different ages in relation to propagation of mouse encephalomyelitis virus. *Science* **112,** 231–232.

Rafelson, M. E., Jr., Pearson, H. E., and Winzler, R. J. (1950b). The effects of certain amino acids and metabolic antagonists on propagation of Theiler's GDVII virus and P^{32} uptake by minced one-day-old mouse brain. *Arch. Biochem.* **29,** 69–74.

Rafelson, M. E., Jr., Pearson, H. E., and Winzler, R. J. (1951a). *In vitro* inhibition of radiophosphorus uptake and growth of a neurotropic virus by 5-chlorouridine. *Proc. Soc. Exp. Biol. Med.* **76,** 689–692.

Rafelson, M. E., Jr., Winzler, R. J., and Pearson, H. E. (1951b). A virus effect on the uptake of C^{14} from glucose *in vitro* by amino acids in mouse brain. *J. Biol. Chem.* **193,** 205–217.

Shaw, M. (1953). Cultivation of egg-adapted Theiler's mouse encephalomyelitis (TO) virus in chick embryo tissue culture. *Proc. Soc. Exp. Biol. Med.* **82,** 547–550.

Sturman, L. S., and Tamm, I. (1966). Host dependence of GDVII virus: Complete or abortive multiplication in various cell types. *J. Immunol.* **97,** 885–896.

Tamm, I., and Tyrrell, A. J. (1954). Separation of hemagglutination-inhibitors for GDVII and influenza viruses in normal allantoic fluid and human urine. *J. Immunol.* **72,** 424–432.

Theiler, M. (1934). Spontaneous encephalomyelitis of mice—a new virus disease. *Science* **80,** 122.

Theiler, M. (1937). Spontaneous encephalomyelitis of mice, a new virus disease. *J. Exp. Med.* **65,** 705–719.

Theiler, M. (1941). Studies on poliomyelitis. *Medicine (Baltimore)* **20,** 443–460.

Theiler, M. (1957). Action of sodium desoxycholate on arthropod-borne viruses. *Proc. Soc. Exp. Biol. Med.* **96,** 380–382.

Theiler, M., and Gard, S. (1940a). Encephalomyelitis of mice: I. Characteristics and pathogenesis of the virus. *J. Exp. Med.* **72,** 49–67.

Theiler, M., and Gard, S. (1940b). Encephalomyelitis of mice: III. Epidemiology. *J. Exp. Med.* **72,** 79–90.

Thompson, R., Harrison, V. M., and Myers, F. P. (1951). A spontaneous epizootic of mouse encephalomyelitis. *Proc. Soc. Exp. Biol. Med.* **77,** 262–266.

von Magnus, H. (1951). Studies on mouse encephalomyelitis virus (TO strain): IV. Neutralization tests by the intraperitoneal and intracerebral routes. *Acta Pathol. Microbiol. Scand.* **29,** 243–250.

von Magnus, H., and von Magnus, P. (1948). Breeding of a colony of white mice free of encephalomyelitis virus. *Acta Pathol. Microbiol. Scand.* **26,** 175–177.

Chapter 18

Encephalomyocarditis Virus

Thomas G. Murnane

I.	Introduction	353
II.	History	353
III.	Etiology	354
	Classification	354
IV.	Epidemiology	354
	A. Occurrence	354
	B. Reservoir	355
	C. Transmission	355
V.	Laboratory Animal Hosts	355
	A. Diagnostic and Preventive Measures	356
	B. Research Modeling	356
	References	356

I. INTRODUCTION

Encephalomyocarditis (EMC) viruses are ubiquitous and have been isolated from a variety of mammals, birds, and insects. The viruses have been recovered from human patients with encephalitis and meningitis and are clearly capable of naturally infecting and causing fatal illness in swine, nonhuman primates, and some captive wild animals, but they have not been known to cause disease spontaneously or to produce latent infection in laboratory mice. The EMC viruses are, however, highly infectious when experimentally inoculated into the laboratory mouse.

EMC viruses are not regarded as a significant disease threat to laboratory mice. Nevertheless, laboratory animal specialists should be cognizant of the history of the EMC viruses and their potential for accidental infection of experimental mouse colonies. EMC virus is popularly used in biophysical investigations and experimental pathogenicity studies. The virus is periodically recovered in diagnostic and epidemiological investigations as the etiological agent of disease in domestic swine, captive wild animals, or nonhuman primates.

II. HISTORY

The encephalomycarditis group of viruses is composed of several strains of viruses. The EMC, Columbia-SK, MM, and Mengo viruses are the strains most frequently associated with

ISBN 0-12-262502-1

the group. These strains are antigenically similar and, collectively, are referred to as the *EMC viruses,* the *EMC group of viruses,* or the *Columbia-SK group*. The name *cardioviruses* has been applied more recently (Fenner, 1968; Melnick *et al.*, 1975).

The Columbia-SK (Jungeblut and Sanders, 1940) and MM (Jungeblut and Dalldorf, 1943) viruses were intially of interest because of their possible relationship to human paralytic diseases; however, the EMC viruses have not been shown to have any relationship to a number of neurotropic and pantropic viruses, including several strains of poliomyelitis virus (Warren and Smadel, 1946). The EMC group of viruses are immunologically indistinguishable (Dick, 1949; Warren *et al.*, 1949b) and antigenically similar (Craighead, 1965) but not identical. Now it is generally believed that these early virus isolations were of latent viruses in the laboratory rodents (cotten rats, *Sigmodon hispidus*). The original EMC virus was isolated from a chimpanzee dying of an acutely fatal disease in a Florida zoological park. Initially, the isolate was thought to be related to lymphocytic choriomeningitis virus. It was designated *encephalomyocarditis virus* because of its distinctive ability to produce encephalomyelitis and myocarditis in laboratory animals (Helwig and Schmidt, 1945). The Mengo virus was recovered from a captive monkey in Uganda (Dick *et al.*, 1948b).

Members of the EMC group of viruses have been reportedly isolated from human patients with encephalitis and meningitis (Dick *et al.*, 1948a; Gajdusek, 1955). The EMC virus has been implicated as the cause of acutely fatal disease in swine (Acland and Littlejohns, 1975; Acland *et al.*, 1970; Gainer, 1961; Gainer and Murchison, 1961; Gainer *et al.*, 1968; Murnane *et al.*, 1960), nonhuman primates (Dick, 1948; Helwig and Schmidt, 1945; Roca-Garcia and Sanmartin-Barberi, 1957), and some species of captive wild animals (Simpson *et al.*, 1970). The virus has been recovered from a variety of mammals and birds as well as from blood-sucking arthropods in diverse geographical areas (Tesh and Wallace, 1977).

EMC viruses have been increasingly used in biophysical investigations and experimental pathogenicity studies. The EMC viruses are highly infectious for a variety of small laboratory animals and are suitable as models for the investigation of the pathogenesis of myocarditis (Gainer, 1974), vasculitis (Burch and Rayburn, 1977), and virus-induced diabetes mellitus (Craighead and McLane, 1968).

III. ETIOLOGY

Classification

The EMC viruses are strains of an RNA virus of the family *Picornaviradae,* genus *Enterovirus* (Melnick *et al.*, 1975; Wildy, 1971), and species *cardiovirus*. The single-stranded RNA has a molecular weight of 2.6×10^6 and comprises 31% of the virion mass (Wildy, 1971). The isometric particle has a diameter of 30 μm in wet preparations when measured by X-ray diffraction (Faulkner *et al.*, 1961). Virions are not retained by Berkefeld and Seitz filters or by gradocol membranes with average pore diameters of 30 μm (Warren, 1965). The virus is ether resistant, acid stable to pH 3, and stores well at −70°C. EMC virus usually loses its infectivity following lyophilization or desiccation (Warren, 1965).

Members of the EMC group of viruses have been shown to be immunologically indistinguishable by mouse neutralization (Dick, 1949), cross-protection, and complement fixation tests (Warren *et al.*, 1949b). Hemagglutination inhibition tests indicate that the strains are antigenically similar but not identical (Craighead, 1965; Craighead and Shelokov, 1961).

IV. EPIDEMIOLOGY

A. Occurrence

The geographical area of the most frequently reported EMC virus activity is Florida, where the virus has been isolated from eight different animal species, i.e., baboon, chimpanzee, raccoon, cotton rat, squirrel, swine, calves, and African elephants. Swine are apparently the most seriously affected of any animal species. Disease outbreaks have occurred in swine of different age groups in large and small herds on commercial farms in Panama, Florida, and New South Wales, Australia. Serological surveys indicate the presence of EMC virus infection of swine in England (Sanger *et al.*, 1977) and Hawaii (Tesh and Wallace, 1977). The fatal EMC infections of captive monkeys, apes, and elephants in Florida suggest that these animals are highly susceptible to infection from extraneous sources (Helwig and Schmidt, 1945; Kissling *et al.*, 1956; Simpson *et al.*, 1970). In all instances in which EMC virus was isolated from apes or monkeys, the affected animals were either moribund, found dead, or partially paralyzed shortly before death (Helwig and Schmidt, 1945; Kissling *et al.*, 1956; Roca-Garcia and Sanmartin-Barberi, 1957). Interstitial myocarditis and pulmonary edema have been observed in histological sections of tissue from dead monkeys and apes (Kissling *et al.*, 1956; Schmidt, 1948). Fatal infections in four African elephants at each of two widely separated zoological gardens in Florida attest to the infectivity of the EMC group of viruses for a broad range of distinctive mammals from the small rodent to the large African elephant. The disease in African elephants was characterized by a fulminating illness, as has been observed in swine and subhuman primates.

The significance of EMC virus as a disease-producing agent

in domestic animals other than swine must await further investigations. The virus has, however, been recovered in Florida from two dead calves with mild lymphocytic myocarditis (Gainer, 1974). Antibodies to EMC viruses have been reportedly found in cattle and horses in Queensland, Australia (Spadbrown *et al.*, 1970), and horses in Canada (Gainer, 1974). Disease in the horse is reported as an ascending spinalitis.

EMC viruses have reportedly been recovered from natural cases of human illness only in Europe (Gajdusek, 1955). The single human infection with the Mengo strain is believed to have been a laboratory-acquired infection (Dick *et al.*, 1948a). Serological studies and surveys have, however, confirmed one human epidemic in the Philippines (Smadel and Warren, 1947) and have revealed a more widespread distribution of human infections in the United States, Mexico, Central America, Panama, and some countries of South America, Africa and Southern Asia, Hawaii, and other Pacific islands (Craighead *et al.*, 1963b; Gajdusek, 1955; Tesh, 1977). The results of these studies indicate that EMC infection occurs primarily during childhood or adolescence. Whereas there have been continuous reports of disease in animals, reports of human illness have virtually ceased since early 1950. Authors of a recent study believe that EMC infection in man is fairly common, but most human cases are probably asymptomatic and/or unrecognized (Tesh, 1977).

B. Reservoir

Considerable attention has been focused on rats as the primary source of EMC infections without conclusively implicating rats in outbreaks of disease in animals or cases of human infection (Acland and Littlejohns, 1975; Gainer, 1969, 1974; Tesh *et al.*, 1977; Warren, 1965; Warren *et al.*, 1949a). EMC virus has been recovered from other wild rodents, e.g., the raccoon, cotton rat, squirrel, mongoose, and water rat, suggesting other possible sources or links in the transmission chain (Gainer, 1969; Pope, 1959).

C. Transmission

The primary source of EMC infection in outbreaks of animal disease is uncertain. Animals may acquire the infection by consuming feed or water contaminated with EMC virus (Dick, 1948) from rats or other rodents or by consuming diseased rodent carcasses. The virus titer in animals is much higher than the concentration in feces, and animals feeding on carcasses or viscera of other animals dying of EMC infection are more likely to become infected. Several captive lions died shortly after being fed viscera and meat of two African elephants that died of EMC infection at a zoological park in Florida (Simpson *et al.*, 1970). The cause of death in the lions was never confirmed, but circumstantial evidence points to infection with EMC virus. The cause of the elephants' deaths had not been diagnosed at the time of the disposition of their carcasses. Infected swine may perpetuate the infection within a herd through excretion of the virus and contamination of their own premises. An experimental attempt to transmit disease to contact pigs under unhygienic conditions was unsuccessful (Littlejohns and Acland, 1975); however, transmission of disease to swine has been accomplished under experimental conditions of feeding the virus (Craighead *et al.*, 1963a). EMC virus has been recovered from two batches of wild-caught mosquitoes, *Taeniorhynchus* sp., in Africa (Dick, 1948) and has also been isolated from mosquitoes in Brazil and the United States and from ticks in India (Tesh and Wallace, 1977). Attempts to infect experimentally or to transmit disease experimentally via mosquitoes have been unsuccessful. Persistent high levels of viremia in animals should offer an exceptional opportunity for insect vectors to transmit disease. However, epidemiological studies of the natural disease in swine and captive wild animals do not indicate that outbreaks of disease are vector-borne or sustained.

V. LABORATORY ANIMAL HOSTS

Manifestations of disease vary among the laboratory animals which are susceptible to infection with EMC viruses (Acland and Littlejohns, 1975; Dick, 1948; Roca-Garcia and Sanmartin-Barberi, 1957). The virus is readily adapted to various routes of passage. A fatal encephalitis is most often produced in mice of any age following inoculation by parenteral, oral, or nasal routes with EMC virus. Central nervous system disease is usually accompanied by myocarditis. Mice inoculated by any of the usual routes with a dilute suspension of EMC-infected tissue begin to show signs of lethargy, ruffled fur, and flaccid paralysis within 72–96 hr, followed shortly by prostration and death. A rapidly fatal encephalitis follows inoculation with concentrated suspensions, and fulminating neurologic disease may overshadow the occurrence of myocarditis. There is evidence, however, to indicate that the virus on initial isolation may not exhibit the encephalotropism that is observed with laboratory-adapted strains.

The experimental disease in hamsters is similar. EMC virus is sporadically fatal for adult guinea pigs as a consequence of myocarditis. It is uniformly fatal in at least some strains of white rats inoculated during the newborn period (J. E. Craighead, 1978 unpublished observations). EMC virus is usually not pathogenic for rabbits and causes only an inapparent infection. Experimentally infected rhesus monkeys do not manifest any clinical signs of infection despite a high concentration of circulating virus. The owl or night monkey (*Aotus trivirgatus*) is highly susceptible to infection and if inoculated with a low passage of EMC virus develops a fulminating disease which is always fatal (Roca-Garcia and Sanmartin-Barberi, 1957). The *Macaca fascicularis* monkey also sustains a nonfatal infection associated with myocarditis. Marmosets are readily infected, developing a fulminating myocarditis with high titers of virus in the heart or a diffuse acinar pancreatitis and hyperglycemia (J. E. Craighead, 1978 unpublished observations).

EMC virus replicates in embryonated hens' eggs when inoculated by any of the customary routes and causes death of the embryo in 72–96 hr. The virus produces cytopathic effects in embryonic and other tissue cultures of the chick, mouse, monkey, hamster, swine, and cattle (Schmidt, 1948). Plaques form when the virus is cultivated on cell monolayers overlayed with agar. Individual mouse strains of the EMC group of viruses may differ from one another in plaque size. It has been observed that mouse-adapted strains are apparently made up of a homogeneous population of plague-forming particles (Craighead, 1965). Plaque varients of EMC virus (r and r^+) have been isolated which differ in sensitivity to a sulfated polysaccharide inhibitor in agar (Takemoto and Liebhaber, 1961).

A. Diagnostic and Preventive Measure

Naturally occurring overt or latent infections of colonies of laboratory mice with EMC virus have not been reported. If infection with EMC virus is suspected, virus recovery and identification must be made to confirm the presence of the disease agent. The virus is readily recoverable from affected mice. EMC virus is sufficiently antigenic so that its presence in rodents or a mouse colony can be determined through serological procedures. Laboratory mouse colonies are not, however, serologically tested for evidence of EMC virus infection, nor is it necessary, since EMC virus is not known or suspected to be an indigenous virus of laboratory mouse colonies. Prominent pathological signs most commonly associated with mice experimentally infected with EMC virus are myocarditis and encephalitis. EMC virus appears to predominate among viruses which may induce myocarditis in laboratory mice. There are several other classes of viruses (Gainer, 1974), i.e., adenoviruses, reoviruses, and vaccinia, which have been shown naturally or by experimental inoculation to induce myocarditis in laboratory mice. Similarly, there are other groups of viruses which naturally or experimentally may cause encephalitis in mice. The pathological lesions in mice infected with EMC virus are indicative but not pathognomonic for the disease. Prevention of EMC virus infection of a colony of laboratory mice is primarily dependent upon basic vermin control procedures which preclude contact with wild mice or other rodents which might have gained entry into the laboratory animal facilities or contaminated laboratory animal feed, bedding, and supplies.

B. Research Modeling

The EMC viruses are a ubiquitous group of antigenically related picornaviruses and have a known broad range of natural hosts. Experimentally, EMC viruses have been shown to be highly infective by peripheral routes in a wide variety of laboratory animal hosts. The pathogenicity of all strains of EMC viruses can be modified by laboratory manipulations. Variants of the EMC virus have been derived which differ in organ tropism and pathogenicity for adult mice. One variant, designated the *E variant,* causes a rapidly fatal encephalomyelitis; the second variant, designated *M,* produces widespread myocardial damage but few, if any, neurological signs (Craighead, 1966). The latter variant has also been shown to produce lesions in the islets of Langerhans and to induce diabetes mellitus in several strains of adult mice (Craighead and McLane, 1968). Occurrence of disease will vary among strains of mice (Craighead and Higgins, 1974). Biochemically, the disease induced in mice strikingly resembles juvenile-type diabetes in man (Craighead, 1976). EMC virus may also provide a model systemic virus infection which permits investigation of the modification of host resistance to viral infection and the enhancement in susceptibility of the myocardium during pregnancy (Farber, 1970).

Increasing interest in human health research is being focused on the role of viral agents in atherosclerosis and arteriosclerosis. Experimental studies employing EMC virus demonstrated that a virus can invade the heart valves and coronary blood vessels in mice producing valvular disease and vasculitis (Burch and Rayburn, 1977). These findings lend support to the concept that patchy atherosclerosis and arteriosclerosis lesions found in coronary vessels of man may represent healed lesions of previous viral damage.

These examples are illustrative of the value of EMC virus to comparative research of human disease. In one study (Tesh, 1977), sera from diabetic, suspected encephalitis, and myocarditis patients were examined for EMC-neutralizing antibodies. The prevalence of antibodies among these groups was not significantly different from that of the control populations, and no association could be made between EMC infection and these diseases.

ACKNOWLEDGMENT

Portions of this chapter are reproduced with permission from the "Handbook Series on Zoonoses," Section B, Viral Zoonoses, Volume 2. Copyright © 1981 The Chemical Rubber Co., CRC Press, Inc.

REFERENCES

Acland, H. M., and Littlejohns, I. R. (1975). Encephalomyocarditis virus infection of pigs, an outbreak in New South Wales, Australia. *Aust. Vet. J.* **51,** 409–415.

Acland, H. M., Littlejohns, I. R., and Walker, R. I. (1970). Suspected encephalomyocarditis virus infection of pigs. *Aust. Vet. J.* **46,** 348.

Burch, G. E., and Rayburn, P. (1977). EMC viral infection of the coronary blood vessels in newborn mice: viral vasculitis. *Br. J. Exp. Pathol.* **58,** 567-571.

Craighead, J. E. (1965). Some properties of the encephalomyocarditis, Columbia SK and Mengo viruses. *Proc. Soc. Exp. Biol. Med.* **119,** 408-412.

Craighead, J. E. (1966). Pathogenicity of M and E varients of the encephalomyocarditis (EMC) virus. I. Myocardiotropic and neurotropic properties. *Am. J. Pathol.* **48,** 333-345.

Craighead, J. E. (1976). Diabetes mellitus (juvenile and maturity onset types), Model No. 70. *In* "Handbook: Animal Models of Human Disease" (T. C. Jones, D. B. Hackel, and G. Migaki, eds.), Fasc. 5. Registry of Comparative Pathology, Armed Forces Inst. Pathol., Washington, D.C.

Craighead, J. E., and Higgins, D. A. (1974). Genetic influences affecting the occurrence of a diabetes mellitus-like disease in mice infected with encephalomyocarditis virus. *J. Exp. Med.* **139,** 414-426.

Craighead, J. E., and McLane, M. F. (1968). Diabetes mellitus: induction in mice by encephalomyocarditis virus. *Science* **162,** 913-914.

Craighead, J. E., and Shelokov, A. (1961). Encephalomyocarditis virus hemagglutination-inhibition test using antigens prepared in HeLa cell cultures. *Proc. Soc. Exp. Biol. Med.* **108,** 823-826.

Craighead, J. E., Peralta, P. H., Murnane, T. G., and Shelokov, A. (1963a). Oral infection of swine with the encephalomyocarditis virus. *J. Infect. Dis.* **112,** 205-212.

Craighead, J. E., Peralta, P. H., and Shelokov, A. (1963b). Demonstration of encephalomyocarditis virus antibody in human serum from Panama. *Proc. Soc. Exp. Biol. Med.* **114,** 500-503.

Dick, G. W. A. (1948). Mengo encephalomyelitis virus: pathogenicity for animals and physical properties. *Br. J. Exp. Pathol.* **29,** 559-577.

Dick, G. W. A. (1949). The relationship of Mengo encephalomyelitis, encephalomyocarditis, Columbia-SK, and M.M. viruses. *J. Immunol.* **62,** 375-386.

Dick, G. W. A., Best, A. M., Haddow, A. J., and Smithburn, K. E. (1948a). Mengo encephalomyelitis. *Lancet* **ii,** 286-289.

Dick, G. W. A., Smithburn, K. C., and Haddow, A. J. (1948b). Mengo encephalomyelitis virus: isolation and immunological properties. *Br. J. Exp. Pathol.* **29,** 547-588.

Farber, P. A. (1970). Viral myocarditis during pregnancy: encephalomyocarditis virus infection in mice. *Am. Heart J.* **80,** 96-102.

Faulkner, P., Martin, E. M., Sved, S., Valentine, R. C., and Work, T. S. Studies on the protein and nucleic acid metabolism in virus infected mammalian cells. *Biochem. J.* **80,** 597-605.

Fenner, F. (1968). "Biology of Animal Viruses," Vol. 1, pp. 20-21. Academic Press, New York.

Gainer, J. H. (1961). Studies on the natural and experimental infections of animals in Florida with the encephalomyocarditis virus. *Proc. U.S. Livest. Sanit. Assoc.* **65,** 556.

Gainer, J. H. (1969). Encephalomyocarditis virus infections in Florida, 1960-1966. *J. Am. Vet. Med. Assoc.* **151,** 421.

Gainer, J. H. (1974). Viral myocarditis in animals. *Adv. Cardiol.* **13,** 94-105.

Gainer, J. H., and Murchison, T. E. (1961). Encephalomyocarditis virus infection of swine. *Vet. Med.* **56,** 173-175.

Gainer, J. H., Sandefur, J. R., and Bigler, W. J. (1968). High mortality in a Florida swine herd infected with the encephalomyocarditis virus. An accompanying epizootic survey. *Cornell Vet.* **58,** 31-47.

Gajdusek, C. (1955). Encephalomyocarditis infection in childhood. *Pediatrics* **16,** 819-834.

Helwig, F. C., and Schmidt, E. D. H. (1945). A filter-passing agent producing interstitial myocarditis in anthropoid apes and small animals. *Science* **102,** 31-33.

Jungeblut, C. W., and Dalldorf, G. (1943). Epidemiological and experimental observations on the possible significance of rodents in a suburban epidemic of poliomyelitis. *Am. J. Public Health* **33,** 169-172.

Jungeblut, C. W., and Sanders, M. (1940). Studies of a murine strain of poliomyelitis virus in cotton rats and white mice. *J. Exp. Med.* **72,** 407-436.

Kissling, R. E., Vanella, J. M., and Schaeffer, M. (1956). Recent isolations of encephalomyocarditis virus. *Proc. Soc. Exp. Biol. Med.* **91,** 148-150.

Littlejohns, I. R., and Acland, H. M. (1975). Encephalomyocarditis virus infection of pigs. 2. Experimental disease. *Aust. Vet. J.* **51,** 416-422.

Melnick, J. L., Agol, V. I., and Bachrach, H. (1975). Picornaviradae. *Intervirology* **4,** 303-316.

Murnane, T. G., Craighead, J. E., Mondragon, J., and Shelokov, A. (1960). Fatal disease of swine due to encephalomyocarditis virus. *Science* **131,** 498-499.

Pope, J. H. (1959). A virus of the encephalomyocarditis virus group from a water rat, Hydromys chrysogaster, in North Queensland, Australia. *Aust. J. Exp. Biol. Med. Sci.* **37,** 117-124.

Roca-Garcia, M., and Sanmartin-Barberi, C. (1957). The isolation of encephalomyocarditis virus from Aotus monkeys. *Am. J. Trop. Med. Hyg.* **6,** 840-852.

Sanger, D. V., Rowlands, D. J., and Brown, F. (1977). Antibodies in sera from apparently normal pigs. *Vet. Rec.* **100,** 240-241.

Schmidt, E. C. H. (1948). Virus myocarditis. Pathological and experimental studies. *Am. J. Pathol.* **24,** 97-117.

Simpson, C. F., Lewis, A. L., and Gasking, J. M. (1970). Encephalomyocarditis virus infection of captive elephants. *J. Am. Vet. Med. Assoc.* **171,** 902-905.

Smadel, J. E., and Warren, J. (1947). The virus of encephalomyocarditis and its apparent causation of disease in man. *J. Clin. Invest.* **26,** 1197.

Spadbrown, P. B., and Chung, Y. S. (1970). Hemagglutinatrin-inhibition antibodies to encephalomyocarditis virus in Queensland cattle. *Aust. Vet. J.* **46,** 126-128.

Takemoto, K. K., and Liebhaber, H. (1961). Virus polysaccharide interactions. I. Agar polysaccharide determining plaque morphology of EMC virus. *Virology* **14,** 456-462.

Tesh, R. B. (1977). The prevalence of encephalomyocarditis virus neutralizing antibodies among various human populations. *Am. J. Trop. Med. Hyg.* **27,** 144-149.

Tesh, R. B., and Wallace, G. D. (1977). Observations on the natural history of encephalomyocarditis virus. *Am. J. Trop. Med. Hyg.* **27,** 133-143.

Warren, J. (1965). Encephalomyocarditis viruses. *In* "Viral and Rickettsial Infections of Man" (F. L. Horsfall and I. Tamm, eds.), pp. 562-568. Lippincott, Philadelphia, Pennsylvania.

Warren, J., and Smadel, J. E. (1946). Further observations on the virus of encephalomyocarditis. *J. Bacteriol.* **51,** 615-616. (Abstr.)

Warren, J., Russ, S. B., and Jeffries, H. (1949a). Neutralizing antibody against viruses of the encephalomyocarditis group in the sera of wild rats. *Proc. Soc. Exp. Biol. Med.* **71,** 376-378.

Warren, J., Smadel, J. E., and Russ, S. B. (1949b). The family relationship of encephalomyocarditis, Columbia, SK, M.M., and Mengo encephalomyelitis viruses. *J. Immunol.* **62,** 387-398.

Wildy, P. (1971). Classification and nomenclature of viruses. First report of the International Committee on Nomenclature of Viruses. *Monogr. Virol.* **5,** 56, 57, 75.

Chapter 19

Protozoa

Chao-Kuang Hsu

I. Introduction 359
II. Parasites of Parenteral Systems 359
A. Flagellates 359
B. Sporozoa 360
III. Parasites of the Alimentary System 366
A. Flagellates 366
B. Sporozoa 370
C. Amoebae 370
References 370

I. INTRODUCTION

As a result of recent advances in laboratory animal science, many parasitic diseases have been minimized or eradicated from laboratory rodents, particularly those diseases that require arthropods as vectors or intermediate hosts. A number of parasitic diseases, however, are still present in many conventional laboratory mouse colonies.

This chapter will deal with those protozoan diseases that occur spontaneously in laboratory mice. This information is organized according to the major organ systems affected and the taxonomic classification of the protozoan.

A brief description of each parasite is presented. The structure, life cycle, pathogenesis, diagnosis, treatment, prevention, and control are discussed. For further information, readers are referred to other review articles (Flynn, 1973; Frenkel, 1971; Hsu, 1979; Levine, 1974).

II. PARASITES OF PARENTERAL SYSTEMS

A. Flagellates

Trypanosoma musculi

Trypanosoma musculi is a hemoflagellate that occurs in wild mice and house mice and is worldwide in distribution. Although *T. musculi,* a blood parasite, closely resembles *T. lewisi,* the rat trypanosome, in both shape and size, it is not known to infect rats. Its life cycle is similar to that of *T. lewisi.*

ISBN 0-12-262502-1

It is transmitted by the flea, *Nosopsyllus fasciatus*. In the absence of fleas, the parasite should not be present in mouse colonies. Infection due to experimental inoculation of laboratory mice with wild mouse blood or tissues may occur.

Trypanosoma musculi infection is not very pathogenic, but it will induce weight loss and gastrointestinal hemorrhages in mice when they are subjected to stress such as inadequate nutrition and low environmental temperatures (Sheppe and Adams, 1957).

B. Sporozoa

1. *Toxoplasma gondii* (Nicolee and Manceaux, 1908)

a. Description and Life Cycle. *Toxoplasma gondii* is a ubiquitous coccidian parasite with the mouse as its principal intermediate host. It occurs in the intestinal epithelium of the definitive hosts, domestic cats, and wild felids. It also infects a variety of organs of the intermediate host.

The cat can be infected with *T. gondii* via three routes or mechanisms: congenital, carnivorism, and ingestion of contaminated tissues or feces containing tachyzoites, bradyzoites, or sporozoites. The prepatent period varies depending upon the infecting stage. The prepatent period is 20–24 days after ingestion of bradyzoites and 5–10 days after ingestion of tachyzoites (Jones, 1973).

The term *tachyzoite* is used in the recent literature to replace the terms *trophozoite* and *endodyozoite,* which were often used in the past. Tachyzoites are the rapidly dividing forms which proliferate in the cells of the intermediate hosts and in the nonintestinal epithelium of the definite host. *Toxoplasma* tachyzoites are crescent-shaped and measure about 2 × 6 μm, with a pointed anterior end and a blunt posterior end. Host cells containing tachyzoites are called *pseudocysts* because no well-defined, true parasitic membrane surrounds a group of tachyzoites. When pseudocysts are ingested by animals, the tachyzoites enter the host cells either by active penetration of cell walls or by phagocytosis. The host cell then isolates the tachyzoite by forming a parasitophorous vacuole. The tachyzoite multiplies asexually by repeated endodyogeny within the host cell; in endodyogeny two daughter cells are formed within the mother cell, which is later destroyed. A group of many tachyzoites in a parasitophorous vacuole is termed a *clone*.

In contrast to the tachyzoite, the bradyzoite is the form that multiplies slowly within a cyst. Bradyzoites are structurally similar to tachyzoites. However, they are less susceptible to proteolytic enzymes. A cyst is the parasitic form which has a group of bradyzoites confined within a well-defined elastic parasitic membrane. The cyst occurs in the tissues and grows by endodyogeny, is a resting stage of *Toxoplasma* within the host, and is usually subspherical and varies in size. Cysts may contain numerous crescent-shaped bradyzoites and usually persist for a long time in the muscular and neural tissues of the host. The cyst stage appears to be an essential part of the life cycle of *T. gondii*.

The life cycle of *T. gondii* in the cat has been reviewed and described (Dubey, 1977). It involves asexual stages and sexual (gametogony) stages in the feline intestinal epithelial cells as well as the sporogony stage outside the definitive host. The asexual stage in the cat intestine begins with the ingestion of tissue cysts (containing bradyzoites), oocysts (containing sporozoites), or tachyzoites. After ingestion by the cat, the tissue cyst wall is dissolved by the proteolytic enzymes in the stomach and the upper part of the small intestine. The released bradyzoites then actively penetrate the epithelial cells of the small intestine and form numerous *Toxoplasma* merozoites that are confined in a schizont.

Certain merozoites are believed to initiate gametocyte formation, which occurs throughout the small intestine but most commonly in the ileum about 3–15 days after infection. Female gametocytes are subspherical and male gametocytes are ovoid to ellipsoidal. During microgametogenesis, the nucleus of the male gametocyte divides to produce many microgametes, which swim to and fertilize a mature female macrogamete. After penetration, an oocyst wall starts to form around the fertilized macrogamete. When they are mature (Fig. 1), oocysts (10 × 13 μm) are discharged into the intestinal lumen by the rupture of the intestinal epithelial cells.

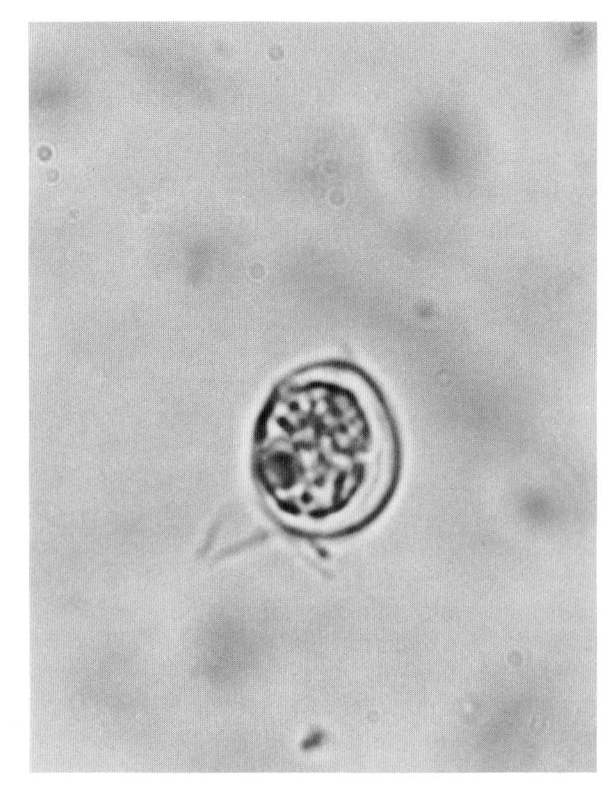

Fig. 1. *Toxoplasma gondii,* unsporulated oocyst, ×2000.

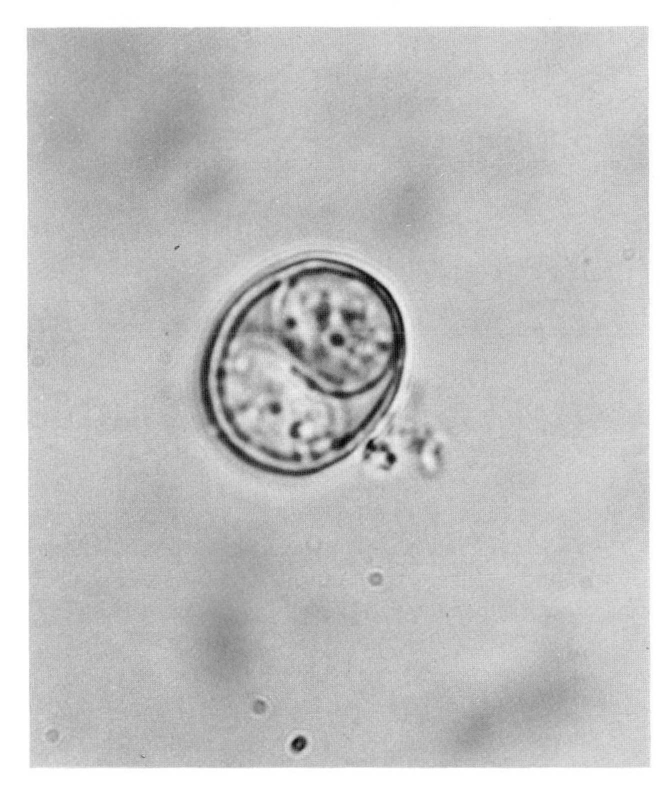

Fig. 2. *Toxoplasma gondii,* sporulated oocyst, ×2500.

Sporulation of the oocyst occurs on the ground. A sporulated oocyst (Fig. 2) contains two sporocysts, which in turn contain four sporozoites each. Oocysts are very resistant to environmental conditions and chemicals. *Toxoplasma* infection may persist in the intestinal and extraintestinal tissues of cats for at least several months.

Vertical infection of *T. gondii* by crossing the placenta occurs in either natural or experimental infections in man, mice, rats, hamsters, guinea pigs, dogs, sheep, cattle, and, rarely, cats.

b. Host. *Toxoplasma gondii* occurs in about 200 species of mammals and birds. The domestic cat and wild felids are the only definitive hosts in which oocysts are produced (Levine, 1977). Intermediate hosts are man, mice, rats, hamsters, guinea pigs, other rodents, dogs, sheep, cattle, nonhuman primates, and zoo animals.

c. Pathobiology. The pathogenicity of *Toxoplasma* infection in animals is dependent upon the number of infecting organisms, the route of infection, and the virulence of *T. gondii* strains. Strains M-7741 and Aldrin are very pathogenic, S-1 is less pathogenic, and BWM is least pathogenic (Dubey, 1977). However, the pathogenicity of all stages of *Toxoplasma* (oocysts, cysts, tachyzoites) may be enhanced by frequent passage in a host. Mice have been commonly used for laboratory testing and maintenance of virulent *Toxoplasma*. Mice of any age are susceptible to *Toxoplasma* infection. Inoculation of *T. gondii* into mice can usually produce either latent infection or acute symptomatic infection depending upon the size of the inoculum and the route of inoculation. Experimental inoculation with oocysts, cysts, or tachyzoites (Fig. 3) intraperitoneally, subcutaneously, or intracerebrally usually will induce acute fulminating infection and death in mice. Animals under stress caused by pregnancy, lactation, immunosuppression, or

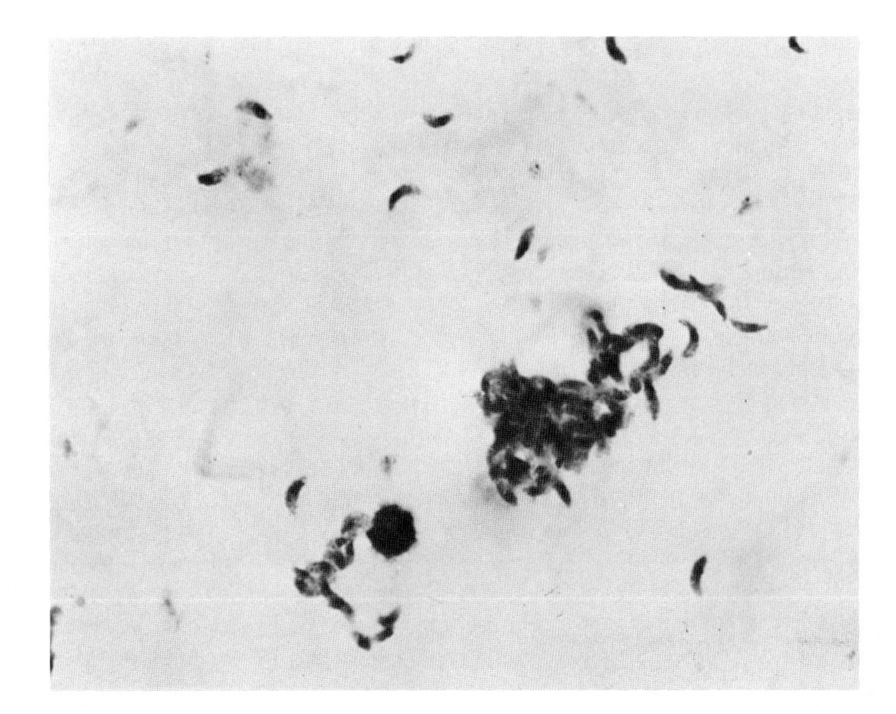

Fig. 3. *Toxoplasma gondii* tachyzoites (trophozoites). Giemsa stain, ×800.

concurrent infection with other organisms are more susceptible than normal animals.

Natural infections of laboratory mice may be acquired by ingestion of food or water that is contaminated with sporulated oocysts from cats. Congenital transmission can occur over several generations in mice (Beverly, 1959). The infection may produce necrosis or lesions in intestinal mucosa, mesentery lymph nodes, eyes, heart, adrenals, spleen, brain, lung, liver, placenta, and muscles. Animals usually recover from natural infection and demonstrate both humoral and cellular immune responses, with cysts being formed in various organs. Cysts remain and will induce inflammation only if ruptured. The acquired immunity is probably lifelong, but it is not a solid immunity, nor can it eradicate infection. Interferon induced by other agents, i.e., viruses or bacteria, enhances host resistance to *T. gondii* infection in mice and in cell cultures (Hsu, 1979).

Research complications induced by *T. gondii* in mice include the long-lasting immunosuppression of T and B lymphocyte function (Huldt *et al.*, 1973; Strickland *et al.*, 1975), nonspecific activation of macrophages (Swartzberg *et al.*, 1975) or enhanced protection of mice against certain unrelated pathogens (Mahmoud *et al.*, 1976; Ruskin and Remington, 1969), and viral (Remington and Merigan, 1969), bacterial (Ruskin and Remington, 1968), or fungal (Gentry and Remington, 1971) infections. *Toxoplasma gondii* infection also alters the development of unrelated leukemia, mammary tumors, or liver tumors in response to a chemical carcinogen (Hibbs *et al.*, 1971; Frenkel and Reddy, 1977).

d. Diagnosis. The diagnosis of *T. gondii* infection is based on immunoserological methods (complement fixation, direct or indirect hemagglutination, Sabin-Feldman dye test, immunofluorescent antibody, enzyme-linked immunosorbent assay or ELISA, toxoplasmin skin test, lymphocyte transformation, etc.), histological examination, and animal inoculation. Detailed methods are described elsewhere (Dubey, 1977).

e. Treatment. Pyrimethamine (30 mg/100 gm in food) plus sulfonamides (i.e., sulfadiazine, 60 mg/100 mg drinking water or 100 gm in food) are two drugs of choice for the chemotherapy of toxoplasmosis (Frenkel, 1971). They act synergistically by blocking the metabolic pathway involving *p*-aminobenzoic acid and the folic-folinic acid cycle. Supplementation with folinic acid and Baker's yeast is necessary in prolonged treatment regimens. These two drugs effectively inhibit the multiplication of *T. gondii,* but they are not effective against established cyst stages.

f. Prevention and Control. The prevalence of *T. gondii* infection in laboratory mice, wild mice, and rats is probably low, as determined by serological surveys (Wallace, 1973). Laboratory mice may acquire infection by the ingestion of food or water contaminated with sporulated oocysts from feline feces. This may happen in a research facility where cats, mice, and other animals are housed in the same room or in close proximity. Furthermore, *Toxoplasma* oocysts are very resistant to adverse temperatures, drying, and chemical disinfectants (Hsu, 1979; Dubey, 1977) and remain infective for other laboratory animals.

2. *Sarocystis muris*

Sarcocystis muris occurs as cysts in the muscles of mice, wild Norway rats, and black rats. It is now extremely rare or nonexistent in surgically derived and barrier-maintained laboratory rodent colonies (Flynn, 1973). This is compared to an expected prevalence of 14% reported in laboratory mice used for cancer research in 1932 (Twort and Twort, 1932). In mice and rats, the elongated cysts of *S. muris* have a thin, smooth wall, are not compartmented, and are several millimeters long. Trophozoites are slightly curved and measure 9–15 × 2.5–3 μm. The life cycle of *S. muris* is similar to that of *Toxoplasma*. Transmission of cats is presumably by ingestion of *Sarcocystis*-infected meat or fecal oocysts. *Sarcocystis muris* is nonpathogenic. Frenkel (1971) has reviewed the biology of this parasite.

3. *Klossiella muris*

a. Description and Life Cycle. *Klossiella muris* is a coccidium that commonly occurs in the kidneys of wild and conventional laboratory mice throughout the world. In a 1932 study, a *K. muris* was found in 90% of the kidneys, 7% of the brains, 15% of the adrenals, 5% of the lungs, 2% of the spleens and 15% of the thyroids examined (Twort and Twort, 1932).

Mice acquire infection by ingestion of sporulated sporocysts, which each contain about 30–35 sporozoites. Sporozoites released from sporocysts enter the blood and are transported throughout the body. From the blood, sporozoites pass into the endothelial cells lining the kidney arterioles and capillaries of the kidney glomeruli, where schizogony takes place. Some sporozoites may enter the capillary endothelial cells of the brain, lung, thyroid, adrenal, spleen, and lymph nodes.

Two types of schizogony have been described. One is the earlier stage that occurs in the endothelial cells of the kidney arterioles, in which each schizont produces 8–12 merozoites. The other type of schizogony is more common and occurs in the endothelial cells of the kidney capillaries and arterioles, with a formation of 40–60 merozoites in each schizont. The mature schizonts are large and eventually rupture into Bowman's capsule, releasing merozoites into the lumen of the renal tubules. Merozoites then enter the epithelial cells lining the convoluted renal tubules, where both gametogony and sporogony take place. Within the tubular epithelial cells, a macrogamete and two microgametes are formed in the same vacuole, and the fertilized macrogamete becomes a sporont. By means of multiple fission, 12–16 sporoblasts are formed. Each sporoblast undergoes further multiple fission and becomes a sporocyst containing 25–35 sporozoites. The sporocysts have a thin wall and measure 16 × 13 μm. The sporocysts eventually rupture the host cell and pass out of the host in the urine. It appears that no oocysts are formed in the life cycle of *K. muris*.

b. Host. The host is mice, species specific.

c. Pathobiology. *Klossiella muris* infection is usually nonpathogenic, causing asymptomatic infections in mice. However, in a heavy infection, significant lesions can be anticipated. Grossly, the heavily infected kidneys are enlarged and have many small, pale gray spots scattered over the cortical surface (Otto, 1957; Yang and Grice, 1964). On the cut surface, the lesions are primarily in the region of the corticomedullary junction. Microscopically, the gray foci are areas of necrosis with marked cellular proliferation. Perivascular and focal lymphocyte and macrophage infiltration is observed in the corticomedullary junction. *Klossiella muris* also causes destruction of tubular epithelium and obstruction of convoluted tubules and may impair the metabolism of mice under stress conditions (Rosenmann and Morrison, 1975).

d. Diagnosis. Diagnosis of *K. muris* infection is based on finding the organisms in the tissues.

e. Prevention and Control. Proper sanitation in the animal room, plus management techniques such as wild mouse control, caesarian rederivation, or housing mice in suspended cages apart from contact with bedding and feces, is necessary to prevent the infection. No treatment is known.

4. *Encephalitozoon cuniculi*

a. Description and Life Cycle. *Encephalitozoon cuniculi* is a microsporidian parasite. It is frequently found in the brain and kidneys of rabbits, mice, rats, guinea pigs, and dogs (Yost, 1958).

The life cycle is direct. Animals are infected by the ingestion of spores (shed in the urine) or by carnivorism. The spores are oval, about 1.5 × 2.5 μm, and are probably ingested by wandering cells in the intestine and enter other organs. *Encephalitozoon cuniculi* proliferates by asexual binary fission (schizogony) and sporogony in peritoneal macrophages and results in the formation of a parasitophorous vacuole. The spores are formed by repeated sporogony and are present in clusters in macrophages, brain tissues, and renal tubular cells. The spores have a capsule which is well stained with Giemsa, gram, and Goodpasture's carbol fuchsin stains but is poorly stained by hematoxylin. A spore contains a nucleus at one end and a polar filament forming 5-6 coils at the other end (Pakes *et al.*, 1975). Spore cells are disseminated to the brain, kidney, liver, spleen, heart muscle, etc. via the bloodstream. *Encephalitozoon cuniculi* infection lasts more than 1 year in the brain and kidney of mice, rats, rabbits, and hamsters. In infected hosts, spores are shed in the urine, which serves as a source of infection. Vertical transmission has also been postulated (Hunt *et al.*, 1972).

b. Host. *Encephalitozoon cuniculi* infections have been reported in mice, rats, hamsters, guinea pigs, *Mastomys*, rabbits, cottontails, cats, dogs, squirrel monkeys, and man. Bywater (1979) has extensively reviewed the available literature and concluded that *E. cuniculi* is not a pathogen of man.

c. Pathobiology. *Encephalitozoon cunniculi* is an obligate intracellur parasite that affects many species of laboratory animals. It usually causes no clinical signs of disease. However, susceptibility, lesions, and tissue distribution are significantly different among mice, rats, rabbits, and monkeys (Shadduck *et al.*, 1979).

Young mice are more susceptible than adult mice. Germfree mice inoculated intraperitoneally show no clinical symptoms (Shadduck *et al.*, 1979). Shadduck *et al.* (1979) observed focal gliosis with lymphocyte and plasma cell accumulation as well as lymphocytic perivascular affinity in the gray matter of the cerebral cortex (Fig. 4). Lymphocytic meningitis and focal necrotizing hepatitis were present in mice with encephalitozoonosis. No lesions were seen in the kidneys. These observations are similar to those found in spontaneous encephalitozoonosis in mice (Innes *et al.*, 1962). In contrast to rabbit encephalitozoonosis, mice do not develop interstitial nephritis. Ascites, hepatosplenomegaly, and hyperplasia of the reticuloendothelial system were observed in mice inoculated intraperitoneally with the parasite. *Encephalitozoon cuniculi* infection causes various biomedical research complications in the areas of cancer research and infectious diseases. For example, mice with inapparent infection had high morbidity or mortality when they received cortisone, irradiation, or 0.1% urethane treatments (Frenkel, 1971).

d. Diagnosis. *Encephalitozoon cuniculi* infection in mice can be diagnosed by: (1) examination of ascites fluid smears (Fig. 4a) for spores in macrophages; (2) histopathological examination of brain tissues; (3) skin test (Pakes *et al.*, 1972); and (4) serological methods, including the immunofluorescence test (Cox and Pye, 1975), immunoperoxidase test (Gannon, 1978), immuno-Indian ink test (Kellett and Bywater,

(a)

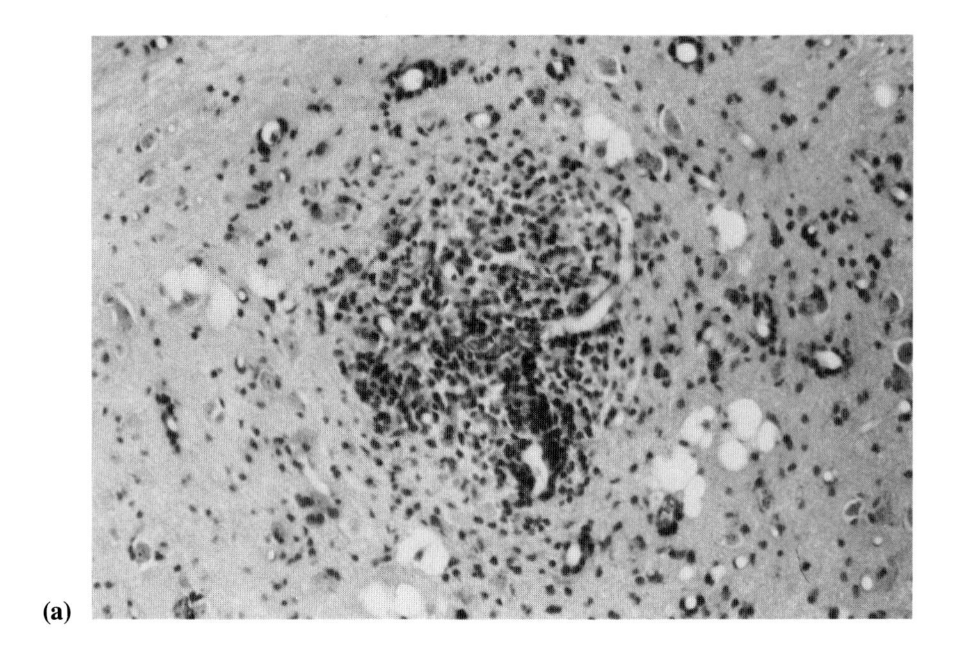

(b)

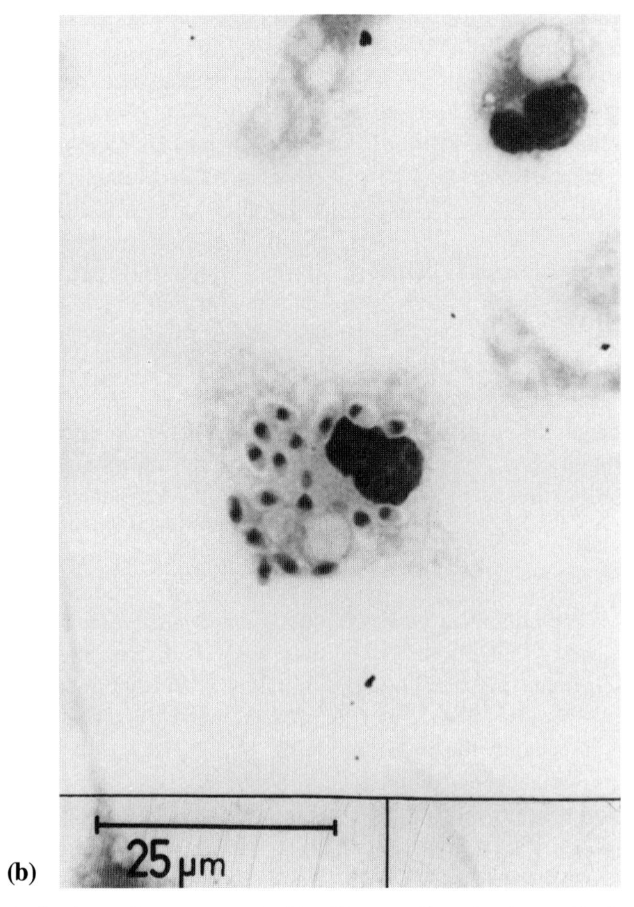

Fig. 4. (a) *Encephalitozoon cuniculi,* focal granulomatous encephalitis. Hematoxylin and eosin, ×150. (b) *Encephalitozoon cuniculi* in a transplantable Yoshida tumor cell. Giemsa stain.

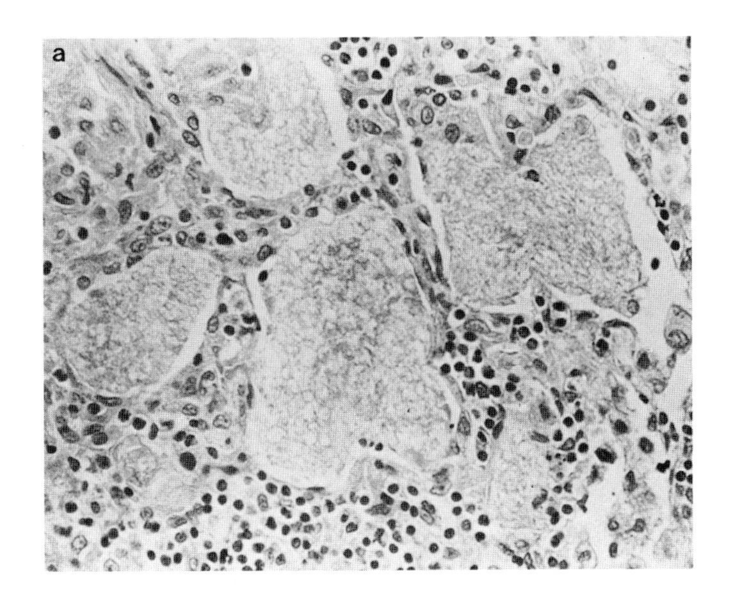

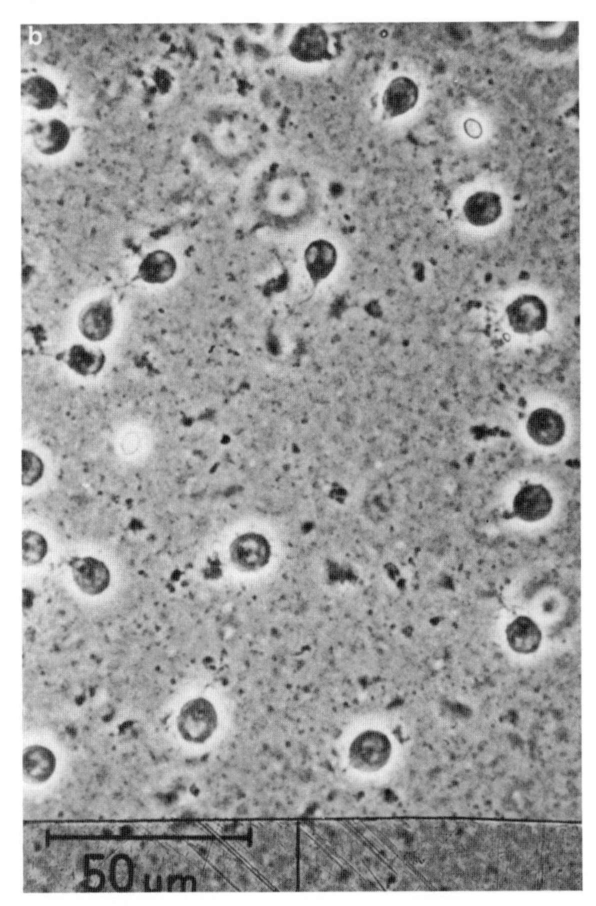

Fig. 5. (a) *Pneumocystis carinii* in lung. Hematoxylin and eosin, ×250. (b) *Giardia muris* trophozoites from a small intestinal wet preparation. Unstained; live trophozoites. (c) *G. muris* trophozoite from the small intestine of a mouse. Giemsa stain. (d) *G. muris* cysts from the feces of a mouse. Unstained, phase contrast. [(a) Courtesy of Dr. J. B. Indsey; (b–d) courtesy of Dr. Ivo Kunstyr.]

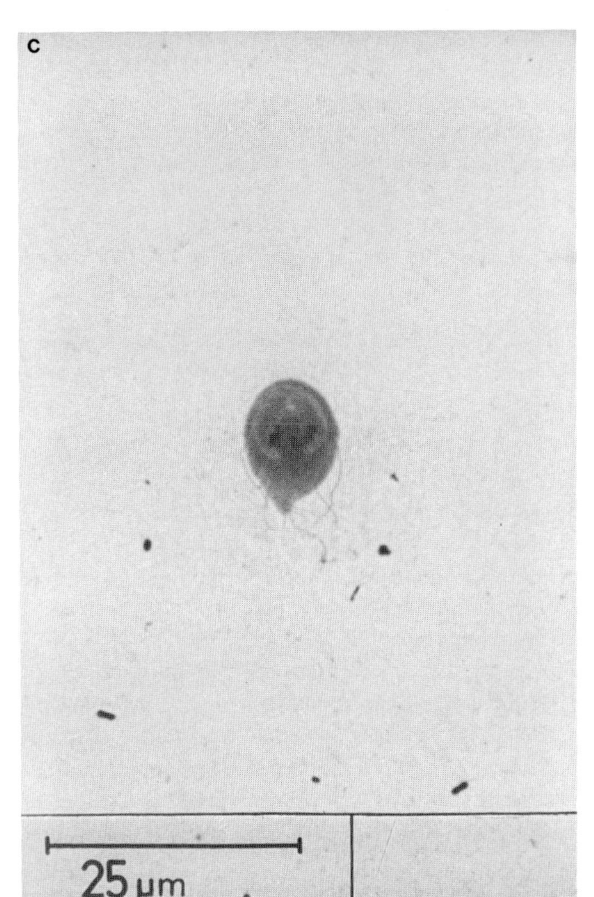

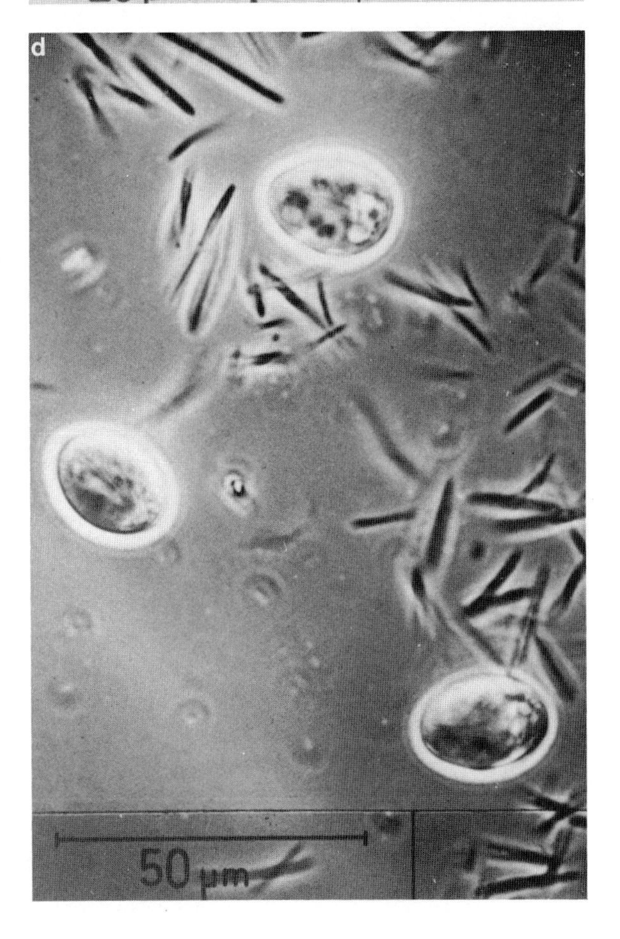

1978), complement fixation test (Wosu, 1975), and indirect microagglutination test (Shadduck and Geroulo, 1979; Shadduck *et al.*, 1979).

e. Treatment, Prevention, and Control. No effective treatment has been reported. The effective prevention and control is rigid testing and elimination of infected colonies and cell lines.

5. *Pneumocystis carinii*

a. Description and Life Cycle. *Pneumocystis carinii* is a ubiquitous, opportunistic microorganism and is present in a wide variety of animal species as latent infections. The organism affects primarily the lung and is found in the foamy, honeycomb materials filling the alveolar space, causing pulmonary insufficiency.

The life cycle of *P. carinii* is simple and occurs within the alveoli of the lung. Four structurally distinguishable entities can be found in the infected lungs. They are trophozoites, precysts, cysts, and intracystic bodies. The structural features of these parasitic forms have been reviewed elsewhere (Hsu, 1979; Seed and Aikawa, 1977).

b. Host. A wide variety of mammalian species serve as natural hosts of *P. carinii* infection, including man, chimpanzees, monkeys, dogs, cats, horses, sheep, goats, pigs, rabbits, hares, guinea pigs, rats, and mice.

c. Pathobiology. *Pneumocystis carinii* occurs widely in many animal species. Normally, it is not pathogenic for the host. Clinical *Pneumocystis* pneumonia usually is associated with patients with neoplasms, immunodeficiencies, immature immune responses, debilitation, or poor nutrition, or with those undergoing immunosuppressive treatment.

Walzer *et al.* (1979) found spontaneous *P. carinii* infection in common strains of mice, including, among others C3H/Hen, BALB/cAnN, C57BL/6N, AKR/J, and Swiss Webster mice that had been treated with corticosteroids, low-protein (8%) diet, or dexamethasone for 8–12 weeks. These mice were from various commercial vendors.

The lung lesions of mouse *Pneumocystis* pneumonia are similar to those in infected rats. The infected lungs are usually enlarged and solid, with a firm rubber consistency. The cut surface is homogeneous and contains multiple irregular gray, brownish, or pink areas of consolidation. Histologically, the pulmonary alveoli are extended and filled with a homogeneous, honeycombed, foamy material (Fig. 5) containing the organisms which stain best with methenamine silver stain. The foamy exudates are thought to be a mixture of disintegrating parasites and antigen–antibody complexes. There is minimal or no cellular response in the alveoli. The alveolar septa are thickened and infiltrated predominantly by plasma cells, with some lymphocytes, histiocytes, and fewer numbers of neutrophils and eosinophils.

Pneumocystis cuniculi grows slowly and extracellularly in animals and man. In the alveolar space, the filopodial surface structures of trophozoites facilitate their attachment to the epithelial cells of the alveolar wall and inhibit their phagocytosis by macrophages. Consequently, the adhering clusters of *Pneumocystis* on the alveolar wall cause alveolar capillary blockage, resulting in pulmonary insufficiency. Immunological factors in the host determine the clinical manifestations of *P. carinii* infection. These factors include IgG, IgA, complement, T lymphocytes, and macrophages. A deficiency in any one of the above factors may result in clinical pneumocystosis. Successful experimental infection (Walzer *et al.*, 1977) and spontaneous pneumocystosis in nude mice (Tamura *et al.*, 1978) have been reported.

d. Diagnosis. The diagnosis of murine pneumocystosis is based on (1) histological examination of lung tissues or lung impression smears stained with methanamine silver stain to reveal cysts or trophozoites, and (2) serological tests, including the complement fixation test and the immunofluorescent antibody test.

e. Treatment, Prevention, and Control. The combined use of sulfadiazine (100–200 mg in 100 gm of food) and pyrimethamine (30 mg in 100 gm of food) is an effective treatment (Hsu, 1979). This regimen results in arrested maturation of granulocytes in the bone marrow. Sulfadiazine inhibits the biosynthesis of folic acid, and pyrimethamine interfers with the conversion of folic acid to folinic acid. This toxicity could be antagonized with a folinic acid supplement.

The control of pneumocystosis is based on surgical rederivation of infected colonies, the maintenance of *Pneumocystis*-free mice behind barriers, and the selection of *Pneumocystis*-free animals as breeders.

III. PARASITES OF THE ALIMENTARY SYSTEM

A. Flagellates

1. *Giardia muris*

a. Description and Life Cycle. *Giardia muris* commonly occurs in the anterior small intestine of mice, rats, hamsters, and various wild rodents. Like *G. lamblia* of man, *G. muris* residers as a trophozoite in the lumen of the upper small intestine. The trophozoites are pear-shaped, bilaterally symmetrical, and approximately 7–13 × 5–10 μm in size (Fig. 5b). The anterior end of trophozoites is broadly rounded with a large sucking disc, and the posterior end is drawn out. There are two anterior nuclei, two blepharoplasts, and four pairs of flagella (Fig. 5c). Trophozoites attach themselves to intestinal epithelial cells by means of concave sucking discs. Trophozoites can be recognized in a wet preparation of small intestinal contents by their characteristic rolling and tumbling movements. The cyst is ellipsoidal (15 × 17 μm), has a thick wall and four nuclei, and is found in the large intestine and feces (Fig. 5d). *Giardia muris* proliferates by binary and multiple fission. Transmission is usually by ingestion of cysts.

b. Host. Mice, rats, hamsters, and other wild rodents are hosts.

c. Pathobiology. *Giardia muris* can infect both young and adult mice and normally exhibits a low degree of pathogenicity. Infected mice usually have a rough hair coat, sluggish movement, and distended abdomen. There is no diarrhea even though at necropsy the small intestines may contain yellow or white watery fluid (Sebesteny, 1969). Athymic (*nu/nu*) and C3H/He mice are more susceptible and have a more prolonged infection with *G. muris* than BALB/c mice (Roberts-Thomas and Mitchell, 1979).

Mice infected with 100–10,000 *G. muris* cysts excrete increasing numbers of cysts in their feces 4–14 days postinfection; thereafter these decline and cannot be detected 32 days after infection (Roberts-Thomas *et al.*, 1976). Trophozoites proliferate in the small intestine. They colonize the proximal 25% of the small intestines, adhere to microvilli of columnar cells near the base of villi, wedge into furrows in the epithelial surface, or lodge in mucus within the unstirred layer (Owen *et al.*, 1979). Over Peyer's patches, *Giardia* adhere to columnar cells and not to M cells, which transport soluble antigens and particulate material from the lumen into the lymphoid system.

Infected mice have significant impairment of weight gain and a significant reduction in the villus:crypt ratio of jejunal mucosa, with an apparent increase in chronic inflammatory cells in the lamina propria (Owen *et al.*, 1979). Impairment of weight gain might reflect anorexia associated with infection or competition for or malabsorption of dietary nutrients.

Giardiasis in mice and the clearance of *Giardia* appear to be related immunological processes of the host. Giardiasis may cause significant mortality and morbidity in athymic (*nu/nu*) or thymectomized mice (Boorman *et al.*, 1973) and in hypogammaglobulinemic patients (Ament and Rubin, 1972). Tissue or intestinal invasion of *Giardia* can occur in heavily infected mice, in irradiated mice, or in mice fed a protein-deficient diet in which the intestinal epithelial barrier may have been damaged (Owen *et al.*, 1979).

In *Giardia*-infected mice, intestinal intraluminal lymphocytes (which appear to be thymus dependent) increase in number (MacDonald and Ferguson, 1978). Intraepithelial lymphocytes are thought to be immunologically primed cells capable of blocking antigen penetration or of differentiating into immunoglobulin-secreting cells (Ferguson, 1977). The in-

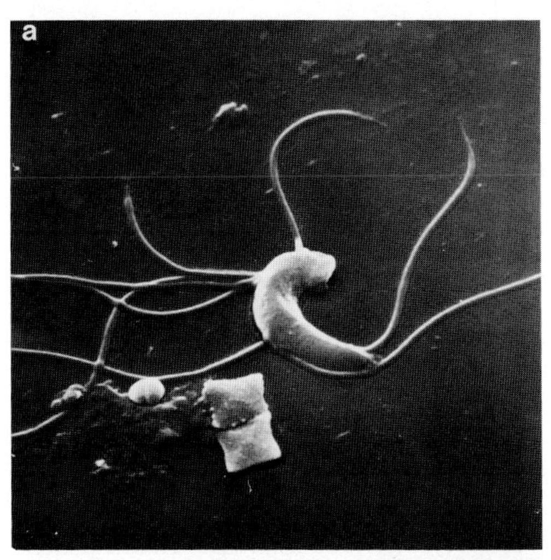

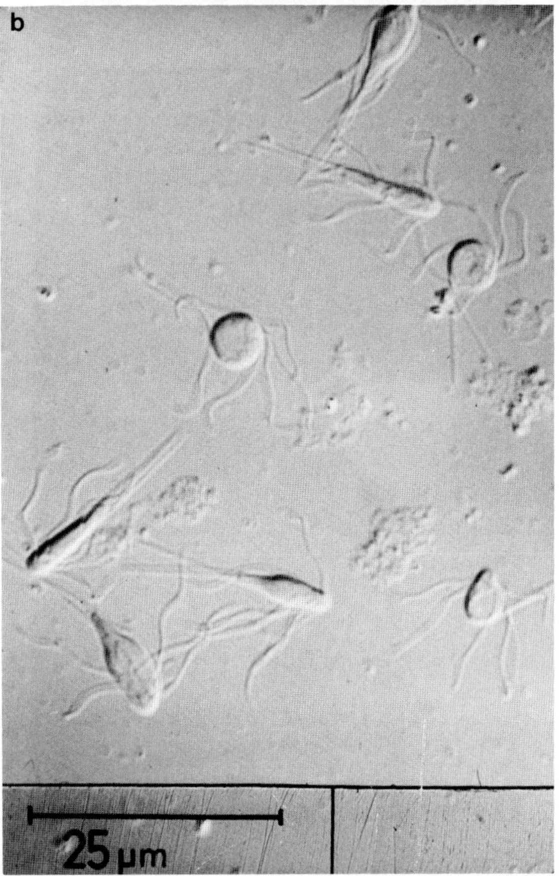

Fig. 6. *Spironucleus muris*. Scanning electronmicrograph, ×5000. (b) *Spironucleus muris* trophozoites from the small intestine of a mouse. Wet mounting, unstained, differential interference contrast. [(a) Courtesy of J. E. Wagner; (b) courtesy of Dr. Ivo Kunstyr.]

traluminal cellular immune response appears to be associated with the clearance of *Giardia* in mice (Owen *et al.*, 1979).

d. Diagnosis. The diagnosis of giardiasis in mice is based on the demonstration of trophozoites in the small intestine or cysts in feces, as well as the histological identification of parasites.

e. Treatment, Prevention, and Control. Treatment of murine giardiasis with 0.1% dimetridazole in drinking water for 14 days (Sebesteny, 1969) or with chloroquine, quinacrine, or amodiaquin is effective. Cysts in feces or the environment can be killed by 2.5% phenol or Lysol or by dry temperatures above 50°C (Hsu, 1979). Prevention and control depend on proper sanitation, colony management, and surgical derivation of infected strains and stocks.

2. *Spironucleus (Hexamita) muris*

a. Description and Life Cycle. Spironucleus (Hexamita) muris is a flagellated protozoon that commonly occurs in the small intestine of conventional mice, rats, hamsters, and other wild rodents. It is usually present in the crypts of Lieberkühn. *Spironucleus muris* is elongated, pear-shaped and bilaterally symmetrical; it measures 7–9 × 2–3 μm (Fig. 6b). It has two anterior nuclei, two longitudinal axostyles, and four pairs of flagella arising from a pair of blepharoplasts situated between the paired nuclei (Fig. 6). *Spironucleus muris* multiplies by longitudinal fission. Infectious cysts have been described and are transmitted by ingestion of trophozoites or cysts (Kunstyr, 1977).

b. Pathobiology. Spironucleus muris is more pathogenic for stressed or young than for adult animals; the former may develop a chronic infection. Infected mice have a roughened hair coat, depression, weight loss or failure to gain weight, listlessness, dehydration, hunched posture, distended abdomen, diarrhea, and sometimes in young mice (3–6 weeks of age) high mortality (20–50%) (Van Kruiningen *et al.*, 1978; Flatt *et al.*, 1978). At necropsy the small intestine is dark, brownish red and distended with watery fluid and gas. Intestinal content smears or preparations contain numerous trophozoites with fast, straight, or zigzag movements. Histologically, the parasite may cause duodenitis accompanied by an accumulation of catarrhal fluid and cellular infiltration in the small intestine. The crypts of Lieberkühn and the lumina between villi are distended by their accumulation of *Spironucleus* parasites (Figs. 7, 8). Athymic mice infected with *S. muris* excrete cysts throughout life (Kunstyr *et al.*, 1977). Clinical hexamitiasis is usually associated with environmental stress or immunosuppression. Concurrent infections with other intestinal pathogens may result in a synergistic effect.

Spironucleosis (hexamitiasis) may cause a number of research complications, including increased macrophage activity (Ruitenberg and Kruyt, 1975), interference with the immune response of the host (Ruitenberg and Kruyt, 1975), changes of

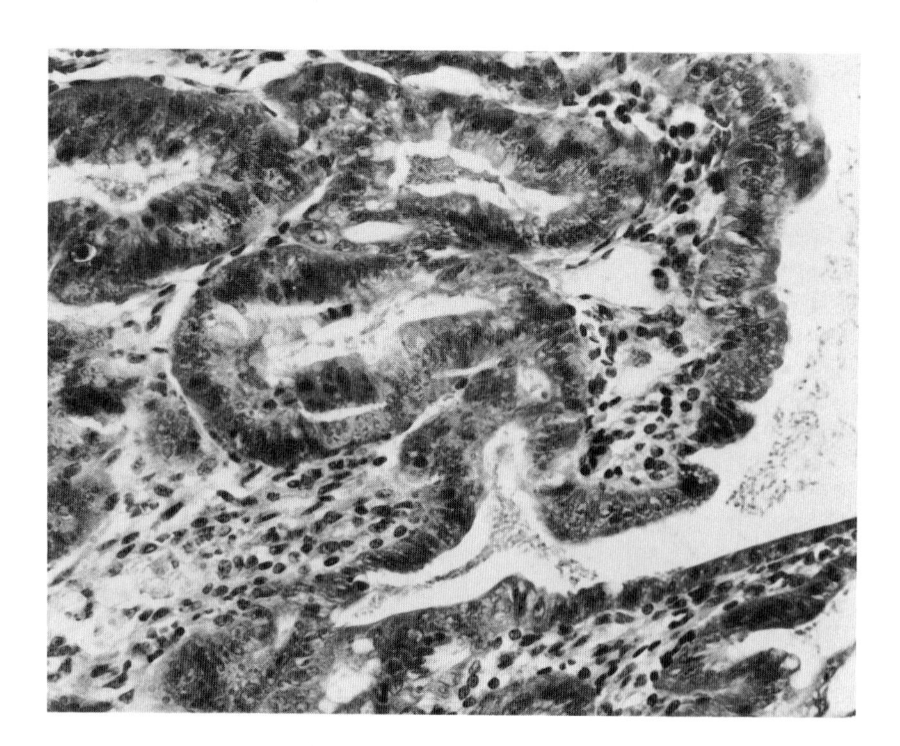

Fig. 7. Spironucleus (Hexamita) muris organisms in the crypt and lumen of the small intestine of a DBA/2 mouse. Hematoxylin and eosin, ×320.

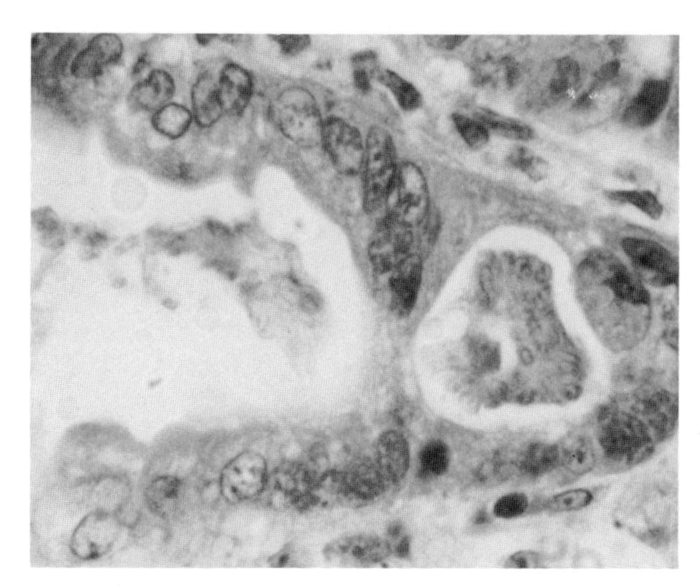

Fig. 8. Higher magnification of *Spironucleus muris* in the crypt. Hematoxylin and eosin, ×1000.

macrophage metabolism (Keast and Chesteman, 1972), increased sersitivity of mice to irradiation (Myers, 1973), increased mortality in cadmium-exposed mice (Exon *et al.*, 1975), and shortened life span of athymic mice. Infected mice have enlarged lymph nodes containing activated macrophages which kill tumor cells nonspecifically in *in vitro* assays (Sebesteny, 1969; Keller, 1973), thus making assessment of specific immunity to tumor cell antigens impossible. Sebesteny (1974) stated that mice infected with *S. muris* are unsuitable for immunological studies in general, since they respond poorly to a wide variety of soluble and particulate antigens in comparison to uninfected animals.

c. Diagnosis. The diagnosis of spironucleosis is based on finding trophozoites in the lumen or crypts of Lieberkühn or the presence of cysts in the feces. *Spironucleus muris* can be distinguished from *Giardia muris* and *Tritribhomonas muris* by its small size, fast horizontal or zigzag movement, and the absence of a sucking disc or undulating membrane.

d. Treatment, Prevention, and Control. Treatment of spironucleosis in mice with 0.04–0.1% dimetridazole in drinking water for 14 consecutive days can control clinical disease but does not eliminate parasites from infected animals (Sebesteny, 1969; Flatt *et al.*, 1978). *Spironucleus muris* cysts are sensitive to most disinfectantants and high temperatures but are resistant to low temperatures (Kunstyr and Ammerpohl, 1978). The prevention and control of spironucleosis depend on proper sanitation and good colony management. Without proper control, cross-transmission of spironucleosis among mice, rats, and hamsters in the animal facility is common.

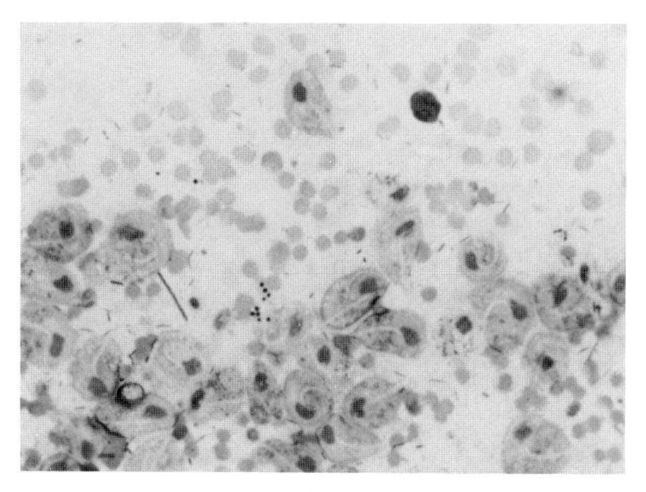

Fig. 9. *Trichomonas* spp. Trophozoites from the intestinal content smear. Giemsa–Wright stain, ×500.

3. *Tritrichomonas muris*

Tritrichomonas muris is nonpathogenic and occurs in the cecum, colon, and small intestine of mice, rats, hamsters, and wild rodents. It is pear-shaped and measures 16–26 × 10–14 μm (Fig. 9). It has an anterior vesicular nucleus. Anterior to the nucleus is the blepharoplast, from which arise three anterior flagella and a posterior flagellum which is attached to the body by means of an undulating membrane that continues posteriorly as a free flagellum. The trichomonad is supported by a stiff, rodlike axostyle which protrudes terminally like a tail (Fig. 10). It has a large, oval-shaped nucleus and a sausage-shaped parabasal body in the anterior region. Reproduction is asexual by binary fission. The parasite swims with a characteristic wobbly movement. Cysts are not formed, and

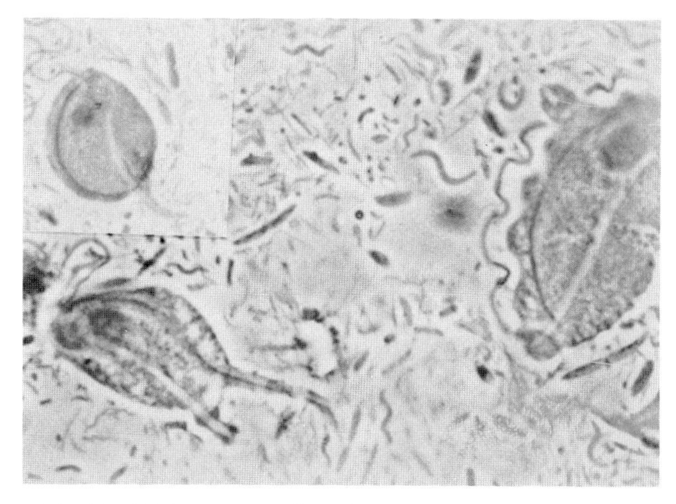

Fig. 10. *Trichomonas* spp. from the intestinal content smear. Giemsa–Wright stain, ×800–1000.

transmission is by ingestion of trophozoites passed out in feces.

B. Sporozoa

1. *Eimeria falciformis*

Eimeria falciformis is a pathogenic coccidium that occurs in the epithelial cells of the large intestine of mice. It is host specific. It has been commonly reported in mice from Europe but is seldom observed in laboratory mice in the United States. The oocyst measures 14–26 × 11–24 μm; they have neither micropyle nor residuum. Sporulated oocysts contain four ovoid sporocysts each with two sporozoites. The sporulation time is 3 days. Heavy infection may cause diarrhea and catarrhal enteritis together with desquamation of the intestinal epithelium and hemorrhage. Levine and Ivens (1965a) have described this parasite in detail.

2. *Cryptosporidium muris*

Cryptosporidium muris occurs on the surface of the epithelium of the mouse stomach. It is uncommon in laboratory mice. Oocysts contain four naked sporozoites which measure 7 × 5 μm and have a residuum, but no sporocysts are micropyle (Levine and Ivens, 1965b). This parasite is only slightly pathogenic. Its life cycle has been well detailed and illustrated (Levine and Ivens, 1965b).

3. *Cryptosporidium parvum*

Cryptosporidium parvum inhabits the small intestine, is nonpathogenic, and is uncommon in laboratory mice. It is attached to the surface and either causes an indentation or is buried in the striated border of the epithelial cells of the villi (Levine and Ivens, 1965b). The oocysts are ovoid or spherical and measure 4–5 × 3 μm.

C. Amoebae

Entamoeba muris

Entamoeba muris is common in the cecum and colon of mice, rats, hamsters, and many wild rodents throughout the world. It occurs in high incidence in conventional laboratory mouse, rat, and hamster colonies. *Entamoeba muris* resembles *E. coli*. Trophozoites are large, 8–30 μm long, and cysts are 9–20 μm in diameter and contain up to eight nuclei (Fig. 11). A food vacuole is often present in the cyst. The cyst produces eight amoebae when ingested by a new host. Trophozoites multiply by binary fission. Normally, the amoebae live in the lumen of the intestine, where they feed on particles of food and bacteria. Though nonpathogenic, *E. muris* can be cross-transmitted among rats, mice, and hamsters.

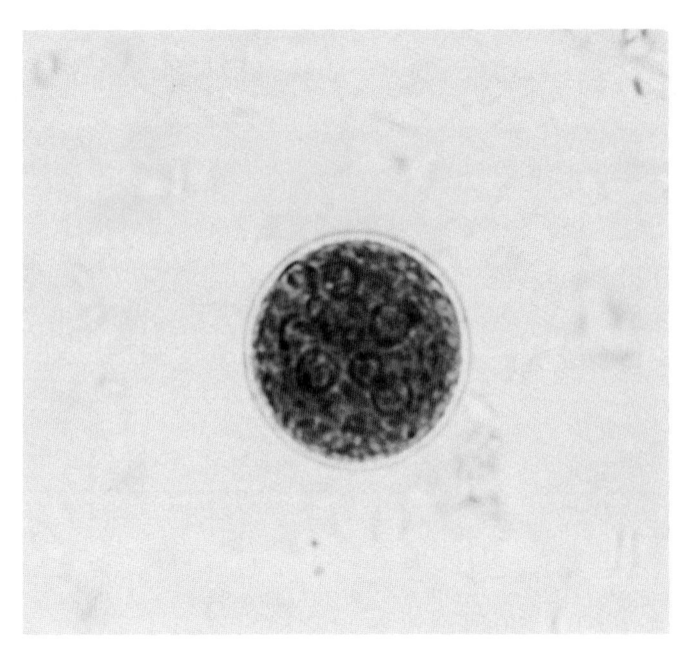

Fig. 11. *Entaomeba muris* cyst, ×1500.

REFERENCES

Ament, M. E., and Rubin, C. E. (1972). Relation of giardiosis to abnormal intestinal structure and function in gastrointestinal immune-deficiency syndromes. *Gastroenterology* **62,** 216–226.

Beverley, J. K. A. (1959). Congenital transmission of toxoplasmosis through successive generations of mice. *Nature (London)* **183,** 1348.

Boorman, G. A., Linda, P. H. C., Zurcher, C., and Nieuwerkerk, H. T. N. (1973). *Hexamita* and *Giardia* as a cause of mortality in congenitally thymus-less (nude) mice. *Clin. Exp. Immunol.* **13,** 623–627.

Bywater, J. E. C. (1979). Is encephalitozoonosis a zoonosis? *Lab. Anim.* **13,** 149–151.

Cox, J. C., and Pye, D. (1975). Serodiagnosis of nosematosis by immunofluorescence using cell-culture-grown organisms. *Lab. Anim.* **9,** 297–304.

Dubey, J. P. (1977). Toxoplasma, hammondia, besnoitia, sarcocystis and other tissue cyst-forming coccidia of man and animals. *In* "Parasitic Protozoa" (J. P. Kreier, ed.), Vol. 3, pp. 101–237. Academic Press, New York.

Exon, J. H., Patton, N. M., and Koller, L. D. (1975). Hexamitiasis in cadmium-exposed mice. *Arch. Environ. Health* **30,** 463–464.

Ferguson, A. (1977). Intraepithelial lymphocytes of the small intestine. *Gut* **18,** 921–937.

Flatt, R. E., Halvorsen, J. A., and Kemp, R. L. (1978). Hexamitiasis in a laboratory mouse colony. *Lab. Anim. Sci.* **28,** 62–65.

Flynn, R. J. (1973). "Parasites of Laboratory Animals," pp. 3–119. Iowa State Univ. Press, Ames.

Frenkel, J. K. (1971). Protozoal diseases of laboratory animals. *In* "Pathology of Protozoal and Helminthic Diseases" (R. A. Marcial-Rojas, ed.), pp. 319–369. Williams & Wilkins, Baltimore, Maryland.

Frenkel, J. K., and Reddy, J. K. (1977). Induction of liver tumor by 3′-methyl-dimethylaminoazobenzene (3′-Me′DAB) in rats chronically infected with *Toxoplasma* or *Besnoitia*. *J. Reticuloendothel. Soc.* **21,** 61–68.

Gannon, J. (1978). The immunoperoxidase test diagnosis of *Encephalitozoon cuniculi* in rabbits. *Lab. Anim.* **12,** 125–127.

Gentry, L. O., and Remington, G. S. (1971). Resistance against *Crytococcus* conferred by intracellular bacteria and protozoa. *J. Infect. Dis.* **123,** 22–31.

Hibbs, J. B., Jr., Lewis, H. L., Jr., and Remington, J. S. (1971). Resistance to murine tumors conferred by chronic infection with intracellular protozoa, *Toxoplasma gondii* and *Besnoitia jellisoni*. *J. Infect. Dis.* **124,** 587–592.

Hsu, C.-K. (1979). Parasitic diseases. *In* "The Laboratory Rat" (H. Baker, J. Lindsey, and S. Weisbroth, eds.), pp. 307–331. Academic Press, New York.

Huldt, G., Gard, S., and Olovson, S. G. (1973). Effect of *Toxoplasma gondii* on the thymus. *Nature (London)* **244,** 301–303.

Hunt, R. D., King, N. W., and Foster, H. L. (1972). Encephalitozoonosis: evidence for vertical transmission. *J. Infect. Dis.* **126,** 212–214.

Innes, J. R. M., Zeman, W., Frenkel, J. K., and Borner, G. (1962). Occult encephalitozoonosis of the central nervous system of mice (Swiss-Bagg-O'Grady strain). *J. Neuropathol. Exp. Neurol.* **21,** 519–533.

Jones, S. R. (1973). Toxoplasmosis: a review. *J. Am. Vet. Med. Assoc.* **163,** 1038–1042.

Keast, D., and Chesteman, F. (1972). Changes in macrophage metabolism in mice heavily infected with *Hexamita muris*. *Lab. Anim.* **6,** 33–39.

Keller, R. (1973). Cytostatic elimination of syngeneic rat tumor cells *in vitro* by nonspecifically activated macrophages. *J. Exp. Med.* **183,** 625.

Kellett, B. S., and Bywater, J. E. C. (1978). A modified India-ink immunoreaction for the detection of encephalitozoonosis. *Lab. Anim.* **12,** 59–60.

Kunstyr, I. (1977). Infectious form of *Spironucleus muris:* banded cysts. *Lab. Anim.* **11,** 185–188.

Kunstyr, I., and Ammerpohl, E. (1978). Resistance of fecal cysts of *Spironucleus muris* to some physical factors and chemical substances. *Lab. Anim.* **12,** 95–97.

Kunstyr, I., Meyer, B., and Ammerpohl, E. (1977). Spironucleosis in nude mice: an animal model for immuno-parasitologic studies. *Proc. Int. Workshop Nude Mice* **2,** 17–27.

Levine, N. D. (1974). Diseases of laboratory animals—parasitic. *In* "CRC Handbook of Laboratory Animal Science" (E. C. Melby and N. H. Altman, eds.), Vol. 2, pp. 289–327. CRC Press, Cleveland, Ohio.

Levine, N. D. (1977). Taxonomy of *Toxoplasma*. *J. Protozool.* **24,** 36–41.

Levine, N. D., and Ivens, V. (1965a). "The Coccidian Parasites (Protozoa, Sporozoa) of Rodents," Illinois Biology Monographs, No. 33, pp. 132–135. Univ. of Illinois Press, Urbana.

Levine, N. D., and Ivens, V. (1965b). "The Coccidian Parasites (Protozoa, Sporozoa) of Rodents," Illinois Biology Monographs, No. 33; pp. 176–179. Univ. of Illinois Press, Urbana.

MacDonald, T. T., and Ferguson, A. (1978). Small intestinal epithelial cell kinetics and protozoal infection in mice. *Gastroenterology* **74,** 496–500.

Mahmoud, A. A., Warren, K. D., and Strickland, G. T. (1976). Acquired resistance to infection with *Schistosoma mansoni* induced by *Toxoplasma gondii*. *Nature (London)* **263,** 56–57.

Myers, D. D. (1973). Sensitivity to x-irradiation of mice infected with *Hexamita muris*. *Annu. Meet., 24th, Am. Assoc. Lab. Anim. Sci.*, Abst. No. 22.

Otto, H. (1957). Kidney lesions in mice with *Klossiella muris* infection. *Frankf. Z. Pathol.* **68,** 41–48.

Owen, R. L., Nemanic, P. C., and Stevens, D. P. (1979). Ultrastructural observations on giardiosis in a murine model, 1. Intestinal distribution, attachment, and relationship to the immune system of *Giardia muris*. *Gastroenterology* **76,** 751–769.

Pakes, S. P., Shadduck, J. A., and Olsen, R. G. (1972). A diagnostic skin test for encephalitozoonosis (nosematosis) in rabbits. *Lab. Anim. Sci.* **22,** 870–876.

Pakes, S. P., Shadduck, J. A., and Cali, A. (1975). Fine structure of *Encephalitozoon cuniculi* from rabbits, mice and hamsters. *J. Protozool.* **22,** 481–488.

Remington, J. S., and Merigan, T. C. (1969). Resistance to virus challenge in mice infected with protozoa or bacteria. *Proc. Soc. Exp. Biol. Med.* **131,** 1184–1188.

Roberts-Thomas, I. C., and Mitchell, G. R. (1979). Protection of mice against *Giardia muris* infection. *Infect. Immun.* **24,** 971–973.

Roberts-Thomas, I. C., Stevens, D. P., Mahmoud, A. A. F., and Warren, K. S. (1976). Giardiosis in the mouse: an animal model. *Gastroenterology* **71,** 57–61.

Rosenmann, M., and Morrison, P. R. (1975). Impairment of metabolic capability in feral house mice by *Klossiella muris* infection. *Lab. Anim. Sci.* **25,** 62–64.

Ruitenberg, E. J., and Kruyt, B. C. (1975). Effect of intestinal flagellates on immune response in mice. *Parisitology* **71,** xxx.

Ruskin, J., and Remington, J. S. (1968). Immunity and intracellular infection: resistance to bacteria in mice infected with a protozoan. *Science* **160,** 72–74.

Ruskin, J., and Remington, J. S. (1969). A role for the macrophage in acquired immunity to phylogenetically unrelated intracellular organisms. *Antimicrob. Agents Chemother.* pp. 474–477.

Sebesteny, A. (1969). Pathogenicity of intestinal flagellates in mice. *Lab. Anim.* **3,** 71–77.

Sebesteny, A. (1974). The transmission of intestinal flagellates between mice and rats. *Lab. Anim.* **8,** 79–81.

Seed, T. M., and Aikawa, M. (1977). Pneumocystis. *In* "Parasitic Protozoa" (J. P. Kreier, ed.), Vol. 4, pp. 329–357. Academic Press, New York.

Shadduck, J. A., and Geroulo, M. J. (1979). A simple method for the detection of antibodies to *Encephalitozoon cuniculi* in rabbits. *Lab. Anim. Sci.* **29,** 330–334.

Shadduck, J. A., Watson, W. T., Pakes, S. P., and Cali, A. (1979). Animal infectivity of *Encephalitozoon cuniculi*. *J. Parasitol.* **65,** 123–129.

Sheppe, W. A., and Adams, J. R. (1957). The pathogenic effect of *Trypanosoma duttoni* on hosts under stress conditions. *J. Parasitol.* **43,** 55–59.

Strickland, G. T., Ahmed, A., and Sells, K. W. (1975). Blastogenic response of *Toxoplasma*-infected mouse spleen cells to T- and B-cell mitogens. *Clin. Exp. Immunol.* **22,** 167–176.

Swartzberg, J. E., Krahenbuhl, J. L., and Remington, J. S. (1975). Dichotomy between macrophage activation and degree of protection against *Listeria monocytogenes* and Toxoplasma gondii in mice stimulated with *Corynebacterium parvum*. *Infect. Immun.* **12,** 1037–1043.

Tamura, T., Ueda, K., Furuta, T., Goto, Y., and Fujiwara, K. (1978). Electron microscopy of spontaneous model for *Pneumocystis carinii* infection. *Science* **197,** 177–179.

Twort, J. M., and Twort, C. C. (1932). Disease in relation to carcinogenic agents among 60,000 experimental mice. *J. Pathol. Bacteriol.* **35,** 219–242.

Van Kruiningen, J. H., Knibbs, D. R., and Burke, L. N. (1978). Hexamitiasis in laboratory mice. *J. Am. Vet. Med. Assoc.* **173,** 1202–1204.

Wallace, G. D. (1973). Intermediate and transport hosts in the natural history of *Toxoplasma gondii*. *Am. J. Trop. Med. Hyg.* **22,** 456–464.

Walzer, P. D., Schnelle, V., Armstrong, D., and Rosen, P. P. (1977). Nude mice: a new experimental model for *Pneumocystis carinii* infection. *Science* **197,** 177–179.

Walzer, P. D., Powell, R. D., Jr., and Yoneda, K. (1979). Experimental *Pneumocystis carinii* pneumonia in different strains of cortisonized mice. *Infect. Immun.* **24,** 939–947.
Wosu, N. J. (1975). Studies in serodiagnosis on encephalitozoonosis of rabbits. *Diss. Abstr. Int. B* **35,** 4277–4278.
Yang, Y. H., and Grice, H. C. (1964). *Klossiella muris* parasitism in laboratory mice. *Can. J. Comp. Med.* **28,** 63–66.
Yost, D. H. (1958). Encephalitozoon infection in laboratory animals. *J. Natl. Cancer Inst.* **20,** 957–963.

Chapter 20

Helminths

Richard B. Wescott

I. Introduction 373
II. Helminths of Major Importance 374
A. *Syphacia obvelata,* Mouse Pinworm 374
B. *Aspiculuris tetraptera,* Mouse Pinworm 376
C. *Hymenolepis nana,* Dwarf Tapeworm 378
III. Helminths of Minor Importance 379
A. Nematodes 379
B. Cestodes 381
References 382

I. INTRODUCTION

Various aspects of the problem of helminth parasites in laboratory mice have been reviewed by a number of authors (Blair *et al.,* 1968; Flynn, 1973; Flynn *et al.,* 1965; Habermann and Williams, 1958; Hoag, 1961; Hussey, 1957; Oldham, 1967; Sasa *et al.,* 1962; Stone and Manwell, 1966; Taffs, 1976a) since Heston (1941) discussed these parasites as a group in the first edition of "Biology of the Laboratory Mouse." Interestingly, many of the same parasites mentioned as important or common in 1941, namely *Syphacia obvelata, Aspiculuris tetraptera,* and *Hymenolepis nana,* appear to have survived rather well and are found in many mouse colonies today. This dilemma has probably been caused both by the resourcefulness of the parasites and by incomplete understanding of their life cycles and control measures. The technology is essentially available to eliminate helminths from laboratory mice, and the main objective of this chapter is to suggest how it should be used most expeditiously.

The hypothesis that moderate nematode infections do not constitute a variable in biomedical research is no longer tenable. Parasites of mice have been reported to produce or potentiate disease (Bieniek and Tober-Meyer, 1976; Bisseru, 1967; Blair *et al.,* 1968; Harwell and Boyd, 1968; Hoag, 1961; Kyaw and Oo, 1976; Mayer and Pappas, 1976; Sasa *et al.,* 1962; Simmons *et al.,* 1967), decrease weight gains (Kyaw and Oo, 1976), change behavior (McNair and Timmons, 1977; Patterson and Vessey, 1973), and alter immune responses (Chowaniec *et al.,* 1972; Goodall, 1973; Pearson and Taylor, 1975; Shimp *et al.,* 1975). In addition, marked host responses have been described for most common helminths of mice (Andreassen *et al.,* 1978; Befus and Featherston, 1974; Befus,

ISBN 0-12-262502-1

1975, 1977; Behnke, 1974b, 1975a, 1976; Hopkins *et al.*, 1972, 1973, 1977; Turner and McKeever, 1976). Consequently, parasitized mice are not considered normal experimental subjects (Jirina, 1952). Parasitized hosts alter the results in blood, nutritional, immunological, and radiation studies (Habermann and Williams, 1957) and are not suitable for cancer and disease research or helminth biology experimentation (Voge, 1958).

Since the number of different helminths commonly encountered in laboratory mice is small, each species will be covered separately and in detail. Other helminths which appear in wild mice and rarely in laboratory mice will receive less attention. The more important of these are described briefly at the conclusion of the chapter, with references provided should further information be desired. The remaining helminths of mice may be found in Table II.

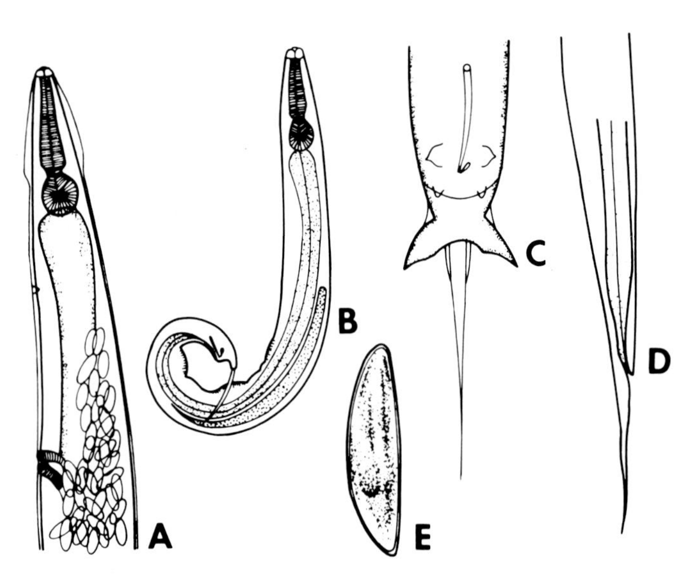

Fig. 1. *Syphacia obvelata.* (A) Anterior extremity of a female, lateral view. (B) Male, lateral view (note the single spicule at the posterior end and three mamelons or enlargements on the cuticle just anterior to it). (C) Posterior extremity of male, ventral view. (D) Posterior extremity of a female, lateral view. (E) Egg. (After Yorke and Maplestone, 1926.)

II. HELMINTHS OF MAJOR IMPORTANCE

A. *Syphacia obvelata*, Mouse Pinworm

This oxyurid is a cosmopolitan parasite of wild and laboratory mice (*Mus musculus*). Surveys indicate that most colonies are involved and that a high percentage of mice within these colonies are infected (Flynn *et al.*, 1965; Fratta and Slanetz, 1958; Habermann and Williams, 1958; Hasslinger and Bosse, 1966; Hoag, 1961; King and Cosgrove, 1963; Sasa *et al.*, 1962; Simmons *et al.*, 1964, 1965; Thompson and Reinertson, 1952; Weber, 1976). Unfortunately, the protocols used in these surveys varied considerably, so it is difficult to compare results and detect changes in prevalence that may have occurred in recent years. However, it does appear that *S. obvelata* often occurs in combination with *A. tetraptera,* and it is not unusual to observe *H. nana* in these same hosts.

Younger mice are most apt to be infected with *S. obvelata* (Sasa *et al.*, 1962). Infection is common in wild mice and will reappear in laboratory mice after treatment unless extreme precautions are taken to prevent reexposure (Simmons *et al.*, 1967; Wescott *et al.*, 1976). It appears that treatment has reduced the magnitude of infections in individual mice, but many colonies still harbor the parasite. The host range of *S. obvelata* includes the house mouse and, on rare occasions, the Norway rat and the Mongolian gerbil (Hussey, 1957; Wightman *et al.*, 1978).

1. Morphology

Important morphologic details of *S. obvelata* are shown in Fig. 1, and complete descriptions are available elsewhere (Flynn, 1973; Levine, 1968; Oldham, 1967; Sasa *et al.*, 1962). Females are much larger than males, ranging from 3.4 to 5.8 mm in length, compared to 1.1–1.5 mm for males. Both sexes have small cervical alae, three broad lips, a simple mouth cavity, an esophagus ending in a prominent bulb, and long, pointed tails. The male has an obvious single spicule and three mamelons on the ventral surface. The vulva of the female is near the anterior end of the body. Eggs are typically flattened on one side and pointed on both ends. The nucleus fills the shell and frequently has progressed to a larval stage when laid. Eggs measure approximately 134 × 36 μm (see Fig. 2).

2. Life Cycle

The cycle is direct and completed in 11–15 days in susceptible hosts (Brown *et al.*, 1954; Chan, 1952; Lawler, 1939; Sasa *et al.*, 1962). Gravid females leave the large intestine and deposit their eggs on the skin and hairs in the perianal region. The eggs develop to an infective stage in 6 hr at 37°C. When ingested by a suitable host, the eggs hatch, liberating larvae in the small intestine which move to the cecum in 24 hr. The parasites remain for 10–11 days in the cecum, where they mature and mate. The females then migrate down the large intestine and deposit their eggs in the perianal area as they leave the host some 12–15 days after exposure.

There is some speculation (Chan, 1952) that retroinfection, in which eggs hatch and larvae enter the rectum, may occur, but no experimental evidence supports this contention.

Other factors influencing the life cycle are the fact that young mice are most susceptible (Sasa *et al.*, 1962) and that hosts develop a definite age resistance to this infection. Also, since the life cycle is much shorter than that of *A. tetraptera*, younger animals (3-4 weeks old) are more apt to be infected with *S. obvelata* than with *A. tetraptera*, which first appears in mice at 5-6 weeks of age (Sasa *et al.*, 1962).

3. Pathology

No gross lesions directly attributable to *S. obvelata* have been produced experimentally. However, rectal prolapse, intussusception, enteritis, fecal impaction, liver granulomas, and poor weight gains have all been observed in heavily infected mice, and most workers consider pinworms to be partially responsible (Bieniek and Tober-Meyer, 1976; Cook, 1969; Harwell and Boyd, 1968; Hoag, 1961; Oldham, 1967). The difficulty in confirming these observations lies in both the universality of *S. obvelata* infection and the life cycle of the parasite. Most colonies are infected without showing overt clinical signs. Hence, in clinical reports it is hard to confirm that the parasites observed in affected mice are causally related to the condition or are relatively harmless inhabitants of the intestine. Furthermore, experimentation with *S. obvelata* is difficult. Even if mice free of the parasite were found for such studies, the life cycle is so rapid (2 weeks) and host resistance so well defined that natural infection would be very difficult to simulate.

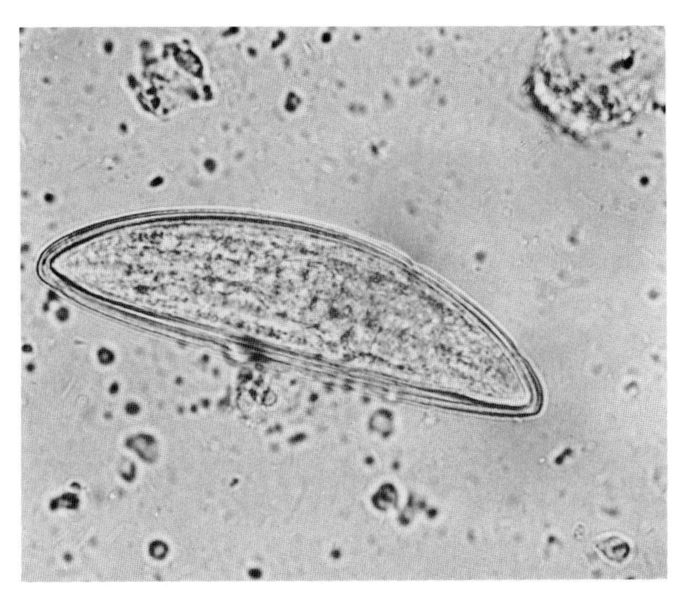

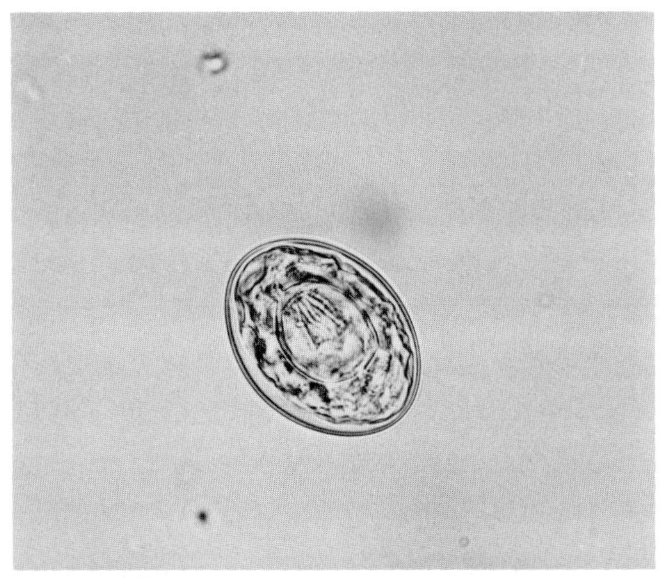

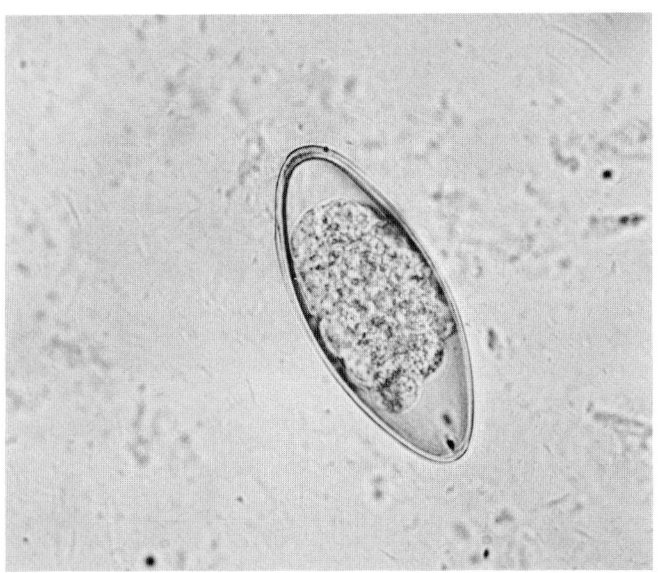

Fig. 2. Eggs of (A) *Syphacia obvelata*, (B) *Aspiculuris tetraptera*, and (C) *Hymenolepis nana* in fecal flotation. (D) Cellophane tape preparation of *Syphacia* eggs.

Another problem is that *S. obvelata* often occurs with *A. tetraptera* and *H. nana*. The former has been shown to invade the crypts of Lieberkühn (Behnke, 1974a) and the latter the intestinal villi (Oldham, 1967) in their development. In such mixed infections, it is impossible to distinguish between the damage directly attributable to *S. obvelata* and that due to the other parasites. Nevertheless, sufficient circumstantial evidence exists to warrant elimination of this parasite from research animals.

4. Diagnosis

Syphacia obvelata can be diagnosed by demonstration of the presence of eggs in the perianal area or adult worms in the cecum and large intestine. Since most eggs are deposited outside the gastrointestinal tract of the host, fecal examination for eggs is not reliable. It is also imperative that some 4- to 5-week-old mice be examined because a much higher percentage of subjects in this age group will be infected than in younger or older hosts (Sasa *et al.*, 1962).

The presence of eggs is best demonstrated in the perianal area by pressing clear cellophane tape to this area and then examining it on a glass slide with a microscope (Fig. 2). Differentiation from eggs of *S. muris* can be done on the basis of size. *Syphacia obvelata* eggs are considerably larger (134 × 36 μm) than those of *S. muris* (75 × 29 μm) (Hussey, 1957) and differ somewhat in shape. Eggs of *A. tetraptera* are not ordinarily found in tape preparations, and if they are, they differ sufficiently in appearance so that there should be no confusion.

Demonstration of adult parasites is readily accomplished by opening the cecum and large intestine in a petri dish filled with warm tap water. After the opened organs have been allowed to stand for a few minutes and the debris has settled, parasites are readily observed with the naked eye or by using a dissecting microscope. For positive identification, parasites should be removed and examined microscopically.

Syphacia obvelata is most easily distinguished from *A. tetraptera* by the eggs, cervical alae, shape of the tail, location of the vulva in females, and the presence of mamelons and spicules in males (Figs. 1–3; see also Table I). Adults of *S. obvelata* and *S. muris* are most readily differentiated by measurement of eggs taken from adult females.

5. Treatment

Effective treatment of *S. obvelata* is not a problem. Many drugs or caesarian derivation can be used to remove the parasites (Flynn, 1973; Foster, 1963). Control is quite a different matter. Unless stringent measures are taken to prevent reexposure, this infection will reappear. Hence, a combination of factors must be considered if progress is to be made in eliminating this parasite from laboratory mice.

Of the drugs available, piperazine salts have been used most extensively. The administration of piperazine citrate at 200–400 mg/kg in drinking water for a week, followed by no treatment for a week and retreatment the third week, as described by Hoag (1961), has been used extensively. It is effective, readily done, and will reduce the number of *S. obvelata* present in mouse colonies. However, the commonly used piperazines have been shown to be less effective than many other drugs (Brody and Elward, 1971), and if greater efficacy is desired, the use of pyrvinium pamoate (Blair *et al.*, 1968), stilbazium iodide (Hunt and Burrows, 1963), newer phosphate anthelmintics (Pilgram and Hass, 1975), metronidazole (Seth and Louekar, 1973), mebendazole (Sharp and Wescott, 1976), trichlorfon (Simmons *et al.*, 1965), newer piperazines (Singhal, 1975), thiabendazole (Taffs, 1975, 1976b), or dichlorvos (Wagner, 1970) can be investigated.

Other means of elimination of *S. obvelata* are caesarian derivation (Foster, 1963), use of purified diets (DeWitt and Weinstein, 1964), and alteration of bacterial flora (Przyjalkowski, 1974).

Prevention of reinfection requires strict isolation (Simmons *et al.*, 1967; Wescott and Rabstein, 1962; Wescott *et al.*, 1976). *Syphacia obvelata* eggs are attached to the host, become infective in as little as 6 hr, survive for weeks, and are common in wild or uncaged mice (Chan, 1952; Owen, 1976). These factors, together with the incomplete efficacy of treatment against immature worms (Cook, 1969), practically ensures a source for reinfection in a colony unless mice are kept under filter tops or in isolators. If it is not possible to employ this degree of isolation, heroic efforts must be made to eliminate uncaged mice and contamination of feed and bedding to achieve total control. In most instances this is not obtained, and routine periodic anthelmintic treatment is required to minimize the effect of this parasite.

B. *Aspiculuris tetraptera*, Mouse Pinworm

Aspiculuris tetraptera is a cosmopolitan oxyurid parasite of the house mouse (*M. musculus*). Surveys suggest that many laboratory and wild mice are infected, that this nematode may be somewhat less prevalent than *S. obvelata*, and that both parasites frequently appear concurrently in the same host. (Behnke, 1975b, 1976; Flynn *et al.*, 1965; Owen, 1976; Sasa *et al.*, 1962; Weber, 1976).

The host range of *A. tetraptera* includes the house mouse and on rare occasions the Norway rat (Sasa *et al.*, 1962). Human infection has not been reported.

Most of the general information presented for *S. obvelata* also applies to *A. tetraptera* and will not be repeated here. The following describes only those features of *A. tetraptera* that are unique or sufficiently different from *S. obvelata* to be noteworthy.

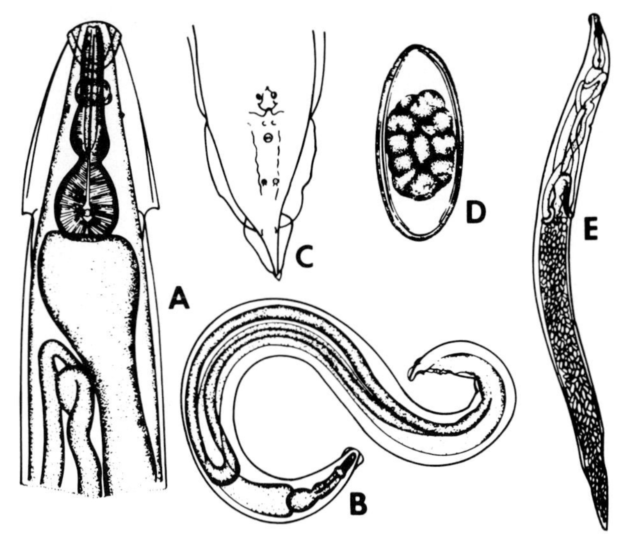

Fig. 3. *Aspiculuris tetraptera.* (A) Anterior extremity, ventral view (note prominent cervical alae). (B) Male. (C) Tail of a male, ventral view. (D) Egg. (E) Female, lateral view. (From Yorke and Maplestone, 1926, after Schulz.)

1. Morphology

Important morphologic features of *A. tetraptera* are shown in Fig. 3 and Table I. More detailed descriptions are available elsewhere (Flynn, 1973; Levine, 1968; Oldham, 1967; Sasa *et al.*, 1962). Females are somewhat larger than males, ranging from 2.6 to 4.7 mm in length, as opposed to 2.0–2.6 mm for males. Both sexes have braod cervical alae, three lips, a simple mouth cavity, an esophagus ending in an elongate oval, and a conical tail. The male has neither spicule nor mamelons. The vulva of the female is near the center of the body. Eggs are spindle-shaped with thin shells. They measure approximately 90×41 μm and are in the morula stage when laid. The nucleus does not fill the shell (Fig. 2).

2. Life Cycle

The life cycle is direct and takes approximately 23–25 days in susceptible hosts (Anya, 1966b; Brown *et al.*, 1954; Sasa *et al.*, 1962). Mature females reside in the large intestine, where they can survive for 45–50 days (Hsieh, 1952). They are usually located in the terminal colon, when they lay eggs, but do not leave the host for this purpose, as does *S. obvelata* (Chan, 1955). The eggs are laid at night (Phillipson, 1974) and leave the host in the mucous layer of fecal pellets. They require incubation for 6–7 days at 24°C to become infective (Anya, 1966a; Sasa *et al.*, 1962). The infective egg is resistant and can survive for weeks outside the host (Sasa *et al.*, 1962). When ingested by a susceptible host, the eggs hatch and larvae reach the middle of the colon, where they enter the crypts of Lieberkühn and remain for 4–5 days (Behnke, 1974a). They then move to the proximal colon in about 7 days, but never enter the cecum, and then gradually migrate to the distal colon about 3 weeks after exposure (Behnke, 1974a; Chan, 1955). Notable features of the host response to *A. tetraptera* are that there is an immune expulsion of parasites and resistance to reinfection (Behnke, 1974b; Stahl, 1966a), male mice are more susceptible than females (Stahl, 1961), and exposure to *S. obvelata* may provide some protection against infection with *A. tetraptera* (Stahl, 1966b).

The length of the life cycle is some 10–12 days longer than that of *S. obvelata,* which means that patent *A. tetraptera* infections appear in somewhat older hosts. The heaviest infections can be expected in 5- to 6-week-old mice, although the presence of *A. tetraptera* can be demonstrated in hosts of a wide age range (Sasa *et al.*, 1962).

3. Pathology

Generally, light to moderate *A. tetraptera* infections do not produce clinical disease. Heavy infections could be expected to produce the same types of conditions discussed for *S. obvelata,* namely, rectal prolapse, intussusception, enteritis, and fecal impaction. Reproduction of these conditions has not been accomplished experimentally.

4. Diagnosis

Aspiculuris tetraptera can be diagnosed by demonstration of eggs by fecal examination of adult worms in the large intestine at necropsy. Eggs are not found in the perianal area, so cellophane tape techniques used for *S. obvelata* are of no diagnostic value for this parasite. Also, since *A. tetraptera* does not inhabit the cecum, examination of that organ has limited diagnostic value. Eggs are sufficiently distinctive (Fig. 2) to diagnose the infection, and adult worms can be differentiated from *S. obvelata* with little difficulty (Table I). Techniques used for recovery and examination of adults from the large intestine are the same as those described for *S. obvelata*.

Table I

Differentiation of *Syphacia obvelata* and *Aspiculuris tetraptera*

Characteristic	*Syphacia obvelata*	*Aspiculuris tetraptera*
Cervical alae	Small	Prominent
Shape of tail of female	Long, pointed	Conical
Location of vulva	Anterior of body	Middle of body
Mamelons in male	Present	Absent
Spicule	Present	Absent
Location in host	Cecum and large intestine	Large intestine only

5. Treatment

Control measures used for *S. obvelata* ordinarily will work for *A. tetraptera*. In fact, the life cycle of the latter probably is more amenable to treatment. *Aspiculuris tetraptera* takes longer to mature, its eggs take longer to develop to the infective stage, and they are not carried on the host but deposited in fecal material. This means that once the adult parasites are removed by treatment and mice are placed in clean cages, chances for recontamination are less than for *S. obvelata*. If reinfection does occur, it will also take longer for large populations of parasites to develop. Although some differences have been noted (Brody and Elward, 1971) in the efficacy of drugs against *A. tetraptera* and *S. obvelata*, they should not influence control. Hence procedures discussed for *S. obvelata* should be adequate.

C. *Hymenolepis nana*, Dwarf Tapeworm

This cestode is a cosmopolitan parasite of the house mouse, the Norway rat, and man. Surveys indicate that its prevalence in laboratory and wild mice at times approaches that of the oxyurids, *S. obvelata* and *A. tetraptera* (Flynn *et al.*, 1965; Owen, 1976; Sasa *et al.*, 1962; Simmons *et al.*, 1964; Weber, 1976). However, most investigators feel that the prevalence of this parasite has decreased in well-managed commercial colonies (Wescott and Van Hoosier, 1974). Apparently, once the infection is eliminated from a colony, chances for reexposure are less than for the oxyurids, and good management and sanitation discourage this cestode. The fact that *H. nana* is less common in laboratory mice than pinworms is fortunate because this cestode presents a human health problem (Bisseru, 1967; Stone and Manwell, 1966; Chandler and Read, 1961) and is more pathogenic (Oldham, 1967; Sasa *et al.*, 1962; Simmons *et al.*, 1967).

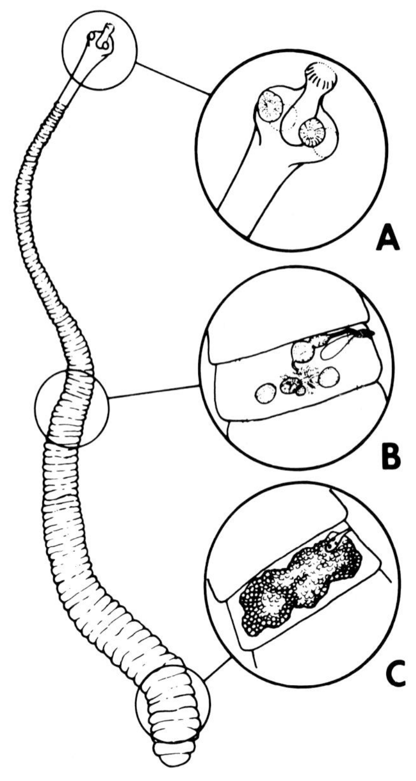

Fig. 4. *Hymenolepis nana*. (A) Scolex with rostellum exserted. (B) Mature segment. (C) Ripe segment. (From Heston, 1941, and Chandler and Read, 1961, after Leuckart and Bailey.)

1. Morphology

Important details of *H. nana* are shown in Fig. 4. Detailed morphologic descriptions are available elsewhere (Flynn, 1973; Oldham, 1967; Wardle and McLeod, 1952). Adults range from 25 to 40 mm in length and 0.5–1.0 mm in width. The scolex has four circular suckers and a short rostellum with 20–27 hooks in a single ring. Mature segments are wider (0.8 mm) than they are long (0.2 mm). They contain three testes, a single ovary, and a compact vitelline gland. Gravid segments are about 1.0 mm wide and contain 100–200 eggs which measure 50–53 × 37–41 μm. Eggs have prominent polar filaments and hooks (Figs. 2, 5).

2. Life Cycle

The life cycle of *H. nana* is unique in that it may or may not utilize an intermediate host (Flynn, 1973; Oldham, 1967). In the direct cycle, eggs are ingested by the host and hatch in the intestine, and the liberated onchospheres penetrate the villi of the duodenum or intestine. There they develop into a cercocystis stage and emerge into the lumen after 10–12 days. The scolex evaginates and attaches to the mucosa and the parasite grows to adult size in 2 weeks, making the entire cycle from ingestion to patency about 20–30 days (Oldham, 1967; Przyjalkowski, 1977: Sasa *et al.*, 1962).

The indirect cycle can utilize a variety of arthropod intermediate hosts (Oldham, 1967; Wardle and McLeod, 1952). When necessary, the parasite develops to a cercocystis in the arthropod and completes development in about 10 more days when ingested by the definitive host (Sasa, *et al.*, 1962). Adult worms survive only a few weeks in most mice before they are expelled (Hunninen, 1935), and eggs survive only a short time outside the host under normal conditions (Oldham, 1967).

3. Pathology

The early stages of parasite development within the intestinal villi have been reported to produce mild to moderately severe lesions (Oldham, 1967; Sasa *et al.*, 1962; Simmons *et al.*, 1967). In addition, this infection has been reported to reduce weight gains (Cook, 1969). The mature worms are less

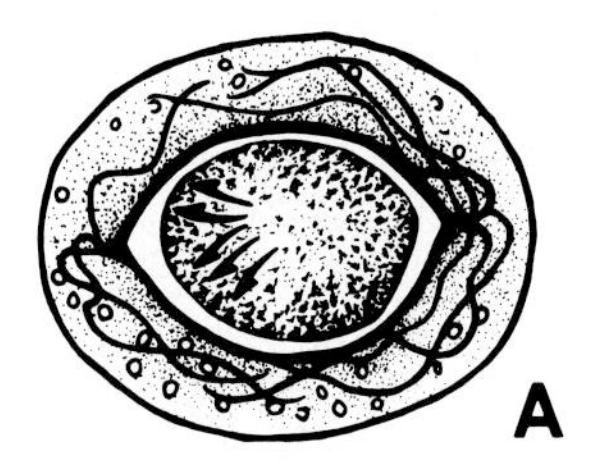

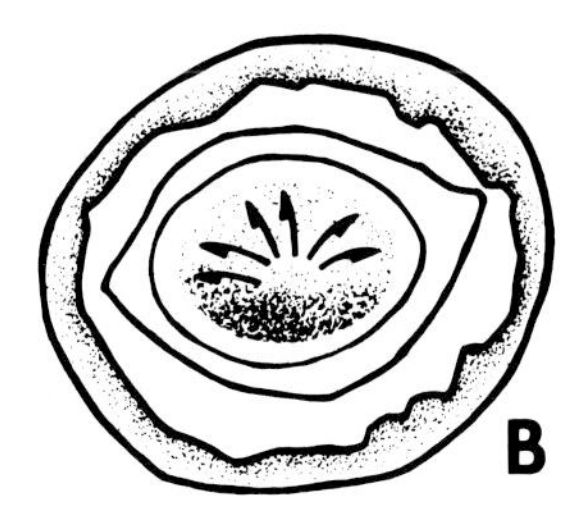

Fig. 5. Egg of (A) *Hymenolepis nana* and (B) *Hymenolepis diminuta*. (From Heston, 1941, after Augustine.)

pathogenic and readily rejected by the host response (Hunninen, 1935).

4. Diagnosis

Hymenolepis nana can be diagnosed by demonstration of the presence of eggs in routine fecal flotations or adult worms in the small intestine. Examinations should include some hosts in the 5- to 7-week age group, which are most frequently infected (Sasa *et al.*, 1962).

Eggs are characteristic and not readily confused with those of mouse pinworms. Differentiation from a similar cestode, *Hymenolepis diminuta,* is based on the presence of polar filaments in *H. nana* and the absence of them in *H. diminuta* (Fig. 5).

The presence of adult worms is readily demonstrated by opening the small intestine in a petri dish filled with tap water. Numerous small cestodes, 25–40 mm in length and <1 mm in width, can be observed attached to the mucosa of infected mice. Measurements of these parasites will help differentiate them from the larger *H. diminuta,* which measures 20–60 mm long by 4 mm wide. However, to confirm the diagnosis, morphologic features such as the lack of hooks on the rostellum of *H. diminuta* and the presence of such hooks in *H. nana* must be compared (Wardle and McLeod, 1952).

5. Treatment

Effective control for *H. nana* is less difficult to achieve than for the oxyurids, *S. obvelata* and *A. tetraptera*. This is probably due to the inability of *H. nana* eggs to survive for long periods outside the host (Oldham, 1967) and to the generally lower incidence of *H. nana* in uncaged mice (Owen, 1976). Once mice are treated and the parasites eliminated by drug therapy or caesarian derivation, there is less chance for reexposure. Undoubtedly, the strict isolation procedures applied to control pinworms have indirectly reduced the prevalence of *H. nana*.

A number of drugs have been recommended for treatment of *H. nana,* including quinacrine hydrochloride (Balazs *et al.*, 1962), bunamidine (Brody and Elward, 1971), niclosamide (Flynn, 1973), and thiabendazole (Taffs, 1976b). However, the compounds which have most promise appear to be the newer benzimidazoles, which have excellent efficacy against both cestrodes and nematodes. The advantages of using drugs with broad-spectrum activity, such as parbendazole (Brody and Elward, 1971, mebendazole (Borgens *et al.*, 1975; Sharp and Wescott, 1976), albendazole (Wescott *et al.,* 1979a), and oxfendazole (Wescott *et al.,* 1979b), would be greatest in hosts with more than one type of parasite. Unfortunately, there are no reports on the efficacy of these compounds on mixed helminth infections in mice. The treatment of choice at present appears to be bunamidine administered for 18 days in the diet at 1000 ppm (Brody and Elward, 1971).

The other effective method of control is caesarian derivation (Foster, 1963). Needless to say, the same precautions should be observed after these procedures as for pinworms, together with control of potential intermediate hosts, to ensure that reinfection with *H. nana* does not occur.

III. HELMINTHS OF MINOR IMPORTANCE

A. Nematodes

1. *Capillaria hepatica*

Slender nematodes, *C. hepatica* are 17–100 mm in length which are found in the liver of rodents, including the house mouse (Levine, 1968). They may produce visible yellow streaks on livers but are relatively nonpathogenic (Oldham, 1967) and have not been reported as spontaneous infections in laboratory mice.

2. *Heterakis spumosa*

Ordinarily this is a parasite of wild and occasionally laboratory rats (Oldham, 1967). Experimental infections have been produced in mice (Smith, 1953). Should infection occur in mice, the parasites would be found in the cecum and large intestine and could be mistaken for pinworms. Differentiation could be made by the appearance of the eggs, which have thick mamillated shells, and the paired spicules of the male. No lesions have been associated with this infection (Oldham, 1967).

Table II

Helminths of Mice Not Described in Text[a]

Parasite	Geographic distribution	Endothermal host	Location in host	Incidence: In nature	Incidence: In laboratory	Reference
Nematodes						
Capillaria bacillata	Europe	Mouse, rat, European field mouse	Esophagus	Unknown	Absent except in specimens obtained from their natural habitat	Levine (1968)
Capillaria muris-musculi	Europe	Mouse	Intestine	Unknown	Absent except in specimens obtained from their natural habitat	Levine (1968)
Gongylonema neoplasticum	Worldwide	Mouse, rat, black rat, hamsters, voles, other rodents, rabbit, hares	Esophagus, stomach	Common	Absent except in specimens obtained from their natural habitat	Ash (1962); Bacigalupo (1934); Chabaud (1954, 1955); Kirschenblatt (1949); Lapage (1968); de Léon (1964); Levine (1968); Oldham (1967)
Gongylonema musculi	Europe	Mouse	Esophagus, stomach	Unknown	Unknown	Yamaguti (1961)
Heligmonoides murina	West Africa	Mouse	Small intestine	Unknown	Unknown	Baylis (1928); Yamaguti (1961)
Longistriata musculi	Southern United States	Mouse	Small intestine	Unknown	Unknown	Dikmans (1935); Morgan and Hawkins (1949); Schwartz and Alicata (1935)
Mastophorus muricola	Europe, Africa, Asia, Central America, West Indies, Philippine Islands	Mouse, rat, black rat, hamster, other rodents, monkey	Stomach	Common	Absent except in specimens obtained from their natural habitat	Chabaud (1954, 1955); Foster and Johnson (1939); de Léon (1964); Myers *et al.* (1962); Yamaguti (1961)
Nematospiroides dubius	North America, Europe	Mouse, deer mice, voles, European field mouse, other rodents	Small intestine	Common	Absent	Dobson (1962); Ehrenford (1954); Fahmy (1956); Lepak *et al.* (1962); Levine (1968); Lewis (1968a,b); Thompson *et al.* (1962)
Nematodes						
Nippostrongylus brasiliensis	Worldwide	Mouse, rat, black rat	Small intestine	Common in rat; rare in mouse	Experimental only	Habermann and Williams (1958); Haley (1961, 1962); Levine (1968); Sasa *et al.* (1962)
Physaloptera massino	North America, Europe	Mouse, ground squirrels, other rodents	Stomach	Unknown	Absent except in specimens obtained from their natural habitat	Morgan (1943); Yamaguti (1961)

Table II—*Continued*

Parasite	Geographic distribution	Endothermal host	Location in host	Incidence		Reference
				In nature	In laboratory	
Rictularia muris	Europe	Mouse	Small intestine	Unknown	Unknown	Yamaguti (1961)
Rictularia magna	Europe	Mouse	Small intestine	Unknown	Unknown	Yamaguti (1961)
Rictularia baicalensis	USSR	Mouse, European field mouse	Small intestine	Unknown	Unknown	Chabaud (1955); Yamaguti (1961)
Rictularia amurensis	USSR	Mouse	Small intestine	Unknown	Unknown	Yamaguti (1961)
Trematodes						
Plagiorchis muris	North America, Hawaii, Asia	Mouse, rat, man	Intestine	Common in some areas	Absent	Ash (1962); Bisseru (1967); Olsen (1967); Sasa *et al.* (1962)
Plagiorchis philippinensis	Philippine Islands	Mouse, rat, man	Intestine	Common in some areas	Absent	Bisseru (1967)
Plagiorchis javensis	Indonesia	Mouse, rat, man	Intestine	Common in some areas	Absent	Bisseru (1967)
Cestodes						
Oochoristica ratti	United States, Japan	Mouse, rat, black rat	Small intestine	Rare	Unknown; probably absent	Oldham (1967); Rendtorff (1948); Wardle and McLeod (1952)

[a] Reprinted by permission from "Parasites of Laboratory Animals" by Robert J. Flynn © 1973 by Iowa State University Press.

3. *Syphacia muris*

This is the common oxyurid parasite of the Norway rat (Levine, 1968), and infection is ordinarily restricted to that host, although it can occur in the house mouse (Hussey, 1957). Differentiation of *S. muris* from *S. obvelata* can be done most readily on the basis of egg size (see Section II, A, 4). Treatment would be as recommended for pinworms of mice.

4. *Trichinella spiralis*

Spontaneous *T. spiralis* infections of laboratory mice have not been reported. However, the prevalence of this parasite in wild rodents is such that infection in laboratory hosts is not impossible. Diagnosis would be made on demonstration of the presence of encysted larvae in striated muscle (Levine, 1968). Control would be based on removal of encysted infective larvae from feed sources.

5. *Trichuris muris*

The whipworm, *T. muris,* is common in wild mice (Behnke and Wakelin, 1973; Fahmy, 1954) but has not been reported as a spontaneous infection in laboratory mice. The relatively long (30 days) period required for eggs to become infective (Fahmy, 1954) precludes laboratory infection when good husbandry practices are used. Should *T muris* be suspected, these worms are relatively large (16–25 mm long), possess the distinctive morphology of whipworms (Morgan and Hawkins, 1949; Flynn, 1973), and are found as adults in the cecum and large intestine.

B. Cestodes

1. *Hymenolepis diminuta*

This tapeworm infects the house mouse, the Norway rat, a number of wild rodents, the monkey, and man (Oldham, 1967; Flynn, 1973). No reports of spontaneous infection in laboratory mice have appeared recently, indicating that its prevalence currently is less than that of *H. nana*. The requirement of an arthropod intermediate host in the life cycle of *H. diminuta* (Wardle and McLeod, 1952) undoubtedly has limited its incidence in laboratory mice where good management practices are observed. Differentiation from *H. nana* can be done on the basis of the size of the mature worm and the appearance of the scolex and eggs. *Hymenolepis diminuta* is larger (20-60 mm long by 4 mm wide) than *H. nana* (25-40 mm long by 1 mm wide). The scolex of H. diminuta has an unarmed rostellum, whereas that of *H. nana* has a prominent row of hooks. The eggs of *H. diminuta* do not possess the polar filaments present in *H. nana* (Fig. 5). Should infection be diagnosed, treatment would be the same as for *H. nana*, with more attention given to removal of potential arthropod intermediate hosts.

2. *Hymenolepis microstoma*

This is a cosmopolitan cestode found in the bile duct of rodents (Jones, 1966). The adult parasite could be confused with *H. nana* should it be found in laboratory mice. However, the location of *H. microstoma* in the bile duct as well as differences in the size of eggs (*H. microstoma* eggs are approximately twice as large as those of *H. nana*) make the diagnosis relatively simple.

3. *Taenia taeniaeformis*

The house mouse and Norway rat ordinarily serve as intermediate hosts for this parasite of the cat. The infective stage (*Cysticercus fasciolaris*) is a strobilocercus in the liver of the rodent host. These are large (up to 1 cm in diameter) white cysts which contain a distinctive larva composed of a scolex, a segmented strobila, and a small bladder giving it the appearance of a miniature cestode (Oldham, 1967). Should this infection be found in laboratory mice, control would consist of removal of the source of *T. taeniaeformis* eggs from the diet (Balk and Jones, 1970). Ordinarily, this would be accomplished by storing the diet where it would not become contaminated with feces of infected cats. Benzimidazole compounds have been shown to damage cysts in mice (Campbell and Blair, 1974; Thienpont *et al.*, 1974) and could be considered if treatment is desired.

ACKNOWLEDGMENTS

The author would like to thank Dr. Robert Flynn of the Argonne National Laboratory, Argonne, Illinois, for his assistance in editing this chapter and graciously granting permission for the use of a great deal of information from his book entitled "Parasites of Laboratory Animals" (1973). Special thanks are also due to Dr. Charles Leathers of Washington State University, who did much of the photographic work supporting the text.

REFERENCES

Anya, A. O. (1966a). Studies on the biology of some oxyurid nematodes. I. Factors in the development of eggs of *Aspiculuris tetraptera* Schulz. *J. Helminthol.* **40,** 253-260.

Anya, A. O. (1966b). Studies on the biology of some oxyurid nematodes. II. The hatching of eggs and development of *Aspiculuris tetraptera* Schulz within the host. *J. Helminthol.* **40,** 261-268.

Andreassen, J., Hindsbo, O., and Ruitenberg, E. J. (1978). *Hymenolepis diminuta* infections in congenitally athymic (nude) mice: worm kinetics and intestinal histopathology. *Immunology* **34,** 105-113.

Ash, L. R. (1962). The helminth parasites of rats in Hawaii and the description of *Capillaria traverae* sp. n. *J. Parasitol.* **48,** 66-68.

Bacigalupo, J. (1934). El ciclo evolutivo del *Gonylonema neoplasticum* (Fibiger-Ditlevsen) en la Argentina. *Actas Trab. Congr. Nac. Med., 5th* **3,** 947-949.

Balazs, T., Hatch, A. M., Gregory, E. R. W., and Grice, H. C. (1962). A comparative study of hymenolepicides in *Hymenolepis nana* infestation of rats. *Can. J. Comp. Med. Vet. Sci.* **26,** 160-162.

Balk, M. W., and Jones, S. R. (1970). Hepatic cysticercosis in a mouse colony. *J. Am. Vet. Med. Assoc.* **157,** 678-679.

Baylis, H. A. (1928). On a collection of nematodes from Nigerian mammals (chiefly rodents). *Parasitology* **20,** 280-304.

Befus, A. D. (1975). Secondary infections of *Hymenolepis diminuta* in mice: effects of varying worm burdens in primary and secondary infections. *Parasitology* **71,** 61-75.

Befus, A. D. (1977). *Hymenolepis diminuta* and *H. microstoma:* mouse immunoglobulins binding to the tegumental surface. *Exp. Parasitol.* **41,** 242-251.

Befus, A. D., and Featherston, D. W. (1974). Delayed rejection of single *Hymenolepis diminuta* in primary infections of young mice. *Parasitology* **69,** 77-85.

Behnke, J. M. (1974a). The distribution of larval *Aspiculuris tetraptera* Schultz during a primary infection in *Mus musculus, Rattus norvegicus,* and *Apodemus sylvaticus. Parasitology* **69,** 391-402.

Behnke, J. M. (1974b). The effect of hydrocortisone upon infection with *Aspiculuris tetraptera* in laboratory mice. *Parasitology* **69,** xviii.

Behnke, J. M. (1975a). Immune expulsion of the nematode *Aspiculuris tetraptera* from mice given primary and challenge infections. *Int. J. Parasitol.* **5,** 511-515.

Behnke, J. M. (1975b). *Aspiculuris tetraptera* in wild *Mus musculus.* The prevalence of infection in male and female mice. *J. Helminthol.* **49,** 85-90.

Behnke, J. M. (1976). *Aspiculuris tetraptera* in wild *Mus musculus.* Age resistance and acquired immunity. *J. Helminthol.* **50,** 197-202.

Behnke, J. M., and Wakelin, D. (1973). The survival of *Trichuris muris* in wild populations of its natural hosts. *Parasitology* **67,** 157-164.

Bieniek, Von H., and Tober-Myer, B. (1976). Zur ätiologie der colitis und des *prolapsus recti* beider mäus. *Z. Versuchstierkd.* **18,** 337-348.

Bisseru, B. (1967). "Diseases of Man Acquired From His Pets." Lippincott, Philadelphia, Pennsylvania.

Blair, L. S., Thompson, P. E., and Vandenbelt, J. M. (1968). Effects of pyrvinium pamoate in the ration or drinking water of mice against pinworms *Syphacia obvelata* and *Aspiculuris tetraptera. Lab. Anim. Care* **18,** 314-327.

Borgens, M., De Nollin, S., Verheyen, A., Vanparijs, O., and Thienpont, D. (1975). Morphological changes in cysticerci of *Taenia taeniaeformis* after mebendazole treatment. *J. Parasitol.* **61,** 830-843.

Brody, G., and Elward, T. E. (1971). Comparative activity of 29 known anthelmintics under standardized drug-diet gavage medication regimens against four helminth species in mice. *J. Parasitol.* **57,** 1068-1077.

Brown, H. W., Chan, K. F., and Ferrell, B. D. (1954). A study of the activity of chemotherapetic agents on infections of *Syphacia obvelata* and *Aspiculuris tetraptera*. *Exp. Parasitol.* **3,** 45-51.

Campbell, W. C., and Blair, L. S. (1974). Prevention and cure of hepatic cysticercosis in mice. *J. Parasitol.* **60,** 1049-1052.

Chabaud, A. G. (1954). Sur le cycle évolutif des spirurides et de nématodes ayant une biologie comparable. Valeur systématique des caractéres biologiques. *Ann. Parasitol. Hum. Comp.* **29,** 42-88, 206-249, 358-425.

Chabaud, A. G. (1955). Essai d'interprétation phylétique des cycles evolutifs chez les nématodes parasites de vertèbres. Conclusions taxomoniques. *Ann. Parasitol. Hum. Comp.* **30,** 83-126.

Chan, K. F. (1952). Life cycle studies on the nematode *Syphacia obvelata*. *Am. J. Hyg.* **56,** 14-21.

Chan, K. F. (1955). The distribution of larval stages of *Aspiculuris tetraptera* in the intestine of mice. *J. Parasitol.* **41,** 529-532.

Chandler, A. C., and Read, C. P. (1961). "Introduction to Parasitology," 10th ed. Wiley, New York.

Chowaniec, W., Wescott, R. B., and Congdon, L. L. (1972). Interaction of *Nematospiroides dubius* and influenza virus in mice. *Exp. Parasitol.* **32,** 33-44.

Cook, R. (1969). Common diseases of laboratory animals. *In* "The I.A.T. Manual of Laboratory Animal Practice and Techniques" (D. J. Short and D. P. Woodnott, eds.), 2nd ed., pp. 160-215. Thomas Springfield, Illinois.

de Léon, D. D. (1964). Helminth parasites of rats in San Juan, Puerto Rico. *J. Parasitol.* **50,** 478-479.

DeWitt, W. B., and Weinstein, P. P. (1964). Elimination of intestinal helminths of mice by feeding purified diets. *J. Parasitol.* **50,** 429-434.

Dikmans, G. (1935). New nematodes of the genus *Longistriata* in rodents. *J. Wash. Acad. Sci.* **25,** 72-81.

Dobson, C. (1962). Certain aspects of the host-parasite relationship of *Nematospiroides dubius* (Baylis): V Host specificity. *Parasitology* **52,** 41-48.

Ehrenford, F. A. (1954). The life cycle of *Nematospiroides dubius* (Baylis) (Nematoda: Heligmosomidae). *J. Parasitol.* **40,** 480-481.

Fahmy, M. A. M. (1954). An investigation on the life cycle of *Trichuris muris*. *Parasitology* **44,** 50-57.

Fahmy, M. A. M. (1956). An investigation on the life cycle of *Nematospiroides dubius* (Mematoda: Heligmosomidae) with special reference to the free living stages. *Z. Parasitenkd.* **17,** 394-399.

Flynn, R. J. (1973). "Parasites of Laboratory Animals." Iowa State Univ. Press, Ames.

Flynn, R. J., Brennan, P. C., and Fritz, T. E. (1965). Pathogen status of commercially produced laboratory mice. *Lab. Anim. Care* **15,** 440-447.

Foster, A. O., and Johnson, C. M. (1939). A preliminary note on the identity, life cycle, and pathogenicity of an important nematode parasite of captive monkeys. *Am. J. Trop. Med.* **19,** 265-277.

Foster, H. L. (1963). Specific pathogen-free Animals. *In* "Animals for Research: Principles of Breeding and Management" (W. Lane-Petter, ed.), pp. 110-137. Academic Press, New York.

Fratta, I. D., and Slanetz, C. A. (1958). The treatment of oxyurid infested mice with piperazine citrate, stylomycin, caricide, phthalofyne and gibberellic acid. *Proc. Anim. Care Panel* **8,** 141-146.

Goodall, R. I. (1973). *Hymenolepis diminuta* in the mouse: effects of heterologous antigens on worm rejection—a possible biological model for antigenic competition. *Parasitology* **67,** xviii.

Habermann, R. T., and Williams, F. P. (1957). The efficacy of some piperazine compounds and stylomycin in drinking water for the removal of oxyurids from mice and rats and a method of critical testing of anthelmintics. *Proc. Anim. Care Panel* **7,** 89-97.

Habermann, R. T., and Williams, F. P. (1958). The identification and control of helminths in laboratory animals. *J. Natl. Cancer Inst.* **20,** 979-1009.

Haley, A. J. (1961). Biology of the rat nematode, *Nippostrongylus brasiliensis* (Travassos, 1914): I. Systematics, hosts and geographic distribution. *J. Parasitol.* **47,** 727-732.

Haley, A. J. (1962). Biology of the rat nematode, *Nippostrongylus brasiliensis* (Travassos, 1914): II. Preparasitic stages and development in the laboratory rat. *J. Parasitol.* **48,** 13-23.

Harwell, J. F., and Boyd, D. D. (1968). Naturally occurring oxyuriasis in mice. *J. Am. Vet. Med. Assoc.* **153,** 950-953.

Hasslinger, M. A., and Bosse, K. (1966). Intestinal parasites in small research animals. *Muench. Med. Wochenschr.* **108,** 2394-2396.

Heston, W. E. (1941). Parasites. *In* "Biology of the Laboratory Mouse" (C. Snell, ed.), pp. 359-369. Blakiston, Philadelphia, Pennsylvania.

Hoag, W. G. (1961). Oxyuriasis in laboratory mouse colonies. *Am. J. Vet. Res.* **22,** 150-153.

Hoag, W. G. (1961). Oxyuriasis in laboratory mouse colonies. *Am. J. Vet. Res.* **22,** 150-153.

Hopkins, C. A., Subramanian, G., and Stallard, H. (1972). The development of *Hymenolepis diminuta* in primary and secondary infections in mice. *Parasitology* **64,** 401-412.

Hopkins, C. A., Stallard, H., and Befus, A. D. (1973). Immunological rejection of *Hymenolepis diminuta* by mice. *Parasitology* **67,** xvii.

Hopkins, C. A., Goodall, R. I., and Zajac, A. (1977). The longevity of *Hymenolepis microstoma* in mice, and its immunological cross-reaction with *Hymenolepis diminuta*. *Parasitology* **74,** 175-183.

Hsieh, K. Y. N. (1952). The effect of standard pinworm agents on the mouse pinworm *Aspiculuris tetraptera*. *Am. J. Hyg.* **56,** 287-293.

Hunninen, A. V. (1935). Studies on life history of and host-parasite relations of *Hymenolepis fraterna* (*H. nana* var *fraterna,* Stiles) in white mice. *Am. J. Hyg.* **22,** 414-443.

Hunt, G. R., and Burrows, R. B. (1963). Stilbazium iodide in the mass treatment of mice infected with *Syphacia obvelata*. *J. Parasitol.* **49,** 1019-1020.

Hussey, K. L. (1957). *Syphacia muris* vs. *S. obvelata* in laboratory rats and mice. *J. Parasitol.* **43,** 555-559.

Jirina, K. (1952). Die bedeutung der mäusecysticercose für laboratoriums versuche. *Z. Tropenmed. Parasitol.* **4,** 510-512.

Jones, A. W. (1966). *Hymenolepis microstoma* (Dujardin, 1845) a cosmopolitan cestode of the bile duct of rodents. *Proc. Int. Congr. Parasitol. 1st, Rome, 1964* **1,** 478-479.

King, V. M., and Cosgrove, G. E. (1963). Intestinal helminths in various strains of laboratory mice. *Proc. Anim. Care Panel* **13,** 46-48.

Kirschenblatt, I. D. (1949). On the helminth fauna of *Mesocricetus auratus brandii* Nehr. (In Russ.) *Uch. Zap. Leningr. Gos. Univ. im. A. A. Zhdanova, Ser. Biol. Nauk* No. 19, 110-127.

Kyaw, A., and Oo, M. (1976). Increased hepatic lysosomal enzyme levels in mice infected with *Hymenolepis diminuta*. *Jpn. J. Med. Sci. Biol.* **29,** 105-108.

Lapage, G. (1968). "Veterinary Parasitology." Oliver & Boyd, Edinburgh.

Lawler, H. J. (1939). Demonstration of the life history of the nematode *Syphacia obvelata* (Rudolphi, 1802). *J. Parasitol.* **25,** 442.

Lepak, J. W., Thatcher, V. E., and Scott, J. A. (1962). The development of *Nematospiroides dubius* in an abnormal site in the white mouse and some apparent effect on the serum proteins of the host. *Tex. Rep. Biol. Med.* **20,** 374-383.

Levine, N. D. (1968). "Nematode Parasites of Domestic Animals and Man." Burgess, Minneapolis, Minnesota.

Lewis, J. W. (1968a). Studies on the helminth parasites of the longtailed field mouse *Apodemus sylvaticus sylvaticus* from Wales. *J. Zool.* **154,** 287-312.

Lewis, J. W. (1968b). Studies on the helminth parasites of voles and shrews from Wales. *J. Zool.* **154,** 313–331.

McNair, D. M., and Timmons, E. H. (1977). Effects of *Aspiculuris tetraptera* and *Syphacia obvelata* on exploratory behavior of an inbred mouse strain. *Lab. Anim. Sci.* **27,** 38–42.

Mayer, L. P., and Pappas, P. W. (1976). *Hymenolepis microstoma:* effect of the mouse bile duct tapeworm on the metabolic rate of CF-1 mice. *Exp. Parasitol.* **40,** 48–51.

Morgan, B. B. (1943). The *Physaloptera* (Nematoda) of rodents. *Wasmann Collect.* **5,** 99–107.

Morgan, B. B., and Hawkins, P. A. (1949). "Veterinary Helminthology." Burgess, Minneapolis, Minnesota.

Myers, B. J., Kuntz, R. E., and Wells, W. H. (1962). Helminth parasites of reptiles, birds, and mammals in Egypt: VII. Check list of nematodes collected from 1948 to 1955. *Can. J. Zool.* **40,** 531–538.

Oldham, J. N. (1967). Helminths, ectoparasites, and protozoa in rats and mice. *In* "Pathology of Laboratory Rats and Mice" (E. Cotchin and F. J. Roe, eds.), pp. 641–679. Blackwell, Oxford.

Olsen, O. W. (1967). "Animal Parasites: Their Biology and Life Cycles," 2nd ed. Burgess, Minneapolis, Minnesota.

Owen, D. (1976). Some parasites and other organisms of wild rodents in the vicinity of an SPF unit. *Lab. Anim.* **10,** 271–278.

Patterson, M. A., and Vessey, S. H. (1973). Tapeworm (*Hymenolepis nana*) infection in male albino house mice: effect of fighting among the hosts. *J. Mammal.* **54,** 784–786.

Pearson, D. T., and Taylor, G. (1975). The influence of the nematode *Syphacia obvelata* on adjuvant arthritis in the rat. *Immunology* **29,** 391–396.

Phillipson, R. F. (1974). Intermittent egg release by *Aspiculuris tetraptera* in mice. *Parasitology* **69,** 207–213.

Pilgram, K., and Hass, D. K. (1975). Novel phosphate anthelmintics. 3. Alkyl and aryl 1-methyleneallyl phosphates, phosphonates, and phosphinates. *J. Med. Chem.* **18,** 1204–1211.

Przyjalkowski, Z. (1974). *Aspiculuris tetraptera* Nitzsch, 1821 (*Nematoda, Oxyuridae*) establishment in gnotobiotic mice. I. Experiments with mice affected by *Escherichia coli. Acta Parasitol. Pol.* **22,** 345–349.

Przjalkowski, Z. (1977). Establishment, growth and rate of expulsion of the cestode *Hymneolepis nana* Siebold, 1882 in germfree and conventional mice. *Acta Parasitol.* **25,** 63–68.

Rendtorff, R. C. (1948). Investigations on the life cycle of *Oochoristica ratti,* a cestode from rats and mice. *J. Parasitol.* **34,** 243–252.

Sasa, M., Tanaka, H., Fukui, M., and Takata, A. (1962). Internal parasites of laboratory animals. *In* "The Problem of Laboratory Animal Disease" (R. J. C. Harris, ed.), pp. 195–214. Academic Press, New York.

Schwartz, B., and Alicata, J. E. (1935). Life history of *Longistriata musculi,* a nematode parasite in mice. *J. Wash. Acad. Sci.* **25,** 128–146.

Seth, D., and Louekar, C. D. (1973). A new chemotherapeutic property of metronidazole: effect against oxyurids in mice. *J. Pharm. Pharmacol.* **25,** 1015.

Sharp, J. W., and Wescott, R. B. (1976). Anthelmintic efficacy of mebendazole for pinworm infection of mice. *Lab. Anim. Sci.* **26,** 222–223.

Shimp, R. G., Crandall, R. B., and Crandall, C. A. (1975). *Heligmosomoides polygyrus* (= *Nematospiroides dubius*): suppression of antibody response to orally administered sheep erythrocytes in infected mice. *Exp. Parasitol.* **38,** 257–269.

Simmons, M. L., Williams, H. E., and Wright, E. B. (1964). Parasite screening and therapeutic value of organic phosphates in inbred mice. *Lab. Anim. Care* **14,** 326.

Simmons, M. L., Williams, H. E., and Wright, E. B. (1965). Therapeutic value of the organic phosphate trichlofon against *Syphacia obvelata* in inbred mice. *Lab. Anim. Care* **15,** 382–385.

Simmons, M. L., Richter, C. B., Franklin, J. A., and Tennant, R. W. (1967). Prevention of infectious diseases in experimental mice. *Proc. Soc. Exp. Biol. Med.* **126,** 830–837.

Singhal, K. C. (1975). A new piperazinobiguanide series effective against *Syphacia obvelata* in mice. *Jpn. J. Pharmacol.* **25,** 352–354.

Smith, P. E. (1953). Host specificity of *Heterakis spumosa* Schneider, 1866 (Nematoda: Heterakidae). *Proc. Helminthol. Soc. Wash.* **20,** 19–21.

Stahl, W. (1961). Influences of age and sex on the susceptibility of albino mice to infection with *Aspiculuris tetraptera. J. Parasitol.* **47,** 939–941.

Stahl, W. (1966a). Experimental aspiculuriasis. I. Resistance to superinfection. *Exp. Parasitol.* **18,** 109–115.

Stahl, W. (1966b). Experimental aspiculuriasis. II. Effects of concurrent helminth infections. *Exp. Parasitol.* **18,** 116–123.

Stone, W. B., and Manwell, R. D. (1966). Potential helminth infections in humans from pet or laboratory mice and hamsters. *Public Health Rep.* **81,** 647–653.

Taffs, L. F. (1975). Continuous feed medication with thiabendazole for the removal of *Hymenolepis nana, Syphacia obvelata* and *Aspiculuris tetraptera* in naturally infected mice. *J. Helminthol.* **49,** 173–177.

Taffs, L. F. (1976a). Pinworm infections in laboratory rodents: A review. *Lab. Anim.* **10,** 1–13.

Taffs, L. F. (1976b). Further studies on the efficacy of thiabendazole given in the diet of mice infected with *H. nana, S. obvelata* and *A. tetraptera. Vet. Rec.* **99,** 143–144.

Thienpont, D., Vanparijs, O., and Hermans, L. (1974). Anthelminthic activity of mebendazole against *Cysticercus fasciolaris. J. Parasitol.* **60,** 1052–1053.

Thompson, P. E., and Reinertson, J. W. (1952). Chemotherapeutic studies of natural pinworm infections in mice. *Exp. Parasitol.* **1,** 384–391.

Thompson, P. E., Worley, D. E., and McClay, P. (1962). Effects of bis (2, 4, 5 -trichlorophenol) piperazine salt against intestinal nematodes in laboratory animals. *J. Parasitol.* **48,** 572–577.

Turner, H. M., and McKeever, S. (1976). The refractory responses of the White Swiss strain of *Mus musculus* to infection with *Taenia taeniaeformis. Int. J. Parasitol.* **6,** 483–487.

Voge, M. (1958). Helminth infections in small laboratory mammals and methods of obtaining a helminth-free colony. *Proc. Anim. Care Panel* **8,** 107–112.

Wagner, J. E. (1970). Control of mouse pinworms, *Syphacia obvelata,* utilizing dichlorvos. *Lab. Anim. Care* **20,** 39–44.

Wardle, R. A., and McLeod, J. A. (1952). "The Zoology of Tapeworms." Univ. of Minnesota Press, Minneapolis.

Weber, A. (1976). Zur parasitenfauna der labortiere: 1. Die parasitenfauna der weissen mäus (*Mus musculus* var. *albinus*). *Wien. Tieraerztl. Monatsschr.* **63,** 30.

Wescott, R. B., and Rabstein, M. M. (1962). Unpublished data. Animal Farm Div., Ft. Deterick, Maryland.

Wescott, R. B., and Van Hoosier, G. L. (1974). Unpublished data. Prime Genet. Cent., Washington State Univ., Pullman, Washington.

Wescott, R. B., Malczewski, A., and Van Hoosier, G. L. (1976). The influence of filter top caging on the transmission of pinworm infections in mice. *Lab. Anim. Sci.* **26,** 742–745.

Wescott, R. B., Farrell, C. J., Gallina, A. M., and Foreyt, W. J. (1979a). Efficacy of albendazole for treatment of naturally acquired nematode infections in Washington cattle. *Am. J. Vet. Res.* **40,** 369–371.

Wescott, R. B., Shelton, T. A., and Gates, N. L. (1979b). Anthelmentic efficacy of oxfendazole in sheep. *West. Vet.* **17,** 22–23.

Wightman, S. R., Wagner, J. E., and Corwin, R. M. (1978). *Syphacia obvelata* in the mongolian gerbil (*Meriones unguiculatus*): natural occurrence and experimental transmission. *Lab. Anim. Sci.* **28,** 51–54.

Yamaguti, S. (1961). The nematodes of vertebrates. *In* "Systema Helminthum," Vol.3, Parts 1 and 2. Wiley (Interscience), New York.

Yorke, W., and Maplestone, P. A. (1926). "The Nematode Parasites of Vertebrates." Blakiston, Philadelphia, Pennsylvania.

Chapter 21

Arthropods

Steven H. Weisbroth

I. Introduction 385
II. Ectoparasitic Ecology 387
III. Diagnosis 387
IV. Identification 388
V. Arthropod Parasites of Laboratory Mice 388
A. *Polyplax serrata* (Burmeister) 388
B. *Myobia musculi* (Schrank) 390
C. *Radfordia affinis* (Poppe) 395
D. *Psorergates simplex* (Tyrell) 395
E. *Myocoptes musculinus* (Koch) 397
F. *Trichoecius romboutsi* (van Eyndhoven) 398
VI. Treatment, Control, and Prevention 398
References 400

I. INTRODUCTION

Arthropod parasites remain as important pathogens of laboratory mice. Looked at from the perspective of time, however, progressive elevation of animal care standards in mouse colonies over the years has had the effect of reducing both the incidence and the diversity of parasitic species. Some indication of this trend may be gained from a survey taken in 1955 (in the United States), in which 100% of commercial offerings were found to be parasitized by arthropod species (Flynn, 1955), compared to 56% in a similar survey a decade later (Flynn *et al.*, 1965) and 7.5% (in the United Kingdom) in 1975 (Sparrow, 1976).

Access to the host by a number of parasitic species has been interdicted by at least two major factors acting in concert. These are (1) commercial and institutional applications of gnotobiotic principles for the development of parasite-free production colonies, with the consequently common availability of arthropod-free mouse stocks, and (2) general acceptance of the principles of modern colony management as recognized by the American Association for Accreditation of Laboratory Animal Care (AAALAC), which are summarized in the ILAR "Guide" (ILAR, 1978). As a practical matter, these factors have essentially limited arthropod parasitism of laboratory colonies to a handful of relatively species-specific forms which must live their entire life cycle on the mouse host:

ISBN 0-12-262502-1

Table I

Insect Parasites of the Laboratory Mouse

Order	Genus	Species	Brand name	References
Hemiptera	*Cimex*	*lectularis*	Bedbug	Flynn (1973); Yunker (1964)
Siphonaptera	*Xenopsylla*	*cheopis*	Oriental rat flea	Flynn (1973); Pratt and Wiseman (1962); Yunker (1964)
	Nosopsyllus	*fasciatus*	Northern rat flea	Flynn (1973); Pratt and Wiseman (1962); Yunker (1974)
	Nosopsyllus	*londsniensis*		Flynn (1973)
	Leptosylla	*segnis*	Mouse flea	Flynn (1973); Pratt and Wiseman (1962); Yunker (1974)
Anoplura	*Polyplax*	*serrata*	Mouse louse	Flynn (1954, 1973); Nelson *et al.* (1972)
	Hoplopleura	*acanthopus*		Flynn (1973); Pratt and Littig (1961)
	Hoplopleura	*captiosa*		Flynn (1973)

Myobia, Radfordia, Trichoecius, Psorergates, Myocoptes, and *Polyplax*. These forms will be emphasized in this chapter for reasons of greatest current significance.

The older literature makes frequent reference to a diversity of arthropod forms. Many of these species are now only rarely encountered as parasites of laboratory mice. Such parasites include the fleas, true bugs (e.g., *Cimex*), and mesostigmatic mites, all of which are forms that must spend part of their life cycle off the host. These have been the parasitic species most vulnerable to the pathogen barriers and sanitization standards characteristic of modern animal facilities. The few citations of parasitism with such species in recent years have occurred as surveys of wild mouse populations or as reports of infestations in colonies with deficient care standards. They will be briefly noted in this chapter for purposes of reference (see Tables I and II).

Table II

Ectoparasites of Laboratory Mice of the Order Acarina

Suborder	Genus	Species	Common name	References
Mesostigmata	*Ornithonyssus*	*bacoti*	Tropical rat mite	Harrison and Daykin (1965); Keefe *et al.* (1964); Strandtmann and Wharton (1958)
	Ornithonyssus	*sylviarum*	Northern fowl mite	Baker *et al.* (1956); Flynn (1973)
	Liponyssoides	*sanguineous*	House mouse mite	Baker *et al.* (1956); Flynn (1973)
	Haemogamasus	*pontiger*		Baker *et al.* (1956); Strandtmann and Wharton (1958)
	Eulaelaps	*stabularis*		Baker *et al.* (1956); Strandtmann and Wharton (1958)
	Laelaps	*echidnus*	Spiny rat mite	Flynn (1973); Strandtmann and Wharton (1958)
	Haemolaelaps	*glasgowi*		Baker *et al.* (1956)
	Haemolaelaps	*casalis*		Baker *et al.* (1956); Strandtmann and Wharton (1958)
Prostigmata				
Family Myobiidae				
Subfamily Myobiinae	*Myobia*	*musculi*	Fur mite	Flynn (1955, 1973)
	Radfordia	*affinis*	Fur mite	Ewing (1938); Flynn (1955)
Family Psorergatidae	*Psorergates*	*simplex*	Hair follicle mite	Beresford-Jones (1965); Flynn (1955)
Family Sarcoptidae	*Notoedres*	*musculi*		Fain (1965)
Family Demodicidae	*Demodex*	*musculi*		Hopkins (1949)
Astigmata				
Family Myocoptidae	*Myocoptes*	*musculinus*		Flynn (1955, 1973)
	Trichoecius	*romboutsi*		Van Eyndhoven (1946); Fain *et al.* (1970); Flynn (1955)

II. ECTOPARASITIC ECOLOGY

Several factors have been found to play roles in regulating population densities of ectoparasites, particularly those that must spend their entire life cycle on the host. Indeed, the mouse has been a useful model for investigations of the ecology of ectoparasitic populations. That such populations are in fact regulated is suggested by the simple mathematic progression in which a mouse, seeded with a single male/female parasitic pair having a life cycle of 20 days, would be predicted to have a parasitic load in the millions by the end of 1 year. Since parasitic populations never approach such magnitudes, it follows that the mouse host effectively limits their growth. It is useful, therefore, to think of the arthropod population on the host at any given moment as representing the net balance of the parasite's propensity to multiply and extend colonization versus the host's defensive mechanisms, which act to limit or even approach eradication of ectoparasitism.

Systematic exploration of some of these factors was initiated by Murray, who established the role of self-grooming competence in regulating population numbers and anatomic distribution of the louse, *Polyplax serrata* (Murray, 1961). Murray observed that mice commonly employ two types of self-grooming: (1) scratching with the hindlimbs followed by licking of the toes, and (2) combing with the grooved lower incisors through the hair with upward motions of the head (Murray, 1961). These studies demonstrated that if the mouse's ability to groom itself with the mouth (and teeth) were restricted by Elizabethan collars, or by spreading the incisor teeth with a small ring, louse populations would increase rapidly and seemingly without limitation. These studies were confirmed and extended by Bell and colleagues, also working with *Polyplax* (Bell and Clifford, 1964; Bell *et al.*, 1962, 1966; Clifford *et al.*, 1967; Stewart *et al.*, 1976). Bell started with the unplanned observation that mice in paralytic phases of rabies virus infection became heavily infested with *Polyplax* (Bell *et al.*, 1962). Whereas Murray (1961) established the role of the paired incisors in grooming of both lice and eggs, the initial experiments by Bell *et al.* (1962) established the significance of the feet for the same purpose. Amputation of the limb at various levels and in varying combinations of laterality was used to study the effect of self-grooming competence. An initial suggestion by Scott that mutual grooming by cagemates also played a role in regulating louse populations (Scott, 1958) was further examined by Bell and Clifford (1964), who confirmed the observation and explored the grooming effectiveness of littermate pairs, intersex and same sex pairs, and symbiotic or mutual grooming effectiveness in larger groupings. It was concluded that mutual grooming was a well-established social function with considerable survival value. Subsequent studies with the *Polyplax* model, in which louse populations were tracked for long periods of time, demonstrated that, even in single-caged mice disabled by amputation, louse populations would initially increase but then decrease due to acquired resistance (Bell *et al.*, 1966), that inbred mouse strains similarly infested would vary in louse population levels and the susceptibility was heritable (Clifford *et al.*, 1967), and that early neonatal infestation did not induce tolerance (Stewart *et al.*, 1976).

A similar series of experiments was conducted by Weisbroth and Friedman and their colleagues with the mite, *Myobia musculi* (Friedman and Weisbroth, 1975; Weisbroth *et al.*, 1972, 1974b, 1976). In these experiments, it was shown that both self-grooming and mutual grooming played effective roles in limiting equilibrium *Myobia* populations (Weisbroth *et al.*, 1974b), that inbred strains characteristically varied in equilibrium mite populations (Friedman and Weisbroth, 1975), that certain strains, especially C57BL and congenic sublines, developed allergic manifestations to acariasis, and that mice developed precipitating antibodies to mite antigens (Weisbroth *et al.*, 1972).

The presence of hair, i.e., fur, appears to be necessary for ectoparasitism of mice, at least with mites. Letscher showed that with both *Myobia* and *Myocoptes*, neonates of infested breeders would remain free of mites until 5–7 days after birth (Letscher, 1970). Infestation occurred first on the heads and backs (the heaviest-haired areas) but then extended to the ventrum by the eighth to ninth day as the latter became haired. Similar observations were made by Weisbroth and Friedman (unreported elsewhere) in their failure to infest nude athymic mice but not their phenotypically normal (i.e., haired) heterozygotic counterparts.

In summation, the following points may be made. Ectoparasites of the mouse generally have short life cycles, on the order of 20–25 days, and may reproduce almost geometrically. That they do not, is a function of immunologic, innate, and societal behavioral patterns that collectively form the resistance mechanisms of the mouse. These mechanisms have been found to include mutual and individual grooming maneuvers by both the feet and teeth, immunologic resistance as manifested by antibodies, allergy, and general lowering of parasite populations over long periods of time, and genetic factors of resistance and susceptibility. The relative emphasis and interplay of these factors, although variably effective in limiting parasitic populations, largely determines the clinical intensity of infestation in a given individual or grouping of laboratory mice.

III. DIAGNOSIS

The methodology of diagnostic sample preparation and/or examination for all surface-dwelling forms spending their entire life cycle on the host is identical. In general, all forms of lice and mites parasitic on mice favor the dorsal anterior region of the body, particularly the dorsum and nuchal crest of the skull, the cervical dorsum, and the area between the shoulder blades. From the foregoing section, it will be appreciated that these anatomic locations are the ones most shielded from grooming maneuvers of the feet and teeth. When infestations become more intensive, all areas of the skin may be parasitized. Also, in the author's experience, the level of parasitic populations is not often directly predictable by the superficial appearance of the mouse, i.e., large populations may be colonized on grossly normal-looking hosts, and conversely, a mangelike appearance may be developed in the presence of relatively few parasites. More will be said on this point when the separate parasitic species are described.

Various techniques have been advocated for diagnosis of ectoparasitism in laboratory mice. Specialized methods have been developed for collection of the entire host ectoparasite population that involve KOH maceration of the pelt and $ZnSO_4$ flotation of parasitic forms (Hilton, 1970; Hopkins, 1949; Sheather, 1915, 1923). Such methods are useful for careful survey collections but are rarely required for clinical diagnosis or diagnostic screening. For the latter purposes, the following techniques have been used or advocated, and are listed here in decreasing order of accuracy and reliability.

1. Direct observation of the hair and skin under low power (20–30×) magnification (Galton, 1963; Weisbroth *et al.*, 1974b). Either dead dead or anesthetized mice may be examined. The body is directly illuminated with a high-intensity lamp, and the hairs are parted with pins or sticks as the observer examines through the dissecting microscope with the principal focus on the epidermal surface. Using this method, quickly moving species (lice) may be seen and collected, as well as hair-clasping and stationary forms and viable or hatched egg cases. The principal advantage of this technique is that small populations may be easily apprehended. Arthropods may be collected for identification by removal with pins or forceps, or by plucking particular hairs.

2. Use may be made of the thermotactic propensity of most parasitic arthropods to crawl out on the hairs and leave the host as the carcass cools after death. Mice may be killed and, 1–4 hr later, parasitic forms collected for mounting and direct identification. As a variation, clear cellophane tape may be dabbed on the pelt and mounted with adhering hairs and parasites, sitcky side down on a glass side, for microscopic examination. Flynn described a modification of this procedure for survey collections in which mice were killed and placed on black paper (Flynn, 1955). The paper was rimmed on the perimeter with cellophane tape, sitcky side up, to prevent parasites from leaving. Use of black paper facilitated the observation of small arthropods. Another variation of this principle was used by Wagner, who placed dead mice in tape-sealed petri dishes, which were refrigerated for about an hour, and then examined the plates under the dissection microscope (Wagner, 1969). The disadvantages of this method and its variations include the fact that the host must be killed, that it may limit other examinations to be made on the same carcass (although the pelts may be removed and placed on black paper), and that small populations may be missed. It is used to best advantage with moderate to high parasitic populations.

3. Scrapings of the skin of dead mice may be made in a manner analogous to that for diagnosis of dermatologic conditions in other species (Cook, 1953; Gambles, 1952). A scalpel blade is used to scrape the skin, and the scrapings are mounted under a cover slip in 10% KOH, glycerine, or immersion oil and examined microscopically. This method is the least reliable and is almost certain to miss quickly moving species (lice) and low-level populations of slower-moving forms. Interestingly, the laboratory mouse (*Mus musculus*) has no reported parasitic representatives of *Demodex, Sarcoptes,* or *Notoedres,* the species for which scrapings are most useful.

IV. IDENTIFICATION

The diagnosis of ectoparasitism should be accompanied by identification of the parasites. Permanent preparations for type collections can be made by any of several methods from the following references (Baker and Wharton, 1952; Baker *et al.*, 1956; Owen, 1972). Permanent preparations are not usually required for diagnostic identification of the common parasitic species. They may be collected as described previously with pins, a fine hair brush, or fine forceps and temporary mounts made by placing the mites in glycerine, 10% KOH, or even immersion oil under a cover slip and examined microscopically. With some experience, common species may be identified under clear cellophane tape. Parasites thus mounted may be compared with those illustrated in this chapter. Parasitic species other than these may require a trained parasitologist for identification.

V. ARTHROPOD PARASITES OF LABORATORY MICE

A. *Polyplax serrata* (Burmeister)

1. Description and Life Cycle

The louse encountered on laboratory mice is *Polyplax serrata* (Fig. 1A,B). It has been reported from both wild mice and rats (Flynn, 1973; Owen, 1972, 1976), as well as from laboratory mice in most parts of the world (Compton, 1926; Flynn, 1955; Flynn *et al.*, 1965; Heine, 1962; Jancke, 1930; Sasa, 1950; Sparrow, 1976). The genus *Polyplax* is in the family

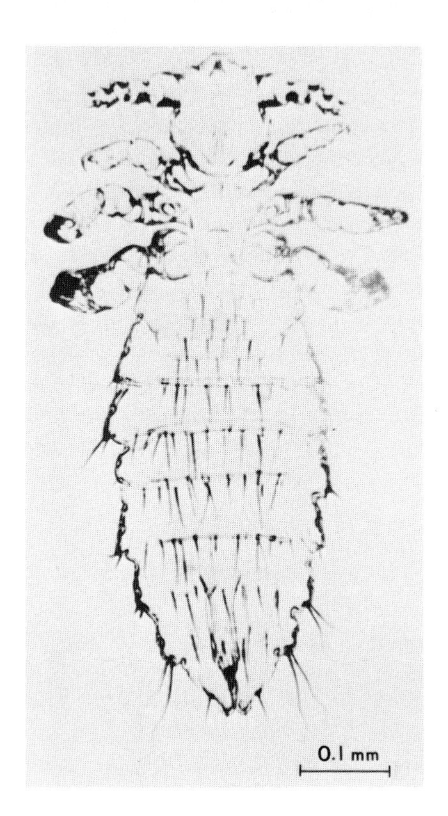

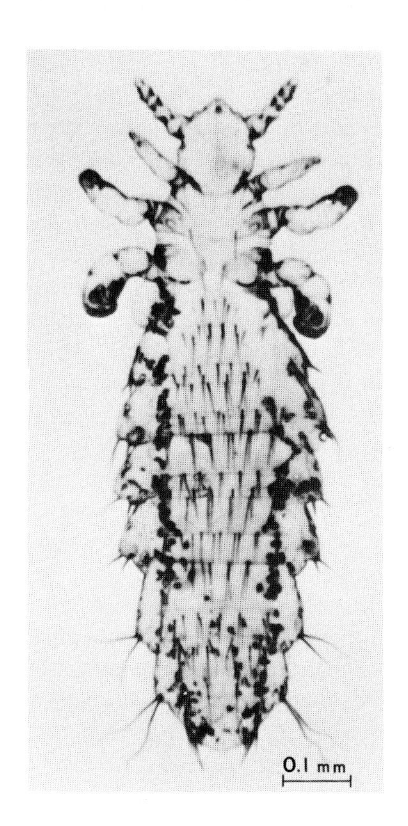

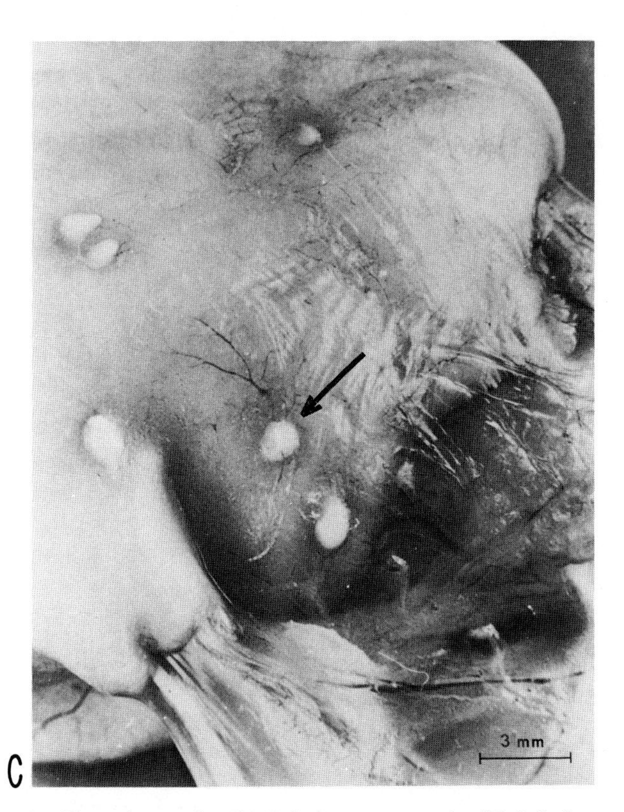

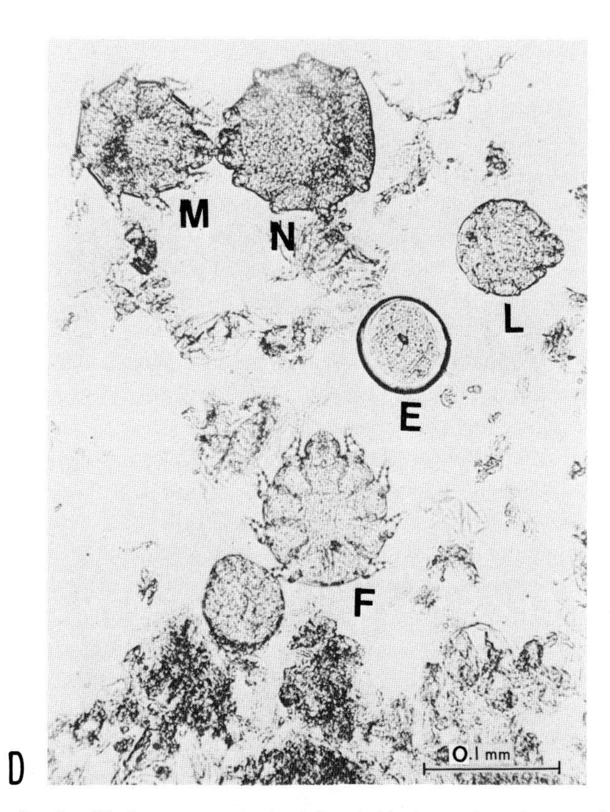

Fig. 1. Photomicrographs. (A) *Polyplax serrata*, male. (B) *Polyplax serrata*, female. (C) *Psorergates simplex*-infested skin inverted to show pouches (arrows). (D) *Psorergates simplex*, pouch contents: egg (E), adult female (F), adult male (M), larva (L), nymph (N). (Photograph courtesy of Robert J. Flynn and *Lab. Anim. Sci.*)

Hoplopleuridae, most of whose members are parasites of rodents and lagomomorphs. Adult females are slender and about 1.5 mm long, whereas the males are thicker and shorter (1.0 mm); however, both sexes are large enough to be seen with the unaided eye. *Polyplax serrata* may be differentiated from *P. spinulosa* (the *Polyplax* of rats) on the basis of both the sternal plate, which is triangular in *P. serrata* and pentagonal in *P. spinulosa,* and the fourth paratergal plate, which has setae of unequal length in *P. serrata* (the dorsal seta is longer than the ventral), whereas these setae are of equal length in *P. spinulosa* (Flynn, 1973).

The five stages recognized in the life cycle are the egg, three nymphal stages, and the adult (Murray, 1961). The eggs are attached near the base of hair shafts. The young hatch by lifting the operculated cap in the dorsum of the egg. A row of pores may be noted along the top of the operculum (Owen, 1972). Stage I nymphs may be found in wide distribution over the body surface; however, more mature stages generally favor the anterior dorsum of the host (Murray, 1961). Nymphal stages may be differentiated by setal arrangements, as described by Murray (1961). Eggs of *P. serrata* hatch in 5-6 days, and nymphs develop into adults in 7 days, giving an average minimum life cycle of 13 days (Murray, 1961). Transmission from host to host is by direct contact.

2. Pathobiology

The Anoplura are (blood) sucking lice, and the effects on the mouse host are mainly related to this feature. In heavy infestations, *P. serrata* is thought to cause anemia and debilitation (Flynn, 1973), as well as causing mice to become irritable through frequent scratching of painful or pruritic bites. Heavy infestations can be fatal (Oldham, 1967). Arthropod-borne infections transmitted by *P. serrata* are known to include *Eperythrozoon coccoides* (Eliot, 1936) and *Francisella tularensis* (Shaughnessy, 1963). Thus, *P. serrata* should be regarded as a potential vector of any septicemic or bacteremic microorganism.

The histopathology of *P. serrata* infestations has been studied in a timed sequence (Nelson *et al.,* 1972). The initial stages induce, over a period of about 4 weeks, an increase in epidermal thickness with underlying hyperemia and infiltration of neutrophils, eosinophils, and lymphocytes in the subcutis. The hyperemia decreases and superficial arterioles undergo constriction toward the end of this period. Later stages include epithelial hyperplasia (acanthosis) and the presence of monocytes and increased numbers of mast cells, some of which are degranulated, in the subcutaneous tissues (Nelson *et al.,* 1972). Localized anaphylactic reactions, as demonstrated by bluing of *P. serrata* antigen-injected skin sites in Evans blue-injected mice, were suggestive of antibody-mediated reactions.

Although a direct proof of induction of immune reactions by *P. serrata* has not been reported, the demonstrated reduction in parasitic populations after initial peak levels, the seemingly pruritic nature of louse bites, the presence of cells with immunologic activity in infested skin, and the demonstration of anaphylactic reactivity of the skin of infested mice all argue for an immunologic factor, in addition to epithelial hyperplasia, as among host defensive mechanisms. As discussed earlier, these mechanisms are also known to include self-grooming and mutual grooming effectiveness and heritable resistance and susceptibility (Bell and Clifford, 1964; Bell *et al.,* 1962, 1966; Clifford *et al.,* 1967; Stewart *et al.,* 1976).

B. *Myobia musculi* (Schrank)

1. Description and Life Cycle

Myobia musculi was originally described as a louse in 1781 (von Schrank, 1781) but was later recognized as a fur mite of the mouse by von Heyden (1826), who established the genus *Myobia.* Early work with the mite was largely descriptive and taxonomic (Claparede, 1868; Guthrie *et al.*, 1971; Trovessart, 1895; von Heyden, 1826; Poppe, 1896), with the family Myobiidae being established in 1877 (Radford, 1934). Ewing placed the genus *Myobia* into a new subfamily, Myobiinae, with four genera, two of which were ectoparasitic on mice (*M. musculus*): *Myobia* and *Radfordia* (Ewing, 1938). This relatively late reclassification (1938) undoubtedly explains many earlier misidentifications of *Radfordia* as *Myobia* in both mice and rats. *Myobia* was recognized in the United States as early as 1915 (Banks, 1915) and may now be regarded as of nearly global distribution on laboratory mice (Boceia, 1942; Flynn, 1963, 1973; Fukui *et al.*, 1961; Heine, 1962; Sparrow, 1976; Wormersley, 1942). The mite is small and about twice as long as wide, with females 0.40-0.50 mm in length and males about 0.28-0.30 mm. The sexes are quite similar in appearance, differing in size, setation, and genitalia (Fig. 2A,B). *Myobia* may be easily differentiated from other mites on the mouse, except *Radfordia,* by the characteristic shape, especially the lateral margins of the idiosoma, which form bulges between the legs (Fig. 2A,B). Differentiation from *Radfordia* may be accomplished by examination of the tarsus of the second pair of legs. In *Myobia,* a true claw is lacking; however, the terminal tarsal appendage (empodium) ends with a single clawlike structure—the empodial claw. In contrast, the terminal tarsal structure (of leg II only) in *Radfordia* ends with two terminal tarsal claws (see arrows, Fig. 2B,D). Stages in the life cycle include the egg, first- and second-stage larvae, protonymph, deutonymph, and adult male and female. Subdivisions of nymphal stages based on chaetotaxy have been described and illustrated (Ama, 1955; Gambles, 1952; Grant, 1942; Radford,

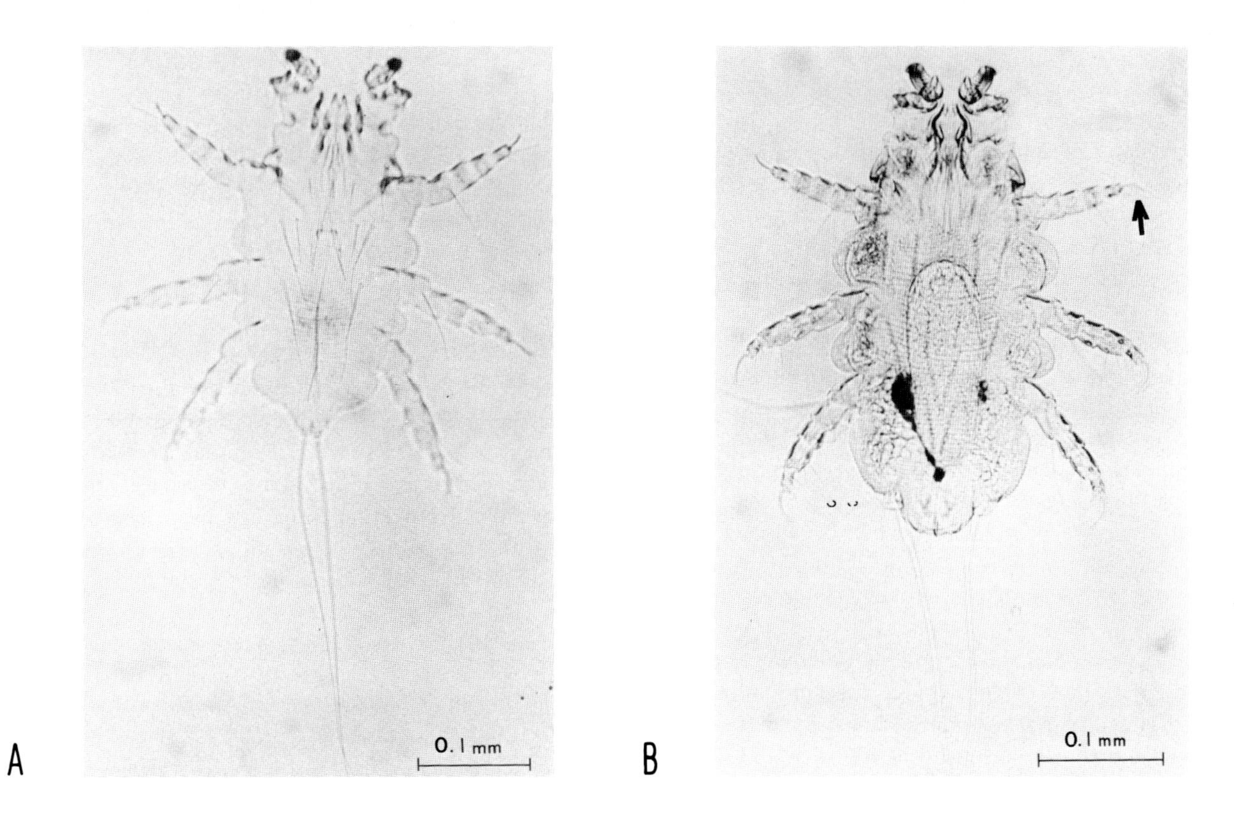

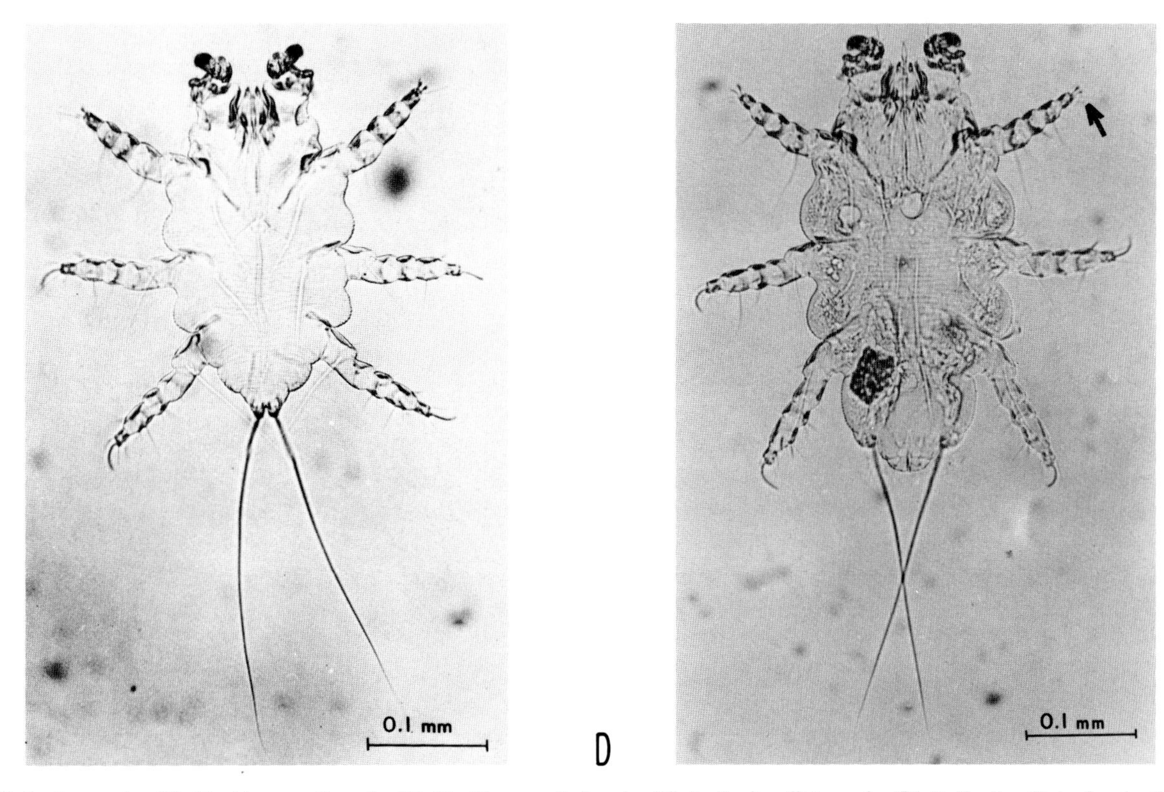

Fig. 2. Photomicrographs. (A) *Myobia musculi*, male. (B) *Myobia musculi*, female. (C) *Radfordia affinis*, male. (D) *Radfordia affinis*, female. (Photographs courtesy of Robert J. Flynn and *Lab. Anim. Sci.*)

1934). The eggs are oval, about 0.20 mm long (smaller than *Polyplax* eggs), and attached at the lower pole to hair shafts near the base. The eggs hatch in about 7–8 days, as determined either *in vivo* or *in vitro* (Friedman and Weisbroth, 1977; von Haakh, 1958; Letscher, 1970). The larval period (forms with three pairs of legs) lasts 10 days, with nymphal forms appearing on the eleventh day after emergence from the egg (Hilton, 1970). Adult forms are seen as early as the fifteenth day and are capable of laying eggs within 24 hr (Friedman and Weisbroth, 1977). The complete life cycle thus may be 23 days. Insofar as treatment programs are concerned, it is important to note that if an acaricide is ineffective on *in ovo* forms, a second application should take place sometime after the eighth day (when all residual eggs must have hatched, but sometime before the sixteenth day, when new adults may have laid new eggs), ideally on the tenth to twelfth day (Friedman and Weisbroth, 1977).

The mouthparts of *Myobia* are small, with simple palpi and styletlike chelicerae (Flynn, 1973). According to Banks (1915), the myobic mites were not bloodsucking, but rather were thought to subsist on matter secreted by the skin. This concept appears to have originated from the common microscopic observation of various forms of *Myobia* clasping a hair shaft with the head down and stylets embedded in the hair follicle, although Grant (1942) recognized that feeding could take place in nonhaired areas of skin. The mode of feeding was clarified by Wharton (1960) using Evans blue dye injection of the host as a means of studying mite feeding. Due to the rapid ingestion of dye by mites (within 15–20 min from intravenously injected hosts), Wharton concluded that inasmuch as host erythrocytes were not seen in the mites, interstitial fluid (presumably lymph) was the most probable food source. This interface, the lymph, or interstitial fluid is also important as a route for immunization of the host with parasitic antigens.

Transmission of *Myobia* from host to host is primarily by direct transfer. However, the question was studied in detail both by Wharton (1960) and by Letscher (1970). These studies have shown that *Myobia* infestations are spread primarily by adult female mites within 24 hr to mice caged with parasitized hosts and to neonates from infested parents by the seventh to eighth day as the hair coats appear. The mites are thermotactic and crawl out to the ends of the hairs of dead hosts, poised to await new hosts, as well as actively crawling out in all directions to contaminate the environment. *Myobia* may live as long as 4 days upon dead hosts but only about 2 days *in vitro* (Letscher, 1970). New infestations are characterized by an initial mite population increase which is followed by a decrease from this peak value as host defense mechanisms are mobilized (Friedman and Weisbroth, 1975; Weisbroth *et al.*, 1974b). Mite populations decrease until an equilibrium level is reached after 8–10 weeks, where the host cannot eradicate the parasite, but parasitic populations cannot significantly expand. The equilibrium population level may be carried for long periods of time, even years. Cyclic fluctuations of 20–25 days in the equilibrium population level may represent waves of egg hatches (Friedman and Weisbroth, 1977; Weisbroth *et al.*, 1974b).

2. Pathobiology

In Section II, it was pointed out that such factors as caging modalities which influence self-grooming and mutual grooming behavior, genetic factors of resistance and susceptibility, effectiveness of the immunologic response, and still others were known to condition the clinical expression and mite population burden of *Myobia* infestation. Indeed, due to inapparent infestation, some have questioned the pathogenicity of *Myobia* (Constantin, 1972; Gambles, 1952; Green and Neeham, 1974; Oldham, 1967; Owen, 1972). These factors underlie the variability of reported clinical presentations, which range from essentially nil (Friedman and Weisbroth, 1975; Weisbroth *et al.*, 1974b; Whitely and Horton, 1962, 1965) to severe, with extensive integumental lesions and systemic effects. An additional factor is that the host genetic constitution also regulates the equilibrium level of mite populations. There appears to be no direct relationship between mite numbers and the severity of resultant lesions, although different stocks and strains vary widely in resident mite populations (Friedman and Weisbroth, 1975; Weisbroth *et al.*, 1976).

Integumental effects include pruritis and self-inflicted trauma, with gross skin changes that vary from inapparent, to a

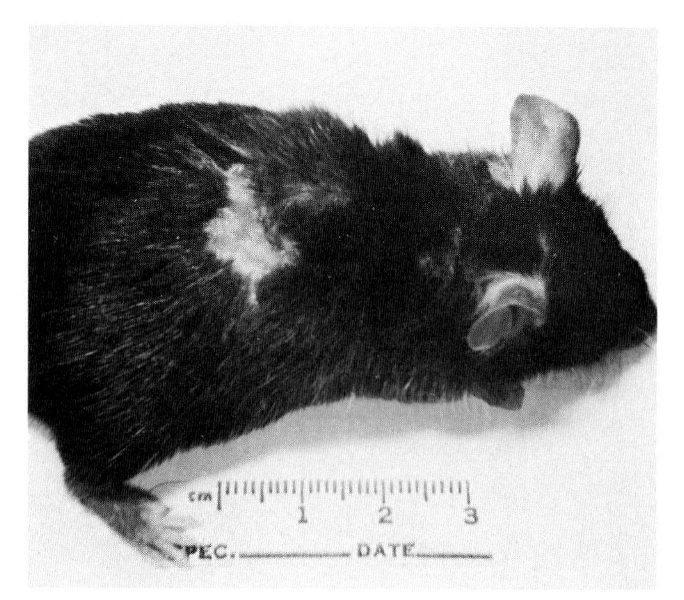

Fig. 3. Ulcerative lesion of the skin in myobic acariasis of a C57BL strain mouse.

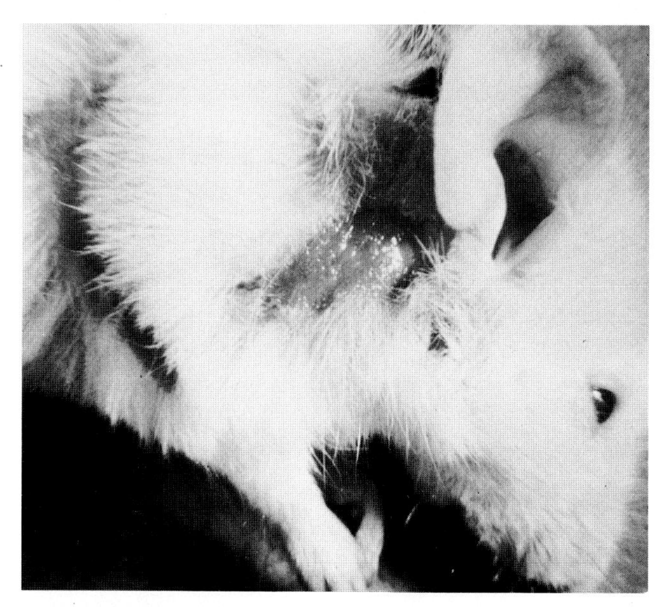

Fig. 4. Ulcerative lesion of the skin of the neck in myobic acariasis of a random-bred Swiss-Webster mouse.

mild scruffiness with patchy depilation of the head, to more severe lesions that involve ulceration and pyoderma. Whitely and Horton have described the histology of the skin (in CBA mice) with uncomplicated chronic infestations that were grossly inapparent or with mild degrees of depilation and loss of hair luster (Whitely and Horton, 1962, 1965). The epithelium was observed to be "thickened," i.e., mildly hyperkeratotic, with an increase from the normal of two to three cells in depth to six to eight cells. They also noted an increase in the presence of chronic inflammatory cells underlying the epithelium and an increased mitotic rate of epithelial cells, but no change in the hair regrowth cycle (Whitely and Horton, 1962, 1965). These changes should be regarded as characteristic of resistant stocks and strains.

The histology of more extensive, ulcerating lesions has also been described and illustrated (Galton, 1963; Weisbroth *et al.*, 1976), as well as observed grossly in mice of the C57BL strain and some of its congenic sublines, e.g., [C57BL/10 Sn, B10.LP, and C57BL/10-H-3^{b} (Constantin, 1972; Csiza and McMartin, 1976; Galton, 1963; Heston and Deringer, 1948; von Heyden, 1826; Sokoloff *et al.*, 1960; Weisbroth *et al.*, 1976)]. Figs. 3 and 4 illustrate ulcerative skin lesions in C57BL and Swiss-Webster mice. The "normal" haired skin adjacent to ulcerative zones resembles that described above for chronic, uncomplicated acariasis (Fig. 5). The transitional zone between the normal skin and the ulcerated surface is characterized by an increase in the depth of the epithelium to 8–20 cells, hyperplasia of the basal cell layer with deeply reaching rete pegs, hyperkeratosis, inspissated hairs, and infiltration of subcuticular layers underlying the epidermis with chronic inflammatory cells, including lymphocytes, macrophages, and mast cells as the chief components (Fig. 6). Ulcerative zones are characterized by complete loss of epithelial elements, with sharp demarcation at the acanthotic periphery (Fig. 6). Fibrinous exudate with leukocytic debris and coccal colonization covers the ulcered surface. The dermis below the fibrinous surface is infiltrated by both acute and chronic inflammatory cells, including plasma cells. The panniculus adiposus and panniculus carnosus underlying the dermis are involved in a fibrotic process best described as granulation tissue with orderly rows of fibroblasts and capillaries and chronic inflammatory cells as above, with abundant mast cells (Figs. 5, 6).

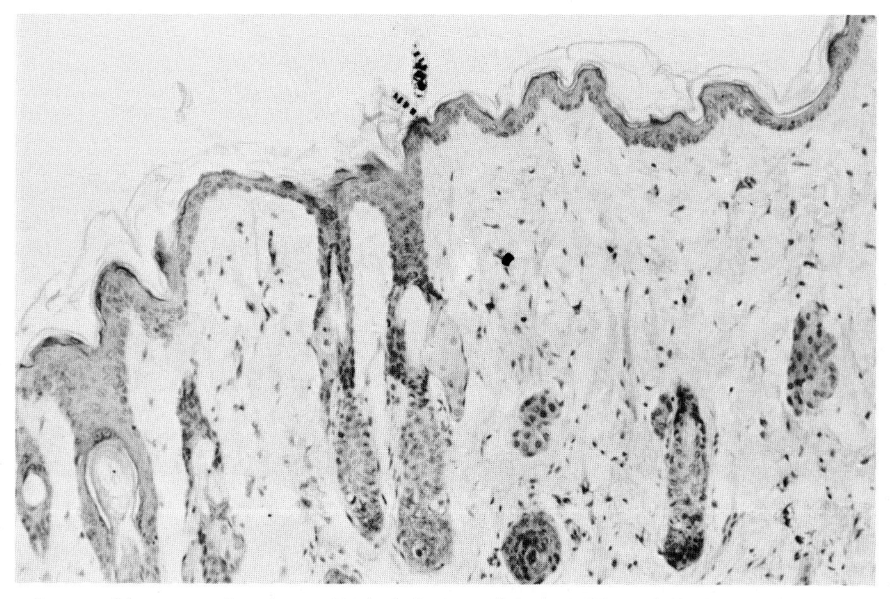

Fig. 5. Seemingly normal mouse skin near an ulcered zone. Typical of uncomplicated myobic acariasis. Note the two- to three-cell depth of the epidermis and mildly increased numbers of chronic inflammatory cells in the dermis. Hematoxylin and eosin. ×100.

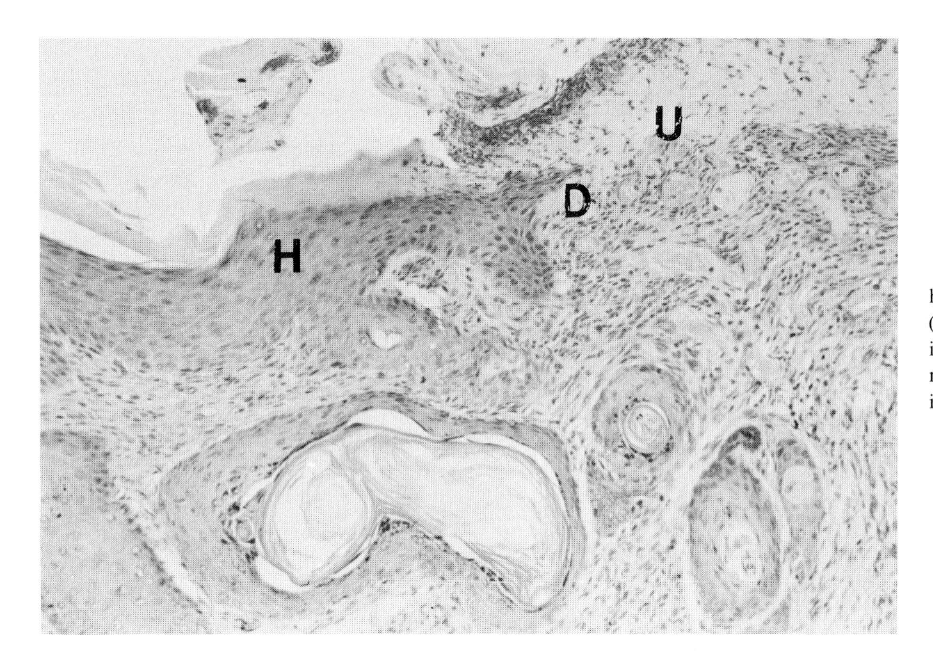

Fig. 6. Photomicrograph at the junction of a hyperkeratotic periphery (H) and an ulcerative zone (U). Note hyperplasia of the epidermis (10 to 15 cells in depth), sharp demarcation of the ulcer (D), and markedly increased numbers of inflammatory cells in the dermis. Hematoxylin and eosin. ×100.

The evidence supporting the interpretation of ulcerative lesions commonly seen in, but by no means restricted to, C57BL and derived substrains as allergic in character may be summarized as follows. The lesion architecture is similar to that of allergic acariasis in other species (Weisbroth *et al.,* 1974a) and has abundant numbers of the IgE effector mast cells (Ishizaka and Ishizaka, 1970; Ishizaka *et al.*, 1971). Such lesions are pruritic due to kinins released by degranulation of mast cells, and the frantic scratching by the host leads to self-inflicted ulceration. Lesion redevelopment in healed, mite-free mice may be induced in a matter of days following experimental reinfestation, which is suggestive of an anamnestic response. Other features characteristic of allergic states are the heritable tendency, the observation that other (nonsusceptible) stocks and strains faced with the same challenge do not develop such lesions, and the minute amounts of antigen required to evoke dramatic lesions. As few as five mites could evoke lesion redevelopment within a few days (Weisbroth *et al.,* 1976).

Systemic effects of *Myobia* infestations in mice have also been investigated. They have been shown to include decreased life span (Whitely and Horton, 1962), decreased body weight (Galton, 1963), development of an immunologic response (Claparede, 1868; Fig. 7), and decreased reproductive indices (Weisbroth *et al.,* 1976). An increase in the size of lymph nodes draining ulcerative skin lesions, hypergammaglobulinemia, and hyperactivity of the spleen in such animals have also been reported (Csiza and McMartin, 1976; Weisbroth *et al.,* 1976). Secondary amyloidosis has been described as attributable to chronic mite (*Myobia*) infestation (Fukui *et al.*, 1961; Galton, 1963; Heston and Deringer, 1948). The critical cause-and-effect experiments to sustain the amyloidosis concept as secondary to chronic mite infestation have not been

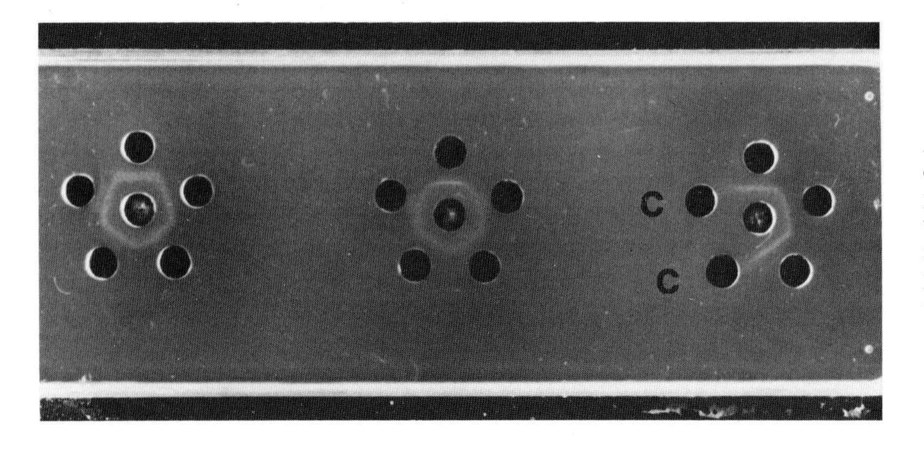

Fig. 7. Immunodiffusion with whole-mite *Myobia* antigen (each center well) against undiluted sera from 13 *Myobia*-infested mice (all peripheral wells except the last two, clockwise, in the group on the right). Note the precipitin lines in the agar gel from infested mice. There are no precipitins in the agar gel from two noninfested control mice (C).

done, and the latter should be regarded as circumstantial at present. Against it is the observation that amyloidosis in mice occurs in the absence of mites (Dunn, 1944, 1949; Thung, 1957), and chronic mite infestations occur in the absence of amyloid.

C. *Radfordia affinis* (Poppe)

Description

Radfordia affinis was first described by Poppe (1896). It was transferred to the Myobiidae family in 1938 by Ewing (1938), who included the mite in the subfamily Myobiinae and renamed it *Radfordia affinis* (syn. *Myobia affinis*). It was seen in laboratory mice as early as 1942 by Gambles, who mentioned a second species of *Myobia,* "probably *M. affinis*" (Gambles, 1952). The mite has subsequently been identified from laboratory mice in many parts of the world (Flynn, 1955, 1963; Flynn *et al.,* 1965; Fukui *et al.,* 1961; Heine, 1962; Seamer and Chesterman, 1967).

Radfordia affinis is a myobiid mite similar in appearance to *M. musculi,* with which it is often confused. Scrutiny of the tarsal terminus of the second pair of legs enables differentiation of these two species (Fig. 2B,D). *Radfordia affinis* has two tarsal claws of unequal length, whereas *M. musculi* has a single empodial claw. In turn, *R. affinis* (of mice) may be differentiated from *R. ensifera* (of rats) again by reference to the tarsal claws of the second pair of legs. The claws are paired and unequal in length in *R. affinis* and paired and equal in length in *R. ensifera* (Flynn, 1973).

Virtually nothing is known of the life cycle or pathobiology of *Radfordia,* and there are no reports of investigations dealing with it beyond a morphologic description. Treatments effective against *M. musculi* have been reported as also effective against *R. affinis* in cases of dual infestation (Flynn, 1955; Heine, 1962).

D. *Psorergates simplex* (Tyrell)

1. Description and Life Cycle

The trombidiform family Psorergatidae includes at least 37 species in the genus *Psorergates* (Ah *et al.,* 1973). The type species *Psorergates simplex* was initially described by Tyrell from a wild house mouse in 1883 (Tyrell, 1883). Interestingly, it had been reported as a species of *Notoedres* in the mouse, both earlier by Gerlach (cited in Beresford-Jones, 1965) and later by Low (1911). It has long been known as a follicle-inhabiting mite of the skin of the mouse (Newmann, 1893; Jambon, 1928). The mite was placed in the family Cheyletidae by Ewing in 1929 (Ewing, 1929). However, the group has been since reorganized by Fain and co-workers into the family Psorergatidae (Fain, 1939a,b; Fain *et al.,* 1966).

Psorergates simplex is quite small and rounded, averaging about 0.10 mm in length (Figs. 7,8). Nymphs and adults have four pairs of legs, whereas larvae have three pairs. The legs have five short segments which terminate in a pair of tarsal claws and a padlike empodium. The chelicerae are minute and styletlike. The life cycle is not known. However, all stages (egg, larva, protonymph, deutonymph, adult male and female) may be found within a single dermal nodule (Cook, 1956; Flynn and Jaroslow, 1956). Transmission is by direct contact (Beresford-Jones, 1967).

2. Pathobiology

Only some 30 years ago, the mite was commonly seen in laboratory mice in many parts of the world (Beresford-Jones, 1965; Cook, 1956; Flynn, 1955). However, more recent surveys (Flynn *et al.,* 1965; Sparrow, 1976) indicate that *P. simplex* may currently be regarded as rare. It has not been reported as a parasite of laboratory mice in the last decade.

There is general agreement that infestation is initiated by invasion of the hair follicle (Beresford-Jones, 1965; Flynn and Jaroslow, 1956; Newmann, 1893) and that development of the cystlike nodule occurs as a result of epidermal growth to accommodate internal pressure of space-occupying mites. Infestation of the follicle results in dermal cysts or pouches which appear as small, whitish nodules on the inverted or subcuticular surface of the skin (Fig. 1C). The nodules are well described and illustrated in both gross and microscopic aspects by Flynn (1955; Flynn and Jaroslow, 1956). Histologically, the pouches resemble comedones and are seen to consist of invaginated sacs of squamous epithelium filled with mites and their products and keratinaceous debris. Typically, the epidermal lining is intact and without an inflammatory reaction (Flynn, 1955). Healing of nests appears to take place by encystment and granuloma formation (Flynn, 1955; Flynn and Jaroslow, 1956).

Lesions produced by this mite are often inapparent in live hosts. However, the auricular form may have a mangelike appearance in the pinna (Cook, 1956; Low, 1911). Nodule formation tends to be more common in the skin of the head, shoulders, and lumbar areas but may occur anywhere. Nodules are larger in loose skin of the trunk (2 mm) and smaller where the skin is tighter (1 mm), as in the face and legs. Diagnosis cannot be established by the procedures outlined in the Section III, which are oriented to surface-inhabiting arthropod forms. *Psorergates* should be diagnosed by examination of the inverted subcuticular surface of the pelt for characteristic nodules. Pouch contents may be expressed by pressure of a scalpel blade or by scraping and the contents mounted under a

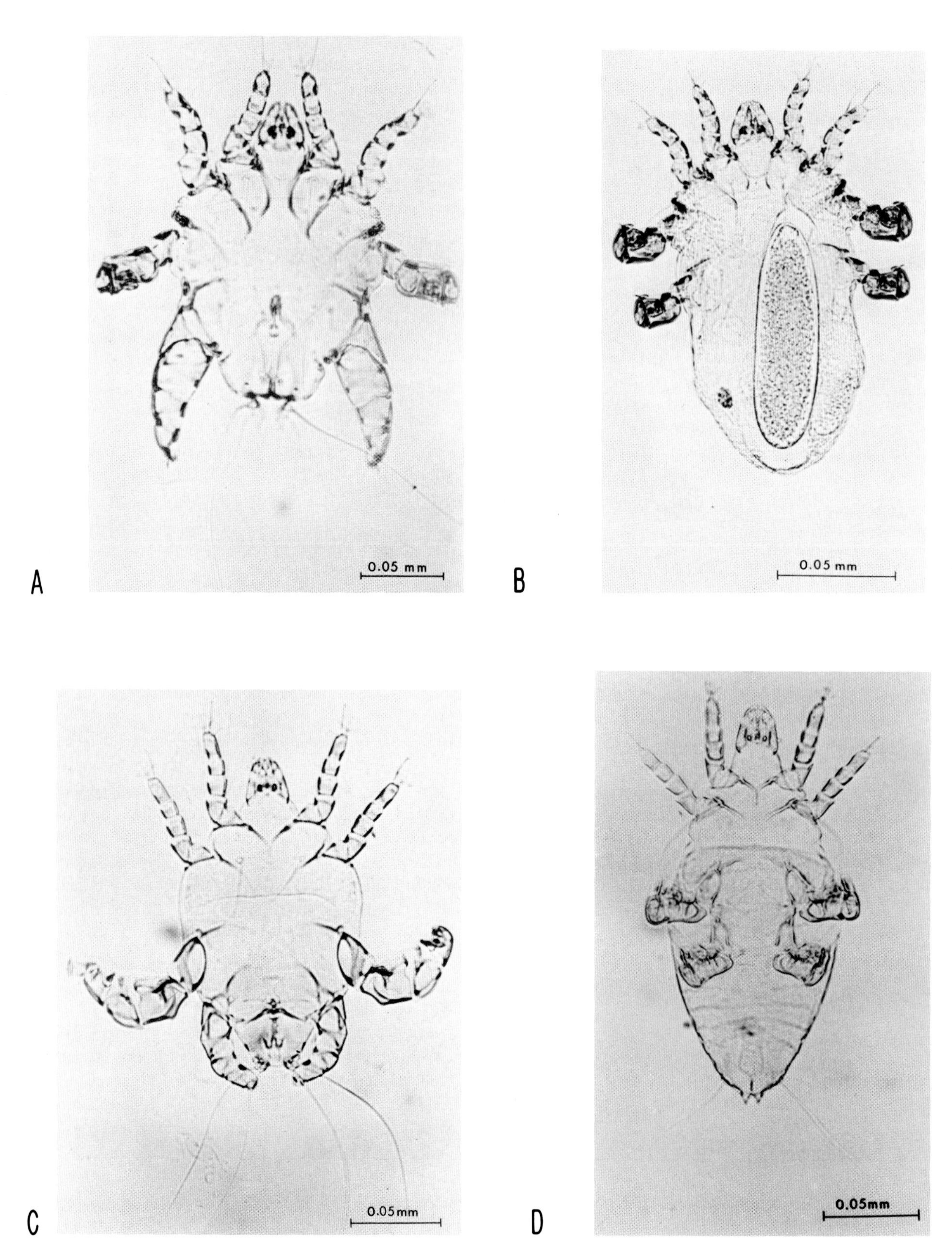

Fig. 8. Photomicrographs. (A) *Myocoptes musculinus*, male. (B) *Myocoptes musculinus*, female. (C) *Trichoecius romboutsi*, male. (D) *Trichoecius romboutsi*, female. (Photographs courtesy of Robert J. Flynn and *Lab. Anim. Sci.*)

cover slip in water, glycerine, or 10% KOH for microscopic examination (Fig. 1D).

E. *Myocoptes musculinus* (Koch)

1. Description and Life Cycle

The myocoptid mite, *Myocoptes musculinus,* was originally described by Koch in 1840 (Koch, 1840), who illustrated the female. Claparede (1868) described the male mite in 1868. Various other descriptions were made by Tiraboschi (1904) and by Gambles (1952), who diagrammed certain of the immature stages. *Myocoptes* is regarded as the most common arthropod parasite of the laboratory mouse (Flynn, 1973), although it frequently occurs in mixed infestation, particularly with *Myobia musculi.* Surveys of laboratory mice from commercial breeders and of conventional mouse colonies have shown that the mite is almost globally distributed (Flynn, 1955, 1963; Flynn *et al.,* 1965; Fukui *et al.,* 1961; Heine, 1962; Sparrow, 1976) even in recent times.

The life cycle has been studied by Watson (1960), who illustrated the immature stages, and more recently, by Letscher (1970). Stages include the egg, larva, protonymph, tritonymph, and adult male and female (Watson, 1960). The eggs average 0.20×0.05 mm in size and hatch within 5 days (Watson, 1960). *Myocoptes* eggs are usually attached to the middle third of the hair shaft, in contrast to *Myobia* eggs, which are attached close to the base of the shaft (Letscher, 1970). The larval and protonymph stages are three-legged (bilaterally), and the tritonymph and adult stages have four pairs of legs. The adult male is approximately 0.19–0.20 mm in length, and the female 0.30 mm (Watson, 1960). The heavily chitinized, dark brown third and fourth pairs of legs, which are modified in the female for hair clasping, are distinctive features. The fourth pair in the male are enlarged but not modified for hair clasping (Fig. 8A,B).

Some controversy exists over the length of the life cycle, being reported as 14 days by Watson (although he recognized the inaccuracy of the estimate) and as 8 days by Letscher (1970). The latter determination was made *in vivo* by placing adult females on a mite-free host, determining the next day that eggs were attached to hairs, and finding adult males 8 days later (the determination assumes that females mature and become fecund when adult males emerge).

Transmission has been shown to require close, direct contact. Although transfer to mite-free hosts can occur in 24 hr or less when they are caged with infested mice, when mite free-hosts are separated in the same cage from parasitized mice, by a wire screen, transfer does not occur (Letscher, 1970). For similar reasons, both Letscher and Watson concluded that bedding of infested mice was not effective in vectoring infestation (Letscher, 1970; Watson, 1961). Infestation of neonates may occur within 4–5 days, as the hair coat appears (Letscher, 1970; Watson, 1961), and all life stages of *Myocoptes* appear to be active in migrations to new hosts (in contrast to *Myobia,* which is transferred primarily by the adult female). *Myocoptes* may live as long as 8–9 days on dead hosts and 3–4 days off the host *in vitro* (Letscher, 1970).

2. Pathology

It has been generally recognized that *Myocoptes* is a more ambulatory species (*in vivo*) than *Myobia* and tends to spread out over greater areas of the body. Particularly in mixed infestations with *Myobia, Myocoptes* has some predilection for the skin of the inguinal areas, abdominal ventrum, and back. In monospecific infestation, it will occupy the neck and head as well. In heavy mixed infestations, there is some tendency for *Myocoptes* to crowd out populations of *Myobia.* A single report of *M. musculinus* infestation of guinea pigs has been described (Sengbusch, 1960). However, there appear to be no other reports for this mite in laboratory hosts other than *Mus.*

Myocoptes has been described as a surface dweller that feeds on superficial epidermal layers and has not been observed to ingest blood or lymph (Watson, 1961). The signs of infestation may be inapparent (Cook, 1953) or may involve patchy thinning of the hair (Fig. 9), irregular alopecia, and erythema of the skin (Cook, 1953; Gambles, 1952; Harrison and Daykin, 1965; Watson, 1963). The lesions can be pruritic (Flynn, 1973), but the ulcerative lesions ascribed to *Myobia* have not been reported in myocoptic infestation.

The histology of infested skin has been studied (Watson, 1961). Chronic infestations induce epidermal hyperplasia (increase in the depth of malpighian layers) and an increase in the mitotic rate of adult skin, but not that of adolescent mice. An

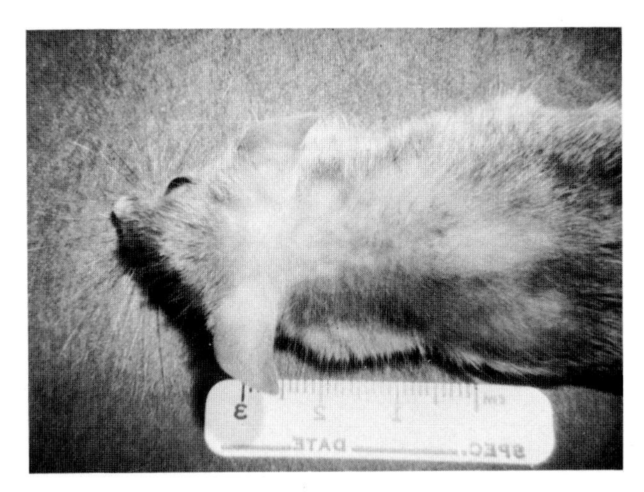

Fig. 9. Mouse with typical, noncomplicated *Myocoptes* infestation. Note thinning of the hair over the thoracic dorsum with some irregular, patchy alopecia.

inflammatory infiltrate of macrophages and lymphocytes was seen in the subcuticular tissues underlying infested epidermis. Systematic studies utilizing monospecific infestations with *Myocoptes* to investigate sequential histologic changes, genetic variability of host strains, or immunologic responses have not been reported.

F. *Trichoecius romboutsi* (van Eyndhoven)

Description

Trichoecius romboutsi was originally described from Holland in 1946 by van Eyndhoven and named *Myocoptes romboutsi* (Van Eyndhoven, 1946). It was encountered in a survey of commercial mouse breeders in the United States in 1955 (Flynn, 1955). However, in a similar survey 10 years later (Flynn *et al.*, 1965), the mite was not observed. The mite was redescribed by Fain *et al.* in 1970 and reclassified in one of the other four genera, *Trichoecius,* of the family Myocoptidae (Fain *et al.*, 1970).

The life cycle and other pertinent aspects of its biology are unknown. Males are approximately 0.16–0.17 mm in length, females 0.20–0.25 mm (see Fig. 8C,D). *Trichoecius romboutsi* is slightly smaller than *M. musculinus,* which it resembles. Indeed, it has been pointed out that this resemblance may account for *T. romboutsi* being overlooked in diagnosis (Flynn, 1973). Nothing is known of the pathobiology of *T. romboutsi*. However, its taxonomic proximity to *Myocoptes* suggests that they may be quite similar.

VI. TREATMENT, CONTROL, AND PREVENTION

It has been pointed out that treatment of arthropod parasitic infestations is less satisfactory than prevention (Owen, 1972). Treatment and control programs are costly in terms of materials and labor and are never 100% effective. The latter observation is a statistical statement. For example, if a particular treatment is 99% effective (a level rarely achieved in practice), a mouse carrying a population of 1000 parasites would be expected to retain about 9 or 10 viable parasites as a posttreatment residuum. The residuum is sufficient to act as a reservoir for infestation rebound (Weisbroth *et al.*, 1976). Particularly for breeding colonies and mice in chronic studies, once infestation has been diagnosed, one must think in terms of ongoing and periodic control programs rather than erodication.

The only effective means of eradicating arthropod infestations is by gnotobiotic rederivation (Clark and Yunker, 1964; Flynn *et al.*, 1965; Foster, 1963) and the repopulation of liquidated colonies by mice obtained from such sources. As discussed earlier, such programs have been of immense value in reducing both the incidence and the diversity of ectoparasitism in laboratory mouse colonies over the last 2–3 decades.

The foregoing is not to say, however, that arthropod parasitism is rarely encountered or that treatment programs are futile. To the contrary, ectoparsitic enzootics remain quite common and are frequently of a clinical intensity sufficient to impede research and breeding programs. A rich body of literature amply demonstrates the continuing significance of ectoparasitism and the importance of treatment and control programs. Control programs have tended to emphasize strategies for mass treatment of all individuals in the colony, rather than only of infested individuals. Such strategies, therefore have had to contend with components such as forms of administration, cost, labor, accumulative contamination of the environment with (possibly) toxic insecticides, and (possibly) intoxication of laboratory personnel during treatment phases.

As successive generations of insecticidal and acaricidal agents have been developed, the approach has been to apply such agents to the treatment of afflicted mouse colonies. The main body of reports dealing with control programs for ectoparasitism of laboratory mice is summarized in Table III. The ideal treatment mode is yet to be developed, and each of the reported modes has certain advantages and disadvantages.

Experiences with dusts, either as desiccating sorptives (Cwilich and Binhar, 1964; Tarshis, 1962, 1967) or as insecticidal powders, have been generally disappointing because of variable efficacy, toxicity of active components, and the high labor component in administration (Cook, 1953; Cwilich and Binhar, 1964; Madden *et al.*, 1954). Various compounds have been evaluated as dips. The use of malathion as a dip (as well as a dust) has had mixed results in terms of efficacy and toxicity (Constantin, 1972; Davis, 1957; Green and Neeham, 1974; Jenkins and Fletcher, 1964; Softly, 1968). Malathion is not recommended for treatment of parasitized mouse colonies. Excellent results have been obtained against *Myocoptes* with DMC or dimite alone (Stoner and Hale, 1953), or in combination with tetmosol (Bateman, 1961). Tetmosol alone has been only variably effective against *Myobia* (Cook, 1953; Green and Neeham, 1974). However, it was reported as effective against *Myobia* in combination with DMC (Bateman, 1961). There have been no reports of the use of DMC or tetmosol from United States laboratories. In contrast, Aramite has been widely used in the United States as a dip with generally excellent results (Clark and Yunker, 1964; Flynn, 1954, 1959; Galton, 1963). Promising results have been obtained with Pyractone, particularly against *Myobia,* and this agent may be of increasing importance due to its lack of mammalian toxicity (Constantin, 1972; Green and Neeham, 1974). Excellent results have been obtained with vegetable oil alone in *Polyplax*-infested rats (Olewine, 1963). As a group, the dips are labor intensive, which has tended to discourage their use.

Table III

Reported Treatment for Ectoparasitism of Laboratory Mice

Chemical or generic name	Trade or common name	Method of administration	Parasite treated and efficacy	Reference
Di-(*p*-chlorophenyl)methyl carbinol and tetraethyl-thiuram monosulfide	DMC or Dimite (2 gm/liter) and tetmosol (67 mg/liter of 25% solution)	Two dippings 1 hr apart by immersion and swimming, 30 sec; repeated 1–3 weeks later	*Myobia, Myocoptes, Psorergates, Polyplax;* eradicated	Bateman (1961)
Di-(*p*-chlorophenyl)methyl carbinol	DMC (Dimite), 2 gm/liter	Single dipping	*Myocoptes;* eradicated	Clark and Yunker (1964)
Tetraethylthiuram monosulfide	Tetmosol (67 gm/liter) of 25% solution	Single dip with swimming	*Myobia;* "usually sufficient"	Cook (1953)
	Tetmosol, 2.5%	Two dippings with immersion at 14-day intervals	*Myobia;* not effective	Green and Neeham (1974)
2-(*p-tert*-butylphenoxy)iso-propyl-2-chloroethyl sul-fite plus Nacconal	Aramite-15W, 2% plus wetting agent, 0.1%	Single dip by immersion and swimming, 17 sec	*Myobia, Myocoptes, Psorergates, Polyplax, Trichoecius, Radfordia;* eradicated	Clark and Yunker (1964); Flynn (1954, 1959); Galton (1963)
Benzene hexachloride	BHC, 0.625%	Two dippings with immersion at 14-day intervals	*Myobia;* not effective	Green and Neeham (1974)
Benzene hexachloride	BHC	Dusting	*Myocoptes,* "cures condition; *Myobia,* ineffective	Cook (1953)
O-1-dimethyl-5-(dicarbeth-oxyethyl)dithiophosphate	Malathion, 2.5%	Two dippings with immersion at 14-day intervals	*Myobia;* not effective	Green and Neeham (1974)
	Malathion, 1% and 4%	Five dippings at 3-day intervals	*Myobia;* not effective	Constantin (1972)
	Malathion, 2%	Single dipping	*Myocoptes;* eradicated	Davis (1957)
	Malathion, 2%	Single dip by immersion and swimming, 17 sec	*Myobia, Myocoptes;* effective	Softly (1968); Webb and Shepherd (1959)
	Malathion, 0.125% (Malastan E-C50, emulsifiable)	Single dip by immersion of rats	*Polyplax;* eradicated	Jenkins and Fletcher (1964)
Pyrethrins 5% with 2.5% bucarpolate	Pyractone, 1%	Five dippings at 3-day intervals	*Myobia;* eradicated	Constantin (1972)
Pyrethrins 5% with 25% bucarpolate	Super Pyractone M.429, diluted to give 0.06% pyrethrin, 0.3% bucarpolate	Two dippings with immersion at 14-day intervals	*Myobia;* effective	Green and Neeham (1974)
SG-67	Dri-Die (0.5 gm per mouse)	Applied individually as a dust	*Myobia;* not effective	Cwilich and Binhar (1964)
Silica sorptive dust	Dri-Die 67	Applied individually	*Polyplax*	Tarshis (1962, 1967)
Silica dust with pyrethrin	Dri-One	Applied individually	*Polyplax*	
Formel 5-brommethyl-1,2,3,4,7,7-hexachlorobi-cyclo-(2,2,1) hepten-(2)	Alugan, 0.6% solution	Two dippings at 11-day intervals	*Myobia;* eradicated	Heine (1966)
Sulfur	325 mesh dusting sulfur	One pinch dusted onto animal	*Myobia, Myocoptes,* 95–99% effective	Madden *et al.* (1954)
	Chlordane, 2 gm 5% dust	Sprinkled over contact bedding	*Polyplax;* eradicated	Madden *et al.* (1954)
	Lindane, 10 gm 1% dust	Sprinkled over contact bedding; several applications	*Myobia, Myocoptes;* 90–98% effective	Madden *et al.* (1954)
	Kelthane, 5% wettable powder	Used as dust, 0.5 gm per mouse	*Myobia;* effective	Cwilich and Binhar (1964)
Gamma HCH	Jacutin spray (20% in oil)	Individually with small brush; twice at 12-day intervals	*Myobia;* effective	Harris and Stockton (1960)
Dibutyl phthalate		Applied topically to affected areas	*Psorergates;* effective	Cook (1956)
Dichlorvos (DDVP)	Vapona (18% in resin strip)	1 × 2.5-in resin strip on cage top, under filter cap	*Myocoptes;* effective	Wagner (1969); Weisbroth *et al.* (1976); Wharton (1940)
		2 × 2.5-in resin strip on cage top, under filter	*Myobia;* effective	
	Atgard (anthelmintic) pellets	2 gm sprinkled in bedding; on day 8 cage changed, 2 gm added (14-day exposure)	*Myobia;* effective	Fraser *et al.* (1974)
	Vaporal (4.65% DDVP, 5% ronnel) liquid dip	2 ml per cage on contact litter	*Myobia;* effective	Csiza and McMartin (1976)

Dichlorvos (DDVP or Vapona) has come into recent and widespread use for treatment of murine ectoparasitism. Its popularity is related both to its value as an exceptional acaricide and to the convenience (low labor component) with which it may be administered. Several formulations have been investigated, including resin strips placed in or on cages (Friedman and Weisbroth, 1977; Wagner, 1969; Weisbroth *et al.*, 1976), pellets formulated for use as oral anthelmintics (Fraser *et al.*, 1974), and liquid placed in contact litter (Csiza and McMartin, 1976). Essentially similar results have been obtained with 2-4-in^2 resin strips, 2-gm anthelmintic pellets, and 2-ml liquid concentrates used for individual cage treatments. Due to the resistance of *in ovo* forms (*Myobia*), recommended regimens require two treatments at 10- to 12-day intervals. Although no reports of use of dichlorvos in the treatment of *Polyplax* have appeared, its effectiveness against avian lice (Kunz and Hogan, 1970) suggests that it would be efficacious in mice. Dichlorvos has been shown to result in transient lowering of serum cholinesterase during exposure phases (Wagner, 1969; Wagner and Johnston, 1970). However, it is rapidly cleared and metabolized (Blair *et al.*, 1975) and is neither teratogenic (Guthrie *et al.*, 1971) nor carcinogenic (Blair *et al.*, 1976). Dichlorvos has been reported to delay breeding transiently during intensive exposures (Wagner, 1969).

REFERENCES

Ah, H.-S., Peckham, J. C., and Atyeo, W. T. (1973). *Psorergates glaucomys* Sp.N. (Acari: Psorergatidae), a cystogenous mite from the Southern flying squirrel (*Glaucomys* v. *volans*), with histopathologic notes on a mite-induced dermal cyst. *J. Parasitol.* **59,** 369–374.

Ama, M. (1955). Chaetotaxy of the stages, in the life cycle of *Myobia musculi* (Schrank). M.S. Thesis, Univ. of Maryland, College Park, Maryland.

Baker, W. E., and Wharton, G. H. (1952). "An Introduction to Acarology." Macmillan, New York.

Baker, E. W., Evans, T. M., Gould, D. J., Hull, W. B., and Keegan, H. L. (1956). "A Manual of Parasitic Mites of Medical or Economic Importance," Tech. Publ. Natl. Pest Control Assoc., New York.

Banks, N. (1915). "The Acarina or Mites," U.S. Dept. Agric., Rep. No. 108. U.S. Gov. Print. Off., Washington, D.C.

Bateman, N. (1961). Simultaneous eradication of three ectoparasitic species from a colony of laboratory mice. *Nature* (*London*) **191,** 721–722.

Bell, J. F., and Clifford, C. (1964). Effects of limb disability on lousiness in mice. II. Intersex grooming relationships. *Exp. Parasitol.* **15,** 340–349.

Bell, J. F., Jellison, W. I., and Owen, C. R. (1962). Effects of limb disability on lousiness in mice. I. Preliminary studies. *Exp. Parasitol.* **12,** 176–183.

Bell, J. F., Clifford, C. M., Moore, G. J., *et al.* (1966). Effects of limb disability on lousiness in mice. III. Gross aspects of acquired resistance. *Exp. Parasitol.* **18,** 49–60.

Beresford-Jones, W. P. (1965). Occurrence of the mite *Psorergates simplex* in mice. *Aust. Vet. J.* **41,** 289–290.

Beresford-Jones, W. P. (1967). Observations on the transmission of the mite *Psorergates simplex* Tyrell, 1883, in laboratory mice. *In* "The Reaction of the Host to Parasitism" (E. J. L. Soulsby, ed.), pp. 227–280. Vet. Med. Rev. Elwert Univ. Verlagsbuchhandl., Marburg/Lahn.

Blair, D., Hoadley, E. C., and Hutson, D. H. (1975). The distribution of dichlorvos in the tissues of mammals after its inhalation or intravenous administration. *Toxicol. Appl. Pharmacol.* **31,** 243–253.

Blair, D., Dix, K. M., Hunt, P. F., *et al.* (1976). Dichlorvous—a 2 year inhalation carcinogenesis study in rats. *Arch. Toxicol.* **35,** 281–294.

Boccia, M. (1942). Sulla dermatosi parassitarie del *Mus musculus albus. G. Batteriol. Immunol.* **29,** 678–685.

Claparede, E. (1868). Studien an Acariden. *Z. Wiss. Zool.* **18,** 445–546.

Clark, G. M., and Yunker, C. E. (1964). Control of fur-mites on mice in entomological laboratories. *Proc. Int. Congr. Acarol., 1st, Ft. Collins, Colo.* pp. 235–236.

Clifford, C. M., Bell, J. F., Moore, J., *et al.* (1967). Effects of limb disability on lousiness in mice. IV. Evidence of genetic factors in susceptibility to *Polyplax serrata. Exp. Parasitol.* **20,** 56–67.

Compton, A. (1926). Phthiriasis of the mouse: *Hematopinus muris* with observations on treatment by salicylidene compounds. *Vet. J.* **82,** 255–258.

Constantin, M. L. (1972). Effects of insecticides on acariasis in mice. *Lab. Anim.* **6,** 279–286.

Cook, R. (1953). Murine mange: The control of *Myocoptes musculinus* and *Myobia musculi* infestations. *Br. Vet. J.* **109,** 113–116.

Cook, R. (1956). Murine ear mange: The control of *Psorergates simplex* infestation. *Br. Vet. J.* **112,** 22–25.

Csiza, C. K., and McMartin, D. N. (1976). Apparent acaridal dermatitis in a C57B1/6 Nya mouse colony. *Lab. Anim. Sci.* **26,** 781–787.

Cwilich, R., and Binhar, A. (1964). The control of mites, *Myobia musculi* (Myobiidae: Prostigmata) on laboratory mice. *Refu. Vet.* **21,** 39–40.

Davis, R. (1957). Control of the myocoptic mange mite, *Myocoptes musculinus* (Koch) on laboratory mice. *J. Econ. Entomol.* **50,** 5.

Dunn, T. B. (1944). Relationship of amyloid infiltration and renal disease in mice. *J. Natl. Cancer Inst.* **5,** 17–28.

Dunn, T. B. (1949). Some observations on the normal and pathologic anatomy of the kidney of the mouse. *J. Natl. Cancer Inst.* **9,** 285–301.

Eliot, C. P. (1936). The insect vector for the natural transmission of *Eperythrozoon coccoides* in mice. *Science* **84,** 397.

Ewing, H. E. (1929). "Manual of External Parasites." Thomas, Springfield, Illinois.

Ewing, H. E. (1938). North American mites of the subfamily Myobiinae, new subfamily (Arachnida). *Proc. Entomol. Soc.* Wash. **40,** 180–197.

Fain, A. (1959a). Les acariens psoriques des chauves-souris. III. Le Genre *Psorergates* Tyrell (Trombidiformes: Psorergatidae). *Bull. Ann. Soc. R. Entomol. Belg.* **95,** 54–69.

Fain, A. (1959b). Les acariens psoriques des chauves-souris. IX. Nouvelles observations sur le genre *Psorergates* Tyrell. *Bull. Ann. Soc. R. Entomol. Belg.* **95,** 232–248.

Fain, A. (1965). Notes sur le genre *Notoedres* Railliet, 1893 (Sarcoptidae: Sarcoptiformes). *Acarologia* **7,** 321–342.

Fain, A., Lukoschus, F., and Hallman, P. (1966). Le genre *Psorergates* chez les Murides. Description de trois especes nouvelles (Psorergatidae: Trombidiformes). *Acarologia* **8,** 251–274.

Fain, A., Munting, A. J., and Lukoschus, A. (1970). Les Myocoptidae parasites des rongeurs en Hollande et en Belgique. *Acta Zool. Pathol. Antverp.* **50,** 67–172.

Flynn, R. J. (1954). Mouse mange. *Proc. Anim. Care Panel* **5,** 96–105.

Flynn, R. J. (1955). Control of ectoparasites in mice. *Proc. Anim. Care Panel* **6,** 75–91.

Flynn, R. J. (1959). Follicular acariasis of mice caused by *Psorergates simplex* successfully treated with Aramite. *Am. J. Vet. Res.* **20,** 198–200.

Flynn, R. J. (1963). The diagnosis of some forms of ectoparasitism of mice. *Proc. Anim. Care Panel* **13,** 111–125.

Flynn, R. J. (1973). "Parasites of Laboratory Animals." Iowa State Univ. Press, Ames.

Flynn, R. J., and Jaroslow, B. N. (1956). Nidification of a mite (*Psorergates simplex,* Tyrell. 1883: Myobiidae) in the skin of mice. *J. Parasitol.* **42,** 49-52.

Flynn, R. J., Brennan, P. C., and Fritz, T. E. (1965). Pathogen status of commercially produced mice. *Lab. Anim. Care* **15,** 440-447.

Foster, H. L. (1963). Specific pathogen-free animals. *In* "Animals for Research: Principles of Breeding and Management" (W. Lane-Petter, ed.), pp. 110-138. Academic Press, New York.

Fraser, J., Joiner, G. N., Jardine, J. H., *et al.* (1974). The use of pelleted dichlorvos in the control of murine acariasis. *Lab. Anim.* **8,** 271-274.

Friedman, S., and Weisbroth, S. H. (1975). The parasitic ecology of the rodent mite, *Myobia musculi.* II Genetic factors. *Lab. Anim. Sci.* **25,** 440-445.

Friedman, S., and Weisbroth, S. H. (1977). The parasitic ecology of the rodent mite *Myobia musculi.* IV. Life cycle. *Lab. Anim. Sci.* **27,** 34-37.

Fukui, M., Matsuzaki, H., Tanaka, T., *et al.* (1961). Studies on the acaric dermatitis of albino mice in Japan. (In Jpn. *Exp. Anim.* **10,** 83-90.

Galton, M. (1963). Myobic mange in the mouse leading to skin ulceration and amyloidosis. *Am. J. Pathol.* **43,** 855-865.

Gambles, R. M. (1952). *Myocoptes musculinus* (Koch) and *Myobia musculi* (Schrank), two species of mite commonly parasitizing the laboratory mouse. *Br. Vet. J.* **108,** 194-203.

Grant, C. D. (1942). Observations on *Myobia musculi* (Schrank) (Arachnida: Acarina: Cheyletidae). *Microentomology* **7,** 64-76.

Green, C. J., and Neeham, J. R. (1974). Control of mange mites in a large mouse colony. *Lab. Anim.* **8,** 245-251.

Guthrie, F. I., Monroe, R. J., and Abernathy, C. O. (1971). Response of the laboratory mouse to selection for resistance to insecticides. *Toxicol. Appl. Pharmacol.* **18,** 92-101.

Harris, J. M., and Stockton, J. J. (1960). Eradication of the tropical rat mite *Ornithonyssus bacoti* (Hirst, 1913) from a cology of mice. *Am. J. Vet. Res.* **21,** 316-318.

Harrison, I. R., and Daykin, M. M. (1965). The biology and control of ectoparasites of laboratory animals with special reference to poultry parasites. *J. Anim. Tech. Assoc.* **16,** 69-73.

Heine, W. (1962). Zur Ektoparasitenbekämpfung bei Maus and Ratte. *Z. Versuchstierkd.* **2,** 1-22.

Heine, W. (1966). Alugan zur Bekämpfung von Ektoparasiten bei Muriden. *Dtsch. Tierartzl. Wochenschr.* **73,** 474-475.

Heston, W. E. (1941). Parasites. *In* "Biology of the Laboratory Mouse" (G. D. Snell, ed.), p. 375. Blakiston, Philadelphia, Pennsylvania.

Heston, W. E., and Deringer, M. K. (1948). Hereditary renal disease and amyloidosis in mice. *Arch. Pathol.* **46,** 49-58.

Hilton, D. F. J. (1970). A technique for collecting ectoparasites from small birds and mammals. *Can. J. Zool.* **48,** 1445-1446.

Hirst, S. (1917). Remarks on certain species of the genus *Demodex* Owen (the *Demodex* of man, the horse, dog, rat, and mouse). *Ann. Mag. Nat. Hist.* **20,** 232-235.

Hopkins, G. H. E. (1949). The host associations of the lice of mammals. *Proc. Zool. Soc. London* **119,** 388-604.

Institute of Laboratory Animal Resources (ILAR) (revised 1978). "Guide for the Care and Use of Laboratory Animals," Nat. Res. Counc., DHEW Publ. (NIH) 78-23. U.S. Dep. Health, Educ. Welfare, Natl. Inst. Health, Bethesda, Maryland.

Ishizaka, K., and Ishizaka, T. (1970). Biological function of γE antibodies and mechanisms of reaginic hypersensitivity. *Clin. Exp. Immunol.* **6,** 25-42.

Ishizaka, T., Tomoika, H., and Ishizaka, K. (1971). Degranulation of human basophil leucocytes by anti-γE antibody. *J. Immunol.* **160,** 705-710.

Jambon, W. (1928). *Psorergates simplex* as a possible factor in epitheliuma of the skin and mammary glands in murids. *Ann. Trop. Med. Parasitol.* **22,** 133-136.

Jancke, O. (1930). *Polyplax serrata* (Burmeister) 1839. *Zool. Anz.* **92,** 105-109.

Jenkins, J. I., and Fletcher, F. J. (1964). The eradication of lice from a rat colony by means of a malathion dipping routine. *J. Anim. Tech. Assoc.* **15,** 1-6.

Keefe, T. J., Scanlon, J. E., and Wetherald, L. D. (1964). *Ornithonyssus bacoti* (Hirst) infestation in mouse and hamster colonies. *Lab. Anim. Care* **14,** 366-369.

Koch, C. L. (1840). C. M. A. Deutschl, fasc. 5, fig. 13 (*Sarcoptes musculinus*).

Kunz, J. E., and Hogan, B. F. (1970). Dichlorvos-impregnated resin strands for control of chicken lice on laying hens. *J. Econ. Entomol.* **63,** 263-266.

Letscher, R. M. (1970). Observations concerning the life cycle and biology of *Myobia musculi* (Schranck) and *Myocoptes musculinus* (Koch). M.S. Thesis, Texas A&M Univ., College Station.

Low, R. C. (1911) An investigation into scabies in laboratory animals. *J. Pathol.* **15,** 333-348.

Madden, A. H., Tozloski, A. H., and Sweetman, H. L. (1954). Control of *Myobia musculinus* (sic) (Koch) and *Myocoptes musculinus* (Koch) in laboratory mice. *J. Econ. Entomol.* **47,** 442-444.

Murray, M. D. (1961). The ecology of the louse *Polyplax serrata* (Burm.) on the mouse, *Mus musculus* L. *Aust. J. Zool.* **9,** 1-13.

Nelson, W. A., Clifford, C. M., Bell, J. F., and Hestekin, B. (1972). *Polyplax serrata:* Histopathology of the skin of louse infested mice. *Exp. Parasitol.* **31,** 194-202.

Neumann, M. G. (1893). Sur un acarien (*Psorergates simplex* Tyrell) de la souris. *Bull. Soc. Hist. Nat. Toulouse* **27,** 13-22.

Oldham, J. N. (1967). Helminths, ectoparasites and protozoa in rats and mice. *In* "Pathology of Laboratory Rats and Mice" (E. Cotchin and F. J. C. Roe, eds.), pp. 641-679. Blackwell, Oxford.

Olewine, D. A. (1963). An effective control of *Polyplax* infestation without the use of insecticide. *Lab. Anim. Care* **13,** 750-751.

Owen, D. M. (1972). "Common Parasites of Laboratory Rodents and Lagomorphs," Handbook No. 1. Med. Res. Counc., Lab. Anim. Cent. HM Stationery Off., London.

Owen, D. (1976). Some parasites and other organisms of wild rodents in the vicinity of an SPF unit. *Lab. Anim.* **10,** 271-278.

Poppe, S. A. (1896). Beitrag zur Kenntnis der Gattung *Myobia. Zool. Anz.* **19,** 327-333, 337-349.

Pratt, H. D., and Littig, K. S. (1961). "Lice of Public Health Importance and Their Control," PHS Publ. No. 772, Insect Control Series: Part VIII. U.S. Gov. Print. Off., Washington, D.C.

Pratt, H. D., and Wiseman, J. S. (1962). "Fleas of Public Health Importance and Their Control," PHS Publ. No. 772, Insect Control Series: Part VIII. U.S. Gov. Print. Off., Washington, D.C.

Radford, C. D. (1934). Notes on mites of the genus *Myobia. Northwest. Nat.* **9,** 356-364.

Sasa, M. (1950). Note on the blood-sucking lice (Anoplura) of rodents in Japan (Part I). *Jpn. J. Exp. Med.* **20,** 715-717.

Scott, J. P. (1958). "Animal Behavior." Univ. of Chicago Press, Chicago, Illinois.

Seamer, J., and Chesterman, F. C. (1967). A survey of disease in laboratory animals. *Lab. Anim.* **1,** 117-139.

Sengbusch, H. G. (1960). Control of *Myocoptes musculinus* on guinea pigs. *J. Econ. Entomol.* **53,** 168.

Shaughnessy, H. J. (1963). Tularemia. *In* "Diseases Transmitted From Animals to Man" (T. G. Hull, ed.), 5th ed., pp. 588-604. Thomas, Springfield, Illinois.

Sheather, A. L. (1915). An improved method for the detection of mange acari. *J. Comp. Pathol. Ther.* **28,** 64–66.

Sheather, A. L. (1923). The detection of intestinal protozoa and mange parasites by a flotation technique. *J. Comp. Pathol. Ther.* **36,** 266–275.

Softly, A. (1968). Production of specific mite free rats and mice by two methods. *J. Inst. Anim. Tech.* **19,** 17–21.

Sokoloff, L., Mickelsen, O., and Silverstein, E., *et al.* (1960). Experimental obesity and osteoarthritis. *Am. J. Physiol.* **198,** 765–770.

Sparrow, S. (1976). The microbiological and parasitological status of laboratory animals from accredited breeders in the United Kingdom. *Lab. Anim.* **10,** 365–376.

Stewart, S. J., Bell, F. J., Hestekin, B., *et al.* (1976). *Polyplax serrata:* Effects of limb disability on lousiness in mice. VI. Lack of tolerance after neonatal exposure. *Exp. Parasitol.* **40,** 373–379.

Stoner, R. D., and Hale, W. M. (1953). A method for eradication of the mite, *Myocoptes musculinus,* from laboratory mice. *J. Econ. Entomol.* **46,** 692–693.

Strandtmann, R. W., and Wharton, G. W. (1958). "A Manual of Mesostigmatid Mites Parasitic on Vertebrates," Contrib. No. 4. Inst. Acarol., Univ. of Maryland, College Park, Maryland.

Tarshis, I. B. (1962). The use of silica aerogel compounds for the control of ectoparasites. *Proc. Anim. Care Panel* **12,** 217–258.

Tarshis, I. B. (1967). Silica aerogel insecticides for the prevention and control of arthropods of medical and veterinary importance. *Angew. Parasitol.* **8,** 210–237.

Thung, P. J. (1957). Senile amyloidosis in mice. *Gerontologia* **1,** 259–279.

Tiraboschi, G. (1904). *Myocoptes musculinus* Claparede. *Arch. Parasitol.* **8,** 327.

Trouessart, E. (1895). Sur les metamorphoses du genre Myobia et diagnoses d'especes nouvelles d'Acariens. *Bull. Soc. Entomol. Fr.* **8,** 213–214.

Tyrell, J. B. (1883). On the occurrence in Canada of two species of parasitic mites. *Proc. Can. Inst.* **1,** 332–342.

Van Eyndhoven, G. L. (1946). Diagnoses of two epizootic mites. *Entomol. Ber.* **12,** 30–31.

von Haakh, U. (1958). Ektoparasitenfreie Laboratoriumsmäuse: die *Myobia-Räude der weissen Mäuse und ihre Bekämpfung. Z. Tropenmed. Parasitol.* **9,** 75–87.

von Heyden, C. H. G. (1826). Versuch einer systematischen Eintheilung der Acariden. *Oken. Isis* **19,** 608–613.

von Schrank, F. P. (1781). Pediculis Muris musculi. *Enumeratio Insect Aust. Indig.* 501–520.

Wagner, J. E. (1969). Control of mouse ectoparasites with resin vaporizer strips containing vapona. *Lab. Anim. Care* **19,** 804–807.

Wagner, J. E., and Johnston, D. C. (1970). Toxicity of dichlorvos for laboratory mice LD50 and effect on serum chlolinesterase. *Lab. Anim. Care* **20,** 45–47.

Watson, D. P. (1960). On the adult and immature stages of *Myocoptes musculinus* (Koch) with notes on its biology and classification. *Acarologia* **2,** 335–344.

Watson, D. P. (1961). The effect of the mite *Myocoptes musculinus* (C. L. Koch, 1840) on the skin of the white laboratory mouse and its control. *Parasitology* **51,** 373–378.

Watson, D. P. (1963). The effect of cysteine and B. group vitamins on the skin of the white laboratory mouse infested with the mite *Myocoptes musculinus* (Koch). *Parasitology* **53,** 265–272.

Webb, J. E., Jr., and Shepherd, E. L. (1959). Notes on control of mange mites in laboratory white mice with malathion. *J. Econ. Entomol.* **52,** 790.

Weisbroth, S. H., Wang, R., and Scher, S. (1972). Antibodies in mouse acariasis detected with antigens from *Psoroptes cuniculi. Annu. Meet., 23rd, Am. Assoc. Lab. Anim. Sci., St. Louis, Mo.* Abstr. No. ■ ■.

Weisbroth, S. H., Powell, M. B., Roth, L., *et al.* (1974a). Immunopathology of naturally occurring otodectic otoacariasis in the domestic cat. *J. Am. Vet. Med. Assoc.* **165,** 1088–1093.

Weisbroth, S. H., Friedman, S., Powell, M., *et al.* (1974b). The parasitic ecology of the rodent mite *Myobia musculi.* I. Grooming factors. *Lab. Anim. Sci.* **24,** 510–516.

Weisbroth, S. H., Friedman, S., and Scher, S. (1976). The parasitic ecology of the rodent mite, *Myobia musculi.* III. Lesions in certain host strains. *Lab. Anim. Sci.* **26,** 725–735.

Wharton, G. W. (1940). Life cycle and feeding habits of *Myobia musculi. J. Parasitol.* **40,** 29.

Wharton, G. W. (1960). Host-parasite relationships between *Myobia musculi* (Schrank, 1781) and *Mus musculus* Linnaeus, 1758. *In* "Libro Homenaje al Dr Eduardo Caballero y Caballero, Jubileo 1930–1960," pp. 571–575. Inst. Politec. Nac., Mexico, City.

Whitely, H. J., and Horton, D. L. (1962). The effect of *Myobia musculi* on the epidermis and hair regrowth cycle in the aging CBA mouse. *J. Pathol.* **83,** 509–514.

Whitely, H. J., and Horton, D. L. (1965). Further observations on the effect of *Myobia musculi* on the skin of the mouse. *J. Pathol.* **89,** 331–335.

Wormersley, H. (1942). Miscellaneous additions to the Acarine fauna of Australia. *Trans. R. Soc. South Aust.* **66,** 85–92.

Yunker, C. E. (1964). Infections of laboratory animals potentially dangerous to man: Ectoparasites and other arthropods, wtih emphasis on mites. *Lab. Anim. Care* **14,** 455–465.

Chapter 22

Zoonoses and Other Human Health Hazards

James G. Fox and James B. Brayton

I.	Introduction	404
II.	Viral Diseases	404
	A. Lymphocytic Choriomeningitis Virus (LCM)	404
	B. Rabies	405
	C. Other Viruses	405
III.	Rickettsial Diseases	406
	A. Rickettsialpox	406
	B. Murine Typhus	406
IV.	Bacterial Diseases	407
	A. Leptospirosis	407
	B. Pasteurellosis	408
	C. Rat-bite Fever	409
	D. *Salmonella*	410
	E. Other Potential Bacterial Diseases	410
V.	Mycoses (Ringworm)	411
	A. Reservoir and Incidence	411
	B. Mode of Transmission	412
	C. Clinical Signs	412
VI.	Protozoan Diseases (*Entamoeba coli*)	413
VII.	Helminth Diseases	413
	A. Tapeworms	413
	B. Roundworms (*Syphacia obvelata*)	414
VIII.	Arthropod Infestations	414
IX.	Bites	416
X.	Allergic Sensitivities	416
	A. Incidence and Clinical Signs	416
	B. Pathogenesis	417
	C. Diagnosis	418
	D. Treatment and Prevention	419
XI.	Conclusion	419
	References	419

ISBN 0-12-262502-1

I. INTRODUCTION

Derived from the Greek words *zoon,* meaning animals, and *noses,* meaning disease, *zoonoses* literally refers to diseases transmitted directly to man by animals. This chapter reviews known or potential zoonotic agents and the disease manifestations produced in man by exposure to infected mice. We discuss other health hazards that may be encountered when working with mice, such as bites and allergies. Selected transmission of human infectious agents to mice is also briefly mentioned.

Although the mouse, either feral, laboratory, or pet, is not commonly considered a reservoir for human pathogens, a review of the literature contests that notion. It should also be noted that many of the zoonotic diseases affecting mice also occur in rats (Geller, 1979). However, with the advent of modern laboratory animal production and management, zoonotic diseases are being curtailed and are nonexistent in many laboratories.

II. VIRAL DISEASES

A. Lymphocytic Choriomeningitis Virus (LCM)

Of the many latent viruses present in the mouse, only LCM naturally infects man. A review of the literature attests to the ease with which LCM can be transmitted from animals to man (Lehmann-Grube, 1971; see also Chapter 12, this volume).

1. Reservoir and Incidence

The natural association of LCM virus and the mouse provides for mutual survival in a symbiotic relationship. Neither the virus nor the host significantly suppresses the other, though each can do so. LCM exists in the wild mouse population throughout the United States, Europe, Asia, Africa, and probably the world (although it has not been isolated from mice in Australia). Wild mice are the ultimate reservoir of infection for laboratory mice and other susceptible hosts (Maurer, 1964). Mice, and hamsters to a lesser extent, are the only species in which a long-term, asymptomatic infection is known to exist (Hotchin, 1971; Parker *et al.*, 1976). In an early study, 21.5% of mice surveyed in the Washington, D.C., area were infected (Armstrong *et al.*, 1940). In a more recent survey (1967-1970) in the United States, LCM infection was detected in only 2 of 22 production or research colonies (Poiley, 1970). It was present at a low-level incidence for at least 2 years in one colony. However, this survey was conducted only in retired breeding stock, and the monitoring technique detected only nontolerant infections. LCM has also been reported in other mouse colonies used for research in the United States. (Soave and Van Allen, 1958). Early investigations in the United Kingdom demonstrated infection in 1 of 18 mouse-breeding colonies (Findlay *et al.*, 1936) and in "many strains" surveyed at a later date (MacCallum, 1949). LCM still existed in colonies in selected institutions in England in 1970 (Skinner and Knight, 1971) and undoubtedly persists in some colonies maintained in the United States. Infection has been eradicated in almost all colonies, however, by surgical derivation, routine serological monitoring, culling, and prevention of entry of wild mice into laboratory colonies.

Another source of infection for man is the presence of LCM virus in experimental tumors induced in mice. This source was first recognized in a much used, transplantable leukemia of C58 mice, line I, in which inoculation of the tumor produced mild clinical illness in mice. It had been assumed that the sickness was due to a toxic substance produced by the leukemia cells; it was discovered, however, that the etiologic agent was LCM virus (Lindorfer and Syverton, 1953; Taylor and MacDowell, 1949). Subsequently, LCM virus has been found in other commonly used tumor lines (Collins and Parker, 1972; Stewart and Haas, 1956). LCM virus has also been found as a contaminant of mycoplasma and murine poliovirus (Findlay *et al.*, 1938; Wenner, 1948).

2. Mode of Transmission

Diagnosis and control of this infection in mouse colonies has been described in Chapter 12. Mice that are congenitally infected are born normal and appear normal for most of their life span, even though they are persistently viremic and viruric. Virtually all cells can be infected with the virus. Most human laboratory infections have been associated with improper handling of infected murine tissues (Baum *et al.*, 1966; Tobin, 1968). Before manipulative procedures begin, all murine tumor lines should be screened for this virus. Man can also be infected with LCM virus either directly from feces or urine of mice or indirectly by inhaling the dried excreta carried on aerosolized dust originating from the animal cage or room. The wild house mouse plays an important role in the incidence of human disease from LCM virus (Dalldorf *et al.*, 1946; MacCallum, 1949). The original description of human infection with LCM was associated with a reservoir of the virus in the form of persistent latent infections in the wild house mice, *Mus musculus* (Armstrong and Lillie, 1934). Although LCM infection can cause death in man, none of these cases was fatal, nor was there evidence of transmission by human contact. Several authors have emphasized that acutal handling of LCM-infected mice appeared to be important in causing the disease in humans (Havens, 1948; Smithard and Macrae, 1951). The bite of an infected mouse can also cause human infection (Scheid *et al.*, 1964).

In general terms, control of LCM is related directly to sanitary conditions in homes and laboratories; infestation of the premises with LCM-infected mice may increase the likelihood of LCM infection (Armstrong and Sweet, 1939). Careful washing of hands or using disposable gloves reduces the chance of infection in man.

LCM can be considered an arthropod-borne virus, having been transmitted experimentally by various bloodsucking insects, including mosquitoes (*Aedes aegypti*), Rocky Mountain wood ticks (*Dermacentor andersoni*), and fleas; all of these organisms could conceivably gain entrance into a laboratory animal facility (Hotchin and Benson, 1973). LCM virus can also occur spontaneously in cockroaches (Armstrong, 1963).

3. Clinical Signs, Susceptibility, and Resistance in Man

Although its expression can vary greatly, LCM virus infection appears most frequently as a mild influenza-like syndrome, with or without apparent involvement of the central nervous system (Duncan *et al.*, 1951).

In one epidemic of nonmeningitic LCM virus infection, caused by exposure to infected hamsters, an influenza-like illness was described with typical symptoms of retro-orbital headache, severe myalgia, malaise, anorexia, and aching pain in the chest (Baum *et al.*, 1966). Fever was a consistent symptom. The author compared this illness to the disease in two other meningitic human cases in his laboratory, caused by contact with infected mice. Sequelae to the initial infection can consist of arthritis, orchitis, parotitis, and a mild generalized alopecia of the scalp (Baum *et al.*, 1966; Lewis and Utz, 1961).

B. Rabies

Rabies virus, a rhabdovirus, has been recognized since ancient times in Europe and Asia. This virus probably produces fatal disease, by inoculation, in all warm-blooded animals; mice must therefore be considered a potential source of rabies virus. In fact, laboratory diagnosis of rabies can be aided by intracerebral inoculation of mice with test suspensions.

1. Reservoir and Incidence

Rabies occurs on all continents of the world except Australia; islands such as New Zealand, Hawaii, and Great Britain are also free of the disease. It has been successfully excluded by rigid quarantine requirements. Rabies is uncommon in man, and its natural reservoirs are wild carnivora, bats, and rarely certain rodents, such as squirrels (Benenson, 1975).

The incidence of rabies varies within select populations and geographic locations. No cases of human rabies in the United States have been associated with the bite of rabid mice or rats. However, in the Federal Republic of Germany, from 1961 to 1967, three mice, one rat (species unspecified), nine Norway rats, and eight muskrats were reportedly infected with rabies and had bitten humans (Scholz and Weinhold, 1969).

Rabies is transmitted via virus-laden saliva and is inoculated by a bite of a rabid animal or contamination of a wound with saliva. Most rabid animals transmit virus 3–5 days before the appearance of clinical signs and during the course of clinical disease.

2. Clinical Signs

Rabies appears nearly the same in man and animals, with both furious and paralytic signs being presented. The incubation period can range from 12 days to 6 months or more.

In experimentally inoculated mice, paralysis of the hindlimbs occurs as early as the seventh day or as late as the twenty-fifth day; death follows paralysis within 24 hr. Convulsions may be observed just before paralysis begins (Bruner and Gillespie, 1973). Extreme caution should be used when working with experimentally infected mice, as with other infectious agents.

Other than in experimental settings involving rabies research, routine antirabies prophylaxis is not practiced for individuals bitten by laboratory-reared mice. Though the likelihood of rabid wild mice biting man is slim, the possibility does exist (Scholz and Weinhold, 1969).

C. Other Viruses

Three other viruses commonly associated with disease in mice have been implicated as being infective to man. Complement-fixing and neutralizing antibody titers to mouse hepatitis virus, a coronavirus, have been found in human sera (Hartley *et al.*, 1964). The titer's presence is most likely due to cross-reactivity with antibody from infections with human coronaviruses, such as OC 38–OC 43 (McIntosh *et al.*, 1967, 1969) and HCV 229B (Bradburne, 1970), rather than to indicators of zoonotic disease.

Another prevalent agent in mouse colonies, Sendai virus (parainfluenza virus), was originally isolated during an epidemic of fatal pneumonitis is Japanese children (Kuroya *et al.*, 1953a; Sano *et al.*, 1953). Lung suspensions from fatal cases were inoculated intranasally into laboratory mice, and Sendai virus was isolated from diseased lungs of the mice. A year later, another investigator demonstrated the indigenous nature of the virus in mice (Fukumi *et al.*, 1954). Rising antibody titers were demonstrated in patients, and the virus was supposedly capable of producing disease in human volunteers (see Parker and Richter, Chapter 8, this volume, for a review). Others have also reported isolating the virus from cases of human respiratory illness (Gerngross, 1957; Kuroya *et al.*, 1953b; Zhdanoff *et al.*, 1957). In a survey to detect the presence of antibody to murine viruses, antibody to Sendai virus was noted in personnel working with laboratory animals; significant titers were also present in personnel with no laboratory animal exposure (Tennant *et al.*, 1967). Though a definitive answer to the question remains debatable, the antibody titer is probably due to cross reactions with antigenically related parainfluenzea viruses (Heath *et al.*, 1962; see also Chapter 8, this volume).

Reovirus 3, a prevalent virus in mouse colonies, was first isolated in 1953 from the feces of a clinically ill child (Stanley *et al.*, 1953). Since then, the presence of antibody to reovirus 3 in human sera has been reported, although no human clinical syndrome has been well defined. The occurrence of reovirus 3 in mice and humans suggests possible natural transmission between these species and others that harbor these viruses. Such transmission has yet to be demonstrated, but it may occur occasionally (Rosen, 1968).

III. RICKETTSIAL DISEASES

A. Rickettsialpox

1. Reservoir and Incidence

A variety of rodent hosts are included in the transmission cycle of rickettsial disease in nature. The house mouse is the natural host of *Rickettsia akari,* the causative agent of rickettsialpox and a member of the spotted fever group of rickettsiae. The organism has also been isolated from rats (*Rattus*) and voles (*Microtus*). Rickettsialpox in humans was first described by two physicians in New York City. The causative agent was isolated from the patient, the mite vector *Liponyssoides* (*Allodermanyssus*) *sanguineus,* and the wild house mouse (Huebner *et al.*, 1946a,b, 1947). Subsequent clinical cases in New York City have been reported, principally among residents of buildings where mice, mites, and rickettsia maintain a cyclic infection (Nichols *et al.*, 1953). Other cases in eastern United States cities, Korea, and the USSR have been reported. In man, the disease has not been associated with naturally infected laboratory mice. The mite vector *L. sanguineus* occurs in many parts of the world but has not been reported in conventional rodent colonies. This may be due to confusion with other species of mites, such as *Ornithonyssus,* which can appear in laboratory mice and rats, or *Dermanyssus,* which may infest rats (Flynn, 1973). The tropical rat mite *Ornithonyssus* (*Liponyssus*) *bacoti,* which also infests mice, has been infected experimentally but is not known to be involved in the natural cycle of rickettsialpox.

2. Clinical Signs

Rickettsialpox is initially characterized by skin papules, chills, fever, and a rash; the clinical manifestations range from mild to severe. Headache and general malaise, with muscular pain, are frequent. Clinical diagnosis is confirmed serologically by a positive complement fixation test between the second and third week of the illness (Benenson, 1975). Because many rickettsial infections mimic each other and occur in varying frequencies, rickettsialpox is difficult to diagnose either clinically or anatomically. Also, other bacterial and viral diseases, such as typhoid fever, chickenpox, or measles, can produce similar febrile reactions with an accompanying rash (Robbins, 1974). Specific serologic tests (complement fixation and agglutination) are extremely important in making proper diagnoses. Skin biopsies may be helpful for early specific diagnosis (Dolgopol, 1948). The blood of febrile patients can also be inoculated into mice and the organism recovered.

Laboratory mice are susceptible to *R. akari;* intranasal inoculation causes fatal pneumonia, and intraperitoneal injection of the organism produces severe illness and death in most animals. Anorexia, depression, and dypsnea are marked. Necropsy findings include peritonitis, splenomegaly, and lymphadenitis. Subcutaneous inoculation of *R. akari* causes active infection for 1 month, with organisms being recovered from the spleen but not from urine or feces. The nature of the natural infection in the mouse is not known (Bell, 1970).

Control and eradication of the disease depend on preventing wild mice and the mite vector from entering animal research facilities and human dwellings.

B. Murine Typhus

Another rickettsial disease, murine typhus or endemic typhus, is transmitted to man by rat fleas (*Xenopsylla cheopis* and *Nasopsyllus fasciatus*); rats and mice are its natural reservoirs. *Rickettsia mooseri,* the causative agent, has not been isolated from natural infections in laboratory mice. Clinical signs, diagnosis, and control in man are similar to those described for rickettsialpox. A total of 18 laboratory workers were infected with *R. mooseri* while performing intranasal inoculations with this agent and while handling infected mice (Loffler and Mooser, 1942; Van den Ende *et al.*, 1943).

IV. BACTERIAL DISEASES

A. Leptospirosis

Leptospira microorganisms were discovered in 1914, when isolated from jaundiced patients (Inada *et al.*, 1916), and after further study were named in 1917 (Noguchi, 1918).

1. Reservoir and Incidence

Reservoir hosts of leptospirosis include rats, mice, field moles, hedgehogs, gerbils, squirrels, rabbits, hamsters, other mammals, and reptiles. A particular species of animal will usually act as the primary host of a particular serotype, but most serotypes can be carried by several hosts. Leptospira are well adapted to a variety of mammals, particularly wild animals and rodents; clinical manifestations in the chronic form are inconspicuous, with the organism being carried and shed in the urine for long periods of time. Rodents and perhaps hedgehogs are the only animal species that can shed leptospires throughout their life span without clinical manifestations (Babudieri, 1958; Faine, 1963). Active shedding of leptospires by laboratory mice can go unrecognized until personnel handling the animals become clinically affected. Many leptospira prototypes, including *L. australis, bataviae, grippotyphosa, hebdomidis icterohaemorrhagiae, pomona,* and *pyrogenes,* are found in the house mouse (Torten, 1979). *Leptospira ballum* has also been reported from mice and is most commonly associated with zoonotic outbreaks (Borst *et al.*, 1948; Friedmann *et al.*, 1973; Stoenner and Maclean, 1958).

Rats and mice are common animal hosts for *L. ballum,* although it has been found in other wildlife, including skunks, rabbits, oppossums, and wild cats (Mailloux, 1975). The infection in mice is inapparent and can persist for the animal's lifetime (Torten, 1979). Although earlier reports indicated that several colonies of laboratory mice harbor the organism (Wolf *et al.*, 1949; Yager *et al.*, 1953), no current estimates of the carrier rate among laboratory rodents in the United States are available. In several European laboratories, transmission of leptospires from laboratory rats to laboratory personnel has been reported (Geller, 1979). In a study of leptospiral infections in feral rodents, 2673 rodents of 10 species were collected in Georgia. Of the 933 tested for leptospires (by kidney culture), *L. ballum* was the only serotype cultured. It was isolated from 22% of the house mice and 0.8% of the old-field mice *Peromyscus polionotus* (Brown and Gorman, 1960).

Since leptospirosis in humans is often difficult to diagnose, the low incidence of reported *L. ballum* infection in man may be misleading. Between 1947 and 1973, only 17 cases of *L. ballum* infection in man were recorded in the United States (Boak *et al.*, 1960; CDC, 1965, 1966; Friedmann *et al.*, 1973; Stoenner and Maclean, 1958). Outbreaks in personnel working with laboratory mice in the United States have been documented (Barkin *et al.*, 1974; Boak *et al.*, 1960; Stoenner and Maclean, 1958. In one study, 8 of 58 employees handling the infected laboratory mice (80% of breeding females were excreting *L. ballum* in their urine) experienced leptospirosis. Humans have also contracted leptospiral infection by handling infected pet mice (Friedmann *et al.*, 1973).

2. Mode of Transmission

Infection with *L. ballum* most frequently results from handling the infected mice (contaminating the hands with urine) or from aerosol exposure during cage cleaning. Skin abrasions may serve as the portal of entry, since *L. ballum* presumably does not penetrate intact skin. In one instance, it was speculated that a father was infected after his daughter, because of an argument, used his toothbrush to clean the contaminated pet mouse cage (Friedmann *et al.*, 1973). Also, laboratory or wild mice that are to be used for primary kidney tissue cultures should be ascertained to be free of leptospires (Turner, 1970).

3. Clinical Signs, Susceptibility, and Resistance in Man

Infected individuals experience a biphasic disease (Heath and Alexander, 1970). They become suddenly ill with weakness, headache, myalgia, malaise, chills, and fever. Leukocytosis, usually associated with leptospirosis, is found inconsistently with *L. ballum* infection. During the second phase of the disease, a common finding is painful orchitis. Unlike the orchitis associated with mumps, leptospirosis caused enlarged testes in only one patient (Friedmann *et al.*, 1973). Two infected personnel in a laboratory mice-associated outbreak required more than a month for recovery (Stoenner and Maclean, 1958). Renal, liver, pulmonary, gastrointestinal, and conjunctival findings may be abnormal (Barkin *et al.*, 1974).

4. Diagnosis

Because of variability in the clinical symptoms and lack of pathognomonic pathological findings in man and animals, it is essential that serologic diagnosis or actual isolation of leptospires be undertaken to establish a correct diagnosis (Torten, 1979). As an aid to diagnosis, leptospires can sometimes be observed by examination or direct staining of body fluids or fresh tissue suspensions. A definitive diagnosis in man or mouse is made by culturing the organisms from tissue or fluid samples or by animal inoculation (particularly in 3- to 4-week-old hamsters) and subsequent culture and isolation. Serologic assessment of host infection is accomplished by indirect hemagglutination, agglutination/analysis, complement fixation, microscopic and macroscopic agglutination, and fluorescent antibody techniques (Stoenner, 1954; Torten, 1979).

In a survey of trapped wild urban rats, diagnosis of leptospirosis was more accurate by urine or kidney culture, rather than by either indirect fluorescent antibody or macroscopic slide agglutination (Sulzer *et al.*, 1968). Another survey of wild rats confirmed that culture techniques identified more positive rats than did macroscopic slide agglutination (Higa and Fujinaka, 1976).

5. Epidemiology and Control

In mouse colonies infected with *L. ballum,* antibodies against *L. ballum* were detected in sera of mice of all ages, but leptospires could be recovered only from mature mice. Progeny of seropositive females had detectable serum antibodies at 51 days of age, but not at 65 days. It was also reported that progeny of seropositive female mice, which possessed antibody at birth and acquired additional antibody from colostrum, remained free of leptospires if isolated from their mothers at 21 days of age, despite exposure during the nursing period (Stoenner, 1957).

Studies in mice experimentally infected with *L. grippotyphosa* demonstrated that maternal antibodies, whether passed through milk or placental transfer, conferred protection of long duration against the carrier state and shedding of leptospirae. Thus, serologically positive immune mothers do not transmit the disease to their offspring. However, mice born to nonimmune mothers, if infected at 1 day postpartum, become carriers with no trace of antibodies. Thus a population of carrier pregnant mice without antibody could serve as a precipitator in outbreaks among susceptible mouse populations (Birnbaum *et al.*, 1972). Field surveys have supported this data in that the percentage of carrier mice that do not have antibodies is significant. This led to the diagnostic approach, which specifies that both serologic and isolation methods must be utilized to determine the rate of leptospiral infection in rodents (Galton *et al.*, 1962).

Leptospira ballum is found frequently in the common house mouse (*M. musculus*) (Brown and Gorman, 1960; Yager *et al.*, 1953). Therefore, eradication of infected colonies, use of surgically derived and barrier-maintained mice or of conventional laboratory mice free of leptospira infection, coupled with the prevention of ingress of wild rodents, should effectively preclude introduction of the organism into research and commercial laboratories (Loosli, 1967). *Leptospira ballum* has been elminated from a mouse colony by administration of feed containing 1000 gm chlorotetracycline hydrochloride per ton for 10 days. After 7 days of antibiotic therapy, mice were transferred to clean containers and administered clean water, both having been sterilized by steam. Mouse traps and DDT were used to destroy escaped mice and to prevent reintroduction of *L. ballum* by the common house mouse (Stoenner *et al.*, 1958).

B. Pasteurellosis

Pasteurella pneumotropica, P. multocida, and *Yersini pseudotuberculosis,* causes of infection in man, have also been recovered from mice. Although *Y.* (*Pasteurella*) *pestis,* usually transmitted to man by a flea bite, occurs endemically in wild rodent populations (Hudson *et al.*, 1964), it should not be found in established mouse colonies if these mice have no contact with wild rodents.

1. Incidence and Reservoir

Pasteurella pneumotropica, first identified and studied in 1948 (Jawetz, 1948, 1950 (and rarely *P. multocida* or *Y. pseudotuberculosis*), usually occur as latent infections, though they can be a primary pathogen in laboratory mice (Brennan *et al.*, 1965, 1969; Hoag *et al.*, 1962). However, these organisms are rarely associated with human disease. Although direct transmission of *P. pneumotropica* from mice to man has not been reported, this organism has been transmitted via the bites of other animals (Miller, 1966; Olson and Meadows, 1969; Winton and Mair, 1969). Because mice harbor this organism in the upper respiratory system and pharynx, exposure could result from mouse bites.

2. Mode of Transmission

Reported cases in man are usually attributed to animal bites or exposure to ill animals. Also, *P. pneumotropica* was reportedly introduced into a barrier-maintained, specific pathogen-free (SPF) rat and mouse colony by personnel working in the area, who carried this organism in their upper respiratory system (Wheater, 1967). *Pasteurella pneumotropica* with similar biochemical characteristics was isolated from sputum and sinus infections from humans (Henriksen and Jyssum, 1961; Henriksen, 1962). An organism closely related to *P. pneumotropica* was recovered from 1% of sputum samples obtained during an 8-month period at a public health laboratory in England (Jones, 1962). It is suspected, however, that transmission of pasteurella infection from man to mice is rare, and despite the lack of confirmatory literature, transmission of pasteurella organisms from mouse to man probably is also rare.

3. Clinical Signs, Susceptibility, and Resistance in Man

Pasteurellosis is seldom reported in man, possibly because *Pasteurella* is usually an opportunistic pathogen with low pathogenicity for man or because the organism may be confused with *Hemophilus influenzae* or *Acinetobacter* sp. (e.g., *Mima polymorpha*) (Freigang and Elliott, 1963; Schipper, 1947). Deaths from *P. pneumotropica* infection have been recorded: One 51-year-old man died 48 hr after being bitten by a dog (Miller, 1966). Local inflammation, purulent discharge, pyrexia, and pain have been caused by bite wounds from which *P. pneumotropica* has been isolated (Olson and Meadows, 1969). Similarly, *P. multocida* was isolated from a wound after a bite from a laboratory rat, although the organism was not found in subsequent cultures from the rat (Bergogne-Berezin *et al.*, 1972). Septicemia and meningitis due to pasteurella have also been reported (Cooper *et al.*, 1973; Rogers *et al.*, 1973). Hubbert and Rosen (1970) listed 316 cases of *P. multocida* in man, usually associated with animal exposure. *Yersinia pseudotuberculosis* in man is reported rarely, but man can develop severe systemic infections from it.

C. Rat-bite Fever

Rat-bite fever can be caused by either of two microorganisms: *Streptobacillus moniliformis* (*Actinomyces muris*) or *Spirillum minus* (*Spirillum minor*) (*Spirillum morsus muris*); synonym (sodoku).

1. Reservoir and Incidence

These organisms are present in the oral cavity and upper respiratory passages of asymptomatic rodents. Reported incidences of mice as asymptomatic carriers of *S. moniliformis* or *Sp. minus* were not found. Nearly 50% of the asymptomatic laboratory rats cultured in an early study harbored *S. moniliformis* as normal oral flora (Strangeways, 1933). In a more recent study in laboratory Sprague-Dawley rats, *S. moniliformis* was the predominant microorganism isolated from the upper trachea of control animals (Paegle *et al.*, 1976). The lack of reported carrier rates in mice is attributed partly to the usual asymptomatic carrier state and partly to the difficulty in isolating the organisms. *Spirillum minus* cannot be cultured *in vitro* and requires inoculation of culture specimens into laboratory animals and subsequent identification of the organism by dark field microscopy. *Streptobacillus moniliformis* grows slowly on artificial media, but only in the presence of sera, usually 10–20% rabbit or horse serum incubated at reduced partial pressures of oxygen (Holmgren and Tunevall, 1970; Rogosa, 1974).

Arkless (1970) and Gilbert *et al.* (1971) described infection caused by the bite of a laboratory mouse and a pet mouse, respectively. The disease is not commonly reported in man but has been reported in personnel engaged in research involving laboratory rodents, particularly rats (Cole *et al.*, 1969; Gilbert *et al.*, 1971; Holden and MacKay, 1964). Historically, however, wild rat bites and subsequent illness have been associated with social conditions of poor sanitation and overcrowding, and almost 50% of all cases have involved children under the age of 12 (Brown and Nunemaker, 1942; Raffin and Freemark, 1979; Richter, 1945; Roughgarden, 1965).

Rat-bite fever is not a reportable disease; thus, its incidence, geographic location, racial data, or source of infection in humans is difficult to assess. Acute febrile diseases, especially if associated with animal bites, are routinely treated with penicillin or other antibiotics without prior culturing of the bite wound. This therapeutic approach, though successful in aborting cases of potential rat-bite fever, does not allow accurate recording of the disease in humans. One would suspect, therefore, because of the high number of rodent bites suffered by humans, that the incidence of rat-bite fever is low.

2. Mode of Transmission

The bite of an infected rodent, usually a wild rat but occasionally a laboratory rat or mouse, is the usual source of infection. In some reported cases, infection was attributed to dog, cat, or other animal bites and rarely to traumatic injuries unassociated with animal contact (Richter, 1945; Roughgarden, 1965). Outbreaks of febrile illness in humans have been associated with *Streptobacillus*-contaminated milk or food. The disease gained the synonym *Haverhill fever* after a 1926 outbreak in Haverhill, Massachusetts, attributed to contaminated milk (Place and Sutton, 1934).

3. Clinical Signs, Susceptibility, and Resistance in Man

Incubation varies from a few hours to 1–3 days in infection with *S. moniliformis* and may range from 1 to 6 weeks with *Sp. minus*. Fever is almost always present during the illness, whether it is caused by *Sp. minus* or *S. moniliformis*. In *Sp. minus* infection, inflammation at the site of the bite wound, as well as regional lymphadenopathy, often occurs; both these signs may coincide with the onset of fever. Inflammation and lymphadenopathy are infrequently documented with *S. moniliformis* infection. The fever and local signs may be accompanied by headache, general malaise, myalgia, and chills (Gilbert *et al.*, 1971; Raffin and Freemark, 1979). In most cases, a discrete macular rash appears on the extremities, frequently involving the palms and soles; it may become generalized, with pustular or petechial sequelae.

Arthritis reported in 50% of the cases of *S. moniliformis* infection usually affects larger joints; it also occurs in *Sp. minus* infection, but less commonly. Prolonged and recurrent joint involvement is noted in untreated patients. Serous to purulent effusion can be recovered from affected joints, with *S. moniliformis* being cultured from the fluid. Complications of the primary infection can result if antibiotic treatment is not instituted early. Pneumonia, heptatitis, pyelonephritis, enteritis, and endocarditis have been reported (McGill *et al.*, 1966). Deaths from *S. moniliformis* endocarditis have occurred, usually in cases with preexisting valvular disease.

D. *Salmonella*

Although there are 1600 recognized serotypes, *Salmonella typhimurium* and *S. enteritidis* have been associated most commonly with infections in laboratory mouse colonies (Haberman and Williams, 1958; Hoag and Rogers, 1961). Other serotypes have also been reported in mice (Ganaway, Chapter 1, this volume). From 1974 to 1978, the most frequently isolated serotype in the United States was *S. typhimurium* (CDC, 1976; MMWR, 1980). Other frequently isolated serotypes were *S. newport, S. enteritidis,* and *S. heidelberg*.

1. Reservoir and Incidence

Salmonella infection in man and animals, including mice, occurs worldwide. The organism is an enteric bacterium inhabiting the intestinal tract of many animals. Salmonella are routinely associated with food-borne disease outbreaks, are contaminants of sewage, and are found in many environmental water sources.

Although the reported incidence of salmonella in laboratory mice has decreased in the last several years because of management practices, environmental contamination with salmonella continues to be a potential source of infection for laboratory animals and, secondarily, for personnel handling these animals. Animal feed containing animal by-products continues to be a source of salmonella, especially if diets consist of raw meal and have not undergone a pelleting process (Hoag *et al.*, 1964; Stott *et al.*, 1975; Williams *et al.*, 1969). Until rodent feeds in the United States and Europe are salmonella-free, laboratory rodent-associated cases of salmonellosis will remain a distinct possibility.

Salmonella infections in rodents have also caused food-borne outbreaks of salmonellosis. In an interesting epidemiological study performed in England, *S. enteritidis* var. *danyz* was isolated from two adults living 4 miles apart. The source of infection was contaminated cakes from a local bakery. Mice from the bakery were trapped and cultured; they had acquired the infection from living *S. danyz* cultures in rodenticide baits and had infected food in the bakery (Brown and Parker, 1957). In another study, salmonella serotypes were isolated from 17% of 170 wild house mice. The authors concluded that house mice are a reservoir of infection and play an important role in human and animal salmonellosis (Shimi *et al.*, 1979). Undoubtedly, rodent excreta is the source of other food-forne outbreaks.

Both man and animals are carriers and periodic shedders of salmonella; they may have mild, unrecognized cases or they may be completely asymptomatic. Asymptomatic animals that shed salmonella are particularly important in biomedical research because they are a potential source of infection for other animals, animal technicians, and investigators (Fox and Beaucage, 1979). The incidence of carrier mice in the colony may vary from 1% to 20% (Haberman and Williams, 1958); indeed, one investigator suggested that clinically apparent salmonellosis is rare in infected mice (Margard *et al.*, 1963). In a survey of 19,137 nonhuman-origin salmonella isolations from pet-type animals, conducted in the United States from 1962 to 1965, a total of 227 isolates were recovered from rodents (Kaufman, 1966). The incidence of salmonellosis in man acquired from mice or vice versa is unknown; however, a treatise on diseases of laboratory mice (Hoag and Meier, 1966) states, "Occasional paratyphoid carriers are found among mouse handlers, but are not important sources of animal infection" (p. 594). No further expansion of this statement was provided.

2. Clinical Signs

The most common clinical sign of salmonellosis in man is acute gastroenteritis with sudden onset, abdominal pain, diarrhea, nausea, and fever. Loose bowels and anorexia may persist for several days. When organisms invade the bowel wall, some cases can lead to febrile septicemia without severe intestinal involvement; in these cases, most clinical signs are attributed to hematogenous spread of the organisms (Robbins, 1974). As with other microbial infections, severity of this disease is related to the organism's serotype, the number of bacteria ingested, and host susceptibility. Inapparent infections are encountered frequently; laboratories maintaining mice should consider screening animal technicians for subclinical salmonella infections.

E. Other Potential Bacterial Diseases

1. Erysipelas

Erysipelas, caused by *Erysipelothrix rhusiopathiae,* which affects a variety of fishes and mammals, including man, was first recognized when Koch discovered an organism he called *bacillus of mouse septicemia*. The disease in man, called *erysipeloid* (not human erysipelas), was recognized by Rosenbach in 1887. The first report of natural infection in wild mammals appears to be an epizootic among migrating meadow mice and house mice in California (Wayson, 1927). Although the laboratory mouse is susceptible to experimental infection, neither natural disease nor human infection from handling diseased mice has been reported.

2. *Pseudomonas*

Pseudomonas aeruginosa, a virulent pathogen in immunosuppressed mice and a potential pathogen in conventional or SPF mice, may be acquired from human carriers. Van der Waaij *et al*. (1963) documented two outbreaks of pseudomonas infection spreading within the colony from infected animal caretakers. Infection was curtailed by transferring infected personnel who served as reservoirs and by instituting rigid sanitation and hygienic precautions.

3. *Staphylococcus*

Staphylococcal organisms are ubiquitous; there is abundant evidence that humans can carry phage types of *Staphylococcus aureus* that can cause clinical disease in mice. Pathogenic *S. aureus* of the human phage type have been introduced into SPF barrier-maintained mouse colonies where they had previously been absent; the same phage type was isolated from animal caretakers working in the area (Blackmore and Francis, 1970; Davey, 1962; Shults *et al*., 1973). The lack of indigenous strains of staphylococci in SPF barrier-maintained animals may account for colonization of and resulting disease produced by human phage-type staphylococci (Blackmore and Francis, 1970). A similar phenomenon has been demonstrated in humans where the nasopharynx resists colonization by virulent strains of staphylococci due to the presence of avirulent *S. aureus* strains (Eichenwald, 1965). Pathogenic *S. aureus* of various human phage types have also been isolated from other laboratory animal species (Fox *et al*., 1977; Renquist and Soave, 1969; Rountree *et al*., 1956). Animals, including mice, therefore conceivably could serve as sources of potentially pathogenic staphylococci for humans.

4. *Streptococcus*

Streptococci of Lancefield groups A, B, C, and D have been isolated from disease outbreaks in mice (Besch and Williford, Chapter 4, this volume). Groups A, B, C, and D streptococci have also been isolated from man; group D streptococci are routinely isolated as common bacteria of human and animal intestinal flora. Group A streptococci (the most common species causing disease in man is *S. pyogenes*) and group B streptococci are responsible for a variety of clinical diseases in man. Although group C streptococci are primarily pathogens in various animals, including mice, strains can be isolated from the human respiratory tract as well (Davis *et al*., 1969). However, we found no documented evidence that streptococci isolated from mice are transmitted to or acquired from man.

V. MYCOSES (RINGWORM)

Ringworm (favus) in the mouse was first recorded in England in 1850. A common house mouse had ringworm lesions identical to those on man (cited in Buchanan, 1919) Classical murine ringworm, reportedly caused by *Trichophyton quinckeanum*, is usually restricted to feral rodents. However, successful crossing of *T. quinckeanum* cultures with tester strains of the perfect state of *T. mentagrophytes*, *Arthroderma benhamiae*, demonstrates that *T. quickeanum* is not a distinct species and is indistinguishable from *T. mentagrophytes* (Ajello *et al*., 1968). Many other zoophilic dermatophytes associated with infections of mice can cause ringworm in man, including *Epidermophyton floccosum*, *Microsporum gallinae*, *M. gypseum*, *T. erinacai*, *T. schoenleini*, and *T. (keratinomyces) ajello* (Dvorak and Otechenasek, 1964; Krempl Lamprecht and Bosse, 1964; Marples, 1967; Refai and Ali, 1970). In almost all mouse-associated ringworm infections in man, *T. mentagrophytes* has been isolated as the etiological agent (Table I).

A. Reservoir and Incidence

Dermatophytes are distributed throughout the world, with some species being reported more commonly in certain geographic locations. For example, in a study of small mammals in their natural habitat, *T. mentagrophytes* was isolated from 57 of 1288 animals representing 15 different species. The dermatophyte was isolated most commonly from the bank vole (*Clethrionomys glariolus*), followed by the common shrew (*Sorex araneus*) and the common house mouse (*M. musculus*) (Chmel *et al*., 1975). In this survey, agricultural workers, exposed to these mammals in granaries and barns, risked contracting *T. mentagrophytes* infection. *Trichophyton mentagrophytes* was isolated from 77% of the 137 agricultural workers infected with ringworm, whereas *T. verrucosum* was isolated from only 23% of the cases. In the same study, of 445 ringworm-infected personnel working with farm animals, 75% were infected with *T. verrucosum* and 28% with *T. mentagrophytes*. Human infection with *T. mentagrophytes* has also followed handling of bags of grain in which mice had been living (Alteras, 1965; Blank, 1957). Thus, specific exposure to

Table I

Trichophyton mentagrophytes Infections Associated with Laboratory or Pet Mice

Probable source of infection	Number of persons infected	Lesions appearing on infected mice	References
Pet white mice	7 children	2 of 104, diffuse hair loss	MacKenzie (1961)
Inbred albino laboratory mice (VSBS, A2G)	2 laboratory technicians		
Laboratory mice	6 laboratory technicians		Alteras (1965)
Laboratory mice	2 laboratory technicians	0 of 96 (222 cultured), survey of commercial stock	Dolan *et al.* (1958)
BALB/c C3H/Bi mice	6 laboratory technicians	<1% of all mice, carrier rate 90%	Davies and Shewell (1964)
White mice	1 laboratory worker	% ND, loss of hair, increased scaling on head and back, 10 mice	Booth (1952)
White mice	1 bacteriologist	60 of 400, crusted or crustless plaques, circular with prominent periphery; general alopecia; mortality in some mice	Cetin *et al.* (1965)

reservoir hosts harboring different dermatophytes determines the type and incidence of infection in man.

Ringworm infection in laboratory mice is often asymptomatic and is not recognized until personnel become infected. A prevalence of *T. mentagrophytes* among laboratory mouse stocks as high as 80–90% has been recorded (Davies and Shewell, 1964). During an 8-month period, before infected mice were treated, 6 of 13 people handling the mice developed ringworm, although less than 1% of the mice showed any clinical signs of the disease. Dipping the animals in an aqueous solution of an acaricide containing a fungicide to remove mites reduced mice carriers from 90 to 21%. Other authors have reported that personnel were infected when handling mice clinically affected with *T. mentagrophytes* infection (Cetin *et al.*, 1965).

B. Mode of Transmission

Transmission of dermatophytes from animal to man is a well-known and serious public health concern. Transmission from laboratory animals, including mice, to personnel is often unsuspected, because laboratory animals usually have a paucity of visible skin lesions. Various surveys have indicated that as few as 1% of infected laboratory rodents show clinical evidence of dermatophyte infection (Cotchin and Roe, 1967). Dolan *et al.* (1958) reported that 43% of healthy mice received from various breeders in the United States were infected with *T. mentagrophytes,* whereas only 2% displayed clinical lesions. Transmission occurs via direct or indirect contact with asymptomatic carrier animals or with skin lesions of infected mice, contaminated animal bedding, equipment, or causal fungi present in air, dust, or on surfaces of animal-holding rooms (MacKenzie, 1961). The infection is communicable as long as infected animals are maintained and viable spores persist on contaminated materials. To maintain laboratory mice free of infection, and thus preclude personnel exposure, newly purchased mice should be screened for dermatophytes, and the animal facility environment and caging should be sanitized regularly.

C. Clinical Signs

The disease, dermatomycosis or ringworm in man, is nonfatal, often self-limiting, sometimes asymptomatic, and thus often ignored by the affected person. In general, the dermatophytes cause scaling, erythema, and occasionally vesicles and fissures. The fungi cause thickening and discoloration of the nails. On the skin of the trunk and extremities, the lesion may consist of one or more circular lesions with a central clearing, forming a ring (Fig. 1) (Mescon and Grots, 1974). Fungal infections in man are catagorized clinically according to location; some examples are tinea capitis, a fungus infection

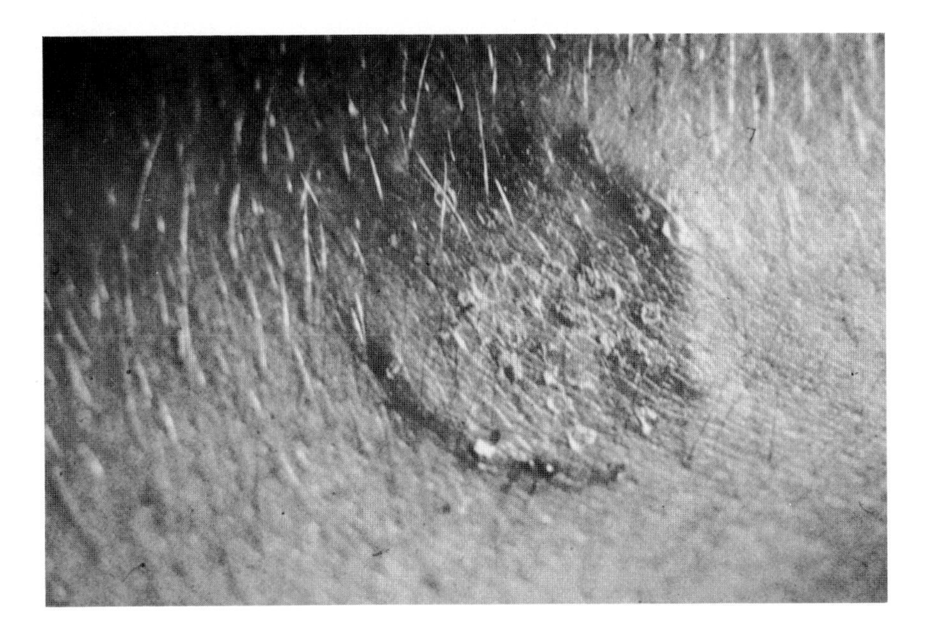

Fig. 1. A circular ringworm lesion on the arm of a man; contracted from a rodent infected with *Trichophyton mentagrophytes*. (Courtesy of Dr. William Kaplan.)

of the scalp and hair; tinea corporis, ringworm of the body; tinea pedis, of the foot; and tinea unguim, of the nails. Of the dermatophytes recovered from mice, when man is infected, the fungus is usually localized on the extremities or body, infection of the arm and hand being the most common. For instance, the zoophilic form of *T. mentagrophytes* is highly inflammatory and often undergoes rapid resolution. The infection may also produce furunculosis, deep involvement of the hair follicles, or a widespread tinea corporis, which is also seen in infections of *E. floccosum*. In a laboratory-acquired infection with *T.* (*keratinomyces*) *ajelloi,* a technician working with mice developed small, grayish-white, scaly lesions on both hands. The organism was isolated from the hand lesions and from 2 of 250 apparently healthy mice (Refai and Ali, 1970).

VI. PROTOZOAN DISEASES (*Entamoeba coli*)

Entamoeba coli, a nonpathogenic protozoan in man, is morphologically similar to *E. muris* seen in mice and rats. It is not definitely known whether transmission between humans and rodents is possible. An early report described the establishment of *E. coli* infection in rats; however, little attempt was made to reduce the possibility of cross-infection with *E. muris* (Kessel, 1923). In a later thorough study, *E. coli* was not transmitted to or established in either mice or rats (Neal, 1950). More recently, an attempt was made to establish human *E. coli* in laboratory rodents. Cysts of *E. coli* from 15 human and four primate stools were passaged in SPF guinea pigs, rats, and mice. In only one instance were cysts established in mice and eventually in rats (Owen, 1978). Because of the difficulty of differentiating between *E. coli* and *E. muris* morphologically, it is not possible to determine whether the *Entamoebe* that became established in rodents, originally isolated from an Ethiopian individual, was a true transmission of *E. coli* in the rodent or a contaminant of the human feces with *E. muris*. The author concluded that human–rodent contact is not responsible for the introduction of *Entamoeba* sp. (most likely *E. muris*) in SPF barrier-maintained rodent colonies. There is no evidence in the literature that humans are *E. muris* carriers.

The mouse can be infected experimentally with *E. histolytica,* but natural infections with this parasite have not been reported (Flynn, 1973).

VII. HELMINTH DISEASES

A. Tapeworms

1. *Hymenolepis diminuta*, The Rat Tapeworm

a. Reservoir and Incidence. Although this parasite occurs in the mouse intestine, it is more commonly associated with rats and is especially common in wild Norway (*Rattus norvegicus*) and black (*Rattus rattus*) rats throughout the world (Wardle and McLeod, 1952); however, it is rarely encountered in man (Faust and Russell, 1970; Stone and Manwell, 1966).

b. Mode of Transmission. Like other tapeworms, *H. diminuta* requires an intermediate host, usually a flour beetle (*Tribolium* sp.), moth, or flea (Voge and Heyneman, 1957). Larval development in *Tribolium* sp. at 30°C requires 8 days.

Therefore, man becomes infected only by ingestion of infected insects, such as flour beetles, which may contaminate rodent food or cereal marketed for human consumption.

c. Clinical Signs. The infection in man is usually asymptomatic, but in moderate to heavy infections it may cause headaches, dizziness, and abdominal discomfort.

2. *Hymenolepis nana,* The Dwarf Tapeworm of Man

a. Reservoir and Incidence. The dwarf tapeworm is a common parasite of both the wild house mouse and the laboratory mouse. As indicated earlier in the text, in most well-managed mouse colonies, *H. nana* incidence is low compared to earlier reports of its high incidence in rodent colonies (Wescott, Chapter 20, this volume).

Stoll (1947) listed *H. nana* as infecting 100,000 persons in North America and 20 million in the world. Surveys conducted in Central Europe report that this tapeworm in man is more prevalent in warm than in temperate regions. An incidence of 10% has been noted in some South American countries (Jelliffe and Stanfield, 1978).

b. Mode of Transmission. *Hymenolepis nana* is unique among tapeworms in that the adult worm develops after the egg is ingested. The hooked oncosphere then invades the intestinal mucosa and develops into a cysticeroid larva. *Hymenolepis nana* eggs can contaminate hands, be trapped on particulate matter, or be aerosolized, and then accidentally ingested. Since no intermediate host is required, the eggs are readily infective for the reciprocal hosts (Faust and Russell, 1970). Precautions against infection include strict personal hygiene, appropriate laboratory uniforms, and use of disposable gloves and face masks when handling contaminated bedding and feces.

c. Clinical Signs. The clinical picture of *H. nana* infection is quite cosmopolitan. In well-nourished persons, essentially no symptoms occur; the infection is noted when the proglottids or ova are seen in the stool. In other persons, the symptoms include headaches, dizziness, anorexia, inanition, pruritis of the nose and anus, periodic diarrhea, and abdominal distress. Convulsions have also occurred. A tapeworm identified as *H. nana* was found in a tumor removed from the chest wall (Jelliffe and Stanfield, 1978). The diagnosis was based on identification of the characteristic eggs or proglottids in the stool.

B. Roundworms (*Syphacia obvelata*)

1. Reservoir and Incidence

Syphacia obvelata is a ubiquitous parasite in both wild and laboratory mice. Although parasitology texts report that *Syphacia* is infectious to man, this citation originates from a publication in 1919, in which two *S. obvelata* adult worms and eggs reportedly were found in the formalin-preserved feces of a Filipino child whose entire family of five was infected with *H. nana* (Riley, 1919). No mention is made of the method of collection of the feces, whether the feces could have been contaminated with murine feces or with the parasite and/or eggs. The only other report is an unpublished finding of *S. muris* eggs in the feces of two children and two rhesus monkeys, cited in a personal letter from Dr. E. C. Faust of Tulane University, dated January 6, 1965 (Stone and Manwell, 1966). Both of these cases may therefore be examples of spurious parasitism, but definitive information for that conclusion is lacking. Regardless, no published information indicates that laboratory personnel have become infected by working with *Syphacia*-infected mice.

2. Mode of Transmission

Contamination of food or utensils, or accidental ingestion of *Syphacia* ova (e.g., via contaminated hands) could result in infection of man. People working with infected mice probably ingest ova occasionally, but there is no evidence that active infection results from this exposure.

3. Clinical Signs

Because *Syphacia* infection in man has not been described, clinical signs have not been noted.

4. Diagnosis

There are striking differences in size between specimens of female. *S. obvelata* and those of *Enterobius vermicularis,* the pinworm, in man (Markell and Voge, 1965). *Syphacia* is 3.5–5.8 mm long, whereas the *Enterobius* female reaches a length of 8–13 mm. The male *Syphacia* measures 1.1–1.5 mm, compared to 2.5 mm for *Enterobius*. The size difference between the eggs of the two species is also marked: *Syphacia* eggs are more than twice as long (125 μm versus 52 μm as those of *Enterobius*. It is unlikely therefore that *Syphacia* would be misdiagnosed as *Enterobius,* assuming, of course, that the observer was aware of the size difference and measured the eggs.

VIII. ARTHROPOD INFESTATIONS

Though many species of mites are found on laboratory mice, only *Ornithonyssus bacoti,* the tropical rat mite, and *Liponyssoides sanguineus,* the house mouse mite, are vectors of human disease. *Ornithonyssus bacoti* is seen in laboratory

Table II

Mouse Ectoparasites of Public Health Significance[a]

Vector	Disease	Reservoir
Mites		
Acarina (mites)	Allergic dermatitis	House mouse, laboratory mice, wild rodents
Ornithonyssus bacoti (tropical rat mite)	Biologic vector of murine typhus	
	Rickettsialpox	
	Q fever	
	Plague	
	Eastern equine encephalitis	
Liponyssoides (Allodermanyssus) sanguineus	Allergic dermatitis	House mouse
(house mouse mite)	Biologic vector of rickettsialpox	Wild rodents
	Rash	
Haemalaelaps casalis	Allergic dermatitis	House mouse, laboratory mouse, wild rodents
Fleas		
Leptopsylla segnis	Biologic vector of plague and typhus	
	Intermediate host of *H. nana* and *H. diminuta*	
	Allergic dermatitis	
Xenopsylla cheopis	Allergic dermatitis	
	Biologic vector of murine typhus	
Nasopsyllus fasciatus	Allergic dermatitis	
	Biologic vector of murine typhus	

[a] For more information regarding life cycles, pathogenicity, and host range, see Flynn (1973).

mice (Fox, 1982); *L. sanguineus* has been identified only on wild mice (Table II). Bites from these mites, as well as from another mouse mite, *Haemalaelaps casalis*, are responsible for allergic dermatitis, or local inflammation, in man (Fig. 2).

Fleas are seldom found in laboratory mice but are common parasites of feral rodents. The Oriental rat flea, *Xenopsylla cheopis*, and *Nasopsyllus fasciatus*, another flea, both naturally infest mice and rats; they are vectors for murine typhus. Apparently, *X. cheopis* is easily established in animal facilities. At a Midwestern United States university, it inhabited rooms housing laboratory mice where, on two separate occasions, the flea caused distress by biting students (Yunker,

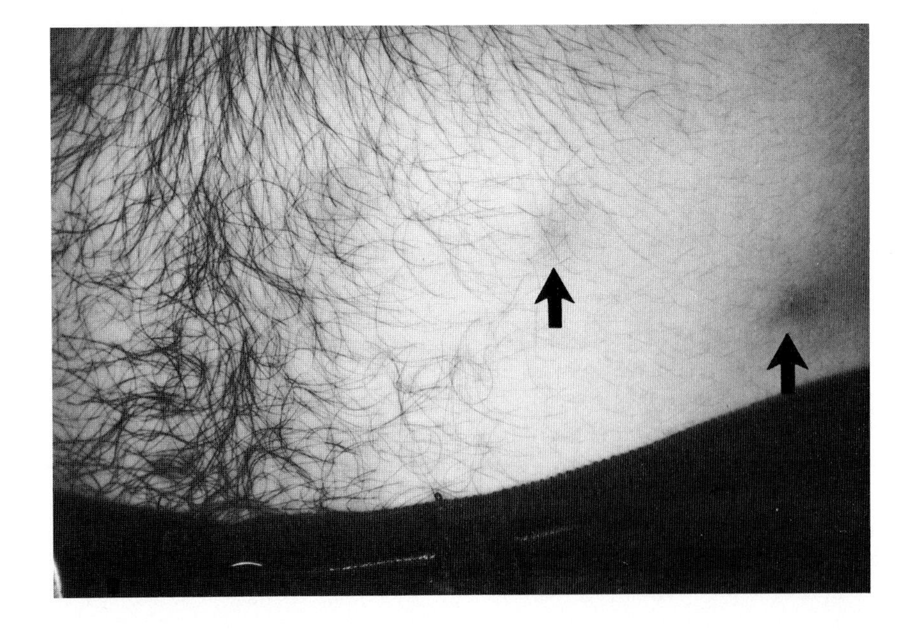

Fig. 2. Severe papular dermatitis with excoriations (arrows) on a man, caused by the bite of *Ornithonyssus bacoti*. Lesions are often noted on the stomach, above the belt line.

1964). *Leptopsylla segnis,* the mouse flea, bites man and is a vector for plague and typhus, serious diseases in man. Also, *L. segnis* can serve as an intermediate host for the rodent tapeworms *H. nana* and *H. diminuta,* which can infect man. The flea's bite can also be irritating and cause allergic dermatitis.

IX. BITES

During a 2-year surveillance period (1971-1972), 196,684 animal bite cases were reported from the 15 reporting areas in the United States (Moore *et al.,* 1977). The type of biting animal was reported for 196,117 persons bitten; 4% were rodents, type unspecified. By tradition, and public emotion, rabies has been the primary reason for investigating animal bite cases. Rodent bites (especially from wild rats), however, present other serious public health hazards, particularly in impoverished areas, where feral rodents are plentiful. Important effects of animal bites that must be considered are pain, anxiety, disfigurement, and infections caused by bacteria such as *Pasteurella* spp., *Clostridium tetani, S. moniliformis,* and *Sp. minus.* Reported incidences and severity of laboratory rodent-associated bites are few, except for published cases of rat-bite fever and *Pasteurella* infections (Hubbert and Rosen, 1970). Depending on the nature of the wound and the health status of the animal inflicting the bite, medical attention may be required. Minimally, for minor mice bites, the wound should be cleaned thoroughly and treated topically as necessary. Current tetanus immunizations should be maintained for personnel working with animals (ILAR, 1978).

X. ALLERGIC SENSITIVITIES

A. Incidence and Clinical Signs

Allergic skin and respiratory reactions are quite common in laboratory workers working with mice. Hypersensitivity reactions to mouse dander and urine are serious occupational health problems. Because of the large number of mice used in biomedical fields, numerous people are constantly exposed to laboratory mouse allergens. Historically, it was believed that only rabbit and cat danders produced laboratory animal-related asthma. This notion has been disproven and the mouse has been incriminated in producing asthma (Rajka, 1961; Newman-Taylor *et al.,* 1977). A biologist developed a typical mild anaphylactic reaction with hypotension, asthma, and giant urticaria after a mouse bite (Lincoln *et al.,* 1974). Hypersensitivity reactions include nasal congestion, rhinorrhea, sneezing, itching of the eyes, angioedema, and asthma. These would also include various skin manifestations such as localized urticaria and eczema (atopic dermatitis). Skin wheal and flare reactions were demonstrated in eight subjects sensitive to mice when tested with mouse pelt extracts (Ohman *et al.,* 1975). Maximal allergenic activity was demonstrated when the skin testing was done using the extract fraction which had the electrophoretic mobility of albumin. Figs. 3 and 4 illustrate a typical wheal and flare reaction on the skin of a patient who is hypersensitive to mouse urine. The patient developed hypersensitivity after working with mice for several years. A mouse whose feet were contaminated with urine walked across the patient's arm and produced these lesions. Often the complaint is manifested by intense itching when the mouse urine or serum has touched the skin (Ohman, 1978). Delayed reactions are also seen. Asthma may develop during the night after an exposure during the working day.

Some allergic disorders exhibit a familial prevalence. This familial predisposition to respond to allergies is called *atopy* and suggests that inheritance plays a role in the pathogenesis of atopic diseases (Gupta and Good, 1979). Members of the same family may manifest their atopy in different ways, with some having asthma and others eczema. These clinical manifestations of atopy are determined by the location of the shock organ, i.e., the skin, mucous membranes, respiratory or gastrointestinal tracts (Criep, 1976).

It appears that the major sources of antigen for personnel working with mice would be mouse dander (Sorrell and Gottesman, 1957; Lincoln *et al.,* 1974) and mouse urine (Newman-Taylor *et al.,* 1977). Newman-Taylor evaluated five patients, all of whom had a history of hay fever or asthma. Four of these patients handled mice. Symptoms appeared within 1 year in four patients and after 4 years in the fifth patient. Initial symptoms were rhinitis and conjunctivitis of rapid onset after animal exposure. Urticaria developed when the animal's feet, contaminated with urine, touched the patient's skin. Asthma developed in the five patients after a few weeks to 2 years of exposure to the animals. Initially, the asthmatic episodes would occur several hours after exposure to the animals. Within a year after the first asthma attack, all five patients developed asthma after a few minutes of exposure. After being separated from the animals for a few days, the patients were clinically normal. All four patients handling mice had immediate skin test reactions to mouse urine, mouse hair extract, and the prealbumin urine protein fraction. The four patients exhibited asthmatic reactions to inhalation tests using mouse urine. No asthmatic reactions were produced by mouse hair extract. Three of the four mouse handlers had immediate skin test reactions to mouse serum. The mouse prealbumin allergen in serum is in concentrations 200-400 times less than in urine.

Levy (1974) studied and quantitated allergic activity of pro-

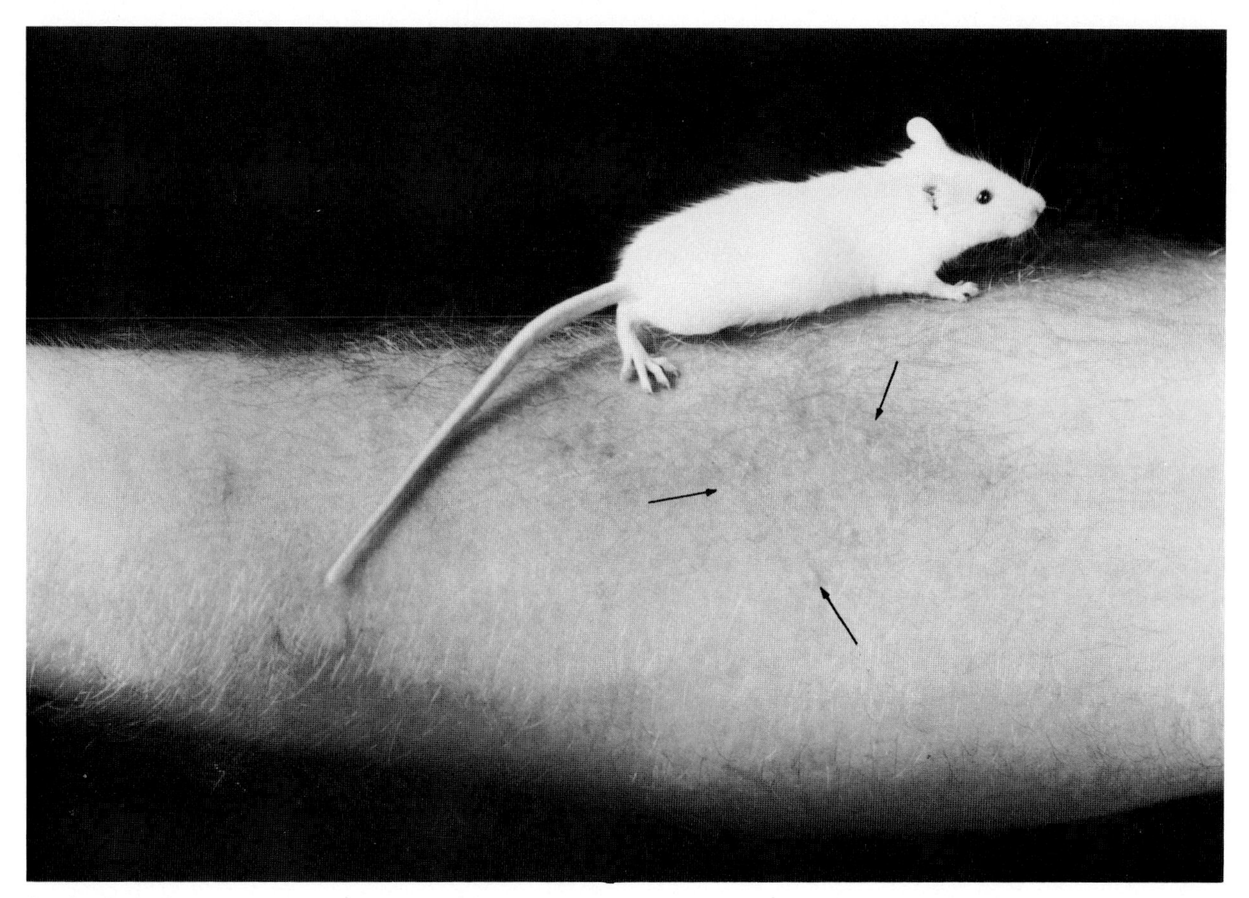

Fig. 3. Small wheals on the mid-forearm (arrows) and a large wheal adjacent to the tip of the tail of the mouse in the skin of a patient sensitive to mice.

teins from mice. He found that albumin was the major component of mouse skin extracts and that it was highly allergenic in some patients who were allergic to mice. Furthermore, Siraganian and Sandberg (1979) demonstrated the presence of at least two major allergens in mouse skin, serum, and urine. Patients sensitive to mice may react predominantly with either or both of these allergens. Further characterization of allergens from urine and animal pelts of inbred laboratory mice identified potent allergens in the mouse urine within the major urinary protein complex (MUP). In the three mouse strains studied, purified MUP proteins that bound to IgE antibodies cross-reacted extensively with each other and with allergens in dust from a mouse room. In addition, allergens from mouse pelts cross-reacted with MUP protein, suggesting that part of the allergenicity of pelt material may result from its content of components of the MUP complex. Other allergens with a high molecular weight were also found in the mouse pelt preparations. This study, which demonstrates cross-reactivity between urinary proteins and antigens in dust from a mouse room, suggests that a possible cause of sensitization in laboratory personnel is the dispersal of urinary protein from litter in mouse cages (Schumacher, 1980).

Other antigens in laboratory animal quarters may cause allergic reactions; these include mold spores and proteins in food that might be aerosolized (Patterson, 1964).

B. Pathogenesis

Because allergic sensitivities are the most common significant health hazard in mouse-associated employee activity, the pathogenesis will be discussed in some detail.

The pathogenesis of immediate hypersensitivity is initiated by an interaction of the mast cell and/or basophils. This degranulation releases preformed chemical mediators such as histamine, serotonin, and heparin. Other mediators are generated by the basophil or mast cell after it has been appropriately sensitized. An example of this type of mediator is the slow-reacting substance of anaphylaxis (SRS-A) (Gupta and Good, 1979). These mediators of immediate hypersensitivity act as messengers between the primary target cell population and populations of secondary effector cells or tissues. Effector cells, such as eosinophils, platelets, T lymphocytes, and monocytes, in turn amplify or modulate the inflammatory host

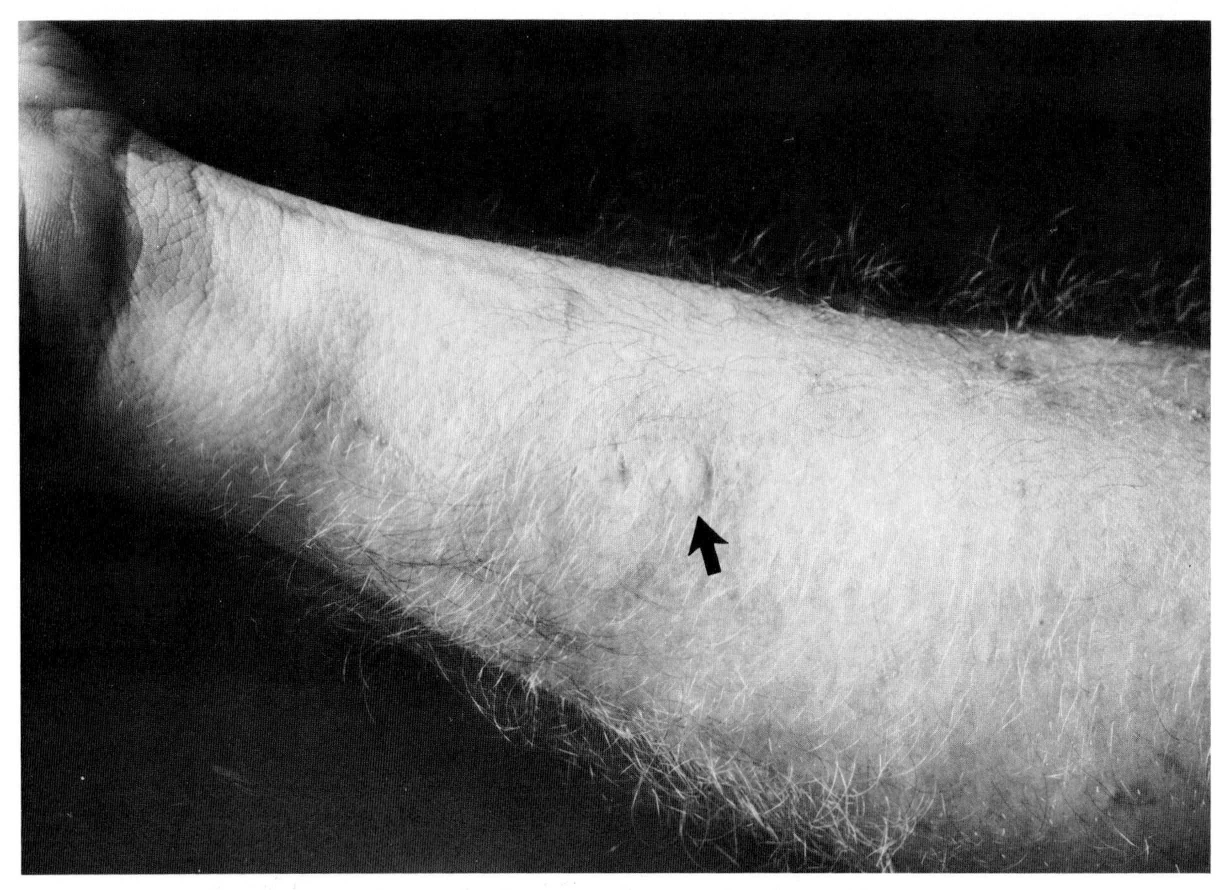

Fig. 4. Large wheal and flare in the skin (arrow) of a patient sensitive to mice.

responses, including smooth muscle contraction, vascular dilatation, and increased vascular permeability. The allergic reaction described above is often classified as type I.

Type II reactions occur when an IgG or IgM antibody reacts with an antigen on target cells. This reaction activates complement which causes cell lysis (Lutsky and Toshner, 1978). This type of reaction is most often seen when drugs act as the antigen and therefore is relatively unimportant in people working with laboratory mice.

Type III allergic reactions are characterized by damage initiated by immune complexes. These complexes activate the complement system which enhances the inflammatory response. This arthus reaction is a vasculitis initiated by the deposition of immune complexes of antigen and immunoglobulin (IgG) on the vessel endothelium. Examples include serum sickness, delayed tuberculin hypersensitivity, and hypersensitivity pneumonitis. Type III reactions may be involved in people working with laboratory animals who develop asthma several hours after being exposed to mice.

Type IV allergic reactions are cell mediated. Antigens are deposited locally and react with sensitized T lymphocytes which release certain cell-free factors (lymphokines). Examples of this type of reaction include homograph rejection, graft versus host reactions, and allergic-contact dermatitis. Type IV reactions may be involved in cases in which mouse dander, urine, or serum produces erythema and pruritis locally on the skin (Sorrell and Gottesman, 1957; Lincoln *et al., 1974).*

C. Diagnosis

To diagnose occupational allergic disease resulting from working with mice, one must establish a clinical diagnosis and incriminate the etiologic factors. A careful, detailed history, including patients' complaints as well as clinical symptoms, must be evaluated. The history of the appearance of clinical symptoms concomitant with or following environmental exposure often helps to narrow the number of allergens considered in the differential diagnosis. Nonoccupational exposure to potential allergens must also be considered. The family history of allergy is also important, since atopy predisposes a person to type I allergic reactions.

The physical examination must be thorough and well documented; often a repeat physical examination is helpful if performed when the patient is not having an acute allergic attack. Repeated pulmonary function tests, especially when the lungs are the target organ, are often helpful; radiological examinations are occasionally used. Bronchial challenge tests with suspected allergens, though rarely indicated and difficult to evaluate, together with pulmonary function tests may detect the etiologic allergen. Skin testing with suspected antigens often identifies the hypersensitivity. Skin tests are almost always positive when properly done on a patient who has type I sensitivity to animal dander. Useful laboratory tests include a complete blood count; immunoglobulins and IgE antibody specific to one allergen, as measured by the radio-allergosorbent test (RAST); nasal smears for eosinophilia; and serum precipitants to specific allergens. The *in vitro* RAST, however, is less sensitive and no more specific than the skin test. The direct eosinophil count is another useful laboratory test; it is often elevated in the presence of nasal allergy and is almost always elevated in patients with asthma.

D. Treatment and Prevention

After an allergic disorder associated with exposure to mice has been diagnosed definitely, pharmacological agents are often used to relieve the acute attack. Useful agents include antihistamines, sympathomimetic agents, corticosteroid, and bronchodilators. Some pharmacological agents are somewhat useful on a long-term basis; these include antihistamines for allergic rhinitis, allergic conjunctivitis, and allergic skin reactions. β-Adrenergic agonists, cromolyn sulfate, and xanthines are sometimes useful in asthmatic patients.

Immunotherapy has been employed to reduce symptomatology of a laboratory worker sensitive to mice (Sorrell and Gottesman, 1957). Allergen immunotherapy is the systemic administration of etiological antigens in increasing dosages to produce hyposensitivity to animal proteins in the patient. This type of therapy may not be recommended because in highly sensitive individuals, it can be accompanied by uncomfortable local and systemic reactions. There is also a serious risk of inducing anaphylaxis in the patient (Gupta and Good, 1979). The risk of treatment of patients with animal dander extract, however, is probably no greater than that of treatment with pollen extracts. Newer forms of immunotherapy are still experimental, but some may prove clinically useful. Complete avoidance of the offending antigen is the method of choice for preventing an allergic reaction to mice. However, when complete avoidance of the allergen is unfeasible for socioeconomic reasons such as earning a living, other avenues of treatment and control must be considered (Lutsky and Toshner, 1978). Reduction of the contact intensity of the offending allergen is frequently used. Methods include reduction of direct animal contact time, increasing the room ventilation, employing filter caps on animal cages, using exhaust hoods when working with mice, and using protective clothing, masks, or respirators when working with mice.

XI. CONCLUSION

With the adoption of modern laboratory animal management, which includes routine disease surveillance, proper sanitary regimens, acceptable personal hygiene, and personnel health monitoring, laboratory mice usually do not present a zoonotic or health hazard. Well-designed animal facilities to prevent ingress of wild rodents and other vermin help preclude the introduction of animal and human pathogens. Careful attention to design of caging and air-flow dynamics within animal rooms is necessary to minimize exposure to allergens.

REFERENCES

Ajello, L., Bostick, L., and Cheng, S. (1968). The relationship of Trichophyton quinckeanum to Trichophyton mentagrophytes. *Mycologia* **60,** 1185–1189.

Alteras, I. (1965). Human infection from laboratory animals. *Sabouraudia* **3,** 143–145.

Arkless, H. A. (1970). Rat-bite fever at Albert Einstein Medical Center. *Penn. Med. J.* **73,** 49.

Armstrong, C. (1963). Lymphocytic choriomenigitis. *In* "Diseases Transmitted from Animals to Man" (T. G. Hull, ed.), 5th ed., pp. 723–730. Thomas, Springfield, Illinois.

Armstrong, C., and Lillie, R. D. (1934). Experimental lymphocytic choriomeningitis of monkeys and mice produced by a virus encountered in studies of the 1933 St. Louis encephalitis epidemic. *Public Health Rep.* **49,** 1019–1027.

Armstrong, C., and Sweet, L. K. (1939). Lymphocytic choriomeningitis; report of 2 cases, with recovery of the virus from gray mice (*Mus musculus*) trapped in 2 infected households. *Public Health Rep.* **54,** 673–684.

Armstrong, C., Wallace, J. J., and Ross, L. (1940). Lymphocytic choriomeningitis: gray mice, *Mus musculus,* a reservoir for the infection. *Public Health Rep.* **55,** 1222.

Babudieri, B. (1958). Animal reservoirs of leptospires. *Ann. N.Y. Acad. Sci.* **70,** 393–413.

Barkin, R. M., Guckian, J. C., and Glosser, J. W. (1974). Infection by *Leptospira ballum:* a laboratory-associated case. *South. Med. J.* **67,** 155–176.

Baum, S. G., Lewis, A. M., Jr., Wallace, P. R., and Huebner, R. J. (1966). Epidemic nonmeningitic lymphocytic-choriomeningitis-virus infection—an outbreak in a population of laboratory personnel. *N. Engl. J. Med.* **17,** 934–936.

Bell, J. F. (1970). Tick-borne (pasture) fever and rickettsial pox. *In* "Infectious Diseases of Wild Mammals" (J. W. Davis, J. H. Karstad, and D. O. Trainer, eds.), pp. 358–362. Iowa State Univ. Press, Ames.

Benenson, A. S., ed. (1975). "Control of Communicable Diseases in Man," 12th ed. Am. Public Health Assoc., Washington, D.C.

Bergogne-Berezin, E., Christol, D., Zechorsky, N., and Bonfils, S. (1972). Human *Pasteurella* infections from bites: epidemiologic survey in a laboratory environment. *Nouv. Presse Med.* **44,** 2953-2957.

Birnbaum, S., Shenberg, E., and Torten, M. (1972). The influence of maternal antibodies on the epidemiology of leptospiral carrier state in mice. *Am. J. Epidemiol.* **96,** 313-317.

Blackmore, D. K., and Francis, R. A. (1970). The apparent transmission of staphylococci of human origin to laboratory animals. *J. Comp. Pathol.* **80,** 645-651.

Blank, F. (1957). Favus of mice. *Can. J. Microbiol.* **3,** 885-896.

Boak, R. A., Linscott, W. D., and Bodfish, R. E. (1960). A case of Leptospirosis ballum in California. *Calif. Med.* **93,** 163-165.

Booth, B. H. (1952). Mouse ringworm. *Arch. Dermatol. Syphilol.* **66,** 65-69.

Borst, J. G. G., Ruys, A. C., and Wolff, J. W. (1948). Eeen Geval van Leptospirosis Ballum. *Ned. Tijdschr. Geneeskd.* **92,** 2920-2922.

Bradburne, A. F. (1970). Antigenic relationships amongst coronavirus. *Arch. Gesamte Virusforsch.* **31,** 352-364.

Brennan, P. C., Fritz, T. E., and Flynn, R. J. (1965). *Pasteurella pneumotropica:* cultural and biochemical characteristics and its association with disease in laboratory animals. *Lab. Anim. Care* **15,** 307-312.

Brennan, P. C., Fritz, T. E., and Flynn, R. J. (1969). Murine pneumonia: a review of the etiologic agents. *Lab. Anim. Care* **19,** 360-371.

Brown, C. M., and Parker, M. T. (1957). Salmonella infection in rodents in Manchester. *Lancet No. 273,* 1277-1279.

Brown, R. Z., and Gorman, G. W. (1960). The occurrence of leptospirosis in feral rodents in southwestern Georgia. *Am. J. Publ. Health* **50,** 682-688.

Brown, T. M., and Nunemaker, J. C. (1942). Rat-bite fever: review of American cases with re-evaluation of its etiology and report of cases. *Johns Hopkins Hosp. Bull.* **70,** 201.

Bruner, D. W., and Gillespie, J. H. (1973). Rabies and other rhabdoviruses. *In* "Hagen's Infectious Diseases of Domestic Animals," 6th ed. Cornell Univ. Press, Ithaca, New York.

Buchanan, R. E. (1919). Favus herpeticus, or mouse favus: possibility of production of favus in man from Australian wheat. *J. Am. Med. Assoc.* **72,** 97.

Cetin, E. T., Tahsinoglu, M., and Volkan, S. (1965). Epizootic of *Trichophyton mentagrophytes* (*interdigitale*) in white mice. *Pathol. Microbiol.* **28,** 839-846.

Chmel, L., Buchvald, J., and Valentova, M. (1975). Spread of *Trichophyton mentagrophytes* Var. Gran. infection to man. *Int. J. Dermatol.* **14,** 269-272.

Cole, J. S., Stroll, R. W., and Bulger, R. J. (1969). Rat-bite fever. *Ann. Intern. Med.* **71,** 979.

Collins, M. J., and Parker, J. C. (1972). Murine virus contaminants of leukemia viruses and transplantable tumors. *J. Natl. Cancer Inst.* **49,** 1139-1143.

Communicable Disease Center (CDC) (1965). "Zoonoses Surveillance: Leptospirosis," Rep. No. 7, p. 25. U.S. Public Health Serv. Atlanta, Georgia.

Communicable Disease Center (CDC) (1966). "Zoonoses Surveillance: Leptospirosis Annual Summary," p. 3. U.S. Public Health Serv. Atlanta, Georgia.

Communicable Disease Center (CDC) (1976). "Salmonella Surveillance: Annual Summary 1975," Rep. No. 126. U.S. Dep. Health, Educ. Welfare, Cent. Dis. Control, Atlanta, Georgia.

Cooper, A., Martin, R., and Tibnles, J. A. R. (1973). Pasteurella meningitis. *Neurology* **23,** 1097-1100.

Cotchin, E., and Roe, F. J. C. (1967). Fungal diseases of rats and mice. *In* "Pathology 9f Laboratory Rats and Mice," pp. 681-732. Davis, Philadelphia, Pennsylvania.

Criep, L. H. (1976). "Allergy and Clinical Immunology." Grune & Stratton, New York.

Dalldorf, G., Tungeblut, C. W., and Umphlet, M. D. (1946). Multiple cases of choriomeningitis in an apartment harboring infected mice. *J. Am. Med. Assoc.* **193,** 25.

Davey, D. G. (1962). The use of pathogen free animals. *Proc. R. Soc. Med.* **55,** 256-262.

Davies, R. R., and Shewell, J. (1964). Control of mouse ringworm. *Nature (London)* **202,** 406-407.

Davis, B. D., Dulbecco, R., Eisen, H. N., Ginsberg, H. S., and Wood, W. B. (1969). Streptococci. *In* "Microbiology," pp. 702-725. Harper, new York.

Dolan, M. M., Kligman, A. M., Kobylinski, P. G., and Motsavage, M. A. (1958). Ringworm epizootics in laboratory mice and rats; experimental and accidental transmission of infection. *J. Invest. Dermatol.* **30,** 23-25.

Dolgopol, B. B. (1948). Histologic changes in rickettsial pox. *Am. J. Pathol.* **24,** 119-133.

Duncan, P. R., Thomas, A. E., and Tobin, J. O. (1951). Lymphocytic choriomeningitis. Review of ten cases. *Lancet* **i,** 956-959.

Dvorak, S., and Otechenasek, M. (1964). Geophites zoophilic and anthropohilic dermatophytes: a review. *Mycopathol. Mycol. Appl.* **23,** 294-296.

Eichenwald, H. F. (1965). Bacterial interference and staphylococcal colonization in infants and adults. *Ann. N.Y. Acad. Sci.* **128,** 365-380.

Faine, S. (1963). Antibody in the renal tubules and urine of mice. *Aust. J. Exp. Biol. Med. Sci.* **41,** 811-814.

Faust, E. C., and Russell, P. F. (1970). "Craig and Faust's Clinical Parasitology," 8th ed., pp. 528-529. Lea & Febiger, Philadelphia, Pennsylvania.

Findlay, G. M., Alcock, N. S., and Stern, R. O. (1936). The virus aetiology of one form of lymphocytic meningitis. *Lancet i,* 650.

Findlay, G. M., Klieneberger, E., MacCallum, R. O., and Mackenzie, R. D. (1938). Rolling disease-new syndrome in mice associated with a pleuropneumonia like organism. *Lancet* **ii,** 1511-1513.

Flynn, R. J. (1973). Mites. *In* "Parasites of Laboratory Animals," pp. 425-492. Iowa State Univ. Press, Ames.

Fox, J. G., and Beaucage, C. M. (1979). The incidence of salmonella in random-source cats purchased for use in research. *J. Infect. Dis.* **319,** 362-365.

Fox, J. G., Niemi, S. M., Murphy, J. C., and Quimby, F. W. (1977). Ulcerative dermatitis in the rat. *Lab. Anim. Sci.* **27,** 671-678.

Fox, J. G. (1982). Outbreak of tropical rat mite dermatitis in laboratory personnel. *Arch. Dermatol.* (In press).

Freigang, B., and Elliott, G. B. (1963). *Pasteurella septica* infections in humans. *Can. Med. Assoc. J.* **89,** 702-704.

Friedmann, C. T. H., Spiegel, E. L., Aaron, E., and McIntyre, R. (1973). Leptospirosis ballum contracted from pet mice *Calif. Med.* **118,** 51-52.

Fukumi, H., Nishikawa, F., and Kitayama, T. (1954). A pneumotropic virus from mice causing hemagglutination. *Jpn. J. Med. Sci. Biol.* **7,** 345-363.

Galton, M. M., Menges, R. W., Shotts, E. B., Nahmias, A. J., and Heath, C. (1962). "Leptospirosis," Publ. No. 951. U.S. Dep. Health, Educ. Welfare, Public Health Serv. Cent. Dis. Control, Atlanta, Georgia.

Geller, E. H. (1979). Health hazards for man. *In* "The Laboratory Rat" (H. J. Baker, J. R. Lindsey, and S. H. Weisbroth, eds.), Vol. 1, pp. 402-407. Academic Press, New York.

Gerngross, O. G. (1957). Peculiarities of the 1956 influenza outbreak in Vladivostok due to D. virus. *Probl. Virol.* (*USSR*) **2,** 71-75.

Gilbert, G. L., Cassidy, J. F., and Bennett, N. M. (1971). Rat-bite fever. *Med. J. Aust.* **2,** 1131-1134.

Gupta, S., and Good, R. A. eds. (1979). "Cellular, Molecular and Clinical Aspects of Allergic Disorders." Plenum, New York.

Haberman, R. T., and Williams, F. P. (1958). Salmonellosis in laboratory animals. *J. Natl. Cancer Inst.* **20,** 933-941.

Hartley, J. W., Rowe, W. P., Bloom, H. H., and Turner, H. C. (1964). Antibodies to mouse hepatitis viruses in human sera. *Proc. Soc. Exp. Biol. Med.* **115,** 414-418.

Havens, W. P. (1948). LCM: Report of a case occurring in a granary harboring infected mice. *J. Am. Med. Assoc.* **137,** 857-858.

Heath, C. W., Jr., and Alexander, A. D. (1970). Leptospirosis. *In* "Joseph Brenneman's Practice of Pediatrics" (U. C. Kelley, ed.), Vol. 2, Cap. 26B. Harper, New York.

Heath, R. B., Tyrrell, D. A. J., and Peto, S. (1962). Serological studies with Sendai virus. *Br. J. Exp. Pathol.* **43,** 444-450.

Henriksen, S. D. (1962). Some pasteurella strains from the human respiratory tract. *Acta Pathol. Microbiol. Scand.* **55,** 355-356.

Henriksen, S. D., and Jyssum, E. (1961). A study of some pasteurella strains from the human respiratory tract. *Acta Pathol. Microbiol. Scand.* **51,** 354-368.

Higa, H. H., and Fujinaka, I. T. (1976). Prevalence of rodent and mongoose leptospirosis on the island of Oahu. *Public Health Rep.* **91,** 171-177.

Hoag, W. G., and Meier, H. (1966). Infectious diseases. *In* "Biology of the Laboratory Mouse," 2nd ed., pp. 589-600. McGraw-Hill, New York.

Hoag, W. G., and Rogers, J. (1961). Techniques for the isolation of *Salmonella typhimurium* from laboratory mice. *J. Bacteriol.* **82,** 153-154.

Hoag, W. G., Wetmore, P. W., Rogers, J., and Meier, H. (1962). A study of latent pasteurella infection in a mouse colony. *J. Infect. Dis.* **111,** 135-140.

Hoag, W. G., Strout, J., and Meier, H. (1964). Isolation of *Salmonella spp.* from laboratory mice and from diet supplements. *J. Bacteriol.* **88,** 534-536.

Holden, F. A., and MacKay, J. C. (1964). Rat-bite fever—an occupational hazard. *Can. Med. Assoc. J.* **91,** 78.

Holmgren, E. B., and Tunevall, G. (1970). Case report: rat-bite fever. *Scand. J. Infect. Dis.* **2,** 71.

Hotchin, J. (1971). The contamination of laboratory animals with lymphocytic choriomeningitis virus. *Am. J. Pathol.* **64,** 747-769.

Hotchin, J. E., and Benson, L. M. (1973). Lymphocytic choriomeningitis. *In* "Infectious Diseases of Wild Mammals" (J. W. Davis, L. N. Karstad, and D. O. Trainer, eds.), pp. 153-165. Iowa State Univ. Press, Ames.

Hubbert, W. T., and Rosen, M. N. (1970). *Pasteurella multocida* infection due to animal bite. *Am. J. Public Health* **60,** 1103-1117.

Hudson, B. W., Quan, S. F., and Goldenberg, M. I. (1964). Serum antibody response in a population of *Microtus californicus* and associated species during and after *Pasteurella pestis* epizootics in the San Francisco Bay area. *Zoonoses Res.* **2,** 15-23.

Huebner, R. J., Stamps, P., and Armstrong, A. (1946a). Rickettsialpox—a newly recognized rickettsial disease. I. Isolation of the etiological agent. *Public Health Rep.* **61,** 1605-1614.

Huebner, R. J., Jellison, W. L., and Pomerantz, D. (1946b). Rickettsialpox—a newly recognized rickettsial disease. IV. Isolation of a rickettsia apparently identical with the causative agent of rickettsialpox from *Allodermanyssus sanguineus,* a rodent mite. *Public Health Rep.* **61,** 1677-1682.

Huebner, R. J., Jellison, W. L., and Armstrong, A. (1947). Rickettsialpox—a newly recognized rickettsial disease. V. Recovery of *Rickettsia akari* from a house mouse (*Mus musculus*). *Public Health Rep.* **62,** 777-780.

Inada, R., Ido, Y., Hoki, R., Kaneko, R., and Ito, H. (1916). The etiology, mode of infection and specific therapy of Weil's disease. *J. Exp. Med.* **23,** 377.

Institute of Laboratory Animal Resources (ILAR), National Research Council (1978). "NIH Guide for the Care and Use of Laboratory Animals," DHEW publ. No. (NIH) 78-23. U.S. Dep. Health, Educ. Welfare, Natl. Inst. Health, Bethesada, Maryland.

Jawetz, E. (1948). A latent pneumotropic pasteurella of laboratory animals. *Proc. Soc. Exp. Biol. Med.* **68,** 46-48.

Jawetz, E. (1950). A pneumotropic pasteurella of laboratory animals. I. Bacteriological and serological characteristics of the organism. *J. Infect. Dis.* **86,** 172-183.

Jelliffe, D. B., and Stanfield, J. P. (1978). "Diseases of Children in the Subtropics and Tropics," pp. 532-533. Arnold, London.

Jones, D. M. (1962). A pasteurella-like organism from the human respiratory tract. *J. Pathol. Bacteriol.* **83,** 143.

Kaufman, A. F. (1966). Pets and salmonella infection. *J. Am. Vet. Med. Assoc.* **149,** 1655-1661.

Kessel, J. F. (1923). Experimental infection of rats and mice with the common intestional amoebae of man. *Univ. Calif, Berkeley, Publ. Zool.* **20,** 409-430.

Krempl-Lamprecht, L., and Bosse, K. (1964). Epidermophyton floccosum (Harz) Langeron und Milochevitsch als spontanin fektion bei Mausen. *Kleintier-Prax.* **9,** 203-207.

Kuroya, M., Ishida, N., and Shiratori, T. (1953a). Newborn virus pneumonitis (type Sendai) II. Report: the isolation of a new virus possessing hemagglutinating activity. *Yokohama Med. Bull.* **4,** 217-233.

Kuroya, M., Ishida, N., and Shiratori, T. (1953b). Newborn virus pneumonitis (type Sendai). II. The isolation of a new virus. *Tohoku J. Exp. Med.* **58,** 62.

Lehmann-Grube, F. (1971). Lymphocytic choriomeningitis virus. *Virol. Monogr.* **10,** 88-118.

Levy, D. A. (1974). Allergic activity of proteins from mice. *Int. Arch. Appl. Immunol.* **49,** 219-221.

Lewis, J. M., and Utz, J. P. (1961). Orchitis, parotitis, and meningoencephalitis due to lymphocytic-chorio-meningitis virus. *N. Engl. J. Med.* **265,** 776-780.

Lincoln, T. A., Bolton, N. E., and Garrett, A. W. (1974). Occupational allergy to animal dander and sera. *JOM, J. Occup. Med.* **16,** 465-469.

Lindorfer, R. K., and Syverton, J. T. (1953). The characterization of an unidentified virus found in association with Line 1 Leukemia. *Proc. Am. Assoc. Cancer Res.* **1,** 33-34.

Loffler, W., and Mooser, H. (1942). Mode of transmission of typhus fever. Study based on infection of group of laboratory workers. *Schweiz. Med. Wochenschr.* **72,** 755-761.

Loosli, R. (1967). Zoonoses in common laboratory animals. *In* "Husbandry of Laboratory Animals" (M. L. Conalty, ed.), Sect. 4. Academic Press, New York.

Lutsky, I., and Toshner, D. (1978). A review of allergic respiratory disease in laboratory animal workers. *Lab. Anim. Sci.* **28,** 751-756.

MacCallum, F. O. (1949). The virus of lymphocytic choriomeningitis (LCM) as a cause of benign aseptic meningitis. Laboratory diagnosis of five cases. *Mon. Bull. Minist. Health Public Health Lab. Serv.* **8,** 177.

McGill, R. C., Martin, A. M., and Edmunds, P. N. (1966). Ratbite fever due to *Streptobacillus moniliformis. Br. Med. J.* **I,** 1213-1214.

McIntosh, K., Becker, W. B., and Chanock, R. M. (1967). Growth in suckling-mouse brain of "IBV-like" viruses from patients with upper respiratory tract disease. *Proc. Natl. Acad. Sci. U.S.A.* **58,** 2268-2273.

McIntosh, K., Kapikian, A. Z., Hardison, K. A., Hartley, J. W., and Chanock, R. M. (1969). Antigenic relationships among the coronaviruses of man and between human and animal coronaviruses. *J. Immunol.* **102,** 1109-1118.

MacKenzie, D. W. R. (1961). *Trichophyton mentagrophytes* in mice: infections of humans and incidence amongst laboratory animals. *Sabouraudia* **1,** 178-182.

Mailloux, M. (1975). Leptospiroses-Zoonoses. *Int. J. Zoonoses* **2,** 45-54.

Margard, W. L., Peters, A. C., Dorko, N., Litchfield, J. H., and Davidson, R. S. (1963). Salmonellosis in mice-diagnostic procedures. *Lab. Anim. Care* **13,** 144-165.

Markell, E. K., and Voge, M. (1965). "Medical Parasitology," 2nd ed. Saunders, Philadelphia, Pennsylvania.

Marples, M. J. (1967). Nondomestic animals in New Zealand and in Rarotonza as a reservoir of the agents of ringworm. *N. Z. Med. J.* **66,** 299-302.

Maurer, F. D. (1964). Lymphocytic choriomeningitis. *Lab. Anim. Care* **14,** 415-419.

Mescon, H., and Grots, I. A. (1974). The skin. *In* "pathologic Basis of Disease" (S. L. Robbins, ed.), pp. 1374-1419. Saunders, Philadelphia, Pennsylvania.

Miller, J. K. (1966). Human pasteurellosis in New York State. *N.Y. State J. Med.* **66,** 2527-2531.

Moore, R. M., Jr., Zehmer, B. R., Moultrop, J. I., and Parker, R. L. (1977). Surveillance of animal-bite cases in the United States, 1971-1972. *Arch. Environ. Health* **32,** 267-270.

Morbidity and Mortality Weekly Report (MMWR) (1980). Human Salmonella isolates—United States, 1978, surveillance summary. **28,** 618-619.

Neal, R. A. (1950). An experimental study of *Entamoeba muris* (Grassi 1879); its morphology affinities and host parasite relationship. *Parisitology* **40,** 343-365.

Newman-Taylor, A., Longbottom, J. L., and Pepys, J. (1977). Respiratory allergy to urine proteins of rats and mice. *Lancet* **ii,** 847-848.

Nichols, E., Rindze, M. E., and Russell, G. G. (1953). Relationship of habits of house mouse and mouse mite (*Allodermanyssus sanguineus*) to spread of rickettsialpox. *Ann. Intern. Med.* **39,** 92-102.

Noguchi, H. (1918). Morphological characteristics and nomenclature of *Leptospira* (Spirochaeta) *icterohemorrhagiae* (Inada and Ido). *J. Exp. Med.* **27,** 575-592.

Ohman, J. L. (1978). Allergy in man caused by exposure to mammals. *J. Am. Vet. Med. Assoc.* **172,** 1403-1406.

Ohman, J. L., Lowell, F. C., and Bloch, K. J. (1975). Allergens of mammalian origin. *J. Allergy Clin. Immunol.* **55,** 16-24.

Olson, J. R., and Meadows, T. R. (1969). *Pasteurella pneumotropica* infection resulting from a cat bite. *Am. J. Clin. Pathol.* **51,** 709-710.

Owen, D. (1978). Attempted transmission of *Entamoeba coli* to specified-pathogen-free rodents. *Lab. Anim.* **12,** 79-80.

Paegle, R. D., Tewari, R. P., Bernhard, W. H., and Peters, E. (1976). Microbial flora of the larynx, trachea, and large intestine of the rat after long-term inhalation of 100 percent oxygen. *Anesthesiology* **44,** 287-290.

Parker, J. C., Igel, H. J., Reynolds, R. K., Lewis, A. M., Jr., and Rowe, W. P. (1976). Lymphocytic choriomeningitis virus infection in fetal, newborn, and young adult Syrian hamsters. *Infect. Immun.* **13,** 967-981.

Patterson, R. (1964). The problem of allergy to laboratory animals, *Lab. Anim. Care* **14,** 466-469.

Place, E. H., and Sutton, L. E. (1934). Erythema Arthriticum Epidemicum (Haverhill Fever). *Arch. Intern. Med.* **54,** 659-684.

Poiley, S. M. (1970). A survey of indigenous murine viruses in a variety of production and research animal facilities *Lab. Anim. Care* **20,** 643-650.

Raffin, B. J., and Freemark, M. (1979). Streptobacillary rat-bite fever: a pediatric problem. *Pediatrics* **64,** 214-217.

Rajka, G. (1961). Ten cases of occupational hypersensitivity to laboratory animals. *Acta Allergol.* **16,** 168-176.

Refai, M., and Ali, A. H. (1970). Laboratory acquired infection with Keratinomyces ajelloi. *Mykosen* **13,** 317-318.

Renquist, D., and Soave, O. (1969). Staphylococcal pneumonia in a laboratory rabbit: an epidemiologic follow-up study. *J. Am. Vet. Med. Assoc.* **155,** 1221-1223.

Richter, C. P. (1945). Incidence of ratbites and ratbite fever in Baltimore, *J. Am. Med. Assoc.* **128,** 324-326.

Riley, W. A. (1919). A mouse oxyurid, *Syphacia obvelata,* as a parasite of man. *J. Parasitol.* **6,** 89-92.

Robbins, S. L. (1974). Infectious disease. *In* "Pathologic Basis of Disease," pp. 396-400. Saunders, Philadelphia, Pennsylvania.

Rogers, B. T., Anderson, J. C., Palmer, C. A., and Henderson, W. G. (1973). Septicaemia due to *Pasteurella pneumotropica. J. Clin. Pathol.* **26,** 396-398.

Rogosa, M. (1974). *Streptobacillus moniliformis* and *Spirillum minor. In* "Manual of Clinical Microbiology" (E. H. Lennette, E. H. Spaulding, and J. P. Truant, eds.), 2nd ed., pp. 326-332. Am. Soc. Microbiol., Washington, D.C.

Rosen, L. (1968). "Reoviruses," Virology Monographs, Vol. 1, pp. 73-107. Springer-Verlag, Berlin and New York.

Roughgarden, J. W. (1965). Antimicrobial therapy of rat-bite fever. *Arch. Intern. Med.* **116,** 39-54.

Rountree, P. M., Freeman, B. M., and Johnston, K. G. (1956). Nasal carriers of *Staphylococcus aureus* by various domestic and laboratory animals. *J. Pathol. Bacteriol.* **72,** 319-321.

Sano, T., Nitsu, I., Nakagawa, I., and Ando, T. (1953). Newborn virus pneumonitis (type Sendai). I. Report: Clinical observation of a new virus pneumonitis of the newborn. *Yokohama Med. Bull.* **4,** 199-216.

Scheid, W., Ackermann, R., Bloedhorn, H., and Kuepper, B. (1964). Distribution of the virus of LCM in Western Germany. *Ger. Med. Mon.* **9,** 157-161.

Schipper, G. J. (1947). Unusual pathogenicity of *Pasteurella multocida* isolation from the throats of common wild rats. *Johns Hopkins Med. J.* **81,** 333-356.

Scholz, M., and Weinhold, E. (1969). Epidemiology of rabies in rats and mice. *Berl. Muench. Tieraerztl. Wochenschr.* **82,** 255-257.

Schumacher, M. J. (1980). Characterization of allergens from urine and pelts of laboratory mice. *Mol. Immunol.* **17,** 1087-1095.

Shimi, A., Keyhani, M., and Hedayati, K. (1979). Studies on salmonellosis in the house mouse *Mus musculus. Lab. Anim.* **13,** 33-34.

Shults, F. S., Estes, P. C., Franklin, J. A., and Richter, C. N. (1973). Staphylococcal botryomycosis in a specific-pathogen-free mouse colony. *Lab. Anim. Sci.* **26,** 36-42.

Siraganian, R. P., and Sandberg, A. L. (1979). *J. Allergy Clin. Immunol.* **63,** 435-442.

Skinner, H. H., and Knight, E. H. (1971). Monitoring mouse stocks for lymphocytic choriomeningitis virus—a human pathogen. *Lab. Anim.* **5,** 73-87.

Smithard, E. H. E., and Macrae, A. D. (1951). LCM: Associated human and mouse infections. *Br. Med. J.* **1,** 1298-1300.

Soave, O. A., and Van Allen, A. (1958). LCM in a mouse breeding colony. *Proc. Anim. Care Panel* **8,** 135-140.

Sorrell, A. H., and Gottesman, J. (1957). Mouse allergy: case report. *Ann. Allergy Nov.-Dec.* 662-663.

Stanley, N. F., Dorman, D. C., and Ponsford, J. (1953). Studies on the pathogenesis of a hitherto undescribed virus (Hepatoencephalomyelitis) producing unusual symptoms in suckling mice. *Aus. J. Exp. Biol. Med. Sci.* **31,** 147-160.

Stewart, S. E., and Haas, V. A. (1956). Lymphocytic choriomeningitis in mouse neoplasms. *J. Natl. Cancer Inst.* **17,** 233-245.

Stoenner, H. G. (1954). Application of the capillary tube test and a newly developed plate test to the serodiagnosis of bovine leptospirosis. *Am. J. Vet. Res.* **15,** 434-439.

Stoenner, H. G. (1957). The laboratory diagnosis of leptospirosis. *Vet. Med.* **52,** 540-542.

Stoenner, H. G., and Maclean, D. (1958). Leptospirosis (ballum) contracted from Swiss albino mice. *Arch. Intern. Med.* **101,** 706-710.

Stoenner, H. G., Grimes, E. F., Thrailkill, F. B., and Davis, E. (1958). Elimination of *Leptospira ballum* from a colony of Swiss albino mice by use of chlortetracycline hydrochloride. *Am. J. Trop. Med. Hyg.* **7,** 423-426.

Stoll, N. E. (1947). This wormy world. *J. Parasitol.* **33,** 1-18.

Stone, W. B., and Manwell, R. D. (1966). Potential helminth infections in humans from pet or laboratory mice and hamsters. *Public Health Rep.* **31,** 647-653.

Stott, J. A., Hodgson, J. E., and Chaney, J. C. (1975). Incidence of Salmonellae in animal feed and the effect of pelleting on content of enterobacteriaceae. *J. Appl. Bacteriol.* **39,** 41-46.

Strangeways, W. I. (1933). Rats as carriers of *Streptobacillus moniliformis*. *J. Pathol. Bacteriol.* **37,** 45-51.

Sulzer, C. R., Harvey, T. W., and Galton, M. M. (1968). Compaison of diagnostic technics for the detection of leptospirosis in rats. *Health Lab. Sci.* **5,** 171-173.

Taylor, M. J., and MacDowell, E. C. (1949). Mouse leukemia, XIV. Freeing Line I from a contaminating virus. *J. Natl. Cancer Inst.* **17,** 233-245.

Tennant, R. W., Reynolds, R. K., and Layman, K. R. (1967). Incidence of murine virus antibody in humans in contact with experimental animals. *Exp. Hematol.* (*Oak Ridge, Tenn.*) **14,** 76.

Tobin, J. O. (1968). Viruses transmissable from laboratory animals to man. *Lab. Anim.* **3,** 19.

Torten, M. (1979). Leptospirosis. *In* "CRC Handbook Series in Zoonoses" (J. H. Steele, ed.), pp. 363-421. CRC Press, Cleveland, Ohio.

Turner, L. H. (1970). Leptospirosis III. *Trans. R. Soc. Trop. Med. Hyg.* **64,** 623-646.

Van den Ende, M., Harries, E. H. R., Stewart-Harris, C. H., Steigman, A. J., and Cruckshank, R. (1943). Laboratory infection with murine typhus. *Lancet* **i,** 328-332.

Van der Waaij, D., Zimmerman, W. M. T., and Van Bekkum, D. W. (1963). An outbreak of *Pseudomonas aeruginosa* infection in a colony previously free of this infection. *Lab. Anim. Care* **13,** 46-51.

Voge, M., and Heyneman, D. (1957). Development of *Hymenolepis nana* and *Hymenolepis diminuta* (Cestoda: Hymenolepididae) in the intermediate host *Tribolium confusum*. *Univ. Calif., Berkeley, Publ. Zool.* **59,** 549-580.

Wardle, R. A., and McLeod, J. A. (1952). "The Zoology of Tapeworms." Univ. of Minnesota Press, Minneapolis.

Wayson, N. E. (1927). An epizootic among meadow mice in California, caused by the bacillus of mouse septicemia or of swine erysipelas. *Public Health Rep.* **42,** 1489-1493.

Wenner, H. A. (1948). Isolation of LCM virus in an effort to adapt poliomyelitis virus to rodents. *J. Infect. Dis.* **83,** 155-163.

Wheater, D. R. W. (1967). The bacterial flora of an SPF colony of mice, rats and guinea-pigs. *In* "Husbandry of Laboratory Animals" (M. L. Conalty, ed.), pp. 343-360. Academic Press, New York.

Williams, L. P., Vaughn, J. B., Scott, A., and Blanton, V. (1969). A ten-month study of Salmonella contamination in animal protein meals. *J. Am. Vet. Med. Assoc.* **115,** 167-174.

Winton, F. W., and Mair, N. S. (1969). *Pasteurella pneumotropica* isolated from a dog-bite wound. *Microbios* **2,** 155-162.

Wolf, F. W., Bohlander, H., and Ruys, A. C. (1949). Researches on *Leptospirosis ballum*—The detection of urinary carriers in laboratory mice. *Antonie van Leeuwenhoek* **15,** 1-13.

Yager, R. H., Gochenour, W. S., Jr., Alexander, A. D., and Wetmore, P. W. (1953). Natural occurrence of *Leptospira ballum* in rural house mice and in an opossum. *Proc. Soc. Exp. Biol. Med.* **84,** 589-590.

Yunker, C. E. (1964). Infections of laboratory animals potentially dangerous to man: ectoparisites and other arthropods, with emphasis on mites. *Lab. Anim. Care* **14,** 455-465.

Zhdanoff, V. M., Ritova, V. V., and Golygina, L. A. (1957). Influenza D in early infancy. *Acta Virol.* (*Engl. Ed.*) **1,** 216-219.

Chapter 23

Selected Nonneoplastic Diseases

J. D. Burek, J. A. Molello, and S. D. Warner

I. Introduction 425
II. Occurrence 426
A. Liver 426
B. Digestive System 428
C. Pancreas 428
D. Cardiovascular System 429
E. Genital System 430
F. Urinary System 431
G. Endocrine System 432
H. Integumentary System 434
I. Musculoskeletal System 435
J. Respiratory System 436
K. Nervous System 436
L. Hematopoietic (Lymphoreticular) System 438
M. Eyes 438
References 438

I. INTRODUCTION

The mouse is a diminutive creature that has evoked terror as well as feelings of untidiness throughout history. It has also exerted an unparalleled impact on virtually every phase of our temporal existence. Medical history is replete with accounts of the benefits provided mankind by virtue of experiments using laboratory mice, but only recently have results from animal studies received such wide publicity. Judgments on the relationship of man and his environment are often dogmatically predicated on the basis of data from animal studies. Therefore, a thorough understanding of laboratory animals, and currently of the mouse, is essential if evaluation of laboratory data is to be meaningful.

Any discussion of the nonneoplastic pathological anatomy of mice must begin with the studies of Thelma Dunn and Margaret Deringer. It was their early work on comparative and rodent pathology that opened the door to the study of the pathology of laboratory mice. Thelma Dunn's paper in 1954 on "The importance of differences in morphology in inbred strains" touched on several issues that are only now being fully appreciated. Issues were raised such as the differences between

ISBN 0-12-262502-1

inbred mouse strains, the importance of the development of pathologic changes with aging in mice, and the critical need for information on each specific mouse strain used in research (Dunn, 1954a).

Most mouse strains were developed because of one or more unique tumors or diseases that were present in these strains. As a result, tumors in mice have been well documented, but only rarely have complete gross and histopathologic examinations been done to characterize both the neoplastic and nonneoplastic lesions in aging strains of mice. Age-associated incidence patterns of nonneoplastic lesions have seldom been published. A literature review of the nonneoplastic lesions of mice revealed only a limited number of publications on various mouse strains.

As a result, the information in this chapter has been derived from a review of the literature and from personal observations. Generalizations do not refer to all mice because genetic background, age, sex, nutrition, and laboratory conditions may influence the development or existence of the observations. These and other variables must be considered in the interpretation of data. Furthermore, no attempt was made to cover all spontaneous diseases and their associated nonneoplastic lesions that are unique to specific mouse strains. Rather, the chapter is intended to cover lesions commonly observed in several species and likely to be encountered by investigators using mice.

The incidence of many lesions can vary dramatically depending on the completeness of the gross necropsy examination. If a lesion is not observed, it may be missed when slides are prepared. A complete gross necropsy examination involves the careful examination of the external appearance of the animal, including the skin, extremities, and inside the mouth. The lower jaw needs to be removed in order to see the tongue, teeth, and palate. Body cavities should be opened and the internal organs and tissues examined. This includes the thoracic, abdominal, and cranial cavities. Such as investigation requires removal of each organ from the mouse's body and careful examination. All surfaces should be examined and all organs carefully identified. The brain should be removed from the cranial cavity and the pituitary examined. Organs such as the adrenal glands, pancreas, and thyroid should be examined. The intestinal tract should be removed and spread out and the entire length evaluated. The male and female reproductive organs need to be removed and individually examined, and ideally, hollow organs should be opened.

Another factor affecting the incidence of lesions is the age of the animal. Therefore, any consideration of nonneoplastic lesions in mice must be tempered with an evaluation of the age of the animals examined. For most of the nonneoplastic, age-associated lesions, the older the mouse the greater the incidence of the lesion and the greater the size or extent of the age-associated lesion. Therefore, a background knowledge of the survival curves for the given strain to be evaluated is of critical importance. Animals that are only 12 months of age will clearly have different age-associated lesions than animals older than 30 months.

Several reports are available on the longevity of aging mice (Blankwater, 1978; Storer, 1966; Abbey, 1979). From these reviews, it is clear that mice have a variable longevity. However, most strains of mice have a 50% survival of 24–30 months of age. In some strains males outlive females, whereas in other strains the female is longer-lived. Storer (1966) found that males outlived females in eight mouse strains, males and females had similar life spans in four strains, and females outlived males in 10 strains. Occasionally, a strain such as the AKR/J is short-lived. In such cases, it is important to know the cause of death and what has influenced and shaped the survival curve. Animals such as those of the AKR strain that die with a high incidence of leukemia early in life have a different background of spontaneous, nonneoplastic lesions than animals that live for more than 24–30 months. Similarly, mice that die because of some infectious disease will also have different nonneoplastic lesions than mice that die naturally.

This chapter focuses primarily on nonneoplastic morphological alterations in aging mice. Since only limited published data are available on the age-associated incidences of nonenoplastic lesions in mice, no attempt was made to list all age-associated, nonneoplastic lesions present in each mouse strain or stock. The important lesions are discussed and, where appropriate, specific strains are mentioned to illustrate the incidence patterns that may be seen with specific lesions.

II. OCCURRENCE

A. Liver

Nonneoplastic, age-associated alterations are common in the liver of mice. One of the most common findings is the presence of cellular and nuclear pleomorphism in mice that are 6 months of age or older. Individual hepatocytes vary in both size and shape. However, their nuclei often have the most striking variability. It is common to find two or three nuclei within a single hepatocyte. Heptatocellular nuclei can vary in shape and size, with some being two or three times the size of nuclei in adjacent cells (polypoidy). Some are elongated, others indented, others slightly lobulated, and still others with varying numbers of prominent nucleoli. Hyperchromatic nuclei can be numerous. Furthermore, mitotic figures may be seen. Both nuclear and cellular pleomorphism appear to be common in older mice and may continue to increase in degree with age (Andrew *et al.*, 1943). However, Van Bezooijen *et al.* (1974) demonstrated that polypoidy in the rat liver increases up to 6–12 months of age and remains relatively constant beyond 12 months. A similar event may or may not be true for polypoidy in mice.

Another common finding is the presence of eosinophilic, homogeneous intranuclear inclusion bodies (Jones, 1967; Hollander, 1975). They are spherical bodies that vary in size and number and are actually cytoplasmic invaginations into the nucleus.

Spontaneously occurring foci of necrosis are common. They are frequently infiltrated with inflammatory cells that may be predominantly polymorphonuclear cells or may contain predominantly mononuclear cells and even form small granulomata. Grossly, such lesions are usually pale and are pinpoint or larger in size. The areas of necrosis vary considerably in size, distribution, and degree (Olitsky and Casals, 1945). Infectious diseases must first be excluded, but infectious agents are not usually observed in these areas and their etiology is unknown.

Small foci of altered hepatocytes are seen in mice, and many are similar to those described in rats (Squire and Levitt, 1975). They consist of aggregates of hepatocytes that vary from surrounding hepatocytes in their tinctorial or morphological appearance (Fig. 1). Most are composed of vacuolated hepatocytes, but others may be eosinophilic or basophilic. The cells in such foci are usually larger than surrounding hepatocytes, but they may be smaller. The importance of these foci in long-term studies has been discussed by Butler and Newberne (1975).

Hyperplasia of bile ducts is not uncommon. It is a focal or multifocal lesion characterized by nests of small bile ducts in the portal area. Varying degrees of inflammatory cells and fibrous connective tissue may also be seen with the biliary tract lesion.

Hepatocellular vacuolization is also common. It is usually present as a few randomly distributed cytoplasmic vacuoles but may be very numerous and is usually suggestive of fat vacuoles. The severity may increase or decrease depending on disease processes elsewhere in the animal's body.

Similarly, glycogen is usually prominent in the mouse liver if the mouse is killed. However, mice that are ill usually do not eat and can deplete their hepatocellular glycogen within a few hours. Depending on the degree of glycogen depletion, there is variable staining in the central and portal areas of the liver lobule, resulting in altered tinctorial properties in various portions of the liver lobule.

The liver is an important site for amyloid deposition in several mouse strains (Dunn, 1967b; Scheinberg *et al.*, 1976; Soret *et al.*, 1977). The earliest deposits are in the periportal region and start as focal or multifocal lesions which may progress to become diffuse and severe.

Strangulated segments of the liver and diaphragmatic herniation, either piercing the diaphragm or resting in a cul-de-sac, are commonly observed. Such lesions may have areas of necrosis that are focal or widespread. Furthermore, they are usually nodular and may be rather large. As a result, they must be differentiated from tumors of hepatocellular origin.

A number of miscellaneous alterations occur in livers of aging mice, some having been reviewed by Jones (1967). For example, pigments may be present. Hemosiderin can occur in the cytoplasm of hepatocytes as well as within reticuloendothelial cells in the liver. Likewise, lipofuscin can be demonstrated using special stains in hepatocytes of many aging mice, and

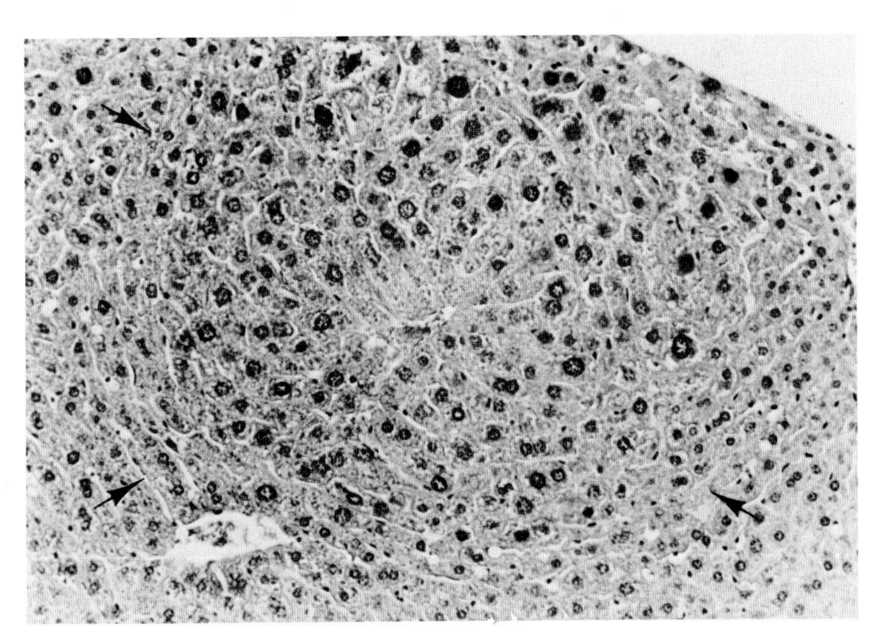

Fig. 1. Focus of altered hepatocytes (↑) in the liver.

bile pigment may be seen with obstructive disease processes. In addition, aggregates of lymphoid cells, areas of fibrosis, or areas of nonspecific inflammation may occur. Cytoplasmic inclusions, vacuolization, and hyaline droplets may be found. Extramedullary hematopoiesis is common in young mice and in older mice as a result of disease processes that result in anemia. Finally, sinusoidal dilatation may occur as a focal, multifocal, or diffuse lesion, with the last possibly related to hypervolemia (Jones, 1967).

B. Digestive System

Spontaneous nonneoplastic lesions of the lips and oral cavity are most often inflammatory or traumatic. Abscesses and necrosis may occur on the lips because of fighting or an injury from wire cages. Incisor teeth may become overgrown and result in malocclusion. They may also break off and develop localized inflammation. Molar teeth may become eroded and dark because of dental carieslike alterations. The tongue may have focal traumatic areas, ulcers, and even abscess formation. Mice that are severely cachectic may show degenerative changes in the muscles of the tongue, similar to those seen in the skeletal muscles of other areas of the body, with atrophy secondary to severe emaciation. The salivary glands are a site of interstitial amyloid deposition and frequently have localized aggregates of mononuclear (predominantly lymphoid) cells.

Spontaneous nonneoplastic gastric lesions in mice are common but are often nonspecific or secondary to something else. Dilatation of crypts, mild submucosal fibrosis, mineralization, and erosions or ulcerations of the mucosa may be seen. Many mice that die as a result of a prolonged illness have small erosions on the nonglandular gastric mucosa that are probably secondary to food deprivation and stress. Also, hyperkeratosis of the nonglandular gastric mucosa can occur. A spontaneously occurring lesion of the glandular mucosa of the stomach of strain I mice has been reported by Stewart and Andervont (1938) which was characterized by an adenomatous, hypertrophic, hyperplastic overgrowth of the gastric rugae. Suntzeff and Angeletti (1961) have reported that the intestinal lamina propria of aging mice becomes replaced with coarse, fibrous tissue and that alterations in alkaline phosphatase activity also occur. The significance of the structural and enzymatic changes is unknown. Of more importance is the deposition of amyloid in the stomach and small intestine, which may be accompanied by the development of fetid, yellow-green bowel contents and by diarrhea. The nutritional state of affected mice undoubtedly is compromised, and since the condition is irreversible, death results.

Amyloidosis in the gastrointestinal tract is common in some strains of mice. Morphologically, it is similar in appearance to the amyloid observed in other tissues (Fig. 2). It is usually deposited in the lamina propria and submucosa of the small intestine, stomach, or colon. It may be localized to one portion of the gastrointestinal tract or diffuse throughout all regions.

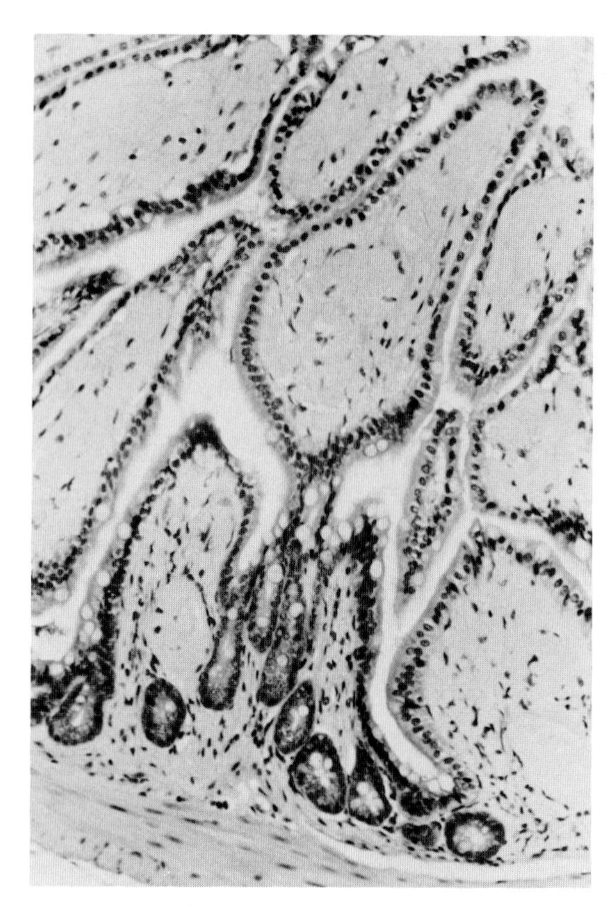

Fig. 2. Amyloid in the lamina propria of the small intestine.

Hyperplastic Peyer's patches or increased aggregations of lymphocytes in the gastrointestinal tissues are seen and should not always be considered neoplastic unless confirmation can be made microscopically. These proliferations probably are associated with some type of stress. Furthermore, Meckel's diverticula and other anomalies have been documented (Kelsall, 1946; Hummel and Chapman, 1959).

C. Pancreas

Exocrine pancreatic insufficiency due to an absence of exocrine tissue resulting in emaciation has been reported in CBA/J mice (Leiter and Cunliffe-Beamer, 1977). Pancreatitis and focal necrosis, seen occasionally, probably represent part of an infectious process and, unless severe, do not affect the well-being of the mouse. Fat cells are commonly interspersed within areas of the pancreatic lobules but may be related to lobular atrophy.

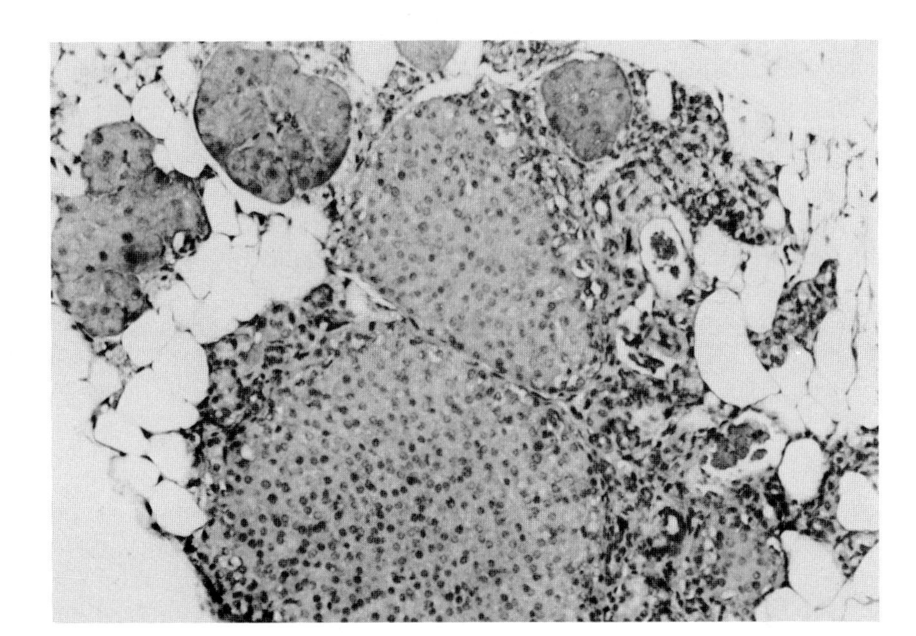

Fig. 3. Severe atrophy of pancreatic lobules. Two islets appear prominent in the center, with only a few acinar structures remaining.

Pancreatic acinar cell atrophy is common (Russfield, 1967), but the incidence is strain and sex dependent. For example, RFM males may have an incidence of 14%, but BALB/c mice have an incidence of less than 1%. In general, it is a common lesion that is present in many strains of mice (Dunn, 1954a) and is characterized by the loss of acinar tissue, often with replacement by fat cells (Fig. 3).

Diabetes mellitus has been documented in several strains of mice (McClure *et al.,* 1978). For example, Coleman and Hummel (1967) determined that an autosomal recessive gene in C57BL/Ks mice led to a degranulation of B cells in the pancreatic islets. This was accompanied by obesity, hyperglycemia, polyuria, and glycosuria. The KK strain of mice have hypertrophy and hyperplasia of the islets and degranulation of B cells (Nakamura and Yamada, 1967). Hyperplastic islets also occur in the pancreas of some mice (Bielschowsky and Bielschowsky, 1956), as do small solitary or multiloculated cystic ducts.

D. Cardiovascular System

Reports on cardiovascular diseases in mice are not as numerous as for other laboratory animals, perhaps because other species provide better accessibility to the cardiovascular system and because only recently has the number of chronic studies been of such magnitude as to permit a greater insight into spontaneous alterations related to aging. Many cardiovascular lesions of mice seem to be age and strain related and have not been reported to contribute significantly to morbidity and mortality.

Thrombosis of either atrium, rarely with extension into a ventricle, is encountered in both sexes. The incidence may be low or as high as 35% in RFM mice (Cosgrove *et al.,* 1978). This lesion may be characterized by enlargement of the atrium, which usually appears pale. Less obtrusive lesions will not be seen unless the atria are opened or tissue sections examined microscopically. The flow of blood through the chamber is sometimes greatly restricted. Consequences may be pulmonary congestion, hydrothorax, enlarged heart, ascites, subcutaneous edema, and unthriftiness (Meier and Hoag, 1975). Although this condition is reported to occur more frequently in older breeding females, we have observed this in virgin males and females of three strains: CDF_1, CD_1, and $B6C3F_1$.

Suppurative pericarditis has been seen in mice occasionally, usually secondary to septic conditions in the thoracic cavity. Hemopericardium has been found and, although unconfirmed, rupture of an aneurysm or atrium as described by Angevine and Furth (1943) has been considered the cause. Dystrophic calcification, seen as pale focal or multifocal areas of white patches or striations of the epicardium, occurs in many strains of mice. A higher incidence has been reported for DBA/2, C, and C3H strains (Hare and Stewart, 1956; DiPaolo *et al.,* 1964), and the possible genetic, hormonal, and dietary factors that may influence the condition have been documented (Eaton *et al.,* 1978). Also, the incidence of mineralization in other soft tissues in DBA/2 mice has been reported (Rings and Wagner, 1972). Replacement of muscle fibers by connective tissue without calcification may be confused grossly with dystrophic calcification. Whereas the first condition may be the last phase of focal myocarditis, the latter is related to genetic susceptibility and to environmental factors (Bellini *et al.,*

1976). A cartilaginous focus may be found at the base of the aortic valve and is similar to that reported in rats (Hollander, 1968).

Vascular lesions such as the severe necrotizing arteritis reported in BL/De mice (Deringer, 1959a) are uncommon, although mild degenerative and inflammatory vascular changes may be found. In mice with lupuslike disease, the vascular degenerative changes may become severe enough to result in myocardial infarction (Accinni and Dixon, 1979). Periarteritis affecting any organ of the body may be found in older mice (Upton *et al.,* 1967). The changes generally involve the outer portions of the vessel wall, although transmural lesions have been reported to occur. Cardiac valvular chondrification, occasionally associated with inflammation, may occur in young adults and older mice (Helper, 1939; Clapp, 1973).

E. Genital System

Evaluation of ovaries in mice requires familiarization with the differences that exist among the various strains since dissimilarities in morphology and function have been reported (Russfield, 1967). Spontaneous ovarian and uterine lesions occur frequently in mice, whereas spontaneous tumors of these organs are reported to occur with different incidences according to the strain of mouse (Deringer, 1959b; Malinin and Malinin, 1972).

Paraovarian cysts filled with straw-colored or red fluid have been observed frequently, and their occurrence may be related to the fact that the mouse ovaries are enclosed in membranous pouches (Marchant, 1977). Ovarian amyloid deposits that characterize some strains, CD_1 for instance, may almost obscure the identity of the ovary. Brown degeneration (ceroid) of cells is common and has been reported (Russfield, 1967; Deane and Fawcett, 1952). Fig. 4 illustrates the appearance of an ovary from a $B6C3F_1$ mouse with atrophy and a small cyst.

Cystic endometrial hyperplasia is frequently encountered (Sheldon and Greenman, 1979; Cosgrove *et al.,* 1978; Malinin and Malinin, 1972; Clapp, 1973). It occurs in one or both uterine horns and may be discernible as a segmental lesion of the uterus. In many strains of mice, nearly every female older than 18 months will have some degree of endometrial hyperplasia. Severe cases are apparent grossly. Microscopically, the glands are dilated and show an irregular cystic proliferation (Fig. 5). Increased mucus and some inflammation may be present in a few mice with this endometrial lesion. Furthermore, the endometrial proliferation is often associated with ovarian atrophy. Hydrometra or pyometra, although reported in mice, has not been frequently observed in our experience. The same is true for uterine atrophy.

Preputial and clitoral gland abscessation occurs in mice from about 1 month of age and older. These nodules are usually visible externally and may be confused with tumors. The health of the mouse is unaffected unless extension to the exterior, urethra, prostate, seminal vesicles, or other adjacent structures occurs (Hong and Ediger, 1978). Although the functional significance of preputial glands is unknown, Chipman and Albrecht (1974) have reported that glandular secretion may be related to synchronization of estrus.

Spontaneous nonneoplastic lesions affecting the mouse testes consist of atrophy (Fig. 6), seminiferous tubular degeneration, sperm granulomas, and tubular calcification.

Prostatitis and seminal vesiculitis do occur and are usually mild. These inflammations are generally secondary to infecti-

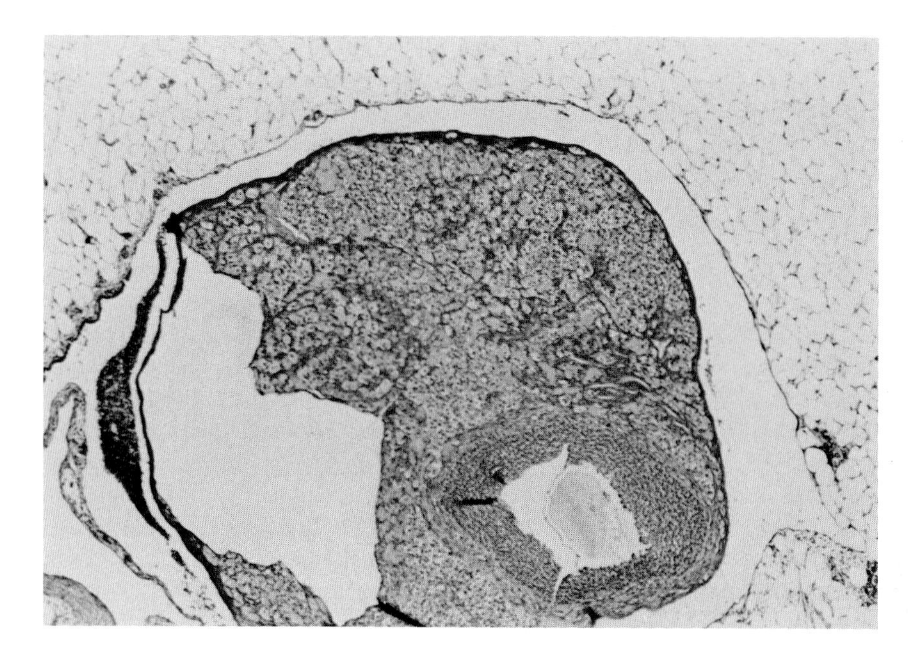

Fig. 4. Atrophic ovary containing a small cyst.

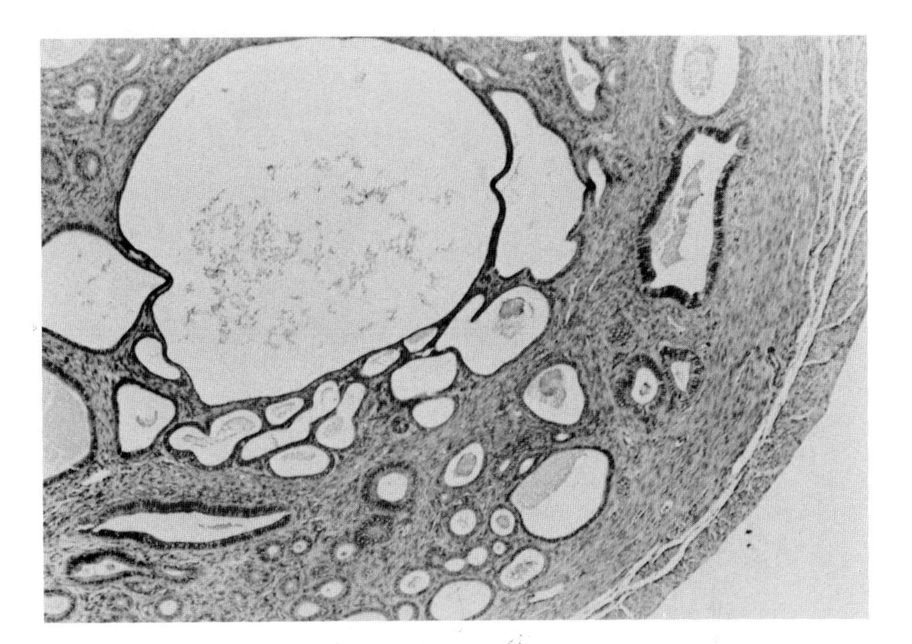

Fig. 5. Cystic endometrial hyperplasia in the uterus. Uterine glands are dilated, cystic, and often fluid-filled.

ous conditions of the urethra or to lesions in the inguinal-perineal tissue and frequently secondary to fighting or to an infectious agent such as *Proteus* sp. or *Pseudomonas* sp. We have seen hyperplasia and a neoplasm of the prostate in the CDF_1 strain, the latter a rare spontaneous alteration. Unilateral or bilateral atrophy or hypertrophy occurs in seminal vesicles of young adult and older mice in both poor and good physical condition.

F. Urinary System

Spontaneous renal disease in mice appears to be relatively infrequent in the mouse strains used for most toxicity and carcinogenicity testing, and the alterations most frequently reported seem related to strain. However, some strains of mice routinely develop varying degrees of glomerular, tubular, and interstitial lesions. For example, NZB (Kelley and Winkelstein, 1980) and C57BL/6 (Linder *et al.*, 1972) mice develop immune complex glomerulonephritis. NZB/NZW hybrid mice also have a high incidence of interstitial nephritis (Hurd and Ziff, 1978). Also, Rudofsky (1978) has reported renal tubulointerstitial lesions in CBA/J mice that did not resemble known spontaneous or induced renal lesions. Spontaneous glomerulonephritis, particularly in RF mice, has been reported (Gude and Upton, 1960). It has also been reported in other

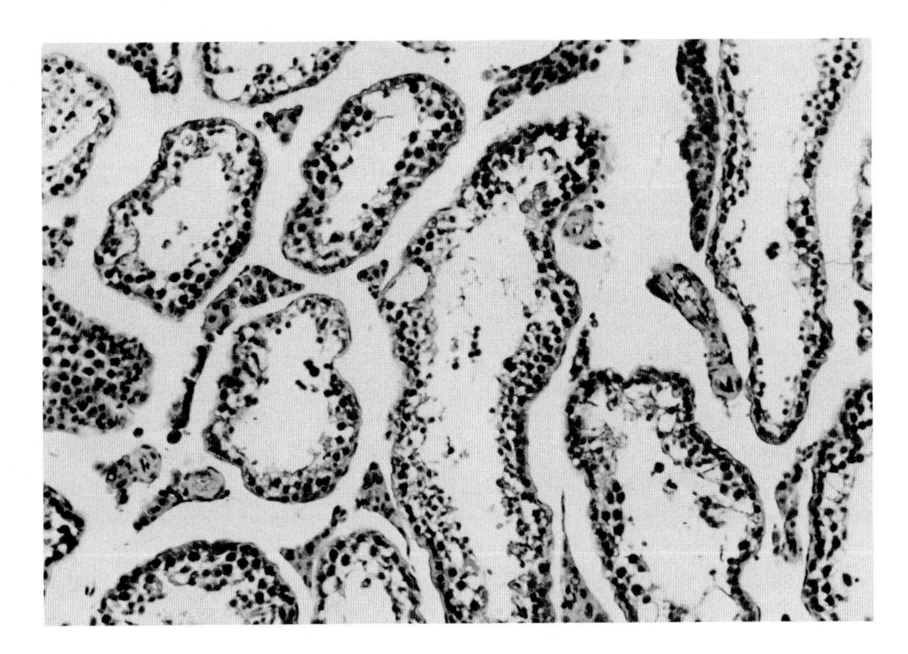

Fig. 6. Severe testicular atrophy.

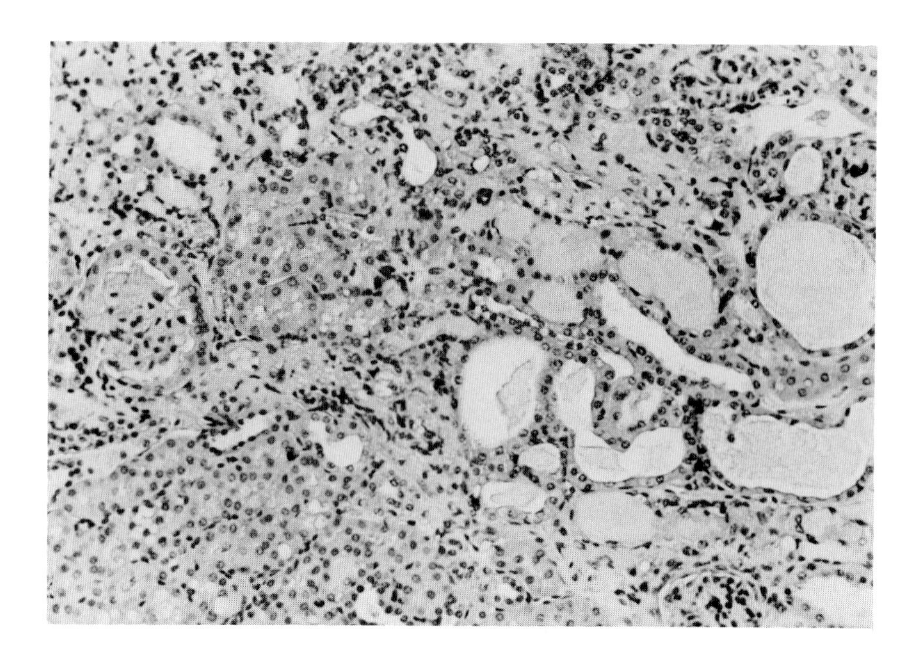

Fig. 7. Kidney with moderate chronic glomerulonephropathy. Renal tubules are dilated, often with eosinophilic material in their lumina. Fibrous connective tissue and a few lymphocytes are present in the interstitial tissue.

strains (Dunn, 1967b), but often at a very low incidence (Ward *et al.,* 1979). Fig. 7 illustrates a moderate degree of glomerulonephropathy in a B6C3F$_1$ mouse. Ascending pyelitis may occur in male mice affected with lower urinary tract infections resulting from fight wounds (Tuffery, 1966), and occasionally renal abscessation may be found in mice with systemic bacterial infections. The most severe glomerular lesions are those associated with amyloidosis, although equally severe changes may be found in the tubules, papilla, or both (Heston and Deringer, 1948; Dunn, 1967a). In some studies, nearly all LLC mice over 1 year of age have been found to have renal amyloidosis (Chai, 1978), as have KK mice (Soret *et al.,* 1977). Perinephritic lesions, not uncommon, are generally related to an infectious process or to malignant lymphoma. Focal interstitial nephritis and a leukemoid reaction are not infrequent findings, and the latter must be recognized as a nonneoplastic condition.

Obstructive uropathy occurs in various strains of mice. We have observed a high incidence in the CDF$_1$ strain. The condition may be characterized by extreme dilatation of the bladder, to a degree that may result in rupture of the organ. Tightly wedged, pale plugs may be found in the neck of the bladder or more distal in the urethra, and back pressure may result in dilatation of the renal pelvis and tubules. Secondary inflammation may involve the urethra and periurethral tissue with extension to adjacent tissues (Babcock and Southan, 1965).

Hydronephrosis may occur as a unilateral or bilateral disease. The incidence can vary from $<1\%$ in B6C3F$_1$ mice (Ward *et al.,* 1979) to 32% in other strains (Collins *et al.,* 1972). It is characterized by slight, moderate, or severe dilatation of one or both kidneys.

G. Endocrine System

1. Pituitary

Enlargement, degranulation, and vacuolation of basophils of the pituitary is seen in the mouse after castration (Anderson and Capen, 1978) and after thyroidectomy (Turner and Bagnara, 1971). Microscopic cysts may be seen between the anterior and posterior lobes of the pituitary of old mice (Dunn, 1944). In addition, focal or multifocal, small, hyperplastic foci may occur that are characterized by cells that vary from surrounding cells in their morphological and tinctorial appearance (Fig. 8).

2. Adrenal

Accessory adrenocortical nodules are commonly found in the periadrenal or perirenal fat of the mouse, the incidence being higher in females of all strains and varying considerably among strains (Russfield, 1967). Hummel (1958) found nodules in more than half of the mice of strains C57BL, C58, and BALB/c and in about one-half of the mice of strains A and C3H. The accessory adrenocortical nodules were generally found on the left side and resembled normal adrenal cortex but did not contain medullary tissue. Aging changes characteristic of the adrenal cortex proper, such as ceroid deposition or brown degeneration, may be found in the nodules. The accessory nodules have little practical significance other than the fact that their presence may account for an apparent failure surgically to adrenalectomize some mice.

Various degenerative changes may occur in the adrenals of

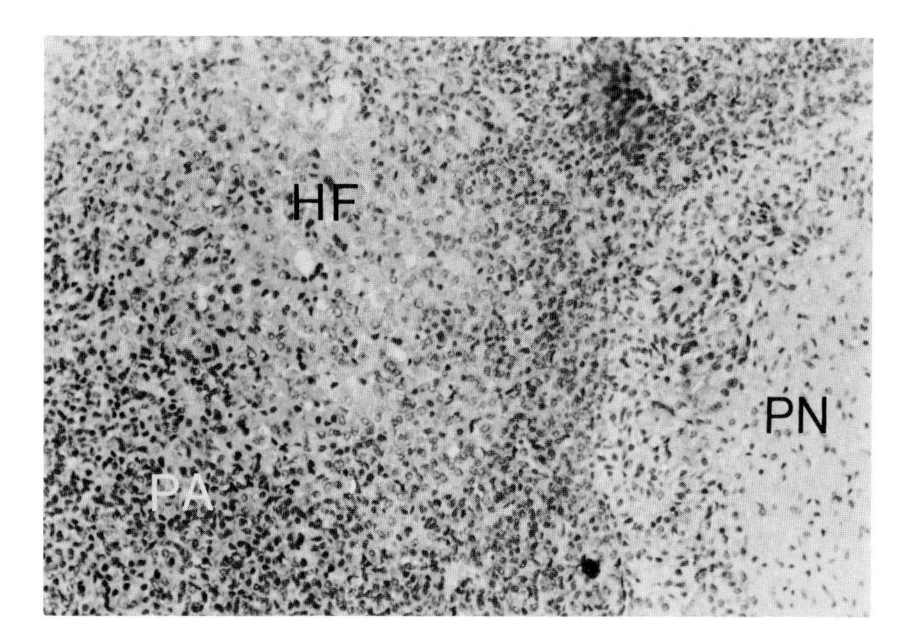

Fig. 8. Small hyperplastic focus in the pars anterior of the pituitary. PA, pars anterior; PN, pars nervosa; HF; hyperplastic focus.

mice. Hypoplasia or atrophy of the fetal adrenal can be produced by maternal corticosteroids crossing the placenta and causing a negative feedback inhibition of ACTH secretion by the fetal pituitary (Anderson and Capen, 1978). Brown degeneration or ceroid deposition is often found in the adrenals of old mice as a bank of pigment-containing cells between the medulla and cortex (Dunn, 1970). The brownish pigment is strongly fuchsinophilic. Chemically, the pigment is believed to be ceroid, which is formed by polymerization and peroxidation of unsaturated fatty acids released within degenerating adrenocortical cells (Bern *et al.*, 1958). Ceroid deposition is especially prominent in BALB/c mice and is increased if the mice recieve estrogens. High-fat diets may also increase its frequency, whereas administration of vitamin E decreases its occurrence. Similar deposits of ceroid may be seen in degenerating cells of the corpus luteum of the ovary or the interstitial cells of the testis (Russfield, 1967).

Cystic dilatation of the cortical sinusoids of the adrenal is frequently observed in old mice as unevenly colored, gross foci of the adrenal surface (Furth and Sobel, 1946; Dunn, 1965). The dilated sinusoids contain red blood cells or a coagulum of red cells, fibrin, and plasma. This alteration and similar changes in the liver, spleen, and kidneys were attributed by Dunn (1965) to increase blood volume or hypervolemia.

Subcapsular spindle cell hyperplasia is a common finding in the adrenal cortex of old mice of various strains. Spindle cells which are fibroblastic in appearance are found radiating from the adrenal capsule for varying distances into the layers of the cortex (Fig. 9). The condition is more prominent in females than males in the age group of 300–499 days, probably reflecting a diminution in ovarian activity. In older mice (over 600

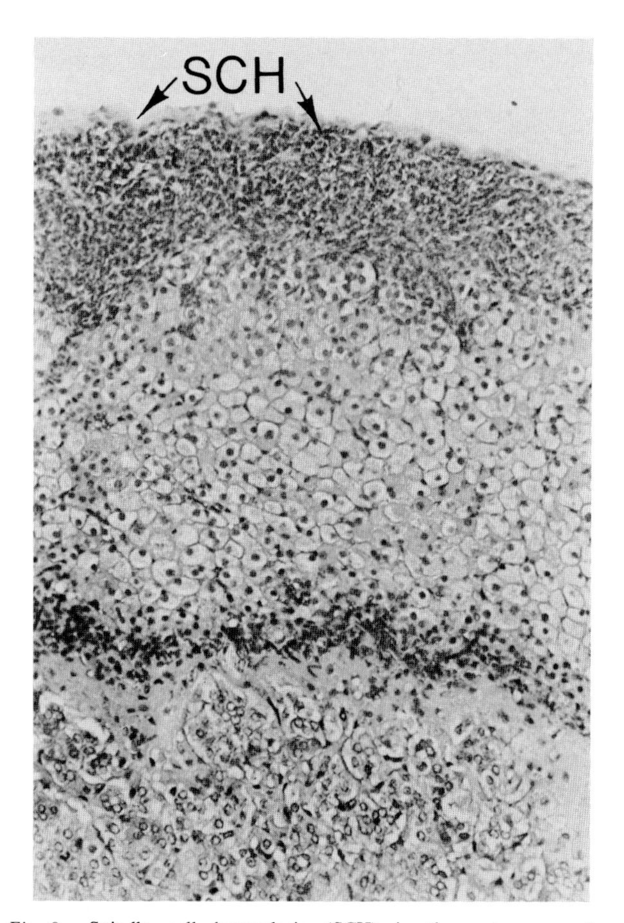

Fig. 9. Spindle cell hyperplasia (SCH) in the outer zone (zona glomerulosa) of the adrenal cortex.

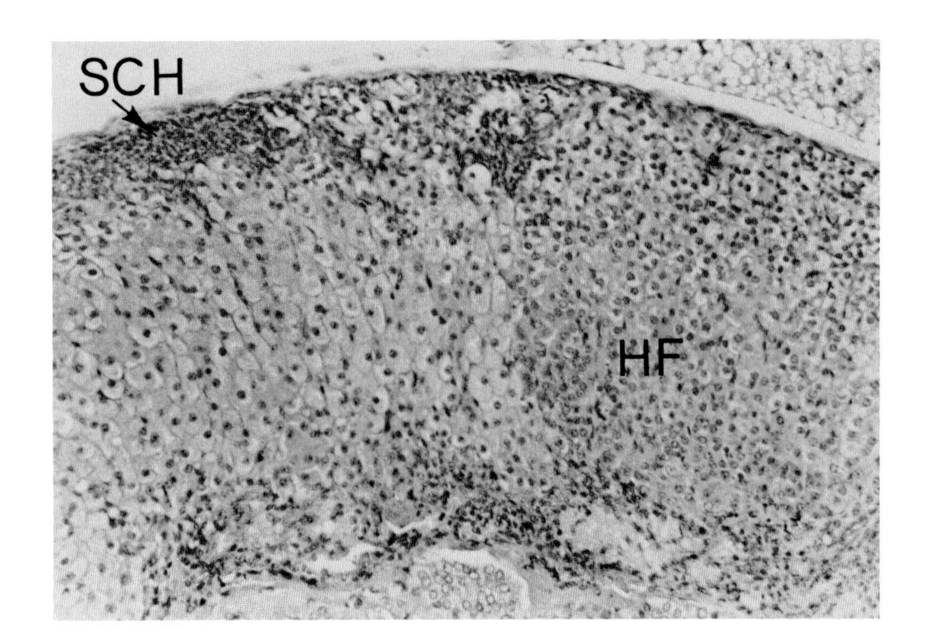

Fig. 10. Hyperplastic focus (HF) in the cortex of the adrenal cortex. A slight degree of spindle cell hyperplasia (SCH) is also present in the zona glomerulosa.

days), the difference between the sexes is less notable (Dunn, 1970). Gonadectomy has been shown to cause hyperplasia in both sexes, and in the BALB/c, HH, and C3H mouse strains, adenomas or carcinomas appear to result from proliferation of these cells. In addition, hyperplastic foci occur in the other zones of the adrenal cortex. These are characterized by aggregates of cortical cells that vary in size and tinctorial characteristics from surrounding tissue (Fig. 10).

3. Thyroid

Atrophy of the thyroid associated with a decrease in colloid may be seen in Snell's (dw/dw) and Ames (df/df) dwarf mouse strains which have a deficiency of thyrotropic hormone (Anderson and Capen, 1978).

Cysts lined by ciliated epithelial cells and containing a pale, foamy cytoplasm (Dunn, 1944) or, alternatively, cystic structures lined by stratified squamous epithelium and generally of ultimobranchial origin (Russfield, 1967) may be seen in old mice.

Birefringent crystals typical of calcium oxalate may occur in the thyroid follicles, particularly in mice of strain C3H (Dunn, 1965).

4. Parathyroid

Microscopic cysts lined by flattened or cuboidal cells, sometimes ciliated, have been found in the parathyroid (Dunn and Andervont, 1963).

Dendritic melanocytes, not to be confused with abnormal pigment, may be present in the stroma of the parathyroid of heavily pigmented mouse strains such as the C58 strain (Dunn, 1965).

Amyloid deposits in various endocrine glands are not an uncommon finding in various strains of mice. Homburger *et al.* (1975) reported involvement of the adrenal in 34% of males and 41% of females of CD-1 HaM/ICR mice. Deposits of amyloid in the adrenal are generally restricted to the inner cortex. Amyloid deposits in the thyroid and parathyroid involve the interstitium in a diffuse fashion. Amyloid deposition in other organs, particularly the intestine, kidney, stomach, spleen, liver, heart, and lungs, has frequently been found in conjunction with deposits in endocrine glands.

H. Integumentary System

Nonneoplastic lesions of the skin and associated structures are rare except for infectious agents, such as mites and bacteria, and trauma. For example, focal superficial skin lesions in the form of pustules or crusts may be seen in mice, particularly in males, as a result of fighting. These lesions are generally found over the back or in the perineal region and lower abdomen (Tucker and DeC. Baker, 1967).

Abscesses of the preputial glands, which are collections of specialized sebaceous gland cells in the subcutaneous tissue on either side of the prepuce or the vagina, are commonly seen in male mice of various strains. The abscesses are generally observed externally as unilateral or bilateral nodular enlargements at the base of the penis. *Staphlococcus aureus* has been isolated from a majority of the affected animals (Hong and Ediger, 1978), but *Proteus* sp. and *Pseudomonas* sp. may also be factors. Fighting among males may be a contributory factor to this condition (Tucker and DeC. Baker, 1967).

Bilaterally uniform alopecia of the skin of the face and back as a result of friction contact with feeding jars and bars on the cages should not be confused with an infectious or toxic manifestation of hair loss. Also, alopecia of unknown etiology may occur in the abdominal and thoracic regions of $B6C3F_1$ mice.

Because the mouse has been used extensively for cancer induction by painting the skin with potential carcinogens, the normal skin and integument have been studied in detail. For example, the distribution of mast cells has been described (Simpson and Hayashi, 1960) and the patterns of hair growth and hair cycles have been reported (Argyris, 1963).

I. Musculoskeletal System

Dystrophic calcification of soft tissue occurs frequently in old mice, a high incidence being reported in DBA/2JN mice (Sokoloff, 1967). Focal calcification is seen most often in the myocardium of the left ventricle and interventricular system but may affect other tissues as well, including skeletal muscle, kidneys, arteries, and the lung. Fibrosis is associated with calcification in the heart, whereas fibrosis is minimal in the skeletal muscle and an inflammatory response with local phagocytosis predominates instead. Feeding diets high in calcium have increased the incidence of dystrophic calcification (Highman and Daft, 1951).

Bone necrosis may occur in mice in association with fractures. Necrosis of the bone of the femur, tibia, or sternum may also occur in the absence of fractures. Sokoloff and Habermann (1958) reported focal necrosis of the femur or tibia in 4.6% of mice in one study, the condition being more prevalent in females and showing considerable variation among mouse strains. Aseptic necrosis in the region of the epiphysis may result in fracture of the head of the femur in old female mice. Necrosis of the knee bones also occurred in a large proportion of older mice. Necrosis of the sternum has been reported in RF/Un mice 15 months of age and older (Clapp, 1973), and we have observed it in CD_1 mice of similar age. The lesion is characterized grossly by nodular, grayish-yellow enlargement of intersternebral areas and microscopically by aseptic coagulation necrosis of cartilage and adjoining bone.

Age-associated osteoporosis or senile osteodystrophy occurs in the bones of some aging mice (Krishna Rao and Draper, 1969; Silberberg and Silberberg, 1956, 1962; Sass and Montali, 1980). It resembles osteoporosis in humans, with a decrease in the amount of bone present, loss of bone marrow in the affected area, osteoclastic activity, and fibrosis (Fig. 11). Superficially, this lesion may be confused with osteodystrophy fibrosis caused by severe renal disease. However, this lesion is not associated with parathyroid hyperplasia and is not correlated with any other gross of histopathologic alteration. It is observed in both males and females but is more often found in females.

Nearly all strains of mice develop some form of osteoarthrosis. The disease is a noninflammatory degenerative disease of articulating surfaces with secondary bone degeneration and thickening. It is more common in males, although the incidence varies greatly from strain to strain (Sokoloff and Jay, 1955a,b). Furthermore, the pathogenesis has been described

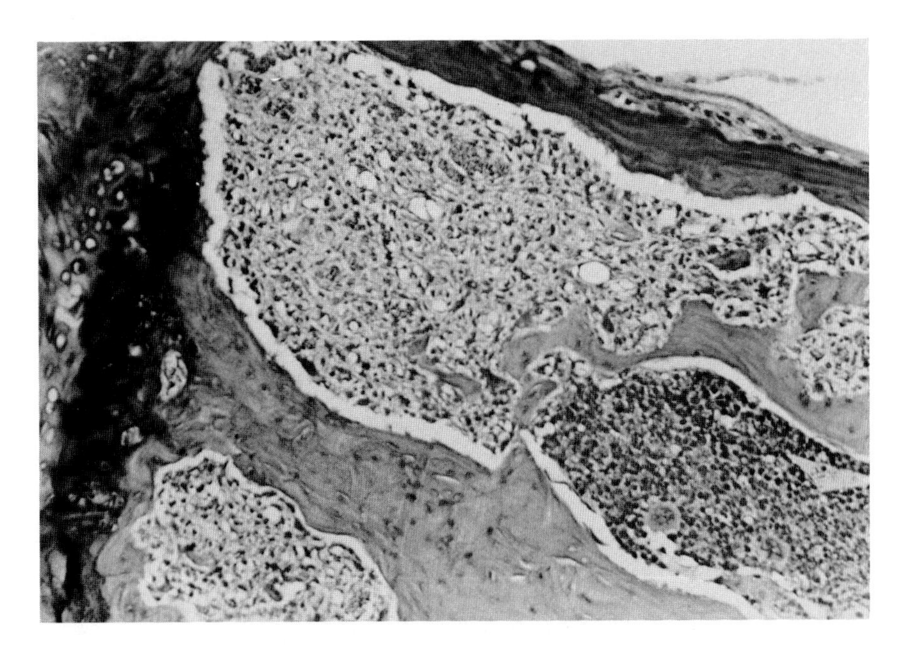

Fig. 11. Section of vertebral bone with age-associated osteoporosis (senile osteodystrophy). It is characterized by the loss of bone marrow, evidence of osteoelastic activity, and replacement of the marrow by fibrous connective tissue.

and the knee joint was found to be most commonly affected (Sokoloff, 1956). Walton (1977a,b, 1978, 1979) studied the joints of STR/ORT mice with this disease and reported the radiological and pathological features of the disease.

Fractures of the tibia and fibula, particularly of the midportion, occur in a small percentage of mice, and housing mice in cages with wire-meshed screen floors may increase the incidence of such fractures (Sokoloff, 1958).

J. Respiratory System

Focal or multifocal pulmonary adenomatosis, a hyperplasia of alveolar or bronchiolar epithelium characterized microscopically by the presence of mucus-containing, ciliated cuboidal or columnar cells lining the alveolar septae (Fig. 12), is a lesion that must be distinguished from alveologenic tumors. Grossly, the pleural surface overlying a lesion of adenomatosis is generally retracted, whereas alveologenic tumors, when superficial, elevate the pleural surface (Horn *et al.*, 1952); Stewart *et al.*, 1970). Microscopically, the presence of mucus can be demonstrated in the adenomatoid lesion but not in similar sites of alveologenic tumors. Adenomatosis can occur spontaneously or be induced experimentally by chemicals and spontaneously by certain viruses (Zurcher *et al.*, 1977).

Crystalline deposits have been observed in the lungs of normal and cancerous Swiss mice (Green, 1942). Dunn and Andervont (1963) reported crystals to be particularly prevalent in the lungs of wild mice. Furthermore, Ward (1978) described pneumonia in moth-eaten mice with needlelike crystals in alveolar macrophages.

Pulmonary edema in mice can be produced experimentally by subcutaneous administration of adrenalin (Wang *et al.*, 1970). It may also occur spontaneously as a result of congestive heart failure, as seen with atrial thrombosis in mice, as a result of capillary damage by anoxia or various irritating gases or chemicals, or pulmonary edema may simply represent an agonal or postmortem occurrence. Gross observations, including the relative weight of the lung, are probably more valid in substantiating a diagnosis of pulmonary edema than microscopic findings (Stookey and Moe, 1978). Amyloid deposits are found in the alveolar septae as one of the first sites of deposition in strain A mice (West and Murphy, 1965). We have observed the deposition of amyloid in the wall of the pulmonary arterial vessels as a frequent finding in CD_1 mice with systemic amyloidosis.

Netteschein and Martin (1970) have reported on glandlike structures in the trachea of aging mice.

K. Nervous System

One of the most common nonneoplastic lesions in mice is mineral deposits in the brain. These deposits are most common as focal lesions in the thalamic region and are usually bilaterally symmetrical; however, they may be multifocal and occur in the midbrain, cerebellum, and cerebrum (Fig. 13). They occur in the A/J mouse (Coburn *et al.*, 1971) and others as well (Tucker and DeC. Baker, 1967). Often, these small mineral deposits appear to be associated with blood vessels, but their pathogenesis is unknown, although they appear to increase with age.

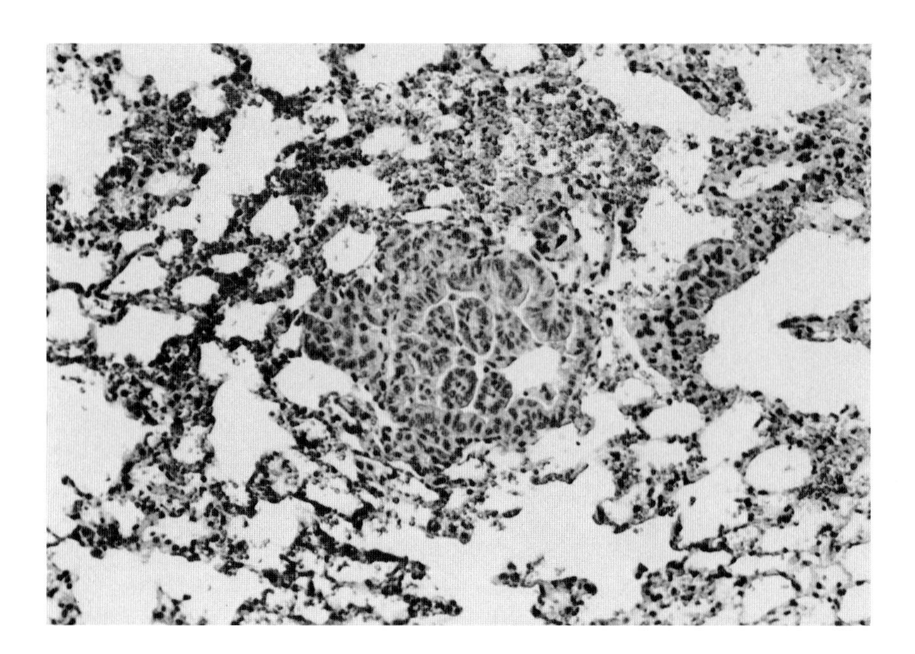

Fig. 12. Focal hyperplasia of the bronchiolar epithelium in the lung.

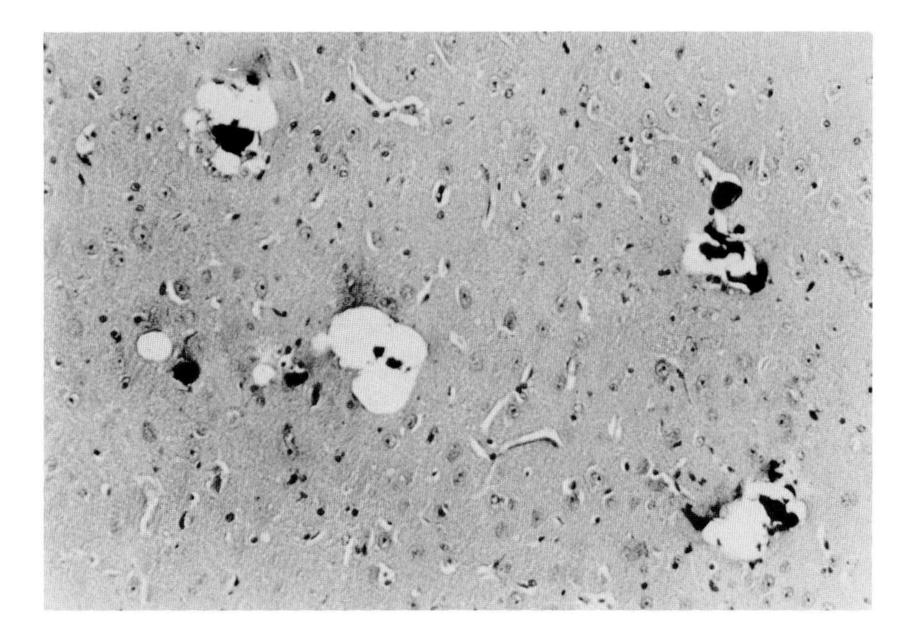

Fig. 13. Multifocal mineral deposits in the brain. Such lesions are usually bilaterally symmetrical.

Lipofuscin pigment appears as a golden-brown to yellowish pigment within the cytoplasm of neurons. The distribution in B6 mice has been documented (Reichel *et al.,* 1967, 1968), but it is very common and can be found in neurons of most aged mice. Melanin pigment may also be seen in pigmented mice. The distribution varies, but it is often found in the meninges and may extend into the brain, especially in the frontal lobes.

A number of miscellaneous conditions or lesions have been described in various mouse strains. They are often unique to a specific strain or stock, or occur as incidental findings. A few such findings are listed here. Neurological mutant mice have also been described (Garner *et al.,* 1967). Absence of the corpus callosum, pseudoencephaly (anencephaly), congenital and acquired hydrocephalus, ataxic ''paralytic'' mice, myelin deficiencies (DBA/2J and C3H/DI mice), leukoencephalosis, vacuolization, and epidermal (dermoid) cysts have all been documented (Garner *et al.,* 1967). Observations of epidermal (dermoid) cysts in the spinal cord (Fig. 14) are not unusual if multiple sections are examined. Age-associated peripheral

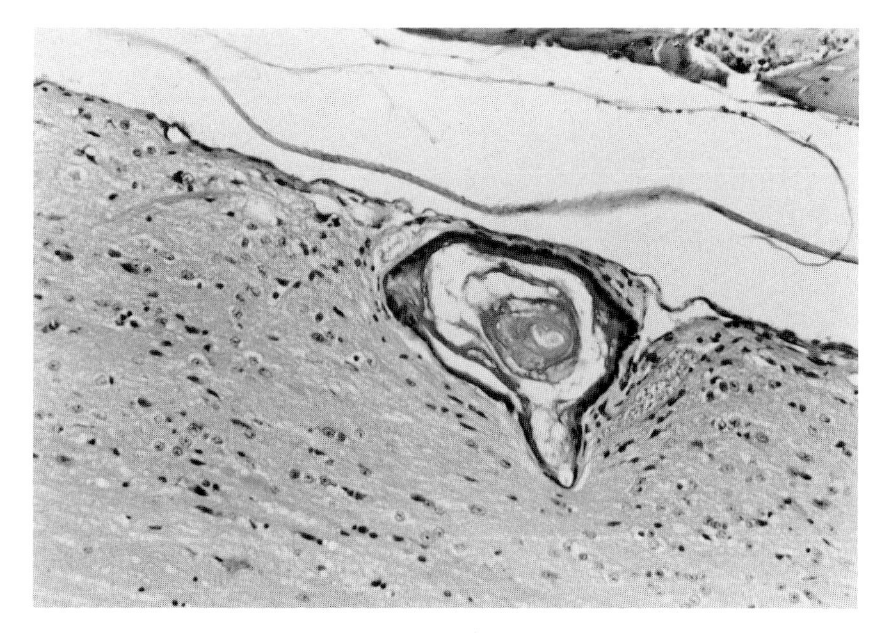

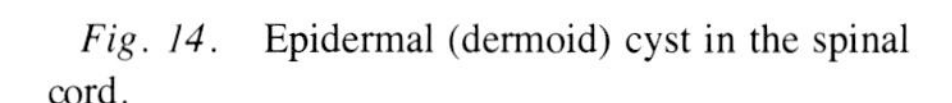

Fig. 14. Epidermal (dermoid) cyst in the spinal cord.

neuropathy may be found in the nerves of the hindlimbs in B6 mice. It is characterized by demyelinization and nerve axon degeneration, varying from very slight to severe.

L. Hematopoietic (Lymphoreticular) System

A number of nonneoplastic lesions are observed in the mouse hematopoietic system, including hyperplasia, lympoid depletions, and inflammatory changes. Many are secondary to infectious or inflammatory disease, whereas others are due to unknown factors. Most were alluded to by Dunn (1954b) in her paper on the mouse reticular tissue.

One of the most common lesions is the presence of focal or multifocal aggregates or nodules of mononuclear (predominantly lymphoid) cells in many tissues. They may be perivascular or may appear as a cluster of cells unassociated with a particular site. They occur commonly in or next to organs such as the salivary glands, thymus, ovaries, uterus, mesentery, mediastinum, urinary bladder, and gastrointestinal tract. At times, these aggregates become very prominent and may resemble lymphosarcomas. As a result, these aggregates often lead to a difficult diagnostic challenge in order to determine where hyperplasia ends and when the proliferation is extensive enough to warrant a diagnosis of lymphosarcoma.

In some lymph nodes (often those in the cervical region), there is an increase in the plasma cells in the medullary cords. Lymph nodes may show cystic sinusoidal dilatation which may be so extensive that the nodal architecture is destroyed and all that remains is a cyst. Crystal deposits, similar to those described in the lungs, may be formed in the lymph nodes of mice with a predisposition for such crystal formation. As in the lungs, they are usually associated with macrophages (Ward, 1978). Pigment (especially hemosiderin) is also common inside and outside macrophages in lymph nodes or the spleen.

"Mesenteric disease" (Dunn, 1954b) is common in C3H mice but may occur in many strains and is characterized by mesenteric lymph nodes that contain large, cystic, sinusoidal spaces filled with blood.

The bone marrow of mice remains active even in older animals. As a result, depletion of marrow elements is not common but is usually easier to diagnose than hyperplasia. However, age-associated osteoporosis (see Section II, I) results in portions of the bone marrow being replaced by fibrous connective tissue. As a result, less marrow is present, but that which remains appears normal.

The spleen, like the lymph nodes, may have a number of nonneoplastic lesions. Lesions include lymphoid depletion, lymphoid hyperplasia, increased extramedullary hematopoiesis, hemosiderin pigment, and necrosis, as well as inflammatory changes. Amyloid or amyloid-like material is often deposited in the spleen first, in the marginal zone of the white pulp, and in these cases it may be nodular or diffuse. In addition, lipofuscin has been reported in C57BL mice, resulting in discoloration in the anterior region of the spleen (Crichton *et al.*, 1978).

The thymus undergoes an age-associated atrophy with a decrease in lymphoid elements. Small cysts may remain in the atrophic gland.

M. Eyes

Cataracts occur infrequently as unilateral or bilateral lesions (Saunders, 1967), except in certain mutant strains with dominant or recessive cataracts which are utilized in studies on models of cataractogenesis (see Chapter 62, Section IV). Several strains of mice have an inherited retinal degeneration resulting from the *rd* gene (Chapter 62, Section III). Otherwise, retinal lesions are rare in most strains, and inherited ocular lesions are well documented (Saunders, 1967, Chapter 62).

REFERENCES

Abbey, H. (1979). Survival characteristics of mouse strains. *In* "Development of the Rodent as a Model System of Aging" (D. C. Gibson, C. C. Adelman, and C. Finch, eds.), DHEW Pub. No. (NIH) 79-161, pp. 1-17. U.S. Dep. Health, Educ., Welfare, Natl. Inst. Health, Bethesda, Maryland.

Accinni, L., and Dixon, F. J. (1979). Degenerative vascular disease and myocardial infarction in mice with lupus-like syndrome. *Am. J. Pathol.* **66,** 477-492.

Anderson, M. P., and Capen, C. C. (1978). The endocrine system. *In* "Pathology of Laboratory Animals" (K. Benirschke, F. M. Garner, and T. C. Jones, eds.), pp. 423-508. Springer-Verlag, Berlin and New York.

Andrew, W., Brown, H. M., and Johnson, J. B. (1943). Senile changes in the liver of the mouse and man with special reference in similarity of the nuclear alterations. *Am. J. Anat.* **72,** 199-221.

Angevine, D. M., and Furth, J. (1943). A fatal disease of middle-aged mice characterized by myocarditis associated with hemorrhage in the pleural cavity. *Am. J. Pathol.* **19,** 187-196.

Argyris, T. S. (1963). Hair growth cycles and skin neoplasia. Conference on Biology of Cutaneous Cancer. *Natl. Cancer Inst. Monogr.* **10,** 33-41.

Babcock, V. J., and Southan, C. M. (1965). Obstructive uropathy in laboratory mice. *Proc. Soc. Exp. Biol. Med.* **120,** 580-581.

Bellini, O., Casazza, A. M., and DiMarco, A. (1976). Histological and histochemical studies of myocardial lesions in BALB/c/Cr mice. *Lab. Anim. Sci.* **26,** 329-333.

Bern, H. A., Nandi, S., Campbell, R. A., and Rissoti, L. E. (1958). The effects of hormones and other agents on weight changes and on ceroid deposition induced by estrogen administration and by hypophysectomy on the adrenal glands of BALB/c rgl mice. *Acta Endocrinol. (Copenhagen)* **31,** 349-383.

Bielschowsky, M., and Bielschowsky, F. (1956). The New Zealand strain of obese mice: Their response to stilbeisterol and to insulin. *Aust. J. Exp. Biol. Med. Sci.* **34,** 181-198.

Blankwater, M. J. (1978). "Ageing and the Humoral Immune Response in Mice." Publ. Inst. Exp. Gerontol. Organ. Health Res. TNO, Rijswijk, Netherlands.

Butler, W. H., and Newberne, P. M. (1975). "Mouse Hepatic Neoplasia." Elsevier, Amsterdam.

Chai, C. K. (1978). Spontaneous amyloidosis in LLC mice. *Am. J. Pathol.* **90,** 381-398.

Chipman, R. K., and Albrecht, E. D. (1974). The relationship of the male preputial gland to the acceleration of oestrus in the laboratory mouse. *J. Reprod. Fertil.* **38,** 91-96.

Clapp, N. K. (1973). "An Atlas of RF Mouse Pathology: Disease Descriptions and Incidences." USAEC, Tech. Inf. Cent., Oak Ridge, Tennessee.

Coburn, A. F., Galy, R. M., and Rivers, S. M. (1971). Observations of the relation of heart rate, lifespan, weight and mineralization in the digoxin-treated A/J mouse. *Johns Hopkins Med. J.* **128,** 169-193.

Coleman, D. C., and Hummel, K. P. (1967). Studies with the mutation diabetes in the mouse. *Diabetologia* **3,** 238-248.

Collins, G. R., Goodheart, C. R., and Hanson, D. (1972). Spontaneous heritable hydronephrosis in inbred mice. 1. Description, incidence, and distribution of lesions. *Lab. Anim. Sci.* **22,** 333-338.

Cosgrove, G. E., Satterfield, L. C., Bowles, N. D., and Klima, W. C. (1978). Diseases of aging untreated virgin female RFM and BALB/c mice. *J. Gerontol.* **33,** 178-183.

Crichton, D. N., Busuttil, A., and Price, W. H. (1978). Splenic lipofuscinosis in mice. *J. Pathol.* **126,** 113-120.

Deane, W. H., and Fawcett, D. W. (1952). Pigmented interstitial cells showing "brown degeneration" in the ovaries of old mice. *Anat. Rec.* **113,** 239-243.

Deringer, M. K. (1959a). Necrotizing arteritis in strain BL/De mice. *Lab. Invest.* **8,** 461-465.

Deringer, M. K. (1959b). Occurrence of tumors, particularly mammary tumors in agent-free strain C_3 HeB mice. *J. Natl. Cancer Inst.* **22,** 995-1002.

DiPaolo, J., Stron, L., and Moore, G. (1964). Calcareous pericarditis in mice of several genetically related strains. *Proc. Soc. Exp. Biol. Med.* **115,** 496-497.

Dunn, T. B. (1944). Ciliated cells of the thyroid of the mouse. *J. Natl. Cancer Inst.* **4,** 555-557.

Dunn, T. B. (1954a). The importance of differences in morphology in inbred strains. *J. Natl. Cancer Inst.* **15,** 573-585.

Dunn, T. B. (1954b). Normal and pathologic anatomy of the reticular tissue in laboratory mice with a classification and discussion of neoplasms. *J. Natl. Cancer Inst.* **14,** 1281-1433.

Dunn, T. B. (1965). Spontaneous lesions of mice. *In* "Pathology of Laboratory Animals" (W. E. Ribelin and J. R. McCoy, eds.), pp. 303-320. Thomas, Springfield, Illinois.

Dunn, T. B. (1967a). Renal disease of the mouse. *In* "Pathology of Laboratory Rats and Mice" (E. Cotchin and F. J. C. Roe, eds.), pp. 149-179. Blackwell, Oxford.

Dunn, T. B. (1967b). Amyloidosis in mice. *In* Pathology of Laboratory Rats and Mice" (E. Cotchin and F. J. C. Roe, eds.), pp. 181-212. Blackwell, Oxford.

Dunn, T. B. (1970). Normal and pathologic anatomy of the adrenal gland of the mouse, including neoplasms. *J. Natl. Cancer Inst.* **44,** 1323-1389.

Dunn, T. B., and Andervont, H. B. (1963). Histology of neoplasms and some nonneoplstic lesions found in wild mice. *J. Natl. Cancer Inst.* **31,** 873-901.

Eaton, G. J., Custer, R. P., Johnson, F. N., and Stabenow, K. T. (1978). Dystrophic cardiac calcinosis in mice. Genetic hormonal and dietary influence. *Am. J. Pathol.* **90,** 173-186.

Furth, J., and Sobel, H. (1946). Hypervolemia secondary to grafted granulosa cell tumor. *J. Natl. Cancer Inst.* **7,** 103-113.

Garner, F. M., Immes, J. R. M., and Nelson, D. H. (1967). Murine neuropathology. *In* "Pathology of Laboratory Rats and Mice" (E. Cotchin and F. J. C. Roe, eds.), pp. 295-348. Blackwell, Oxford.

Green, E. U. (1942). On the occurrence of crystalline material in lungs of normal and cancerous swiss mice. *Cancer Res.* **2,** 210-217.

Gude, W. D., and Upton, A. C. (1960). Histological study of spontaneous glomerular lesions in aging RF mice. *Am. J. Pathol.* **40,** 699-709.

Hare, W., and Stewart, H. (1956). Chronic gastritis of the glandular stomach, adenomatous polyp of the duodenum, and calcareous pericarditis in strain DBA mice. *J. Natl. Cancer Inst.* **16,** 889-911.

Helper, W. C. (1939). Cartilaginous foci in the hearts of white rats and of mice. *Arch. Pathol.* **27,** 466-468.

Heston, W. E., and Deringer, M. K. (1948). Hereditary renal disease and amyloidosis in mice. *Arch. Pathol.* **46,** 49-58.

Highman, B., and Daft, F. S. (1951). Calcified lesions in C3H mice given purified low protein diets. Tissues involved: Heart, skeletal muscle, arteries and lungs. *AMA Arch. Pathol.* **52,** 221-229.

Hollander, C. F. (1968). Cartilaginous focus at the base of the noncoronary semilunar valve of the aorta in rats of different ages. *Exp. Gerontol.* **3,** 303-307.

Hollander, C. F. (1975). Embryology and aging effects. *In* "House Hepatic Neoplasia" (W. H. Butler and P. M. Newberne, eds.), pp. 7-19. Elsevier, Amsterdam.

Homburger, F., Russfield, A. B., Weisburger, J. H., Lim, S., Chak, S. P., and Weisburger, E. K. (1975). Aging changes in CD-1 HaM/ICR mice reared under standard laboratory conditions. *J. Natl. Cancer Inst.* **55,** 37-45.

Hong, C. C., and Ediger, R. D. (1978). Preputial gland abscess in mice. *Lab. Anim. Sci.* **28,** 153-156.

Horn, H., Congdon, C., and Stewart, H. L. (1952). Pulmonary adenomatosis in mice. *J. Natl. Cancer Inst.* **12,** 1297-1315.

Hummel, K. P. (1958). Accessory adrenal cortical nodules in the mouse. *Anat. Rec.* **132,** 291-295.

Hummel, K. P., and Chapman, D. B. (1959). Visceral inversion and associated anomalies in the mouse. *J. Hered.* **50,** 9-13.

Hurd, E. R., and Ziff, M. (1978). Association of interstitial nephritis with tubule cell injury and proliferation in NZB/NZW mice. *Clin. Exp. Immunol.* **32,** 1-11.

Jones, T. C. (1967). Pathology of the liver of rats and mice. *In* "Pathology of Laboratory Rats and Mice" (E. Cotchin and F. J. C. Roe, eds.), pp. 1-23. Blackwell, Oxford.

Kelley, V. E., and Winkelstein, A. (1980). Age- and sex-related glomerulonephritis in New Zealand white mice. *Clin. Immunol. Immunopathol.* **16,** 142-150.

Kelsall, M. A. (1946). Anomalies in the small intestine and cecum of inbred strains of mice. *Anat. Rec.* **95,** 1-6.

Krishna Rao, G. V. G., and Draper, H. H. (1969). Age-related changes in the bones of adult mice. *J. Gerontol.* **24,** 149-151.

Leiter, E. H., and Cunliffe-Beamer, T. (1977). Exocrine pancreatic insufficiency syndrome in CBA/J mice. *Gastroenterology* **73,** 260-266.

Linder, E., Pasternack, A., and Edgington, T. S. (1972). Pathology and Immunology of age-associated disease of mice and evidence for an autologous immune complex pathogenesis of the associated renal disease. *Clin. Immunol. Immunopathol.* **1,** 104-121.

McClure, H. M., Chapman, W. L., Jr., Hooper, B. E., Smith, F. G., and Fletcher, O. J. (1978). The Digestive System. *In* "Pathology of Laboratory Animals" (K. Benirschke, F. M. Garne, and T. C. Jones, eds.), Vol. 1, pp. 280-286. Springer-Verlag, Berlin and New York.

Malinin, G. S., and Malinin, I. M. (1972). Age-related spontaneous uterine lesions in mice. *J. Gerontol.* **27,** 193-196.

Marchant, J. (1977). *In* "Animal Tumors of the Female Reproductive Tract—Spontaneous and Experimental" (E. Cotchin and J. Marchant, eds.), Springer-Verlag, Berlin and New York.

Meier, H., and Hoag, W. G. (1975). Blood coagulation. *In* "Biology of the

Laboratory Mouse" (E. L. Green, ed.), Ch. 18, pp. 373-376. Dover, New York.

Nakamura, M., and Yamada, K. (1967). Studies on a diabetic (KK) strain of the mouse. *Diabetologia* **3,** 212-221.

Nettescheim, P., and Martin, D. H. (1970). Appearance of gland-like structures in the tracheobronchial tree of aging mice. *J. Natl. Cancer Inst.* **44,** 687-693.

Olitsky, P. K., and Casals, J. (1945). Certain affections of the liver that arise spontaneously in so-called normal stock albino mice. *Proc. Soc. Exp. Biol. Med.* **60,** 48-51.

Reichel, W., Hollander, J., and Clark, J. H. (1967). Lipofuscin pigment accumulations and distribution in mouse brain as a function of age. *Gerontologist* **7,** 15. (Abstr.)

Reichel, W., Hollander, J., and Clark, J. H. (1968). Lipofuscin pigment accumulations as a function of age and distribution in rodent brain. *J. Gerontol.* **23,** 71-78.

Rings, R. W., and Wagner, J. E. (1972). Incidence of cardiac and other soft tissue mineralized lesions in DBA/2 mice. *Lab. Anim. Sci.* **22,** 344-352.

Rudofsky, U. H. (1978). Renal tubulointerstitial lesions in CBA/J mice. *Am. J. Pathol.* **92,** 333-348.

Russfield, A. B. (1967). Pathology of the endocrine glands, ovary and testis of rats and mice. *In* "Pathology of Laboratory Rats and Mice" (E. Cotchin and F. J. C. Roe, eds.), pp. 391-468. Blackwell, Oxford.

Sass, B., and Montali, R. J. (1980). Spontaneous fibro-osseous lesions in aging female mice. *Lab. Anim. Sci.* **30,** 907-909.

Saunders, L. F. (1967). Opthalmic pathology in rats and mice. *In* "Pathology of Laboratory Rats and Mice" (E. Cotchin and F. J. C. Roe, eds.), pp. 349-371. Blackwell, Oxford.

Scheinberg, M. A., Cathcart, E. S., Eastcott, J. W., Skinner, M., Benson, M., Shirahama, T., and Bennett, M. (1976). The SJL/J mouse: A new model for spontaneous age-associated amyloidosis. I. Morphologic and immunochemical aspects. *Lab. Invest.* **35,** 47-54.

Sheldon, W. G., and Greenman, D. L. (1979). Spontaneous lesions in control BALB/c female mice. *J. Environ. Pathol. Toxicol.* **3,** 155-167.

Silberberg, M., and Silberberg, R. (1956). The skeletal aging of mice. *Anat. Rec.* **91,** 89-105.

Silberberg, M., and Silberberg, R. (1962). Osteoarthritis and osteoporosis in senile mice. *Gerontologia* **6,** 91-101.

Simpson, W. L., and Hayashi, F. (1960). Distribution of mast cells in the skin and mesentery of BALB/c and C57BL mice. *Anat. Rec.* **138,** 193-202.

Sokoloff, L. (1956). Natural history of degenerative joint disease in small laboratory animals. 1. Pathologic anatomy of degenerative joint disease in mice. *Arch. Pathol.* **62,** 118-128.

Sokoloff, L. (1958). Joint diseases of laboratory animals. *J. Natl. Cancer Inst.* **20,** 965-969.

Sokoloff, L. (1967). Musculoskeletal lesions in experimental animals. *In* "Pathology of Laboratory Rats and Mice" (E. Cotchin and F. J. C. Roe, eds.), pp. 391-467. Blackwell, Oxford.

Sokoloff, L., and Habermann, R. T. (1958). Idiopathic necrosis of bone in small laboratory animals. *Arch. Pathol.* **65,** 323-330.

Sokoloff, L., and Jay, G. E. (1956a). Natural history of degenerative joint disease in small laboratory animals. 2. Epiphyseal maturation and osteoarthritis of the knee of mice of inbred strains. *Arch. Pathol.* **62,** 129-135.

Sokoloff, L., and Jay, G. E. (1956b). Natural history of degenerative joint disease in small laboratory animals. 3. Variations in epiphyseal maturation in the head of the femur among inbred strains of mice. *Arch. Pathol.* **62,** 136-139.

Soret, M. G., Peterson, T., Wyse, B., Block, E. M., and Dulin, W. E. (1977). Renal amyloidosis in KK mice that may be misinterpreted as diabetic glomerulosclerosis. *Arch. Pathol. Lab. Med.* **101,** 464-468.

Squire, R. A., and Levitt, M. H. (1975). Classification of specific hepatocellular lesions in rats: Report of a workshop. *Cancer Res.* **35,** 3214-3233.

Stewart, H. L., and Andervont, H. B. (1938). Pathologic observations on the adenomatous lesion of the stomach in mice of strain I. *Arch. Pathol.* **26,** 1009-1022.

Stewart, H. L., Dunn, T. B., and Snell, K. C. (1970). Pathology of tumors and nonneoplastic proliferative lesions of the lungs of mice. *Proc. Oak Ridge Natl. Lab.*, pp. 161-181.

Stookey, J. L., and Moe, J. B. (1978). The respiratory system. *In* "Pathology of Laboratory Animals" (K. Benirschke, F. M. Garner, and T. C. Jones, eds.), pp. 71-113. Springer-Verlag, Berlin and New York.

Storer, J. B. (1966). Longevity and gross pathology at death in 22 inbred mouse strains. *J. Gerontol.* **21,** 404-409.

Suntzeff, V., and Angeletti, P. (1961). Histological and histochemical changes in intestines of mice with aging. *J. Gerontol.* **16,** 225-229.

Tucker, M. J., and DeC. Baker, S. B. (1967). Diseases of specific pathogen-free mice. *In* "Pathology of Laboratory Rats and Mice" (E. Cotchin and R. J. C. Roe, eds.), pp. 787-824. Blackwell, Oxford.

Tuffery, A. A. (1966). Urogenital lesion in laboratory mice. *J. Pathol. Bacteriol.* **91,** 301-309.

Turner, C. D., and Bagnara, J. T. (1971). "General Endocrinology" 5th ed. Saunders, Philadelphia, Pennsylvania.

Upton, A. C., Conklin, J. W., Cosgrove, G. E., Gude, W. D., and Darden, E. B. (1967). Necrotizing polyarteritis in aging RF mice. *Lab. Invest.* **16,** 483-487.

Van Bezooijen, C. F. A., Von Noord, M. J., and Knook, D. L. (1974). The viability of parenchymal liver cells isolated from young and old rats. *Mech. Ageing Dev.* **3,** 107-119.

Walton, M. (1977a). Degenerative joint disease in the mouse knee; histological observations. *J. Pathol.* **123,** 109-122.

Walton, M. (1977b). Degenerative joint disease in the mouse knee; radiological and morphological observations. *J. Pathol.* **123,** 97-107.

Walton, M. (1978). A spontaneous ankle deformity in an inbred strain of mouse. *J. Pathol.* **124,** 189-194.

Walton, M. (1979). Patella displacement and osteoarthrosis of the knee joint in mice. *J. Pathol.* **127,** 165-172.

Wang, N. S., Huang, S. N., Sheldon, H., and Thrulbeck, W. M. (1970). Ultrastructural changes of Clara and Type II alveolar cells in adrenal-induced pulmonary edema in mice. *Am. J. Pathol.* **62,** 237-252.

Ward, J. M. (1978). Pulmonary pathology of the motheaten mouse. *Vet. Pathol.* **15,** 170-178.

Ward, J. M., Goodman, D. G., Squire, R. A., Chu, K. C., and Linhart, M. S. (1979). Neoplastic and nonneoplastic lesions in aging (C57BL/6N and C3H/HeN)F_1 (B6C3F1) mice. *J. Natl. Cancer Inst.* **63,** 849-854.

West, W. T., and Murphy, E. D. (1965). Sequence of deposition of amyloid in strain A mice and relationship to renal disease. *J. Natl. Cancer Inst.* **35,** 167-174.

Zurcher, C., Burek, J. D., Van Nunen, M. C. J., and Meihuizen, S. P. (1977). A naturally occurring epizootic caused by sendai virus in breeding and aging rodent colonies. I. Infection in the mouse. *Lab. Anim. Sci.* **27,** 955-962.

Index

A

Abortion, 49
Absidia corymbifera, 66
Acariasis, 393-394
Acarina, 415
Achorion, 65
Acinetobacter, 409
Actinobacillus equuli, 60
Actinobacillus lignieresii, 60
Actinobacillus moniliformis, 84
Actinomyces muris, 409
Actinomycosis, 94
Addison's disease, 337
Adenine arabinoside, 288
Adenomatosis, 436
Adenovirus, 304, 335-340
  characteristics, 337-338
  epidemiology, 338-339
  pathology, 336-337
  serology, 338
Adrenal gland, 279
  accessory nodules, 432
  adenovirus, 337
  medullary carcinoma, 302
  nonneoplastic lesion, 432-434
Aedes aegypti, 405
Aedes vigilax, 172
Aging, nonneoplastic lesions, 425-440
Albendazole, 379
Alcaligenes, 47
Alimentary canal, amoebae, 370
  flagellate parasite, 366-370
  nonneoplastic lesion, 428
  sporozoa, 370
Allergic sensitivity, 416-419
  clinical sign, 416-417
  diagnosis, 418-419
  pathogenesis, 417-418
  treatment, 419
Allodermanyssus sanguineus, 100, 406, 415
Alopecia, 69-72, 397
Alugan, 399
Alveolar bronchialization, 125
Ameloblastoma, 300
Amodiaquin, 368
Amyloidosis, 394-395
  renal, 432
Amyloid deposition, 427, 428, 434, 436
Anaphylaxis, 417
Anaplastic carcinoma, 301
Anemia, 277, 302
Anesthetized injury, 11
Antibody, complement-fixing, 243-245
  lactate dehydrogenase-elevating virus, 196-197
  neutralizing, 243-245
  Sendai virus, 118-119, 130-132
Antithymocyte serum, 274
Aotus trivirgatus, 355
Apodemus, 222
Aramite, 398, 399
Arenovirus, 234
  immune reaction, 243-249
  properties, 234-235
Arteritis, necrotizing, 430
Arthritis, 23, 49, 85
Arthroderma benhamiae, 65, 411
Arthropod parasites, 385-402, 414-416
  diagnosis, 387-388
  ecology, 387
  identification, 388
  treatment, 398-400
Aspiculuris tetraptera, 373, 374, 376-378
Asthma, 416, 419
Atgard, 399
Aviadenovirus, 335

B

BHC, 399
Bacillus piliformis, 7-9, 88
Bacillus pseudotuberculosis, 57

Bacterial infection, digestive system, 2–13
Bacterium putscheri, 57
Bartonellosis, 103
Bedbug, 386
Benzene hexachloride, 399
Bile duct, hyperplasia, 427
Bite, 416
Bone necrosis, 435
  tumor, 301, 302
Brain, nonneoplastic lesion, 436–437
5-Bromodeoxyuridine, 329
Bronchitis, infectious, 174
Brown fat, 279
Bronchopneumonia, 27
Bucarpolate, 399
Bulbourethral gland, 47
  infection, 34
Bunamidine, 379
2-(*p-tert*-Butylphenoxy) isopropyl-2-chloroethyl sulfite, 399

C

Calcification, dystrophic, 435
Camelpox, 210
Candida albicans, 13, 282
Candida tropicalis, 50, 94
Capillaria bacillata, 380
Capillaria hepatica, 379
Capillaria muris-musculi, 380
Cardiac valvular chondrification, 430
Cardiomyopathy, 174
Cardiovascular disease, nonneoplastic, 429–430
Castration, 432
Cataract, 438
Catarrh, infectious, 23
Catarrhal enterocolitis, 9
Central nervous system, diseases, 77–81
  encephalomyelitis virus, 344–346
Cerebellar hypoplasia, 317
Ceroid deposition, 433
Chattering, 93
Chlamydia psittaci, 105, 106
Chlamydia trachomatis, 34, 35–36, 105–106
Chlamydial infection, 105–106
Chlordane, 399
Chloroquine, 368
Chlortetracycline hydrochloride, 408
Choriomengitis virus, lymphocytic, 51, 231–266, 404–405
  carrier state, 249–254
  cerebral form, 237–238, 239
  clinical disease, 237–239
  diagnosis, 256–257
  epizootiology, 254–256
  immunity, 242–249
  late onset disease, 238, 242
  neonatal infection, 238–239, 242
  pathology, 239–242
  prevention, 257–258
  virus, culture, 235–237
    properties, 233–237
  visceral form, 238, 239–242
  zoonose, 404–405
Chorioretinitis, 277
Chronic immune complex disease, 242
Cimex lectularis, 386
Circling, 79
Citrobacter freundii, 9
Clethrionomyces glariolus, 411
Cletrionomys, 222
Clitoral gland abscessation, 430
Clostridium perfringeus, 12–13
Clostridium tertium, 163
Clostridium tetani, 416
Cockroach, vector, 217
Colitis, 9–11
  cystica, 9
Colonic hyperplasia, transmissible murine, 9–11
Colostrum, 273, 281
Conjunctivitis, 4, 34, 35, 49, 59, 86, 87, 92
Convulsion, 345
Coronavirus, 174
Coronavirus, culture, 175
  diagnosis, 182–183
  inactivation, 176
Corticosteroid treatment, 366
Cortisone, 178, 294
Corynebacterium, 47
Corynebacterium kutscheri, 36, 44, 80, 86
  skin lesion, 57–58
Corynebacterium murium, 57
Corynethrix pseudotuberculosis, 57
Cowpox, 210
Coxiella burnettii, 100
Cryptococcus neoformans, 80, 94
Cryptosporidium muris, 370
Cryptosporidium parvum, 370
Culex fatigans, 172
Culture, coronavirus, 175
  cytomegalovirus, 270–273
  ectromelia virus, 211
  encephalomyelitis virus, 343–345
  K virus, 142–144
  lactate dehydrogenase-elevating virus, 200–202
  mouse pneumonia virus, 136
  polyomavirus, 298–300
  reovirus, 169–170
  Sendai virus, 113–115
Cyclophosphamide treatment, 11, 179, 336
Cyst, *Toxoplasma,* 362
Cysticerus fasciofaris, 382
Cystitis, 46
Cytomegalovirus, 267–288
  epizootiology, 275, 285
  handling, 286–287
  identification, 285–286
  pathogenesis, 273–285
    genetic factor, 275
    immune response, 280
    inoculation route, 273–275
    latent, 277, 284–285
  model role, 275–278
  overt disease, 273
  pathology, 278–280
  resistance, 273
  transmission, 280
  prevention, 287, 288

properties, 269–273
vacine, 277–278
Cytosine arabinoside, 288

D

DMC, 398, 399
Dacryoadenitis, 34
Deafness, 277
Demodex musculi, 386
Depilation, 393
Dermacentor andersoni, 405
Dermanyssus, 406
Dermatitis, 55, 62, 64, 90, 415
Dermatophytoses, 65
Dermis, biological function, 56–57
Dermophyte infection, 411–413
Diabetes mellitus, 429
Diagnosis, coronavirus, 182, 183
ectoparasites, 387–388
leptospirosis, 407, 408
lymphocytic choriomeningitis, 254, 255
mouse pneumoniae virus, 132, 140–141
mousepox, 224–226
pinworm, 376, 377
tapeworm, 379
rotavirus, 166
Diarrhea, 4, 368, 370
epidemic, 160–167
epizootic, 160–167
control, 166–167
diagnosis, 166
epizootiology, 165–166
pathogenesis, 163–165
properties, 160–163
Dibutyl phthalate, 399
Di-(*p*-chlorophenyl)methyl carbinol, 399
Dichlorvos (DDVP), 399
Digestive system, bacterial infection, 2–13
epizootic diarrhea, 160–167
hepatoencephalomyelitis, 167–173
murine hepatitis, 173–183
mycoplasmal infection, 13
mycotic infection, 13
nonneoplastic lesion, 428
reovirus 3 infection, 167–173
O-1-Dimethyl-5-(dicarbethoxyethyl) dithiophosphate, 399
Dimetridazole, 368, 369
Dimite, 398, 399
Dri-Die, 399
Dri-One, 399
Duodenitis, 368
Duodenum, hemorrhagic lesion, 337
Duovirus, 160
Dyspnea, 23

E

Echo 10 virus, 167–173
Ectromelia virus, 86
Eimeria falciformis, 370
Embryo abnormality, 129
Encephalitis, 277, 355
focal granulomatous, 364
necrotizing, 273, 274
Encephalitozoon cuniculi, 363–366
Encephalomyelitis, 174, 341–352
culture, 343–345
diagnosis, 350
epizootiology, 349–350
immune response, 349
pathogenesis, 345–349
properties, 342–345
Encephalomyocarditis virus, 353–357
epidemiology, 354–355
host, 355–356
Endocarditis, 336
Endocrine system, nonneoplastic lesion, 432–434
Endrometrial hyperplasia, cystic, 430
Entamoeba coli, 413
Entamoeba muris, 370, 413
Enteritis, 375
catarrhal, 370
Enterobacter cloaca, 49
Enterobius vermicularis, 414
Enterovirus cardiovirus, 354
Eperythrozoon, 100–103
Eperythrozoon coccoides, 100–105, 175–179, 197, 390
Epidermophyton floccosum, 411, 413
Epizootiology, K virus, 143–144
Sendai virus, 116–123
Erysipelas, 86, 410
Erysipelothrix insidiosa, 92–93
Erysipelothrix rhusiopathiae, 410–411
Erythrocyte, agglutination, 314
Escherichia coli, 12, 37, 49
Eulaelaps stabularis, 386

F

Favus, 65–66, 411
Fecal impaction, 375
Feces, virus transmission, 316
Fibroblast, minute virus infection, 320–321, 326–332
Filter, air, 166
Flea, 360, 386, 405, 406
Formel 5-brommethyl-1,2,3,4,7,7-hexachlorobiocyclo-(2,2,1)hepten-(2), 399
Francisella tularensis, 390

G

Gamma HCH, 399
Gastroenteritis, 174
Genetics, cytomegalovirus response, 275
salmonellosis resistance, 3
Genital infection, 47–49
nonneoplastic lesion, 430–431
George's disease, 341
Giardia muris, 365, 366–368
Giardiasis, 367–368
Globulin, antilymphocyte, 281
Glomerulonephritis, 50–52, 252, 280, 302
acute, 274
immune complex, 431
Glomerulosclerosis, 50

Glycosuria, 429
Gonadectomy, 434
Gongylonema musculi, 380
Gongylonema neoplasticum, 380
Granulocytopenia, 11
Granuloma formation, 395
Guinea pig, mousepox, 221

H

Haemogamasus pontiger, 386
Haemalaelaps, 415
Haemolaelaps casalis, 386
Haemolaelaps glasgowi, 386
Hair chewing, 69, 72
Hamster, 221, 255-256, 303
Heligmonoides murina, 380
Helminth parasite, 373-384
  cestode, 381-382
  nematode, 379
  pinworm, 373-376
  tapeworm, 376-379
Hemangioendothelioma, 302
Hematopoietic system, nonneoplastic lesion, 438
Hemobartonella muris, 100, 103-105
Hemopericardium, 429
Hemophilus influenzae, 121, 409
Hemosiderin, 427
Hepatitis, murine, 120, 173-183, 279, 405
  control, 183
  diagnosis, 182-183
  epizootiology, 182
  focal necrotizing, 363
  pathogenesis, 177-182
  properties, 174-177
Hepatocyte, altered, 426-427
Hepatoencephalomyelitis, 167-173
Herpesvirus, 288
Heterapis spumosa, 379
Hexamita muris, 368
Hexamitiasis, 368-369
Histoplasma capsulatum, 93-94
Hoplopleura acanthopus, 386
Hoplopleura captiosa, 386
Human, Sendai virus, 116
  zoonoses, 403-423
Hydrocephalus, 129, 303
Hydronephrosis, 432
Hymenolepsis diminuta, 379, 381-382, 413-414
Hymenolepsis nana, 373, 374, 378-379, 414
Hymenolepsis microstoma, 382
Hyperexcitability, 345
Hyperammaglobulinemia, 394
Hyperglycemia, 429

I

Immune response, cytomegalovirus, 280-283
  mousepox, 219-220
Immunity, encephalomyelitis, 349
  H-2 restriction, 247-249
  humoral, 280-281
  lymphocytic choriomeningitis, 242-249
Immunofluorescence, 31-32
Immunoglobulin, antirotaviral, 165
Immunoperoxidase method, 31-32
Immunological tolerance, 250-252
Immunology, adenovirus, 338
Immunosuppression, 36, 282, 362
Immunosuppressive virus, 317
Infection, restrictive interaction, 319-320
Influenza virus, 315
Integumentary system, 55-75
  bacterial disease, 57-65
  hairy, 56-57
  mycotic disease, 65-68
  noninfectious disorder, 68-72
  nonneoplastic lesion, 434-435
Interferon, 172, 204-205, 220, 283
Interuterine infection
  cytomegalovirus, 275-276
  mousepox, 218
Intestine, mousepox, 216-217
  mucosa, encephalomyelitis, 346-347, 349
    necrosis, 362
    ulceration, 428
Intussusception, 375
5-Iodo-2′-deoxyuridine, 288
Iron deficiency, 4

J

Jaundice, 171, 399

K

K virus, 142-150, 178
  control, 150
  diagnosis, 132, 148-150
  epizootiology, 143-144
  pathogenesis, 144
  properties, 142-143
Kelthane, 399
Kidney, gray spot, 363
  infection, 44-46, 50-52, 274, 280
  K virus, 146
  tumor, 300, 302
Kilham virus, 142
Klebsiella pneumoniae, 13, 33, 37, 49, 91-92
Klebsiellosis, 33
Klossiella muris, 362-363
Kupffer cell, 284

L

Lactate dehydrogenase-elevating virus, 193-108
  detection, 200
  infection characteristics, 194-200
  properties, 203-205
  replication, 195-196, 200-203
Laelaps echidnus, 386
Lassa virus, 233
Leprosy, 88
Leptospira australis, 407
Leptospira ballum, 44, 407-408
Leptospira bataviae, 407
Leptospira grippotyphosa, 407, 408
Leptospira hebdomidis icterohaemorrhagiae, 407

Leptospira icterohaemorrhagiae, 44
Leptospira pomona, 407
Leptospira pyrogenes, 407
Leptospirosis, 44, 407–408
Leptosylla segnis, 386, 415–416
Leukemia, 426
  virus, 313, 315
    Moloney, 174, 178
Lindane, 399
Lipofuscin, 427, 437
Liponyssoides sanguineous, 386, 406, 414–415
Liponyssus bacotii, 406
Liver, age-associated lesions, 426–428
  granuloma, 375
  hemangioma, 302
  hepatitis, 179–181, 279
    lobular, 240
  K virus, 146, 147
  mousepox lesion, 215–216, 217
Longistriata musculi, 380
Louse, mouse, 101, 386, 388–390
  rat, 105
Lung pathology
  K virus, 145–148
  Pneumocystis carinii, 366
  pneumonia virus, 138–140
  pneumonitis, 280
  Sendai infection, 123–129
Lymph node, cytomegalovirus infection, 278
  enlargement, 369
  mousepox, 216
  nonneoplastic lesion, 438
Lymphadenitis, 64, 90, 91
Lymphocyte, cytotoxic T, 247–248, 250

M

MOPC-315 tumor, 197
Macaca fascicularis, 356
Machupo virus, 233
Macrophage, cytomegalovirus effect, 283–284
Malathion, 398, 399
Mammary adenocarcinoma, 302
Mast cell, 56
Mastadenovirus, 335
Mastophorus muricola, 380
Mebendazole, 379
Mengo virus, 354, 355
Meningitis, 180
  lymphocytic, 363
Meningoencephalitis, necrotizing, 279
  subclinical, 274, 279
Mesenteric disease, 438
Mesocricetus auratus, 255
Mesothelioma, 302
Micrococcus, 47
Microsporum gallinae, 411
Microsporum gypseum, 66, 411
Microsporum quinckeanum, 65
Microtus, 143, 232
Mima polymorpha, 409
Minute virus, 313–334
  antigenic relationship, 316, 321
  cell interaction, 318–321
  DNA properties, 324–326
  detection, 314–316
  epidemiology, 316–317
  hamster passage, 317–318
  pathogenesis, 317–318
  replicative life cycle, 326–332
  structure, 322–326
Mite, 386, 387, 406
  infestation, 391–398
  rat, vector, 217
  zoonoses, 414–416
Mitogen, response to, 282
Monkeypox, 210
Mosquitoe, 355, 405
Mouse dander, hypersensitivity, 416
Mouse strain, 129/J, 117
  129/ReJ, 117
  A, 303, 432, 436
  A/HeJ, 117
  A/J, 117, 181
  AKR, 199, 222, 366, 426
  AKR/J, 117
  B6C3F1, 429, 432
  BALB/c, 174, 222, 275, 366, 429, 432, 434
  BALB/cJ, 117
  BDF, 222
  C, 429
  C3H, 181, 222, 275, 349, 366, 367, 429, 432, 434
  C3H/Bi, 117
  C3H/HeJms, 182
  C3HeB/FeJ, 117
  C57, 303
  C57BL/6, 117, 181, 222, 275, 366, 393, 431
  C57BL/10Sn, 117
  C57BL/K, 429
  C57L, 432
  C57L/J, 117
  C58, 199, 432
  C58/J, 117
  CAF, 222
  CBA, 222, 275, 349
  CBA/J, 431
  CD, 429, 430
  CD-1, 435, 436
  CDF, 429
  CF1, 222
  DBA/1, 222
  DBA/1J, 117
  DBA/2, 117, 222, 429
  DBA/2J, 117
  DK-black, 275
  HH, 434
  KK, 429
  LLC, 432
  MA, 222
  MA/Nd, 222
  N2B/N2W, 431
  NIH Swiss, 275
  New Zealand (NZ), 117, 196–197
  rude, 115, 117, 128–129, 174, 178–179, 183, 303, 337
  PHI, 275
  RF, 431

Mouse strain (*continued*)
RF/J, 117
RF/Un, 435
RFM, 429
SJL/J, 117, 198, 349
STR/ORT, 436
SWR, 222
SWR/J, 117
Swiss, 117
Sy D, 222
Sy N, 222
wild house, 406
Mousepox, 209–230
control, 226–227
diagnosis, 224–226
epizootiology, 220–224
pathogenesis, 211–220
properties of virus, 210–211
transmission, 217–218
Mus caroli, 220
Mus musculus, 143, 165, 220, 254, 255, 336, 343, 374, 376, 408, 411
Mus musculus domesticus, 255
Mus musculus musculus, 255
Muscle calcification, 435
necrosis, 279
Mutatis mutandis, 253
Muzzle alopecia, 69
Mycobacterium lepraemurium, 88–89
Mycobacterium pseudotuberculosis, 57
Mycoplasm, digestive system, 13
Mycoplasma arthritidis, 93
Mycoplasma neurolyticum, 31, 78–79, 93
Mycoplasma pulmonis 22–33, 34, 49, 79, 93
Mycoplasmosis, respiratory, 22–33
control, 32–33
diagnosis, 30–32
disease, 22–29
epizootiology, 32
immune response, 29–30
Mycoses, 411–413
Mycotic infection, 93–94
central nervous system, 80
digestive tract, 13
integumentary, 65–68
Mycotoxicosis, 93
Myobia, 399
Myobia musculi, 386, 387
life cycle, 390–392
pathobiology, 392–395
Myocarditis, 273, 336, 355–356
focal, 429
lymphocytic, 355
Myocoptes, 399
Myocoptes musculinus, 386, 396–398
life cycle, 397
pathology, 397–398
Myocoptes romboutsi, 398

N

Nacconal, 399
Nematode infection, 373, 379–381
Nematospiroides dubius, 380
Neonatal death, 273
Neoplasia, 9
Nephritis, 44–46
interstitial, 431
Nephrosis, 174
Nervous system, nonneoplastic lesion, 436–438
Neuramindase, 327
Newcastle disease virus, 315
Niclosamide, 379
Nigg agent, 35
Nippostrongylus brasiliensis 380
Nonneoplastic disease, selected, 425–440
Nosopsyllus fasciatus, 360, 386, 415
Nosopsyllus londsniensis, 386
Notoedres, 395
Notoedres musculi, 386

O

Oncornavirus, murine, 51
Oochoristica ratti, 381
Oophoritis, 23
Oochitis, acute, 48
Ornithonyssus, 400
Ornithonyssus bacoti, 217, 386, 406, 414–415
Ornithonyssus sylviarum, 386
Orthopoxvirus, 210
Orthoreovirus, 167
Osteoarthrosis, 435
Osteodystrophy, senile, 435
Osteoporosis, 435
Otitis media, suppurative, 23, 90
Ovary lesion, 430
Oxfendazole, 379

P

Pancreas, necrosis, 171–172
nonneoplastic lesion, 428–429
Pancreatitis, 428
Panophthalmitis, 34, 59, 60, 62
Pantothenate deficiency, 36
Papovavirus, 178, 295
K, 142
Parrigofavosa, 66
Parainfluenza type I virus, 116, 405
Paralysis, 345, 348
Paramyxovirus, 112
Parathyroid, lesion, 434
Parbendazole, 379
Parotid tumor, 293, 301
Parvovirus, autonomous, 314
DNA replication, 325–326, 328–331
Pasteurella, 47, 416
Pasteurella multocida, 93, 408
Pasteurella pestis, 408
Pasteurella pneumotropica, 28, 33–34, 47, 49, 58–60, 80, 91, 408–409
cutaneous infections, 59–60
Pasteurellosis, 33–34, 408–409
Pathology, cytomegalovirus, 278–280
ectromelia virus, 212–216
lactate dehydrogenase-elevating virus, 199
Myobia infestation, 392–395

murine hepatitis, 179–181
Myocoptes, 397–398
pinworm, 375–376, 377
Psorergates simplex, 395–397
reovirus, 171–173
rotavirus, 164–165
tapeworm, 378–379
Periarteritis, 430
Pericarditis, suppurative, 429
Periodontal disease, 317
Peromyscus, 143
Peyer's patch, 367, 428
Pharyngitis, 90
Phosphonoacetic acid, 288
Phycomycetes, 80
Physaloptera massino, 380
Pinworm, 374–378, 414
Piperazine citrate, 376
Pituitary, lesion, 432
Plagiorchis javensis, 381
Plagiorchis muris, 381
Plagiorchis phillippinensis, 381
Pneumocystis carinii, 365, 366
Pneumonia, 174, 406, 436
immunization, 121
Pneumocystis, 366
virus, 134–141
control, 141
diagnosis, 140–141
epizootiology, 136–138
pathogenesis, 138–140
properties, 134–136
Pneumonitis, 105
chlamydial, 35
interstitial, 276, 280
virus, mouse, 142
Pneumovirus, 112, 134
Poliomeylitis-like infection, 345, 347
Polyarthritis, 85, 93
Polyomavirus, 112, 142, 293–311, 323
cell transformation, 294–295, 298, 299
cultivation, 298–300
DNA resistance, 297–298
hosts, 303
pathology, 300–304
prevention, 304–305
structure, 295–297
Polyplax, 399
Polyplax serratus, 101, 386, 387, 388–390
life cycle, 389–390
pathobiology, 390
Polyplax spinulosa, 104
Polyuria, 429
Prednisolone, 281
Preputial gland abscessation, 430, 434
infection, 47–49
Prostatitis, 430
Proteus, 37, 47, 431, 434
Proteus mirabilis, 44, 93
Protozoa parasites, 359–372
alimentary system, 366–370
parenteral system, 359–366
Pruritis, 392
Pseudocyst, 360
Pseudomonas, 431, 434
Pseudomonas aeruginosa, 37, 44, 49, 87–88, 282, 411
central nervous system, 79–80
digestive system, 11–12
Pseudomoniasis, 87
Pseudotuberculosis, 36, 44, 58
Psittacosis, 104
Psorergates, 399
Psorergates simplex, 386, 395–397
life cycle, 395
pathobiology, 395–397
Pulmonary adenomatosis, 436
edema, 436
endothelium, 145
Pyelonephritis, 44–46
Pyoderma, 393
Pyractone, 399
Pyrethrin, 399
Pyridoxine-deficient diet, 139
Pyrimethamine, 362, 366

Q

Quinacrine, 368, 379

R

Rabbit, mousepox, 221
Rabies, 405
Raccoonpox, 210
Radfordia, 390, 399
Radfordia affinis, 386, 391, 395
Radiation, 303–304
injury, acute, 11
Rat, ectromelia virus, 220–221
polyomavirus infection, 303
Rat-bite fever, 409–410
Rattus, 136, 406
Rattus norvegicus, 413
Rattus rattus, 413
Rectal prolapse, 375
Renal infection, 274
Reovirus, 167
control, 173
diagnosis, 173
epizootiology, 172–173
inactivation, 168–169
pathogenesis, 170–172
properties, 167–170
Reo-1, 169
Reo-3, 167–173, 406
Respiratory disease
chlamydial pneumonitis, 35–36
corynebacteriosis, 36
klebsiellosis, 33
mycoplasmosis, 22–33
nonneoplastic lesion, 436
pasteurellosis, 33–34
viral diseases, 109–157
K virus, 142–150

Respiratory disease (*continued*)
pneumonia virus, 134–141
sendai virus, 110–134
Reticuloendothelial system, 198
Ribavirin, 288
Rickettsia akari, 100, 105, 406
Rickettsia australia, 104
Rickettsia mooseri, 406
Rickettsia prowaseki, 220
Rickettsia siberica, 103
Rickettsia tsutsugamushi, 100, 104
Rickettsial infection, 99–105
Rickettsialpox, 100, 406
Rictularia amurensis, 381
Rictularia baicalensis, 381
Rictularia magna, 381
Rictularia muris, 381
Riley virus, 194
Ringworm, 411–413
Rolling disease, 78
Rotavirus, cutlure, 163
diagnosis, 166
enteritis, 160–167
inactivation, 162
Roundworm, 414
Runt disease, 178, 273, 302, 317

S

SG-67, 399
SV40, 295
Saliva, vector, 273
Salivary gland, 428
infection, 274, 279
tumor, 300, 301
virus, 268
Salmonella, 2–7, 86–87, 410
Salmonella cholerae-suis, 2
Salmonella enteritidis, 2–4, 86–87, 410
Salmonella gallinarium, 3
Salmonella heidelberg, 410
Salmonella newport, 410
Salmonella paratyphi B, 3
Salmonella paratyphi C, 3
Salmonella typhi, 2, 5
Salmonella typhi-typhimurium, 3
Salmonella typhimurium, 3, 86, 410
Salmonella typhosa, 178
Salmonellosis, 2–7, 86, 410
diagnosis, 6–7
immunity, 5–6
Sarocystis muris, 362
Sebaceous gland, 57
Sebum, 57
Self-grooming, parasite control, 387
Seminal vesiculitis, 430
Sendai infection, 28, 30
Sendai virus, 405
control, 133–134
diagnosis, 132–134
epizootiology, 116–123
pathogenesis, 123–131
properties, 111–116
Seroconversion, 315, 316
Serodiagnostic test, 132, 149
Sialodacryoadenitis, 174
Sigmodon, 136
Sigmodon hispidus, 354
Silica dust, 399
Skin graft, 282
heterogenization, 246
lesion, mousepox, 212–215
nonneoplastic, 434–435
ulcerative, 393–395
Smallpox, 210
Social dominance, 69, 72
Sorex araneus, 411
Sound stress, 304
Spinal cord, epidermal cyst, 437
lesion, 345–346
Spinalitis, ascending, 355
Spirillum minor, 409
Spirillum minus, 409, 416
Spirillum morsus muris, 409
Spironucleosis, 368–369
Spironucleus muris, 368
Spleen, cytomegalovirus infection, 278
K virus, 146
killer cell, 281
mousepox lesion, 216, 217
necrosis, 275
nonneoplastic lesion, 438
Splenectomy, 178
Splenomegaly, 197
Stachybotrys chartarum, 94
Stachbotrytoxicosis, 94
Staphylococcus aureus, 44, 47, 49, 60–62, 91, 282, 411, 434
Staphyloccus, cutaneous infection, 60–62
Staphyococcus epidermidis, 49, 60
Streptobacillus moniliformis, 49, 84–86, 409, 416
Streptococcus, 89–91
cutaneous infection, 62–65
pneumonia virus interaction, 138–139
Streptococcus anginosa, 64
Streptococcus pyogenes, 63–64, 411
Streptococcus zooepidermicus, 64
Sulfadiazine, 362, 366
Sulfur, 399
Sweat gland adenocarcinoma, 302
Syphacia muris, 376, 381, 414
Syphacia obvelata, 373, 374–376, 377, 414

T

Tachyzoite, 360
Taenia taeniaeformis, 382
Taeniorhynchus, 355
Tapeworm, 381–382, 413–414
dwarf, 378–379
Tatera kempi, 210
Taterapox, 210
Teratogenesis, 317, 319
Testes, atrophy, 430
Testudo graeca, 113
Tetmosol, 398, 399

Tetraethylthiuram monosulfide, 399
Thiabendazole, 379
Thrombocytopenia, 277
Thrombosis, 429
Thymic virus, 288–289
Thymectomy, 303
  neonatal, 178
Thymosin, 173
Thymus, atropy, 273, 279
  necrosis, 289
  tumor, 301
Thyroid lesion, 302, 434
Thyroidectomy, 432
Tick, Rock Mountain wood, 405
Tooth germ tumor, 300
Torticollis, 79
Tortoise, 113
Toxoplasma gondii, 360–362
Toxoplasmosis, treatment, 362
Transplant, organ, 276
Trematode infection, 381
Tribolium, 413
Trichinella spiralis, 381
Trichoecius, 399
Trichoecius romboutsi, 386, 396, 398
Trichophytom ajello, 411, 413
Trichophyton benhamiae, 411
Trichophyton erinacai, 411
Trichophyton keratinomyces, 411, 413
Trichophyton mentagrophytes, 65–68, 411–413
Trichophyton quinckeamum, 65, 411
Trichophyton schoenleini, 65, 411
Trichophyton terrestre, 66
Trichophyton verrucosum, 411
Trichophyton violaceum, 65
Trichuris muris, 381
Tritrichomonas muris, 369–370
Trypanosoma lewisi, 359
Trypanosoma musculi, 359–360
Tumor induction, polyomavirus, 300–304
  transplanted, 313, 315
Tumorigenesis, 199–200
Typhus, murine, 406
  schrub, 100
Tyzzer's disease, 7–9, 86, 88

U

Uremia, 174
Urethritis, 47
Urinary system, nonneoplastic lesion, 431–432
Urine, hypersensitivity, 416
  virus transmission, 316
Urogenital disease, 43–53
  cystitis, 47
  genital, female, 49
    male, 47–49
  glomerulonephritis, 50–52
  mycoplasmal, 49
  mycotic, 49–50
  nephritis, 44–46
  pyelonephritis, 44–46
  urethritis, 47
Uropathy, obstructive, 432
Uterine lesion, 430

V

Vaccination, mousepox, 226–227
Vaccine, cytomegalovirus, 277–278
Vaccinia, 210
Vapona, 399
Vaporal, 399
Variola virus, 210
Vibris cholerae, 169, 315
Virus, capsid protein synthesis, 323–324
  isolation, 132–133, 141, 150, 173
  screening, 315
  transcription, 331
Vitamin E, 433

W

Wasting syndrome, 50, 174
Whipworm, 381

X

X-irradiation, 36, 179
Xenopsylla cheopis, 386, 406, 415

Y

Yersini pestis, 408
Yersini pseudotuberculosis, 408, 409

Z

Zoonoses, 403–423
  allergic sensitivity, 416–419
  arthropod, 414–416
  bacterial, 407–411
  bite, 416
  helminth, 413–414
  mycoses, 411–413
  protozoan, 413
  rickettsial, 406
  viral, 404–406